WELDING
SKILLS AND TECHNOLOGY

DAVE SMITH

Member American Welding Society and American Society for Metals

CONSULTING EDITORS
WILLIAM H. MERCER
Training Center Manager • Airco Welding Products
RALPH E. BARNETT
Safety Technical Coordinator • Airco Educational Services

GREGG DIVISION • McGRAW-HILL BOOK COMPANY

New York   Atlanta   Dallas   St. Louis   San Francisco   Auckland   Bogotá   Guatemala
Hamburg   Johannesburg   Lisbon   London   Madrid   Mexico   Montreal   New Delhi
Panama   Paris   San Juan   São Paulo   Singapore   Sydney   Tokyo   Toronto

Sponsoring Editor: Myrna Breskin
Editing Supervisor: Iris Wheat
Design Supervisor/Interior Design: Karen Tureck
Production Supervisor: Frank Bellantoni

Cover Photographer: Dave Smith
Cover Designer: Renée Kilbride

Library of Congress Cataloging in Publication Data

Smith, Dave (date)
    Welding skills and technology.

    Includes index.
    1. Welding.   I. Title.
TS227.S597 1984   671.5′2   83-931
ISBN 0-07-000757-8

Welding Skills and Technology

  2 3 4 5 6 7 8 9 0 VNHVNH 8 9 0 9 8 7 6 5 4

ISBN 0-07-000757-8

Contents

PREFACE   vi

PART 1

INTRODUCTION   1

CHAPTER 1

Why Welding Is a Good Trade   2
1-1   Profile of a Welder   2
1-2   The Welder Shortage   4
1-3   Where to Find Welding Jobs   6
1-4   Welders and Labor Unions   15
1-5   Welder Qualification Tests   16
1-6   Your First Look at Metallurgy   16

CHAPTER 2

Introducing 98 Kinds of Welding and Cutting   18
2-1   Brazing and Soldering   18
2-2   Solid-State Welding   19
2-3   Fusion Welding   20
2-4   What Welders Need to Know   20

PART 2

FUEL-GAS PROCESSES   25

CHAPTER 3

Fuel Gases   26
3-1   Important Facts About Fuel Gases   26
3-2   The Uses of Fuel Gases   27
3-3   Fuel-Gas Physics   28
3-4   Making Gas Calculations   35
3-5   Other Important Numbers   39
3-6   Heat Transfer   40
3-7   Combustion   42
3-8   Backfires and Flashbacks   46
3-9   Maximum Regulator Working Pressure   50
3-10   Your Nose—The Perfect Leak Detector   50
Color Plates   51

CHAPTER 4

The Tools You Use with Gases   56
4-1   Gas Cylinders   56
4-2   Handling and Storing Cylinders Safely   65
4-3   Gas Regulators   68
4-4   Flowmeters   76
4-5   Gas Manifolds   78
4-6   Hoses   79
4-7   Check Valves   83
4-8   Welding and Heating Torches   84
4-9   Cutting Torches   88
4-10   Packaged Welding Outfits   90
4-11   Necessary Accessories   90
4-12   Protective Outerwear   92
4-13   Welding Distributors   94
Shop Practice: Shop Experience with Oxyfuel Equipment   95

CHAPTER 5

Flame-Cutting Shapes, Edges, and Holes   99
5-1   Where Flame-Cutting Is Used   99
5-2   How Oxyfuel Cutting Works   100
5-3   Cutting Tips   101
5-4   Getting Ready to Cut   106
5-5   Setting Your Cutting Oxygen Pressure   107
5-6   Adjusting Your Flame   107
5-7   Starting the Cut   110
5-8   Making a High-Quality Cut   111
5-9   Making a Steady Cut   118
5-10   Troubleshooting Your Cuts   119
5-11   Other Cutting Methods   122
5-12   Guiding Devices   130
5-13   Cutting Machines   130
Shop Practice: Practice with Flame-Cutting   136

CHAPTER 6

Using a Heating Tool   138
6-1   Heating Torches and Tips   138
6-2   Expansion and Contraction   139
6-3   Wrinkle-Bending Pipe, Tube, and Bars   142
6-4   Hot-Bending Pipe, Tubing, and Bars   143
6-5   Straightening Dents and Bending Structurals   145
6-6   Other Processes Using Heating   147
Shop Practice: Learning to Use Heat   149

CHAPTER 7

Brazing and Soldering Can Join Anything   151
7-1   A Few Good Things to Know   151
7-2   Basic Torch Methods   153
7-3   Working Safely   161
7-4   Filler Metals   161
7-5   How Filler Metals Melt and Solidify   165
7-6   Joints That Hold Tight   168
7-7   Braze-Welding   169
Shop Practice: Soldering, Brazing, and Braze-Welding Practice   175

CHAPTER 8

Strength and Hardness of Welded Metals   178
8-1   Metal Properties   178
8-2   The Strength of Metals   179
8-3   Ductility   184
8-4   Elasticity   185
8-5   Hardness   186
8-6   Toughness   190
8-7   Bend Tests   195
8-8   Metal Fatigue   197
8-9   Creep Rupture   198
8-10   Metal Properties and Welding   199

CHAPTER 9

Making Fusion Welds with Oxyfuel   200
9-1   Why Gas Welding Is Important   200
9-2   Selecting Gas-Welding Rods   202
9-3   Welding Techniques   206
9-4   Multipass Welding Procedures   208
9-5   Welding Fluxes   209
9-6   Gas-Welding   209
Shop Practice: Gas Welding Practice   225

CHAPTER 10

Oxyfuel Pipe Welding   233
10-1   The Difference Between Pipe and Tubing   233
10-2   Pipe and Tubing Specifications   235
10-3   Pipe Positions   237

10-4 Pipe Joints 238
10-5 Pipe-Welding Rods 240
10-6 Consumable Inserts 240
10-7 Welding Small Sections 241
10-8 Welding Positions 242
10-9 Large Pipe Sections 244
10-10 Multipass Pipe Welding 246
10-11 Stress-Relieving Finished Welds 247
10-12 Pipe-Welding Procedures 248

Shop Practice: Oxyfuel Pipe-
Welding Practice 252

PART 3

ELECTRICAL PROCESSES 255

CHAPTER 11
Learn Electricity before Striking an Arc 256
11-1 What Electricity Is 256
11-2 Why Electricity Flows 258
11-3 Charged Ions 262
11-4 Measuring Electricity 264
11-5 Ohm's Law 266
11-6 Power and Energy 266
11-7 Resistivity, Heat, and Power 268
11-8 Magnetism 270
11-9 Understanding Electricity 275
11-10 Duty Cycles and Volt-Amp Output
Relationships 281
11-11 Using What You've Learned 285

CHAPTER 12
Arc Welding Safely 290
12-1 Welding Machines 290
12-2 Electrical Hazards 291
12-3 Arc Voltages and Machine Ratings 292
12-4 Multiple Machines 294
12-5 Machine Duty Cycles 295
12-6 Power-Supply Wiring 296
12-7 Selecting Welding Cable 296
12-8 Electrode Holders 298
12-9 Grounding the Workpiece 298
12-10 Water Is Hazardous 299
12-11 Safe Maintenance 299
12-12 Working in Confined Spaces 299
12-13 Hot Metal and High Places 300
12-14 Protecting Your Eyes 300
12-15 Welding Curtains and Blankets 301
12-16 Protective Clothing 303

CHAPTER 13
Cutting Metal with Arc and Air 304
13-1 How AAC Works 304
13-2 Gouging Equipment 305
13-3 Gouging Techniques 308
13-4 Solving Gouging Problems 309
13-5 Arc-Gouging Safety 310

Shop Practice: Getting
Experience with Gouging 311

CHAPTER 14
TIG Welding 313
14-1 What is TIG? 313
14-2 An Overview of TIG 315
14-3 GTAW Power Sources 318
14-4 Starting Your Arc 321
14-5 Alternating versus Direct Current 322
14-6 Types of Machines for TIG Welding 324
14-7 Selecting the Right Tungsten 326
14-8 Care of Electrodes 329

14-9 TIG Holders 330
14-10 Welding Cable 332
14-11 Hose for GTAW 332
14-12 Selecting a Shielding Gas 332
14-13 Filler Metals for GTAW 334
14-14 Manual GTAW Techniques 335
14-15 Welding Schedules 341

Shop Practice: Learning to Weld
with TIG 343

CHAPTER 15
Welding and Cutting with a Plasma Arc 347
15-1 What's a Plasma? 348
15-2 Two Types of Arcs 348
15-3 Plasma Welding's Advantages 348
15-4 Plasma Welding's One Drawback 349
15-5 Plasma Gases 349
15-6 PAW Outfits 350
15-7 Plasma-Welding Power Sources 351
15-8 Plasma-Welding Torches 353
15-9 Plasma-Welding Techniques 354
15-10 Plasma-Welding Schedules 355
15-11 Plasma-Arc Cutting 356
15-12 Plasma-Arc Equipment 358
15-13 Plasma-Arc Operations 359
15-14 Safety 360
15-15 Operating Procedures 360

CHAPTER 16
Welding with Stick Electrodes 364
16-1 What Is SMAW? 364
16-2 SMAW Electrodes 366
16-3 AWS SMAW Electrodes 368
16-4 Mild-Steel Electrodes 371
16-5 Low-Alloy-Steel Electrodes 380
16-6 Stainless Steels 382
16-7 Copper-Alloy Electrodes 388
16-8 Nickel Electrodes 391
16-9 Aluminum and Aluminum-Alloy
Electrodes 391
16-10 Welding with Stick Electrodes 392
16-11 Welding Flat 394
16-12 Horizontal Welding 397
16-13 Vertical Welding 399
16-14 Overhead Welds 402
16-15 Solving Problems with SMAW 404
16-16 Preheating and Postheating 406
16-17 Pipe Welding 411

Shop Practice: Welding Steel
with Stick Electrodes 415

CHAPTER 17
MIG and Cored-Wire Welding 420
17-1 The GMAW Processes 420
17-2 Moving Metal Through an Arc 423
17-3 Dip-Transfer Welding 424
17-4 Straight CO_2-Shielded Buried-Arc
Welding 427
17-5 Globular and Spray Transfer 427
17-6 Spray-Transfer Welding 430
17-7 Pulsed-Spray Welding 431
17-8 Narrow-Gap Welding 432
17-9 Flux-Cored Arc Welding (FCAW) 432
17-10 GMAW Power Sources 434
17-11 Other GMAW Equipment 438
17-12 GMAW Wire 442
17-13 Shielding Gases 445
17-14 GMAW Techniques 454

Shop Practice: Becoming Skilled
at GMAW 458

CHAPTER 18

Submerged-Arc Welding 462
18-1 The SAW Process 462
18-2 SAW Equipment 463
18-3 Weld-Metal Backing 469
18-4 Preparing to Weld 472
18-5 Types of SAW Joints 473
18-6 Positioning Your Work 478
18-7 Starting the Welding Arc 480
18-8 Variables Affecting Weld Quality 483
18-9 Multipass Welding 485
Shop Practice: Welding with
Submerged Arc 493

CHAPTER 19

Applying Metal Surfaces 494
19-1 Hardfacing Terminology 494
19-2 Which Weld-Surfacing Process? 503
19-3 Filler-Metal Forms 503
19-4 Using Stick Electrodes 503
19-5 Surfacing with GMAW, GTAW,
and SAW 508
19-6 Spraying Molten Metal 508
19-7 Applications and Service Conditions 517
Shop Practice: Learning
Cladding, Hardfacing, and
Buildup 524

PART 4

SKILLS FOR ALL PROCESSES 525

CHAPTER 20

Welding Defects and NDT 526
20-1 Welding Defects 526
20-2 Proof Tests 537
20-3 Leak Testing 538
20-4 Nondestructive Testing 539
Shop Practice: Developing an
Eye for Defects 547

CHAPTER 21

Reading Blueprints 548
21-1 Why Welding Symbols are Used 548
21-2 Taking Symbols Apart 549
21-3 Putting the Symbols Back Together 555
21-4 Groove Welds 558
21-5 Plug and Slot Welds 560
21-6 Flange Welds 563
21-7 Melt-Through and Weld-
Backing Symbols 564
21-8 Seam, Spot, and Projection Welds 566
21-9 Surfacing Symbols 569
21-10 Brazing Symbols 571
21-11 NDT Symbols 571
21-12 Summing Up 571

CHAPTER 22

Other Equipment Large and Small 577
22-1 Why Welding Positioners are Used 577
22-2 Welding Robots 589
22-3 Press-Brake Forming 596
22-4 Shearing and Slitting 598
22-5 Bending Rolls 598
22-6 Shot Peening 599
22-7 Determining Weld Temperature 600
22-8 Scales, Tapes, and Markers 601
22-9 Angle-Measuring Devices 602
22-10 Hacksaws 602
22-11 Chisels and Chipping Hammers 603
22-12 Grinding Wheels and Discs 604

CHAPTER 23

Welding Codes and Getting Certified 609
23-1 The Perfect Welder Doesn't Exist 609
23-2 Rely on Your Welding School 612
23-3 Getting Certified 612
23-4 Jobs Without Certification 614
23-5 The ASME Code 614
23-6 Qualification Tests 617
23-7 What's It Worth to You? 627

APPENDIX 630

INDEX 637

Preface

I wrote this book for many reasons. Two of the most important are to help train you as quickly as possible to become a skilled welder and to give you enough facts so you can quickly learn new welding and cutting procedures and solve problems on your first welding job. That often means reading other books, welding specifications, magazine articles and welding procedures, and this book gives you the background to do that. The words and ideas you will learn from this book will open up a huge resource of other people's experience which you will be able to understand and use. Several of the chapters in this book and the Appendix include references to some of these other resources, and I have listed them just when and where you are most likely to need them in the future.

As much as you want to light up a torch the first day you open this book, we don't want you to. Once you have learned what torches are, how to select and use the tips, hoses, regulators, cylinders, and all the other things that go with them, including oxygen and fuel gases, and even the kind of clothes you should wear, you will be ready for your first hands-on work.

The first chapter of this book is about welding jobs. The last chapter is about getting welding jobs. Everything in between is about learning the skills needed to land a job as a welder. Review Questions follow nearly every chapter. You can use these to review some of the highlights of the chapter or to test newly acquired knowledge

without the aid of the instructor. Wherever they apply, Shop Practice sections are included at the end of chapters. These sections give detailed step-by-step instructions on various kinds of shop work.

The chapters in this book are organized in such a way that you can progressively develop your manual skills. For example, a discussion of mechanical properties of metals, often left out of other welding textbooks, is presented in Chap. 8. This prepares you to read the rest of the book, because you can't select filler metals without understanding mechanical properties. In the same way, oxyfuel welding techniques are introduced (in Chap. 9) prior to a discussion of electric-arc welding. The oxyfuel welding skills that you develop will make electric-arc welding much easier to do.

Safety is a subject we do not let you forget. Chapter 3 on oxyfuel safety begins with a discussion of safety procedures, and this subject is repeated throughout the book in various ways. Special symbols have been used to call out specific safety procedures. A symbol ⚡ followed by the word *caution* means the possibility of bodily injury. The word *careful* preceded by ⚡ means the possibility of equipment, tool, or material damage. Safety is your responsibility and ultimately yours alone.

The latest technology is covered extensively at the end of the book. There you will find discussions of laser and electron-beam welding; welding manipulators, positioners, and turning rolls; temperature-in-

dicating crayons; and pneumatic (air-driven) grinders and chipping hammers. Various types of welding robots are illustrated in Chap. 22, because by that point you should be skilled enough in welding to be able to program welding robots to weld.

Like the Chinese who knew that a journey of 1000 miles begins with a single step, take your first step and turn to the first page of Chap. 1.

I owe deep thanks to Walter Goerg, Communications Director of Airco Inc.; Gene Wolfe, Senior Editor of *Plant Engineering;* Bob Irving, Technical Editor of *Iron Age,* and Herb Hinkle, publisher; and Rosalie Brosalow, chief editor of *Welding Design and Fabrication.*

I owe special thanks to the staff and committee members of the American Welding Society who made their enormous resources, operating data, and literature available to me.

I owe the most to my wife, Renée, who gave invaluable support during the three years while I researched and wrote this book.

Finally, my thanks to every skilled welder, burner, and welding instructor who stopped work to talk to someone who knew so much less than they. I have tried to put down all I could remember of what they told me because these people gained their knowledge the hardest way of all . . . by learning their skills at the side of other men and women who, in turn, teach each new generation.

Dave Smith

PART 1

Introduction

The following two chapters will introduce you to welding. Of the two, the first chapter is by far the most important. It tells you where the welding jobs are most likely to be, what industries hire the largest number of skilled welders, and what industries the U.S. Government projects will need the most welders over the next few years.

Not counting recessions (when skilled people in almost any trade have trouble finding work), there is a long-term and growing demand for skilled welding labor, as will be shown in this chapter. What's more, you will learn in Chap. 1 that once you have basic welding skills, you can branch out into many other fields including welding sales, inspection, supervision, or welding engineering. The type of work welders do in different industries is also described in this chapter.

Chapter 2 will introduce you to welding technology. You will soon learn that there are almost 100 different welding processes, and that the world's best welding engineers have never learned them all. Welding is simply too big a subject for one person to ever completely master. But in Chap. 2 you will learn about the key welding processes in which the great majority of skilled welders are employed. Together, these two chapters put the rest of this textbook in perspective.

CHAPTER 1

Why Welding Is a Good Trade

Skilled machinists are fast being replaced by punched paper tape controllers and computers. Many tool- and diemakers are dependent upon the ups and downs of the automotive industry for their jobs. Residential construction workers' jobs are tied to the interest rates on home mortgages, inflation, and the weather. At present, there are more long-haul truck drivers than there are trucks to haul.

But the outlook for skilled welders looks good through the year 2000 and probably will remain so through the beginning of the twenty-first century. Beyond that nobody can make a good forecast about any kind of work. But if you want to learn a highly skilled trade that gives you the freedom to work where you want to, to advance or go somewhere else when you want to, to make significant products, to earn the respect of other workers around you, and to earn good pay, too, learn welding.

1-1
PROFILE OF A WELDER

Welders are the last of the great industrial craftspeople— the men and women who work with their heads as well as their hands to create useful products. Unlike the average assembly-line worker who does the same thing day in and day out, welders have to think everything through and their work varies enormously.

The objects welders make are important, often dramatic, and seldom dull. A welder has skills to sell, and industry wants to buy them. A good welder also cares about the quality of the work done in return for wages earned. You answer first to yourself, and only then to a company manager if a job is wrong and a weld is no good.

There are three reasons that welders care about doing quality work. First, somebody's life may depend upon the weld produced. Second, a single bad weld may cost from $400 to $4000 to repair (that includes the original weld-

ing costs, testing, weld-metal removal, rewelding, and retesting). The third reason is simple pride in doing good work.

Table 1-1 lists typical products made by welders. Every one of these products is essentially a handcrafted item in terms of welding. And the welder responsible for each product is a lot more than a machine operator. He or she puts pride into each weld. Some welders even sign their own names to the paperwork that goes out with any big job. Some welders leave their names behind inside the job itself, because each job is a personal achievement.

Welding is not always fun. It can be very hard work. It can be hot, tiring, sometimes dirty and it can be dangerous if you are not careful to obey all safety rules. Neverthless, according to the National Safety Council, the injury rate for welders on the job is far lower than that of their families at home, because good welders work safely. If they don't, they don't stay employed very long no matter how severe the welder shortage may be, or how good they think they are.

Job satisfaction for welders is higher than for most industrial or clerical workers. The welder knows more than the average worker, and is better paid for it. And a good welder can continue what he or she does for years. In addition, there are chances to move into many other kinds of work, including management and sales.

If you prefer your own companionship and you would rather deal with things than with people, welding is still a good job because you work in a world behind that mask that is all your own. Pride in what you are doing is an important part of that world.

Walk through any plant that uses welders and ask other plant workers about the welders that work with them. You'll soon find out that other people think welders are a breed apart. They're right. Welders are respected for what they do and what they know. A good welder is part electrician, metallurgist, chemist, physicist, and design and mechanical engineer. But mostly a welder is a skilled craftsperson who has spent several thousand hours just learning the basics of his or her craft. The great welders spend a lot more time than that.

It takes several years in Local 798 of the United Association of Journeymen and Apprentices of the Plumbing and Pipefitting Industry of the United States and Canada, in Tulsa, Oklahoma, to make the journeyman level. You'll understand why if you know that labor union by its common name, *the pipeliners*. They built the Alaska pipeline literally by hand in summer and winter, in extraordinary weather conditions, hauling pipe across ice bridges on the frozen Yukon River and fighting mosquitoes in the mud under a preheated pipe joint while lying on their backs with hot sparks falling on them. That's not normal for anybody . . . except the pipeliners. And knowing that every inch of what you weld will be visually inspected, then x-rayed, and torn out if not good enough, is not normal work for most welders or anybody else. For pipeliners, it's routine (Fig. 1-1).

What's not routine is the money they made on the first big Alaska pipeline job. Some pipeliners say they made over $80,000 a year with overtime, working up to 3 weeks straight without a day off (and nowhere to spend the money if they did take off). But pipeliners, and a small band of boilermakers who produce welds that must meet nuclear pressure-vessel power-plant codes, are not just skilled welders—they are the great ones. You won't become a great welder in a year or two or five, even if you have the right skills and intelligence. It takes years of work and study to be very good. But someday you might be one of the great ones yourself.

The first step in that direction is simply to make it through your first welding course. If you do that, work hard enough, and can prove that you have learned the

TABLE 1-1 Just a few of the things America wouldn't have if America didn't have welders

air compressors	cultivators	industrial process furnaces and ovens	pressure and nonpressure tanks
air-conditioning units	dairy products machinery	industrial scales	printing trades machinery
air purifiers and dust collectors	dehydrating equipment	internal combustion engines	pulverizers
airplanes	derricks	jet propulsion engines	radiators
ambulances	desalination equipment	leatherworking machinery	radio and television equipment
anchors	diesel engines	lift trucks	railroad tracks
anvils	docks	locomotives and cars	railway motor cars
arbor presses	electrostatic precipitators	locomotive wheels	rapid-transit cars
armored cars	elevators	marine engines	reconnaissance cars
automobiles	excavators	meat grinders	refrigeration machinery
backhoes	exhaust fans and air-moving equipment	military carriers	rocket casings
bakery machinery	extruding machines	mining machinery	rolling mill machinery
barges	fans	mobile lounges	safes and vault doors
bending and forming machines	farm buildings	monorail systems	septic tanks
blowers	farm machinery	motor buses	sewage purification equipment
bookmaking machines	fences	motor homes	smokestacks and liners
bottling machinery	ferris wheels	motor truck scales	snowblowers
breadslicing and wrapping machines	ferryboats	motor vehicle chassis	snowmobiles
brewers' and malters' machinery	filters	motor vehicles	snow plows
brickmaking machinery	fire escapes	motors and generators	space simulation chambers
bridges	fire department vehicles	motorcycles, bicycles	specialty transformers
buildings	fireboats	nuclear reactors	spreaders
canning and packing machinery	flagpoles	oil-field machinery	steam turbines
car-washing machinery	floating drilling platforms	outboard motors	stills
carousels (merry-go-rounds)	forging machines	paint-making machinery	street sprinklers and sweepers
carports	foundry machinery	paper mill machinery	switchgear and switchboard apparatus
cement-making machinery	furnace blowers	patrol cars	textile machinery
coal chutes	furnaces	pavers	tractors
commercial laundry, dry cleaning, and pressing machines	gas turbines	petroleum refineries and equipment	truck trailers
compactors	gears	personnel carriers	tugboats
conveyor belts	glass-making machinery	piledrivers	ventilators
cotton-ginning machines	grocery carts	piping and valves	washing machines
crankshafts	gutters and downspouts	pharmaceutical machinery	wheelbarrows
crushers	hearses	plastics-working machinery	woodworking machinery
	helicopters	plows	
	hoists	power and marine boilers	
	hoppers and chutes	power cranes	
	house trailers	power distribution and specialty transformers	
	hydrofoil vessels		

FIGURE 1-1 The last weld goes in place on the first trans-Alaska pipeline from above the Arctic Circle in Prudoe Bay to the all-weather port of Valdez in Southern Alaska. Pipeline welders travel the world and make more money than they can spend, but few people ever develop the skills to qualify as one of these super-welders.

basic skills, there's probably a job waiting for you outside. One reason is that our nation faces a serious welder shortage that will last for decades.

1-2
THE WELDER SHORTAGE

The United States is rapidly running out of skilled welders and flame cutters (who are called "burners" in the trade). Meanwhile machines (including industrial robots) are taking over other production jobs. Robots are replacing a few welders, too, but very few and only on jobs that are relatively simple, repetitive, and boring. Even then, skilled welders are required to set up these robots so they won't make more mess than metal. The U.S. Department of Labor, Bureau of Labor Statis-

tics, estimates that the welder shortage will probably last for the rest of your career.

According to the labor bureau, American industry needs about 833,000 welders and flame cutters in 1985. That is 55 percent more welders than industry needed in 1970, and 22 percent more than were needed in 1976. Industry can't find them or train them fast enough to fill the demand. Table 1-2 gives you some details.

The Bureau of Labor Statistics and the publisher of the trade magazine *Welding Design & Fabrication* also estimate that U.S. industry must train almost 20,000 new welders and flame cutters every year to meet demand for new welders through 1985. After that, predictions are that industry will need 35,000 new welders each

year, every year into the twenty-first century.

Neither the 20,000 welders-per-year nor the 35,000 welders-per-year figure includes replacing welders and burners that have moved to other kinds of work ranging from welding engineering and production superintendents to welding equipment salespeople. Nor do the figures include jobs lost due to men and women who simply retire and leave their positions open.

These figures also do not include about 1 million people (maybe a lot more) that do some welding and flame-cutting as part of another job or special interest. Examples are steamfitters, plumbers, plant maintenance people, even tool- and diemakers, some machinists, and many mechanics. A great deal of welding also is done on farms.

The total of 1 million additional people who may do welding but aren't called welders does not count a growing number of artists, sculptors, home handicrafters, auto customizers, and racing-car builders who do their own work with welding and flame cutting. The government statistical people would never count somebody who builds his or her own wrought-iron fence and customizes cars as a welder. They're not. But these hobbyists have a lot of fun doing what they can with welding, and some of them make extra money doing it.

There is another aspect of the welder shortage that you should think about. Looking at Table 1-2 more closely, you'll see that the very smallest overall increase in welder demand over the 15 years from 1970 to 1985 is in New York State. But that state alone needs at least 4000 more welders over those 15 years. The highest welder demand is in Wyoming where the need for welders is expected to jump 250 percent by 1985. And that estimate was made before big oil exploration action began in the overthrust belt in Wyoming. Of course, there were not very many welders to start with in Wyoming, and even after adding 250 percent more there still won't be a crowd of welders there.

If you like the sun-belt states, look at the number of welders in Texas. Texas industry will need 88 percent more welders in 1985 than it needed in 1970. Or try Arizona. That desert and sun state must double its number of welders by 1985 just to stay even with the demand in 1970.

You want to work in Hawaii? The islands will need 93 percent more welders in 1985 than they needed in 1970. One of the world's largest welding distributor companies is located there, too. But if you prefer the deep south, look at welder demand in Mississippi or Louisiana. They need 145 percent and 85 percent more welders in 1985 than they had in 1970.

Look over the entire state-by-state list in Table 1-2. There is not one state that will need fewer welders. The average for all states is a 55 percent increase in welder demand by 1985. The growth rate of the U.S. population for the same period is expected to be one-fourth to one-fifth the growth rate in welder demand every year from now on, discounting an occasional recession.

Welding has one more advantage over other skilled trades. When business is good, welders can't keep up with demand. Overtime pay is not unusual. When business is bad, industry tends to repair the equipment it has instead of buying new equipment. Welders can be kept pretty busy maintaining production equipment and plants until better times come back. As a result, welders, overall, tend to be less affected by swings in the business cycle and by layoffs than other workers.

Of course, there are no guarantees for anybody, including management, but welding tends to be more recession resistant (if not recession-proof) than most kinds of work.

TABLE 1-2 Where welders and flame-cutters work

State	Number in 1985	Number in 1970	Increase from 1970–1985	
			Number	Percent
Alabama	18,372	11,358	7,014	61.8
Alaska	1,142	383	759	198.2
Arizona	8,167	3,813	4,354	114.2
Arkansas	9,744	4,773	4,971	104.1
California	61,833	41,231	20,602	50.0
Colorado	9,726	4,215	5,511	130.7
Connecticut	8,147	5,732	2,415	42.1
Delaware	2,082	1,277	805	63.0
District of Columbia	455	212	243	114.6
Florida	22,775	10,903	11,872	108.9
Georgia	15,935	10,233	5,702	55.7
Hawaii	3,319	1,722	1,597	92.7
Idaho	3,788	1,543	2,245	145.5
Illinois	46,225	34,651	11,574	33.4
Indiana	31,641	22,176	9,465	42.7
Iowa	17,404	9,371	8,033	85.7
Kansas	14,867	6,532	8,335	127.6
Kentucky	12,861	9,571	3,290	34.4
Louisiana	27,039	14,627	12,412	84.9
Maine	3,114	2,095	1,019	48.6
Maryland	9,894	7,756	2,138	27.6
Massachusetts	10,785	9,560	1,225	12.8
Michigan	44,804	37,303	7,501	20.1
Minnesota	14,685	9,498	5,187	54.6
Mississippi	17,054	6,966	10,088	144.8
Missouri	15,915	12,757	3,158	24.8
Montana	1,770	1,151	619	53.8
Nebraska	6,207	3,447	2,760	80.1
Nevada	1,789	754	1,035	137.3
New Hampshire	1,693	1,142	551	48.2
New Jersey	19,528	13,424	6,104	45.5
New Mexico	5,372	1,851	3,521	190.2
New York	29,736	25,723	4,013	15.6
North Carolina	17,014	8,608	8,406	97.7
North Dakota	1,448	741	707	95.4
Ohio	55,188	41,812	13,376	32.0
Oklahoma	16,784	9,249	7,535	81.5
Oregon	10,390	5,442	4,948	90.9
Pennsylvania	54,362	43,895	10,467	23.8
Rhode Island	3,340	1,600	1,740	108.8
South Carolina	12,341	6,052	6,289	103.9
South Dakota	1,653	1,050	603	57.4
Tennessee	22,175	10,941	11,234	102.7
Texas	72,832	38,715	34,117	88.1
Utah	6,370	2,517	3,853	153.1
Vermont	548	383	165	43.1
Virginia	15,749	9,000	6,749	75.0
Washington	11,810	7,460	4,350	58.3
West Virginia	7,754	5,229	2,525	48.3
Wisconsin	22,545	16,953	5,592	33.0
Wyoming	2,826	802	2,024	252.4
Total	832,997	538,199	294,798	Average: 54.8%

Source: U.S. Bureau of Labor Statistics, unpublished data.
Courtesy *Welding Design and Fabrication.*

1-3
WHERE TO FIND WELDING JOBS

Some industries employ many more welders than others, even though you can find welding work almost anywhere. Table 1-3 shows you where the largest number of welders' jobs were when this book was printed, and that is a clue to where the new jobs will be when you are ready to go find one.

Out of the 36 categories, only 6 industries hire over three-quarters of all the welders and flame cutters (let's call them burners). The table, compiled by *Welding Design & Fabrication* magazine, also lists the number of welders and burners working in those industries in 1970 versus 1985 expectations.

Table 1-3 also shows you whether welder demand probably will be up or down in specific industries, and most importantly, that overall demand for welders between 1970 and 1985 will be up over 50 percent. It also shows that only 5 industries will need fewer welders, while 31 will need more.

Let's look at the first eight industries that hire well over 80 percent of all the welders and burners and see what those industries do.

Fabricated Metal Products

More than 17,100 plants in the United States make things out of ferrous metals (irons and steels) and nonferrous metals (aluminum, copper, zinc, magnesium, titanium and other metals and alloys not based primarily on iron). Even when industrial machinery and electrical and transportation machinery and equipment are not included in this category (we'll describe them separately), you can see from Table

TABLE 1-3 Where the jobs will be

Industry	1985	% All jobs, 1985	1970	% Change, 1970–1985
1. Fabricated metal products	151,049	18	79,826	+ 89.3
2. Construction	133,074	16	60,135	+ 121.3
3. General industrial machinery	129,332	15	77,804	+ 66.3
4. Transportation equipment	108,959	13	97,053	+ 12.3
5. Primary metals	49,256	6.0	36,121	+ 36.4
6. Repair services	46,198	5.5	35,322	+ 30.8
7. Electrical and electronic equipment	32,380	4.0	30,415	+ 6.5
8. Wholesale trade	29,525	3.5	14,437	+ 104.5
(subtotal)		(81%)		
9. Mining	22,484	2.7	11,807	+ 90.5
10. Wood and forest products	20,019	2.4	9,737	+ 105.6
11. Electric, gas, and sanitary services	13,476	1.6	9,598	+ 40.5
12. Stone, clay, cement, and glass products	13,368	1.6	7,340	+ 82.2
13. Instruments and miscellaneous manufacturing	12,407	1.5	9,339	+ 32.9
14. Chemical and related products	10,341	1.2	6,932	+ 49.2
15. Public administration (government)	8,746	1.1	5,824	+ 50.2
16. Food and related products	7,363	0.9	3,799	+ 93.9
17. Retail trade (e.g., welding distributors)	6,943	0.8	5,438	+ 27.7
18. Railroad transportation	6,344	0.8	9,569	− 33.8
19. Pulp and paper mills	4,252	0.5	3,158	+ 34.7
20. Rubber and plastic products	3,821	0.5	1,930	+ 98.0
21. Transportation services, except trucks and RR's	3,737	0.5	2,609	+ 43.3
22. Petroleum and coal products, other than mining	3,329	0.4	5,612	− 40.7
23. Trucking and warehousing	2,336	0.3	1,684	+ 38.8
24. Business services	2,521	0.3	1,582	+ 59.4
25. Agriculture, forestry, fishing	2,465	0.3	2,216	+ 11.2
26. Other services not listed separately	1,931	0.2	1,479	+ 30.6
27. Textile mill products	1,833	0.2	1,182	+ 55.1
28. Educational services	1,709	0.2	1,571	+ 8.8
29. Apparel and other textile products	749	0.1	334	+ 124.3
30. Entertainment and recreation services	714	0.1	409	+ 152.3
31. Health services	663	0.1	409	+ 62.2
32. Finance, insurance, real estate	472	0.1	440	+ 7.3
33. Printing and publishing	457	0.1	1,165	− 60.8
34. Tobacco and leather products	335	—	624	− 46.4
35. Communications	265	—	163	+ 57.1
36. Other professional and related services	153	—	1,262	− 87.9
Total	832,997	100%	538,199	+ 54.8 average

Source: U.S. Department of Labor, Bureau of Labor Statistics.
Courtesy *Welding Design & Fabrication.*

1-3 that there are a lot of welding jobs in fabricated metal products plants.

Of the 17,000 fabricated metal products plants, 4240 or 25 percent of them are considered to be large plants with 100 or more people working in them. Some of the biggest of these plants hire thousands or even tens of thousands of workers.

Plants that make fabricated metal products fall into roughly 42 product categories ranging from plants producing metal cans to those making hacksaw blades. In between these extremes (neither of which need many welders) are the big plants that produce heating equipment, including industrial furnaces, pipe fittings, fabricated pipe and pipe sections, and the very important category of fabricated structural metal products.

Structural metal products

Fabricated structural metal products include fabricated sections for bridges and buildings, structural components for ships, barges, boats, transmission lines, and TV towers; and dams and the huge water pipes that go with them, called *penstocks*. Fabricated structural metal products also include everything made by boiler and pressure-vessel shops. *Shops* is an understatement for many of these very large plants, but that's what most people call them. (See Fig. 1-2.)

Boiler and pressure-vessel fabricators make everything from gas storage tanks under a filling station to nuclear-reactor containment vessels for atomic power plants. More welders work in this one segment of industry than in any other except construction. They build heat exchangers; farm storage tanks and silos; big oil and gas tanks; water towers for cities; blast furnaces for steel mills; pressure vessels for liquefied oxygen, nitrogen, and natural gas; missile silos; petroleum refinery and chemical-distillation columns; steel smokestacks; foundry

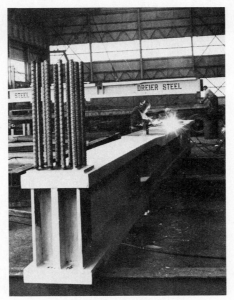

FIGURE 1-2 This special beam is being prefabricated in a structural steel shop. It will soon be shipped to the construction site where it will span a wide space over a well-known New York restaurant. The long beam eliminates columns in the dining area so that more people can be served.

FIGURE 1-3 Ornamental iron is an industry that can be a hobby too. After you complete this course, you can get the equipment you need from almost any local welding distributor to make your own fancy balconies, gates, chairs and tables, plant hangers, room dividers, and barbecue racks. Some people even go on to become metal sculptors.

furnaces called *cupolas;* welded crane hooks; the steel forms used for tunnel liners, steel mill ladles; and hundreds of other large fabricated metal products.

Pipe fittings and valves Closely related to the fabricated structural metal industries are plants that specialize in manufacturing pipe fittings and valves. These are not the kind of pipes and valves you have in your sink. They can easily be taller than you are. They are industrial hydraulic and pressure-vessel valves that are used in the oil, gas, and chemical industries; in large water mains; and in some large hydraulically operated machinery and construction equipment. Even valves that initially are produced as castings usually require some welding and flame cutting for completion.

Architectural and ornamental ironwork An interesting group of smaller companies in fabricated metal products makes architectural and ornamental ironwork (Fig. 1-3). These products include fire escapes, flagpoles, open grillwork for factory flooring and stairs, lamp posts, elevator guide rails, and stationary steel ladders of the kind that rise up the side of storage tanks. They also make the fancy "wrought iron" railings (actually steel and not wrought iron) for porches and balconies.

Prefabricated structures Prefabricated metal buildings also belong to fabricated metal-products people, before construction workers finally erect them. Prefabricated metal buildings are produced for everything from storage sheds to good-sized manufacturing plants and for farm buildings. Many large shopping centers also are prefabricated buildings.

Many of the modern prefabricated buildings win architectural awards of excellence. These are no longer the tin shacks and quonset huts of World War II that looked like half-round silos that fell over. Welded steel is sometimes even used to replace the common two-by-four as the price of wood rises faster than the price of steel. Some better homes now have structural members of welded steel in them.

Construction Industries

There are at least 28 separate and clearly defined types of construction contractors (although many of them work in more than one kind of construction). These diverse groups of contractors range from people working on single-family homes to specialists at bridge, tunnel, and underwater or high-pressure caisson work (they call themselves *sand hogs*). All these companies can be grouped under three basic headings: general contractors working in building construction, general contractors working on everything else, and specialty trade contractors who usually subcontract work from the two kinds of general contractors.

Welders have the least chance of finding jobs with building contractors, especially those working on single-family homes and other structures such as motels, hotels, and dormitories. There are many more welding jobs available with contractors working on large apartment buildings and industrial buildings and plants, warehouses, and nonresidential structures such as high-rise office buildings.

Heavy construction The largest number of welders and burners working in construction work for contractors who specialize in highways and highway bridges, tunnels, and elevated roadways and for firms specializing in water, sewer, pipeline, communications tower, and power-line construction (Fig. 1-4).

Another important group of contractors who hire a lot of welders are those working on heavy construction jobs. Heavy construction is generally defined as anything from dam construction to hydroelectric plants, blast furnace builders, missile silos, drydocks, canal construction, mining and

FIGURE 1-4 A massive curved beam was discovered to have a blow hole in the steel when it was flame-cut. This construction welder, using stick electrodes, is repairing the defect so that the project can continue.

mine development and facilities contractors, and builders of oil refineries and chemical and power plants. Almost any company that makes things that are big, long-term, and that makes heavy use of metals and other structural materials is a prospect for welding jobs.

Specialty contractors The third category of contractor, the specialty contractor, usually subcontracts work from the general contractor. These firms include everything from people who work in plumbing, heating and air con-

ditioning trades (who need welding skills), to painters, paperhangers, and decorators (who don't).

The specialty contractors most likely to have jobs for welders and flame-cutters are the industrial plumbing and heating contractors (often called *steamfitters*) and the specialty contractors who work with sheet metals, including those who work with ductwork, metal floors, and steel ceilings and siding, tinsmiths, metal roofing firms, and preengineered building erectors. The final major category of specialty contractors of interest to welders is the structural-steel erectors. They are something special.

The ironworkers Most people who work as welders for structural-steel erectors are called *ironworkers*. Their work ranges from placing concrete reinforcing bars to field-erecting water storage tanks. A big part of that work is welding structural steel in place on high-rise buildings. The famous Mohawk Indian ironworkers of New York City, who are specialists at steel erection on things like the 110-story twin towers of the World Trade Center, subcontract this work from general contractors.

While construction work offers a welder good to excellent pay and a chance to work outdoors, it also has (depending upon your point of view) some disadvantages, like being 110 stories up on open steelwork. Some people like ironwork, either for the money (which can be substantial) or for the "challenge" (read as overcoming a fear of heights if you like). If it's any consolation, most ironworkers say that beginners literally work their way up from the ground level and that they soon get used to it (Fig. 1-5). But some people don't even want to watch Mohawk Indians work on high-rise structurals, let alone do what they do. So if you consider construction, consider first whether you will want to work off the ground or not.

Construction, incidentally, has become far safer to work in than it was many years ago. In fact, all large construction firms have full-time safety directors, and the smaller companies have at least one person watching safety as part of the job. All firms, with very few exceptions, have safety regulations that you must follow if you want to stay employed. Most of these regulations are set down by OSHA. OSHA stands for "Occupational Safety and Health Act." It's both a law established by Congress and a U.S. Government agency whose job is to make

FIGURE 1-5 Ironworkers combine welding and burning with other talents, only one of which is a good sense of balance.

FIGURE 1-6 A welder is depositing layers of wear-resistant manganese steel on the worn surface of construction equipment. The new metal surface work-hardens, actually getting harder and more wear-resistant the longer it is used. This work is being done with the gas-shielded continuous-wire process called MIG welding.

sure that workers work in safe conditions. OSHA regulations dominate construction industry work.

Another important job for both welders and burners in construction is maintenance welding (Fig. 1-6). Every bulldozer, drag line, power shovel, crane, scraper, off-highway truck, and backhoe takes a terrible beating. Large construction companies keep maintenance welders very busy repairing this equipment, usually in the field, because no contractor can afford to have idle equipment. There is a special chapter in this book on weld overlay and hardfacing techniques that are frequently used to repair the cutting edges of construction equipment (from bulldozer blades to bucket teeth) right where the construction work is being done.

There is a final group of specialty contractors that a beginning welder or burner should consider. They don't build things, they tear them down. If you like to handle a cutting torch and to smash things up, you may have fun working with the demolition contractors. Don't think that work is unskilled. One of the most important skills is to keep from getting hurt while doing your work. Some of the finest torch specialists outside of prison work for demolition companies.

General Industrial Machinery

This major group of companies with over 16,950 plants (of which 26 percent or 4,345, are firms with 100 or more employees) hire a substantial number of welders. Plants in this category produce engines and turbines, farm and garden machinery, construction and mining and materials-handling equipment, oil field machinery and equipment, metalworking machinery from heavy-duty lathes to forging presses, and special industrial machinery such as textile machines, woodworking machin-

ery, paper industry machinery, and printing presses (Fig. 1-7). Also included in the long list of general industrial machinery and equipment built by welders are pumps and air compressors, blowers and exhaust fans, and industrial process furnaces and ovens.

Some of the firms in this category, especially those making farm and construction equipment, know more about welding and flame cutting as production tools (and about welders as productive people) than any other large industrial corporations. One of the firms, Caterpillar Tractor, in Peoria, Illinois, probably has more full-time welding engineers on its staff than any other company in the United States.

Welding is a critical part of heavy construction equipment, farm equipment, and mining machinery. Skilled welders, who want to work with the best-quality welding equipment available, will

FIGURE 1-7 Part of a new materials-handling device requires special modifications during installation in a user's plant. This stick-electrode welder was called in to make modifications on-site, as part of the equipment manufacturer's installation and customer-service team.

like working in these companies, and their work will be appreciated by management.

Transportation Equipment

This group of companies includes 4460 plants with 42 percent of them employing over 100 workers (actually several of these companies employ tens of thousands of workers). They include the major automotive producers, shipyards, aircraft, and truck- and bus-body builders (Fig. 1-8).

When people think of welding, they often think of automobile manufacturing. They are correct in that auto bodies are almost totally welded and brazed together. However, much of the sheet-metal work is spot welding, which usually is a machine operation requiring only an operator to set up and run the equipment. Many of the automotive spot-welding lines are now fully automated and the work is done by computers or industrial robots. Nevertheless, there are jobs for highly skilled welders in the automotive industry. But most of the skilled jobs for welders will be found in transmission and axle departments, or with frame builders, and not with the body assembly plants.

On the other hand, truck and bus manufacturers who do not have to perform much spot welding, and build their products with much thicker sections than auto bodies, have a strong demand for highly skilled welders. Truck fabricators also require special welding skills for the vans or "boxes" as well as the truck cabs or dump bodies. A great deal of aluminum welding is required on most trucks, often on the rather thin-gauge sheets of the vans or boxes. Another important material in trucks is high-strength low-alloy steels. Welding these materials requires both special skills and specialized welding knowledge.

The same is true of aerospace fabrication, but even more so (Fig. 1-9). Aircraft, missiles, and space vehicles use a wide range of exotic and often difficult-to-weld met-

FIGURE 1-8 The aluminum boxes for new truck trailers will be given a final touchup by this MIG welder, who also is responsible for final inspection before the equipment is shipped to trucking companies.

FIGURE 1-9 TIG welding, plasma-arc welding and cutting, and other processes you will learn to use in this textbook are widely used in the aerospace industry.

als. Welders working in these industries have very interesting jobs.

Shipbuilding requires extensive welding fabrication, often on thick sections of steel or aluminum plate. Shipyards produce far more than ships, and not all of them are on the oceans or the Gulf Coast. In fact, the United States has more shipyards on its inland lakes and waterways and rivers than it has on its coasts. Follow any major river or the Great Lakes and you will find shipyards and boat yards looking for skilled welders.

Shipyards produce drydocks, drilling platforms for the oil exploration industry, floating radar towers, marine rigging, hydrofoil vessels, pontoons, fishing vessels, dories, tugboats, lighters (floating flatboats), ferries, and barges galore (you will find barge yards any time you get near the water). Shipyards produce other kinds of work boats as well as big freighters, passenger liners, naval ships, and tankers (Fig. 1-10).

Railroad shops are found in specific centers in the United States, such as around Bettendorf, Iowa; Omaha, Nebraska; the Chicago area; New York City; and Denver. All are principal rail centers. Car repair shops are found over much of the country, tucked away on sidings and on rail spurs, because the damage sustained by rolling stock is substantial. It is a constant maintenance problem requiring extensive flame cutting and welding. Welding jobs on the railroads also include repairing track, signals, signs, fences, and buildings.

Omaha, Nebraska, is both a major rail center and an important aerospace center, too. Airlines employ welders for maintenance work. They also either subcontract or build their own aluminum shipping containers which usually are welded. The major primary fabrication centers for the aerospace industry are in Seattle, Long Island near New York City, St. Louis, Denver, and

FIGURE 1-10 A large oil tanker under construction in a shipyard drydock. This tanker is slightly longer than New York City's Empire State Building is high. When the tanker is ready to go, the dry dock will be flooded with water and the tanker will float into the river. Shipyards often employ thousands of welders and burners.

Houston, with smaller firms of one kind or another located almost anywhere in the United States.

One final transportation products area that requires a lot of welding, or more specifically brazing, is motorcycle and bicycle manufacturing. There are not too many bicycle plants left in the United States, but if you ever visit one you will see more production brazing in one location than you might ever see anywhere else.

Primary Metals

Big plants dominate the industries that produce metals. Primary-metals plants produce iron and steel, aluminum and copper, and all the other metals that industry uses. There are 5115 primary-metals plants with 2030, or 40 percent of them having more than 100 employees. Some of these plants have 10,000 to 20,000 employees and operate on three shifts around the clock. However, minimills specializing in only one kind of product (such

as bar stock) are becoming more important.

There are five principal kinds of jobs for welders and burners in this area. Most are in plant and equipment maintenance. A large steel mill can have more employees than some small cities. They have their own railroads, truck fleets, buildings, and shops, and all these things must be kept in good working condition along with the production equipment. By the very nature of the processes used to make metals, the equipment takes a terrific beating. Welders are in constant demand to repair it.

The second type of work in both steel and aluminum mills (and some iron foundries) is flame cutting, or a similar arc cutting process called *air-carbon-arc gouging*. Many burners in mills and foundries cut up heavy scrap for use in the various kinds of furnaces that make iron that will later be processed into steel (Fig. 1-11). Flame-cutting is also needed to shape and square flat-rolled steel products such as plate

FIGURE 1-11 It takes a highly skilled torch operator to burn through a round section this thick. The worker, in a steel mill scrap yard, is preparing material to be fed to electric arc furnaces. The structure in the background is a blast furnace.

FIGURE 1-12 A burner with a hand-held cutting torch removes the damaged section of a locomotive in an Omaha railroad repair shop.

The third kind of work for welders and flame cutters in primary-metals companies is building very large structural metal parts from welded plate and structural shapes. Several large steel producers not only make the plate and the structurals that their customers need but also have separate companies, usually located close to the primary mill, that fabricate large construction equipment such as the bases of coal-stripping drag lines and the large hulls of dredges, and parts for large cranes and sections for bridges.

The fourth kind of work in iron and steel foundries involves more cutting than welding. When castings are poured and harden in molds, there is always some excess metal that must be removed. Sometimes that metal is too thick to be ground off the casting with a grinding belt or a wheel. It must be cut off with cutting torches, plasma torches, or by the air-carbon-arc cutting method.

The fifth and most interesting kind of work for welders and burners is customer service and work in research and development (R&D) labs. These jobs also are the most difficult to get. Only the best welders get to work on customer problems or to help develop methods for fabricating new alloys. However, no metals-producing company can sell a new alloy without knowing how to fabricate it. Development welders work with the metallurgists who invent the new materials.

Repair Services

People and companies who repair things hire welders and burners to do the work. These establishments range from local shops handling general welding jobs and boiler-repair work to large industrial companies that handle special repair jobs on contract from other businesses.

Repair shops can range from small firms that fix bicycles, motorcycles, farm machinery, musical instruments, and gunsmith-

and sheet into forms acceptable to customers. Even structural steel and bar stock often are flame-cut to length as they come off the rolling lines. But not everything is flame-cut. Thinner sections are cut by shears (they work like huge mechanical scissors).

In aluminum mills, cut-off work is done either by mechanical shears or by a special electrical cutting process called *plasma-arc cutting*. Both steel and aluminum producers also cut finished parts for customers such as round disks cut from thick plate for use in forming pressure-vessel heads. This shape-cutting work keeps a lot of burners busy.

A related kind of mill work is removing as-rolled surface imperfections from semifinished products. Very large cutting torches with special attachments are used in a process called *scarfing* to remove these imperfections and cut the product down to good sound metal before it goes into the final finishing mills.

ing, to firms that repair ships in drydocks and marinas, and large industrial machinery either in the user's plant or their own (Fig. 1-12). But the shops probably most familiar to you are the ones that repair automobile bodies.

Auto and truck body repair shops make up a large group of companies that need people familiar with flame cutting and some kinds of welding. They also are more likely to hire beginning welders and burners than large industrial companies. There are four kinds of companies doing automotive repair work. The first kind of company is the neighborhood body repair shop. Similar work is done by new car dealers, but less frequently.

More active, and probably more interesting for the welder, are the farm and construction equipment dealers who (just like new-car dealers) provide extensive repair services for their customers. The fourth kind of automotive repair shop is the car or truck fleet shop. These are maintained by companies that have large fleets of vehicles and must keep them in good repair. Large trucking firms and big companies that deliver packages are only examples. A big bus company and a rental-car company are others.

The smaller repair shops (blacksmiths, local boiler repair, auto shops) have both advantages and disadvantages. They pay less than most large corporations, on the average. On the other hand, they are good places to get a starting job without top skills, and there is always the possibility that you will get good enough so that someday you can open your own repair service.

Some repair shops and services in special fields make a fortune for their owners. In oil fields, independent workers called *rig welders* own their own equipment and often are highly skilled welders who work on a day or hourly rate for whoever needs them. They load everything on a pick-up truck, for example, and do maintenance work for the oil drilling and extraction industry. They also may be on call 24 hours a day. But they make more money than most of the oil field workers they help.

Electrical Equipment Builders

Among the largest employers of welding and flame-cutting operators are very large companies that make heavy industrial transformers and generators, heat exchangers for power plants, and big motors. There are many excellent welding jobs in these companies, if you develop top skills (Fig. 1-13).

Welders often have to take special welding tests called *qualification tests* (more about them later) to prove what they can do. Welder qualification tests almost always are required in companies that make equipment for electrical power plants (and the contractors who build them).

The electrical products industries also include appliance builders (who often are different divisions of the same large electrical apparatus companies mentioned above), and even the manufacturers of welding equipment. However, most large appliance builders, such as companies who make refrigerators, freezers, and stoves, generally use brazing or spot welding. This work is done by automated equipment. Skilled welders are more likely to find produc-

tion jobs in the heavy equipment end of the electrical equipment business.

Wholesale Trade

Wholesale trade seems like an odd category for welders and burners to be working in. However, one part of this business category includes metal service centers (Fig. 1-14). To give you an idea of their importance, metal service centers buy more steel from steel mills than the entire automotive industry. They process and warehouse metals (especially steel, aluminum, and copper) for anybody who cannot buy in mill-size quantities, or who needs fast deliveries or special processing services. Even steel mills get deliveries from metal service centers when they need special metal shapes for maintenance or repair work. Special processing is a very important part of the metal service center business.

For example, one steel service center in Seattle does more than warehouse steel plates for resale to a furniture manufacturer; it also fabricates the steel into table legs by flame-cutting the parts on very large and efficient 10-torch flame-cutting machines. When the furniture manufacturing company wants steel, the company buys precut table legs made at far lower cost than the furniture company could do by itself. That's a modern way to warehouse metals until you need them. Buy metals almost ready to use, and only when you need them. Preprocessing benefits the furniture company and the extra sales benefit the metal service center.

Therefore metal service centers are heavily service oriented and often perform initial fabricating work as well as warehousing metals. You may be surprised to learn that metal service centers own more large flame-cutting machines than shipyards or contruction equipment makers. There are 2000 metal service centers in the United States. Some specialize in steel or aluminum or copper.

FIGURE 1-13 Sections for large electrical apparatus are being cut by a numerically controlled cutting machine. The operator, in the background, monitors the controls.

(A)

(B)

FIGURE 1-14 (A) Metal service centers often cut parts to size for customers, as well as simply stocking steel, aluminum, copper, and other materials. (B) Various standard shapes are dialed into the computer control on this cutting machine, and the four (or more) torches produce circles, rectangles, squares, or triangles of any size. More complex, custom shapes are produced by an electric eye that follows a paper and ink drawing of the desired part.

Some specialize in plate or structural mill forms, or flat-rolled sheets and strip, or pipe and tubing. The ones to look for when you look for a job are the specialists in steel and aluminum, preferably those working with plate and structurals or pipe and tubing because these materials are most often flame-cut.

Now that you know a little more about where welders and burners work, what they do, and why they do it, let's look at the unions that some of them join.

1-4
WELDERS AND LABOR UNIONS

There is no special labor union for welders. Nor do all welders or burners join unions, because welding and cutting are tools of many different trades. But those that do join a union are likely to join any 1 of 20 to 30 different unions that have substantial numbers of welders or burners as members.

Table 1-4 is a partial list of unions with signficant percentages of skilled welders and burners in their membership. Many of these unions actively sponsor welder training programs after the student has developed the basic skills needed to meet apprentice requirements. These union training programs can be demanding if you don't learn your basic work now. Unions are intolerant of people who don't "do their homework" and then expect to get a union job. There usually are enough people trying to enter a union that the union stewards can pick and choose between really motivated people and drop those who won't make the grade on the job.

One of the toughest training programs of all is that of the pipeliner's union, whose Tulsa local we've already mentioned. It takes 7 years of solid, hard work to become a journeyman pipeliner, about as long as it takes for a doctor to learn his or her profession. A master pipeliner has worked at it even longer. That's one of the reasons why the pipeliners are some of the best welders in the world.

1-5
WELDER QUALIFICATION TESTS

You will be asked to perform and pass special welding qualification tests as part of getting most welding jobs. You will find shop exercises at the end of many of the chapters in this book. They are there to help you learn how to do what you've read about in each chapter. You can never learn to weld just by reading this book. You have to practice hard until you are able to produce good welds or do superior cutting work. Those exercises will help you make the grade.

The exercises are there for another reason too. Where possible, each shop exercise section includes real examples of the kind of work you will be expected to do to pass a welder qualification test. You'll know before the end of this course whether you are going to make it as a skilled welder or

burner. If you can't do good shop work, you had better find another way of making a living.

1-6
YOUR FIRST LOOK AT METALLURGY

The primary-metals industries include steel and aluminum producers; iron and steel foundries; copper, zinc, and lead smelters; and plants that produce or refine primary antimony, beryllium, bismuth, cadmium, calcium, chromium, cobalt, columbium (also called niobium), copper, germanium, gold, iron, magnesium, manganese, molybdenum, nickel, silver, sodium, tantalum, tellurium, tin, titanium, tungsten, vanadium, uranium, and zirconium. We wrote all these metals' names for a reason. Many are the alloying elements used in weld metal as well as the base metals that you will join.

Some elements that you will find in the metals you cut or join are called *nonmetallic elements*. Three very important ones are carbon, sulfur, and phosphorus.

You also will use (or try to get rid of) certain chemicals that are gases in many different welding processes. They include oxygen, argon, helium, carbon dioxide, and hydrogen.

Chemists, physicists, metallurgists, and welders use a shorthand notation to talk about metals and other elements that are in the materials they work with. You will often see these chemical symbols used in welding filler-metal specifications and in books and job instructions that you will read later on. These materials can be chemical elements or elements joined together to make compounds.

For example, the symbol for a hydrogen atom is H. The symbol for an oxygen atom is O. If you put two hydrogen atoms together with one oxygen atom and make them combine, you get water. Water is an example of a chemical compound. The chemical symbol for water is H_2O.

The chemical symbol for carbon is C. Combine carbon with oxygen (burn it in oxygen) and you get carbon dioxide, which is CO_2. That is, two atoms of oxygen combine with one atom of carbon when the carbon is completely burned. If the carbon is only partially burned, you might get carbon monoxide (CO). Acetylene is made up of equal parts of carbon and hydrogen atoms. It's formula is C_2H_2.

Table 1-5 lists the chemical symbols you are most likely to run into either in this book or in a welding shop. Let's call it your first look at welding metallurgy. Every time we use one of these symbols in the book, we will remind you of what it stands for. But it's to your advantage to learn this list now and never forget it.

You may be wondering why hydrogen is listed with the metals. Maybe you also noticed that argon and helium are not listed either with the metals or the nonmetals, while we listed water as a gas. Some of the chemical ele-

---

TABLE 1-4. Some unions that have larger welder memberships.

- Blasters, Drillers and Miners Union
- Building, Concrete, Excavation and Common Laborers Union
- Civil Service Employees Union
- Communications Workers of America
- Compressed Air Workers Union
- Factory and Building Employees Union
- Highway Road and Street Construction Laborers Union
- House Wreckers Union
- International Association of Bridge, Structural and Ornamental Iron Workers Union
- International Association of Machinists and Aerospace Workers
- International Brotherhood of Electrical Workers
- International Brotherhood of Teamsters
- International Union of Electrical, Radio and Machine Workers
- International Union of Elevator Constructors
- Electrical Production and Industrial Workers Union
- International Brotherhood of Pulp, Sulfite and Paper Mill Workers Union
- National Maritime Union of America
- Production, Industrial, Technical Amalgamated Workers Union
- Scrap Iron Demolishers and Handlers Union
- Sheet Metal Workers International Association
- Steamfitters Union
- Timbermen, Hod, Hoist, Carpenters, Core Drillers and United Brotherhood of Journeymen's Helpers
- United Auto Workers
- United Association of Journeymen and Apprentices of the Plumbing and Pipefitting Industry
- United Auto, Aerospace, Agricultural Implement Workers
- United Steelworkers Union
- Utility Workers Union of America

ments have obvious abbreviations (like O for oxygen, N for nitrogen, and C for carbon), while lead's symbol is Pb and iron's is Fe, tin is Sn, gold is Au, and silver is Ag. You'll find out the reasons for other things on the list later on,

TABLE 1-5. Important chemical symbols for welders

Chemical Name	Chemical Symbol
Metals	
aluminum	Al
antimony	Sb
calcium	Ca
cadmium	Cd
chromium	Cr
cobalt	Co
columbium	Cb
copper	Cu
gold	Au
hydrogen	H
iron	Fe
lead	Pb
magnesium	Mg
manganese	Mn
molybdenum	Mo
nickel	Ni
potassium	K
silver	Ag
sodium	Na
tin	Sn
titanium	Ti
tungsten	W
vanadium	V
zirconium	Zr
Nonmetals	
boron	B
carbon	C
nitrogen	N
phosphorus	P
oxygen	O
selenium	Se
silicon*	Si
sulfur	S
tellurium	Te
Gases	
acetylene	C_2H_2
argon	Ar
carbon dioxide	CO_2
carbon monoxide	CO
helium	He
hydrogen	H_2
nitrogen	N_2
oxygen	O_2
water	H_2O

*Silicon is an example of a semiconductor. It's half-way between a metal and a nonmetal in its properties. For example, silicon conducts heat and electricity better than nonmetals but not as well as metals. Selenium also is a semiconductor. It is used in arc-welding equipment.

probably while you work with this book in one hand and a torch in the other.

We'll only answer the last question here. There are two reasons why not all chemical symbols sound like the chemicals they describe. For one, these symbols are used worldwide in all languages. And two, the older elements such as iron and tin and gold and silver were discovered so long ago that they had chemical symbols when most scientists spoke and wrote in Latin. Most odd chemical symbols simply reflect the old Latin names. *Plumbum* is the Latin name for lead (Pb). *Ferrum* is the Latin name for iron (Fe). *Stannum* is the Latin name for tin (Sn). *Cuprum* is Latin for copper (even earlier it was called *cyprium,* the metal from the island of Cyprus) so the symbol Cu for copper is understandable.

On the other hand, wolfram is the name of one of the primary minerals that make up the ore from which tungsten (W) is extracted, and the W is handy because there are all sorts of chemical elements (including some we didn't list here) whose names begin with T plus something else, like titanium (Ti), tellurium (Te), tantalum (Ta), thallium (Tl), thulium (Tm), and thorium (Th). So forgive the world's chemists if they occasionally stick in another letter or two that may not make sense to you.

While it's not a beginning welder's job to worry about alloying elements directly (let alone Latin), you have to know that they exist. Later on, you will develop a good idea of what some of them will do for you (and even to you if you are not careful). Some of these metals are used to coat other metals, and metal coatings not only affect the metals a welder works on, they can affect the welder. That's why we include a detailed chapter every once in a while, where it will help you, on welding safety as well as welding techniques.

Now let's get to work.

REVIEW QUESTIONS

1. Name 10 different things, each weighing over 50 lb [22.7 kg], that are made by welding or flame-cutting.

2. Name five major industries that employ many welders and burners.

3. List 10 companies in your area that might hire you after you complete this course.

4. What segment of the construction industry is *least* likely to hire welders and burners?

5. Name five unions likely to be interested in you after you complete this course.

6. What are the chemical symbols for each of the *elements* listed below (any welder would know all of them by heart)?

7. Which of these elements are metals and which are nonmetals or semiconductors?

aluminum	nickel
argon	oxygen
carbon	phosphorus
chromium	silicon
copper	silver
columbium	sulfur
helium	tin
hydrogen	titanium
iron	tungsten
lead	vanadium
magnesium	zinc
manganese	zirconium
molybdenum	

8. What is the chemical formula for these *compounds?*

acetylene	nitrogen
carbon dioxide	oxygen
hydrogen	water

9. Methane (natural gas) is sometimes used as a fuel gas for flame-cutting. Its chemical formula is CH_4. What two chemical elements is it made of?

10. Acetylene burns completely in oxygen to produce two other gases, as follows:

$$2\ C_2H_2 + 5\ O_2 \rightarrow 4\ CO_2 + 2\ H_2O$$

What are the two gases on the right-hand side of this chemical equation?

11. What color is the chemical compound Fe_2O_3? What's a common name for this material?

CHAPTER 2

Introducing 98 Kinds of Welding and Cutting

Imagine learning 98 separately named welding and cutting processes. That's how many the American Welding Society (AWS) lists, complete with distinct official abbreviations, definitions, equipment, and process techniques. Figure 2-1 shows you the list.

Take HFRW for example. That's high-frequency resistance welding. HFRW is an electric welding process using the resistance of the base metal to a high-frequency electric current to produce heat for welding. It requires special machines, special electric circuits, and very few operator skills to make it work.

Then consider GMAW-S. This welding process is called *gas metal-arc welding with a short-circuiting filler-metal transfer into the molten weld puddle* (what a mouthful). You will learn a lot about this process. It requires special equipment and gases and a skilled operator. The words that make no sense to you now will soon be as clear and understandable as reading a newspaper.

You don't have to learn anything about most of the welding processes that exist to be a skilled welder. Even welding engineers don't know all about all possible welding processes. They don't have to. What they do know are the basic facts about welding. With these facts they can look up any of the process details to use as the need arises. You need to acquire the same basic welding knowledge, basic welding skills, and basic know-how so that you can look up and understand any welding process you don't know, just like a welding engineer.

But the fact that the list of official welding and cutting processes now extends to 98 items gives you an idea of the scope of modern welding technology. Ten years from now the list will be longer because welding science is as hot as its subject matter. New methods are constantly being developed to weld or cut new materials while solving old maintenance and production problems.

Actually it is very easy to dip into this alphabet soup of processes and scoop up an understanding of the most basic ideas about welding and cutting. Let's start by assuming that you want to join two things together. They need not be made of metal. They might even be made of plastic. Figure 2-2 gives you the possibilities. You can try anything from library paste to rubber cement to high strength epoxy glues, which are adhesive bonding materials, to join the two pieces. Or you could try for a very strong joint and use welding, brazing, or soldering.

2-1 BRAZING AND SOLDERING

If you join the materials with something like a filler rod that melts at a lower temperature than the base metal or plastic, that would be either brazing or soldering. The molten filler rod melts and when it hardens sticks the base materials together with a metallurgical bond that resembles what glue does when it hardens.

What's the difference between soldering and brazing? If the filler rod melts at less than 840°F (Fahrenheit) [450°C (Celsius)], the process is soldering. If the filler material melts at a temperature higher than that, but lower than the melting point of the base metal, the process is officially brazing. (The cutoff point of 800°F between brazing and soldering also is often used as a rule of thumb. The difference is not important, but the AWS prefers 840°F for certain technical reasons, for one because it's a round number when converted to degrees Celsius.)

With this definition, we already have simplified the 37 separate soldering and brazing processes shown in Fig. 2-3. You will learn a lot more about some of them. The rest are basically variations on the same theme.

2-2 SOLID-STATE WELDING

You might like to try solid-state welding. The name sounds advanced, but it's actually a very old technique. Instead of using a filler rod that melts and sticks two pieces of material together after it hardens, try the technique used by early blacksmiths and Japanese sword makers to join two pieces of metal. Heat up the two pieces of metal until they glow orange to white-hot. Put them together and beat on them with a hammer. The hot atoms on either side of the joint (with the help of the hammering) mix up with each other at the joint surface until the separate surfaces no longer exist. You can actually form a solid, continuous piece of metal this way by forcing atoms to diffuse (move slowly) across the joint barrier. This also gives us our first technical word: *faying*.

The faying surfaces of a joint are where the metals will be joined together. It's where all the action is in soldering and brazing. The word *faying* is a very old English word meaning "to fit or join tightly." Since the base metals do not actually melt in this process (they slowly diffuse into each other), the base metal re-

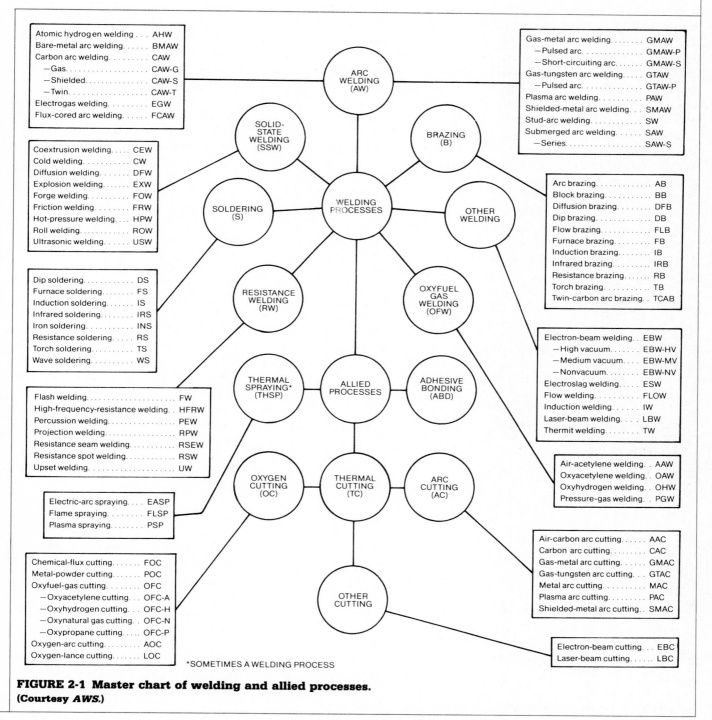

FIGURE 2-1 Master chart of welding and allied processes. (Courtesy AWS.)

mains in the solid state. Thus you have a solid-state welding process. This particular solid-state heat-and-hammer process is called *forge welding.* FOW is its AWS symbol. (You might see that symbol used on a welding blueprint.) If you have nothing else to work with but your base metal, a heat source, and a heavy hammer, you can make a solid weld with this method.

There are 22 solid-state welding methods (see Fig. 2-4), one of which is FOW, or forge welding. But you already have the basic idea behind all solid-state welding methods. The base metals do not melt. The joint is made by diffusion of atoms across the faying surface. The diffusion process is speeded by heating. When atoms get hot, they move faster, and therefore they diffuse faster. Additional force, such as pressing or hammering, is usually needed to make the process work because few surfaces are mirror flat, and that's necessary for the atoms to jump quickly across the faying surface barrier.

2-3
FUSION WELDING
What if you join the two faying surfaces of the base metal by ac-

tually causing them to melt and flow together to harden into a continuous, solid piece. You might add a filler metal to the joint, or you might not. Either way, that's fusion welding. *Fusion* means "melting and flowing together" and refers to the melting and flowing of the base metal (or plastic). There are 31 separately named fusion-welding processes shown in Fig. 2-5, but this basic unifying idea connects them all.

As you look at Fig. 2-5, you'll see that the first distinction among fusion-welding processes is based on the *source of heat.* The heat is produced either by electricity (as in electric-arc welding and electric-resistance welding) or by a chemical reaction (burning a fuel gas in oxygen or in air is a good example, but there are other possibilities).

The next distinction in the structure of welding definitions is *how the heat gets to the workpiece.* It may be directly from the electric arc through a process called *radiation,* or the heat may be created inside the workpiece through electrical resistance, or by convection, conduction, and radiation all at once as you find in a fuel-gas flame when it burns in pure oxygen. (You'll also find more details on radiation, convection, and con-

duction when we tell you about fuel gases.)

Some of these heat-producing methods also require pressure or force to make the weld by squeezing the molten metal together. Some methods don't. Either way, *fusion welding processes require some kind of shielding to keep oxygen in the air away from the hot weld metal.* That shielding can be a gas that won't hurt the hot weld metal or it can be flux, which is a chemical that melts and protects the molten weld metal from oxidation. Fluxes do other things in soldering, brazing, and welding, too, such as tying up dirt, oil, and oxides. This helps make the base metal easier to wet by the molten weld metal, brazing metal, or solder, as you'll find out over and over again before you finish this course.

The welding process may even require a vacuum to prevent atmospheric oxygen and nitrogen from touching the weld metal. Even a bare welding flame can sometimes be used as a special type of "gaseous" atmosphere that will shield the base metal and your weld from air until your molten weld metal turns solid. That's a common practice used in oxyfuel gas welding. You'll soon learn how to do it by setting your torch flame the right way.

Then the individual processes are listed, all 31 of them. But you don't have to learn 31 fusion-welding processes; only a handful are used for over 95 percent of all the fusion welding done by skilled welders.

2-4
WHAT WELDERS NEED TO KNOW
If you learn the basics of soldering and brazing, flame cutting, oxyfuel welding and cutting, about four kinds of electric-arc welding processes, and maybe one kind of arc cutting, you will know as much about welding as 95 percent of all the skilled welders in the world, and most welding engi-

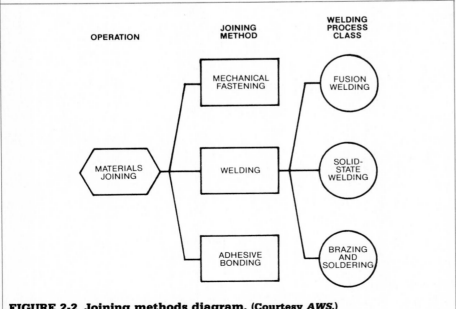

FIGURE 2-2 Joining methods diagram. (Courtesy *AWS*.)

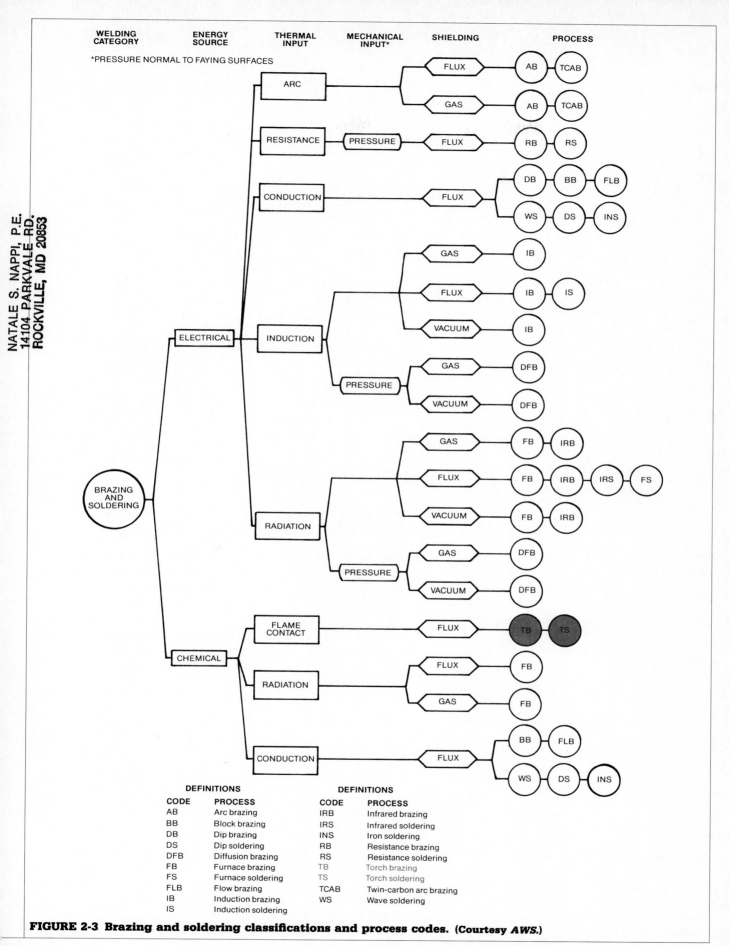

WELDING CATEGORY | ENERGY SOURCE | THERMAL INPUT | MECHANICAL INPUT* | SHIELDING | PROCESS

*PRESSURE NORMAL TO FAYING SURFACES

NATALE S. NAPPI, P.E.
14104 PARKVALE RD.
ROCKVILLE, MD 20853

DEFINITIONS

CODE	PROCESS
AB	Arc brazing
BB	Block brazing
DB	Dip brazing
DS	Dip soldering
DFB	Diffusion brazing
FB	Furnace brazing
FS	Furnace soldering
FLB	Flow brazing
IB	Induction brazing
IS	Induction soldering

DEFINITIONS

CODE	PROCESS
IRB	Infrared brazing
IRS	Infrared soldering
INS	Iron soldering
RB	Resistance brazing
RS	Resistance soldering
TB	Torch brazing
TS	Torch soldering
TCAB	Twin-carbon arc brazing
WS	Wave soldering

FIGURE 2-3 Brazing and soldering classifications and process codes. (Courtesy AWS.)

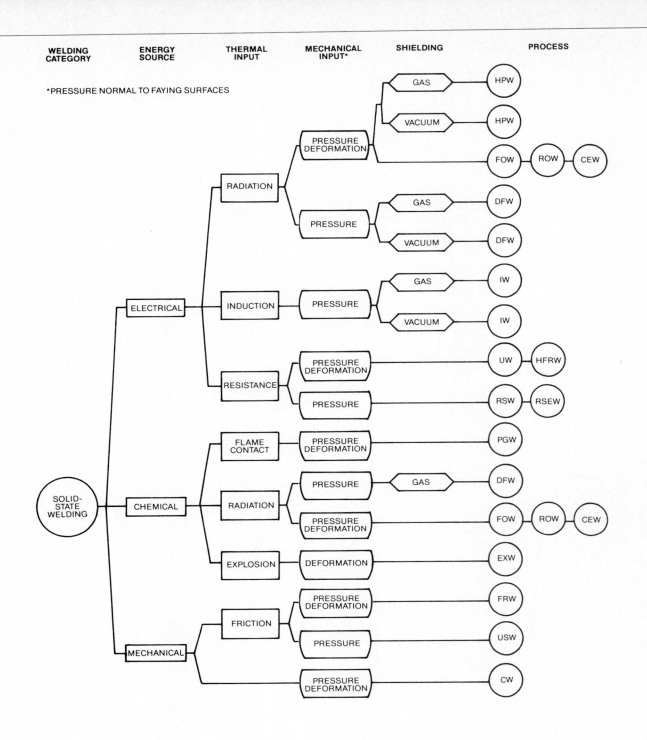

FIGURE 2-4 Solid-state welding classifications and process codes.
(Courtesy *AWS*.)

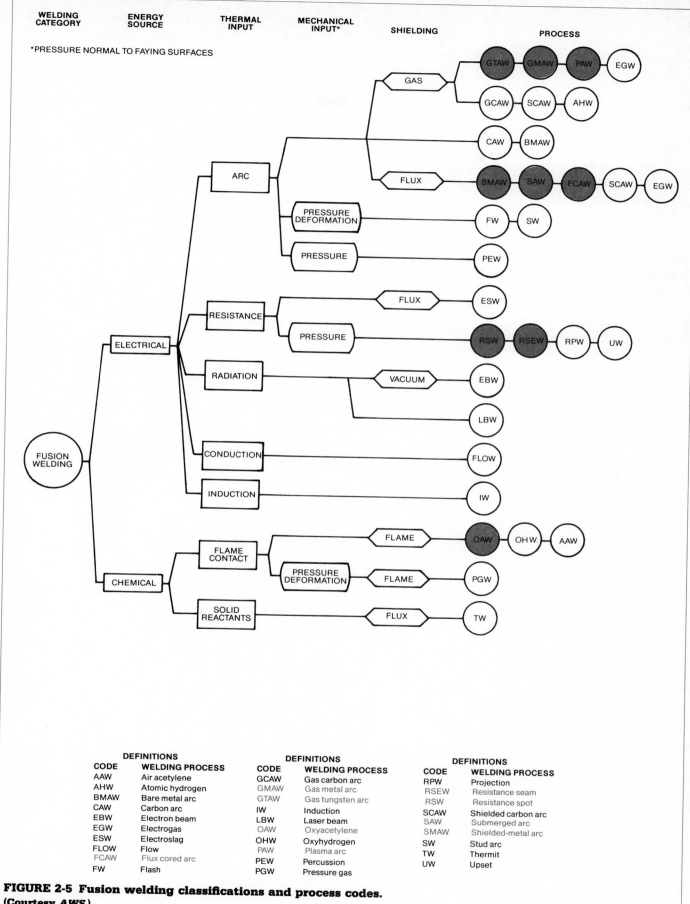

FIGURE 2-5 Fusion welding classifications and process codes.
(Courtesy *AWS*.)

DEFINITIONS

CODE	WELDING PROCESS
AAW	Air acetylene
AHW	Atomic hydrogen
BMAW	Bare metal arc
CAW	Carbon arc
EBW	Electron beam
EGW	Electrogas
ESW	Electroslag
FLOW	Flow
FCAW	Flux cored arc
FW	Flash

CODE	WELDING PROCESS
GCAW	Gas carbon arc
GMAW	Gas metal arc
GTAW	Gas tungsten arc
IW	Induction
LBW	Laser beam
OAW	Oxyacetylene
OHW	Oxyhydrogen
PAW	Plasma arc
PEW	Percussion
PGW	Pressure gas

CODE	WELDING PROCESS
RPW	Projection
RSEW	Resistance seam
RSW	Resistance spot
SCAW	Shielded carbon arc
SAW	Submerged arc
SMAW	Shielded-metal arc
SW	Stud arc
TW	Thermit
UW	Upset

neers, too. You also will have your hands full just learning these basic welding and cutting processes.

To make it easier for you to find the most important welding processes in Fig. 2-3, Fig. 2-4, and Fig. 2-5 if you need to refer to them later on, we have shaded the diagrams with a light-colored tint. These are the processes that you must learn to become an all-purpose skilled welder. The logical place to start learning the skills and knowledge you need to do brazing, soldering, and fusion welding is to follow the ideas behind the development of the welding charts.

Start with the method used to provide the heat to the base metal. The best place to do that is with the fuel gases used for soldering, brazing, welding, and flame-cutting and then learn about the oxygen used to burn them to produce heat. Once you know about these basic welding gases and techniques, you will know a great deal about OFW, OHW, AAW, TB, TS, FB, BB, FLB, WS, DS, INS, PAW, . . . and you will know how to work with them *safely*.

REVIEW QUESTIONS

1. How does fusion welding differ from soldering and brazing?

2. How does brazing differ from soldering?

3. What is a faying surface?

4. Does welding always require a filler metal?

5. What is a brazing filler rod?

6. Name three different sources of heat (thermal) energy input for fusion welding.

7. Name two kinds of soldering processes.

8. Give two examples of methods for joining metals that are *not* welding, brazing, or soldering.

9. The following are standard American Welding Society abbreviations for important welding, brazing, or soldering processes. What do the abbreviations stand for? (This textbook tells you how to do all of these processes.)

GMAW	PAW
TB	FCAW
OAW	SAW
TS	SMAW
GTAW	FOW

PART 2

Fuel-Gas Processes

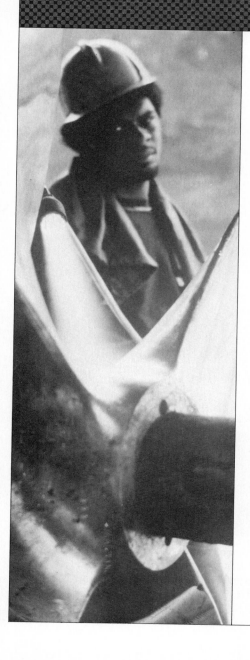

Major welding and cutting processes in this book have been divided (with the exception of hardfacing and surfacing) between processes that use fuel gases to produce heat and processes that use electricity to produce heat. We'll begin with safety methods. After you learn how to take care of yourself, then you will learn about your basic gas-using equipment.

When you are ready to begin real welding and cutting work, we will start you with flame-cutting because most products must be cut to size and shape before they are welded together.

Flame-cutting leads logically to other uses for flames and heat. You will learn how to use that heat to do jobs that many modern welders have completely forgotten how to do.

Next you will learn about soldering and brazing. Then you will learn how to do oxyfuel welding.

You also will learn in this section about the mechanical properties of metals. You will soon realize that the real job of a welder is to make metal parts stick together, and stay together when they are put to use.

The final chapter in this section, Chap. 10, which is on oxyfuel pipe welding, will begin a series on pipe work that will continue in the following section on electric welding methods.

CHAPTER 3
Fuel Gases

Welders and burners don't always get to choose the fuel gas for their work, but knowing the differences among the fuels you use is part of your job. Knowing how different fuel gases work also directly affects the quality of a joint. This fuel-gas know-how can be critically important to your personal safety as well as your job.

There is quite a lot to know about the fuel gases used in welding and cutting, brazing, and soldering. In this chapter you will first learn about the different fuel gases and how they are used. You also will find out about the cylinders they come in and how the pressure in the cylinders changes as you use the gas.

In Chap. 4 you will then learn about the other equipment that makes it possible to work with gases. This equipment includes regulators to control the gas pressure, the hoses used between your regulator and your torch, and the torches and tips that control the way the gas burns. You will also learn more about acetylene in Chap. 4 because the acetylene and the cylinder that holds it are different from other fuels and fuel containers and relate closely to how acetylene is used. After that, you'll learn how to turn your equipment on safely, how to light your torch, and how to turn it off again without an unpleasant bang.

3-1
IMPORTANT FACTS ABOUT FUEL GASES

About 75 percent of all high-energy industrial fuel gases are used for flame-cutting (Fig. 3-1). These fuel gases are acetylene, MAPP Gas® (a trademark, but MAPP stands for methylacetylene propadiene stabilized), propane, propylene, and natural gas. Although rarely used for flame-cutting, hydrogen also is a high-energy fuel gas. These gases are called high-energy fuel gases because, when burned in pure oxygen, they produce large amounts of heat. As you'll soon learn, however, these gases are not always interchangeable. Certain fuel gases perform certain kinds of work better than others.

To put things in perspective for you, while flame-cutting uses 75 percent of all fuel gases, only 10 percent of these gases are used for heating metals to bend,

FIGURE 3-1 This welder is cutting steel plate in the field for an oil tank. He only has to set up and monitor the job while the small track-mounted cutting-torch carriage makes a perfectly straight cut. Chapter 5 will tell you how to do flame-cutting and how to operate cutting machines.

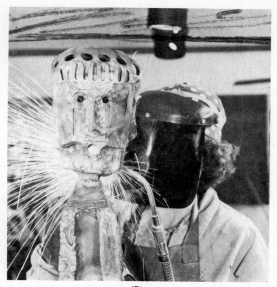

FIGURE 3-2 (A) Auto repair shops often hire beginning welders. Many people who learn oxyfuel welding and cutting use what they know to customize their own cars. It's much less expensive than having a custom shop do it. Chapter 9 tells you how to do oxyfuel welding. (B) Cars aren't the only things that can be customized with welding and brazing. This professional artist creates welded metal sculpture with her torch. She also makes a good living making brazed jewelry.

straighten, and preheat them prior to gas or arc welding. Brazing and soldering use up about 5 percent of all industrial fuel gases. Other processes such as metal spraying consume the rest.

Although only a very small amount of conventional gas welding is done in industry (consuming about 1 percent of the fuel gases used), gas welding is a very important process for you to learn first. There are four good reasons.

First, gas welding (Fig. 3-2) is the best way to learn how to handle a welding torch in one hand and a filler rod in the other while at the same time controlling a puddle of molten weld metal. These are exactly the same manual skills you have to learn to do GTAW (gas tungsten arc welding, commonly called TIG) welding work that often pays a premium wage.

Second, it's much easier to see what happens to molten metal when you're behind gas-welding goggles than behind the very dark face plate of an arc welding helmet (where you can't see anything except the inside of the helmet

until you strike an arc). You may not realize it when you melt your first metal with your torch, but what you'll be gaining is a lot of firsthand knowledge about welding metallurgy. This will help you later on when you start working with more complex welding methods and base metals.

Third, gas welding and brazing are important welding processes in certain plants and any complete welder must know them. In fact, gas welding and brazing still are used extensively in plant maintenance work.

Finally, gas welding, brazing, and soldering are fun. They also are the processes you are most likely to use at home to make things for yourself because the equipment usually costs less than arc-welding equipment. Before we look at the fuel gases you'll use, let's start by learning what they're used for.

3-2
THE USES OF FUEL GASES

Flame-cutting is the most important user of fuel and oxygen gas.

Before a welder welds or brazes anything, he or she will most likely need to cut some metal to the right shape for the part. That's what oxyfuel flame-cutting does, but only for carbon steels. Other arc-cutting processes have to be used for stainless steels, aluminum, copper, and other nonferrous metals and their alloys. You'll learn about the other cutting processes later on.

The skilled use of a heating torch (Fig. 3-3) has almost been lost. We will bring it back by teaching it to you. It could mean a job for you if you have to compete with more "sophisticated" or more experienced welders. For example, you can straighten a thick steel shaft that's been bent out of shape without using an ounce of muscle power if you master the use of your heating torch. You also can make very fancy wrought-iron furniture with a heating torch, braze it together, and maybe make money on the side.

In fact, we've devoted all of Chap. 5 to using the heating torch to give you a competitive edge

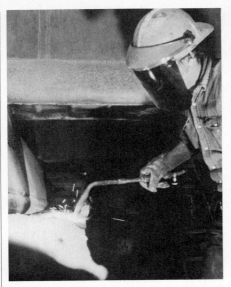

FIGURE 3-3 This welder is using a large heating torch to straighten the bent side of a railcar. When he is finished, the steel section will be perfectly straight, without the use of a sledgehammer or any other force except his skill with the torch. Chapter 6 tells you how to do it.

over other workers. But to use a heating torch with skill, you have to understand fuel gases, and heat, and torches and tips, and a lot more about metals than has been discussed till now.

Oxyfuel brazing and soldering (Fig. 3-4) are entirely different skills. Their successful use is dependent on your knowledge of fuel gases and oxygen. We mentioned that bicycles are brazed together. So are delicate electronic parts, jewelry, and many of the smaller "welded" sculptures you see in wealthy homes and museums.

If you complete the first part of this book on fuel gases and equipment, and oxyfuel welding, brazing, and soldering, you will be equipped for jobs in many different industries. (But we want you to complete the entire course, including electric-arc welding as well as oxyfuel work for a special reason . . . it could mean an even larger paycheck at the end of each week.)

We want you to learn some things about fuel gases and oxy-

gen for another reason: to keep you out of trouble. Fuel gases and oxygen can be dangerous if you don't know what you're doing. If you understand the physics of fuel gases, you'll be able to take better care of yourself.

3-3
FUEL-GAS PHYSICS
We're going to introduce you to some new ideas. They are very practical ideas as well as basic physics ideas. Many of them will not only tell you *how* to handle gases safely, but *why*.

In a real shop situation, the welder is often the plant's leading authority on gases, including gas safety, cylinder handling, and often fuel-gas and oxygen purchasing. You may not do the actual buying at first, but this chapter will help you know what you're talking about. It won't be long before management will notice what you know, as well as what you can do.

Gas Pressure
All gases are made up of atoms or molecules. Pressure is the pushing force of these atoms or molecules against the wall of a container, or even against another gas such as air. The container can be a thick steel cylinder, a gas regulator, a length of hose, your torch tip, or your entire work area, including you. Even the air you breathe is under pressure.

Gas pressure is special because it is a force that operates equally in every direction (Fig. 3-5C). The force of a wrench lying on a work table only goes straight down against the table. Gravity causes that. We don't usually think of the wrench's weight as pressure, but in a general sense it is because gravity causes the surface of the wrench that is in contact with the table to press against it. A physicist would lump all the kinds of pressures we are going to talk about into one word, *force*. Weight "force" works only in one direction, down (Fig. 3-5A).

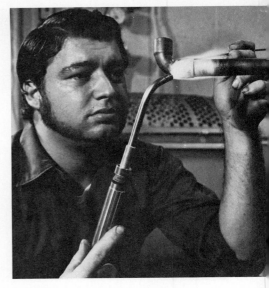

FIGURE 3-4 Plumbers have to know how to braze pipe joints. The same heating torch used in this job also could be used to bend or straighten the pipe. Chapter 7 will tell you all about brazing and soldering. When you complete this course, you will be able to do your own plumbing.

The pressure of water in a glass or in the ocean *pushes equally in every direction* at any given depth under the surface. But water pressure increases as you get deeper (Fig. 3-5B). At the surface of the water, the pressure of the liquid is zero. Underwater, the pressure increases by 0.45 lb/in.² for every foot of depth. At 10,000 ft depth the water pressure is 4500 lb/in.² In the deepest part of the ocean the pressure will be almost 12,000 lb/in.² or about 6 tons/in.²

Gas pressure also works in every direction. In that sense it's like water pressure. However, gas equalizes its pressure by expanding or contracting to fill its container (Fig. 3-5C). For all practical purposes the pressure of the gas in the container is uniform throughout. The difference due to gravity is so small that you would need a gas container several miles high before it became noticeable. So let's assume that the gas pressure in any container is the same throughout the container.

If the container has a hole in it, the gas will escape as long as its pressure is greater than the pressure of the outside air.

Gases also will slowly diffuse through other gases even when they are under equal pressure. That's why you smell the pilot light on your gas stove after the flame goes out. The gas from the flame has an odorizer that diffuses through the air to reach your nose. Industrial fuel gases don't always have odorizers.

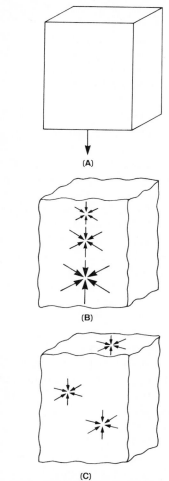

FIGURE 3-5 (A) The force of gravity produces weight. Gravity is a one-way force. (B) The force of water pressure is equal in all directions at any one depth, but the force increases with depth and disappears at the top surface of the water. (C) Gas pressure is equal in all directions at any one point in a closed container. The shape of the container has no effect.

Therefore, you should stop fuel-gas leaks as soon as you detect them. The longer you wait, the more gas will get out of the container and the more dangerous the leak becomes.

All of this probably sounds rather obvious to you. However, welders who didn't remember these simple facts about gases have been killed or injured by "empty" cylinders (cylinders are never completely empty). And they have been smothered in enclosed areas where a little otherwise harmless inert shielding gas such as argon or helium has diffused into the work area until there wasn't enough good air left to breathe. Sometimes totally "harmless" inert shielding gases can be more dangerous than explosive fuel gases. In Chap. 4 we'll tell you how to handle fuel gases and inert gases safely. We'll also tell you a lot more about the safe handling of oxygen.

Now you can see that there are reasons for the things you will learn next, even if some of them are not immediately obvious. Sometimes you won't clearly see the reason for some of the "scientific" facts we'll give you until much later on in this book. Be assured that there is a solid practical reason for everything we're going to tell you.

How pressure is measured In principle, measuring gas pressure is very simple. Since gas pressure is a force that acts uniformly over any surface, you simply have to select some measure for the force, such as pounds (lb), and some equally simple measure for the unit of surface area, such as square inches (in.^{2}). The ratio of pounds to square inches (lb/in.^{2}) is a good basic measure of the gas pressure in a container, no matter what its size or shape. Using a unit gas pressure, you don't even have to know about the container's size or dimensions.

The gauges on gas regulators attached to gas cylinders will give you these unit gas pressure readings. In the United States, they are almost invariably calibrated in pounds per square inch (lb/in.^{2}). Since lb/in.^{2} is still a bit wordy, most people simply say "psi." That's not a word. You say each letter one after the other, "p-s-i."

In science labs, and in every other country in the world (except one African nation), the metric system is used and the units are different, but the idea is the same. The metric system is much easier to use than the so-called English system. However, most shops in this country still use the English system. We will use the English system of units in this book most of the time because that will be the system you will use in most of your shop work. We would prefer to use metric units throughout the book, but the long strings of conversion factors that would have to be inserted in many of the tables and much of the text would make it difficult for you to read the lessons. If you do work with metric units in the shop, you will soon become familiar with them and most likely will learn to prefer them to English units.

If you're interested, the most commonly used metric unit for gas pressure is kilograms (kg) of force per square meter (m^2), which is called the *pascal*. (A kilogram equals about 2.2 lb and a pound is 0.45 kg, while a meter is about 39.37 in., or just 3.37 in. longer than a yardstick.) The abbreviation for the pascal is simply Pa. Since the metric system is based on powers of ten (kilo means 1000 and mega means 1,000,000), you sometimes will see the units written as kilopascals or megapascals, which are simply abbreviated kPa and MPa. The metric ton is 1000 kg which is 2205 lb (very close to a long ton of 2200 lb).

The regulator on your cylinder probably will have two separate dial gauges (Fig. 3-6). One tells you what the gas pressure is in your cylinder. The other tells you the pressure of the gas coming out of the regulator and into your

FIGURE 3-6 This is an oxygen cylinder regulator. The cylinder pressure gauge is in back and the delivery pressure gauge is in front. The delivery-pressure gauge dial reads up to 200 psig, while the cylinder pressure gauge dial reads up to 4000 psig. The screw-in handle on the right is used to adjust the delivery pressure.

hose. The first pressure is called your *cylinder* pressure. The smaller pressure out your hose is called the *working* or *delivery* pressure. It's the pressure setting that you make when you start to work on a job. Working or delivery pressures often are around 10 psi. Cylinder pressures are as much as 220 times higher for high pressure gases.

If you have a gas in a cylinder at a pressure of 2200 psi, the gauge dial on the cylinder-pressure regulator will be pointing at 2200 psi, which is a very common cylinder pressure for full high-pressure gas cylinders. When you think about it, though, *2200 psi is equivalent to a pressure of one long ton per square inch on every square inch of the inside of the cylinder.* That is enough gas pressure, if suddenly released all at once, to turn a 120-lb steel cylin-

der into a rocket with enough thrust to crunch its way through a concrete-block wall. Now you understand why there is a gas cylinder regulator between your torch and you and the gas in the cylinder. The regulator reduces the pressure of the gas in the cylinder to a delivery pressure that you can safely use.

Psi and psia There are several different units used to describe gas pressure. You already know about psi, the most frequently used unit. Therefore, if your supervisor asks you to "get a full cylinder of oxygen," you will know that you are looking for one with about 2200 psi cylinder pressure reading on the cylinder-pressure regulator gauge (Fig. 3-6). If the regulator delivery valve is shut (as it should be), the delivery-pressure gauge needle will point to zero.

However, out in the cylinder storage shed you might only find cylinders with MT marked on their sides with chalk. One or two may have some gas in them, but then they'll probably be tagged with a note such as "leaker" or a comment about a stuck valve. These cylinders are waiting to be picked up for a refill and, where necessary, a repair job. But are the MT cylinders really empty?

You know they're not. Every cylinder in that shed whose cylinder-pressure gauge is reading zero is actually full of gas. However, that gas is at the same pressure as the air outside the cylinder. This shows you there must be two entirely different kinds of psi gas-pressure units. The difference is very important.

There are *psia* units, and *psig* units. Psia stands for *pounds per square inch absolute* pressure. Psig stands for *pounds per square inch gauge* pressure. The "empty" cylinders you looked at were at 0 psig. The real or total or absolute pressure of the gas in the cylinder is the same as the air pressure outside the cylinder. It's measured in psia or psi (absolute).

If your cylinder shed were at sea level, that absolute pressure would be equal to 1 atmosphere (1 atm), which is another handy pressure-measuring unit, especially when doing certain kinds of work like deep-sea diving (which some welders working for oil and gas companies do a lot of), or working in the weather bureau, or in certain kinds of science labs.

How much pressure is 1 atm in psia units? A pressure of 1 atm, if you were standing with your empty cylinder at sea level on a nice, comfortably warm day without any weather pressure fronts coming your way, would be 14.7 psia. One way of looking at 14.7 psia is that it's the gas pressure in the cylinder above a "hard vacuum," or outer space. A pressure of 1 atm is equal to 14.7 psia, so that is the real gas pressure inside your supposedly empty cylinder.

If you took the same cylinder (keeping it shut) and put it on the top of one of the lower mountains in the Colorado Rockies at, say, 10,000 ft, the cylinder pressure gauge would no longer read 0 psig. It would read a few psig higher simply because of the difference between the atmospheric pressure at that high altitude and the atmospheric pressure at sea level. At only 1 mile (mi) high, the difference in atmospheric pressure between Denver, Colorado, and New York City is 2.5 psig. At 10,000 ft the difference is even greater, and your empty cylinder at sea level would have a gauge-pressure reading of 9.5 psig. If you opened the cylinder at high altitude, gas would escape even though nothing would come out if the same cylinder were at sea level.

The cylinder pressure gauge actually measures the *difference* between the pressure of the gas inside the cylinder and the pressure of the outside air. It's a big differential barometer. That's why it reads 0 psig (cylinder gauge pressure) when the gas pressure inside and outside the cylinder is the same.

The only truly empty cylinder is one with a vacuum pump working on it. *All cylinders you will work with will always be full of gas.* If the gas is dangerous, flammable, or something else, then so is the empty cylinder.

Incidentally, you should never wait until a high-pressure gas cylinder (2200 psi when full) gets down to 0 psig on the cylinder gauge before you change it and put a full cylinder on line. *Always change cylinders when the cylinder-pressure gauge of a high-pressure cylinder gets down to about 25 psig.* Since atmospheric air pressure can change during the day (you'd see it change if you had a meteorologist's barometer), you actually can suck some air into your cylinder if you leave the regulator or cylinder valve even slightly open and the cylinder's inside pressure gets too close to 0 psig.

One reason, in addition to changes in barometric pressure, that the pressure inside a cylinder also can change, is that if the gas gets hot (like when you leave it in the sun) it expands. If the gas gets cold (when the sun goes down) it contracts. That expansion and contraction can suck air into your nearly empty cylinder if you forget to seal it tight or if it has a slight leak. Oxygen and nitrogen in the air can contaminate most of the high-pressure gases you will be using. At 25 psig you aren't throwing away very much gas. You could be saving some very expensive weld metal.

Now that you know the difference between psia or absolute pressure, and psig or gauge pressure, keep it in mind. We will return to those empty cylinders in the storage shed after we look further at the effects of temperature on gas pressure and watch one of them burn the shed down.

Density
The inert welding gases helium and argon, symbolized by He and Ar, are atoms and not molecules. Oxygen, which is O_2, or hydrogen,

which is H_2, and carbon dioxide, which is CO_2, are all molecules. The inert gases are so chemically unreactive that their atoms won't join up with each other like oxygen's atoms will (as in O_2). Each of these atoms or molecules has its own characteristic weight. Weight and mass, by the way, are also different ideas because weight depends on gravity. You (or the gas) weigh six times as much on the earth with its higher gravity than you would on the moon. But your mass remains the same.

A tankful of gas will have a different weight depending on the kind of gas in the cylinder, but for an entirely different reason. The weight of the gas depends on the kind of gas in the cylinder, the pressure of the gas, and its temperature, which makes things a little complicated. Therefore we'll introduce another idea, gas density.

Gas density is normally measured in pounds per cubic foot (lb/ft^3). The density of gases, however, even if 1 ft^3 of each different gas were weighed under the same temperature and pressure, varies with the weight of the molecules or atoms that the gas is made of (Fig. 3-7). Oxygen, for example, has a higher density than hydrogen. In fact, the density of molecular oxygen gas is 16 times higher than the density of molecular hydrogen. A chemist would

say that the molecular weight of O_2 divided by H_2 equals 16. Helium is only one-eighth as dense as oxygen because the ratio of the atomic weight (there are no helium molecules) of helium to the molecular weight of oxygen is $32/4$ = 8.

Argon, a very common welding shielding gas, has a density 2.5 times that of oxygen. The difference in density between oxygen and argon (let alone oxygen and hydrogen or helium) is so large that a truck that can haul a full load of oxygen often has to haul much less argon to keep from exceeding its weight limit and possibly blowing out its tires. For example, a large 36-ft-long high-pressure tube trailer (a giant gas cylinder truck used to make large deliveries of pressurized gas) can carry 175,000 standard cubic feet (scf) of helium, but only 49,900 scf of argon at 70°F, with the gas pressurized to 2640 psig. We will explain standard cubic feet in a moment.

Just because most of the gases you will work with are tasteless, colorless, and odorless, don't think they are weightless. Density (in pounds per cubic foot or lb/ft^3) is a handy way to measure the weight differences between gases.

Gas densities are always reported under standard conditions of temperature and absolute pressure so everybody knows what

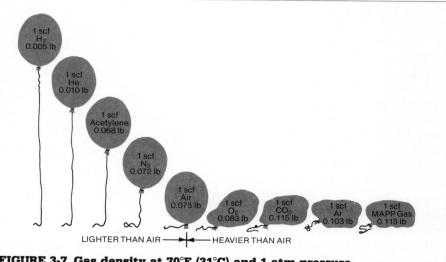

FIGURE 3-7 Gas density at 70°F (21°C) and 1 atm pressure.

you mean. The pressure is always reported as sea-level pressure which is the same as 1 atm pressure, or 14.7 psia. The temperature almost always used in making density measurements is 70°F. If you are a metric enthusiast, and some people who use gases are because metric units make gas calculations much easier to do, 70°F equals 21.111111...°C. Therefore the metric equivalent most often used is 21°C. And 70°F or 21°C are used most often simply because they are very close to what people normally think of as a comfortable room (ambient) temperature.

Sometimes gas densities are compared directly to the density of air. That is called the *specific gravity* of the gas. It's a useful way of thinking about gases that are heavier or lighter than air. For example, the specific gravity of helium is only 0.13796. The specific gravity of argon is 1.380 (38 percent heavier than air). The specific gravity of carbon dioxide gas is 1.53, so CO_2 is about 50 percent heavier than air.

Why should a welder have to know anything about gas densities and specific gravities when working on a weldment? The density of a gas directly affects the flow rate of the gas through a hose and a regulator. It is one of several important reasons why different kinds of regulators are used with different gases (as we'll discuss in great detail in Chap. 4). Gas density also is very important in arc welding where different kinds of gases, with different densities, are used to shield molten weld metal.

Some shielding gases are so dense (carbon dioxide or argon) that they sink down over the weld metal. Others have such a low density (helium, for example) that they rise up (which is very good for welders working on welds over their heads). The difference in shielding-gas densities is one of many factors you use to select the shielding gas for welding different kinds of work.

Purity

The purity of the gases used in oxyfuel welding, brazing, soldering, and flame cutting, as well as in gas shielding for arc welding, is very important. In fact, it's one of the most important factors in making good welds on almost any metal and smooth flame-cut edges on steel plates.

As you can see from Fig. 3-8, a mere 0.5 percent difference in oxygen purity can make a big difference in cutting speed when cutting plate with an oxyfuel torch. Gas purity is even more critical if you are using a shielding gas for arc welding. That's why we strongly recommend that you change your shielding gas cylinder when the cylinder regulator gauge gets down to 25 psig.

Purity in parts-per-million One way to measure gas purity is simply to state the percent purity of the gas, like we did in the oxygen example. The good oxygen was at least 99.5 percent pure, which is slightly purer than Ivory soap. Percent purity is accurate enough for most of the oxygen you will use, but minute (and we really mean small) percentages of impurities in some shielding gases can affect a weld so badly that it may cost up to $4000 to repair. Shielding gas at 99.5 percent purity is so dirty you can't possibly use it. Most of the shielding gases you will use will be 99.999 percent pure, or better.

Instead of stringing many 9s together on the right side of the decimal point, another way of expressing gas purity is used when dealing with very pure gases. The amounts of impurities are measured in parts per million, which often is abbreviated ppm. Sometimes you will read a label that says that the gas you are working with has "5 ppm hydrocarbons maximum," or "10 ppm oxygen in argon." The expression parts per million or ppm is related to percentages very simply:

$$1 \text{ ppm} = 0.0001\%$$
$$10 \text{ ppm} = 0.001\%$$
$$100 \text{ ppm} = 0.01\%$$
$$1000 \text{ ppm} = 0.1\%$$
$$10,000 \text{ ppm} = 1\%$$

The oxygen that you will be using for flame cutting will be around 99.999 percent pure (it contains no more than 0.001 percent or 10 ppm impurities) simply because of the way modern oxygen is made. It starts as a liquid at temperatures colder than $-297°F$ [$-183°C$] and most of the other impurities (including many other gases) are actually frozen out of it before it is warmed up back to a gas and pumped into cylinders.

Dew point and purity A different way to talk about gas purity when water vapor might be in the gas is in terms of the *dew point* of the gas. Dew point is a specific measure of how dry the gas is. It only considers possible water content. Since even very tiny amounts of water vapor can make strong weld metal very brittle (you'll find out why when you get

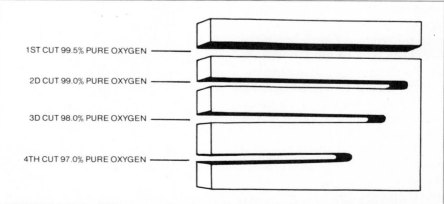

FIGURE 3-8 A difference of only 0.5 percent in oxygen purity makes flame-cutting difficult, or impossible.

1ST CUT 99.5% PURE OXYGEN

2D CUT 99.0% PURE OXYGEN

3D CUT 98.0% PURE OXYGEN

4TH CUT 97.0% PURE OXYGEN

to the arc-welding processes), the dew point of a shielding gas like argon or helium often is reported along with its purity.

You wouldn't believe what just a little moisture in your shielding gas can do to a high-strength steel weld. Among other things, it can cause the steel to crack along a line in the base metal right next to the weld metal. It is safe to say that a minute amount of moisture can help bring down a building or a bridge.

The dew-point measurement is based on the same principle that causes a cold mirror to fog up when you breathe on it while a hot mirror won't. Dew point tells you at what temperature water vapor will condense out of the gas onto a very cold, very smooth surface. The lower the dew point (in degrees Fahrenheit, for example) the less moisture the gas contains. It's not unusual to have welding shielding gases with dew points of less than $-60°F$ to $-80°F$ [$-51°$ to $-60°C$]. A dew point of $-60°F$ equals only 34 ppm of water vapor, or only 0.0034 percent moisture in the gas. A dew point of $-80°F$ equals 7.8 ppm water vapor, or only 0.00078 percent moisture.

Table 3-1 shows the relationship between percent purity, parts-per-million purity, and the dew point of gases. Liquefied oxygen, or nitrogen, and argon usually have dew points of around $-90°F$ [$-68°C$], and welding-grade compressed gases in cylinders almost always have dew points of $-60°F$ [$-51°C$] or lower.

The reason that cylinder gases may have a few parts per million more impurities in them than a liquefied gas is that the liquefied gas is first vaporized and then pumped into your cylinder to make a cylinder gas. In the process, it's necessary to pump the cylinder empty to a fairly high vacuum to remove everything that was in it. It is much harder to pull a hard vacuum on a cylinder than it is to purify gases initially by liquefying them and then vaporizing them.

You never have to worry about the dew point of fuel gases like acetylene, MAPP Gas, propane, natural gas, propylene, or other gases that are burned to generate heat instead of to shield your weld metal (gases like argon or helium or carbon dioxide). All gases that burn in oxygen and that contain carbon and hydrogen (all hydrocarbon fuel gases), or hydrogen alone when used as a fuel gas, will produce water when they burn. The reason is simple. Oxygen combines with hydrogen to produce water, and that's one of the primary chemical reactions that occurs when any fuel gas is burned. Therefore, you don't care whether your fuel gases are dry to start with because they won't remain that way when you burn them (make them combine with oxygen) to produce heat for your work. (When we get to the gas-shielded arc-welding process you will learn why moisture is no problem in a flame even though moisture makes high-strength weld metal brittle and lower-strength welds porous when it passes through an electric arc.)

Heat

There is a big difference between heat and temperature. Heat measures the *amount* of thermal energy. Flame temperature measures the *intensity* of that energy.

Heat is caused by atoms moving around. The more atoms in the material that move around (actually vibrate or jump around), the more heat the material contains. But heat doesn't tell you how fast the atoms are jumping around. That's one reason why heat and temperature have entirely different units. Temperature is measured in degrees Fahrenheit or degrees Celsius. (There are even two other "absolute" temperature scales we'll tell you about shortly called *degrees Rankine* and *Kelvin*). Heat is measured in British thermal units (Btu's) in our awkward English system and calories (cal) or kilocalories (kcal), which equal 1000 calories in the nice base-ten metric system. Let's look at temperature and heat and see why they are different.

Temperature

Temperature is a measure of heat intensity. While a lot of heat tells you that a lot of atoms or molecules are moving around, *temperature tells you how fast they are moving.* Something can have a very high temperature and a very low total energy or heat content. Outer space gives us the best example. There is a very rare, thin "atmosphere" of hydrogen and helium atoms (and even some hydrogen and other molecules) in outer space between the galaxies. This space may contain no more than one atom or molecule of hydrogen per cubic yard. There obviously isn't very much heat in a cubic yard of anything that only con-

TABLE 3-1 Moisture content of welding gases

Dew Point, °F	H₂O, ppm	H₂O, %	
-130	0.10	0.00001	
-120	0.25	0.000025	
-110	0.63	0.000063	
-100	1.53	0.000153	
-90	3.53	0.000353	Typical of liquefied gases such as argon, helium, oxygen
-80	7.8	0.00078	
-70	16.6	0.0017	
-60	34.0	0.0034	Typical of good high-pressure shielding gases
-50	67.0	0.0067	
-40	128.0	0.0128	
-30	235.0	0.0235	
-20	422.0	0.0422	Bad welds can occur in high-strength steels
-10	740.0	0.0740	

tains one hydrogen atom in it. However, if that one hydrogen atom is vibrating at very high speed, equivalent to 100,000,000°F, you'll have to admit that the single hydrogen atom is at a very high temperature.

Now if you took a cubic mile of that nearly empty space with all those fast, hot, vibrating hydrogen atoms in it, you would say that matter in that region of space has a very high temperature, 100,000,000°F. It also would contain 545,180,000 hydrogen gas atoms, which sounds like a lot of hydrogen, but 545,180,000 hydrogen atoms only weigh 0.00000000000002 lb. Even at 100,000,000°F the heat content of the atoms is so small that an astronaut wouldn't feel it at all, even if all the 100,000,000°F hydrogen atoms were crammed together and stuffed in the astronaut's ear. That shows you what we mean when we say that heat measures the amount of thermal energy and temperature measures its intensity. This fact has a lot of very important uses in welding, brazing, and soldering that will come clear when we look at the flame temperature versus the heating value of different fuel gases.

Btu's and calories The heating value of a fuel gas is obviously important, especially since different fuel gases have widely different heat values (and therefore differing abilities to do preheating, welding, brazing, flame-hardening, and flame-cutting). The Btu's of heat energy produced by a fuel gas is one measure of the heat energy produced by that gas when it is burned with oxygen. It's usually expressed in units of Btu/ft^3 of gas. Similarly, kilocalories per cubic metre ($kcal/m^3$) is the most common metric unit when talking about a fuel gas's heating values.

One Btu is the amount of heat energy required to raise one pound of water one degree Fahrenheit. Similarly, one kilocalorie (or 1000 "small" calories) is the

amount of heat required to raise one kilogram of water (about 2.2 lb) one degree Celsius.

Natural gas produces only 1000 Btu/ft^3 when completely burned in pure oxygen. Compare that with 2406 Btu/ft^3 for MAPP Gas, 2371 Btu/ft^3 for propylene, and 1470 Btu/ft^3 for acetylene and you will immediately see why these three gases are used much more frequently than natural gas for most welding, heating, and cutting jobs. That's why they are considered to be high-energy fuel gases, and natural gas is not.

Celsius and Fahrenheit temperature scales No doubt you've been noticing the little (°C) symbols that we have been using after the Fahrenheit temperatures. You know that this is the metric temperature scale based on degrees Celsius, especially now that many weather reporters on TV are giving the daily temperatures in both degrees Fahrenheit and Celsius. You also may know that it is a much more logical scale to use than the Fahrenheit scale. Here's why.

212°F	100°C	Water boils
32°F	0°C	Water freezes

In the Celsius system, all you need for a thermometer is pure water at 1 atm (sea level) pressure. You measure the temperature at which water boils, the temperature at which it freezes, and then you arbitrarily call the higher one 100°C and the lower one 0°C. If you then divide the "distance" between these two temperatures into 100 equal parts, each part on your new temperature scale is defined as 1°C.

The Fahrenheit system, on the other hand, makes no sense at all, except that we are used to it. The point at which water boils at 14.7 psia or 1 atm pressure is 212°F. Water freezes at 32°F. Thus there are 180 temperature-scale units between them, and the zero point corresponds to nothing special at all. It is far more difficult to make temperature calculations with the Fahrenheit scale than with the

Celsius scale. Interestingly, the two scales are exactly equal at at least one point; −40°F = −40°C. Also, your body temperature of 98.6°F = 37.0°C (exactly). Knowing that, and combining it with the fact that 32°F = 0°C, you have a handy way for "feeling" what different Celsius temperatures are like. If the TV weather reporter says it's 30°C outside, it's about 86°F, or a pretty warm day.

How do you get from one temperature scale to another? There are two simple formulas that will do the job for you. They are:

$°C = 5/9 (°F − 32°)$ Celsius when Fahrenheit is known

$°F = 9/5 (°C) + 32°$ Fahrenheit when Celsius is known

For example, if you have 1000°F and want to know what that is on the Celsius scale:

$°C = 5/9(1000°F − 32°) = 5/9(968°F) = 538°C$

Now let's say that you have 728°C and haven't the slightest idea what that is on the Fahrenheit scale.

$°F = 9/5 (728°C) + 32° = 1310°F + 32° = 1342°F$

We are bothering you with the metric system for temperatures on purpose. It can make your life a lot easier when figuring out certain practical things about gases.

The absolute temperature scales You may be wondering why we are stuffing all this math into a shop course. It's for a very practical reason, because the ideas (even if you hate math) will help you in your work.

Table 3-2 shows you not two but four temperature scales. Why do you need two additional temperature scales? Remember, we said that temperature is a measure of how fast atoms or molecules are vibrating or jumping around? Imagine gas molecules. Then imagine that they slow down, becoming slower and slower until they stop moving. Does that mean that there's no more temperature? Yes, that's exactly what it means. There is a lower limit to how cold anything, gas, liquid, or solid, can get. No motion. No temperature. This point is called *absolute zero*.

There are two scales that start at absolute zero and go up from there. One scale, the Kelvin scale, uses units equal to those of the Celsius scale. The other, the Rankine scale (which is seldom used), starts at absolute zero and goes up in units equal to degrees Fahrenheit.

Very low temperatures, from $-100°F$ all the way down to the point where the temperature scale runs out ($-459.7°$ on the Fahrenheit scale), have become very important to people who use gases, including welders. The reason is that many gases such as oxygen, nitrogen, argon, helium, and hydrogen are now being supplied as liquids in big bulk tanks. Not everyone would agree with the cutoff point, but any gas that becomes a liquid at a temperature below $-100°F$ can be considered to be a cryogenically liquefied gas. Cryogenics is the study and use of very low temperatures. It's a very practical science.

Liquefied gases take up much less space than pressurized gases. Their containers cost less for equal amounts of gas. The shipping costs for liquefied gases are much less than those for equal

volumes of compressed gases. Even on remote construction sites, the oxygen you will use may be delivered as liquid oxygen.

These liquefied cryogenic gases can be at incredibly low temperatures. But when you need the gases as gases and not as liquids, it's very easy to warm them up to room temperature and then use them. The temperatures of liquefied gases often are measured in terms of absolute temperature scales. One reason these absolute scales are used is that they make gas calculations much easier. The other reason they are used is that a gas couldn't care less where some human being decided the zero point on some arbitrary temperature scale would be placed, either at $0°F$ or $0°C$. *Gases expand and contract when heated and cooled in proportion to the absolute temperature.*

The Rankine scale The scale that uses degree units equal to those of the Fahrenheit temperature scale but starts at absolute zero is called the *Rankine temperature scale*. It's named for the physicist who invented it. Rankine also did most of the research on how internal combustion engines work. But his scale is not often used, even in the United States, where almost everything

else is in the English rather than the metric system.

Here's how to convert from degrees Fahrenheit to degrees Rankine. It's very easy.

$$°R = °F + 459.7°$$

That is,

$$32°F = 32°F + 459.7° = 491.7°R$$

Absolute zero on the Rankine scale ($0°R$) is an unbelievably cold $-459.7°F$. Nevertheless, helium, one of the arc-welding shielding gases, only becomes a liquid if it is cooled below $-452.1°F$, or $7.6°R$ above absolute zero.

The Kelvin scale The most commonly used low-temperature scale is based on Celsius-sized temperature units but starts at absolute zero and goes up. It is named for Lord Kelvin, a British scientist who was the first person to realize that there must be a lower limit to temperature.

On the Kelvin scale, the ice-water temperature of $32°F$ or $0°C$ is equal to $273.15°K$. Therefore, absolute zero on the Celsius scale is $-273.15°C$. On the Fahrenheit scale, as we've already mentioned, absolute zero is $-459.7°F$. You can convert from degrees Celsius to Kelvin and back again very simply:

$$°K = °C + 273.1°$$

and

$$°C = °K - 273.1°$$

For example,

$$37°C = 37° + 273.1° = 310.1°K$$

and

$$500°K = 500° - 273.1° = 226.9°K$$

3-4
MAKING GAS CALCULATIONS

Sometimes it's necessary to calculate the way the pressure of a gas changes at different temperatures. It's also useful to know how the pressure changes in a cylinder when you remove gas from it even though the temperature remains the same.

That may sound more like a job for the engineering department than the welder, but any welder

TABLE 3-2 Four temperature scales and how they compare

Fahrenheit, °F	Rankine, °R	Celsius, °C	Kelvin, °K	Comment
5589	6049	3087	3360	Acetylene flame temperature
2800	3260	1540	1813	Low-carbon steel melts
1981	2441	1082	1355	Pure copper melts
1220	1680	660	933	Pure aluminum melts
212	671.69	100	373.15	Water boils
32	491.69	0	273.15	Water freezes
−109	351	−78.3	194.85	CO_2 becomes dry ice
−297.33	162.37	−182.96	90.19	Liquid oxygen boils
−302.55	157.15	−185.86	87.29	Liquid argon boils
−320.36	139.34	−195.75	77.36	Liquid nitrogen boils
−361.1	98.6	−218.4	54.75	Oxygen freezes
−423.0	36.7	−252.8	20.35	Hydrogen liquid boils
−434.6	25.1	−259.2	13.95	Hydrogen freezes
−452.1	7.6	−268.9	4.25	Liquid helium boils
−459.7	0	−273.15	0	Absolute zero

can easily do it. Even if you don't make any of these calculations and forget completely how to do it, the *ideas* behind the calculations are very important. They are so important, in fact, that they can make the difference between your getting injured or not. These ideas also will help you remember, understand, and use a lot of safety rules about different kinds of gases.

Temperature-Volume Changes

When a gas is not tightly sealed in its cylinder (so it can expand and contract unhindered by the cylinder walls), the gas volume increases as the temperature gets hotter and contracts as the gas cools (Fig. 3-9). The only tricky part is that this expansion and contraction are proportional to the absolute temperature (°K or °R) of the gas. It is *not* proportional to the regular Fahrenheit or Celsius temperature so forget them when you make gas calculations or else all your numbers will come out wrong. Let's go through a series of quick calculations to show you how to do it. It will also show you something about gas safety.

An empty acetylene cylinder (at least as indicated by its cylinder regulator gauge pressure dial,

FIGURE 3-9 Temperature-volume changes in an enclosed gas that is allowed to expand and contract. The total *amount* of gas remains the same.

which points to 0 psig) is full of acetylene gas, as you already know. If the volume of the fat steel cylinder is 300 ft³ then there are exactly 300 ft³ of acetylene in this supposedly empty cylinder. Let's also assume that it's an early summer morning and the outside temperature in the cylinder storage shed already is 70°F. It will get a lot hotter before the day's finished.

What's more, some careless welder has left the cylinder valve open and forgot to close the regulator outlet valve. Therefore, the cylinder is not shut tight like it should be whenever it is not in use. How much acetylene can leak out of the cylinder and into the shed on a hot day when the temperature in the shed might get as high as 100°F?

Before we go to work on this problem, let's review the process. We have the volume of gas that we are going to work with. It's 300 ft³. We have the initial temperature of the gas, 70°F. We have the temperature of the gas after it will be heated, 100°F. We have something else, too, the pressure of the gas in the cylinder even though its gauge pressure is zero.

To make things simple, assume that the cylinder shed is close to sea level. (Even if it isn't, it won't make too much difference in the calculation.) Therefore the gas in the cylinder is initially at 14.7 psia. (We told you we were giving you information that was practical; you probably didn't believe that the psia absolute pressure was very important, did you?) We won't need the absolute pressure data now, because we have an open cylinder. We will need the pressure data later on when we calculate the effect of heat on a closed cylinder. That's because, since the gas can't leak out when the cylinder is heated, the only other thing that can happen is for the pressure to rise.

We have all sorts of data and all we need is one more item, the final volume of gas after the acet-

ylene has been heated. That final volume minus the cylinder volume of 300 ft³ will tell us how much acetylene will leak out at 100°F.

The change in gas volume is directly proportional to the change in *absolute temperature*. If the absolute temperature doubles, the gas volume doubles. If the absolute temperature is cut in half, the gas volume is cut in half.

We could even set up a simple equation:

$$\frac{\text{new vol.}}{\text{old vol.}} = \frac{\text{new absolute temp.}}{\text{old absolute temp.}}$$

or, in the case of doubling the temperature, which doubles the volume:

$$\frac{X \text{ volume}}{1 \text{ volume}} = \frac{2 \times \text{absolute temperature}}{1 \times \text{absolute temperature}}$$

That could have just as easily been written:

$$\frac{X}{1} = \frac{2}{1}, \qquad \text{therefore,} \qquad X = 2$$

It's pretty obvious that X also has to equal 2, since both sides of a complete equation have to be equal. That is, the new volume is twice as large as the old volume.

Direct-proportion problems are always quite straightforward. Watch how we do it, and let's see what kind of a problem that careless welder made for you.

First, of course, we have to convert our Fahrenheit temperature into one of the absolute temperature scales. Let's use the Rankine scale because the conversion from degrees Fahrenheit is so easy.

$$°R = °F + 459.7°$$

so that

$$70°F + 459.7° = 530°R$$

(we rounded off the number).

Now we can set up our simple direct proportion equation that says that the old volume of acetylene (300 ft³ at 70°F) and the new volume of acetylene (X ft³ at 100°F) are directly proportional to their absolute temperatures:

$$\frac{X \text{ ft}^3}{300 \text{ ft}^3} = \frac{560°R}{530°R}$$

or

$$X \text{ ft}^3 = \frac{(300)(560)}{530} = 317 \text{ ft}^3$$

The new volume of acetylene will be 317 ft^3 instead of the 300 ft^3 volume of the acetylene cylinder. We wind up with 17 ft^3 of "excess" acetylene. That's how much acetylene will leak out of the accidentally open cylinder when the gas gets to 100°F around noon with a hot sun shining on the cylinder shed. That's enough acetylene to give you quite a bang if a spark happened to occur in or near the shed. It might even be enough to blow the shed's door off its hinges.

Pressure-Volume Changes

Gas volume also changes with pressure (Fig. 3-10). If you have a gas in a cylinder and you double the cylinder pressure, you reduce the volume of gas originally in the cylinder by half. Of course, since you can't change the volume of the steel cylinder (unless you run over it with a bulldozer), what this really means is that you can cram twice as much gas in a cylinder simply by doubling the cylinder pressure as you pump more gas into it.

Similarly, if you open the valve on the cylinder and let gas out (allow the gas volume to expand), the gas pressure inside the cylinder drops. In fact it drops in what is called the *inverse proportion* to the increase in the volume of the escaping gas.

An inverse proportion problem is just like a direct-proportion problem except that it's partly upside down. Gas volume and gas pressure are inversely proportional to each other. As one goes up, the other goes down. As one is doubled, the other is cut in half. If you increase one of them by 10, the other is cut to one-tenth of its original value. The way you'd write that mathematically is

$$\frac{\text{New volume}}{\text{Old volume}} = \frac{\text{Old pressure (psia)}}{\text{New pressure (psia)}}$$

or, in the case of doubling the pressure, which cuts the volume in half (or vise versa),

$$\frac{X \text{ volume}}{1 \text{ volume}} = \frac{1 \text{ absolute pressure}}{2 \text{ absolute pressure}}$$

That could have just as easily been written:

$$\frac{X}{1} = \frac{1}{2}, \qquad \text{therefore,} \qquad X = \frac{1}{2}$$

Be careful in making these calculations. Don't use your cylinder regulator pressure (psig) without adding on your atmospheric pressure (14.7 psia at sea level, also use 14.7 psia whenever you're not sure what the atmospheric pressure might be) to convert the pressure into absolute or psia units.

We already calculated that the gas in our overheated acetylene cylinder would increase in volume from 300 ft^3 to 317 ft^3 if it were open, and heated from 70°F up to 100°F. What would happen to the pressure inside the cylinder if the acetylene cylinder were closed? That is, if the volume didn't change so that the cylinder pressure was forced to increase?

For starters, the gas in the "empty" acetylene cylinder was at atmospheric pressure, or 14.7 psia (even though the cylinder pressure gauge read 0 psig). That's the only additional fact we need to complete the calculation. Here it is:

$$\frac{\substack{\text{New pressure} \\ \text{(psia)}}}{\substack{\text{Old pressure} \\ \text{(psia)}}} = \frac{\text{New absolute temp.}}{\text{Old absolute temp.}}$$

$$\frac{X \text{ psia}}{14.7 \text{ psia}} = \frac{560°\text{R}}{530°\text{R}}$$

That could have just as easily been written:

$$X = \frac{(14.7 \text{ psia})(560°\text{R})}{530°\text{R}} = 15.53 \text{ psia}$$

We're not quite through yet. Let's convert 15.53 psia back into something more familiar, a conventional cylinder regulator pressure, by subtracting the atmospheric pressure (remembering that cylinder regulators measure the *difference* in pressure between the inside and outside of the cylinder):

15.53 psia − 14.7 psia = 0.83 psig

Not much change at all. Now let's burn down the cylinder shed. The hot cylinder is going to rise

from 70 to 1500°F. Watch what happens to the cylinder pressure:

70°F = 70 + 459.7 = 530°R

1500°F = 1500 + 459.7 = 1960°R

so that:

$$X = \frac{(14.7 \text{ psia})(1960°\text{R})}{530°\text{R}} = 54 \text{ psia}$$

Now

54 psia − 14.7 psia = 39.3 psig

It just so happens that the gas in our empty acetylene cylinder has just increased in pressure by 40 psig. Do you know what the cylinder pressure would have been if it had been a full acetylene cylinder at about 250 psig? The gas in the cylinder would have risen to about 925 psig, which is about three times the allowable pressure for any acetylene cylinder.

Would it explode? Not quite. All gas cylinders have safety valves on them for overpressure situations just like this one. But if the safety valve blew on a full acetylene cylinder in a fire it would dump about 300 ft^3 of acetylene. Of course that is the lesser of two evils. Without a safety pressure relief valve, the cylinder is a bomb. It might have done more than fuel the shed fire. It might have blown the shed away.

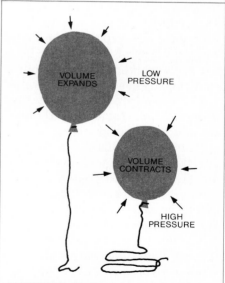

FIGURE 3-10 Pressure-volume changes in an enclosed gas that is allowed to expand and contract. The total *amount* of gas remains the same.

Acetylene is a moderately low pressure cylinder gas. There may have been some high-pressure oxygen cylinders in the shed, too. Want to see what happens to the pressure inside a full 2200-psig cylinder when it's in a fire if it weren't for the safety relief pressure valve?

$$X = \frac{(2200 \text{ psig} + 14.7 \text{ psia})(1960°R)}{530°R}$$
$$= 8190 \text{ psia}$$

It's not even worth subtracting 14.7 psia from that to get the cylinder pressure in psig. We've got a pressure of almost 4 tons/in.². It's like having a pickup truck sitting on every square inch of the cylinder wall. Now you know why *all* gas cylinders have safety relief valves. They also have fusible plugs in the bottom that will melt away and release the pressure in case of a fire if the safety relief valve doesn't work. Never, ever tamper with any safety device on a cylinder. Never overheat any cylinder. Don't even store them in the direct sun. If you can't do anything else, put a canvas shade over them. Never store oxygen and fuel gases in the same location.

Now let's see why some welders have been asphyxiated. How much argon shielding gas will escape from a full 270 ft³ argon cylinder? The cylinder's gauge pressure is 2200 psig. The 270 ft³ is the total volume of gas in the cylinder if the gas were allowed to leak out over the weekend because a welding gun was carelessly left on the floor with the trigger slightly depressed (which causes the shielding gas to come out of the cylinder, through the gun, and into the air). There are various sizes of high-pressure gas cylinders, but 270 ft³ cylinders and 244 ft³ cylinders are common sizes.

You don't have to make any detailed calculations because the rated gas volume of the cylinder, 270 ft³, is in a measure called *standard cubic feet*. What that means is that the contents of the cylinder would take up 270 standard cubic feet or 270 scf of space if the gas were released at 14.7 psia (sea level) and at a temperature of 70°F. Most gas volumes for cylinders and other gas containers are precalculated in standard cubic feet. The gas volume is adjusted for standard conditions. Whenever you see the letters *scf* after a gas volume, you know that the numbers are in standard cubic feet, that is, they are preadjusted for sea-level pressure and 70°F [21°C].

The oxygen you breathe is only 21 percent of the air volume. If that oxygen level drops to 16 percent, you have less than 5 min left before you die. You only have to dilute the oxygen in the air from 21 percent to 16 percent to get in serious trouble. How much dilution of the air is required? Another direct proportion will solve it easily.

$$\frac{X\% \text{ bad air}}{100\% \text{ good air}} = \frac{16\% \text{ oxygen}}{21\% \text{ oxygen}}$$
$$X\% \text{ dilution} = \frac{(16)(100)}{21}$$
$$= 76\% \text{ bad air}$$

In other words, if you replace about 24 percent of your air with any gas other than oxygen, it will kill you. Let's use another direct proportion to find out how much space the leaking argon cylinder will fill to reduce the breathing air to dangerous levels. It's another direct proportion.

$$\frac{X \text{ ft}^3 \text{ of space}}{270 \text{ ft}^3 \text{ of argon}} = \frac{100\%}{24\%}$$
$$X \text{ ft}^3 = \frac{270 \times 100}{24}$$
$$= 1125 \text{ ft}^3$$

Now a 1125-ft³ space can be a box-shaped room just about 10 ft on a side. That's just about the size of a small compartment in the hull of a ship, or the volume of a small pressure vessel or a manhole in the street.

If you see somebody (in the hold of a ship or inside a water tank) suddenly collapse, think before helping. Don't go after them to pull them out unless you can very quickly get a rope around yourself and get a crew member to pull you out, too. If you ever suddenly start to get a headache, dizzy, or feel short of breath for no apparent reason while you are working in any enclosed area, run for the nearest fresh air.

Temperature-Pressure-Volume Changes

Let's review what we've learned so far. The absolute pressure of a gas increases in *direct proportion* to the absolute temperature if you hold the volume constant. If you heat up or cool down a closed cylinder, the cylinder pressure will also go up or down exactly the same amount.

The absolute pressure of a gas increases in *inverse proportion* to a change in the volume. As a gas expands out of a cylinder (increases its volume) the cylinder pressure decreases.

Standard cubic feet means the volume of a gas adjusted to standard conditions of pressure and temperature (sea-level pressure and 70°F).

Now let's put it all together in one all-purpose formula.

$$\frac{\text{New vol.}}{\text{Old vol.}} = \frac{\text{new temp.}}{\text{old temp.}} \times \frac{\text{old pressure}}{\text{new pressure}}$$

This one simple equation combines our two direct and inverse proportions. Using the proper absolute temperature and pressure units, if you know five of the terms, all you have to do to calculate the remaining one is fill in the numbers and do the math. It often is simpler than that.

If the pressures in the problem remain the same, cancel them out and you'll have the previous equation that relates gas volume changes with temperature. If the volumes don't change (the cylinder is closed), cancel out the old and new volume values and the left-hand side of the equation will equal one. Then you'll have the equation that shows how gas pressure varies with temperature. What's a lot more important, though, is *understanding the ideas* behind this equation.

FIGURE 3-11 The big fat liquid oxygen cylinder behind the civil engineer on this construction site feeds vaporized oxygen to workers for days at a time without changing cylinders. The small cylinder of MAPP Gas just to the right of the big liquid oxygen cylinder is the fuel gas called MAPP Gas. Note the length of the big cutting torch. Gas welding and cutting equipment comes in many sizes. The next chapter tells you all about it.

Using Standard Cubic Feet

If you already have your volume data in terms of standard cubic feet, most of the time it's not worth adjusting it for your particular atmospheric pressure and temperature (unless you're working in the Peruvian Andes or at −60°F in the dead of a Montana winter) because the numbers will be close enough for most quick calculations.

We put these gas calculations in terms of your safety, because we think that's important. But it should be obvious that the very same calculations are useful in buying gases and calculating how much gas you need for a job, how much gas is left in your cylinder, and how many cylinders you'll need next week.

Using standard cubic feet, you can even compare a liquefied gas to the same gas supplied in high-pressure cylinders. For example, 1 gallon (gal) of liquid oxygen equals 115.1 scf of oxygen gas. Therefore, it's easy to see that about 2 gal of liquid oxygen equals one 244-scf high-pressure

gas cylinder because 244/115.1 = 2.1 gal.

Liquid gas cylinders (commonly called *liquid cylinders*) contain cryogenic liquid oxygen, nitrogen, or argon at very cold temperatures and are now commonly used both in plants and even on construction sites (Fig. 3-11). Liquid cylinders normally operate at only 125 psig. They have vaporizers that will warm up the ultracold cryogenic liquid so that you can use it.

The capacities of these small cryogenic or ultracold liquid tanks range from 3630 scf for nitrogen to 4500 scf for oxygen. One liquid oxygen cylinder standing about 4 ft high holds about 45 gal of liquid oxygen or as much oxygen gas as twenty-eight 244-scf high-pressure cylinders. You can carry more oxygen to a remote construction site on one pickup truck using liquid cylinders than you could deliver with several large delivery trucks hauling compressed-gas cylinders. We'll tell you more about cryogenic storage tanks in the next chapter.

If you are not sure of the proper way of using or storing any gas, ask somebody. If nobody in your plant knows, call the welding distributor who delivers the cylinders. Welding distributors who handle gases are specialists in their safe handling and use. Whenever you have a problem or need advice, they will be glad to help you.

3-5
OTHER IMPORTANT NUMBERS

Continuing our adventures with gas physics, we're going to look at some other important data on gases. But this time there won't be any calculations. What we'll tell you about are some additional basic facts about gases in general, and then we'll finish this chapter with some facts that are specific to the fuel gases that you burn to get your work done.

Boiling Range

Very few gases are like hydrogen. Hydrogen is seldom used as a fuel gas and never as a shielding gas. Hydrogen (like oxygen, argon, and helium) is delivered in high-pressure gas cylinders (slightly over 2200 psig when full). The other fuel gases are delivered in medium- to low-pressure cylinders, and most of them are liquids under pressure. All fuel gases contain hydrogen atoms as part of the fuel-gas molecule; this includes acetylene, MAPP Gas, propylene, propane, natural gas, and other gaseous fuels like the butane in butane cigarette lighters. Gases like butane are not used in welding because they don't produce enough heat and temperature. Gases that are made out of hydrogen and carbon atoms are called *hydrocarbons*. All fuel gases, except hydrogen itself, are hydrocarbon gases because they are combinations of carbon and hydrogen, which makes them organic compounds.

Most hydrocarbon gases become liquid when subjected to moderate pressure (typically in the range of about 100 to 300 psig). Therefore, they also are called *liquefied fuel gases*. Their cylinders are not like the thick, heavy-walled cylinders used for high-pressure gases. They are designed for use at only moderate pressures.

When the pressure on a liquefied fuel gas in the cylinder is released (by opening the cylinder valve), the liquid fuel gas boils, even at room temperature. When the liquefied fuel boils, it turns into a gas and that gas is what you burn (using either air or oxygen to support the fuel-gas combustion) to produce heat. Liquefied fuel gases include MAPP Gas, propylene, propane, and, in fact, all the hydrocarbon gases *except* acetylene. Acetylene is a unique fuel gas and we are going to talk about it separately in Chap. 4, although we'll include some information on it here.

If you happen to be working with a gas whose liquid form has a higher boiling point than the outside temperature, the gas won't come out of the cylinder. It happens in very cold weather (it doesn't have to be Alaska, any northern state during winter can get cold enough). Even if the outside temperature around the cylinder is slightly below the boiling point of the liquid fuel gas, the gas will only come out of your cylinder very slowly. What do you do?

What you don't do, ever, is heat the cylinder with an open flame. You can increase the gas flow by putting the cylinder in a warm shed. You can even use electric heating blankets like many construction workers do in Alaska. Or you can quit work and wait for warm weather.

Table 3-3 lists the temperatures that can cause problems for you.

Vapor Pressure

Another fact worth knowing is a fuel-gas liquid's vapor pressure. This is the pressure of the gas that has evaporated and now sits above the liquid inside the cylinder (Fig. 3-12). It determines the actual gas pressure in the cylinder at a given temperature.

For example, at room temperature the vapor pressure of MAPP Gas is 94 psig, which means that the cylinder pressure of a full MAPP Gas cylinder at 70°F is 94 psig. That's handy to know because if the cylinder pressure gauge is much below that figure, you know you don't have a full cylinder. Of course, the vapor pressure will be lower if the full cylinder is cold. That goes right back to the boiling point of liquefied fuel gases.

The vapor pressure of propane at room temperature is 120 psig and the vapor pressure of propylene at room temperature is about 220 psig. Those also, of course, are the cylinder pressures for these gases when the cylinder is full and the temperature around it is 70°F.

As you can see, these "normal" cylinder pressures are much lower than the standard 2200 psig cylinder pressures for oxygen, nitrogen, argon, helium, and hydrogen gases. That's why these gases are called *high-pressure* gases and that's why their cylinders look so different; tall and skinny instead of short and fat. In contrast, natural gas out of a pipeline is usually delivered at less than 1 psig. That's an example of a low-pressure fuel gas.

3-6
HEAT TRANSFER

It should be obvious by now that a welder's job involves controlling heat to get useful work done. Heat moves in three different ways from your flame (or even a welding arc) to your workpiece. These ways are radiation, conduction, and convection. Knowing how heat moves around will have a direct bearing on your safety as a welder and on how well you do your job. Figure 3-13 shows you how these three modes of heat transfer work.

Radiation

Radiation (Fig. 3-13A) is the same process that light uses to travel from one place to another and that radio and TV waves use to get from a broadcast station to your receiver. All bodies radiate heat unless they are at absolute zero temperature. Even a cake of ice placed near a steam radiator sends heat to the radiator. Meanwhile, of course, the radiator sends a lot more heat back to the ice.

Figuring how radiation works gets a little complicated, but it is worth understanding because radiation is the primary (but not the only) reason you wear welding goggles if you are gas welding or cutting, and a dark face plate in your welding mask if you are arc welding. (In fact, radiation from the welding arc is the reason you have the mask covering all of your face, whereas you don't need full face coverage for gas welding.)

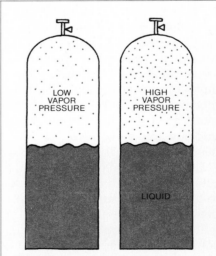

FIGURE 3-12 Vapor pressure over a liquefied gas in a cylinder.

TABLE 3-3 Boiling temperatures of fuel gases

Fuel Gas	Boiling Point, °F at 1 atm
MAPP Gas	−4 to −36 (MAPP Gas is a mixture of several gases, so it has a range over which it boils.)
Propane	−50
Propylene	−54
Acetylene	−84

The basic law for radiation is that any object radiates heat in direct proportion to the fourth power of its absolute temperature. (There's those absolute Rankine and Kelvin temperature scales again.) Thus, if the absolute temperature is doubled, the object's radiation will be 16 times as great, or $2 \times 2 \times 2 \times 2 = 16$.

This means that hot weld metal will radiate 16 times as much heat as colder weld metal that is at only half the absolute temperature. This also explains (as you'll learn later on) why higher arc-welding currents require a much darker shielding glass in your face plate than lower arc-welding currents. Higher arc-welding currents produce more energy, which results in much higher radiation.

Another useful law about how intense radiation will be (whether it's from a light bulb or white-hot steel) as you walk toward it or back away from it is that the intensity of the radiation varies as the inverse square of the distance you are from it. That sounds pretty complicated but it's not. If you are getting "cooked" by the heat from a batch of molten steel, get back from it twice as far as you were before and the heat will be only one-fourth as intense because $(\frac{1}{2})^2$ equals $\frac{1}{4}$. Get back three times as far, and the radiation intensity will be only one-ninth as great because $(\frac{1}{3})^2$ equals $\frac{1}{9}$.

It also works for light bulbs and welding arcs. You have to get four times as close to a 60-watt bulb to get the same amount of light on the page of this book as you get from a 120-watt bulb (which produces twice as much light).

Practically speaking, this means that things radiating heat, or light, or even x-rays, will get "hot" a lot faster than you think they will when you get close to them (the radiation gets more intense in proportion to the *inverse square* of your distance from the radiating source), and the heat or radiation will fall off a lot faster than you might expect as you

back away (as the inverse square of the distance).

It's not necessary to be able to state this fact as an equation, but it's why holding your bare hand in front of you as you get near a lot of hot metal or a large heat treating furnace will keep you from walking into something unpleasant and unexpected.

Radiant energy can travel through a vacuum. Sunlight does. So do x-rays. Both are examples of radiant energy. The sun's heat and light get to the earth from 98 million miles away in space by radiation. The ultraviolet radiation from an electric arc can hurt your eyes when you weld unless you use special very dark glass in the lens of your helmet. The arc-welding helmet also is bigger than your entire face because the same ultraviolet radiation can give you a very bad sunburn, just like the rays from the sun do when you are on the beach.

Since the welding arc is much hotter than a gas flame, the heat radiation (heat radiation is called *infrared radiation*) also is much more intense than a gas flame. With a gas flame, you don't have to worry about being sunburned or overheated as much as you do when arc welding, and goggles are enough protection.

Interestingly, light waves, infrared radiation, and ultraviolet radiation don't penetrate solid materials, except glass. They simply aren't powerful enough. The only type of radiation that can penetrate solids is very high energy radiation, such as x-rays and gamma rays, both of which are used in welding inspection. Special precautions are always taken when working with x-radiation and gamma radiation (which is of the same basic wavelike nature as radio waves, infra-red radiation, visible light, ultraviolet radiation and x-rays, but they are simply more energetic).

Most kinds of radiation travel well through the air. There isn't enough matter in the gas in the air to stop it. Any welder knows

that and uses this fact to advantage before picking up a piece of metal that might just be too hot to handle. Simply wave your hand near the metal. If you feel it radiating heat, it's too hot to pick up without gloves even though it's cooled down enough so that it is no longer radiating light (glowing).

Conduction

Heat moves through metals and other solids by conduction (Fig. 3-13B). Metals conduct heat much better than nonmetals such as wood or rubber. High conductivity is a characteristic of any metal. Some metals such as silver, or copper, or aluminum (in that order) have much higher thermal conductivity (as it's called) than carbon steel or stainless steel.

If you have one end of a copper bar that is several feet long, and you heat the other end of the bar with a torch, the cold end soon gets very hot, too. This is thermal conduction down the length of the bar from the hot end to the cold end.

You can control radiant heat by blocking it out with a solid workpiece (or even the fiberglass cover of your arc welding mask). But a material will still conduct heat through itself, even fiberglass, if one side gets hotter than the other.

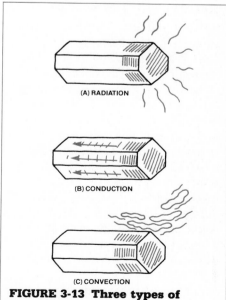

(A) RADIATION

(B) CONDUCTION

(C) CONVECTION

FIGURE 3-13 Three types of heat transfer.

You can and will use this fact to advantage. Lots of times when you are brazing or soldering you don't want to put the torch directly on the solder or brazing filler metal because the flame may overheat the material, or maybe the pressure of burning gas will push it away. The simple solution is to heat the workpiece from the other side. Heat will be conducted very quickly through the workpiece and it will melt your brazing filler metal or solder in a nice, controlled manner. That's exactly what you do when you "sweat solder" a pipe connection or an electric circuit.

Convection

Let's look at the radiator again. The cast iron in an old fashioned radiator is heated by steam or hot water. The heat gets from the inside of the radiator to the outside through the solid cast iron by conduction. Then a combination of radiation and conduction heats up a layer of air near the surface of the radiator. The warm swirling air moves away from the radiator because gases expand as they get hotter (as you know after calculating that fact for hot open cylinders). Since the air expands, it gets lighter. It has lower density than colder air. The hotter, less dense air rises and colder, denser air takes its place next to the radiator and the heating process continues. That's heating by convection (Fig. 3-13C).

Convection is how heat moves in liquids and gases. Think of it as boiling, swirling, and mixing, and you get the idea. You can hold a torch on molten weld metal and watch it boil, too. The top of the weld puddle is being heated by radiation from your torch flame or welding arc. But the metal inside the molten puddle is being heated by conduction and convection; conduction because it's a metal (molten metal still conducts quite nicely) and convection because it's a liquid.

Hot gases produced around your welding torch flame or welding arc also heat up the workpiece and weld metal by convection. In fact, your welding flame itself is really a kind of high-temperature gas mixture moving at a very high velocity toward your workpiece.

3-7
COMBUSTION

Combustion, or burning, is a chemical process. When things burn, no matter what they are, it means that a very rapid chemical reaction is going on that releases a lot of heat while raising the temperature of the material being burned. Wood burns in air only because the air has 21 percent oxygen by volume (or 23 percent by weight) in it. The burning wood is combining with the oxygen.

Metals also burn in oxygen. Iron burns very rapidly in pure oxygen to produce iron oxides. Zinc burns in oxygen to form white zinc oxide. And, of course, fuel gases burn in oxygen. Since all the fuel gases we are considering are made up of hydrogen and carbon (hydrocarbons), the hydrogen combines with the oxygen to form water (H_2O) and the carbon burns in the oxygen to form carbon dioxide (CO_2). All hydrocarbon fuels have the same end products, water vapor and carbon dioxide, when they are completely burned up in oxygen. If there is not enough oxygen to burn these fuel gases completely, you can get water plus carbon monoxide (CO), or even carbon (C), which looks bright yellow in your flame but becomes black when it cools.

Oxygen, incidentally, is not the only chemical that makes things burn. Hydrogen burns in oxygen to produce water and lots of heat and light energy. But hydrogen also will burn in chlorine gas (releasing even more energy) to produce hydrochloric acid (HCl). But as far as we, as welders, are concerned, oxygen is the only gas that makes things burn to produce heat.

Never make the common mistake of thinking that oxygen burns. Oxygen is not a fuel gas. Oxygen is not flammable. It only supports combustion (with certain odd chemical exceptions like ozone, or O_3, where oxygen "burns" itself). It may take a moment for you to agree with that idea, but when you think about it, it will become clear that oxygen doesn't burn. Oxygen is not a flammable gas. It only helps other things burn.

If you have a spark source or even a lit match and put it inside pure oxygen, the match will burn, but the oxygen won't. If you had some way to eliminate the match stick and leave just the flame, the flame would go out because there is nothing to burn. Oxygen does not burn. It only supports combustion. It makes other things burn, often violently.

Kindling Temperatures and Flash Points

Things need more than oxygen to burn. Other special conditions also are required. Hydrogen and oxygen are very dangerous when mixed and ignited. They burn explosively if not burned under carefully controlled conditions in the correct kind of torch and torch tip. However, a hydrogen-oxygen gas mixture won't do anything without some kind of heat input to start the reaction going.

Steel won't burn in oxygen unless the steel is heated red hot (1600°F or 871°C). Even paper won't burn in oxygen unless the paper is heated to about 453°F [234°C]. The equivalent of flash point when you talk about materials is called the *kindling temperature*. The high kindling temperature of steel in oxygen at 1600°F is going to be a problem if you try to flame-cut steel before it's hot enough.

In other words, any material must be heated to a certain temperature before burning starts. You must preheat the steel before it will burn. Only when the steel gets heated enough by the preheat flames in your cutting tip will the jet of pure oxygen coming out of the middle of the torch tip make the steel burn.

Fuel gases have something similar to kindling temperatures, called *flash points*. These flash points are very low. Because of their low flash points, fuel gases combine with oxygen very easily and the chemical reaction is very fast and produces a lot of heat. The heat produced in the reaction heats up more of the gas to its flash point, and the process feeds on itself in a chain reaction.

Nitrogen, argon, helium, and carbon dioxide are not flammable gases. Argon and helium won't combine with oxygen at all. They are totally inert gases. They won't combine with themselves, either. Therefore their liquids and gases are made of atoms and not molecules. Nitrogen won't combine with oxygen unless the temperature is very high. Nitrogen often is used as if it were an inert gas, even though it isn't. Carbon dioxide won't combine with oxygen because it already has. Carbon dioxide, nitrogen, argon, and helium will put out fires, not start them.

Carbon dioxide is used in the common red wall-mounted fire extinguishers you see in building hallways. Large volumes of nitrogen have been pumped into burning coal mines to put them out. Nitrogen is also used to blanket chemicals inside tanks that might otherwise react with oxygen or air. It's sometimes used to purge these tanks of explosive vapors before cutting or welding them.

Don't ever try this. If you make a mistake, the results will incinerate you, but even gasoline has a flash point—you can put out a small fire by dumping gasoline on it . . . if the gasoline cools the burning material below its flash point before the burning material raises the gasoline to its flash point.

Flame Temperature
The flames produced by burning fuel gases in air or in oxygen have temperatures as well as heat content. While the heat energy (Btu's) produced by a flame tells you how much more (or less) of the metal the flame can melt in a given time, the temperature of the flame tells you whether a fuel gas can ever heat the metal enough to melt it. That can be a very important consideration when you select a fuel gas.

Acetylene has a flame temperature that can go as high as 6000°F [3316°C] if you use excess oxygen to burn it. However, the typical temperature for an acetylene flame is more like 5589°F [3087°C]. Propylene's flame temperature is 5189°F [2865°C] and MAPP Gas has a flame temperature of 5301°F [2927°C]. Natural gas only has a flame temperature of 4600°F [2538°C]. Propane has a flame temperature of 4579°F [2526°C].

All these flame temperatures assume that you are using pure oxygen with the fuel gas. You will get lower flame temperatures if you burn the fuel gas in air (which only contains 21 percent oxygen). Later on you will learn about air-fuel torches used for small brazing jobs and by plumbers. These torches produce lower flame temperatures than oxyfuel torches.

If any fuel gas you use has a high enough temperature to melt a metal, then the heat content of the flame is what tells you how productive that fuel gas will be at melting the metal. Some fuel gases can take hours to do a job that another fuel gas can do in a very short time. Using lower temperature or lower-heat-content fuel gases is like trying to saw wood with a dull saw blade.

A few high-melting-point alloys that contain cobalt can only be worked with acetylene. At the other extreme, natural gas is never used for oxyfuel welding. Its low flame temperature combined with its low heat output (4600°F plus 1000 Btu/ft³) make natural gas almost impossible to use to weld even common mate-

rials. In fact, natural gas is seldom used even for brazing or soldering. However, it is sometimes used for flame-cutting work on very large flame-cutting machines because when steel is burned in oxygen it adds its own heat to the reaction.

By far the two most commonly used general-purpose fuel gases for welding, brazing, soldering, heating, and flame-cutting are acetylene (5589°F and 1470 Btu/ft³) and MAPP Gas (5301°F and 2406 Btu/ft³). They both have high flame temperatures combined with high heat output. Other fuels either are short on temperature, have low heat output, or both. Or they have other drawbacks such as higher cost or limitations in handling them. Natural gas, for example, is delivered in very low pressure pipelines, often at less than 1 psig gas pressure, and isn't normally supplied in cylinders of any kind, portable or otherwise.

Heat Distribution
Temperature and heat value are not the whole story behind the differences among fuel gases. The way heat is distributed inside the flame is even more important as a clue to what the flame will do for you. This fact is so critical to operating a welding torch or doing flame-cutting or heating that many production companies give their welders or burners special lessons on flame settings for doing different kinds of work.

Later on we'll show you what different flame settings look like. Right now we'll do something more basic and show you that flames actually have a structure and that you can look into the flame (with goggles on) and see that structure. Every flame you use in welding, flame-cutting, or heating has at least two parts. Inside the flame you will see a little blue flame cone. This is called the *primary* flame cone. Sometimes you will see a little yellow or white cap on the tiny primary flame called a *feather*. We'll talk about that later. It's one of the important clues you have to making correct flame settings.

But first, let's look at a drawing of an imaginary flame in Table 3-4 that could be made by almost any fuel gas. Along with the drawing is a table of different fuel gases and their different temperatures and heat contents. You'll note that the numbers vary for different parts of the flame.

This imaginary flame and the very real data in the table show that different fuel gases produce their heat in distinctly different parts of the flame. Most of the heat in an acetylene flame is concentrated in the primary flame. The acetylene primary flame, in fact, has just a smidgen more heat than MAPP Gas. Both of these fuels have much hotter primary flames than any of the other fuels. Now look at the heat content of the primary flame of natural gas. See how low it is?

Most gas welding is done by putting the primary flame very near to the filler metal that you will supply by hand, or the base metal that you are joining. Acetylene has the hottest primary flame, which explains why it is the best fuel gas for general-purpose welding, especially when working on sections less than an inch thick. Natural gas is worst; in fact you can't weld steel with it at all. Propane, with all its high total heat value in the secondary flame, can't be used for welding most metals.

The secondary flame cone is far more important for flame-cutting, brazing, and heating than the primary flame cone. If propane didn't have such a low flame temperature, it would be a great flame-cutting fuel. Unfortunately, even with all of its heat content, the flame temperature of propane is so low that it doesn't raise steel to a temperature high enough to do good flame-cutting.

Acetylene and MAPP Gas, on the other hand, are excellent flame-cutting fuel gases. For thick sections, MAPP Gas flame cuts faster than acetylene because it can preheat the steel ahead of the oxygen stream that does the cutting much faster than acetylene can. Propylene also is a good flame cutting fuel. All three fuels have flame temperatures high enough to do the job, and they all have enough heat output in the secondary flame envelope (as the flame cone often is called), as well as in the primary flame cone, to make the cutting job go quickly.

These fuels also are very good for welding thick pieces of iron and steel because big blocks of metal absorb a lot of heat before you can ever get the material up to a good working temperature. (Some large castings are heated for half a day or more before they can be welded.) MAPP Gas and propylene are better than acetylene for heavy heating jobs, too, because MAPP Gas and propylene will heat up thick chunks of aluminum, copper, or gray cast iron. These materials absorb heat like a sponge.

You can see the reason for the difference in the fuels by looking at the heat distribution in their flames. While acetylene has a very high temperature and a lot of heat concentrated in a very small area (and that's excellent for welding thin sheet), the same fact makes it a slow fuel for welding, heating, or flame-cutting thick plate or big cast iron sections. MAPP Gas combines a lot of the best qualities of acetylene with those of good heating fuels. It has a high temperature and high heat.

Amounts of Oxygen

Different fuel gases need different amounts of oxygen to burn completely, because each fuel gas has different amounts of hydrogen and carbon that must be burned up. This "welder's chemistry" is very important, because it explains why different fuel gases require different oxygen and fuel regulator settings and torch adjustments.

Not all of the oxygen you use to make a flame burn will have to be supplied through the torch. In fact, a lot of it comes directly from the air, and only a certain amount of the combustion-promoting oxygen is supplied through the torch hose. This fact also explains some things you'll learn later on about oxyfuel versus air-

TABLE 3-4 The flame properties of five major fuel gases

Fuel Gas	Neutral Flame Temperature, °F	Primary Flame, Btu/ft³	Secondary Flame, Btu/ft³	Total Heat, Btu/ft³
Acetylene	5589	507	963	1470
MAPP Gas	5301	517	1889	2406
Propylene	5193	438	1916	2371
Propane	4579	255	2243	2498
Natural gas	4600	11	989	1000

fuel soldering and brazing torches.

Table 3-5 shows how different fuel gases use different amounts of oxygen. To burn 1 scf of acetylene you have to have 2.5 scf of oxygen. The air around the flame supplies 50 percent of the oxygen needed (assuming you are not working at a very high altitude), so only the other 50 percent of the oxygen needed to burn acetylene completely to form water and carbon dioxide must come from the oxygen cylinder.

Table 3-5 also shows what welders call the *supplied oxygen* (cylinder oxygen, for example) required for other fuels. As you can see, the supplied oxygen requirements for fuel gases range from a low of 50 percent for acetylene to a high of 95 percent for natural gas. This is one of many reasons why natural gas is not often used in joining or cutting metals. Its supplied oxygen requirements are too costly.

What happens to the fuel when you don't supply quite enough oxygen from the cylinder to burn it completely? If you are just a bit short on the oxygen requirement, you will see a flame with an inner cone that has a tiny feather on it. Turn down the oxygen a tiny bit more and the feather gets lighter blue to white. Keep turning down the oxygen and the flame gets yellow and pretty soon, if you keep turning the oxygen down, there is so little supplied oxygen going into the reaction for the available amount of fuel that the flame produces clouds of black soot or unburned carbon. What has all that to do with welding? Plenty.

Carburizing flames The flame that is slightly short of oxygen is called a *reducing* or *carburizing* flame. Certain metals that are sensitive to oxidation are often welded with slightly carburizing flames to protect the molten metal from being exposed to any excess oxygen.

Certain copper alloys, for example, have very good electrical properties simply because they are very low in oxygen content. They actually are called *oxygen-free coppers* or *OFC coppers*. A careless welder can add a lot of oxygen to a molten puddle of oxygen-free copper and ruin the metal's electrical conductivity in the process, simply by using the wrong flame setting.

Oxidizing flames An oxidizing flame is one with slightly more oxygen than is needed to burn the fuel gas completely. Oxidizing flames also have some uses in welding and brazing. Among other things, they tend to be slightly hotter than reducing flames. Certain metals, most notably cast irons, are a lot easier to weld with slightly oxidizing flames.

Neutral flames In between a carburizing and an oxidizing flame is the neutral gas flame setting. It has just the right amount of oxygen to burn the fuel gas completely, no more or less. Most welding and brazing is done with a neutral flame. If you are not sure what flame setting is correct, a neutral flame is the best choice.

Making Flame Settings
The plates on pages 51–54 show you exactly what we are talking about. These color photos show carburizing, neutral, and oxidizing flames for acetylene and for a typical liquefied fuel gas, in this instance MAPP Gas. We chose these two gases for comparison because their different flame settings and appearance cause even the most experienced welders and burners many problems. Flame settings can affect things like cutting speed and welding or brazing results. In fact, flame settings are one of the most important first lessons you can learn in the welding shop. The following will explain why.

Acetylene An acetylene flame burning in air picks up 1.5 times its own volume of oxygen from the atmosphere to burn itself. If the flame needs more oxygen, it must be supplied through your torch. The following comments only cover that extra, cylinder-supplied oxygen.

CARBURIZING FLAME When the supplied oxygen is less than the amount of acetylene fuel gas burned, the acetylene has a strongly carburizing flame. The flame has a yellow or white luminescent inner flame cone and may even be smoky. Molten steel will boil when heated by this flame. Strongly carburizing flames are not used for welding because they add carbon to sensitive metals,

TABLE 3-5 Oxygen consumption for fuels (neutral flames only)

	Total Oxygen Required, ft³/ft³	Oxygen Supplied by Torch, ft³/ft³	% Total Supplied Oxygen, ft³/ft³	Supplied Oxygen, ft³/lb of fuel
Acetylene 5589°F 1470 Btu/ft³	2.5	1.3	50	18.9
MAPP Gas 5301°F 2406 Btu/ft³	4.0	2.5	62.5	22.1
Propylene 5193°F 2371 Btu/ft³	4.5	3.5	77.0	31.0
Propane 4579°F 2498 Btu/ft³	5.0	4.3	85.0	37.2
Natural Gas 4600°F 1000 Btu/ft³	2.0	1.9	95.0	44.9

even low-carbon steels, producing a more brittle high-carbon steel weld metal that will easily crack when the weld puddle hardens.

A slightly carburizing oxyacetylene flame is produced by oxygen-to-acetylene flow ratios (controlled by you at the cylinder regulators and by a valve at the torch handle) of slightly less than a 1:1 oxygen-to-acetylene ratio by volume.

Numbers like 1:1 simply mean that the volume of oxygen equals the volume of acetylene. It doesn't matter what that volume is. It could be 1 ft^3 of each, or one cylinder of each (if their cylinders contained the same amount of gas, which they don't). The practical use is that if you set your acetylene regulator's delivery pressure for 15 psig and you want a 1:1 oxygen-to-acetylene ratio, you set your oxygen regulator's delivery pressure for 15 psig. If the oxyfuel volume ratios were 2:1, then if your oxygen delivery pressure is set for 10 psig, your acetylene delivery pressure should be half that, or 5 psig.

The slightly carburizing acetylene flame has an inner flame cone that is brighter than that of the strongly carburizing flame. It also has pale-green streamers or feathers of partially burned acetylene trailing from the inner cone tip. Slightly carburizing flames are frequently used for welding with low-alloy steel welding rods. Flame temperature at the tip of the inner cone is about 5300°F [2900°C].

NEUTRAL FLAME A neutral acetylene flame is produced by slightly more oxygen than a 1:1 oxygen-to-acetylene ratio. This mildly reducing flame produces a flame temperature of around 5480°F [3027°C] and 1469 Btu/ft^3 of fuel burned. The neutral flame is the most commonly used flame setting for welding or brazing with acetylene and is also often used as the preheat setting for flame-cutting steel.

OXIDIZING FLAME Oxidizing acet-ylene flames are produced by oxygen-to-fuel-gas ratios from greater than 1:1 up to 1.7:1. These are the hottest flame settings (but not necessarily the best for your work). The flame temperature may be almost as high as 6000°F [3300°C], but the flame will oxidize molten weld metal, the molten puddle will spark and foam as the fuel-gas flame literally starts to burn the molten metal, and the welds that will be produced will be full of particles of slag (mostly iron oxide).

Oxidizing flames are only used to weld certain brasses and bronzes or to work glass, for example, when making laboratory glassware. The reason the oxidizing flame is needed for glass is its high temperature combined with the fact that the glass won't be oxidized or burned as easily as metals.

MAPP Gas® Now we'll describe a typical liquefied hydrocarbon fuel like MAPP Gas. You will see that it is very easy to confuse the neutral acetylene flame with the oxidizing MAPP Gas flame (and you've already learned that oxidizing flames are not likely to be what you want at all).

CARBURIZING FLAME The car-burizing MAPP Gas flame is produced by oxygen-to-fuel-gas ratios of around 2.2:1 or lower. Slightly carburizing, or reducing, flames are used to weld or braze easily oxidized alloys such as aluminum. It's also good for oxygen-free copper alloys that must not be contaminated with oxygen.

NEUTRAL FLAME The neutral MAPP Gas flame has an oxygen–fuel-gas ratio of about 2.3:1. Like acetylene, the neutral MAPP Gas flame is the most commonly used flame setting for welding or brazing. The flame will remain neutral up to about a 2.5:1 oxyfuel ratio.

OXIDIZING FLAME Oxidizing flames with MAPP Gas are produced above 2.5:1 oxygen-to-fuel-gas ratios. Here's where you can do a lot of damage or slow down production if you are not careful. *The oxidizing MAPP Gas flame looks very much like the neutral oxyacetylene flame.*

Many welders who grew up with acetylene will try hard to make MAPP Gas oxidizing flames because they look "correct" for acetylene. Then nobody can figure out why this very productive fuel gas is doing such a terrible job. The neutral MAPP Gas flame has a primary flame cone about 1.5 to 2 times as long as the acetylene neutral flame cone. The oxidizing MAPP Gas flame looks like the neutral acetylene flame. As simple as that sounds, it's probably the cause for more human error and lost production than any other factor in using fuel gases. Flame settings are very important.

Other fuels While other fuel gases also have different oxyfuel ratios, the principles are very similar to those of acetylene and MAPP Gas. Carburizing flames are white to bright yellow. Slightly carburizing flames have a little feather of partially unburned fuel at the tip of the inner flame cone. Neutral flames are just right. They have a nice blue color and the inner flame cone will be slightly bluer than the outer flame cone. Oxidizing flames are too hot and too full of oxygen for most uses. They will burn your metal and make steel boil and spark.

3-8
BACKFIRES AND FLASHBACKS
When a candle is lit, the wick takes fire and the wax beneath it melts and forms a pool of liquid that saturates the wick and feeds the flame. The burning or combustion of the wax and the wick are progressive and constant. Similarly, when an oxyfuel gas jet is lit at the end of your welding, heating, or cutting torch tip, the rate of combustion of the combination fuel-oxygen mixture is

FIGURE 3-14 Guess what's missing? The torch operator. He disappeared about 30 seconds after his torch flashed back into the hose. The flame could burn right up the fuel-gas hose and into the black acetylene cylinder on the left. Hopefully, he turned off the fuel-gas cylinder before exiting. This book tells you how to avoid flashbacks.

cutting torches purposely keep the oxygen and fuel gas in separate tubes until just before they are ready to burn. Only at that point are the gases mixed together, just before they leave the torch. The only point where the oxygen-fuel mixture is burning is in a very thin layer at the end of the torch tip.

Any kind of combustion can occur either slowly under carefully controlled conditions or all at once, which means explosively. Finely milled wheat flour, for example, has enough explosive energy to blast down a grain elevator if it is combined with the right amount of oxygen under the right conditions, and somebody adds a spark.

If a substantial amount of a combustible fuel gas mixes with air or oxygen, it will explode if you ignite it. Welders and burners who do not handle their equipment carefully learn this fact the hard way.

Figure 3-14 shows you what can happen. This welder has let a little air remain in the fuel-gas hose when attaching the hose to the cylinder regulator. The welder should have purged both the oxygen hose and the fuel-gas hose before lighting the torch. Now the welder has a flaming hose and a hole blown in the heavy rubber-and-cloth hoses running to the cylinders.

These welding fuel and oxygen hoses have a burst strength of more than 300 psig. You know what a tire blowout sounds like, and auto tires are only filled to around 30 psig pressure. Imagine the sound of a heavy-walled hose bursting at 10 times the force of a tire blowing out and you will always follow your welding instructor's safety directions.

One reason the flame-cutting operator is no longer in sight in the photo is that this kind of accident may not stop with a burning hose. Especially with acetylene, the explosion and fire can continue up the hose and into the regulator, blowing it apart. The

steady and constant. That's why it doesn't blow up. The oxygen-fuel mixture burns back just as fast as the gas escapes from the single hole in the welding torch tip, or from the many preheat holes around the central oxygen bore in the cutting torch tip.

The speed at which a combustible gas burns is constant only for a given set of conditions. This combustion rate changes as the gas-oxygen mixture is changed

and as the flow rates of these gases are increased or decreased. If a container (the cylinder shed with the leaking acetylene cylinder in it) is full of a large amount of oxygen and fuel gas and the mixture is ignited all at once, you'll get an explosion.

Welding and cutting torches are designed to control and compensate for these variables so that you get a steady flame and not an explosion. Welding, heating, and flame and explosion may even travel into the acetylene cylinder and blow everything up. Accidents like this should never happen; they are completely avoidable.

In addition to purging your oxygen and fuel-gas hoses properly before using them, there also are devices you will read about later called *check valves* that should be put between the hose and your regulators and the hose and your torch to stop a reaction like this.

Always purge your oxygen and fuel gas lines both before you use them and after you finish. All you have to do is run a little of the correct gas through each line and it will clear the hose and your equipment of any combustible mixture that may have been sucked up the wrong way.

The following will describe the two kinds of explosions you can get. One is not serious and happens every once in a while. The other kind is very serious and Fig. 3-14 is only a mild example.

Backfires

The terms *backfires* and *flashbacks* are often discussed in welding shops. They should not be confused. They are entirely different.

The backfire, a loud pop-pop-pop sound, is an annoyance caused by incorrect manipulation of your torch, which overheats the tip. Sometimes a dirty tip will do the same thing. Either stop overheating your torch tip or clean it.

When a backfire occurs, the flame momentarily burns back into the torch and immediately pops back out with sounds ranging from a mild pop to a loud bang. Backfires are caused by touching your torch tip to the work (which not only cuts off the gas flow but also overheats the tip and limits the amount of oxygen from the air that is needed to complete combustion in the flame). A little slag or dirt clogging the tip holes will also overheat the tip and cause backfires.

Shoving the tip face down flat against your work will cause a backfire. Overheating the tip by holding it too long inside a large, hot metal joint will do the same.

Sometimes the flame will go out when the torch backfires. If the tip is overheated, the torch will pop-pop-pop-pop like a string of firecrackers until the gas coming through it is turned off or the tip cools.

If the torch goes out and does not relight itself (a hot workpiece often will relight your torch):

■ Close the oxygen torch valve immediately
■ Then close the torch fuel valve
■ Then clean your tip
■ *Remember, turn off the oxygen first, then the fuel. Don't do it the other way around.*
■ Just think "O for Oxygen Off," and do it first.

If you turn off the fuel valve on your torch first, the flame will go out. But you may not have turned the valve completely closed. If the oxygen valve is still slightly open it is not as dangerous as the fuel valve being slightly open, which creates conditions for a possible explosion.

After you have correctly turned the torch gas valves off, check the torch tip with a wire cleaner to make sure that the holes in it are free of dirt or slag. All welders and burners have special wire tip cleaners in their pockets for this purpose. These wires are specially designed for tip cleaning. Do not use any old wire that is lying around. Chapter 4 will tell you more about tip cleaners.

Flashbacks

A flashback is serious. If one occurs, all work should stop to find out what caused it. A flashback means that the flame burns back inside the torch itself, not just the tip. The flame is burning backward faster than the gas is flowing out. The flashback may burn the torch handle, blow the hose

up, and burn or even blow a hole in your heavy-walled brass gas regulator (Fig. 3-14).

CAUTION: If you get a flashback, don't run. First turn off the torch oxygen valves, then the fuel valves. Then, if you think you have time, shut off the valves of the oxygen and fuel-gas cylinders, turning off the oxygen cylinder first, unless you see that the hose flame is heading right for the fuel-gas cylinder. Then quickly make up your mind to turn off the cylinder valve itself or run!

Flashbacks are caused by several basic errors that inexperienced welders are most likely to make. They don't use the proper gas pressures for their equipment. They don't screw their torch tips on tight. They may not have been watching the hose lines and let them get kinked, cutting off the gas flow. The torch tip may be plugged with dirt (often because the careless operator never bothered to clean the tip before starting work). Or the torch tip has been shoved deep into a hot joint and held there too long, or up against molten weld metal where it doesn't belong.

A flashback means that something is radically wrong. Before relighting your torch, check your gas pressures. Are they near the recommended pressures for the job? Then remove both hoses from the torch and the gas regulators and inspect the hoses for damage. Replace them if necessary. If the flame has burned back into the regulator, get a new regulator from the supply room. Then purge your old or new regulator, reconnect your hoses to the two regulators, and purge both of them with gas. After that, connect the torch and tip and purge them with oxygen and fuel gas, too. Only after that should you light up. These are the same procedures you should always follow when first setting up your equipment.

If you have check valves (which are special little one-directional flow valves that you can connect between the hoses and the torch or the hoses and the regulator) use them. Insist that your company use them, too. If you have to, go out and buy some check valves for your equipment rather than wait for the company to supply them.

Under normal conditions, the flame cannot enter the torch tip, torch, or hose because the velocity of the gas flowing out of the equipment is at least equal to the speed that the flame is burning back along the column of mixed fuel gas and oxygen. The outflowing gas often pushes the flame slightly off the end of the tip. That's why backfires and flashbacks won't happen if you use the correct flow or pressure settings for the type of equipment you are using and job you are doing.

Burning velocity Experiments have shown that flame spreads very quickly in oxyfuel gas mixtures. The speed of flame propagation can be extremely fast. It depends upon the ratio of the fuel gas to oxygen in the mixture, the temperature of the gas mixture, and other things. The burning velocity of acetylene in pure oxygen, for example, is 22.7 ft/s. That's a lot faster than you can run; it equals 15.5 mi/h.

MAPP Gas burns in oxygen at a speed of 15.4 ft/s, almost the same as propane's 15.2 ft/s burning velocity and about 30 percent slower than acetylene. Propylene burns at about the same speed in oxygen as MAPP Gas, around 15.1 ft/s. Natural gas burns more slowly in oxygen, at a speed of about 12.2 ft/s. It's not a coincidence that acetylene, which has the highest flame propagation speed in oxygen, is the most likely fuel gas to backfire or flashback.

Explosive limits of fuel gases There is a very good reason for turning off the oxygen first before turning off the fuel gas, not only if you get a flashback, but also any time you turn off your torch. Different fuel gases form explosive or combustible mixtures in oxygen (or in air). It is possible to have so much fuel gas in the oxygen or air that the gas won't ignite. It also is possible to have so little fuel gas in the oxygen or air that you won't get an explosive mixture. This is true of all combustible gases.

If you turn off the fuel valve first on your torch, the mixture of gases coming out of the tip is likely to remain within the explosive limits of the fuel you are using for quite a while. If you turn off the oxygen valve first, the gas mixture passes through the explosive range so quickly that it doesn't matter, meanwhile the flame continues to burn on the end of the tip so that it uses up any explosive mixture by burning it safely. If you turn off the fuel first, the flame may go out, but the gas mixture will be reignited by the hot tip with one very loud bang.

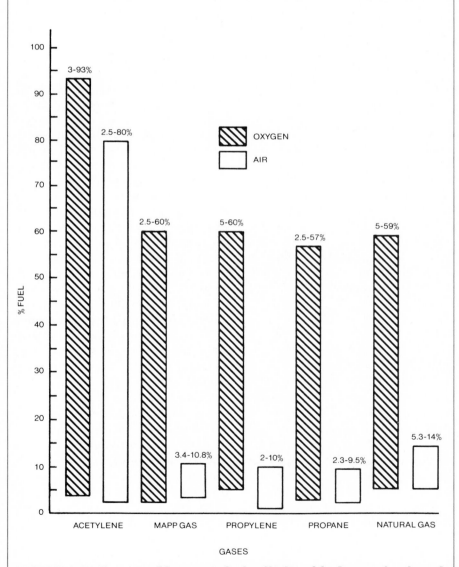

FIGURE 3-15 Upper and lower explosive limits of fuel gases in air and in pure oxygen. The vertical bars show the range of explosive mixtures.

Figure 3-15 shows you the explosive ranges for the major fuel gases in oxygen and in air. The explosive proportions of fuel mixed in oxygen for propane, propylene, natural gas, and MAPP Gas are just about the same and nearly equal when mixed in air. However, acetylene has much wider explosive limits than the other fuels. In fact, the explosive limits for acetylene are just about the same as those for hydrogen. This is another reason why acetylene is more likely to cause a flashback than the other fuels.

Some welders and burners worry more about what happens when a dense gas such as MAPP Gas leaks versus a less-dense gas such as acetylene. Will the MAPP Gas accumulate in low spots over a weekend, for example, and give you a very high concentration of fuel and air that could make a very serious explosive mixture? No. They overlook a fact about gases. Gases diffuse through other gases. They don't remain concentrated in one spot.

The diffusion rate of a gas is inversely proportional to the square root of its density. That means that hydrogen, a very low-density gas, will diffuse very fast through air. Heavier gases like acetylene, MAPP Gas, or other liquefied petroleum gases will diffuse through air more slowly. But the relative diffusion rate of two hydrocarbon gases is not that much different, because it's a function of the square root of their densities, which smooths over minor density differences.

3-9
MAXIMUM REGULATOR WORKING PRESSURE

When you read about regulators in Chap. 4 you will see that the delivery or working pressure of acetylene regulators is red-lined above 15 psig on the cylinder gauge dial. There is a special reason for that. Acetylene is a very unstable molecule. Acetylene gas can burn or, more accurately, decompose all by itself, producing heat even when no oxygen is around. Acetylene under high pressure is a very dangerous substance. Therefore, all acetylene regulators have red lines on them that indicate that you should never, ever turn your acetylene delivery pressure (the working pressure of the gas coming out of the cylinder and through your hose and torch) above 15 psig.

CAUTION: Do not use acetylene at pressures over 15 psig.

Other fuel gases have delivery pressures that are limited only by the size of the regulator or the cylinder pressure, or line pressure if you are using a pipelined gas. For example, you could turn a MAPP Gas cylinder regulator up to 94 psig delivery pressure (which equals the full cylinder pressure at 70°F). Of course you would get one very big flame and you would quickly empty the cylinder, but there is no fundamental safety reason why you shouldn't do it. Similarly, propylene, propane, or even natural gas are only limited by the capacity of the regulator, hose, torch, and tip combination, or the line pressure at which they are supplied.

3-10
YOUR NOSE—THE PERFECT LEAK DETECTOR

You can't see, smell, or taste gaseous oxygen. Like most other gases, oxygen can be liquefied. At −297.2°F [−183°C] oxygen becomes a beautiful robin's-egg-blue liquid. That rather low temperature is the boiling point of liquid oxygen at atmospheric pressure (sea level, 14.7 psia). An oxygen leak is a problem only if it is likely to increase the amount of oxygen in the air around you above 21 percent, which will make things burn faster. Since you can't see or smell oxygen, you can't tell that you are working in a high-oxygen atmosphere. For that reason, a pure oxygen atmosphere obviously is a serious problem. But an oxygen gas leak is not as much of a problem as a fuel-gas leak. (A liquid oxygen leak is more serious simply because so much more oxygen is involved.) A fuel-gas leak can lead to an immediate explosion. Nevertheless, your nose will help you detect a leaking fuel gas. (You can find the leak in your equipment or cylinder by using soapy water, just as you would to locate a leak in a tire. Putting your torch or hose underwater in a bucket will also locate the leak if it's there.)

Acetylene has a peculiar odor. The smell is not overpoweringly strong, but it is pungent and distinct. After you smell acetylene, you will always remember that smell. Any time you smell acetylene in the shop, look for a leak. You can smell acetylene at about one-fifteenth of the concentration of the gas in air at its lower explosive limit.

MAPP Gas also has a strong, distinct odor. Most welders and burners can only describe the smell as "awful" or "it stinks." That's good, because again, if you smell MAPP Gas, something is leaking. MAPP Gas can be smelled at only 100 ppm (100 ppm = 0.01 percent), which is only one-three hundred and fortieth of the concentration of the gas at its lower explosive limit in air.

Neither natural gas, nor propane, nor propylene have a natural odor. However, they often have an artificially added odor put into them just for the purpose of leak detection. Natural gas, propane, and propylene with an odorizer are only detectable at a concentration of one-fifth their lower explosive limit in air. The odorizer most often used is the same one that you smell when you turn on the gas stove at home and don't light it.

These four-color photographs will help you adjust your flame correctly when you use either acetylene or MAPP Gas. The flame adjustments are different for each of these gases, as well as for natural gas or propane which are not illustrated but will resemble MAPP Gas flames more than acetylene flames. Correct flame adjustment makes an enormous difference in your ability to weld, braze, solder, or flame-cut. Note that all the acetylene *cutting* flames are produced by one-piece tips (all acetylene tips are one-piece tips) while MAPP Gas uses either one-piece or two-piece tips for flame-cutting. Propane, propylene, and natural gas also use two-piece tips for cutting. When using two-piece tips for flame-cutting, the primary flame cones of the MAPP Gas preheat flames should be shorter than those used for natural gas or propane. The star-pattern method of cutting-flame adjustment used for MAPP Gas (Plates T, U, and V) should *never* be used for acetylene. Acetylene will backfire or even flash back under these conditions. Also note that the oxidizing MAPP Gas flame looks very much like the neutral acetylene flame setting. Don't mistake the two or you will have trouble flame-cutting because your MAPP Gas cutting speed will be much lower and your cut quality not as good. Flame adjustments for welding with MAPP Gas are not shown because acetylene is generally a better fuel gas for welding. MAPP Gas, however, is excellent for brazing and soldering (as is acetylene).

ACETYLENE FLAME SETTINGS
Welding Torches

A. Acetylene burning in air. Not suitable for welding. The yellow flames are hot, unburned carbon particles.

B. Strongly carburizing flame. Weld metal boils and is not clear.

C. Neutral flame. Weld metal is clean and clear, flowing easily. Correct for most welding. Used on steel.

D. Oxidizing flame. Weld metal will foam and spark. You are burning the weld metal.

Heating Torches

E. Carburizing flame. Not recommended for rapid heating.

F. Neutral flame. Most commonly used.

G. Oxidizing flame. Not usually recommended.

Cutting Torches

H. Strongly carburizing with flame-cutting oxygen flowing. Especially suitable for cutting cast iron.

I. Neutral flame with cutting-oxygen flowing. Standard adjustment for cutting steel.

J. Neutral flame with preheat flames only. Standard adjustment for cutting steel.

K. Oxidizing flame with cutting-oxygen flowing. Not recommended for cutting.

MAPP GAS CUTTING-FLAME SETTINGS
One-Piece Tips

L. Carburizing flame. For stack cutting only.

M. Neutral flame. For machine cutting.

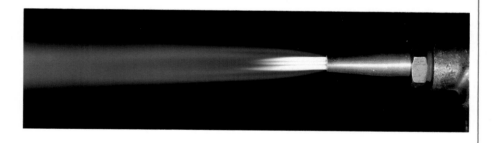

N. Slightly oxidizing flames. For hand cutting or fast starts and for making bevel cuts.

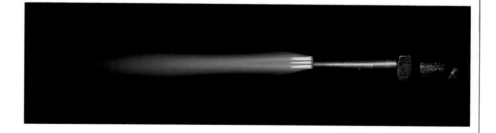

O. Oxidizing preheat flames. Not recommended for cutting.

Two-Piece Tips

P. Carburizing preheat flames. For stacking cutting only.

Q. Neutral preheat flames. For machine cutting.

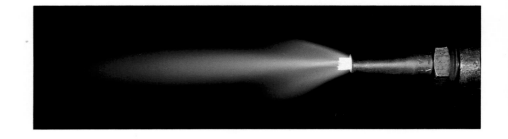

R. Slightly oxidizing preheat flames. For hand cutting or fast starts and for beveling.

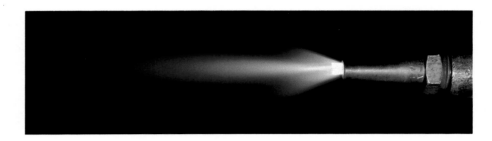

S. Oxidizing preheat flames. Not recommended for cutting.

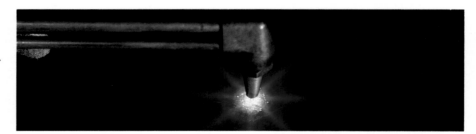

Star Patterns for Setting MAPP Gas Flames

Hold the tip flush against a plate *without* turning on the cutting oxygen. Do *not* use this procedure for acetylene.

T. Neutral preheat flames. For general cutting.

U. Very slightly oxidizing preheat flames. Will result in comfortable operation if the neutral flame setting makes your fingers (in your gloved hand) get too hot.

V. Slightly oxidizing flames. Tight star pattern produces preheat flames that concentrate heat for piercing.

REVIEW QUESTIONS

1. What does psi stand for?

2. What's the difference between psia and psig?

3. A cylinder gauge reads 10 psig. What is the absolute pressure (psia) of the gas in the cylinder?

4. How much pressure is "one atmosphere" at sea level and 70°F, in psia units?

5. At what cylinder pressure (in psig) should you change a high-pressure gas cylinder because the cylinder is getting too low (the gas in the cylinder might become contaminated)?

6. How much heavier is argon than oxygen?

7. A welding shielding gas has 10-ppm impurities. What does ppm stand for, and what is the purity of the gas in percent?

8. A shielding gas has a dew point of −80°F. How much water vapor (moisture) is in the gas?

9. A gas is at 32°F. What is the temperature of the gas in degrees Celsius?

10. Do gases expand or contract when heated if they are not enclosed? If they can't expand (they are enclosed in a container) what happens?

11. What does scf mean? What does scfh mean? Under what conditions of temperature and pressure are they measured (using English, not metric units)?

12. You are working in Siberia and you can't get the acetylene to come out of your cylinder. How cold is the weather where you're working?

13. One kind of radiation will hurt your eyes much more than other kinds of radiation when you are arc welding. What is this kind of radiation called?

14. You stuck a copper rod in a small oxyfuel gas flame on the end of your torch, and even though you are holding the far end of the copper rod 2 ft from the flame, you soon burned your hand. Why? Was it convection, radiation, or conduction of the heat to the cold end of the copper rod that hurt you?

15. Does oxygen burn? What does oxygen do when something else burns?

16. What does Btu stand for? What does it measure? What will one Btu do to the temperature of one pound of water?

17. Fahrenheit and Celsius are not absolute temperature scales (they don't have a zero point that starts at absolute zero). What are the equivalent absolute temperature scales that do start at absolute zero (they are used to calculate the expansion and contraction or changes in the pressure of gases when they are heated or cooled)?

18. What is the pressure inside an empty oxygen cylinder?

19. If the absolute temperature of a gas in a closed cylinder doubles, what will happen to the pressure inside the closed cylinder?

20. What common fuel gas has the highest heat content in the primary flame cone, in Btu/ft³? What common fuel gas has the highest flame temperature? What common fuel gas has the lowest heat content in both the primary and secondary flame cone?

21. Which common fuel gas requires the most supplied oxygen (from a cylinder, for example) per pound of fuel gas when burned? Which common fuel gas requires the least amount of supplied oxygen per pound of fuel gas when burned?

22. Pick one answer from the following. A carburizing flame has:

    a. Excess oxygen.

    b. Just the right amount of oxygen to completely burn the fuel.

    c. Not enough oxygen to completely burn the fuel.

23. What single flame setting is most commonly used for welding, brazing, soldering and flame-cutting? Is it the carburizing flame, the neutral flame or the oxidizing flame?

24. What's the first thing you should do if you ever get a flashback?

25. Which of the common fuel gases has the widest range of explosive limits in air? Is it acetylene, MAPP Gas, propylene, propane, or natural gas?

The Tools You Use with Gases

This chapter will tell you about the tools used with gases. It covers the tools for flame-cutting, welding, brazing, soldering, and heating. It also includes gas tools and equipment you will use when you learn about metal-inert-gas (MIG) and tungsten-inert-gas (TIG) arc welding.

This chapter is long. It's also very important. As you learn about each type of equipment, we'll also tell you how to use it safely, which will let you know what to do if you ever get into trouble. Safety rules are good only if you remember them. Understanding the reasons for the rules is better, because you will remember the reasons longer than rules.

After finishing this chapter you will know how to set up any kind of gas-using equipment you are ever likely to use. You will also know enough to be able to buy your own tools, and you will have some of the background needed to track down welding and flame-cutting problems that result from your equipment or how it's set up.

We'll start with gas cylinders and work down your gas lines, and wind up with torches and tips. That's the order in which any welding or flame-cutting job is normally set up.

4-1
GAS CYLINDERS

There are different kinds of gas cylinders simply because there are different kinds of gases. Some gases can be supplied in compressed-gas cylinders under pressures of 2200 psig [15.2 MPa], or even higher (Fig. 4-1). Examples are the oxygen used for flame-cutting, welding, and heating; the argon and helium gases used for shielding molten weld metal in certain arc-welding processes; and hydrogen (which is rarely used as a fuel gas today). Another gas in this category is nitrogen, which is almost never used in welding (with the exception of small amounts added to argon when welding copper by one of the gas-shielded welding processes). But

FIGURE 4-1 Compressed-gas cylinders on a welding distributor's loading dock. On the cylinder cart is an acetylene cylinder (the dark one) and a high-pressure oxygen cylinder (the taller, light-colored one).

nitrogen is used for testing tanks, pressure vessels, and pipelines that welders produce because nitrogen is safe and inexpensive.

Some gases are not supplied in high-pressure cylinders because they turn into liquids under moderately high pressure. More pressure will not reduce the volume of the liquefied gas very much. Examples are carbon dioxide and the liquefied petroleum-based gases (LPG) such as MAPP Gas, propane, and propylene.

Some of these medium-pressure gases have very peculiar properties. A good example is acetylene, which has a lot of characteristics all its own. The cylinders used for acetylene look like no other gas cylinders.

Carbon dioxide also is odd. It can be supplied as a pressurized gas or as a liquid. It all depends on the combination of pressure and temperature used to store it. What's more, if you release the pressure on carbon dioxide very quickly, it will turn into a solid.

You know that solid as dry ice. Carbon dioxide becomes a solid at $-94°F$ [$-70°C$]. Not surprisingly, CO_2 gas regulators have electrical heaters attached that keep the gas from freezing the equipment solid when the pressure is rapidly reduced.

All of the high-pressure gases also come in ultracold cryogenic versions that are liquid instead of gas and don't need much pressure to keep them that way. Liquid oxygen, for example, is supplied in special containers at $-270°F$ [$-169°C$]. The liquid has to be vaporized into a gas and then warmed up before you can use it.

Finally, natural gas is a low-pressure fuel gas. It is never supplied in cylinders (at least not to welders). It almost always is supplied in low-pressure pipelines as a gas, even though it can be liquefied.

You can learn a lot about all high-pressure gas cylinders (and pick up the most important safety rules, too) by studying oxygen. We'll do that first.

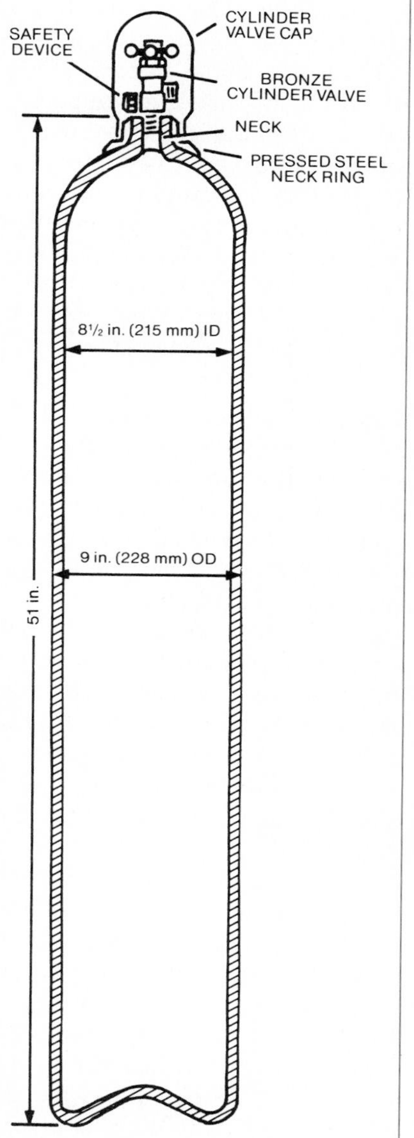

High-Pressure Gas Cylinders

High-pressure gas cylinders come in different sizes and pressure ratings. But most of the high-pressure gas cylinders you will use will be 2200-psig [15.2 MPa] oxygen cylinders (that's the cylinder pressure when full of gas at 70°F or 21°C). The most common cylinder of this type contains 244 scf [6.9 m^3] of oxygen, which is the volume of the oxygen in the cylinder if it were measured at room temperature and pressure, not at 2200 psig [15.2 MPa]. Oxygen cylinders are typical of all high-pressure gas cylinders.

The weight of a standard oxygen cylinder, complete with valve and cap (Fig. 4-2) ranges from 104 to 139 lb [47 to 63 kg], depending upon the kind of steel used to make the cylinder. Add another 20 lb [9 kg] for the oxygen it contains when full. If you put an oxygen regulator on the full cylinder and turned the cylinder on, the cylinder gauge would read about 2200 psig [15.2 MPa].

If the gas in the cylinder were hydrogen, 244 scf [6.9 m^3] of it would weigh only 2.5 lb [1.1 kg] instead of 20 lb [9 kg]. If the cylinder contained argon instead of oxygen, the gas would weigh about 25 lb [11 kg].

Smaller and larger oxygen cylinders are also used. For example, high-pressure 122-scf [3.5 m^3] oxygen cylinders (at 14.7 psig [101 kPa] and 70°F [21°C]) are supplied by most oxygen producers for people who only need small volumes of oxygen, or where very portable cylinders are used. One example is flame-cutting reinforcing bars (rebars) in a mine. Miners who cut the rebars often carry the cutting oxygen cylinders on their backs.

We learned in Chap. 3 that the pressure in any unopened gas cylinder varies with the temperature of the gas in the cylinder. Table 4-1 shows you how the pressure in an oxygen cylinder changes as the gas gets hotter or colder.

FIGURE 4-2 A 244-scf oxygen cylinder in cross section. The oxygen capacity of this cylinder is 244 scf at 2200 psig and 71°F [21°C].

TABLE 4-1 How oxygen cylinder-gauge pressure changes with temperature

Temperature		Gauge Pressure	
°F	°C	psig	MPa
120	49	2500	17.2
100	38	2380	16.4
80	27	2260	15.6
70*	21	2216*	15.3
60	16	2140	14.7
50	10	2080	14.3
40	4	2020	13.9
30	−1	1960	13.5
20	−7	1900	13.1
10	−12	1840	12.7
0	−18	1780	12.4
−10	−23	1720	11.9
−20	−29	1660	11.4
−30	−34	1600	11.0

* The standard cylinder-filling temperature and pressure for the most commonly used high-pressure cylinder gases.

The hotter the gas, the higher the cylinder pressure, which is one reason why high-pressure gas cylinders should never be stored in direct sun or near furnaces (no matter whether the gas in the cylinder is oxygen or some inert gas such as helium or argon).

⚠ **CAUTION: Never leave gas cylinders in the sun.**

Table 4-2 and the graph that accompanies it show you how the pressure changes in a 244-scf [6.9-m³] oxygen cylinder initially filled to 2216-psig [15.3-MPa] cylinder pressure at 70°F [21°C]. If the temperature remains the same but the gas in the cylinder is used, the pressure on the gas remaining in the cylinder drops. Therefore you can tell how much gas (in this instance oxygen) is left in the cylinder by reading the cylinder gauge and making a quick mental estimate. If the cylinder pressure-gauge dial is half way between 2200 psig [15.2 MPa] and 0 psig (at 1100 psig or 7.6 MPa), the cylinder is half full. If the needle on the cylinder gauge is three-quarters of the way from 2200 psig [15.2 MPa] to 0 psig [0 MPa] (at 550 psig [3.8 MPa]), the cylinder must be three-quarters empty.

The available gas contents figure in Table 4-2 applies only to 244 scf [6.9 m³] gas cylinders (gas volume corrected to 70°F [21°C] and 14.7 psia [101 kPa] sea level pressure). But the pressure readings will apply to any size cylinder. High-pressure gas is high-pressure gas, whether there's 244 scf [6.9 m³] of it, or 24.4 scf [0.7 m³] of it.

As you learned in Chap. 3, when the cylinder is shut the pressure goes up as the remaining gas gets hotter and goes down as it gets cooler. The gas pressure and absolute temperature are directly proportional to each other. Doubling the absolute temperature doubles the pressure inside a sealed container. If you bothered to calibrate the cylinder pressure in terms of degrees Fahrenheit (or degrees Celsius) like we did in Table 4-1, but in greater detail, you'd have a very large, accurate gas operated thermometer.

Oxygen Cylinder Safety

Oxygen cylinder valves (Fig. 4-3) are similar to the valves used on other types of high-pressure gas cylinders. There are several types of cylinder valves but the details are not important for your purposes. The valve shown here is a ball-type packed oxygen-cylinder valve. All oxygen cylinder valves should be tightly closed and protected by a cylinder valve cap when the cylinder is not in use, whether the cylinder has any gas in it or not.

There is a very good reason for the cylinder cap over the valve. A forklift truck, a swinging pallet, a crane hook, or anything else just might hit the uncovered valve and break it off. If the valve stem is suddenly pulled out of a cylinder with a full load of oxygen at 2200 psig [15.2 MPa], it can turn the cylinder into a 160-lb [73-kg] rocket with enough force to drive itself through a concrete block wall. That's also the reason why all high-pressure gas cylinders

TABLE 4-2 How a 244-scf [6.9-m³] oxygen cylinder empties as the cylinder pressure drops*

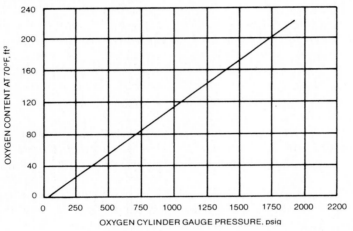

Gauge Pressure		Available† Cylinder Contents		Gauge Pressure		Available† Cylinder Contents	
psig	MPa	scf	m³	psig	MPa	scf	m³
2216	15.3	244	6.9	1110	7.6	120	3.4
2090	14.4	230	6.5	1020	7.0	110	3.1
2000	13.8	220	6.3	930	6.4	100	2.8
1910	13.2	210	5.9	840	5.8	90	2.5
1820	12.5	200	5.7	745	5.1	80	2.3
1730	11.9	190	5.4	655	4.5	70	2.0
1640	11.3	180	5.1	565	3.9	60	1.7
1550	10.7	170	4.8	475	3.3	50	1.4
1465	10.1	160	4.5	380	2.6	40	1.1
1375	9.5	150	4.2	285	1.9	30	0.9
1285	8.9	140	4.0	190	1.3	20	0.6
1200	8.3	130	3.7	95	0.7	10	0.3
				0	0†	0	0†

* A 122-ft³ [3.5-m³] oxygen cylinder would contain about one-half of these volumes at the given temperatures and gauge pressures. The volume figures are standard cubic feet (and standard cubic meters), not, of course, the volume of the gas cylinder, which remains constant.
† There will always be oxygen left inside the cylinder, even at a pressure reading of 0 psig. But you can't get that extra oxygen out without a vacuum pump. That's why we called the gas volume *available cylinder contents*.

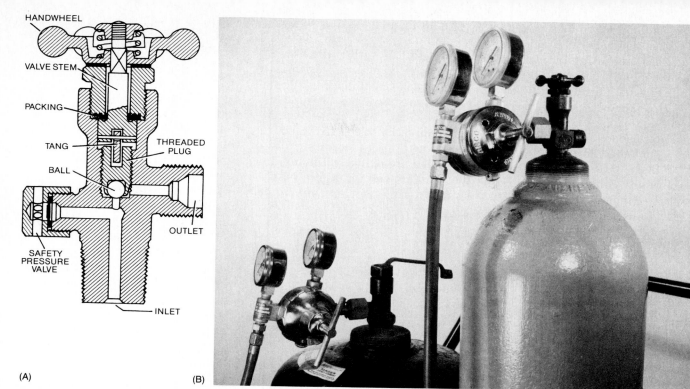

FIGURE 4-3 (A) A cross section of a typical ball-type oxygen cylinder valve. Other types of valves are also used, but they are similar in operation to this one. **(B)** An oxygen cylinder (on the right) with its hand valve. Regulator and hose are attached to it. On the left is an acetylene cylinder. It does not have a hand valve. It is turned on by a special wrench (which you see on the acetylene cylinder).

should be chained upright to a wall or post when not in storage. If the cylinder is not tied down and you pull on the hose while you work, the cylinder will tip over.

The cylinder stem in the valve is connected to the lower part of the valve by a tang. This tang is strong enough to allow manual operation of the valve in normal opening and closing. But the tang may break if excess force is applied to the hand wheel when closing the valve with a wrench or lever. Never try to force the cylinder valve. If the valve is stuck, call the distributor who delivered the cylinder. They have the equipment to handle the problem. You don't.

> 🏃 **CAUTION: Never force a stuck valve.**

Oil, grease, and other easily oxidizable (burnable) materials must be kept away from oxygen cylinders, valves, and regulators, or anything else that will be used to contact or work with oxygen, including your hands and clothes. The smallest spot of grease inside an oxygen valve can start a fire and cause an explosion. One reason is that materials that are combustible in air are five times more combustible in pure oxygen. A hot piece of steel wool will not do anything in the air except rust faster. In pure oxygen the hot steel wool will burst into flame.

If you suddenly release high-pressure oxygen into your cylinder regulator by turning the valve all the way on with one sharp twist instead of opening the cylinder valve slowly, the outrushing oxygen will drop from 2200 psig [15.2 MPa] to atmospheric pressure inside the regulator in milliseconds. Then it will slam into parts of the valve or regulator at the speed of sound, suddenly recompress, and heat up like a tire being pumped up very fast. The recompressed oxygen gets hot enough to set steel on fire. It can easily ignite the valve seats in your regulator, which are made of neoprene rubber, nylon, or Teflon. That's called a *regulator seat fire*.

High-pressure oxygen cylinder valves and gas regulators are made of brass or bronze because these materials do not burn easily in oxygen. Brass and bronze also draw off heat more quickly than steel. The nonmetallic parts of oxygen regulators are also made of materials that are resistant to oxygen, but not superhot oxygen. Open the valve on an oxygen cylinder rapidly and you may wind up with a hole melted in your cylinder regulator (Fig. 4-4).

> 🏃 **CAUTION: Open oxygen cylinder valves slowly.**

FIGURE 4-4 This is what can happen if you turn on a high-pressure oxygen cylinder all the way when it has a regulator attached to it. This was an oxygen regulator before the fire. The operator neglected to "crack" the oxygen cylinder valve first, to bring the regulator up to pressure slowly.

Oxygen cylinder valves have special CGA (Compressed Gas Association) threads on the outlet where the gas regulator is attached. So do all other high-pressure gas cylinders. But valve connections on oxygen cylinders are threaded only with right-hand threads. Fuel-gas connections are threaded only with left-hand threads. That prevents you from attaching fuel-gas regulators and hoses to oxygen cylinders.

The various possible CGA connections for different kinds of gases and devices have different numbers. For example, acetylene uses CGA #510 connections (or #200 or #520 for small valves). Oxygen valves and fittings use another set of numbers. The complete list of CGA fittings with their code numbers is quite long. Don't worry about the CGA numbering system right now, as long as you know that it exists. Your welding distributor will help you hook up to the right connections when you or your company buys the equipment.

Never force any threaded connection on any gas-using device. If it doesn't fit, either you are forcing the wrong connection or the threads are damaged. If the threads are damaged the connection will leak gas. Whenever a fitting cannot be connected hand-tight, something is wrong. Stop what you are doing and figure out what the problem is before tightening anything with a wrench.

CAUTION: Never force threaded connections.

Liquid Cylinders

One reason that gases are liquefied is that liquefied gas takes up less room than the gas under high pressure. Let's take oxygen, again, as our example. A cubic foot [0.02 m^3] of liquid oxygen under normal atmospheric pressure equals 865 scf [24.5 m^3] of oxygen gas at room temperature and pressure. Even if you took all 865-scf [24.5 m^3] of the room-temperature oxygen gas and pressurized it to 2200 psig [15.2 MPa] in several 244-scf [6.9-m^3] high-pressure cylinders, you still would need 3.54 high-pressure oxygen cylinders to equal 1 ft^3 [0.02 m^3] of liquid oxygen. That isn't much. One portable liquid oxygen cylinder will hold enough oxygen to equal about 20 high-pressure gas cylinders.

Another advantage of a liquid container is that it weighs less than 100 lb [45 kg], empty, whereas the five or more high-pressure gas cylinders that it replaces weigh over 750 lb [340 kg]. That is why liquid oxygen often is used on construction jobs. Liquid containers are much more portable and they last up to 20 times longer than high-pressure gas cylinders between refills or cylinder changes. Liquid argon and liquid nitrogen also are supplied in these special liquid containers.

Liquid-gas containers are sealed to the atmosphere, although they have a pressure-relief valve. Liquid-gas "dewars" (named for an early scientist, Dr. Dewar, who

studied low temperature gases) have tops open to the atmosphere. Liquid helium comes in dewars, but it is seldom used that way by welders. Helium must be incredibly cold before it liquefies, −452°F [−269°C]. Liquid helium is so cold that it freezes air into solid ice. That creates special problems that welders don't want to worry about.

There are very few problems handling liquid-oxygen containers. The same general safety requirements that apply to compressed oxygen also apply to liquid oxygen. Of course, you don't want to splash liquid oxygen on yourself, but you are very unlikely to ever see the liquid. It remains sealed inside the cylinder and is vaporized into a gas before you ever use it. Don't subject dewars to rough handling, and never put them on their sides, or you just might see liquid oxygen leaking on the ground.

CAUTION: Never tip liquefied-gas cylinders over.

Liquefied-gas dewars and liquefied-gas cylinders are actually big Thermos bottles with one or more containers inside each other. The inside container is separated by a vacuum and special insulation from the outside container. Rough handling can break the vacuum seal between the double containers. It also can shake the insulation loose. Always handle them gently.

Very cold gases like liquid helium may use three separate concentric containers along with a refrigerant such as liquid nitrogen between the inside and the next outside container. These helium dewars are more fragile than liquid oxygen containers. You also have to watch out for ice plugs of solid air frozen by the liquid helium. These air-ice plugs can plug up outlets and vent lines. As the liquid helium continues to warm up, the gas will vaporize and the pressure will increase until it blows out the ice plug or cracks

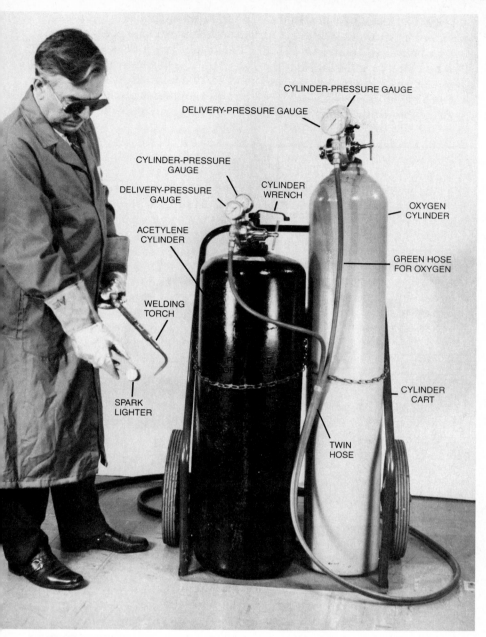

CYLINDER-PRESSURE GAUGE

DELIVERY-PRESSURE GAUGE

CYLINDER-PRESSURE GAUGE

DELIVERY-PRESSURE GAUGE

CYLINDER WRENCH

ACETYLENE CYLINDER

OXYGEN CYLINDER

GREEN HOSE FOR OXYGEN

WELDING TORCH

CYLINDER CART

SPARK LIGHTER

TWIN HOSE

FIGURE 4-5 A basic oxyacetylene welding setup. A flame-cutting setup is made just like this except that a different type of torch and tip is used.

the dewar. Liquid helium is rarely used by welders.

Cylinders for handling cryogenic (ultracold) gases in industry always come equipped with vaporizers. These are steam or electrical devices for warming up the liquid and turning it into a warm gas that you can use. That's why you are not likely to see any liquid oxygen, even if your oxygen supply comes from a liquid container.

Acetylene Cylinders

Figure 4-5 shows you how to connect a basic oxyfuel welding outfit with an acetylene cylinder. Other fuel-gas cylinders could be used in place of this type of fuel gas with minor changes such as a different kind of regulator and different torch tip. A cutting outfit would be set up the same way, except that a different kind of torch and tip would be used.

Acetylene is a colorless gas at

normal temperatures and pressures. Every molecule contains two atoms of carbon (chemical symbol C) and two atoms of hydrogen (symbol H). Its chemical formula is:

$$H - C \equiv C - H$$

The horizontal lines are called *chemical valence bonds*. The three bonds between the two carbon atoms contain enormous energy. They release that energy when the bonds are broken. Since triple bonds are very unstable, they are easily broken. Combining acetylene with oxygen and heat (to burn it) is one way to break the bonds and release enormous amounts of extra energy in the form of heat. That is why acetylene is a good fuel gas.

These chemical bonds or forces that hold the atoms together in the acetylene molecule also make acetylene a very sensitive fuel gas. The triple bonds can also be broken simply by heating the gas, even though there is no oxygen around. The bonds also can be broken with a heavy impact, like dropping an acetylene cylinder off a loading dock or building. Moderately high pressure can also cause the bonds to break. This explains why acetylene is handled the way it is.

Acetylene gas is easy to detect. You can smell it. It has a characteristic pungent odor. Acetylene, however, is not something you want to breathe too much of. This gas is an anesthetic in large concentrations. Old-time "painless Potter" dentists used it when pulling teeth.

Acetylene has other properties that set it apart from all the other gases you will use in welding and cutting. For example, you can make acetylene yourself. Acetylene is made simply by dripping water onto a gray lumpy or granular material called *calcium carbide* (CaC_2). Calcium carbide is made in electric arc furnaces by smelting limestone and coal or petroleum coke. Companies that generate their own acetylene make it the same way that acetylene suppliers do. They put the crushed and screened calcium carbide in acetylene generators and react the carbide with water. The reaction goes like this (we wrote the formula for water as HOH instead of H_2O to make it clear how the hydrogen, H^+, and hydroxyl ion, OH^-, act separately in the reaction; this is how water molecules most often break up in chemical reactions):

$$CaC_2 + 2HOH \rightarrow C_2H_2 + Ca(OH)_2$$

$$\underset{\text{carbide}}{\text{calcium}} + \text{water yields} \underset{\text{lene}}{\text{acety-}} + \underset{\text{lime}}{\text{slack}}$$

Most people don't generate their own acetylene because the process can be hazardous; it requires a lot of money to buy the special generating equipment, compressors, and other things needed to get the acetylene into cylinders, and the reaction produces calcium hydroxide (slack lime) sludge that you have to get rid of after making the gas. Acetylene cylinders also are very expensive. No other gas cylinder resembles them, inside or out.

The acetylene cylinder (Fig. 4-6) is a lot more than a simple steel container. It contains materials that hold the acetylene gas in a stable condition at moderately high cylinder pressures until you are ready to use it.

The maximum safe operating pressure for acetylene outside its special cylinder is 15 psig [103 kPa]. The acetylene inside the cylinder is usually around 250 psig [1720 kPa] when full. Such a pressure is possible only because most of the acetylene in the cylinder is not present as a gas. It is actually dissolved in a liquid called *acetone* (a common cleaning fluid). When the acetylene cylinder valve is opened, the acetylene bubbles out of the acetone solution like carbon dioxide fizzing out of a carbonated soft drink or champagne bottle.

When the acetylene is pumped into the cylinder, 1 volume of acetone will dissolve 25 volumes of acetylene gas for each 14.7 psi pressure increase (that is, each in-

FIGURE 4-6 Acetylene cylinders look like no other gas cylinder you will use. They usually are painted black. But you can recognize them in any color simply by their square-shouldered shape.

crease in pressure equal to 1 atm). At 30 psig [207 kPa] the acetone will dissolve 50 volumes of acetylene, and at 250 psig [1720 kPa] or about 17 atm of pressure, 1 volume of acetone will dissolve 420 volumes of acetylene. The higher the cylinder pressure, the more acetylene will dissolve in the acetone liquid. However, you don't want to get the cylinder pressure too high because, even in acetone, the acetylene gets very sensitive at high pressures. Acetylene cylinders usually are filled up to 250 psig [1720 kPa].

The acetone solution only takes up 36 percent of the void space in the porous filler material before the cylinder is filled with gas. A lot of extra space is needed for the liquid acetone to expand when it fills up with dissolved acetylene.

Most welders have little to do with filling acetylene cylinders. But they occasionally find out what's inside them when they don't treat the cylinders properly. You can draw acetylene gas out of the cylinder too fast. If you exceed the withdrawal rate of the cylinder (for instance because the cylinder is cold and the molecules

slow down so that the gas pressure drops), the gas can't get out of solution fast enough and you start pulling acetone along with the acetylene. You also will get a torch that spits acetone if you put the acetylene cylinder on its side.

CAUTION: Always use acetylene cylinders upright.

One way to draw too much acetylene from the cylinder is to attach a gas regulator that is too big to the cylinder. A large flow-rate regulator, when wide open, is just like punching a hole in the acetylene cylinder. The gas won't come out of solution fast enough to keep up with the drop in cylinder pressure, so you'll suck acetone liquid out with your fuel gas. That is why acetylene cylinders have maximum withdrawal rates. Acetylene's maximum withdrawal rate is 50 ft³/h [1.4 m³] from a 275-scf [7.78-m³] cylinder at 70°F [21°C]. If you need more fuel gas than that, use a cylinder manifold. We will describe manifolds later on in this chapter.

Acetylene cylinders are peculiar for other reasons. In addition to liquid acetone, they also contain a highly porous inert mixture of materials (Fig. 4-7) such as lime, diatomaceous earth, charcoal, shredded asbestos, and cement. These inert fillers act like shock absorbers to make the acetylene-acetone solution less sensitive to impact and heat.

The mixture of acetylene gas, acetone liquid, and inert porous filler material is packaged inside a squat, black, heavy-walled steel cylinder that looks like no other gas cylinder. The cylinder also has fusible plugs on the top and bottom that will melt and vent the gas and acetone solution into the air if the cylinder temperature gets above 220°F [104°C]. That's only 8°F [4°C] higher than the temperature of boiling water at sea level.

The fuse plugs prevent a cylinder explosion if the cylinders are in a fire. Its better to release the contents of the cylinder than to have it blow up. However, you don't want to release the contents of any acetylene cylinder while you are next to it. Don't try heating an acetylene cylinder to warm it up and make it produce gas faster. Don't even use boiling water.

The valve stem on the top of an acetylene cylinder is squared off for a special cylinder wrench. The acetylene cylinder does not have a hand wheel like an oxygen, argon, or helium cylinder. Instead, it has a hand wrench. (That wrench must be left on the cylinder whenever it is in use, or else you can't turn off the cylinder quickly in an emergency.) The outlet valve of the cylinder also is threaded to connect only with acetylene regulators.

Small acetylene cylinders may have smaller outlet connections (called *A-size connections*) than large B-size or even larger C-size cylinders. The small connection prevents you from putting too big

a regulator on a small cylinder and getting into trouble with acetylene withdrawal rates. Very small acetylene cylinders are called *MC cylinders*. The MC stands for *motor coach* because these small cylinders were used years ago to light the lamps on early automobiles. The MC designation remains to this day.

Sometimes the threads on a cylinder or regulator are chipped or scratched. They won't fit very well or else they require a lot of force to make the connection. If so, don't use them. Mark defective regulators and turn them in for repair. Label cylinders with chalk and a tag and put them outside in a designated storage area for the supplier to pick up. If the threads are chipped or scratched, don't use them. Never try to fix a cylinder or regulator yourself.

Using Acetylene Safely
Now that you know something about acetylene cylinders, you can see why there are so many rules

FIGURE 4-8 Every acetylene regulator is red-zoned above the 15 psig delivery pressure. Don't exceed that pressure when you use the fuel gas.

about using this gas. The most important rule is: *Never use acetylene at a delivery pressure on your regulator gauge higher than 15 psig [103 kPa].*

⚠ **CAUTION: Never exceed 15-psig acetylene delivery pressure. Keep the cylinder wrench on when using the cylinder.**

Every acetylene gas regulator is marked with a red zone above 15 psig [103 kPa] (Fig. 4-8) to remind you not to exceed this delivery pressure. Some production people actually do set the delivery pressure higher than this because they are doing heavy heating work or making sprayed metal coatings (called *metallizing*) that require high volumes of acetylene. This is a very unsafe practice. When more acetylene is needed than can be delivered by a single cylinder at 15 psig [103 kPa] an acetylene manifold should be used (also at 15 psig [103 kPa]), which will gang up a number of acetylene cylinders to increase your gas flow rate.

Another odd thing about acetylene is that you can't judge the remaining volume of an acetylene cylinder simply by reading the cylinder gauge pressure. The reason, of course, is that a lot of the gas is in solution in acetone and is not a gas above the acetone liq-

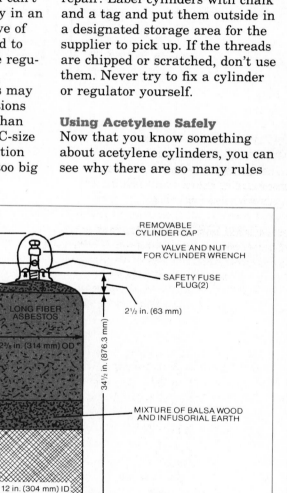

1. Porous filler fills cylinder
2. Acetone added to make up tare weight
3. Acetylene dissolved in acetone

REMOVABLE CYLINDER CAP

VALVE AND NUT FOR CYLINDER WRENCH

SAFETY FUSE PLUG (2)

ASBESTOS CLOTH

LONG FIBER ASBESTOS

2½ in. (63 mm)

12⅜ in. (314 mm) OD

34½ in. (876.3 mm)

45⅜ in. (1150 mm)

42 in. (1050 mm)

MIXTURE OF BALSA WOOD AND INFUSORIAL EARTH

12 in. (304 mm) ID

12²¹/₃₂ in. (321 mm) OD

FINE ASBESTOS

SAFETY FUSE PLUG (3)

FIGURE 4-7 This drawing shows the inside of an acetylene cylinder. As you can see, it contains a lot more than just acetylene.

uid where you can detect its pressure (which for a full acetylene cylinder is 250 psig [1720 kPa] at 70°F [21°C], at sea-level pressure).

The only way to know accurately how much acetylene is in a cylinder is to weigh the entire cylinder, then calculate how much gas is in it. You do that the same way people calculate the weight of a loaded truck. First you weigh the truck when it's empty. Then you weigh the truck when it's full. Subtracting the empty weight (the tare weight) from the full (gross) weight gives you the net weight of the load.

You do exactly the same thing with acetylene cylinders, except that the empty cylinder weight includes the acetone and filler materials inside the cylinder without any acetylene. The gross weight is the same cylinder filled with acetylene. The net weight (gross weight minus tare weight) is the weight of the acetylene gas, whether that acetylene is dissolved in acetone, or is a gas in the cylinder. You will see a tare weight stamped on the acetylene cylinder's shoulder. Don't use that weight, it's the tare weight of the cylinder from the manufacturer *before* it has been filled with acetone and inert filler materials.

Once you have the weight of acetylene in the cylinder, it's easy to convert that figure into standard cubic feet of acetylene. Multiply the pounds of acetylene by the cubic feet per pound density of acetylene (14.74 lb/ft³ at 70°F). If you use metric units, multiply by 236 kg/m³ at 21°C. This gives you the volume of the acetylene gas left in the cylinder in standard cubic feet or in standard metric units.

For example, an acetylene cylinder with a tare weight of 200 lb [90.6 kg] when empty and 220 lb [99.6 kg] gross weight when full contains 20 lb [9 kg] of acetylene. Therefore 20 × 14.74 = 294 ft³ of acetylene at standard conditions of sea-level pressure and room temperature. If you want a more accurate reading, you would have

to adjust this standard gas volume by the pressure and temperature of the place you work, using the equations we gave you in Chap. 3. A much simpler fact to help you is that 14.7 scf of acetylene weighs one pound.

How do you know when you are running low on acetylene in the cylinder? When the cylinder regulator pressure gets down around 10 psig [69 kPa] it's time to change cylinders. (Actually, it can go lower than 10 psig before changeover, but that's a good point to start thinking about changing cylinders.) When you take the cylinder off line, mark your cylinder EMPTY or MT with chalk and put it outside in a shaded, protected cylinder storage area for the distributor to pick up.

Acetylene is the only gas that cannot be supplied as a bulk liquid. It is available only in cylinders because of its special safety requirements.

LPG Cylinders

Compared with acetylene, other fuel gases come in very simple cylinders. With the one exception of hydrogen (which comes in 2200-psig [15.2 MPa] high-pressure cylinders) almost all other fuel gases are either medium- or low-pressure gases and come in light-walled steel tanks or even in pipelines like natural gas. With the two exceptions of hydrogen and acetylene, all other common welding and cutting fuel gases are LP or *liquefied petroleum gases*.

The designation LPG includes liquids in the cylinder that turn into a gas only when the cylinder valve is opened so that the moderate cylinder pressure on the liquid is relieved. Examples of LPG fuels are butane, propane, propylene, and MAPP Gas. MAPP Gas is the most important, and we'll use it as the example for the group.

MAPP Gas cylinders come in half a dozen different sizes. You can buy 1-lb [0.45-kg] capacity MAPP Gas cylinders in many hardware stores. Propane also comes in these tiny cylinders.

(Even smaller cylinders of butane sometimes are used in cigarette lighters.)

The most common industrial LPG cylinder is a 70-lb [32 kg] capacity MAPP Gas cylinder (Fig. 4-9A). It weighs 120 lb [54 kg] when full of 70 lb [32 kg] of MAPP Gas, so that the empty cylinder weighs 50 lb [23 kg] (compared with 220 lb [99.6 kg] for an empty acetylene cylinder). Therefore the net weight of a full MAPP Gas cylinder is 70 lb [32 kg] of fuel gas; its tare weight is 50 lb [23 kg], and its gross weight is 120 lb [54 kg].

There are even bigger MAPP Gas cylinders—up to 500 lb [227 kg] net weight. MAPP Gas also is stored on site by high-volume users in 1000-gal [3785-liter] storage tanks (Fig. 4-9B).

The pressure inside a typical 70-lb [32 kg] cylinder of MAPP Gas at 70°F [21°C] is 94 psig [650 kPa]. When the cylinder is opened, the liquid vaporizes and the fuel comes out as a gas. Propane and propylene cylinders are very similar to MAPP Gas cylinders, although these gases are not as efficient at flame-cutting.

Natural gas, on the other hand, doesn't come in cylinders at all. It is a low-pressure gas normally supplied only in pipelines. The line pressure of a natural-gas pipeline may be very low so that special *injector* torches and tips are needed to use it. We will explain how injector and the related positive-pressure torches work later on in this chapter.

The pressure in a natural-gas pipeline also will vary with demand. As more people use the gas at any given time, the pressure will drop, which can cause some unexpected problems with flame-cutting, even though a regulator is attached between the pipeline and your work.

Natural gas also can be supplied as a liquid at very low temperatures, just like argon, oxygen, helium, and hydrogen. However, gas utilities never supply liquid natural gas to users.

(A) (B)

FIGURE 4-9 MAPP Gas cylinders come in many sizes from 1 lb up to 500 lb, as well as bulk storage tanks. (A) Two 70-lb MAPP Gas cylinders. (B) Two 1000-gal storage tanks.

LPG cylinders have a lot in common with all other gas cylinders. They have a valve on top of the cylinder and a hand wheel to operate the valve. The valve should be protected by a valve cap when the cylinder is not in use. You always should chain or tie the cylinders upright when you use them.

Like acetylene, using any LPG cylinder on its side will produce liquefied gas out your torch. You don't want a large volume of liquefied fuel gas squirting out of your torch because it will immediately vaporize and form a cloud of fuel gas around your work area.

LPG cylinders also have safety relief plugs to vent the gas to the atmosphere if the cylinder is in a fire. Never tamper with the safety relief valves or devices on any cylinder.

Just like acetylene cylinders, LPG cylinders have maximum withdrawal rates that depend upon the gas and the outside temperature and inside pressure of the cylinder. You won't run into problems with gas withdrawal rates from LPG cylinders under normal temperatures on routine work unless your withdrawal rate is too high. But if you need large volumes of gas or are working outdoors during the winter when the gas vaporizes more slowly,

you can easily exceed the withdrawal rate for a liquefied-gas cylinder.

It is possible to use electric heating pads to warm the gas inside a liquid cylinder to increase the amount of fuel you can get out of it. That was done frequently on pipeline jobs in Alaska. It is not something that a beginning welder should do without expert advice. Whatever else you do, *do not try heating the cylinder yourself.* Get advice from someone who knows a lot about the fuel gas you are using. Ask a welding engineer, your welding foreman, or the gas distributor or supplier about what to do on very cold days.

4-2
HANDLING AND STORING CYLINDERS SAFELY

All compressed-gas cylinders carry markings showing that they comply with the requirements of the Interstate Commerce Commission (ICC). In addition, any local or state ordinances should be followed. ICC regulations cover transporting gases in trucks and rail cars. Once the cylinders are at your work site, they are your responsibility. Here are some things you should (and should not) do.

Never tamper with or alter the safety devices on any cylinders. It's not only illegal, it's stupid. Only owners or suppliers of the cylinders are permitted to adjust, alter, or repair safety devices.

Never fill your own cylinders. It takes special training and often special licenses to fill cylinders. It also takes special equipment.

Never put cylinders close to heat sources such as open fires or radiators. Even out of doors, cylinders should not be left exposed to the direct rays of the sun. The hotter the cylinders get, the higher the gas pressure rises.

Protect cylinder valves and safety devices from snow and ice by storing cylinders under cover in a warm building and by protecting them with a wind screen in severe weather. A safety device might not work when you need it if it is frozen solid.

Never keep cylinders in unventilated rooms or sheds. They might leak gas. If they are fuel-gas cylinders, you can quickly get enough fuel gas into the air to form an explosive mixture. If they are inert gases, you can die from asphyxiation.

Store cylinders in well-protected, well-ventilated, dry storage areas. Store them at least 20 ft [6 m] away from highly combustible materials such as oil, grease, and paper products. Also keep cylinders away from elevators, stairways, or passageways where moving equipment can damage them. Avoid locations where passing trucks or falling objects can hit the cylinders. Never put cylinders in unventilated buildings, or work areas, or in lockers or cupboards.

⚠ CAUTION: Always store oxygen and fuel gases separately.

Always store fuel-gas cylinders and oxygen cylinders in separate locations. Oxygen cylinders should be stored at least 50 ft [15 m] away from fuel-gas cylinders. Remember the problem of leaking cylinders? You could get a leaking oxygen cylinder and a leaking fuel-gas cylinder in the same place. Use fire resistant partitions between cylinder storage areas with a fire-resistance rating of at least ½ h. Your safety director or welding foreman will give you details on how to build an approved barrier.

When a cylinder is empty, or simply too low to continue in use, close the cylinder valve. Make sure that all the equipment is removed from the cylinder such as hoses and regulators. Put the valve caps back on. Then mark the cylinder EMPTY or MT and store it in a cool, dry location ready to be picked up by the sup-

FIGURE 4-10 Always chain cylinders upright when they are in use, even while moving them on a hand truck.

plier. Even in storage, chain the cylinders upright.

Store cylinders with different gases in different areas that are clearly marked with the name of the gas stored there. Don't mix up oxygen cylinders with argon or helium cylinders.

Even though the different cylinders are painted different colors, the paint may be worn off. If it is, ask the distributor to repaint it since it's difficult to tell high-pressure cylinders apart. They are exactly alike except for the color coding. Repainting will help the distributor as much as it helps you.

When you move cylinders around a plant, tilt and roll them on their bottom edges. Do not drag cylinders or slide them around. If at all possible, use a cylinder hand truck or dolly and chain the cylinders to it (Fig. 4-10). Always put the cylinder valve caps back on the cylinders before you move them. Keeping the cylinders on the truck at all times is often a convenient, as well as a safe, idea. You can always operate your cylinders directly from the hand truck.

⚠ CAUTION: Handle cylinders with care.

Never use slings or magnets to carry cylinders. If a cylinder falls from a height, the valve stem or a safety device may be damaged. Do not attach hooks to the valve cap to lift a cylinder when using a crane. If cylinders are lifted as a group, use a sturdy pallet with a rail or center post to which the cylinders can be chained.

When cylinders have been stored outdoors, they can become frozen to the ground. Use special care to remove frozen cylinders. Warm (not boiling) water is the best to thaw them loose. Do not use boiling water or steam because the safety relief devices often are set very close to the temperature of boiling water.

If you have to pry frozen cylinders loose from the ground, do not use bars under the valve caps or

valves. If you hold a cylinder by its cap to lift it to a vertical position, first make sure that the cylinder cap is screwed on tight. A loose cap may come off, not only dropping the cylinder on your foot but also bending or snapping off the valve stem.

⚠ CAUTION: Never heat cylinders with an open flame.

Never let a welding or cutting torch flame get near the cylinders. Before you start a job, place your cylinders away from sparks, slag, or molten metal. Always remove the regulators, shut the cylinder valves, and put the cap on the cylinder before you move it to a new location.

Always shut off the cylinder valves at the end of the day. Several welders have died from asphyxiation because they left inert-gas cylinder valves on and only closed the gas at the TIG torch or MIG welding gun. Even a small inert-gas leak can push all the air out of an enclosure overnight or over a weekend.

Don't put gas cylinders near third rails, trolley wires, and circuits used for grounding arc-welding equipment. Cylinders are metal. They conduct electricity. Never strike an arc or even tap a welding electrode on a gas cylinder. Any contact by the gas cylinder with an electric circuit may burn a portion of the cylinder wall and weaken it. Even if you don't get hurt immediately, the weak cylinder can burst later on.

Always refer to fuel gases or high-pressure gases by their correct names. Don't call them "gas" or "air." Someday you may want compressed gas and somebody will hand you a pure oxygen line.

Before connecting a fuel-gas cylinder, except in the case of hydrogen cylinders, always "crack" or open the cylinder valve slightly, then close it. Cracking the valve serves to blow out any dirt or foreign material that may have become lodged in the valve. Be careful, however, not to crack

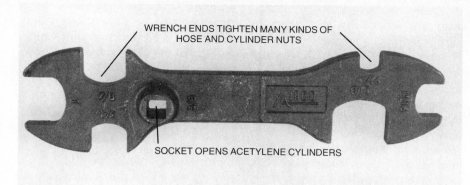

WRENCH ENDS TIGHTEN MANY KINDS OF HOSE AND CYLINDER NUTS

SOCKET OPENS ACETYLENE CYLINDERS

FIGURE 4-11 The multipurpose cylinder wrench is used to turn acetylene cylinders on and off and to attach gas regulators and hose. Always leave this wrench on the acetylene cylinder. You might have to turn it off in a hurry, and you can't if you can't find the wrench.

a fuel-gas cylinder valve near an open fire, sparks, hot metal, or any other possible ignition source. Stand to one side of the cylinder outlet when you crack the valve. The pressure of the gas can blow dirt at you with great force.

After cracking the valve, wipe the outlet with a clean cloth as a further precaution against getting dirt into the gas regulator. Attach the regulator and then purge it and the hose lines with a second spurt of gas from the cylinder. That will clean any oxygen or air out of a fuel-gas line. It also will assure you that your oxygen lines are free of fuel gas.

Never crack a hydrogen cylinder before attaching a regulator and hose. Hydrogen gas burns with an almost colorless flame, and the product of combustion is water vapor. You cannot see the hydrogen flame in daylight and might walk right into it. Therefore you should never crack a hydrogen cylinder until the regulator and hose are attached.

A special cylinder wrench or key (Fig. 4-11) is used to operate acetylene cylinders. Always keep the wrench or key on the valve at all times while the cylinder is in use. If you have an emergency and want to turn the cylinder off fast, you also want to know exactly where the cylinder wrench is. It always belongs on the cylinder valve when the cylinder is in

use, not in your pocket or toolbox.

Careless welders use cylinders to hang clothes on. Don't do it. Clothes, tools, and welding equipment can damage safety devices on the cylinder. Or they can make it difficult for you to turn the cylinder off quickly if you have an emergency.

If you ever do find a leak around the valve stem of a cylinder when you turn it on, tighten the valve glad nut with a wrench. If that does not stop the leak, do not use the cylinder. Instead, move the cylinder outdoors, tag it as a leaker, and immediately notify the supplier.

If the cylinder is leaking through the valve seat (closing the valve does not shut off the gas flow), install a regulator and shut the gas off at the regulator. Then move the cylinder with its attached regulator outdoors very carefully. Get at least 500 ft [150 m] away from all sources of ignition. Put up warning signs, and then slowly empty all the gas you can out of the cylinder. After the cylinder has been emptied, tag it as a leaker and call the supplier.

Treat cylinders that leak around a fuse plug or some other safety device the same way, but open the cylinder very slowly to let all the gas out. Be careful. The safety device may be leaking because the gas is at higher than normal pressure and temperature.

Whenever you call a cylinder-gas supplier, always give them all the details of any cylinder problems you have. They will appreciate your help. They also want to know what cylinders are creating problems so that they can handle them separately from the rest of their fuel or oxygen cylinder pickups.

Special Oxygen Precautions
Because of oxygen's special properties, special precautions must be taken when handling it.

Cylinder oxygen Oxygen supports combustion, but it will not burn. Therefore there is no legal limit on how much cylinder oxygen you can store in a room. However, always keep oxygen cylinders away from combustible materials, especially oil or grease and gasoline. Also store oxygen cylinders at least 20 ft [6 m] away from calcium carbide containers.

⚠ CAUTION: Oil and oxygen start fires.

Do not handle oxygen cylinders with oily hands or gloves. Do not place oxygen cylinders where oil from overhead cranes, shafts, or belts can fall on the cylinders and regulators. Never let a jet of oxygen strike an oily or greasy surface. Never let oxygen enter a fuel oil or other storage container unless the container has previously been thoroughly purged and cleaned. Above all else, never dust yourself off with an oxygen hose. The next spark that drops on your shirt or pants will set them on fire.

Never turn an oxygen cylinder valve on full force without first cracking the valve to build up oxygen pressure inside the regulator. If the full pressure of an oxygen cylinder suddenly enters an empty regulator, you can start a regulator fire.

Storing liquid oxygen Liquid oxygen can be stored indoors if the quantity does not exceed 100 gal [380 liters] (about the equivalent of two liquid-oxygen containers). When the quantity of liquid oxygen to be stored is greater than 100 gal [380 liters], a separate storage building, not for any other purpose, must be used. Liquid-oxygen containers also can be located outdoors in cool, dry, shaded areas. However, never store liquid-oxygen containers on gravel, asphalt, or bituminous blacktop. Dripping liquid oxygen can set a road on fire.

✴ CAUTION: Use special handling for liquid oxygen.

Liquid-oxygen containers are best stored on concrete pads. If there is a big liquid-oxygen storage tank at your workplace, it should be sitting on a steel-reinforced concrete pad. The pad also should have a spill apron about 12 × 12 ft [4 × 4 m] and about 6 in. [150 mm] thick, made of reinforced concrete. That gives the driver of the delivery truck enough room to pull up to the tank and make deliveries safely. Liquid-oxygen truck drivers won't make deliveries on asphalt or gravel.

4-3
GAS REGULATORS

All gas regulators have one basic job. They take in high-pressure gas from a supply system and reduce the pressure to a lower, safer, useful level. They do that while also controlling the flow rate (volume of gas per hour) to your equipment, so that you have enough gas to get work done.

The regulated supply system is usually a gas cylinder. It also can be a pipeline, a liquid-gas container, or a manifold that combines a group of cylinders.

If that's all there is to gas regulation, why are there from 150 to 200 different kinds of regulators (Fig. 4-12) in a typical welding equipment manufacturer's catalog? Because welders need them for different gases, flow rates, pressures, cylinder sizes, and work to be done.

There are heavy-duty, medium-duty, and light-duty regulators for high, medium, and low flow rates. There are regulators with flowmeters that tell you exactly how much gas is coming out of the cylinder, and regulators with no meters because flow rates are already preset by the manufacturer.

There are regulators for 2200 psig [15.2 MPa] cylinder gases like oxygen, and regulators for 100 psig [0.7 MPa] cylinder gases like propane, propylene, and MAPP Gas. There are special regulators just for acetylene with dials that tell you when you are over the safe 15 psi [103 kPa] working pressure. There also are gas regulators for low-pressure natural gas (which may be under 5 psig when it comes from the pipeline). Since some gases like carbon dioxide and nitrous oxide freeze when the pressure is reduced, there even are electrically heated gas regulators to keep them warm.

What's more, not all high-pressure, heavy-duty gas regulators can be used interchangeably. For example, cylinder oxygen used for flame-cutting and cylinder helium or argon (used for shielding weld metal during certain types of arc welding) all come in 2200-psig [15.2-MPa] gas cylinders. But oxygen, argon, and helium gases have entirely different densities.

In fact, a cubic foot [0.03 m^3] of oxygen weighs about eight times as much as a cubic foot [0.03 m^3] of helium. Since flow rates depend on both pressure and gas density, a helium regulator is completely wrong for use with oxygen. The helium regulator would lower the pressure of the oxygen gas from the cylinder, but it also might catch on fire or even blow up. Inert-gas regulators are not designed to work safely with oxygen.

Fortunately, you don't have to be a regulator expert to figure out which one to use. Most plants only use a limited number of regulator models. All regulator manufacturers also specifically state what gases their regulators are designed to handle. So once you

FIGURE 4-12 Just a few of more than one hundred and fifty different kinds of gas regulators that are used with fuel gases, shielding gases, and oxygen.

use the correct regulator for a given gas, don't switch it for another gas unless you know exactly what you are doing.

Never use a high-pressure argon or helium regulator on an oxygen cylinder. Oxygen regulators are specially built to prevent regulator fires caused by recompressed oxygen. Just like fuel-gas regulators, they have a FOR USE WITH OXYGEN or some sign printed right on them. Just read the label and you'll stay out of trouble.

There is no problem selecting a regulator for a given gas and cylinder. If you aren't sure, tell your welding distributor the size of your cylinder and the gas it contains and you will get the correct regulator for that job. Most of the errors that occur are made by operators, so read the following section very carefully.

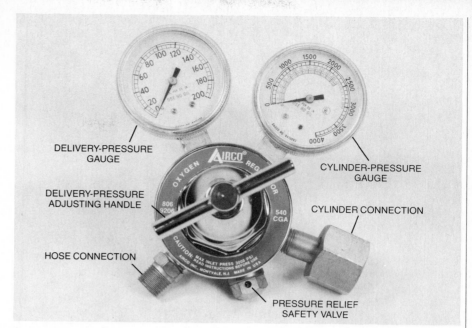

DELIVERY-PRESSURE GAUGE

DELIVERY-PRESSURE ADJUSTING HANDLE

HOSE CONNECTION

CYLINDER-PRESSURE GAUGE

CYLINDER CONNECTION

PRESSURE RELIEF SAFETY VALVE

FIGURE 4-13 Oxygen regulators like this one have most of the important features found on regulators used for other types of gases. As the cylinder gauge pressure dial indicates, this regulator is used for high-pressure gas service.

How Regulators Differ

Most of the gas regulators used by welders are high-pressure regulators. Most of them are used on 2200-psig [15.2-MPa] gas cylinders. Some 6000-psig [41 MPa] ultra-high-pressure gas cylinders are used, but it's very unlikely that you will ever see one. If you do, make sure that it has an ultra-high-pressure gas regulator to go with it.

Most of the regulators you use will have two gauges (Fig. 4-13). The first gauge tells you what the cylinder pressure is when the cylinder valve is opened. The second gauge tells you the pressure of the gas coming out of the regulator. You have to open the regulator before you get a reading on the second gauge dial. This is the delivery pressure and you set it yourself before you start to work.

Both the cylinder-pressure and delivery-pressure gauges give you readings in gauge pressure (psig [MPa])—a measure of the pressure above atmospheric pressure. The needle on the cylinder gauge dial of a full high-pressure cylinder will point to about 2200 psig [15 MPa], even though the dial scale may go all the way up to 4000

psig [27.5 MPa]. A cylinder that is used up will show a cylinder gauge pressure of 0 psig. At sea level that is the same as 14.7 psia [101 kPa]. High in the Rocky Mountains that might be equal to 12 psia [83 kPa].

But even though the dial on the regulator gauge of an empty gas cylinder reads 0 psig, the cylinder is still full of gas. The difference is that the gas is only at 14.7-psia [101 kPa] pressure (if you are at sea level), and you would need a vacuum pump to get the rest of the gas out. Remember, no gas cylinder is ever really empty—and if you don't remember why this fact is so important, go back and reread Chap. 3, pages 30–31 and 35–39.

High-pressure regulators The delivery-pressure gauge on most high-pressure gas regulators only goes up to 200 psig [1.4 MPa]. That's the maximum delivery pressure you can set and use with that regulator. For most work, you will set the delivery pressure a lot lower than that.

For example, the typical delivery pressure of oxygen for flame-

cutting with a big cutting or heating tip and torch might be as high as 80 psig [550 kPa]. The oxygen pressure used for gas welding will seldom be set at more than 35 psig to 40 psig [240 to 275 kPa]. The oxygen delivery pressure for a small welding torch can be as low as 10 psig [69 kPa] when working on thin material. The delivery pressure for fuel gases will usually be 15 psig [103 kPa] or less. That shows you something about delivery pressures. They depend upon the equipment as well as the job you are going to do.

The second most common type of regulator is a medium-pressure regulator used for acetylene (see Fig. 4-14). You will recognize it immediately. For one thing, the regulator label will tell you it's for use only with acetylene. For another, the delivery-pressure gauge only goes up to 30 psig, and half that range, from 15 to 30 psig has a red band on it. That red warning band tells you not to set the delivery pressure any higher than the safe 15-psig working pressure for acetylene.

The cylinder-pressure gauge for a typical acetylene cylinder only

FIGURE 4-14 Not all regulators have two dial gauges. This one is an acetylene regulator which is used on pipelined acetylene. Only the delivery or working pressure will vary, so that this regulator only has one gauge for the working-pressure setting. The pipeline pressure remains constant.

goes up to 400 psig [2.8 MPa] (one-tenth the level of a high-pressure gas-cylinder gauge). On a full acetylene cylinder, the cylinder gauge seldom will read above 250 to 300 psig, or about 75 percent of the full gauge scale.

Other medium-pressure cylinder regulators for LPG fuels will have maximum delivery pressures of around 40 to 60 psig and maximum cylinder pressures of around 200 psig, depending on the fuel gas for which the regulator is to be used and the outside temperature. In almost all cases, the fuel gas will be printed on the regulator label.

Low-pressure regulators Low-pressure gas regulators are usually used on natural-gas pipelines (Fig. 4-15). Very likely, they will only have one gauge because the line pressure is also the delivery pressure. It can easily be less than 5 psig [35 kPa].

FIGURE 4-15 Low-pressure gas regulators for gases such as pipelined natural gas often have only one pressure gauge. The note on the regulator tells the operator to never use this regulator on gas supplies over 300 psig.

Some regulators have no gauges at all (Fig. 4-16). If you go to a hardware store and buy a 1-lb [0.45-kg] cylinder of propane or MAPP Gas to burn old paint off your garage, the cylinder will have a tiny regulator with a delivery pressure that is preset at the factory. The only way you

know what the cylinder pressure is, is when it runs out of gas. But for small jobs like that, you don't want to pay extra for a fancy cylinder regulator that you will soon throw away.

One- and Two-Stage Regulators

An important aspect of cylinder regulators is whether they work in one step or two steps to lower the cylinder gas pressure to a usable level. The difference is important. Both kinds of regulators are used frequently on high-pressure cylinders, and even on some special high-pressure gas pipelines. They can be used interchangeably (as long as the regulator model is correct for the gas being used). But two-stage regulators can make a big difference in the quality of work you do.

As you already know, the working pressure inside a cylinder that contains nothing but gas drops as the gas is used up. However, if the cylinder contains a liquid that vaporizes to become additional gas as the gas over the liquid is used up, the cylinder pressure remains fairly steady until all the liquid in the cylinder has vaporized. Only then does the cylinder pressure begin to fall. That's characteristic of MAPP Gas and propane cylinders, and other LPG

FIGURE 4-16 Some regulators don't have any pressure gauge. These regulators are used on small MAPP Gas and propane cylinders that only weigh a pound.

cylinders such as butane. The cylinder pressure changes only if the outside temperature changes (unless the gas is almost all used up).

The full cylinder pressure of MAPP Gas is 94 psig [648 kPa] at 70°F [21°C] and about 250 psig [1720 kPa] at 70°F [21°C] for acetylene. The difference between the highest cylinder pressure and the lowest delivery pressure is not very much. The pressure drop from a full cylinder load to the delivery pressure you set on the regulator can easily be handled in one step.

But the difference between the 2200 psig [15.2 MPa] full cylinder pressure for argon shielding gas and the delivery pressure of argon (say 5 psig [35 kPa] at the welding gun) is enormous. What's more, you don't want the shielding-gas flow rate to vary while you work, even though the cylinder pressure is dropping as you use the gas. Keeping the delivery-gas flow rate constant is simply too much to expect any regulator to do if it has to drop the pressure of a full cylinder from 2200 psig to 5 psig [15.2 MPa to 35 kPa] in one step, or stage, as it's called.

Two-stage regulators solve that problem by reducing the pressure in two stages; first to an intermediate stage (maybe around 250 psig [1.7 MPa]) which is preset by the manufacturer, and then to the low delivery pressure you set yourself on your delivery-pressure gauge.

Figure 4-17A graphs the results of using three typical high-pressure gas regulators; two of them are single-stage regulators and the other one is a two-stage regulator. The cylinder pressure and delivery pressure are plotted against time as the gas inside the cylinder is used up.

The delivery pressure that you set on the single-stage regulator either rises or sags as the cylinder pressure drops. The two-stage regulator keeps the preset delivery pressure nearly level right up to the point where there is almost no

more gas left to extract from the cylinder.

Welders call that final usable cylinder pressure the *end point* of the cylinder (Fig. 4-17B). In science labs doing very critical work where no variation in the gas flow rate can be allowed, and when using ultra-high-pressure cylinders, three-stage regulators are some-

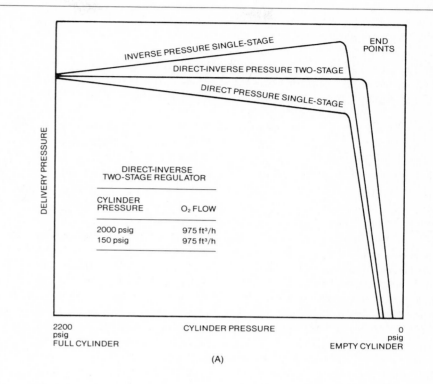

(A)

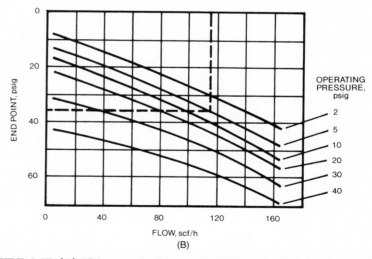

(B)

FIGURE 4-17 (A) This graph shows the difference between regulator valve designs. It also shows why two-stage regulators are better than single-stage regulators for precision work. Single-stage regulators have to be adjusted as the cylinder pressure falls to maintain a constant delivery pressure. (B) This graph shows how cylinder regulators act at the end point of an almost empty cylinder. Regulator end points depend on the flow rate and operating pressure. This regulator has an end point of about 37 psig cylinder pressure when working at 5 psig with a gas flow rate of 115 scf/h, which is indicated by reading the graph data along the dashed lines.

times used. The flow rates from these regulators, however, are quite low.

A single-stage regulator might be fine if you don't mind stopping every once in a while to readjust the working pressure back to where you want it. If you are cutting scrap in a junkyard, or rebar on a construction site, you don't really mind whether your oxygen pressure varies. You can take time to adjust your oxygen pressure (and thus your gas flow rate) at your torch to correct the flame setting.

But let's assume that the cutting machine hoses are too long, too small in diameter, or both too long and too small. This will cause a drop in the cubic feet flow rate of both your oxygen and your fuel gas. This situation will cause a flashback with certain fuel gases, especially acetylene, or it will give you insufficient oxygen for cutting the part.

Let's further assume that you are working on a big flame-cutting machine with 12 torches and that each torch has its own regulator. If the oxygen flow varies from torch to torch, the quality of cut you produce will also vary. One torch might give you a beautiful, sharp cut in ½-in. thick [25-mm] steel plate. Another torch might start losing the cut because it's not getting enough oxygen. The single big piece of steel plate you are cutting may be worth more than an automobile. You can't afford to make mistakes. You can either run all over the steel plate trying to adjust a dozen oxygen regulators, or you can use two-stage regulators that keep the oxygen flow rate perfectly even and balanced at each torch. (Of course, if you're smart, you set the correct torch entry pressure for the torch farthest from the gas supply, and all the other torches will fall in line.)

Actually, a big cutting machine would have a manifold that does the same job as a bunch of cylinders and regulators. But the manifold would still use at least one two-stage regulator.

Two-stage regulators cost more than single-stage regulators; therefore, the decision to use single-stage or two-stage regulators depends on how critical the job is, your skill, how much downtime you can afford while adjusting things, and even how close you are to the regulators while you work (whether you have to walk a long distance to adjust them).

Regulator Ratings

Gas flow, gauge pressure, delivery, and final working pressure are all different. The delivery pressure (in psig or kPa) is the push behind the gas coming out of the regulator as measured on the delivery-pressure gauge dial. Final working pressure is the gas pressure after it has traveled through a long hose and a torch. The final working pressure will be less than the delivery pressure on your regulator gauge. So will the gas flow rate, measured in cubic feet per hour (or cubic metres per hour). The flow rate out the end of your torch or welding gun, not the gas pressure at the regulator, is really what you want. It just happens that pressure is easier to measure than flow is. It also can mislead you.

Let's assume that your welding hoses are far too long, so that the gas has a very long way to go to get to the work. That reduces the flow rate at your torch. Let's also assume that the hose is full of metal connectors you used to patch it. That also slows down the amount of gas that comes out the other end of the hose.

Let's further assume that you are a beginning welder and don't remember to straighten out your hoses before you start to work. The hoses have a couple of kinks in them. None of these will affect the delivery-pressure reading on your regulator. They all will reduce your gas flow rate and final working pressure. What you read on the regulator delivery-pressure

gauge, and what you get out of the torch, are two entirely different things.

We have just told you three of the most common problems that occur in shops: extra long hoses; hoses full of repairs; and kinks. More than one welder has stopped production because he or she thought the equipment was failing. The only thing that was failing was the welder's common sense and the gas flow rate.

How, then, are gas regulators rated in terms of gas flow? Regulator manufacturers give you a lot more than vague words like "low," "medium," and "high" flow (or "light," "medium," and "heavy" duty) to rate their regulator models. If you look in a regulator catalog, you will see one of two kinds of very specific gas-flow ratings for each model.

The first rating is *maximum free flow,* or MFF. An example for a given regulator model is "425 scf/h oxygen, MFF." This is the flow rate measured in standard cubic feet of gas per hour at 70°F [21°C], right at the regulator gas outlet.

A better rating system is *maximum conventional application* or MCA. That is the flow rate for the regulator and the gas at some distance from the regulator, measured through a standard 12 ft [3.6 m] or 25 ft [7.6 m] of ⅜-in. diameter welding hose, for example. The same regulator that has a 425-scf/h oxygen rating under the MFF system might only have a 400 scf/h flow rating under the MCA system. The MCA rating is more realistic, however, because you don't use a regulator without some hose attached. Both ratings are perfectly valid, of course, as long as you understand what they mean and how to use that information to get your work done.

Some regulator manufacturers will give you both kinds of data. Each regulator model in a catalog will list the regulator with MFF data for the gas being rated. Then a separate table or chart will tell you how to calculate the pressure drop and flow rates for this gas, depending upon the diameter and length of the hose you use. A few regulator manufacturers even give MCA flow data from a regulator model for nearly full and nearly empty cylinders. That's really being thorough.

Gas Flow Rates

There are other facts about gases and regulators that will help explain why you have to be very careful when selecting a regulator for a given gas and job. One of them is the actual flow rates of different gases through the same regulator model. We already mentioned that gas flow rates depend upon the gas density as well as the delivery pressure, even before the gas gets into a hose. The MCA and MFF ratings given in regulator catalogs are almost always measured with compressed air. Compressed air is cheaper than argon, helium, oxygen, or carbon dioxide.

If the manufacturer's catalog doesn't tell you whether the data are for MCA or MFF conditions, and what gas was used for the test, assume that you are reading a MFF value for air. It makes the regulator model sound better in the manufacturer's catalog.

However, the flow rate for helium at the same pressure as the compressed air used to rate a helium regulator is 2.7 times higher than air. Hydrogen's flow rate is 3.79 times higher than air at the same pressure. The flow rate of argon gas is only 0.87 times that of air. Only nitrogen and oxygen are close to the flow rate of air, and that's because air is composed largely of those two gases.

One time we watched a beginning welder set up a gas-shielded arc-welding job. She had no flow meter (see pages 76 through 78), and she took it for granted that the argon gas she was using had the same flow rate as air through 25 ft [7.6 m] of ⅜-in. diameter hose. Later on, she couldn't figure

out why her welds were getting so porous. She checked her hoses. They weren't tangled, kinked, or full of repair connections.

Next she checked the regulator. The delivery-pressure gauge indicated that she was getting enough argon shielding gas, based on the pressure reading. But the appearance of her weld metal (and soon after that, her welding supervisor) told her that she was not getting enough argon shielding gas on the hot weld metal to protect it from oxidation until it cooled off.

Her error was that she set up the job using regulator and flow-meter readings based on air, not argon, and a very long hose. Since the flow rate of argon is only 87 percent that of air, she was getting 13 percent less argon shielding gas than the regulator reading seemed to indicate. She lost another 30 percent of her preset shielding-gas flow through her 50-ft long [15-m] hose. It happens every day, and not just to beginning welders.

We are not going to tell you right now how her welding supervisor helped her solve the problem. We'll leave the solution for you to find when you read about using welding hose.

Regulator Problems

Gas regulators are complicated, sensitive instruments. A careless welder can do more to ruin a regulator than any other gas-using instrument he or she works with. One of the fastest ways to ruin any regulator is to forget to wipe the regulator, cylinder, and hose connections clean with a dust-free cloth before connecting them.

When you turn on the cylinder, high-pressure gas blasts into the regulator at the speed of sound. Any dirt in the connection is carried along with the gas. If you have seen what sand blasting will do to the side of a granite wall, think what a little grit will do blasted into the delicate parts inside a regulator. The first thing the dust particles do is wear away

part of the precision-fitted high-pressure valve seats inside the regulator. The next thing that happens (at the least) is that the regulator starts to "creep."

You see regulator creep when you shut off the gas at the regulator but keep it on at the cylinder. The delivery-pressure gauge should drop to zero. If the delivery pressure does not stay at zero but tends to creep up the dial, some gas is leaking from the open cylinder, through the high-pressure end of the regulator into the low-pressure stage. That's regulator creep and most of the time it's caused by poor care and handling of regulators.

If the gas is explosive or toxic and continues to creep out of the regulator while you are not around, be careful or you may learn about the final results of regulator creep the hard way when you return to your equipment.

Safety Devices

Regulators are built with at least two relief devices that are intended to protect you in case things like regulator creep get out of hand and the regulator gets a full surge of high-pressure gas dumped into the device all at once. All regulator gauges have blow-out backs. The pressure in the gauge is released from behind, away from you, before the gauge glass is blown at you. In fact, most regulator gauges don't use glass. They use very tough, shatterproof plastic cover plates.

The regulator body also is protected from sudden overpressure. A variety of safety relief devices are used. Blow-out disks are one kind. Spring-loaded pressure-relief valves are another. Either one is designed so that when the overpressure reaches a certain level, the pressure is reduced suddenly before the regulator blows up. The burst disk can be unpleasant. It sounds like a cannon when it goes off. Spring-loaded relief valves make a lot of noise, but it's more like a shriek than a big bang.

If you hear either one, you'll know it. Immediately turn the gas off at the cylinder valve. Then remove the regulator and tag it for inspection and repair. It's probably faulty. Never clamp down the relief device or weld it shut, or the next noise you hear will be a lot more serious.

As a general rule, always open the cylinder valve slightly before turning it on full both before and after you attach a regulator and hose. Also stand to one side when you do, just in case the thing throws some dirt at you or the regulator is defective. In addition to preventing dust from sandblasting the regulator insides and overstraining the regulator parts, there is a special reason for opening a regulator slowly when the gas you use is oxygen.

If you suddenly turn the oxygen cylinder valve full on, instead of gradually opening or "cracking" the valve first, pure oxygen at up to 2200 psig [15.2 MPa] will slam into the inside of the nearly empty high-pressure chamber of the regulator. Part of the gas will expand for a microsecond. And

then the in-rushing gas will slam into parts inside the regulator. When it hits something, the oxygen molecules immediately recompress and get very hot. The recompression reaction occurs in milliseconds. The gas heats up to 1800°F [980°C]. Oxygen that hot will burn steel and set the valve seats on fire.

You'll know immediately because your regulator may either catch on fire or blow up. For the same reason, never oil or grease an oxygen regulator, or handle the regulator with dirty hands. Oil or grease in an oxygen regulator almost guarantees a regulator valve-seat fire or an explosion.

How Regulators Work

Now that you understand a lot more about regulators, it will be a lot easier for you to understand how they work. Single-stage regulators have a high-pressure chamber next to the gas inlet that is attached to the cylinder, and a low-pressure outlet that hooks up to the welding hose. In between these two chambers is a valve (Fig. 4-18) that has a tiny nozzle

and a valve seat or plug that fits into the nozzle and lifts up or down. A given amount of gas entering the high-pressure chamber will expand when it goes through the nozzle, enters the low-pressure chamber, and goes out through the hose. When the gas expands, the pressure drops.

One side of the low-pressure chamber is a flexible neoprene rubber (or even stainless steel) disk called a diaphragm. When you adjust the regulator delivery pressure by turning the big screw on the outside of the regulator, you put more (or less) spring pressure on the diaphragm. The higher the spring pressure on the diaphragm, the less it can expand. That makes the delivery pressure in the second stage higher. The lower the spring pressure, the more the diaphragm can expand. The low-pressure chamber can get bigger so that the gas expands more. That lowers the delivery pressure as the gas leaves the regulator.

A real regulator has more parts than shown in the drawing. For example, there is a second small

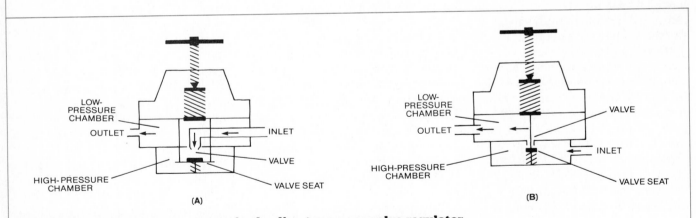

FIGURE 4-18 (A) High-pressure gas in the direct-pressure valve regulator pushes the valve seat down, tending to open the high-pressure chamber for gas flowing into the low-pressure chamber. (B) High-pressure gas in the inverse-pressure valve pushes the valve seat up, tending to close the flow of gas from the high-pressure chamber into the low-pressure chamber. The small valve-seat springs oppose the motion of the valve seats. The force of the big spring on top of each regulator diaphragm is adjusted by the threaded handle. These devices are all connected by mechanical links. Therefore, adjusting the regulator with the hand screw determines how far each direct-pressure or inverse-pressure valve seat can open or close the regulator.

spring in the high-pressure gas chamber that pulls the valve stem and valve seat back down.

When you turn the adjusting screw in, the screw presses down against the big coil spring inside the regulator. The spring pushes against the top of the regulator diaphragm. Since the diaphragm is attached to the regulator valve seat (which is on the high-pressure side) by the valve stem, the screw-and-spring force pushes the valve seat away from the nozzle. That unseats the valve and lets high-pressure gas from the cylinder rush through the nozzle and into the low-pressure chamber.

The gas flowing into the low-pressure chamber pushes back against the diaphragm so that the opposing forces of spring pressure on one side of the diaphragm and gas pressure on the other side try to reach a balance. More high-pressure gas from the cylinder rushes into the first chamber, decompresses as it enters the second chamber, and escapes from the regulator into your hose and torch.

When you turn off the gas flow at the low-pressure end of the regulator (by shutting off the gas at your torch, for example), gas pressure builds up in the low-pressure chamber. That forces the diaphragm up against the big coil spring, which pulls the valve stem seat up with it against the regulator nozzle. That shuts off the regulator, but it leaves your delivery-pressure gauge with a reading that tells you what the working pressure is.

The more you turn in the adjusting screw, the more high-pressure gas rushes into the second chamber and then into the delivery-pressure chamber of your regulator. If the cylinder supply pressure did not vary, this would produce a constant pressure and flow rate to your torch and work.

When you turn the adjusting screw out, it pulls the diaphragm up, which also pulls the valve seat up against the nozzle. The

valve seat fits into the nozzle and shuts off the flow of gas to the low-pressure chamber in the regulator. Since the regulator outlet is still open (and your torch valves are, too), any remaining gas rushes out of the low-pressure chamber and into the outside air. This produces a zero reading on your delivery-pressure gauge. If, however, you shut the torch off, the gas stays in the regulator.

Both ways of shutting off the regulator, at the torch or with the regulator adjusting screw, also leave high-pressure gas in the first stage of the regulator. This high-pressure gas is at the same pressure as the gas in the cylinder. Therefore you know what the cylinder pressure is by reading the cylinder-pressure gauge.

Now you know why you should first turn the gas off at the cylinder valve when you finish a job, or when you want to remove the regulator. Then you should open up the regulator to let all the high-pressure gas out of it. Only after that can you safely remove the regulator from the cylinder.

Two-stage regulators When a cylinder is full of gas at 2200 psig and the delivery pressure is set at 22 psi (for example), then the ratio between the two pressures is 100:1. When the cylinder is nearly empty (say 44 psig), and the working pressure is still 22 psig, the ratio between the cylinder pressure and delivery pressure is only 2:1.

The big coil spring that has to be stiff enough to withstand a 100:1 pressure ratio is simply too stiff to detect a much smaller pressure ratio of 2:1. That's why two-stage regulators (Fig. 4-19) work better than single-stage regulators.

Two-stage regulators have two opposing springs and two diaphragms. They really are two regulators in one case. One spring and diaphragm handle high pressures best; the other spring and diaphragm handle low pressures

best. Therefore, two-stage regulators work well at both high and low cylinder pressures.

Regulator Care and Handling
From the discussion of the mechanics involved, you can see why you should not try to fix a regulator yourself. Your welding distributor has specially trained people to do that work. You, however, can do a lot to keep the regulators in your shop in good working order by following the proper procedures.

Always purge all your regulators, hoses, and your torch before you use them and after you are finished. Even if you are a safe worker, you don't know who else has been using your equipment. They might not be as careful as you are.

Check valves are designed to keep gas from backing up into the hose or regulator, but you may not have them on your equipment. Even if you do, it only takes a minute to purge your equipment before and after use, just to be sure. Otherwise, if some fuel gas backed up into your oxygen lines (or oxygen into your fuel-gas lines), you'll get an explosive mixture that will cause a flashback or even a regulator fire or explosion.

CAUTION: Always purge your gas lines before and after work

You should purge your regulators, lines, and torch or welding gun even if you are using inert shielding gases during arc welding. A little air, or even moisture that can condense inside the lines during a humid day, can ruin a good metal-inert-gas (MIG) or tungsten-inert-gas (TIG) weld. Shielding gas from the cylinder is very dry. It will not only remove any air in the lines, but it also will dry them out.

Be sure to purge each line separately. Simply turn the cylinders on a little, one at a time, after

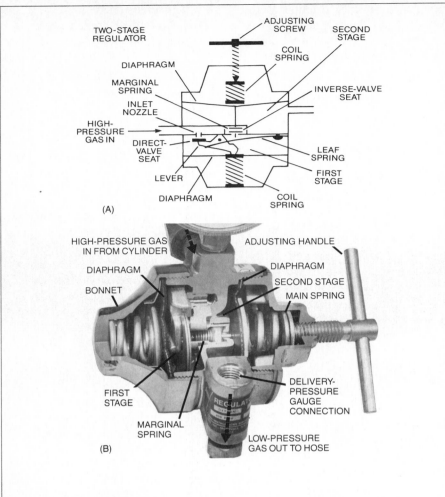

(A)

(B)

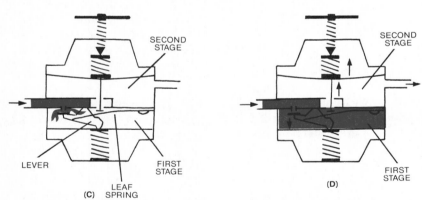

(C)   (D)

FIGURE 4-19 (A) A good two-stage regulator combines both direct and inverse valves in one case. That is why they have a steady flow rate that will not change even though the cylinder pressure drops as the gas is used. (B) The insides of a real two-stage oxygen cylinder. It has two diaphragms and two springs. The first-stage pressure is set by the manufacturer. The operator adjusts the second, delivery-pressure stage with the threaded screw on the right. (C) High-pressure gas enters a direct-inverse valve, two-stage regulator by pushing open the first-stage valve. When the pressure in the first stage reaches the present value, the valve closes. (D) Gas at the present intermediate pressure flows through the second-stage valve and exerts force on the second-stage diaphragm. The second-stage valve then closes at the set pressure as gas flows out the hose. The process in (C) and (D) repeats at high speed, producing a constant gas flow and pressure out of the regulator and into the hose and torch.

each regulator, hose, and torch is hooked up. Open the torch valve and let some oxygen out. Then open the fuel-gas valve and let fuel gas out. If you happen to have a mixture of fuel gas and oxygen in either the fuel gas or oxygen regulator, or in the hoses or the torch, it will be swept out and not give you trouble.

If you are working with very dry welding shielding gases, any moist air or gas that backed up into the hose or regulator also will be swept out before you start welding. Then you won't have problems such as starting porosity in the first few feet of weld caused by moisture or oxygen.

Finally, when you stop work always remove your cylinder regulators and put them in a drawer or tool box. They are delicate precision instruments and should be treated with great care.

4-4
FLOWMETERS

Flowmeters are used to regulate precisely the flow rates of shielding gases in MIG and TIG arc-welding processes. Flowmeters are special valves with a cylindrical glass gauge and a little steel ball that rides up on the gas column flowing into the glass metering tube.

How Flowmeters Work

The higher the gas flow rate, the higher the steel ball rises inside the tube. A scale on the tube glass can be calibrated so that you can read the gas flow rate in standard cubic feet per hour [cubic meters per hour]. The reading is seldom more than 90% accurate, but that's close enough because the flow rates for shielding gases are much less than 100 scf/h [2.8 m³/h] and often around 5 to 10 scf/h [0.1 to 0.3 m³/h].

Flowmeters can be purchased separately or already attached to two-stage gas regulators (Fig. 4-20). They can be used on low-pressure gas pipelines as well as cylinders. Some of them are adjust-

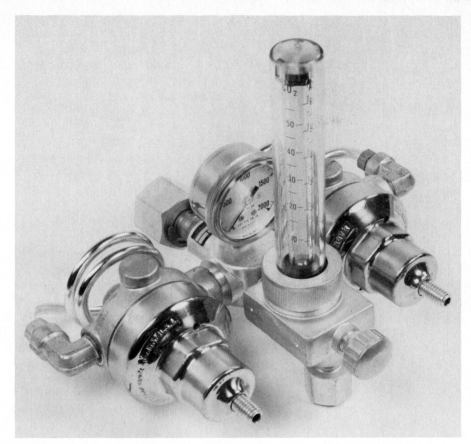

FIGURE 4-20 A typical gas flowmeter. You may use gas flowmeters for GTAW and sometimes MIG welding and for metal spraying jobs. They also are used to help you make mixed shielding gases.

able so that you can regulate your shielding-gas flow rate. Some flowmeters are used with gaugeless regulators that are preset by the manufacturer to produce a single flow rate and gas pressure. Still other flowmeters have a dual range; one for low flow rates, the other range for high flow rates.

Shielding-gas flowmeters are designed to regulate the flow of shielding gases such as argon, helium, and carbon dioxide. Most flowmeters used in welding operate at either 15 psig or 30 psig [103 or 207 kPa]. The adjustable meters will give you gas flows as low as 2.9 scf/h [0.1 m³/h] up to 60 scf/h [1.7 m³/h] for argon and argon-carbon dioxide mixtures; from 4 to 150 scf/h [0.1 to 1.4 m³/h] for helium gas, and from 2.6 to 50 scf/h for straight carbon dioxide. Sometimes other gases such as oxygen or nitrogen are used, but not very often.

How to Read a Flowmeter
The welder reads the flowmeter by looking straight ahead at the very center of the little steel ball inside the glass gauge tube. The nearest mark on the glass tube will be a number between 1 and 100. Be careful. That number is *not* a gas flow rate.

The numbers on the glass gauge, from 0 to 100, have no direct meaning because different gases have different flow rates under a given pressure. Also, more than one inlet gas pressure may be used. The little steel ball rides up the gas stream in the tube a different distance, depending upon the gas being monitored and the pressure of the gas entering the flowmeter. That's why you have to calibrate the flowmeter scale.

Calibration Charts
Flowmeters come equipped with special charts. These charts show how to read the flowmeter when different gases and gas pressures are used. These are called *calibration charts* because they relate the glass tube-and-ball readings to actual flow rates. Different flowmeters will have different calibration charts, so don't make quick judgments without looking at the chart for your own flowmeter and inlet pressure.

A specific model of flowmeter often will have two charts with lines on them for different gases. One chart calibrates the numbers on the glass tube with the flow rates for different gases at 15-psig [103-kPa] inlet pressure. The second chart might give the same data for a 30-psig [207-kPa] inlet pressure.

Let's assume that your flowmeter is running on argon and that the inlet pressure is 15 psig [103-kPa]. The calibration chart for your flowmeter might look like Fig. 4-21. Look dead on at the center of the little steel ball to get the correct reading, not up or down or at an angle. Let's assume that the ball is very close to the number 30 on the glass gauge scale. According to the calibration chart for this flowmeter, a 30 is equivalent to about 12 scf/h [0.34 m³] of argon. If you had helium running through the flowmeter, the same 30 reading would equal 48 scf/h [1.4 m³] of helium.

There's a lesson here far more important than how to run a flowmeter (they always come with instructions, anyway). Two separate gases, a heavy one like argon and a light one like helium, will have widely different flow rates even though they have the same inlet pressure. These gas flow rates are critical, especially for TIG shielding gases.

The rule is very simple. Everything else being equal (such as the pressure, the size of hole the gas is flowing through, and so on), heavy gases will have lower flow rates than light gases. It's almost as if heavy gases act like pancake syrup and light gases act like water. This holds true whether the gases are shielding gases or fuel gases. See Sec. 4-6 on welding hose and the footnote to Table 4-3.

4-5
GAS MANIFOLDS

Large shops often will have a lot of welders working from a single oxygen or fuel supply. Large cutting machines with anywhere from 2 to 12 or more torches also need a large volume of oxygen and fuel gas. Even a single welder sometimes needs more oxygen or fuel than a single cylinder can provide. Two examples are very heavy cutting and using a heavy heating torch that would empty the contents of a single oxygen or fuel cylinder in a short time. Whenever you start spending a lot of time changing cylinders instead of working, you have a job for a manifold.

What Manifolds Do

A manifold is nothing more than special plumbing that connects two or more cylinders into one gas supply line (Fig. 4-22). In addition to supplying large quantities of gas, manifolds also solve the problem of exceeding the maximum removal rate of gas from an acetylene or liquefied-gas cylinder. For example, a manifold will help you if you are extracting acetylene from one cylinder so fast that you start to draw the acetone out along with the fuel gas.

Manifolds also empty all the cylinders on line at a uniform rate. When they near the end

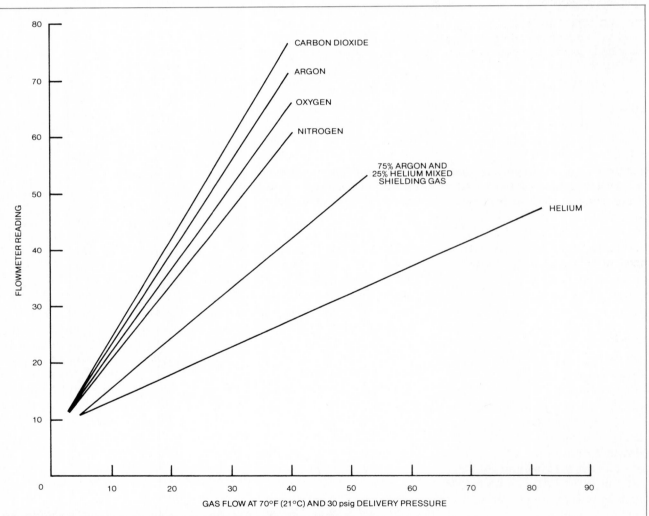

FIGURE 4-21 Different gas flowmeters have different calibration charts. However, this chart shows some important general facts about gases. For the same flowmeter reading (50 in this particular example), six different gases will have different flow rates under the same temperature and regulator delivery pressure. Helium always has the highest flow rates (87 scf/h has a meter reading of 50 in this example), while heavier gases such as argon and carbon dioxide will have the lowest flow rates. If the flowmeter had a reading of 50 scale units for CO$_2$, it would mean a flow rate of about 23 scf/h.

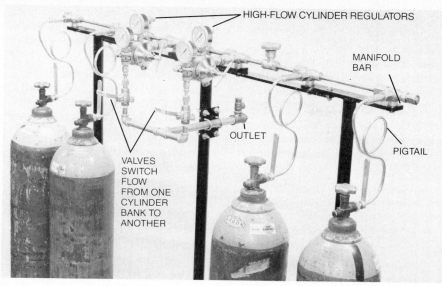

FIGURE 4-22 This large cylinder manifold is used so that you won't have to change cylinders very often. They also can be used to give you very high gas flow rates that you can't get from single cylinders.

point, you can easily switch to a second bank of manifold-connected cylinders and keep working while a laborer changes the cylinders on the first manifold. Bigger manifolds have a separate valve that turns one entire bank of cylinders on while shutting off the other bank of cylinders. These duplex manifolds (cylinders on both sides in two banks) often are used in large plants to supply a group of welders, or even in-plant oxygen or fuel pipelines.

Manifolds come in different sizes. There are small, portable ones for 2 to 5 cylinders, and big, stationary manifolds that can take a bank of 26 cylinders in two rows. Manifolds are also used "in reverse" in gas filling stations run by welding distributors. The distributors use the manifolds to fill dozens of cylinders at once.

The portable oxygen manifold in Fig. 4-23 is a very simple one. It has four "pigtails" leading from connections to four cylinders into

a central bronze block. If a single large-flow oxygen manifold regulator were attached to the threaded connection on the block, you could draw gas from four cylinders instead of one.

The plugs that hang down on chains from the block are used to stopper up any pigtail outlet that doesn't have a cylinder attached. That's absolutely necessary, of course, or else gas would flow out of the open pigtails instead of through the regulator.

Both stationary and portable gas manifolds can be supplied for oxygen, nitrogen, argon, hydrogen, helium, carbon dioxide, or even shielding-gas mixtures. If you don't know if you have the correct manifold for the gas and cylinder pressure you will use, ask somebody before you use it.

Manifold Regulators
Everything we've said about cylinder regulators applies to regulators for manifolds, with one additional fact. Regulators for manifolds have to be bigger than cylinder regulators. They have to have much higher flow rates even though their pressure ratings will be the same.

A typical oxygen manifold regulator, for example, might be able to deliver oxygen at flow rates up to 4000 scf/h [113 m³/h] at pressures up to 200 psig [1.4 MPa], whereas (at the same delivery pressure) a good heavy-duty cylinder regulator might deliver only 1000 scf/h [28.3 m³/h]. You can see why you mustn't put a cylinder-oxygen regulator on your oxygen manifold.

4-6
HOSES
All gas-welding processes use special welding hose to get the gas from the cylinder or pipeline regulator to the welding torch. Welding hose causes more problems for welders and flame cutters than any other tool they use. Even many experienced welders don't know why.

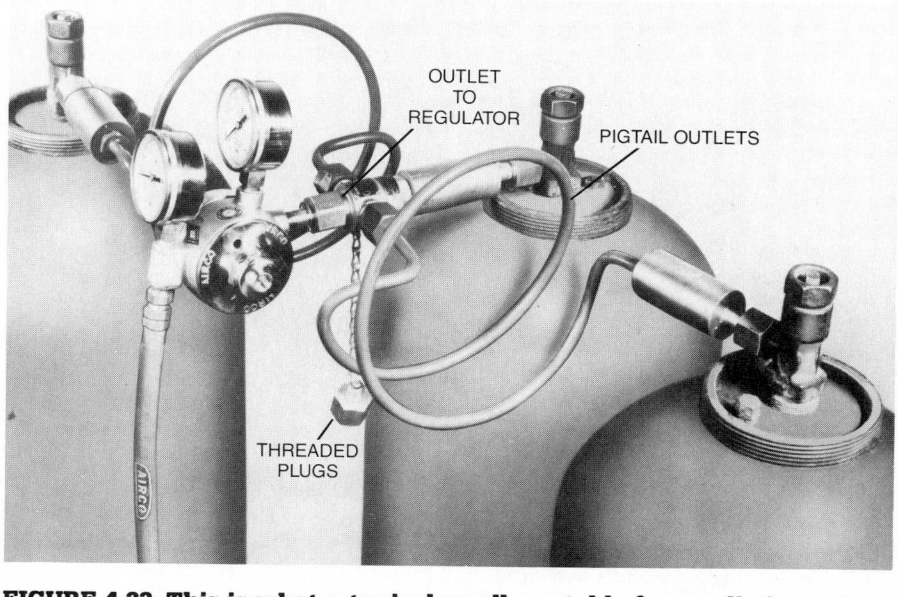

FIGURE 4-23 This is what a typical small, portable form cylinder oxygen manifold looks like.

People drive forklift trucks over welding hose. Trucks drive over it, people step on it, welders drag it through puddles of oil, kink it, pull it too hard, and welding hose still survives most of the time. When it wears it can even be patched and put back into service.

People often buy the cheapest hose they can get. Then the troubles start. The hose wears out. It has to be patched. And every time you patch a welding hose with special high-pressure fittings and connections, you increase the friction inside the hose.

Don't think of welding hose as what you use to water your lawn. You certainly can't patch it with black tape. The pressure inside a welding hose can easily exceed 100 psig [690 kPa] in use. That's about three times higher than the air pressure in an automobile tire.

If you ever stood next to an auto tire when it blew out, you'll get a good idea of what a welding hose sounds like when it blows out. Only it can be three times as loud. A good welding hose is built like a good tire, only better. The maximum recommended working pressure for good-quality welding hose is between 200 to 250 psig [1.38 to 1.72 MPa]. The maximum burst strength may be close to 1000 psig [6.89 MPa].

How Welding Hose Is Constructed

If you take a knife and cut a welding hose in half and look at the cut ends, you will see at least three separate layers (Fig. 4-24).

The outside layer is a rubber or neoprene compound designed to resist combustion, sparks, hot slag, hot metal, weathering, water, ozone (a special type of oxygen molecule produced in electric arcs), and wear from being dragged around on the floor. The cover also is designed not to break loose or blister where hose couplings are attached, even after severe flexing.

The cover is usually either red for fuel gases or green for oxygen. Color coding helps you avoid mixing up fuel and oxygen lines so you won't create an explosive mixture in your hose.

The middle part of the hose is called the *carcass*. It is usually woven of twisted nylon or rayon cord. Its job is not only to add strength to the hose but also to help it resist impact, crushing, and sharp objects. The carcass also helps keep the hose from kinking and shutting (like a garden hose will if you bend it and then pull on it hard). Sometimes several plys of braided cords are wrapped around the hose in between layers of rubber to strengthen the carcass and to make sure it won't leak.

The inside of the hose is either rubber or neoprene. This inner tube should be absolutely free of porosity and as smooth as possible. Hose manufacturers vulcanize the inside tube by injecting live steam under pressure into it, after the carcass is added but before the outside cover is put on. Any porosity in the inner tube is sealed shut by this process. A rubber inner lining is acceptable for acetylene. You always should specify a neoprene inner hose lining for LPG fuels such as MAPP Gas because the neoprene is more resistant to the liquefied fuels.

Welding hose is supplied either cut to length with all connections already attached to it or in 500-ft [152-m] reels without connections. (Fig. 4-25). When you use hose from a reel, it's your responsibility to cut the lengths you want perfectly square and add the correct connectors so you can attach your hose to your regulators and torch. (If you make up your own hose from single-line hose on a reel, we strongly suggest that you tape the two fuel and oxygen lines together every 8 or 10 in. [200 to 250 mm] with black tape. That will keep the separate lines from tangling up under your feet.)

In addition to single-line hose, welding hose is available in prepared lengths or on reels as double or twin hose (one red fuel line and one green oxygen line). The twin hose is handy because the hoses are already attached side-by-side along their full length. Most welding hose in use is the twin-hose variety.

Inside Diameter (ID)

Twin welding hose is most often available in common inside diameters of 1/8, 3/16, 1/4, 5/16, and 3/8-in. That's the inside diameter of the inner tube, not the outside diameter of the hose cover. The 1/8-in.

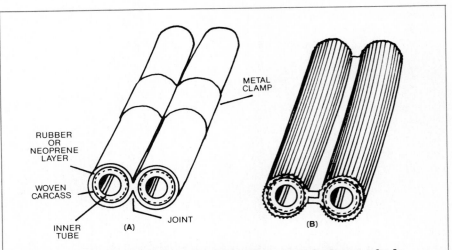

FIGURE 4-24 Twin welding hose for oxygen and fuel gases is designed for high pressures and rough handling. The two hoses (green for oxygen and red for fuel gas) are joined together by clamps or by a special rib to keep them from becoming tangled up. If you do use two separate hoses, tape them together or they will tangle.

FIGURE 4-25 Welding hose can be purchased complete with connections in short lengths such as 15 ft, or in long sections which you cut to length and then add your own connections.

ID hose is only used on very small welding torches. The ³⁄₈-in. hose is most often used on heavy-duty manual cutting torches.

Single-line hose is available in these diameters as well as ½, ¾, and 1-in. ID hose. The larger hose sizes are usually only used on very heavy duty cutting and heating jobs, and on large flame-cutting machines.

The inner dimensions of a welding hose are very important. That's where our beginning welder had her argon flow problem. Her hose was too long and the diameter was too small.

Hose Length and Flow Rate

A ³⁄₁₆-in. hose might be just fine for a job if the hose is only 20 ft [6.1 m] long. The same ID in a 50-ft-long [15.2-m] hose will not deliver the same amount of gas. For a given hose diameter, the pressure drops as the hose gets longer, which means that the flow rate also drops because friction inside the hose slows the gas flow. If you double the length of the hose, you double the friction and also double the pressure drop.

If the hoses are short and the inside diameter is adequate for

the job, the pressure drop from one end of the hose to the other won't be noticeable. Our beginning welder, however, was using a 50-ft [15.2-m] hose when a 20-ft [6.1-m] hose would have been enough for the job.

The problem becomes even more noticeable when you are doing work requiring large gas flows. Examples are heavy cutting or heating. There are two things you can do.

Choosing Hose to Fit the Job

First, you should use a hose whose internal diameter is large enough for the required flow rate and working pressure on the job. Only then, if you need to, should you adjust your regulator delivery pressure higher to get more gas into the hose and out the other end.

How do you know what hose diameter you need? Most welders "eyeball" it. They adjust the regulator flow rate until the torch flame looks correct. An experienced MIG or TIG welder will do something very similar. He or she will listen to the shielding gas flowing out of the gun. Experienced welders estimate the flow rate of a shielding gas literally by holding the gun to their ear.

But you have no experience. What can you do? When you set up the job, use a set of tables that give you the desired output flow rate (scf/h or m³/h) and regulator pressure settings (psig or kPa) for different hose diameters and different lengths within each diameter. Table 4-3 will help you a lot. This table is based on oxygen flow rates. See the note at the bottom for using other gases—you can approximate for most work. Only hydrogen, helium, and natural gas will seriously throw you off if you don't correct the oxygen data for the low density of these gases.

How to Use the Pressure-Drop Chart

Look at Table 4-3 for 25 ft of ½-in. ID hose. Look under the "torch entry pressure" column at 70 psig.

If you want 500 scf/h of gas, you have to set your regulator at 72 psig. That is, the pressure drop through 25 ft of ½-in. ID hose is only 72 psig at the regulator, minus a 2-psig pressure drop, equals 70 psig at the torch for a 500-scf/h flow rate.

Now look at the same data for 25 ft of ³⁄₁₆-in. ID hose. To get 500 scf/h of gas at 70 psig out of the torch, you have to set your regulator delivery pressure at 105 psig. That's a pressure loss of 35 psig (105 psig at the regulator minus 70 psig at the torch equals 35 psig pressure drop through the small-diameter hose). That's a 33 percent pressure drop from the regulator to the torch, assuming that the hose is in perfect working condition (that is, it doesn't have repair connections, and there are no kinks or twists to cut down the flow rate.)

What if you wanted 500 scf/h of gas through 50 ft of hose instead of 25 ft of hose because the job is far away from the cylinders? (You might be working on the roof of a building.) The pressure drop for 50 ft of hose will be twice as much as for 25 ft. The pressure drop would be 70 psig instead of 35 psig between the regulator and the torch if you were using ³⁄₁₆ ID hose, so you will have to set your regulator another 70 psig higher than you need to for a 25-ft hose, or 105 psig plus 70 psig equals 175 psig for a gas flow rate of 500 scf/h. It would be better to get a larger-diameter hose.

If you switched to ³⁄₈-in. ID hose (the even larger ½-in. ID hose wasn't available for this job) 50 ft from your cylinder regulator, and still want 500 scf/h of gas with 70 psig at the torch, read it from the table for ³⁄₈-in. ID hose.

At 70 psig into your torch, with 500 scf/h of gas, 25 ft of this ³⁄₈-in. ID hose requires a regulator setting of 74 psig. That's a pressure drop of 4 psig. But your hose is twice as long, 50 ft, not 25 ft as in the table. Therefore the pressure drop will be twice as large, 8 psig instead of 4 psig. Instead of set-

TABLE 4-3 Torch entry pressure versus regulator outlet pressure*

Flows, scf/h	Torch Entry Pressure, psig								
	20	30	40	50	60	70	80	90	100
Hose Diameter—3/16 in. Length—25 ft									
50	23	33	43	53	63	73			
75	25	35	45	55	65	75			
100	28	37	46	56	66	76			
150	33	41	48	57	67	77			
250	46	52	58	67	74	84			
350	59	63	68	75	84	91			
500	82	85	89	94	99	105			
Hose Diameter—1/4 in. Length—25 ft									
100	22	32	42	52	63	73	83	94	103
250	27	36	45	55	65	75	84	95	104
500	36	44	53	61	70	79	89	98	108
1000	76	78	82	86	91	97	104	111	118
Hose Diameter—5/16 in. Length—25 ft									
100	20	30	41	52	62	72	84	94	105
250	23	34	43	54	64	74	84	94	105
500	30	37	47	56	66	75	85	95	106
1000	47	53	60	67	75	83	92	101	112
1500	68	73	76	82	88	95	103	111	120
2000	94	95	96	100	104	108			
Hose Diameter—3/8 in. Length—25 ft									
250	21	31	41	51	62	72	82	92	
500	24	33	44	53	64	74	83	93	
1000	33	39	48	58	67	77	86	95	
2000	57	60	67	74	84	89	96	104	
3000	92	94	96	100	103	110	112	125	
Hose Diameter—1/2 in. Length—25 ft									
500	21	31	42	52	62	72	82	93	104
1000	23	34	44	54	64	74	83	94	105
2000	32	41	50	59	68	78	87	97	107
3000	47	51	57	66	75	84	93	102	113
4000	66	69	73	77	83	90	100	110	119
5000	82	84	86	88	92	98	107	117	126

*Adjustments for Other Gases: These charts are measured using oxygen as the gas. If you need more exact flow rates for other gases (which you probably won't), multiply your known flow rate in standard cubic feet on the left side of the chart by a correction factor to get the equivalent oxygen flow rate in standard cubic feet. *The lighter the gas, the greater the volume flow rate for the same pressure difference across a long hose.* Here are the multiplying factors for other commonly used gases:

Helium gas in scf/h	= Oxygen flow rate, scf/h × 0.35
Natural gas in scf/h	= Oxygen flow rate, scf/h × 0.71
Acetylene in scf/h	= Oxygen flow rate, scf/h × 0.91
Nitrogen gas in scf/h	= Oxygen flow rate, scf/h × 0.93
Air in scf/h	= Oxygen flow rate, scf/h × 0.95
Argon in scf/h	= Oxygen flow rate, scf/h × 1.12
Propylene in scf/h	= Oxygen flow rate, scf/h × 1.15
MAPP Gas in scf/h	= Oxygen flow rate, scf/h × 1.16
Propane in scf/h	= Oxygen flow rate, scf/h × 1.17

These multiplying factors are based on the following formulas (insert the number for the gas, divide it and take the square root)—if you ever need them for hose or flowmeters:

$$\text{Multiplying factor} = \sqrt{\text{Molecular weight of gas}/32}$$
$$\text{Multiplying factor} = \sqrt{\text{Density of scf of gas}/0.83}$$
$$\text{Multiplying factor} = \sqrt{\text{Specific gravity of gas}/1.105}$$

ting your regulator at 74 psig, you will have to add another 4 psig for the 50-ft hose and set your regulator at 78 psig.

Notice the difference in regulator pressure settings at 50 ft for very small diameter hose (3/16-in. ID hose requires 175 psig at the regulator to get 70 scf/h of gas at your torch) and for large-diameter hose (3/8-in. ID hose requires 78 psig at the regulator to get 70 scf/h of gas at your torch). That's a huge difference of almost 100 psig between the two regulator pressure settings. Unfortunately, large-diameter hose is heavy to carry around.

A good trick in field welding, when very long hoses are absolutely necessary, is to use large-diameter hoses from the regulator to about 12 ft [3.6 m] from the welder or burner. The last 10 to 12 ft [3 to 3.6 m] are handled by a 1/4 or 3/16-in. ID "lead hose" connection. That way you don't have a heavy hose to drag around and you still get a high gas flow rate from a distant cylinder.

You won't believe how many experienced welders have problems getting enough gas through their torch and don't know why. We have even seen people turn in three good torches in a row for repairs, when all they needed was a one-size-larger welding hose. Or worse, they will use a beat-up hose full of repair connections and wonder why their torch won't give the same flow rate that it did when the hose was new.

Checking Pressures

You can check the initial flow rate from a gas regulator with a simple test stand. Use a deep bucket or container of water. Here are some useful facts to help you. The pressure of 27.7 in. [704 mm] of water equals 1 psig [6.9 kPa] (or 2.0 ft [0.6 m] of water equals 0.866 psig [6 kPa]). Water at the bottom of a tank 277 in. [7 m] or 23.1 ft deep, for example, is under a pressure of exactly 10 psig [69 kPa]. If you set your regulator for 3 psig [18 kPa] with 7 ft [2.1 m] of a large (½-in. diameter) hose on it (too short to make a noticeable pressure drop through the hose), the water pressure at the bottom of a 6.925-ft [2.11-m] deep tank should just stop the gas from coming out of the hose.

If you have pressure-testing equipment that uses a column of mercury instead of a column of water, 1 in. [25.4 mm] of mercury almost exactly equals ½ psig [3.4 kPa] (0.4913 psig to be precise). The weight of a column of mercury an inch high [25.4 mm] also equals 1.13 ft [0.34 m] of water pressure.

Fixing Hose Correctly

All the connections for welding hose have been standardized by the Compressed Gas Association. Letter grades (A, B, C, D, E) plus the type of gas to be used correlate directly with the corresponding connections on regulators and torches. A-, B-, and C-size connections are the most common. A-size connections are for low flow rates; B sizes are for medium flow rates, and C-size connections are for maximum flow rates. Sizes D and E are used for very large heating or welding torches that require manifolds. Details on the thread size, ID, and other information are in all equipment catalogs and often are stamped right on the connections.

For example, you need A-size nuts and *glands* to hook up hose to regulators with A-size outlet regulators and to a torch with an A-size inlet. A torch and regulator with A-size connections would

probably be a small brazing outfit. B-size connections are typical of most industrial production units.

The reason why the type of gas also has to be specified (oxygen, acetylene, LPG, or inert gas) is that an A connection, for example, will be threaded differently for acetylene than for oxygen, or for inert fuel gases. Some A- or B-size connectors will have right-hand threads instead of left-hand threads, or even inside versus outside threads.

Right-hand threads are always used for oxygen; left-hand threads are for acetylene and other fuel gases. The number of threads per inch (or per millimeter) even though the threads run the same way and the ID is identical, also can be different. The reason is not to make things complicated. It's to protect you. By using right- and left-hand threads that don't match, you are prevented from hooking up oxygen equipment to fuel-gas cylinders and regulators. And you can't hook up inert-gas equipment to oxygen. Certainly equipment manufacturers and the CGA have done everything in their power to keep you from making these mistakes. But would you believe that there are cases where somebody went out of the way to force (maybe even weld or braze) the wrong connectors onto the wrong equipment?

The basic hose connection is a nut and a gland (Fig. 4-26). The nut is threaded on the inside and connects to the threaded inlets and outlets of your torch or regulators. The grooved gland fits in-

side the hose very tightly. The nut is loose and can be turned by hand or by a small wrench to tighten the threaded nut onto the equipment. A metal strap called a *brace* usually is attached about 2 ft [0.6 m] back from the hose end connections, allowing enough freedom between the lines so they can be attached to regulators or torch connections.

Two lengths of hose can be connected with hose couplings. Hose-splicing nipples insert between two pieces of hose where a leaking or blown-out section has been removed. Hose splicers provide a strong, nonslip connection, especially when they are combined with a hose clamp wrapped around a hose ferrule. The ferrule is usually crimped onto the outside of the hose. If a crimping machine is not available, the hose clamp is tightened up over the ferrule. Binding a hose splicer by wrapping the hose ends with wire will not work. The joint will not be tight enough to withstand high gas pressures.

4-7
CHECK VALVES

We have been following your gas system from the cylinder to the regulator to the hoses and connections. If you are a safe welder, the next item in your equipment list will be check valves installed on both the fuel and oxygen lines between the hose and the torch.

Check valves (Fig. 4-27) are very small, simple one-way valves with a tiny spring inside them. They only allow gas to flow one

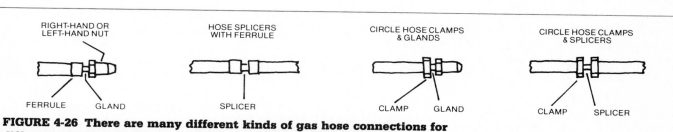

FIGURE 4-26 **There are many different kinds of gas hose connections for different gases and sizes of torches and gas regulators. Some connections are used to splice lengths of hose together and to repair broken sections. Do not use too many splices. If a hose has many repairs, throw it away and get a new hose. Each connection you add can lower the delivery-gas pressure.**

FIGURE 4-27 Check valves are one-way valves between your torch and hose, or hose and gas regulator that help prevent flashbacks. You need at least two (one threaded for oxygen and one threaded for fuel-gas connections) if you use them.

direction, from the cylinders into the torch. If gas tries to back up into either the fuel or gas line, the check valve on that line closes and seals it off. Check valves, while optional, are used by all safe welders. If you ever do get a serious flashback, it will stop at your torch check valves before it can travel up the hose and into the regulators, or even into the fuel-gas cylinder.

You have to specify both the type of gas (fuel or oxygen) and the connection size and threads (⅜-in. right-hand—threaded for oxygen, A-size connection, for example) and whether the check valve is for the torch end or the hose end that connects to the regulator. The reason for specifying which end of the hose the check valve is designed to attach to is simply that some welders are extra careful and attach check valves between the regulator and the hose as well as between the hose and the torch connections.

Make sure that the check valves have the same inside diameter as the hose. While they are very valuable to you (and very inexpensive to buy), you don't want them to restrict the gas flow to your torch.

4-8
WELDING AND HEATING TORCHES

The torch provides your final control point for handling and using gases. We will only describe gas welding, flame-cutting, and heat-

ing torches in this chapter. Other gas-using "torches" and "guns" are used in TIG and MIG arc welding and are discussed when we get to these important processes.

There are two basic types of oxyfuel gas welding and flame-cutting torches. Both types of torches have different ways of mixing the fuel and oxygen before it comes out of the tip to be burned.

Welding Torches

The basic welding torch (Fig. 4-28) used to be called a *blow pipe* and some old-timers still call it that. Welding torches come in a number of different sizes for a variety of tasks ranging from making tiny welds on jewelry to big torches used primarily for heating large castings such as machine bases and frames.

Gas-flow capacity Like regulators, all torches are rated by their gas-flow capacity. The gas-flow ca-

FIGURE 4-28 The basic oxyfuel welding torch. They come in many sizes from miniature to huge torches used for heating large sections instead of welding. The pointer shows you where to connect check valves between the hose and the torch.

pacity directly correlates with the thickness of work they will weld, heat, or cut. The classes are miniature torches, light-duty torches, medium- and heavy-duty torches, and very heavy duty heating torches. Cutting torches often can be used as heating torches simply by changing the tip, although a large heating torch is better to use for a big job. The following flow rates are only approximate guidelines for telling torches apart.

Miniature torches are used by model makers, sculptors, jewelers, and precision instrument makers. They have gas-flow rates less than 20 scf/h [0.6 m/h] and are used to join material from 0.001- to ⅛-in. [0.25-mm to 3.2-mm] thick.

Light-duty torches with more than 20 to 40 scf/h [0.6 to 1.1 m³/h] flow capacity are used for gas welding and brazing metal from 24 gauge to ⅜-in. [9.6-mm] thick. For example, plumbing, heating, and ventilating contractors use them to make air conditioning and heating ductwork.

Medium-duty industrial production torches are the most common tools used by oxyfuel gas welders. They are used to join metal from 1/32-in. up to ⅝-in. [0.8- to 16-mm] thick. They have gas flow rates of 80 to 100 scf/h [2.26 to 2.83 m³/h] or more.

Heavy-duty torches also are commonly used by welders. They have gas-flow capacities up to 180 or even 200 scf/h [5.1 to 5.7 m³/h]. They are used to weld metal from 1/32- to 1- or 2-in. thick. The very heavy duty heating and welding torches may have flow rates up to 400 scf/h [11.3 m³/h]. That's a lot more gas per hour than you can get out of a single acetylene or MAPP Gas cylinder. Obviously very large torches are used with manifold-connected cylinders.

Miniature and light-duty torches use A-size hose connectors. A medium-duty or heavy-duty torch will use B-size or sometimes C-size hose fittings, depending on the flow rate. The very big heating torches use D- or

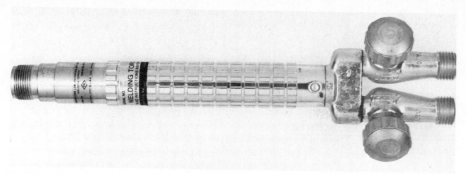

FIGURE 4-29 The basic oxyfuel torch handle. Different sizes of welding tips can be connected to it. It also can be connected to a small cutting torch with a screw-on cutting attachment.

Almost all modern welding and heating torches have separate tips that screw into the front end of the torch handle, just ahead of the mixer. The welder selects the tip size and style (or shape) that he or she wants for a given job and screws it into the front end of the torch (Fig. 4-30).

The two basic types of welding and heating torches are based on the type of gas mixer used. The difference is important. You can think of a torch gas mixer as the carburetor on a car, except that the mixer combines oxygen with fuel gas instead of air with vaporized gasoline.

The two types of welding and heating torches use either positive-pressure torch (or tip) mixers or injection torch mixers (Fig. 4-31). Therefore, all welding and heating torches are described as either positive-pressure or injector torches.

Positive-pressure torches are

E-size hose fittings and connections.

When you buy a torch or get one from stock, the package and instruction literature will tell you what thickness of material the torch is designed to weld (or flame-cut in the case of a cutting torch). They don't always give the flow rate, but you usually can find that data in the instruction manual or the manufacturer's catalog. The correct hose and fitting sizes also will be given, so you don't have to guess.

How welding torches work

Oxygen and fuel gas enter the torch handle (Fig. 4-29) through two separate gas lines from their respective hoses. The torch inlets have needle valves with round knurled knobs designed to be operated with gloves. The valve knobs usually are at the torch inlet, but sometimes they are up near the tip end of the torch handle. These simple valves control the oxygen or fuel-gas flow rate. The gases also can be shut off completely by these two separate little valves. The welder uses them primarily to adjust the oxygen and fuel-gas ratios and flow rates to produce the kind of flame—oxidizing, neutral, or reducing—that is best for the job.

The separate gas tubes run through the torch handle to a mixer either in the torch handle or sometimes attached to the separate screw-on tip. The latter type of torch is called a *tip-mix torch*.

The mixer has two jobs. Most important, it thoroughly mixes the oxygen and fuel gas so that the mixed gases produce the right flame characteristics and burn uniformly. Secondarily, the mixer also helps prevent flashbacks or backfires from traveling back into the torch handle. (Check valves still should be used between the hose and torch inlet connections.)

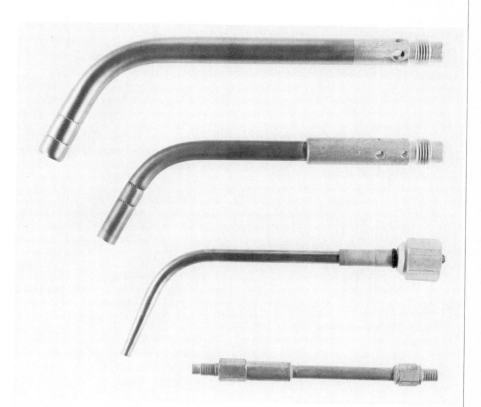

FIGURE 4-30 Here are a few of the dozens of different types of welding, brazing, and soldering tips that you can use. The two tips on the top with holes in the base are air-fuel torch tips. They go with single-hose air-fuel torches. Second from the bottom is a typical oxyfuel torch tip. On the bottom is a flexible copper tip extension.

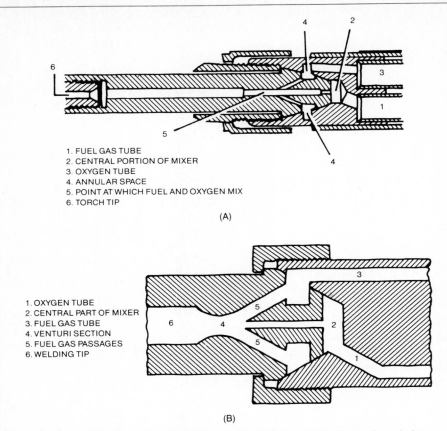

1. FUEL GAS TUBE
2. CENTRAL PORTION OF MIXER
3. OXYGEN TUBE
4. ANNULAR SPACE
5. POINT AT WHICH FUEL AND OXYGEN MIX
6. TORCH TIP

(A)

1. OXYGEN TUBE
2. CENTRAL PART OF MIXER
3. FUEL GAS TUBE
4. VENTURI SECTION
5. FUEL GAS PASSAGES
6. WELDING TIP

(B)

FIGURE 4-31 (A) The drawing shows the mixing chamber of an injector-type welding torch. It sucks (injects) low-pressure fuel gas into the oxygen stream. (B) The drawing is the mixing chamber for a positive-pressure tip-mix welding torch (again similar in principle to positive-pressure tip-mix cutting torches). The fuel pressure is high enough to be mixed directly with the oxygen in the base of the tip.

used mostly with medium-pressure fuels such as acetylene, MAPP Gas, propylene, and propane. The injector torches (Fig. 4-31A) are used mostly for low-pressure gases such as natural gas, or occasionally with small acetylene cylinders that have low withdrawal rates.

Positive pressure mixers Figure 4-31B shows a cross section of a typical positive pressure tip-mix torch mixer. We won't show the rest of the torch because it looks the same as an injector torch and there's really very little else to see in the diagram. The tip is simply a hollow copper tube with threads on one end. The back end only has two fuel and oxygen in-

let tubes and the needle valves that control fuel and oxygen pressure and flow rates.

Fuel gas from the fuel-gas tube (1) in the torch handle enters the mixer at (2) while oxygen enters from the oxygen tube (3) and spirals around a ring-shaped space (4). The fuel is under a positive pressure greater than 3 psig [6.9 kPa] and usually a lot more than that. The oxygen draws the fuel gas into the single tube (5) where it mixes with the fuel. The pressure for both gases is set on their respective regulator delivery-pressure gauges, less pressure losses in the hose.

The oxygen and fuel-gas pressure are often set equal to each other when acetylene is used, up to its safe maximum operating

pressure of 15 psig [103 kPa]. Oxygen pressures may go as high as 25 psig [170 kPa] on a typical positive-pressure welding torch. Other fuel gases use different fuel-to-oxygen ratios. The larger the size of the tip opening on the torch, the higher the regulator gas pressure should be set.

Tip-mix torches have several advantages over injection-mix torches. As long as you use the correct tip for the fuel gas, you can use any standard welding fuel gas in a tip-mix torch. You don't have to own different torches for different fuel gases. The most serious thing that can happen with a tip-mix torch is using a propane tip with acetylene fuel gas. About all that will happen is that you will burn out the tip. Do that with an injector torch and you can burn out the torch handle and possibly burn your hand quite badly.

Since tip-mix torches burn all fuel gases, only a small number of torches have to be in inventory or in your tool box. You may easily get away with one light-duty torch, one heavy-duty torch, and an air-fuel torch (to be described shortly) for almost all the work you will ever do.

Another advantage of tip-mix torches is that they are far less likely to go into flashback than an injector torch. Every tip-mix torch is designed to withstand repetitive backfiring without the occurrence of a flashback. The consequences of a flashback are far less serious with a tip-mixer than with an injector torch.

If a tip-mixer cutting tip, for example, goes into a flashback due to a manufacturing defect or because the mixer is partially clogged, damage is normally limited to the tip because that's as far back as the explosive mixture of oxygen and fuel gas extends into the torch. On the other hand, when an injector torch goes into flashback, the torch is usually damaged, and you can get hurt. Injector torches are best used for very low pressure fuel gases such as natural gas.

Injection mixers In injector torches oxygen flows from the hose and torch inlet and handle through the tube and enters the mixer through a central chamber. Fuel enters from the handle through the tube and flows into a circular chamber around the edge of the mixer. Then the fuel gas flows through small ports or passages into a special mixing area known as the *venturi,* where the gases mix and then enter the welding-tip tube.

The venturi works much like an airplane wing. It narrows down at the throat, which increases the flow rate. When the oxygen passes through the venturi at a higher pressure than the fuel gas, the oxygen actually sucks the low-pressure fuel gas along with it. The fuel gas is injected into the oxygen stream and that's where this type of torch gets its name. That also is why injector-type torches are used with fuel gases that have low line or delivery pressures.

Oxygen and fuel-gas pressures are never equal in an injector torch. Oxygen delivery pressures often are set around 5 psig [34 kPa] for small tips and up to 15 psig [103 kPa] for larger tips. The fuel gas often will have a line pressure of 1 psig [6.9 kPa] or less.

Different injector torches are required for different fuels because the injector system in the torch handle or near the head must be designed for the particular fuel gas you are using. If you use an injector torch, make sure that you have the right mixer for the fuel gas you will use. If the mixer in the torch has been changed, you may not know what you have unless you disassemble the torch and take the mixer out and look for a part number. If you are not familiar with the equipment, ask someone who has experience with

the torch model you will use.

Also be sure that your torch has a high enough flow rate for the job you will do. Injector torches often have lower flow rates than positive-pressure tip-mix torches.

The mixers in injector torches have to be cleaned from time to time, especially if you have had a series of backfires. Carbon deposits will clog the mixer. Some mixers can't be removed from the torch without difficulty. However, if you use a positive pressure tip-mix torch, you get a new mixer every time you change tips.

Heating Torches
Heating torches are almost always very big versions of positive-pressure torches. Low-pressure fuel gases simply don't deliver enough fuel flow to be used in heavy heating. Heating torches even look like big welding torches. The real difference (besides high gas flow rates) is the type of tips used. Tip selection is a big part of the skill involved in heating and bending heavy sections, which you will learn in Chap. 6. Figure 6-1 shows what typical heating tips look like.

Air-Fuel Torches
Special air-fuel torches (Fig. 4-32) are even simpler than oxyfuel torches. They do not use supplied oxygen. Instead, they only have one torch inlet for the fuel gas and one needle valve to control the fuel-gas flow and turn the torch on and off. The oxygen is provided by the air. It gets mixed with the fuel gas because the screw-on air-fuel tips have holes at their base or mixer section that suck the air in. Fuel gas flowing from the torch handle sucks the air into the tip-mixer section the same way a venturi section sucks oxygen into the mixer in an injector torch. The mixer section of an air-fuel torch then swirls the air-fuel mixture around before it flows out of the torch tip.

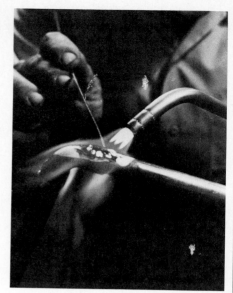

FIGURE 4-32 Air-fuel torches have one hose for fuel only. The oxygen is supplied by air sucked into the base of the tip. They are excellent for brazing because they have a wide multiflame pattern.

Air-fuel torches do not make the hot flames that you get from oxyfuel torches because the air is only 21 percent oxygen. The rest of the air is 89 percent nitrogen and about 1 percent argon. Neither of these other gases adds any heat to the combustion reaction. In fact, they tend to cool the flame.

Nevertheless, air-fuel torches are ideal for low-temperature brazing, sweat soldering, and other jobs done by plumbers, air conditioning and ventilating contractors, and some jewelry and model makers (Fig. 4-33). Air-fuel torches are so simple that sometimes you don't even need the torch.

For example, an air-fuel tip with its mixer can be screwed right onto the gaugeless regulator of a small 1-lb [0.45-kg] MAPP Gas or propane cylinder. All you need to do is select the right air-fuel tip style and you can get a nice big bushy flame that will wrap around a copper water-pipe joint, or a flat-ended tip that will give a broad flat flame for burn-

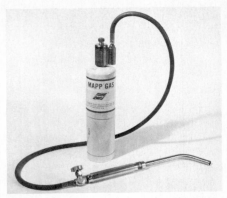

FIGURE 4-33 An air-fuel torch attached to a 1-lb MAPP Gas cylinder. You can't get more portable than that.

FIGURE 4-34 Wall Colmonoy, Detroit, makes metallizing torches and guns, and the special wires and powders you need to spray molten metal to hardface, surface, or built-up worn parts. This is a Wirespray metallizing gun. Note the gas flowmeters behind the operator. The company also makes more conventional-looking oxyfuel torches with a funnel attached to feed powdered metals into the oxyfuel flame. (See Chap. 19.)

ing paint off a wall, or a small narrow tip for pinpoint soldering.

There are many different kinds of air-fuel tips and torches, but all of them are low-flow torches used for soldering, low-temperature brazing, and sometimes for melting and welding lead. Telephone linemen use these portable cylinders with an air-fuel tip to repair lead cable covers. The entire outfit only weighs a few pounds.

Specialized Torches

If you think we've covered all the kinds of oxyfuel welding torches there are, we haven't. But we have covered all the commonly used ones. What's left?

There are special oxyfuel torches that suck powdered metals into the flame, or melt a wire, to make a flame that sprays molten-metal droplets to coat or *metallize* special alloys onto metal parts (Fig. 4-34).

These powder-spray or metallizing torches are used by maintenance departments to build up worn sections of parts such as gears and shafts with steel or even harder, wear-resistant alloys. After the part is built up, the excess weld metal is machined off and the part is as good as new. Metallizing torches (also sometimes called *guns*) also will spray corrosion-resistant alloys, such as aluminum on steel tanks, for ex-

FIGURE 4-35 The business end of an oxyfuel cutting torch. Note the four holes around the end of the tip. These are for the preheat flames. The hole in the middle is for the cutting oxygen. Different cutting tips for different types of work may have more than four preheat flame ports.

ample. They literally spray the molten aluminum on like a paint gun sprays paint.

There also are special large heating heads that are really huge tips with a giant mixer and dozens of flame ports, but no torch handle. They are used in industry. For example, a heating head can be designed to fit over the top of a railroad rail. The rail passes under the heating head and is flame-hardened by a water quench also produced by the heating head just after the rail passes through the flame.

These are only examples of other specialized oxyfuel equipment that you may see or even be asked to operate. With what you now know about fuel gases and oxygen, their operation will be easy.

4-9
CUTTING TORCHES

Oxyfuel cutting torches are used much more frequently than welding and heating torches. Since they also are more complicated than welding torches, we covered the basic welding torch first. Most of what you have already learned applies to cutting torches, too.

An oxyfuel cutting torch is a precision instrument. It looks a lot different than a welding torch. For one thing, it has a big trigger on the top or bottom of the handle. It has either two or three long tubes running from the torch handle to the tip head. A cutting tip also looks quite different from a welding or heating tip (Figs. 4-35 and 4-36).

The typical cutting tip has preheat holes around the outside and a single big cutting-oxygen bore hole in the middle. The preheat flames are operated similarly to the single flame from a welding torch. They are used to bring steel up to a cherry-red temperature.

When the steel is hot enough, the operator presses the oxygen trigger on the torch and a big blast of pure oxygen starts a cut that will burn a narrow slice right through the hot steel. The

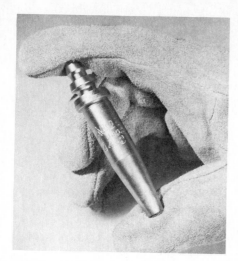

FIGURE 4-36 This is a typical one-piece cutting tip that you might use for acetylene. Propane or MAPP Gas often use two-piece cutting tips. Burners carry many different tip sizes and styles in their tool boxes to do different kinds of work.

next chapter will tell you a lot more about flame cutting and the special tips that are used for this process. Nevertheless, cutting torches are selected the same way as welding and heating torches, that is, by gas-flow capacity.

For example, a medium-sized hand-held cutting torch with a total gas-flow capacity of 140 to 175 scf/h [3.9 to 4.9 m³/h] will cut steel up to 1 in. [25 mm] thick. A bigger torch with a flow rate of about 760 scf/h [21.5 m³/h] will cut steel up to 12 in. [0.3 m] thick.

Oxyfuel cutting torches are designed either as injector torches for low-pressure fuel gas like natural gas, or as positive-pressure torches for acetylene, MAPP Gas, propylene, and other cutting fuels. The mixer is located inside the torch handle in these models. A third kind of cutting torch also is often used. It is a positive-pressure torch except that the tip mixer is located in the torch head and uses the cutting tip to do part of the fuel and oxygen mixing job. These are called *tip-mix* torches and they are very popular because a backfire or flashback will stop

at the torch tip and not travel back into the torch handle.

Figure 4-37A shows cross sections of a positive pressure torch head with the tip screwed into it. The torch has 1 fuel-gas tube (7) and 2 oxygen tubes (5 and 6). Tube (5) is for the cutting oxygen, and tube (6) is for the preheat oxygen. The preheat oxygen spins around inside a ring machined into the cutting tip and torch (4) while the fuel gas enters a similar ring (2). The fuel gas and oxygen are mixed in the tip and then exit the tip through the small holes fed by rings (2 and 4). The cutting oxygen is entirely separate. It enters the middle of the cutting tip (1) and is squeezed into a jet that exits the tip when the torch's cutting-oxygen lever is pressed.

Compare Fig. 4-37A with Fig. 4-37B, a cross section of the torch head and tip for an injector cutting torch. Since the oxygen and fuel gas have already been combined in the mixer in the handle,

they enter the cutting tip preheat holes (2) directly through a single groove machined in the tip (1). The cutting oxygen enters the middle hole in the tip (3) and exits the tip in a jet when the cutting-oxygen lever is pressed.

Figure 4-38 shows how some welding torch handles can be converted into cutting torches by adding a cutting attachment that screws into the hole where the welding tip otherwise would be placed. When you have a welding torch that converts into a cutting torch (not all of them do), you will have two needle valves on the handle. One needle valve controls the preheating-oxygen; the other controls the fuel gas. A third needle valve attached to a lever arm controls the high-pressure cutting oxygen. This is the lever that you press when you are ready to cut metal.

Some cutting torches (usually heavy-duty models) are dedicated only to flame cutting. They cannot be converted into welding torches

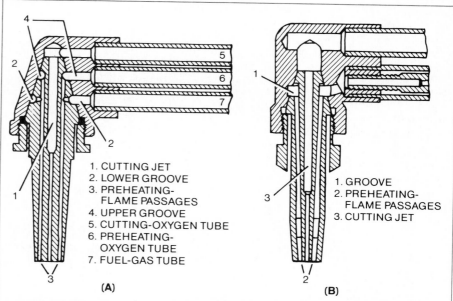

1. CUTTING JET
2. LOWER GROOVE
3. PREHEATING-FLAME PASSAGES
4. UPPER GROOVE
5. CUTTING-OXYGEN TUBE
6. PREHEATING-OXYGEN TUBE
7. FUEL-GAS TUBE

(A)

1. GROOVE
2. PREHEATING-FLAME PASSAGES
3. CUTTING JET

(B)

FIGURE 4-37 (A) Schematic cross section of a typical tip-mix cutting torch. Cutting torches have one fuel-gas tube but two oxygen tubes. One oxygen tube supplies oxygen to burn the fuel gas. The other one provides high-pressure oxygen for cutting. The cutting oxygen always exits from a center hole in the cutting tip while the preheat fuel gas flames are in a ring around the central cutting-oxygen orifice. (B) Since injector cutting torches use separate gas mixers in the handle, they need different types of cutting tips than tip-mix torches that mix the fuel and oxygen in the torch head.

and back again into cutting torches by using different attachments and tips.

In the next chapter we will look at cutting torches and tips in more detail. We'll even look at an oxygen lance that has no flame at all, yet it can cut a hole or slice through a piece of steel 18 ft [5.5 m] or thicker as long as you preheat a small area before you start the cut. The burning steel, itself, will provide the rest of the heat to keep the reaction going.

We'll also give you some details on machine flame-cutting torches. But with what you already know, the rest of the details will come easily.

4-10
PACKAGED WELDING OUTFITS

Because of their convenience, whoever invented instant cake mixes must have invented prepacked welding outfits next. Welding equipment suppliers now package fully compatible, matched welding and cutting torches, mixers, hose, and regulators; spark lighters; a cylinder wrench; a selection of welding, heating, and cutting tips; and sometimes even portable fuel and oxygen cylinders into Styrofoam-filled boxes packaged to look good on any supermarket shelf (Fig. 4-39). Some portable outfits with the small oxygen and fuel cylinders even have their own fiberglass carrying cases. You can buy a good small welding outfit for about $150. A big maintenance outfit might only cost $250 to $300, but what you get is a complete welding shop in a box.

There are welding and cutting outfits for large maintenance shops, farm repair, contractors, sculptors, plumbers (even air-fuel outfits now come prepackaged), production welders, and auto body shops. A trip to a good local welding distributor will turn up more special kinds of welding outfits than you will ever need.

Many of these specialized welding and cutting outfits have the detachable cutting torch sections that screw onto the handle of the basic welding torch. That way you get two torches in one and can convert quickly from welding to cutting.

4-11
NECESSARY ACCESSORIES

Certain accessories are vital to a welder or burner and your welding and cutting outfit or tool kit is not complete without these items.

Spark Lighters
Never light your torch with matches or a cigarette lighter. That's a sure way to get a severely burned hand. Welders and cutting-torch operators either use spark lighters (Fig. 4-40) or various types of pilot lights. All welders and burners have their spark lighters close at hand. They often hang them from their belts.

A spark lighter has a spring handle and pan to catch the mixed oxyfuel gas coming out of the torch tip. (They also are used to light air-fuel torches.) A flint on one end of the spring handle, over the gas pan, strikes a metal scratch bar that throws sparks into the gas in the pan. You get an instant light almost every time. Spark lighters even work well on windy days. The flints are removable and many welders and burners carry extra flints.

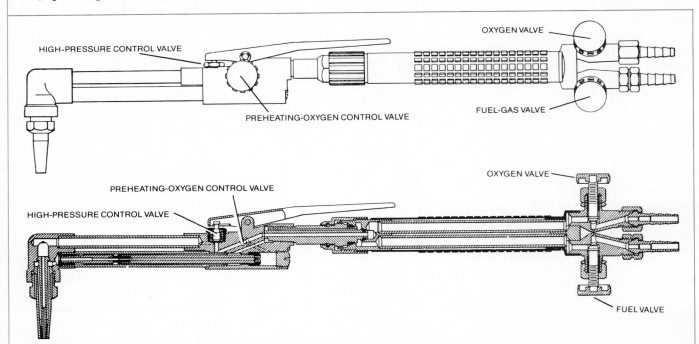

FIGURE 4-38 Although some large cutting torches do only one job, many welding torches can be converted into cutting torches by removing the welding tip and screwing on a separate cutting attachment to the welding-torch handle.

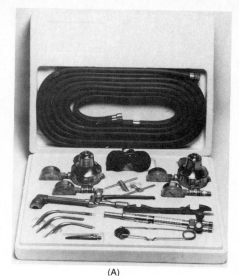

(A)

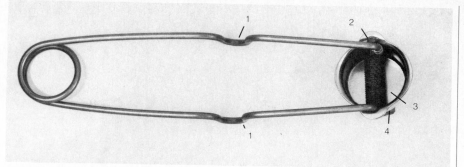

FIGURE 4-40 Every oxyfuel welder carries a spark lighter. Never light a torch with matches. Your hand will be badly burned.

Tip Cleaners

Every good welder and flame-cutting operator carries a set of tip cleaners (Fig. 4-41) in his or her pocket or tool box. Welding and cutting tips in use become covered with oxide and slag on the outside and in the gas bore holes. The dirt in the gas ports distorts the flame and reduces the gas-flow rate. That can lead not only to bad cuts and poor welds but also to overheated tips and a backfire, or even worse, a flashback.

Clean your tips frequently. A set of tip cleaners (high-carbon stainless steel wires of different diameters with tiny filelike teeth on their central sections) is the best way to do it. Be careful to use the correct diameter wire for the holes you are cleaning. Also be careful to keep the cleaning wire aligned with the tip bore holes. Do not ream out the exit holes of a tip to make them bigger. Never use a hard wire or rod or piece of scrap to clean a tip. You will get it stuck in the torch tip. You also might ruin an expensive tip.

Other Tools

Many welders carry a handy all-purpose cylinder wrench that looks something like a flat box wrench with extra squares cut out for different sizes of cylinder and hose nuts, torch, regulator, and cylinder glands and connections. The handle of the torch also may have a square or triangular hole cut out for use as a key to turn on

(B)

FIGURE 4-39 (A) Many companies now buy complete, prepackaged welding outfits like the one shown. It has all the basic components you would need to get started, except for the cylinder oxygen and fuel gas and a pair of good gloves. (B) Not all prepackaged welding outfits are big. This one is extremely portable and even includes both a MAPP Gas cylinder and an oxygen cylinder that together are small enough to carry on your back.

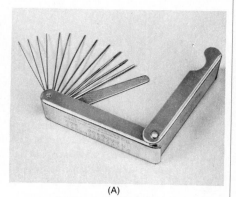

(A)

(B)

FIGURE 4-41 (A) Never clean your tip with a piece of wire or a nail. Always carry these special tip cleaners with you. Clean your tips frequently and reduce your chances of getting a flashback. (B) Do not let your tips get as dirty as the one in the photo.

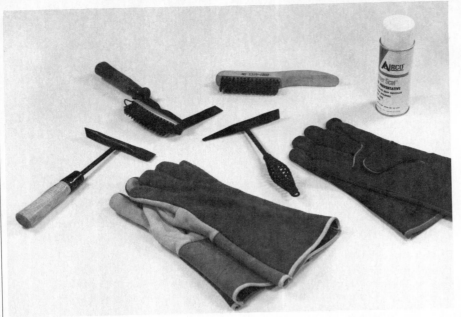

FIGURE 4-42 There are many different kinds of tomahawk chipping hammers in use. They are used to remove solidified slag from a weld, and to clean weld surfaces between passes. Grinding wheels and air-operated or electrically operated wire brushes also are used. This photo shows a can of antispatter compound that you can spray on your work before welding to make droplets of weld spatter brush off easily. Most welders also carry two sets of gloves in case one pair gets soaked with perspiration. That's especially important when doing arc welding.

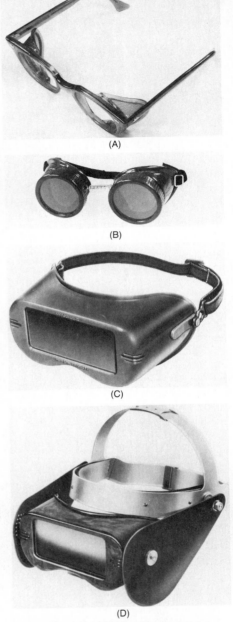

FIGURE 4-43 (A) An example of prescription safety glasses. (B) Conventional oxyfuel burner's or welder's goggles. (C) These glasses are also used for oxyfuel work, but they allow a wider field of view and are more comfortable. (D) A burner's goggles with side shields and an attachment to fit over a hard hat.

acetylene cylinders. If you use this wrench on acetylene cylinders, don't carry it with you. Leave it on the cylinder. Otherwise use a separate cylinder wrench for the acetylene and leave it on the cylinder.

You also should have a stiff wire brush, clean cloths for wiping dirt out of cylinder, regulator, and hose connections, and a good tomahawk-type chipping hammer (Fig. 4-42) to knock off slag and spatter from your work. A few pieces of emery paper also are good for cleaning soot from the outside and bottom surface of cutting tips. From there on in, your job and personal needs will dictate what you carry with you. What you wear is something else.

4-12 PROTECTIVE OUTERWEAR

Your gloves, goggles, welding cap, protective leatherwear to keep sparks from hurting you, and your shoes should be selected with great care. They are your most important safety equipment. And they had better fit and be comfortable since you have to live in them every day.

Some of your safety wear will be supplied by the shop where you work. Some things you'll probably prefer to buy yourself. While most tool rooms will supply goggles, gloves, caps, and leatherwear, many welders prefer to buy their own. That's especially true of welding gloves.

Goggles

Safety goggles for gas welding (Fig. 4-43) come in dozens of sizes and shapes. There are very cheap goggles with a simple elastic band that goes around your head; goggles with a single wide faceplate, and even goggles that attach to hard hats.

But oxyfuel welding goggles should never be used for anything except gas welding and cutting. Never use them for any kind of arc welding. They are not usually dark enough to stop the danger-ous ultraviolet radiation from an electric arc that can damage your eyes. Arc-welding helmets also cover your whole face because the ultraviolet radiation from the arc not only will hurt your eyes, it also will give you a very bad sunburn.

Gas-welding goggles all have dark glass lenses. Each lens has a shade number, indicating how dark the glass is. Most welding goggles use a #5 or a #6 lens shade. The bigger the lens-shade number, the darker the light filter or lens is. For example, #5 lenses are used for light-duty oxygen cutting and gas welding. A #6 lens is for medium-duty oxygen cutting and welding. A #8 lens shade is for heavy-duty gas welding and cutting. The lens-shade numbers continue up to a #14 lens, but these darker shades are used for arc welding.

Goggles do a lot more than just shield your eyes from the bright light of hot slag and molten weld metal. They also protect your eyes from flying sparks and slag. That's why all welding goggles are enclosed on the sides. Because the eye cups are closed, ventilation ports are also included on the side of each cup to keep the glasses from fogging up.

The dark lens in the goggles is covered by a clear shatterproof cover plate. You don't want to scratch the lens and it's a lot cheaper to replace the cover plate when you need to because it gets spots of metal on it. All you have to do is unscrew the cap on the eye cup and drop out the round cover plates (or the lens if you are changing shades). Put new lenses or cover plates back in, and screw the cap back on.

Goggles should have an adjustable headband and an adjustable link between the eye caps over the bridge of your nose. Some people have wider faces than others, so these glasses really have to be adjustable two ways. They also should be comfortable. If you ever get a rash around the eye caps where the goggles fit onto your head, try putting a little petroleum jelly on your face. That usually works well. The rash is caused because you are sweating under the goggles and that might irritate your face.

Gloves

The single most important item you wear is a good pair of gloves. Conventional work gloves are not welding gloves. If you go to a welding distributor, buy at least two pairs of the best welding gloves you can afford. When one set gets wet if you perspire, switch to the second set of gloves.

Welders' gloves (Fig. 4-44) cover your lower arm as well as your hands. These gauntlet gloves also have seams that are turned inside to keep from trapping slag, sparks, and even molten metal. They are always made of leather except for some asbestos gloves we'll mention in a minute.

Get a pair of gloves that fit comfortably, but loosely enough so that you can throw them off if they get too hot, or if a spark gets inside. Some gloves have wool linings to reduce heat flowing through the leather. Other special gloves may even have aluminized palms and backs for additional resistance to high heat. Fiberglass gloves usually are worn over a regular pair of welders' gloves for very hot work. Some lightweight gloves are worn by TIG welders who don't have too much problem with the heat and want to feel what they are doing.

Most welders wear chrome-tanned split cowhide gloves. Welding gloves should be easily removable in case they get too hot.

Leatherwear

Chrome-tanned fire-resistant leather aprons, hoods, capes, sleeves, jackets that button up to the neck, leggings that also cover your boot tops, and other kinds of leather goods are used by welders and burners who get close to their work (Fig. 4-45). One example is a pipeline welder who spends a lot of time right under the weld he or she is making. There's nothing like living in a shower of sparks

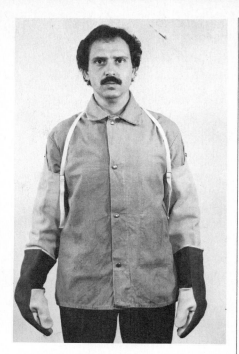

FIGURE 4-44 Welder's gloves come in many styles, including leather, on up to special gloves with heat-reflecting surfaces for very hot work. This man also has leather sleeves to protect his upper arms.

FIGURE 4-45 This dummy in arc-welder's clothes is dressed complete with a leather jacket, apron, leggings, and even leather covers over the boots.

FIGURE 4-46 These special leather-covered steel foot guards and leather anklets will protect you from dropping heavy pieces on your own foot, or from splashing slag up your pants legs.

and hot slag while working in the snow at −60°F to explain why these people earn the kind of money they do.

You won't often need very much protective clothing. But if you do need it, get it. Companies who hire welders often buy it for them because leatherwear is expensive.

Welding Caps

You know a welder because he or she will probably have a skullcap with anything from pink flowers to a pattern that looks like wallpaper selected in poor taste. Some caps have short bills. The operator probably will be wearing it with the bill down over one ear to make room for a helmet.

Welding caps are important as well as inexpensive. They cover your hair and keep sparks off. Always wear them. Pick a pretty one (or an ugly one) to your taste. Your welding distributor will have a bin full of them to select from.

An ex-mayor of Cleveland opened a big metalworking convention and refused to use a welding cap while on TV, when he cut a chain to open the show. He had used hair spray for the TV cameras. One tiny spark landed on him and his hair spray made a nice little cone of blue flame right on top of his head. The cameras caught it all and he was on na-tionwide TV the next day—looking startled while someone hit him over the head to put the fire out while sparks melted holes in his double-knit suit. Don't make the mayor's mistake; always wear your cap.

Work Clothes

Wear a long-sleeved shirt that you can button up at the collar and the wrists. It keeps sparks from going down your back or up your sleeves. Button your pockets. Don't believe the wild stories about welders getting hurt by exploding butane lighters. They are totally untrue. But if you smoke, don't carry loose kitchen or paper matches in your pockets. Use that lighter. Don't light your cigarettes from the torch. Among other things, it's hard on the tip of your nose.

Cotton work shirts are OK for gas welding and cutting if they are treated for flame retardance. Remember that the treatment may wash out with the next laundry. Never wear any synthetics such as nylon or rayon. They melt.

A wool shirt is far better for arc welding. We don't know why, but ultraviolet radiation does strange things to cotton. The next time you wash a cotton shirt after arc welding it may simply fall apart.

Blue jeans are in for welders, like everybody else. Never wear pants with cuffs. Cuffs are the first place the sparks go to.

Use high-topped work boots so that sparks won't fall in your shoes. Don't tuck your pants in. Let them hang straight down. Use steel-capped toes on your boots if you work with heavy parts. In addition to safety boots, you can also order steel arch protectors that fit over the entire top of your foot and strap on (Fig. 4-46). Use them if there's any chance something very heavy will drop on your feet.

In rare instances where you have to work on very hot plate you can even get wooden shoes.

4-13 WELDING DISTRIBUTORS

Now that we've described many of the tools you use for welding with gases, let's tell you where you or the company you work for can get them.

All major welding equipment and gas producers sell through welding distributors. Some large-volume sales are made direct from the manufacturer to the user, but the great bulk of welding equipment, gases, supplies, and accessories are sold through these specialized retail outlets (Fig. 4-47).

A good welding distributor can sell you everything else you need,

FIGURE 4-47 Welding distributors supply almost everything you will ever need for welding. This distributor has set up his store like a welding-products supermarket.

from your chipping hammer and welding cap to the filler metals you will use for welding, the shielding gases, fuel and oxygen, your torches and tips, along with all the arc-welding equipment we will describe in later chapters of this book.

Welding distributors can give you good advice on using your equipment, too. They often even know from day to day what companies in your area are hiring welders, and what kinds of welders they need. Most of the time the person behind the distributor's counter is an ex-welder who will be happy to help a new welder.

Almost all welding distributors carry specific brands of equipment and supplies. If you prefer one manufacturer's brand over another, find the welding distributor near you that handles that product line. Simply look in the Business Yellow Pages under "Welding Equipment and Supplies," and you will see the name of every welding distributor in your area, along with display ads featuring the major product brands that they carry. If you can't find a local welding distributor that handles a product you are interested in, call the manufacturer for the nearest sales office or retail sales outlet.

You can buy a detailed directory of all welding equipment, supplies, and gas manufacturers serving the entire United States from *Welding Design & Fabrication* magazine. Every year their January issue (usually running around 200 pages) is devoted entirely to a "Welding and Fabricating Buyers' Guide." It contains the names, addresses, and telephone numbers of every manufacturer and contains detailed lists on the availability of every welding product you will ever need. The single-issue copy price is only a couple of dollars. Write to *Welding Design & Fabrication*, 1111 Chester Avenue, Cleveland, Ohio 44114. Their number is (216) 696-7000.

REVIEW QUESTIONS

1. Why should you always put a cylinder cap back on the cylinder after you finish using it?

2. What can happen when you open an oxygen cylinder valve very fast when a regulator is attached to the cylinder?

3. What is the CGA?

4. Why should all cylinders be chained upright when you use them?

5. Is it OK to use acetylene, MAPP Gas, or propylene cylinders when they are lying on their sides?

6. Can you use high-pressure gas cylinders such as oxygen, argon, or helium while they are lying on their sides?

7. Can you judge the contents of an acetylene cylinder by reading the cylinder-gauge pressure?

8. What is the highest working pressure you can safely set on an acetylene regulator?

9. There are 15 acetylene cylinders and 23 oxygen cylinders in the back right-hand corner of the storage shed. What's wrong with that?

10. What can happen if you add oil or grease to an oxygen regulator?

11. Is it OK to use any handy regulator on any gas cylinder as long as the gas and the regulator are both designed for use at the same maximum pressure?

12. What's the basic difference between a single-stage and a two-stage regulator? Is it the number of gauges that are on them (one or two)?

13. Why should you always purge your gas lines before and after you finish working?

14. Why are cylinder manifolds used?

15. Are special regulators used on cylinder manifolds?

16. What color is an oxygen hose? What color is a fuel-gas hose?

17. If you are working a long distance from your cylinders, will you increase the delivery pressure at your torch by using larger diameter hoses, or by using smaller diameter hoses?

18. What does the mixer mix in a welding or cutting torch? Why does it do that?

19. What's the best kind of cutting torch to use for low-pressure natural gas, a positive-pressure torch or an injector torch?

20. What kind of torch, an injector torch, or a positive-pressure tip-mix torch, is more likely to produce a flashback?

21. Can an injector torch be used with any common fuel gas without doing anything special to the torch? What about a positive-pressure tip-mix torch?

22. What do gas-welding goggles do for you besides protect your eyes from flying sparks and slag?

23. What lens shade should you use in your goggles for heavy-duty gas welding and cutting?

24. What lens shade should you use in your goggles for light-duty oxygen cutting and gas welding?

25. What should you use to light your torch?

Shop Experience with Oxyfuel Equipment

Never do any work until your welding instructor has briefed you and has given you permission to start.

4-1-S
YOUR EQUIPMENT
You will need the following equipment for this shop work.

- Fuel and oxygen cylinders with cylinder- and delivery-pressure gauges
- Cylinder cart or post and chain
- Correct oxygen and fuel regulators for your cylinders
- Fuel and oxygen hoses with connections sized for your equipment
- Oxygen and fuel-gas check valves sized for your hose and torch connections
- A cutting torch, or welding torch with cutting attachment
- Cutting tips made for use with your torch
- Spark lighter
- Set of tip cleaners
- Clean, dry, lint-free cloth
- Cutting goggles (use light lens shade)
- Pair of shatterproof safety glasses
- Universal cylinder wrench
- Gloves
- Proper clothing
- A hard hat is optional unless required for safety

4-2-S
SETTING UP

We will conduct this shop work a step at a time. We won't go into so much detail in future shop work. In fact, the instructions in each shop section will get shorter and shorter as you learn more and more, so we can give you more work to do and let you figure out how to do it.

What follows is basic, *but you have to know it before you can be allowed to do any shop work safely*. If you don't understand how to do anything, or why you should do it, ask your instructor first. You will follow the same general procedures we give you here every time you do any shop work, in a welding school or on the job. Here's the first thing to do.

1. Make sure you have the proper clothing on. You only need to wear the goggles when you light up. Put on your safety glasses right now. Wear them whenever you don't have your goggles on. Most shops require that all workers wear them at all times. Put your gloves on right now. Learn to work with your gloves on all the time. Also make sure your work area is clear of combustible materials. If you don't work in a welding booth, you should have a welding screen or curtain, or sheet metal deflector around you to stop flying sparks (even though you will not be cutting anything until later). Learn to do things right the first time.

2. Chain your cylinders to a wall, post, or cylinder cart. Remove the cylinder caps. Put them close by the cylinders so they won't get lost.

3. Stand to the side of each cylinder. Quickly open and close the valve a little (crack each cylinder) to blow out any dirt that may be deep in the valve outlets. Clean the outlets with the cloth.

4. Inspect and then clean the regulator inlets and outlets with the cloth.

5. Make sure that your regulator-adjusting threads are both turned clockwise (in) to open the regulators.

6. Attach your oxygen and fuel-gas regulators, making the connections hand tight. Do not force them. Be sure that your fuel-gas regulator is correct for the cylinder fuel you will be using.

7. Now tighten the regulator inlet connection nuts firmly onto the cylinders with a close-fitting wrench, preferably the universal cylinder wrench. Do not force any connections that don't seem to fit well. Do not overtighten the connections. You want a leakproof connection, not a broken one.

8. Inspect your hoses for cuts and wear spots. We assume that the hose is not new. (If it is, see the next section on how to purge it of dust before you connect your torch to it.) Clean the hose connections with the cloth.

9. Attach the green hose to the oxygen regulator and red hose to the fuel-gas regulator. If your hoses are black, ask your instructor for advice on which hose to attach to each regulator, or look for a notch in the connections (this is the fuel-gas hose). The fuel-gas hose also will have left-hand threaded connections. The oxygen hose will have right-hand threaded connections. Tighten the hose connections at the regulator with the wrench.

10. Clean the check valves and torch inlets with the cloth. Attach the check valves to your torch hand tight. Note that each check valve is marked either for oxygen or fuel gas. Attach them to the right inlets on the torch.

11. Attach the hoses to the check valves only hand tight. Then use the wrench to tighten all connections near the torch inlets. Don't overtighten the connections.

12. Close the hand-operated torch fuel and oxygen valves with your fingers. (Never use a wrench on them.)

13. Clean the cutting torch threads on the head and then insert the cutting tip seat into the torch head and tighten the tip nut with a wrench.

14. Check both of your regulators, and your torch, to make sure that all the valves are closed.

15. Close both your regulators by turning the adjusting screws counterclockwise (out).

16. Open the oxygen cylinder valve very slowly, and only a little. Allow pressure to build up in the high pressure chamber of the regulator. Gas won't enter the low-pressure chamber because the regulator is shut. Do not stand in front or in back of the regulators, but to the side when you work with them.

17. After the pressure has built up in the high-pressure chamber of the oxygen regulator, fully open the oxygen cylinder valve. Read your cylinder gauge pressure. Is your oxygen cylinder nearly full or is it nearly empty? If it's almost empty, now is the time to change cylinders. Do you hear any leaks around the regulator connections? What about around the cylinder valve? If you do hear a leak (you can also test for one with soapy water), try tightening the glands or nuts a little more. If the leak continues, take the equipment out of service.

18. Open the fuel-gas cylinder valve *only partway.* You should never open a fuel-gas cylinder valve as far as it will go (even though you just did so for oxygen). About 1½ turns is usually good enough if you are using a liquefied petroleum gas such as MAPP Gas. If you are using acetylene, don't turn the wrench more than three-quarters of a turn. Leave the wrench on the acetylene cylinder. You may want to turn it off quickly in the event of a fire or flashback. Do you hear any leaks from the cylinder gland, valve, or regulator connection? If you are operating off a natural-gas pipeline, turn the valve open all the way because your line pressure will be so low that you will need all the gas pressure you can get.

19. Inspect your work area. Is there anything that will ignite fuel or oxygen when you purge your regulator, torch, and lines? Don't point your torch at anybody or anything that will burn. Don't light your flame if the torch leaks.

20. Turn the adjusting screw on the oxygen cylinder regulator clockwise (left to right) until the recommended oxygen pressure for the torch tip is recorded on the delivery-pressure gauge. Your instructor will give you an oxygen delivery-pressure setting to aim for. In the next chapter we will tell you how to find a good oxygen pressure setting for any tip you use.

21. Now turn on the preheat-oxygen valve on your torch. Allow the oxygen to flow for about 5 to 10 s for each 50 ft [15.2 m] (or less) of hose you have. Then shut off the torch oxygen valve.

22. Next, open the cutting torch fuel-gas valve and let the fuel gas flow for about 5 to 10 s for each 50 ft [15.2 m] (or less) of hose you have. Shut the torch fuel-gas valve. Now you have separately purged each of your lines from the regulators down through the

torch. You are ready to light your torch as soon as you put on your cutting goggles and gloves.

23. Put on your goggles. Put on your gloves (if you are not already wearing them). Pick up your spark lighter and torch. Don't aim the torch at anybody or anything that can be set on fire.

24. The method you will use to light your torch depends on the type of cutting torch you have.

If you have a positive-pressure torch . . . open the torch fuel-gas valve slightly and light the gas with your spark lighter, holding the tip pointed away from you.

Open the torch oxygen valve slowly and adjust the fuel gas and oxygen volume until you have a stable flame maintained on the end of your cutting tip. Then depress the cutting-oxygen lever and continue to increase the oxygen flow until you have the desired type of flame (which we will discuss in detail in Chap. 5).

You can adjust your flame by varying both the oxygen and fuel-gas torch valves. Experiment with setting different flames on your tip. Always make final flame adjustments with your cutting oxygen on. (You'll follow the same procedures adjusting the flame on a welding or heating torch, but you won't have any cutting oxygen to worry about.)

Look for the inner and outer flame cones. Watch for the feather as you change the preheat oxygen. Look at the change in color of the inner and outer flame cones as you change your oxygen and fuel-gas flow rates. Listen to the change in the sound of the flame as you make changes in gas settings.

We also suggest that you read the sections on flame settings in the next chapter (see pages 107 through 110). Then try to set a carburizing flame, a slightly carburizing flame, a neutral flame, a slightly oxidizing flame, and a highly oxidizing flame on your tip (as described in the next chapter).

Now depress your cutting-oxygen lever. Watch how the appear-

ance of the various flame settings changes with cutting oxygen on and off. Get used to handling the torch, cutting-oxygen lever, and torch valves all at the same time while wearing your gloves.

Here's an alternate method for lighting a positive pressure torch:

Open the torch fuel-gas valve *slightly* to make sure fuel gas is flowing from the tip. Then open the oxygen valve *slightly.*

Light your torch with your spark lighter.

Depress the cutting-oxygen lever if you're working with a cutting torch. Adjust the oxygen and fuel-gas flow with the appropriate torch valves in successive steps to obtain your desired flame. (Certain fuel gases are more easily adjusted by this method.)

If you have a low-pressure injector torch and low pressure fuel gas like natural gas, here's how to light the torch:

Open the torch fuel-gas valve about three-quarters of a turn. Then open the oxygen valve *slightly* and quickly light the oxygen and fuel-gas mixture with your spark lighter.

Press the cutting-oxygen lever if you are adjusting the flame on a cutting torch. Adjust the preheat oxygen valve on the torch to obtain the desired flame.

Experiment with flame settings and cutting oxygen as described in the positive-pressure torch lighting procedure, above.

4-3-S
PURGING NEW HOSE

After connecting the hose to regulators on the cylinder, hold the outlet end of each hose to prevent it from whipping around when you turn on the gas. Adjust your oxygen regulator to deliver about 5 psig [35 kPa] while holding the hose. Let the oxygen flow for about 10 s for each 50 ft [15.2 m] of hose. Do not direct the oxygen at your skin, your clothing, or any highly combustible material.

Shut off the oxygen at the regulator by turning the adjusting screw counterclockwise (out). On three-hose cutting machines, make sure that both oxygen hoses are purged before attaching them to the machine torch. You won't have to do that in this lab because you have a hand-held torch, but remember to do it whenever you work on large flame-cutting machines.

Repeat the procedure with the fuel-gas regulator and hose, again making sure that you vent the gas into a safe place. Then back out the fuel-gas regulator adjusting screw to shut off the flow of fuel gas.

Now you can connect up the rest of your equipment and light your torch.

4-4-S
SHUTTING DOWN

1. Rapidly close the torch valves in the sequence recommended by the manufacturer. If you don't know that procedure, first turn off the oxygen. Then turn off the fuel gas. The torch may make a loud pop. That's OK.

2. This is a good time to practice cleaning your torch tip. Be careful, the tip is hot. Use your gloves. Use only the tip cleaners that fit your tip preheat holes and cutting-oxygen hole in the center. Don't jam the cleaner into the hole or change the shape of the holes.

3. After extinguishing the torch, close both the oxygen and fuel-gas cylinder valves or your pipeline valves. This procedure is safe as long as your work area is well ventilated.

4. Open the torch fuel-gas valve and bleed off fuel gas from the regulator, hose, and torch. Be sure not to point the gas at anything that might ignite it.

5. Back out the regulator adjusting screw on the fuel gas regulator to shut off the regulator.

6. Then close the torch fuel-gas valve. Be sure that the torch valve is closed. You don't want any oxygen backflowing into your fuel-gas lines.

7. Open the torch oxygen valve and bleed oxygen from the oxygen regulator, hose, and torch lines.

8. Back out the regulator adjusting screw on the oxygen regulator to shut off the regulator.

9. Then close the torch oxygen valve.

10. Now check all your equipment. *Are you sure that the two cylinder or pipeline valves are fully shut?* (That was step 2.) Are you sure that both your oxygen and fuel-gas lines and regulators are depressurized? Are all your torch valves shut?

11. Remove the cutting tip. Be careful. It may still be hot. Use your gloves. Put the tip away in a tool chest or drawer where it can't be lost or damaged.

12. Take off your torch check valves and hose. Roll up the hose and put it away. Never put the hose in an unventilated cabinet or leave hose in a confined space. It still has fuel and oxygen in it and a spark could land in the space later on.

13. Take off the fuel and oxygen regulators. Put them away in a safe, clean, dry place. A cabinet or toolbox is OK, but it's better to put them in a box in the storage place so that the gauge glasses won't be broken if you throw in something else later on.

Put the cylinder caps back on your cylinders.

15. If you use a cylinder cart, wheel it out of the way. Put your universal cylinder wrench, torch, check valves, goggles, safety glasses, and gloves in a cabinet or your toolbox. Lock the box. These are your tools. Somebody else might steal them. Put the box where it's safe from theft. Now clean up your area before you leave.

4-5-S
USING EQUIPMENT ALREADY SET UP

If you start to work with equipment already connected up, check everything for signs of damage.

Make sure all cylinder valves are closed hand tight. Make sure all gases have been drained from the regulators, hose, and torch. Unpurged hoses can contain an explosive mixture of residual fuel gas and oxygen, or fuel gas and air. Somebody also may have fiddled with the regulator settings.

4-6-S
HANDLING FLASHBACKS AND BACKFIRES

A flashback can give you a severe burn. The flame backs up into the mixing chamber of the torch and could burn right through it. The tip-mix torch is less likely to give you trouble. But either torch can cause problems.

If you get a flashback with a torch-mix torch instead of a tip-mix torch, you will hear a high-pitched whistle or shrill humming sound. The torch will start to get hot. It may burn right through the handle and direct flames at you and anybody else standing near it. Here's what to do.

Immediately close the torch oxygen valves. Next, shut off the torch fuel-gas valve. Let the torch cool down.

Purge the fuel and oxygen lines from the regulator through the torch.

Relight the torch. If it flashes back again, the torch should be removed from service.

Don't confuse a backfire with a flashback. The backfire usually makes more noise and does nothing except startle you. For example, when you turn off a hot torch it probably will backfire. That's not abnormal.

However, if you get repeated backfires, you probably have a dirty tip. Clean it with your tip cleaner. You also can make your torch backfire when you overheat the tip by getting it too close to hot metal, especially when flame-cutting. If cleaning the torch tip or changing your procedure doesn't help, turn the torch and tip in for cleaning and possible repairs.

CHAPTER 5

Flame-Cutting Shapes, Edges, and Holes

After you have mastered this chapter you will know enough to work as a torch or cutting-machine operator. But that's only your first step to becoming a fully skilled welder. This chapter teaches you how to flame-cut carbon and low-alloy steels. We show you everything you can do with one of a welder's most versatile tools, the hand-held cutting torch. Then we give you some background on using cutting machines.

We also describe some special cutting processes that can cut anything from aluminum and stainless steel to solid concrete. Other chapters on electric arc welding and cutting will give more details on some of these special cutting processes, but we want you to have an overview of how welders cut metal. Most cutting in industry is done with the oxyfuel torch. The big changes have not been in the basic oxyfuel cutting process, but in the many new fuels besides acetylene that are now in use in industry. There is a good chance that your "acetylene torch" won't use acetylene, and may not even be designed for it.

5-1
WHERE FLAME-CUTTING IS USED

It would be nice if all the parts for everything you make were supplied precut to precise shapes and accurate dimensions so that you could simply weld them together. Reality doesn't work that way. Welders often have to cut their own shapes before they join them into finished products (Fig. 5-1). That's one of the jobs done with a cutting torch.

With one hand-held cutting torch and the right fuel gas you could cut a precision slice through 1-in.-thick [25-mm] steel plate at 18 to 24 in./min [460 to 610 mm/min], leaving a gap (called the *kerf*) of only 0.065 in. [1.65 mm] between the pieces, while holding the cut tolerance to within ±0.016 in. [±0.4 mm].

That cutting speed is fast enough to cut 800 ft [24.4 m] of 1-in. [25-mm] plate in one 8-h shift (if you didn't stop to eat or rest). However, there are flame-cutting machines that can do the same thing at 20 percent faster cutting speed (using a different kind of tip), while mounting anywhere from two to ten or more torches. In one 8-h shift an eight-torch machine could cut 9600 ft [290 m] of the same 1-in. [25-mm] plate (almost 2 miles or 3.2 km), if it could work continuously without running out of steel.

The hand-held torch and the cutting machine are not limited to cutting straight lines. You can cut your name out of plate, in script, with a hand torch if you want to. In fact, that's one of the exercises in the shop session that follows this chapter.

Flame-cutting also is used to take things apart. Every police and fire company emergency vehicle carries a cutting torch and oxygen and fuel-gas cylinders to cut people out of wrecked cars. When a train derails, the first workers on the scene, after the emergency vehicles arrive, have cutting torches to remove bent rail and broken wheels and damaged couplings before a crane removes the train.

When a high-rise building is demolished, the steel is cut to size with torches to be resold to foundries and steel mills as high-grade scrap. (Cutting torches also were one of the main tools used to cut the steel to shape when the building was erected.)

Auto scrap yards pull engines from junk cars with torches because it's faster than sawing or unbolting. For the same reason, cutting torches remove fenders, bumpers, and bent frame sections from auto bodies in body repair shops. A good cutting torch is one of the key tools needed to turn an auto body into a funny car or a hot rod.

A farmer without a cutting torch and a welding outfit is in trouble when equipment fails in

FIGURE 5-1 Oxyfuel cutting is not the only way to cut metals. It only works on steel. It is, however, the most common cutting method, and you can get a job based only on the skills you'll learn in this chapter. Look at the shapes you can cut.

the field. Cutting torches are primary tools for plant maintenance departments, millwrights, steamfitters, drilling rig operators, mine repair shops, pressure-vessel builders, pipeliners, construction contractors, farm and construction equipment builders, shipyards, and anybody else who works with steel . . . except safecrackers.

Safecrackers and safe builders are constantly trying to outwit each other. They both know that all good safes are built of high-alloy steels and other metals that you can't cut with an oxyfuel flame. That's the first lesson to be learned. Oxyfuel flame-cutting has limitations just like everything else. For example, you can't cut aluminum, titanium, copper alloys, or many types of stainless steel with it without difficulty (you might be able to melt your way through aluminum . . . but that's not oxyfuel cutting).

5-2
HOW OXYFUEL CUTTING WORKS
Carbon steel rusts very slowly in the air. Air is only 21 percent oxygen and rust is made up mostly of iron oxide. Rusting

works a little faster if there is some water vapor present (water is a catalyst that speeds the rusting process). However, steel at room temperature rusts slowly mostly because it is too cold.

Oxygen reacts faster with iron (and therefore steel) when the metal is hot. In fact, all materials that react with oxygen (and that's what burning as well as rusting really is) have to reach a *kindling temperature* before the burning or oxidation reaction starts going very fast. The kindling temperature of a solid is roughly the same idea as the flash point of a combustible gas or liquid vapor.

Low-carbon steel has a kindling temperature of about 1600°F [870°C] in pure oxygen. If you heat carbon steel to that temperature (or just a little higher) and add a jet of pure oxygen, the metal will literally catch on fire. The high-pressure oxygen jet will burn a slice right through the steel, which is precisely what an oxyfuel cutting torch and tip are designed to do. First the preheat flames from the tip bring the steel up to the kindling temperature. Then, when you press the cutting-oxygen lever on the torch, the

oxygen jet through the center orifice of the tip burns the steel away.

When you burn paper, the products of combustion are carbon or carbon dioxide and water vapor with a little bit of mineral ash left over. The product of combustion after burning steel is the molten slag that is sprayed out of the cut by the high-pressure oxygen jet.

A little chemistry will help you understand how flame-cutting works, and why, sometimes, it doesn't.

The chemical symbol for iron is Fe. The symbol for oxygen is O. A molecule of oxygen contains two oxygen atoms. Therefore, molecular oxygen is O_2. Iron combines with oxygen to produce several different oxides depending upon the temperature of the reaction and how much iron and oxygen are available. The symbols for the iron oxides show the difference. One of the oxides is FeO (ferrous oxide). Another is Fe_2O_3 (ferric oxide). The third is Fe_3O_4 (or if you wish, $FeO + Fe_2O_3$). This last iron oxide is especially magnetic.

Table 5-1 shows the chemical reactions that produce iron oxides when you cut steel. These reactions show you something very important about using a cutting torch. Iron burns in oxygen to produce a lot of heat. (If you haven't had a chemistry course before, you will soon discover that chemical equations are no big mystery. Like any other equations, you have to wind up with as much finished product on the right-hand side as you start with on the left-hand side. Chemists can't create or destroy matter. They can only rearrange it. For example, the first reaction in the table shows that one iron atom combines with half of every molecule of oxygen to make one molecule of a certain kind of iron oxide. (It just so happens that iron has three kinds of oxides, so we have three separate equations.)

Table 5-1 explains why some cutting jobs can continue using only an oxygen lance once you have the steel up to its kindling temperature. The reactions also give you a clue as to why some other metals such as aluminum, stainless steel, bronze, and highly alloyed steels and cast irons can't be flame-cut (at least not with an oxyfuel flame). They resist combining with oxygen at the temperatures you can produce with your torch. Their kindling temperatures are too high.

These same metals "rust" much more slowly at room temperature, too. That's part of the reason why stainless steel is stainless. You may be able to melt your way through aluminum, but you can't cut it without a much hotter "flame" than you can get with fuel gas and oxygen.

However, a process that combines a high-voltage and high-current electric arc with certain gases (not fuel gases) does cut aluminum, stainless steel and carbon steel, too. It's called *plasma-arc cutting*. It does not oxidize its way through the metal at all. It literally vaporizes the metal away because the temperature of the gas-plasma arc is about the same as the visible surface of the sun. If you think cutting 1-in. [25-mm] thick plate at 20 in./min [508 mm/min] is fast, a plasma torch can probably do the same job at over 100 in./min [2.54 m/min].

Still another process called *powder cutting* combines an oxyfuel flame with powdered iron. The combination of a hot flame, a pure oxygen jet, the heat released from burning powdered iron, and the fact that the iron powder also dilutes the alloying elements in steel, produces a process that not only can burn through a high-alloy steel casting but can cut concrete. It's also one of the most unpleasant processes to work with because of the thick cloud of red iron-oxide smoke that is produced. That smoke is the Fe_2O_3 in the second chemical equation in Table 5-1.

We will stick to more conventional oxyfuel flame-cutting in this chapter. It's the cutting process that does by far the most work in industry.

Your first problem as a torch operator is deciding when your steel has reached its kindling temperature. You hardly are expected to have a high-temperature thermometer in your hip pocket. How do you know when to press the cutting-oxygen lever on your torch?

You can easily recognize the kindling temperature of steel (1600°F [870°C] or higher) by sight. As the steel gets hot it changes color from gray to black, then to a deep purple to dark red, and then it will begin to glow cherry red (reddish-yellow) under your preheat flame. It takes only about a minute to preheat 0.5-in. [13-mm] thick plate. The thicker the section the longer you have to preheat it to bring the steel up to temperature. When the steel reaches the cherry red color, squeeze the oxygen lever on your torch and the steel will start burning away.

But flame-cutting isn't quite that simple. To illustrate, let's do a mental experiment. Pretend you've just been asked to cut some 4-in. [102-mm] thick carbon steel plate. Pick up your mental torch and get to work.

Let's say that the first thing you do is put a two-piece propane tip on your torch and then use acetylene for your fuel gas. You start your cut in the middle of the plate, preheat to the correct temperature, and squeeze the cutting-oxygen trigger hard to produce a blast of oxygen. Several things will happen rather quickly.

First, instead of making a nice

sharp slice into the plate when you press the oxygen lever, the cutting-oxygen jet will gouge out an ugly pit in the steel. The oxygen cut finally will break through the back of the plate and spray more molten slag on the floor and into your shoes. Since your socks are on fire you probably won't notice that the hot slag also has ignited the wooden floor of your work area. That's about the time that your tip will start to melt because you have the wrong tip model for the fuel gas. You may not notice your melting tip because the overheated tip just caused your torch to flash back. The flame has backed up into the mixing chamber and is beginning to melt through the torch handle, after which it will spit flame all over your arm.

Obviously, there's a little more to flame-cutting than we've told you about so far. (Flame-cutting actually is a very safe process in spite of our hypothetical, but very real, mental experiment—as long as you do it correctly.) Therefore, it's time to back away from the plate, drop what's left of your torch, and learn a few more things about flame cutting before you pick up another torch and really ruin your steel, clothes, torch, skin, and building.

5-3 CUTTING TIPS

The tip you use depends upon the fuel you use. Match your tip to the metal thickness being cut. Make sure that your tip is suitable for the conditions of the steel surface. For example, electroplated, metal-coated, plastic-coated, or painted steel is more trouble than clean steel. Steel

TABLE 5-1 How oxygen combines with iron

Iron	plus	Oxygen	yields	Slag	plus	Heat
Fe	+	$\frac{1}{2}O_2$	→	FeO	+	127,600 cal [534 kilojoules (kJ)]
2Fe	+	$\frac{3}{2}O_2$	→	Fe_2O_3	+	196,800 cal [824 kJ]
3Fe	+	$2O_2$	→	Fe_3O_4	+	267,800 cal [1,121 kJ]

with a heavy mill scale (the oxide produced when the steel is hot-rolled at the mill) also is more trouble to cut than clean steel.

Do you want a high-speed cut simply to turn one piece of steel into two, or do you want a high-precision cut with a shiny smooth surface produced to close tolerances? They require different cutting speeds and sometimes different tips.

How hard is it to reach the work? Do you need a tip extension? There are tips that help you hold the torch several feet away from the cut. Not all cutting tips are straight and round. Some are flat to make wide grooves. Others have a cross section that looks like the letter D to slice the heads off rivets. These D-shaped cutting tips are bent to lie flat on a plate and point preheat flames and cutting oxygen right at the rivet head, parallel to the plate. They are often used to repair things

like rail cars, to recover old structural steel for reuse, and to wash out bad rivets and remove tack welds.

Let's look at the process of selecting the right tip for the job one factor at a time. First, let's see what kinds of tips are available.

Types of Tips

Cutting tips made by one manufacturer cannot be used directly with another manufacturer's torch (unless the tips are specifically made for another torch model). One reason is that the back ends of the tips that seat into the torch head have different shapes (Fig. 5-2A) that only fit specific torch models. For example, the tip on a tip-mix torch also is part of the mixing chamber in the torch head. Therefore, tip-mix torch tips work only with tip-mix cutting torches and injector torches that premix fuel gases require tips for

this kind of torch. Tips also come in different shapes for different kinds of work. They are not all round and straight (Fig. 5-2B).

Another reason that cutting tips are not standardized is that torch manufacturers make more money selling tips than they make selling torches. It's much like the razor blade business, where razor manufacturers sell their razors at cost to get the blade business for a profit. Just like razor blades, there also are "bootleg" tip manufacturers who do not make torches at all. They simply copy the design of another manufacturer's torch tip and then make and sell them at lower cost.

Beware of bootlegged tips. Some of them aren't as good as original-equipment tips. You also can bet that the original torch manufacturer has established a reputation with a good-quality product. There's really no choice but to keep up the quality control if they are to get your torch and tip business. A bootlegger doesn't have such problems. What's more, cutting tips are something you should never gamble with just to save a buck.

Cutting tips are made to work only with specific gases. Solid or single-piece cutting tips are used for acetylene and hydrogen. With modified, bigger ports, solid tips also can be used for MAPP Gas and propylene. They have square or flat faces on the flame end. In comparison, the tips used for propane and natural gas are two-piece tips.

Cutting tips can be one-piece tips or two-piece tips, and two-piece tips can be either *skirted tips* or *flush tips* (Fig. 5-3). The outside shell of a skirted tip extends down below the face of the tip insert. In a flush two-piece tip, the insert and the shell are flush at the tip face. Skirted two-piece tips are used with slow-burning fuels like natural gas, propane, butane, and Acetogen (propane inoculated with propylene).

The flames of these slow-burning fuels simply can't burn back

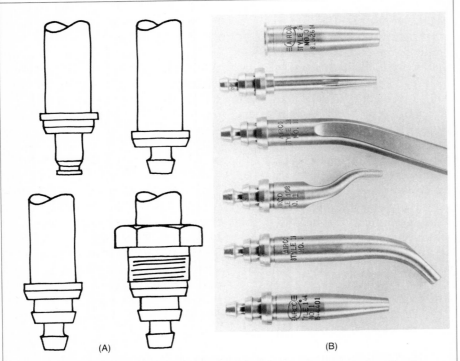

FIGURE 5-2 (A) The back end of a welding or cutting tip that hooks up to the torch is called the seat. Each torch manufacturer has different tip-seat designs; therefore, different tips are needed for different torch models. The front end of cutting, welding, and heating tips also come in many styles for different kinds of work. (B) These are only a few of the hundreds of shapes, styles, and tip extensions that exist.

(A)

(B)

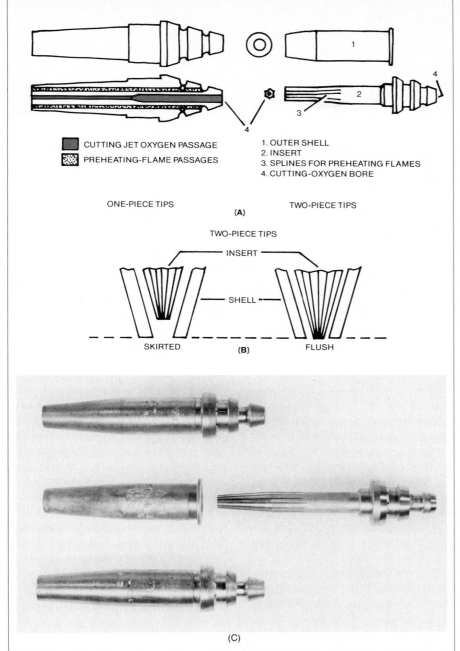

CUTTING JET OXYGEN PASSAGE
PREHEATING-FLAME PASSAGES

1. OUTER SHELL
2. INSERT
3. SPLINES FOR PREHEATING FLAMES
4. CUTTING-OXYGEN BORE

ONE-PIECE TIPS
TWO-PIECE TIPS
(A)

TWO-PIECE TIPS
INSERT
SHELL
SKIRTED
(B)
FLUSH

(C)

FIGURE 5-3 Two kinds of cutting tips are used for different kinds of fuel gases. Some cutting tips are solid, single-piece tips (most often used for acetylene and MAPP Gas). Two-piece tips are used for MAPP Gas, propane, and natural gas.

as fast as the gas flows out of the tip. It will be blown off the end of the tip. The flush two-piece tips in Fig. 5-3B are used for fast-burning fuels such as acetylene, MAPP Gas, and propylene. These fuels burn so fast that the flame stays anchored on the end of the tip.

The two-piece tips in Fig. 5-3A have an outer shell and an insert. The insert has grooves milled down its lower end. The ridges between the grooves are called *splines*. When the shell is put on the tip, the milled grooves become the flame ports for the preheat flames. The cutting-oxygen bore is located in the center of the spline insert just like any other cutting-torch tip.

The milled splines and grooves in the tip insert also are sized so that the flow rate of fuel and oxygen will be correct to produce good preheat flames for the particular fuel and oxygen combination being used. As you already know, different gases require different amounts of supplied oxygen to give complete combustion. Tips for individual gases take that into consideration.

Table 5-2 shows the amount of supplied oxygen needed by different fuels to produce a neutral flame. This isn't all the oxygen they need. As we described in Chap. 4, different fuels pick up the rest of their oxygen requirements from the air. Therefore, these ratios will vary a little with altitude. The higher the altitude, the more supplied oxygen is required because the less atmospheric oxygen there is. The difference won't be noticeable for most work in the United States but it stresses another important point—for underwater cutting, all the oxygen must be supplied.

Tips for Different Fuel Gases
MAPP Gas and propylene can be burned either in solid tips or in two-piece flush tips. If acetylene were used in a skirted tip, it would melt the tip or cause a flashback. The reverse is not nec-

TABLE 5-2 Oxyfuel ratios for neutral flame settings

Fuel	Perheat Oxyfuel Ratio	Flame Temperature, °F [°C]
Acetylene	1.1:1	5600 [3093.3]
Natural gas	2.0:1	5000 [2760]
MAPP Gas	3.5:1	5300 [2926.6]
Propane	4.5:1	5100 [2815.5]
Propylene	4.5:1	5193 [2867.2]

essarily true. MAPP Gas and propylene can be used in acetylene tips. However, they won't operate very well because the preheat holes are not large enough.

If MAPP Gas or propylene is used in a two-piece natural-gas (skirted) tip, there is a tendency for these fuels to overheat the tip. That can lead to backfires or a flashback, just as acetylene will if you burn it in a two-piece (skirted) propane tip.

As you can see, there are a lot of reasons why tips have to be engineered for each gas. The subject is even more complex than we've outlined. Tip designs also take into consideration the burning velocity of the gas in the flame and the flow rate of gases through various small hole diameters while making sure that you get enough oxygen and fuel gas to produce a neutral flame for cutting different metal thicknesses.

Tips for Different Thicknesses

Every tip manufacturer lists the range of metal thicknesses for which different tip sizes should be used. You use a much larger tip to cut 2-in.-thick [51-mm] plate than you would select for cutting ¼-in.-thick [6.4-mm] plate. That also means that the oxygen and fuel flow rates that you set on your delivery gauges are a little different for each fuel and tip size you use.

For example, if you are cutting ³⁄₁₆-in.-thick [4.8-mm] plate with MAPP Gas using machine flame-cutting, you will need a #72 (0.025-in. [0.6-mm] oxygen bore hole) one-piece or two-piece MAPP tip. Your cutting speed will be 24 to 30 in./min [610 to 760 mm/min] if you set your cutting oxygen at 30 to 40 psig to give you a flow rate of 20 to 30 scf/h [5.7 to 8.5 m³/h]. Your preheat oxygen should be set at about 5 to 10 psig for a flow rate of 5 to 20 scf/h [0.14 to 0.57 m³/h]. Your MAPP Gas should be at 2 to 6 psig [14 to 41 kPa] delivery pressure at the torch for a flow rate of 2 to 8 scf/h [0.06 to 0.23

m³/h]. That will give you a kerf width of 0.03 in. [0.76 mm].

Sound complicated? It is. But you don't have to remember any of these details. Just understand the reasons. Tip and torch manufacturers and fuel-gas suppliers (which usually are the same people) supply "cutting cards" that you can carry around in your pocket, hang from your cylinder or put in your tool box. These data also are in their product catalogs.

What's more, these data tables are very useful for setting up optimum production conditions, but they are not necessary if you just want to chop up some plate . . . as long as you understand the basic principles. Table 5-3 is a guide to selecting tip sizes for cutting different thicknesses. You don't have to memorize it, but you must understand it. Note how the tip dimensions increase with the thickness of steel to be cut.

Use the right tip for the fuel gas. Use small tips for thinner material, bigger tips for thicker material. If you can make your flame settings by eye and adjust your torch to get a neutral flame, you will come very close to the operating conditions given in tip manufacturers' and fuel-gas suppliers' cutting cards.

What's more, every tip comes packaged with recommendations for its use. If you keep the tip in

the little box that it comes in, you will know what size it is. The metal thickness range and fuel gas that the tip is designed to work with is usually printed right on the label of the box. All you have to do is keep your tips in their boxes in your tool chest and you won't forget what they're for.

Tips for Difficult Conditions

What about tips for difficult surface conditions? In general, the larger the cutting-oxygen bore, the larger the flow rate of cutting oxygen you'll get from the tip when you depress the cutting lever on your torch. Bigger tips also will have bigger preheat flame ports since the size of the flame port increases with the thickness of the metal to be cut, just as the cutting-oxygen bore does. You can use that fact to advantage in handling problems with dirty or coated steel surfaces.

Larger preheat flames are necessary for cutting painted, dirty, rusty or mill-scale-coated steel (no matter what the optimum tip size to cut the clean material may be). Some small single-piece special tips only have a couple of preheat holes. Most standard acetylene cutting tips have four to six preheat-flame holes. Some acetylene cutting tips have eight preheat holes for cutting very thick material where the extra preheat is needed. Single-piece MAPP Gas

TABLE 5-3 Tip sizes for hand-cutting carbon steel*

Plate Thickness, in. [cm]	Oxygen Orifice Wire Gauge Drill Size Number	Diameter of Oxygen Orifice, in. [mm]
Up to ⅛ [0.32]	#72	0.025 [0.64]
⅛ to ¼ [0.32 to 0.64]	#65	0.035 [0.89]
¼ to 1 [0.64 to 2.54]	#56	0.0465 [1.18]
1 to 2 [2.54 to 5.08]	#52	0.0635 [1.61]
2 to 4 [5.08 to 10.16]	#44	0.086 [2.18]
4 to 8 [10.16 to 20.32]	#31	0.120 [3.05]
8 to 12 [20.32 to 30.5]	#19	0.166 [4.22]
12 to 16 [30.5 to 40.6]	#10	0.1935 [4.91]
16 to 20 [40.6 to 25.4]	# 2	0.221 [5.61]

* Sizes are based on the diameter of the cutting-oxygen orifice necessary to produce a good kerf for a given thickness.

tips have as many as eight preheat holes. Use tips with a lot of preheat flames for steel with a bad surface.

The larger, more numerous preheat flames will flake off mill scale, heavy rust, and burn paint ahead of your cutting-oxygen jet. The worse the condition of the surface, the more preheating it needs. Materials that are difficult to flame-cut, such as cast iron, also need a maximum number of preheat flames.

Sometimes the surface will be so bad that you will have to use your cutting torch as a heating torch to clean the surface. Simply turn up the preheat flames to maximum flow and run the torch along the line to be cut without depressing the cutting-oxygen lever. Burn the surface clean where you intend to cut it. Then return to the beginning, bring the steel up to heat, and start flame-cutting.

Fast and Slow Tips

Assuming you have the correct tip for the fuel gas and torch, the right tip size for the material, and good preheating flames on your tip, you can still choose tips to cut things quickly or tips to cut them slowly. If you do a good job with a slightly slower tip, you will get what is called a *high-quality cut*. It has a very smooth surface, sharp edges, and no slag hanging down from the bottom of the parts.

A *high-speed cut* is designed to chop things up quickly when you don't care what the results look like. Some tips and fuel gases can give you a little of both quality and speed, but you can't get both quality and speed without a high-speed cutting machine.

The shape of your cutting tip's oxygen orifice largely determines how fast you can cut. The operating pressure and length of the cutting-oxygen jet also helps speed things.

There are two kinds of cutting-oxygen tip bore designs. One is the straight-bore tip and the other is the divergent-bore tip (Fig. 5-4).

You will use straight-bore tips on hand-held cutting torches and divergent-bore tips on high-speed cutting machines because they increase your cutting speeds by about 25 percent. That's an excellent increase for machine cutting, but it's too fast for you to follow with a hand torch.

Divergent-bore tips for cutting machines have cutting oxygen bores that flare out at the bottom on the flame end of the tip. These high-speed tips can operate with cutting-oxygen pressures of 60 to 100 psig [400 to 690 kPa] while maintaining a uniform cutting-oxygen jet running at supersonic gas speeds.

Straight-bore tips for hand-held torches operate at cutting-oxygen pressures of 30 to 60 psig [200 to 400 kPa]. You can cut quickly enough with one to not want a faster tip on a hand torch.

Procedures for Dangerous Coatings

Be careful of the fumes created by burning off certain coatings.

ZINC-COATED STEELS The coated steel you are most likely to encounter is galvanized (zinc-coated) steel. Galvanized steel looks like the metal on a metal garbage can. Of course not all galvanized steels have the big "spangles" of zinc crystals on them that produce the characteristic pattern on a galvanized garbage can; some newer galvanized-steel coatings are very smooth. Some zinc coatings are made only on one side instead of two. But they all have the same gray or gray-white color unless they are painted over.

When zinc is burned it produces fumes of zinc oxide that look like white smoke. Zinc-oxide fumes can give you a fever, chills, and a splitting headache for a day. Then you will recover.

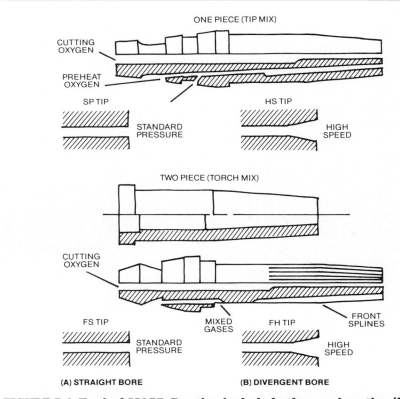

FIGURE 5-4 Typical MAPP Gas tips include both one-piece tips (for tip-mix torches) and two-piece tips (for torch-mix torches). Both kinds of tips may have cutting oxygen orifices that are straight-bore or divergent-bore, which are used for high-speed cutting on flame-cutting machines.

Galvanized steels are used to make air conditioning and heating ductwork, rain spouts and gutters, prefabricated farm and factory buildings, the underparts of automobiles to stop rust, highway guard rails and culverts, and some types of corrosion-resistant steel wire and fence posts. If you have to cut a lot of zinc-coated steel, use a filter mask to stop the fumes and work in a well-ventilated area. A related product, aluminum-coated steels, won't cause you problems.

LEAD-COATED STEELS Lead-coated steels are more dangerous. The lead-oxide fumes can give you lead poisoning. Use a filter mask (like painters use), and work in a well-ventilated area when cutting lead-coated steels. A supplied-oxygen mask would be even better.

Lead- or lead-tin-coated sheet steels are used as roofing materials on some commercial buildings and churches and as gutters and downspouts. Lead-coated steels include lead-tin alloy coatings called *terne plate*. Terne plate is used to make auto and tractor gas tanks and electric equipment chassis because the lead coating is easy to solder things to, while the steel underneath makes the product strong and rigid.

A more likely lead-poisoning hazard for torch operators is cutting structural steel in demolition work. Some structural steels are coated with a red-lead primer. You've probably seen this deep purple-red paint. Remember that an old structural column may be coated with a lead primer and then finished with cement so that you can't immediately see the red-lead surface on the steel. (Check for it with a chipping hammer.)

ELECTROPLATED COATINGS The third group of surfaces to be especially careful with are electroplated coatings. These include chrome-plated steel bumpers which have a layer of copper under a layer of nickel under a layer of chrome. You could cut them while wearing a filter mask, but

you're better off not cutting them at all.

Never cut anything that is coated with cadmium or mercury. You aren't likely to find mercury coating anything you will cut, but cadmium can occur. It's a highly corrosion resistant silvery electroplated coating sometimes used on nuts, bolts, and other kinds of fasteners. Cadmium fumes are more poisonous than lead. Most fastener manufacturers have stopped using it, but you can still find it in scrap and old parts.

Although it's not a coated metal, beryllium is a material you are never likely to flame-cut (at least not with oxyfuel). Its fumes are highly poisonous. The only place you are likely to find beryllium used is in beryllium-copper alloys. These high-strength, corrosion-resistant alloys are not very common and can't be flame-cut with the oxyfuel process. They are reddish-orange to red-brown-colored alloys.

PREPAINTED STEELS Some sheet steels now come prepainted or coated with plastics by the steel producer before they arrive at the plant. Plastic-coated steel pipe also is in use; the plastic is usually a liner on the inside diameter. Whenever you flame-cut these materials, remove the plastic far enough on either side of the cut so that you won't burn the coating or produce harmful fumes.

Don't breath fumes from burning paint or plastic in any instance. Besides smelling awful, paint fumes also may contain lead, while the plastic smoke can have all sorts of strange things in it. Structural steels often are primed with lead-containing paints. These lead primers usually are dark red. Other colors, including gray, also are used now.

DEGREASED STEELS Plates sometimes are cleaned of oil with degreasers or solvents before you get them. Don't breath fumes from these cleaning compounds. Many of these coatings can be removed with a damp cloth. If you

are not sure how to clean the plate, ask somebody. Burning solvent fumes can hurt your lungs and nervous system. Some of the fumes will damage you permanently.

USE A MASK In general, wear a filter mask for dust and fumes, and a supplied-oxygen mask for gases or for where you can run out of air. Remember that a filter mask or a gas mask is not a supplied-oxygen mask. The filter mask stops particles down to a certain size, depending upon the filter. The gas mask adsorbs gases. Neither one supplies oxygen if you run out while working inside an enclosed tank or in a sewer or excavation. Only a mask with a tank of air or a long breathing air supply line is safe when the oxygen in the work area is reduced. Most work in plants won't involve flame-cutting coated steels. But the safety director will be happy to give you the right mask for your work if you ever need it.

If you have doubts about cutting any material with which are are not familiar, find out first . . . or don't do it at all. Most of the time you will be cutting familiar materials. We don't want you to think that you will run into these flame-cutting risks every day, but we do want you to know about them for those rare instances when you might have a problem.

5-4
GETTING READY TO CUT
The Shop Practice in Chap. 4 told you how to set up your equipment and light your cutting torch. There are some other things you must (or mustn't) do in a real work situation. These are problems you won't face in a welding school, so remember them.

Don't flame-cut on a wooden floor. Lay down sheet metal, a fireproof asbestos or woven glass fiber blanket on the floor before you light up. Have a fire extinguisher nearby. Get a safety permit to use your torch if local ordinances or plant regulations require it. Always find out what is on the other side of any wall, barrier, or container you will cut into. Remember that hot slag sprays out the other side of your cut. If you cut through a wall or bulkhead there could be anything from gasoline to paper to people on the other side.

Never work in an unventilated area. Most welding and cutting stations have fume hoods, air suction ducts or blowers, fume collectors, or directed ventilation such as large floor fans.

Don't cut into a pipeline without prior permission and inspection. There are methods for sealing off pipelines containing explosive or flammable gases or liquids. One method used on oil lines is to freeze the liquid solid with dry ice placed on either side of the area to be cut. Training for this work is required and a detailed description of how to do it belongs in an advanced cutting class.

Don't cut into any enclosed container unless you know what it contained and have completely cleaned it. It might still contain gasoline vapors, explosive gases or something else dangerous or even merely unpleasant. There are special procedures for purging containers like auto gas tanks before welding and cutting them. The procedures are either to fill the tank with water after a thorough cleaning, or to clean the tank with industrial detergents and then purge it completely with nitrogen. After either procedure the atmosphere in the vessel is tested for flammability with a special flame or hazardous-gas detector. You're not ready yet for that kind of work. *Don't try it.*

5-5
SETTING YOUR CUTTING-OXYGEN PRESSURE
For correct adjustments of the working pressure of your oxygen or fuel-gas regulator, the needle valve on the torch handle must be open and the gas flowing. The cylinder adjustment screw is then turned until the working-pressure gauge on the cylinder shows the desired pressure.

If the pressure is adjusted with the needle valve on the torch handle closed, your working pressure will not be maintained when you open the torch valve and start to use it. Instead, the gas pressure at your torch will fall below the pressure you want and that will make further adjustment necessary. Especially when cutting, where heavier oxygen flows are involved, it is essential that the cutting-oxygen pressure be adjusted with the high-pressure oxygen valve (usually a lever or trigger) on the cutting torch momentarily held open to let cutting-oxygen flow. You won't get satisfactory results any other way.

When you don't have recommended oxygen and fuel-gas regulator pressure settings to work with (from a cutting card, operator instructions, or tip data sheets from a catalog), there still is a simple way to quickly determine the best cutting-oxygen pressure to use for any tip.

Put a low-volume, soft (short) preheat flame on your tip. Then depress the cutting-oxygen lever to turn on the cutting-oxygen jet and vary the pressure to find the best looking *stinger* (the visible cutting-oxygen stream). Low cutting-oxygen pressures give very short stingers, maybe 2 or 3 in. [50 to 76 mm] long. Low-pressure cutting-oxygen stingers will break up at the front end. As cutting-oxygen pressure is increased, the stinger will suddenly become longer. This point is the correct cutting-oxygen pressure for the tip on your torch.

Beyond that pressure, the long cutting-oxygen stinger will remain relatively unchanged over a fairly wide range. But as the cutting-oxygen pressure is increased too much, the stinger returns to the short, broken form it had under low pressure.

5-6
ADJUSTING YOUR FLAME
Preheat-flame adjustment is critical. The procedures are the same whether you are flame-cutting, heating, welding, brazing, soldering, or metallizing.

There are three basic kinds of flame settings. They are carburizing, neutral, and oxidizing (Fig. 5-5). You adjust them right on your torch using your torch gas-control valves while making the settings by the flame's appearance and sometimes by its sound. You can use these settings creatively to do different kinds of work.

Be sure to reread the Shop Practice section at the end of Chap. 4 so that you know how to set up your torch safely. Then recall what you saw in the flames.

The Color Plates on pages 51–54 show the different types of flames with the inner flame cone, the outer flame cone, feathers on the inner flame cone, and the oxygen stinger you get when you press the cutting-oxygen lever.

Remember how, when you snapped your spark lighter under

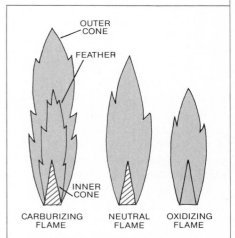

FIGURE 5-5 What MAPP Gas cutting-flames should look like. The carburizing flame on the left is what you usually get when you light your torch. It is very yellow. The neutral flame setting is the most common setting used for flame-cutting or welding. Oxidizing flame settings are only used for special jobs. See Color Plates on pages 51–54.

your tip, you got a long, bushy, yellow smoky flame from the fuel gas burning only in air? You had not yet turned on your oxygen supply, and the flame did not have enough oxygen to burn efficiently. The color was due to particles of unburned carbon and hot soot that result from the partial oxidation of the fuel gas. This flame is not very hot. Its temperature is only about 1500°F [800°C].

When you turn up the preheat-oxygen valve on the torch, you add more oxygen to the combustion reaction. The yellow flames get a lot shorter and start to change color from yellow to blue. Now you can see the inner and outer flame cones. The shape and size of the inner and outer flame cones will depend upon the fuel gas and tip you are using.

One of the reasons is that acetylene preheat flame ports are not as large as propylene and MAPP Gas flame ports. Look back at the footnote in Table 4-3 in Chap. 4 on gas flow through hoses (or any other restriction) and you'll see that the gas flow rate (velocity) depends on the density of the gas. Acetylene is much lighter than other common fuel gases. For that reason, it needs smaller tip holes to get the correct velocity. Also, if larger acetylene holes in the tip were used, it would cause flashbacks, as well as too much heat which would melt the edge of the steel.

Acetylene preheat flames, in general, are shorter than MAPP Gas and propylene flames. Natural gas and propane flames are longer and sometimes more rounded. (There even are special welding torch tips that can produce bulbous or very rounded flame cones with natural gas or propane.)

Experienced welders and torch operators have trouble switching from one fuel to another because they often don't know that flames from different fuels look very different. For example, if they've only had experience with acetylene and switch to MAPP Gas,

they will turn the fuel and oxygen pressure down until the MAPP flame looks exactly like an acetylene flame. Then they can't understand why a fuel that produces so much more heat than acetylene acts like it's cold. Any fuel gas runs cold if you don't use enough of it.

If they are used to natural gas or propane and switch to acetylene, they probably will want to set big flames on the acetylene tip that will soon overheat it. Instead of making nice sharp cut edges, the edge will be overheated and melt off to a round shoulder. That's another reason for using tips of the right size.

We can't tell here exactly what every possible flame should look like. Even photographs won't help much. But after an hour or so in the shop with a torch in your hand, you will easily learn how to master your preheat-flame settings to do different kinds of work. We can, however, give you some guidelines.

The first guideline is: Always make your final cutting-torch preheat-flame setting with the cutting oxygen on. You can make your initial adjustments without the cutting oxygen, but as soon as you add cutting oxygen, you will change the flame setting, so you must make final adjustments with the cutting oxygen operating.

Strongly Carburizing Flames
When you turn up the preheat oxygen a little, the blue inner flame cone still has a trace of yellow in it. Therefore the blue flame may have a green tint. The inner flame cone will have small yellow or white feathers on it. You now have a carburizing flame.

The carburizing flame can either be strongly carburizing (longer, with yellow or white feathers on the inner cone) or slightly carburizing (the feather on the inner cone gets shorter and turns white). The strongly carburizing acetylene flame has a flame temperature of about 5400°F [2990°C]. The slightly carburizing

flame is at 5500°F [3040°C]. MAPP Gas and propylene flames will be only a couple of hundred degrees cooler—still very high temperatures. Their flames, however, will be bigger than the acetylene flame and will produce more heat. A natural gas flame will be even larger, but it will produce less heat and will have a lower flame temperature.

If you had a welding tip instead of a cutting tip, the single carburizing flame would make a molten-steel weld puddle boil. The molten metal would look murky, not clear. This flame adds extra carbon to a low-carbon steel, which reduces its melting temperature but makes the metal harder and more brittle when it solidifies and then cools to room temperature.

Strongly carburizing flames are used in flame-cutting, heating, and welding work on thick cast-iron sections because cast iron already has much more carbon in it than steel. A little extra carbon won't do any damage. Strongly carburizing flames also look much bigger than neutral flames and they pump more total heat into the iron because of their large size. But the flame temperature of a carburizing flame is lower than a neutral flame, or the even higher-temperature oxidizing flame.

Burners often call a strongly carburizing flame a "cold" flame. Part of the technique of successfully cutting cast iron (which is difficult to do) is to use the heat of a strongly carburizing preheat flame to bring a massive piece of iron up to a temperature where the material in the kerf can be melted or "washed" out of the cut because cast iron can't be oxidized or "burned" out like steel.

Slightly Carburizing Flames
Slightly carburizing flames are good for flame-cutting steels and other ferrous metals that tend to produce a lot of slag. The slightly carburizing flame has no free oxygen and produces chemically reducing conditions (the opposite of

oxidizing conditions). Therefore, it makes fairly smooth cut surfaces and reduces the amount of slag clinging to the bottom of the cut. While a neutral flame setting is best for almost all cutting, a slightly carburizing flame may be better on special cutting jobs when you want to get a lot of heat down inside the kerf.

The best example is the use of the slightly carburizing flame setting for stack cutting (cutting a layer of sheets or plates all at once), where high heat input is needed. Since the carburizing flame produces less slag, it keeps the layers of a stack from welding themselves together with their own slag. However, any kind of carburizing preheat flame has several drawbacks. In particular, the cutting speed is lower than a neutral flame. Therefore, the slightly carburizing flame is seldom used in production flame-cutting.

Neutral Flames

The neutral preheat flame setting is by far the most common flame setting for oxyfuel cutting. It's also the most commonly used oxyfuel welding flame setting.

A slight adjustment of your preheat-oxygen valve will add very little extra oxygen to the slightly carburizing flame, turning it into a neutral flame. The feather will disappear completely from the inner flame cone. All you will have left is the dark-blue inner flame cone and the lighter-blue outer cone. This is the neutral preheat flame. If you had a welding tip on your torch, you would most likely use the neutral flame for gas welding. Its flame temperature is about 5600°F [3090°C] for acetylene, 5300°F [2920°C] for MAPP Gas, 5190°F [2870°C] for propylene, 5100°F [2820°C] for propane, and 5100°F [2820°C] for natural gas.

The neutral flame will neither add carbon to steel nor oxidize it. In fact, the neutral flame protects molten steel from oxidation. Carburizing, neutral, and oxidizing flames will all exclude oxygen from the air, but the neutral flame also avoids the problem of a carburizing flame, which adds carbon to the steel, and an oxidizing flame, which adds its own oxygen even when pushing away the air around the hot metal.

Properly set neutral welding, heating, or cutting flames act very much like the inert gases used in MIG and TIG arc welding. That is why the neutral oxyfuel flame is the most commonly set flame for welding, as well as for flame-cutting, flame-hardening, heating, brazing, and soldering.

If you hold a neutral flame on one spot on the steel until it melts the metal, the molten puddle will look clear and lie very quietly under the flame. That's one of the ways for testing for a neutral flame setting.

Slightly Oxidizing Flames

Add a little more oxygen at the oxygen preheat valve on your torch and the flames will suddenly get shorter. They will start to neck down at the base next to the flame ports. The color of the inner flame cone will change from dark blue to light blue. You'll hear an increase in sound. This is the slightly oxidizing flame. Its temperature is about 6000°F [3300°C] for acetylene, or 5800°F [3200°C] for MAPP Gas.

Oxidizing flames are easier to look at because they are less radiant than neutral flames. Molten steel under an oxidizing flame will foam and spark. The excess oxygen is burning the carbon out of the steel.

The oxidizing flame is almost never used for welding. The oxide particles that it produces make harmful, brittle inclusions or particles in the weld metal after it has hardened, reducing the strength of the weld. The oxidizing flame also is almost never used in conventional flame-cutting (except to speed the preheating of very large steel or iron sections, and that only rarely). The oxidizing flame produces excess slag which makes a rough surface on the cut edge. The high temperature also rounds off square-cut edges.

There are several very important special uses for slightly to moderately oxidizing preheat flames in flame-cutting and that's why many large cutting machines have HIGH-LOW oxygen switch positions. These oxidizing flames can give you fast starts using a high-speed cutting machine. They'll also make a piercing start with a cutting machine, and you can use them for bevel cutting the edges of thicker sections either while working on a flame-cutting machine or working with a hand torch.

The most important machine use for the oxidizing flame produced by the HIGH setting is to make more difficult piercing starts. Without the hot oxidizing flame, the piercing start would be much more difficult to make. You can do the same thing with your torch.

An oxidizing flame will give a fast start when using a cutting machine with the special high-low preheat oxygen gas controls needed for this technique. The torch is positioned at the edge of the plate with the preheat oxygen switch on HIGH. As soon as the edge is hot enough to burn, the cutting oxygen is switched on and the cutting-machine operator immediately drops the preheat oxygen switch to LOW. The cutting machine automatically changes the preheat-oxygen flow and creates a neutral flame as the cut continues. But in fact edge starts are not that hard to make.

Slightly oxidizing flames speed preheating. Because of their higher temperatures, they often are needed for bevel cuts made on thicker plates. When the cutting tip is angled to produce a bevel cut on an angled slice instead of making a perpendicular cut through the surface, the flame has to cut what's equivalent to a much thicker section of metal.

Bevel cuts are made on thicker

FIGURE 5-6 When you can't get an edge start, you must start your cut in the scrap to avoid digging a hole in the workpiece. Note how the operator used a piercing start in the scrap, and then followed a continuous "lead-in" line to the part, guiding the kerf tangent to each of the circles being cut.

sections (over ½ in. [13 mm] thick and up) to allow complete penetration of weld metal through the joint and often to allow room for the welding rod or electrode to get into the joint.

Very Oxidizing Flames

Very oxidizing flames are never used in welding and almost never in flame-cutting—on land. There is one place where a very oxidizing flame setting is set on the land, but it isn't used there. A diver's tender will light the torch, adjust the flame to a highly oxidizing condition (just short of the point where the flame will be blown off the torch), and then lower the torch to the diver. The diver will probably turn the oxygen up even higher before starting to flame-cut.

On land, the torch flame gets part of its oxygen from the cylinder gas and part of it from the air (which is 21 percent oxygen, 78 percent nitrogen, 0.94 percent argon and 0.6 percent other rare gases such as krypton, xenon, and neon). Underwater, all of the oxygen for the flame must be supplied through the torch. If a neutral flame (above water) is set by a diver or an assistant, then low-

ered into the water, hot unburned gases will rise to the surface, meet atmospheric oxygen, and finish burning as a sheet of flame on top of the water.

5-7
STARTING THE CUT

You have to use slightly different methods for starting a cut depending on whether you are starting from the edge of the plate or have to pierce a hole somewhere on the plate's surface.

Edge Starts

Whenever possible, start a cut on the edge of a plate or workpiece. This avoids the problems of a piercing start. Edge starts also leave any ragged cut surfaces or starting imperfections outside the workpiece, on the scrap area of the steel, when precision parts are being made.

A "lead-in" line (Fig. 5-6) can then be used to bring the cut to the area of the part by guiding the kerf tangent to the edge of the shapes being produced. Then you can continue the cut around the workpiece. Similarly, you can finish the cut in the scrap area outside the workpiece and assure

that the piece will drop out of the scrap. Flame-cutters always try to produce *drop cuts* because it's a lot of extra work when the workpieces are hung up in the scrap.

There are two good ways to start a cut on the edge of a steel section.

■ The most common method is to hold the torch tip halfway over the edge of the steel, keeping the inner preheat-flame cones about ¹⁄₁₆ to ⅛ in. [1.6 to 3.2 mm] above the surface of the material to be cut. When the top corner of the plate or section reaches a reddish-yellow (cherry yellow) color, *slowly* squeeze the cutting-oxygen lever and you will produce a sharp-edged starting cut. Move the torch and the cutting-oxygen jet into the plate at the appropriate cutting speed for the fuel and the section thickness, while holding the preheat flames a short distance above the steel (still about ¹⁄₁₆ to ⅛ in. [1.6 to 3.2 mm]), and you will produce a precision drop cut.

■ The second method is to put the tip entirely over the edge of the material to be cut, and then move the preheat flames back and forth along the line of the cut for a distance about equal in length to the plate thickness. Be sure to leave a small ¹⁄₁₆-in. to ⅛-in. [1.6- to 3.2-mm] gap between the inner preheat-flame-cone tip and the surface of the steel. When the edge reaches the kindling temperature, back the tip off the edge, squeeze the cutting-oxygen lever, and slowly move the torch into the steel to begin the cut. This method produces sharper corners at the beginning of the cut. It takes a little longer to start the cut, but this is the best method to use when the edge of the part also is the edge of the steel.

Piercing Starts

Sometimes you have no choice but to start a cut somewhere besides the edge of the steel. One way to do it (especially on very thick sections such as 2-ft [0.6-m] slabs) is to drill a hole all the way through the metal with a power drill and use this hole as your starting edge. However, it usually is much faster to use the cutting torch to produce the starting hole. This technique is known as *piercing*.

Piercing anything except very light plate requires somewhat larger preheat flames than are used for an edge start. The preheat flames also should be slightly oxidizing to create more heat.

Locate a spot in the scrap area outside the part to be cut, if possible. Let the preheat-flame cones touch the steel. That will rapidly bring the steel up to its kindling temperature. Short up-and-down motions of the torch tip will soon indicate the proper distance for you. Move the tip slowly around the hot spot to bring the entire area, including the center of the hot spot, up to temperature. When a few sparks appear, the steel is hot enough to burn.

For a section up to 1 in. [25.4 mm] thick, squeeze the cutting-oxygen lever slowly and simultaneously lift the tip straight up from the plate to keep slag from being blown into the the torch tip. Stand to one side where you will not be cutting directly away from yourself. If you were to turn the cutting oxygen on quickly, you would blow slag into the tip holes and probably back on yourself.

As soon as the hole is completely pierced and breaks through the bottom side, turn the cutting-oxygen lever on all the way and start making your cut. Avoid cutting through slag on the top of the section. It will cause more slag and sparks to fly in all directions and will deflect your cutting oxygen.

When cutting a section over 1 in. [25 mm] thick, you will have difficulty keeping slag from piling up or spraying into the torch tip as the pierced hole gets deeper. One solution is to hold the torch close to the plate until you get a good preheated spot and then *slowly* press the cutting-oxygen lever as you pull the torch back

from the plate. As the hole begins to penetrate the plate, bring the torch head back close to the surface as the flame sinks in, and start cutting as soon as the hole breaks through.

If you are getting too much slag sprayed back at you, you can angle the torch head left and right and wash the slag out of the hole. (Watch out that you don't spray slag on yourself or somebody else in the area.) While you do that, move the torch a short distance in the direction of the cut, during which time the metal will continue to be oxidized by the oxidizing preheat flames, while you use very short bursts of the cutting oxygen. Continue to blow the slag out of the deepening pierced hole while angling the torch back and forth, watching out where you spray the molten slag.

As you continue cutting deeper into the plate, you will finally break through the other side. When you have penetrated the plate or structural section, angle the tip up to the vertical position with the cutting oxygen on and you will straighten out the pierced hole. Then start your cut in the usual way.

For very thick sections you can use an oxygen lance, which we describe later on.

5-8
MAKING A HIGH-QUALITY CUT

Now you know how to start your cut, but there are many things you must do to make your cut a good one. Let's discuss them one at a time.

Coupling Distance

In all the procedures we've described, we say that you should hold your torch tip a "little distance" above the surface of the steel. This stand-off distance is called the *coupling distance*. The reason for the coupling distance is to hold the part of the preheat flame with the highest amount of heat close to the surface of the plate. As you know, the acetylene

flame has most of its heat in the inner flame cone. MAPP Gas has more heat in the outer flame cone. Therefore, the correct coupling distance depends on the fuel gas you use.

Remember that the coupling distance is not the distance from the tip of the outer flame cone to the plate, or the distance from the torch tip to the plate, but it's usually quite close to or even inside the inner flame cone.

If you stand off too far, you won't preheat fast enough and your cutting speed will slow down. If you get too close, you will overheat your cutting tip, start to melt your plate, and your tip will start backfiring. It might even overheat enough to give you a flashback.

Many operators think coupling distance is the distance from the torch tip to the plate—but it's not. It is approximately the distance from the inner preheat-flame-cone tip to the plate (Fig. 5-7). Since different fuel gases and flame settings produce flames with different inner and outer cone sizes, measuring the distance by eye from the torch tip to the plate makes no sense. Measuring by eye (behind a goggle's lenses of course) and holding the tip of the *inner* flame cone close to the plate makes lots of sense.

When cutting ordinary plate or structurals from $\frac{1}{4}$ in. to 2 in. or 3 in. thick, keep the inner preheat flame cones about $\frac{1}{16}$ in. to $\frac{1}{8}$ in. off the surface of the work to keep from melting the surface of the steel. When piercing or making very fast starts, let the preheat cones just touch the surface. This will give you very fast preheating.

As plate or structural thickness increases above 6 in. [152 mm], use a very big preheat flame. As one burner said, "build a bonfire" on the tip of the torch. Increase the coupling distance to get more heat from the secondary flame cone when using MAPP Gas or propane fuel (because they have a lot of their heat in the secondary flame).

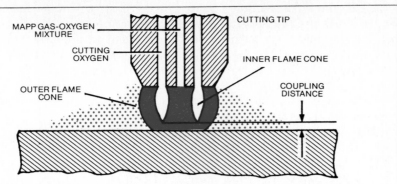

FIGURE 5-7 The coupling distance is very important when you are flame-cutting. The inner flame cone of the preheat cutting-torch flames should not touch the plate. A small coupling distance puts the maximum heat area of the flame on the plate while avoiding oxidation or carburization of the steel. The right coupling distance produces maximum cutting speeds.

Keep the coupling distance smaller even on heavy plate when using acetylene, which has most of its heat in the primary flame cone. For example, the secondary preheat flame of MAPP Gas will rapidly preheat 12-in.-thick [305-mm] steel well ahead of the cutting oxygen using a coupling distance of ¾ to 1¾ in. [18 to 43 mm]. The coupling distance for acetylene would be half as much or less.

Cutting Speed

Once the kerf has been started by the oxygen stream, move the torch along the line of cut. It is important that the forward cutting speed of your torch be correct for the job. It must be just fast enough so that the cutting-oxygen jet passes completely through the plate thickness to make a clean cut on the top *and* the bottom of your steel. The slag will be thrown out the bottom of the plate in the *same direction* you move the torch if you are using the correct travel speed (Fig. 5-8).

If you move the torch too slowly, you will create a lot of oxide slag. The slag will be thrown out of the cut straight down or even opposite the direction you are traveling. This sticky slag will hang from the bottom of the cut like icicles, and the surface of

the kerf will be rough and irregular. Excess slag can also "glue" a part into the scrap so you won't get a perfect drop cut and will have to climb on the plate to hammer parts loose.

If you cut too fast, the high-pressure cutting-oxygen jet will not make it all the way through the plate and you will either get parts that don't drop out and may have to be recut, or you can move so fast that you lose the cut completely. When that happens, the torch continues moving but the

cutting stops and all you do is either heat up a lot of steel or make burn marks on the surface. When you lose the cut, you also have problems getting it going again without gouging a hole in the part where you restart.

When you do lose the cut, immediately release the cutting-oxygen lever. Return the preheat flames to the point where the cut was lost and restart the cut. Otherwise you can go back onto the scrap to restart the cut, then lead the cut into the kerf and continue from there.

Travel Speed and Drag Lines

There is one best cutting speed for each fuel, plate, or structural thickness and job. On material up to about 2 in. [50.8 mm] thick, a high-quality cut will be obtained when there is a steady "purring" sound from the torch, and the spark stream under the metal has a 15° lead angle (Fig. 5-9). Note in Fig. 5-9 how the sparks point in the direction you are cutting.

Using this 15° lead angle for the sparks is a good way to determine the correct cutting speed when hand cutting, especially when you don't have a cutting card. Just like setting your oxygen and preheat fuel-gas flows for

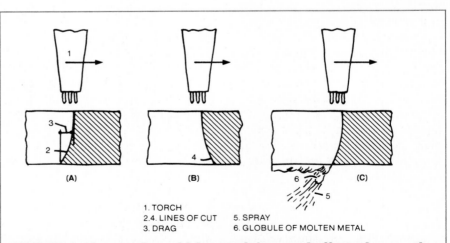

1. TORCH
2,4. LINES OF CUT
3. DRAG
5. SPRAY
6. GLOBULE OF MOLTEN METAL

FIGURE 5-8 The speed at which a torch is moved affects the way the steel is cut. (A) The top of the cut should be slightly ahead of the bottom of the cut. (B) If you move too slowly, the bottom of the cut will get ahead of you. (C) You can tell if you have the right cutting speed by the angle of the sparks spraying out the bottom of the cut. If you cut very fast, the cut will not pass all the way through the metal.

FIGURE 5-9 The correct cutting speed will give you a 15° drag angle. This operator is cutting up and to the left, in the same direction that the sparks fly.

the job and the tip, there are methods for determining just about every cutting parameter by eye, even when you work without data.

The reason that the sparks go out the bottom of the cut ahead of the torch in the direction you are traveling is that the bottom of the cutting-oxygen jet always trails the top. The flame cutting through the metal leaves fine lines on the side of the kerf at the same angle as the lead angle. These lines are called *drag lines* (Fig. 5-10).

The angle the lines on the face of the kerf make with the bottom of the plate is called the *drag angle* of the cut. Obviously this angle is the same as the lead angle of the sparks you see coming out of the bottom of the plate. If you cut the material at the correct speed and produce a 15° lead angle with your sparks, you'll also see a 15° drag angle on the kerf when you are finished. You can actually estimate approximately how fast a cut has been made by looking at the angle of the drag lines on the surface of the kerf after the job is finished. That's important in troubleshooting cutting problems.

You also can experiment with a piece of scrap of the same thickness as your work to determine the best speeds and cutting methods before you start a big cutting job. When you are producing a 15°

lead angle, your sparks and the kerf show the same 15° drag angle on the scrap, you have a good cutting speed for the material, and you can start cutting the plate.

The cutting pattern on the kerf will help you troubleshoot a cutting operation with problems. For example, a slow-speed, high-quality cut will have a small drag angle, right around 15°. Cutting too fast will produce a large drag angle, up to 45° when you go very fast. There are many more things you can learn from just looking at the cut pattern on the kerf and the edge of the cut steel.

There are different ways of moving your torch, depending on the type of metal you are cutting (Fig. 5-11). Carbon steel requires less preheating than cast iron, and therefore the torch moves straight ahead at a steady, slow speed.

Since cast iron needs so much more preheat than carbon steel, you have to move the torch in an arc that varies with thickness as

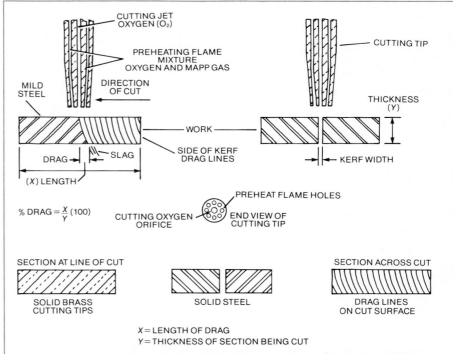

FIGURE 5-10 Drag lines on the finished cut surface indicate whether you were cutting too fast or too slow. This diagram gives you useful names for the parts of the cut, including the kerf (width of cut), the percentage of drag, the location of preheat and cutting oxygen orifices in the tip, and other information.

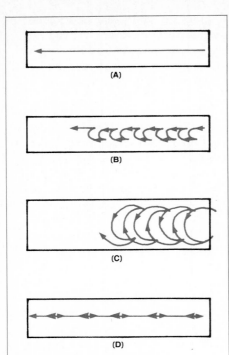

FIGURE 5-11 Different materials require different torch motions when you cut them. (A) For carbon steel use a straight line torch motion. (B, C) You need to weave your torch to successfully cut cast iron. (D) Sometimes you can melt your way through thinner sections of stainless steel (which can't be oxidized) with the back and forth motion shown here. This last motion is something like sawing wood. Other cutting processes usually are used on stainless steels.

you cut. That keeps the preheat flame in one area longer. When cutting stainless steel, the torch motion is forward, then back slightly, then forward, then back slightly. Both of these materials require special techniques because you can't really burn (oxidize) metal out of either of them with an oxyfuel torch. You have to melt your way through them. See the special section on "Cutting Cast Iron," page 125, for more details.

Tip Angles
Your torch head and cutting-oxygen jet must be in line with the cut. But for straight up-and-down cuts (as opposed to beveled edges) you can angle the torch head so that the cutting oxygen slices through a larger thickness of the steel when you are cutting thin plate. Figure 5-12 shows you how.

This angle is called the *lead angle* of the torch and tip. It will help you cut thinner material (under ½ in. [12.7 mm] thick without melting the edges, overheating the base metal, or creating excessive distortion while you work. Angling the torch tip forward in the direction you are cutting thin plate also will help prevent slag from welding itself to the back of the plate, across the kerf, which would keep your parts from dropping out when you finish cutting. If you are cutting material that is ½ in. [12.7 mm] thick or thicker, hold your torch tip straight up and down, perpendicular to the face of the horizontal surface.

When cutting steel plate or structurals that are in the vertical position, start your cut on the lower edge of the material and cut upward along the section, not down. That will help you avoid cutting through slag that otherwise would trickle down the back of the plate if you had started your cut at the top and worked down.

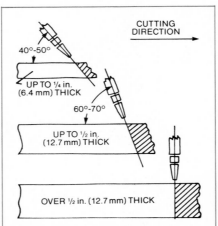

FIGURE 5-12 Thin sections are cut with the torch held at an angle. As the section thickness increases, hold the torch more toward the vertical. To cut steel over ½ in. thick, hold the torch perpendicular to the steel surface.

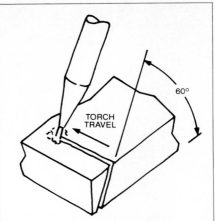

FIGURE 5-13 Bevel cuts are made by slanting the torch to one side. This torch is making a 60° bevel cut. Bevel cuts are often used to make joints in thicker plates for subsequent welding.

Bevel cuts (Fig. 5-13) are made on thicker sections because you can't get a complete weld melted through the metal section in one pass. If you only weld the top of a thicker section and the weld does not pass all the way through, you get a weld defect called *lack of fusion,* which is one of the two reasons why the edges of thicker sections are always beveled before they are jointed for welding.

Without edge beveling on thicker sections, usually ¾ in. [18 mm] thick or over, the welder would create what amounts to a built-in crack in the weld joint where the two pieces of metal are not completely joined. Lack of fusion is a bad defect because it often cannot be seen. A very good-looking weld on the surface can be no weld at all in the center of the joint.

The other reason for edge beveling joints before welding is to make room for the welder to get the filler metal into the joint. If the angle of the beveled edge is too narrow, the edges of two narrowly beveled sections, when butted together, would leave no room in the joint for the welder to get at the bottom part of the weld.

You make a bevel cut by tipping the angle of your torch to the right or left of the cut (Fig. 5-14) instead of forward or backward as you do on thinner metal. The angle of the bevel depends on the thickness of the steel and the type of weld joint you will be preparing. Since you will be cutting through what amounts to a thicker section, you will have to use slower cutting speeds and plenty of preheat flame to get the job done. If you are using a cutting card to set up the job, don't read the data for the vertical section thickness of the material you cut. Use the data for the actual thickness of metal that your flame will pass through. For example, a 4-in. [102-mm] thick plate cut on a 45° bevel is the same as cutting a section 5.7 in. [145 mm] thick.

Also, rotate your tip's heating ports as shown to get a good bevel cut. Use a slightly oxidizing preheat flame to get more heat into the larger section you are cutting at this angle. *Remember that any bevel cut is equivalent to cutting thicker plate.*

Providing for Kerf Width

We've taken you from cutting very thin material to very thick material. So far, we haven't talked about a most important factor when you are cutting steel to precision shapes . . . the width of the kerf.

If you are cutting up steel on a wrecking job or to reduce the size of scrap to manageable chunks, the width of the kerf is insignificant. If you are cutting precision parts, the kerf width makes the difference between creating usable parts to the correct dimensions and producing a pile of undersized scrap.

The width of the kerf is determined by the size of the cutting-oxygen orifice and to a lesser extent by the cutting-oxygen pressure used. Charts of kerf width usually are included in tip manufacturers' cutting cards, listed along with tip size and oxygen and fuel gas pressures. Tables 5-4 through 5-10 show typical data

for different fuel gases and cutting tips. On those tables that don't list kerf width as such, the cutting-oxygen orifice diameter is approximately equal to the kerf width.

These tables give you good starting data for setting up a job. You may want to alter the conditions (after you become an experienced burner) to suit your particular job and your "feeling" of what looks best for you in terms of cutting speed and in terms of cut quality.

Don't put the kerf inside the part

When you lay out a job on a plate or structural, you must allow for the kerf width on the *outside of the part* (or on the inside, too, if the part is ring-shaped and has a center cut-out). As a general rule, thicker plate requires a wider kerf (because it also requires larger tips, and larger tips have larger oxygen bores). Never make the horrible error of putting the kerf on the inside of the part. You will wind up with dozens of parts all cut too small.

Remember that the actual kerf width you get depends on the tip style and model you use. And since tips are not standardized, neither are kerf widths. Tables 5-4 through 5-10 show how the typical kerf gets wider as the material gets thicker. They show that oxyfuel cutting does a pretty good job of slicing some rather

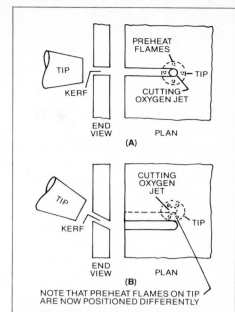

FIGURE 5-14 This diagram will help you position your torch for making straight cuts and bevel cuts. Watch where the preheat flames are on the top of the plate. The preheat flames for a bevel cut should heat the thickest part of the plate edge, not the thinnest part of the plate edge.

thick material without taking much of the material with it.

Sometimes the dimensions you get for the part already will include an allowance for the kerf width, in which case you can cut directly on the centerline of the part edge. *If someone else laid out the pattern, find out whether the kerf should be outside the cut or on the centerline.*

TABLE 5-4 MAPP Gas for hand-cutting using Type SP (one-piece) and FS (two-piece) standard cutting tips

Material Thickness, in.	MAPP Tip Size No.	Hose ID Size, in.	MAPP pressure, psig*	Oxygen Pressure, psig*	Kerf Width, in.
Up to 1/8	72	3/16	2–4	20–40	0.03
1/8–1/4	65	3/16	2–6	30–70	0.04
1/4–1	56	1/4	2–8	40–80	0.06
1–2	52	1/4	2–10	50–90	0.09
2–4	44	1/4	4–12	60–100	0.14
4–8	31	5/16	6–15	70–110	0.19
8–12	19†	5/16	10–25	80–120	0.25

* Pressures given are at regulator and based upon up to 25 ft of hose.
† Type SP MAPP tip is not available in size 19.

TABLE 5-5 MAPP Gas for machine flame-cutting using standard-speed cutting tips

| Plate Thickness, in. | MAPP Tip Size No. | Cutting Speed, in./min | Oxygen | | | | | MAPP Gas | | |
| | | | Cutting | | Preheat | | | | | Kerf Width, in. |
			Pressure, psig	Flow, scf/h	Pressure, psig	Flow, scf/h		Pressure, psig	Flow, scf/h	
3/16	72	24–30	30–40	20–30	5–10	5–20		2–6	2–8	0.03
1/4	68	22–28	35–45	30–40	5–10	5–20		2–6	2–8	0.035
3/8	65	21–27	35–45	40–55	5–10	5–25		2–8	2–10	0.04
1/2	60	20–26	40–50	55–65	5–10	10–25		2–8	4–10	0.045
3/4	56	16–21	40–50	60–75	5–10	15–25		2–10	4–10	0.06
1	56	14–19	40–50	60–75	5–10	20–38		2–10	6–10	0.06
1 1/4	54	13–18	40–60	105–120	10–20	20–38		2–10	8–15	0.08
1 1/2	54	12–16	40–60	105–120	10–20	20–38		2–10	8–15	0.08
2	52	10–14	40–60	145–190	10–20	20–50		2–10	8–15	0.09
2 1/2	52	9–13	40–60	150–200	10–30	20–50		6–10	8–20	0.09
3	49	8–11	40–70	200–250	10–30	20–50		6–10	8–20	0.10
4	44	7–10	40–70	300–360	10–30	20–50		6–10	10–20	0.14
6	44	5–8	40–70	300–360	10–30	50–100		6–15	10–20	0.14
8	38	4–6	40–80	415–515	30–50	50–100		10–15	20–40	0.17
10	31	3–5	40–90	550–750	30–50	50–100		10–15	20–40	0.19
12	28	3–5	40–100	750–950	30–50	75–150		10–15	30–60	0.22
14	19	2–4	40–120	1000–1250	30–50	75–150		10–15	30–60	0.25

Conditions: Guidelines based on average operation. Data are for straight cutting using a three-hose torch perpendicular to the plate in all axes. Preheat pressures given are at regulator and based on 25 ft or less of 5/16-in. ID hose supplying a single torch. MAPP Gas: 2 to 15 psig; oxygen: 5 to 50 psig (positive-pressure torch) or 30 to 50 psig (injector torch). Data for MAPP tips: Type SP (one-piece) and Type FS (two-piece) (SP = single piece, FS = fine spline).

TABLE 5-6 MAPP Gas for machine flame-cutting using high-speed cutting tips

| Plate Thickness, in. | MAPP Tip Size No. | Cutting Speed, in./min | Oxygen | | | | | MAPP Gas | | |
| | | | Cutting | | Preheat | | | | | Kerf Width, in. |
			Pressure, psig	Flow, scf/h	Pressure, psig	Flow, scf/h		Pressure, psig	Flow, scf/h	
1/4	68	24–31	60–70	55–65	5–10	8–25		2–10	3–10	0.05
3/8	65	23–30	70–80	60–80	5–10	10–25		2–10	4–10	0.055
1/2	60	22–29	80–90	75–95	5–10	12–25		2–10	5–10	0.06
3/4	56	20–26	80–90	115–130	5–10	12–25		2–10	5–10	0.065
1	56	18–24	80–90	115–130	5–10	12–25		2–10	5–10	0.065
1 1/4	54	16–22	80–90	155–170	10–20	20–38		2–10	8–15	0.08
1 1/2	54	15–20	80–90	170–180	10–20	20–38		2–10	8–15	0.08
2	52	14–19	80–90	215–255	10–20	20–38		2–10	8–15	0.09
2 1/2	52	12–17	80–90	215–255	10–30	20–38		4–10	8–15	0.09
3	49	10–14	80–90	310–365	10–30	20–38		6–10	8–15	0.10
4	44	9–13	80–90	420–510	10–30	30–38		6–10	10–15	0.12
6	44	7–11	80–90	420–510	10–30	30–38		10–15	10–15	0.12
8	38	6–9	80–90	590–720	15–30	30–50		10–15	15–20	0.15

Conditions: Guidelines based on average operation. Data are for straight cutting using a three-hose torch perpendicular to the plate in all axes. Preheat pressures given are at regulator and based on 25 ft or less of 5/16-in. ID hose supplying a single torch. MAPP Gas: 2 to 15 psig; oxygen: 5 to 50 psig (positive-pressure torch) or 30 to 50 psig (injector torch). Data for MAPP tips: Type HS (one-piece) and Type FH (two-piece) high-speed cutting tips (HS = high speed, FH = fine spline high speed).

TABLE 5-7 Acetylene for hand-cutting

Material Thickness, in.	Acetylene Tip Size No.	Hose ID Size, in.	Acetylene Pressure, psig*	Oxygen Pressure, psig*	Kerf Width, in.
⅛	00	5/16	3	30	0.06
¼	0	5/16	3	30	0.75
⅜	1	5/16	3	30	0.95
½	1	5/16	3	40	0.95
¾	2	5/16	3	40	0.110
1	2	5/16	3	40	0.110
1¼	3	5/16	3	40	0.110
1½	3	5/16	3	45	0.130
2	3	5/16	3	45	0.130
2½	4	5/16	3	50	0.145
3	4	5/16	3	50	0.165
4	5	5/16	4	50	0.165
5	6	5/16	5	55	0.190
6	7	5/16	6	55	0.190
7	7	5/16	6	55	0.190
8	8	5/16	6	55	0.220
10	8	5/16	6	65	0.220
12	8	5/16	6	75	0.220

* Pressures given are at regulator and based upon up to 25 ft of hose.
Data based on Airco style 124 Acetyle tip for light preheat and clean, low-carbon steel surfaces. Other manufacturers provide similar data but different tip size designations since tip size numbers have never been standardized.
Preheat-gas ratios are preheat-oxygen/preheat-acetylene = 1.1:1.

If you are not familiar with the part you are making, make sure which is the outside of the part and which is the inside. There are many instances where scrap was made to precise dimensions and the part was cut undersized. The operators thought that the scrap was the part and the part was the scrap, and put the kerf on the wrong side.

Most larger cutting machines designed to produce parts from a pattern of some kind (they are called *shape cutters* as opposed to *ripping and squaring machines*) have built-in controls to adjust for kerf width. If you are using a hand-held torch and have no data table for your tips and fuel gas to make kerf corrections, you can still find out what your kerf will be. Simply cut a section through the steel you will cut (along an edge or on a piece of scrap). Do it under the same conditions you will use to cut the parts. Then

TABLE 5-8 Acetylene for machine flame-cutting

Plate Thickness, in.	Acetylene Tip Size No.	Cutting Speed, in./min	Cutting-Oxygen Pressure, psig	Preheat-Acetylene Pressure, psig	Cutting-Oxygen Flow, scf/h	Preheat-Oxygen Flow, scf/h	Total Oxygen Flow, scf/h	Preheat-Acetylene Flow, scf/h	Kerf Width, in.
⅛	00	23	30	1½	25	12	33	11	0.060
¼	0	20	30	2	42	15	57	14	0.075
⅜	1	19	40	3	62	22	84	20	0.095
½	1	17	40	3	78	22	100	20	0.095
¾	2	15	40	3	110	24	134	22	0.110
1	2	14	50	3	130	24	154	22	0.110
1½	3	12	50	3	165	25	190	23	0.130
2	3	10	50	3	180	25	205	23	0.130
2½	4	9	50	3	235	28	263	25	0.145
3	5	8	55	4	280	33	313	31	0.165
4	5	7	60	4	350	33	383	31	0.165
5	6	6	60	5	410	47	457	43	0.190
6	6	5	60	5	445	47	492	43	0.190
7	6	4½	65	6	505	47	552	43	0.190
8	7	4	70	6	600	54	654	49	0.220
10	7	3½	70	6	680	54	734	49	0.220
12	8	3	75	6	830	62	892	56	0.260
14	9	3	85	6	1135	72	1207	65	0.295
15	10	3	85	6	1230	72	1302	65	0.345

Conditions: Guidelines based on average operation. Data are for straight cutting using a three-hose torch perpendicular to the plate in all axes. Preheat pressures given are at regulator and based on 25 ft or less of ⅜-in. ID hose supplying a single torch. Acetylene: 1 to 15 psig, oxygen: 30 to 85 psig (positive-pressure torch). Preheat-gas ratios: Preheat-oxygen/preheat-acetylene = 1.1:1. Data for tip: Airco style 144 medium-preheat for rusty or painted surfaces on low-carbon steels, for straightline or shape cutting. Style 144 cutting tips are also used for cutting steel castings. Sizes 11 through 13 have straight-bore cutting holes for rivet washing and gouging.

TABLE 5-9 Natural gas for machine flame-cutting

Plate Thickness, in.	Natural Gas Tip Size Drill No.	Cutting Speed, in./min	Cutting-Oxygen Pressure, psig	Cutting-Oxygen Flow, scf/h	Preheat-Oxygen Flow, scf/h	Preheat–Natural Gas Flow, scf/h	Kerf Width, in.
⅛	68	15	20	20	15	10	0.060
¼	60	15	25	45	15	10	0.075
⅜	60	15	30	45	15	10	0.095
½	60	15	35	55	15	15	0.095
¾	53	12	35	113	15	15	0.110
1	53	11	35	113	15	15	0.110
1½	53	8	35	113	15	15	0.130
2	53	8	35	113	15	15	0.130
2½	46	8	35	235	15	15	0.145
3	46	8	35	285	15	15	0.165
4	46	7	40	335	15	15	0.165
5	39	5	40	335	15	15	0.190
6	39	4	40	360	15	15	0.190
7	39	3	40	480	20	20	0.190
8	30	3	45	580	20	20	0.220
10	30	3	45	650	25	25	0.220
12	18	2	40	700	25	25	0.260

Conditions: Guidelines based on average operation. Data are for straight cutting using a three-hose torch perpendicular to the plate in all axes. Preheat pressures given are at regulator and based on 25 ft or less of ⅜-in. ID hose supplying a single torch. Natural gas: less than 1 psig; oxygen: 20 to 55 psig (injector torch). Data for tip: Tip suitable for low-pressure natural gas and injector torch.

TABLE 5-10 MAPP Gas for very heavy cutting

Thickness of Steel, in.	Oxygen Flow, scf/h	Diameter of Cutting Orifice, in.	Cutting-Oxygen Pressure at Torch, psig
12	1000–1500	0.147 –0.221	26–56
16	1300–2000	0.1935–0.290	25–49
20	1600–2500	0.200 –0.221	25–48
24	1900–3000	0.221 –0.332	22–46
28	2200–3500	0.230 –0.400	18–44
32	2500–4000	0.250 –0.450	16–40
36	2900–4500	0.290 –0.500	12–38
40	3400–5000	0.330 –0.550	10–33
44	3800–5500	0.375 –0.550	9–29
48	4500–6000	0.422 –0.600	7–27

measure the kerf width with a machinist's rule.

5-9
MAKING A STEADY CUT

Once the cut has been started, move the torch along the line of cut with a smooth, steady motion. Maintain a constant flame-tip-to-workpiece (coupling) distance. Move the torch at a speed that produces a light ripping sound and a smooth spark stream. Be sure that your hoses are not under your sparks or your cutting oxygen (burners have been known to cut through their own hoses).

With a little practice you can learn to cut reasonably straight lines and simple curves freehand. There are some aids to cutting lines and circles with great precision even when you use a hand-held torch. We'll describe them later on.

A lot of welders and burners drape their hoses over their shoulder. That's a good way to get a tired shoulder, and if anybody jerks on the hose you will get pulled over. The ideal place to keep hose is hung overhead on rods with sliding hangers very much like big curtain rods and hangers. This is precisely what is done with the hoses on larger flame-cutting machines. (You can easily make your own hose hangers.) Barring that, throw the hoses back and let them trail behind you. Don't keep them under your workpiece or your feet.

Hold the torch with both hands (Fig. 5-15) and make sure that your arms are free to move the torch along the line of cut without difficulty. The arm and hand controlling the cutting-oxygen lever should be held close at your side with your forearm and torch extending straight out. The other hand should hold the torch further toward the tip, with the back of your wrist resting on the workpiece for stability. It is this arm and hand that acts as a pivot point and guide for the torch.

The cutting line should be marked with chalk or a scribed line. You can use a sharp tool-steel punch like a pencil to scribe or draw the line on steel plate or structurals. Some large cutting machines include similar punches to locate hole positions and do other marking jobs in addition to cutting the steel.

When cutting long sections you will have to stop from time to time and reposition yourself. If the cut is stopped and then restarted, a blowhole will occur each time you pick up the stopped cut. You can avoid this ugly defect by moving the torch a short distance into the scrap area perpendicular to the line of cut. The cut can then be restarted by swinging the torch back into the line of cut or kerf at a point just behind the location where the cut was stopped.

If a long cut is made to separate two pieces of steel, both of which

FIGURE 5-15 Use both hands to control the torch. When possible, steady your hands with your forearms.

will be used (and thus you have no scrap area), blowholes can be avoided by stopping the torch travel before releasing the cutting-oxygen lever. That also will straighten out the front edge of the cut inside the kerf. What you are doing is turning the drag angle into a straight vertical line. It helps produce drop cuts.

5-10 TROUBLESHOOTING YOUR CUTS

The surface of the cut holds clues to most of the problems you can have when flame-cutting. Figs. 5-16 through 5-28 are useful for troubleshooting your own work. If you want more help, write for the American Welding Society's publication AWS C.4, 1-77, *Criteria for Describing Oxygen-Cut Surfaces,* American Welding Society, 2501 N.W. Seventh Street, Miami, FL 33125, or call (305) 642-7090. Every burner should buy a personal copy of this useful AWS publication.

The AWS brochure illustrates flatness, drag lines, notches, angularity, top-edge rounding, and slag problems. A three-dimensional surface-roughness guide comes with the brochure so that you cannot only see but even feel the difference between good and bad cuts.

Using Kerf Surfaces
Following are some illustrations (Figs. 5-16 through 5-28) from the AWS roughness guide for the cut kerf surface, courtesy of the American Welding Society.

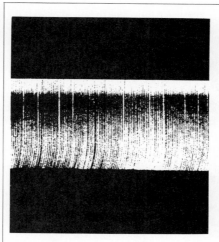

FIGURE 5-16 The perfect cut. The perfect cut has a regular surface with slightly sloping drag lines. A slight amount of scale at the top of the cut is caused by the preheat flames and is easily removed. This surface can be used for many purposes without finish-machining.

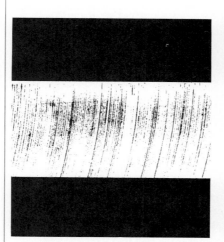

FIGURE 5-17 Production cut. The production cut has moderately sloping drag lines and a reasonably smooth surface. For production purposes, a cut of this type represents the best combination of quality and economy.

Cracking Problems
A problem that you will run across from time to time is cracks in the edges of the steel after you have flame-cut it. Or sometimes the flame-cut edges will be so hard that it is difficult to machine them with anything except a grinder. Both problems have the

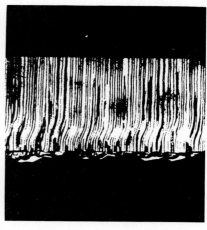

FIGURE 5-18 Dirty tip. Dirt or scale in the tip will deflect the oxygen stream and cause one of the following problems: excess slag on the steel, an irregular cut surface, pitting, or undercutting.

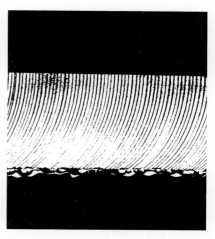

FIGURE 5-19 The angle of the drag lines illustrated here indicates an extremely fast cutting speed. The top edge is good and the cut face is smooth. However, slag adheres to the bottom edge, and there is danger of losing the cut. Not enough time is allowed for the slag to blow out of the kerf. The cut face is often slightly concave.

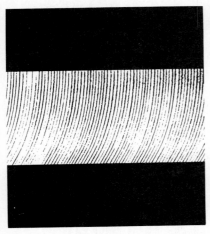

FIGURE 5-20 Slightly too fast cutting speed. The drag lines on this cut incline backwards, but a "drop cut" is still attained. The top edge is good; the cut face is smooth and slag-free. This quality is satisfactory for most production work.

FIGURE 5-22 Slightly too slow cutting speed. The cut is high-quality although there is some surface roughness caused by vertical drag lines. The top edge is usually slightly beaded. This quality is generally acceptable, but faster speeds are more desirable because the labor cost for this cut is too high.

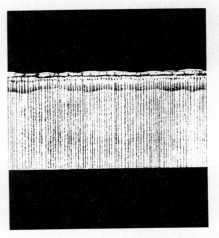

FIGURE 5-24 Tip too high off the steel. The top edge is beaded or rounded, the cut face is not smooth, and often the face is slightly beveled where preheat effectiveness is partially lost due to the tip being held so high. The cutting speed must be reduced because of the danger of losing the cut.

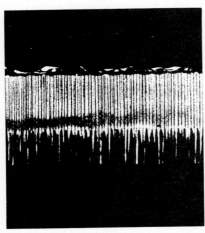

FIGURE 5-21 Extremely slow cutting speed. Pressure marks on the cut face indicate too much oxygen for the cutting conditions. Either the tip is to big, the cutting-oxygen pressure is too high, or the speed is too slow as shown by the rounded or beaded top edge. On reducing the cutting-oxygen volume to the correct proportions for the thickness of the cut, the pressure marks will recede toward the bottom edge until they finally disappear.

FIGURE 5-23 Tip too close to the steel. The cut shows grooves and deep drag lines, caused by an unstable cutting action. Part of the preheat cones burned inside the kerf, where normal gas expansion deflected the oxygen cutting stream.

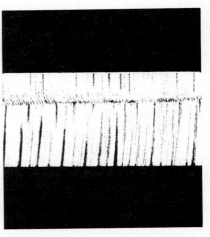

FIGURE 5-25 Too much cutting oxygen. The cut shows pressure marks caused by too much cutting oxygen. When more oxygen is supplied than can be consumed in oxidation, the remainder flows around the slag, creating gouges or pressure marks. Correct this fault by reducing the cutting-oxygen pressure, increasing the speed, or using a smaller tip. As the cutting-oxygen volume nears the correct proportion for the section thickness and tip, the pressure marks recede closer to the bottom edge until they finally disappear.

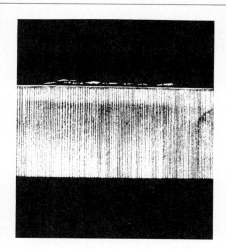

FIGURE 5-26 To much preheating. The cut shows a rounded top edge caused by too much preheat. Excess preheating does not increase cutting speed. it only wastes gases.

FIGURE 5-27 Good-quality bevel cut. The top edge of the cut is excellent and the cut face is extremely smooth. Slag will be easy to remove and the cut part will be dimensionally accurate. This cutting speed is slower than square cutting because the preheat flames are working through a thicker steel section.

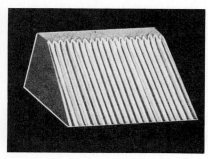

FIGURE 5-28 Poor-quality bevel cut. The most common fault is gouging, caused by either excessive speed or inadequate preheat flames. Another fault is a rounded top edge, caused by too much preheat, indicating excessive gas consumption.

same source, high-carbon steels. It's a metallurgical problem, and with a little help from a plant metallurgist, you can fix it.

Carbon is the primary element in steel that makes it stronger than wrought (nearly pure) iron. As the carbon content goes up, the strength and hardness go up, too—up to a certain point at least. One of the basic ways of classifying steels is by their carbon content into low-carbon steels, medium-carbon steels and high-carbon steels. The steel sheet that auto producers use to form into fenders, for example, is low-carbon steel. A hardened shaft or axle might be made out of medium-carbon steel. Auto leaf springs are made out of high-carbon steels.

A low-carbon steel has between 0.05 percent and 0.20 percent carbon in it, depending upon the grade. A medium-carbon steel might have somewhere between 0.25 percent to 0.35 percent carbon. A high-carbon steel would have 0.35 percent carbon up to about 1.20 percent carbon, the practical upper limit for carbon in high-carbon tool steels or very hard bearing steels. (These definitions are broad and will vary with what you do, but for flame-cutting they are adequate.)

When you get more than 2.0 percent carbon in the metal, graphite starts appearing and you no longer have a steel at all, but a cast iron, with entirely different properties. (Cast irons typically have carbon contents around 3.5 percent up to 4.0 percent.)

The more carbon there is in the steel, the faster it will harden if it is heated and quickly cooled. Sometimes when high-carbon steels cool too quickly they crack. At the worst they get a lot harder.

Your preheat torch heats the steel, which quickly cools after the torch passes on. You are heat-treating the edge of the steel you cut. The edge-hardening effect won't be noticeable unless you are cutting a medium- to high-carbon

steel or an alloy steel that is designed to get much harder when heated and air-cooled.

You seldom will have troubles with edge hardening or cracking when flame-cutting steel with a carbon content of 0.25 percent or less. But if you cut steels with a carbon content of 0.35 percent or more, you should preheat and postheat the steel to prevent hardening and cracking of the steel along the cut surface.

You will run into the same problem when welding high-carbon steels (which we'll discuss in several of the welding chapters later on in this book). The same solution is used in welding as in flame-cutting. The higher the carbon content, the more important it is to use preheating and possibly postheating to avoid cracking.

One way of avoiding the problem right from the beginning is to heat up all the steel uniformly before you cut it. That lets the plate cool down more slowly because not only the edge but the whole piece has more heat in it that it must get rid of to cool down. That takes more time. Sometimes the cooling rate of a steel is slowed even more by continuing to heat it a little while it is cooling. That lets it cool very slowly.

Elements other than carbon, especially in alloy steels, will give similar problems, but to a lesser extent. For example, some wear- and abrasion-resistant steels have a very high percentage of manganese in them. Some of them have 14 percent manganese and 1.5 percent carbon. Other alloy steels may have less than 1 percent alloying elements, but the hardening effect of these elements can still be severe. What's more, different steels (and other metals, too) have different recommended preheat and postheat temperatures. The time and temperature you select for preheating and postheating an alloy steel is critical.

Welders have exactly the same problem because they also dump a lot of intense heat into steels in a very small volume, followed by a rapid cooling rate. We will talk more about pre- and postheating when we get to other chapters. See Chap. 16, Table 16-19, for further details.

Fortunately, most of the steels you will cut (and most of the steels used in industry) are low-carbon steels. (More ship hulls and auto fenders are made than coil springs and tool-steel dies.) Right now, we simply want you to know about the possible problem so you can recognize it if it occurs and go get the right kind of help, usually a production manager or the plant metallurgist.

5-11 OTHER CUTTING METHODS

Now that you've learned the basics of flame-cutting, let's look at some of the methods that use flame-cutting to get special jobs done on different kinds and shapes of iron and steel. Several of these processes use special tips. A few don't even use the conventional cutting torch, but instead use special equipment (one even uses a straight piece of plumber's pipe). Nevertheless, the principles that you've learned still hold for these other methods.

Flame Gouging

Gouging is used extensively to remove deep defects in steel revealed by flame-scarfing plate in a steel mill, or by radiographic, magnetic, and other nondestructive inspection methods.

The most common gouging job is to remove tack welds that initially hold work together until the welder can make the final weld. Gouging is also used for removing weld defects, blowholes, or sand inclusions in castings; for welds made on temporary brackets or supports; for removing flanges from piping and pressure-vessel heads; and for removing old tubes from boilers. Flame gouging is used in demolition work and preparing specially shaped plate edges for welding.

Manual gouging with an oxy-fuel cutting torch (Fig. 5-29) is a rapid metal-removal method. Cutting speeds from 12 to 60 in./min [5.1 to 25.4 mm/s] are possible. Using MAPP Gas, a speed of 24 in./min [10 mm/s] across ¾-in. [6-

mm] thick plate will make a gouge 5/16 to 3/8 in. [7.9 to 9.5 mm] wide. Using acetylene, a speed of 12 to 18 in./min [5.1 to 7.6 mm/s] will produce a gouged channel 5/16 to 3/8 in. [8 to 10 mm] wide. A gouge made at this speed would be made with an oxygen-bore diameter of about 1/8 in. [3.2 mm]. The tip is angled at about 25° to 30° to the surface of the plate. The oxygen pressure should be set at about 25 psig [170 kPa] and the acetylene pressure for the preheat flames would be set at 7 psig [48 kPa].

For a gouged groove ranging from 9/16 to 5/8 in. [14 to 16 mm] wide, the tip used would be slightly greater than 7/32 in. [5.5 mm] in diameter at the exit end of the cutting-oxygen orifice. Oxygen pressure would be set at 25 psig [170 kPa] and the acetylene pressure at 10 psig [69 kPa]. The forward cutting speed would be about 20 to 30 in./min [8.5 to 12.5 mm/s].

Flame gouging also is used to prepare J-grooved and U-grooved joints in thicker plate for welding on both sides. The first step in preparing a weld joint by flame gouging is to butt two square-edge plates together and completely weld one side. Then turn the plate over with a forklift or overhead crane and use a gouging tip to groove out the opposite side of the joint.

The gouge will be taken down to a depth such that the original square edge of the butted plates is removed and the sound weld metal on the other side is exposed. The groove is then welded up, joining sound weld metal on both sides of the plate.

This procedure has the benefit of permitting you to align precisely and hold two plates with the first weld on the back while you flame gouge the other side. If you did not do it this way, the heat of the gouging operation could easily warp the two plates and you would have a terrible time getting them back into alignment.

FIGURE 5-29 Flame gouging is being used to remove the top skin of a steel billet prior to finish-rolling to check for metal soundness (no porosity or cracks) underneath. You also can remove bad weld metal with flame gouging. This torch is a special scarfing torch.

Depending on your skill, gouging accuracy can be controlled to width and depth tolerances of about 1/16 in. [1.6 mm]. You can buy flattened cutting tips specially designed for removing metal by flame gouging. Tips designed to remove rivets also work well.

Flame gouging is a quiet process compared with using a pneumatic chipping hammer or an electric-arc process called air-carbon-arc cutting. Maintenance costs are much less than those for electric-arc cutting processes or for machine tools that require repeated sharpening. Flame gouging also has the advantage that the groove produced is smooth at the entrance and exit as well as at the edge contours.

Flame Drilling

Flame drilling is used to make relatively shallow holes in bars or heavy steel sections. Such holes might be used for centering massive pieces of steel in a lathe or inserting tools for making seamless pipe. The operation is similar to making a piercing start in a plate. When the steel is preheated to its kindling temperature, withdraw the tip about 1/2 to 1 in. [13 to 25 mm] from the surface and turn on the cutting oxygen.

Rotate the tip to make the desired hole size. Use straight-bore tips. Continue the hole until you think it's deep enough. It is very difficult to drill a blind hole to a precise depth using this method.

If you need a precision hole, you should burn out a smaller-diameter hole and finish it with a larger mechanical drill of the precise bit size.

Removing Rivets

You can easily remove rivets and rusted bolt heads with an oxyfuel cutting torch. It's possible to do this without damaging the holes or the plate through which the rivets pass.

A torch with a conventional tip can be used to remove rivets, or you can use a special rivet-washing tip that will make the work easier. When using a conventional tip (Fig. 5-30), cut a slot in the rivet head from the top to the underside of the rivet head. As this cut nears completion, withdraw the torch about 1½ in. [40 mm] from the rivet to prevent burning the plate under the rivet head.

Next, just as the cut slot reaches the plate, rotate the torch through a small arc so that the high-pressure cutting oxygen will strike the rivet shank to cut off one half of the rivet head. Immediately rotate the torch again in the opposite direction and cut off the other half of the rivet head. Doing it this way will not damage the plate. In fact, the plate will not even become very hot.

You can then hammer out the rivet shank or even use the same technique used for flame drilling to burn out the body of the rivet. This last step requires a lot of skill. Be careful of flying slag.

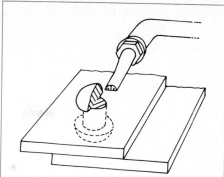

FIGURE 5-31 Rivets are much easier to cut away if you have a special cutting tip designed for this work. It is a bent tip with a half-moon cross section.

Another way to remove rivets or bolt heads is to use a tip specially designed for the job (Fig. 5-31). Put the flat part of the tip directly on the plate next to the rivet head and, after preheating it, use the high-pressure cutting oxygen to slice off the entire head in one step.

There's still another way to remove rivets and bolt heads, called

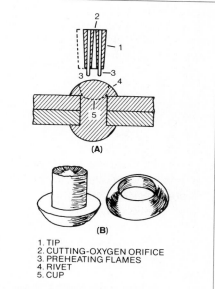

1. TIP
2. CUTTING-OXYGEN ORIFICE
3. PREHEATING FLAMES
4. RIVET
5. CUP

FIGURE 5-32 Rivets can be washed out of a plate by first burning a hole through the rivet head. A hammer and punch will then knock the rest of the rivet out of its hole. Very skilled burners can wash rivets out of a hole using only a torch, without widening the hole itself.

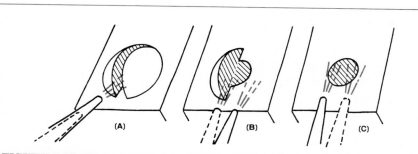

FIGURE 5-30 Rivet cutting technique is illustrated. (A) Slice the rivet in half. (B) Then wash away one side. (C) Then wash away the remaining side. Hold the tip almost parallel to the surface of the plate.

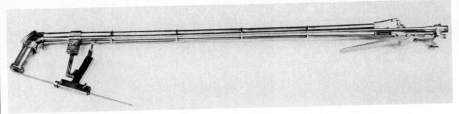

FIGURE 5-33 A scarfing torch is a special type of cutting torch that feeds a steel wire into the flame to increase the cutting speed to skim off the surface of the metal very quickly.

washing. As you can see in Fig. 5-32, a special tip with a divergent cutting-oxygen orifice is used. This tip operates at relatively low pressures and removes large amounts of metal rapidly. In the operation, the preheat flames are played on the rivet head, and when it reaches the kindling temperature the cutting oxygen is turned on slowly. As the cut starts, the tip is moved so that it washes out a cup shape in the top of the rivet head.

A sharp blow with a punch driven into the cup will shear the remaining metal and drive out the rivet. The plate will not be damaged by this process unless the rivet-washing tip is too big or the cutting-oxygen pressure is too high, or you preheat the rivet too long or move the torch too slowly.

Flame Scarfing

Flame scarfing is a very special cutting process used to flame-machine a thin layer of metal from a piece of thicker steel. It usually is done in steel mills, often on thick billets that will subsequently be rolled into plate or sheet.

The scarfing operation removes surface defects that otherwise would show up on the finish-rolled steel. Before the process was used, a crew of men had to remove hot billet defects with pneumatic chipping hammers. The flame-scarfing process removes them at speeds greater than 125 ft/min [0.63 m/s] and requires only one operator. Sometimes scarfing is done automatically by a machine that smooths off all four sides at once.

Scarfing requires a special torch (Fig. 5-33). Manual scarfing usually uses straight-bore cutting tips. Machine scarfing uses divergent-bore tips. The oxygen bores on machine-scarfing tips are usually rectangular instead of round to cover a wider section of the billet or slab. The same technique is used in steel foundries to remove risers and other projections from cast steel left by the moulding process.

The flame-scarfing torch is held so that the high-pressure cutting oxygen strikes the steel at an angle of from 15° to 30°, depending upon the depth to which the scarf is being cut. A sharper angle produces a deeper scarf. The depth of metal removed varies between 1/32 amd 1/8 in. [0.8 to 3.2 mm]. The torch tip is held at an angle of about 25° to the line of motion of the scarfing operation. This blows the slag resulting from the scarfing operation to one side.

Starting the scarfing cut is the most difficult part of the job. When the high-pressure cutting oxygen is first turned on, it sometimes gouges the surface instead of washing off a wide, thin layer. A flat surface is washed on the steel by using a steel starting rod with the special torch. This steel rod is usually 3/16 in. [4.8 mm] in diameter and is held in the path of the preheat flames. The end of the rod is quickly heated to the kindling temperature and when the cutting oxygen is turned on, the rod is immediately burned. Manipulation of the cutting-oxygen lever on the torch also advances the steel starting rod into the cut area.

The burning steel rod transfers heat to the surface being scarfed and the plate or billet surface continues to burn under the cutting oxygen. The start of the scarfing operation is so quick that there is no time for the flame to gouge the surface.

Cutting Bars and Pipes

Round bars and pipe can easily be cut, but you must know what you're doing.

Making a start The lack of sharp edges makes it difficult to start the cut, but any of the three methods below will do the trick.

■ Get the cut started by raising a chip on the side of the round section with a hammer and cold chisel. This gives you an edge notch where the cut will start quickly.

■ Play the preheat flames back and forth along the line of cut in a small arc on the circumference of the bar or pipe. Preheat the area longer than you would cutting other shapes, so that the cut starts quickly when you squeeze the cutting-oxygen lever.

■ Use a carbon-steel starting rod to assist the cut start. Start by preheating the round section along the line of cut, then preheat the mild-steel starting rod while it's held in line with the cut, just touching the bar or pipe. When you depress the cutting-oxygen lever, the starting rod will immediately burn and carry part of the bar or pipe wall along with it, making a notch you can continue as you slice into the section.

Types of cuts The shape of the edge you create when cutting pipe lengths will vary, depending on how you point your torch.

SQUARE-EDGE CUTS The simplest pipe-cutting job is to make a square-edged cut around the circumference of a pipe, as you would to cut off a length for use. Using chalk or a center punch, make a line around the pipe that you will cut. Start the cut the same way you would start it on a solid round bar.

The trick after that is to keep the torch tip pointed at the interior centerline of the pipe at all times (Fig. 5-34). Simply think while you work that the cutting oxygen is always an extension of the pipe radius.

It may be necessary to stop the cut several times and rotate the pipe to bring an uncut surface to the top. There are special pipe-cutting machines that avoid that problem and rotate completely around the section in one pass. Pipe-fabricating shops also have motorized turning rolls that rotate the pipe at a predetermined speed while the operator holds the torch in one position to cut (or weld) in a single down-hand position. However, a welding school is not likely to have this equipment. What's more, you won't always have these fancy machines when doing pipe work in the field.

BEVELED EDGES Bevel cutting a pipe with a hand-held torch will give you excellent general practice in free-hand control of a cutting torch. After a section has been cut to length, the torch is simply held at the angle that will create the degree of bevel called for and the outside edge is beveled. The difficulty is that the cutting position constantly changes. You should practice this job until you get good at it. If you can cut and bevel pipe, you can use a torch on almost any shape.

MITERED EDGES Some pipe connections require an angled instead of a square-cut pipe end. This is called a *miter* cut (Fig. 5-35). Miter cuts are made by holding the torch tip pointed at the centerline of the pipe. The difference is that the cut is made along a plane that intersects the pipe at an angle, such as 30° or 45°.

SPECIAL CUTS Other special cuts are made in pipe to join it to larger or smaller pipe connections. These special cuts use patterns called *templates* that have

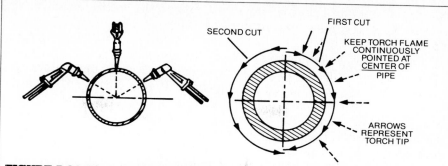

FIGURE 5-34 Pipe cutting is difficult for beginners. You have to constantly change the angle of your torch, keeping it pointed at the centerline of the pipe.

already been calculated so that, for example a 4-in. branch connection can enter an 8-in. pipe at an angle. (The hole to be cut in the 8-in. pipe is not a circle, it's a rather complex shape). Pipe-fabricating and welding is a special category of very skilled work and we'll give you a complete chapter on it later on.

Cutting Cast Iron

Much higher preheat is needed to cut cast iron than an equivalent steel section. Thinner cast-iron sections are even more difficult to keep heated than thicker sections.

Increase the oxygen pressure about 25 percent higher for cast-iron sections than for an equivalent thickness of steel section. Adjust the preheat flames to a carburizing flame setting, making the setting with the cutting-oxygen lever open. Then close the cutting-oxygen lever and start preheating.

Begin preheating the cast-iron section along the widest possible area through which your cut will pass. You often can hold the torch head so that the preheat flames lie flat along the surface, along the line of cut. In general, the hotter you get the section, the faster you will start cutting.

When your preheat line is up to temperature, hold the cutting tip at a 45° angle in line with the cut, pointing the tip backward (away from the uncut section). The reason is that cast iron will produce a very large drag angle and you want to keep the bottom

of the cut preheated while the top of the cut moves ahead of it. Figure 5-36 shows you how.

Heat a semicircular area about ½ to ¾ in. [13 to 20 mm] in diameter on the edge of the section until the cast iron is actually molten. Do it by swinging the flames in a tiny arc back and forth over the semicircle. When the cast iron begins to bubble, press the cutting-oxygen lever for an instant and blow off the slag. That will carve out a tiny starting notch for you.

Keep preheating the tiny notch, and when you get it hot enough, press the cutting-oxygen lever wide open and begin moving the torch along the line of cut with the tip at about a 45° angle to the surface, still pointing the tip backward away from the uncut section. Continue moving the tip from side to side in a small arcs

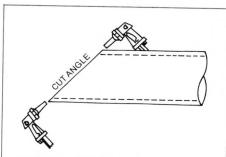

FIGURE 5-35 Miter cutting makes pipe work even more difficult. The torch not only has to point at the centerline of the pipe, but you also have to hold a constant cut angle as the torch is moved.

to concentrate more heat in the section, just as you did to start the cut.

When cutting has progressed all the way through the top of the section, don't stop cutting. The bottom of the cut will be lagging behind your top surface. Hold the tip at the finish of the top cut and gradually turn the tip up until it is at an angle of about 75°. That will draw the bottom cut along to catch up with the top section. Finally, finish the bottom of the cut and then withdraw your torch.

When cutting very thin cast-iron sections (up to about 1¼ to 2 in. [32 to 51 mm] thick), you

should do the cutting in the following sequence to keep the cut going. Start the cut as we've already described. As the cut progresses, watch the top surface of iron. If it goes dark, or if the preheat flame does not seem to progress along the line of cut, advance the tip about ¼ in. [6 mm] or so ahead of the cut to preheat another semicircular area. Then cut off the molten surface just as in starting the cut, keeping the cutting oxygen on.

Then move the tip backward a little to the kerf, slowly tilting the flame downward in a more vertical direction to cut through

the lower section, and the cut will progress. As long as the cut keeps moving and the metal remains cherry red, keep going. If the cast iron begins to cool, perform the same preheating steps that you used to start the cut.

Heavy cast-iron sections are a little easier to cut. The heat of the reaction will help preheat the cast iron. Just like cutting a thin section, there always will be a considerable amount of drag when cutting cast iron, even with your tip at 90° to the surface. Therefore, when you get to the opposite edge of the cut, hold the tip at about the drag angle so that the lower end of the cut continues and ultimately breaks through the opposite bottom side.

If you lose a cut while cutting cast iron, release the cutting oxygen, move back about ½ in. [13 mm] along the line of cut and preheat overlapping semicircles with the preheat flame at the point where you stopped cutting. That will bring the section back up to temperature so that you can start cutting again.

You must maintain plenty of heat in cast iron whenever you cut it and preheat the entire volume of metal evenly around the area that you will cut. Very large sections (several feet thick) sometimes may have to be preheated with a large heating torch for several hours before you start cutting.

Molten cast iron normally flows faster than water, but some highly alloyed irons produce a very sticky slag. You can increase the fluidity of cast-iron slag and make the section easier to cut by feeding a carbon-steel welding rod or even an iron bar into the cut underneath the preheat flames and cutting oxygen. The iron will dilute the alloy cast iron, add more molten iron to the slag, and also add more heat to the section.

Oxygen Lancing

Lancing is a special piercing technique used to make holes in very thick plate or steel billets. The

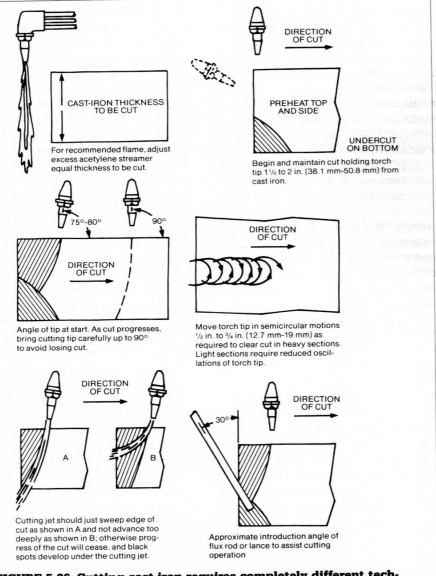

FIGURE 5-36 Cutting cast iron requires completely different techniques than those used for steel. Follow these directions carefully and you will get good results.

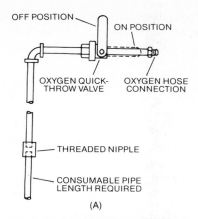

OFF POSITION — ON POSITION

OXYGEN QUICK-THROW VALVE

OXYGEN HOSE CONNECTION

THREADED NIPPLE

CONSUMABLE PIPE LENGTH REQUIRED

(A)

(B)

FIGURE 5-37 An oxygen lance will cut very thick sections, even steel over 10 ft thick. (A) This drawing shows how to fabricate a lance. (B) This photo shows one in use, in combination with a cutting torch. The lance is cutting a starter hole and the torch is completing the cut.

oxygen lance is made of a length of small-diameter black-iron pipe (Fig. 5-37A). The ¼ to ½ in. [6 to 13 mm] diameter pipe is connected with fittings and nipples and a quick-throw oxygen-valve switch. Oxygen is fed through hose to the pipe with a regulated pressure of 80 to 90 psig [550 to 620 kPa]. The lance pipe will be burned up in the process.

A good rule of thumb to follow for figuring out how much lance pipe is needed is to use 2.5 in. [63.5 mm] of pipe for every 1 in. [25.4 mm] of metal thickness to be pierced. For example, if you pierce a 3-ft-thick [0.91-m] billet you will be piercing 3 × 12 in. [305 mm] = 36 in. [92 mm] of

steel, so you will need 36 × 2.5 in. = 90 in. [230 mm] of lance pipe, plus another 6 to 8 in. [152 to 203 mm] of pipe so that the threaded pipe connections will not be damaged.

A spatter-protection shield will be required to shield you from splashing slag. A very handy shield can be made out of a steel bucket. Cut a ½-in. [13 mm] hole in the bottom of the bucket. Turn the bucket upside down, slide the lance pipe through the bottom of the bucket so that open bucket faces toward the steel. All you have to do after that is hold the bucket down over the surface to be lanced and the bucket will catch the slag and sparks.

When you are ready to start piercing, use a hand-held cutting torch to preheat a spot on the billet and the lower end of the pipe to near melting. As soon as the spot on the billet is very hot, pull away the torch, slap the lance bucket on the hot spot, and flip the quick-action oxygen valve. Then push the pipe firmly and smoothly through the thick section as it burns the steel (Fig. 5-37B).

A slight twisting motion applied to the lance pipe will produce a larger hole than the pipe diameter to permit slag to be blown out. As long as you keep the bucket over the hole you will be protected from flying slag. As soon as the hole is completely through the plate, close the oxygen valve on the lance. Remove what is left of the pipe. Remove the bucket shield and clean the slag off the top surface of the billet with a chipping hammer. Now the thick section is ready for oxygen cutting.

You will need a high-flow cutting tip for this job, and a regulator, fuel gas, and oxygen supply to match. You should set up the job with fuel-gas and oxygen manifolds. Another thing to keep in mind is the amount of slag you will produce cutting such a thick section. A thick layer of sand under the billet, built to dam up

flowing molten metal, is a good idea. Table 5-10 lists data for cutting very thick sections with MAPP Gas.

If you lose the cut when lancing a deep hole, it is very difficult to start cutting again. The solution is to stuff a handful of steel wool down the hole. Steel wool ignites and burns like crazy in pure oxygen. Steel wool will start the rest of the hole burning like kindling starts a log fire. You can substitute commercial steel-wool pot-cleaning pads in a pinch.

Powder Cutting

A variety of powder-burning processes are used to cut, scarf, gouge, wash, lance, or pierce metals such as stainless steels and highly alloyed cast irons. When the metal-powder-laden oxygen stream comes into contact with the preheat flames and the reaction zone, it rapidly oxidizes, producing enormous amounts of heat.

The powder is usually iron. Iron powder will dilute the refractory slag to speed the burning reaction. Techniques for powder cutting are virtually identical to conventional oxygen lancing or scarfing methods. Powder scarfing, for example, uses a conventional scarfing torch and nozzle with an external powder tube. Or else the torch may be designed so that the powder feed is an integral part of the torch.

Materials such as concrete, cast iron, blast furnace pig iron, and steel mill ladle spills resist lancing by an ordinary oxygen lance. However, they can be lanced by mixing the oxygen stream with iron powder, or even mixtures of iron and aluminum. The powder-lancing technique is identical to conventional lancing, except that deeper holes can be cut at much faster speeds.

Materials that are highly resistant to other cutting methods can be lanced by this technique, with lower consumption of the steel lance pipe. it is possible to restart a lance hole at any time if the end of the lance pipe is heated to the ignition temperature before the mixture of oyxgen and powder is turned on.

Underwater Cutting

Underwater cutting techniques are unique. The problems are much different from those of all other cutting methods. Underwater cutting is used in construction, maintenance, repair, and salvage work on ships, drydocks, pipelines, dams, piers, pilings and caissons, tunnels, offshore drilling rigs, and, of course, to salvage sunken ships.

The work is frequently critical and very often hazardous. If you can skin dive professionally, handle a cutting torch, and live along the Gulf Coast, you have a good chance of making a lot of money working for the offshore drilling industry.

Fuels Fuel gases such as propane and natural gas are not used underwater. They lack sufficient heat. In deep water most fuel gases will liquefy due to the pressure (about ½ psig [3.3 kPa] is added to the operating pressure of your torch for each foot [0.3 m] of depth).

Acetylene cannot be used very far underwater because it is limited to a 15-psig [103 kPa] operating pressure. At 30 ft [9.1 m] depth, for example, the water pressure equals 15 psig [103 kPa] so that you would have no operating pressure and acetylene won't come out of the torch.

Until the development of MAPP Gas, the only other underwater fuel was hydrogen which won't liquefy at the extremely high pressure found deep in the ocean. In addition to its obvious explosive hazard (it has very wide explosive limits in oxygen or air), hydrogen's low density meant that you needed a barge full of manifolded cylinders to supply a single diver on a single cutting job. The only alternatives were electric-arc cutting methods, which were worse. Would you like to work underwater with a 600-volt dc power line?

The U.S. Navy has standardized on MAPP Gas for most of its salvage work. So have a lot of contractors that do underwater work. Hydrogen is still used, but only at relatively great depths, usually over 150 ft [45.7 m] that exceed the 94-psig [647 kPa] cylinder pressure of MAPP Gas under standard conditions of 70°F [21°C] and 1 atm pressure.

Torches Standard cutting torches can be used underwater to depths of 30 ft [9 m] or less. Special underwater torches are used below that depth. Some of the problems you will run into in underwater work include the fact that light steel sections are much more difficult to cut than heavier sections. They actually require more preheating than thicker steel. In addition to rapid heat transfer from thin sections to the water caused by a high surface area and small steel volume, water has a cooling action about 40 times greater than air.

When making regulator pressure settings, you have to allow for an additional 0.445 psig/ft [10 kPa/m] of depth to your operating pressures for the same job at the surface. Work should be stopped at once when the cylinder pressures get close to the operating pressure at the torch. Otherwise water will back up into the hoses.

The milled splines of a two-piece skirted tip for MAPP Gas will help because the water is displaced more evenly than from a solid single-piece tip. Therefore, the skirted tip will preheat faster. Since it can be very difficult to see the steel through the bubbling water, a special device known as a *tip spacer* (Fig. 5-38) is sold. It is simply a round cylinder with holes in it that fastens over the cutting tip on the underwater torch and holds the flame tip about ¹⁄₁₆ in. [1.6 mm] from the plate surface so that the proper coupling distance is maintained.

Several special torches are manufactured for underwater oxy-fuel cutting (Fig. 5-39 and Table 5-11). These torches are equipped with *air hoods* (which also are called *air caps*). Compressed air is supplied to the hood to form an air bubble at the end of the torch around the preheat flame. The purpose of the air bubble is not to keep the flame dry but to speed preheating by keeping water away from the surface. Torches with air caps can use single-piece cutting tips. A needle valve on the torch regulates the volume of

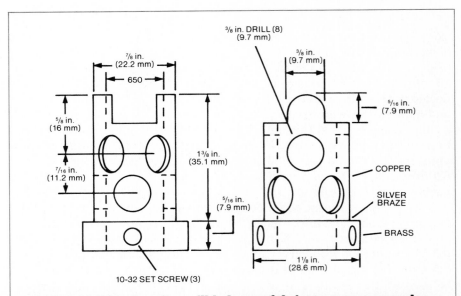

FIGURE 5-38 This drawing will help you fabricate your own underwater cutting tip spacer if you can't buy one. You can use it with a standard cutting torch for shallow underwater work.

compressed air. If the compressed-air volume is too great, it is nearly impossible to light the torch.

When possible the torch should be lighted and adjusted before being lowered to the diver. This was discussed in our earlier section on "Very Oxidizing Flames" (page 110). It is the responsibility of the diver's tender, who also is responsible for the maintenance of correct regulator pressures on the fuel, oxygen, and compressed-air lines.

Since underwater cutting pressures exceed those of work on shore, the diver will make final adjustments to the preheat flames to obtain maximum cutting performance. Simultaneously, the diver will be adjusting the compressed-air bubble in the air hood. As you can see, divers have their hands full underwater.

Several methods are used to light torches underwater. If underwater flares are used to illuminate the work area, they also can be used to light the torch. Otherwise an electric arc igniter can be used.

Underwater cutting practice

You can practice underwater cutting techniques without getting all wet. Get a 50-gal drum and cut it in half the long way to make two dish-shaped containers. Fill one of them within 2 in. [50

mm] of the top with water. Place a section of ½-in.-thick [13 mm] steel scrap in the tank at an angle to the water's surface. If your welding school doesn't have a tip spacer, you can make one yourself by following the specifications in Fig. 5-38. Light a conventional torch and shove it about 10 to 12 in. [254 to 305 mm] under the water and make final flame adjustments there.

Begin preheating the steel by pressing the legs of the tip spacer lightly against the steel. Start preheating. When you see a bright red-yellow spot form on the preheat area, slowly open the cutting-oxygen lever and start your cut. When you have a good cut started, slowly advance the torch

across the plate. The cut will proceed at about the same speed as cutting ½-in.-thick [13-mm] plate on the surface. If you hear loud popping sounds, other than those made by hot slag forming in the water, adjust the flame to neutral and then add a little more preheat oxygen to the flame.

Stack Cutting

Stack cutting is a special type of high-quality (as opposed to high-speed) cutting method used on multiple layers of flat material. Sheet or light plate (less than ½ in. [13 mm] thick) is stacked and cut in one operation (Fig. 5-40). The stack must be rigidly clamped to eliminate air gaps, especially along the line of cut. A

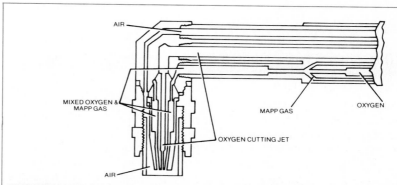

FIGURE 5-39 Special underwater cutting torches also are used, especially in the offshore oil industry and marine construction and salvage work. This torch has been designed for use with MAPP Gas and includes the underwater tip spacer as an integral part of the torch.

TABLE 5-11 Operating data for underwater cutting with MAPP Gas

Water Depth		MAPP Gas			Compressed Air			Oxygen		MAPP Cutting Tips			
										#52	#49	#44	#38
		Pressure,	Hose		Pressure,	Hose		Hose			Plate Thickness, in.		
			ID,	Length,		ID,	Length,	ID,	Length,	¼–1	1–2	3–4	5–6
ft	psig	psig	in.	ft	psig	in.	ft	in.	ft	Cutting Oxygen, psig			
10	4.5	25	¼	50	20	¼	50	⁵⁄₁₆	50	60	70	80	90
20	9.0	25	¼	50	20	¼	50	⁵⁄₁₆	50	60	70	80	90
30	13.5	25	¼	50	20	¼	50	⁵⁄₁₆	50	60	70	80	90
40	18.0	30	¼	50	30	¼	50	⁵⁄₁₆	50	65	75	85	95
50	22.5	35	⁵⁄₁₆	75	35	¼	75	⁵⁄₁₆	75	70	80	90	100
60	27.0	40	⁵⁄₁₆	75	40	¼	75	³⁄₈	75	75	85	95	105
70	31.0	45	⁵⁄₁₆	100	45	⁵⁄₁₆	100	³⁄₈	100	80	90	100	110
80	35.5	50	⁵⁄₁₆	100	50	⁵⁄₁₆	100	³⁄₈	100	85	95	105	115
90	40.0	55	³⁄₈	125	55	⁵⁄₁₆	125	³⁄₈	125	90	100	110	120
100	45.0	60	³⁄₈	125	60	⁵⁄₁₆	125	³⁄₈	125	95	105	115	125

FIGURE 5-40 Stack cutting can cut complex shapes in many layered pieces of thin steel in one pass. The stack must be rigidly clamped or welded tight to eliminate air gaps between the layers.

weld bead often is run down the sides of the stack to hold it together. The material in the stack should be as flat as possible.

Stack cuts on flat stock have square edges that are free from burrs, slivers, and drag marks. Fuel consumption is somewhat less than proportional to the increased stack thickness—the addition of another unit thickness to the stack does not increase the amount of fuel required by the amount needed to cut the same unit individually.

The total thickness of a stack is determined by required dimensional tolerances of each cut piece, with a gradual decline in tolerances up to stacks about 6 in. [152 mm] thick. Considerable operator skill is required to make good stack cuts.

5-12
GUIDING DEVICES

The simplest "machine" you can use to cut straight lines accurately is a long straight strip of steel placed as a guide along the side of the line of cut (don't cut it). If you put the steel straightedge just to the right or left of your tip and preheat flames after marking the line you will cut, your unaided cut will be much straighter.

If you are not cutting downhand, sometimes you can use clamps to hold the strip in position on the work. Be sure to allow for your kerf correction when you lay the strip down. Another piece of plate works just as well as a steel straightedge but it is harder to move around.

You also can buy a torch holder on wheels. This miniature buggy (Fig. 5-41) lets you walk or crawl along the plate guiding the torch. The tip is placed in the little carriage and positioned to give you the correct coupling distance between your preheat flames and the steel to be cut. You also can guide the wheels by putting them up against a second piece of plate aligned with your cut. This produces an even more accurate cut.

Very good freehand circles or rings are easy to cut if you have a simple rotating-radius rod and a center post pointed to fit into a punch mark on the center of the circle (Fig. 5-42). The rod should have a bracket and clamp for your tip and a screw to tighten the rod extension to the center post. Again, keep in mind that you must include your kerf correction when you extend the rod.

Irregular shapes are more diffi-

cult to follow. One solution is to produce one good part, then use it as a template to guide your torch tip to produce duplicates. Using the same steel template for every new part will quickly produce quite a number of uniform parts with excellent cut tolerances. If you use each part to make the next part, all you need is one error and it will be carried along through the rest of the job. When you use a part for a template, always use the same part and check it for correct measurements and tolerances before you use it to make more parts.

5-13
CUTTING MACHINES

Flame-cutting machines range from tiny one-torch track- and wheel-mounted units to big monster machines that fill the bay of a shipyard. Let's look at how they operate. They can actually be easier to use than a hand-held torch.

Machine Torches

Torches for cutting machines are straight-headed torches with the tip in line with the torch axis (Fig. 5-43). They don't have cutting-oxygen levers because the cutting oxygen is controlled elsewhere on the machine. A special type of control valve called a *solenoid* is used on cutting machines. A conventional gas valve of the kind you already are familiar with is used on the small one-torch track-mounted machines to turn on the cutting oxygen.

FIGURE 5-41 A cutting-torch holder on wheels lets you walk along a line, cutting as you go. The cutting-torch tip clamps into the holder.

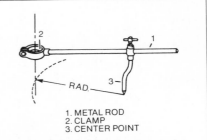

1. METAL ROD
2. CLAMP
3. CENTER POINT

FIGURE 5-42 Radius rods and a clamp to hold the cutting torch make circle cutting simple. The radius rod can be extended or retracted to cut large or small circles.

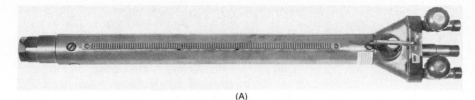

(A)

(B)

FIGURE 5-43 Special torches are used on flame-cutting machines. (A) This torch is a simple one with a ratchet face used to manually lift and lower the torch. (B) A machine torch with one vertical tip and two bevel-cutting tips and a motor that automatically lifts and lowers the entire torch under the control of the machine operator.

warped or wavy plate. The operator can adjust the torch height only slightly and keep the coupling distance between the preheat flames and the plate correct without either losing the cut, overheating the plate, or stopping the cutting machine to adjust torch height.

New devices have now been developed that actually sense the height of the torch above the plate automatically and adjust it for a preset coupling distance. Some torches also include automatic electric spark lighters so that the operator can light all the torches at once without climbing out on the plate.

Another common option selected for torches used to cut thinner material is a water spray just above the torch tip. The water spray cools thinner plate while it is being cut, preventing the material from warping.

Each torch also has some method for clamping it to the machine. All cutting machines have a strong, stiff beam by which the cutting torches attach to the machine. The beam must be very rigid because a group of torches add a lot of weight. When the machine changes direction to cut a pattern, the torches must react instantly without any delay to the mass of the torches or the beam on which they are mounted.

Almost all hand-held torches are two-hose models (oxygen and fuel). The cutting and preheat oxygen come from the same line. Many larger machine torches have three hose lines and inlet connections. One line is for fuel gas, one is for preheat oxygen, and third is for a separate cutting-oxygen supply. Three-hose torches are used for cutting thicker plates that require very high cutting-oxygen gas flows.

Each of the hoses is equipped with its own regulators so that the gas supplied is individually controlled for each line on each torch. Many of the larger machines provide automatic gas controls using a series of solenoid

Instead of a cutting-oxygen lever, the barrel of the cutting-machine torch (with the fuel and oxygen tubes inside) has ratchet gear teeth running up the side. This is used to raise and lower the torches on the machine. The operator does it manually when setting up the machine.

Some torches come equipped with motors that lift them up and down on command by the machine's program or by the operator at the machine's control center. The motorized torches look like a regular machine torch with a small box on the side. Motorized torches are useful when cutting

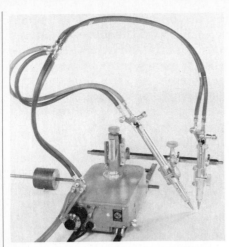

FIGURE 5-44 The simplest flame-cutting machine is a motorized torch holder on a track. It can cut straight lines, bevel cuts, and even circles when the radius arm is used.

valves for each gas line. All flow rates are set in advance and the machine controls just about everything, except when a manual override is desired to handle special problems.

The torch size for a cutting machine is determined by a maximum flow rate for oxygen and fuel gas, which in turn is determined by the tip sizes which will be needed on the machine. The tip size, of course, is determined by the thickness of the material to be cut.

Divergent-bore tips are almost always used on flame-cutting machines to get increases in cutting speeds of about 25 percent over manual cutting tips.

Plasma-arc cutting torches and electrical and gas controls also can be mounted on some cutting machines. These always are very high speed machines. If a cutting machine has plasma capability, the literature on the machine will tell you.

Small Machines

The simplest oxyfuel cutting machine is a one-torch, track-mounted, motor-driven torch carriage (Fig. 5-44). It cuts very straight lines at speeds from a few inches per minute (or milli-

meters per second) up to about 60 in./min [2.54 mm/s]. Faster models for cutting with plasma-arc torches may have cutting speeds above 120 in./min [5.08 mm/s].

These machines can carry one or two torches, or duplex tips to make right-angled edge cuts followed by a bevel cut on the squared plate. Circle-cutting attachments also can be used by adding a radius arm and pivot mounted on the carriage.

These little track-mounted, wheeled buggies are a common sight in plate-fabricating shops. They are so portable that you can carry them from one job to another under your arm. The track is modular so that any length of it can be laid on a plate. The machines operate off conventional 115-V ac wall plugs.

These small cutting machines take quite a beating in the shop. Their motors are so close to the cut that they often are exposed to a considerable amount of heat that can affect them. The cutting speed is set by a simple numbered dial. Although the motors are well shielded from the heat, it's a good idea to occasionally calibrate their speed to make sure that what you dial in is what you get out.

Calibrating the torch carriage

If you don't trust a machine's cutting-speed control, make a test. Lay several lengths of track on the plate or even on the floor. Put the machine on the track. Load the machine with the torch or torches, tips, and the hose you will use to do the work. Then turn on the machine and time it across a known distance that you have paced off along the track. From the time the machine takes to travel the known distance with a given speed-dial setting, you can calibrate its true travel speed for that dial setting.

For example, if you set the machine dial on #5 which is supposed to give you 25 in./min, then the machine should run the 25-ft

track at 25×12 in. = 300 in. or in 300 in./25 in./min = 12 min. Unfortunately, your fully loaded machine just ran the course in 15 min. What is the machine's true present cutting speed at a dial setting of #5, and what dial setting will give you the 25 in./min cutting speed you want? You can use a simple relationship to find out.

$$\frac{\text{real distance (in.)}}{\text{real time (min)}}$$
= number of inches per minute

Don't forget to multiply the number of feet in the real distance by 12 to get inches, and measure your time in minutes, converting seconds to a fraction of a minute—to get your dial calibration for *inches* per minute.

The real cutting speed in the above example is:

$$\frac{25 \text{ ft}}{15 \text{ min}} = \frac{5}{3} \times 12 \text{ in.} = 20 \text{ in./min}$$

Since a dial setting of #5 gives you 20 in./min, you can now easily calculate the dial setting to give you 25 in./min.

$$\frac{\text{unknown dial setting}}{\text{known dial setting}} = \frac{\text{desired cutting speed}}{\text{known cutting speed}}$$

$$\frac{\#X}{\#5} = \frac{25}{20}$$

$$\#X = \frac{\#5(25)}{20} = \#6.25 \text{ dial setting for 25 in./min}$$

You have just calibrated your machine. Sometimes you will need to calibrate electric meters and other equipment used in welding. The procedure is exactly the same. Get an initial measurement to determine what the machine really is doing. Then set up a proportion to calculate what you would like it to do and make corrections accordingly.

Torch angles The torches on all cutting machines, even the small ones, can be angled right or left to make bevel cuts as well as being set vertically to produce square-edged cuts. There also are special calibrated tips that you can use for bevel cutting. Figure 5-45

FIGURE 5-45 Here are some of the special tips that are available for use on flame-cutting machines. A tip with a calibrated swivel is ideal for making bevel cuts to a precise bevel angle. Dual tips can cut plate and bevel edges in one pass. Some dual tips can simultaneously cut the inner and outer edges of rings in one pass.

(right) shows a dual-cutting tip that cuts edges and then makes a 45° bevel. Figure 5-45 (left) shows a calibrated bevel-cutting tip that can be preset at any desired angle, and a double-headed cutting tip that can be operated by one cutting machine to make two parallel cuts (or if it's rotated, the inside and outside of a ring).

Larger Cutting Machines

Bigger cutting machines are usually multitorch machines ranging from two-torch models that work nicely in a small tool room up to monster machines that carry a dozen or more torches and fill the hull-fabricating shop of a large shipyard. But cutting machines are not classified by the number of torches they carry. They are classified by their function.

Bridge machines Some very big machines may only have one or two torches. They are designed for stripping and squaring the edges of as-rolled plates shipped from steel mills without further processing. (Steel mills also will supply squared plate with prepared edges, but they charge more for it. If a lot of plate is processed, it is more economical to cut incoming plate to the required size.) Mill-edged plates are not fully square and their edges are rounded and sometimes wavy. If you have ever seen bread dough rolled by a rolling pin, you'll be able to visualize a mill-edge plate.

Stripping and squaring machines (Fig. 5-46) usually are *bridge machines*. The torches are mounted on a long beam spanning the plate and supported on either side by wheeled carriages. Sometimes the torches can move right or left along the bridge while it moves forward or backward. That gives these machines limited shape-cutting capability in that they can make diagonal cuts as well as cuts at right angles to the edge of the plate.

Many of these shapes are suitable for making large structural sections for hulls, bulkheads, floor plate, truck beds, rail car sides and bottoms, and plate to be subsequently rolled into cylinders and other shapes for water tanks, pressure vessels, and steel silos. Some of these large machines have a second or even a third auxiliary bridge to carry additional torches that cut the ends of plates while other torches on the main bridge cut the long plate edges, but the shape-cutting capability of large bridge machines is limited.

Shape cutters The second machine category is the *shape cutters*. These are usually multitorch machines (from two torches to half a dozen or more). Most of them are tracer-controlled units, although the microprocessor-controlled machines are becoming far more common. Tracer-controlled machines have two separate sections. One area contains the cutting torches and on the opposite side is a second wide area for the tracer template. When the torches are on the right and the tracing area is on the left, the machine is called a *right-handed* machine. When the tracer is on the right and the torches are on the left, the unit is a *left-handed* machine.

Many of these machines have a bridge on one side to hold the torches. A cantilevered beam (supported on only one end) on

FIGURE 5-46 This large bridge machine is used in a shipyard. The bridge rides on rails the full length of the cutting bay. Torches can be moved along the length of the bridge to be positioned over plate anywhere in the cutting area.

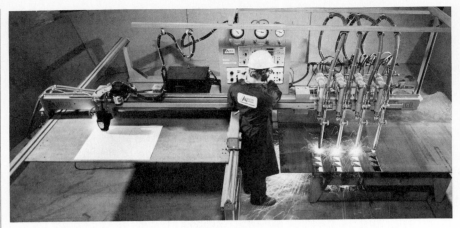

FIGURE 5-47 An optical tracer follows the paper-and-ink template on this shape-cutting machine. Finished parts are produced by each torch on the cutting side of the machine. This tracer-controlled machine also can be controlled by a dial-in shape microprocessor.

the other side holds the lighter-weight optical tracing head.

CAM CONTROLS Some old machine models are still used that follow a wooden template or cam cut to the same size and shape as the part to be produced. These cam-controlled machines have mostly been replaced by the far more versatile tracer or microprocessor-controlled cutting machines.

TRACER CONTROLS An optical tracer is an electronic "eye" with a light beam that can "see" a pattern drawn on a piece of paper. The light beam follows an ink or pencil drawing or template of a part to be cut while the torches burn the same shape into the plate (Fig. 5-47).

The paper-and-ink pattern or template usually is the same size as the finished part. There are two kinds of patterns used. One is the silhouette in which the light beam follows the edge of a solid pattern. The other is the line tracer that follows a narrow line defining the cut edge. The line cannot be too thin or else the electronic eye will not see it and will lose the cut. Line tracers normally follow rather narrow pencil lines not much heavier than those on a conventional engineering drawing.

The optical tracer also can lose the cut if its travel speed is too high when it comes to a sharp corner on the paper template. Modern tracers are now designed to sense an upcoming sharp change in direction on the template and will actually slow the machine down to make the corner. Nevertheless, a very sharp angle will be difficult or impossible for the tracer to follow. The tracer patterns must have a small radius at corners and angles to help the electronic eye (and the

heavy torches) make it around the corner.

The newer optical tracers will automatically turn off the cutting oxygen if the light beam loses the pattern. Manual override is always available so that the machine may operate under tracer or under the guidance of the machine operator. Obviously, tracer-controlled machines can cut straight lines as well as shapes, and they are used as stripping and squaring machines when a larger, higher-production strip-and-square machine is not available.

MICROPROCESSOR CONTROLS Another type of shape-cutting machine control is the microprocessor dial-in shape cutter. Machine instructions are dialed in on a control panel that resembles a pocket calculator (Fig. 5-48). The control requests each step in order (this is called *interactive control*) including type of shape (square, circle, rectangle, triangle, ring), its dimensions, and position on the plate (parts often are nested to reduce the amount of scrap left over). Dial-in microprocessor controllers are very easy to use but the number of shapes available is limited.

FIGURE 5-48 This flame-cutting machine is controlled by a computer. The operator dials in the shapes and their position on the plate and the machine does the rest.

The difference between a microprocessor control and a computer is that the microprocessor has been preprogrammed (like a small pocket calculator). The programs are a permanent part of the device. A computer can have new programs written for it by you or a computer programmer.

Newer machines now incorporate an option of both microprocessor and tracer control on one machine, combined with manual guidance when desired. All three control modes can be used, depending on which one is more efficient for a given part.

NUMERICAL CONTROLS The next level up in shape-cutting sophistication is the numerically controlled (NC) cutting machine. Very complex shapes (but no more complex than can be produced by an optical tracer) can be programmed and entered on punched paper tape. The paper tape is fed into a tape reader connected with the machine. The punched paper tape includes all cutting instructions, even when to turn on or off the preheat flames and the cutting oxygen. These machines are programmed like other NC machine tools.

FULL COMPUTER CONTROL Fully programmable computer numerical control (CNC) for cutting machines is becoming more common. A central computer may control not only one or more cutting machines, but also other machine tools. In addition to controlling a group of machines the computer also keeps records of part production and calculates when to reorder more plate and other bookkeeping tasks.

NC and CNC machines do more than cut plate. They can order punch marks to locate where holes should be drilled later on by other machine tools. They can calculate a complete part layout and mark it on the plate as well as preparing the edges for welding and cutting the steel to shape.

The cost of a cutting machine ranges from less than $1000 to more than $0.5 million, depending upon its control sophistication, size, number of torches, and whether oxyfuel or plasma torches and power sources are mounted on the machine. But the man or woman with the hand-held cutting torch remains by far the most versatile production cutting "machine" of all. High-quality plate must be delivered to a cutting machine. A human being can go anywhere and cut anything that can be sliced up with the oxyfuel cutting process.

REVIEW QUESTIONS

1. What is the kindling temperature of carbon steel in pure oxygen? What color does steel become when it reaches that temperature?

2. What is the slag produced by flame-cutting mostly made of?

3. When steel and pure oxygen react together, does the reaction produce heat or use up heat? Why is that important for flame-cutting?

4. Can any cutting tip be used with any fuel gas?

5. Do you use flush two-piece tips for acetylene and MAPP Gas? What fuels would typically use a skirted two-piece cutting tip?

6. You are going to burn a piece of plate covered with heavy mill scale, rust, and paint. Should you use:

(a) A tip with a heavy preheat flame?

(b) A tip with a light preheat flame?

7. What do you have to watch out for when flame-cutting zinc-coated (galvanized) steels?

8. When are divergent-bore (oxygen orifice) cutting tips used?

9. Why is it a bad idea to cut into a gasoline or oil drum without cleaning it out very carefully first?

10. Do you adjust your preheat flames with the cutting oxygen on or off? Do the preheat flames of all fuel gases look alike when properly adjusted?

11. Most flame-cutting is done with a neutral preheat flame. What kind of metal would you cut with a strongly carburizing preheat flame? What kind of cutting is best done with a slightly carburizing preheat flame adjustment?

12. What one place would you use a very oxidizing preheat flame adjustment for flame-cutting?

13. What is a kerf?

14. What's wrong with making a piercing start, after preheating a section to the kindling temperature, by turning the cutting oxygen on full force, all at once?

15. When you are flame-cutting, what does coupling distance mean?

16. What are drag lines? What can they tell you? What is the best drag angle for getting an optimum cutting speed?

17. At about what metal thickness should you consider making a bevel cut, instead of a perpendicular edged cut, when you will weld the material later on?

18. When adjusting for kerf width, should you put the kerf inside or outside of the part you are cutting?

19. Why can't you flame cut stainless steel, copper, and aluminum with an oxyfuel flame?

20. What process is most often used to cut stainless steel, copper, aluminum, and other nonferrous metals?

21. Name three special flame-cutting operations other than conventional straight-line cutting or shape cutting?

22. Can you flame-cut cast iron, or do you just melt it away when using an oxyfuel flame?

23. What is the biggest difference, in your opinion, between a manual cutting torch and a machine cutting torch?

24. Do all flame-cutting machines do essentially the same kind of work?

25. What kind of part pattern is most often used for making parts with an optical-tracer-controlled shape-cutting machine?

Practice with Flame-Cutting

Set up your equipment safely. Make sure that you are properly dressed. If you don't succeed with any of these assignments the first time (which you won't), continue practicing until you get good at it. You can only learn to do this work with repeated practice. You cannot learn to flame-cut simply by reading about it.

5-1-S
MATERIALS AND EQUIPMENT

You will need the same equipment as in Shop Practice 4, plus suitable materials to be flame-cut. Use a selection of ¼-in. to 1-in. plate.

5-2-S
STRAIGHT CUTS

1. Your first assignment is to cut a straight line in ¼-in.-thick carbon-steel plate using an edge start. Work with the plate in the flat, downhand position (with the plate lying on the cutting table, not held vertically or in the overhead position). Practice making different kinds of edge starts. Make the cut as straight as possible.

2. Now make horizontal cuts on the plate. Using the horizontal position the plate is clamped standing upright on its longest edge, but your cut will still be parallel to the bottom of the plate, or horizontal. Use various edge starts. Make the cut as straight as possible.

3. Clamp the plate so that it is standing vertically with at least half of it off the edge of the cutting table. Make a vertical-up cut. Starting at the bottom edge of a vertical plate, cut upward to avoid running through slag. Make the cut as straight as possible. Then make the same cut down. Which is easier, working vertical-up or vertical-down? Did you have trouble with slag?

4. Learn to cut off a piece of angle iron in the downhand, horizontal, and vertical positions. Angle iron varies in section thickness and cut direction, but you should cut the material in one continuous pass.

5-3-S
EDGE BEVELING

5. Make a 45° bevel cut on the edges of 1-in.-thick plate. Then make a 45° chamfer edge cut (it does not intersect the bottom of the plate). Make the bevel and chamfer cuts in the downhand position. Leave a ¼-in. [6-mm] "nose" at the bottom of the chamfer-cut edge. These are common joint designs for welding. Flat, horizontal, and vertical positions also are important positions you must learn for welding. You will see a lot more of them in the chapters that follow.

6. Repeat the bevel and chamfer edge cuts while working in the horizontal and vertical positions.

5-4-S
THIN MATERIAL

7. For comparison, now cut a line or pattern of any kind on thin carbon-steel sheet (the thinner the better), working in the down-hand position. How will you handle the distortion caused by the preheating? How will you keep the sheet from warping while you work? Try using clamps, hold-down dogs, or anything else you can think of to keep the sheet from moving around and warping while you work. Strong magnetic clamps can be very useful if you have them.

5-5-S
STARTING AND STOPPING

8. Start a piercing cut in 1-in.-thick plate and continue the cut to the farthest edge, while work-ing in the downhand position. Try different ways of starting the piercing cut.

9. Start a cut in ¼- or ½-in. plate. Cut part way across the steel and then purposely lose the cut (lift your torch). Restart your cut without creating a blowhole.

5-6-S
CIRCLE CUTTING

10. Cut an 8-in. [200 mm] free-hand circle in ¼-in.-thick steel, using a lead-in line from the outside of the circle. Allow for your kerf so that the circle, when cut, is a little oversized for final grinding to the finished diameter. That is, you can be a little over 8 in. [200 mm], but not under. Nevertheless, try to keep within ¹⁄₁₆ in. [1.5 mm] of the circle circumference so that very little finish grinding is required. Remember, if you cut into the circle it will become a piece of useless scrap. Mark your plate with chalk, a punch, or a scribe before cutting it.

5-7-S
PIPE CUTTING

11. This will be difficult. Try to make a circumferential cut around a piece of large-diameter round steel pipe or tubing, say about 6 to 8 in. [180 to 200 mm] in diameter. Cut the pipe off while it is first in the flat position (the pipe's central axis is horizontal), and then while the pipe axis is in the vertical position. After that, set up the pipe so that the axis is at a 45° angle and then make a square-edged cut across the pipe.

12. Now make a miter cut at a 45° angle in the pipe, cutting it into two pieces each of which has a 45° angle cut on the end. Repeat the cut in the horizontal and vertical positions and when the pipe axis is at a 45° angle.

13. Now make a 30° bevel all the way around the outer lip of the pipe edges. Any position will do.

5-8-S
CUTTING COATED METAL

14. Cut a series of 2- × 6-in. [50- × 150-mm] slots in a galvanized steel sheet. Work in a well-ventilated area. Allow for kerf and make the cuts as accurate and square-cornered as you can. Also control the sheet so that it won't warp out of shape while you cut it.

5-9-S
CUTTING THICK SECTIONS

15. Turn a 2- or 3-in. solid round bar into a pile of 1-in.-thick disks by slicing the bar like a piece of bologna. Each disk should be 1 in. thick (or as close as you can get to that). The tighter your cutting tolerances, the better, but if you cut the disk thinner than 1 in., including grinding, you will produce scrap. Now grind them to exactly 1 in. thick.

16. Clamp two pieces of ½- or 1-in. plate side-by-side to form a butt joint. Gouge a ½-in.-thick deep groove along the butt joint to make a U-groove joint. Grind the joint smooth.

17. Make a J-groove edge on a piece of 1-in.-thick plate (half of a U-groove joint). But you cannot use a second plate (duplicate the previous exercise). You must cut the groove using a single plate. Grind the joint smooth.

5-10-S
STACK CUTTING

18. Clamp 5 to 10 pieces of steel sheet tightly together. Then make a stack cut, cutting them all in one pass.

5-11-S
FLAME DRILLING

19. Drill a small-diameter hole through a thicker piece of steel. Make the hole just large enough for a bolt.

20. Put a bolt through the hole you just drilled. Put a nut on the bolt and fasten it as tight as you can. Do not use a bolt or nut that looks like it has been electroplated or has a bright silvery metallic finish—it could be plated with cadmium. Now burn off the bolt head without damaging the plate.

5-12-S
CAST IRON CUTTING

21. Practice cutting (melting) a slice through a piece of cast iron. Use any cast-iron scrap that you can find. But try to find some that offers both thin (less than 1½ in.) and thick (over 2 in.) sections. Engine blocks from an automotive scrap yard are good to practice on (you can break one up into pieces with a sledge hammer).

22. Continue working with cast iron, using different preheat control patterns as you manipulate your torch. Try cutting very thick sections of iron. Use a long preheat and a dam to stop the flow of molten iron.

5-13-S
UNDERWATER CUTTING

23. *Do this only under your instructor's supervision.* Wash out an oil drum with detergent and rinse it until it is *completely* clean. Fill it with water well above the line you will cut so that it's safe to cut in case there are any flammable gases or vapors left in it. Leaving the top open for expanding hot gas, cut the container in half (either horizontally

or vertically) to make a tank for underwater flame-cutting.

24. Fill your new container three-quarters or more full of water. Put some ½- and 1-in.-thick plate in it, underwater, and practice cutting the plate. Make sure you have the correct flame adjustment for underwater cutting. Note the longer preheat time required to make a cut underwater.

5-14-S
SMALL MACHINE CUTTING

25. Set up a portable machine-torch carriage on its tracks. Calibrate the machine's speed-dial indicator.

26. Practice cutting straight lines with the torch carriage.

27. Practice cutting bevel cuts with the torch carriage, with the torch angled at 30° or 45°.

28. If your torch carriage has a circle cutter, practice cutting circles with the machine.

5-15-S
CUTTING WITH A TEMPLATE

29. Prepare a paper and ink template of a part (a gear, for example), and cut it out of ½-in.-thick plate, transferring the template pattern to the plate with whatever method you think works best.

30. Prepare a paper and ink template with your name on it. Transfer the pattern to a piece of ½-in.-thick plate by tracing it with carbon paper or marking it with a punch. Allow for the fact that the insides of some letters will drop out if not connected to the plate. Cut your own name plate. Use either block letters or script (which is harder to do).

Using a Heating Torch

A skilled operator with a heating torch can bend pipe or straighten it, take dents out of a fender, turn a heavy straight I beam into a curved girder, shape the plate on a ship to conform with the curvature of the hull, make wavy plate edges line up into a straight butt joint for welding, bend large-diameter pipe with nothing but a torch, flame-harden steel parts so that they resist wear much longer, and even weld metals together without melting them. This chapter will show you how to do these jobs.

This chapter is a lot easier and shorter than the previous one on flame-cutting. But this is an important chapter because what you'll learn here leads logically into the next step in your training: oxyfuel brazing and soldering. And it will prepare you for handling a lot of work you'll do later on in both gas and arc welding.

6-1 HEATING TORCHES AND TIPS

A heating torch is like a very big gas welding torch. It's rated by the fuel gas it will use and usually by the maximum gas-flow rates it can handle. A production-sized welding torch can easily be converted into a heating torch simply by changing the tips. Sometimes a regular, large-diameter "single-hole" welding tip can be used for part-time heating work instead of using a multihole (multiflame) heating tip. But you will have to be careful how long you hold the welding-tip flame on the work to make it heat, but not melt, the base metal.

A cutting torch also can be used as a temporary heating torch without changing tips. In that case, you simply set a neutral flame on the torch tip *without turning on the cutting oxygen* when you make the flame adjustment. In fact, no matter what type of torch you use, almost all torch heating is done with a neutral flame setting.

Heating Tips

You can spot a heating tip immediately. Unlike the welding tip with its one hole, or the cutting tip with a mixture of two to four or more preheat holes and a central oxygen bore, the heating tip has nothing but heating ports on its face. It probably will have more of these holes in the tip than other torch tips, too. It also will tend to be a little more squat and solid looking. Heating tips often come equipped with your choice of tip extensions. We've shown some of them together, tips and tip extensions, in Fig. 6-1.

Just like cutting tips, different heating tips are used for different fuel gases, and for the same reasons. The gas-flow rate, heating values, mixing requirements, oxygen and fuel ratios, and other factors vary with different fuels and different torches. For example, acetylene would soon burn up a heating tip designed for propane. It also would flash back into the tip because the holes are too large for the high burning velocity of acetylene gas.

Heating very big workpieces usually requires very high volumes of fuel and oxygen, so you will need large-capacity torches and tips for some work, and you probably will have to manifold your cylinders.

Recently, "universal" heating tips have been developed that will burn a lot of different fuels in any one of a large variety of tip sizes and shapes. The tips come equipped with recommended fuel and oxygen flow rates and also are rated by their heat output in British Thermal Units (Btu) for different gases. (Recall that one Btu equals the amount of heat to increase one pound of water by one degree Fahrenheit.) Nevertheless, they are not available from all tip manufacturers. Be sure that you have the right heating tip for the fuel gas you will use. Read the instructions supplied with each tip. Follow them carefully. The manufacturer's operating data were developed to help

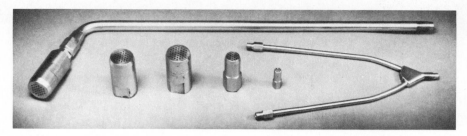

FIGURE 6-1 Oxyfuel heating uses all the equipment you've already learned about, except that special heating tips are needed. This photo shows some tips and tip extensions.

you get work done faster, and to keep you safe while doing it.

⚠ CAUTION: Always use the right heating tip for your fuel gas, and always follow the manufacturer's recommendations.

The same basic procedures for making flame settings for welding or cutting tips also are used for heating tips (except that the final flame setting is made without any cutting oxygen on). You can choose a carburizing, neutral, or oxidizing flame setting. The neutral flame is almost always used because you don't want to burn your workpiece with the big heating flame. An oxidizing flame would do that. You also don't want to add any more carbon to the steel (most of the time), and a reducing flame would do that.

6-2 EXPANSION AND CONTRACTION

Metals expand when heated and contract when cooled. These facts cause a lot of trouble when you weld or flame-cut thin-gauge metals. Cooling weld metal, for example, can distort a solid piece of steel as shown in Fig. 6-2, which is one of the reasons why a lot of welding is done with the work clamped tightly in special jigs or fixtures. It's also the main reason why many weld joints are first "tack" welded to hold them in place before the finish weld is made, and why stresses sometimes are relieved by reheating a workpiece after it is welded. Re-

heating reduces built-up stresses caused by expansion and contraction during welding. But you also can use the forces of expansion and contraction to advantage, if you know how.

There are only a few important things to remember about expansion and contraction in metals.

■ *Thermal expansion and contraction work in all directions* when a metal is heated or cooled. Not only will the surface dimensions of a workpiece change, but the thickness will change, too.

■ The total amount of expansion and contraction *in any direction* is proportional to the metal thickness in that direction (unless the metal is clamped down very tight). A 2-ft-long [0.6-m] steel or copper rod, when heated or cooled, will expand or contract along its length twice as much as a 1-ft [0.3-m] steel or copper rod under the same change in temperature. Of course it also will expand twice as much in thickness as a piece only half as thick to start with.

■ Thermal expansion and contraction are approximately proportional to the *change in temperature*. If you double the temperature change, you double the expansion or contraction. For example, if the 1-ft-long [0.3 m] copper rod is heated from 500 to 1000°F [260 to 539°C] (a 500°F [279°C] *temperature difference),* the rod's length will expand twice as much as the same rod heated from 50 to 300°F [10 to 149°C] (a 250°F [139°C] *temperature difference).* Of course, the diameter of the rod also will expand twice as

much when heated 500°F [260°C] than when heated a difference of only 250°F [139°C], but the actual amount of expansion of the diameter will be much less than that of the length of the rod because there is much less metal across the diameter than across the length.

■ *Different metals expand and contract at different rates.* The actual amount of expansion or contraction may not look very large per inch (or millimeter) of material, or per degree Fahrenheit or

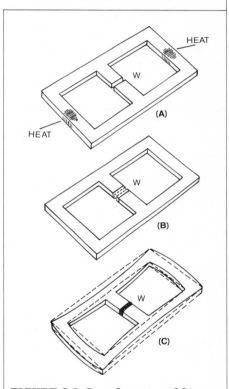

FIGURE 6-2 Step by step, this figure illustrates the effect of expansion and contraction caused by heating a part and welding it. If the two ends are first heated (A), the gap for the weld will expand (B). When the weld metal cools, along with the heated ends of the part, the entire piece shrinks evenly and distortion is avoided. However, if the weld were made in the middle section without any other heating, the weld metal would shrink when it solidified and this would distort the part (C).

Celsius. For example, 1 in. of aluminum will expand only 0.000013 in./°F, 1 in. of copper will expand 0.0000098 in./°F, and 1 in. of gray cast iron will expand only 0.000006 in./°F. But if you have a piece of each of these metals 8 ft long (96 in.) and heat each one of them up from room temperature to 1800°F (1730°F change in temperature), the three pieces of metal will expand 2.2 in., 1.6 in., and 1.0 in., respectively.

That's easily calculated because (using aluminum as the example) 0.000013 in./°F × 1730°F × 96 in. = 2.2 in.

The numbers that tell you how much a given unit length of metal (inch or centimeter or yard or mile) will expand or contract when heated or cooled a unit temperature degree (°F or °C) have a name. They are called *coefficients of linear thermal expansion.* Linear means that they measure the amount of expansion or contraction in any *one* direction that you choose. Table 6-1 lists a few of them to give you a feeling for which metals expand and contract more than others.

Detailed lists of these coefficients are published in standard handbooks for almost any metal or alloy you will ever use. The best of these handbooks is Vol. 1, *Metals Handbook,* "Properties and Selection." It is 1300 pages long and available from the American Society for Metals in Metals Park, Ohio. You can get a copy along with membership in ASM for the price of the book alone.

The Forces of Expansion

If each of the 8-foot-long metal pieces in our last example is clamped on the edges so that it can't expand front to back or side to side, you will wind up with a big warp in the middle about a quarter to half a foot long, depending on the metal. The forces that develop when materials are heated or cooled are enormous.

You've probably seen what a hot summer day can do to pavement. It can make a blacktop road surface curl up, change the length of a bridge by several feet, and break concrete. Imagine the effects you can get when big temperature changes occur, or when very long pieces of metal are used. Most of the roads in Alaska, including those in the cities, are gravel roads and not asphalt-coated for the opposite reason that roads in the southern U.S. and especially in tropical areas are often gravel-topped—the cold weather in Alaska winters makes the surfacing shrink too much and the blacktop cracks.

The forces produced when metals expand and contract are strong enough to permanently change the shape of the metal. Sometimes they are strong enough to tear a weld right out of the base metal, or cause it to crack. But by controlling thermal expansion and contraction, you can use these forces to your advantage to make metals take shapes you want.

Besides expanding when they get hot, metals also lose strength when heated. That's why a blacksmith hammers a hot horseshoe rather than a cold one. The strength of the steel in the hot horseshoe is much lower than that of cold steel. The hot steel is more "plastic." It has a lower "tensile" (or pull-apart) strength. It also has a lower "yield" strength (which means that less force is required to permanently change its shape). The hot metal also gets softer (it has lower surface hardness).

In Chap. 8 on the mechanical properties of metals, we will tell you more about yield strength, tensile strength, and hardness. They are mechanical and physical properties of metals that are very important to welders. Right now we're simply going to put them to work.

Expansion in Sheet Metal

Figure 6-3 shows drawings of sheet metal. The first drawing (A) shows that heat has been applied only along one edge, the rest of the sheet remaining cooler. The second drawing (B) shows what will happen to that sheet. The metal in the hot edge expands while the cooler areas try to stay the same shape, fighting the expansion of the hotter metal. The third drawing (C) shows what the same piece of sheet metal will look like after it cools again (with the edge thickening magnified for clarity).

If the forces of expansion do not exceed the metal's yield point or yield strength, the hot metal will contract back to its original shape when it cools again. But if the

TABLE 6-1 How much metals can expand

Metal	Coefficient of Thermal Expansion, in./in./°F	An 8-ft (96-in.) Piece Would Expand This Much in Length if Heated from 0 to 1000°F
Tungsten	2.5 × 10⁻⁶ = 0.0000025	1.92 in.
Ferritic stainless steel (type 430)	5.8 × 10⁻⁶ = 0.0000058	4.45 in.
Gray cast iron	6.0 × 10⁻⁶ = 0.0000060	4.61 in.
Martensitic stainless steel (type 410)	6.5 × 10⁻⁶ = 0.0000065	5.00 in.
Carbon steel (0.10% carbon)	8.4 × 10⁻⁶ = 0.0000084	6.45 in.
Austenitic stainless steel (type 304)	9.6 × 10⁻⁶ = 0.0000096	7.73 in.
Pure copper (oxygen-free)	9.8 × 10⁻⁶ = 0.0000098	7.52 in.
Bronze (phosphor-bronze)	10 × 10⁻⁶ = 0.000010	7.68 in.
Aluminum (type 1100)	13 × 10⁻⁶ = 0.000013	9.98 in.
Magnesium (AZ31B)	16 × 10⁻⁶ = 0.000016	12.3 in.
Zinc	18 × 10⁻⁶ = 0.000018	13.8 in.

metal is not uniformly heated and cooled, especially when the material is heated to high temperatures, the expansion and contraction forces will be too great and the metal will have a permanent set or distortion in it when it cools.

Something else also will happen to the hot spots in the metal. The cold areas act like clamps holding the expanding hot metal, because the cold metal is much stronger and doesn't want to expand. The hot metal cannot expand as much as it would like to because the cold metal won't let it. The only other way for it to expand is to get thicker.

When the hot metal cools down it will begin to contract. Now, however, the hot metal not only is cooling, but it's getting harder, stronger, and more rigid as it cools. As a result it is more diffi-

cult for the previously hot metal to contract back to its original shape, especially when surrounded by even colder areas that are still stronger.

Therefore, it is easier for the thicker contracting metal to pull the whole sheet out of shape than it is for the sheet to thin out the thickened sections. As a result, the heated areas remain thicker after heating than before, while the sheet is warped out of shape.

When metal is made thicker this way it is said to be "upset." "Bunched up" is a better description.

Figure 6-4 shows a common welding problem. A box shape has been welded to a flat sheet base. Weld metal not only heats the sheet locally, but it also shrinks a lot when it cools and solidifies, which creates very large shrinkage stresses that will distort the base sheet, tending to make it bulge at the bottom. If the base plate inside the frame is heated, the resulting upsetting action, as the metal becomes hot and then cools, will draw the parts back into fairly close alignment. The bulge at the bottom also will nearly disappear. This is exactly the same method used to take a dent out of sheet metal if it is solidly clamped first on all sides.

Expansion in Thick Metal

Look at another piece of metal (Fig. 6-5). Instead of thin sheet, this time it's a narrow piece of thick plate. First we will heat the plate along a centerline only on one surface (Fig. 6-5A).

Let's first assume that the metal is steel so that we can get the line very hot, cherry red or better. While we're heating one side, cool the edges and the other side of the plate with water while we heat the center strip on one surface. That makes the permanent deformation larger because of the big temperature difference between the heated flame and water-cooled areas.

The result will be a lump or ridge that is upset on the plate

(shown in Fig. 6-5B). If the plate is not too thick, the expansion of the hot ridge on one side will pull the opposite surface along with it, giving the plate a slight bow.

The amount of distortion outward from the free surface of the hot ridge (arrow 3 in Fig. 6-5B) is greater because there is no steel to stop it. The metal distorts less in the opposite direction (arrow 1) because the mass of the steel block "clamps" or holds back the distortion. That doesn't mean that there is no effect in that direction. It simply means that high stresses are built in inside the metal block even if you can't see the change in the shape of the metal.

The heated surface (arrows 2, 4, 5, and 6) also expands more than the unheated surface on the opposite side of the plate, forcing the plate into a "bow" shape parallel to the long direction of heating (arrows 5 and 6).

If these stresses are just a little below the yield point of the metal,

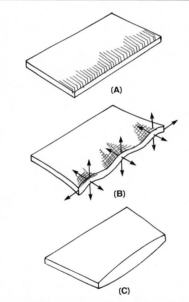

FIGURE 6-3 (A) When the edge of a piece of sheet metal is heated, the hot metal expands while the cold metal does not. (B) That produces a warped edge. (C) If the section is too thick to buckle, the section will become thicker where it has been heated. Since no new material has been created, the thickened central section will cause the edges to be pulled in and they no longer will be parallel.

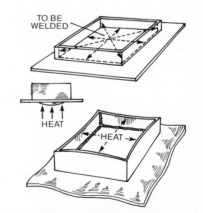

FIGURE 6-4 Box shapes made of thin material often create severe distortion problems. By carefully sequencing the welding with intermittent welds, and by welding opposite sides simultaneously, some distortion can be controlled. Applying heat to the bottom of the box to make the metal shrink and thicken when it is cooled also will correct initial welding distortion. Dents in fenders can be reduced using the same technique.

you don't need to add much more force to the metal block to make it bend or even crack. That is why certain types of weld joints and certain metals are heated both before and after welding and why some weld joints look perfectly good but crack when the part is used.

Heating before welding makes the metal more plastic so that stresses don't build up high enough to crack the metal while you weld it. The postheating is done to reduce or relax any more stresses that may still remain in the metal after welding.

The metal block in Fig. 6-5C shows what the original piece of plate in Fig. 6-5B will look like after it has cooled. It now has bowed back the other way, in the direction of the upset ridge while cooling, but you also can see that heating has reduced the metal thickness on the heated side of the plate, unless the metal block is very thick.

If the plate is thick (as in Fig. 6-5D), the top and bottom edges of the plate on the heated side will be pulled in toward the ridge.

If the entire piece of steel is very thick in all directions (Fig. 6-5E), you can't see much change in the overall shape of the steel block apart from the visible ridge. What you also can't see are the built-up stresses inside the block. These stresses can be very high, even though the thickness of the metal is strong enough to resist them.

The bowing and thickening effects could have been reduced or eliminated by clamping the metal *very tightly* while you were heating it. Stresses would build up in the metal, but they wouldn't cause it to warp or change shape nearly as much.

Another way to avoid these effects is to uniformly heat the entire piece so that there is very little temperature difference (and very little stress or distortion) between different sections of the part. Since heating effects depend on the temperature differences in a part, making the part uniformly hot reduces those differences from one place in the metal to another. That reduces the chances of accumulating built-in stresses.

A third way to eliminate built-in stresses is to reheat the part very soon after fabrication to a temperature hot enough to make the metal "give" plastically. You do that by taking all clamps off and making the metal hot all over, heating it to a rather high temperature (the temperature depends on the metal you are working with; we'll tell you more about it in Chap. 16). That reduces the strength of the metal holding the built-in stresses in place. The stresses will then be relieved.

These are all very important ideas that you will see used time and time again when welding, especially when using arc welding, which concentrates much more heat in a very small area than gas welding does. Before you have

finished with this book, you will be very familiar with these ideas. They are called *preheating, post-weld heating,* and *stress relief heating.*

6-3
WRINKLE-BENDING PIPE, TUBE, AND BARS

Most of the time pipe fabricators try very hard to keep pipe free of dents, wrinkles, and kinks when they bend it. Among other reasons, these "bending defects" decrease the inside pipe diameter or increase the roughness of the inner wall so that fluid flow is not as smooth. It's something like getting a kink in an oxygen hose or reducing the diameter of the hose at a connection. The pipe will have a reduced flow rate wherever the inside diameter changes, gets rougher, or even when the fluid flow changes direction at an elbow or bend.

But there is one technique for pipe bending that is often used in

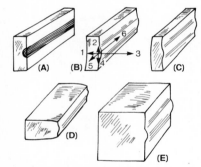

FIGURE 6-5 Since heating a section tends to make it thicker, a skilled worker with a heating torch can use this to advantage. If a strip is heated along the side of a block (A), the metal will expand in six directions (B). When the metal cools, the heated section will not contract back to a fully flat position. (C) Instead, it will leave a bulge and produce some warping. (D) Thicker sections will warp less. (E) Very thick sections will show almost no distortion, but they will have a bulge after cooling.

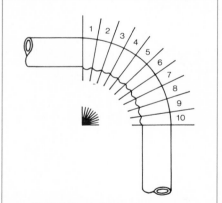

FIGURE 6-6 The section thickening (upsetting) effect of heating a section can be put to work. Here a single worker with a heating torch can bend a thick pipe, bar, or shaft simply by heating the section in closely spaced steps on the inside of the bend. In this drawing, a 90° bend has been produced in ten steps, each step producing a separate wrinkle on the outside surface of the section. If this were a pipe, the inside surface would still be fairly smooth.

the field where there is no special machinery big enough to bend pipe. The process actually creates wrinkles. It's called (appropriately enough) *wrinkle bending* (Fig. 6-6). Surprisingly, these outside wrinkles don't affect the inside surface of the pipe very much. About all it takes to do the job is one skilled person with a good heating torch, a tip big enough to do the job, a high-Btu fuel such as acetylene or MAPP Gas, and the usual oxygen cylinder, hose, and regulators.

The thick-walled metal pipe is heated by a large-capacity hand-held heating torch. When the metal becomes fairly plastic (bright red color), the pipe is bent *very slightly,* either by hand or by block and tackle.

The pipe is never bent very much in any one place (a maximum of 12° per wrinkle) to avoid buckling the pipe. After the first wrinkle bend is made, another one is made slightly forward of the first one along the pipe. Then a third wrinkle is created, and a fourth and and so on until the pipe is bent around the desired angle. The pipe in Fig. 6-6 was bent 10 times to make it turn a 90° angle. That's an average of only 9° bend per wrinkle.

Wrinkle bending has been used successfully on pipe more than 20 in. in outside diameter (OD). Experience shows that heating control is better for pipe over 6 in. OD if two people with two torches work together. In any event, the heating procedure is the same whether done by one person or more than one (see Fig. 6-7). The torch or torches heat about two-thirds of the pipe's circumference and leave the other one-third relatively cool. The heated portion will become the outside of the bend because the softer, lower-strength hot metal will stretch more easily than the cooler section.

The pipe section that is not so hot will upset when it is bent. It will pucker and wrinkle as the hot section yields and stretches. The wall thickness of the cool section will increase a little at the upset. The wall thickness of the hot section will get a little thinner. If the job is done a step at a time around the bend, the wrinkle inside and outside the pipe will not be as noticeable as it is in the drawing in Fig. 6-6, and the changes in wall thickness at each point will not be very important for most jobs.

The method for bending each heated section is so straightforward that it's hard to believe. Instead of hiring Superman to make the wrinkle bend, the torch operator simply rolls the pipe over and lifts it up at one end (Fig. 6-8). The cool section will face up and the hot section will face down. If the pipe is more than a couple of yards long, the lever arm of its length combined with the pipe's weight will cause the pipe to bend itself. The larger two-thirds hot section will stretch and the smaller one-third cold section will thicken and wrinkle. Then the next wrinkle bend is made.

Such is the simple power of a heating torch in skilled hands.

The problem is not making the pipe bend, but keeping it from bending too much at each step. Other than that, wrinkle bending

works for both large- and small-diameter pipe. You can try it yourself with steel or copper water pipe when we come to the shop work for this chapter.

6-4
HOT BENDING PIPE, TUBING, AND BARS

Any piping system of significant size will have bends in it. When pipe is fabricated for such a system, bends can be made either hot or cold. Cold bends are usually made on special pipe-bending machines. This equipment ranges from portable benders used by plumbers to large hydraulically driven machines used in pipe shops that can cold bend steel pipe up to 16 in. in diameter. As a torch operator, you will be involved mostly with hot bending jobs.

Whatever method you use to bend pipe, make a bend template first. The template represents the desired shape of the bend. Pipe-bending templates can be as simple as a wire or small, flexible tube measured and bent to the desired shape by hand. The template will represent the centerline of the pipe. You can lay the template over the actual pipe while you work.

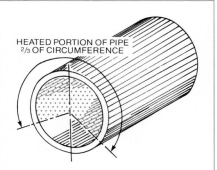

FIGURE 6-7 This shows how much of the pipe should be heated. About two-thirds of the circumference is heated. The pipe will bend toward the heated side when it cools. The nonheated side can even be further cooled with water spray or wet rags to increase the bending effect.

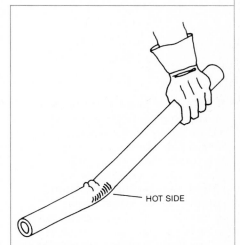

FIGURE 6-8 You can make a wrinkle-bend on smaller sections all by yourself simply by heating one side, rolling it over (hot side down), and then lifting one end.

Hot bends in pipe, tubing, and solid round bars are made about the same way. A bending slab or cast-iron table is used in most shops. The table has square or rectangular holes in it to stick in hold-down clamps or stakes called *dogs*. Most welding shops have similar tables for making temporary jigs and fixtures to fabricate welded products. A typical table and the hot-bending method are shown together in Fig. 6-9.

The first step in hot bending pipe or tubing is to pack the workpiece tightly with dry sand. That prevents the "heel" or outside of the bend from flattening. If flattening occurs, it will reduce the cross-sectional area of the pipe and restrict the flow of fluid

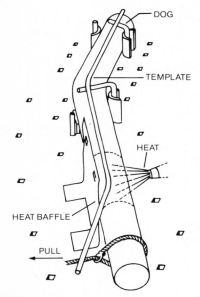

FIGURE 6-9 Some sections are simply too thick to use wrinkle-bending without extra force. This shows how a heating torch can be combined with pulleys or winches to soften and bend a very heavy section. Since metal loses strength when it's heated, the heat is applied to the outside of the bend and much less tension is needed to pull the section. The work is held on a cast-iron table and dogged down. The wire or rod template on top is the guide for the desired final shape.

through the system. Sometimes flattening will be followed by the complete collapse of the section. It will fold up like a hairpin.

Drive a tapered wooden plug into one end of the pipe or tube. Place the pipe or tubing in the vertical position with the plugged end down. Fill it with dry sand. Leave just enough space at the upper end to take a second plug.

Tap the pipe continuously with a wooden or rawhide-covered mallet while you fill it to make sure that the sand is tightly packed inside. A good rule of thumb is to tap *one hour* for each inch of pipe diameter. Pneumatic hammers or jolters also are used.

Next a second wooden plug, identical with the first one except for a small vent hole drilled through its length, is hammered into the other end. The vent hole in the second plug permits air and gases (mostly steam) to escape from the sand when the pipe is heated. No matter how dry the sand may appear, it will "out-gas" or produce a lot of steam. If you don't provide a vent, you will almost certainly blow out one of the plugs before you bend the pipe.

When you have packed the pipe with sand and plugged it at both ends, the next step is to heat it and make the bend. Mark the bend area with chalk or soapstone and heat the area to an even red heat along the distance indicated from A to B in Fig. 6-10.

Preventing Problems

Apply heat to the bend area, first on the outside of the bend and then on the inside. When you have an even heat, bend the pipe to conform to the wire template. The main problem you will have in bending copper tubing is preventing wrinkles and flat spots. Wrinkles are caused by compression of the pipe wall at the throat (inside) of the bend. Flat spots are caused by lack of support for the pipe or tube wall, by excess stretching in the heel (outside) of the bend, or by improper, nonuniform heating.

If the pipe is properly packed, plugged, and heated, wrinkles and flat spots can be prevented by bending the pipe or tube in segments (very much like the wrinkle bending idea, but without the wrinkles) so that the stretch is spread evenly over the whole bend.

When a pipe is bent, the stretch tends to occur at the middle of the bend. If the bend area is divided into a number of segments and then bent one segment at a time, the stretch will occur at the center of each segment. That will spread the stretching evenly over the whole bend area. Another advantage of bending in segments is that it is about the only way you can accurately follow a wire template.

When bending steel and some other pipe and tubing materials, you can control wrinkles and flat spots by first overbending the pipe slightly and then pulling the end back, as shown in Fig. 6-11.

When a bending slab or hold-down table is used, pull the pipe, tube, or solid bar in a direction parallel to the top of the table. The leverage for forming the bend is obtained by using chain falls or block and tackle, or by using a

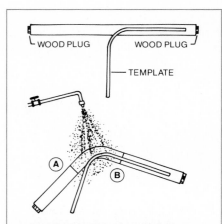

FIGURE 6-10 Pipe and hollow tubular sections can be bent through large angles without the walls collapsing if they are filled first with dry sand and plugged in the ends. Again, the wire template on top is used to indicate the final, desired bend angle.

length of pipe that has a large enough diameter to slip over the end of the shape to be bent.

Be sure to wear heat-resistant gloves when working on hot-bending jobs. Pins, clamps, and hold-down dogs will have to be moved during bending. These devices absorb heat radiated from the hot section as well as from the torch flame itself. You cannot safely handle these bending accessories without gloves.

Bending Specific Materials
Each type of pipe, tube, or bar has its own bending characteristics. Since there are a lot of jobs available for torch operators skilled at pipe work (not to mention plumbers, steamfitters, plant maintenance people, and others who make their living with this material), we'll give you a little more detail on how to handle different types of pipes, tubes, and bars.

Wrought iron It becomes brittle when hot, so always use a large

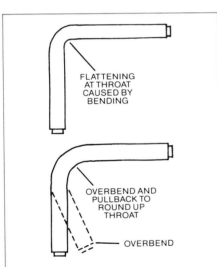

FIGURE 6-11 Sometimes a section will start to kink or flatten at the throat when bending. The solution is a more gradual bend angle, which will produce overbending. Once the bend angle has been made, the section can then be pulled back slightly to correct for the overbend and the kink in the throat of the section will be eliminated.

bend radius. Apply the torch to the throat of the bend instead of to the heel. Wrought-iron pipe is rarely used now. Carbon steel pipe is far more common. Carbon steel in "plumbing" sizes can be bent cold or hot. Larger steel pipe diameters are bent hot.

Brass Do not overbend brass. Brass is likely to break when the bend direction is reversed.

Copper Hot bends may be made in copper although the copper alloys are more adaptable to cold bending. This material is one that is not likely to give you any trouble.

Aluminum Overbending and reverse bending do not harm aluminum, but because there is only a small range between the bending and melting temperatures of aluminum and its alloys, you will have to work with care. Keep the heat in the throat at all times. You will not be able to see any change in heat color (aluminum looks silver colored even when it melts), so you must depend on "feel" to tell you when the heat is right for bending. You can do this by keeping a strain on the pipe, tube, or bar while the bend area is being heated. As soon as bending starts, flick the flame away from the area. Play it back and forth onto the work as needed to maintain the bending temperature and, at the same time, to avoid overheating.

Molybdenum-alloy and chromium-molybdenum alloy steels These materials are commonly used for high-temperature steam pipes, chemical reactor piping, and nuclear power-plant piping. They may be heated for bending, if necessary, and if the fabricating codes permit it. However, caution must be exercised. These steels are easily recrystallized when extreme heat is applied. If possible, pipes or tubes made from these steels should be bent cold with manual or power-bending machines.

6-5
STRAIGHTENING DENTS AND BENDING STRUCTURALS
Spot heating is frequently used by steel fabricators and erectors and in shipyards to straighten bent structural sections and bend straight plates into useful shapes. You will find spot heating is easy to do as long as you use the right amount of heat in the right place. The expanding and contracting metal will then do most of your work for you.

A 12-in. bar of steel, for example, when uniformly heated to a cherry-red temperature (about 1800°F [980°C] or higher), expands about ⅛ in./ft. When it is cooled to room temperature the bar will contract to its original length unless it is solidly clamped down.

When the ends of the bar are held by a die, jig, welding fixture, or hold-down clamps on either end (or put lengthwise into a vise), it can't get longer. The only thing it can do is get thicker and shorter after cooling. If the foot-long bar is restricted from expansion at the ends, and cools back down to room temperature and there is nothing to keep it from contracting, it will be about ⅛ in. shorter than when originally heated.

When the bar is heated only on a spot on one side while it is clamped on the ends, it has nowhere to go in the lengthwise direction, so it expands to the side. However, the side that is heated will shrink when the bar is cooled. The result is shown in Fig. 6-12A. The heated side of the bar shrinks on cooling and the bar is warped. To straighten a bar that already is warped, you reverse the process. Heat it on the outside or long side of the bend to make the metal shrink and upset on that side when it cools (Fig. 6-12B). That will straighten out the bent bar.

Figure 6-12C also shows the same fundamental idea being applied to straighten a bent structural-steel channel. The torch is played on the inside of the bend

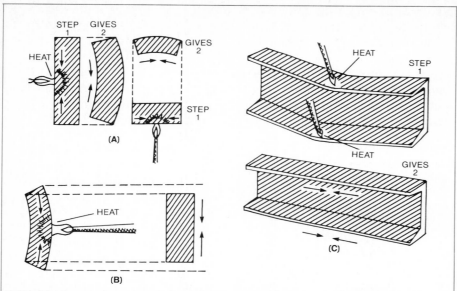

FIGURE 6-12 Even structural sections can be bent, or straightened, using nothing but a heating torch. These drawings show where to apply the heat, and then illustrate for you how the section shape will change.

so that it will shrink and upset when it cools. The shrinking, cooling metal will pull the channel back into a straight line.

You can't completely control heat, of course. When heat is applied at one spot, it rapidly flows to other parts of the metal section by thermal conduction. But this heat flow can be partially controlled so that the metal will upset just where you want it to.

You control the heat by manipulating your torch to keep the hot spot as hot as possible and the heat flow to other parts of the section as small as possible. Remember that temperature differences, not absolute temperatures, are what matter.

NOTE: Only the difference in temperature, not the absolute temperature, is important.

Figure 6-13 shows a good way to control your torch and the heat and temperature differences it produces in the metal. First mark the area to be heated with soapstone. The heated area will be V-shaped or triangular because more heat is needed at the outer part of the bend and less is needed toward the center.

Start heating at the tip of the triangle and work toward its outer edge, as shown in Fig. 6-13. When you start, wait for the the small area beneath the heating tip to reach the proper temperature, then *slowly* move the tip to one side so that the circular area of heated metal moves with it. Continue to work the torch so that the entire triangular area is brought up to temperature. Then let it cool and the metal will upset and the section will bend.

In spot heating you must hold the heating tip very steady. It's somewhat like flame-cutting. Do not use the weaving motion that you will learn later when you start welding. Watch the work closely and keep the hot spot moving and make sure that you don't start melting the base metal. Otherwise there will be no upsetting and the metal may be "burned." Under no circumstances should the metal be heated to the melting point.

✦ **CAREFUL: Never heat your metal to the melting point.**

You can cool the unheated side of the bend by using water, ice,

wet rags, or a water spray. Continue heating until the little circle of hot metal has covered the entire area within the triangle. Then let the metal section cool. With the proper procedure you'll get an appreciable bend. If the first bend is not sufficient, repeat the process. If you make too big a bend the first time, correct it by working on the other side of the section.

It's much better to apply a little heat several times than too much heat all at once. Long bends are best straightened by heating several times at intervals along the length of the bend. Remember that heating must be carried out on both sides of the flange or centerline of the section when the metal is particularly thick. This is to provide even heating followed by even contraction, upsetting, and bending.

Figure 6-14 shows how different kinds of structural shapes can be curved by spot heating. Don't try this on real structural sections (they're very expensive) until you know what you are doing and have had plenty of practice on scrap sections. If the section is a medium- to high-carbon heat-treated steel or a tool-steel part, leave it alone until the entire section can be cut out or removed and replaced with a separate curved section.

As you already know, the physical and mechanical properties such as the strength and hardness of medium- to high-carbon steels, alloy steels, and tool steels will be

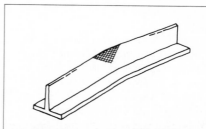

FIGURE 6-13 Distorted parts can be straightened by heating and cooling, using upsetting and expansion and contraction to advantage.

strongly affected by being heated and cooled even in air.

6-6
OTHER PROCESSES USING HEATING

Bending pipe and fixing dents aren't the only jobs you can do with a heating torch. You can harden certain kinds of steel, make tight parts fit in too-tight holes, loosen up parts that are stuck, and even make welds.

Blacksmith Welding

An important and little-practiced capability of a heating torch is that you can weld metal with a bond as strong as the base metal, without even melting it. The process is sometimes called *blacksmith welding*. It's one of a variety of solid-state welding processes. Blacksmith welding also is the oldest welding process in the world.

When two pieces of the same kind of metal are put face-to-face and heated hot enough and long enough while being pressed or hammered, the atoms of the metal in each piece will travel or diffuse across the weld joint. Pretty soon the joint itself becomes hard to

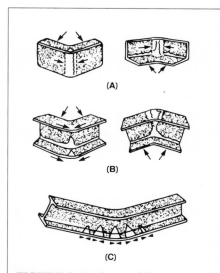

FIGURE 6-14 Curved beams often are used in bridges. Curved sections are commonly found in farm and machinery frames. Heat the light areas to make the metal move in the direction of the arrows.

define as the two metals blend into one.

Most of the older tools made by blacksmiths were welded together this way because the smith's forge could not heat the metals high enough to melt them. Most people think that the common picture of a blacksmith hitting a workpiece with a hammer is supposed to mean that the smith is simply shaping heated metal. Actually, in many cases the blacksmith is welding pieces of metal together; putting the tang on a hayfork, adding a handle to a cooking pot, or closing the links of a chain.

Four things are required for blacksmith welding. They are:
■ A high enough temperature to let the bond form quickly
■ Pressure (or hammering) to make the two surfaces (called the *faying* surfaces) press as closely as possible to each other
■ Very smooth, clean metal surfaces
■ Time to let it happen. (The higher the temperature and pressure, the shorter the time required.)

You would be surprised at how fast a blacksmith weld can be made. Two pieces of carbon steel can be joined in about 10 minutes with a pressure of 1000 psi [6.9 MPa] at a temperature of about 1800°F [980°C] or higher, but below the melting point of steel.

Preparing the work If the work is done properly the resulting weld joint will be as strong as the steel. The biggest part of the job is simply making sure that the faying surfaces are perfectly clean of oxides (let alone dirt and grease).

⚙ CAREFUL: The faying surfaces must be absolutely clean of oxides, grease, and dirt, and be very smooth.

The faying surfaces usually are prepared by grinding and even polishing them to as flat a surface as you can get. Fine emery cloth will do a nice polishing job after

you have removed just the surface metal oxide with a fine-grained grinding wheel. Be sure not to burn the metal when you grind it.

Testing your weld After the two faying surfaces are prepared, they are put together to make a "lap joint" (Fig. 6-15). Then the joint is uniformly heated on all sides to a cherry-red temperature or to a higher yellow temperature. After that you can use a heavy hammer to beat on the joint until it is formed. There is, however, an easier way.

Put the two cold pieces of metal into a vise with the cleaned and polished faying surfaces in contact. Put a firebrick on either side of the joint to prevent the jaws of the vise from being overheated. (You may also want to put some leather or asbestos padding between the brick and the vise jaws to keep the abrasive brick from marring the jaw surfaces.) Then tighten the vise as tight as you can by hand.

Now heat the lap joint on all sides as evenly as possible (but don't heat the vise in the process). As the metal is heated it will expand against the firebrick and increase the pressure on the lap joint, pushing the mating surfaces very tightly together. In about 15 min, stop heating the joint. Let it cool until it is black (does not glow). Then pull the joint out of the vise and quench it in water (that will simply cool it off for you because the metal is already too cold to be affected by a rapid quench), or simply let it cool to room temperature.

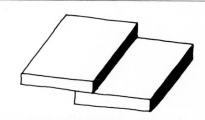

FIGURE 6-15 This is a lap joint. The sections should overlap by at least three times the thickness of the material being joined.

If you want to see how good a weld you have created, you can test it. Bend the joint backward (Fig. 6-16) to try to open up the lap joint. Continue to bend the metal around the joint (this is called a *reverse-bend test*) until you have it bent into a hairpin shape. If the weld still holds, give it a real tough test. Hammer the hairpin bend flat. If the weld still does not tear, you know you have created an extremely sound weld.

Lap joints are commonly used in soldering, brazing, and in both gas and arc welding. Reverse-bend tests are one of the most important "destructive" weld tests used. Reverse-bend tests also are one of the most difficult tests you will have to pass when you qualify as a certified welder. Before we go on to the shop practice, where you will have your first chance to actually make a weld, let's mention just a few more uses of heat on metals.

Flame Hardening

The procedure of heating the workpiece to a special high temperature is called the *austenitizing temperature* (a temperature where ferrite iron crystals turn into higher-temperature austenite crystals, another crystal form of iron as explained in Chap. 8). After the austenitic workpiece has

been quenched to form martensite, it is then tempered or reheated to a lower temperature to reduce the hardness only slightly while increasing the part's ductility. If the part passes mechanical property tests, it meets specifications.

Heat-treating specifications are complicated and usually worked out by metallurgists. Many parts are heat-treated in special furnaces under controlled atmospheres that eliminate oxygen and sometimes add carbon to the metal (or take it out) if the parts are made of steel.

But many products don't fit into a furnace. And some parts are simply easier to heat-treat with an open torch flame. Railroad rail is a typical example. The top part of a rail can be hardened to prevent wear, even after the rail is in place in the road bed, by using a heating torch. The edges of bulldozer blades, the teeth on large earth-moving shovels, wear plates on ore chutes in mines, gear teeth on tractor drives, and many other parts often are *surface hardened* with heating torches. (There also are very hard metals that can be put on the surface by either gas- or arc-welding methods. This is called *hardfacing* and it's the topic of Chap. 19.)

Shrink Fitting

Wagon wheels were made with wooden spokes and rims held tightly together by a steel band called a *tire*. It tied the assembly together. When the tire broke or slipped off, a blacksmith put a new one on using thermal expansion and contraction to make a tight fit. When the steel tire was uniformly heated, it expanded so that its diameter was larger than the wheel rim and the tire could easily be slipped over the rim. When the tire iron cooled, it shrank. The diameter grew smaller and the tire clamped so tightly onto the wheel that it held the rim and spokes together.

This invention made it possible to design wagon wheels much

lighter than the solid wood wheels previously used. The iron band around the rim held the rim and spokes in compression against the wheel hub (just the opposite of bicycle wheel spokes, which are in tension). The blacksmith called the process *shrink fitting*. Shrink fitting is still used today to attach gears onto shafts and unthreaded pipe connections onto pipes. Heating torches are still used to do it.

Sometimes you can use a combination of heat and cold to do the job (because the result is a function of the *temperature difference*, not the actual temperatures involved). For example, while the outside part is heated, the inside part can be frozen. Dry ice does a nice job. Its temperature is −109°F [−79°C]. Ice water also works, but not as quickly.

You can use the reverse of shrink fitting to loosen stuck parts. If you have a threaded connection on a shaft or fastener and it won't come loose, shrink the shaft by cooling it (or heat the connection, the nut on a bolt, for example) and you can take them apart a lot more easily. You do exactly the same thing when you heat the lid of a jar in hot water to get it off. The metal lid expands faster than the glass jar, so when heated it's easy to take it off.

REVIEW QUESTIONS

1. Can you use a cutting torch for heating instead of cutting? If so, how?

2. Can you use any heating tip with any fuel gas?

3. Do metals expand, or contract when heated?

4. Do all metals expand or contract equal amounts when heated or cooled an equal amount? If not, which of these three metals will expand or contract the most, an intermediate amount, and the least when equally heated or cooled: gray cast iron, carbon steel, aluminum?

5. If a metal has a coefficient of linear thermal expansion of 0.00001 in./in./°F, is 100,000 in.

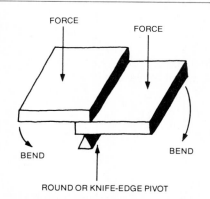

FIGURE 6-16 To find out if you made a good blacksmith weld, try to break it. Set up a test like this: If your plate bends at the weld, but does not separate at the welded seam, you have succeeded.

long, and is heated up 1°F, how much will it expand?

6. Do metals get stronger or weaker when heated?

7. If you heat a flat sheet of galvanized steel, lying loose on your workbench, in one spot with a torch until the spot gets red hot, then you let the entire sheet cool off, will it still be flat when you finish?

8. Can you make a raised bump appear on a 12-in. cube of copper? If so, how?

9. How would you avoid changing the shape of the 12-in. cube of copper if you wanted to heat it up, then let it cool, again?

10. How would you make the bump on the copper cube even bigger, if you could only heat it so hot without melting it?

11. What process would you use to bend large-diameter pipe, tube, or solid bar, if you could lift one end by yourself, and you had no one else around to help you.

12. Why should you try hard *not* to overbend brass and then bend it back to where you want it? Can you overbend aluminum and then bend it back to where you want it without any problem?

13. What is the single biggest problem when heating aluminum?

14. Describe how you would straighten a section of structural steel H beam, I beam, L- or Z-shaped section with a big dent in one of the flanges. (That's a typical problem when fixing anything from an auto frame to a railroad car that's been in an accident.)

15. Do you use a filler metal to make a blacksmith weld?

16. How much of the base metal should you melt to make a good blacksmith weld?

17. How do you get a rod into a hole in a metal part when the hole is slightly too small for the rod? Do you heat the rod or do you heat the metal with the hole in it?

Learning to Use Heat

You will gain valuable experience in this Shop Practice by learning how to predict the effects of heat on thin and thick sections, round bars and structural shapes. You will learn how to bend pipe and tube and straighten dents out of fenders. You will also learn how to make parts for "wrought-iron" furniture, trellises, balconies, and other decorative iron work. And you will make your first weld, followed by a reverse-bend test to see how good your weld is. The reverse bend test is an important test that you will have to pass for all the welds you make if you want to become a certified welder.

6-1-S
WHAT YOU'LL NEED

You'll need a heating torch (or a large-capacity welding torch) and good selection of both large and small heating tips. You'll also need thin-gauge steel sheet; thin-gauge aluminum sheet; steel, copper and aluminum rods; a small block of 1-in.-thick carbon-steel plate; any round steel pipe or tube from 1 to 2 in. OD, about 4 ft to 6

ft long; a heavy angle iron (2 × 2 in. or larger); hot-rolled carbon-steel strip 1 in. wide by ⅛ to ¼ in. thick, and heavy-gauge mild-steel wire.

6-2-S
HEATING THIN MATERIAL

1. Select a small heating tip or a small-diameter welding tip, set a neutral flame on it, and heat several separate hot spots on the edge of a piece of thin-gauge carbon-steel sheet. Watch how the sheet distorts and changes thickness as it cools.

2. Heat a piece of thin-gauge aluminum sheet. Can you heat the edge of the aluminum sheet high enough to make it upset and warp without melting it? Learn to judge how long you can heat thin-gauge aluminum without melting it. It won't glow red to tell you when it's hot.

3. Now take a steel rod, a copper rod, and an aluminum rod, each about 2 ft [0.6 m] long, and clamp them side-by-side on both ends on

a welder's worktable or a layer of firebrick. Be sure that the ends are clamped tightly so they cannot stretch and contract in the long direction. Heat the rods uniformly, all at once, with the same flame. Watch how each type of metal reacts when heated and cooled. That will show you an example of the difference in coefficients of thermal expansion and contraction between these three different metals.

6-3-S
STRAIGHTENING DENTS

4. Press or hammer a shallow dent in the middle of a piece of thin-gauge carbon-steel sheet about as thick as an auto fender. Clamp the sheet tightly on all four edges and heat it to make the dent go away. You may want to use a hammer to smooth out the remaining wrinkle or bump. It will be a lot easier to hammer the metal when it is hot than when it is cold. The metal also is much less likely to crack when hammered hot. When sheet metal is stretched cold it tends to "work harden" and grow stronger, which also makes it more liable to crack. Heating removes the effect.

5. Cut yourself a piece of thick carbon-steel plate (about 1 in. or thicker if you can get it). Put a narrow ridge on one face and then on the edges using a small heating or welding tip. Note how the plate reacts when heated and cooled.

6-4-S
BENDING SECTIONS

6. Use your heating torch to wrinkle bend a piece of 1 or 2 in. OD steel water pipe, preferably about 4 to 6 ft long so you can get more leverage on it when you bend it. Make a 45° wrinkle bend in one direction and using another part of the pipe make a 90° wrinkle bend in the opposite direction. Also try making hot bends without a wrinkle. If you had time and plenty of pipe you could continue this job to make anything from a complete "jungle gym" to the frame for a race car.

7. Bend a piece of angle iron, using only your torch if possible. Put a second bend in the iron at another angle, preferably bending it in another direction. Then try straightening the first bend, again using only your torch if possible.

8. Would you like to make "wrought-iron" furniture, fancy porch railings and New Orleans–style balconies? Most of the modern stuff is not made of wrought iron at all. It's made of low-carbon steel strip or slit sheet.

Using a heating torch, a mandrel, or a set of metal spikes or dogs in your worktable, and occasionally a vise or a small anvil, you can bend by torch and by hand very fancy curves, spirals, and other doodads in steel strip about 1 in. wide and ⅛ in. thick, or heavy carbon-steel wire between ¹⁄₁₆ and ¼ in. in diameter. After you join the sections (which you'll learn to do in the next lesson), a pot of white paint will make the objects very pretty.

9. Get some experience bending steel strip and heavy wire. It will also help you learn how to make your own pipe templates and welding fixtures. The only thing that's missing is a method to join the parts together. We'll try making a blacksmith weld, but the next chapter on soldering, brazing, and braze-welding will show you a much easier way to join your work.

6-5-S
BLACKSMITH WELDING

10. Select two pieces of carbon steel strip about 1 in. wide and ⅛ to ¼ in. thick. Grind off a thin layer on the faying surfaces of each piece, and then polish the faying surfaces with emery cloth to make them as shiny as you can. Be sure to keep the faying surfaces as clean as possible.

Put the faying surfaces together and clamp them in a vise between two pieces of firebrick. Tighten the vise by hand as tight as you can get it and then heat the lap joint that you have created to a uniform high temperature without melting the steel. (Be sure not to heat the vise.) Continue to heat the lap joint for about 15 min, then let the work cool.

6-6-S
TESTING YOUR FIRST WELD

11. Take your blacksmith-welded metal out of the vise. Does it stick together? Try bending the two pieces backward to separate the joint at the weld. If the weld still holds, bend the two pieces into a hairpin shape with the weld in the middle, facing out (to put maximum stress on the weld joint).

12. If your weld still holds together, put it to a real test. Hammer the bend as flat as you can. If your blacksmith weld can withstand this treatment you have successfully completed a very severe reverse-bend test. Even if the base metal cracks in the middle of the sharp bend, as long as the weld holds you have passed the test.

Brazing and Soldering Can Join Anything

This is the first chapter that explains in any real detail how joints are made with molten metal. There are four good reasons why you should learn soldering and brazing before you learn to weld. First, soldering and brazing require fewer manual skills than welding, but they do require a good knowledge of how metals melt. This knowledge will help you understand how welding works when you study it later on.

Second, both soldering and brazing are important skills that can help you get a job, and neither subject is covered in most welding courses. Third, soldering and brazing can join dissimilar metals that can't be welded. For example, it's almost impossible to weld steel to aluminum. It's very easy to braze steel to aluminum. Fourth, soldering and brazing are fun. Now that you can cut and shape parts, you will be able to join metals to make things for yourself.

7-1 A FEW GOOD THINGS TO KNOW

Gas and arc welding melt the base metal to complete the joint. Brazing and soldering do not melt the base metal, they wet it, and by doing so combine with the base metal in a totally different way. When they harden, brazing alloys and solders make a thin alloy surface film with the base metal that can be very strong.

The word *wet* means precisely the same thing for molten metals that it does for water. Pure water will not wet or stick to a greasy glass. Instead of running free, the water rolls up into balls or droplets. Water also won't wet many plastics and stainless steels. But if you add a detergent to it, the water-detergent mixture will wet a glass, a plastic cup, or a stainless-steel knife. Fluxes help solders and brazing alloys wet the surfaces of metals.

Welding may or may not need a filler metal or a flux. Soldering and brazing always use filler metals and usually need some flux. (The exception is certain kinds of furnace brazing where the controlled furnace atmosphere does the same job a liquid, paste, solid, or powder flux will do in manual brazing.) The flux protects the hot metal joint from oxygen in air and makes the faying surface of the joint more likely to wet, or react with, the molten filler metal on one side and the base metal on the other side of the brazing alloy-base metal interface.

The joint clearance or *fit-up* between two pieces to be welded can range from zero up to an inch or more, depending on the material thickness and the process. In contrast, brazing and soldering joint fit-ups are always very narrow. The optimum clearance between two parts to be joined by brazing or soldering usually is between 0.001 and 0.005 in. [0.025 to 0.13 mm], or about the thickness of this paper (Fig. 7-1).

The welding operator usually controls the molten weld metal as the joint is formed. The brazing or soldering operator has very little control over the molten metal, but that's not much of a problem because brazing and soldering filler metals flow right into a properly prepared joint, no matter what its shape (Fig. 7-2).

A third process, braze-welding, is also covered in this chapter. Braze-welding is not welding. But it's not exactly brazing, either.

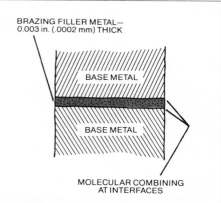

BRAZING FILLER METAL— 0.003 in. (.0002 mm) THICK

BASE METAL

BASE METAL

MOLECULAR COMBINING AT INTERFACES

FIGURE 7-1 Brazed joints must be extremely thin in order for capillary action to work.

The base metal does not melt during braze-welding, but you can use joint fit-ups as wide as those used in gas welding. The filler metal used in braze-welding is usually a high-strength bronze, and the base metal most often is steel or one of several kinds of cast iron.

We mention the braze-welding process for several reasons. Braze-welding is commonly used and is adaptable to plant maintenance because it often is easier to braze-weld cast irons than to fusion-weld them. Braze-welding also is very handy for do-it-yourself work. You can use it to make the ornamental-iron railings we talked about in the last shop section in this book.

You also can make beautiful braze-welded steel and copper sculpture with nothing more than a selection of different-colored copper, bronze, and silver filler metals, some scrap steel and copper sheet, and some wire (Fig. 7-3). Braze-welding is often used by metal sculptors. We won't make metal sculpture a required shop exercise, but you should try it at home.

Soldering versus Brazing

The difference between soldering and brazing is mostly a difference in temperature. By the American Welding Society (AWS) definition, soldering uses filler metals that melt below 840°F [450°C]. Brazing uses filler metals that melt above 840°F [450°C] up to, but not including, the melting temperature of whatever base metal you are working with.

The definition is obviously arbitrary. In fact, 840°F was chosen by AWS not only for technical reasons but also because this temperature is a practical intermediate between soldering and brazing alloy filler-metal systems. To make it simple, let's call the cut-off point between soldering and brazing 800°F [427°C]. It's what most shop people use; it's a lot easier to remember, and it won't make a bit of difference in what you do.

NOTE: As a rough rule of thumb, you solder metals below 800°F [427°C] and braze above 800°F [427°C].

Don't think that all welding is done at very high temperatures. For example, you can weld two sheets of lead together at their melting point of 621°F [327°C]. Since you melt both sheets to make them join together, the process is welding, not brazing. You could also solder them together with a lead-tin filler metal, being careful not to melt the sheets. But you can never, by definition, braze pure lead; it melts at less than 840°F [450°C] (or, if you prefer, 800°F [427°C]). Therefore, some metals can be fusion-welded at temperatures lower than the limit that separates soldering from brazing.

Filler Metals

The filler-metal alloys used for solders (all filler metals that melt below 800°F [427°C]) tend to be softer and of lower strength than brazing alloys. Most of the solders that are used are lead-tin alloys, but some solders contain tin and antimony and no lead at all. Others are made up of tin and silver, tin and zinc, lead and silver, cadmium and silver, or zinc and aluminum. Some special low-melting-point alloys have more than two elements in them.

For example, tin-indium-bismuth-lead-cadmium alloy system forms a whole range of alloys that

melt below the boiling point of water. You probably are familiar with at least one of them, which is used in auto body shops to fill fender dents, door dings, and hood humps. Another one of these alloys is used in automatic sprinkler systems. It melts before the fire gets going and sets off the sprinkler system.

Before we go any further, let's correct a common mistake in naming brazing alloys and solders. Most "silver solders" are not solders—they are brazing alloys. There are true silver-bearing solders which are lead-silver alloys, but most silver-based filler metals that are called silver solders are higher-strength brazing alloys that melt above 800°F [427°C].

Capillary Force

The final thing to understand about brazing and soldering is how the filler metal gets into the joint. Brazing and soldering filler metals flow into tight joints almost faster than you can see them go, once they become completely liquid.

The force that pulls them into the joint is called *capillary force* or *attraction* (Fig. 7-4). Capillary attraction is the same force that draws soda higher up a narrow straw than up a wide straw. It's also the force that pulls water through fine soil or blood plasma through the very narrow blood vessels in your body (which, in turn, are called capillaries).

The same capillary force pulls mercury up a thin thermometer tube and liquid metal into a tight joint. If you looked at the surface of metals through a microscope, you would see peaks and valleys instead of a shiny smooth surface. These minute pathways, plus the very slight gap that they create between any two pieces of metal, make the ultra-thin capillaries through which capillary action can operate. But the pathways must be very tiny.

If the fit-up gap is over about 0.010 in. [0.25 mm], the capillary force will be lost and the molten

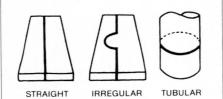

STRAIGHT   IRREGULAR   TUBULAR

FIGURE 7-2 Because of the capillary attraction of molten brazing metal into a tight joint, complex shapes can be joined with equal ease.

FIGURE 7-3 This red, white, and brown eagle is made of brazing alloys. Brazing and braze-welding alloys give you many colors from the gold or brass to the rich browns and reds of copper and bronze and the white of silver. Metal sculpture often is brazed.

metal will not flow completely around your brazed or soldered joint. The gap will be too big and you won't get a complete joint. This is one reason why the fit-up of a brazed or soldered joint is so critical.

NOTE: Tight joint fit-up is absolutely essential.

Heat Sources

Brazing and soldering processes are defined mostly by the source of the heat (torch brazing, torch

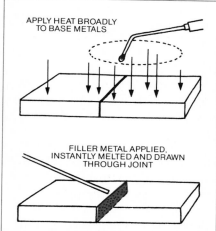

APPLY HEAT BROADLY TO BASE METALS

FILLER METAL APPLIED, INSTANTLY MELTED AND DRAWN THROUGH JOINT

FIGURE 7-4 Capillary force draws molten brazing metal deep into tight joints.

soldering, infrared soldering, furnace brazing, electrical-induction brazing, electrical-resistance brazing) or by the way the molten filler metal is applied (dip soldering, dip brazing, iron soldering). Most of these processes don't require much skill. It doesn't take much know-how to dunk a leaking radiator in a tub of molten solder (dip soldering). You don't need much more skill to operate a soldering iron.

Some processes for high-volume production, especially furnace brazing and soldering where thousands of parts may be processed in one batch, only require that you be able to lift the metal parts basket and shove it into the furnace. The skill in furnace brazing is in assembling the parts and selecting the furnace atmosphere, brazing temperature, and the correct brazing alloy, not in doing the brazing.

Instead of putting flux on each part and then adding filler metal by hand, furnace-brazed parts are preassembled with the filler metal already added. The gas atmosphere of the furnace itself is the flux. The furnace atmosphere has a controlled composition designed

to do what you will do with your flux when you learn manual brazing. The work requires high technology, but not high manual skills.

We won't tell you much about this type of work. We want you to learn the hard stuff—how to make superior-quality brazed and soldered joints with a torch in one hand and the filler metal in the other. Not many people have ever learned to do it well.

Process symbols Just like gas and arc welding, brazing and soldering have brief code symbols you may see on blueprints. The AWS (American Welding Society) codes for these processes are very simple. We'll give you some examples and you'll quickly get the idea. Torch brazing is TB, torch soldering is TS, furnace brazing is FB, furnace soldering is FS, induction brazing is IB, resistance brazing is RB, dip brazing is DB, and dip soldering is DS.

About the only time you will ever see these codes is at the end of a little arrow pointing to a joint in a blueprint. We won't go into blueprint reading here. We save that for Chap. 21. But we'll give you a few examples in Fig. 7-5.

While it's important to learn to braze and solder, you can learn something even more important from this lesson; that is, what happens to molten metals when they melt and harden. In the long run this knowledge will help you in all kinds of shop work (not just welding, brazing, and soldering). But first let's look at the basic steps you'll take whenever you do either brazing or soldering.

7-2
BASIC TORCH METHODS

There are nine simple steps to know for soldering and brazing. Although the particulars of any given torch-brazing or torch-soldering job will vary with the parts, the temperature at which the filler metal melts, and a few other things you'll learn about shortly, what follows applies to any kind of manual brazing and soldering job, and to most high-production furnace work. The process must begin with the cleanest possible metal surface you can get.

Precleaning

Clean faying surfaces are critical to the bond strength of your soldered or brazed joint. Capillary action works well on metals only when their surfaces are very clean. Problems will occur if you don't properly clean the faying surfaces of your joint before you solder or braze it. All grease, oil, dirt, and metal oxides must be removed from the two surfaces to be joined.

Oil and grease An oily base metal will repel the flux, leaving bare spots that oxidize under heat and result in "voids," or spaces of unbrazed, open joint space. Oil and grease will turn into carbon when heated, forming a film to which the filler metal will not bond.

CAREFUL: You can't braze anything without cleaning it first.

Commercial degreasers will get rid of oil. Rust or mill scale is a more difficult problem. If you have both a rusty or mill-scale-coated surface that also is oily, remove the oil first with a degreaser. Don't try using an abrasive cleaner on an oily surface. You will just grind the oil in. You can get rid of oil or grease simply by dipping the parts in a degreasing solvent like trichloroethane, by vapor degreasing, or by using an alkaline cleaner. If you don't have anything else, try laundry detergent.

Rust and mill scale Brazed parts should not be made with rusty or mill-scaled surfaces because your work won't stick together. Brazing and soldering alloys require as clean a surface as possible. Just looking at the surface isn't good enough. Metals like aluminum and stainless steel have a natural layer of oxide on the surface that you can't see because it's only a few hundred atoms thick. These thin oxide coatings are what make aluminum, stainless steel, or titanium so corrosion resistant. The oxide coating also makes them difficult to braze unless special fluxes are used, or the metal is specially treated in advance.

Oxide or scale can be cleaned from metal surfaces chemically or mechanically. Chemical cleaning means using strong acids. When you use acids, be very careful so that no acid gets on you. Clean the metal with water afterward so that no acid is left on the metal. You can use one of several acids

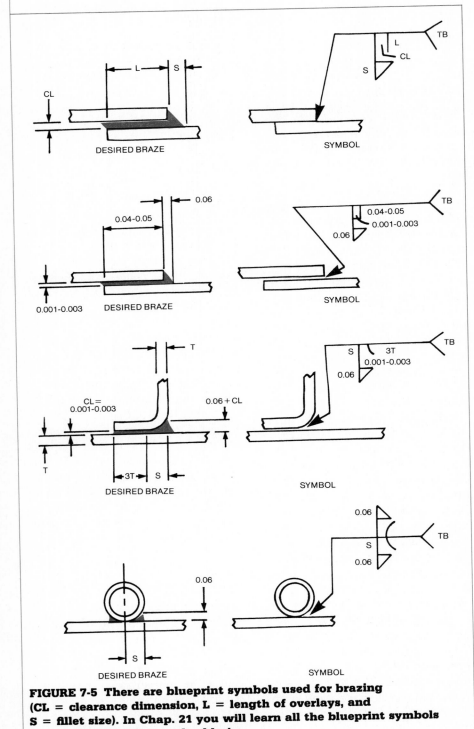

FIGURE 7-5 There are blueprint symbols used for brazing (CL = clearance dimension, L = length of overlays, and S = fillet size). In Chap. 21 you will learn all the blueprint symbols for welding, brazing, and soldering.

depending on the base metal. For example, hydrochloric acid (also commonly called muriatic acid) will treat a steel surface. Sulfuric acid, nitric acid, orthophosphoric acid, and even hydrofluoric acid are used for certain kinds of metals such as stainless steel, aluminum, and some cast irons.

Unless you are working in production conditions and know exactly what you are doing, we recommend that you use mechanical methods rather than using acids for cleaning rust or mill scale. Mechanical cleaning methods include grit or shot blasting, disk sanding, hand sanding, hand filing, grinding, cleaning with dry (not wet) steel wool, using a wire brush, or even scraping the surface off with a knife.

After using a grinding wheel, emery cloth, or grit blasting, follow up by washing the dirty residue off with water. Dry the part. Use a clean, dry, lint-free cloth or hot air. (A hair dryer will do a nice job.) Then apply the flux immediately to prevent the part from reoxidizing or tarnishing in the air.

Fit-up

Joint clearances between parts are critical. They should be small enough that molten filler metal will fill the space by capillary attraction, and not so large that the molten metal cannot fill up the gap. A paper-thin clearance of 0.001 to 0.005 in. [0.025 to 0.13 mm] is required for most work and most filler metals. When joining coated steels such as galvanized steel, use the lower end of this range.

Figure 7-6 shows the effect of fit-up on the strength of a typical silver-brazed joint. The joint has its maximum strength with about a 0.0015-in. [0.04-mm] clearance gap. If the fit-up is tighter than that the joint loses a little strength. If the fit-up is much more than 0.005 in. [0.13 mm], the joint loses too much strength.

The ideal joint fit-up of 0.0015 in. [0.04 mm] is too hard to main-

tain in ordinary day-to-day brazing, so you have to assume that you will work with a little leeway in strength to make a good joint under real shop conditions. Figure 7-6 also shows that it's quite possible to do that. As long as your fit-up clearance is somewhere between 0.001 and 0.005 in. [0.025 to 0.13 mm], the silver brazing alloy will still give you a very strong bond of 100,000 psi [689 MPa] tensile (pull-apart) strength.

When you consider that the tensile strength of low-carbon steel is around 35,000 psi [241 MPa], this silver-brazed joint will be about three times stronger, or about the strength of a quenched and tempered high-strength steel. Silver brazing alloys are both the easiest filler metals to use and generally the best ones. Although they are very expensive when purchased by the pound, so little metal is used in a typical silver-brazed joint that silver brazing alloys are usually very economical choices.

Let's take a closer look at the required fit-up gap. Translating these requirements into everyday shop language, an "easy slip-fit" will give you a perfectly adequate brazed joint between two tubular steel parts. And if you're joining two flat parts, you can simply rest one on top of the other. The metal-to-metal contact is all the clearance you'll usually need, since the average "mill finish" of metals gives you enough surface roughness to create capillary paths for the flow of molten filler metal.

Highly polished surfaces tend to restrict solder or brazing-alloy flow. There are not enough pathways to give you capillary action. If you have to join two highly polished pieces of metal, rough up the faying surfaces a little first with very fine emery paper or steel wool.

There's a special factor that you do have to consider carefully in planning your joint clearances. Brazed joints are made at brazing

temperatures (usually over 1100°F [593°C] for silver brazing alloys), not at room temperature. So you have to take into account the coefficient of thermal expansion of the metals being joined, just like you did when working with a heating torch. Table 7-1 lists some of these data for you.

You will see names of metals and nonmetals you may never have heard of on the list in Table 7-1. Many of them, like silicon carbide and titanium carbide, are used as small brazed inserts on cutting tools for metal-working lathes and other kinds of cutters, including rock drills, or for milling cutters used on metals. Some of them are very expensive precious metals that are soldered or brazed when making electronic parts. Examples range from silver and gold to metals called osmium and iridium, which are worth more than platinum.

Ceramics are usually metal oxides or metal-silicon materials. Cermets are combinations of ceramics and metals. They are used as metal cutting tools and for other jobs requiring extremely high hardness.

Carbides are a metal or nonmetal combined with carbon. Two important examples are tungsten carbide and boron carbide, which are extremely hard materials used in cutting tools. Carbon and graphite are on the list because, while they are not brazed or sol-

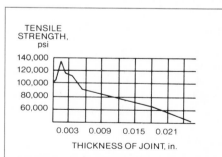

FIGURE 7-6 This graph shows the effect of joint thickness on the tensile strength of a brazed joint. Strength is maximum at about a 0.0015-in. gap. Wide gaps between brazed parts produce low-strength joints.

dered, they often are used for molten-metal crucibles and molds to handle or cast all these other materials.

Materials that expand and contract a lot when heated and cooled are at the top of the list in Table 7-1. Materials that expand and contract very little are at the bottom of the list.

Besides providing information on the characteristic thermal expansion (or contraction) rates of materials that you would have trouble finding anywhere else, Table 7-1 also gives you some idea of the scope of materials that can be joined by either brazing or soldering. The table includes just about every metal used in industry.

Here's how to use data from Table 7-1. Let's say that you're brazing a brass bushing into a steel sleeve. When heated, brass expands faster than steel. So if you machine the parts to have room-temperature clearance of 0.002 to 0.003 in. [0.05 to 0.075 mm], by the time you've heated the parts to brazing temperatures the gap may have closed completely! What do you do then? Simply allow a greater *initial* clearance (Fig. 7-7) so that the gap at brazing temperature will be about 0.002 to 0.003 in. [0.05 to 0.075 mm].

Of course, the same principle holds in reverse. If the outer part were brass and the inner part steel, you would start with a light force fit at room temperature. By the time you reached brazing temperature, the more rapid expansion of the brass would create a suitable clearance (Fig. 7-8).

Flux

Fluxes are chemical compounds that are applied to the joint's faying surfaces before brazing. They are essential in most soldering and brazing processes. Why? Heating a metal surface accelerates the formation of oxides. Oxides result from the chemical combination of hot metal with oxygen in the air. These oxides will prevent the brazing filler metal from wetting and bonding to the faying surfaces. Flux prevents the formation of these oxides by picking up excess oxygen from your torch flame.

A coating of flux on the joint area will shield the molten metal surfaces from the air, preventing new oxide formation. The flux also will dissolve and absorb any oxides that may already be present on the metal and not completely removed in the cleaning.

How to apply flux You apply the flux to the joint any way you can. Dip it on. Brush it on. Rub it on. Spray it on. Brushing is the most common method; an old but clean paintbrush will do fine. Put the flux on just before brazing, if possible. That way the flux has no time to dry out, flake off, or get knocked off the parts you are handling.

TABLE 7-1 Coefficients of thermal expansion: How different materials compare[a]

Material	10^{-6} in./in./°F		10^{-6} in./in./°C	
	High	Low	High	Low
Zinc and its alloys[c]	19.3	10.8	3.5	1.9
Lead and its alloys[c]	16.3	14.4	2.9	2.6
Magnesium alloys[b]	16	14	2.8	2.5
Aluminum and its alloy[c]	13.7	11.7	2.5	2.1
Tin and its alloys[c]	13	—	2.3	—
Tin and aluminum brasses[c]	11.8	10.3	2.1	1.8
Plain and leaded brasses[c]	11.6	10	2.1	1.8
Silver[c]	10.9	—	2.0	—
Cr-Ni-Fe superalloys[d]	10.5	9.2	1.9	1.7
Heat-resistant iron alloys[d]	10.5	6.4	1.9	1.1
Nodular and ductile irons[d]	10.4	6.6	1.9	1.2
Cast stainless steels[d]	10.4	6.4	1.9	1.1
Cast tin–bronzes[c]	10.3	10	1.8	1.8
Austenitic stainless steels[c]	10.2	9	1.8	1.6
Phosphor-silicon bronzes[c]	10.2	9.6	1.8	1.7
Coppers[c]	9.8	—	1.8	—
Nickel-base superalloys[d]	9.8	7.7	1.8	1.4
Cast aluminum–bronzes[c]	9.5	9	1.7	1.6
Cobalt-base superalloys[d]	9.4	6.8	1.7	1.2
Beryllium copper[c]	9.3	—	1.7	—
Cupro-nickels and nickel silvers[c]	9.5	9	1.7	1.6
Nickel and its alloys[d]	9.2	6.8	1.7	1.2
Cr-Ni-Co-Fe superalloys[d]	9.1	8	1.6	1.4
Alloy steels[d]	8.6	6.3	1.5	1.1
Carbon-free machining steels[d]	8.4	8.1	1.5	1.5
Cast alloy steels[d]	8.3	8	1.5	1.4
Age-hardenable stainless steels[c]	8.2	5.5	1.5	1.0
Gold[c]	7.9	—	1.4	—
High-temperature steels[d]	7.9	6.3	1.4	1.0
Ultra-high-strength steels[d]	7.6	5.7	1.3	1.1
Malleable irons[c]	7.5	5.9	1.3	0.8
Titanium carbide cermets[d]	7.5	4.3	1.3	—
Wrought irons[c]	7.4	—	1.3	0.9
Titanium and its alloys[d]	7.1	4.9	1.2	—
Cobalt[d]	6.8	—	1.2	1.0
Martensitic stainless steels[c]	6.5	5.5	1.2	—

[a] Values represent high and low sides of a range of typical values.
[b] Value at room temperature only.
[c] Value for a temperature range between room temperature and 212°F to 750°F [100°C to 390°C].
[d] Value for a temperature range between room temperature and 1000°F to 1800°F [540°C to 980°C].
[e] Value for a temperature range between room temperature and 2200°F to 2875°F [1205°C to 1580°C].

Choosing flux Which of the many possible fluxes do you use? First you should use one formulated for the base metal. Beyond that you can select fluxes for the expected soldering or brazing temperature and the application method in your shop (dip or brush, or do you need something that sprays from a paint gun for production work?). You also should select a flux that becomes liquid at the brazing temperature you want, because when it melts, you will know that it's time to add the brazing alloy filler metal.

There are fluxes formulated for just about every need. For example, there are fluxes for very high temperatures, fluxes for metals with refractory oxides such as aluminum, stainless steel, and ti-

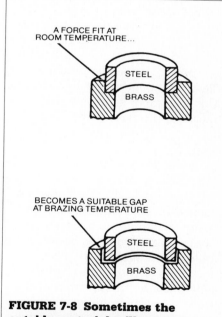

FIGURE 7-7 When materials with different expansion rates are used, you must take that into consideration. Brass, for example, expands more than steel when heated. What might be a loose brazing joint at room temperature will be a perfect joint gap at brazing temperatures.

FIGURE 7-8 Sometimes the outside material will expand at a faster rate than the inside part of a joint when heated. In this drawing, a force fit (too tight) becomes a good brazing joint at high temperatures because the outside brass section expands at a faster rate than the steel.

Material	10⁻⁶ in./in./°F		10⁻⁶ in./in./°C	
	High	Low	High	Low
Nitriding steels[d]	6.5	—	1.2	—
Palladium[c]	6.5	—	1.2	—
Beryllium[b]	6.4	—	1.2	—
Chromium carbide cermets[e]	6.3	5.8	1.1	1.0
Thorium[b]	6.2	—	1.1	—
Ferritic stainless steels[c]	6	5.8	1.1	1.0
Gray cast irons[c]	5.8	—	1.1	—
Beryllium carbide[d]	5.8	—	1.0	—
Low-expansion nickel alloys[c]	5.5	1.5	1.0	0.3
Beryllia and thoria[e]	5.3	—	0.9	—
Alumina cermets[d]	5.2	4.7	0.9	0.8
Molybdenum disilicide[e]	5.1	—	0.9	—
Ruthenium[b]	5.1	—	0.9	—
Platinum[e]	4.9	—	0.9	—
Vanadium[b]	4.8	—	0.9	—
Rhodium[b]	4.6	—	0.8	—
Tantalum carbide[d]	4.6	—	0.8	—
Boron nitride[d]	4.3	—	0.8	—
Columbium and its alloys	4.1	3.8	0.7	0.68
Titanium carbide[d]	4.1	—	0.7	—
Steatite[e]	4	3.3	0.7	0.6
Tungsten carbide cermets[e]	3.9	2.5	0.7	0.4
Iridium[b]	3.8	—	0.7	—
Alumina ceramics[e]	3.7	3.1	0.7	0.6
Zirconium carbide[d]	3.7	—	0.7	—
Osmium and tantalum[b]	3.6	—	0.6	—
Zirconium and its alloys[d]	3.6	3.1	0.6	0.55
Hafnium[b]	3.4	—	0.6	—
Zirconia[e]	3.1	—	0.6	—
Molybdenum and its alloys	3.1	2.7	0.6	0.5
Silicon carbide[e]	2.4	2.2	0.4	0.39
Tungsten[b]	2.2	—	0.4	—
Electrical ceramics[e]	2	—	0.4	—
Zircon[e]	1.8	1.3	0.3	0.2
Boron carbide[e]	1.7	—	0.3	—
Carbon and graphite[e]	1.5	1.3	0.3	0.2

[a] Values represent high and low sides of a range of typical values.
[b] Value at room temperature only.
[c] Value for a temperature range between room temperature and 212°F to 750°F [100°C to 390°C].
[d] Value for a temperature range between room temperature and 1000°F to 1800°F [540°C to 980°C].
[e] Value for a temperature range between room temperature and 2200°F to 2875°F [1205°C to 1580°C].

tanium; fluxes for long heating cycles; and fluxes for dispensing by automated machines. Fortunately, there also are a lot of perfectly good general-purpose fluxes for commonly brazed or soldered metals. Any one of them will do nine out of ten brazing jobs perfectly well. Simply ask your local supplier for a recommendation.

The companies that produce brazing alloys will know a lot about fluxes too.*

How much flux? How much flux you use depends on your work. The rule is simple. Use only enough to get the job done, but never use less. That's not a fair statement. As a general guide, thin metal pieces heat up faster and need less flux than thick metal pieces, which need to be heated longer to bring them up to temperature. Be sure the entire faying surface of each piece is coated with flux.

You can use flux that melts at a slightly lower temperature than the filler metal as a thermometer. When the flux melts and becomes transparent (when ready, fluxes usually look like water on top of orange-hot metal), you have reached brazing temperature and are ready to add the filler metal. Test the temperature by touching the brazing or soldering filler metal to the base metal. If it melts, you are ready to go to work.

Incidentally, some high-production furnace-brazing jobs don't use any flux (as such). They use controlled furnace atmospheres instead. Examples of such atmospheres are hydrogen or relatively inert gases such as nitrogen, or mixtures of them. Inert gases shield the hot metal from oxidation. A hot hydrogen atmosphere turns any oxygen in the furnace into water. Never use any of these

* Handy & Harman, for example, is one of the world's leading suppliers of brazing alloys (they provided a lot of information for this chapter). Handy & Harman has an excellent customer-service department that can help you. Their phone number is (212) 752-3400. Or you can write to them at 850 Third Ave., NY, NY 10022.

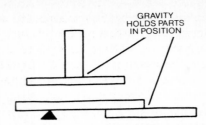

FIGURE 7-9 Special jigs and fixtures are not always needed for brazing. Here's a simple setup where gravity holds the parts in place.

atmospheres, especially one containing hydrogen, without experienced supervision. Some special furnace atmospheres can be explosive if there are any leaks, or if the furnace is not properly purged of air before it is used.

Positioning the Parts
The parts of the assembly have been cleaned and fluxed. Now you have to hold them in position for brazing. You want to be sure they remain in correct alignment during the heating and cooling cycles so that capillary action can do its job.

The simplest way to hold parts together is by gravity, if the shape and weight of the parts permit it (Fig. 7-9). You also can give gravity a helping hand by adding more weight (Fig. 7-10).

If you're brazing just one or a few parts, you can use any clamping or supporting device that holds the parts together long enough to complete the brazing cycle (Fig. 7-11). If you have a

number of assemblies to braze or solder and their configuration is too complex for self-support or clamping, rig up a brazing support fixture or jig.

In planning such a fixture, it should have the least mass (material) possible, and the least possible contact with the parts of the assembly. A massive fixture that contacts most of the part will draw heat away from the joint area. Use pinpoint or knife-edge contacts with the part to reduce heat flow into the jig or fixture (Fig. 7-12).

Many parts are self-jigging. Their very shape, itself, holds them together while you work. Examples are shown in Fig. 7-13. Sometimes parts will even be tack-welded first and then brazed because brazing is easier to do than welding, and it makes complicated joints just as easily as it makes simple ones. Make sure that your joint is properly fitted or you will have trouble making the brazing alloy flow into the joint.

Heating
Where and how the heat is applied, the torch and tip used, the flame setting, and the proper preheating of the different metals are all vital in any successful brazing or soldering operation.

Apply the heat broadly, not in a pinpoint area. Heat up as much of the entire workpiece as you can without overheating it. If you are torch soldering or brazing, use a heating or welding tip that gives

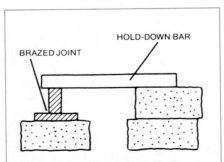

FIGURE 7-10 This is a little more elaborate fixture requiring only fire brick and a heavy hold-down bar.

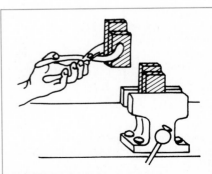

FIGURE 7-11 A vice, or even tongs, can be used to attach many parts for brazing.

you a bushy flame, or use a dual tip that heats the part from both sides. Either an oxyfuel or an air-fuel torch can be used because you won't be needing temperatures high enough to melt the base metal.

Air-fuel torches have the advantage of being a lot simpler. There is no charge for the oxygen, and they provide a good selection of tips that produce nice bushy flames suitable for brazing or soldering. Air-fuel torches are often used by pipefitters and plumbers and by auto body shops. Most of the inexpensive hobby "welding" outfits are really air-fuel brazing outfits with a 1-lb [0.45-kg] bottle of propane or MAPP Gas.

Flame settings Set a slightly reducing flame for steel or aluminum. Use a slightly oxidizing flame setting for copper and copper-base alloys. Copper alloys (except oxygen-free copper) may contain dissolved oxygen, and if you use a reducing flame, excess hydrogen or carbon monoxide in the flame can make water (as steam) or carbon dioxide in the metal. The gas pressure inside the metal can increase so much that the copper will blister.

Heat the workpiece uniformly all around the joint. Make sure both pieces to be joined are heated equally. That's one of the advantages of using a larger heating tip

with multiple flames, or a dual tip with two sets of flames, one for each side of the work. Keep in mind when joining a big part to a smaller part that you will have to heat the big part longer.

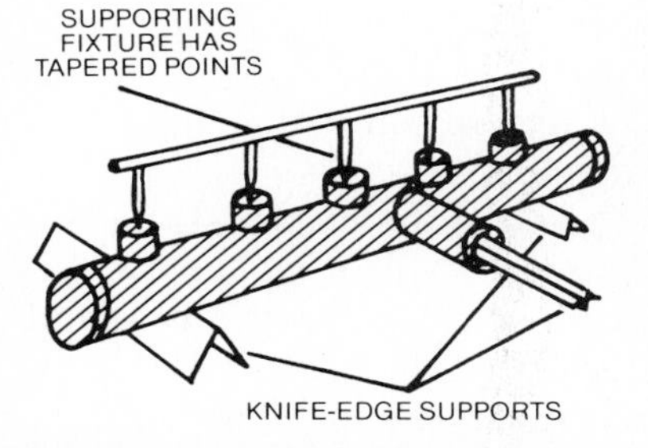

⚙️ **CAREFUL: Remember the basic brazing rule:** *The heat of the work, not the torch, does the brazing.*

Heat conduction Some metals are better heat conductors than others. If you are joining copper to stainless steel, for example, the copper will come up to temperature faster than the stainless steel. So you should heat the stainless steel longer. But be careful not to overheat the workpiece. Remember you don't want to melt it and you don't want to oxidize or "burn" your work. You are soldering or brazing, not welding.

Do not point the flame directly at the filler rod or wire. You will overheat the filler alloy and you don't want a pile of molten brazing alloy or solder on your workbench. You want it in the joint. When the part gets hot enough, the filler metal will melt and take care of itself, filling the joint automatically and quite rapidly because of capillary attraction (as long as you don't accidentally melt the filler metal before the part is hot enough).

Adding Filler Metal
Brazing and soldering filler metals come in wire and rods. They also are available in sheets, strips, rings, and many other special shapes. Experienced workers often will put a ring or strip of brazing or soldering alloy into the joint after fluxing it when they set up the work. Then all they have to do is heat the workpiece to the right temperature and the filler metal does the rest.

A lot of manual soldering and brazing requires you to feed in the wire or rod. *Face-feeding* is the best method to feed wire or rod filler metal into a brazed or soldered joint. Simply touch the end of the wire to the hot joint

right on the gap, and if the work is hot enough and your fit-up is correct, the filler metal will instantly flow into and all around the joint. While you do that, make sure you continue to heat the workpiece, not the filler metal.

There's only one small precaution to watch for. Molten solders or brazing metals flow toward areas of higher temperature. In the heated assembly, the outer base metal surfaces may be slightly hotter than the interior faying surfaces. So you should take care to deposit the filler metal right next to the joint. If you deposit it away from the joint area, it will tend to plate the hot outside surfaces rather than flow into the joint.

In addition, heat the side of the assembly opposite the point where you're going to feed the filler metal. In Fig. 7-14 the work is being heated on the backside while the filler metal is face-fed

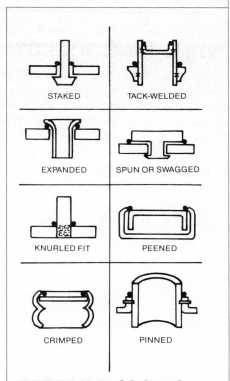

FIGURE 7-13 Prefabricated brazing filler metals in the form of wire rings or strips are often put in place before the part is heated. The black dots are cross sections of brazing filler metal preforms.

FIGURE 7-12 Some brazing fixtures can be rather complex, especially when high-volume production work is done. The supports all have knife edges or sharp points to prevent the fixture from stealing heat from the brazed joint.

from the topside into the joint. The filler metal will flow toward the heat source, exactly where you want it to go. This is an example of *sweat soldering* or *sweat brazing* where the heat is applied on one side of the part and the filler metal is fed on the other. It is a common procedure used in joining copper tubing.

Flux Removal
Flux removal is simple and absolutely necessary. Most flux residues are chemically corrosive and, if not removed, might weaken certain kinds of joints or start long-term corrosion. Therefore, immediately after the joint is made, take away the torch and clean the excess flux from the joint.

Quenching *Quenching* means to cool hot metal very rapidly. Since most brazing fluxes are water soluble, one easy way to remove them is by quenching the assembly in hot water (about 120°F [49°C] or hotter). Your best bet is to quench the finished part while it's still hot. But make sure that the filler metal has solidified completely before quenching.

The glasslike flux residue will usually crack and flake right off. If it's a little stubborn, brush the

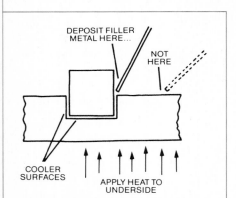

COOLER SURFACES
APPLY HEAT TO UNDERSIDE
DEPOSIT FILLER METAL HERE...
NOT HERE

FIGURE 7-14 Indirect torch heating is much better than direct heating for brazed joints. It avoids oxidation of the brazing filler alloy if the torch flame is not adjusted to the neutral condition. Indirect (sweat) brazing also avoids overheating low-temperature brazing alloys.

area lightly with a wire brush while the completed part is still in the hot water. Or you can tap the flux with a hammer to fracture it and then put the part back into the water.

There's another reason for cooling the joint fast, especially when it is a nonferrous metal joint. Many nonferrous metals such as copper and silver will start growing big crystals or "grains" at brazing temperatures. Your filler metal is much stronger when it is fine-grained.

Production methods There are other, fancier, ways of cleaning flux from parts in production shops. These include ultrasonic cleaning tanks to speed the action of hot water, or live steam using a pulsating steam lance. But that's a frill you won't have in do-it-yourself work or in a welding school shop.

Problems with flux removal
The only time you'll have trouble getting flux off is when you didn't use enough to start with, or if you've overheated the part. Either way, the flux will become totally saturated with oxides. It usually will turn green or black. In that case, the flux has to be removed with a mild acid solution. A 25 percent hydrochloric acid bath heated to 140° to 160°F [60° to 71°C] will usually dissolve the most difficult flux residues. Simply shake the soldered or brazed part in the bath for 30 s to 2 min. There's no need to brush.

A word of caution. Acid solutions are potent, especially when hot. So when you're quenching hot parts in an acid bath, be sure to wear safety glasses and a transparent face shield, rubber gloves, and a rubber apron. Concentrated acids also can release enough heat when water is poured into them to make the water boil, which will splash acid on you. Always remember to mix acid into water and not water into acid. This may help you remember:

There once was a chemist named Ted
Whose job was to analyze bread.
He added water to his acid—
In a flash it reacted—
And took Ted into solution instead . . .
Then from out of the smoke came a sound,
"If you would be a chemist renowned,
Do like you ought'er,
Add acid to water,
And not the other way 'round."

CAUTION: Wear special protective clothing when handling acids.

Inspect the Part
How does the brazed part look? Is all the flux cleaned off? Does it look like the filler metal made a complete joint? If your employer has established visual inspection procedures for you, follow them. Remember that Japanese workers have been successful because they have been making products with few defects. If you get below 5 percent defects (1 percent defects is even better) on anything you make in large quantities, congratulations! If not, try it again. Chapter 20 on testing and inspection will give you more details.

Reviewing Brazing
This was a long discussion of the basic methods of brazing and soldering. We can boil it down into nine simple steps:
- Preclean the faying surfaces of all work
- Choose the correct filler metal alloy
- Make the fit-up tight
- Apply the flux
- Assemble the parts
- Heat the parts properly
- Add the filler metal
- Clean off excess flux
- Inspect your work

7-3
WORKING SAFELY

You already should be familiar with the general safety precautions for brazing and soldering. They are the same as flame-cutting and heating, and for the gas welding processes we cover later on. Wear the right goggles with a light lens shade just as you would for heating. Always use gloves and protect yourself from hot metal. Always observe all gas-handling safety precautions.

Brazing and soldering also have special safety precautions you should follow, especially when using or working with certain types of material. *There is always the possibility of dangerous fumes arising from cadmium-bearing brazing alloys and solders, or even from some cadmium-electroplated base metals. Fumes also can be a serious problem when using fluoride-based fluxes.* The following well-tested precautions should be followed to guard against any hazard from these fumes.

■ Ventilate confined areas. Use ventilating fans and exhaust hoods to carry all fumes away from work, or supplied-air respirators as required, as they are when working with cadmium alloys and fluorine fluxes.

■ Clean base metals thoroughly. A surface contaminant of unknown composition on base metals may add to fume hazards and may cause too-rapid breakdown of the flux, leading to overheating and fuming. One example is commercial degreasing solvents. They should be washed from the work before you apply heat. Another example is pickling acids. Remove them before heating the base metal.

■ Use sufficient flux. Flux protects base metals and filler metals from oxidation during the heating cycle. Full flux coverage reduces the chances of fumes.

■ Heat metals broadly. Heat the base metals uniformly over a wide area because intense local heating

uses up flux and increases the danger of fuming. Apply heat only to the base metal, not to the filler metal.

■ Know your base metal. A cadmium-electroplated or hot-dipped coating may be on the base metal. Question anything that looks like it has a bright white, metallic coating. You will find this material used mostly on fasteners and some printed circuits. Cadmium-based filler metals also are subject to toxic fuming if overheated. *Don't breath the fumes.* Cadmium-bearing filler metals all carry a clear warning notice on the package. Be sure to look at the package label and carefully follow the directions. Work under a fume hood or in a well-ventilated area. If possible, use a supplied-air mask, not a filter mask. Cadmium is normally found in alloys used to make bearings, in special low-melting-point alloys, sometimes as a very bright electroplated coating on fasteners, and often in industrial electrical contacts in large switches and relays. Its use is declining because of its poisonous properties.

7-4
FILLER METALS

Now you know a lot about soldering and brazing. It's time to learn the different kinds of soldering and brazing filler metals, the reasons why there are different families or systems of filler metals, and why their metallurgy will help you select the one best grade of soldering or brazing alloy for any given job.

Choosing a Solder

Specific solders (and brazing alloys) are selected for strength and melting temperature, and sometimes for a color match with the base metal, for corrosion resistance, for electrical properties if you need them, and then finally by cost. Cost considerations have to come last because if any filler

metal can't meet one or more of the other requirements, its cost doesn't matter because the alloy simply can't be used. Let's look closer at some of these filler-metal selection factors.

Melting temperature Solders are frequently selected on the basis of melting temperatures. Solders with higher melting temperatures also tend to have higher strength. That's not a hard and fast rule, but it's a good general guide. There is another reason for giving you a choice of melting temperatures.

Expert solderers and brazers have a special trick that they use when they have to join more than one part together in several separate steps. The first solder used has the highest melting point. It joins the first two pieces together. The second solder has a lower melting point than the first solder. It joins a third part onto the other two pieces at a temperature below that at which the first solder will melt (Fig. 7-15). As many as a dozen connections can be made, one at a time, simply by using a dozen lead-tin solders, for example, each with a slightly lower melting point than the one used before it. The very same idea

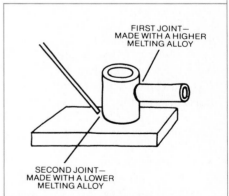

FIRST JOINT—
MADE WITH A HIGHER
MELTING ALLOY

SECOND JOINT—
MADE WITH A LOWER
MELTING ALLOY

FIGURE 7-15 Parts with more than one joint can be made with brazing alloys that have different melting temperatures. Make the high-temperature joint first, or otherwise the tubular arm will fall off when the second joint is made.

is occasionally used when brazing complicated parts.

If all you ever needed was a range of alloys with different melting temperatures, the list of solders would begin and end with lead-tin solders. They are the least expensive and are easy to apply. Any lead-tin solder would make excellent alloy to join the seams of food and beer cans, except that lead is hazardous to your health. Therefore another system of solders must be investigated for possible use for food-service containers.

Let's take another example. Lead-tin solders have fairly good electrical conductivity, but many electronic instruments need more conductive electrical connections than lead-tin solders can give. That's why there also are lead solders with silver in them. All lead-silver solders have high electrical conductivity. They are used to solder printed circuit boards. However, if you look at the family of lead-silver solders you will find that they not only have different melting temperatures but also different electrical properties and different costs, depending on how much silver they contain. The higher the silver content, the higher the electrical conductivity, the melting point, the strength, and the cost. Therefore the selection of a specific grade becomes a trade-off between higher electrical conductivity on one hand and higher cost on the other.

Some solder families have few members, maybe only one, two, or three, separated by different melting temperatures. That's a clue that they exist for some very special reasons. For example, a small family of indium-tin solders with very low melting points includes one grade, a 50 percent indium and 50 percent tin solder, that has the peculiar property of soldering glass-to-glass and glass-to-metal. That's how the metal bottom of a light bulb gets attached to the glass top.

Some solders have such specialized applications that they have

become the standard choice for only one or two kinds of work and often no more than that. Nevertheless, without that one type of solder, its special job (making light bulbs, for example) would be very difficult to do.

Wetting ability No matter what else you are looking for in either a solder or a brazing alloy, it must be able to hold the base metal together. Not all solders and brazing alloys will do that for all metals. The most important factor for any soldering or brazing alloy is that it must wet the base metal.

Lead and iron won't alloy or bond together because pure lead acts much like oil and water when poured into molten steel. It simply separates like cream does from milk. But tin does wet iron, and lead wets tin. That is, lead and tin make alloys. Tin and iron make alloys. Therefore, lead-tin solders will bond to steel (which is mostly iron). Similarly, zinc will alloy with aluminum. Therefore tin-zinc alloys are used to solder or braze aluminum. They often come in the form of a hard stick solder which is rubbed on the aluminum faying surface. The hard solder rubs through the aluminum oxide surface layer, breaking the oxide skin to get at the pure metal underneath.

Tinning
You probably are familiar with the use of a soldering iron. Sometimes the soldered connection isn't easy to make unless you "tin" the faying surfaces of the base metals first. The idea behind tinning is very simple. Using a flux (which acts something like a detergent) and a little molten solder (which is like water), you make sure that a thin layer of solder has wet the faying surfaces of the base metal. Once that thin layer is fully bonded, then you can complete the soldered joint because the rest of the solder easily sticks to the tinned surface, which is usually made of the same filler metal.

Sometimes, however, metals

don't make ductile alloys. Steel and aluminum are an example. They do combine but the result is an "intermetallic" compound of iron-aluminum. Iron-aluminum is extremely brittle. It does not act at all like a ductile metal. That, in fact, is the main reason why steel cannot be welded directly to aluminum. Copper and brass cannot be brazed directly to aluminum either, but steel and copper alloys do make good ductile joints with nickel. So does aluminum. Therefore you can use nickel alloys as a go-between to join these very dissimilar metals.

Aluminum causes all sorts of special soldering and brazing problems. So do stainless steels. The problem, of course, is that the same protective surface oxides that otherwise are so helpful in keeping these metals from oxidizing gets in the way when you braze them. Individual copper and brass alloys also may have special peculiarities, not so much in how they are joined, but what specific filler metals will join them to make the best connection. Some of these alloys become brittle under certain conditions. Some of them change in corrosion resistance depending on how they are joined.

Table 7-2 lists the standard AWS/ASTM solders. Each solder family or system is listed with data on its composition and its solidus and liquidus temperature (which also defines its pasty range, see p. 166). If you want details, the American Welding Society publishes a 170-page book on the subject titled *The AWS Soldering Manual*. There also are many proprietary solders, so call a supplier and see what's available.

Brazing Alloys
What's true of solders is equally true of brazing alloys. There are eight or nine separate groups of them. In most cases they are far simpler alloy systems than solders, even though they melt at higher temperatures. For example, there are silver-based brazing alloys, copper alloys, copper-zinc

(brass) and copper-phosphorus (phosphor-bronze) alloys, aluminum-silicon alloys used to braze aluminum, magnesium alloys for magnesium metals, nickel-brazing alloys for nickel alloy base metals, gold-alloy filler metals often used by jewelers, and even one special cobalt-base alloy. Table 7-3 lists the standard AWS brazing alloys.

Just like solders, most of these brazing alloys have special applications and only a few are for general-purpose use. The silver- and copper-based alloys are used most often on iron and steel because they produce high-strength bonds.

Sometimes nickel-based brazing alloys are tinned separately onto aluminum and steel. Then the joint is completed with more nickel-based brazing alloy. Sometimes the nickel may be put on the steel by electroplating, and then the nickel-alloy brazed joint is completed with the aluminum. It's even possible to make a *transition joint*. Transition joints are used in welding as well as in brazing and soldering. Some metals will make good bonds to a third material when they won't join up with each other. The third material becomes the material in the transition joint that bonds other materials together that normally don't get along well.

Steel and aluminum are examples. Together, iron and aluminum alloy to form extremely brittle compounds. But both steel and aluminum make ductile bonds with copper alloys such as bronze. Therefore the bronze filler metal can make a transition joint between two metals. Transition joints sometimes are used when two metals have expansion and contraction coefficients that are very far apart. Even if they can be joined, unequal thermal expansion and contraction can tear the joint apart. A third alloy with thermal expansion properties intermediate between the other two materials in a transition joint will often solve the problem.

TABLE 7-2 AWS standard solder grades

Lead-Tin Solders

AWS/ASTM Solder Class	Composition, weight %		Temperature, °F		
	Tin	Lead	Solidus	Liquidus	Pasty Range
5A	5	95	572	596	24
10A	10	90	514	573	59
15A	15	85	437	553	116
20A*	20	80	361	535	174*
25A*	25	75	361	511	150*
30A*	30	70	361	491	130*
35A	35	65	361	477	116
40A	40	60	361	455	94
45A	45	55	361	441	80
50A	50	50	361	421	60
60A†	60	40	361	374	13
70A	70	30	361	378	17

* Often called "wiping" solders because of their wide pasty range.
† The lowest-melting-point solder in the group. Often called "60-40" or "soft solder," it's the most common of all solders and the grade you will probably get if you go to a hardware store and say, "please give me some solder."

Tin-Antimony-Lead Solders

AWS/ASTM Solder Class	Composition, weight %			Temperature, °F		
	Tin	Antimony	Lead	Solidus	Liquidus	Pasty Range
20C	20	1.0	79.0	363	517	154
25C	25	1.3	73.7	364	504	140
30C	30	1.6	68.4	364	482	118
35C	35	1.8	63.2	365	470	105
40C	40	2.0	58.0	365	448	83

Tin-Antimony Solder

Composition, weight %		Temperature, °F		
Tin	Antimony	Solidus	Liquidus	Pasty Range
95	5	452	464	12

Tin-Silver Solders

Composition, weight %		Temperature, °F		
Tin	Silver	Solidus	Liquidus	Pasty Range
96.5	3.5	430	430	0
95.0	5.0	430	473	43

Tin-Zinc Solders

Composition, weight %		Temperature, °F		
Tin	Zinc	Solidus	Liquidus	Pasty Range
91	9	390	390	0
80	20	390	518	128
70	30	390	592	202
60	40	390	645	255
30	70	390	708	318

continued on page 164

Table 7-2 (*Continued*)

Lead-Silver Solders

AWS/ASTM Solder Class	Composition, weight %			Temperature, °F		
	Lead	Silver	Tin	Solidus	Liquidus	Pasty Range
2.5S	97.5	2.5	—	579	579	0
5.5S	94.5	5.5	—	579	689	110
1.5S	97.5	1.5	1.0	588	588	0

Cadmium-Silver Solder

Composition, weight %		Temperature, °F		
Cadmium	Silver	Solidus	Liquidus	Pasty Range
95	5	640	740	100

Cadmium-Zinc Solders

Composition, weight %		Temperature, °F		
Cadmium	Zinc	Solidus	Liquidus	Pasty Range
82.5	17.5	509	509	0
40	60	509	635	126
10	90	509	750	241

Zinc-Aluminum Solder

Composition, weight %		Temperature, °F		
Zinc	Aluminum	Solidus	Liquidus	Pasty Range
95	5	720	720	0

Indium Solders

Composition, weight %					Temperature, °F		
Tin	Indium	Bismuth	Lead	Cadmium	Solidus	Liquidus	Pasty Range
8.3	19.1	44.7	22.6	5.3	117	117	0
12	21	49	18	—	136	136	0
12.8	4.0	48.0	25.6	9.6	142	149	7
50	50	—	—	—	243	260	17
48	52	—	—	—	243	243	0

Typical Fusible Solders

Common Alloy Name	Composition, weight %				Temperature, F°		
	Lead	Bismuth	Tin	Other	Solidus	Liquidus	Pasty Range
Lipowitz	26.7	50	13.3	10 Cd	158	158	0
Bending (Wood's metal)	25	50	12.5	12.5 Cd	158	165	7
Eutectic	40	52	—	8 Cd	197	197	0
Eutectic	32	52.5	15.5	—	203	203	0
Rose's	28	50	22	—	204	229	25
Matrix	28.5	48	14.5	9 Sb	217	440	223
Mold and pattern	44.5	55.5	—	—	255	255	0

To make a transition joint, first braze a specially prepared thin area of the faying surface with a nickel electroplated coating, and then braze the other side of the coating to steel or copper. The same general idea is used in welding to make steel-aluminum transition joints although the details are very different.

For example, to make a transition joint by welding, an intermediate alloy such as bronze might be used because it will bond both to aluminum and to steel. In another type of welding transition joint, the steel and aluminum are joined directly together using a process that makes the iron-aluminum layer in between very thin so its low ductility doesn't matter. This is done with explosive bonding. The two pieces are actually blown together by high explosives. Then the bimetal section is used as a transition joint to weld more steel onto the steel side and more aluminum onto the aluminum side of the transition section.

Just like solders, there are dozens of special procedures, little tricks, and a few important precautions that you need to know before getting into brazing. We won't give you all the details. We doubt whether any one human being knows them all without looking some of them up. That's why the American Welding Society has a 300-page book titled *The AWS Brazing Handbook*.

You will also find that there are many proprietary brazing alloys (even more proprietary brazing grades than special solder alloys). These alloys have been developed by the filler-metal manufacturers to give you special properties that exceed those of the standard AWS/ASTM brazing alloys.

In a real shop situation, you will be given detailed brazing instructions when you need them. And, just like soldering, the person who gives you the specific procedure probably got it from the *AWS Brazing Handbook*, or from a company that specializes in producing and selling brazing alloys and fluxes.

7-5
HOW FILLER METALS MELT AND SOLIDIFY

Pure lead melts at 621°F [327°C]. Pure tin melts at 450°F [232°C]. When they are alloyed together, strange things happen when the alloys are heated or cooled. For example, most alloys of two metals do not have a single melting temperature. They have one temperature at which the alloy starts to melt and a higher temperature at which the melting is completed. In between this range is a semimolten slushy mixture partly filled with liquid metal and partly filled with solid metal crystals.

Water acts exactly the same way. Pure water has one specific freezing temperature, 32°F (0°C). But water with salt dissolved in it freezes at a different temperature depending on how much salt is included. The salt acts like an alloying element in the water-ice system. Instead of freezing all at once, the mixture of ice, water,

TABLE 7-3 AWS brazing alloys

AWS Classification	Solidus °F	Solidus °C	Liquidus °F	Liquidus °C	Brazing Temperature Range °F	Brazing Temperature Range °C
Silver						
BAg-1	1125	603	1145	618	1145–1400	618–760
BAg-1a	1160	627	1175	635	1175–1400	632–760
BAg-2	1125	607	1295	702	1295–1550	702–843
BAg-2a	1125	607	1310	710	1310–1550	710–843
BAg-3	1170	632	1270	688	1270–1500	688–816
BAg-4	1240	671	1435	779	1435–1650	779–899
BAg-5	1250	677	1370	743	1370–1550	743–843
BAg-6	1270	688	1425	774	1425–1600	774–871
BAg-7	1145	618	1205	651	1205–1400	651–760
BAg-8	1435	779	1435	779	1435–1650	779–899
BAg-8a	1410	766	1410	766	1410–1600	766–871
BAg-13	1325	718	1575	857	1575–1775	857–968
BAg-13a	1420	771	1640	893	1600–1800	893–982
BAg-18	1115	602	1325	718	1325–1550	718–843
BAg-19	1435	779	1635	891	1610–1800	891–982
BAg-20	1250	677	1410	766	1410–1600	766–871
BAg-21	1275	691	1475	802	1475–1650	802–899
Aluminum-Silicon						
BAlSi-2	1070	577	1135	613	1110–1150	599–621
BAlSi-3	970	521	1085	585	1060–1120	571–604
BAlSi-4	1070	577	1080	582	1080–1120	582–604
BAlSi-5	1070	577	1095	591	1090–1120	588–604
BAlSi-6	1038	559	1125	607	1110–1150	599–621
BAlSi-7	1038	559	1105	591	1090–1120	588–604
BAlSi-8	1038	559	1075	579	1080–1120	582–604
Precious Metals						
BAu-1	1815	991	1860	1016	1860–2000	1016–1093
BAu-2	1635	891	1635	891	1635–1850	891–1010
BAu-3	1785	974	1885	1029	1885–1995	1029–1091
BAu-4	1740	949	1740	949	1740–1840	949–1004
BAu-5	2075	1135	2130	1166	2130–2250	1166–1232
Copper and Copper-Zinc						
BCu-1	1980	1082	1980	1082	2000–2100	1093–1149
BCu-1a	1980	1082	1980	1082	2000–2100	1093–1149
BCu-2	1980	1082	1980	1082	2000–2100	1093–1149
RBCuZn-A	1630	888	1650	899	1670–1750	910–954
RBCuZn-C	1590	866	1630	888	1670–1750	910–954
RBCuZn-D	1690	921	1715	935	1720–1800	938–982
Copper-Phosphorus						
BCuP-1	1310	710	1695	924	1450–1700	788–927
BCuP-2	1310	710	1460	793	1350–1550	732–843
BCuP-3	1190	643	1495	813	1325–1500	718–816
BCuP-4	1190	643	1325	718	1275–1450	691–788
BCuP-5	1190	643	1475	802	1300–1500	704–816
BCuP-6	1190	643	1450	788	1350–1500	732–816
BCuP-7	1190	643	1420	771	1300–1500	704–816

continued on page 166

Table 7-3 (*Continued*)

AWS Classification	Solidus		Liquidus		Brazing Temperature Range	
	°F	°C	°F	°C	°F	°C
Magnesium						
BMg-1	830	443	1110	599	1120–1160	604–627
BMg-2a	770	410	1050	566	1080–1130	582–610
Nickel						
BNi-1	1790	977	1900	1038	1950–2200	1066–1204
BNi-1a	1790	977	1970	1077	1970–2200	1077–1204
BNi-2	1780	971	1830	999	1850–2150	1010–1177
BNi-3	1800	982	1900	1038	1850–2150	1010–1177
BNi-4	1800	982	1950	1066	1850–2150	1010–1177
BNi-5	1975	1079	2075	1135	2100–2200	1149–1204
BNi-6	1610	877	1610	877	1700–2000	927–1093
BNi-7	1630	888	1630	888	1700–2000	927–1093
BNi-8	1800	982	1850	1010	1850–2000	1010–1093
Cobalt						
BCo-1	2050	1121	2100	1449	2100–2250	1149–1232

and salt forms a solution (like a metal alloy) that produces an ice-water-salt system with a range of freezing points. The result is slush. Metal alloy systems that form slush are very useful. Phase diagrams show how to find them.

Phase Diagrams

Every alloy system has a diagram that shows how the various possible alloy combinations melt and solidify, depending upon how much of either of the two (or more) alloying elements there are in it, and the temperature. These are called *phase diagrams* and metallurgists use them a great deal. Geologists use the same type of diagrams to understand rocks, but their phase diagrams usually include three or four elements plus temperature and pressure. A five- or six-dimension phase diagram is very complicated.

Figure 7-16 is the phase diagram for the binary (two-element) lead-tin filler-metal system. It may look complicated at first . . . because it is. (The copper-nickel phase diagram, for example, is ex-tremely simple while the iron–iron carbide diagram that includes steel is even more complicated.) However, it's worthwhile looking at this lead-tin diagram because what it can tell you will be useful when you work with any filler-metal system, including those used in welding.

As you can see, the bottom of the chart, on the left, starts with 100 percent lead. Reading up the left side of the chart, you can see that 100 percent lead simply gets hotter and hotter until it melts at 621°F [327°C] (or point A on the lead-tin alloy phase diagram). But as we move to the right on the bottom of the diagram, adding more tin and less lead up to about 20 percent tin, the temperatures at which lead-tin alloys melt drop off very fast.

Solidus The line drawn through all the temperatures at which all alloys in this system melt is called the *solidus*. Everything below the solidus (running from point A through C and E to point D and up to B in the diagram) is totally solid metal.

Let's see what happens if you heat up an alloy of 20 percent tin and 80 percent lead (one fifth of the way from the left-hand side of the chart). The alloy gets hotter and hotter until it reaches the solidus at 361°F [183°C] (point C on the chart). Then the metal starts to melt . . . but it doesn't finish melting until it reaches about 560°F [293°C], which is the *liquidus*.

Liquidus The liquidus is simply a line drawn through all the temperatures at which all the possible lead-tin alloys that can exist become fully molten. The liquidus in the lead-tin alloy phase diagram that you see here runs down from point A to point E and up to point B.

The region between the solidus and the liquidus lines (where there is a space between them) is the slushy region (called the *pasty range* when you are working with solders or brazing alloys), where the metal will be partly very hot solid crystals and partly molten metal. This region can be useful in selecting and using certain alloys. For example, if you want to use an alloy that won't be too runny, or that you can mold while partially melted, this is the region in which you would work the alloy.

The vertical distance between the solidus and liquidus lines tells you how much room (temperature range) you have to work in. The lead-tin "wiping" solders listed in Table 7-2 are a perfect example of such alloys. Note that the wiping solders have a fairly wide gap between the solidus and the liquidus. Wiping solders are used in auto body shops because they can be molded into shape while semi-molten and still in the pasty range. They are used to fill dents in fenders and hoods and to smooth out hammer marks that otherwise would show up on a repair job after painting.

Eutectic point There is another very important point in this diagram. It is the *eutectic* point.

Not all alloy systems have a eutectic point, but many do. The eutectic point is probably one of the most important ideas in metallurgy. A eutectic point is simply the lowest possible melting point in an alloy system where two liquidus lines and two solidus lines come together.

Many important reactions occur at eutectic points in different alloy systems. For example, cast irons (which contain 2 to 5 percent carbon) include much of their carbon as graphite, not as iron carbides that form in steel. Not only does graphite give entirely different properties to cast irons, but the solidus and eutectic temperatures 2075°F [1135°C] for the eutectic) are much lower than the solidus temperatures of any steel (usually around 2600°F [1427°C]). That explains why cast irons melt at lower temperatures than steels and why cast irons are much easier to cast than steels. Eutectics are important in many alloy systems from aluminum to zinc.

You can see from the diagram (point E) that a solder containing 63 percent tin and 37 percent lead has the lowest possible melting point in the entire lead-tin solder alloy system. (That's what you would select when soldering a thin-gauge part, when heat distortion would pull it out of shape.)

Arrangement of atoms What's all the "alpha," "beta," and "gamma" stuff on the diagram? We won't get into that now, not because it's complicated but because it's not too important for you to know right now. When working with steel it gets very important. For example, the differences between various families of stainless steels are based directly on their position in different phase diagrams. The way that carbon steels are heat-treated is predicted directly from similar phase diagrams, all of which are full of Greek letters identifying different phases that occur at different temperatures.

For the moment we'll simply mention that the atoms in all crystalline solids take up different positions in metal crystals at different temperatures, and those temperatures and atomic arrangements vary with the amount and kind of alloying elements that are in the material. These different atomic arrangements are given Greek-letter names by metallurgists. They are the key to why different alloys have different properties. The same is true of materials that aren't made of metal.

We'll give you an example. The chemical element carbon can exist as black smoke, as graphite, or as diamond. Each material is formed at a progressively higher temperature (and sometimes high pressure). Whether you called them alpha carbon, beta carbon, or gamma carbon does not make things any more complicated than calling them carbon, graphite, or diamond.

Let's look at one more alloy phase diagram, the silver-copper phase diagram in Fig. 7-17. This time we left out the Greek letters completely (although they would be there in a complete phase diagram). This diagram has a eutectic point at 72 percent silver (Ag) and 28 percent copper (Cu). The eutectic temperature is 1435°F [780°C].

The eutectic 72 percent silver and 28 percent copper is a very common brazing alloy. However, a lot of silver-copper brazing alloys also are found to the right of the eutectic composition. The reason is the relatively wide *melting range* of the alloys on the right-hand side. These alloys are very useful when the part fit-up is not perfect and there may be some gaps that are a little wider than you would like to have in a perfect brazed joint. Therefore, the "mushy zone" brazing alloys on the right hand side will help overcome gaps with wider clearances (over 0.005 in. [0.13 mm], for example). On the other hand, the eutectic composition is very nice if you have a perfect fit-up and don't want to heat the workpiece any hotter than necessary, to avoid warping.

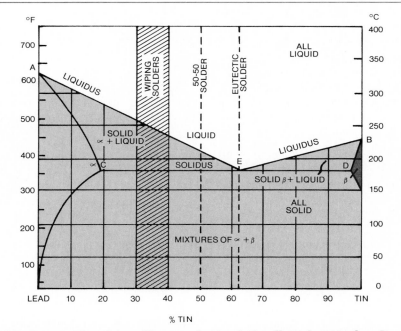

FIGURE 7-16 The phase diagram for lead-tin alloys is complex. Its overall shape (but not the temperatures, compositions, or phases) resembles the phase diagram for iron and carbon: the basis for steel metallurgy. The intermediate phases that are part solid and part liquid are useful in soldering practice and in selecting specific soldering alloys.

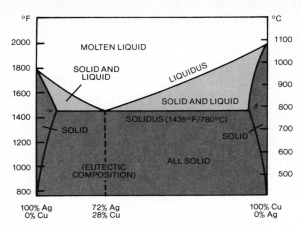

FIGURE 7-17 Another moderately complex phase diagram, this one is for silver-copper alloys. The lowest melting point alloy in the silver-copper system is 72 percent silver and 28 percent copper. This is the eutectic composition for the silver-copper alloy system.

7-6
JOINTS THAT HOLD TIGHT

What is the best kind of joint to hold two pieces of metal together when brazing or soldering? There are many kinds of joints, but only two basic joints are possible for brazing and soldering. These are the *butt joint* and the *lap joint*. All other soldering and brazing joints are essentially modifications of these two basic ways of putting metal together.

The Butt Joint

Let's look first at the butt joint, both for flat and tubular products (Fig. 7-18). As you can see, the butt joint gives you a single thickness at the joint. Preparation of this joint is usually very simple and the joint will have enough tensile strength for many applications. However, this joint is not the strongest way to join metals by brazing or soldering.

The strength of the butt joint depends upon the amount of bonding surface (the area of the faying surface) and there's really not much surface area in this joint. What's more, the butt joint can't be any larger than the cross section of the thinnest member, as shown in Fig. 7-19.

The Lap Joint

Now let's compare this with the lap joint, both for flat and tubular products (Fig. 7-20). The first

thing you'll notice is that for a given thickness of base metal, the bonding area of the lap joint can be larger than that of the butt joint—and it usually is. With larger bonding areas, lap joints can usually carry larger loads.

The lap joint gives you a double thickness at the joint, but in many applications, such as plumbing connections, for example, the double thickness is not objectionable. The lap joint is generally self-supporting during brazing or soldering. Resting one flat member on the other is usually enough to maintain a uniform joint clearance. In tubular joints, nesting one tube inside the other holds them in proper alignment for brazing.

The rule of thumb for a good lap joint is to make the overlap distance of the two faying surfaces three times as long as the thickness of the thinnest joint member (Fig. 7-21). A bigger overlap wastes filler metal and

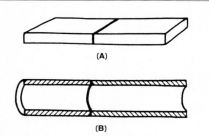

FIGURE 7-18 (A) Brazed butt joints can be made on flat parts. (B) They can also be made on round tubing. Butt joints are the most common of all joints for brazed tubular products because they are easy and inexpensive to make.

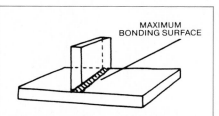

FIGURE 7-19 This is a fillet joint. Its maximum strength depends upon the width of the faying surface between the two parts.

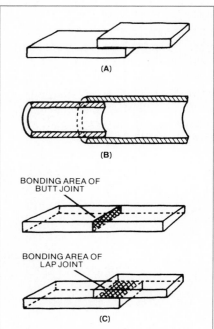

FIGURE 7-20 Lap joints have larger faying surface areas than butt joints. (A) A lap joint for flat parts. (B) A lap joint for tubular parts. (C) Compare the size of a lap joint surface with that of a butt joint.

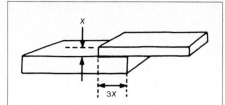

FIGURE 7-21 A lap joint should always have these proportions: The size of the lap should be three times the thickness of the smaller section.

uses more base metal than is needed. A shorter lap will lower the strength of the joint.

NOTE: The overlap distance of a good lap joint will be three times as long as the thickness of the thinnest member.

Butt-Lap Joints

Suppose that you want a joint that has the advantages of both butt and lap joints: a single thickness at the joint combined with maximum tensile strength. You get this combination by using a butt-lap joint (Fig. 7-22). The butt-lap joint is usually a little more work to prepare than straight butt or lap joints, but the extra work will pay off. You wind up with a single-thickness joint of maximum strength, and the joint is usually self-supporting when assembled for brazing.

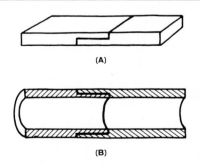

FIGURE 7-22 Butt-lap joints combine the best properties of both joint types. They also are much more expensive to prepare. (A) A butt-lap joint for flat products. (B) A butt-lap joint for tubular products.

Scarf Joints

There's still another way to combine strength with a single thickness of base metal in a brazed joint. That's by making a scarf joint (Fig. 7-23), which actually is a modification of the butt joint where the bonding surfaces are angled to increase the bonding area. The bonding area of a scarf joint must be at least three times larger than that of a butt joint if you want to get maximum strength.

NOTE: The area of a good scarf joint must be at least three times that of the butt-joint area for maximum strength.

Scarf joints are often useful, but they have some drawbacks. The matching angled surfaces are more difficult to prepare, and the joint is usually not self-supporting. Where you want strength and single thickness in the same joint, the butt-lap joint will generally prove your best bet.

You can modify the basic butt and lap joints in many ways to meet special purposes and different kinds of base-metal shapes, or to join different kinds of parts. Figure 7-24 gives you some examples. You also have learned a lot about welding just by looking at these joints. These brazing and soldering joints are also basic welding joints; but in welding, there are even more kinds of joints, as you'll see in later chapters.

7-7
BRAZE-WELDING

As we mentioned in the beginning of this chapter, braze-welding is a method for making strong bonds in conventional welding joints without melting the base metal. Braze-welding will make high-strength joints in most carbon steels, tool steels, some stainless steels, gray cast iron, malleable and ductile iron, and it does a very good job of joining galvanized steel sheets, not to mention things like copper, steel, and cast-iron water pipe. The process does

a beautiful job in repairing cracked cast-iron engine blocks. Braze-welding also is a principal tool of hobby welders and metal sculptors.

Advantages and Limitations
Braze-welding has both advantages and limitations. The advantages far outweigh any problems you may have with this joining process.

Advantages The braze-welding process is used not only to join irons and steels, but also to build up broken sections such as gear teeth that have worn away, or to make corrosion- or abrasion-resistant surface coatings. One example is building up the worn surfaces of pistons. Another is repairing propellers and propeller shafts.

Other advantages of braze-welding concern its relatively low temperature. Distortion will be less when braze-welding than when fusion-welding. For the same reason, massive cast-iron machine parts will only require local heating to bring them up to braze-welding temperatures, while the same big pieces might have to be preheated heavily before they could be gas-welded. Therefore braze-welding often can be done faster than full fusion-welding.

Limitations The most important limitation of braze-welding is that it should never be used to repair

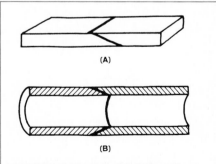

FIGURE 7-23 Scarf joints offer a lower-cost solution than butt-lap joints for increasing the faying surface to increase the strength of the brazed joint.

work that will then be used at temperatures higher than 500°F [260°C]. The reason is simple. Bronze loses strength and hardness above this temperature. Another drawback is that you can't get any color match between the bronze filler metal and iron or steel parts.

A third disadvantage is that the filler metal costs more than iron or steel filler metals used for full fusion-welding. If a very big weld is to be made, the cost of brass or bronze filler metal can be quite high.

Still another drawback to braze-welding is that in certain chemical processes materials may be used that have almost no effect on cast iron but will corrode the bronze filler metal very quickly. For example, hot alkaline (basic as opposed to acid) solutions such as sodium hydroxide, which you know by names such as "lye" or "caustic soda" (used in making soap and many other industrial chemicals), will eat away the braze-welded joints of piping that carries it.

Braze-Welding Filler Metals
Braze-welding filler metals are usually either bronze or brass alloys. They melt at temperatures considerably above 800°F [425°C], usually around 1600 to 1800°F [871 to 982°C], but still below the melting point of irons and steels. For example, the eutectic temperature of the iron-graphite system (in which you find cast irons) is 2075°F [1135°C]. All commercial cast irons melt above that temperature. Low-carbon steels melt at temperatures higher than 2550°F [1399°C]. Therefore, when you are braze-welding iron or steel you still have a range of 500 to 1000°F [260 to 540°C] between the melting point of the nonferrous filler metal and melting point of the ferrous base metal. The exact difference between the melting point of the filler metal and base metal will depend on which alloys are compared, of course.

An important feature of the copper-bearing filler metals most commonly used for braze-welding is that they are very strong. They can make joints with tensile or pull-apart strengths of over 100,000 psi [689 MPa], which is roughly three times stronger than carbon steel. Most of them are bronze alloys.

Metal names Since ancient times it has been known that copper alloyed with zinc to make brass, and copper alloyed with tin to make bronze. Today, the name *bronze* covers many copper alloys (except copper-zinc, which is brass). There are copper-silicon alloys called silicon-bronze alloys, copper-tin (the original bronze), copper-phosphorus (phosphor-bronze) and other bronzes. However, there are some confusing exceptions.

Most of the bronze alloys used in braze-welding are not bronze at all. They are really brasses with roughly 60 percent copper (Cu) and 40 percent zinc (Zn) but with a pinch or two of other metals such as nickel (Ni), manganese (Mn), silicon (Si), and other ingredients. These alloyed brasses are called bronzes because the alloying elements serve to

■ Increase the tensile strength of the alloy.
■ Improve its hardness and wear resistance.
■ Improve its flowing and wetting properties.
■ Make it tin better on hot metal surfaces.
■ Deoxidize the molten filler metal.
■ Help you obtain a slag-free deposit with only a minimal tendency to fume at brazing temperatures.

What's more, the special 60–40 copper-zinc alloys (with a few extra additions) have another name entirely. They are called *Muntz metal*. Muntz metal is used not

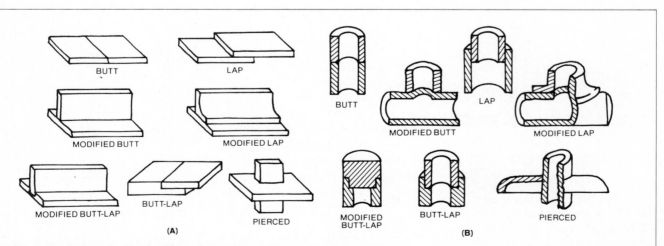

FIGURE 7-24 Here are a wide selection of brazed joints you can use for (A) flat surfaces and (B) tubular products.

only as a braze-welding filler metal but also as one of the most common brazing alloys.

We'll stick to tradition and call our brass braze-welding alloys bronzes. You do the same and at least everybody will know what you are talking about.

Preparing Metals for Braze-Welding

Braze-welding joints, unlike brazed joints, can have plenty of space between them. If the edges to be joined are ¼ in. [0.6 cm] thick or less (sheet metal or thin-walled tubing), you can use a chisel or cutting torch to get down to clean bright metal. Before joining them, it's a good idea to pass a grinding wheel or a hand file over the surface to make sure that you have a clean joint. Also remove any grinding residue with a wire brush.

If the metal to be joined is over ¼ in. [0.6 cm] thick (which is where plate thicknesses begin), the edges should be beveled to give about a 90° V joint. Again, run a grinder or hand file over the surface to be joined to make sure you are down to clean base metal.

Special cast-iron problems If you machine or grind gray, ductile, or malleable cast irons you will run into a little trouble. These materials all contain graphite flakes. The grinding wheel, for example, will tend to smear the graphite out on the faying surface. That means that flux will not react with the surface and you won't be able to tin it (which is a very important step in braze-welding).

Like so many things in welding, there's also more than one way out of the problem. First you can use a cleaning solution (actually a detergent) made of trisodium phosphate to remove surface oil. Then you can heat the cast-iron edges up to a dull-red temperature before you do any work on them. Not only heat the surfaces

to be joined but also heat an area about ½ in. [1.25 cm] on either side or back from the top of the joint. That will burn off the smeared graphite or oil and give you clean base metal to work with.

Sometimes you may have to repair a cast-iron part that has been in a fire, perhaps a boiler heating section. You will have trouble tinning these surfaces with bronze filler metal.

Another common problem is machine parts such as machine-tool bases and frames or an automotive crankcase that have been soaked with oil or cutting lubricants for a long time. Again, work 15 min or more with a heating torch on the surface to clean it up and maintain it at a dull- to bright-red temperature (up to 800°F [425°C] if you have a temperature-indicating crayon) will burn out the graphite, the oil, or what have you, and it will give you clean metal.

Some cast irons subject to sea-water or corrosive chemicals may become "soft" on the surface. What's happened is that the crystals or grains in the metal are "opening up." The area between the grains has been partially dissolved on the surfaces of the part and that will make it difficult to tin the part with bronze. Alternate heating and rapid cooling will "close up" the surface and make it suitable for tinning. Very obstinate cases may make it necessary to machine off some of the surface to get down to unaffected, clean base metal. This sometimes happens with diesel locomotive pistons which are loaded with oil and greasy soot.

After you have prepared the joint, clamp the pieces to be joined in a jig or fixture so they won't move around when heated and cooled. Now you are ready to get to work.

Flame Settings

The tip used for braze-welding should be about one size larger

than the tip you would use for welding steel of the same thickness. The idea is simply to get more heat into the base metal. A slightly oxidizing flame is used for braze-welding steel or cast iron.

As you know, a slightly oxidizing flame setting is obtained by first making a neutral flame and then adjusting it. Remember that the oxidizing flame will look different depending upon the fuel gas you are using. The slightly oxidizing flame setting will help you get a better bond between the bronze and the steel or cast iron when you wet it. It also will keep a slight film of oxide over the surface of the molten filler-metal puddle, which protects the molten braze-weld metal from further oxidation by the oxygen in the air.

Preheating the Base Metal

You won't have to preheat thin steel sections (up to 1 in. [2.54 cm] thick, at least). If you do any preheating at all, it probably will be on a casting. Even then, you won't have to preheat castings up to about 100 lb [45 kg] in weight because the welding torch itself will do the preheating by the way you will hold it. The method of holding the torch is called the *forehand* welding position (Fig. 7-25), which means that the welding flame is pointing in the direction you will be traveling. In backhand welding, you hold the flame so that it points at where you've already been. We'll show you that shortly.

If you do have to preheat the base metal, all you will probably have to do is preheat the joint area along the line you will braze-weld. The only time you need to do extensive preheating is on a very massive casting such as a machine base or frame, because it will soak up heat as fast as you pump it in if you don't preheat, and you will not be able to bring the braze-weld joint area up to temperature. Incidentally, the same is true of fusion-welding.

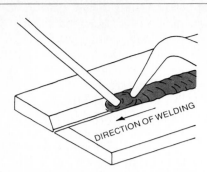

FIGURE 7-25 This is your first welding technique. Use it for braze welding. Learn it well. You will use it again and again in both gas and electric-arc welding. In forehand welding, the flame points forward in the direction you are welding. Alternate the welding flame from side to side to heat the base metal as well as the filler metal rod.

Some experienced maintenance welders simply set up a couple of very large heating torches and let the big casting "cook" for 8 h or so while they do something else. Sometimes they'll even build a brick oven around the big part out of firebrick, or heat a big machine base along the crack with oil or natural-gas burners. When it's very warm, they get back to work.

Braze-welding a massive preheated machine base during midsummer is not the best job in the world, but some people wear air-conditioned heat-reflecting "hot suits" like those worn by airport fire fighters, when doing this very massive work. You won't need to do anything like that when working with small parts.

Fluxing the Work and Rod

You will have to select a flux suitable for working with whatever base metal you will repair or join. If in doubt, ask a welding distributor, a welding engineer, or the flux supplier. If you are not experienced, be liberal with the flux. It's better to use too much than too little when you tin the workpiece faying surfaces even though you should use only as much flux as you need to get good tinning action.

Also heat your bronze filler rod up just enough to make flux stick to it, and dip the rod in the flux container every once in a while as you work.

The molten flux should not be allowed to melt to the point where it is clear. Let it stay slightly cloudy with dissolved oxides. If it starts to get too clear, add more flux. That way you know you have plenty of additional fluxing power to pick up more metal oxides if you need to and you won't load up the flux with oxygen to the point where it turns black. Overoxidized flux is very hard to remove from a joint.

Tinning the workpiece
When the base metal is up to temperature, start dipping or touching the bronze braze-welding rod to the base metal as needed to make the joint. The filler rod should flow smoothly and spread out like a very wet liquid. Keep moving your torch forward along the joint, pointing the flame slightly ahead of the direction you are moving (use the forehand welding position), while you continue to add filler metal to all the exposed faying surfaces of the joint.

One of the nice things about braze-welding is that tinning takes place only when conditions are just right. If the base metal is too cold, the bronze is sticky and refuses to run out and spread over the heated surface. If the base metal is too hot, the bronze will form little balls, like droplets of water on a hot stove. If the temperature of the base metal is correct and the tinning is done properly, the molten bronze resembles water spreading over a clean, moist surface.

Tinning is the most important step in braze-welding. It forms the bond or molecular alloy surface union between the base metal and the rest of the filler metal you will add later on. Do it right and everything else will be easy.

NOTE: Tinning is the most important step in braze-welding. Do it right the first time and everything else is easy.

Don't leave any part of the faying surface uncovered. Bright spots on the molten filler-metal puddle indicate oxides and impurities and should be worked out with the welding flame and the flux. The bronze braze-welding rod should never be melted faster than the progress of the tinning action. The success of the rest of your work depends upon getting a good tinned surface to begin with.

While you do this work, you will see why welding is more difficult than brazing. With one hand you are swirling a torch flame slowly in little circles and with the other you are guiding and dabbing on filler metal. Welders have to be good enough to do two entirely separate jobs at once with two different hands while thinking about where the molten metal is going and keeping the torch flame pointed more or less in the direction of travel.

Starting the Braze-Weld

After tack-welding the joint, make sure that it's clamped down securely unless it's much too big to move around and you know it's not going anywhere. After the base metal is preheated (if necessary), use your slightly oxidizing torch flame to make swirling circular motions over the area where you will start the tinning operation. Start on one or the other end of the open joint. As the base metal gets hot enough to melt the bronze filler metal, the joint will start to glow cherry red.

It's hard to judge exactly the right temperature until you get some experience because your dark goggles will make it difficult for you see this color. But it's about the same temperature as you use to preheat steel before flame-cutting it.

Making More Passes

After the entire faying surface is fully tinned, you are ready to add the *root pass* at the bottom of the V joint (Fig. 7-26). Make a second pass over the first or root pass to fill up the V joint. If you have to, make more passes, even if they have to be side by side, to build up the joint gap.

Go sparingly on the flux when making a root pass in a V joint. You don't want too much flux floating on the top of the molten metal you will deposit in the bottom of the V joint, because you will be making more than one pass. The excess flux might cause problems with making a solid, void-free joint when you make second and third passes over the V joint.

Every time you make a pass, wash the molten metal up against the edge of the tinned base metal.

This will "tie in" your filler metal to the base metal. You will do exactly the same thing when you make fusion welds, whether you use a gas flame or an arc-welding electrode. Learn how to tie in your welds early, and you will pass a lot of welding certification tests later on.

For practice, try using the backhand welding technique when you apply second and third passes over the bronze root pass. In this method, the torch points in the direction you came from, not the direction you are going (Fig. 7-27). It's not necessarily the right way to do this work, but now is a good time to learn backhand as well as forehand welding.

Only on the final pass should you contour the molten weld metal up in the middle to add a little reinforcement on the weld bead. All the other passes should

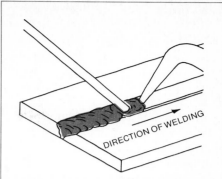

FIGURE 7-27 The backhand welding and braze-welding technique points the flame back over the deposited filler metal. Because the filler rod is between the flame and the weld, the welding action can proceed more rapidly than in forehand welding.

be concave or dipping slightly in the middle and washed up onto the sides of the tinned base metal. Do not let the first root pass have a hump in the middle with its edges sloping down. That condition is called *wagon wheels*. Wagon wheels will very likely make it difficult to get a good joint because the filler metal won't bond completely to the base metal.

Later on, when you are gas or arc fusion-welding, you will find that even worse problems will occur if you don't make the root pass correctly. You will run into such problems as undercutting and cold laps that won't pass any welder certification test.

Removing Flux

Everything we told you about removing soldering fluxes also applies to brazing fluxes, including the methods for removing them. But since the base metal will be steel, you'll often find that the easiest way to remove the flux is with a wire brush. If you are at all concerned about scratching the surface, use a chemical instead of mechanical flux-removal method.

Inspecting Your Work

Figure 7-28 shows you how to determine the quality of your weld,

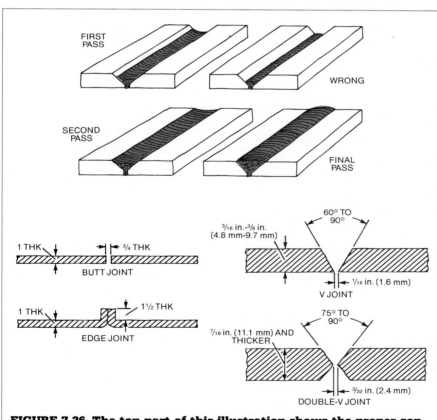

FIGURE 7-26 The top part of this illustration shows the proper contours for bronze alloy braze-welding of grooved joints. The top surface of each pass, except the final pass, should be concave. The convex top pass in the illustration, marked "wrong," should be avoided. The bottom illustrations show typical butt joint and groove joint designs which are used on thicker plates.

assuming your first braze-weld will be on a piece of steel plate rather than a heavy casting that you might actually have to repair later on. (The test won't work on a casting because cast iron is too brittle and will break, so you can't bend it.) Put the sample plate joint over a gap and then put a wide, thin steel bar, about the shape of a steel ruler, on top, right in the middle of your joint. Make sure that the root of the joint is facing down. Now press hard and steadily on the joint. (If your welding shop has a small press brake, this will do very nicely.)

If your work bends but does not break as it is folded up, you have made a pretty good joint. After folding up the joint (assuming it did not crack open), try sticking it into a vise and really folding it flat. You can even try pounding it flat with a hammer. If you can fold your joint in half like folding up a letter and it still holds up, you have learned how to make a very good braze-welded joint and are well on your way to becoming a skilled welder.

Some tips Braze-welding is one skill that you must practice a lot to get down pat. You won't get it

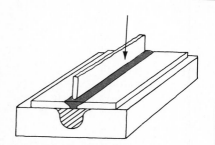

FIGURE 7-28 This is how to test the strength of your first braze-weld on a butt joint. Put the finished braze-weld over an air gap, root side down, and set a steel bar on top. Then hit the vertical bar with a hammer. If the joint breaks instead of bending, you flunk the test.

immediately. However, when you learn to do it correctly, you will have taken a big step in learning to do gas welding, and you also will have developed a lot of the manual skills you will need when you learn to do arc welding. Here are a couple of ideas to help you.

You don't have to use a straight bronze filler rod. Bend it to whatever angle makes you happy and makes it easy to poke the point into the base metal joint. Also use the force of the gases coming out of the torch at the flame to make the molten tinning bronze flow where you want it to go. You can actually blow molten metal uphill if you want. Dip the bronze filler rod in and out of the hot metal joint whenever you want to add more of it.

Try to keep the bronze rod out of the direct flame of the torch, unless you want to melt an extra amount off to work around the surface of the joint. Keep the torch flame moving in small circles or in a weaving pattern, washing the filler metal up the edges of the joint. If you leave the torch flame in one spot, you will start melting the base metal and that's strictly forbidden.

The inner flame cone should be about ¼ to ⅛ in. [0.6 to 0.3 mm] from the surface of the base metal. The flame is pointed ahead of the melting filler metal at an angle of about 45° with the puddle, and held slightly behind the flame. This angle varies, of course, with flat, vertical, and overhead braze-welding. It depends on the thickness of the joint or surface to be built up. Your manipulation of the torch and bronze rod also depend heavily upon the size of the puddle being carried, the nature of the joint or surface being braze-welded, and the speed you work at.

On heavy sections, proper tinning is most easily obtained by positioning the joint so that the braze-weld will progress slightly uphill, about 30° above the horizontal. Not all joints can be positioned like this, but it's nice to try for it.

REVIEW QUESTIONS

1. As a rough rule of thumb, what's the difference between brazing and soldering?

2. What's the most important difference between fusion-welding and brazing (or soldering)?

3. What force causes molten brazing metal or solder to get into very tiny spaces between metal parts?

4. What does the process code FB stand for?

5. Is a joint clearance between two parts of 0.01 in. [0.25 mm] tight enough for brazing?

6. What is the linear coefficient of thermal expansion of copper in units of inch per inch per degree Fahrenheit?

7. When mixing up an acid cleaning solution, would you add the water to the acid, or the acid to the water?

8. What is the solidus in degrees Fahrenheit of an ASTM 50A tin-lead solder?

9. Sketch a lap joint. What is the correct length of the overlapping area in terms of the thickness of the thinnest section in the joint?

10. For full effectiveness, the scarf-joint faying surface should be at least how many times that of the normal butt-joint cross-sectional area?

11. Bronze filler alloys are most often used for which process: soldering, welding, or braze-welding?

12. When you are forehand welding or brazing, which direction does your torch flame point—in the direction you are moving (the flame is pointed at the joint before it is filled)—or in the opposite direction (at the deposited filler metal)?

13. Should you use a highly oxidizing flame, a slightly oxidizing flame, a neutral flame, or a reducing (carburizing) flame when braze-welding gray cast iron?

14. What is the faying surface of a brazed, soldered, or welded joint?

Soldering, Brazing, and Braze-Welding Practice

7-1-S
EQUIPMENT AND SUPPLIES

Since we don't know what materials you have readily available, we won't specify the thickness or type of base metal in most cases. If possible, use a good selection of both flat-rolled and tubular products and wire, selecting from a combination of ferrous and nonferrous metals and several types of solders and brazing alloys.

You will need the same basic equipment used for the prior shop-practice work, plus an air-fuel torch if available. The following are only suggested shop exercises. Your welding instructor will determine what equipment and materials are available for you to use.

In most instances you will be working with steel, not only because it costs less than copper and aluminum, but more importantly, about 94 percent by weight of all engineering metals used are irons and steels. Aluminum, copper, zinc, nickel, titanium, and other metals make up the remaining 6 percent of use. Therefore, most of your actual work in a real shop probably will be spent working on steel products rather than other materials. Nevertheless, gain as much experience as you can with other metals, especially aluminum and copper alloys. They also are very important. Being able to join them is one measure of a highly skilled welder.

We'll give you a long list of shop projects because the next chapter will not have much shop work. We want you to continue these projects while you study the next chapter on mechanical properties of metals.

These projects will prepare you for the fusion-welding chapter that follows Chap. 8. The manual skills you will be learning here apply directly to most forms of gas and arc welding. If you don't learn them now, the manual work will get progressively more difficult. Learn them now and the work to come will get easier.

Don't be discouraged if you have trouble with some of these exercises. We haven't told you yet how to do a lot of it, especially the out-of-position work. (Sometimes its better to figure things out for yourself.) You will spend the rest of this course learning many of the techniques we'll start with here. You will spend the rest of your working life improving those skills.

7-2-S
SOLDERING AND BRAZING

1. Solder two pieces of stranded copper wire together with a lead-tin solder. This is the the most common electrical connection, except that you should learn to do it with a torch instead of a soldering iron. Use an air-fuel torch if available, or a small gas welding torch. Be sure to use a small tip suitable for soldering small parts.

2. Braze or solder a lap joint from two pieces of sheet. Use an overlap three times the thickness of the metal. This is the most common of all brazed joints. It's also a common welding joint. Use a small-capacity air-fuel or oxy-fuel welding torch.

3. Repeat Exercise 2, but join two pieces of galvanized steel. This is a common air conditioning, automotive, and farm job. Use an air-fuel torch if available.

4. Braze or solder a round tube to a flat sheet using a butt joint. This is a common kind of brazed connection to join a tube to a flat base. The flat base often will have a hole in it the same diameter as the ID of the tube. This is a common job in air conditioning and refrigerator work.

5. Braze a T joint. This is another very common brazing or soldering joint. If you want the leg of the T to be at an angle other than 90° to the base, you will have to grind the bottom edge of the upright leg of the T at the desired angle to make a good faying surface against the base. That wouldn't be necessary if you welded or braze-welded it to the base (or top) of the T.

6. Prepare a butt joint from two pieces of sheet and braze them together. This is also a common welding joint.

7. Prepare and braze a scarf joint from sheet or plate thick enough so that you can prepare the joint with a grinder. The biggest problem will be in making the joint so that it fits together smoothly with a uniformly tight fit-up.

8. Braze a copper tube to a flat steel base by using a butt joint. Use an air-fuel torch if available.

9. Join two tubes of equal diameter with a butt joint. This is a common joint used in plumbing, heating, air conditioning, and refrigeration work. Use an air-fuel torch if available.

10. Braze five screws or bolts onto steel sheet or plate with their threaded ends pointing straight up. Grind or file off the heads of the fasteners to get a good flat fit-up before you join them to the base. If the fasteners are so close together that the heat from one job melts the prior joint, use several alloys with different melting points to complete the work. This exercise shows you how to join mechanical fasteners to a part. (Actually, there are many kinds of mechanical fasteners available already prepared for soldering,

brazing, braze-welding, or welding, but you can make your own).

11. Bend one edge on each of two sheets up to a right angle. The edge should be about ½ in. [12.5 mm] deep and as long as the sheets. Turn the sheets around so that the bent-up edges butt against each other. Butt the two edges together to make an edge joint. Braze it. Then fold it over flat to tuck it away so that it doesn't stick up. Then lightly hammer or press the folded-over edge flat. This is a common sheet-metal joint that can make a very flat surface seam on the opposite side of the edge joint.

12. Make a lap-seam joint from sheet and braze or solder it. In this joint, the two edges of the sheet are folded over each other and one edge is tucked in. This is very much like the edge joint you made in Exercise 11, with one extra fold, except that you start with the sheets folded over and then braze them. Get a tight clearance before brazing by hammering the joint uniformly flat. When you do it, put some filler inside the joint when you make it, before you braze it, so that the metal flows without having to add much more filler metal after the joint is brought up to temperature. That's a way to get filler metal to flow into a very long joint. This joint should be water-tight. It's often used on cans and other liquid containers.

13. Braze a pipe or tube connection onto another piece of pipe or tubing. An elbow bend or a straight connection will do fine. First use steel. Then use copper and finally use aluminum materials. This is a commonly performed job in plumbing, heating, air conditioning, refrigeration, steamfitting, and plant maintenance work.

14. Braze or solder two nested tubes together to make a single, continuous hollow section. Make sure the lap joint is three times the distance of the metal thickness. If the tubes do not give you good fit-up, figure out some way

to tighten them up. You may even want to use a hacksaw on the outer one to make a single, short cut so you can squeeze the outer edges at the cut onto the inner tube section, to fit it onto the inner tube. That will make a tightly fitting joint for brazing with a nearly smooth inner surface for good fluid flow. This special joint preparation would not be necessary using braze-welding or welding because they don't require a very tight fit-up.

7-3-S
BRAZE-WELDING

15. Practice running a number of braze-welded "beads-on-plate" (put filler metal directly on the surface of the plate or sheet without a joint). Work only in the flat (down-hand) position. Use a bronze rod about ¹⁄₁₆ to ⅛ in. in diameter. Learn how to control your molten filler metal by using the speed at which you work, the motion of your flame and filler rod. Learn how to control your molten filler metal with either the forehand or the backhand welding methods. Work until you get perfectly uniform beads that run for at least a foot or two without much variation in width, height, or appearance. If the bead has narrow areas, you are going too fast at that point. If the bead is too wide in some places, you are going too slowly or the base metal is getting too hot.

16. When you get good at Exercise 15, repeat it with the sheet or plate sloping up at a 45° angle. Braze-weld the bead-on-plate uphill. You may want to start the job by building a little dam at the lower end to keep the filler metal from running off the plate. Continue to use more solidified filler metal as a dam to hold the new metal you deposit. You also may have to lift the flame off the work from time to time to let the plate cool a little.

17. Repeat Exercise 15 with the base metal in the vertical position with your bead running straight uphill. Use the same filler-metal-

dam technique as you work up from the bottom.

18. Repeat Exercise 15, except that the beads will run in the horizontal position (parallel to the floor) with the plate vertical.

19. Repeat Exercise 15 with the plate in the overhead position (put the bead on the underside of the horizontal plate). If the base metal gets too hot, stop periodically and let it cool a little (lift your torch away) before you continue so that you can control the molten filler metal.

20. Braze-weld both sides of a lap joint while working in the down-hand position. Unlike brazing, braze-welding, and welding joints, even butt joints, have two sides to worry about.

21. Braze-weld a simple butt joint, using material no thicker than ¼ in. [6 mm], while working in the down-hand position. You only need to do one side of the joint. You don't need to prepare a special joint for this job. Leave a ¹⁄₁₆ to ⅛ in. [15 to 30 mm] gap between the butted ends when setting up the job to let filler metal flow to the other side. Tack both ends of the joint together with a dab of bronze brazing alloy first. This will hold the gap open and will avoid distortion while you work.

22. Braze-weld a T joint on both sides of the T, using base metal of equal thickness for both sections. Use sheet no more than ⅛ to ¼ in. [3 to 6 mm] thick so you don't have to prepare any special joints. Tack the vertical section upright and in position using braze-welding, before you complete the work. Leave about a ¹⁄₁₆ in. [1.5 mm] gap at the bottom, between the two pieces of metal, to compensate for shrinkage as the filler metal solidifies and pulls the sections together. That also lets filler metal penetrate the joint. Next to the simple butt joint, this is the most common of all welding joints.

23. Braze-weld the same T joint, after tack-welding, but work in the vertical position, braze-welding uphill.

24. Repeat Exercise 22, working overhead, after first tack-welding the two sections of base metal in place.

25. Braze-weld a corner joint (two pieces of sheet at right angles with just their edges touching), filling in the gap along the parallel edges to make two sides of a four-sided box. Unlike brazing, you need only put filler metal along the two edges of the corner. You will first have to tack-weld the two pieces together at right angles. Then you can put one of them on a piece of firebrick with the edge to be joined on one side of the brick so it's easy to work on. The other side can hang down from the brick. Make sure when you finish that the two sides of your V-shaped trough are at right angles (90°). Check it for accuracy. You will be making a box in the next job with another piece just like this one, and the corner must be square.

26. Repeat Exercise 25 to make two more sides of the box. Then assemble the box by braze-welding the remaining two corner joints. After that, put a bottom on the box. If you work carefully and use the proper jigging or clamps, the bottom will be square with the sides.

27. Flame-cut a V-butt joint from steel plate at least ¾ to 1 in. [18 to 25 mm] thick. Use a 45° angle on each leg of the V. Make a solid, down-hand braze-welded root pass after lightly tacking the two pieces into position at either end. Follow up with as many cover passes as needed to fill the joint, completing the joint with a final, slightly reinforced (slightly convex or humped up) cover or capping pass.

28. Repeat Exercise 27 in the vertical position or a 45° sloping position.

29. Repeat Exercise 27 in the horizontal position with the plate vertical and the joint parallel to the floor.

30. Repeat Exercise 27 in the overhead position, working on the bottom of the joint.

31. Repeat Exercise 27 using cast iron instead of steel. Work downhand. Be sure to prepare the faying surfaces by grinding, and then burn off excess graphite on the surface so that you can tin the faying surfaces of the joint.

32. Pipe welding is very difficult but well-paying work. Prepare a beveled pipe joint from two pieces of steel or cast-iron pipe about 3 in. [75 mm] in diameter. Bevel the outer edges to 45° before you start. Then tack-weld the two sections into position with four tacks spaced equally around the joint (at points about 0°, 90°, 180°, and 270° around the pipe). Then braze-weld the pipe sections together. Keep the pipe centerline in the horizontal position while you work. It will help you to rest the pipe in a piece of channel iron or two angle irons while you work. That way you can roll the pipe when you have to, braze-welding a quarter of the section at a time, while always working in the down-hand position. Pay special attention to getting a good solid root pass. Make sure your cover passes are tied into the tinned sides of the joint and the root pass.

33. Repeat Exercise 32 with the pipe centerline in the vertical position (the pipe is standing upright) after you tack-weld it in line. This time the pipe joint is like making a circular weld in the horizontal position. It's tough to do.

34. Braze-weld a butt joint to join two pieces of heavy-gauge steel wire or small-diameter bar end to end. Grind off excess filler metal to make a smooth joint.

35. Braze-weld a cross joint (using two wires at right angles, only in contact at one point). This joint is the most common connection for making products from steel wire that range from fan-blade covers and shopping baskets to wire

mesh screens and steel wire furniture.

36. Braze-weld the two wires or small-diameter bars side-by-side so that they are parallel and touch from end to end. Exercises 34, 35, and 36 are the three principal joints used in joining wire objects ranging from shopping baskets to welded sculpture.

37. If you can get a scrap-steel or cast-iron gear, brake off one tooth (or melt it off with a torch). Then build up a new tooth on the gear with braze-welding. If you can't get an old gear, build new teeth on a piece of round bar. Make the gear tooth you build up slightly oversized, assuming that it will be machined to the correct size later on. Work downhand. This is a typical maintenance job often using braze-welding.

38. Start with a steel bar. Grind a groove around the circumference, as if it were a worn axle. Then fill up the groove with braze-welding, leaving only a little excess metal to be machined off. Now grind the excess metal smoothly to match the surface of the bar. This is another typical maintenance job for braze-welding.

39. Use a small steel plate or thick sheet, about 12 in. [305 mm] square or less. This is a part that must be surfaced with braze-welding alloy to prevent wear. Put a layer of braze-welding alloy of uniform thickness over the entire surface of the wear plate. A ⅛-in. [3-mm] coating will be sufficient. Lay down parallel, slightly overlapping beads until the plate is covered. Make the deposits as equal in thickness over the entire area as you can. Assume that it will subsequently be used as part of a bulldozer blade. (On a real job another, harder alloy would be used, but the gas-welding technique is precisely the same.) For a whole chapter on surfacing methods using both oxyfuel and arc welding processes, see Chap. 19.

Strength and Hardness of Welded Metals

This chapter will explain words like *strength, elasticity, hardness, ductility,* and *toughness.* These are words that have very special meanings when applied to metals, and they are very important to welders.

Welding filler-metal specifications, especially those used in arc welding, are written in terms of specific measures of the strength, hardness, ductility, and toughness of the weld metal that the electrodes will produce. Even the names for arc-welding electrodes such as AWS E7018 include a code for the strength of the weld metal it will deposit. The detailed specification for this E7018 electrode, for example, has a long list of weld-metal property data. Those data are the basis for selecting an AWS E7018 electrode over some other one. But you can't select an electrode when you don't understand the specifications.

The base metals you will weld are usually specified by many of the same properties as welding filler metals. That's how you know what filler metal to select for a given base metal, fabricated product, or repair job. Welders who can't understand the properties of the metals to be welded are seriously handicapped. They are welders who can only take orders and can't make their own decisions.

The tests you will take to qualify as a certified welder measure whether your work meets minimum acceptable levels of these same base-metal properties. The tests you make on your own work do the same thing. A welder who can't test the results of a proposed new welding procedure is doomed to do the same old work.

You can't begin to talk about metals and welding without talking about the properties of the base metal and the weld metal you produce. A welding inspector, a supervisor, or a production manager doesn't think in terms of welding procedures. They think in terms of results. Most of these results are going to be measured in terms of good (or bad) weld-metal properties.

8-1 METAL PROPERTIES

All metals have different properties. The composition of an alloy steel or aluminum, a gray cast iron or a titanium alloy are examples of *chemical properties.* The fact that brasses are alloys of copper and zinc is a chemical fact. Corrosion resistance is a chemical property of metals. Even the things that happen inside metals and alloys when they melt or solidify, or are strengthened, heat-treated, stress relieved and cooled result from the chemical properties of the weld metal and base metal.

You need to know the electrical conductivity of a welding cable to see if it will carry a given current. The number of pounds per cubic inch [kilograms per cubic centimeter] of a base metal is used to calculate the weight of a huge weldment that will be lifted by a crane. You already have used thermal conductivity and coefficients of expansion to help you judge how much you need to preheat a base metal, or how much distortion to expect when you weld or braze something. These are all examples of *physical properties* of metals.

As you now know, all of these "technical" properties are very practical, and most are not hard to understand. Some of these properties you just accept. Some of them you can modify. You are generally stuck with chemical and physical properties—they come with the material and the only way to change them is to change the material. Most mechanical properties are different.

Strength, elasticity, hardness, ductility, and toughness are called *mechanical properties.* Most of them change as you weld the base metal. Unlike physical and chemical properties, you have some control over them. You might call them the *mechanical engineering properties* of metals.

Mechanical Properties

Mechanical properties are the properties that make things work in use, with the important exception of the chemical property of corrosion, and sometimes physical properties like electrical conductivity. Mechanical properties are the design properties that tell you whether a piece of metal, or the weld that holds it together, will stay together when a product is put into service. They also help explain why products have failed after they were used.

Mechanical properties are the easiest of all the metal properties to understand. When you hardface a bulldozer blade, the wear- and abrasion-resistant filler metal you lay down will have a minimum specified hardness. Whether it actually meets that hardness or not depends on how you do the hardfacing job. To some extent, you can control the final hardness of the deposit. You also can easily measure the result of your work so you know whether it meets specifications.

If a weld cracks right after you finish it, or later on when it's in use, the ultimate strength of the metal has somehow been exceeded. You usually can change the properties by changing the welding procedure so that other parts like it will not fail again.

If you weld certain boiler and pressure-vessel and pipe steels or carbon steels with more than 0.30 percent carbon in them without preheating and post-heat treatment, a narrow region next to the weld metal (called the *heat-affected zone*) will not have enough ductility. The pressure vessel or pipe section you are building won't give under sudden, unexpected high forces. Tensile (pull apart) stresses can suddenly build up in the vessel or pipe and make it crack (or maybe blow up). Your welding procedure will be designed to prevent this type of mechanical failure.

Some metals that are very tough at room temperature become very brittle when cold. This is a problem with construction-equipment used and maintained in cold climates. Filler-metal specifications tell you just how ductile or brittle the weld metal will be when it is cold.

Low-temperature embrittlement is obviously a serious problem in designing containers for ultracold cryogenic gases. It's something you should know about even if all you are doing is using cryogenically stored liquid oxygen or argon for your shielding gases. Low-temperature embrittlement is something you have to know a lot about if you work in a big pressure-vessel fabricating plant or on a pipeline job.

8-2 THE STRENGTH OF METALS

There are many different ways of measuring the strength of metals. Each method gives you a different kind of information on how a metal reacts when a force is applied to it. But first you have to have some definitions.

Defining Some Terms

Stress is the force that acts on a metal. *Strain* is the result. When you are under personal stress, you are heavily loaded with problems. The personal stress produces physical and mental strain. It might even lead to a breakdown. Exactly the same is true of metals.

Load is a word used in metalworking and welding for an overall stress that is put on a part or a machine. Weight is a load on anything. Bending force is a load on a structural beam. A pile of dirt in the back of a truck is a load on the axles. The gas pressure inside a cylinder is a load on the cylinder walls. The outward force of a spinning turbine shaft is a load on itself.

Load usually means the whole

amount of force applied to a part, a bridge, a ship hull, a car body, a tractor frame, or a pressure vessel. *Stress* measures how much of that gross load is spread over a given cross sectional area of an object. That idea makes the force much easier to understand and deal with. For example, the steel bar on the right in Fig. 8-1 has a 2- × 2-in. [51- × 52-mm] cross section. The section obviously has an area of 2 in. × 2 in. = 4 in.² The other bar in Fig. 8-1, on the left, has exactly the same square shape but the cross section is smaller. It only has a section 1 in. on a side. It has a cross-sectional area of 1 in. × 1 in. = 1 in.², or only one-fourth as large.

Let's put a 30,000-lb [13,608-kg] load on the end of each of the two

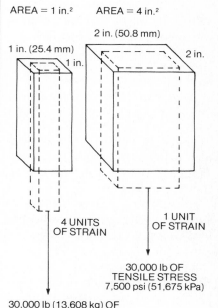

AREA = 1 in.²   AREA = 4 in.²

2 in. (50.8 mm)

1 in. (25.4 mm)
1 in.

4 UNITS OF STRAIN   1 UNIT OF STRAIN

30,000 lb OF TENSILE STRESS 7,500 psi (51,675 kPa)

30,000 lb (13,608 kg) OF TENSILE STRESS 30,000 psi (206,700 kPa)

FIGURE 8-1 Stress causes strain. Stress is the force that distorts a material. Strain is a measure of the distortion. We will generally use English units (with metric equivalent units in parentheses) in illustrations in this chapter. However, metric units are used alone in areas where these are the standard units to use, and not English units. One example is the impact-test bar shown in Fig. 8-13.

bars. If the load is designed to push on the bars, it produces *compressive stress*. If the load pulls on the bars it produces *tensile stress*. The actual stress and strain will be measured in exactly the same way for either load. Only the direction in which they work is opposite.

In the bar on the right, the 30,000-lb [13,608-kg] load is divided up between each square inch [square millimeter] of the bar's 4-in.² [2580 mm²] cross section. The tensile stress on each square inch of the bigger bar is 30,000 lb/4 in.² [13,608/2580 mm²], which equals 7500 lb/in.² [527 kg/mm²]. The same load on the 1-in.² [645-mm²] bar produces a stress of 30,000 lb/in.² because the smaller bar happens to have exactly a 1-in.² [645-mm²] cross section. Therefore the load exactly equals the unit stress.

All this certainly makes sense. While the bigger bar has twice the edge distance, it has four times the cross-sectional area. You would expect it to be able to hold four times the load of the smaller bar if they are made of exactly the same material. And it does.

That's all there really is to the idea behind load and stress. Things only get complicated when the shapes get complicated or the load and stress are not at nice right angles like they are in the first figure, or the load varies with time. You won't have to worry about those problems in this lesson. Those are problems for the engineers who design the products you will fabricate.

The units of load and stress don't have to be pounds, or pounds per square inch, either. The force can be tons, ounces, kilograms, or whatever. Similarly, the area may not be square inches but square feet, or square millimeters, or any other units of area. That's something else we won't worry about in this chapter. We will stick to pound and inch units to make them easy to visualize.

Back to the two bars in Fig. 8-1. What's the unit stress on the

bigger bar? It's 7500 lb/in.². You can also say 7500 psi (pounds per square inch) just like you did with gases. The stress on the smaller bar with the same 30,000-lb load is four times larger, or 30,000 psi, because the cross-sectional area holding the same load is four times smaller.

Any amount of load (stress) will have some kind of effect (strain) on both of the bars. The units of stress are typically pounds per square inch. The units of the strain that results from that stress are in inches (usually a small fraction of an inch) per inch of material length, measured in the direction in which the stress tends to change the shape of the metal. If the stress is compressive, it will squeeze the length of the bar down slightly, making it shorter (and slightly bigger around). If the stress is tensile, it will pull the bar, making it

slightly longer (and the cross-sectional area a little smaller).

Strain is the measure of the change due to stress. In our examples, the strain will be the increase (or decrease) in the length of the bars, measured in decimal fractions of an inch for each inch of bar length (like 0.00033 in./in. or 0.01223 in./in. of the test bar's length). The strain will be less if the metal in the test bar is very strong, and more if it is weak.

That's just saying that the same force will stretch or compress a weak metal more than a strong metal. You can hardly argue with that idea, but let's see it happen. We'll put a known stress on both bars and actually measure the strain that results. That way we can calculate the unknown compressive or tensile strength of the metal that make up the bars (which is exactly how it is done in a metals test lab).

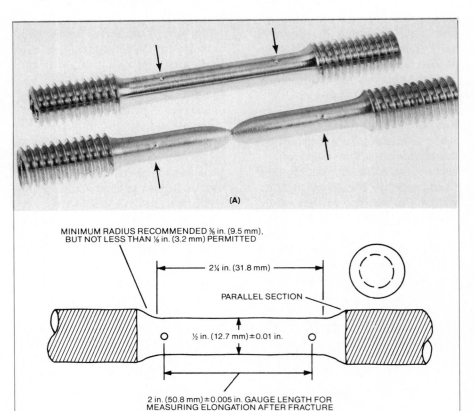

FIGURE 8-2 (A) These drawings show what tensile test bars will look like before and after the tensile test. Note that the broken bar, when put back together, will have a much larger gauge length than before the test (see arrows). (B) Dimensions of the standard round tensile-test bar.

Testing the Strength of Metals

Figure 8-2A shows a round bar with large threaded ends that fit into a tensile testing machine, which is simply a machine designed to pull test bars apart or push them together. The machine has a large dial on top where you can read the total applied load. The threaded ends of the test bar screw into cross arms on the machine. Tensile testing machines can be quite large, but there also are portable machines that pull smaller bars and can be used right in the welding shop.

Let's assume that the test bar's radius is 0.5642 in. That will make the area at the middle of the bar exactly 1 in.² (The area of a circle is 3.1416 × the radius squared, or 3.1416 × 0.5642 in. × 0.5642 in. = 1.0000 in.²) Therefore, if the tensile-testing machine applied a gross pulling load of 30,000 lb on the bar, the gross load reading you get from the machine also is automatically equal to the applied unit stress of 30,000 lb on 1 in.² of the test bar's cross section, or 30,000 psi.

Most real tests are made on bars about 0.5 in. diameter, which is just about equal to a 0.2 in.² cross section, or one-fifth the area of a 1-in.² bar. The reason that these smaller test bars are used is that they won't cause the testing machine to pull itself apart. The testing machine's load data sim-ply can be multiplied by 5 to correct it, after testing, to equal the load on a bar with a cross sectional area of exactly 1 in.².

Round test bars aren't always used. Flat tensile test bars, for example, are made from sheets or thinner plates. Figure 8-3 shows an example, complete with its gauge-length marks. Instead of big threaded ends, it has big flat ends that the tensile-testing machine can grip. The loading, data gathering, and interpretation of the results are the same as if a round test bar were used, with one exception. The gauge length used for sheet metal or thin plate often is 8 in. because there's less metal to pull. It doesn't make much difference in the final test result.

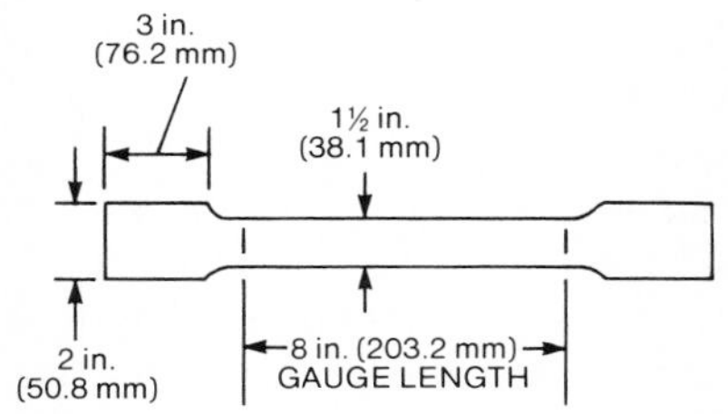

Let's assume that the two lines or dots scribed on the middle of the bar are exactly 1 in. apart before testing. That gives us a unit gauge length of 1 in. Using that figure we don't have to find out how much the entire bar is strained. When the bar is pulled or pushed, any strain that does occur will be easy to measure. It will be the distance between the gauge marks, less their original 1-in. gauge length. A tensile strain will make the distance between the scribed lines longer than 1 in. A compressive strain will make the distance between the gauge marks shorter than 1 in.

Let's make our own tests on a tensile-test bar made of some kind of steel (whose properties we don't know yet) and see what we have. While we are at it, let's list our data (Table 8-1). Let's also adjust the machine for making a tensile rather than a compressive test.

As shown in Table 8-1, a 10,000 psi tensile stress pulls the scribe marks apart by only 0.00033 in./in. of gauge length. Doubling the stress to 20,000 psi doubles the strain to 0.00066 in./in. Five times the initial 10,000 lb stress, or 50,000 psi, separates the 1 in. gauge length by five times as much strain, or 0.00165 in./in. of test bar length. So far, so good. It looks like stress and strain are directly proportional. Double the stress and you double the strain. A bar 100 in. long under a 50,000 psi stress would stretch 100 times as much as the gauge length, or 0.165 in. That's how design engineers can calculate the effect of stress on real parts.

Let's reverse this part of the test and see what happens. When we release the applied stress, the slightly elongated 1-in.-plus gauge marks snap right back to 1 in. even. It's just as if the bar were a very strong spring. Up to a certain point, all metals act just like very stiff springs. When they are loaded they stretch. When the load is taken away they snap right back to the original shape. This is called *elasticity*. All materials from rubber to plastics to high-strength steels initially act elastically under load.

Continuing the test, let's put a 60,000 psi stress load on the bar. For the first time, we don't get the strain data we expected. If stress is proportional to strain, we would expect the gauge-length increase to be six times the original 0.00033-in./in. length, or 0.00198 in./in. It's not. The gauge-length separation beyond the original 1 in. is 0.00260 in./in. What's more, the bar stays slightly pulled apart when the stress is removed. It no longer snaps all the way back to

FIGURE 8-3 Standard dimensions for a flat tensile-test bar. Flat bars are used to test plate. An 8-in. gauge length is commonly used in plate specimens while a 2-in. gauge length is most commonly used for round test bars.

TABLE 8-1 Results of tensile test on steel bar

Load Stress, lb	Strain, in./in. gauge length
10,000	0.00033
20,000	0.00066
30,000	0.00099
40,000	0.00132
50,000	0.00165
60,000	0.00260
65,000	0.00460
72,500	0.00510
80,000	0.00860
88,000	0.01130
86,250	0.01815
86,000	Rupture

shape. It no longer acts elastically. There's much more strain occurring than we expected for the given stress. Something very unusual is happening.

Let's continue to pull on the bar. Suddenly the bar starts to shrink or "neck in" in the middle. After that only a little added stress produces a lot of extra strain. The bar continues to neck in and begins to stretch a lot. You can see it happen.

At a test-load stress of about 86,000 psi, the bar suddenly snaps in half right in the middle at the smallest part of the necked-in area. It actually makes a clanking sound when it breaks.

Now let's plot all these data on a graph. Figure 8-4 shows you what our *stress-strain curve* looks like.

The elastic range The straight upward line on the left side of Fig. 8-4 is the region in which stress and strain are directly proportional. The bar also acted like a spring at that time; it snapped right back to shape when we re-leased the test load. This line represents the elastic range of this test material. The line between 50,000 psi and 60,000 psi, where the bar first started to act oddly, is just beyond the *proportional limit,* the point at which the bar stopped stretching proportionately to the load.

Yield points The first round point on the curve is called the *upper yield point* of the steel. This particular steel also has a *lower yield point*. The metal is starting to *work harden*. Its strength is actually increasing while the ductility is decreasing, because it is being severely strained. Work hardening is what causes a piece of sheet metal to crack if you pound it with a hammer until it gets too hard. It's also the reason why you can break a strong wire by bending it back and forth a lot. In fact, it's also one of the reasons why the curve turns up again toward the ultimate tensile strength of the steel.

Ultimate tensile strength At the very top of the curve is the ultimate tensile strength of the steel bar and the steel from which it is made. That means that the ultimate tensile strength of the steel is slightly less than 90,000 psi, or about 86,000 psi. Just beyond the top peak load on the graph, the bar started to neck in so that more stress was actually acting on less bar. That's why the curve drops. The bar continued to neck in until it broke at the fracture or breaking strength. We aren't really concerned about the breaking strength. Once a piece of metal has gone beyond its ultimate tensile strength it pulls apart very fast. That's what happened when the bar necked in fast and stretched very rapidly until it broke in half.

The tensile strength is a common number used in filler-metal specifications. It tells you the maximum load that the base metal or the weld metal can handle before it starts to fail by rapidly pulling apart and breaking. For example, we already mentioned an arc-welding electrode called AWS E7018. The 70 in its name means that "as-deposited" weld metal will have a tensile strength of at least 70,000 psi. Therefore it must be a good electrode for welding steels in that general strength range. And it is. AWS E7018 electrodes are among the most commonly used arc welding electrodes for a large family of special steels with yield strengths around 50,000 to 65,000 psi and tensile strengths about 10,000 psi higher (60,000 to 75,000 psi). These are called *high-strength low-alloy steels.*

Yield Strength
Actually, not all metals show such clearly defined yield points. Therefore, a more commonly used measure is *yield strength at 0.2 percent offset* from the original 1-in. gauge length. (A 0.2 percent offset along the strain axis of the stress-strain graph is simply 0.002 in. if the gauge length is 1 in.) This figure is measured directly on the stress-strain curve.

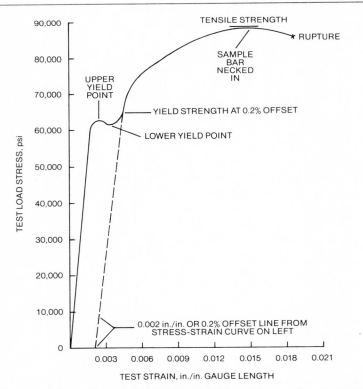

FIGURE 8-4 The results of a typical tensile test on steel are plotted on this graph.

It's a measure that's not easy to explain unless you look at the stress-strain curve.

Figure 8-4 shows how it is measured. An imaginary line is drawn from the strain coordinate of the graph 0.2 percent offset to the right, along the horizontal axis on the bottom of the graph, from the zero point on the left-hand lower corner of the graph. The imaginary line is drawn parallel to the straight part of the real stress-strain line up to the point where the imaginary line intersects the curved part of the stress-strain line. That point is the yield strength at 0.2 percent offset. It is used because in real stress-strain curves, it is usually much easier to measure. The exact yield point frequently is difficult to find on the curve.

You often will see *yield strength* data reported at some offset distance, typically 0.2 percent of the gauge length, but sometimes 0.8 percent of the gauge length. These data most often are reported instead of *yield point* data,

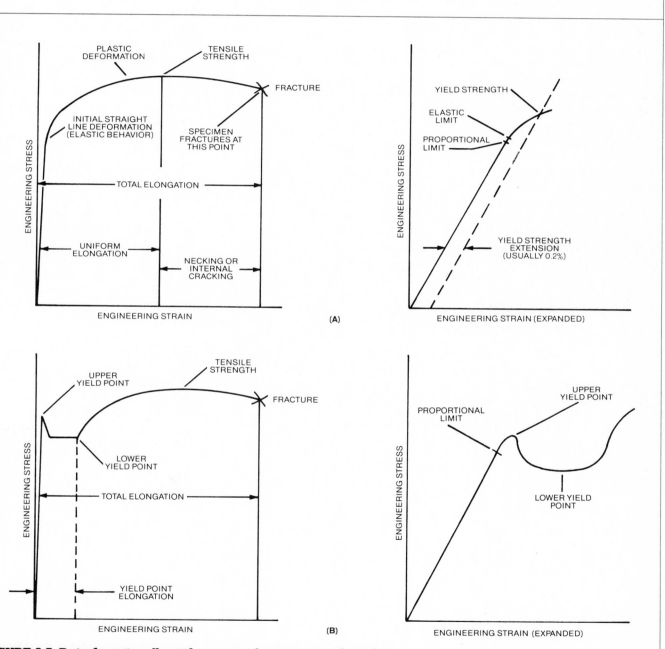

FIGURE 8-5 Data from tensile and compressive tests are plotted to make curves. (A) The left curve shows an entire hypothetical stress-strain curve and tells you its key parts. The curve on the right is only the first part of the curve on the left, expanded to show you how yield strength at 0.2 percent offset is calculated. (B) The left curve is for a metal with a clearly defined yield point. The curve on the right shows the yield point area magnified so you can see it better.

but for your purposes, the idea is precisely the same. That's the load at which the metal starts to change shape markedly or yield under heavy stress. Figure 8-5 shows other facts about stress-strain curves.

Yield Strength and Product Design

Most engineering structures, whether they are buildings and bridges or automobiles and ships, are designed to operate below their yield strength. Yield strength (or yield point) is probably the single most important type of mechanical property data used.

Designers usually don't design to the ultimate tensile strength simply because the metal part will already have pulled out of shape at that point. An out-of-shape gear might just as well be a broken gear for all the use it is in a machine. Things like gears, machine parts, truck bodies, and ship hulls are designed to operate so that all expected loads will be below the yield strength and the proportional limit of the metals used. The metals are used in their *elastic range,* but they are formed (by stamping, for example) by much higher stresses in their *plastic range.* The plastic range is

above the yield point or yield strength, the elastic range, and the proportional limit.

The idea of the plastic range of a metal is very simple. Any time a piece of metal is stretched so much that it permanently changes shape, the metal has been stressed until it reached the plastic range. An automobile fender started out as a flat piece of sheet. It was stretched so much that it couldn't return to its original flat shape. Therefore, it was formed by stresses in the plastic range to make the fender permanently change shape.

8-3 DUCTILITY

Ductility is quite different from strength. To understand ductility, think of its opposite, brittleness. Ductility means the ease with which a metal can be deformed without breaking. The tensile test gives us a way to measure the ductility of a metal.

If we put the broken pieces of our test bar back together tightly and measure the final gauge distance, we will find that it now is just a little over 1.018 in. between the scribe marks. That means that the *total elongation* during the test was 0.018 in. more than

the original 1-in. length, or just about 1.8 percent of the original gauge length. Its name, very simply, is *percent elongation* (in a specified gauge length). That's the amount of stretching the metal did before it broke. A very brittle material (concrete, for example) would hardly stretch at all before it broke if loaded with enough pull-apart force.

In our test, the stretching that occurred before the metal test bar broke would be measured as 1.8 percent elongation in 1 in. If we had used a flat bar with an 8-in. gauge length, the data would be reported as a percent for the excess stretch over 8 in. Percent elongation is a very common number used in both base-metal and filler-metal specifications. It is a clue to the ductility of the metal.

Measuring Ductility

We also need a related number to give us a little more information on ductility. If we measure the diameter of the broken test bar at the necked-in area right where it finally cracked, we will find that the area of the bar at that point is about 25 percent smaller than before the test.

Its area is now around 0.75 in.^{2} instead of 1 in.^{2} in the middle of the bar. This measurement is called *percent reduction of area.* It is used together with percent elongation in the unit gauge length as a combined measure of the metal's ductility. A high percent reduction in area and a high percent elongation means that the metal is very ductile. Low numbers mean the metal is brittle.

Figure 8-6 shows interesting comparison stress-strain curves for several different materials ranging from very ductile but low-strength steels to a very high-strength steel with almost no ductility at all. That's not unusual. Most materials get less ductile as they get stronger and harder. For example, automotive sheet steel to be formed into fenders and complex hood shapes must have

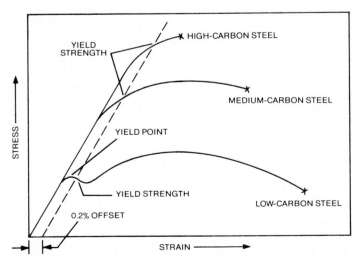

FIGURE 8-6 Stress-strain curves for steel change with the carbon content of the steel. Low-carbon steels are not very strong, but they are more ductile than high-carbon steels. High-carbon steels have high strength, but low ductility.

high ductility or it will crack when formed and stretched by the stamping dies.

Ductility is more important than strength when making formed fenders. Strength is more important than ductility when making machined axles and heavy-duty leaf springs. The function of an object will determine the required tensile strength and ductility.

The stress in which any metal is permanently formed is somewhere between its yield point and its ultimate tensile strength. That's just as true of hammering, forging, and rolling steel or aluminum in a rolling mill as it is in shaping the metal by forming it in an automotive stamping press. Ductility tells how much the metal can be formed without breaking. Yield strength tells how much force is required to form the metal.

The Effect of Temperature

Tensile strength and yield strength drop as the metal gets hotter. The ductility tends to go up (not counting some special brittleness problems at certain temperatures that are used when forging certain metals). That is why a blacksmith can hammer and shape a hot piece of metal on an anvil much easier than a cold piece of metal. That's also why thick ingots or slabs in a steel or aluminum rolling mill are usually rolled hot. It's the same reason why it's easier to bend pipe hot rather than cold. Because the ductility has increased with temperature, much less force (stress) is needed to shape (strain) the metal.

Some special tests are even conducted on very hot test bars just to find out what the stress-strain curve for the hot metal will look like. Figure 8-7 is an example. It's an inexpensive way to find out how well a metal will form before expensive dies are made for a metal that perhaps can't be formed into a required shape when it is cold.

8-4 ELASTICITY

You can tell a lot about a piece of metal from a stress-strain curve. One item will surprise you. If you were only to look at the first part of the curve below the proportional limit, where the stress is proportional to the strain, and then divided the stress by the strain (for example, 10,000 psi/ 0.00033 in./in. = 30,000,000 psi), you would get a very large number. Its units, as you can see from our calculation, are in pounds per square inch.

What's surprising is that any other steel test bar would give you almost exactly the same large number, 30,000,000 psi, and it doesn't matter much what kind of steel you test. All steels have about the same stress/strain ratio in the elastic region (give or take 1,000,000 psi either way, which is very little difference).

If the bar were made of aluminum, you would get a completely different number, about 10,000,000 psi. Again, it wouldn't matter much what kind of aluminum alloy you tested. The number will always be just about 10,000,000 psi for all aluminum alloys. Copper and many copper alloys would give you a number around 17,000,000 psi.

A typical gray cast iron is different. It will give you a number in a wide range between 11,000,000 and 21,000,000 psi, depending on the grade of iron. The reason is that cast iron is a combination of iron and graphite flakes. The more iron and less graphite it contains, the higher the big number gets as it approaches the 30,000,000 psi number for pure iron or steel. A cast iron's properties depend not only on how much graphite it contains, but also on what shape that graphite has (flakes or lumps, for example). So a composite material like gray cast iron or fiberglass will have a range of these big stress-strain numbers. What could these numbers mean?

The Elastic Modulus

The ratio of stress to strain is called the *elastic modulus* of the material. It is a direct measure of how stiff the material is. Engineers and metallurgists use *elasticity* to mean *stiffness,* and they are using their words correctly—the two things have precisely the same meaning. It's the average

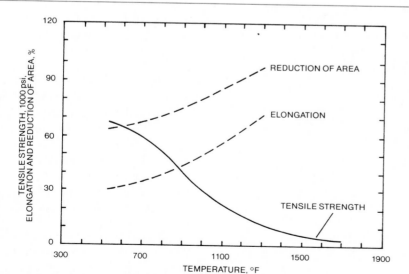

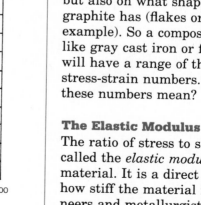

FIGURE 8-7 This graph is typical of most engineering materials, especially metals. As temperature increases, tensile strength drops but elongation and reduction of area increase. That is why blacksmiths (and big steel plants) heat metals before forging or rolling them.

person who confuses elasticity with high stretchability, like the rubber band versus the steel bar. A rubber band actually has very low elasticity and a very low elastic modulus because it has very low stiffness. The stress-strain curve for a rubber band would be almost horizontal. The elastic modulus would be very small. The ductility would be extremely large.

The best way to show what the elastic modulus and stiffness of a metal means is to hang 100-lb weights from the ends of three bars with equal cross sections. Each bar is made of a different metal. They are sticking out an equal distance from the edge of a table (Fig. 8-8). The aluminum bar will bend down the most under the applied load. The copper bar will bend less and the steel bar will bend the least. The steel bar has the highest elastic modulus, the copper bar is next, and the aluminum bar has the lowest elastic modulus.

Even the distance the weights pull the bars down is directly proportional to the elastic modulus of each metal. Steel is three times as stiff as aluminum. The aluminum bar will bend down three times as much as the steel bar. Copper is about 1.7 times stiffer than aluminum and slightly more than half as stiff as steel. The copper

bar will bend down almost twice as much as the steel bar and proportionately less than the aluminum bar.

The elastic modulus is a direct measure of the stiffness of *equal sections* made from different metals. Stiffness has nothing to do with the strength of the metal. It is only a function of the section thickness, its geometry or shape, and the kind of metal used (the metal's elastic modulus).

Elasticity and Product Design
Stiffness is often a problem in making metal parts. A bridge must not only be strong, it also must be stiff enough to keep it from sagging. Thin sections made of very strong material won't hold up unless the section thickness is increased. For example, guard rails of high-strength low-alloy steels are used inside the doors of modern automobiles to keep side collisions from jamming the door into the driver's lap.

Automotive engineers would love to make these rails thinner so they weigh less. They can't, because the rails wouldn't be stiff enough to do the job. What the engineers do, however, is change the section. They don't use thicker metal. That would weigh more. Instead, they use corrugated metal. They increase the section

to get an effect similar to the different stiffnesses of the different metal bars in Fig. 8-8. Corrugated galvanized steel is used in culverts and on the sides and roofs of farm buildings for the same reason. The corrugations don't make the building stronger. They make it stiffer.

A steel I beam is stiffer than a flat steel plate of equal weight and total metal cross-sectional area for the same reason. In fact, you can change the shape of the metal (as in an I beam versus a square plate or bar of equal weight) and have a stiffer section. That explains why a hollow tube is "stronger" (actually stiffer) than a solid steel bar of equal weight. That's why racing car body frames and birds' wing bones are hollow. They are stiffer while weighing less.

8-5
HARDNESS
Strength, whether it's yield or ultimate, tensile or compressive, is obviously a very important property. But hardness is equally important. In fact, you can often use hardness to estimate strength, especially when working with common kinds of carbon steels. What's more, hardness tests are easier to do in the shop than tensile tests. You also don't have to tear the workpiece apart to find out what its strength is (or was).

What Is Hardness?
Hardness is a difficult property to define, partly because the closer you look at a metal, the more variation in hardness you get. Since metals are made of grains (crystals) of different kinds of material, each type of grain will have a different hardness. Therefore, welders mostly use the *average hardness* of a metal, not its *microhardness*. They measure the hardness of a lot of grains of metal rather than one or two grains or crystals.

What, then, is hardness? It has been defined as "the resistance to

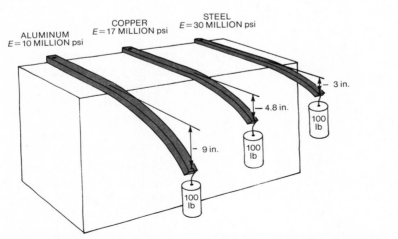

FIGURE 8-8 Modulus of elasticity measure a material's resistance to bending under load, that is, the stiffness of the metal. Steel is three times as stiff as aluminum, given equal sections. Copper has an intermediate stiffness.

penetration." That's as good a definition as any because that's what most hardness tests measure. Hardness also could be defined as resistance to abrasion, or resistance to being cut by a machine tool. As it works out, resistance to penetration works as well for machinists as for welders, metallurgists, and sheet-metal workers.

It even works well as a definition for a property of armor plate that measures resistance to penetration by an artillery shell. The only time the definition runs into trouble is in hardfacing, where abrasion or wear occur that may combine corrosion with penetration by hard particles, the cutting action of sharp particles, and other factors. Even then, the general definition for hardness works quite well about 90 percent of the time.

As we mentioned, there is a direct correlation between hardness and tensile strength for heat-treated carbon steels and low-alloy steels. Hardness (as defined by some kind of penetration test) and tensile strength are very close to being directly proportional (Fig. 8-9). You often can estimate the tensile strength of a carbon steel simply by giving it a hardness test with a simple portable hardness-testing machine. For example, a very useful rule of thumb is that the tensile strength of a carbon steel is 500 times the Brinell hardness number. But don't be fooled into thinking this works for all metals (or all steels). It does only in the sense that almost all very strong metals are very hard metals. You can't predict the ultimate tensile strength of all very strong metals (including certain kinds of stainless steels, for example) just by measuring their hardness. But as a rule of thumb this strength-estimating procedure works just fine for carbon steel.

Measuring Hardness

So how do you measure hardness? There are at least five commonly applied hardness tests. Each has advantages and disadvantages in

use. We'll start by explaining the two tests that you will most often see reported in welding specifications. The first one is called the *Brinell hardness test*. Its data usually are reported as a BHN number without any other dimensions on it. It got its name from its inventor (which is true of three of the four other common hardness tests).

The other common test is the *Rockwell* hardness test. The *Vickers* or *diamond pyramid* hardness test is less important to most welders. The *Schorr scleroscope test* is a fourth test which is very easy to perform in the shop. The fifth test doesn't give an exact number; you simply use a hard file. A metal is either softer than

the file or it's harder. If it is harder it will be very difficult to weld without cracking.

Brinell hardness What Mr. Brinell invented when he developed the Brinell hardness test was a standard method for seeing how deep he could push a tiny, but very hard, round ball into a piece of metal. The harder the base metal, the smaller the dent or indentation made by the ball under a known standard load. The smaller the dent, the more resistance to penetration by the ball, and therefore the harder the base metal must be.

The Brinell hardness test uses either one of two standard loads to press a 10-mm [0.39-in.] di-

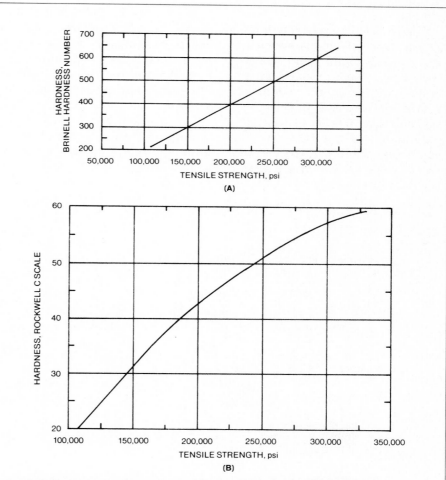

FIGURE 8-9 The relationships between (A) Brinell hardness and tensile strength and (B) Rockwell hardness and tensile strength are shown here. Hardness is a good way to estimate the tensile strength of carbon steels. The correlation between hardness and strength holds for many materials, but these graphs only apply to carbon and high-carbon stainless steels hardened by heat treatment.

ameter, very hard steel or tungsten-carbide ball into the base metal. Ferrous metals are tested with a 3000-kg [6600-lb] load applied for 10 to 15 s to make sure that the ball is pressed into the base metal as far as it will go. Softer nonferrous metals such as aluminum or soft copper are tested with a 500-kg [1100 lb] load applied for 30 s.

The distance across the dent is converted into a depth-of-penetration reading. A complicated formula can be used to calculate the penetration depth, but tables are now provided with each test machine so that a given load and test time can be read directly from a chart to give you a Brinell hardness number. These numbers range from 50 BHN for soft metals to 700 BHN for very hard metals.

The use of the relatively large Brinell test ball compared with the typically small grain size of the metals tested produces an average hardness number for all the grains of base metal under the test ball. The Brinell test is very easy to make if you have the equipment. However, the test has two drawbacks. The first is that the dent produced by the ball

marks the surface of the base metal and can spoil the surface finish of a part.

The other drawback is more serious. Since the ball may sink into the base metal to a distance of almost half the radius of the test ball, you can't get reliable hardness data on thin material. The ball can actually punch a hole through thin strip. You also can't get reliable hardness readings on thin, hard cases over softer metal. Therefore, the Brinell test is usually taken on thicker sections where surface finish is not too important. Examples are steel plate, structurals, and hot-rolled bars.

Rockwell hardness The Rockwell hardness test uses much smaller standard loads and a larger variety of them than the Brinell test. The Rockwell test can be used on very soft to very hard materials and on thinner sections than the Brinnel test. The test procedure is also a little different.

In operation (Fig. 8-10), the metal to be tested is placed on an anvil and raised up against an indenter until a *minor load* is reached. Then a *major load* is ap-

plied on the indenter by a lever. When enough time has passed and the downward motion of the indenter pressing into the test metal stops, the major load is released and the indenter is raised again to the position corresponding with the minor test load. The distance the indenter returns is a measure of its depth of penetration and therefore of the hardness.

This distance is indicated on a dial gauge on the Rockwell test machine. The dial gauge reading converts directly into a Rockwell hardness number on one of several scales that range from A to D. The scales use different-size indenters and different minor loads. For example the Rockwell A scale uses a 60-kg [132-lb] minor load while the Rockwell C scale uses a 150-kg [330-lb] minor load. The Rockwell B and C scales are the most commonly used because the C scale is designed to test the hardness of steel including hardened alloy and tool steels, while the B scale often is used for nonferrous metals.

The Rockwell C scale uses a diamond-point indenter with a 120° angle at the point, and 150-kg [330-lb] major load and a 10-

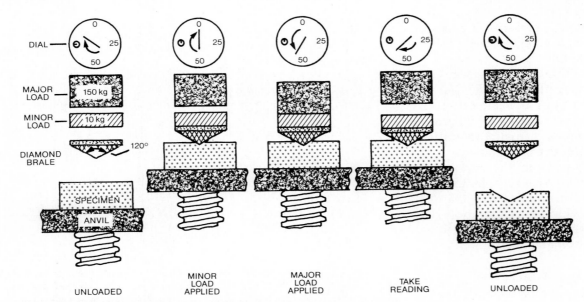

FIGURE 8-10 Rockwell hardness tests require several steps as outlined in this illustration. Metric units for major and minor loads are standard in these tests. English units are not standard.

kg [22-lb] minor load. The Rockwell B scale uses a ¹⁄₁₆ in. [0.16 cm] diameter steel ball with a 100-kg [220-lb] major load and a 10-kg [22-lb] minor load.

The minor load first presses the indenter into the metal until it is solidly in place. Then the major load pushes the indenter farther into the metal until it won't penetrate any further. That is typical of all Rockwell hardness tests. The different Rockwell scales simply use different major loads and different kinds of indenters. The depth the indenter penetrates into the metal is converted into a Rockwell hardness number. The deeper the indentation, the softer the metal, and the lower the hardness number.

There also is a surface Rockwell hardness test with much lighter major loads of 15 to 30 kg [33 to 66 lb] for use when testing metals with a thin but very hard surface, and for thin strip and sheet metal. This surface tester's reading are reported on one of three N scales.

Any Rockwell test number always states the scale used. For example, R_c 65 to R_c 70 is a very hard material, probably a steel with a tensile strength of over 300,000 psi, like a tool steel, for example, or a very hard bearing steel with about 1.20 percent carbon content. A mild steel might have a Rockwell C hardness of only 30 to 40.

The Rockwell test may sound a little more complicated than the Brinell test, but in fact it is very fast and easy to make on typical test specimens. Readings can be converted to other Rockwell scales or to Brinell or other hardness test numbers with conversion tables listed in many materials handbooks. Table 8-2 is an example.

For example, a Rockwell C reading of 50 is equal to a Brinell hardness of 475–481 BHN, and a Rockwell C of 30 equals 286 BHN or a Rockwell A reading of 65. The details of these conversion tables won't matter to you in your

work. What is important is that different hardness tests can be used depending on available test equipment and type of sample. Then the test data can be converted to hardness numbers on other scales for comparison with welding test procedures or welding code specifications.

Vickers hardness The Vickers hardness test is also called the diamond pyramid hardness test, describing the shape of its indenter. The Vickers test combines some of the features of both the Brinell and Rockwell tests. It is used on very small test specimens and often can test the hardness of individual grains or crystals in a metal. The numbers produced are often called *DPH* numbers.

For example, a Rockwell C of 30 is equal to a DPH of 302. Unlike the Rockwell test, the Vickers or diamond pyramid hardness test only has one scale, like the Brinell test. Unlike the Brinell test, Vickers tests are limited only to very small areas on test specimens. The indentation often is measured by looking through a microscope at the individual grains of metal being tested.

The Vickers test uses a square-based diamond-pyramid indenter with test loads varying from 5 to 100 kg [33 to 220 lb] (usually 10, 30, or 50 kg [22, 66, or 110 lb] loads are used). The Vickers test specimen is placed on a tiny anvil and brought up to within a millimeter of the diamond indenter. Then the test load is applied by a lever, after which the indenter returns to its original position. The long diagonal of the diamond-shaped indentation made by the indenter on the test metal is measured. This length is converted into a penetration depth into the metal, which converts to a DPH hardness number.

The advantage of the Vickers test is its ability to measure the microhardness of individual grains of metal. This makes the test very important when troubleshooting welding problems involv-

ing the microstructure of the weld or base metal. Using the Vickers test alone, without a metallurgical microscope and special polished sections of the metal being tested, sometimes can tell you that too rapid postweld cooling of higher-carbon steels has greatly hardened the heat-affected zone next to a weld while reducing the base metal's ductility so much that the base metal in the heat-affected zone cracks right after welding. (See how important postweld and preweld heating can be in medium- and high-carbon steels and in certain alloy steels?)

Vickers hardness tests also are good for measuring thin, hard surfaces such as the chrome plate on a bumper. The mark made by the test is very small (it usually is measured with a low-power microscope). Therefore the mark will not spoil the surface of most finished parts.

Scleroscope hardness The scleroscope test is not really a hardness test at all, but it is the easiest test to make on a large workpiece in the shop. Rather than a direct measure of penetration hardness, the scleroscope test (shown in Fig. 8-11) measures the rebound energy of a standard diamond-tipped tiny hammer falling

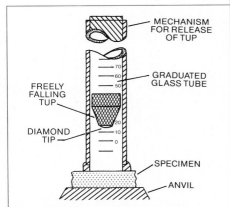

FIGURE 8-11 The Scleroscope hardness test is a very fast way to measure the hardness of a part right in the shop. Like other hardness tests, scleroscope test data can be converted into approximate tensile strength data.

freely onto the surface of a work-piece.

The tip is first raised a measured distance above the test surface inside a glass tube, and then the tip falls and hits the surface. The distance that the tiny diamond hammer bounces back up the tube after hitting the work is measured. This distance measures the *rebound energy* of the test surface, which is closely related to surface hardness. Objects bounce higher off hard things than they bounce off soft things. The tall glass tube is calibrated in scleroscope hardness numbers. A scleroscope hardness of 42 is about equal to a Rockwell C hardness of 30, a Brinell hardness number of 286, or a Vickers diamond pyramid hardness of 302.

Hardness tests are very useful for quality control during welding, for testing new welding procedures, and for estimating the tensile strength of carbon steels without making a tensile test. Hardness tests also are a major clue to telling similar-looking alloys apart. Hardness tests are even used to evaluate the quality of as-deposited weld metal on wear-resistant parts (produced by one of several hardfacing welding processes).

8-6 TOUGHNESS

Toughness is measured by a metal's ability to absorb sudden impact energy at a given temperature. When reading a filler metal's specification, or even the label on a can of welding electrodes, you may see a line of copy about as-deposited weld metal that reads "15 ft-lb Charpy V-notch at $-20°F$." What does that mean? It means that the steel remains "tough" down to at least $-20°F$ [$-28.8°C$]. (Below that temperature the steel weld metal may become very brittle.)

TABLE 8-2 Approximate equivalent hardness numbers to Rockwell C hardness numbers to steel

Rockwell C-scale Hardness No.	Diamond pyramid Hardness No.	Brinell Hardness No., 10-mm ball, 3000-kg load			Rockwell Hardness No.			Rockwell Superficial Hardness No., Superficial Brale Penetrator			Shore Scleroscope Hardness No.	Tensile Strength (approx), 1000 psi	Rockwell C-scale Hardness No.
		Standard Ball	Hultgren Ball	Tungsten Carbide Ball	A-scale, 60-kg load, brale penetrator	B-scale, 100-kg load, 1/16-in. diameter ball	D-scale, 100-kg load, brale penetrator	15-N Scale, 15-kg load	30-N Scale, 30-kg load	45-N Scale, 45-kg load			
68	940	—	—	—	85.6	—	76.9	93.2	84.4	75.4	97	—	68
67	900	—	—	—	85.0	—	76.1	92.9	83.6	74.2	95	—	67
66	865	—	—	—	84.5	—	75.4	92.5	82.8	73.3	92	—	66
65	832	—	—	739	83.9	—	74.5	92.2	81.9	72.0	91	—	65
64	800	—	—	722	83.4	—	73.8	91.8	81.1	71.0	88	—	64
63	772	—	—	705	82.8	—	73.0	91.4	80.1	69.9	87	—	63
62	746	—	—	688	82.3	—	72.2	91.1	79.3	68.8	85	—	62
61	720	—	—	670	81.8	—	71.5	90.7	78.4	67.7	83	—	61
60	697	—	613	654	81.2	—	70.7	90.2	77.5	66.6	81	—	60
59	674	—	599	634	80.7	—	69.9	89.8	76.6	65.5	80	326	59
58	653	—	587	615	80.1	—	69.2	89.3	75.7	64.3	78	315	58
57	633	—	575	595	79.6	—	68.5	88.9	74.8	63.2	76	305	57
56	613	—	561	577	79.0	—	67.7	88.3	73.9	62.0	75	295	56
55	595	—	546	560	78.5	—	66.9	87.9	73.0	60.9	74	287	55
54	577	—	534	543	78.0	—	66.1	87.4	72.0	59.8	72	278	54
53	560	—	519	525	77.4	—	65.4	86.9	71.2	58.6	71	269	53
52	544	500	508	512	76.8	—	64.6	86.4	70.2	57.4	69	262	52
51	528	487	494	496	76.3	—	63.8	85.9	69.4	56.1	68	253	51
50	513	475	481	481	75.9	—	63.1	85.5	68.5	55.0	67	245	50
49	498	464	469	469	75.2	—	62.1	85.0	67.6	53.8	66	239	49
48	484	451	455	455	74.7	—	61.4	84.5	66.7	52.5	64	232	48
47	471	442	443	443	74.1	—	60.8	83.9	65.8	51.4	63	225	47
46	458	432	432	432	73.6	—	60.0	83.5	64.8	50.3	62	219	46
45	446	421	421	421	73.1	—	59.2	83.0	64.0	49.0	60	212	45
44	434	409	409	409	72.5	—	58.5	82.5	63.1	47.8	58	206	44
43	423	400	400	400	72.0	—	57.7	82.0	62.2	46.7	57	201	43
42	412	390	390	390	71.5	—	56.9	81.5	61.3	45.5	56	196	42
41	402	381	381	381	70.9	—	56.2	80.9	60.4	44.3	55	191	41
40	392	371	371	371	70.4	—	55.4	80.4	59.5	43.1	54	186	40

The Effects of Temperature on Toughness

Toughness (or brittleness) is much more sensitive to temperature than are strength and hardness. Sometimes a 10°F [5.5°C] drop in temperature can turn a tough metal into a brittle metal. This can be crucial. For example, the frame of a large bulldozer that is tough in Florida or Houston will become brittle and crack when operated in a Minnesota winter.

Liberty-class troop ships in World War II started splitting in half in heavy weather in the cold North Atlantic. For a long time it was thought that they were sunk by U-boats. Later on it was discovered that poor welding and design did the job more effectively than torpedos. Cracks often would start at a square-cornered welded hatch. During warm weather the crack wouldn't go anywhere. When the weather got very cold, the crack took off and ran around the ship's hull, splitting the ships in half. When the weld metal and hull steel used at that time were tested in the lab at room temperature they were ductile. But at lower temperatures they would suddenly become very brittle. On the ships, small cracks or weld defects would race around the metal nonstop, especially when the steel was stressed on cold stormy days.

At about the same time one of the world's first liquid methane tanks used to store liquefied natural gas at −259°F [−160°C] got hit by something, possibly just a forklift truck. All that is known for sure is that when the accident occurred the steel tank cracked like a brittle egg shell and spilled liquid methane down the sewers of Cleveland. The methane liquid vaporized inside the sewers and then hundreds of thousands of cubic feet of natural gas blew up, eliminating about five city blocks. Pipelines sometimes split when

Rockwell C-scale Hardness No.	Diamond pyramid Hardness No.	Brinell Hardness No., 10-mm ball, 3000-kg load			Rockwell Hardness No.			Rockwell Superficial Hardness No., Superficial Brale Penetrator			Shore Scleroscope Hardness No.	Tensile Strength (approx), 1000 psi	Rockwell C-scale Hardness No.
		Standard Ball	Hultgren Ball	Tungsten Carbide Ball	A-scale, 60-kg load, brale penetrator	B-scale, 100-kg load, 1/16-in. diameter ball	D-scale, 100-kg load, brale penetrator	15-N Scale, 15-kg load	30-N Scale, 30-kg load	45-N Scale, 45-kg load			
39	382	362	362	362	69.9	—	54.6	79.9	58.6	41.9	52	181	39
38	372	353	353	353	69.4	—	53.8	79.4	57.7	40.8	51	176	38
37	363	344	344	344	68.9	—	53.1	78.8	56.8	39.6	50	172	37
36	354	336	336	336	68.4	(109.0)	52.3	78.3	55.9	38.4	49	168	36
35	345	327	327	327	67.9	(108.5)	51.5	77.7	55.0	37.2	48	163	35
34	336	319	319	319	67.4	(108.0)	50.8	77.2	54.2	36.1	47	159	34
33	327	311	311	311	66.8	(107.5)	50.0	76.6	53.3	34.9	46	154	33
32	318	301	301	301	66.3	(107.0)	49.2	76.1	52.1	33.7	44	150	32
31	310	294	294	294	65.8	(106.0)	48.4	75.6	51.3	32.5	43	146	31
30	302	286	286	286	65.3	(105.5)	47.7	75.0	50.4	31.3	42	142	30
29	294	279	279	279	64.7	(104.5)	47.0	74.5	49.5	30.1	41	138	29
28	286	271	271	271	64.3	(104.0)	46.1	73.9	48.6	28.9	41	134	28
27	279	264	264	264	63.8	(103.0)	45.2	73.3	47.7	27.8	40	131	27
26	272	258	258	258	63.3	(102.5)	44.6	72.8	46.8	26.7	38	127	26
25	266	253	253	253	62.8	(101.5)	43.8	72.2	45.9	25.5	38	124	25
24	260	247	247	247	62.4	(101.0)	43.1	71.6	45.0	24.3	37	121	24
23	254	243	243	243	62.0	100.0	42.1	71.0	44.0	23.1	36	118	23
22	248	237	237	237	61.5	99.0	41.6	70.5	43.2	22.0	35	115	22
21	243	231	231	231	61.0	98.5	40.9	69.9	42.3	20.7	35	113	21
20	238	226	226	226	60.5	97.8	40.1	69.4	41.5	19.6	34	110	20
(18)	230	219	219	219	—	96.7	—	—	—	—	33	106	(18)
(16)	222	212	212	212	—	95.5	—	—	—	—	32	102	(16)
(14)	213	203	203	203	—	93.9	—	—	—	—	31	98	(14)
(12)	204	194	194	194	—	92.3	—	—	—	—	29	94	(12)
(10)	196	187	187	187	—	90.7	—	—	—	—	28	90	(10)
(8)	188	179	179	179	—	89.5	—	—	—	—	27	87	(8)
(6)	180	171	171	171	—	87.1	—	—	—	—	26	84	(6)
(4)	173	165	165	165	—	85.5	—	—	—	—	25	80	(4)
(2)	166	158	158	158	—	83.5	—	—	—	—	24	77	(2)
(0)	160	152	152	152	—	81.7	—	—	—	—	24	75	(0)

Note: Values in parentheses are beyond normal range and are given for information only.

they are cold and under stress. The crack can run for thousands of feet at the speed of sound, emptying the complete contents of the pipeline between pumping stations.

A submarine's hull not only has to be strong in cold undersea currents at high pressure and great depth, but the hull metal also must be resistant to sudden impact from hitting the bottom (or even from depth charges).

These are all examples of metal toughness, which is the ability to absorb sudden impact energy at a given temperature. Of course now there are many different kinds of alloy steels, stainless steels, aluminum alloys, titanium, and other materials that can be used at low or even cryogenic temperatures without becoming brittle. But not as much about metal toughness and alloy selection was known 20 or 30 years ago.

Measuring Toughness

The square corners of welded hatch covers on the Liberty ships; a simple small flaw caused by bad welding practice, especially in the root pass of a weld; a machined keyhole slot in a round shaft; bolt threads in bolt holes; and a minor nonmetallic imperfection or inclusion in a weld or the base metal are all examples of notches. What has this to do with toughness? Read on.

Notches are stress raisers. The sharper the notch, the higher the *local* stress at the tip of the notch, even through the overall stress may not be all that large. For example, the overall stress on a part may be well below the yield point of the metal, but the localized stress at the tip of a notch can be several times larger than the metal's tensile strength. If it is, a crack forms.

If a metal is tough and a crack forms at the tip of the notch it is not likely to continue very far into the base metal because the metal will give or yield locally, reducing the stress at the crack

tip. If a metal is brittle it will not yield at the crack tip fast enough and the crack will continue to run through the metal. This *crack* or *notch sensitivity* is a function of the temperature and stress on the metal as well as some base-metal properties. It can be measured by a variety of different tests.

Tensile versus impact tests

The first tests used to measure notch sensitivity were conventional tensile tests on bars with a notch machined into the middle. Later on it was found that impact tests did a better job of testing for toughness or brittleness and that temperature was very important during the test. Notched-bar impact tests were developed to measure what has now become known as the *fracture toughness* of metals. Therefore, impact tests are not only temperature-sensitive, they also are *rate-sensitive* to loading speed.

A slow tensile test might not show the effect while a very rapid whack with a heavy pendulum will break the steel in half. That explains why impact tests are used much more often than notched-bar tensile tests to determine toughness. The extreme example of the difference that speed can make is toy material such as Silly Putty—extremely plastic when pulled apart slowly, but brittle when rapidly deformed.

There are several commonly used impact tests reported in welding filler-metal specifications. They are the *Charpy impact test* (using one of two different types of notched bars), a series of different drop-weight tear tests that are becoming very important for testing high-strength steel plate, and occasionally an older test called the *Izod impact test* that is not used much any more. An entire branch of mechanical and metallurgical engineering has now developed around these tests, their interpretation and use in designing products. It is called *fracture mechanics*.

Charpy impact tests The Charpy impact test uses a rectangular test bar made from the metal to be tested (Fig. 8-12). The standard bar is $10 \times 10 \times 55$ mm [$0.4 \times 0.4 \times 2.2$ in.]. It has either a sharp V notch or a hole and slot machined halfway through one side. These simulate notches. A heavy blow is delivered to the reverse side of the test bar, opposite the notch, by swinging a weighted pendulum against the bar while the bar is held tightly in place.

The units of the test (and of impact energy in general) are foot-pounds (or one of several metric equivalents). Since this is an entirely new test unit for you, we'll explain it before we go on.

FOOT-POUNDS Foot-pounds are units of energy. A 1-lb weight falling off a 4-ft-high table onto your shoe will deliver 4 ft-lb of energy to your toe. A 4-lb weight falling a distance of only 1 ft will deliver exactly the same amount of impact energy, 4 ft-lb. A 40-lb weight falling through 0.1 ft also will deliver 4 ft-lb of impact energy and a 0.04-lb weight falling through 100 ft will deliver the same 4 ft-lb of impact.

Foot-pounds of impact energy are simply the weight multiplied by the distance the weight falls through, *or the equivalent amount of energy applied very rapidly some other way*. The weighted pendulum used in impact tests is a very easy way to impart a known impact to a test bar. The vertical height through which the pendulum "falls" is simply multiplied by the weight of the pendulum-hammer to give the impact energy in foot-pounds.

NOTCH SHAPE Some metals that are fairly tough without notches become rather brittle with notches. They are *notch-sensitive*. Some metals also become notch brittle quite rapidly at lower temperatures while they are not very notch sensitive at higher temperatures. The Charpy test and other impact tests measure these properties.

In all types of notched-bar impact tests, the shape of the notch is very important. If the notch is not accurately made the data will be unreliable. The usual notch in one form of the Charpy test bar is made by drilling a hole with a Number 47 wire size drill 5 mm [0.2 in] from one wall of the bar, right in the bar's center. Then the notch is cut into the hole from one edge of the bar. This bar is used in the Charpy keyhole test.

But another kind of Charpy test bar is now much more commonly used. This is the Charpy V-notch test bar. It has a much sharper tip on the notch and better simulates the effect of real crack or welding defect in a stressed part. The shape of the V-notch is the same as the one used in the Izod test bar, which we'll describe shortly. The Charpy V-notch test is now widely used in filler-metal specifications. Whenever Charpy tests are reported, they include the impact energy at which the test bar suddenly broke, and the test temperature at which it happened.

The most important combination of impact energy and temperature is the temperature at which the metal stops acting tough and starts acting brittle. That brings us back to our "15 ft-lb Charpy V-notch impact at −20°F," printed on the filler metals can label, except that now it has a little more meaning for you. It's kind of a cut-off point at which the metal will become brittle if it gets any colder. The 15 ft-lb of impact energy is simply a commonly used level of impact energy at which to make that cut-off.

In fact, 15 ft-lb of impact energy may or may not be enough toughness for a given design. If it's not adequate, a tougher filler metal will be selected for use, for example, one with 15 ft-lb of impact energy at −40°F [−40°C] (which will have a lot more toughness than 15 ft-lb at −20°F [−29°C]). Later on we will show you the kind of test curve you can get from different impact tests. But first let's look at the older Izod impact test.

Izod impact test The principal difference between the Charpy V-notch and Izod impact tests is the way the test bar is held during the test. Figure 8-12B shows the

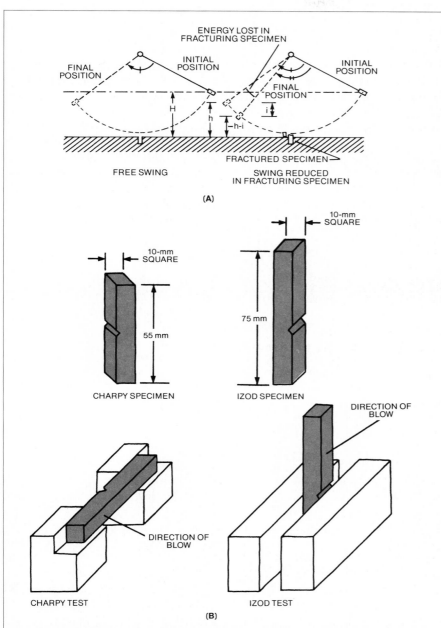

FIGURE 8-12 Impact tests are conducted with a weight on a pendulum. By measuring the difference in maximum pendulum swing before and after breaking a test specimen, the energy lost in breaking the test specimen can be converted to energy units such as foot-pounds. The higher the energy required to break the test specimen, the tougher it is. Impact data almost always includes the test temperature because many tough materials become brittle when they are cold. The two major impact test methods, (A) the Charpy and the (B) Izod test, differ primarily in the size of the test specimen and how it is held ready for impact in the test pendulum. The Charpy V-notch test is by far the most commonly used test.

Izod test set-up. The test bar notch should have a very sharp root or point at the tip. The sharply notched test bar gives you data that show a wider difference between notch-tough and notch-brittle steels, for example. (A similar test sometimes used in welder qualification puts a much bigger notch on the edge of a welded coupon; then a chisel is put into the notch and hit with a hammer to see if the weld splits.)

The weighted pendulum in Izod or Charpy tests hits the face of the bar (Fig. 8-13) and, if the steel is sensitive to impact at the test temperature, a crack will start at the root of the notch and run across the test bar at extremely high speed. If the material's impact strength is too low (the toughness is too high) at the given test temperature, the bar may bend or crack but it won't break in a shattering or brittle fashion.

When the test bar does break in a brittle fashion, the crack runs through the bar at very high speed, starting at the sharp tip of the notch. This is exactly the way a pressure vessel, a pipeline, a ship hull, or even a bulldozer blade or truck frame will fail under high impact or sudden stress, especially when it is cold. You may have heard someone say "it was so cold that metal snapped in two." That's literally true.

Transition Temperature

Many steels that are impact-tough at room temperature become impact-brittle at temperatures not much below the temperature of ice water. Most nonferrous metals such as copper, aluminum, and nickel remain notch-tough to extremely low temperatures. So do certain kinds of stainless steel. That is, they are not embrittled by cold. In fact, they tend to get stronger as they get colder. That explains why a certain group of stainless steels called *austenitic* stainless steels are used to contain liquid oxygen and why certain 9 percent nickel steels and aluminum alloys are now used to safely contain liquid methane.

The 9 percent nickel steels do get brittle, but their transition temperature is below about $-320°F$ [$-195°C$], so they are very useful in building tanks for many cryogenic liquids. Some metals such as the austenitic stainless steels, aluminum, and titanium alloys don't get brittle at all, even down to the temperature of liquid helium at $-452°F$ [$-269°C$]. That's also true of the welding filler metals used to join them.

Figure 8-14 shows you the results of a long series of Charpy V-notch tests on a notch-brittle steel. Each test bar was struck at different temperatures and the impact energy required to break it was recorded. At a certain narrow temperature range the steel suddenly became quite brittle. That's where the curve drops off fast from the high shoulder at higher temperatures to the low energy level at lower temperatures. The middle of the temperature range in which this steel

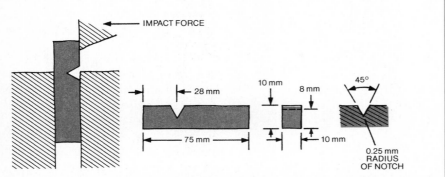

FIGURE 8-13 Dimensions for the Izod test bar must be very precise or the pendulum impact-test data will be useless.

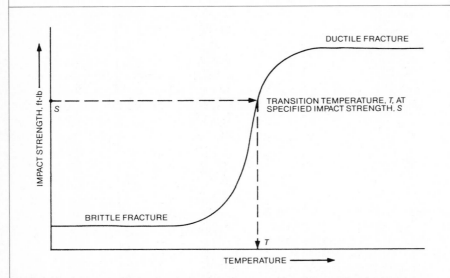

FIGURE 8-14 The difference between ductile and brittle fracture is shown when Charpy impact data are plotted against temperature for a single material. The impact strength drops off rapidly at some point (called the *transition temperature*). The transition temperature of both steels and nonferrous alloys varies considerably. However, the most commonly used impact energy for measuring transition temperature is 15 ft.-lb impact.

changes from ductile to brittle is called the steel's *transition temperature.*

You may have seen the same thing happen to nonmetallic materials. Try putting some rubber bands in the freezer compartment of a refrigerator and see how ductile they are. They will be very brittle and will crack instead of stretching. That's one reason why neoprene rubber is used for welding hoses. It remains ductile at lower temperatures. If you live in Canada or along the Great Lakes, in the Dakotas, or in Alaska you may even know of rubber auto tires that cracked during very cold winter weather.

Now you can see why impact data are important in selecting welding filler metals for certain kinds of products such as construction equipment, ship hulls, crane booms, mining equipment, pipelines, and other products that have to work in cold environments. That also includes spaceships, aircraft, rockets, and cryogenic tanks and pressure vessels for liquefied gases.

We won't give any details on the new fracture mechanics test procedures because you are not likely to use them as beginning welders. The fracture mechanics field is still very new and most of the reliable data produced by fracture mechanics tests only apply to very high strength metals. While some test bars have been standardized for these tests, they are much bigger than Charpy test bars, although Charpy V-notch test data also may be used.

Sometimes large plates several feet on a side are tested by first starting a crack in the plate and then bending it with an explosive charge (Fig. 8-15). The crack is easy to make, by the way. A brittle bead-on-plate weld will do the job nicely. But basically what fracture mechanics tests look for is the maximum allowable notch or welding defect that can be permitted in a given design. Using this information, nondestructive test procedures such as x-rays can

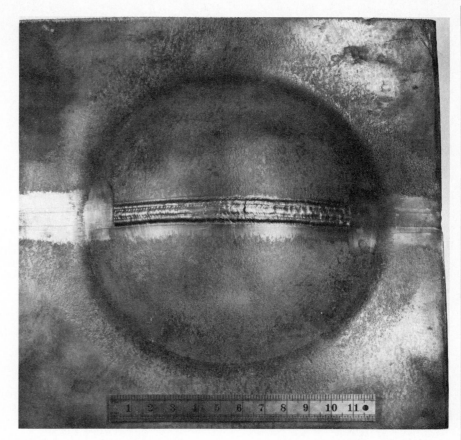

FIGURE 8-15 Explosion bulge tests are used to determine the toughness of heavy plates. A brittle weld bead starts the crack.

be used to inspect welds to make sure that any defects that they do have are too small to initiate brittle fracture under the expected operating conditions.

8-7
BEND TESTS
Bend tests can be used for both ductile and brittle base metals and weld metal. These tests measure the ductility of the sample in terms of its bendability. How does the amount a piece of metal will bend measure the ductility of that metal? It's simple. When a piece of metal bends, the surface on the outside of the bend is stretched. The more that outside surface will stretch without cracking, the higher the ductility of the metal.

Bend tests are particularly important to you because they are the most common destructive tests used to determine whether a welder can do the assigned work

on a job. The same tests often will show up weld defects such as poor tinning and resulting weak base-metal bonds when soldering and braze-welding, and the lack of fusion between weld metal and base metal when testing oxyfuel and electric-arc welds. They are an important part of most welder qualification and welding certification tests.

Free- and Restricted-Bend Tests
Bend tests may be made either *free* or *restricted*. Figure 8-16 shows a *free-bend test* performed on a piece of welded plate. In the first step, the test specimen is loaded at two points and bent anywhere from 30 to 40°. Bending is then completed in the second step when the partially bent specimen is placed upright between two jaws that are forced together until the specimen makes a 180° U bend, or it fractures. If your

weld is at that U bend you had better hope that if the metal does break, it fractures in the base metal well away from your weld and heat-affected zones, or you might not get the job.

The restricted-bend test in Fig. 8-17 uses a series of dies around which the test specimen is bent. Different dies have different bend radii, from a very severe bend of ¹⁄₁₆-in. [1.6-mm] radius to a mild bend of 1½-in. [38-mm] radius. The test specimen is hammered over the die with a wooden mallet covered with leather or one having a hard composition-board head. Sometimes a machine called a *press brake* is used. Again, you had better hope the metal doesn't crack in or around your weld if this test is being used to see how well you can weld.

The American Welding Society has recently standardized some of these tests for use in welder qualification procedures. (See the last

chapter of this book for details.) Two kinds of bend tests, however, measure several different things (besides how well or badly you weld), depending upon how they are performed and how the test specimen is set up.

Minimum bend radius The tests can measure the *minimum bend radius* that a given thickness of metal such as plate or sheet can be bent without forming surface cracks. That's very important when preparing sheet or plate for welding by bending it first in a press break or rolling the cylindrical sections for a tank through bending rolls.

As you might imagine, metals with higher yield strengths are harder to bend because they take more power to do it. Metals with low ductility also don't bend as far without breaking as steels with high ductility, and thicker metals need a larger minimum

bend radius than thinner metals, no matter what the ductility or strength of the material.

Bend elongation A measure of the bendability of a metal is its *bend elongation*. This can be measured by scribing gauge marks on the test sheet or plate just like the marks used on tensile-test bars. The distance the gauge length increases is measured along the outer surface of the bend. The data are plotted on a curve that looks very much like a tensile stress-strain curve except that the bend elongation is plotted against the bend angle.

All bend tests indirectly measure a metal's ductility. Bend tests, in fact, are more realistic measures of actual shop conditions than tensile tests are when the metal is going to be roll-formed or bent in a press break as is often done in large pressure-vessel fabricating shops and in shipyards.

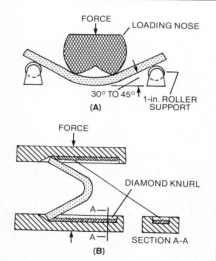

FIGURE 8-16 Two steps are required to make a free-bend test. (A) A partial bend is made with the test specimen in a horizontal position. (B) Then the partially bent test specimen is placed vertically between two knurled jaws which are forced together until the specimen either fractures or is flattened into a 180° U bend.

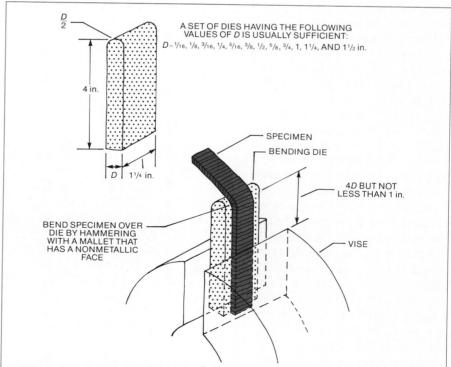

FIGURE 8-17 Restricted-bend tests often are used to test a welder's skill in qualification tests for getting a job. The welded joint is placed right in the middle of the section where the bend will occur. If the metal breaks in the weld joint, or if the joint pulls apart, then the welder fails the test.

Transverse Bend Tests

Figure 8-18 shows you a typical set-up for a transverse-bend test. This type of test is used to evaluate the strength and ductility of relatively brittle materials such as graphite, cast irons, hardened tool steels, and sintered tungsten-carbide cutting-tool inserts used on lathes, milling cutters, drill bits, and other cutting tools. The transverse bend test gives more reliable strength data than a tensile test on these brittle materials because brittle materials will crack in a tensile test before you have enough data to measure anything. Either a round or flat test bar can be used in a transverse bend test. A complete stress-strain diagram similar to the tensile stress-strain curve used for a tensile test can be plotted for the load and amount of bend recorded in the bend test.

8-8
METAL FATIGUE

Under certain conditions welded parts can actually crack under stress well below the tensile strength of the metal. The effect is called *metal fatigue.* Parts subject to cyclic stressing, such as a tensile stress followed by a compressive stress followed by a tensile stress, repeated over and over again, can fail due to metal fatigue.

What kind of part would be subject to that kind of loading? A lot of very familiar parts. Examples are springs, axles on trucks, crankshafts, automobile and truck frames getting bounced around on bad roads, railroad-car wheels, gears of all kinds, power-generation turbines, and many other rotating load-bearing parts in automobiles, trucks, aircraft, and power plants. Other parts that get cyclically stressed are the wire ropes in an elevator and even some of the beams on highway bridges (which are not only cyclically stressed by passing cars and trucks but also by daily changes in the temperature).

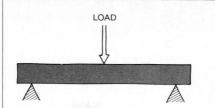

FIGURE 8-18 Transverse-bend tests often are used to measure the tensile strength and ductility of brittle materials such as graphite, ceramics, and high-carbon tool steels that have such low ductility that the conventional tensile test would not provide data.

Fatigue Tests

Fatigue tests are designed to measure the effect of cyclic-stress loading on metals. All metals are subject to the problem. Nothing can completely escape it. However, there are ways around it, as we'll explain shortly.

Specially designed test bars and sometimes complete structures or machine parts and fully welded sections are tested. The number of cycles of loading and unloading necessary for failure to occur by cracking is recorded along with the stress level and the direction of the stress (tensile or compressive). Figure 8-19 shows the results of a typical fatigue test after a number of samples have been tested at different load and stress levels until they failed. This typical fatigue curve shows that the stress at which metals fail, called the *fatigue strength,* drops as the number of changes in the direction of stress or *cycles* increase. The greater the number of cycles, the lower the fatigue strength . . . up to a point.

Even though it may take a lot of cycles under some pretty heavy loads to make a metal crack (often tens of thousands or more cycles), just think how many times a truck axle turns round and round (repeated cycles of tensile stress and compression) during a cross-country trip and you'll see that it's not hard to build up a very large number of cycles pretty quickly.

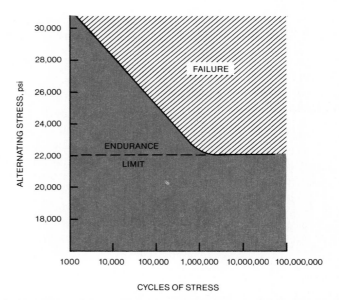

FIGURE 8-19 This is typical of a fatigue curve plotted from many different fatigue strength tests. The cross-hatched area represents the stress-strain combinations where failures can be expected in service. The dark tinted area shows safe stress-strain combinations over many cycles of use. The metals for automobile axles and airplane wings often are fatigue tested.

Endurance Limit

Below a certain low stress level in our fatigue test the metal did not fail by cracking, no matter how many stress cycles occurred. That lowest possible stress level with no failure, no matter how many cycles are repeated, is called the *endurance limit* of the metal.

Figure 8-19 also shows that the endurance limit of the metal being tested is 22,000 psi although the metal's tensile strength is over 30,000 psi. This is a typical fatigue test for a low-carbon steel. What's the fatigue strength of this steel at 10,000 cycles? Simply read directly up the curve from the 10,000-cycle point and you find that it's a little over 28,000 psi. At 100,000 cycles the fatigue strength of the carbon steel is over 25,000 psi.

If a part were expected to be cycled only 10,000 times during its useful life it could be loaded more heavily than a part expected to last for over 100,000 cycles, or even 1 million cycles. Loading the part below the endurance limit of the metal, at 22,000 psi or less, is safe under any number of stress-cycle combinations.

Variables Affecting Metal Fatigue

There are a number of variables that can affect the fatigue properties of a metal part, including the welds that join it.

Stress raisers The presence of notches (bad welds) and other stress raisers can seriously lower a metal's resistance to fatigue failure. Test specimens very similar to the Charpy V-notch or Izod test bars are used in many fatigue tests to determine the notch sensitivity of a metal part or a weld under cyclic loading.

Temperature As you would expect, temperature is a variable in these tests. Cold will decrease the fatigue strength of many metals.

Corrosion Another serious problem that can substantially reduce the fatigue strength of a part is corrosion. A rusty test bar will have a much lower fatigue strength than a clean test bar (or part). This is due mostly to tiny pits or stress raisers on the surface of the part caused by the corrosion.

Surface finish Smooth surfaces have higher fatigue strength than rough surfaces. Finish-machined bars are stronger in fatigue than rough-machined bars, for example. Sometimes bright electroplated surfaces (chrome-plated, for example) have higher fatigue strength than uncoated surfaces.

Fighting Metal Fatigue

There is something you can do to decrease metal fatigue. You may have noted that surface conditions have a lot to do with fatigue strength. You are right. Polished surfaces have higher fatigue strength than rough surfaces. (That's one reason why many shafts used in industry are brightly polished to produce a very smooth surface.)

Another method to combat metal fatigue is to make parts that have surfaces that initially are in compression. When the tensile forces on the surface that produce fatigue react with a surface that already is in compression, all the opposing tensile and surface-compressive forces tend to cancel each other out. That greatly reduces the chance of a fatigue failure. How can you possibly do that? It's done all the time by welders.

It's not unusual to see a written welding procedure end with the comment, "peen all over weld metal" or "shot blast for 10 min," or similar instructions. "Peen all over" means that you are supposed to hammer your finished weld bead, and maybe the heat-affected zone too, to squash or compress the surface of the weld metal and sometimes that of the base metal.

An air hammer often is used to save you extra effort. Sometimes a heavy rotating wire brush with steel balls on it will "beat" the metal surface enough to do the job. And sometimes the parts will be put into a special cabinet and pounded by millions of tiny hard glass balls or steel BBs blown by compressed air (that's shot blasting). This is described in Chap. 22.

The object of shot blasting or peening or hard wire brushing is not just to strengthen the surface metal. The treatment actually puts the metal surface into compression. The compressive force may not go very deep into the metal surface, but it will go deep enough so that when the metal part is bent in tension, the surface actually goes from compression to a very low level of tension, or maybe just to a neutral condition with no stress on the surface at all, even though it is being bent slightly. It's like adding $+5$ (high tension) to -3 (moderate compression) to give $+2$ (low tension), to use an arithmetic "model" of the effect.

The result (looking back at the fatigue-test curve) is to lower the tensile stress on the surface during load cycles. If you were to test such a treated metal surface, you would find that the metal's endurance limit has increased.

8-9
CREEP RUPTURE

Did you ever look closely at a marble park bench? If it is a very old bench it may sag slightly in the middle (and we don't mean the worn areas where people have worn "sit spots" on it). The continued weight of thousands of people sitting on the stone bench has caused it to sag or *creep*, that is, slowly change shape over time, even though one or two people at a time sitting on the bench hardly put the marble slab under very much tensile stress.

Metals creep, too. They deform plastically at stress levels below their yield strength. At room temperature the effect is not noticeable, and you'd have to wait a very long time to see it happen. At high temperatures (like the inside of a jet or diesel engine) creep occurs much faster. If the combination of tensile or compressive load stress and the temperature are both high enough, creep can slowly pull a part out of shape. If that part is a gear or a jet engine compressor vane, it doesn't take very much creep to ruin the entire engine.

If the stress-and-temperature combination is even higher and creep continues for a while, the parts can break or rupture at stress levels significantly lower than their room-temperature tensile strength. That's the "rupture" part of the creep-rupture effect.

Actually creep rupture is a pretty exotic materials problem, and unless you are working in the aircraft industry or maybe making parts for heat-treating furnaces or welding boiler tubes for a fossil fuel or nuclear power plant you won't hear about it. But there are special creep-rupture tests that determine which metals are more resistant to the effect than others.

Creep-rupture tests measure the elongation of test bars or actual parts under different combinations of stress and temperature. Several things usually are measured. One is the total elongation of a test bar before it fractures. The other is data on the time a test takes to change a test bar's shape a given amount as measured by the change in elongation, or until the part ruptures or breaks completely. Thus creep-rupture tests may measure either the amount of stress to produce a specified creep rate (the stress needed to produce a specified elongation in a given time), or the stress to produce a fracture after a long time under a heavy load.

Rupture strength is the name of the maximum stress level needed to cause a piece of metal to fracture in a specified time at a certain high temperature. Results

are usually reported as the stress necessary for rupture in 10, 100, or 1000 h of continuous stress. The temperature always is stated along with the stress level. For example, the rupture strength of an alloy might be 15,000 psi at 1875°F [1024°C] after 100 h of exposure.

Certain metals such as high-alloy nickel, molybdenum, and iron-chromium alloys are much more resistant to creep rupture at high temperatures than are carbon steels. These special superalloys also have much higher tensile and yield strengths than carbon steel at very high temperatures. Collectively they are called "high-strength high-temperature alloys," and, while they can be welded, this work belongs to an advanced welding course.

8-10
METAL PROPERTIES AND WELDING

That's all we'll tell you about mechanical property tests except to point out that we have only scratched the surface when it comes to mechanical property testing. But we scratched it in such a way that you can understand and use what we turned up for you when you are a fully qualified skilled welder. And you will start building those skills soon by making your first oxyfuel fusion welds in the next chapter.

This chapter has no shop practice. It's ideal for shop work, but many welding schools simply can't afford the specialized equipment. Even when they have it, operation of the test equipment varies considerably. So we'll leave it up to your welding instructor to decide what tests, if any, you should run. It's more important that you get hands-on experience joining metals.

Even if you don't run any mechanical property tests yourself, you now have the language you need to understand filler-metal and base-metal specifications, to understand the welder qualifica-

tion tests you'll sooner or later have to pass, and to help your future employer get work done.

Sound weld metal and base metal without problems are what you will be paid to produce as a skilled welder. This chapter has told you how those properties are determined. Welder qualification tests are based on many of the tests you just read about. This chapter will help you understand how do a better job, and it might also show you why you didn't get a job that you applied for.

REVIEW QUESTIONS

1. Name at least two kinds of physical properties of materials.

2. Name two kinds of chemical properties of materials.

3. Name two kinds of mechanical properties of materials.

4. If you hang a 100-lb weight on the end of a dangling rope attached to the limb of a tree, is the 100-lb weight a stress, or a strain?

5. If the rope is 50 ft long before the 100-lb weight is added, and the rope stretches 0.5 ft, is the extra stretch in the rope a stress, or a strain? Quantitatively, how much is it?

6. A steel bar has a square cross section. It is 10 in. on each side. What is the cross-sectional area of the square bar? The bar now has a circular cross section which is 10 in. in diameter. What is the cross-sectional area of the round bar?

7. The 10- × 10-in. square bar is hanging vertically. It has a 1000-lb weight hanging from the bottom end. What is the unit stress (in psi) on the material in the bar due to the 1000-lb weight hanging from the bottom?

8. A tensile stress is applied to the wing spar of an airplane while it is in flight. The stress is 80,000 psi. The aluminum wing spar has a yield point of 70,000 psi. The tensile stress is removed because the pilot was able to land the plane. Is the aluminum wing spar now longer than before, equal to its prior length before the high loading, or shorter than before?

9. The heat-affected zone next to your weld has a hardness of Rockwell C 46, and your welding specification says that the heat-affected zone must not exceed 400 BHN hardness. Are you in trouble? The part is made of carbon steel. What is the base metal in the heat-affected zone? Assuming 400 BHN is the average hardness of the base metal, what is the average tensile strength of the steel in the part? Do you think this is a low-carbon steel or a high-carbon steel? How weldable is this metal?

10. Your welding specification calls for as-deposited welds with a Charpy V-notch of 15 ft-lb at −50°F. Is that a hardness specification, a strength specification, a new way to measure temperatures, or something else?

11. A filler-metal specification shows that a welding rod or arc-welding electrode has all the right mechanical properties for your base metal, except that the data on Charpy V-notch says the as-deposited weld metal will have a Charpy V-notch of 20 ft-lb at +90°F. In your opinion, is the weld metal produced by this rod or electrode tough, or brittle?

12. When a metal gets fatigue, does it:
   a. Stretch.
   b. Crack.
   c. Corrode.

13. Name several variables in a part design, or its operating conditions, that will affect metal fatigue.

14. Why are metal properties important to welding? Give some specific reasons. This is an open-ended question. Talk it over with other people in your class. Do you have any personal experience with metal properties, for example, an automotive engine block that cracked on a cold day, that you now understand but didn't before this lesson?

Making Fusion Welds with Oxyfuel

Gas welding (Fig. 9-1) uses the heat produced by burning oxygen and a fuel gas to melt and weld a base metal (with or without using a filler rod). Its American Welding Society process code is OFW. If you did all the lab exercises in Chap. 7 on brazing and braze-welding, oxyfuel gas welding will be easy to learn.

About 50 years ago, almost all production and maintenance welding was done by oxyfuel. Today, perhaps 90 percent of all production welding is done by one of several arc-welding processes. It's usually not practical to use oxyfuel welding for plate ³⁄₁₆ in. [5 mm] thick or more, or on pipe sections with a wall thickness over ³⁄₁₆ in. [5 mm]. Thicker sections usually are arc-welded because it's faster.

Arc welding has another advantage over gas welding; there is less base-metal distortion. While the heat generated by the electric arc is far more intense than that produced by a flame, it is also concentrated in a much smaller area. The oxyfuel flame pumps much more heat into a much larger area, including the base metal, than an electric arc, resulting in more base metal distortion.

It also means that proper jigging and clamping will be even more important in oxyfuel welding than in arc welding.

The other reason that arc-welding processes are now far more common than gas welding is that they tend to have much higher weld-metal deposition rates. You can lay down more pounds of weld metal per hour, and your employer's labor costs are much lower as a result.

9-1
WHY GAS WELDING IS IMPORTANT

There are four important reasons why you should become as skilled in oxyfuel gas welding as possible. The first reason is that oxyfuel welding is still very important in maintenance work because arc-welding electrodes don't always exist for everything that has to be fixed.

The second reason is that an oxyfuel welding outfit complete with a cutting attachment, cylinders, regulators, and tips costs much less than one good arc-welding power source. You will still need the gas-welding equipment if

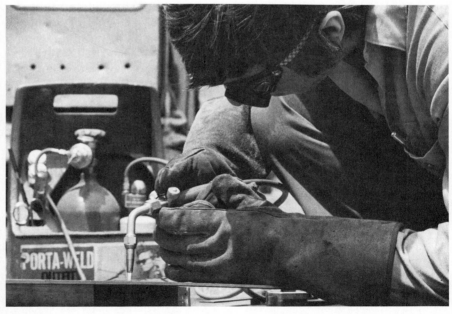

FIGURE 9-1 Oxyfuel gas welding is now used mostly for maintenance work. The equipment for this type of welding costs the least to buy. The process is exceptionally portable.

you are arc welding because of its flame-cutting capability.

The third reason gas welding is still used is that an electrical outlet isn't always handy where you want to work. While there are portable arc-welding machines that use gasoline or diesel engines to generate their own electricity, you're still faced with higher equipment costs, and the portable arc-welding generators or alternators are a lot less portable than an oxyfuel outfit. If you use small oxygen and fuel cylinders, you can strap a complete oxyfuel welding outfit on your back, so anywhere you can climb, crawl, or walk you can do oxyfuel welding and cutting.

The fourth reason this chapter is important is that the manual skills you will learn in oxyfuel gas welding apply directly to the very important (and usually well paid) tungsten inert-gas (TIG) arc-welding process. Tungsten inert-gas or GTA (gas tungsten-arc) welding is used extensively to join stainless steels and aluminum. Tungsten inert-gas welding is used in industries that make nuclear reactor piping, for example. It also is important in making aircraft and missiles. After learning oxyfuel welding, you will probably be able to pick up TIG welding techniques more quickly than an experienced welder who only knows how to do stick-electrode welding.

Uses of Oxyfuel Gas Welding
Most of the commonly used ferrous metals are readily welded by the oxyfuel process. In fact, the art of welding itself was developed primarily with wrought iron, plain carbon steels, and gray-iron castings. The tonnage of iron and steel products manufactured now is many times greater than all other metals combined. Over 90 percent of all the metals used today are iron-based. The relative amount of welding done on irons and steels is equally great.

Nevertheless, the oxyfuel process can weld just about any

metal that is weldable. One of its virtues is that all you need is a welding rod that more or less matches the base metal and you can weld with it. On thinner sections you may not need a filler metal at all. Oxyfuel welding also can join aluminum, magnesium, nickel alloys, titanium, copper, brass and bronze, cast irons, stainless steels—almost any metal that can be welded, which is one reason why oxyfuel welding is still extensively used in maintenance work. It is impossible to get a special flux-coated arc-welding stick electrode or a special gas-shielded bare welding wire for GMAW that will match just about any base metal. Some filler metals can't be drawn into wires.

While there is a wide variety of flux-coated stick electrodes, that doesn't mean it's always easy to get the right one for a small repair job without buying more than you need. But in an emergency, a piece of scrap from a similar broken part with the same chemical makeup can be used as a gas-welding rod. That's how American Sea Bees kept bulldozers in operation on Pacific islands after Japanese shore batteries sank landing craft carrying their welding rods. In a real emergency a gas-welding rod doesn't have to look like a rod. We don't recommend that you try welding with scrap because modern gas-welding

filler metals have a lot of things either in them (or removed from them) that help ensure that you get a good ductile weld, but it illustrates the versatility of gas welding.

Everything you have learned about braze-welding applies to fusion welding: working in different welding positions, using different flame settings from carburizing to oxidizing, the tips and torches you use, and the effect of heat on base-metal distortion. The welding joints you used for braze-welding are the same as those for oxyfuel welding. Some of the fluxes used for braze-welding also are used for gas welding. Braze-welding filler metals are included in the same AWS filler-metal specifications as gas-welding rods because they are the same rods.

Gas-welding rods are even specified on the basis of some of the same mechanical properties you just learned about. In fact, you are very close to knowing most of what you need to start making oxyfuel welds.

9-2 SELECTING GAS-WELDING RODS
Arc welders use electrodes. Gas welders use rods (unless filler metal is not needed to fill a gap in a welding joint, in which case the base metal is melted directly to-

---

TABLE 9-1 American Welding Society gas welding, braze-welding, and brazing rod specifications

AWS A5.2	Iron and Steel Oxyfuel Gas Welding Rods.
AWS A5.7	Copper and Copper-Alloy Welding Rods.
AWS A5.8	Brazing Filler Metals.
AWS A5.9	Corrosion-Resisting Chromium and Chromium-Nickel Steel Welding Rods and Bare Electrodes.
AWS A5.10	Aluminum and Aluminum-Alloy Welding Rods and Bare Electrodes.
AWS A5.13	Surfacing Welding Rods and Electrodes.
AWS A5.14	Nickel and Nickel-Alloy Bare Welding Rods and Electrodes.
AWS A5.15	Welding Rods and Covered Electrodes for Welding Cast Iron.
AWS A5.16	Titanium and Titanium-Alloy Bare Welding Rods and Electrodes.
AWS A5.19	Magnesium and Magnesium-Alloy Welding Rods and Bare Electrodes.
AWS A5.21	Composite Surfacing Welding Rods and Electrodes.

gether without an oxyfuel filler-metal rod).

The American Welding Society publishes a number of different filler-metal specifications for both ferrous and nonferrous gas-welding rods. Table 9-1 lists them, along with one extra AWS specification we haven't mentioned yet, AWS A5.8 for brazing filler metals. The most important gas-welding-rod specification is AWS A5.2, iron and steel oxyfuel gas-welding rods. Here are some important facts about it.

All welding rods (no matter what they are made of) are designated by an R followed by some numbers or letters. The R indicates that the filler metal is a gas-welding rod. In the case of AWS A5.2, you'll see rods designated by RG and followed by some digits (either 45, 60, or 65). The R tells you that you are looking at a specification for a welding rod. The G means it's a gas rod (as opposed to a filler rod for TIG arc welding, for example). The following numbers tell you the approximate tensile strength in 1000 psi of a test weld made with the rod in accordance with test procedures explained in the AWS specification.

Table 9-2 lists the key minimum properties of weld metal meeting AWS A5.2 for mild- and low-alloy-steel gas-welding rods. This table lists the minimum tensile strength that each grade of gas-welding rod will produce, both when you finish a weld (as-welded) and if (or when) the weld metal has been stress-relieved by a follow-up heat treatment. The percent elongation data give you information on the minimum ductility of the weld metal each grade of steel rod will produce when it is turned into weld metal or after the weld has been stress-relieved.

For example, an AWS RG 65 gas-welding rod will produce weld metal that is actually a little stronger than 65,000 psi tensile strength; about 67,000 psi tensile strength according to the specifi-

cation. In fact, the weld metal produced may actually be stronger than that. But what the specification says is "don't count on it, this is the minimum strength that you can expect."

If the filler metal is specially certified to meet exact mechanical requirements, then you can be sure. But certified filler metals are very expensive and are mostly used for code welding on nuclear power plants.

Also note that because of the particular alloying elements used in an AWS RG 65 rod, it will probably be a little more ductile than an RG 60 rod. That's very unusual in welding because increased strength normally means reduced ductility. However, RG 65 rods were specially developed to weld high-strength low-alloy steels.

There is a trade-off, however. RG 65 rods cost more than RG 60 rods, and both RG 60 and RG 65 rods cost more than RG 45 rods. The idea is not to select the strongest welding rod you can, but to select the one that is best for a particular job. It makes no sense to produce more expensive welds that are much stronger than the base metal. Oftentimes, an RG 45 rod will be good enough.

Also, don't be confused by a lot of different trade names. Regardless of their trade names, welding rods that meet all the requirements of a given AWS welding specification will produce major weld-metal characteristics very similar to those of the AWS specification. Certain minor differences

may exist among proprietary brands due to differences in manufacturing methods and chemical analyses, but these usually are minor, especially when using gas-welding rods. Don't ever pay more for a gas rod than you have to simply because it has a fancy brand name.

The situation is quite different with brazing alloys, arc-welding stick electrodes, and hardfacing alloys. While they may meet the minimum requirements of one AWS filler-metal specification or another, many of the proprietary brands will exceed the AWS minimum requirements. You may even find some proprietary brands that don't have an equivalent AWS specification. The reason is that much more research and development is being done on new arc-welding electrodes, for example, than on bare welding rods for gas welding. The markets are larger, so are the profits, and thus the competition. An electrode manufacturer who doesn't constantly improve the product will soon lose business to somebody else who makes better electrodes.

Iron and Steel Gas Rods

In the case of AWS A5.2 specification covering steel welding rods, the weld-metal tensile strengths of the RG 45, RG 60, and RG 65 rods are based on as-welded rather than stress-relieved properties of the deposited weld metal. The AWS A5.2 steel rod specification does not demand much more detailed mechanical properties from steel gas-welding rods than minimum tensile strengths and

TABLE 9-2 Minimum mechanical properties of AWS A5.2 steel gas-welding rods

AWS Class	As-Welded Minimum Tensile Strength, psi		As-Welded Minimum Elongation in 2 in., %	
	As-Welded	Stress-Relieved	As-Welded	Stress-Relieved
RG 45	45,000	40,000	Not reported	Not reported
RG 60	60,000	60,000	16	20
RG 65	67,000	67,000	20	25

minimum percent elongations to assure ductile weld metal.

The specification, however, does put strict limits on how much sulfur, phosphorus, and aluminum can be in the rod. We won't add that information here, but sulfur and phosphorus in particular are very detrimental to iron and steel weld metal. Just a little too much and the weld metal will be brittle, which is just one example of why you should use a standard gas-welding rod and not try to rig your own from scrap.

AWS A5.2 also spells out the dimensions of the rods. They are available in 1/16, 3/32, and 1/8 in. diameters. The standard length is 36 in. for all three diameters. Most of the rest of the specification tells how to conduct weld metal tests to see if an iron or steel rod produces welds that meet the AWS specification, but those details aren't necessary here.

Mixing and matching steel welding rods It is possible to mix the molten weld metal with the molten base metal to control the properties of the final weld. Start with a low-alloy steel rod and a carbon-steel base metal to make weld metal with a chemical composition in between those of the base metal and the filler metal. The mechanical properties of such a weld also will be in between those of the base metal and the filler metal.

For example, if the filler metal is a low-alloy steel (RG 65 rods) and the base metal is mild steel, then the weld metal's chemical composition and mechanical properties will be between those of the mild-steel base metal and low-alloy-steel filler metal.

This idea is sometimes used to gas-weld certain high-alloy steels such as AISI 4130 with any one of the three carbon-steel gas rods. The high-alloy steel has very high tensile and yield strength but its ductility is not all that great. Tubes made from the alloy steel must have weld joints with pretty

good ductility without sacrificing too much tensile or yield strength. Sometimes the job is done simply by using a lower-strength gas-welding rod and mixing it into the weld metal from the full-alloy steel.

For example, AISI 4130 full-alloy steel welded with an AWS RG 65 rod will have an as-deposited weld-metal tensile strength of around 100,000 to 125,000 psi, and a tensile strength as high as 145,000 psi when the joint is heat-treated (stress-relieved) after welding. Using an AWS RG 45 gas rod, the tube joints with mixed weld metal will have a tensile strength consistently around 90,000 to 100,000 psi, which is quite strong, but the ductility of the weld metal will be higher than that of the base metal.

Few modern welding shops would do this. It's much better practice to select an alloy-steel welding rod with the correct weld-metal properties for the job. But this example still shows you what can be done with mixing and matching of filler metals and base metals.

Oxyfuel welding rods (ferrous or nonferrous) are all bare rods. They do not have a previously added flux coating like arc-welding stick electrodes. Thus the ability to weld vertically and overhead is essentially a matter of skill. It is not related to the chemical composition of the flux or welding rod.

In contrast, the flux coating on stick electrodes often is formulated to make out-of-position welding easier by producing fast-freezing weld metal. Later on you will find out that specific grades of arc-welding electrodes are made only for downhand welding, for downhand and some out-of-position work, and even for overhead and horizontal welds.

The out-of-position stick electrodes will work in all positions, but the specifically formulated arc welding electrodes designed for downhand work will have much higher weld-metal deposition

rates, meaning lower labor costs and getting the job done a lot faster. Carbon and low-alloy-steel arc-welding stick electrodes are even coded with information on their most suitable welding positions as well as their tensile strength.

You don't have this complication in selecting an uncoated gas-welding rod. Instead you have the complication of making it work out of position using your talent as a skilled welder.

The RG 65 rods listed in AWS A5.2 are used for oxyfuel gas welding of carbon and high-strength low-alloy steels that have tensile strengths in the range of 65,000 to 75,000 psi. You know because AWS A5.2 tells you so (indirectly by giving the minimum tensile strength of 67,000 psi). You also know because you can simply match (or mix) the tensile strength of the gas welding rod with the base metal. AWS specification A5.2 even suggests how to do it in an appendix.

Matching the filler metal to the base metal often is how you pick the specific class of AWS A5.2 gas rod you want to use. AWS A5.2 RG 65 gas-welding rods are suitable for use on sheets, plates, tubes, and pipes. (Again, the spec tells you so.) The weld-metal deposits made by AWS A5.2 RG 65 rods have the highest tensile strengths of all the rods in that specification.

However, when the base metal has been selected for a specific property such as creep resistance or corrosion resistance, then the filler-metal rod should match the base metal alloy chemical analysis. Class RG 65 rods have a high-strength low-alloy steel analysis. To weld stainless steel or cast iron you would have to go to an entirely different AWS gas welding rod specification—ASW A5.9 covering corrosion-resistant steel filler metals is one example.

RG 60 steel welding rods are used for oxyfuel welding of carbon steels in the tensile strength range of 50,000 to 65,000 psi and for welding wrought iron. Wrought iron is very low carbon iron made in a special way. Not much of it is made anymore. (We'll tell you more about irons and steels like this in a later chapter.)

AWS A5.2 Class RG 60 rods are general-purpose oxyfuel gas-welding rods with medium strength and good ductility. They are most commonly used for welding carbon-steel pipes for power plants, process piping, and other conditions with severe service requirements. These rods provide intermediate strength levels. Class RG 60 rods generally have a high-strength low-alloy-steel composition.

AWS A5.2 Class RG 45 gas-welding rods have a simple low-carbon-steel analysis. Most rods of this AWS class have around 0.07 percent carbon content. This welding rod is used to weld very low carbon wrought iron and low-carbon steels. Like AWS A5.2 Class RG 60 rods, AWS A5.2 Class RG 45 welding rods are general-purpose oxyfuel rods commonly used in maintenance work when a stronger gas rod is not needed.

Cast-Iron Rods
AWS A5.15 "Welding Rods and Covered Electrodes for Welding Cast Iron" lists both oxyfuel welding rods and flux-coated stick electrodes (commonly called *covered electrodes*). The rods begin with an R designation; the electrodes start with an E. A typical cast-iron oxyfuel-welding rod is RCI (R for a gas rod, CI for cast iron). AWS A5.15 not only spells out the chemical composition of the weld metal produced by an RCI rod but also tells you what kinds of cast iron a particular RCI gas-welding rod joins best.

There actually are several RCI rods listed in AWS A5.15. There's a regular-class RCI rod, an RCI-A, and an RCI-B rod. The letters at the tail end of the rod class code simply designate a slightly different chemical composition or a somewhat different set of mechanical properties for as-deposited weld metal. This system of grade designation may include special letters or numbers after the filler-metal class code.

This is especially true in filler-metal specifications for arc-welding electrodes. In our cast-iron rod specification, the A or B after the rod class simply means that the rod has a slightly different chemistry, that the weld metal may have a modified set of mechanical properties for some special kind of work, or it may mean that the weld metal is qualified to meet the specification in a different way than normally used.

An example that you will find on many manufacturers' proprietary filler-metal grades that are identified by an AWS specification number and filler-metal grade and class code is an M after the number, which simply means it is a filler-metal grade suitable for military use. The chemical composition and mechanical properties may be exactly the same as any other filler metal with the same code designation, but the Pentagon has welding specifications and test requirements that sometimes differ from those of the American Welding Society.

You don't have to be a welding engineer to guess from the chemical composition of the as-deposited weld metal which of many possible kinds of cast irons each cast-iron rod should be used for. AWS A5.15 has an extremely useful appendix (as do all AWS filler-metal specifications) that tells you what rod is *usually* best for a specific kind of base metal . . . gray cast iron, malleable cast iron, ductile cast iron, or what have you.

RCI cast-iron rods are far from the only cast-iron oxyfuel rods and electrodes listed in AWS A5.15. It happens that nickel-bearing iron-welding electrodes do a very good job welding many different kinds of cast iron. That's why you also will see another list of filler metals with designations like ENi and ENiFe. Copper-bearing gas rods also are used. They have designations like RCuZn or RCuSn when they contain zinc or tin as well as copper. You are already familiar with some of these rods. You used them for braze-welding brass or bronze.

Stainless-Steel Rods
AWS A5.9, "Corrosion-Resisting Chromium and Chromium-Nickel Steel Welding Rods and Bare Electrodes," covers gas-welding rods and GTAW electrodes that are used mostly for stainless steels. All of the them have ER designations because they can be used both ways, either for gas-welding rods or as filler metals for TIG welding. They have designations like ER 308 (which is used on AISI 308 and a lot of other stainless steels), ER 310 and ER 316 for AISI 310 and 316 stainless-steel base metal.

There are quite a few of these stainless dual-purpose oxyfuel–TIG rods, but in the appendix of AWS A5.9 or in any good welding handbook you'll find tables that suggest which stainless-steel gas-welding rods should be used with what grades of stainless-steel base metal. The single most common rod used for either gas or TIG welding of stainless steel is ER 308.

The equivalent covered stick electrode (in an entirely different AWS specification on SMAW electrodes) has the name E 308. So as you see, once you get the hang of things, filler-metal specifications for either gas or arc welding are less complicated than they appear at first.

Aluminum Rods
AWS A5.10, for "Aluminum and Aluminum-Alloy Bare Welding Rods and Electrodes," covers gas rods for welding aluminum. The letter designation is very similar to iron and steel gas-welding rods except that this specification includes uncoated bare rods that also are used in TIG (GTAW) arc welding.

A rod that is intended only for gas welding has an R in front of the class code. A rod with an ER in front of the class code can be used either for electric arc (GTAW) welding or for gas welding. A rod with only an E in front of the class number is only for arc welding. The E means electrode.

AWS A5.10 includes a long list of aluminum-alloy base metals and the correct rods to use with them, whether you are gas welding or using the GTAW process. Aluminum-alloy welding rods usually are selected first to match the composition of the aluminum base metal, which controls its corrosion resistance, and then try to match its mechanical properties.

We won't list the aluminum alloys covered in AWS A5.10 here. It's best to wait for the chapter on GTAW before we get into the selection of aluminum welding rods versus different aluminum base metals to be joined. At that time, you will find that the very same filler rods and gas-welding specifications are used for TIG welding as well as oxyfuel gas welding.

The actual selection of the correct aluminum gas-welding rod is a little more complicated than steel rods because aluminum has properties other than strength and ductility that also are important. Corrosion resistance is especially important and makes aluminum-alloy rod selection more complicated. You have to know something about aluminum alloys and corrosion resistance and how they both work. In a real shop situation you'll usually be told what the correct rod for a given aluminum base metal is in advance, or you can guess very closely based on information from any standard welding handbook.

For example, if your base metal is 1100-alloy aluminum (nearly pure aluminum), one of the rods the AWS specification tells you to use is a class ER1100 welding rod. The weld metal it produces will have the same chemical and corrosion-resistant properties as 1100-grade aluminum. You also know you can use this rod either for oxyfuel gas welding or for GTAW because of the ER (electrode or rod) designation.

Copper-Alloy Rods

AWS A5.7, "Copper and Copper-Alloy Welding Rods," uses a slightly different designation system than steel rods. The specification also details the actual chemical composition of the weld metal produced by each copper rod or electrode.

The copper rod designations are more like the aluminum rod designations because they are primarily based on the chemical composition of the as-deposited weld metal. They look a little different because the copper rod classes in the AWS A5.7 specification actually tell you (in a shorthand way) the kind of weld metal each rod will produce. The specification then spells out what the weld-metal composition should be.

For example, the basic copper welding rod in AWS A5.7 is RCu (rod, copper). Copper-tin (phosphor-bronze) rods are R CuSn-A. Cu is the chemical symbol for copper, Sn is the chemical symbol for tin (*stannum* is Latin for tin). The last letter, A, refers to a specific grade designation for a specific bronze rod. Further on in the specification you'll find a list of mechanical property data specifically for weld metal produced by class R CuSn-A rods. Copper-nickel rods are designated R CuNi. Therefore, if you look up a rod for copper-nickel base metal in AWS A5.7, you probably would pick the one listed as R CuNi. You weld brass with an R CuZn gas rod because brass is made of copper and zinc.

While there's much more detail in each of these AWS specifications, you can see from our brief description how different gas-rod specifications work. They may use slightly different coding systems on rods or electrodes, but those systems are very practical.

The coding systems are based on the realities of material selection and use. Carbon steels are selected primarily on the basis of strength. So are the gas rods used to join them. The selection of aluminum and copper alloys depends partly on mechanical properties and partly on specific chemical compositions (to get a specific corrosion resistance for example). The rod designations for those materials reflect these facts.

Fortunately, welders don't have to memorize all those facts. They simply look them up when they need them. In a real shop you will only be working with a small number of actual rods. You will very quickly get used to what they are called and what they will do when you use them.

9-3
WELDING TECHNIQUES

The first step in learning oxyfuel welding is learning how to make tack welds on butted plates or heavy sheets. Figure 9-2 shows you how. Support your work on fire brick as shown in the illustration. Leave a 1/16-in. gap between the sections and make tack welds

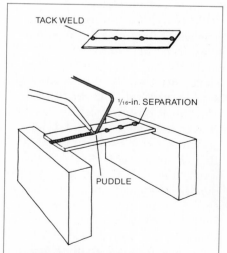

FIGURE 9-2 Learn backhand welding after you have mastered forehand welding. Learn to weld backhand in all positions. Then learn to switch from forehand to backhand welding while making a single pass, without stopping welding.

every few inches along the joint. That will help control distortion later on when you start putting down a continuous bead.

Oxyfuel welding beads are made either with the forehand or the backhand method. We already covered these methods in braze-welding. By now you should know them well. But to remind you, the welding tip and the flame in the forehand technique are pointed ahead of the solidified weld metal, in the direction you are welding. The backhand technique points the flame at the solidified weld metal, opposite from the direction you are welding. Forehand = flame forward, backhand = flame backward.

These welding techniques are so important that we will review them again, and tell you more about what they can and can't do.

Forehand Welding

The forehand welding technique illustrated in Fig. 9-3 is sometimes called *puddle welding* or *ripple welding*. The rod is kept ahead of the flame in the direction in which the weld is being made. You point the flame in the direction you are welding, and you hold the tip at an angle of about 45° to the welding base-metal surfaces. With your other hand, you move the welding rod in the same direction as the tip.

You can melt the end of the rod and the welding surfaces into a uniformly distributed molten puddle if you move the torch tip and

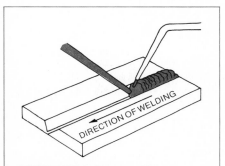

FIGURE 9-3 Learn to do forehand welding. Point the flame in the direction of torch travel.

the welding rod in opposite, semicircular paths. As it passes the rod, the flame will melt a short length of the rod and add it to the puddle.

You will have to practice with the torch and rod until you are able to distribute the molten weld metal evenly into the puddle and along both edges of the welding joint. Since you have to do different things with either hand while also controlling a molten weld-metal puddle, it's harder than simple brazing. That's why we started braze-welding with a filler rod before we took up full oxyfuel fusion welding.

The forehand method can be used in all welding positions (downhand, vertical, overhead, and horizontal) for welding sheets and light-gauge sections up to $\frac{1}{8}$ in. [3 mm] thick because the method gives you better control of a small weld puddle and results in a smoother weld than the backhand technique. However, the forehand technique is not the best method for welding heavy plate.

Backhand Welding

Backhand welding (Fig. 9-4) is a slightly newer technique. In this method the torch tip leads the welding rod in the direction of welding, and the flame is pointed back at the molten puddle and completed weld. The end of the welding rod is placed between the torch tip and the molten weld puddle. The rod is held at an angle of about 60° with the plates or joint surfaces being welded.

The backhand method gives you greater base-metal penetration than forehand welding. It also gives you a narrower weld bead. It's generally better and faster than forehand welding. Less motion is required in the backhand method than in forehand welding.

If you use a straight welding rod (bare rods can easily be bent by hand into any angle that helps you get them into tight joints), the rod should be rotated so that the end will *roll from side to side* and melt off evenly. If you are

using a bent electrode, move the rod and the torch back and forth at a rather rapid rate since you can't easily rotate a bent rod.

If you are making a large weld, you should move the rod in complete circles in the molten puddle instead of in circular arcs. Also move the torch back and forth across the weld while it is advancing slowly and uniformly in the direction of welding.

Most beginners weld too fast. Work slowly until you get the right combination of speed and torch and rod motion. In any case, you must practice these motions until they are second nature. Simply reading about them won't help your hands, arms, and eyes make it happen.

You'll find that the backhand method is best for welding material more than $\frac{1}{8}$ in. [3 mm] thick. You can use a narrower V groove at the joint than is possible with forehand welding (which requires more room). An included angle of 60° between joint edges is enough to get your rod into the joint and make a good weld. It doesn't take as much welding rod or puddling (dipping the rod into the molten metal) for the backhand method as it does for forehand welding.

When welding steel with a backhand technique and a slightly reducing flame, the flame adds extra carbon to a thin surface layer of hot base metal. That reduces the melting point of the steel in that area because the increasing carbon content lowers the melting point of steel, speeding up the welding operation. This same technique also is used when surfacing with chromium-cobalt hardfacing alloys. Be careful with your flame setting. This method works nicely with a *slightly carburizing* setting, but if you give your torch a fully carburizing flame you can add too much carbon and wind up with very hard, brittle welds.

By using the backhand technique on heavier material you get increased welding speeds, better

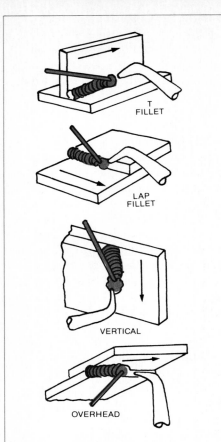

FIGURE 9-4 When you can successfully weld groove joints in the overhead position with a torch and filler rod, the rest of the manual welding techniques in this course, including arc welding, will be easy to learn. Note that tack welds are only one-half the section thickness.

control of the larger puddle, and more complete fusion at the root of the weld joint. Furthermore, by using a slightly reducing flame with the backhand technique (when welding steel), a smaller amount of base metal will be melted while welding the joint.

The somewhat more simple-to-use backhand welding technique combined with lower welding rod consumption per joint and faster welding speeds has produced cost reductions of up to 65 percent in some jobs. That doesn't mean it's the only method you should learn. There are times when you will be working in a difficult position, or on a part that only allows you to

get at the weld with the forehand method. And remember, the forehand technique works better on thin materials. *You need both methods.* You should not only learn both of them but also be able to switch from one to the other right in the middle of making a single weld. If you can do that and do it out of position, too, you'll be able to pass the toughest gas welding job qualification test anybody can dream up.

9-4
MULTIPASS WELDING PROCEDURES

In the preceding section we described two basic oxyfuel welding techniques, the forehand method and the backhand method. Both methods can be used on the same weld when you are joining a thick section that will require more than one pass with your filler metal and oxyfuel gas flame.

Thick plate and pipe joints often require more than one pass to fill the joint. When you make a multilayer weld, the root pass (the first pass at the bottom of a V joint, for example) is critical. This holds true for any welding process, gas or electric. No matter what welding process you use, *always concentrate on getting a good root pass at the bottom of the joint on the first pass.* If you can't get a solid, high-quality root pass without defects, all the rest of the work you do is worthless. There is a good chance that the weld will fail in service.

Root passes in heavy plate, and especially in pipe welding, are so critical that on some arc-welding pipeline jobs entirely different arc-welding electrodes or even different welding processes are used to make them. For example, GTAW may be used to make pipe root passes and stick electrodes or some semiautomatic process may be used to finish the weld.

Very often, even on a pipeline job done entirely with stick electrodes, different electrodes are used for the root pass than the

fill-in and cover or cap passes. In fact, on some pipeline construction jobs the best welders (with the best pay) often work only on root-pass welding. Less skilled people work on the fill-in and cover passes.

Multipass welding is used on thicker joints not only because more than one pass is required to fill the joint, but also because it's much easier to carry a smaller weld puddle than a big one. You'll also find that after you have completed a solid root pass (which inherently requires a smaller weld puddle), the next layers will go more easily.

The big job to do with fill-in passes is to get good fusion with the base metal on the sides of the joint, as well as on the root-pass layer. Welders call it "tying in the weld." The final top pass, often called the *cap pass,* is the easiest one to control to get a smooth surface. It's almost like welding side-by-side bead-on-plate welds.

Multipass welding has an added advantage in that it refines the previous layer of weld metal underneath as the new weld metal is layed on top. *Refines* means that the heat from the new weld bead tends to give the previously cooled older weld metal underneath a more fine-grained structure.

Fine-grained metals (with smaller individual crystals) are tougher than coarse-grained metals. Fine-grained structure improves the ductility and impact strength of the weld metal. If this added quality is desired in the last cover or cap pass, an added layer of weld metal sometimes is deposited and then machined off.

Joint Edges
Sheet metal is easily melted. It doesn't need special edge preparation. However, when you work on heavy-walled structurals, thick plate, or pipe, you cannot melt the entire section in one pass. You have to bevel at least one of the edges so that you get complete penetration of the weld metal throughout the section, and you

have to use multipass welding methods to do it.

You also need room to get into the root of the weld when you make your first pass. If your joint clearance is too small, you may not even have enough room to get your rod (or if you are arc welding, your electrode) into the bottom of the joint. That's when you need to prepare the edges of the base metal for welding.

Joint-edge preparation and the proper spacing between the plates, pipe, tubing, or structural sections are very important in multipass welding. The thickness of the section generally determines the amount of edge preparation and joint clearance, or gap, required between the two pieces of base metal. It also depends on the basic design of the welded joint.

The faces of square edges can be butted together and welded with complete penetration using plate up to about $3/16$ in. [5 mm] thick. For plate or other materials between $3/16$ and $1/4$ in. [5 and 6 mm] thick, a slight root opening or gap between the parts is necessary to obtain complete weld-metal penetration. Plate over $1/4$ in. [6 mm] thick requires beveled edges and a root opening of $1/16$ in. [1.5 mm].

For welding plate or structurals or a pipe section over about $1/4$ in. [6 mm] thick the edges are beveled at an angle of 35 to 45°, making the included angle of the complete joint from 60 to 90°. These edge-joint designs can be prepared by flame-cutting (as you already know), machining, chipping with a chisel on an air hammer, or by grinding. In any case, the edge surfaces must be free of oxides, mill scale, dirt, grease, or any other foreign matter.

Plate from $3/8$ to $1/2$ in. [10 to 12.5 mm] thick can be satisfactorily welded only from one side. Thicker sections should have edges prepared on both sides. In general, butt joints with edges prepared on both sides are easier to weld than edges prepared only

on one side. These double-edged or two-sided joints also produce less heat distortion and give better mechanical properties in weld metal in heavy sections than joints prepared only from one side because of the extra refining action of the heat produced by subsequent weld passes that produce fine-grain weld metal.

In an upcoming chapter on stick-electrode welding (SMAW), we'll show you all the basic joint designs used for welding, along with the proper dimensions, separation, or gap between the sections; the proper bevel angles to use, and other things that you will need. What we've given you above is enough to get you started properly with fusion welding. It's more important right now to learn how to make good welds than how to make good weld joints.

Since the same basic joint designs are used both for oxyfuel gas and most arc-welding processes, you can turn to the chapter on welding with stick electrodes and get this information on edge preparation, shapes, and dimensions right now, if you need it. Most likely you won't yet because Sec. 9-6 of this chapter will give you all the details you'll need for the work you will be doing.

9-5
WELDING FLUXES

You won't need to use welding fluxes on mild steel as long as your base metal is clean, the edges are properly prepared, you don't have oxide or slag on them, *and you set your welding flame to the neutral or slightly carburizing condition.* The main idea behind a neutral flame setting is that the flame envelope acts like the shielding gases used in many gas-metal arc-welding processes. As long as you protect your weld metal with the neutral flame, the hot metal won't be oxidized by the air before it's had time to freeze and cool a little.

If you see that you are getting burned steel, that doesn't mean you need welding flux. It means that your flame is too oxidizing to weld steel base metal. Reset your welding flame for a neutral or slightly carburizing flame.

When you work on nonferrous materials that are easily oxidized, especially copper, aluminum, or magnesium (as you will in the following shop section), you'll need to use welding flux. You also will need flux when working on cast iron. It's easy to know if you have the right flux. Just read the instructions on the label. Get the flux from a reputable welding distributor or your welding instructor. In general, you will use the same fluxes with these metals as when you were braze-welding them. Follow the directions on the flux container. Clean the flux off thoroughly after you finish the weld.

9-6
GAS-WELDING SHOP PRACTICE

You'll have two basic things to learn in your shop work. First, you'll have to master the manual skills required to make a fusion weld. Then you'll have to learn about the welding techniques used on the more common types of base metals, such as the difference between welding carbon steel and gray cast iron, or the difference between welding higher-melting-point metals (steels) and lower-melting-point metals (aluminum, for example). We'll also start teaching you the basic methods for making common welding joints in different positions.

Now you are ready to go into the shop and really get down to the business of learning to weld.

Steel Bead Welds on Sheet
Making bead welds with and without welding rods is a fundamental exercise that you must practice to learn the principles of fusion welding. You must master this procedure before attempting

those that follow. The basic procedure is the same for any thickness of base metal, but a larger tip size is necessary for heavier sheets. See the following Shop Practice lesson for suggestions on selecting

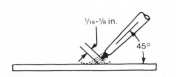

FIGURE 9-5 Hold your torch at a 45° angle with the tip of the inner-flame cone held a short distance off the workpiece.

FIGURE 9-6 This is the weaving motion you must learn to make with your torch and filler rod. You will make much the same motions later on when you learn arc welding.

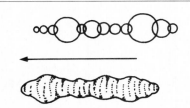

FIGURE 9-7 If you use a circular torch motion that is not uniform in width, your weld will immediately show it.

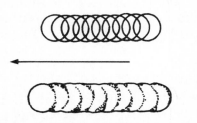

FIGURE 9-8 A uniform torch motion produces uniform weld beads.

tip sizes for different material thicknesses.

Support a sheet of steel on firebricks in such a manner that a portion of the metal where you will make the weld is clear of the bricks. All welding must be performed on areas that are not in contact with the bricks. Hold your torch at an angle of 45° to the plate and point it in the direction of welding (see Fig. 9-5). For right-handed operators weld right to left; for left-handed operators, weld left to right.

The inner cone of the torch flame should be held from $\frac{1}{16}$ to $\frac{1}{8}$ in. [1.5 to 3 mm] from the molten puddle when you use oxyacetylene (also shown in Fig. 9-5). Direct the inner cone of the flame at the starting point and hold it there until the metal becomes molten. Then move forward and at the same time impart a slight oscillating motion of your torch tip as illustrated in Fig. 9-6. Travel speed must be constant to achieve a uniform bead. If the speed is varied, a large molten weld puddle will result at slow speeds, and a small puddle will result when moving too fast. In the former case there is the danger that you will overheat the sheet and burn a hole in it. With the latter condition, poor fusion or no fusion will result with the base metal. The results of incorrect welding speed are shown in Fig. 9-7, while the correct technique is shown in Fig. 9-8.

A correct, satisfactory weld bead must be uniform in width and slightly depressed below the surface of the metal being joined. The under side of the sheet metal must be free of "icicles," that is, solidified drops of molten metal or slag under the joint. A thin film of oxide will cover the top and bottom surfaces of the weld. Obviously, if any of these features are absent, the weld is unsatisfactory. You must practice these beads until you are able to make satisfactory welds consistently. When this degree of proficiency has been acquired, you will pro-

gress to the next step, using a welding rod.

Using a gas rod In running bead welds with a rod the preliminary steps are the same. However, when the molten pool is established at the start of your weld, insert the welding rod and proceed in the direction of welding (see Fig. 9-9A). While oscillating your torch, carry it around the rod to ensure melting the rod and the base metal at the same time (see Fig. 9-9B). Again, it is important to maintain a constant travel speed to obtain a uniform width of bead and depth of fusion. If welding speed is too slow, icicles will be formed by the flame burning through the metal. Welding too fast when using a rod results in lack of fusion of the filler metal to the base metal.

Bead welding must progress at a speed that will ensure you complete fusion on both sides of the base-metal joint and at the same time result in a maximum reinforcement of one-quarter (25 percent) of the base-metal thickness. The contour of your bead must be

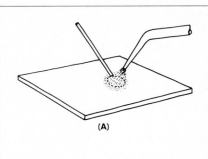
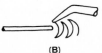

FIGURE 9-9 (A) Start your weld in one spot, making sure your base metal has formed a weld puddle under the torch. (B) Then start the weaving motion and continue to move the puddle along the joint as your weld bead forms. When you first learn to do this, go slowly. Most beginners move too fast.

smooth and the junction with the base metal must not show undercutting or overlapping. (See Fig. 9-10.)

Corner Joints on Sheet

Corner joints will introduce you to the importance of getting complete weld-metal melt-through, or penetration, through a weld joint. Failure to fuse the joint at the root or side walls or failure to provide proper weld-metal reinforcement will result in failures when

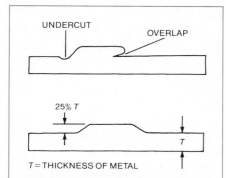

FIGURE 9-10 Typical welding defects such as undercutting and overlapping result from incorrect torch and filler rod motion. The ideal weld should have a reinforcement of no more than 25 percent of the section thickness when thin material is used.

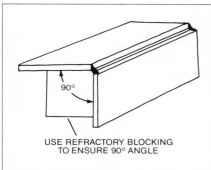

NOTE: ON 1/16 in. THICKNESS, NO SPACING
ON 1/8 in. THICKNESS, 1/16-in. SPACING

FIGURE 9-11 The easiest way to make an edge weld is to tack-weld the two corners while the metal is clamped or held at a 90° angle on a refractory brick. The brick is a simple example of a welding fixture.

you test your weld. Therefore, the principles learned in this lesson will be of great value not only in all future shop work, but also in getting tested for a welding job.

The procedures to be followed for 1/16- and 1/8-in.-thick [1.5 and 3 mm] sheets are essentially the same when making a corner joint. The only difference is that a 1/16-in. [1.5-mm] spacing is used on 1/8-in. [3-mm] thicknesses while no spacing between the sheets is used on 1/16-in.-thick [1.5 mm] metal. Take two sheets of steel of the required size with square-cut or sheared edges, and set them up at a right angle as shown in Fig. 9-11. Then tack-weld them together at each end. Be sure that the sheets are at 90° angles to one another. If welded at less than 90°, the hammer test will impose severe stresses that are likely to cause failure. If welded at greater than 90°, the weld will not be stressed high enough when tested.

After your tack welds have cooled, place your workpiece on your welding table so that the tack-welded corner joint presents a single V-shaped butt joint facing up, as shown in Fig. 9-12.

Hold your torch at an angle of 45° to the vertical as shown in Fig. 9-12. Introduce your welding rod ahead of the torch in the direction of welding.

The position of the inner flame cone of your torch and the way you oscillate your torch and rod are the same as in the preceding welding method (see Fig. 9-9A and 9-9B). A fast travel speed will result in fusing your base metal only across the upper surfaces of the V joint (see Fig. 9-13A). This same result will be obtained if the torch is held at too flat an angle. Abnormally slow travel speed, or holding the torch too near the vertical, or both, are likely to result in excess penetration at the root of your weld that will form icicles under the joint (see Fig. 9-13B). This excess filler metal is frequently not fused to the base metal and is therefore a source of cracking when you test your weld.

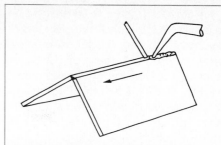

FIGURE 9-12 Once the corners are tack-welded, you can remove the work from the firebrick and finish the edge weld on the shop bench.

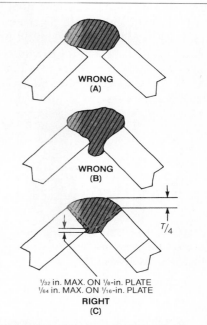

1/32 in. MAX. ON 1/8-in. PLATE
1/64 in. MAX. ON 1/16-in. PLATE
RIGHT
(C)

FIGURE 9-13 Examples of weld defects that often go unseen in real products. (A) This defect is the worst problem. It is an incomplete root penetration and is sure to mean weld failure. (B) This defect can result from burn-through, from traveling too slowly, or from using a filler rod that is too large for the workpiece and joint. (C) The ideal edge weld should have a reinforcement of 25 percent of the section thickness T, and a small reinforcement in the root area.

A correct corner weld will have a cross section similar to that illustrated in Fig. 9-13C. Reinforcement up to 25 percent of the base-metal thickness is OK, as shown in Fig. 9-13C. Lack of adequate reinforcement causes failure through the center of the weld due to insufficient weld-metal cross section to withstand an applied stress. Weld-metal penetration should not exceed the figures given in Fig. 9-13C and there should be no icicles anywhere in the length of your weld. A corner weld of the type shown in Fig. 9-13C will withstand flattening and bending back against itself without any failures.

Square-Groove Butt Joints on Sheet

The first step in making square-groove butt joints on sheet steel is to tack-weld the sheets together. The edges of the sheets should be spaced a distance equal to one-half the sheet thickness (see Fig. 9-14). Tack welds should be deposited at each end of the joint to

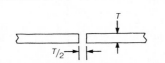

FIGURE 9-14 The best gap for a butt joint is one-half of the section thickness T.

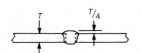

FIGURE 9-15 The joint gap allows for shrinkage of cooling weld metal which will pull the section together. If no space is left for contraction, very high built-in stresses result that can crack or even tear the base metal or weld metal. The correct joint gap also will help you control the height of the weld-metal reinforcement.

maintain this spacing and alignment of the sheets. The tack welds should not exceed one-half of the thickness of the sheets.

Flat butt welding on sheet

Place the tacked sheets on fire-bricks in such a manner that the joint is free of the bricks. Make a forehand butt weld. Use ⅟₁₆-in.-diameter [1.5 mm] rods on your sheet metal. This weld must be thoroughly fused throughout the cross section and built up with a 25 percent reinforcement as shown in Fig. 9-15.

As the end of the weld is approached, note that the accumulation of heat causes the weld puddle to become bigger. When you see this, stop welding and allow the metal to cool down to black heat before finishing the weld. If you attempt to finish the weld without letting it cool, the molten weld metal will run away and it will be impossible for you to build up the required reinforcement.

Vertical butt welding on sheet When welding in the vertical position, you will be confronted with the problem of supporting your weld metal in a vertical plane against the force of gravity. This control of the molten weld metal is achieved first by directing the force of your flame, second through the viscosity or "stiffness" of your molten weld pool, and third by manipulating your torch and rod while occasionally removing the rod and flame to permit the molten weld metal to cool when it gets too hot and too fluid.

The third factor is what controls the stiffness of your weld puddle. The hotter the weld puddle, the more it will want to flow out of the weld. Any time you weld out of position, using any welding process, working on the "cool" side of the temperature range is the main trick to make the weld metal stay where you want it.

To make a vertical weld begin at the bottom and progress up-

ward. Start the weld by building up a "shelf" of weld metal at the bottom of the grooved joint. This shelf will support the molten weld puddle. Hold the torch at an angle of 45° to the sheets as shown in Fig. 9-16 and use the same oscillating motions of your torch and filler rod that you used in making a flat-position weld. If a constant upward speed is maintained, a keyhole-shaped opening will be established at the root of the V joint, ahead of the puddle (see Fig. 9-17). The maintenance of this opening *ahead* of your weld is your assurance that you are securing proper penetration at the root of the joint. In addition, be careful to secure fusion at the face of the weld and to build the weld up to the required reinforced weld-metal contour of 25 percent.

Overhead butt welding on sheet The same factors used to obtain molten metal control in

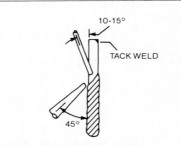

FIGURE 9-16 Use of the forehand technique when welding vertically. Tack welds will help you control distortion while holding your work in place.

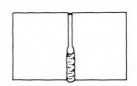

FIGURE 9-17 Another view of a vertical weld shows how the finished weld metal acts as a dam to support new metal. This is called "vertical-up" welding.

vertical welding are used for overhead welding—the difference lies in the position of your torch and welding rod. In overhead welding the flame is directed vertically upward and its force holds the molten metal in place.

To make an overhead weld you stand directly under the weld and in line with it, or slightly to one side. Welding progress should be toward you. Hold the torch tip perpendicular to the sheets and insert the rod from the front at an angle of 45° as shown in Fig. 9-18. Here again, the maintenance of the keyhole ahead of your weld is good practice to ensure that you get complete penetration. The same oscillation of your torch and filler rod used in flat and vertical welding is used when welding overhead to secure proper distribution of heat and weld metal.

Butt Welding Plate Up to ½ in. [12.5 mm] Thick—Forehand

By now you should be familiar with the fundamental techniques necessary for making butt welds. Therefore the detailed instructions given in the previous sections will not be repeated for those joints that follow.

For best results in butt welding plate steel from ³⁄₁₆ in. [5 mm] thick and over, the plates should be beveled before welding. Where forehand welding is to be used, as in this section, the bevel should be 37.5° (45° maximum) with a ¹⁄₃₂- to ¹⁄₁₆-in. [0.7- to 1.5-mm]

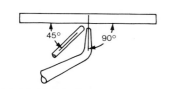

FIGURE 9-18 These are the best positions for torch and filler rod when welding overhead. Control the fluidity of the weld metal by withdrawing the flame from time to time until the molten metal solidifies.

"land" or flat area as shown in Fig. 9-19.

You can make a beveled edge on ³⁄₁₆-in. [0.5-cm] plate with a grinder with fairly satisfactory results. However, for ¼-in. [6-mm] plate and thicker, the bevel should be made accurately by oxyfuel gas cutting with a cutting machine. Weld one poorly beveled set of plates and you'll soon appreciate the importance of accurate edge beveling.

Set up the beveled plates on firebricks in the proper manner previously described, spaced one-half the thickness of the plate for ³⁄₁₆- to ¼-in. [5- to 6-mm] thick material. Then tack-weld them together on either end and in the middle if necessary.

Your tack welds should not penetrate the plate more than half of the thickness of your base metal. An excessively large tack weld will make starting and finishing your weld difficult. Note the similarity of these beveled joints to the corner joint. In both types of connections the weld faces form a 90° V. The principal difference between them is in the distribution of the mass of the base metal as shown Fig. 9-20. Therefore, the technique used in making these butt welds will be substantially the same as that used in making corner joints.

Flat-position plate butt welding—forehand

In starting the flat-position weld on a plate joint it may be necessary to cut the heat down slightly until the proper weld contour is built up. With the correct gas pressures for the tip you use, form a puddle completely enveloping both sidewalls of the V joint and extending all the way to the root of the V. This puddle is carried forward with the oscillating motion of the torch previously described. On these heavier thicknesses you should carry the torch almost completely around the rod to melt the base metal at the bottom of the V joint and to ensure fusion

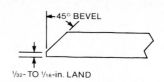

FIGURE 9-19 Thick sections cannot be welded through in one pass. They require beveled joints. This is a typical edge bevel for material over 1 in. [25 mm] thick. The open joint gap also is needed to give you room to insert your welding torch and filler rod.

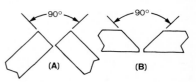

FIGURE 9-20 A corner weld has a built-in joint angle. (A) In this drawing the bevel edge on each piece is equivalent to 45°, making a full 90° double-bevel joint. (B) The single-V butt joint also has two 45° bevels, one on each piece, along with a flat land in the root of the joint.

throughout the entire weld cross section.

Due to the large area of the puddle in these welds, it will be necessary to oscillate or weave both the rod and the torch. As shown in Fig. 9-21A and 9-21B, the rod is carried to the far side of the pool and upward in a clockwise direction while the torch is being carried downward and forward in the same clockwise direction. As the limit of this motion is reached, the process is reversed. This motion must be maintained uniformly. Your welding proceeds at a slow, even rate of speed. Oscillating the rod effects the uniform distribution of your weld metal and aids you in building up the necessary 25 percent reinforcement.

When you are nearing the end of your weld, the same precautions for reducing the heat as explained previously should be observed. Let the metal cool a little before you finish off the weld.

Vertical plate butt welding—forehand It may be necessary to cut the heat down slightly until a shoulder having the proper weld contour is built at the start of your weld. In starting, a minimum amount of heat should be used to ensure proper control of your weld metal. Once the weld is started, your torch and rod are held in the same position as outlined for thinner sections (see Fig. 9-22). Your torch and rod are oscillated from side to side as welding proceeds. In this motion the rod is carried to the opposite side of your weld puddle from the torch, which must be directed back against the pool to keep it from running back onto the cold metal. You should not attempt to carry the puddle back down the

weld to add additional reinforcement because this often causes the beginning welder to lose control of the weld pool. If the puddle gets out of control the molten metal will run down the weld and make a very sloppy joint.

Overhead plate butt welding—forehand Overhead welding is substantially the same as welding in the horizontal position except that the torch is held at right angles to the work to give you better control of your molten pool. You should stand in front of and in line with the joint. Your welding rod is introduced from the front more or less in line with your torch.

Manipulate your torch and rod essentially the same way as you do for light-gauge steel. The increased thickness of plate material, however, means that you need greater rod and torch oscillation.

Butt Welding V-Groove Plate Joints Backhand
Up to this point you have been practicing forehand welding. It's time to practice backhand welding techniques (Fig. 9-23).

Flat plate welding—backhand As in the forehand welding of ³⁄₁₆- and ¼-in. [5- and 6-mm] steel plate it will be necessary to provide the plates with a beveled edge. However, for backhand welding, the bevel should be 30°, as shown in Fig. 9-24.

For ³⁄₁₆- and ¼-in. [5- and 6-mm] thicknesses, the plates are spaced half the thickness of the plate when set up and tack-welded.

To make it easier to start the weld, the first 1½ to 2 in. [37.5 to 50 mm] of the joint should be preheated to a temperature of 800°F to 1000°F [425 to 645°C] (a dull red color). When this has been done, start the weld at the edge and build up to the proper weld contour before proceeding.

Due to the narrower V used in backhand welding, oscillation of your torch is virtually dispensed

with once your weld contour is built up and forward travel begins. However, it is necessary to oscillate or roll your rod. For a while you will experience some difficulty in holding your torch still, and at the proper angle while you oscillate your rod. But you must master this procedure, since oscillation of your torch will tend to slow down your rate of forward progress. This will cause a widening of the weld due to the excess heat imparted as a result of your slow travel speed. Also, the molten metal will drop through the V without proper fusion and result in incomplete fusion at the root.

Vertical plate welding—backhand When making vertical butt welds on ³⁄₁₆- and ¼-in. [5-

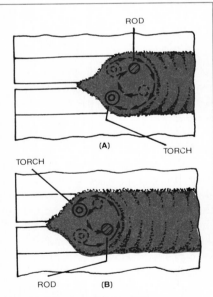

FIGURE 9-21 When filling a wide-angled joint, be sure to tie the weld metal into the sides of the joint. You can do that by alternately rotating the torch and the filler rod as shown as you travel down the length of the joint.

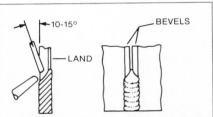

FIGURE 9-22 The technique for making vertical-up welds in a bevel joint is similar to that for a vertical-up butt joint, except that you must also be sure to tie the weld into the walls of the joint, just as you would when making the weld in the flat position.

FIGURE 9-23 Learn to make backhand welds in double-bevel joints in the flat position. This technique will be useful when making V joints in more difficult positions.

and 6-mm] plate with the backhand technique, the same general practices hold true as for backhand welding in the flat position. Here you also must guard against oscillating your torch since overheating the weld will result in loss of control and will permit the molten metal to run down the face of the V joint. It is equally important that your torch be held at the proper angle with relation to the puddle and the base metal (see Fig. 9-25). The force of the properly directed flame will hold the weld metal in the V at the correct angle to prevent running ahead and closing the V, which would result in lack of fusion.

Overhead plate welding— backhand Overhead backhand welding is essentially the same as the other positions for backhand welding. You should stand in line

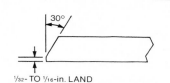

FIGURE 9-24 This 30° bevel joint will use less filler metal than a 45° joint. It is easier to learn on 45° bevel joints, but 30° bevel joints are more economical. They use up less filler metal, labor, and time on large structures.

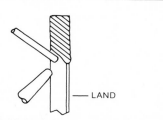

FIGURE 9-25 After you have learned to weld vertically up on groove welds, learn to weld vertically down using the backhand technique.

with the joint as explained previously. However, the torch should not be held at right angles to the plate. The flame must be directed back against the completed weld in a manner similar to that in the horizontal and vertical positions (see Fig. 9-26).

Fillet and Lap Welding Sheet— Forehand

So far you have been concerned only with simple butt welds. In industry, however, it is not always possible to place the pieces in line for the butt joint. At times it is useful to lap the sheets in a manner similar to putting shingles on a roof or making a riveted joint. Also, sheets with surfaces instead of edges meeting at right angles (a fillet weld) present conditions other than those encountered in butt welding. For both these types of connections the fillet weld is used.

Fillet welds are similar to and present many of the same problems as corner welds. The faces of the fillet welds, however, are generally in the horizontal and vertical planes, whereas the corner-weld faces were set at 45° to the horizontal. In addition, the position of the base metal is different, which influences the application of your welding heat.

Lap joints—flat position Set up two sheets so that the top of one overlaps the lower one by half its width (see Fig. 9-27). Then tack-weld the joint at both ends. Tacks are confined to the weld faces. If the overlap merely extends a short distance from one edge the problem is simplified, since the short piece of metal heats up rapidly.

Direct your torch flame at the corner made by the juncture of the two sheets so that the tip will make an angle of 45° with the horizontal and with the direction of travel as shown in Fig. 9-28. Use forehand welding and maintain the puddle so that its center is at the root of the V joint. Use the oscillating movement you

learned for forehand welding. Melt down the edge of the upper sheet for a distance inward of about half the diameter of the puddle.

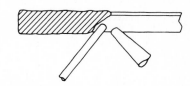

FIGURE 9-26 When you can successfully weld groove joints in the overhead position with a torch and filler rod, the rest of the manual welding skills in this course, including those for arc welding, will be easy to learn.

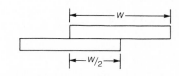

FIGURE 9-27 The lap weld joint is somewhat different from the brazing lap joint. The overlap should be made as shown.

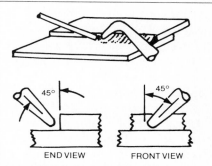

FIGURE 9-28 A welded lap joint really is one form of a fillet weld since weld metal will not flow between the faying surfaces like brazing metal does. Hold the torch as shown while you work.

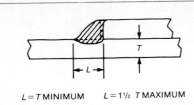

$L = T$ MINIMUM $L = 1\frac{1}{2} T$ MAXIMUM

FIGURE 9-29 The ideal fillet weld in a lap joint will have these dimensions and a smooth weld profile, as shown.

The required weld dimensions are given in Fig. 9-29. In a welding blueprint, actual dimensions would be located where you now see a T (for thickness) and an L (for length). To obtain complete penetration at the root carry fusion forward in this area for a distance about equal to the puddle diameter before building the weld up to the required dimension. The welding rod is fed in front of the torch.

Difficulties in making this weld will result primarily from failing to hold your torch at the correct angle. If your torch is inclined too much toward one or the other of the faces of your weld joint, the other side of the joint is likely not to be fully fused.

Lap joints—vertical position In making lap joints in the vertical position the same techniques you practiced earlier are applied with the obvious modifications necessary for this type of joint. You should set the plates up firmly in an adequate jig. Forehand welding from the bottom upward is best. Stand in front of the sheets so that the weld faces are presented substantially at a 90° V. Hold your torch tip at 45° to the edges of the weld and inclined upward at 45°. The rod is introduced from the top at an angle of about 10 to 15°. Correct positions are illustrated in Fig. 9-30. In starting, build up a shelf of the proper contour to serve as a support for further welding. Momentarily withdrawing your flame will help you control the puddle

temperature and let the molten weld metal solidify as you place it, helping hold the puddle up against the force of gravity.

Lap joints—overhead welding As in making overhead butt welds, making fillet welds is merely a modification of the procedures you used in the flat and vertical positions. The correct positions for your torch are illustrated in Fig. 9-31. It is important in all of these fillet welds to keep the flame directed into the root or corner of the fillet to make sure that you get thorough penetration into the joint. As in the previous overhead weld, progress should be toward you, not away from you.

T joints—all positions In making T joints follow the same general procedures used for lap joints. The principal difference is in setting up the sheets, which are now tack-welded so that one is placed at the middle of the other and at right angles to it (see Fig. 9-32). On T joints it will be found that if too much heat is played on the upstanding leg of the T, the base metal will burn through. To prevent this, direct the torch equally at both plates at an angle of 45°.

Fillet- and Lap-Welding Plate—Forehand
The fundamental principles for making fillet welds on 3/16- and 1/4-in. [5- and 6-mm] steel plate are essentially the same as those for 1/16-in. and 1/8-in. [1.5- and 3-mm] sheet steel. The set-up with regard to the relationship of one piece to another is, in every case, identical to the methods used in the previous welding procedures. The principal difference lies in the need for increased fillet size due to the increased thickness of the material. The oscillation of your torch will cover a larger area and the increased size of your molten puddle will necessitate oscillation of your welding rod as well. Suitable modification of the method of oscillation of your torch and the rod that we previously explained should be used.

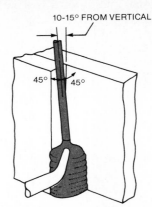

FIGURE 9-30 Learn to make fillet welds working vertical-up with the forehand technique, as shown here. You can bend your bare filler rod to make it easier to insert it into the joint.

FIGURE 9-31 Here is the way to make overhead fillet welds. Note that the torch flame is directed equally toward both pieces of the base metal. Uniform heating avoids welding defects.

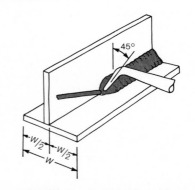

FIGURE 9-32 T welds are the most common fillet welds you will make. Your first samples should be made as shown here, using the forehand technique.

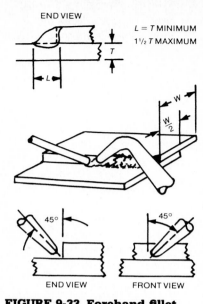

FIGURE 9-33 Forehand fillet welds on lap joints are made as shown here.

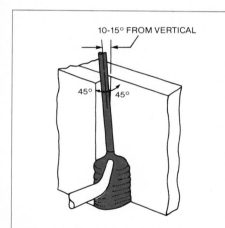

FIGURE 9-34 This is how you make a forehand vertical-up fillet weld.

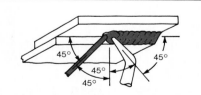

FIGURE 9-35 This is the technique for making an overhead fillet weld lap joint using the forehand technique.

Lap welds in the flat, vertical, and overhead positions are shown in Figs. 9-33 through 9-35. Figure 9-33 shows the setup for making a T weld in the flat position; Fig. 9-34 shows the vertical position, and Fig. 9-35 shows the procedure for overhead lap welds.

The method used in making T joints in the vertical and overhead positions varies from that for lap joints only in the distribution of the mass of the upper plate (Fig. 9-36). As in the thinner sections, you will find that improper heat application in making T joints will cause overheating of the upper plate and may result in burning through it.

If complete penetration of the upper plate of a T joint is desired on 3/16- or 1/4-in. [5- to 6-mm] steel plate, the upper plate should have a 45° bevel and a 1/32- to 1/16-in. [0.7- to 1.5-mm] land. This is a single-bevel groove weld and should be used where maximum strength is required in a T joint (Fig. 9-37).

Fillet and Lap Welding— Backhand

After you have practiced all of the work up to now and mastered both the forehand and backhand welding techniques, you will have little difficulty in making fillet welds on heavy plates (plates over 3/16 or 1/4 in. [5 or 6 mm] thick) using the backhand technique.

In making fillet welds with the backhand procedure, note again that almost no movement of your torch is required. However, your rod should be rolled and oscillated slightly (compared to forehand welding) to get uniform distribution of the molten metal and to build up the required weld contour. You will see that with the backhand method, fusion at the root of the V joint is much easier to get than doing the work with the forehand technique.

The relative positions of your torch and rod for this work are illustrated in Figs. 9-38 through 9-41. Compare these illustrations with the similar one on backhand,

single V-groove butt joints. In effect these fillet welds are V welds, the faces of which have been rotated through 45° placing them horizontally and vertically. Of course the distribution of the mass of base metal is different and calls for some modification of your rod manipulation technique.

Many skilled welders prefer backhand welding for fillet welds, since the procedure simplifies the work and helps make welds with good mechanical properties.

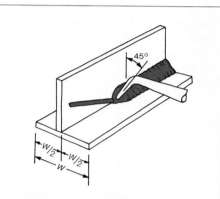

FIGURE 9-36 Procedure for making T joints using the forehand method.

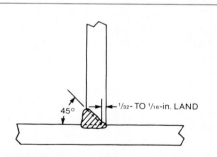

FIGURE 9-37 T joints often require bevels when thicker sections are welded so that you get full penetration into the center of the section. This is a single-bevel fillet T-weld joint. Double-bevel fillet T-weld joints are used for even thicker sections. They are beveled on both sides.

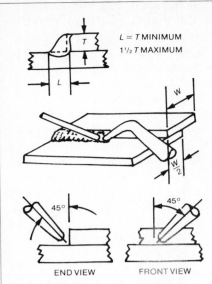

FIGURE 9-38 After you have learned to make fillet welds with the forehand technique, learn to do it with the backhand method.

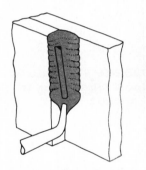

FIGURE 9-39 Learn to make vertical backhand fillet welds. The more methods and welding positions you can work in, the better your chances of getting a good job and higher pay.

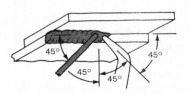

FIGURE 9-40 This is the way you make overhead backhand fillet welds.

Butt Welding Cast Iron—Up to ¼ in. [6 mm] Thick

Before you attempt to weld cast iron at all, you should have a fair idea of what the material you are going to work with is and how it reacts when heated and cooled.

Gray cast iron is one of several different kinds of iron that you will probably have to work on from time to time. It is by far the most common type of cast iron. Gray cast iron contains much more carbon than steel does. Depending upon the grade, it can contain from 3 to 4 percent carbon compared with just fractions of 1 percent carbon in almost all steels. Most of the carbon in gray cast iron is not dissolved in the iron. It is in the metal as free graphite flakes. That's what gives gray cast iron its color.

Gray cast iron has comparatively low ductility and tensile strength compared with steels. The material averages around 15,000 to 20,000 psi [103 to 137 MPa] tensile strength (although higher-strength "alloy" cast irons are produced). However, gray cast irons have much higher compressive strength than most steels. Gray cast iron can take a very large compressive load, and that's one of the main reasons why you will often find it used in the bases of machines and the frames of very large equipment. All of these characteristics must be taken into consideration when cast iron is to be welded, and particularly when a broken casting is being repaired.

Undoubtedly the most important factors to know about when welding gray cast iron are the proper application and control of preheat and post-heat treatments. Proper heating is essential to prevent locked-up stresses that would result in repeated cracking of the base metal (because of its relatively low tensile strength) on rapid cooling.

Preheating in the case of smaller cast-iron parts may be done easily with your welding torch. Slightly larger parts may be preheated with a heating torch, and very large parts (bigger than you are) are usually preheated in a furnace. In any case, the part should be preheated to about 900 to 1100°F [482 to 593°C] (dull red) before welding.

Postweld heating is necessary to keep welded gray-cast-iron parts from cracking. Contact with cool air would permit the face of the weld to cool much more rapidly than the inner or central portion of the weld deposit. Postheating consists of bringing the piece to what is called its *normalizing* temperature (1600°F [871°C]) and cooling it slowly. Heating the part to the normalizing temperature reduces all internal built-up stresses. Controlled (slow) cooling then makes sure that new stresses are not added to the part before it cools down to room temperature. Remember that deposited weld metal shrinks upon cooling and that the base metal expands when heated. These contrary forces must be controlled if you don't want a cracked cast-iron part after you have welded it.

One way to visualize exactly where the most preheating should

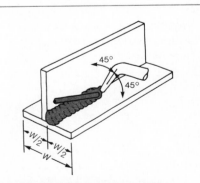

FIGURE 9-41 Backhand fillet welds on a T joint are made this way. T joints often are used to put stiffeners in place to reinforce large panels on ship bulkheads, storage bins, and steel buildings.

be done on a part is to imagine a big wedge driven into the crack you will repair or the joint you will weld. The point or points where the imaginary wedge would cause the most bending or stretching in the base metal are the areas where you should apply the most preheat. You also may have to keep that preheat going all the time while you weld a large cast-iron part.

Cooling off a cast-iron joint
Following the welding operation, your torch flame should be played rather gently over the surface of the cast metal on either side of the joint for a considerable distance. This serves to bring unequally heated sections of the base metal to an even heat. The casting should then be covered with sand or refractory wool to make it cool even more slowly. If the part is small enough, bury it in dry slaked lime or crumbled refractory brick. If you use sand, make sure that the sand is very dry. You don't want to have a steam explosion or similar problem.

Sometimes you can put the repaired casting into an industrial heating oven and set the oven to cool slowly. If you use an oven, make sure that the casting is protected from drafts and cold air, which definitely will cause uneven cooling.

The best rod size to use on cast iron up to about ¼ in. [6 mm] thick is a cast-iron rod (they are square instead of round) about ⅛ in. [3 mm] thick. Common gray iron castings require cast-iron welding rods that are high in silicon, while high-strength iron castings require a cast-iron gas rod with alloying elements in it, such as molybdenum and nickel.

A suitable flux is equally important for welding gray cast iron (unlike working with carbon steel). There are good proprietary cast-iron fluxes available. Details on their use are almost always given on the label. The exact

amount of flux to be used is largely dictated by the type of cast iron to be welded. For example, low-grade (low-strength) cast irons, or irons that are high in impurities, may require a more liberal application of flux than a good grade of clean gray cast iron. The flux should be applied by sprinkling it liberally in and along the V joint before tacking or welding the casting. You also must frequently dip your hot rod in the flux container as you weld.

Cast-iron plates must be preheated even before they are tack-welded. These tack welds on cast iron will differ from the smaller ones you used on steel plate. They should be substantial, with a length equal to twice the thickness of the material you are welding. Nevertheless, your tack welds should be no more than half the thickness of the section when you are working on thin cast-iron parts.

You should exercise care in building up and forming the proper weld contour at the beginning of your weld before attempting to proceed. Proper welding contour throughout the entire length of your weld is important. You want a smoothly flowing, rippled surface that gradually tapers off to either side and fuses smoothly into the base metal on either side of the weld joint, with no sharp edges or very abrupt changes in section. The deposited weld metal must be thoroughly fused with the base metal at the upper and lower corners, and sharp corners or edges should not show anywhere. There must be absolutely no pinholes evident on the surface, as this is an indication that the weld has entrapped oxides and, as a result, poor fusion to the base metal.

Butt Welding Cast Iron ½ to ¾ in. [12 to 20 mm] Thick
The same general procedures as outlined for thinner cast-iron sections apply to thicker material. Preheating and postweld slow

cooling remain especially important. Your biggest problem with thicker cast-iron sections will be in maintaining and carrying along the large molten weld pool you need to join cast iron, especially in material thicker than ½ in. [12.5 mm] when welding a joint with one pass. The reason, of course, is that the thicker the section, the more heat it draws from your weld puddle and the more difficult it will be for you to keep the molten puddle from cooling. It's much easier to turn the job into two passes instead of one. Here's how.

You should progress along the joint for about 1 in. [25 mm] filling one-half of the lower or root part of the V joint. Then return to the start of the weld and fill in the top half of the V (see Fig. 9-42). In other words, make progress in two passes with the top one overlapping (but lagging slightly behind) the lower one, completing an inch or two of both passes at a time before you go on to fill up more of the V joint.

When working on material around ¾ in. [20 mm] thick, you will even have trouble filling up a V joint in two passes. If you do have trouble, use three overlapping passes, following the same progressive method as outlined above. The third pass should be

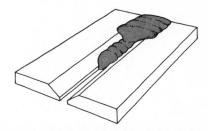

FIGURE 9-42 Groove welds on thick sections become so deep that two or more passes are required to fill them. Make sure that you produce a perfect root pass before you cover it up. Also make sure that the cover pass is tied into the walls of the joint.

added to reinforce and ensure complete fusion at the upper corners of the section, as shown in Fig. 9-43.

Welding a Broken Casting

Before you attempt to repair a broken casting you should refer back to the last two welding procedures and make sure that you have practiced them until you have no trouble. Also reread the section that tells how the properties of cast iron affect your job in welding it. Look again at the idea of driving an imaginary wedge into the broken crack in the casting and think where that wedge would put the most stress on the remaining metal—that's where most of the preheat should go.

Castings that are broken in service are frequently soaked with oil or covered with other foreign material that will affect your welding job. You must clean a broken casting before welding it. If you don't, you will get a porous, weak weld. Grease and oil can be removed with organic solvents, including gasoline. We strongly recommend that you *don't* use gasoline. It's too dangerous. Use a commercial cleaning solvent instead.

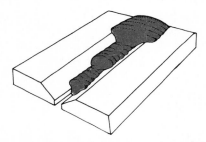

FIGURE 9-43 If the joint angle is sufficiently narrow, you can make up to three passes to fill the joint by using larger-diameter filler rods. Sometimes the gap will be so wide that several passes, made side by side, will be required. Make sure that all weld passes are tied (melted) together as well as to the base metal.

Castings that cannot be easily cleaned with a solvent sometimes can be cleaned with a hot caustic solution of trisodium phosphate. Where the foreign material is extremely persistent or sticks to the casting, you have to try sandblasting to clean the casting before welding it.

The original fracture in your cast-iron part may have occurred due to a weakness in the original casting. Therefore it is important that the cross-sectional area of the repair weld in the casting be larger than that of the original part. This larger cross-sectional area is necessary to increase the strength at the point of the break and to prevent another fracture from occurring in the same spot when the part is put back in use.

Obviously, weld contour, proper build-up at the end of the weld, complete fusion throughout the section, and lack of porosity are vital to getting a sound repair weld.

CAUTION: Don't weld hollow or cored castings without drilling a vent hole to let hot gas, air, or steam escape.

Cored or hollow castings must be ventilated with a drill hole, if necessary, to permit hot gases, air, and steam to escape before you preheat and weld the workpiece. If you apply heat to a hollow casting without a relief vent, it could blow up while you work on it because of the pressure that will build up inside it.

Welding Thin Sections of Yellow Brass

The set-up and welding methods for joining thin sections of yellow brass (copper-zinc alloys) will be almost identical to those outlined for welding steel. There are, however, a few slight changes in your welding procedure.

First and most important is using a different flame setting. Use a slightly oxidizing flame to weld brass. Second, use a filler rod designed for the base material. An

AWS A5.7 R CuZn bronze filler rod will be good. Use a 1/16-in.-diameter rod for 1/16-in.-thick brass, and a 1/8-in.-diameter rod for brass of the same thickness.

Third, use flux. The edges and surrounding area of the joint, top and bottom and all sides, must be thoroughly cleaned and coated with welding flux before tack welding. Apply the flux on the base metal and the rod. You will find that a flux-coated rod suitable for welding brass will work better than periodically dipping a bare rod into dry flux. Some flux-coated rods are sold especially for brass welding.

In your early attempts, you should try to make several joints using a bare rod dipped in flux and then use a flux-coated rod. Compare the results, both in terms of the quality of your weld deposit and in the ease of operating with the different kinds of rod. When you do use a bare rod dipped in dry flux you will get excessive fuming. This fuming tends to give porous welds, usually resulting in low ductility and low strength. Welds made with specially precoated commercial brass-welding rods will make denser, stronger, more ductile welds.

You can use either the forehand or backhand welding procedure on this material. In either event, use a 25 percent weld-metal reinforcement on the section to get a satisfactory weld. Develop skills in welding yellow brass in the downhand, horizontal, vertical, and overhead positions.

Welding Thick Sections of Yellow Brass

Use a slightly oxidizing flame and the proper flux and filler rod. Heavier sections (over 1/4 in. [6 mm] thick) must be provided with a 30 to 45° bevel. You can use an AWS A5.7 R CuZn bronze rod for this work. Use a 1/8-in.-diameter [3 mm] rod for brass sections around 3/16 in. [5 mm] thick. Use a 3/16-in.-diameter [5 mm] rod for brass that is about 1/4 in. [6 mm] thick. As described in the previous exercises, a few welded samples should be made up using a dry flux and a bare rod versus a rod that is precoated with flux.

Develop skills working this material in the downhand, horizontal, vertical, and overhead positions. Use both forehand and backhand welding while you work.

Welding Copper

By now you should be thoroughly familiar with all the fundamental procedures and manipulations required for gas welding. Therefore, it's not necessary to repeat them here. You will find that welding copper is very similar to welding steel. However, because of the much greater heat conductivity of copper, it is necessary to use somewhat larger tip sizes than you would use to weld the same thickness of steel.

Copper also has some peculiar properties that will affect your job. Copper work-hardens very rapidly. That is, it loses its high ductility and becomes brittle when it is hammered when cold. Therefore, when working copper it must either be heat-treated in the *dead-annealed* condition for maximum softness, or, if you are going to hammer it at all, you must work on it at a temperature over 1000°F [538°C] to keep it from becoming brittle.

Dead annealing is accomplished by heating a copper part to a red heat and immediately quenching the part in water. You have to do that frequently when cold-working copper parts. You also might remember that just the reverse would happen when you quenched steel after heating it to a red heat. The steel would become very hard. Nonferrous materials like copper and aluminum are just the opposite from carbon steels. You won't have any trouble with the materials becoming hard and brittle when you cool them very fast.

The analysis or grade of copper you are using also has a marked effect on your welding method and the strength of the welded

joint you will get. For example, electrolytic copper (also called tough-pitch copper) is a type of relatively pure copper, but it contains some oxygen in it from the copper-making process. It has very high electrical conductivity. That's one of the main reasons why it is used. The method you use to join this material can readily affect the electrical conductivity of the joint you make. Since electrical conductivity is the reason the material is used, be very careful joining it.

Another problem with tough-pitch copper is that when it is heated it has a tendency for microscopic particles of copper oxide to precipitate along the grain boundaries in the heat-affected zone of the base metal, right next to your weld. These grain-boundary oxide precipitates will lower the strength of the base metal in a narrow line next to the weld. It's not unusual for welds on electric or tough-pitch copper to have only 60 percent of the normal strength of the base metal prior to welding. That's one of the reasons why another kind of "pure" copper often is used.

A grade of pure copper called oxygen-free copper, or more commonly OFC, has been treated to remove copper oxides. Deoxidized copper will usually produce welds with a strength about equal to the base metal. Therefore, wherever you can, use deoxidized or oxygen-free copper for making welded products.

The welding rods used for most copper products are selected from the filler-metals specification AWS A5.7; a silicon-copper rod usually is used. You also have to apply a welding flux suitable for copper. Also use a flame adjustment that ranges from neutral to slightly oxidizing. For thinner sections (⁵⁄₃₂ in. [4 mm] and less) use the square-groove butt joint. For heavier sections (³⁄₁₆ in. [5 mm] thick and greater) use the single V-groove butt joint.

Deoxidized copper is not suited for all classes of work, particu-

larly when used for electrical equipment. It doesn't have quite as high an electrical conductivity as tough-pitch copper. Hence, it is sometimes necessary to use electrolytic tough-pitch copper whether you like to weld it or not. In that case, you may want to use brazing instead of welding to avoid the joint weakening that is bound to occur when fusion welding electrolytic tough-pitch copper.

Bead Welds on Aluminum

Pure aluminum is a soft, white metallic element somewhat resembling tin in appearance. It is very ductile and malleable, hence it may be readily worked. Its outstanding properties are its lightness and ability to withstand many forms of chemical and atmospheric corrosion. However, pure aluminum is not very strong.

Commercially pure aluminum is often called 1100 series aluminum (the 1100 number is a grade designation by the Aluminum Association, the equivalent of the steel industry's American Iron and Steel Institute). In this grade, the material is about 99.5 percent pure aluminum. It will have a tensile strength of only 13,000 psi [89.6 MPa] unless it has been cold-worked to strengthen it a little.

Alloying elements such as silicon, manganese, magnesium, copper, and chromium can be added to aluminum to make some strong alloys with tensile strengths ranging up to 70,000 psi [482 MPa] or higher after heat treatment. Some of these alloys also are strengthened by cold-rolling (work-hardening). In the aluminum industry, grades that are strengthened by cold-rolling (called temper-rolling) have a special number/letter designation after the alloy number.

Even 1100 series commercially pure aluminum can be strengthened somewhat by cold-rolling; material strengthened to a level about half of the maximum hardness achieved would be called 1100 series ½-H temper aluminum. A fully cold-hardened aluminum would have a 1-H temper designation. (The H stands for hard.)

Welding wrought (rolled as opposed to cast) aluminum is both a production and repair operation. Welding of cast aluminum is most often a repair job. The welding of aluminum alloys depends a lot on whether the material has been strengthened by cold-rolling or by heat treatment.

For example, welding a heat-treated aluminum grade will cause the aluminum in the heat-affected zone to lose most of its added strength. Therefore, welding heat-treated aluminum alloys is not usually recommended when working with oxyfuel welding. The oxyfuel welding process produces a very large heat-affected zone compared with electrical welding methods, such as TIG welding, which we will learn about later on. TIG, or as it's officially called, GTAW, is the major method for joining aluminum alloys.

Aluminum melts fast The biggest problem you will have with welding aluminum, especially in thin sections, is that aluminum not only melts at a much lower temperature than steel (in a range of about 815 to 1220°F [435 to 660°C]), but it also does not change color when it gets hot. Even liquid aluminum has the same silvery color as solid aluminum. When welding a thin section, you can easily apply a lot of heat and suddenly burn or melt a hole right through the section. You have to learn to use your judgment as to how hot the aluminum is and when it's ready to melt. That takes experience, and the only way to get good at it is trial and error (see Chap. 22).

For the bead welds you will first learn to make on aluminum, you should concentrate on learning when the metal is ready to become molten, how to control the molten aluminum weld puddle, and when to add your welding rod.

Clean metal is important

Cleanliness of both your base metal and welding rods is very important in getting sound welds on aluminum. You have to take far more precautions than you do when gas-welding carbon steel.

Welders often will clean the base metal where the weld will be deposited, and also their bare welding rod, with emery cloth or sandpaper, and then immediately weld it before an invisible aluminum-oxide film reforms on the surface.

In production welding of aluminum, the parts are frequently cleaned by immersing them in a 5 to 10 percent solution of sulfuric acid followed by a clear water rinse. Then the part is thoroughly dried with a very clean cloth and immediately covered with a welding flux designed for use with aluminum. After that, the aluminum is welded quickly before the air can reform an invisible oxide coating on the material.

Aluminum welding rods often are heated and dipped directly in the flux container if the container is large enough. Distilled water is recommended for mixing liquid aluminum fluxes since drinking water often contains iron and other chemicals that will adversely affect the strength of the weld. When using dry fluxes for welding aluminum, keep the dry flux in a covered container when not in use. The chemicals used in the dry fluxes will take up moisture from the air and become useless for welding.

Aluminum welding procedures

Use the forehand technique when making aluminum bead welds with a minimum of rod and torch oscillation. The beads should be uniformly built up to 25 percent reinforcement on the top and have a small root bead (see Fig. 9-44). There must be no undercutting or overlapping of the deposited weld metal. A satisfactory bead weld will have smooth-flowing contours and the ripples will not be as pronounced as those encountered in welding steel.

Butt Welding Thin Aluminum

Having mastered the technique for fusing aluminum and running bead welds, you are ready for some really difficult work. The basic techniques you have mastered for steel welding are also used in aluminum welding, modified only to the extent necessary to care for the differences in the two metals.

In making simple, square-edged butt joints on aluminum sheets you will soon discover that it is more difficult to make these welds on thin materials such as 1/16-in. [1.5-mm] sheet than on heavier

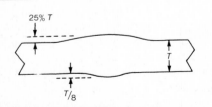

FIGURE 9-44 The weld reinforcement for a thick joint is similar to that for thin joints—no more than 25 percent of section thickness on top, and no more than one-eighth excess weld metal below the root. Use less reinforcement than you would use on very heavy sections.

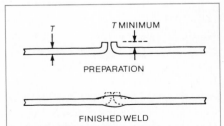

FIGURE 9-45 Thin gauge metal can be jointed without filler metal by using this technique. The final result will be a simple butt weld, which normally is difficult to make in sheet metal.

sections ⅛ in. [3 mm] or thicker. Therefore, you should begin by first making edge welds on flanged edge seams (see Fig. 9-45). That will give you a thicker section to join. After that, work on square-groove butt welds in ⅛-in.-thick [3-mm] material (Fig. 9-46). These welds are all made in the flat and vertical positions. The deep-flanged edge weld (Fig. 9-47) is a joint encountered in production welding of water tanks and other containers.

In butt welding sheet aluminum, you must clean the material and your welding rod before doing the work. Use the methods described in the previous section. Edges to be welded should be thoroughly cleaned with emery cloth, sandpaper, or by dipping in a dilute sulfuric acid solution. After cleaning, do not touch the surfaces with your hands. Even the natural oils on your hands will contaminate the cleaned surface. Use very clean welding gloves. Apply the flux as quickly as possible after cleaning.

Pure aluminum and simpler aluminum alloys require a different flux than the more complex aluminum alloys. Make sure that you use a flux specifically formulated for the grade of aluminum you will be joining.

When applying flux to butt joints, be sure to cover the edges and the top and bottom surfaces. Remember that both sides of your metal will get very hot and require a protective flux coating,

even when you are only joining the material from one side. The top should be flux-coated for a distance of about 1 in. [25 mm] on each side of the area to be welded, while the bottom should be coated at least ½ in. [12.5 mm] from the weld area on each side. Since the function of the flux is to combine with the oxides that will form when the weld metal and base metal are heated, it follows that flux must be applied at all points where that oxide will form as a result of your welding operation. That includes your welding rod.

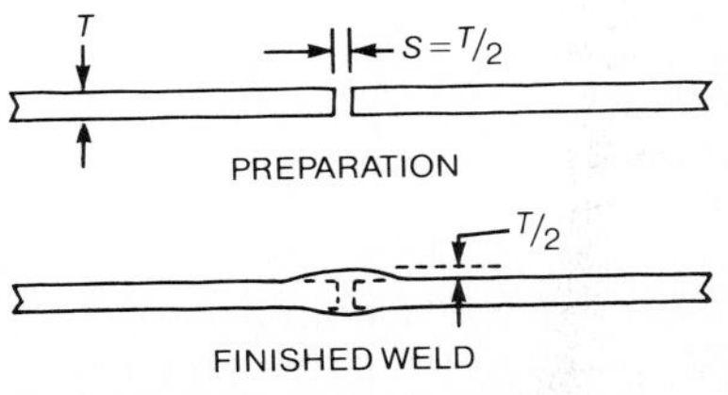

Flanged edge joints Flanged edge joints are set up by bringing the fluxed edges into contact and tack welding as frequently as necessary to preserve alignment. In production work, you will use jigs or fixtures to hold the parts in the correct position. Don't use filler rod. Using the forehand technique, carry the weld forward, melting down the flanges as you go to form your finished weld and reinforcement. Your weld penetration into the joint should be complete right down through the root of the joint. Welds of this type should be capable of withstanding a guided bend test, bending the flange flat against the part without cracking the weld. For vertical welds, weld upward.

Butt joints on thin sheets
These joints are made with the conventional forehand welding technique, exercising due care to secure complete fusion through to the root of the weld. The flux-coated edges of the joint should be spaced apart about one-half the thickness of the sheets and tack-welded as frequently as necessary to maintain alignment. Jigs and fixtures such as clamps would be used for this purpose in production work.

Deep-flanged edge joints This joint is similar to the other flanged edge joints except that no attempt is made to produce a butt weld by melting the flange all the way down to the root of the weld

joint. The depth of the flange is at least 1 in., as shown in Fig. 9-47.

The edges and faces of the flanges should be fluxed for about ½ in. [12.5 mm] and held tightly together with clamps during welding. Filler metal is not necessary in making this joint. For testing, this joint should be cut into strips ¾ to 1 in. [20 to 25 mm] wide. One end of the strip is held in a vise while the other is gripped with pliers. When the joint is pulled apart, a sound weld will tear in the base metal and not in the weld. If the joint does tear in the weld, or especially in the area where the weld and base metal meet, examine the area for thorough fusion and proper weld-metal penetration.

Oxyhydrogen welding The oxyhydrogen flame is sometimes recommended for welding thinner sections of aluminum, but care must be exercised to use a neutral or a reducing flame. Even a slightly oxidizing flame will create aluminum oxide particles in the weld. Since aluminum oxide is a brittle refractory material, the oxide particles will greatly reduce the strength of the weld metal.

Theoretically, 2 volumes of hydrogen are required with 1 volume of oxygen to produce a neutral oxyhydrogen flame. In practice, this is increased to a range of 2½ to 3 volumes of hydrogen to 1 of oxygen in production work. However, oxyhydrogen

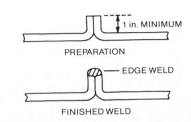

FIGURE 9-47 Flange welds on thin material are often easier to make than butt welds. The extended flange also functions as a stiffener in large-area, thin-gauge panels.

FIGURE 9-46 A square butt joint should have these general dimensions, where the gap, *S*, equals one-half the section thickness (*T*/2).

welding should not be attempted by beginning welders. It's not the difficulty of the work (it's often easier to do than other kinds of oxyfuel welding), but the danger of handling and using hydrogen.

Butt Welding Heavy Aluminum

As with other metals, aluminum sections ³⁄₁₆ in. [5 mm] and thicker must be beveled for welding. In common practice the edge to be welded is provided with a 45° bevel and a land (which makes a 90° V joint when the two edges are butted together) as shown in Fig. 9-48. Some welding specialists recommend welding Vs as large as 120° (two 60° bevels) but this extreme is not normally required. On heavy sections of ½-in. [12.5-mm] thickness and greater, a double V-groove butt joint as shown in Fig. 9-49 can be used if you have to have access to both sides of the joint.

The forehand welding method is the standard technique for welding aluminum and its alloys on heavy-section butt joints as well as thinner sections. Don't forget to clean both your base metal and filler rod first. Select your filler rod from the AWS A5.10 specification, usually matching the analysis of the rod to the base metal. For example, an 1100 series aluminum rod is used with 1100 series aluminum base metal.

You will have to preheat aluminum sections that are ³⁄₈ in. [10 mm] thick and greater, particularly if large assemblies are involved. Preheating should be carried to a temperature of 700 to 800°F [371 to 425°C]. That will serve to relieve heat-induced strains and distortion and also lower the amount of oxygen and fuel gas you need for welding heavy sections.

Butt Welding Cast Aluminum

Cast aluminum is available in any of a large number of chemical compositions or grades. The commercially pure aluminum compositions (equivalent to 1100 series wrought aluminum) are not used to any extent for castings. Therefore, virtually all aluminum castings will be some alloy composition. If the particular alloy has been heat-treated, then welding will destroy the original physical properties built into the casting by heat treatment. If facilities are available for heat-treating, the welded casting can be restored to approximately its original strength by reheat treatment.

Aluminum castings can be made in sand molds, permanent metal molds, or in pressure die-casting machines. Certain cast-aluminum alloys are better adapted to one or another of these casting methods.

Since practically all aluminum castings are of alloy analysis, it is necessary to use an alloy rod in welding them. The rod most commonly used is 5 percent silicon, 95 percent aluminum, selected from AWS specification A5.10; usually ER 4043-grade rods are selected for use. Use a ³⁄₁₆-in.-diameter rod for welding ¼-in.-thick metal. Use a ¼-in. rod for welding ½-in.-thick metal.

Flux is essential when welding cast aluminum. Special aluminum gas-welding fluxes are commercially available. They will help you control the oxides that otherwise would immediately form on the casting when you start to weld it. The action of these fluxes

is similar to that of other fluxes mentioned in this chapter, and in the chapter on brazing. The flux dissolves oxides and floats them to the surface of your molten weld puddle while you weld the part. The flux also helps shield the hot metal from further oxidation. The major difference between aluminum welding fluxes and others is their chemical composition. As previously described, both the base metal and your filler metal must be covered with flux before you start welding.

When you are repairing a broken aluminum casting, it is important to clean the parts thoroughly to remove grease, oil, and other foreign matter. A wire brush and a solvent or degreaser will usually get the job done. Heavy sections of aluminum (½ in. [12.5 mm] and thicker) should be beveled to 45° along the line of your weld.

Some welders recommend welding castings of all but the very heaviest sections without beveling. They recommend melting out the metal on each side of the crack to form a V, and then using a "puddling rod" to assist in removing excess metal. A puddling rod is made from a ⅛- or ³⁄₁₆-in. [3- to 5-mm] diameter steel welding rod by flattening the end of the rod for about 1 in. [25 mm]

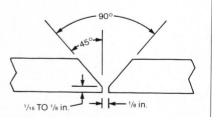

FIGURE 9-48 This is typical of the detail you will see on large double-bevel–welded joints in thick sections. The single-bevel angle may be smaller than 15° in some special narrow-gap, wire-fed arc-welding processes designed to save costly filler metals.

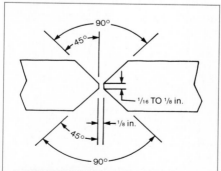

FIGURE 9-49 Exceptionally thick plate requires a double bevel on each side of the plate with a small land and root gap in the middle. The root of the weld joint is now located in the middle of the section.

and bending the flattened end at right angles to the rest of the rod. Puddling rods are used as scrapers to remove excess metal and oxide floating on top of the weld. The puddling rod should be heated in your torch flame before putting it into the weld metal, otherwise the rod will chill the weld puddle.

Large or complicated castings involving combinations of thin and heavy sections should be preheated in a suitable furnace before you start welding them. Preheating is usually carried out at a temperature of about 700 to 800°F [371 to 425°C]. Do not exceed 800°F [425°C] as there is a good chance that you will burn out (oxidize) some of the alloying elements in the aluminum. If the furnace does not have a pyrometer control (a device to record furnace temperature), you can check the temperature of your castings with one of the following methods:

■ Use thermal crayons that melt between 700 and 800°F [371 to 425°C] (they are readily available in most welding shops)—see Chap. 22 for details on thermal indicators used by welders.
■ At the correct welding temperature a pine stick rubbed on the casting will show a "char" mark
■ Marks made with a carpenter's blue chalk will turn white at the correct temperature for welding
■ The cold aluminum casting, when struck a hard blow, gives a metallic sound. As temperature is increased, this sound becomes duller. At the correct welding temperature the casting will no longer give off a metallic sound.

Preheating also can be done with your oxyfuel flame or with oil or gas burners. If you preheat a section of a casting, use the same imaginary wedge experiment we talked about in preheating gray iron castings. Determine where the most stress on the part would be if a wedge were driven into the crack, and put most of the preheat in that area.

When aluminum castings are preheated or where welding is performed on long seams, unsupported from underneath, there is a chance that the part will collapse of its own weight because preheated aluminum does not have the same strength as room-temperature aluminum. To avoid this problem, sheet copper sometimes is fitted around the part and clamped or otherwise held in place to give support to your heated aluminum part.

Welding is best performed using the forehand technique. Flux is applied by dipping your heated welding rod into an open container holding powdered flux. On heavy castings, the flux may be sprinkled on the surface of the V joint of your weld.

The shop section that follows will give you practice in the welding techniques we have described.

REVIEW QUESTIONS

1. What does the process code OFW stand for on a blueprint?
2. What is often the practical and economic *maximum* thickness of sections joined by oxyfuel welding? Can oxyfuel welding be used to join much thicker sections?
3. If gas welding is slower than arc welding, why is it still used?
4. You have a box of filler metals on your shelf. The label says they meet AWS A5.2, Class RG 45. Without looking further, do you know what it is?
5. What does the filler metal specification AWS A5.15 cover?
6. You are going to gas-weld 4043 alloy aluminum. You don't know much about this material. What filler metal specification would you look in to select the correct gas-welding rod?
7. Does your flame point *at* the previously deposited weld metal when you are using the forehand welding technique or *away* from it?
8. What is the most critical pass or layer when making a multipass groove weld (no matter what welding process you use)?
9. What's the best flame setting for welding steel?
10. You just finished a butt weld on steel plate. Looking underneath, you see icicles of weld metal hanging down from the joint. What probably went wrong?
11. When making a butt joint on plate or sheet, or when joining pipe, what kind of welding is done first to hold the base metal in position and reduce distortion?
12. What is the maximum thickness that your tack welds should be on steel?
13. Your weld puddle is getting too big and you are having trouble controlling it. What should you do?
14. Why do you preheat cast-iron sections before gas welding, and slow cool afterward?
15. Why should you avoid welding hollow castings that don't have a vent hole?

SHOP PRACTICE

Gas-Welding Practice

9-1-S
EQUIPMENT AND SUPPLIES
We gave you more information in this chapter than you can remember out in the shop. Now is the time to learn to write your own welding procedures. For each new material or thickness you weld, review our recommended procedures in this textbook (or your instructor's directions). Then write down a brief welding procedure before you do the work. Include all the data you need for the job. Keep your welding procedure in a

notebook if it works well. Save the notebook in your toolbox. It will help you when you get a full-time job.

You will find that this shop section is set up somewhat more formally than the earlier ones, with a format that sets out the purpose, base material, filler rod, torch tip size, and so on. We believe that you will find this information useful since this is the single most important shop section in this entire welding course. The manual skills you will start to learn here will help you quickly learn other welding processes, including arc welding.

Since tip-size designations have never been standardized, we will be using a coding system provided by only one tip manufacturer who uses tip-size numbers ranging from #00 for very small tips up to #10 for very large ones. You often will run into welding instructions based on information from one supplier, but you will have another supplier's equipment. Therefore, you can check the hole diameter in your tips with a wire gauge to determine the actual hole diameter in your tip, then compare it against the drill size number in the tables to make sure that you have the right tip for each job.

We also have included some real-life operator qualification tests in this shop. These tests are very similar to real tests that you may be required to take when you apply for a job.

9-2-S
EQUIPMENT GUIDELINES

When you set up your equipment, use Table 9S-1 and Table 9S-2 as guidelines. Always use the correct welding tip size for the material you will weld. We'll code these tip size numbers from #00 through #10 in the following tables.

Table 9S-2 gives you the approximate regulator settings and minimum hose diameters for good flow rates for each welding tip, once you have selected the correct tip for the material thickness that you will weld and the fuel gas you will use.

For example, our hypothetical tip supplier has a #00 tip listed in Table 9S-1. This tip has a 0.020-in. hole according to the manufacturer, which is equivalent to a #76 standard drill size. Their #10 tip size has a hole that is produced by a #30 drill, which makes a 0.128-in. hole. If you select a #10 tip, Table 9S-2 shows that the regulator settings for this tip (using a $\frac{5}{16}$-in.-diameter hose) should be 6 psig with a 36 scf/h flow rate for MAPP Gas or 3 to 10 psig with a flow rate somewhere between 39 and 140 scf/h if acetylene is used as the fuel gas.

Now let's look in detail at some specific jobs you can do to practice what you've learned. If you don't have time to do all of these jobs, do as many as you can. Remember that the more work you do, the easier the rest of this course will be. Select at least one kind of work from each type of welding

situation so that you get broad practice in working in different positions, and with different types of metal and metal thicknesses. This shop experience is vital to learning the manual skills you need to make good welds—no matter what welding process you use.

9-3-S
SHOP PRACTICE WELDING STEEL

The purpose of this work is to help you learn how to make fusion welds on low-carbon steel with the oxyfuel process, first without using a filler rod and then with a filler rod.

Materials
You will need $\frac{1}{16}$- and $\frac{1}{8}$-in.-thick sheet steel for this work. You also will need $\frac{1}{4}$-, $\frac{3}{16}$-, and $\frac{3}{4}$-in.-thick low-carbon steel plate. Small 2- × 4-in. coupons (small metal samples) are OK to use when you are practicing this work, although your final welder-qualification tests are best conducted on 6- × 6-in. coupons.

When you use a welding rod, you should select a steel rod from AWS A5.2, using an RG 45 or RG 60 grade. Use a $\frac{1}{16}$-in.-diameter rod for both thicknesses of steel you will be working on. Be sure to select the correct tip size for your torch, fuel gas, and the material you will be welding.

Making bead welds
1. Make bead welds on the $\frac{1}{16}$-in.-thick steel without filler metal. (Simply melt through the material, making a single good weld bead on the metal surface.)

2. Make bead welds on the $\frac{1}{8}$-in.-thick steel without filler metal.

3. Make bead welds on the $\frac{1}{16}$-in. steel with filler metal.

4. Make bead welds on the $\frac{1}{8}$-in. steel with filler metal.

Bead weld test There is no physical test for a bead weld. The only test is the appearance of the weld bead that you produce. The best weld bead will have a uniform width and an even, smoothly

---

TABLE 9S-1 Welding tip size and fuel gas versus metal thickness welded.*

Some Common Tip Size Numbers												
Tip Size	#00	#0	#1	#2	#3	#4	#5	#6	#7	#8	#9	#10

Material Thickness Commonly Welded with Each Tip Size and Fuel Gas												
Acetylene	$\frac{1}{64}$	$\frac{1}{32}$	$\frac{1}{16}$	$\frac{3}{32}$	$\frac{1}{8}$	$\frac{3}{16}$	$\frac{1}{4}$	$\frac{5}{16}$	$\frac{3}{8}$	$\frac{1}{2}$	$\frac{5}{8}$	$\frac{3}{4}$ up
MAPP Gas	—	$\frac{1}{64}$	$\frac{1}{32}$	$\frac{1}{16}$	$\frac{3}{32}$	$\frac{1}{8}$	$\frac{3}{16}$	$\frac{1}{4}$	—	—	—	—

* Use table by reading down the column from tip size to fuel gas and then across that line to the metal thickness to be joined.

rippled surface with no pin holes, cracks, gaps where you missed making the weld, and no holes burned through the material. The more you practice making good-looking bead-on-sheet or bead-on-plate welds, the easier it will be to make the more difficult welds you will do next.

Corner welds on steel

5. Make a weld on a corner joint on two pieces of $\frac{1}{16}$-in.-thick steel sheet.

6. Make a weld on a corner joint on two pieces of $\frac{1}{8}$-in.-thick steel.

Corner weld test

7. Subject your completed corner welds to a hammer test. If you produced a satisfactory corner joint, you will be able to bend flat and then hammer the joint without cracking or breaking your weld.

Square-groove butt joints on sheet steel

8. Make a weld on a square-groove butt joint on $\frac{1}{16}$-in.-thick steel in the flat position.

9. Make a weld on a square-groove butt joint on $\frac{1}{8}$-in.-thick steel in the flat position.

10. Make a weld on a square-groove butt joint on $\frac{1}{16}$-in.-thick steel in the vertical position.

11. Make a weld on a square-groove butt joint on $\frac{1}{8}$-in.-thick steel in the vertical position.

12. Make a weld on a square-groove butt joint on $\frac{1}{16}$-in.-thick steel in the overhead position.

13. Make a weld on a square-groove butt joint on $\frac{1}{8}$-in.-thick steel in the overhead position.

14. Make a weld on a butt joint on $\frac{3}{16}$-in.-thick steel plate in the flat position. Use a 45° edge-beveled joint for this and the other plate jobs we list next. First do this and the following jobs with the forehand technique. Then repeat Exercises 14 through 19 using the backhand welding technique.

15. Weld a butt joint on $\frac{1}{4}$-in. steel plate in the flat position.

16. Weld a butt joint on $\frac{3}{16}$-in. steel plate in the vertical position.

17. Weld a butt joint on $\frac{1}{4}$-in. steel plate in the vertical position.

18. Weld a butt joint on $\frac{3}{16}$-in. steel plate in the overhead position.

19. Weld a butt joint on $\frac{1}{4}$-in. steel plate in the overhead position.

Steel butt-joint test All butt welds should be tested by a guided root-bend test. If you are not sure how to do it, see the section at the end of this shop practice on "Operator-Qualification Tests."

Making fillet and lap joints

20. Make a fillet-welded lap joint on $\frac{1}{16}$-in.-thick steel in the flat position.

21. Make a fillet-welded lap joint on $\frac{1}{8}$-in.-thick steel in the flat position.

22. Make a fillet-welded lap joint on $\frac{1}{16}$-in.-thick steel in the vertical position.

23. Make a fillet-welded lap joint on $\frac{1}{8}$-in.-thick steel in the vertical position.

24. Make a fillet-welded lap joint on $\frac{1}{16}$-in.-thick steel in the overhead position.

25. Make a fillet-welded lap joint on $\frac{1}{8}$-in.-thick steel in the overhead position.

26. Make a fillet-welded T joint on $\frac{1}{16}$-in.-thick steel in the flat position.

27. Make a fillet-welded T joint on $\frac{1}{8}$-in.-thick steel in the flat position.

28. Make a fillet-welded T joint on $\frac{1}{16}$-in.-thick steel in the vertical position.

29. Make a fillet-welded T joint on $\frac{1}{8}$-in.-thick steel in the vertical position.

30. Make a fillet-welded T joint on $\frac{1}{16}$-in.-thick steel in the overhead position.

31. Make a fillet-welded T joint on $\frac{1}{8}$-in.-thick steel sheet in the overhead position.

32. Make a forehand fillet-welded lap joint on $\frac{3}{16}$-in.-thick steel plate in the flat position. Use a 45° edge bevel. Repeat the job using backhand welding.

33. Make a forehand fillet weld on a lap joint on $\frac{1}{4}$-in.-thick plate in the flat position. Use a 45° edge bevel. Repeat the job using the backhand welding method.

34. Make a forehand fillet weld on a lap joint on $\frac{3}{16}$-in. steel plate in the vertical position. Use a 45° edge bevel. Repeat the job using the backhand welding method.

35. Make a forehand fillet weld on a lap joint on $\frac{1}{4}$-in.-thick steel plate in the vertical position. Use a 45° edge bevel. Repeat the job using the backhand welding method.

36. Make a forehand fillet weld on a lap joint on $\frac{3}{16}$-in. steel plate in the overhead position. Use a 45° edge bevel. Repeat the job using the backhand welding method.

TABLE 9S-2 Oxyfuel regulator settings for given tip sizes

Tip Sizes	Drill Sizes, in.	Minimum Hose Diameter, in.	Acetylene		MAPP Gas	
			Regulator Pressure Range, psig	Flow, scf/h	Regulator Pressure Range, psig	Flow, scf/h
#00	#76 (0.020)	$\frac{3}{16}$	2–10	1–3	—	—
#0	#72 (0.025)	$\frac{3}{16}$	2–10	2–4	—	—
#1	#68 (0.031)	$\frac{3}{16}$	2–10	3–6	1	1
#2	#62 (0.038)	$\frac{3}{16}$	2–10	4–8	1.5	1.5
#3	#56 (0.047)	$\frac{3}{16}$	2.5–10	5–13	1.5	1.5
#4	#54 (0.055)	$\frac{3}{16}$	2.5–10	6–15	2	4
#5	#51 (0.067)	$\frac{3}{16}$	2.5–10	11–29	2.25	8
#6	#48 (0.076)	$\frac{1}{4}$	3–10	21–62	2.5	12
#7	#44 (0.086)	$\frac{1}{4}$	3–10	25–74	3	16
#8	#40 (0.098)	$\frac{1}{4}$	3–10	29–86	4	22
#9	#35 (0.110)	$\frac{1}{4}$	3–10	31–99	6	28
#10	#30 (0.128)	$\frac{5}{16}$	3–10	39–140	6	36

37. Make a forehand fillet weld on a lap joint on ¼-in. steel plate in the overhead position. Use a 45° edge bevel. Repeat the job using the backhand welding method.

38. Make a forehand fillet weld on a T joint on ³⁄₁₆-in. steel plate in the flat position. Use a 45° edge bevel. Repeat the job using the backhand welding method.

39. Make a forehand fillet weld on a T joint on ¼-in. steel plate in the flat position. Use a 45° edge bevel. Repeat the job using the backhand welding method.

40. Make a forehand fillet weld on a T joint on ³⁄₁₆-in. steel plate in the vertical position. Use a 45° edge bevel. Repeat the job using the backhand welding method.

41. Make a forehand fillet weld on a T joint on ¼-in. steel plate in the vertical position. Use a 45° edge bevel. Repeat the job using the backhand welding method.

42. Make a forehand fillet weld on a T joint on ³⁄₁₆-in. steel plate in the overhead position. Use a 45° edge bevel. Repeat the job using the backhand welding method.

43. Make a forehand fillet weld on a T joint on ¼-in. steel plate in the overhead position. Use a 45° edge bevel. Repeat the job using the backhand welding method.

Testing steel fillet and lap joints

44. All of your lap joints are to be subjected to a hammer test, folding the joint 180° along the weld to make a flat sandwich with the welded joint running along one side. Then pound the welded joint flat without any weld-metal cracking.

45. All of your fillet joints are to be subjected to a hammer test for T joints, folding the 90° leg of the tee flat to one side of the joint, followed by hammering the joint without producing any weld-metal cracking.

9-4-S
WELDING CAST IRON
Welding cast iron is a very common and very important repair job. Learn how to do it well.

Cast-iron-welding materials You should have some samples of gray cast iron about ¼, ½, and ¾ in. thick for this work. You also should have a small cast-iron part to repair. A quick trip to a junkyard with a sledge hammer could soon get you all the cast-iron samples you will need for this work. A local gray-iron foundry also would be glad to donate leftover risers and other scrap castings for you to work on. Select your cast-iron filler rod from AWS A5.15. Rod that is ⅛ in. square will be a good size to work with.

Making cast-iron welds

46. Make a weld on a butt joint on ¼-in.-thick cast iron in the flat position. Repeat the job welding in the vertical position.

47. Make a weld on a butt joint on ½-in.-thick cast iron in the flat position. Repeat the job welding in the vertical position.

48. Make a weld on a butt joint on ¾-in.-thick cast iron in the flat position. Repeat the job welding in the vertical position.

49. Repair a fractured gray cast-iron part by welding. The bigger the part the better. A worn-out engine block from a junkyard is useful. You can do a nice job of cracking it with a sledge hammer. Several student welders can work together on a badly cracked engine block. That will help you learn to coordinate your work with that of other people.

Testing your cast-iron welds Although in commercial practice a repair weld cannot be broken and tested to see if the weld metal is sound, you should know by the way you made your weld whether or not you got sound weld metal. However, if your practice castings are small enough, they should be broken to determine the quality of the welds you made. Each weld and the adjacent base metal should be tested with a file for hardness. If the metal is too hard to be cut by the file, you let the casting cool too fast after welding or you didn't

preheat it enough beforehand, or you made both errors. Cast-iron welds are often machined to a smooth finish after welding. But a file-hard weld in a casting is a machinist's nightmare.

9-5-S
WELDING YELLOW BRASS
This work will help you learn how to weld yellow brass. Be sure to use plenty of flux.

Brass-welding materials You will need samples of yellow brass about 2 × 4 × ¹⁄₁₆ in. and ⅛ in. thick. You also will need welding rod selected from AWS A5.7. Use a grade R CuZn bronze rod. Use a ¹⁄₁₆-in.-diameter bronze rod for ¹⁄₁₆-in. brass material. Use a ⅛-in.-diameter bronze rod for welding ⅛-in.-thick brass. Select a flux that is suitable for use on brass.

Making welds on brass

50. Make a weld on a butt joint of ¹⁄₁₆-in.-thick brass in the flat position.

51. Make a weld on a butt joint of ⅛-in.-thick brass in the flat position.

52. Make a weld on a butt joint of ³⁄₁₆-in.-thick brass in the flat position.

53. Make a weld on a butt joint of ¼-in.-thick brass in the flat position. Use a 45° beveled edge.

54. Make a weld on a butt joint of ¹⁄₁₆-in.-thick brass in the vertical position.

55. Make a weld on a butt joint of ⅛-in.-thick brass in the vertical position.

56. Make a weld on a butt joint of ³⁄₁₆-in.-thick brass in the vertical position.

57. Make a weld on a butt joint of ¼-in.-thick brass in the vertical position. Use a 45° beveled edge.

58. Make a weld on a butt joint of ¹⁄₁₆-in.-thick brass in the overhead position.

59. Make a weld on a butt joint of ⅛-in.-thick brass in the overhead position.

60. Make a weld on a butt joint of

³/₁₆-in.-thick brass in the overhead position.

61. Make a weld on a butt joint of ¼-in.-thick brass in the overhead position. Use a 45° beveled edge.

Testing brass welds All your welds should be subjected to a guided bend test. See the end of this shop lesson for more details.

9-6-S
WELDING COPPER

This is your chance to learn how to weld a material that has high thermal conductivity. The practice you get with copper will make it much easier to work with aluminum, which you will learn to weld next.

Copper-welding materials You will need sheet copper samples, about 2 × 4 × ⅛ in. thick, and ¼ in. thick. You also should get some standard-weight, 2-in.-diameter copper water pipe to work on. The sheet copper preferably should be a good selection of both deoxidized (OFC or oxygen-free copper) and electrolytic copper. Scrap copper used in electrical parts will probably be made of the latter material.

Select your gas rods from AWS A5.7, using a silicon-copper rod about ⅛ in. in diameter. Use larger-sized tips for the thicker copper sections. You are going to need extra heat to work with this material, so keep that in mind when selecting your welding tips. Select a tip that is about one size larger than the tip size you would use for the same thickness of steel or brass.

Making welds on copper

62. Make a weld on a butt joint of ⅛-in.-thick copper in the flat position.

63. Make a weld on a butt joint of ¼-in.-thick copper in the flat position.

64. Make a forehand weld on a butt joint on 2-in. copper pipe in the horizontal rolled position.

65. Make a forehand weld on a butt joint on 2-in. copper pipe in the horizontal fixed position.

66. Make a forehand weld on a butt joint on 2-in. copper pipe in the vertical fixed position.

Testing your copper welds Successful completion of each of these copper-welding exercises will constitute qualification for this work. There are many tests for copper welding, but they vary so much with the material to be tested and the application for the product, that simply making a decent weld on this highly conductive material is proof enough that you can weld this material as far as your shop course is concerned.

9-7-S
WELDING ALUMINUM

Aluminum is not quite as thermally conductive as copper, but you will find that it does soak up heat very fast. Aluminum, however, presents another problem that we described in the lesson. It doesn't change color when heated, and if you are not careful it will suddenly melt without warning. We described several methods for testing for the approach of the melting point of this material. The best method is simply getting experience so you can judge when the stuff is about to melt. And that means doing as much aluminum welding as you can.

Aluminum welding materials You should have 2- × 4-in. sheet and plate samples of aluminum to work on that are ¹/₁₆, ⅛, ³/₁₆, and ¼ in. thick. If you can obtain them, get some aluminum castings to work on. They should be about ¼ to ½ in. thick. If possible, you should practice welding on commercially pure 1100 series aluminum, and on at least one weldable aluminum alloy such as a silicon-aluminum grade.

You should select your aluminum gas rod from AWS A5.10, using ER 1100 rods for the pure aluminum and ER 4043 rods for the silicon-aluminum alloy material. After you have experience working on wrought aluminum, try welding your aluminum castings.

Be sure on all jobs to only use a flux that is suitable for aluminum welding. Aluminum requires special fluxes, so read the label on the can to make sure that it's the right stuff before you use it.

Making welds on aluminum

67. Make bead welds on ¹/₁₆-in.-thick aluminum sheets without filler metal. First try it with 1100 series aluminum, and then with a silicon-aluminum alloy such as 4043, using the correct filler rod for this material.

68. Make bead welds on ⅛-in.-thick aluminum sheets without filler metal. First try it on commercially pure aluminum, and then on the alloyed material.

69. Make welds on flanged-edge joints in the flat position on ¹/₁₆-in.-thick aluminum sheets—no filler metal. Work with both 1100 and 4043 aluminum grades.

70. Make the same flanged-edge joints in the flat position, working on ¹/₁₆-in. sheets, but this time use a filler rod. Work with pure aluminum sheet and then with a weldable alloy grade.

71. Make welds on butt joints in the flat position on ⅛-in.-thick 1100 series aluminum sheets. This time, use your filler rod.

72. Make welds on butt joints in the flat position on ⅛-in.-thick silicon-aluminum sheets. Use filler rod.

73. Make welds on flanged-edge joints in the vertical position on ¹/₁₆-in.-thick aluminum sheets—no filler rod.

74. Make welds on flanged-edge joints in the vertical position on ⅛-in.-thick silicon-alloy aluminum sheets—no filler rod.

75. Make welds on butt joints in the vertical position on ⅛-in.-thick 1100 aluminum sheets. Use filler rod.

76. Make welds on butt joints in the vertical position on ⅛-in.-thick 4043 aluminum sheets. Use a silicon-aluminum filler rod.

77. Make welds on deep flanged-edge joints in the flat position on ¹/₁₆-in.-thick aluminum sheets—no filler rod.

78. Make welds on deep flanged-edge joints in the flat position on $\frac{1}{16}$-in.-thick silicon-alloy aluminum sheets—no filler rod.

79. Make welds on butt joints in the flat position on $\frac{3}{16}$-in.-thick 1100 grade aluminum sheets.

80. Make welds on butt joints in the flat position on $\frac{3}{16}$-in.-thick 4043 silicon-aluminum alloy sheets.

81. Make welds on butt joints in the flat position on $\frac{1}{4}$-in.-thick 1100 grade aluminum sheets.

82. Make the same welds on butt joints in the flat position on $\frac{1}{4}$-in.-thick 4043 silicon-aluminum alloy plates.

83. Repeat the butt joints on $\frac{1}{4}$-in.-thick plate but this time work in the vertical position.

84. You may be ready to handle overhead work with aluminum. Try it. Make a butt joint on $\frac{1}{4}$-in.-thick plate in the overhead position. Watch out for the melting aluminum.

85. Not all cast aluminum alloys are readily weldable. With luck, you'll get a chance to work on one that is. Use an AWS A5.10 ER 4043 gas rod. Use a #3 tip for the $\frac{1}{4}$-in.-thick sections, and a #4 tip for the $\frac{1}{2}$-in.-thick sections. The larger the casting, the more problems you will have getting the material up to welding temperature. Work downhand where possible. But if you think you are good enough, repair the casting by working out of position.

Testing your aluminum welds All aluminum butt welds on material $\frac{1}{4}$ in. thick and thicker should be submitted to the nick-break test. Test your flange welds and butt welds on thinner material to the guided bend tests. There is no test for bead-on-plate or bead-on-sheet except weld-bead appearance.

9-8-S
WELDING STEEL PIPE

This work is optional, but very important if you can possibly get it done. Review your work on pipe and tube brazing in Chap. 7. Using 4-in. steel pipe, repeat as much of the brazing work as you can, except weld the pipe instead of brazing it. Use an AWS A5.2 RG 45 or RG 60 gas welding rod about $\frac{1}{16}$ in. in diameter. Work in as many different pipe-welding positions as you can. Save the best samples for your pipe-welder operator-qualification test (Test IV), which follows.

Here are some jobs to try if you want to get some advance work done on pipe welding before you go on to the next chapter.

86. Pipe welding is very difficult but well-paying work. Prepare a beveled pipe joint from two pieces of steel pipe about 4 in. in diameter. Bevel the outer edges to 45° before you start. Then tack-weld the two sections into position with four tacks spaced equally around the joint (at points about 0, 90, 180, and 270° around the pipe). Then weld the pipe sections together. Keep the pipe centerline in the horizontal position while you work. It will help you to rest the pipe in a piece of channel iron or two angle irons. That way you can roll the pipe when you have to, welding a quarter of the section at a time, while always working in the downhand position. Pay special attention to getting a good solid root pass. Make sure your cover passes are tied into the tinned sides of the joint and the root pass.

87. Repeat Exercise 86 with the pipe centerline in the vertical position (the pipe is standing upright) after you tack-weld it in line. This time the pipe joint is like making a circular weld in the horizontal position. It's tough to do.

Testing your pipe welds Subject strip cuttings from your pipe welds to guided root- and face-bend tests. See the following operator-qualification Test IV—Pipe, for more details. You'll find even more details in the next chapter on pipe welding.

9-9-S
OPERATOR-QUALIFICATION TESTS

The purpose of Chap. 9 and this shop practice has been to develop your ability to make butt and fillet welds under conditions approximating those you will find in industry. Your work has progressed to the point where it is necessary to find out how good a welder you are with the oxyfuel process.

We'll do that by giving you some operator-qualification tests very similar to those you will have to take to get a job. In these tests, you will select your own tip size, rod diameter, and procedure (forehand or backhand), and you will conduct your own test on your own material. In a real operator-qualification test, you would make the welds but someone else would test your work.

Some of these tests are more detailed than we indicate here. You can get more details on how they are actually conducted by looking back to Chap. 8, Sec. 8-7 under "Free- and Restricted-Bend Tests," where we gave some detailed procedures for root-bend tests, face-bend tests, nick-break tests, and other test methods. For the purposes of this work, your welding instructor will help you make the tests. The objective of these tests is to show you where you are having problems.

If you don't pass one of these tests, it shows that you need more work. For example, you may be able to make very good welds in the downhand and horizontal positions, but you may be having trouble making sound welds in the vertical and overhead positions. Or maybe you can make good welds in any position on steel, but aluminum is giving you trouble. Or you may not have any difficulty with wrought materials, but cast iron is giving you a hard time. These tests will show you where you need extra shop practice.

We strongly recommend that you practice enough oxyfuel welding with different materials and in different positions until you can consistently pass an oxyfuel operator-qualification test. If you develop good skills with oxyfuel, you will have no trouble at all learning TIG welding. In fact, you'll probably find that TIG is easier to do.

After you master both oxyfuel and TIG welding, you will find that learning the manual skills you need for other welding processes will become progressively easier as you work your way through this textbook. That's why we gave you so many oxyfuel shop assignments to do. On the other hand, if you don't spend enough time developing good skills with gas, you will continue to have problems through the rest of this course.

Test I. Butt welds These tests will require six plates of metal (low-carbon steel, brass, or aluminum, but not cast iron, which requires a different kind of test). You will have to decide whether or not the material thickness requires edge beveling, and at what angle to bevel the material. You also will do the required edge preparation before you start to weld.

The first test consists of making:

■ Single V-groove butt joints in the flat position.
■ Single V-groove butt joints in the vertical position.
■ Single V-groove butt joints in the overhead position.

Each of the welded plates you make will be cut into four strips as shown in Fig. 9-S1. The two specimens that are retained are machined flush all over as shown in Fig. 9-S2. One test specimen from each of the four welding positions (flat, vertical, horizontal, and overhead) will be submitted to a guided face-bend test, and the other center strip will be submitted to a guided root-bend test (as described in Chap. 8, Sec. 8-7 under "Free- and Restricted-Bend Tests") If any defects greater than 1/8 in. [3 mm] in any dimension are developed in these tests, your weld will be considered as failing. In addition to the guided face-bend and guided root-bend test, your weld will be examined for appearance. Sloppy-looking welds, whether they pass the mechanical tests or not, will be rejected.

Test II. Fillet and lap welds These tests require six plates each of which is 6 × 6 × 1/4 in. thick.

FIGURE 9-S2 You must produce a V-groove butt joint that can be folded flat along the weld without cracking or tears occurring in the weld metal.

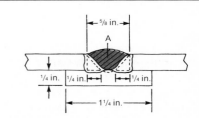

FIGURE 9-S3 Dimensions of a V-groove butt weld made with a 1¼-in.-wide (32-mm) backup bar.

One edge of each of the plates will be ground square. Also, one backup strip 6 × 1¼ × ¼ in. will be required for each joint.

Set up the test plates for a double fillet weld T joint and make:

■ One set of fillet welds in the flat position.
■ One set of fillet welds in the vertical position.
■ One set of fillet welds in the horizontal position.
■ One set of fillet welds in the overhead position.

After the fillet welds have cooled, cut off the leg of the T flush to the base and then make a third weld in the area marked A in Fig. 9-S3, in the position where the leg of the T joint previously stood. This third weld may be done in any position you want, regardless of the position in which the fillet welds were made. However, you must use the 1¼-in.-wide by ¼-in.-thick backup strip behind the final weld in area A, when you make it.

When welding is completed, the backup strip shown on the bottom

FIGURE 9-S1 Samples are cut from butt welds during welder-qualification tests as shown here. The samples are tested to failure (or folded flat in a 180° bend without failure) to show whether the welder can make solid welds and pass the qualification test.

Weld, 1 1/8 in., 1 1/2 in., 1/4 in., 1 1/2 in., 1/4 in., 1 1/8 in., 1/4 in., ALLOWANCE FOR CUTTING, DISCARD, FACE-BEND TEST, ROOT-BEND TEST, DISCARD, 6 in., 12 in.

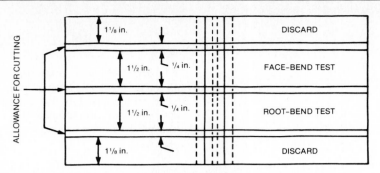

FIGURE 9-S4 Sample preparation for bend tests on vertical T-joint fillet welds.

of Fig. 9-S3 should be removed flush with the underside of the plates. For best results, this backup strip should be machined off. However, if a suitable grinding machine or other machine tool is not available, this strip can be removed by a hand cutting torch. You must in this case be very careful not to cut into the root of the weld. That would tend to fuse in or cover up any root defects and would not permit a true or fair test of your ability.

The plates are then cut into four strips as shown in Fig. 9-S4. The two specimens retained are machined flush all over as shown in Fig. 9-S2. One specimen from each of the three welding positions is submitted to the guided face-bend test and the others to the guided root-bend test.

If any defects greater than ⅛ in. [3 mm] in any dimension are developed in this test, the weld will be considered as failing.

Test III. Cast-iron welding The following tests will use two pieces of cast iron about 5 in. long × 2 in. wide × ½ in. thick. These pieces should be in the "as cast"

condition, and not heat-treated prior to welding. If the metal is very light colored and cannot be scratched with a file, it is cast iron that has been rapidly quenched and it is called *white iron*. It cannot be welded without cracking, so don't use it. Your test iron should be gray cast iron, not an alloy iron grade.

First, you should verbally demonstrate to your welding instructor your knowledge of how to prepare a casting properly to avoid welding defects such as shrinkage cracks, blow holes, sand holes, and excess hardness after welding.

Then, when you have completed your cast-iron butt weld and properly cooled it, test the weld area and the base metal around it with a file. If the metal is too hard, you fail the test.

Next, prepare the sample for a nick-break test. Machine or grind sharp notches or "nicks" on either side of the casting, on directly opposite edges of the weld. Then, using a cold chisel and hammer, split the weld along the line of the nick marks. Examine the broken surface. Look for zones of

poor fusion and gas bubbles on the joint surface. Any weld metal defects larger than ⅛ in. across will mean that you failed the test.

Finally, test the inside surface of the broken joint with a file to make sure that nowhere in the joint area is there exceptionally hard metal. As you know, hard spots mean that the casting was not properly preheated before welding or not slowly cooled afterwards.

Test IV. Pipe welding Just to see if you are ready to learn pipe welding, try this test. (We'll really get deep into pipe welding in the next chapter.)

Make several samples by welding two pieces of 4-in. standard steel pipe approximately 6 in. long. The two pieces of pipe will be placed in a horizontal plane and position welded as a butt joint. This is a horizontal fixed-position pipe weld.

A minimum of four specimens should be taken from the completed weld, spaced at 90° intervals around the weld. These specimens should be taken from the top, bottom, and opposite sides of the completed pipe weld. The top and one side specimen should be submitted to the guided face-bend test, and the bottom and other side specimen should be subjected to the root-bend test.

All of the above specimens should bend to 180° or the full capacity of the guided-bend-test jig without any cracks greater than ⅛ in. in any dimension resulting from bending or existing prior to welding.

If you pass this pipe-welding test, you are well on your way to becoming a skilled welder.

CHAPTER 10

Oxyfuel Pipe Welding

You will start learning to weld pipe with the oxyfuel process in this lesson. Much of what you learn here also will help you weld pipe with stick electrodes or TIG (tungsten inert-gas) welding. SMAW (stick-electrode) welding of pipe is even more difficult than gas welding. Welding pipe and tubing with either oxyfuel or arc welding is difficult for several reasons.

■ When you can't roll the pipe and tube as you work, the welding position changes as you work from downhand (when you are on top of the section) to vertical (on the side of the pipe), to overhead (when you are below it). When the pipe or tube is vertical, the weld joint probably will be horizontal and you must weld continously around a curved surface. Pipe connections make things even more complicated because you may have to start with a butt weld which becomes a fillet weld and finally returns to a butt weld . . . all in one continuous pass while your welding position changes constantly.

■ Pipes and tubes are made of many different kinds of steel alloys, cast irons, and nonferrous metals; many of them are difficult to weld unless you know exactly how to do it.

■ Pipe joints are always critically important. There's no room for error or weld imperfections. This is especially true when the pipe or tube must carry high-pressure fluids at very high or very low temperatures or under high stress. And those are just the environments in which most kinds of pipe and tubing are used. Examples are boiler tubes, chemical plant piping, structural tubing (whether it's used in the frame of a farm tractor or to hold up a building), or the hollow shaft used in a power drive system.

The most important thing for you to learn right now is how to weld pipe and tubing as the welding position continuously changes. That skill, plus a lot of the other things you will learn here, will help you prepare for the electric arc-welding processes that follow.

10-1
THE DIFFERENCE BETWEEN PIPE AND TUBING

Pipe is almost always round. It's specified first by material, then by a "nominal" diameter, often followed by a wall thickness, and, of course, by the desired length. Pipe specifications are described in detail in many different engineering manuals and handbooks. You'll find that there is no one materials standard for all pipe or all tubing. There is even more than one system used to describe the most common types of steel pipes. The two most important ones are listed in Table 10-1.

The older system that's still frequently used includes designations such as "standard-weight pipe," "extra-strong pipe," and "double-extra-strong pipe." In that system, a standard 1-in.-diameter pipe has a wall thickness of 0.133 in. and an inside diameter of 1.049 in., very close to 1 in.

An extra-strong 1-in.-diameter pipe has a wall thickness of 0.179 and an inside diameter of 0.957 in., still very close to 1 in. (therefore, the outside diameter must be larger than standard-weight pipe).

Finally, a double-extra-strong 1-in.-diameter pipe has an even larger wall thickness, about 0.358, and an inside diameter of only 0.599 in., or a little more than 0.5 in. even though it's classified nominally as 1-in.-diameter pipe.

The newer system designates pipe by *schedule numbers*. The larger the schedule number, the thicker the pipe wall. Pipe schedule numbers run from schedule No. 5 pipe to schedule No. 160 pipe (which has very thick walls and, therefore, is much stronger). Just like standard-weight, extra-strong, and double-extra-strong pipe, this schedule number material is usually designated by a nominal pipe size ranging from ⅛ to 30 in. diameter.

Nominal in engineering jargon means "more or less," "somewhere around," or "approximately." Its meaning in nontechnical English is "in name only," and that's close to what it means technically. However, all basic pipe specifications do give exact dimensions for inside and outside diameter and wall thickness along with the nominal diameters. Pipe usually is used to carry fluids such as gases or water and precise dimensions aren't as important as the ability to handle these fluids at different pressures and not leak or burst. A pipe's strength (measured by its wall thickness) is usually the most important factor.

Tubing is quite different. First, tubing need not be round. Square or rectangular tubing is quite common. Other special shapes are also available (Fig. 10-1). Second,

FIGURE 10-1 Tubing comes in a much wider variety of shapes, sizes, and metal grades than pipe. Tubing isn't always used to carry pressured fluids, either. This array of tubular shapes (much of it extruded from solid steel) is an example of the tubular shapes produced by the Babcock & Wilcox Tubular Products Group, Beaver Falls, Pennsylvania, a McDermott company. They also make welded tubing. Many of these tubes will be used for machine parts or as hollow shafts. When produced for those kinds of applications, the material is called *mechanical tubing.*

TABLE 10-1 Pipe size and nominal wall thickness

Nominal Pipe Size	Outside Dia.	Schedule 5	Schedule 10	Schedule 20	Schedule 30	Standard +	Schedule 40	Schedule 60	Extra Strong	Schedule 80	Schedule 100	Schedule 120	Schedule 140	Schedule 160	Double Extra Strong
⅛	0.405	—	0.049	—	—	0.068	0.068	—	0.095	0.095	—	—	—	—	—
¼	0.540	—	0.065	—	—	0.088	0.088	—	0.119	0.119	—	—	—	—	—
⅜	0.675	—	0.065	—	—	0.091	0.091	—	0.126	0.126	—	—	—	—	—
½	0.840	—	0.083	—	—	0.109	0.109	—	0.147	0.147	—	—	—	0.187	0.294
¾	1.050	0.065	0.083	—	—	0.113	0.113	—	0.154	0.154	—	—	—	0.218	0.308
1	1.315	0.065	0.109	—	—	0.133	0.133	—	0.179	0.179	—	—	—	0.250	0.358
1¼	1.660	0.065	0.109	—	—	0.140	0.140	—	0.191	0.191	—	—	—	0.250	0.382
1½	1.900	0.065	0.109	—	—	0.145	0.145	—	0.200	0.200	—	—	—	0.281	0.400
2	2.375	0.065	0.109	—	—	0.154	0.154	—	0.218	0.218	—	—	—	0.343	0.436
2½	2.875	0.083	0.120	—	—	0.203	0.203	—	0.276	0.276	—	—	—	0.375	0.552
3	3.5	0.083	0.120	—	—	0.216	0.216	—	0.300	0.300	—	—	—	0.438	0.600
3½	4.0	0.083	0.120	—	—	0.226	0.226	—	0.318	0.318	—	—	—	—	—
4	4.5	0.083	0.120	—	—	0.237	0.237	—	0.337	0.337	—	0.438	—	0.531	0.674
5	5.563	0.109	0.134	—	—	0.258	0.258	—	0.375	0.375	—	0.500	—	0.625	0.750
6	6.625	0.109	0.134	—	—	0.280	0.280	—	0.432	0.432	—	0.562	—	0.718	0.864
8	8.625	0.109	0.148	0.250	0.277	0.322	0.322	0.406	0.500	0.500	0.593	0.718	0.812	0.906	0.875
10	10.75	0.134	0.165	0.250	0.307	0.365	0.365	0.500	0.500	0.593	0.718	0.843	1.000	1.125	—
12	12.75	0.156	0.180	0.250	0.330	0.375	0.406	0.562	0.500	0.687	0.843	1.000	1.125	1.312	—
14 OD	14.0	—	0.250	0.312	0.375	0.375	0.438	0.593	0.500	0.750	0.937	1.093	1.250	1.406	—
16 OD	16.0	—	0.250	0.312	0.375	0.375	0.500	0.656	0.500	0.843	1.031	1.218	1.438	1.593	—
18 OD	18.0	—	0.250	0.312	0.438	0.375	0.562	0.750	0.500	0.937	1.156	1.375	1.562	1.781	—
20 OD	20.0	—	0.250	0.375	0.500	0.375	0.593	0.812	0.500	1.031	1.281	1.500	1.750	1.968	—
22 OD	22.0	—	0.250			0.375			0.500						—
24 OD	24.0	—	0.250	0.375	0.562	0.375	0.687	0.968	0.500	1.218	1.531	1.812	2.062	2.343	—
26 OD	26.0	—	—	—	—	0.375	—	—	0.500	—	—	—	—	—	—
30 OD	30.0	—	0.312	0.500	0.625	0.375	—	—	0.500	—	—	—	—	—	—
34 OD	34.0	—	—	—	—	0.375	—	—	0.500	—	—	—	—	—	—
36 OD	36.0	—	—	—	—	0.375	—	—	0.500	—	—	—	—	—	—
42 OD	42.0	—	—	—	—	0.375	—	—	0.500	—	—	—	—	—	—

tubing is used a lot of different ways besides handling fluids, and the different kinds of tubing reflect this fact. For example, structural-steel tubing is used to hold up buildings and is more often than not square or rectangular in cross section. The strength of the material is its most important feature. Therefore the steel grade used and the tube wall thickness are the most critical dimensions once an overall size such as a 4- × 4-in. or an 8- × 12-in. section has been selected.

Mechanical tubing usually is specified by very precise inside and outside diameters and wall thicknesses. This material is often used to make machines, shafts, and other "mechanical" products. Even the surface finish is important. Either a hot-rolled or one of several smoother cold-finished surfaces may be specified.

Of course, tubing also can be used to carry liquids or gases. The copper tubing in an air conditioner is one example. Many of the tubular products used in a chemical plant are tubing, not pipe. One reason is that tubing is widely available in almost any metal or alloy.

Pipe materials are usually limited to ferrous alloys, most frequently either carbon steel or one of several kinds of cast iron, with ductile cast iron being the most important because this material has properties much closer to steel than does gray cast iron. Ductile cast iron also is often called nodular iron because the graphite is in a round or nodular shape instead of the graphite flakes that you find in gray cast iron when you look at it under a microscope.

Tubing can be made of any metal and often is. You can order stainless-steel tubing, titanium tubing, low-carbon-steel tubing or high-alloy heat-resisting steel tubing, copper tubing, or aluminum tubing in any of hundreds of different alloys.

Pipe and tubing also are selected by how they are made.

Seamless tubing is drawn or extruded and (as the name implies) it has a hole in the middle but no welded seam running along its length. Welded tubing is made from precisely rolled strip or plate which is roll-formed into a cylinder which is then seam-welded along the entire length. The surface of a tube may be either hot-rolled (like steel plate is), or cold-finished (like some some kinds of very smooth bar stock and some types of steel sheets).

10-2
PIPE AND TUBING SPECIFICATIONS

Pipe and tubing are used in so many different kinds of work that many special types of pipes and tubes have their own categories, specifications, and even materials-specifying organizations. Therefore, many pipe and tubing specifications are based on the end use of the material.

Drill pipe used in oil wells is one example of a large class of products generally lumped together under the category of *oil-country goods*. Line pipe is a special product that transmits the oil or gas that the well produces. Line pipe can be up to 48 in. in diameter, or even larger. The American Petroleum Institute writes the most important specifications for these materials.

Another example of a tube specification based on use is *aircraft hydraulic tubing*. This material has to have its own set of specifications because it is used under high pressures to operate the control surfaces of an aircraft. Both aircraft manufacturers and the airlines take a special interest in specifications for aircraft hydraulic tubing.

The structural-steel tubing we already mentioned is a third example of a use-oriented description for a tubular product. Structural-steel tubing comes complete with its own specifications, dimensions, tolerances, and materials. The American Society of Civil En-

gineers has a lot to say about the specifications for this material.

Pipe and tubing can be coated. Galvanized steel pipe, which is coated with zinc, is only one example. There's also plastic-coated steel pipe and tubing, aluminum-coated steel tubing, and many other kinds of coated pipes and tubes. Therefore, on top of everything else, a pipe or tubing specification may include information on the coating weight and quality on the outside (or the inside) of the section.

Corrugated galvanized-steel culvert pipe is an example of steel pipe that is not very thick-walled, is coated with zinc, and is even crinkled or corrugated to make it stiffer (because changing the section changes the section stiffness of a piece of material). So you won't always find that pipe is something that is always round and thick.

The American Petroleum Institute (API), The American Iron and Steel Institute (AISI), the American Society for Testing and Materials (ASTM), the Society of Automotive Engineers (SAE), and the American Society of Mechanical Engineers (ASME) all have pipe and tubing specifications for different kinds of materials or special jobs. These materials specifications often are coordinated with each other.

AISI writes pipe or tubing descriptions (technically they are not specifications) from the point of view of the steel producers. API writes oil-country-goods specifications for the oil and gas industry and also from their own special point of view. API also writes many very important pipe-welding specifications. The American Society of Mechanical Engineers' (ASME) well-known Boiler and Pressure Vessel Code documents in detail pipe- and tube-welding specifications for tubular products used in both fired and unfired boilers and pressure vessels.

The SAE writes tubing specifications for materials to be used in cars, trucks, and aircraft. The

ASTM will often coordinate all these different points of view into a single materials specification that describes the material from the point of view of producers, users, and even national and local governments. That's why ASTM specifications are so widely used. They will stand up in a court of law, including the U.S. Supreme Court, where a case against a manufacturer was once won on the basis of a single ASTM specification.

Since the details can be found in almost any engineering handbook, we won't fill up this book with pipe and tubing data. We'll only include a little information in Table 10-2. This lists the common ASTM steel pipe and tubing specifications that you are most likely to work with. Although long, this list is far from complete. Also note that most of these pipe and tubing materials are cited by the ASME Boiler and Pressure Vessel Code (not shown here). We won't get into details on ASME requirements because they can be very complicated.

There is an equally long list of ASTM specifications for nonferrous pipe and tubing that always begin with the letter B plus a number. Even if you had both lists of A (ferrous) and B (nonferrous) ASTM materials specifications, you would still need the user-oriented specifications from the API and other groups on how to use the pipe or tubing you select.

You can learn two important facts from this table of ASTM pipe and tubing specifications. The first is that any mill form you select, from bar to plate to wire to structurals or sheet metal, will have an equally detailed list of ASTM materials specifications. As a welder, you are more likely to see ASTM specifications referenced on your work orders than any others because these are materials-oriented and joining materials is the business a welder is in.

Finally, Table 10-2 hints how much more there is to know about

TABLE 10-2 ASTM steel pipe and tubing specifications

Specification	Description
Wrought-Steel Pipe	
ASTM A-53	Welded and seamless steel pipe
ASTM A-120	Black and hot-dipped zinc-coated welded and seamless steel pipe for ordinary use
ASTM A-139	Electric-fusion-welded (arc-welded) steel pipe (sizes 4 in. and over)
ASTM A-134	Electric-fusion-welded (arc-welded) steel pipe (sizes 16 in. and over)
ASTM A-135	Electric-resistance-welded steel pipe
ASTM A-155	Electric-fusion-welded (arc-welded) pipe for high-temperature service
ASTM A-530	General requirements for specialized carbon steel and alloy pipe
ASTM A-369	Ferritic alloy steel forged and bored pipe for high-temperature service
ASTM A-430	Austenitic steel forged and bored pipe for high-temperature service
ASTM A-381	Metal-arc-welded steel pipe for high-pressure transmission service
ASTM A-106	Seamless carbon steel pipe for high-temperature service
ASTM A-312	Seamless and welded austenitic stainless-steel pipe
ASTM A-333	Seamless and welded steel pipe for low-temperature service
ASTM A-376	Seamless austenitic steel pipe for high-temperature central station service
ASTM A-524	Seamless carbon steel for process piping
ASTM A-335	Seamless ferritic alloy-steel pipe specially heat-treated for high-temperature service
ASTM A-405	Seamless ferritic alloy-steel pipe specially heat-treated for high-temperature service
ASTM A-211	Spiral-welded steel or iron pipe
ASTM A-409	Welded large outside diameter light-wall austenitic chromium-nickel alloy steel pipe for corrosive or high-temperature service
ASTM A-252	Welded and seamless steel pipe piles
Steel Tubes	
ASTM A-556	Cold-drawn carbon steel feedwater heater tubes
ASTM A-254	Copper-brazed steel tubing
ASTM A-250	Electric-resistance-welded carbon-molybdenum alloy-steel boiler and superheater tubes
ASTM A-226	Electric-resistance-welded carbon steel boiler and superheater tubes for high-pressure service
ASTM A-178	Electric-resistance-welded carbon steel boiler tubes
ASTM A-214	Electric-resistance-welded carbon steel heat-exchanger and condenser tubes
ASTM A-498	Seamless and welded carbon ferritic and austenitic alloy steel heat-exchanger tubes with integral fins
ASTM A-199	Seamless cold-drawn intermediate-alloy steel heat-exchanger and condenser tubes
ASTM A-179	Seamless cold-drawn low-carbon steel heat-exchanger and condenser tubes
ASTM A-213	Seamless ferritic and austenitic alloy-steel boiler, superheater, and heat-exchanger tubes
ASTM A-249	Welded austenitic steel boiler, superheater, heat-exchanger, and condenser tubes
Mechanical Steel Tubing	
ASTM A-512	Cold-drawn butt-welded carbon-steel mechanical tubing
ASTM A-513	Electric-resistance-welded carbon and alloy steel mechanical tubing
ASTM A-519	Seamless carbon steel mechanical tubing
ASTM A-511	Seamless stainless-steel mechanical tubing
ASTM A-554	Welded stainless-steel mechanical tubing

Structural Steel Tubing

ASTM A-500	Cold-formed welded and seamless carbon steel structural tubing in rounds and shapes
ASTM A-501	Hot-formed welded and seamless carbon steel structural tubing

Still Tubes

ASTM A-422	Butt welds in still tubes for refinery service
ASTM A-271	Seamless austenitic chromium-nickel steel still tubes for refinery service
ASTM A-200	Seamless intermediate alloy-steel carbon-molybdenum steel still tubes for refinery service
ASTM A-161	Seamless low-carbon and carbon-molybdenum steel still tubes for refinery service

Fuel Line Tubing

ASTM A-539	Electric-resistance-welded coiled steel tubing for gas and fuel-oil lines

Welded Fittings

ASTM A-403	Factory-made austenitic steel welding fittings
ASTM A-420	Factory-made wrought steel and alloy steel welding fittings for low-temperature use
ASTM A-234	Factory-made wrought carbon steel and ferritic alloy steel welding fittings

Cast-Iron Pipe

ASTM A-451	Centrifugally cast austenitic steel pipe for high-temperature use
ASTM A-452	Centrifugally cast austenitic cold-wrought steel pipe for high-temperature use
ASTM A-426	Centrifugally cast ferritic alloy steel pipe for high temperature use
ASTM A-216	Carbon steel castings suitable for fusion welding for high-temperature service
ASTM A-217	Alloy steel cast pressure-containing parts suitable for high-temperature service
ASTM A-389	Alloy steel castings specially heat-treated for pressure-containing parts suitable for high-temperature service
ASTM A-351	Ferritic and austenitic steel castings for high-temperature service
ASTM A-352	Ferritic steel castings for pressure-containing parts suitable for low-temperature service
ASTM A-487	Low-alloy steel castings suitable for pressure service

Line Pipe

American Petroleum Institute, API 5L	Line pipe
American Petroleum Institute, API 5LX	High-test line pipe

pipe and tubing. Since welders who have enough skill to join pipe and tubing usually make very good money, you also can see why it's worth your trouble to at least learn the basic welding skills you need to join pipe and tubing.

If you get a job in a pipe shop or work as a steamfitter or on a pipeline, your employer will take care of teaching you the extra details you will need to do work on the particular ASTM pipe or tubing grade you will be joining. What you need to know to get that job in the first place is the different pipe-welding positions and the more common welded pipe and tube joint designs. You also need to learn more about pipe and tubing welding techniques.

Later on we'll tell you more about welding pipe and tubing, including how to handle special pipe joints, T connections, Y connections, K connections, and diameter-reducing couplings. We'll cover those details after you learn to do shielded-metal arc (stick electrode) welding.

10-3
PIPE POSITIONS

The welding positions for pipe and tubing depend a lot on the direction or orientation of the long axis that runs through the middle of the pipe or tube and also on whether the pipe or tube can be rolled while you work. The ideal situation for any welder is pipe mounted horizontally on devices called *turning rolls* that rotate the pipe while the welding is done. Using them, the welder can always work downhand because the pipe weld is always below the torch or welding electrode. Figure 10-2 shows you an example.

Another pipe or tube welding position that's more difficult to work in, but not the hardest, is when the pipe or tube axis is vertical. That usually makes the pipe weld horizontal; it wraps around the pipe on a curved horizontal surface.

All other pipe-welding positions range between these two orienta-

FIGURE 10-2 Turning rolls are used to rotate round work such as pipe or tubing so that welding is always done in the downhand position. Downhand welding is not only easier to do, but the weld-metal deposition rates are higher than for out-of-position work.

tions of the long pipe or tubing axis falling somewhere between a horizontal pipe and a vertical pipe, either rolled while you weld or not moving while you work. Figure 10-3 shows what these different pipe and tubing welding positions look like. It also gives you some welding position names you will often hear in welder qualification tests, welding specifications, and in pipe-fabricating shops. Since a very similar naming system is used for fillet welds, and since fillet welds are used to weld connections onto pipes and tubing, we've also included examples of these fillet-welding positions.

In this welding-position naming system, a groove weld is signified by a G, as in "1G." A fillet weld is signified by an F, as in "1F". You should stop and look at these pictures for a while. It's much easier to say, "The welder qualification test requires me to make one T joint in the 1F position and one pipe butt joint in the 6G position," than taking five times as many words to explain what these positions look like.

Figure 10-4 shows the limits of these welding positions. The fig-

ure shows that a vertical weld may actually be 10° off the vertical (90°) in any direction. A flat or horizontal weld can tip up to 15° from the horizontal plane and still be classed as a flat or horizontal weld. We show you this diagram because it's included in technical literature and in welder qualification tests. It may look a little complicated at first, but if you study it you'll find that it's easy to understand.

10-4
PIPE JOINTS

Pipe and tube welding can only be done from the outside of the section unless very large diameter pipes are joined. Therefore almost all pipe joints are single-groove (one side only) joints, as shown in Fig. 10-5. Thicker pipe walls are welded by one of several electric-arc processes because oxyfuel welding is too slow for thick material, and you won't necessarily get the best weld penetration on thicker pipe walls. That's why you will mostly work with the joints shown at the top of Fig. 10-5 when you do oxyfuel pipe welding. Later on, when you learn to use

GTAW (TIG) and SMAW (stick electrodes), you will need to use the other pipe joints for welding thicker-walled pipe. You also might use *consumable inserts* in the pipe joint (Fig. 10-5G) when joining pipe with one of the arc-welding processes such as stick electrode welding. We show these inserts here, but you will learn more about them when you learn to weld pipe with arc-welding electrodes.

For pipe and tube walls up to 5/32 in. [4 mm] thick you will use the square-groove butt joint. If a three-wheeled pipe cutter is used to cut the pipe to length, the two butted ends won't be exactly square. The pipe cutter's wheels leave a slight bevel on the pipe ends of about 10°. When two such cut ends are butted together you'll have a butt-weld V joint with a 20° included angle. This joint works nicely for small-diameter pipe and you should get good weld penetration through the complete pipe section with it.

As the pipe wall gets thicker than 5/32 in. [4 mm], you should prepare a 37.5° beveled edge on each section of pipe or tube to be welded together. That will give you a 75° V-groove butt joint. This angle has been standardized by the pipe-fabricating industry and pipe producers. All pipe shipped from steel mills routinely will have a 37.5° machine-beveled edge unless you specially order it some other way.

When you weld pipe or tubing with a wall thickness of more than 3/4 in. [20 mm], you are moving out of the practical range of oxyfuel welding and into sections that have to be arc-welded. (However, small-diameter pipe up to 2 in. [50 mm] is gas-welded with acetylene because this fuel gas offers better control of the weld puddle than some arc-welding processes.) Sections over 3/4 in. [20 mm] thick often use a U-groove butt joint to allow room for the movement of stick electrodes or a GTAW welding torch and filler rod. The same U-joint design is

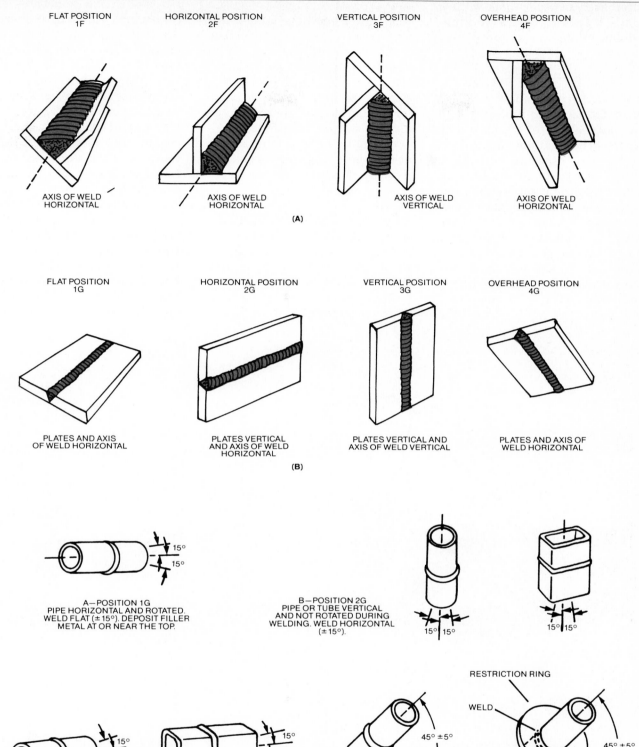

FLAT POSITION 1F · AXIS OF WELD HORIZONTAL

HORIZONTAL POSITION 2F · AXIS OF WELD HORIZONTAL

VERTICAL POSITION 3F · AXIS OF WELD VERTICAL

OVERHEAD POSITION 4F · AXIS OF WELD HORIZONTAL

(A)

FLAT POSITION 1G · PLATES AND AXIS OF WELD HORIZONTAL

HORIZONTAL POSITION 2G · PLATES VERTICAL AND AXIS OF WELD HORIZONTAL

VERTICAL POSITION 3G · PLATES VERTICAL AND AXIS OF WELD VERTICAL

OVERHEAD POSITION 4G · PLATES AND AXIS OF WELD HORIZONTAL

(B)

A—POSITION 1G
PIPE HORIZONTAL AND ROTATED.
WELD FLAT (±15°). DEPOSIT FILLER
METAL AT OR NEAR THE TOP.

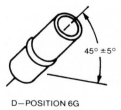

B—POSITION 2G
PIPE OR TUBE VERTICAL
AND NOT ROTATED DURING
WELDING. WELD HORIZONTAL
(±15°).

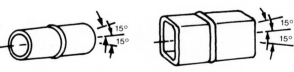

C—POSITION 5G
PIPE OR TUBE HORIZONTAL FIXED (±15°).
WELD FLAT, VERTICAL, OVERHEAD

D—POSITION 6G

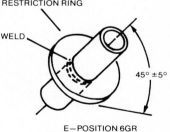

RESTRICTION RING

WELD

E—POSITION 6GR
(T, K, OR Y CONNECTIONS)

PIPE INCLINED FIXED (45° ±5°) AND NOT ROTATED DURING WELDING.

(C)

FIGURE 10-3 Welding positions. (A) Plate fillet welds. (B) Plate groove welds. (C) Pipe groove welds.

often used on other thick-walled joints, whether you are working on pipe, plate, or structural steel.

10-5
PIPE-WELDING RODS
You already glanced at the table of ASTM pipe and tubing specifications (Table 10-2). It should be obvious that you may be dealing with a wide range of different materials and that you will have to select a suitable gas-welding rod with good ductility to match the mechanical properties of the base metal in whatever pipe you work on. For ordinary standard steel pipe, you can always rely on AWS A5.2 grade RG 45 gas-welding rods. For carbon-steel pipe that will operate at higher pressures,

use AWS A5.2 RG 60 gas-welding rods unless you have special instructions to use something else.

If you weld other materials, you will find the correct rod in one of the AWS filler-metal specifications listed in the previous chapter. On most jobs, the rod you use already will be determined for you, usually by a welding engineer or an experienced welding foreman or production manager.

10-6
CONSUMABLE INSERTS
Consumable welding backup strips and specially shaped root-pass inserts occasionally are used in arc welding to improve root-pass fusion and to give a smooth backside to the pipe weld. Backup

strips often can be nothing more than a straight strip of metal of the same analysis as the base metal. It is tack-welded onto the back of a one-sided weld. Backing strips sometimes are inserted into the root of the joint and tack-welded into place. The welding process will melt through the backup strip and make it part of the weld metal. Any excess material is then arc-gouged or ground off. The 3M Company, among others, supplies adhesive weld-backing strips for this purpose that are something like heat-resistant adhesive tape.

Backing strips and consumable inserts that fit into the root of the joint are used mostly in arc welding. Consumable backing strips

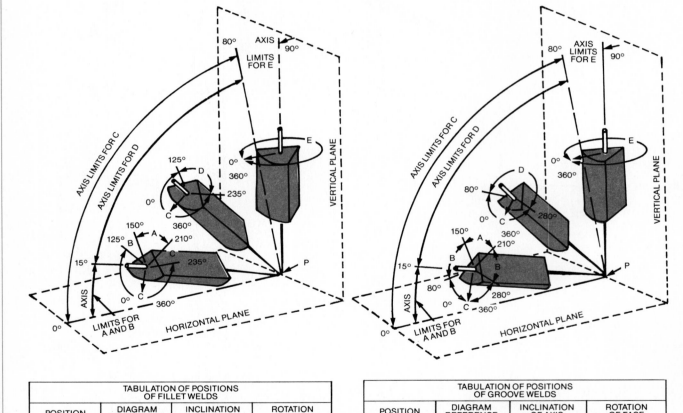

TABULATION OF POSITIONS OF FILLET WELDS			
POSITION	DIAGRAM REFERENCE	INCLINATION OF AXIS	ROTATION OF FACE
FLAT	A	0 TO 15°	150 TO 210°
HORIZONTAL	B	0 TO 15°	125 TO 150°
			210 TO 235°
OVERHEAD	C	0 TO 80°	0 TO 125°
			235 TO 360°
VERTICAL	D	15 TO 80°	125 TO 235°
	E	80 TO 90°	0 TO 360°

TABULATION OF POSITIONS OF GROOVE WELDS			
POSITION	DIAGRAM REFERENCE	INCLINATION OF AXIS	ROTATION OF FACE
FLAT	A	0 TO 15°	150 TO 210°
HORIZONTAL	B	0 TO 15°	80 TO 150°
			210 TO 280°
OVERHEAD	C	0 TO 80°	0 TO 80°
			280 TO 360°
VERTICAL	D	15 TO 80°	80 TO 280°
	E	80 TO 90°	0 TO 360°

FIGURE 10-4 Another way to define welding positions for groove, fillet, or pipe welding.

don't work well with oxyfuel welding. Neverthless, it's important that you know about them. But it's far more important that you learn how to control molten weld metal first with your torch and filler rod or welding electrode and leave these advanced techniques for later when you have mastered your basic skills.

10-7 WELDING SMALL SECTIONS

Oxyfuel welding is done mostly on small-diameter, standard-weight pipe and tubing up through sched-

ule 40 wall thicknesses, on diameters not larger than about 4 in. [100 mm]. Oxyfuel welding can join pipe as large as 6 in. [150 mm] in diameter in schedule 40 pipe, but 2-in.-diameter [50 mm] pipe is about the practical limit. Up to that wall thickness you don't need more edge preparation than the 20° V-groove butt joint created by a plumber's three-wheeled pipe cutter.

Joint Preparation

Figure 10-6 shows a typical oxyfuel V-groove butt joint. It also shows the 3/32-in. [2.4-mm] gap

you should leave between the two pipe sections after the pipes have been tack-welded together in line, and the tack-weld metal has cooled. This 3/32-in. [2.4-mm] gap is necessary to make sure that you get complete fusion all the way through the pipe wall. You will see similar gaps between sections used in many welding joint designs including butt joints and fillet welds. They are there for the same purpose; to get complete fusion in the root of the weld.

The gap (shown in Fig. 10-6) not only serves the purpose of improving root penetration but also helps adapt the section to the effect of cooling, shrinking weld metal. When weld metal cools it shrinks. If two fairly thick sections of metal are butted solidly together and welded, the cooling weld metal will pull the base metal together even tighter. Since the two pieces of base metal already are in contact, the only thing that can happen is a buildup of tensile stress in the weld and base metal. The stress of cooling weld metal can get so high that it will exceed the yield or tensile strength of the base metal and cause it to bend or crack. The gaps between the sections will close up as the weld metal solidifies, cools, and shrinks, without building up excess stresses in the base metal.

You'll only need one tack weld to hold 1-in. [25-mm] pipe in position if you are immediately going to weld it. You should use two or three tack welds spaced equally around the pipe if you will not weld this small section immediately. These tack welds should only be about half the wall thickness of the pipe. They should be made very carefully and finished off with a grinder to thin them before your final pass so that they can become part of the finished weld.

Start the weld on the opposite side of the pipe from the tack weld and work around the section toward the tack weld, first on one side and then on the other. When

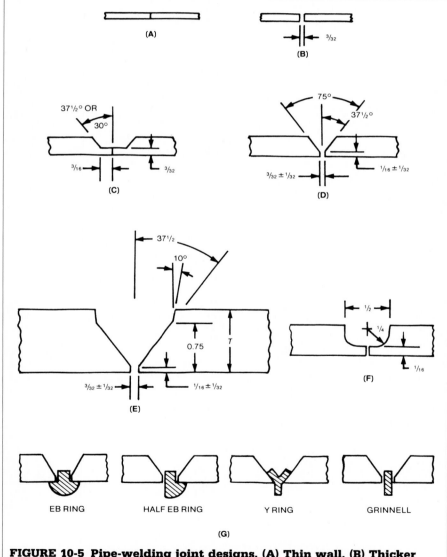

FIGURE 10-5 Pipe-welding joint designs. (A) Thin wall. (B) Thicker wall. (C) Thicker wall—GTAW. (D) Standard wall—uphill welding, steel. (E) Heavy wall—electrical arc welding, steel. (F) Heavy wall—aluminum. (G) Standard consumable welding inserts (used primarily for arc welding). The insert becomes part of the finished weld metal.

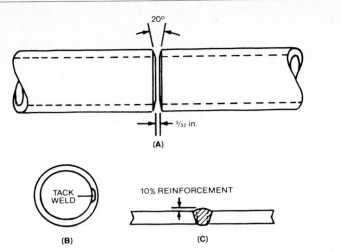

FIGURE 10-6 Small-diameter pipe joint preparation for 0.154-in. (3.9-mm) maximum wall thickness section. (A) Joint preparation and alignment of pipe. (B) Tack-weld pipe joint. (Note that tack welds are never full-section welds.) (C) Cross section of completed weld.

Manufactured fittings One possibility is to use manufactured pipe fittings such as 30, 45, and 90° elbows, or reducer sections that help you connect two pipes of different diameter either in a straight line or a bend. You'll find that they are all supplied machine-beveled and ready for use. They look just like the letters that describe them. A T connection is a right-angle takeoff from a main pipe. A K connection gives you two branches off a main. A Y connection splits the main into two branches.

Cutting templates One problem with joining pipe will be cutting the right hole to connect a smaller-diameter pipe section with a larger main pipe or tube at an angle. Since both surfaces are curved, they will intersect in another three-dimensional curved line. The actual shape that is cut in the side of the pipe or on the end of the other pipe has to match it. The shape of the cut can be quite complex. It won't be a simple circle.

Pipe shops don't have to use calculus to figure out how to lay out the job. (You could learn to make your own layouts if you took a mechanical drawing course and learned to project curves from two or three different views of an object onto a piece of paper, pain-

you come to the now-cold tack weld, be sure to pause momentarily and preheat it before you complete the weld, otherwise you will have a weld defect at that point. You should always do that with any previous tack weld that will become part of the final weldment, no matter whether you are working on pipe or plate or structural steel.

10-8
WELDING POSITIONS
Most welders prefer to make small-diameter pipe welds with the forehand method because forehand welding makes it much easier to manipulate the torch and filler rod around the small pipe.

The 1GR Position
Rolled horizontal pipe welds (the 1GR position) in small diameters are made the same way as any flat or fully downhand-position welds are made. You can use either the forehand or backhand technique for them.

The 5G Position
If you weld small pipe in the horizontal fixed position (the 5G position), use the forehand technique. Start the weld at the bottom of the pipe and carry the weld metal

up one side of the pipe to the top before starting the other side of the joint. The tack weld should be on top.

The 2G Position
For a fixed-position vertical pipe (the 2G position), use the backhand welding technique. It will give you the best results for this position.

Changing Flow Directions
If you weld T-, K-, or Y-pipe connections onto a pipe in the fixed 6G position, you will want to switch from backhand to forehand welding depending on the shape of the connection and its position as you work on it. In some ways welding branch connections to pipe is more difficult than making a simple V-groove butt weld to connect two pipe sections in a straight line. The work may require that you pass from a butt joint to a fillet weld all in one pass (Fig. 10-7). After you have mastered the basics of pipe welding, pipe connections won't give you much trouble.

You can change the direction of the pipeline by bending the pipe with a heating torch, but you also have a couple of other alternatives.

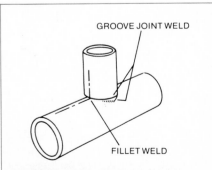

FIGURE 10-7 How to change the direction of a pipe with a T joint using fillet welds which become groove welds. Passing from one type of welding joint to another in a continuous sequence is one of the things that makes pipe welding difficult to do.

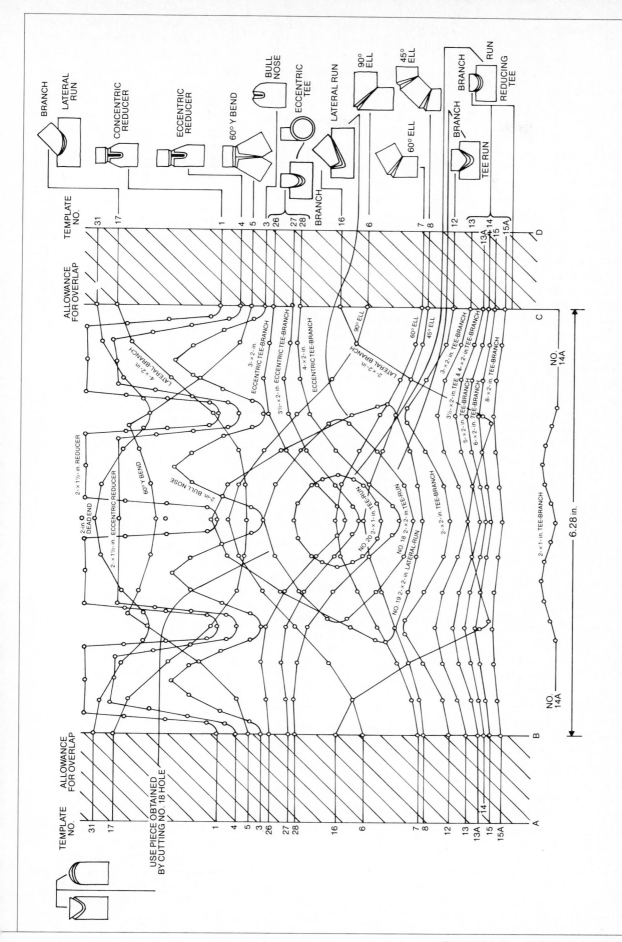

FIGURE 10-8 This is a typical set of pipe-welding templates. Each line is a separate template for a cut in a pipe to produce a different pipe joint. This set of templates was originally drawn for 2-in [51-mm] diameter pipe; therefore, the distance between the vertical lines B and C (where the left and right edge of the template meet when wrapped around the pipe) is the circumference of the pipe, or 2.00 in. × 3.1416 which equals 6.28 in. [160 mm]. The distances AB and CD are used for overlapping the template when it is wrapped around the pipe. You will be given a complete library of separate pipe-cutting templates to work with if you work in a pipe shop.

lessly, without advanced mathematics.) Standard pipe-fabricating templates have been developed for different sizes and diameters of pipe connecting into each other at many different angles. All you have to do is lay one of these templates on your pipe, trace the cut from the template, and then cut the hole. Figure 10-8 is a pipe template for cutting 2-in.-diameter pipe into a variety of joints. To use the template, trace one of the lines for the illustrated joint that you want (see the right side of the template) onto a separate piece of paper. Then cut the temporary paper tracing (not the template, which should be saved for future work) into a long strip.

Lay the single traced-line strip around the pipe circumference until the cross-hatched lines (between A–B and C–D on the left and right side of the master template) overlap. Then use a hard steel marker or carbon paper laid face down under the single-line temporary tracing template and transfer the cut line onto the pipe. That will give you the pattern you need to cut the pipe with a torch.

Figure 10-8 looks complicated, but if you study it carefully you will figure out how to use it for cutting 2-in. OD pipe. To turn the template into a master for cutting 4-in. OD pipe, double the linear width of the drawing (working with a draftsman or using a photocopy machine that can expand drawings). To prepare a template for 1-in. OD pipe, reduce the linear dimensions of the 2-in. OD pipe template by one-half. It may look difficult but you'll get good at it with a little practice.

An alternate method to flame-cutting for making small holes is to use a rotary hacksaw blade attached to a drill to cut out the correct shape. Chamfer the edge after cutting to give you the correct edge bevel.

Mitering Another way to make pipe go around a bend is to make a miter cut in two pipe or tubing sections at a 45° angle, as shown in Fig. 10-9. The 45° miter joint is simple to make, but it does a bad job of passing fluids around the sharp corner. A more gradual bend is better. Therefore, you can make the three-piece miter cut shown in Fig. 10-10 with a 30° angle on each section. The drawback is that it increases the amount of welding and cutting you have to do.

The best solution is to use a one-piece manufactured seamless steel welding fitting that gives you better, smoother fluid flow through the pipe while turning the corner. Of course, if you can bend the pipe section you'll get the same effect without having to cut and weld the pipe at all. You already know how to do that with a heating torch.

10-9
LARGE PIPE SECTIONS

One of the reasons why we said that most oxyfuel welding is done on standard-weight or schedule 40 pipe is that the wall thickness of 12-in.-diameter [300-mm] pipe is ⅜ in. [10 mm]. This is a maximum cutoff point between oxyfuel welding that you can do with one pass and welds that require multipass techniques. Arc welding is better and faster than oxyfuel when doing most multipass work.

Pipe Beveling

Most pipe beveling is done by small cutting machines. It takes a skilled torch operator to cut pipe to the required angles and tolerances needed to ensure accurate fit-up. Conversely, one of the biggest welding problems with pipe (and, in fact, a lot of other kinds of welding) is poor edge preparation during flame-cutting.

If you can't handle a cutting torch with accuracy yet, spend some more time and practice. It will save you a lot of trouble later when you don't have to weld your own poorly beveled edges.

In the early days, welders who worked with the forehand technique usually wanted large-diameter pipe that was beveled at 45° to make a 90° V-groove butt joint. Welders who used the backhand technique wanted pipe beveled to 30° to make a 60° V-groove joint.

The pipe mills compromised by coming up with the standard 37.5° bevel, and that's why this has become the standard bevel for almost all larger-diameter pipes, tubes, and welded fittings. When two 37.5° edges are butted together, you get a V joint with an included angle of 75°. This joint is suitable for both the forehand and backhand welding techniques. It

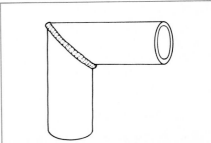

FIGURE 10-9 Change in the direction of flow through a pipe using a 45° miter cut on the ends.

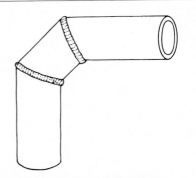

FIGURE 10-10 Change in the direction of flow through a pipe with a three-piece miter joint. Each bend is a 30° turn in the pipe. The result is less turbulent flow of fluid through the finished pipe section, but the joint is more expensive to make.

is shown in Fig. 10-11 along with the positions for tack welding larger-diameter pipe using three or four tack welds.

Remember that the tack welds should not be very thick. They should be less than half the wall thickness of the pipe in larger pipe sections. You must make absolutely sure that the tack welds flow smoothly into the pipe section. You don't want any sudden change in section or a lump on the inside of the pipe. Also remember that the tack weld is there only to hold the sections until you weld them. On some critical jobs, tack welds are even removed ahead of the welder as the final weld progresses around the section. One solution is to grind the tack weld flat prior to welding, blending the tack into the base metal to make a smooth, thin transition before you make the final welded joint. Larger pipe, up to 6 in. [150 mm] in diameter, should have at least three tack welds. Use four tack welds on pipe from 6 in. [150 mm] on up.

You'll also see in Fig. 10-11 that the joint has a small $\frac{1}{16}$- to $\frac{1}{32}$-in. [1.5- to 7-mm] nose or flat on the bottom of the V bevel cut. Also note that the forehand welding technique requires more rod and torch-tip manipulation than the backhand technique. You may have to open up the root of the joint to about $\frac{5}{32}$ in. [4 mm] instead of using the $\frac{1}{8}$-in. [3-mm] gap shown in the diagram if you use the forehand technique.

The 1G position If your pipe will be rolled in the 1G position (which makes it the 1GR position—the R stands for rotation), you can always work in the flat downhand position. This is by far the easiest way to weld pipe. It's also the best way to make sure that there are no welding defects.

Large-diameter pipe often is rolled into this position with special weldment turning rolls. One set of these rolls will be motorized and the other set (or sets if you

need several along the length of a long pipe) will be idler rolls. You will be able to control the speed of the drive motor so that the pipe will rotate at whatever speed you want it to. A foot treadle is used so that you can control the speed of your pipe's rotation without taking your eyes off what you are doing.

You can use either the forehand or the backhand welding technique for this downhand flat-position work. You probably will get the best results, however, using the backhand technique. If you need to make a third layer after completing the root pass and the first filler pass (on top of a deep V groove, for example), the forehand technique is better adapted to handling the wide opening at the top of the groove.

The 5G position To weld pipe in the 5G position (fixed position, horizontal groove), you should start the weld in the bottom centerline of the pipe (the root of the joint) and work up each side in turn to the top. You can use either forehand or backhand welding to do it, although the forehand method is easier in this position.

Be sure to get complete fusion whenever you come to a tack weld. Use a grinder to feather the tack back into the base metal for a smooth transition. Thin it down with the grinder if necessary.

When you complete the second side and are approaching the top where you are about ready to meet your first bead, move the torch up a little and preheat the first bead you made before you complete the second part of the weld. That will ensure that your two weld beads meet without a cold lap or other welding defect.

The 2G position When you weld a fixed pipe in the vertical 2G position, you are making the equivalent of a horizontal weld. Use the backhand technique. Again, preheat your previous bead as you turn the corner and approach the starting point. Also make sure that all tack welds are completely mixed with and fused into the final weld bead.

Changing Flow Directions
We already pointed out under directions for welding smaller pipe diameters that you have four ways to change positions. These are bending the pipe, using manufactured fittings, using templates

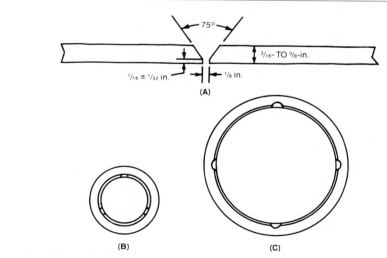

FIGURE 10-11 Large-diameter pipe joint preparation. (A) Single-V-groove joint design. (B) Three tack welds are placed 120° apart for pipe up to 6 in. in diameter. (C) Four tack welds are placed 90° apart on pipe that is from 8 to 12 in. [200 to 300 mm] in diameter. Note, again, that tack welds must not be full-penetration welds.

to cut your own openings in the pipe to change the direction of flow out of a section, and mitering to produce a pipe connection with a different flow direction. The same is true of large-diameter pipe. But you also know simply by observation that your choices are much more limited. Large-diameter pipe is harder to bend and more difficult to cut with a template, and manufactured fittings for large pipe are very expensive. Mitering, however, is easy to do and you will find that many larger pipe sections are mitered and then welded.

10-10 MULTIPASS PIPE WELDING

When you weld pipe or tubing with a wall thickness from ⅜ in. [10 mm] (two passes) to ¾ in. [20 mm] (three or four passes), the joints are best made with multipass welding. You put down a root pass first and then follow up with a cover pass. On very deep V joints you may need to make a root pass, several filler passes, and finish with cover pass layers. You may even want to make a capping pass to refine the weld metal in the cover pass, after which you can grind off the extra weld metal. (Remember from previous chapters that the heat from weld metal makes the weld bead underneath it finer-grained, and therefore tougher—this technique is commonly used in pipeline work.)

When making multipass welds on heavy-walled pipe, a good rule to follow is to put down one layer of weld metal for each ¼ in. [6 mm] of wall thickness. If you want even higher quality welding, lay down one layer of weld metal for each ³⁄₁₆-in. [5 mm] thickness of pipe or tube wall to refine the metal underneath. In general, you will probably use two or three layers to weld pipe with walls about ½ in. [12.5 mm] thick, and three or four layers on pipe with walls about ¾ in. [20 mm] thick.

Multipasses in Cascade

You can make a perfectly good weld by using the procedure described above. However, there's another method you should know about for multipass welding of pipe and tube.

Figure 10-12 shows the cascade method for laying down a multipass weld. Using this method, you start by laying down pass 1 as shown in the figure. That is followed by pass 2, pass 3, and so on until the section is complete all the way around the joint. Note how each pass overlaps the one before it and starts a new layer.

The idea behind the cascade method is to preserve more heat in the weld metal and base metal as you work, which makes the job go more quickly. Therefore, it's more productive. But if you use this method, you must be absolutely sure that you are getting complete fusion at each point where the weld is restarted. For this reason multipass cascade welding is not for beginners. The three most difficult problems you will have are getting complete fusion in the root pass, tying in the filler and cover passes onto underlying weld beads, and restarting a weld when you have stopped without leaving a welding defect such as a cold lap behind.

NOTE: In multipass cascade welding, you must get complete fusion at each weld restart point.

This is as true for arc welding as it is for gas welding. Therefore, *every time you change welding rods or stop for any other reason, be careful to get the weld restarted and thoroughly tied into the previous weld metal. Always reheat your starting point before continuing to weld, even if you stop for only a minute.* Take a little extra time to weave your weld bead into the walls of the joint so that you get total fusion between new weld metal and the base metal next to it. Then continue welding.

Multipasses on 1G Welds

The first layer of a multipass weld on horizontal pipe rolls (the 1G position) should be made continuously around the pipe. In actual pipe shops, this critical root pass often is made by TIG welding even though some other arc-welding process will be used for subsequent passes. One of the characteristics of GTAW welding is that it makes very good quality welds in narrow spaces. However, TIG welding is not used for root passes when you are using the oxyfuel process.

The first root pass is followed by each filler pass deposited completely without stopping (except to change welding rods). If you do have to stop for more time than it takes to pick up a new rod, you should reheat the new starting point of the weld before making more new weld metal. (If you were arc welding, you probably would use TIG welding for the root pass and stick electrodes for the cover and capping passes.)

If you do not reheat previous weld metal and start welding on a cold bead, you will have a serious lack-of-fusion defect at the starting point and for some distance beyond it until the underlying weld metal is brought up to temperature. Of course you also can use the cascade method in the 1G

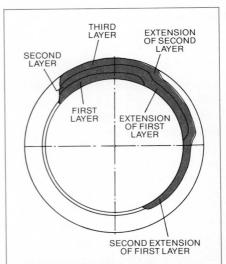

FIGURE 10-12 **Schematic view of cascade welding method for large-diameter pipe.**

position and carry the full weld around the joint as you work.

You can use either the forehand or the backhand welding technique for 1G-position pipe welding. If the gap between the walls of the joint gets too wide for backhand welding, switch to the forehand technique, which is better for handling a wider weld. You also can mix forehand and backhand welding as suits your fancy and your work.

Multipasses in the 5G Position

For working in the 5G position, you'll get the best results with forehand welding. But the backhand technique should be used for the root pass. To use the backhand technique for the root pass, start each half of the root layer at the centerline of the pipe, halfway up the joint, and carry the weld up the sides of the pipe to the top centerline. That's called *vertical up* welding. Each layer must be completed around the pipe before the next layer is started.

Multipasses in the 2G Position

Use the backhand technique exclusively for the vertical 2G fixed position. This joint is pure horizontal welding, even though it goes around a curve. Depending on how easy it is to get at the joint, the weld can be carried continuously around the section for each layer, or the weld may be broken up and welded in segments around the joint. However, all segments should be deposited exclusively by the backhand technique. Cascade welding may be helpful in making a vertical pipe weld in the 2G fixed position.

Another method is to use a series of *stringer beads* on the joint, laying them down over each other and tying each pass into the previous weld metal and the base metal until the joint is completed. Stringer beads are narrow, straight, parallel weld beads similar to bead-on-plate welds, but thoroughly tied into the base metal and into each other. Typi-cally five or six stringer beads will be required on a moderately thick pipe section.

10-11
STRESS-RELIEVING FINISHED WELDS

As you already know, stresses build up in a section as the weld metal cools and shrinks. These forces can exceed the yield or tensile strength of the base metal. When the section is loaded in service, it doesn't take much more added force to cause the section to fail. Postwelding stress relief reduces built-in stresses in the section. That's why preweld and postweld heating are specified on some jobs. All critical welds in thicker sections should be stress-relieved before the product is used.

Stress-Relief Methods

Most small-diameter pipe welds don't require stress relief. The section wall thickness is too small to develop significant built-in residual stresses. Most welds in thick-walled pipe or tubing (which usually means larger pipe or tube diameters) are stress-relieved in large furnaces or by using electric heating elements because that ensures complete stress relief. It also costs less.

But weld joints also can be stress-relieved with a heating torch. One method is to heat the entire pipe joint to the correct stress-relief temperature while making sure that the pipe wall in the area of the weld is at the correct temperature on both sides of the wall (and the wall is therefore uniformly heated). Then the pipe joint is wrapped in an insulating blanket or paper to make the section cool slowly. The open ends of the pipe should be plugged to stop any drafts. This method gives very effective stress-relief treatment without using special furnaces. Pipe-welding codes usually spell out exactly what holding temperatures and times are required, whatever heating method is used.

Time at Temperature

The usual requirement for stress-relieving thick pipe joints with a torch is to heat the weld area and part of the base metal on each side to a distance about six times the thickness of the pipe wall. This area should be brought up to temperature for 1 h for each inch of wall thickness, with a minimum heating time of 1 h at the specified temperature.

This means that for pipe walls below 1 in. [25 mm] thick, the heating time must still be 1 h. Welding codes or your work order will spell out the details of the stress-relief process and what the minimum section thickness is that should be stress-relieved after welding. For example, you won't have to stress-relieve small-diameter pipe used for water or low-pressure gas, unless there is some very special requirement that makes it necessary. If the same material is tubing used in high-pressure hydraulic lines, stress relief makes a lot of sense after welding, even on small-diameter material.

Stress-Relief Temperatures

For ordinary carbon-steel pipe, the stress-relief temperature is usually around 1150°F [620°C], plus or minus 50°F [27°C]. This temperature should be checked with proper temperature-measuring equipment such as a thermocouple or temperature-indicating crayons that melt or change color at a specified temperature. (See Chap. 22 for more details.) Other stress-relief heating and cooling temperatures and times may be required for materials other than carbon steel.

The proper stress-relief temperatures for many different kinds of welds and materials other than just carbon-steel pipe are spelled out in various materials specifications. The Boiler and Pressure Vessel Code of the ASME gives many details for these procedures for products covered by the code. Many people follow the ASME requirements even for products not specifically covered by the code.

We heard of one pressure vessel that was so large that it couldn't fit into any existing heat-treating furnace. The schedule also required a water quench after the vessel had cooled for a certain length of time. But there was no tank large enough to quench it in. The manufacturer simply built a huge fire inside the vessel and let it heat up to the required temperature, cooled the tank for several days, and then quenched it by rolling the vessel into Lake Michigan. However, 16 welders were arrested by the game warden for catching and eating boiled lake trout without a license.

10-12
PIPE-WELDING PROCEDURES

Now it's time to stop reading about how to weld pipe and start learning how to do it. The best place to learn is in the shop. Therefore, the rest of this chapter should be studied while you work with a torch in one hand, a gas rod in the other, and a pipe section in front of you.

In the shop section following this chapter we will give you some specific jobs to do. Follow the instructions in this chapter while you do the work outlined in the shop section. Repeat each job over and over until you get it down perfectly before going on to the next assignment. This is your best chance to get good at oxyfuel welding because in the next chapter you will start learning about the electric arc-welding processes.

Butt Welding 3-in. Standard Pipe

Pipe and other tubular products are used for a wide variety of applications including water, steam, air, gas, oil, gasoline, and so on. Operating conditions vary from high vacuums to pressures of several thousand pounds per square inch. The piping for modern steam-electrical generating plants not only operates at high pres-

sures (typically 900 to 2400 psig), but also at high temperatures (900 to 1100°F [480 to 590°C]). These temperatures are so high that they make steel pipe glow a dull red; these are severe conditions. Many piping service conditions cannot be handled by mechanical joints; welding is the only answer. Furthermore, welded joints reduce weight compared with mechanical joints; they also increase strength and lower the cost of piping installations. There is great need for pipe welds that produce a high yield strength and corrosion resistance at high temperatures.

Many of these joints are made to API or ASME Boiler and Pressure Vessel Code requirements. While much modern pipe welding is done with electrical processes such as the use of TIG welding for the root pass followed by stick electrodes for cover passes, the best way to learn to weld pipe is to start with oxyfuel procedures. If you can't make a good pipe weld with an oxyfuel torch, you won't be able to make one with arc welding. Arc-welded pipe work is even more difficult than oxyfuel welding, but there are a lot of similarities.

Pipe cutting Pipe cutting is essentially the same as plate cutting when starting away from an edge. Flame cutting is used for both oxyfuel and arc-welded pipe joints. In early stages of practice, you will find it easier first to cut a square edge and then bevel the edge to produce the correct joint geometry. As you gain more skill, you will be able to do this in one operation.

To ensure your first leakproof joint, cut the pipe in a true circle lying in a plane perpendicular to the centerline of the pipe. This job can be set up very easily. Take a strip of heavy paper or leather belting to make a straightedge longer than the circumference of the pipe to be marked. Wrap the belt tightly around the pipe and line up the overlapping edges. A

line marked against the straightedge of the paper or belt will have the required accuracy.

After the pipe has been cut, the ends must be beveled on the outside edges. For all pipes with a wall thickness greater than ⅛ in. [3 mm], the bevel should be 37.5°, or half of a 75° angle. (Make a 37.5° protractor from two pieces of steel strip to check your work, and keep this angle template in your toolbox.) On a pipe wall thickness up to 3/16 in. [4.6 mm], beveling can be done on a grinding wheel. Over that thickness, it's better to use your cutting torch.

Before welding, both pipe sections should be set in a jig and tack-welded so that their centerlines coincide and the spacing is uniform throughout (see Fig. 10-13).

Horizontal rolled-pipe (1GR) welding In the horizontal rolled-pipe welding position, mount your pipe on suitable rollers as shown in Fig. 10-14. In a pipe-fabricating shop, special turning rolls are commonly used to rotate pipe for welding. In your shop work, you may have to rotate the pipe by hand. Start welding at point C in Fig. 10-14, building up a proper weld contour and progressing upward to point B. When point B is reached, stop welding and rotate the pipe to bring the weld down until the stopping point of the weld is at point A. Then weld from A to B.

If you had a helper, or pipe rolls, the pipe would be slowly ro-

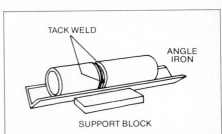

FIGURE 10-13 A simple angle iron with back supports makes a good welding jig for joining a few pipe sections together.

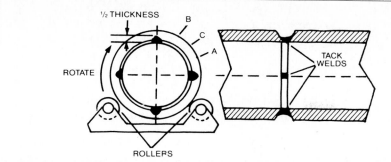

FIGURE 10-14 Welding turning rolls (shown in cross section) are used in production pipe welding.

tated as you worked and all your welding would be done within the small zone between point B and point C as the pipe joint rotated under your welding torch (or arc-welding electrode).

The position of your torch and rod at point A is similar to that for a vertical weld. As B is approached, the weld assumes a nearly horizontal position and the angles of application of your welding torch and rod are alternated slightly to compensate for this change.

You will encounter a new condition when finishing a weld on pipe. The weld should be stopped just before the root of the advancing weld puddle reaches the root of the starting point, so that a small opening remains in the joint. The starting point is then preheated so that the area surrounding the junction point between your last deposited weld metal and the place where you started welding is at uniform temperature to complete fusion of the advancing weld. Since one of the test specimens will be removed from this zone when you do a welder qualification test, you should be very careful to obtain complete fusion.

Horizontal fixed-pipe (5G) welding
After the pipe is tack-welded it is set up so that the centerline of the pipe is horizontal. Once welding has started, you must not move the pipe in any direction.

Welding starts at the bottom point A (see Fig. 10-15), which must be midway between two tack welds. This weld is made in two stages; start at the bottom, point A, and work up one side to the top; then return to the bottom and work up the other side to the top. Again you must take care to ensure complete fusion at the finishing point of your weld. You must return to the original starting point to begin the second half of your weld. The exposed face of this original starting point should be tapered so that the puddle for the second half may be conveniently started.

Vertical fixed-pipe (2G) welding
In this position, the tack-welded pipe is set up so that the centerline is vertical. Therefore this is a horizontal weld from the point of view of your welded joint position. Since this is not a vertical weld, the force of your flame is not beneath the molten weld puddle as it is in the case of a vertical weld. Therefore, greater skill is required to control the weld puddle and produce a satisfactory weld. Years of experience have proved that the backhand technique gives the best results. Since this is a backhand weld, the pipe should be beveled, preferably to a 37.5° angle. The weld is started at a tack and is carried continuously around the pipe. Relative positions for your torch and rod are shown in Fig. 10-16.

Butt Welding 6-in. Standard Pipe
You should use the backhand method to weld 6-in. pipe, just as you did with 3-in.-diameter pipe. Welding this larger-diameter pipe is a natural progression up from the 3-in.-diameter material you now should know how to weld. In the 3-in. standard-weight pipe, the wall was 0.216 in. thick. In this 6-in. standard-weight pipe, the wall thickness will be 0.280 in. thick. Therefore, the principal difference between these jobs lies in the heavier pipe wall thickness. Of course the length of weld on 6-in. pipe should be about twice that for the 3-in. pipe.

Experience again shows that the backhand technique is faster and produces better-quality welds on pipe diameters from 4 in. up. For backhand welding, the pipe should be beveled to 37.5°. Four tack welds, half the pipe thick-

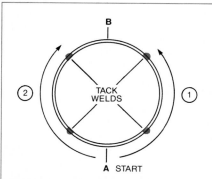

FIGURE 10-15 Welding sequence for vertical-up pipe welding. Two welders often work on opposite sides of pipe, starting at (A) and finishing at (B), when marking large-diameter pipe. This helps balance welding stresses.

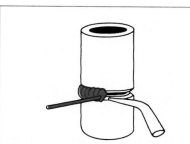

FIGURE 10-16 Backhand welding method used on pipe joint in the 2G position.

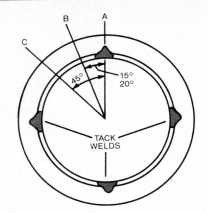

FIGURE 10-17 Procedure for horizontal fixed-position pipe welding.

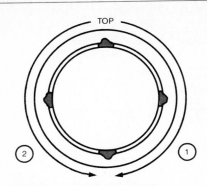

FIGURE 10-18 Multiple-layer welding is required to fill up the wall joint of pipe heavier than ½ in. [13 mm] thick.

ness, are equally spaced around the circumference of the pipe. Welding should be started at a tack weld, the top one if the pipe is in the horizontal position.

Horizontal rolled-pipe (1GR) welding On horizontal rolling welds, after starting at the top tack (A in Fig. 10-17), lower the weld puddle to B by rotating the pipe. If you have a helper or turning rolls to rotate the pipe, maintain the weld puddle at this point, which is 15 to 20° off the vertical centerline.

If you must turn the pipe yourself, you will find it convenient to rotate the pipe further, to point C in Fig. 10-17, so that you can make longer welds before stopping to turn the pipe. You then proceed to weld from this point toward the top. This is called welding *vertically up*.

The weld is carried around in a continuous operation, stopping only when necessary to change rods or rotate the pipe. As the starting point of the weld is approached, toward the finish, stop welding to leave a slight opening at the root. Preheat the starting point to a dull red and proceed to weld up the opening, taking care to get thorough fusion at the root.

Horizontal fixed-pipe (5G) welding Horizontal pipe welding in the fixed position is done with the backhand technique starting

at the top, at a tack, progressing down one side to the bottom and restarting at the top (that preheats previously deposited weld metal as you go). Progress down the other side to the bottom as shown in Fig. 10-18. Be sure to properly finish off the juncture of the two welds at the bottom, for this is the location of a specimen that will be taken for test welds in any pipe welder qualification test.

Vertical fixed-pipe (2G) welding Vertical pipe fixed-position welds are made exactly as described in making horizontal fixed-position welds. The heavier wall thickness obviously will require a larger tip size and welding rod.

Multilayer Butt Welds
Heavy-walled pipe (extra-heavy and double-extra-heavy pipe

grades) is used for power, industrial, and process piping systems in which the pressures or temperatures or both are very high, or on installations where corrosion is particularly active. Heavy-walled pipe is used under severe conditions. It's usually 6 in. [150 mm] in diameter or more, and the weld joints must be of a high quality consistent with the service. Multilayer welding is recommended for use on all pipe with a wall thickness of ⅜ in. [10 mm] or greater.

Multilayer welding simplifies the welding of heavy-walled pipe, lowers the cost of the operation by making the weld in layers rather than in one complete operation, and produces welds of the highest quality. Depending upon the thickness of the pipe wall, two or more layers are deposited. Thus, the welding operation is, in effect, thin-section welding which is repeated until the full-section weld is completed. This method requires less time than it takes to make a single-pass full-section weld.

The recommended procedure for multilayer welding is to deposit two layers on pipe with about ½-in. [12.5-mm] wall thickness; three layers on pipe with ¾-in. [20-mm] walls, and four layers on pipe 1 in. [25 mm] in diameter and heavier (see Fig. 10-19). Even more layers will be required on pipe or tubing that has walls substantially thicker than 1 in.

Multilayer welds are superior to single-layer, full-section welds because of the simplification of the

RANGE OF PIPE WALL THICKNESS	NUMBER OF LAYERS	CROSS SECTION OF WELD
⅜ TO ⅝ in.	2	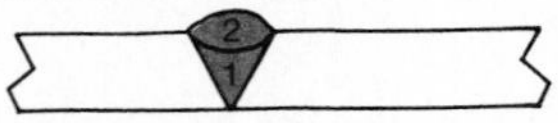
⅝ TO ⅞ in.	3	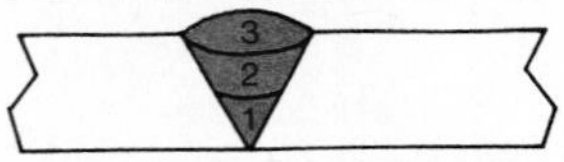
⅞ TO 1⅛ in.	4	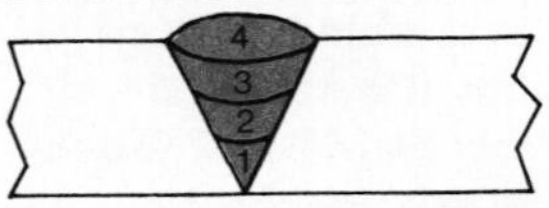

FIGURE 10-19 Number of weld-metal layers increases with the pipe wall thickness.

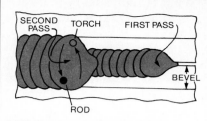

FIGURE 10-20 Torch and rod manipulation for second pass in large-groove pipe joint.

welding procedure. There also is another reason. Follow-up passes reheat the prior weld and tend to refine the grain structure of the prior pass, increasing the toughness of the weld deposit and improving its ductility.

As previously explained, there is practically no oscillatory motion of your torch in backhand welding. This holds true in the first pass of multilayer welding (the root pass). In subsequent cover and capping passes, the widening of the V joint and the resultant larger weld puddle make it necessary to move the rod to get even distribution of the weld metal. To aid in this operation, the torch must also be oscillated. However, instead of the circular motion used in forehand welding, the torch and rod are carried in a very slightly curved path as shown in Fig. 10-20.

Welding Galvanized Steel Pipe
Joining mill lengths of galvanized (zinc-coated) steel pipe with threaded fasteners is a serious problem. The threads are unsatisfactory because they cut through the protective zinc coating, exposing the unprotected bare base metal. That means that a threaded mechanical pipe joint will be subjected to severe corrosion in the threaded area, often where it cannot be seen until the pipe joint fails.

Various methods have been tried to solve the problem. Each one has created more problems. For example, brazing tends to damage the zinc coating on steel pipe. Welding with a nonoxidizing

flux that protects the zinc coating on the pipe during welding or brazing will help solve the problem. The flux prevents atmospheric oxygen from attacking the zinc while it is in the molten state due to the heat developed in the operation.

To weld zinc-coated steel pipe, the pipe ends are provided with a standard mill bevel of 37.5° (to make a 75° V). Before the pipe is set up and tacked as shown in Fig. 10-21, it is coated on both the inside and the outside with liquid flux for a distance of about 3 in. [75 mm] back from the end of the pipe where you will be welding. When the welding operation is completed, the contraction that results from the cooling of the deposited weld metal will draw the ends of the pipe nearly together.

Figure 10-22 is a diagram of a

completed braze-welded joint. This method has been used successfully on pipe diameters up to 12 in. [300 mm]. In such a piping system, galvanized welding fittings are naturally used throughout. Also note that galvanized pipe should be torch-cut by first fluxing the area to be cut for a substantial distance on either side of the cut area.

For ease in lining up the pipe prior to tacking, a piece of angle iron or channel iron should be used as shown in Fig. 10-13. Once the pipe is tacked, it can be set up on pipe rollers as shown in Fig. 10-23, which will make the welding go much faster. The horizontal rolled and horizontal fixed pipe-welding positions may be done with either the forehand or the backhand welding method. The vertical fixed-position joint

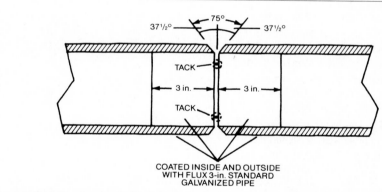

FIGURE 10-21 Galvanized steel pipe must be coated inside as well as outside with flux before welding begins.

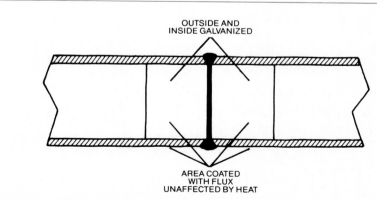

FIGURE 10-22 Finished galvanized-steel braze weld. The flux protects the galvanized (zinc) coating from being burned off during welding.

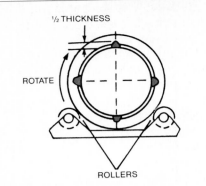

FIGURE 10-23 Use a welding fixture to line up the centerline of pipe before tack-welding it.

should be done with the backhand method as shown in Fig. 10-16.

Welding Yellow Brass Pipe

The procedure for welding yellow brass pipe is essentially the same as for welding steel pipe except that you must be sure to use the correct flux (suitable for brass) and flux the pipe inside and out. This must be done before the pipe is tack-welded. Flux-coated welding rod (a manganese bronze rod is good) is preferable for welding brass pipe but bare rod can be fluxed in the field or shop.

You should make at least one joint using dry flux and compare the soundness of your weld deposits with that made when using a commercially precoated fluxed rod.

In making these pipe welds, either the backhand or forehand welding method can be used. The backhand welding method is preferred for vertical fixed-position welds. Use the forehand welding method on horizontal rolled-pipe joints.

REVIEW QUESTIONS

1. In the expression, the nominal pipe diameter is 3 in., what does the word "nominal" imply?

2. Name one application of mechanical tubing.

3. What organization writes most of the materials specifica-

tions for petroleum line pipe and drill pipe?

4. What material does ASTM A-120 cover?

5. Name one grade of line pipe used to carry petroleum from a well to a refinery.

6. You have been asked to weld some pipe in the 1G position. What kind of edge preparation on the pipe joint is required?

7. The welding test asks that you make a weld in the 3F position. What kind of weld joint is it, and what welding position will you work in?

8. You will weld some pipe in the 6GR position. What does the R stand for?

9. What welding technique do you use for steel pipe to be welded in the 2G position? Do you use the forehand or the backhand technique?

10. What is the primary purpose of T, K, and Y connections in a pipeline?

11. What does a mechanical fitting called a reducer do in a pipe?

12. What is the standard angle of the edge bevel provided on pipe from a steel supplier when you don't specify something special?

13. How many tack welds would you use when setting up a 4-in.-diameter pipe for welding? How many tack welds would you use for a 6-in.-diameter pipe?

14. Which welding technique, forehand or backhand, is most commonly used when welding steel pipe? That is, which one will probably give you the best weld on most jobs?

15. When is the forehand technique most likely to be useful when making a groove-welded pipe joint?

Oxyfuel Pipe-Welding Practice

By now you should be able to make correct flame and regulator pressure settings and select the right tip size for any common material and section thickness you will weld with the oxyfuel process. You also should be able to select the correct filler rod for each material. Nevertheless, we'll still give you some guidelines from time to time. The first assignments will be to weld steel pipe in progressively larger diameters and wall thicknesses. We'll also work with you as you develop skills in welding pipe in progressively more difficult positions. See the operator-qualifications tests at the end of the shop session for details on testing your welds.

10-1-S
3-IN. STANDARD PIPE

In this job you will learn tech-

niques for butt welding steel pipe in the horizontal rolled, horizontal fixed, and vertical positions.

Materials

You will need 3-in.-diameter standard weight black steel pipe (steel pipe that has *not* been galvanized) in approximately 6-in. lengths. Select your gas-welding rods from AWS A5.2; use an RG 45 or RG 60 grade rod. If you use acetylene for fixed-position pipe welds, select a tip with a 0.047-in.-diameter drill size hole and use a ³/₁₆-in.-diameter hose. For MAPP Gas, use a tip one size larger (one with a 0.055-in. drill size hole) and the same ³/₁₆-in.-diameter hose.

For rolled welds, use a tip with a 0.055-in. drill size hole for acetylene, and one size larger (a tip with a 0.067-in. drill size hole) for MAPP Gas. The reason is that you will need a little more heat from the tip when welding a rolled section because, unlike working with a pipe in the fixed position, the rolled pipe will keep moving while you work. Since the welding will progress faster, you need a hotter flame to compensate for the fact that you can't adjust your welding speed as easily.

Welding 3-in. Pipe

1. Make a forehand weld on a butt joint on 3-in. pipe in the horizontal rolled position.
2. Make a forehand weld on a butt joint on 3-in. pipe in the horizontal fixed position.
3. Make a backhand weld on a butt joint on 3-in. pipe in the vertical fixed position.

10-2-S
6-IN. PIPE

This work will help you develop the ability to make butt welds on pipe in the horizontal fixed and vertical fixed positions using backhand welding.

Materials

You will need lengths of 6-in.-diameter black steel pipe at least 12 in. long. Select your welding rods from AWS A5.2, using either RG 45 or RG 60 grade welding rods. Since the wall thickness is greater than that in the previous exercise, use larger tip sizes. Use a tip with a 0.076-in.-diameter drill hole size for acetylene, and one size larger (a 0.086-in.-diameter drill hole size) for MAPP Gas for welding pipe in fixed positions. Use a 0.086-in.-diameter drill hole tip size for acetylene, and a 0.098-in. drill size tip hole for MAPP Gas. Use ¼-in.-diameter welding hose for the larger tips to get a higher gas flow volume.

Welding 6-in. Pipe

4. Make a backhand weld on a butt joint on 6-in.-diameter pipe in the horizontal rolled position.
5. Make a backhand weld on a butt joint on 6-in.-diameter pipe in the horizontal fixed position.
6. Make a backhand weld on a

butt joint on 6-in.-diameter pipe in the vertical fixed position.

10-3-S
6-IN. HEAVY-WALLED PIPE

Now you will learn to weld heavy-walled pipe using the backhand technique while making multi-layer butt welds on pipe in the horizontal rolled, horizontal fixed, and vertical fixed positions.

Materials

Use a tip size with a 0.086-in.-diameter drill hole and ¼-in.-diameter hose for acetylene, and one size larger (0.098 in.) with a ¼-in.-diameter hose for MAPP Gas. Use AWS A5.2, grade RG 65 welding rod this time. You will need grade A extra-heavy-walled steel pipe about 6 in. in diameter.

Welding Heavy-Walled Pipe

7. Make a backhand multilayer butt weld on 6-in.-diameter extra-heavy-grade steel pipe in the horizontal rolled position.
8. Make a backhand multilayer butt weld on 6-in.-diameter extra-heavy-grade steel pipe in the horizontal fixed position.
9. Make a backhand multilayer butt weld on 6-in.-diameter extra-heavy-grade steel pipe in the vertical fixed position.

10-4-S
GALVANIZED PIPE

This work will help you learn how to weld galvanized steel pipe in the horizontal rolled, horizontal fixed, and vertical positions.

Materials

Use 4-in.-diameter galvanized steel pipe. Select your welding rod from AWS A5.2 grade RG 45 or RG 60, using ³⁄₁₆-in.-diameter welding rods. Your flux should be suitable for use on galvanized steel. Use a smaller tip size than before to reduce heat input into the steel to prevent burning off too much of the zinc coating. Use a 0.047-in. drill size tip for acetylene or one size larger (0.055-in. drill size tip) for MAPP Gas when

working on the fixed-position welds. Use a 0.055-in. drill size tip for acetylene or a 0.067-in. drill size tip (one size larger) for MAPP Gas when working on rolled welds.

Welding Galvanized Steel Pipe

10. Make a forehand or backhand welded butt joint on 4-in.-diameter galvanized steel pipe in the horizontal rolled position.
11. Make a forehand or backhand welded butt joint on 4-in.-diameter galvanized steel pipe in the horizontal fixed position.
12. Make a backhand welded butt joint on 4-in.-diameter galvanized steel pipe in the vertical fixed position.

10-5-S
YELLOW BRASS PIPE

Now you are ready to try welding yellow brass pipe or tubing.

Materials

You will need 4-in.-diameter yellow brass pipe or tube and a manganese-bronze filler rod about ³⁄₁₆ in. in diameter. Because of the filler metal you are using, this actually is a braze-welding job, and not a full fusion-welding job. You will need flux suitable for brass. Your welding tip should be fairly large, about 0.076-in.-diameter drill hole size for acetylene and one size larger, 0.086-in.-diameter drill hole size, for MAPP Gas because this material is more thermally conductive than steel.

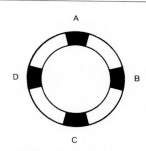

FIGURE 10-S1 Location on pipe joint from which test specimens that are taken in school welder qualification tests depend on your instructor's suggestions and your ability to make the joints.

Braze-Welding Brass Pipe or Tubing

13. Make a forehand braze-weld on a butt joint on 4-in.-diameter brass pipe in the horizontal rolled position.

14. Make a forehand braze-weld on a butt joint on 4-in.-diameter brass pipe in the horizontal fixed position.

15. Make a backhand braze-weld on a butt joint on 4-in.-diameter pipe in the vertical fixed position.

10-6-S OPERATOR-QUALIFICATION TESTS

These tests closely follow those you would have to take to get a small-diameter pipe-welding job (as opposed to a job joining line pipe, which is much larger). In an actual test, you may have to make the welds with an arc-welding process such as TIG or stick electrodes (or both), with gas welding, or with both gas and arc welding. It all depends on what kind of job you are qualifying for. Nevertheless, the qualification test procedure will probably be the same for any process.

Operator-Qualification Test

Make two connections on grade A extra-heavy steel pipe 6 in. in diameter. One connection will be made in the horizontal fixed position and the other one will be made in the vertical fixed position. The length of the completed piece must be not less than 6 in., and preferably 10 to 12 in., particularly if tensile tests are to be made on your welded joint.

At least four specimens taken at 90° around the pipe should be cut by milling or sawing, or by flame-cutting from each welded pipe. Specimens A and B, Fig. 10-S1, should be subjected to guided root-bend tests and specimens C and D should be subjected to the face-bend test.

Figure 10-S2 indicates the points from which the various test coupons are to be cut—a minimum of four must be tested. These are the two face-bend and two root-bend test specimens. If tensile-testing facilities are available in your welding school, tensile tests should be made. In the case of the vertical fixed-position welds, the root-bend test in Fig. 10-S2 should include the starting-finishing point; the other test locations should be distributed about the pipe in the relationship shown in Fig. 10-S2.

The face- and root-bend test specimens should conform to the specifications shown in Fig. 10-S3. It will be necessary to provide a plunger and die that will accommodate a ½-in. specimen when making the bend tests. The plunger should have a 2-in. diameter while that of the die should have a radius of 3⅛ in.

Tensile-test specimens should be prepared as in Fig. 10-S4.

Failure to pass the test simply means that you don't have the skills, yet, for pipe welding.

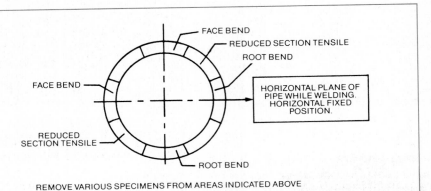

FIGURE 10-S2 Locations on pipe weld from which test samples are cut for many AWS and API pipe welder qualification tests are described in great detail in their welder qualification procedures.

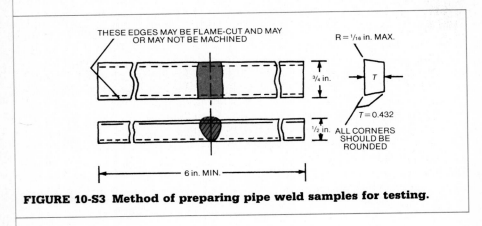

FIGURE 10-S3 Method of preparing pipe weld samples for testing.

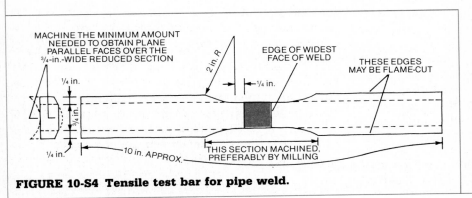

FIGURE 10-S4 Tensile test bar for pipe weld.

PART 3

Electrical Processes

Before learning arc welding, you must learn about electricity. We won't turn you into an electrician, but any skilled arc welder must have a good working knowledge of electricity, just as a skilled oxyfuel welder must know a lot about welding fuel gases and oxygen.

After your introduction to electricity, we will teach you how to use it safely. Then we will show you how to cut and remove metals with electricity.

Most welding textbooks start you with stick electrodes. While your instructor may also want to do that, we firmly believe that you should capitalize on your present gas welding skills. And there's no better way to do that than to learn tungsten-inert-gas (TIG) welding first. The process shares many manual techniques with oxyfuel welding, and TIG welders also earn excellent incomes.

In many ways plasma welding is an extension of TIG, so we recommend that you learn that process next. Only then turn to stick electrodes which require entirely new manual skills.

After stick electrodes, you will find that the techniques and technology in the rest of this section are easy to learn. Enjoy learning the rest of the major arc-welding processes. You are on your way to becoming not only a skilled welder, but a highly skilled one.

Learn Electricity before Striking an Arc

Most arc welders have a good practical understanding of electricity for two reasons. They need it to operate their equipment and to keep from hurting themselves. Understanding what electricity is and how it works isn't all that difficult. Let's start with what it is.

11-1
WHAT ELECTRICITY IS

To understand electricity, you have to understand electrons, and that means knowing a little about atoms.

The Atom

All atoms are made of a tiny nucleus with a positive electrical charge. Since the complete atom has no electrical charge (unless it is ionized), something must balance the positive charge of the atomic nucleus. That something is negatively charged *electrons* (Fig. 11-1).

For every positive unit charge on the nucleus of every electri-cally neutral atom, there is one electron with an equal but opposite electric charge. Naturally occurring elements can have from 1 electron (hydrogen) to 92 electrons (uranium) spinning around each positive atomic nucleus. Each negative electron flies around the positive nucleus in an orbit, something like the way the planets of our solar system orbit the sun. The difference is that electrons are not solid objects like planets. They act more like a gas cloud. A lot of things happen as a result of that.

One is that the electrons in electrical conductors such as copper and aluminum, and even in less-conductive metals such as iron or chromium, are not very tightly attached to their positive nuclei (Fig. 11-2). The outermost electrons of metal atoms can be knocked loose without using much energy. When electrons are turned loose in this way, an electric current flows.

How Metals Conduct Electricity

Metals conduct electricity because they have relatively loose outer electrons. Some metals such as copper or silver have outer electrons that are easier to detach from their atoms than other metals. That makes them even better conductors of electricity than iron or chromium. However, all metals conduct electricity better than nonmetals such as sulfur, phosphorus, or oxygen.

Insulators and semiconductors

The opposite of a conductor is an *insulator*. Insulators such as rubber, plastic, and ceramic are made mostly of nonmetallic atoms, or compounds of metals and nonmetals such as aluminum oxide, with electrons that are more tightly attached than those of metals. It takes much more energy to remove electrons from their atoms. That is why insulators don't carry an electric current as easily as metallic conductors do. When the outside electrons of an insulator atom or molecule are finally

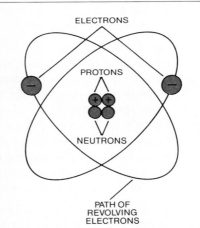

FIGURE 11-1 Structure of a helium atom in stylized form. Two electrons with negative charges are in orbit around nucleus which has two positive charges and an atomic weight of 4, provided by the two positive protons and two chargeless neutrons inside the atomic core. Real electrons in actual atoms act more like a diffuse cloud than little pinpoint particles.

stripped off by very high energy, the insulator is said to "break down" or "leak."

Semiconductors are intermediate in conductivity between metals and nonmetals. Silicon is a very important example of a semiconductor. Computers use thin chips made out of silicon with other elements laid over the material or added to it as controlled impurities. Welding power sources also use semiconductors to turn alternating current (ac) into direct current (dc). These devices are called silicon-controlled rectifiers. You will soon learn how rectifiers work.

Atoms in solids Metal atoms in a solid usually are held together in neat geometric arrays. One of the most common arrangements for iron atoms, for example, looks like a pile of cannonballs (Fig. 11-3). This pattern is repeated throughout each iron crystal. It's typical of a crystal-lattice struc-

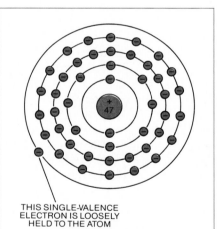

THIS SINGLE-VALENCE ELECTRON IS LOOSELY HELD TO THE ATOM

FIGURE 11-2 Simplified drawing of silver atom has 47 negatively charged planetary electrons, each with a negative charge. The nucleus has a plus-47 charge, giving the atom an overall neutral charge. The very outermost electron on silver (and other good-conductor) atoms is held loosely. When stripped away, the mobile electron becomes part of an electric current. The more electrons that are on the move, the higher the current.

ture for the common low-temperature variety of iron called *alpha iron*. Alpha iron makes soft, ductile iron crystals. When iron or steel gets very hot, the iron atoms rearrange themselves into another pattern (the face-centered cubic crystal lattice of austenite), but they are still close together.

Atoms don't hold still in the crystal lattice. They always vibrate more or less in place. That's what heat is and what temperature measures. The hotter the atoms, the more they vibrate and the higher the temperature. But as long as the temperature is not too high, the atoms remain in place in the lattice. Therefore they still stay close together and it's easy for them to pass loose electrons back and forth. That's exactly what happens when electricity flows through a metal conductor.

If the atoms weren't stuck together in the crystal lattice of the metal, they would flow and the metal would melt (high heat energy means that atoms are vibrating faster than they would at low heat energy). That explains why metals get less conductive when they get hotter and more conductive when they get colder and stay closer together. When the atoms get very hot, they vibrate so fast and far that they can't stick together in the crystal lattice any more and they start to flow. That's what happens when the metals melt and become liquids.

Since the metal atoms in a liquid are not as close together as metal atoms in a solid, the liquid takes up more room than the solid and doesn't conduct electricity as well. The reverse occurs when metals cool and become solid. That's why cooling weld metal shrinks and builds up stresses and why metals expand when they get hot and contract when they get colder.

Conductors

Picture in your mind a wire conductor that looks like a big tube

packed full of loose sand (Fig. 11-4). The sand grains are the hard atomic nuclei of atoms in a crystal lattice. The spaces between the sand grains are filled with loose electrons. Being inside the tube, the electron gas won't leave the sand but can still move around. The electrons can move around inside the metal crystal lattice but they won't jump out of the metal conductor unless they are pushed very hard. If the electrons do jump out of the conductor, you see a spark.

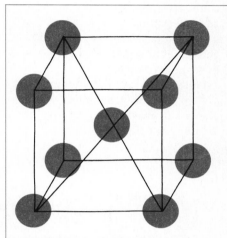

FIGURE 11-3 Typical crystal lattice for low-carbon steel or pure iron. If the iron atoms at the corners of the cube were drawn close together, they would stack up neatly like cannon balls, leaving very little space between them. This "close-packing" structure is actually what happens in some real metals. The close-packed cubic crystal lattice is only one of several atomic arrangements that are possible in the iron system, depending upon the temperature and alloying elements used. The atomic arrangement of the iron lattice helps explain many of the properties of ferritic iron alloys and low-carbon steels. For example, carbon atoms are much smaller than iron atoms. They will jam up this lattice, making the steel harder and stronger (like putting sand in gears). That is why even low-carbon steels can be made to be much stronger than pure iron.

If we pump more negatively charged electron "gas" into one end of the conductor tube, an equal amount of the electron gas will flow out the other end if it has somewhere to go, even though the atomic nuclei "sand grains" in the crystal lattice stay in place. The total amount of electrons still inside the tube always remains the same because electrons that flow out one end are always replaced by electrons flowing in at the other end, so that the overall electrical charge produced by all the atoms and electrons in the conductor stays neutral, or has a zero electrical charge. That is why a conductor can have no electrical charge even when electricity is flowing through it.

Gases also can conduct current if they are ionized (under enough electrical pressure to remove some of the outer electrons from the normally neutrally charged gas atoms or molecules). Fig. 11-5 shows how positive and negative gas ions (electrically charged gas atoms) and negatively charged electrons can conduct current in an ionized gas.

11-2
WHY ELECTRICITY FLOWS

Who pumps the electrons into a conductor? You do when you turn on a light switch or strike a welding arc. Where do the electrons come from? They usually come from the local electrical power company, unless you generate your own electricity. In either case, the electrons are produced by a generator, an alternator, or a battery at the other end of the conductor. The other end could be only a few feet from you if you

FIGURE 11-6 Triple the cross-sectional area (not the linear diameter or radius) of an electrical conductor and three times as many electrons (and three times the current) can flow through the conductor.

are using a gas-driven welding generator or alternator, or miles away if you buy your electricity from the local utility.

Electrical Resistance

Flowing electricity has a lot of other characteristics that are similar to a gas flowing in a tube or a hose. You already know that if a gas-welding hose is too narrow, the flow rate of the fuel gas or oxygen through it is reduced. The flow rate is reduced not only because the hose is smaller, but also because the frictional resistance in a long hose is more than in a short hose. You also should know that the resistance doubles if the length of the hose doubles.

The same is true of electricity flowing through a conductor (Fig. 11-6). If you double the area of a conductor, you can pump twice as much electricity through it. If you double the length of an electrical conductor, you double the resistance to current flowing through it. If you triple the conductor's cross-sectional area, you triple the amount of current (electrons) that can flow through it. If you cut the area of the conductor in half, you cut the amount of electricity that can easily flow through it. If you cut the length of the conductor in half, you cut the electrical friction or resistance in half and the same number of electrons can flow through it under less electrical pressure.

Very important electrical cir-

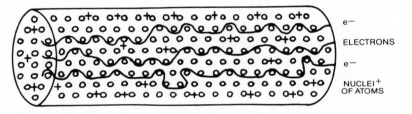

FIGURE 11-4 Think of electrons flowing through the crystal lattice of a conductor as a gas flowing through a sand-filled pipe. Each electron moves only a very short distance before changing direction, but the net flow of electrical current is substantial when billions of electrons are moving.

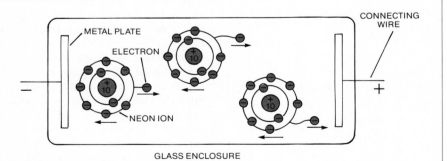

FIGURE 11-5 Electric current in a gas is carried by both negatively charged electrons and positively charged ions. The negative electrons are attracted to the positively charged plate and the positive ions are attracted to the negatively charged plate. This electrified gas is called a *plasma*. A welding arc is a typical plasma.

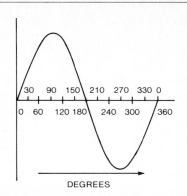

FIGURE 11-7 A waveform can be measured with a simple 360° scale. One complete cycle of an alternating-current waveform is often thought of as having 360 electrical degrees. This idea makes it possible to calculate a lot of the characteristics of alternating current using trigonometry (such as sine, cosine, and tangent functions of the electrical angles). Electricians often use trigonometric formulas in their work with ac current.

cuit elements called *resistors* actually add electrical "friction" to circuits. They are important because resistors keep circuit conductors from being overloaded with current which otherwise would flow too much and make the circuit get hot and maybe even melt. That's what happens when you have a short circuit.

Resistors have another even more important use besides preventing short circuits. You will see what it is when we explain the relationship between voltage,

amperage, and resistance in electrical circuits.

The only way to increase the flow rate of a gas through a narrow hose is to turn up the pressure at the other end. Again, exactly the same is true of electricity. If a gas hose is too long (too much resistance), you have to increase the pressure at the regulator to get more of the gas through the longer hose to maintain a given flow rate. The same is true of electrical circuits. To maintain a given amount of electricity (amperes) through a conductor as its resistance increases, you have to increase the electrical pressure, that is, the voltage.

Electrical current flow is measured in *amperes* (A). Electrical pressure is measured in *volts* (V). Electrical resistance is measured in *ohms* (Ω). We'll tell you about these and other units used to measure electricity later on in this chapter.

Alternating and Direct Current

The gas or water flowing through hoses or pipes normally flows in one direction, from the high-pressure end to the low-pressure end. That's also true of direct-current electricity, which often is simply called *dc*. Direct current always flows in one direction. The electrons travel from a negative source of extra electrons to a positive area that's short of electrons and needs them to stay electrically neutral.

If you ever sat on a beach and watched the waves hit the shore,

you know that the water first flows in, then flows back out. It also rises and falls as waves. When electrons flow first in one direction and then in the opposite direction in a conductor, it's called *alternating current* electricity, or simply *ac*. First the electrons flow in one direction, and then they reverse their direction and flow the other way. As a result, the direction of flow of positive and negative changes with each cycle of the wavelike electrical flow.

The reversal is not instantaneous. The number of electrons flowing in any one direction at a given time in ac builds up to a peak and then declines. Then they build up, peak, and decline again. This produces a water-wave-shaped curve of positive and negative electrical flow when plotted on graph paper against time (Fig. 11-7).

You are quite familiar with the number of cycles of alternation that occurs 60 times a second as the ac in your house flows into a light bulb, a toaster, or a TV. That's an example of 60-cycle ac. The electrical name for cycles is *hertz* (Hz), so the technical way to describe it would be 60-Hz ac.

Electrical Charge

Electricity has one characteristic that gases (except gas ions, which we'll describe later) don't have. As you already know, the electrons have an electrical charge and by convention that charge is called *negative*. If something has too many electrons it also has a negative charge. If the same object has a shortage of electrons, it has a positive charge (Fig. 11-8). The larger the surplus of electrons, the higher the negative charge. The larger the deficiency of electrons, the larger the positive charge.

Benjamin Franklin is responsible for the negative and positive labels put on electricity. When he was flying kites in thunderstorms to collect electricity, he decided that the positive charge was the electricity and the negative

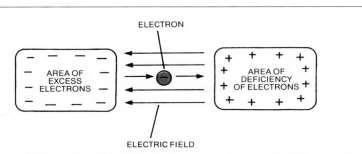

FIGURE 11-8 Negatively charged electron moving through an electric field. Part of the electron's energy is converted into other forms of energy, mostly heat and light, when you are arc welding.

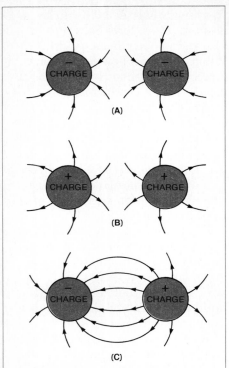

FIGURE 11-9 Electrical fields between charges. (A) and (B) are like charges. They repel each other. (C) The dissimilar charges attract each other. These are examples of static electricity (there is no net current flow).

charge was the lack of it. It turned out that electricity was just the other way around. People have been using the terminology so long that everybody understands it.

What is confusing is that the arrows that show the direction of current flow on many electrical-circuit drawings still point the opposite direction from the actual electron flow. This holdover from our Revolutionary War times shouldn't bother you because you are not learning to be electricians. You don't have to make the distinction between current flow and real electron flow.

Static electricity The discussion of electrical charge leads us directly to a third kind of electricity—static electricity (Fig. 11-9). Static electricity is either negatively charged (too many electrons) or positively charged (not

enough electrons), and it doesn't flow in a current. That's why it's called *static* electricity.

Static electricity is used by photocopy machines to make charged particles (graphite) produce images of your documents. An electrical spark is made of static electricity. Ball lightning is a static electric spark on a grand scale. Static electricity is even used in circuits with flowing electric current when the electricity needs to be stored for a short while. (That's the job of a *capacitor,* one of many basic elements in electrical circuits.)

You know static electricity personally from rubbing your feet on a carpet and then touching a metal doorknob. What you actually did was rub electrons off the rug and built up a static electric charge (excess electrons) in your body. That slight negative charge produced a tiny tingle on your finger when you touched a conductor like a doorknob. A tiny static electric spark actually jumped across to the knob and then redistributed itself in the metal. You felt it happen when your finger got a shock.

Static electricity can make part of an object negative and part of it positive, even though the object as a whole has no charge (Fig. 11-

10). If you hold a positively charged rod next to an object, the positive charge will pull the electrons in the object toward the positive charge. That leaves the other side of the object temporarily short of electrons. While holding the positive rod near the object, you can even make the object (like an insulated doorknob) positive. First you force the electrons to one side, and then you bleed them off with a conductor leading to a ground. Then you take away the positive rod that moved the electrons around in the doorknob, and the knob will remain with a positive charge.

Capacitors
A capacitor has a tiny gap filled with a nonconductor, or even air, between two conducting metal plates (the symbol for a capacitor in a circuit actually looks like that; it has two facing plates and a gap between them, with current leading into one plate and out the other). The electrons stop at the incoming plate and pile up, making it negative (which forces electrons away from the plate on the opposite side, making the second plate positive).

The voltage, or electrical pressure, on the incoming side of the capacitor builds up until it is

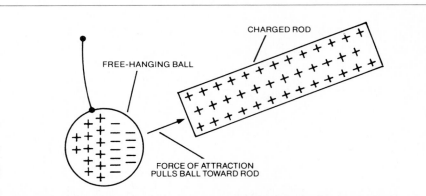

FIGURE 11-10 Inducing a static electrical charge. When the ball touches the rod, electrons will be transferred (flow) to the rod, leaving the ball with a net positive charge when the rod and ball are separated again. Very large static electric charges can develop in welding power sources that are not properly installed or protected. This static electricity becomes high-amperage direct current flowing through you if your equipment is not properly designed, maintained, and grounded. Making sure that your power source is properly grounded is your responsibility.

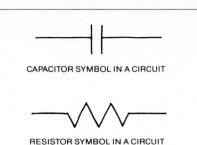

CAPACITOR SYMBOL IN A CIRCUIT

RESISTOR SYMBOL IN A CIRCUIT

FIGURE 11-11 Capacitors in an electric circuit act very much like valves. They won't let direct current flow until the voltage or electrical pressure is high enough. Resistors in an electric circuit control the amount of current that the circuit can carry. We will tell you about them later on.

strong enough to force electrons to jump the gap in the capacitor. Then the current starts flowing in the circuit again. Capacitors are used for many reasons. One is using a capacitor as an electrical valve. Current only flows through the circuit when the voltage is high enough to make electrons jump the gap in the device, thus forcing the electrons through the capacitor.

Figure 11-11 shows the electrical symbols for both electrical capacitors and electrical resistors.

Electrical Grounds
Static electricity easily becomes flowing dc as soon as it has some place to flow to. The current that a built-up charge of static electricity can deliver when it's released can be quite large in a lot of electrical equipment. The metal case of a welding machine can build up a static charge that can knock you off your feet if you happen to touch it . . . unless the cabinet is *grounded*. The same is true if you have a fault in the machine's wiring that short-circuits the current (sends current flowing through a path of least resistance . . . which could be you).

✖ CAUTION: Always ground your power source. Otherwise, if a hot wire touches it at the same time you do, you'll be hurt, or killed.

What is grounding? It means providing a conductor that will bleed off any excess current, static

or otherwise, into the ground. The earth can absorb a lot of electricity simply because it is so big. If it can hold a lightning bolt, the earth can take any amount of current you will ever feed it.

The term *ground* as in the "ground cable" used to set up arc-welding equipment also means "providing a path of low resistance through a good conductor" so that the electricity will have a complete circuit to travel through (Fig. 11-12). If a workpiece is not attached to the ground cable, the circuit cannot be completed. However, the ground cable isn't enough for complete safety. The chassis of the welding machine should also be grounded. Welding machines all come with special cables or connectors for this purpose.

If you look at the plug from the power source that goes into the wall behind the welding machine, the plug will have three prongs. The first two provide a round-trip circuit for electrical current. The third prong is the ground wire for the machine. You'll find the same kind of three-pronged plugs on electric drills, saws, and other shop equipment.

The ground wire won't work unless an electrician has hooked up the wall receptacle to a ground lead behind the wall. Never weld from a plug that has not first been inspected and tested for use with welding equipment by a certified electrician. If the ground connection on your power source wall plug is damaged or missing, don't use the machine. Call an electrician.

There is another kind of cable used in welding. It's also called the ground, or ground cable. One end of that cable hooks to your workpiece or the metal table it is clamped onto, and the other end fastens onto your welding machine. It completes the electric circuit from the welding power source to your workpiece and back to the power source. But that circuit is not completed until your electrode touches the work.

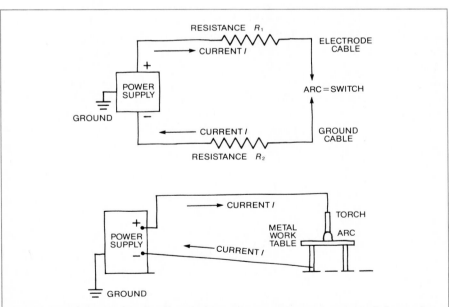

FIGURE 11-12 A complete welding circuit requires a ground from the power source to leak off stray electrical charges or current that might otherwise build up to dangerous levels in the machine cabinet and harm you. The welding electrode acts like a switch. When the electrode is lifted, the circuit is broken. When the electrode touches the work, the circuit is completed.

When the circuit is not complete (the ground cable to the work is not attached, for example), you can turn the juice up as high as you want at the welding machine and nothing will happen when you try to strike an arc. The electricity has no continuous pathway from the the starting point and back again; the circuit is not complete. The ground cable from the workpiece to the power source in Fig. 11-12 completes the circuit that lets the electricity flow when the circuit is closed. When you strike an arc, you close the switch in the circuit. That completes the circuit, the welding current flows, and you can weld.

If you were to measure the voltage between your electrode holder and the ground cable with a voltmeter, without welding, you would get a very high voltage reading. That is the *open-circuit* voltage of the power source, a very important machine rating. It is the highest voltage the machine will produce at a given control setting on the machine panel. It's also the most dangerous voltage if you connected yourself between the ground cable and the power cable, instead of connecting the workpiece to them. While open-circuit voltage ratings for welding power sources may not sound very large compared to your home wall circuit (you will see numbers like 70-V or 80-V open-circuit voltage, or OCV), there is much more electricity (amperes) in that circuit than in the electricity coming from a wall plug at home. That's what makes it dangerous.

Flowing Electricity and Heat
Any break in the circuit will stop the current flow; an open switch can do it. A broken cable can do it. You do it by pulling your welding electrode away from contact with the workpiece. An electrode acts just like an overheated switch—one with poor conductivity and high resistance to the excess welding current flowing through it. When you lift your

electrode off the work, the current stops flowing. When you touch the electrode to the work, again, the electric current flows like crazy.

So much current flows through a welding circuit that the electrode switch gets overheated and melts, along with the base metal near it. That, roughly speaking, is how all kinds of electric welding work, spot welding as well as arc welding. High resistance and high current flow produce high heat energy to melt metals.

The main difference between arc welding and resistance welding is where the heat to make the weld builds up. In resistance welding the electrodes are large and have very high conductivity. The current flows rapidly into the workpiece. But the contact between two pieces of sheet metal has very low conductivity and high resistance. That's where the heat due to resistance forms and where the weld is made. In stick electrode welding, the heat builds up at the tip of the electrode in a very narrow gap between the electrode and the workpiece where the welding arc forms.

Direct current flows one way. Alternating current flows back and forth. Static electricity is much like the gas in a cylinder with the regulator shut—it doesn't go anywhere until you release it. Welding electrodes are like switches. And arc-welding power sources should always be grounded.

11-3
CHARGED IONS
Before we tell you how to work with volts, amperes, ohms, coulombs, and other electrical units we'll describe later, let's take a quick look at what's left of an atom after we strip off (or add to) some of its electrons. An atom that has an electrical charge (positive because it's missing some electrons or negative because it has too many) is called an *ion*. Ions are very active (and reactive) because they have a need for more electrons if they are positive, or need to get rid of extra electrons if they are negative.

Argon, helium, and other atoms that you will use for shielding gases in GMAW and GTAW (MIG and TIG) become ionized by passing through the extremely energetic welding arc that strips off some of their electrons. If the ions are negative they get rid of their ionization energy by immediately dumping their extra electrons, or if they are positive they grab extra ones from the surroundings (Fig. 11-13).

This electron-swapping process produces a lot of heat energy (the same amount of energy that was required to ionize the shielding-gas atoms in the first place). Therefore arc-welding shielding gases not only protect molten metal from the air, but their gas ions help pass along a lot of the heat from the arc into the weld metal. When we get to GMAW and GTAW, you will find that the

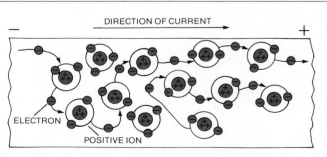

FIGURE 11-13 Current in a plasma. Free electrons flow in one direction while positively charged ions flow in the opposite direction. The result is current flow through an ionized gas, called a *plasma*. The electrical welding arc is an example of a plasma.

depth of penetration of a weld into a workpiece is strongly affected by the type of shielding gas used. It's also strongly affected by the direction of electron current (and positive ion) flow.

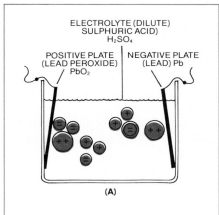

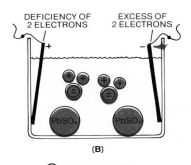

OXIDE ION (O⁻⁻)

LEAD ION (Pb⁺⁺)

HYDROGEN ION (H⁺)

SULPHATE ION (SO₄⁻⁻)

FIGURE 11-14 Lead-acid storage battery cell. (A) Chemical reaction between sulfuric acid and lead produces positive and negative ions so that current flows, charging one battery plate and terminal positive and the other negative. (B) Sooner or later all the lead in the negative battery plate is used up and the current no longer flows. The battery is worn out. Sometimes the reaction is partially blocked as the charge on the lead plates wears down. Adding more water, more sulfuric acid, or even recharging the battery, will help return it to its original, charged state.

You now know why some shielding gases require more energy to become ionized than others, and why they release more heat when they deionize next to the base metal. The gases that require the most energy to become ionized, also give up the most heat as energy of ionization to produce much deeper weld-metal penetration. That's an important factor in selecting a shielding gas for gas-shielded arc-welding processes.

Plasmas

The shielding-gas ions carry current just like solid metal conductors because they have positive charges and are surrounded by a cloud of negative electrons. This ripped-apart, ionized matter is neither a solid nor a liquid nor a gas. It's a fourth state of matter called *plasma*.

The atmosphere of the sun is a plasma. The combination of gas and the high-voltage arc used in plasma flame-cutting is a plasma. That is why plasma welding is so hot and plasma cutting cuts any metal literally by vaporizing it instead of oxidizing it. Since a plasma has both negative and positive particles, it can carry dc. The positive atomic ions will be attracted toward a source of negative charge because they want extra electrons to make them neutral again. The electrons will travel toward a positive charge because they want to neutralize it.

Current-Carrying Ions

Your automobile battery does the same thing, except that the ions are in solution in the water that you have to add to make a car battery work (Fig. 11-14). The ions in an auto battery are the positive and negative parts of sulfuric acid molecules. The negative part of each sulfuric acid molecule is the sulphate ion (SO₄⁻⁻) with a negative charge of 2, because it

has two extra electrons attached to it.

The positive part of the sulfuric acid molecule is simply the two positively charged nuclei of the hydrogen ion (H⁺ and H⁺). Each positive hydrogen-ion nucleus is nothing more than a single proton with a charge of +1. There must be two hydrogen ions (+2) in each molecule of sulfuric acid to balance the one sulfate ion that has a charge of −2.

These positive and negative ions travel through the battery water to the negative and positive terminals, respectively, of the battery. The positive ions want to borrow electrons from the negative terminal and negative ions want to get rid of their excess electrons at the positive terminal.

That is how electrical current flows through water or gas or any other fluid that contains ions. These ions carry electricity with them that was previously stored up in the battery. They release the energy from the battery terminals. But they only flow and carry current when an outside circuit is completed by joining the two battery terminals to a circuit. When nothing is attached to the outside battery terminals, the circuit is not completed and no current can flow. Batteries only produce dc current.

Actually, there's a lot more to batteries than we have described here, but the point we are making shows two things about all electrical circuits, whether they have ac or dc flowing in them:

■ Current will not flow if both terminals are not connected to each other through some kind of continuous electrical device, because the circuit isn't completed.

■ If the external device or circuit has a very low resistance to the flow of current, a very large amount of current will flow through it. If you put a short copper wire directly across both battery terminals, so lots of electricity (electrons) will flow through the wire all at once from the negatively charged terminal to the positively charged terminal, the copper wire will probably overheat and melt. That's a direct current *short circuit*.

Short Circuits

Any time an electric circuit that is carrying a lot of electricity in it is "short," it means that there is almost no resistance to the flow of electrons (Fig. 11-15). Most likely, a fuse will blow out (if the line or device has a fuse); otherwise the wires and other electrical circuit components will overheat and burn up from the excess current flow. Short circuits start fires.

11-4 MEASURING ELECTRICITY

Volt, ampere, and ohm were first studied respectively by Alessandro Volta, André Ampère, and Georg Ohm, some of the first people to study electricity. These electrical properties have close parallels with the properties of gases and other fluids. The electrical unit that is most unlike gases is the basic unit of electrical charge, called the *coulomb* after Charles de Coulomb.

Coulombs

A coulomb of charge is the amount of electrical charge possessed by 6,250,000,000,000,000 electrons (more or less). The coulomb is not an electrical unit you will use very often, but being the base unit of electric charge, it is very useful for defining the other electrical units like ampere and volt.

NOTE: The coulomb is the base unit for electrical charge.

The abbreviation for electrical charge is *Q*. The coulomb has its own abbreviation, C, because other, smaller units also are used to measure electrical charge.

The coulomb has another very handy use. It's one of several units used to measure the amount of charge that can be stored by capacitors. A capacitor in this instance operates something like a gas cylinder with a spring-loaded relief valve. It will store up a static electric charge and then release it in a burst of electrons as current.

Amperes

If you want to know how much traffic flows on a highway, a simple way is to count the number of cars passing by a spot on the road in a given time, say an hour. On a busy wide-lane highway you might count 850 cars per hour, while on a back country road you could easily count less than one car per hour. The flow of electrical current is measured the same way. Instead of automobiles, electric charge units (coulombs) are counted. Instead of hours, the time unit is a second. Therefore the amount of electricity flowing by a given point is measured by the number of electrical charges per second (coulombs/s) that pass by the measuring instrument. It's exactly the same idea as gal/s to measure water flow or ft^3/s to measure the flow of a gas.

Welders refer to electric current by another name, the *ampere*. An ampere is simply one coulomb of charge flowing for one second. The abbreviation for ampere is A, as in "38 A of welding current." That's something that you set on your power source before you start arc welding. Most people will write A or amperes, but they say "amps."

Direct current and alternating current produce amperes of electricity only because they flow over a period of time. Static electricity has no amps (A) because it's not flowing. That's why the amount of static electricity has to be measured by coulombs (C).

Welding power sources often are rated by the ac or dc amperage that they can produce. You might not want to use a 200-A machine because you will be doing heavy-duty work on thick plate and will need a 400-A or maybe even a 500-A or 600-A machine. Of course when you talk about the machine you want you will call it a "400-amp," "500-amp," or "600-amp" machine.

Some useful equations While amperes (A) measure current, the general symbol for current is *I*, time is *t*, and electrical charge is *Q* [even though it's measured in coulombs (C)]. Since electrical current flow is equal to charge units per time unit, then for any

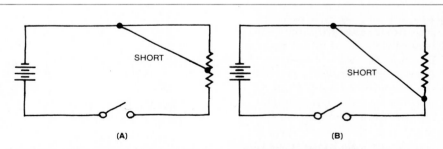

FIGURE 11-15 Short circuits. (A) A partial short circuit that bypasses most, but not all, of the resistance in the circuit. When the switch is closed, high current will flow. (B) A dead short circuit has no resistance. When the switch is closed an extremely high current will flow and overheat the circuit. It may even melt the wires or start a fire if the circuit is not protected by a fast-acting fuse.

electrical circuit, no matter what units are used to measure it:

$$I = \frac{Q}{t} \quad \text{or} \quad Q = It \qquad \text{(Eq. 11-1)}$$

If amps, coulombs, and seconds are used, the equation becomes:

$$A = \frac{C}{s} \qquad \text{(Eq. 11-2)}$$

NOTE:

$$\text{current} = \frac{\text{electrical charge}}{\text{time}}$$

Using Eq. 11-2, one ampere equals one coulomb divided by one second. That's the definition of an ampere. Also, 100 C flowing for 25 s equals 4 A. And 5 A of current flowing for 10 s equals 50 C (using $Q = It$) because 50 C = 5 A × 10 s.

Volts

Amperage measures the amount of electrons flowing in a circuit. Voltage measures the electrical pressure that causes the current to flow.

NOTE: Volts measure electrical pressure.

Voltage also is called *emf* or electromotive force. It's also called the *potential difference* because voltage can only be measured as the difference in electrical pressure between two points (like your ground cable and your electrode holder). Voltage is always a *difference* in pressure just like psig on a cylinder gauge is a difference in pressure between the gas inside and the air outside of the cylinder. All of these terms mean the same thing . . . voltage. The symbol for voltage is V.

Just like the gas in the sealed cylinder, the electrical pressure (voltage) may "potentially" exist, but it's not yet used. The voltage of a static electric charge is an example of potential energy. A 100-lb workpiece sitting on a 5-ft-high workbench has the potential energy to squash your foot. As you already know from the chapter on mechanical properties, the workpiece carries out that potential, it becomes kinetic energy,

only if it drops on your shoe. And the kinetic energy force that it delivers is 100 lb × 5 ft = 500 ft · lb. You've seen figures like that used to measure the toughness (impact energy) of metals. That's an example of mechanical force. Voltage is an example of electrical forces.

Voltage applies both to static electricity and a flowing electrical current just like a sealed cylinder still gives you a pressure reading. The electrical equivalent is static voltage. The gas flowing in a hose also gives you a pressure reading. The electrical equivalent is the voltage of current flowing in a conductor. An example of a static voltage is the potential difference between a source of static electricity and a conductor that hasn't yet been connected to it to drain off the electrical charge. That commonly occurs in capacitors as well as door knobs next to deep-pile rugs.

Another example of a potential difference is a 110-V wall circuit before you plug something into it. The electrical-pressure difference between the 110-V line and the earth (a ground) is 110 V. But since nothing has been plugged in, the voltage is only a potential difference; it hasn't been used yet.

Voltage can be dynamic, as when you measure the electrical pressure in the same 110-V wall circuit *after* you plug something into it and *while* you use the electricity. In this last example, the flowing current has 110 V of electrical pressure pushing it through the device you connected to the wall plug to complete the circuit.

If you look closely at the cord attached to anything you plug into a wall, it will have at least two electrical lines. That shows you that your toaster or light bulb is only one end of a long electrical circuit or loop, the other end of which is at the power company.

Ohms

Coulombs (C) is a measure of static electric charge (Q) when the current is not flowing. Amperes

(A) measure the amount of current (I) flowing per second. Voltage (V) measures the electrical pressure between two points (whether the current is flowing or not). Ohms (Ω) measure electrical resistance (R). The Greek letter omega (Ω) stands for ohms. But just what is this electrical resistance that ohms measure?

Electrical resistance is the opposition to current flow that a device or conducting material produces. It is just like friction in a pipeline. It trys to keep the electricity from flowing. Resistance is directly proportional to the length of the conductor. If you double the length of a welding cable, you double its resistance. Exactly the same is true of a welding hose with gas flowing through it.

The electrical pressure we call voltage has to push against this electrical friction force to make the amps of electrical current flow in the circuit. In the same way, you have to increase the pressure on a gas hose if you want to get gas through a line against increased friction. Welding cable often is rated on the basis of its resistance per 1000 feet, measured in ohms (Ω).

How much resistance is one ohm? A 0/4 American Wire Gauge (AWG) copper welding cable (0.875 in. in diameter) that is 3.5 mi long has a resistance of about one ohm. You can see from that figure that the copper welding cable has a very low resistance. (We'll show you all the AWG wire gauge sizes at the end of this chapter.)

The higher the cable's resistance, the more voltage you need to push a given amount of current through it. Resistance also plays an important role in making welding current produce the heat that makes arc welding possible. Just like a mechanical system, resistance (friction) produces heat.

Since the ohm is a unit of electrical resistance and since resistance to current flow is the opposite of conductivity, electricians call the unit of electrical conductivity the *mho*. That's ohm spelled backward.

11-5
OHM'S LAW

As you might expect, volts and amps and ohms are all very closely related. They are all related to each other by a simple formula called *Ohm's law.* Unlike Eq. 11-1 that relates amperage to coulombs over a period of time, this new formula is very important for you to know about. Memorize it. Learn to work with it. *Ohm's law is the basis for almost all practical work in electricity.*

The formula is:

volts = amperes × ohms

This formula is often written:

$$V = IR \qquad \text{(Eq. 11-3)}$$

**NOTE: Electrical pressure = current × resistance
or
Voltage = current × resistance**

This equation says that one volt equals the electrical pressure required to push one ampere of current through an electrical resistance of one ohm.

■ Let's put the resistance at 10 Ω and the current (*I*) at 15 A. Then $V = 10 \ \Omega \times 15 \ A = 150 \ V$.
■ Here's another example. What is the voltage drop across a stick-electrode current-adjusting resistor box on a welding power source when the box dial is set to 0.2 Ω and the current is 200 A? From Ohm's law, the voltage drop V = 200 A × 0.2 Ω = 40 V.

Whenever you know two of the three basic characteristics of electricity, you can always calculate the third one using Ohm's law. In the above example, we were given current (*I*) measured in amps and resistance (*R*) in ohms, so we can calculate the electrical pressure (*V*) in volts.

From the same simple equation you can solve for an unknown quantity if you know the other two. For example, to calculate amperage when you only know volts and ohms, use:

$$I = \frac{V}{R} \qquad \text{(Eq. 11-4)}$$

NOTE: current = $\dfrac{\text{voltage}}{\text{resistance}}$

■ If $V = 10 \ V$ and $R = 2 \ \Omega$, then $I = 5 \ A$ because 10 V/2 Ω = 5 A.

To calculate resistance when all you know is volts and amps, use:

$$R = \frac{V}{I} \qquad \text{(Eq. 11-5)}$$

NOTE: resistance = $\dfrac{\text{voltage}}{\text{current}}$

■ If $V = 10 \ V$ and $I = 5 \ A$, then $R = 2 \ \Omega$ because 10 V/5 A = 2Ω.

Figure 11-16 is a handy memory device. Put your finger over the unknown quantity and do whatever the rest of the diagram says to do. For example, if you know *I* and *R,* put your finger over the *V* and you'll see the *I* and *R* standing side by side. That means that you should multiply them to get *V*. But if you know *V* and *I*, for example, put your finger over the *R*. What's left is a *V* divided by the *I*. Do that and you immediately get the missing *R* for resistance.

Let's look at what these simple equations (actually one equation in three forms) tell us. Take Eq. 11-3, for example. *V* (voltage) is directly proportional to current (*I*) and resistance (*R*). That means that if the resistance doubles, the voltage doubles. If the resistance is cut in half, the voltage is cut in half. Also, it means that if we hold the resistance constant and double the current (or cut it in half), the voltage also will be doubled (or cut in half).

Similarly, as Eq. 11-4 shows, the current (*I*) is directly proportional to the voltage (*V*) and inversely proportional to the resistance (*R*). If you want to double *I* (the current) on the right-hand side of Eq. 11-3, then you have to double *V* (the voltage) on the left-hand side of the equation, pushing more current through the

given resistance. Otherwise, you have to cut the resistance in half to double the current to keep the voltage unchanged.

Later in this chapter we'll give you tables for selecting welding cable on the basis of the amount of current the cable can carry without requiring too high a voltage or producing too much heat because of resistance. You'll find that the heat produced in a circuit also is related to current and resistance. However, the relationship is slightly different from anything we've shown you so far. By the time we get to the section on resistance and heat, you'll understand why they have a different relationship. Meanwhile, let's look at power.

11-6
POWER AND ENERGY

Power
Power tells you how fast you can convert energy from one form into another. As welders we are interested in how fast an electrical device can convert electrical energy into heat energy. A race-car driver wants a high-powered engine that turns chemical energy (fuel) into mechanical energy pro-

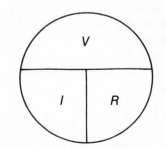

FIGURE 11-16 How to remember Ohm's law. If you want current *I* (amperes for example), put your finger over the *I* and read *V* (volts) divided by (underneath) resistance (*R*). To get *V* (voltage), put your finger over the *V*, and the *I* and *R* are next to each other. Multiply them. Similarly *V* is divided by *I* to get *R*.

ducing speed and acceleration. Some of the electrical energy you will use in arc welding will be turned into light energy, but you can't use that kind of energy for most welding (except laser welding). Most of the energy you produce, however, will be heat energy to melt base metals and filler metals.

The same electrical power units will tell you how fast a machine like a motor can convert electrical energy into mechanical energy for turning a motor shaft or a car axle, or even the amount of power needed to operate plant lighting, where electrical energy is turned into light energy.

Power equations The general expression for power (P) of any kind, electrical or otherwise, is energy (E) used up over time (t) or

$$P = \frac{E}{t} \qquad \text{(Eq. 11-6)}$$

NOTE: power = $\dfrac{\text{energy}}{\text{time}}$

A one-horsepower (1-hp) single-piston gasoline-powered motor will convert a certain amount of chemical energy (burning gasoline) into mechanical energy (a cycling piston) in a certain amount of time (1 s).

You're already familiar with energy units in foot-pounds to measure the impact strength of metals. A horsepower is simply 550 ft·lb of mechanical energy used up in one second (550 ft·lb/s). For example, a 1-hp engine working at maximum capacity could lift a 550-lb weight one foot in one second. That's an example of mechanical power.

A drag racer has to have a high-horsepower engine or lose the race because the object of drag racing is to convert a lot of chemical energy very rapidly into mechanical energy to move the car forward. A tractor might be able to convert the same amount of energy as a dragster to get work done. In fact, the tractor probably can do a lot more work than a high-performance car. But the tractor uses up the energy much more slowly. That's why power must take time as well as energy into consideration.

Measuring power The units of electrical power are watts (W) or kilowatts (kW) (which are 1000-W units). The watt is named after the Scottish scientist James Watt. You probably already know about both watts and kilowatts because a typical light bulb is a 60- or 100-W bulb, and you often hear about kilowatts in electrical company advertising and you may see it on your electrical bill.

How much power is a watt? It takes a little more than 1000 W to raise the temperature of one pound of water one degree Fahrenheit in one second. It takes about 570 times that much power (570 kW) to melt one pound of steel in one second. One horsepower is equal to 746 watts. One watt equals about 0.7 ft·lb/s.

More power equations Electrical power has a very simple formula. Power (P) equals the current times the voltage, or:

$$P = IV \qquad \text{(Eq. 11-7)}$$

NOTE: Power = current × voltage

This equation will help you calculate the power rating of an electrical device, such as an engine-driven arc-welding power source (where it's often used) or even a space heater.

■ What is the power rating of an electric heater that draws 6 A from a 120-V outlet? It's 120 V × 6 A = 720 W or 0.720 kW.

■ If an engine-driven welding power source has a power rating of 10,000 W (10 kW) and you want to run it at 100 A, how much open-circuit voltage (OCV) must the machine be able to deliver at its maximum rating? The machine is rated at 85-V OCV. We already know the power (P) and the voltage (V) so we can solve for the current (I) by using Eq. 11-8 (actually just Eq. 11-7 in another form).

$$I = \frac{P}{V} \qquad \text{(Eq. 11-8)}$$

Substituting our known numbers for the letters in the equation, we have:

$$\frac{10,000 \text{ W}}{85 \text{ V (OCV)}} = 118 \text{ A}$$

So, you can't get more than 118 A out of this machine. But since all you needed were 100 A, the machine will do the job.

Here's another practical example of how to use this basic power formula.

■ Your welding machine is rated at 15,000 W, or 15 kW, and you want to weld with it. You also want to run a bank of lights off the first auxiliary outlet to see what you are doing (it's a night job). Also, you want to use a disk grinder off the other auxiliary output plug. Will the machine be able to carry this power load without shutting down?

You will plug in 10 × 100-W bulbs (power is additive) which is 1,000 W = 1,000 W

You will be welding with 300 A at 35 V, or 300 A × 35 V (from Eq. 11-7) which is 10,500 W = 10,500 W

The disk grinder's name plate says it needs 1,200 W to operate (You also might calculate that figure based on 120 V × 10 A which equals 1,200 W, for example, using other data on the tool's name plate. = 1200 W

Total power demand = 12,700 W or 12.7 kW

Your power source has a rating of 15 kW and you need 12.7 kW for the job, so you can (maybe) just squeeze all this extra equipment onto the power source and have

2.3 kW to spare. The trouble is, this calculation is not very precise because electrical requirements on tools often range above or below what you think by about 10 percent. For example, a grinder may work on 115 V, but that's really 115 V ± 10 percent. In our example, if all the ratings are actually 10 percent higher than shown, you would only have 1 kW left over below the machine's rated capacity. If that rated capacity of 15 kW is actually 10 percent too low, then you wind up needing 0.5 kW that you don't have.

Never push your luck with electrical equipment. Remember that these hints are for estimating purposes only. There are other things that come into play, especially with ac equipment. One example is *power factor*. We won't discuss that here, because you need to know more about ac electricity than you do right now to understand it.

Energy

Going back to Eq. 11-6 and our definition of power as equal to energy divided by time, let's rearrange it to get a simple equation for energy.

$$E = Pt \qquad \text{(Eq. 11-9)}$$

NOTE: energy = power × time

This equation tells us that energy is power multiplied by the amount of time the power is used.

Measuring energy The units most often used for mechanical energy are foot-pounds. The units of heat energy are calories or Btu's (British thermal units). The units of electrical energy are joules (J). The joule is not a unit that you will use much as a welder, but we want you to have a complete feeling for electricity and its most commonly used units. As you'll soon see, we are leading you to something else that is quite practical.

Calculating energy Our simple equation for energy (Eq. 11-9) is

the ultimate basis for your electrical bill. Let's spell it out and see how it works.

Energy (J) = power (W) × time (s)

■ Let's start by seeing how much energy is required to operate a 100-W bulb (P) for 10 min. First, we must remember to convert minutes into seconds so we can fill in the t in the equation: 10 min × 60 s/min = 600 s. Then we can make our substitutions in the formula:

100 W × 600 s = 60,000 J

It's not at all hard to calculate.

In case you are wondering how big a joule of energy is, it's not very big. A joule equals about 4.19 cal and 1 Btu equals 1054 cal. However, 60,000 joules is a pretty big number.

In electrical units, one joule of energy equals one watt-second (W·s). Watt-seconds can produce just as big numbers as joules where energy is involved, as we can see from the number of joules used by our little lamp in only 10 min. The watt-second, then, doesn't seem to be a very useful number, either. Our single light bulb using 60,000 J of energy as both heat and light in just 10 min (or 600 s) would produce 60,000 W·s of energy. Imagine getting a monthly electric utility bill for 100,000,000 J. The joule is a very small energy unit, about like the small calorie used to measure small amounts of heat energy. One kilowatt-hour equals about 3,600,000 J. When dealing with electrical energy, kilowatt-hours is usually a much better unit with which to work unless you are using very small amounts of electricity. Therefore other, larger units for electrical energy often are used to produce smaller energy numbers.

Instead of using watts, use kilowatts (1000 W). Instead of using seconds, use hours (3600 s). The result is kilowatt-hours (kW·h) and there's your electric bill. The power company will bill you for one-sixtieth of a kilowatt-hour for

running your tiny lamp. Power companies charge for supplying energy (not power). Your electric bill is based on the amount of energy they supply to you, no matter how fast you use it.

11-7
RESISTIVITY, HEAT, AND POWER

Georg Ohm discovered something else about electricity. When electrical power is used up by resistance in a conductor, the energy produced by using up the power over time may appear as mechanical energy, light energy, heat energy, chemical energy, or electrical energy. Whatever kind of energy comes out, a certain amount of power is used up or "lost" in the process. Ohm even figured out a way to calculate the amount of power loss and energy production caused by electrical resistance (just like friction produces heat in a mechanical system). The power loss produced is equal to the current squared times the resistance. In equation form, that's:

Power loss = I^2R (Eq. 11-10)

The power lost in any electrical circuit due to resistance to current flow often is called I^2R *loss*. Where does the lost power go? In a light bulb part of it becomes useful light and part becomes waste heat. In electric welding, you want the heat and not the light. Most of the power consumed in electrical welding goes to produce heat and very little actually turns into light. Therefore I^2R power losses are a good way to figure out how much heat can be produced by an electrical circuit.

NOTE: HEAT = I^2R LOSS

If that circuit includes a welding cable, it becomes a very practical problem of finding out how much heat will be produced in the cable. If a small-diameter (or extra-long) cable produces too much heat, you will either have to shorten the cable, use a larger-diameter cable that has less resistance, or else you will have to lower the welding current, which will make it much more difficult (maybe impossible) to do arc-welding. On the other hand, if you ignore the problem, the heat in the cable may get so high that the insulation will burn off. The copper strands in the cable may even melt.

In resistance welding, heating caused by I^2R losses actually does all the work to make the base metal melt. Since energy (E) is power (P) used up (lost) through time (t):

$$E = Pt = I^2Rt \qquad \text{(Eq. 11-11)}$$

or energy = watts × seconds

or joules = amperes × ohms × seconds

This equation also tells you a lot about cable selection. When you increase the amperage in a welding cable, the heat energy produced by the resistance of the cable increases very quickly.

Resistivity and Current

If you double the amperage flowing in a given cable, the heat produced by the resistance of the cable is increased by a factor of four, the *square* of the increase in the current itself. If you made the mistake of increasing the current by a factor of 10 in a cable, you increase the amount of heat produced by a factor of 100 (the I^2 factor).

Here's an example.

■ The current flowing in a 1000-ft-long AWG 1/0-sized welding cable is 300 A. The resistance of the cable is 0.109 Ω. How much heat is produced in the cable?
Using Eq. 11-10:

$(300 \text{ A})^2 \times 0.109 \text{ Ω} =$
　　　　　power lost as heat

$90,000 \text{ A}^2 \times 0.109 \text{ Ω} =$
　　　　　9810 J of heat energy

Since 1 J = 0.223 cal, 9810 J = 2190 cal or 6.3 Btu of heat. That's enough heat to raise a pound of water 6.3°F. The end of this chapter shows you that you also will have a voltage drop of 34 V (enough to weld with). In other words, you can't weld with 1000 ft of AWG I/0 cable at 300 A. You need a much larger cable size.

Resistivity and Cable Size

The amount of current you send through the cable is not the only factor that will determine your power loss (heat production). The dimensions of the cable also are very important. Since resistance increases with the length of the cable, you can double the heating losses simply by doubling the length of the cable. Even worse is using a cable that has too small a radius or diameter. The resulting I^2R heating is one more reason why you have to be careful to select just the right cable size for your current and not use a cable that is longer than you really need. You want to lose power as heat at the point where you are welding. You don't want to create heat in your welding cable.

If the cable in the above example were only 250 ft long instead of 1000 ft, the heat produced in the cable would be 9810/4 = 2453 J. However, our 250-ft-long welding cable with heat losses of 2453 J/s will lose 2453 W of power per second due to resistive heating. That's equal to 2.45 kW of power lost every second. That's a lot of energy lost as heat in the welding cable that is not used in the welding job.

Welding cable, even very large diameter, highly conductive welding cable, should not be any longer than is necessary to get the job done. The longer the cable the more power you lose as heat caused by the resistance of the cable.

Cable Diameter

A far more important fact about the resistivity of a welding cable (or any other conductor) is that when the diameter of the cable is doubled, the resistance of the entire cable gets smaller by a factor of 4. Similarly, if you select a copper cable with half the diameter of a previous copper cable, the resistance of the entire cable will be four times larger. It's not hard to understand why (Fig. 11-17).

The resistance of an electrical conductor is directly related to the *area* of the cross section of the cable, not to the cable diameter or radius. The reason is that the area of a circle increases four times when you simply double the diameter or the radius. The area is four times smaller when you cut the diameter or radius of the cable in half. The reason behind this is that the area of a circle is πr^2, that is, the area is proportional to the square of the radius (or the diameter, or any other linear measurement on the circular cable cross section).

Obviously, when you provide twice as much conductor for the same amount of current to flow, something has to happen. What occurs is that you change the resistance of the entire conductor in inverse proportion to the change in area. Increase the area of the conductor by a factor of 9 (which is the same as increasing its radius or diameter by 3), and you cut the resistance to one-ninth of

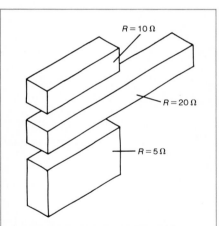

FIGURE 11-17 The effect of conductor length and cross-sectional area on resistance. Doubling the area cuts the resistance in half. Doubling the conductor length doubles the resistance. That is why welding cable must be carefully selected for different jobs. One cable size does not do all work. It's your job to select the correct cable size for every job you do.

its previous value. Use a conductor with one-quarter of the area of a previous conductor (equivalent to cutting the diameter or radius in half) and you increase the resistance by 4. It's easy to understand when you see it as a simple equation:

$$R \text{ varies as } \frac{1}{\text{area}} \qquad \text{(Eq. 11-12)}$$

$$\text{or} \quad R \text{ varies as } \frac{1}{\text{diameter}^2} \qquad \text{(Eq. 11-13)}$$

$$\text{or} \quad R \text{ varies as } \frac{1}{\text{radius}^2} \qquad \text{(Eq. 11-14)}$$

You don't have to worry about the math most of the time, as long as you understand the ideas behind the equations. The welding cable selection section of this chapter will give you tables you can use to select welding cable. Nevertheless, these tables are totally useless if you don't understand the ideas behind them, and the reasons for specifying a certain size conductor for each job listed in the tables. *Don't change electrode cable sizes just because another size is handy.*

11-8
MAGNETISM
Magnetism is an invisible force that acts on some materials and not on others. Only a few materials are strongly magnetic. They are iron, nickel, and cobalt. Why are we going to tell welders about

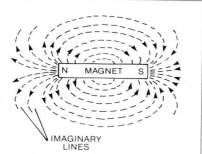

FIGURE 11-18 The magnetic field of a simple bar magnet. The imaginary lines of flux are most dense at the magnetic poles where the magnet has the strongest attraction or repulsion.

magnetism? Because it will help you to understand how electricity is generated in certain kinds of welding power sources. Magnetism is also used by the power company to generate the electricity supplied to welding power sources that plug into a wall. Magnetism helps you understand why alternating current alternates. Magnetism also creates a problem called *arc blow* when welding certain high-nickel-alloy steels and all thicker carbon steels. In Chap. 16 we'll tell you how to deal with the problem of arc blow when you are welding.

Magnetic Fields
The invisible magnetic force is called a *magnetic field*. The field extends out in all directions. Figure 11-18 shows a drawing of a typical bar magnet, only one of many kinds of magnets that are made. The shape of the bar magnet shows you several important things about the magnetic field.

The first thing to know about magnetic fields is that they have two ends, or poles. The magnetic field goes out of one pole and into the other, travels through the body of the magnet and returns to the pole, where it leaves the magnet again. In other words, magnetic fields set up "magnetic circuits" just like electric fields have electric circuits.

If you were to visualize imaginary lines of force running through the magnetic field, you would see the lines get closer together as they get closer to the two poles of the bar magnet. There really are no such things as lines of force, but they help to visualize how magnets work. When the lines of force (the magnetic flux) are close together, the magnetic field is stronger. When they are farther apart, the field is weaker. These flux lines are very close together at the poles of any magnet.

If an imaginary plane cuts the lines of force, it will cut more lines as it gets closer to either one of the poles. The reason is that

the flux lines are more concentrated at the poles of a magnet. That idea is used to create electricity, as you will see shortly.

Magnetic north poles don't like to get too close to other magnetic north poles. South poles don't like south poles. But north and south poles strongly attract each other, as shown in Fig. 11-19. The effect is very much like two positive static electric charges that don't like to get near each other, while two static positive and negative electric charges are strongly attracted to each other.

When north and south magnetic poles get close to each other, the magnetic flux or lines of force get very thick. That means that the magnetic force gets very strong, also as shown in Fig. 11-19.

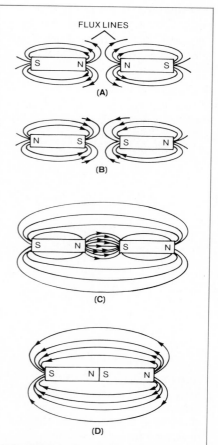

FIGURE 11-19 Like magnetic poles repulse each other. The repelling poles may both be (A) north poles, or (B) south poles. (C) Unlike magnetic poles attract. (D) The force of attraction is greatest when the poles are touching.

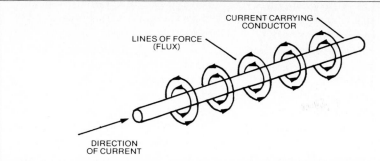

FIGURE 11-20 The magnetic field around a conducting wire is circular. The flux is perpendicular to the direction of current flow.

Induced Current

Now we can tie electricity and magnetism together. Whenever a conductor such as copper wire cuts magnetic lines of force, a direct current (dc) is "induced" into the wire. The more lines of force the wire cuts, the stronger the induced dc that flows through the conductor. Also the faster the conductor cuts the lines of force, the stronger the induced current will be. And finally, if you wind the conductor in a coil so that many wraps or *ampere-turns* of the single conductor wire cut the magnetic lines of force each second,

each segment of the conductor wire adds to the dc induced into the conductor.

You can look at this process both ways. A magnetic field can be thought of as producing dc in a wire, or dc flowing through a wire can be thought of as producing a magnetic field around the wire (Fig. 11-20). The stronger the dc in the wire, the stronger the magnetic field around the conductor.

The direct electric current that is induced into the conductor flows only in one direction, depending on which way the conductor cuts the magnetic force lines. That's the same as saying that when a direct current flows through the conductor, the induced magnetic field around the conductor "wraps" around the wire in one direction.

Hold up your left hand with the thumb and two fingers sticking out as shown in Fig. 11-21. If the current is moving in the direction of your thumb while cutting the flux lines of magnetic force, the magnetic flux lines around the conductor will point in the direction of your index finger, and the current will flow in the direction that your middle finger is pointing.

In Fig. 11-22 you are looking at the round ends of two conductors. The drawing on the top shows that the electrons flowing through the conductor are entering the paper (the middle finger of your left hand is pointed down), so that the induced magnetic field around the

conductor rotates in a counter-clockwise direction.

If the electrons flow through the conductor and out of the paper (left middle finger is pointed up), the induced magnetic field around the conductor flows in a clockwise direction. This often is called the *left-hand rule* for current flow. It shows you which way the dc will flow through a conductor when the conductor cuts magnetic flux lines.

The magnetic field induced around a single conductor is usually too weak to do much. However, a strong magnetic field can be created by wrapping one long wire into a coil and then spinning the coil very fast through a magnetic field. That idea combines a method for cutting a lot of lines of force with one wire, and cutting

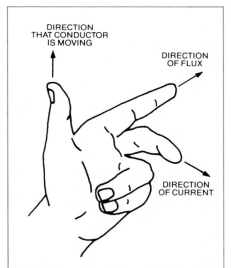

FIGURE 11-21 A magnetic field cut by a moving conductor wire induces an electric current in the wire. The left-hand rule relates the direction of the moving conductor to the direction of flux and current flow.

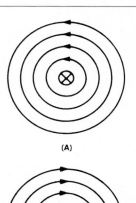

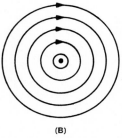

FIGURE 11-22 Another way of viewing current flow and magnetic field around a conductor. (A) Visualize the current as an arrow entering the paper. The X mark is the feathers of the arrow. The magnetic flux flows counter-clockwise, from right to left around the arrow. (B) Visualize the current as an arrow coming out of the page. The dot is the arrow's point. The magnetic flux flows clockwise, from left to right around the arrow.

them very fast. Combine both of the above ideas with a strong magnetic field and you will induce a lot of dc into the conductor wire.

The more wraps of wire the coil has, the stronger the induced electric current will be. The faster the coil spins in the magnetic field, the stronger the induced current will be. The stronger the magnetic field, the more induced dc it will produce. That's how both a big electrical generator in a power plant and a small one in an engine-driven welding power source produce electricity.

Generators and Alternators

A generator has a large electromagnet on the outside with stationary coils that create a strong magnetic field. It's called a *stator*. It does not rotate. A spinning electric rotor on the inside, with

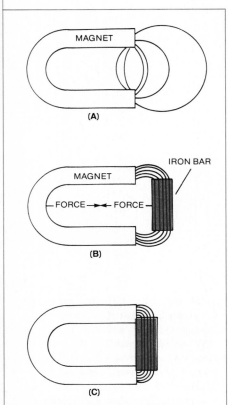

FIGURE 11-23 Magnetic materials concentrate the magnetic flux lines, increasing their density and therefore their strength. The magnetic field force increases in strength from (A) to (B) to (C), where it is strongest.

another coil of wire wrapped around it, cuts the magnetic field lines produced by the magnet. This is the *rotor*. The result is a strong dc flowing through the wire of the rotor coil.

An alternator is designed about the same way, except that the magnet coil is on the inside and spins (it's the rotor) while the static coil (stator) that produces the current is on the outside and doesn't move. Both devices generate electric current from the magnetic field.

Another way to get the magnetic flux lines closer together so that they are stronger and induce more dc in a conductor is to use an iron bar like a magnetic lens to focus the magnetic field lines (Fig. 11-23). In fact, that is why a magnet attracts materials like iron, nickel, and cobalt so strongly. These materials concentrate the magnet's force lines, which produces a much stronger magnetic force and makes you think that the magnet is pulling on the iron, nickel, or cobalt much harder than on other, less magnetic materials. The reality is that the iron, nickel, or cobalt is simply concentrating the magnetic field and making it stronger.

A soft iron bar, often called a *core*, can be slipped inside an electrical coil to concentrate more, or less, of the magnetic field produced by the coil depending on how much of the iron is inside the magnet. If the bar is pulled out of the coil, or pushed into it, the magnetic field produced by the coil decreases or increases in strength. The sliding iron core can actually "tune" the strength of the magnetic field, making it stronger or weaker.

The stronger magnetic field will induce more dc into the electric coil, and the weaker magnetic field will induce less dc. That's exactly how a lot of welding equipment is controlled when you turn the crank wheel or a dial on the machine.

Inducing more dc into a conductor is the same as increasing the

voltage in the conductor so that more dc electricity flows (according to Ohm's law). Therefore the dial or the crank on your welding machine can be calibrated to dial in either more or less current in amps, or more or less voltage in volts. As long as the resistance of the electric circuit doesn't change, increasing the amps is just the

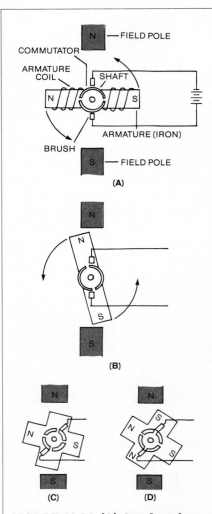

FIGURE 11-24 (A) An electric motor uses currents induced by electromagnets (B) to attract or repel other electromagnets on the spinning armature. In a four-pole commutator, (C) and (D), the same driving principles apply. An electrical generator works much the same way, but in reverse. It has a spinning armature coil that cuts magnetic force lines to induce current flow, instead of using current flow to rotate the armature.

same as increasing the volts, and vice versa.

If you reverse the process and pump more electricity into the coil, it will produce a stronger magnetic field. All you have to do is join several magnetic coils with their north and south poles alternately attracting or repelling the north and south poles of the magnet rotor coil, and you will have produced a dc motor (Fig. 11-24).

Alternating current A simplified view of an electrical generator with a coil having only one loop is shown in Fig. 11-25. The spinning "one-wrap" coil in the center of a magnet cuts magnetic lines of force created by a separate electric magnetic coil. Induced dc starts flowing through the conductor wire as it spins in the magnetic field and cuts the lines of force.

As we know from the left-hand rule of current flow through a conductor, the direction that the electricity flows through the wire loop depends on which direction the loop rotates. In our hypothetical one ampere-turn generator, the single wire normally would cut the lines of force on top making the current flow one way. Then the wire would cut lines of force on the bottom, making the current flow in the opposite direction. Since the current keeps reversing itself, you will produce alternating current (ac). If the rotor turned one full revolution 60 times a second, you would have 60-cycle (60 Hz) ac. Figure 11-26 shows an ac generator.

The induced current is led out of the conducting coil in the generator by current-carrying graphite brushes that rub against copper commutator bars attached to the spinning coil.

Direct current To make dc, the commutator bars are designed with two gaps in them, as are the ones that you see in Fig. 11-27. As the inner conductor loop turns, the commutator turns with it. The commutator bars touch the

brushes and slide against them as the loop rotates. Thus, the graphite brushes make an electrical connection with the rotating loop of the wire through the commutator.

Second, the brushes reverse the connections to the rotating loop every time the polarity of the induced electrical voltage changes. As a result, the top part of the

loop of wire in Fig. 11-27 will first be moving from left to right relative to the magnetic field, as shown in the diagram. After it has spun 90°, it will start rotating from right to left relative to the magnetic field. At the same time, the brushes will change commutator bars. The result is that one brush will always be negative and the other brush will always be

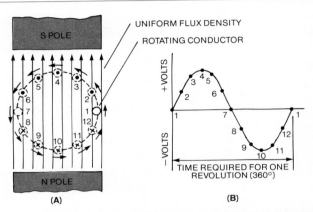

FIGURE 11-25 (A) A conductor rotating in a magnetic field has no current flow in positions (1) and (7) because no lines of force are cut. Current flow is maximum in positions (4) and (10). If the voltage in the circuit is plotted against time for one revolution of the conductor, this graph results. (B) The curve in this graph is called a *sine wave*. It is typical of alternating current. When the voltage is positioned above the zero reference line, the current flows in one direction. When the voltage is below the zero reference line, that means that the current is flowing in the opposite direction. It has alternated its direction. It is ac electricity.

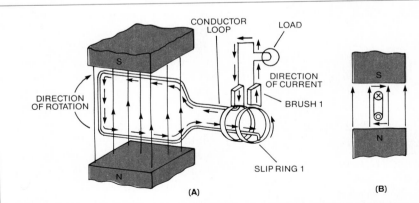

FIGURE 11-26 Alternating current can be generated if a conductor rotates in a magnetic field. (A) Slip rings and graphite brushes take the induced current out of the machine and put it into an electrical circuit. (B) A cross section of the rotating conductor with current flowing. All that is needed to induce more current is to cut more magnetic lines of force with more conducts at the same time. That's what a coil does.

positive. This produces dc flowing out of the generator instead of ac.

As you can see by comparing Figs. 11-26 and 11-27, ac welding power sources are actually simpler to make than dc machines because ac generators are simpler than dc generators. That's why a simple ac machine usually costs less than a dc welding power source of equal rating. We'll tell you a lot more about ac shortly.

Direct current generators are an important type of welding power source, even though they may cost a little more than equally rated ac machines. They often are driven by internal-combustion engines using either gasoline or diesel fuel. These engine-driven welding power sources are used on construction sites, farms, and pipelines where you can't hook up to an existing power supply. Some engine-driven power sources are even designed to give you either ac or dc, depending on which switch you operate or which plug receptacle you use for cable.

There also are electric-motor-driven welding machines that operate with ac input and produce dc output. Other machines oper-ate on high-voltage, low-amperage ac and produce lower-voltage, higher-amperage ac current for welding. Of course, since these machines must be plugged into a wall, they can't be used in remote locations where there are no electric power lines.

Transformers

We just mentioned that some kinds of welding machines can take in high-voltage, low-amperage current and can produce low-voltage, high-amperage current for welding. These are *transformer* welding machines. Here's how transformers work.

Magnetic fields induce voltage, and thus current flow, in conductors that cut through the magnetic flux lines. A voltage also can be induced into a conductor even when it is not moving but the magnetic field is, in the sense that the magnetic field is changing shape.

Figure 11-28 shows a small magnetic field with its flux lines concentrated by an iron core. This time, two separate coils are wrapped around the iron bar. When electricity flows through the top coil wrapped around the iron bar, the current induces a strong magnetic field in the iron core which concentrates the magnetic flux like a glass lens concentrates light.

The induced magnetic field does not change shape as long as the current is running. But if the switch shown in the drawing on the top-core circuit were opened, the magnetic field concentrated in the iron core would collapse.

There is no faster way to change the shape of a magnetic field than to make it fall in on itself and disappear; that very rapid change in the shape of the magnetic field will cause many flux lines to be cut all at once to induce a separate voltage and current to flow through the second coil on the bottom of the magnetic core.

The voltage induced by a changing magnetic flux is the principle on which electrical transformers and other devices operate, including many kinds of welding power sources. The ignition coil of an automobile also is a transformer. The auto ignition coil only has one wrap or ampere-turn of conducting wire wrapped around the top end of the iron core. But the separate circuit on the bottom of the iron core has a lot of ampere-turns of wire around the iron core to make the current output of the second coil much larger. Since the second coil has far more turns than the first coil, the changing shape of the magnetic field cuts many more lines of the conductor (or the conductor cuts many more lines of the magnetic flux field . . . it doesn't matter which way you look at it) in the second coil than in the first coil. The induced voltage and current flowing out of the bottom coil will then be much higher. That's how the low-voltage battery in a car can make a high-voltage spark in the combustion chamber over the piston that will ignite gasoline spraying into it, starting your engine.

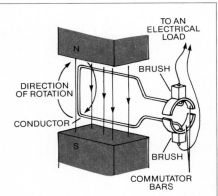

FIGURE 11-27 Direct current is simpler to understand than alternating current. But it is more difficult to produce because more electrical hardware is needed to produce it. That is why ac power sources usually cost less than dc welding power sources of equivalent rating.

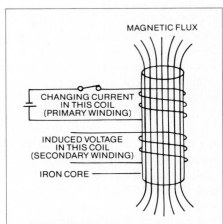

FIGURE 11-28 Voltage can be induced by a changing magnetic flux. The flux is made to change by a changing current in the primary winding of a coil and the induced voltage (and current) is taken off by the secondary-current winding. The flux is tightly concentrated by using an iron core inside the coil.

Turning low-voltage current into high-voltage current is what a *step-up* transformer does. If you reverse the process and put high-voltage, low-amperage current into the transformer and output high-amperage, low-voltage electricity, you will have a welding power-source transformer.

That's how a 120-V ac supply line with just a few amps of current can be turned into a lower voltage welding current with 100 A, or a 240-V dc line with low amperage can make 300 A of welding current at, maybe, only 75 V OCV. Different types of welding power sources will do that job in different ways. For example, the OCV on many stick-electrode welders is fixed by the machine's design.

As you know from the power and energy equations, the heat you produce increases as the square of the current (I^2R heating), so current rather than voltage is what you need most to produce heat for welding. If the input

voltage to the transformer coil is ac which changes direction 120 times a second, the "switch" in the transformer circuit that turns the magnetic field on and off isn't needed. Sixty-cycle ac automatically turns itself on and off 120 times every second. The output current from this kind of transformer also would be ac, except that the voltage would be lower and the current would be higher if the transformer is a step-down model. That's perfect for welding jobs.

An ac transformer is much simpler to make than a dc transformer because the high-speed switch is built into ac itself. That's another reason why pure ac welding power sources tend to cost less than dc welding power sources . . . both of which need some kind of transformer to boost the current so you can get more I^2R heat energy for welding.

11-9
UNDERSTANDING ELECTRICITY

Alternating current is electrical current that periodically changes the direction in which it flows through a circuit. The amount of current also changes, as does the voltage. Since the electrons no longer flow in one direction, but instead flow back and forth, even the direction that is positive and negative (the polarity) reverses periodically.

You can plot any of these electrical units on a graph to see what they look like. The resulting line on the graph is called a *waveform*. Pure dc will produce a straight line on the graph; ac will produce a line for the current or the voltage that varies up and down when plotted against time.

In Fig. 11-29 you see a simple drawing called a *circuit diagram*. The horizontal lines on the left are the symbol for a battery. The battery in this drawing is a 4-V battery. You also can see a switch on the top of the drawing. The switch

is closed. On the right is a jagged line. That is the symbol for a *resistor,* or in the case of our drawing, all the electrical resistance in the entire circuit. The total circuit resistance is 1 Ω.

Since voltage is a potential difference measured between two points, we've shown two possible points on the diagram. In this case we are measuring the voltage difference between the battery output and the resistance in the circuit. If one lead from a voltmeter were placed on the reference spot and the other lead touched the point on the other side of the circuit resistor, you could plot the change of voltage in this dc circuit with time, from the moment that you closed the switch to start electricity flowing until you opened the switch again to turn off the electricity.

We show that plot in Fig. 11-30. Except for a momentary build-up in current or voltage (it doesn't matter which we plot on the left side of the graph), the result is a straight line until we open the switch at 14 units of time (seconds for example). The graph is typical of a pure dc waveform.

DC Waveforms

Direct current is not always flat. It's possible for dc to wander around as shown in the top drawing in Fig. 11-31; dc also can be pulsing, as shown in the middle drawing in the figure. Pulsing dc sometimes is purposely used to in-

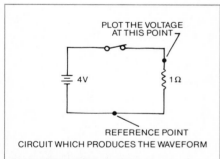

PLOT THE VOLTAGE AT THIS POINT

4V   1Ω

REFERENCE POINT
CIRCUIT WHICH PRODUCES THE WAVEFORM

FIGURE 11-29 A circuit can be described by a drawing showing its elements. Here we have a 4-V battery and a switch (which is closed). The circuit has a total resistance of 1 ohm. The voltage is measured between two points in a circuit, one of which is a reference point. The voltage in a circuit can vary depending on which point in the circuit it is measured relative to a reference point, even though the current flow remains the same, if the resistance between the different voltage measurement points differ. See the waveform in Fig. 11-30.

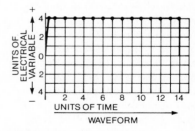

FIGURE 11-30 When the switch is closed, the voltage rapidly rises from 0 to 4 V and remains there until the switch is opened 14 time units later. The result of this plot is a direct-current voltage waveform.

crease the efficiency of TIG and MIG welding. A typical ac waveform is shown on the bottom of Fig. 11-31. Note that the difference between ac and dc is not that one is constant and the other varies, but that dc never changes polarity (it stays above or below the zero reference line) while ac alternately changes polarity (from + to −, from − to +, etc.).

One modified welding process using pulsed dc is called *pulsed-current* MIG welding. One of the most important trade names for the process is Pulsed Arc MIG welding. Among many other things, the pulsing dc helps you

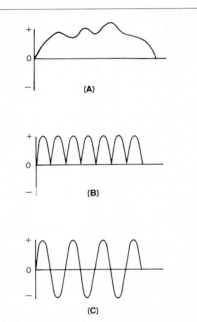

FIGURE 11-31 Not all electrical waveforms are flat like that of a pure direct current, or a sine wave like that of a pure alternating current. (A) This waveform is typical of a fluctuating direct current. (B) This waveform alternates, but never becomes negative; therefore it is dc, and not ac, current. Waveform (C) is typical of ac welding current. Waveform (A) can be a source of trouble when direct-current welding and your power company's input current into your power source varies from what it should be. (C) A waveform of a pure alternating current. See Fig. 11-33.

weld while holding down the total heat build-up in the weld metal. That substantially reduces welding distortion. In TIG welding, a pulsed "square-wave" current can help the welding arc break through the high-resistance, refractory aluminum-oxide skin that forms over molten aluminum weld metal. There are many other uses for pulsed-current dc welding and we'll cover them in later chapters on specific welding processes.

Sometimes variations in dc can cause you trouble. If the power company varies the line voltage or current in the electricity they supply you, you won't see it affect a light bulb, but it can affect your welding-machine operating settings.

The very fact that you now know that dc is not always dead-flat and constant, and that you occasionally may have to adjust your welding power-source machine settings to compensate, will help you. When trying to find out what's wrong with your welding technique or welding power source, sometimes the answer to your problem is "nothing is wrong, except the quality of the electricity." The fact is that electrical power companies allow a variation in supplied voltage of as much as 10 percent either way. When they are having problems, the variation can be even greater, as you may have noticed when your lights momentarily dim.

What's important right now is that while the dc amps or voltage may rise or fall, they never change direction. The polarity of a dc circuit remains the same.

Welding with DCSP and DCRP Current

While the polarity of dc remains the same, you can connect your ground cable and electrode cable two different ways, depending upon which terminal on a dc power source you use; the positive terminal or the negative terminal. In other words, you can't change the polarity of the dc, but you can

change the polarity of the workpiece and your electrode as shown in Fig. 11-32. That reverses the direction of electron flow.

Every time you set up dc welding equipment, you have to pay strict attention to how you set up the polarity of the electrode and ground cables. Many welding electrodes will only give you good results when they are positive. Others only work properly when they are negative. The instructions on the electrode box, as well as the code designation for the electrode coating and even the type of welding process you use, tell you which way to set up your equipment.

The first way is to set up the electrode so that it is attached to the negative lead from the power source. The ground cable is then attached to the work and hooked up to the positive lead from the power source. When the electrode is negative the work setup is *dc straight polarity*. It's almost always abbreviated DCSP. When the electrode is positive and the workpiece ground is negative, it's called *dc reverse polarity*. It's abbreviated DCRP.

If you don't set your dc welding job up correctly, you probably

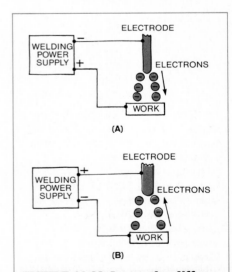

FIGURE 11-32 Learn the difference between (A) direct-current straight polarity and (B) direct-current reverse polarity. You will need to know it every time you hook up a direct-current welding machine to a workpiece.

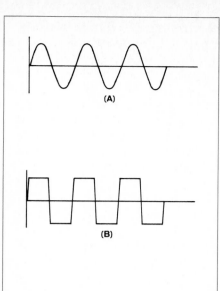

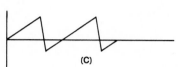

FIGURE 11-33 Not all alternating-current waveforms are simple sine waves (A). (B) This waveform is a square wave. (C) is a sawtooth wave that spends most of its time on the positive side (current flowing in one direction), but sometimes goes on the negative side (current flowing in the opposite direction).

won't be able to make a good weld. How do you remember the difference between DCSP and DCRP? Try this. "Senators' *E*lectrodes are *N*egative and *R*epresentatives' *E*lectrodes are *P*ositive. . . . SEN = straight-electrode-negative. REP = reverse-electrode-positive." Therefore, DCSP electrodes are negative and DCRP electrodes are positive.

AC Waveforms

Alternating current is totally different from dc. In fact the one basic thing we can say about ac is that the polarity of the circuit will constantly be changing from plus to minus and back again. That's the same as saying that the direction of the current flow is constantly reversing.

Most ac flow looks like a water wave. It doesn't always have to look that way. We show you three kinds of common ac waveforms in Fig. 11-33, including the water-wave variety in the top drawing; a square-wave ac waveform in the middle drawing, and a saw-toothed ac waveform in the bottom drawing. The square-wave shape is an important waveform

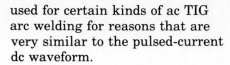

used for certain kinds of ac TIG arc welding for reasons that are very similar to the pulsed-current dc waveform.

Again, the most important thing to learn right now is that ac changes polarity over time. And every time the polarity switches from plus to minus or back again, the direction of the current flow changes. Now we'll give you some handy words you will often see in literature describing ac welding power sources.

Cycles In Fig. 11-34, the *cycle* of the current refers to the total time required to make one complete cycle of ac variation. In this figure, the ac starts at 0 A and builds up to a little more than 1 A in the negative direction, then returns to 0 A. That is half of the cycle. The current continues to build up in the positive direction until it reaches a maximum of a little more than 1 A. Then the current starts to fall back to zero again. When it finally gets back to zero, the first cycle is completed and the sequence will repeat itself in the second cycle, the third cycle, the fourth cycle, and so on.

Period The time it takes the ac to complete one cycle is called the *period* of the current. The period of the ac we have on the graph is 0.25 time units. Note that the ac was at zero twice during the cycle. That's why 60-cycle ac switches on and off 120 times a second.

Frequency If you were to tell someone how many complete cycles of ac alternation occurred in one second, you would call it the *frequency*. The graph plots 4 cycle/s ac. There is a special name for the number of cycles per second. It's called the *Hertz* and is abbreviated Hz.

You are most familiar with 60-cycle/s, or 60-Hz ac because that is the current supplied by power companies to your house or apartment. In many countries, 50-Hz ac is used instead of 60 Hz. If you buy arc-welding equipment for use in another country, check the

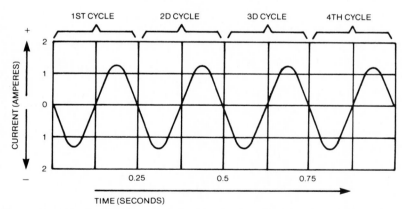

FIGURE 11-34 The sine wave of an alternating current is plotted against time to help you learn three terms: cycle, period, and frequency. This waveform has a cycle of 0.25 seconds and completes 4 cycles per second. Therefore, its frequency is 4 hertz. Hertz means cycles per second and is abbreviated Hz. The ac current used at home in the United States is 60-Hz current. In Europe, 50-Hz current is common. Some arc-welding power sources such as GTAW machines use 60-Hz input current but produce a special and much higher frequency output current to start the arc.

type of ac used and then notify the manufacturer.

What is the period of a 60-Hz-frequency circuit? The period is one divided by the frequency. Similarly, the frequency is one divided by the period. They are reciprocals of each other. *Reciprocal* is the arithmetic equivalent of turning something upside down:

$$\text{Period} = \frac{1}{\text{frequency}} \qquad \text{(Eq. 11-15)}$$

$$\text{Frequency} = \frac{1}{\text{period}} \qquad \text{(Eq. 11-16)}$$

■ How much time is required to complete one cycle of ac if the frequency is 60 Hz?

$$\text{period} = \frac{1}{60\ \text{Hz}} = 0.0167\ \text{s}$$
$$\text{(Eq. 11-17)}$$

■ What is the period of a very high frequency 2 million Hz (2-megahertz, or 2-MHz) ac?

$$\text{period} = \frac{1}{2,000,000\ \text{Hz}} = 0.0000005\ \text{s}$$
$$\text{(Eq. 11-18)}$$

■ If an ac circuit has a period of 0.001 s to complete one cycle, what's the frequency of the current in hertz?

$$\text{frequency} = \frac{1}{0.001\ \text{s}} = 100\ \text{Hz}$$
$$\text{(Eq. 11-19)}$$

AC Amplitude

To better understand ac voltage it helps to know a few more terms. Amplitude is the height of the waveform. On a measuring device called an *oscilloscope,* you not only would see the shape of the ac waveform or signal, you could even measure its amplitude.

Another word for amplitude is *peak height*. Figure 11-35 shows the amplitude or peak height and "peak-to-peak" distance of just one ac cycle. The peak height is exactly 1 V (in either the positive or the negative direction). The total peak-to-peak distance, therefore, is 2 V. Does that mean that, if you looked at this ac circuit and tried to measure or use the voltage, the voltage in the circuit is 1 V, or 2 V?

The answer is neither. The peak ac voltage is only at a peak of +1 V or −1 V for a tiny moment in time. Before and after that, the voltage is different and somewhere between the two upper and lower peak values.

Average voltage To get around this problem we talk about the *average voltage* of the ac circuit. If you could use calculus to calculate the area under the curve above or below the zero voltage line and "turn the area into a square" to get the average size of the area, the square area would be 0.637 times the peak voltage. The average voltage of the current, then, is 0.637 × peak voltage (and the average current is 0.637 × peak amps). The average voltage for any sine waveform ac circuit is 0.637 × peak voltage, no matter what that peak voltage is. The same is true of the average amps.

Effective (rms) voltage Another, more important way to specify the voltage or amperage of an ac circuit is slightly different. You most often will hear about the *effective* voltage or amperage, or *root-mean-square* voltage or amperage, or simply the rms voltage or amperage of the current.

How can you use the rms voltage or amperage data? The rms ac voltage or amperage produces the same amount of heat (measured as I^2R losses) as an equivalent dc circuit. The rms amperage is defined similarly. Root-mean-square data for ac circuits lets you compare the welding work they will do with dc circuits.

NOTE: RMS voltage or amperage in an ac circuit produces the same amount of heat as that produced by an equal dc circuit.

The effective voltage of any ac is 0.707 times peak voltage. The effective amperage also is 0.707 times peak amperage. The number 0.707 is approximately 1 divided by the square root of 2. The square root of 2 is very close to 1.414, so that $\frac{1}{1.414}$ = 0.707. You can remember this number

as a Boeing 707 airplane with a decimal point on its nose. One ampere of effective ac and 1 A of pure dc produce the same amount of heat in a circuit when they flow through the same electrical resistance. Also, equal amounts of dc voltage and effective ac voltage will produce equal power losses across equal-valued resistance in a circuit. That lets you convert dc cable data to ac cable data.

■ The average voltage of a 150-V ac circuit is 0.637 × 150 V = 95.6 V. The effective voltage of the same 150-V ac circuit is 0.707 × 150 V = 106 V. An ac circuit with a 150-V peak current will produce the same amount of I^2R heating as a 106-V dc circuit when run through the same resistance.

Since you are not studying to be an electrician, you don't have to worry much about these details. All you should remember is that the rms voltage of an ac circuit means that the voltage or amperage number is equal to the same dc voltage or amperage, in terms of the heat it will produce for you as a welder.

Three-Phase AC

One of the advantages of ac is that more than one current at a

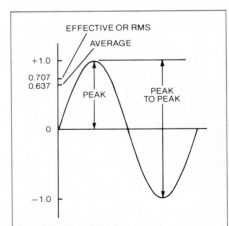

FIGURE 11-35 Three very important ac ideas are shown here. They are peak height, average current, and effective or root-mean-square (rms) current. See the text for details. Learn how to use these ac facts when welding.

time can be run through the same conductor. You can combine three different 60-Hz currents in the same conductor if you put in one, then a second one, and then a third one offset a little in time. That's another way of saying that the three currents are *out-of-phase* with each other. Figure 11-36 shows what such a current would look like if you plotted all three voltages for the three currents separately.

You can add up the three ac voltages (or three ac amps) and actually measure only their overall effect. This addition can result in one single ac curve that has much higher peak voltages than the three separate currents in the line, but only if all three positive and negative peaks nearly coincide. If that happens you have three-phase current that is totally in phase. You also get very large voltages and currents.

If the addition of all three phases gives you almost no ac at all because the positive and negative peaks nearly cancel each other out, you have three-phase current with three completely different phases. The result of adding three alternating currents on one line with their positive and negative peaks just about cancelling each other out is a funny-looking current that is very close to rippling or wavy dc. We'll show it to you when we explain the dc rectifier.

Compared to single-phase ac, three-phase ac makes more efficient use of copper conductors to carry higher voltages because a smaller conductor can handle more three-phase alternating current or voltage than a single-phase ac line. The single-phase line requires about 1.15 times as much copper for power lines as does the three-phase system. When the conductor is small because the current and voltage are small, any saving in the size of the conductor doesn't matter much. For example, a 120-V ac line almost always carries single-phase electricity. That's the kind of current that is sent through the wires in your house.

Many plants are wired for 240-V three-phase ac, or even more. That's the voltage needed to run most arc-welding machines. Another common arc-welding power-supply voltage comes from a 480-V three-phase wall plug.

Even higher supply voltages are sometimes used for very high powered welding processes. Three-phase current makes it possible to use substantially smaller conductors for these power sources.

There are many other facts about three-phase current that are interesting, but not important to you as welder. However, one thing about three-phase current that is very important to you is that it can be turned into dc that only has a slight ripple. If your welding power source only had some sort of filter that could cut off the bottom half or the top half of the ac wave, you'd turn three-phase ac into rippling dc. There is just such a device and all dc welding machines hooked up to an ac supply use them. These devices are called *rectifiers*. Another name for the device commonly used in the shop is a *choke*.

DC Rectifier Power Sources

A *diode* is roughly the electrical equivalent of a check valve in a high-pressure gas line. The check valve lets the gas flow in one direction but not in the other. The electrical diode lets current flow in one direction but not in the other. Diodes are used to make rectifiers.

If the current flowing into the rectifier is ac, the current flowing out of the rectifier will be dc, as shown in Fig. 11-37. If the ac flowing into the rectifier is single-phase, the current flowing out of the rectifier will be a very wavy dc. But if you put three-phase ac into the rectifier, when all three phases are separated (they are each 120° out of phase), the ripple on the dc out of the rectifier will be much less. It will look very much like nearly smooth, flat dc.

Kinds of rectifiers There are many different kinds of rectifiers used in welding power sources. One of the most common is the SCR, or *silicon-controlled rectifier*. Silicon is one of the semiconductor materials so commonly used now in electronics. Before the invention of the SCR, vacuum

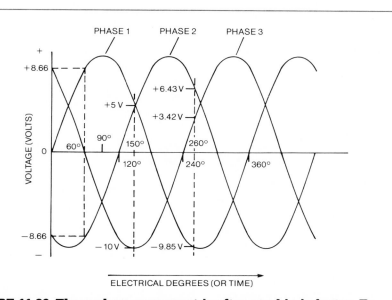

FIGURE 11-36 Three-phase ac current is often used in industry. To understand three-phase current, think of three separate ac sine waves separated by 120 electrical degrees, all combined in one circuit. Three-phase current allows higher voltages and current flow without using extremely large, expensive conductors.

tubes often were needed to rectify ac into dc. SCRs are not the only kind of solid-state rectifiers used in welding power sources. (*Solid state* means that the electrical apparatus does not use vacuum tubes.)

Another kind of rectifier commonly used in welding power sources is a *selenium* rectifier. Selenium is another chemical element that is a good semiconductor material. The selenium rectifier looks like a series of cooling fins on a shaft. They look quite a lot like a steam-pipe radiator with fins several inches long (Fig. 11-38A). Each radiator fin is part of one diode unit that makes up the selenium rectifier. More diode sections are added in series to increase the rectifier's voltage rating.

Silicon diodes look like a big bolt and nut with a braided copper pigtail (Fig. 11-38B). Silicon rectifiers are bolted onto aluminum cooling plates or some specially shaped cooling device with fins. They have a much higher voltage rating than selenium units.

Mercury-pool rectifiers are yet another kind of rectifier used in certain kinds of welding equipment. They have very high current capacity. They are often used where high power is needed, such as in electric locomotives and in resistance-welding equipment.

Rectifier breakdowns When the voltage rating of a diode is exceeded, it will break down. Silicon diodes fail completely, but selenium diodes sometimes "heal" themselves, even though they have lower voltage ratings than silicon-diode rectifiers. In fact, selenium diodes sometimes are connected across silicon diodes to protect the silicon units from damage caused by transient voltage surges from your friendly power company.

It is not possible to repair a selenium rectifier assembly. When it fails, the entire unit must be replaced. When a silicon diode fails, it can be removed just as a bolt can be removed, and then replaced. It is not necessary to discard the cooling plates and other

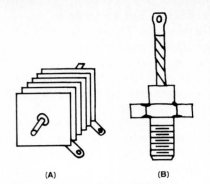

FIGURE 11-38 The rectifiers used in welding machines are mostly solid-state devices. Years ago, they used to be vacuum tubes. Rectifier (A) is made from selenium while rectifier (B) is made from silicon.

things attached to the silicon-diode rectifier. Only the damaged silicon diode needs to be replaced.

Contactors, Relays, and Switches

Welding circuits, control circuits, all other kinds of electrical circuits have to be turned on and off. The easiest way to do that is by opening a set of *contacts*. Contactors, relays, and switches all use contacts to interrupt a circuit. Generally, the term *contactor* is used in arc-welding equipment to mean a heavy-duty relay that interrupts the main-line power.

The term *relay* is usually reserved for things with contacts (switches) that are operated by an electromechanical device. However, solid-state relays are rapidly replacing the electromechanical relays that have been in use for so many years. Relays are most often used to protect electrical circuits from current overloads. For example, a lightning arrestor is a special type of relay that switches off before the very high voltage and current of a lightning bolt can burn out the rest of the electrical circuit.

The term *switch* is usually used for a device with contacts that are hand-operated. All three terms are apt to be used interchange-

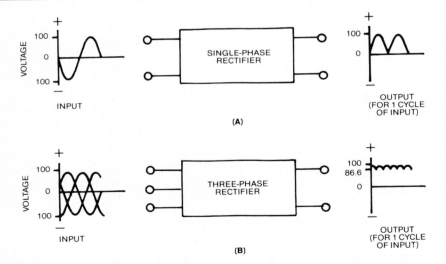

FIGURE 11-37 Alternating current can be converted into pulsing direct current (all flow in one direction) with devices called *rectifiers*. (A) A single-phase rectifier produces a dc current that varies from 0 V to +100 V and back to 0 V (but it never changes from one polarity to another which would make the current flow in the opposite direction, producing alternating current). (B) A three-phase dc rectifier produces a nearly smooth positive-polarity direct current of about 86.6 V. Large production welding machines often use three-phase input current. They require special plant wiring and special wall plugs with three conductors.

ably. An example is saying that "a contactor is a relay that switches current." A switch is a contactor that you operate by hand.

The contact material is one of the most important parts of a switching system. When a circuit is interrupted, the current wants to keep flowing. Some electrons continue to jump the space between the contacts as the switch is opened and the two sides of the contact are separated. When the space between the contacts is made large enough, the electrons cannot jump the gap. Current flow stops. The circuit is turned off.

The higher the voltage, the greater the gap the electrons can jump. This is why low-voltage switching devices should not be used in high-voltage circuits. With high voltage, the electrons will continue to jump the contact gap. Even though the contacts are opened, the circuit won't turn off. Not only that, but the heat generated by the arc between the contacts also will destroy the contacts themselves and may cause a fire. The sustained arc that can jump between high-voltage contacts is just as hot as a welding arc.

The design of switching systems for welding power sources is very complicated. It requires a great deal of know-how. Selection of the correct switch, contactor, or relay is not easy. In addition to all of the electrical problems, there are problems of mechanical life, safety factors, mounting methods, and cost. Even though contacts, relays, and switches are the most common equipment failures that occur in arc-welding equipment, you should never attempt to repair or replace them yourself without expert advice and guidance.

⚙ **CAREFUL: Never try to repair your own arc-welding equipment.**

The best idea is to call the manufacturer of the equipment or the welding distributor who supplied the power source to you and let them repair the equipment. Most large manufacturers and many welding distributors have repair facilities located near most major industrial areas.

Fuses, Circuit Breakers, and Thermal Cutouts

Electrical equipment should be protected from overloads. Overloading welding power sources (asking for more current than the machine is rated to supply) will cause them to overheat because of the extra I^2R heating inside the machine. Most safety devices on welding power sources are based on the effects of I^2R heat. A fuse is nothing more than a conductor with a calibrated resistance and a low melting point. Normal current flow cannot create enough I^2R heat to melt the fuse, but extra current will cause the fuse to blow (melt apart so that current cannot flow through it).

A slight current overload takes much longer to melt a fuse than a severe current overload. The current-time relationship which causes a fuse to operate is dependent upon the size of the fuse and the material from which it is made. All fuses have an I^2t rating. The I^2 part of the rating comes from the I^2R heat, and the t comes from the time it takes the current to melt the fuse. The R is the calibrated fuse resistance. The higher the current, the shorter the time it takes to blow a fuse.

Most circuit breakers are tripped by the I^2R heat generated in a bimetal strip (two thin pieces of different metal sandwiched together). Current is passed through the bimetal strip. The I^2R heat developed in the strip causes it to expand. The two different metals expand at different rates. The different expansion rates cause the bimetal strip to bend and apply pressure to a tripping mechanism.

Just as with fuses, circuit breakers have I^2t ratings. The higher the current, the shorter the time it takes to open the circuit breaker. Whereas fuses must be replaced when they are blown, a circuit breaker can be reset. Some circuit breakers use a magnetic field to trip the mechanism. Whether it is heat or magnetism that trips the breaker, the result is the same—the circuit is interrupted and protected.

Thermal cutoffs are nothing more than bimetal strips (two metals with different coefficients of thermal expansion bonded together so that when they are heated the composite strip bends and either makes or breaks a circuit). The bimetal strips are arranged to operate small contacts. The mechanism is usually sealed in a small, electrically insulated tube with the switch leads brought out of it. The insulated tube can be placed inside a current-carrying coil, in touch with a transformer core, on a rectifier cooling plate, or wherever a current overload will cause a temperature rise. When the thermal cutoff switch contacts open and close, they can be used to operate a relay or contactor to interrupt the main power.

11-10 DUTY CYCLES AND VOLT-AMP OUTPUT RELATIONSHIPS

There are two more practical topics left to cover—duty cycles and volt-amp output relations of power sources. You need to understand both of them before you can start arc-welding.

Power-Source Duty Cycles

Duty cycle is one way all welding power sources are rated and also protected from thermal overloads. Duty cycle tells you how long a power supply can be run at its rated output amperage without overheating. Duty cycle is always expressed as a percent.

Only arc-welding machines with a 100 percent duty cycle can

be run continuously. With anything less than a 100 percent duty cycle, the machine must be turned off part of the time to cool down. The I^2R heating losses inside the machine are what cause power sources to heat up. The machines with 100 percent duty cycle have bigger copper conductors and better fans and other cooling devices to keep them from overheating. That also means that 100 percent duty cycle machines cost more to make and more to buy.

Determining duty-cycle ratings

A 10-min test period usually is used by the manufacturer to determine the duty cycle of a welding power source. A power supply that can be operated from one 10-min period to the next without interruption, for an unlimited time, is said to have a 100 percent duty cycle.

If a machine can only run 5 min out of 10 at its rated amperage capacity, then the machine has a 50 percent duty cycle. If a machine can be run at its rated current output for 8 min but has to be shut down for 2 min out of every 10, the power source has an 80 percent duty cycle. A welding machine with a 60 percent duty cycle can be operated at its rated load for 6 min out of every 10, with a 4-min cooling period.

Since stick-electrode welding is inherently intermittent because you have to stop periodically to get rid of a used electrode stub and replace it with a new electrode, most SMAW welding machines are not 100 percent duty cycle machines. While you are changing electrodes, the machine is cooling down. Machines that operate semiautomatic and automatic welding processes such as MIG, flux-cored wire welding, and submerged arc welding, are almost always 100 percent duty cycle machines. These processes require continuous, nonstop welding capability.

Most machines have built-in safety devices to turn the welding current off if the duty cycle is exceeded. These are the devices we described above as *thermal cutoff* devices. If your welding machine keeps shutting itself down automatically, don't think there is something wrong with the machine. It's doing exactly what it is supposed to do. You may be exceeding the machine's duty cycle.

Calculating a duty cycle

Generally, I^2R heat is the limiting factor in determining the duty cycle. Since the internal resistance (R) of the power source is usually constant, it is sometimes possible to use current-squared

(I^2) ratios to determine the duty cycle at some other current than the maximum rated current for the machine. Less current means less heat produced inside the welding machine, which means that the power source can run for a longer time.

For example, at one-half the rated current, the I^2R heat produced in the machine is only one-fourth of the amount of heat that would be produced if the machine ran at maximum rated output. Operating for four times as long with one-fourth the I^2R heat generation will produce the same total heat inside the welding machine as when the power source is operated at full current output.

The duty cycle of a power source can be estimated by using the following formula:

Percent duty cycle =   (Eq. 11-20)

$$\frac{(\text{rated current})^2}{(\text{load current})^2} \times \text{rated duty cycle}$$

The literature that comes with welding machines almost always gives you the rated duty cycle at different welding currents in amps. Figure 11-39 is an example showing three different machines. In this figure, the three machines are rated at 20 percent, 60 percent, and 100 percent duty cycles. The percent of rated current is plotted against the percent duty cycle to give the three amperage curves.

Here's how to use tables like this one.

■ The duty cycle of a machine rated at 100 A is 60 percent. When the power source is used to produce 150 A of welding current (a 150 percent load), what will the machine's duty cycle actually be? From the duty-cycle equation (Eq. 11-20):

$$\% \text{ duty cycle} = \frac{(100 \text{ A})^2}{(150 \text{ A})^2} \times 60\%$$

$$= 27\% \text{ duty cycle} \quad \text{(Eq. 11-21)}$$

■ Find the output current of a 100-A, 60 percent duty cycle power source when it is operated on a 100 percent duty cycle.

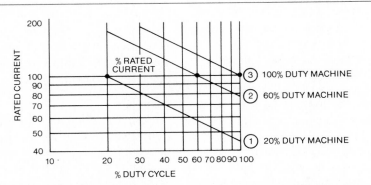

FIGURE 11-39 The duty cycle of a power source is very important to any arc welder who uses it. The rated duty cycle tells you how long the machine can be used before it has to stop and cool off. Curves for three machines are plotted here. Machine (1) can be used at 45 A for 100 percent of the time, but at 100 A 20 percent of the time. If you needed a 100-A machine with a 100 percent duty cycle, you would select machine (3).

A different form of the previous equation will give us the answer to this problem. Rearranging Eq. 11-20 to solve for (load current)², we need then only take the square root of our answer to determine the load current itself.

(Eq. 11-22)

$$(\text{load current})^2 =$$
$$(\text{rated current})^2 \times \frac{\text{rated duty cycle}}{\text{percent duty cycle}}$$

■ Substituting from the above problem into Eq. 11-22, we have

$$(\text{load current})^2 = (100\ \text{A})^2 \times \frac{60\%}{100\%}$$
$$= 10{,}000\ \text{A} \times 0.60$$
$$= 6000\ \text{A}^2$$

Load current is the square root of 6000 A² or approximately 77.5 A.

The results of these calculations also can be found by using the graph in Fig. 11-39, curve 2. It shows a 77 percent rated current (on the right-hand side of the graph) at 100 percent duty cycle for the 60 percent duty cycle machine. And 77 percent of 100 A is 77 A.

Similarly, curve 1 shows a power supply rated for a 20 percent duty cycle at 100 A that may be operated on a 55 percent duty cycle to produce 63 A. The same power supply can operate on an 80 percent duty cycle when the output current is reduced to 50 A.

Curve 3 shows that a machine with a 100 percent duty cycle of 100 A can be operated at 150 percent of load or 150 A if the duty cycle is reduced to 44 percent, and at 200 percent of load or 200 A if the duty cycle is 25 percent (curve projected to 200 A).

While curves like these can be used with individual power sources, it's never a good idea to overload a power source, even for a brief period of time. You run the risk of blowing fuses or permanently damaging the machine. This is the main reason why welding power sources should be purchased with greater rated capacities than you expect to use. Sooner or later you will need the extra capacity.

Volt-Amp Output Relationships

The current-output characteristics of a power source can be represented on a graph of volts versus amperes for the machine. The volt-amp curve of an ideal dc transformer or a battery can be represented by a straight line, as shown in Fig. 11-40. The line shows that no matter what current is drawn from the perfect transformer (or battery) the voltage remains constant.

Theory and reality Theoretically there is no limit to the amount of current that could be drawn from such a perfect electrical device. The same curve even represents the output of the public utility that supplies the ac to your home. No matter how many appliances you use (increased current draw), the voltage produced by the power company and supplied to your home remains the same because the amount of current you are drawing is so small compared with the amount of current they can supply you. In reality, of course, even the public utilities experience voltage reductions—that's what a "brown out" is.

Standard transformers and batteries do not provide constant output voltage at all currents. Real welding power sources are made with wires and materials that have resistance. For example, a real battery (or any other power source) must have an internal resistance or it would short out. A typical internal resistance for a power supply, is 0.02 Ω, which would cause the terminal voltage to drop off 2 V for every 100 A drawn from the unit. It is shown in Fig. 11-41. Welding power sources called *constant-voltage* (CV) machines have output curves that look very much like this figure.

A battery would have a perfectly flat instead of slightly sloping output curve up to to the point where the battery's ability to supply more current becomes limited. But batteries can't provide enough current to be used in welding. (They once were used on low-capacity machines many years ago.)

In transformer terminology, the falloff of voltage with an increase in current is referred to as *regulation*. In welding, *slope* is used to describe voltage regulation. The slope is expressed in voltage drop per 100 A of current flow. In Fig.

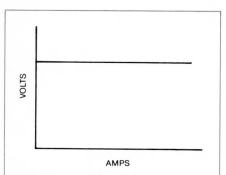

FIGURE 11-40 Volts or amperes can be plotted against time. They also can be plotted against each other. These voltage-amperage graphs are very important when describing the characteristics of any dc power source. In this voltage-amperage curve, the voltage remains unchanged as the amperage increases. This curve is typical of a battery with only a very small resistance in it.

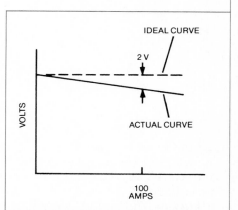

FIGURE 11-41 This is the voltage-amperage curve of a power source with 0.02 ohms of internal resistance. In use at 100 A, the output voltage will drop by 2 V.

11-41, the slope of the line is 2 V/100 A.

The term *slope* is derived from the downhill slant of the volt-amp curve. The slope of a hill is its rise divided by its run. Power-supply slope also is defined in Fig. 11-42 as the voltage rise/run.

Slopes and power sources A power supply with a great deal of internal voltage drop due to resistance or impedance (a type of resistance caused by capacitors) will have a steep slope. A power supply with very little internal voltage drop due to resistance or impedance will have a flat slope.

CV POWER SOURCES A flat-slope power supply is usually called a *constant-voltage* (CV) or *constant potential* (CP) unit. These are names that you often will see used when describing welding power sources. The output voltage of a CV power source remains essentially constant over the normal operating range of currents. Such power sources are usually used for MIG welding because a steady voltage is very important in this process.

CONSTANT-CURRENT POWER SOURCES (DROOPERS) Some power sources have a very steep slope. These units are often called *droopers* because the droop (slope) of the volt-amp curve is very steep. Their output current is essentially constant in the normal operating range. A typical drooper power supply volt-amp curve for a constant-current drooper machine is shown in Fig. 11-43.

Droopers are used for a number of welding processes, but primarily for covered stick-electrode (SMAW) welding and GTAW (TIG) welding because a steady current is very imporant in these processes. These power supplies are also called constant-current or CC machines. The OCV of these machines is fixed by their design so that the power source provides constant current output.

SLOPE-CONTROLLED CV POWER SOURCES These machines are used most frequently for MIG welding. Some MIG welding machines are built with all sorts of slopes that the welder can dial in by using feedback circuits, magnetic amplifiers, solid-state controls, and so on. They are called *slope-controlled CV power sources.* These machines are more expensive than standard MIG power sources, but they are extremely versatile and very productive.

When used in combination with voltage adjustment, it is possible to use the slope adjustment to dial in the slope of the power supply voltage-amperage curve as it passes through a specific arc operating point. This is shown in Fig. 11-44. Note that the no-load voltage (another name for the OCV) of the solid line in this figure is higher than that of the dashed line, and its slope is greater (steeper).

The use of voltage and slope control in various combinations can generate families of power-supply output curves from a single arc-welding power source.

Figure 11-45 shows the voltage-amperage output curve from a CV power source without slope control.

Figure 11-46 shows the voltage-amperage output curve from the same CV machine when some slope is added by the operator. A CV machine will never have a perfectly flat output curve. The best the machine builder can provide is a curve with a very small slope of about 2 V per 100 A. The reason is that all electrical machines have some internal resistance, no matter how big the copper wires in them are.

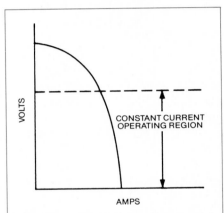

FIGURE 11-43 Constant-current welding machines have drooping voltage-amperage curves. They are most often used for stick-electrode welding.

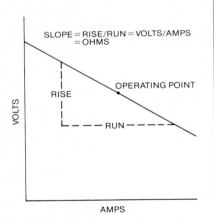

FIGURE 11-42 A drop in voltage with increasing amperage is measured by slope. Slope is the tangent of the angle on this graph. The tangent is calculated as output volts divided by output amperes. Slope is measured in units of ohms. Slope is used to control the welding characteristics of some welding power sources.

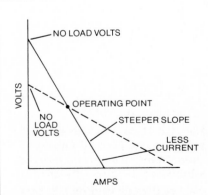

FIGURE 11-44 Some welding power sources allow you to change the slope of the machine. In following chapters of this book, you will learn how to use variable slope to control different arc-welding characteristics.

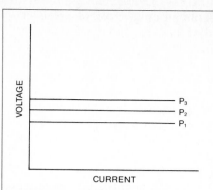

FIGURE 11-45 Voltage-amperage curves for three constant voltage (CV) power sources (alternatively called *constant-potential*, or *CP*, *machines*) are shown here. Constant-potential machines are most often used for GMAW (MIG) welding.

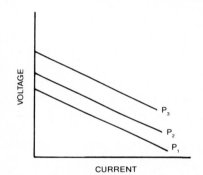

FIGURE 11-46 Slope can be added to CV power sources. More expensive CV machines have dials that allow you to adjust the slope. Other machines have a single built-in slope.

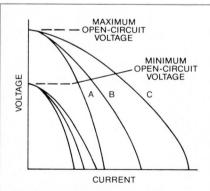

FIGURE 11-47 If you have a slope control on your dc constant-current machine, you can use different welding currents at the same minimum and maximum open-circuit voltages.

Figure 11-47 shows you the voltage-amperage output curve for a drooping type constant-current (CC) machine with different amounts of slope control, when the welding output current is set at two different OCVs. The machine can operate between these extremes. However, the lowest OCV setting would produce very bad welding results with stick electrodes.

At this point you may have decided to give up arc welding as impossible to understand. Don't. You will catch on because we will go over these details again and again as we describe the machines used for each arc-welding process. What's more, you will be making machine settings on your own power sources and you will soon see and hear the difference in the arc. You also will see the results in the weld metal when these different machine settings are made. Everybody is confused with arc-welding power sources at first, and everybody who sticks with them soon begins to understand what it's all about.

The specific use of slope in welding power sources and in different welding processes also will be described later on. All you need to remember right now is that there are two kinds of arc-welding-machine current and voltage controls, either constant-voltage (CV or CP) machines, and constant-current (CC) machines. CV machines are used for MIG welding, while CC machines are used for stick-electrode and TIG welding.

Some welding processes only use ac or dc from these machines and some processes use either type of current, depending on the filler metal used, the type of power source you have on hand, and the welding job.

11-11
USING WHAT YOU'VE LEARNED

The best way to sum up this chapter is to show you that you've already learned more about arc welding and electricity than you may think. We'll also show you how some of the information you've learned in this chapter is actually applied in real arc-welding situations.

A set of tables that you normally would have to find by fishing around in a lot of different welding handbooks is included here. Each table is important to you in setting up welding equipment. Each table also shows you how some part of this lesson is used in welding. Without this lesson, you would have had difficulty understanding what these important tables are all about.

The first one is very simple. Table 11-1 is a listing of American Wire Gauge cable sizes and

TABLE 11-1 American Wire Gauge welding cable sizes and characteristics

American Wire Gauge (AWG) Size	Nominal Overall Diameter, in.	Approximate Weight per Foot, lb/1000 ft	DC Resistance, Ohms/1000 ft at 68°F	Nominal Metric Cross-Sectional Area, mm²
8	0.340	121	0.688	10
6	0.390	137	0.435	16
4	0.440	194	0.272	25
2	0.550	306	0.173	35
1	0.600	376	0.137	50
1/0	0.660	464	0.109	60
2/0	0.715	563	0.087	70
3/0	0.785	708	0.068	95
4/0	0.875	884	0.054	120
250 MCM	0.980	1070	0.045	150
300 MCM	1.060	1260	0.038	185

the equivalent nominal (average) diameters for the cable conductor in inches. It also gives you data on the weight of each size cable, its resistance in ohms/1000 ft and even the metric cross-sectional area of the cable, in case you wanted precalculated cable cross sections to make comparisons of resistance data with other cable sizes. The fact that the areas of the cable are in metric units doesn't matter at all. *It's the relative cross-sectional areas of different cables that count* when comparing them to each other to see which ones will give the least resistance, and how much that resistance will be.

As you know, the current-carrying ability of a cable is increased four times as the diameter is doubled and twice when the area is doubled. Table 11-2 lists the maximum open-circuit welding current in amps that each AWG cable should carry if you want to restrict voltage drops (due to cable resistance) to no more than a certain amount. For example, a 50-A welding current flowing through 100 ft of AWG 2 cable will produce a voltage drop of 1 V due to the cable's resistance. If you increase the cable size, you decrease the cable resistance, so that the same 50 A of current will produce only a 0.3-V drop run through 100 ft of AWG 4/0 cable. A huge 700-A welding current would require at least a 4/0 AWG cable size.

The next table, Table 11-3, lists copper cable sizes for a maximum 4-V line drop due to cable resistance. It's much like the previous table except it is probably more useful to you. A 4-V drop in electrical line pressure is usually about all you want to allow in an electrode or ground cable. This table helps you select the best AWG cable size depending upon the welding current that you intend to use. Also note that the table is split by welding machines with up to 60 percent duty cycles and welding machines with 60 percent duty cycles and more. Bigger electrode and ground cables are used for welding machines with higher duty cycles because the cables, as well as the welding machines, are subject to I^2R heating.

Table 11-4 is yet another way of looking at welding cable, this time specifically for single-phase and three-phase current, for dc

TABLE 11-2 Voltage drop per 100 ft of lead versus AWG cable size

Welding Current, A	AWG 2	AWG 1	AWG 1/0	AWG 2/0	AWG 3/0	AWG 4/0
50	1.0	0.7	0.5	0.4	0.3	0.3
75	1.3	1.0	0.8	0.7	0.5	0.4
100	1.8	1.4	1.2	0.9	0.7	0.6
125	2.3	1.7	1.4	1.1	1.0	0.7
150	2.8	2.1	1.7	1.4	1.1	0.9
175	3.3	2.6	2.0	0.7	1.3	1.0
200	3.7	3.0	2.4	2.0	1.5	1.2
250	4.7	3.6	3.0	2.4	1.8	1.5
300	—	4.4	3.4	2.8	2.2	1.7
350	—	—	4.0	3.2	2.5	2.0
400	—	—	4.6	3.7	2.9	2.3
450	—	—	—	4.2	3.2	2.6
500	—	—	—	4.7	3.6	2.8
550	—	—	—	—	3.9	3.1
600	—	—	—	—	4.3	3.4
650	—	—	—	—	—	3.7
700	—	—	—	—	—	4.0

TABLE 11-3 AWG copper cable sizes for maximum 4-V welding line drop for single cable (double the distance for two cables)

Feet from Welding Machine

Welding Current, A	50	75	100	125	150	175	200	225	250	300	350	400
Welding Machines with Up to 60% Duty Cycle												
100	2	2	2	2	1	1/0	1/0	2/0	2/0	3/0	4/0	4/0
150	2	2	1	1/0	2/0	3/0	3/0	4/0				
200	2	1	1/0	2/0	3/0	4/0	4/0					
250	2	1/0	2/0	3/0	4/0							
300	1	2/0	3/0	4/0								
350	1/0	3/0	4/0									
400	2/0	3/0	4/0									
450	3/0	4/0										
500	3/0	4/0										
550	4/0											
600	4/0											
Welding Machines with 60 to 100% Duty Cycle and Automatic Welding Equipment												
400	4/0	4/0										
800	2–4/0	2–4/0										
1200	3–4/0	3–4/0										
1600	4–4/0	4–4/0										

Note: These recommended cable sizes will produce a voltage drop of no more than 4 V between the welding machine and the workpiece. Note that cable size increases as you read across the chart and amperage increases as you read down the chart.

rectifier or ac transformer-type welding machines, and for motor-driven three-phase welding power source transformers. If you intend to use a three-phase ac transformer, for example, and want to weld with 140 A of current, then look in the third column of the table from the left (three-phase transformers), then look on the last line of this column and your requirement will fall between 130 A and 150 A. Reading over to the right, you will need an AWG 4/0 welding cable.

Now assume that you are going to run with more than one welding operator using one big 1000-A power source with a 60 percent duty cycle rating. Each operator will be working with 200 A when welding. They're all using stick electrodes so that the 60 percent duty cycle rating should be OK for the machine. How many operators can you hook up to this 1000-A, 60 percent duty cycle power source without overloading it. Read down the right-hand "60 percent duty cycle" column of Table 11-5. Then read across the row at 200 A and you will see that you can hook up as many as eight separate operators without overheating the 1000-A power source.

Table 11-6 is a very handy summary of electric welding processes by the type of welding power source they use. TIG welding uses CC drooping power sources, either of the ac or dc type, depending on whether you are TIG welding with ac or dc. GMAW welding with inert-gas shielding (that's MIG welding, of course) sometimes can use either CC drooping dc power sources or CV power sources. The CC machines are very limited for MIG welding, however. They are primarily used for stick electrodes. CV power sources are by far most often used for MIG welding. You don't use ac welding with the MIG process. And so the table continues, welding process by welding process.

Welding power sources aren't the only equipment rated by duty

TABLE 11-4 Copper power cable sizes for welding machines

Motor-Driven Three-Phase, A	Rectifier or Transformer Single-Phase, A	Rectifier or Transformer Three-Phase, A	Three-Phase Power Cable Wire Size, AWG
Up to 24	Up to 30	Up to 24	10
24–32	30–40	24–32	8
32–44	40–55	32–44	6
44–64	55–70	44–64	4
64–76	70–95	64–76	2
76–88	95–110	76–88	1
88–100	110–125	88–100	1/0
100–130	125–165	100–130	2/0
130–155	165–195	130–150	4/0

TABLE 11-5 Maximum welding machines from 1000-A power supply

Amps per Arc	Duty Cycle of Welding Machines on Line							
	25%	30%	35%	40%	45%	50%	55%	60%
50	80	66	56	50	44	40	36	32
75	56	44	38	32	28	26	24	22
100	40	32	28	24	22	20	18	16
125	32	26	22	20	16	16	14	12
150	26	20	18	16	14	12	12	10
175	22	18	16	16	12	10	10	8
200	20	16	14	14	10	10	8	8
225	16	14	12	12	8	8	8	6
250	16	12	10	10	8	8	6	6
275	14	12	10	8	8	6	6	6
300	12	10	8	8	6	6	6	4
325	12	10	8	6	6	6	4	4
350	10	8	8	6	6	4	4	4
375	10	8	6	6	4	4	4	4
400	10	8	6	6	4	4	4	4

TABLE 11-6 Welding processes and recommended types of welding machines

Arc-Welding Process	DC		AC
	CC Drooping	CV Flat	CC Drooping
Gas-tungsten arc welding (GTAW) or TIG	Yes	Not used	Yes
Plasma-arc welding (PAW)	Yes	Not used	No
Shielded-metal arc welding (SMAW) or stick	Yes	Not used	Yes
Gas-metal arc welding (GMAW) (inert gas)	Yes	Yes	Not used
Gas-metal arc welding (GMAW) (spray transfer)	Not very good	Yes	Not used
Gas-metal arc welding (GMAW) (CO_2 shielding)	Not very good	Yes	Not used
Gas-metal arc welding (GMAW) (shorting arc)	Not used	Yes	Not used
Flux-cored arc welding (FCAW)	Yes	Yes	Possible
Carbon-arc welding/cutting/gouging (AAC)	Yes	Possible	Yes
Submerged-arc welding (SAW)	Yes	Yes	Yes
Stud-arc welding (SW)	Yes	Possible	Not used
Electroslag welding (EW)	Possible	Yes	Usable

cycles. As shown in Table 11-7, SMAW electrode holders for stick electrodes also are rated the same way (and with the very same 10-min period equal to a 100 percent duty cycle rating). But since stick-electrode welding is by its very nature intermittent—the operator has to stop periodically and change electrodes—there is no such thing as a stick-electrode holder with a 100 percent duty cycle rating. It's simply not needed.

The biggest and best "extra-large" stick-electrode holder you can buy is one with a 75 percent duty cycle rating for use with ³⁄₈-in.-diameter electrodes operated on a maximum current of 600 A, and that's a very hot welding setting. In all probability you will never weld with stick electrodes at that high a current setting. But if you do, you know what class of electrode holder to order so that it won't overheat in your gloved hand.

Speaking of electrodes, Table 11-8 gives you a table of stick-electrode and uncoated welding-wire diameters versus the average current used to weld with them, by five different welding processes. Also note our polarity settings right in the table; these are DCRP (REPresentative = reverse electrode-positive) and DCSP (SENator = straight-electrode negative).

Finally, in Table 11-9 we give you a list of the types of welding-wire feeders that are used with different kinds of welding power sources. While you haven't studied MIG, flux-cored wire welding, or submerged-arc welding yet (all of which use electrode wire and automatic feeders to feed the wire into the weld metal), it's probably no surprise to you that different welding-machine output currents even affect the type of controls needed to operate different welding-wire feeders. This table gives you a guide to matching the welding process and the power source to the right kind of wire feeder for the job.

TABLE 11-7 Stick electrode holder sizes and ratings

Electrode Holder Class	Maximum Current, A	Duty Cycle, %	Electrode Size, Max. in.	Cable Size, Max. AWG	Nominal Weight, oz
Small	100	50	¹⁄₈	1	10–12
Small	200	50	⁵⁄₃₂	1/0	10–14
Medium	300	60	⁷⁄₃₂	2/0	12–20
Large	400	60	¹⁄₄	3/0	16–26
Extra-large	500	75	⁵⁄₁₆	4/0	22–30
Extra-large	600	75	³⁄₈	4/0	28–36

TABLE 11-8 Average current required versus electrode size and welding process

Diameter Electrode, Carbon or Welding Stud	SMAW with DCSP or DCRP	GTAW with DCSP	Air-Carbon-Arc with DCSP	Stud Welding with Maximum DCSP
0.035	—	50	—	—
0.045	—	75	—	—
¹⁄₁₆	—	110	—	—
³⁄₃₂	75	180	—	—
¹⁄₈	100	235	—	—
⁵⁄₃₂	150	300	150	—
³⁄₁₆	200	400	200	300
⁷⁄₃₂	275	450	300	350
¹⁄₄	350	500	400	400
⁵⁄₁₆	—	—	500	500
³⁄₈	—	—	600	600
¹⁄₂	—	—	1200	900

TABLE 11-9 Selecting a wire feeder for different power sources

Type of Power Source	Welding Voltage Sensing (Wire Speed Varies)	Constant Speed (Wire Speed Preset)
CV direct current	Difficult to adjust, seldom used, self-regulating within limits.	Best for GMAW. Best for flux-cored arc welding. Best for submerged-arc welding when using small-diameter electrode wire. Self-regulating.
CC direct current	Best for submerged-arc when using large-diameter electrode wires. Used for GMAW on aluminum. Self-regulating.	Difficult to control. Not used. Not self-regulating.
CC alternating current	Used for submerged-arc (medium and large electrode diameters). Used for flux-cored arc welding. Self-regulating.	Difficult to control. Not used. Not self-regulating.

Before you started this lesson, none of this information would have made any sense to you at all. With each lesson that follows, electricity, and arc welding, will make more and more sense to you.

REVIEW QUESTIONS

1. What is the difference between an insulator, a semiconductor, and a conductor?

2. Name a material that is a good insulator for welding voltages and another material that is a good conductor of welding current. What is a common semiconductor material?

3. What is the electrical symbol for a resistor? What units are commonly used to measure electrical resistance?

4. What units are most often used to measure electrical pressure?

5. What units are used to measure electrical quantity or current?

6. Describe the difference between ac and dc current.

7. What is the most important difference between static electricity and the electricity used in shielded metal-arc welding, MIG and TIG?

8. What does open circuit voltage mean when applied to a welding power source? Where is it measured in a welding circuit?

9. Describe what happens to the resistance in a circuit when it shorts out. What can happen to a conductor carrying a short circuit?

10. State Ohm's law three different ways for calculating first the current, then the resistance, then the voltage in a dc circuit.

11. Give a formula that describes power in terms of energy and time. What does it mean? What electrical units are used to measure power?

12. Give a formula for power in terms of current and voltage.

13. What is the formula that describes power loss in terms of current and resistance? Where does the lost power go in an electrical circuit; that is, what does it become? Why is that information useful to welders?

14. One welding cable is 1 in. in diameter. Another cable is ½ in. in diameter. Which cable will have the lower electrical resistance? Which cable will heat up most when a current flows through it?

15. You are going to set up two dc power sources for two different welding jobs. One machine setup requires dc straight polarity and the other machine requires dc reverse polarity. Is your welding electrode positive or negative in the DCSP circuit? Is it positive or negative in the DCRP circuit?

16. An ac welding circuit has a peak voltage of 100 V. What is the average voltage of the circuit in volts? What is the effective (RMS) voltage?

17. The peak amperage of an ac circuit is 10 amps. How many amps does an equivalent dc circuit carry if both circuits will produce the same amount of heat? Use the idea of RMS voltage and amperage of the ac circuit to explain your answer.

18. If a rectifier is put into an ac circuit, what kind of current will flow out of the rectifier?

19. What does the duty cycle of a power source mean? If a power source can operate for 6.5 min before overheating, what is its duty cycle?

20. What common welding processes are most likely to use a dc constant current (drooper) power source?

21. What common welding process is most likely to use a dc constant voltage power source?

22. How much voltage is likely to be lost due to the resistance of a 100 ft AWG 1/0 copper welding cable?

23. If you are 50 ft from a 60 percent duty-cycle dc welding power source and you are hooking up the electrode and ground cables, what copper cable size should you use to keep the voltage drop in the line to no more than 4 V?

24. Would you use a constant current or a constant voltage dc power source for GTAW? What about SMAW? PAW? Would you most likely use a constant voltage or constant current dc power source for GMAW welding? Name one welding process that might use constant current dc, constant voltage dc, or an ac machine.

25. What copper cable size and duty cycle should your stick electrode holder be if you plan to use ⅜-in.-diameter covered electrodes?

CHAPTER 12

Arc-Welding Safely

Many welding and cutting processes use electricity instead of oxygen and a fuel gas to make the heat used for welding. In fact, electric arc-welding processes are the dominant production-welding methods used in industry.

There are a variety of arc-welding processes, the most common of which are SMAW or shielded-metal arc welding using flux-covered stick electrodes; gas-metal arc welding [GMAW which includes metal inert-gas-shielded (MIG) solid-wire and CO_2-shielded flux-cored wire welding]; and submerged-arc welding (SAW) which is extensively used for downhand welding of thick plate by running a continuous wire electrode submerged in a pile of flux.

One common arc-cutting process is air-carbon arc (or oxyarc) cutting which often is used for gouging out bad welds, for chamfering plate edges, and for piercing holes. It's used on carbon steels and on metals such as stainless steel, aluminum, and copper which cannot be cut by oxyfuel processes. Two newer processes, plasma-arc welding (PAW) and plasma-arc cutting (PAC), use very high energy arcs to make a gas as hot as the surface of the sun do the work for you. Plasma-cutting, like air-carbon-arc cutting, can slice up any metal.

All of these processes from SMAW to PAC use high-amperage currents that can hurt you. They all require special care and safety precautions that are different from those of the oxyfuel processes. For example, invisible rays (ultraviolet radiation) from any kind of arc can give you a severe sunburn. It also can sunburn the insides of your eyes, and that is why arc welders wear lens shades on their helmets that are much darker than those used by oxyfuel welders and flame-cutters.

This chapter will tell you the common arc-welding hazards and what to do to avoid them. Before we scare you, you should know that arc welders have fewer injuries than the average plant worker. In fact, welders are safer at work than they are at home.

Welders stay safe partly because most arc-welding equipment is designed to be safe as well as easy to use. However, the main reason arc welders have such good safety records is that they have carefully trained themselves not to make mistakes. You won't know what those mistakes are until you understand what the hazards are, which is what this chapter is for.

12-1 WELDING MACHINES

Arc-welding machines (also called power sources) are designed and built to protect you from most electrical hazards as long as you use them correctly. The manufacturers of these machines must comply with many safety requirements specified by the American National Standards Institute (ANSI), the American Welding Society (AWS), and the National Electrical Manufacturers Association (NEMA). Sometimes local city or state ordinances, and often your plant's safety director, will have a say in how this equipment is installed and used.

Hazardous Conditions

Arc-welding machines conforming to these requirements will be safe and will provide good service in normal work environments. For unusual operating conditions welding machines must be specially constructed or modified. What are these unusual conditions?

■ Air temperatures higher than 104°F [41°C] and lower than 32°F [0°C]. High temperatures will overheat the machine and very cold weather will make some types of electrical insulation brittle.
■ Altitudes above 3300 ft [1000 m], especially if the power source is an engine-driven machine. Just like an automotive engine, the carburetor in the engine drive needs a different mixture of fuel and air to work efficiently at

higher altitudes than low ones. Welding machines also cool off more slowly at higher altitudes and their duty cycles (which we'll discuss again shortly) may be different from the machine's rated duty cycle.

■ Exposure to severe weather conditions, to steam, excessive humidity, and unusually corrosive vapors and fumes. These conditions are tough on any kind of electrical equipment. But welding-equipment manufacturers can seal the machine against most of these hazards and problems if they know that's what's needed when the machine is sold.

■ Exposure to abrasive dust, to oil vapors, to flammable gases and explosive powders such as coal, or to wheat or soybean flour in a grain mill, grain elevator, or silo. These are all inherently explosive materials.

■ Abnormal vibration or mechanical shock that can loosen circuits, or exposure to nuclear radiation which rapidly ages electrical insulation.

■ Seacoast or shipboard conditions. Salt spray is very corrosive and will eat up many metals and other materials that you would think are corrosion-resistant. Salt spray will even eat the paint right off a welding-machine cabinet if you don't wash it off periodically.

If you are going to work in any unusual or hazardous conditions, be sure that your arc-welding machine and equipment are designed for such service and that they are safe for welding in these places. If you are in doubt, consult your welding supervisor, a welding engineer, or your welding distributor or equipment supplier. Never strike an arc (or an oxyfuel flame) in any area with a potentially explosive atmosphere. That not only includes things like automobile gas tanks that still have gasoline vapors in them but also places like flour mills or coal mines with fine organic dust or explosive gases.

Get Help from Electricians

Welding machines must be installed and their electrical connections made according to the instructions on the machine or in the user's manual. Only competent electricians or other trained and authorized people are allowed to install and repair them. Welding distributors and machine manufacturers both operate local welding-equipment repair stations for their customers.

As a welder, you obviously are expected to know how to set up and connect a welding machine. But you are not authorized to repair one. Never fool around with the wiring inside a machine, or poke around inside the cabinet unless you know exactly what you are doing, and are *both* qualified and authorized by your company to do the work. Also remember that the output end of the welding machine that you work with actually is the low-voltage end of the machine. The input end (and parts of the inside of the machine) can have voltages anywhere from 220 V on up to 575 V compared with open-circuit output voltages of 70-V dc to 80-V ac.

If you do operate any arc-welding equipment, you must first be properly instructed in its use. Written instructions are available from your equipment manufacturer. These manuals usually are supplied with every machine when it is delivered. Don't expect to find the operating instructions for an old machine still hanging inside the cabinet, but ask for it. You'll probably find the manual on file somewhere in the plant so you can read up on the specific machine model before you use it.

Different machines that are rated the same often will have slightly different limitations, and occasionally special directions and requirements. Be sure that you understand the limitations of your machine. If you have any questions, ask your welding supervisor, engineer, local welding distributor, or the equipment manufacturer. Nobody will think

you are a "green" operator for asking good questions, because nobody wants to work with an unsafe welder around them. If you don't hurt yourself, you can hurt other innocent people in the plant.

Keep unqualified people away from your equipment. People who meddle with your equipment can hurt you as well as themselves.

12-2
ELECTRICAL HAZARDS

Electric shock is the most critical hazard in arc welding. That's pretty obvious. You can better understand and protect yourself from this hazard if you understand how electricity can hurt you.

An electric circuit, as you already know from the last chapter, has four important properties that can do some damage to you. They are the amperage (current flow), the voltage, the resistance, and the type of current (dc or ac) that you will work with.

High voltage is more serious than high amperage, although both are dangerous. Electrical current flow is the "substance" that causes injury in electric shock. The more electricity (amperes) that flows through you, the more likely you are to be hurt. But high voltage is the factor that forces the current to overcome the resistance in your glove or your skin so that the amperage can enter you. That's why you have to be so careful with a power source when you are not welding. The open circuit voltage (OCV) can really do some damage to you. That's why the OCV of most welding power sources is limited to a maximum of 80 V on ac machines, and 70 V on dc machines.

Other factors that affect the degree of injury are the duration of current flow, the frequency when ac is used, and the part of your body that carries the current.

What a Shock Can Do

If a dangerous current goes into your right thumb and out of your right little finger, you will probably survive. In fact, you may not even realize that you were in any danger. If the same current goes into your right thumb and out your left thumb, it might kill you. The electricity probably will pass through your heart while getting from one side of you to the other.

Most electrical injuries occur when current enters a hand and leaves through a foot. Most of these accidents occur when the welder is not welding but is exposed to the full OCV of the machine. Most accidents can be avoided simply by wearing gloves and not standing in a puddle of water while welding.

Human skin acts as a natural resistor to current flow. Calloused or dry skin has a fairly high resistance, but the resistance of skin drops quickly if a person is perspiring or standing in water. Once resistance is lowered, the electric current flows easily through the victim's muscles and blood stream. That is why many arc welders have two sets of gloves when working on a hot day. When one set of gloves becomes wet from perspiration, they simply switch to the dry set of gloves. Working safely can be as simple as that.

As the voltage of an electrical circuit increases, your skin's resistance also drops rapidly. High-voltage ac of 60 Hz (or 50 Hz in some foreign countries) will cause violent muscular contractions, often so severe that the victim is thrown clear of the circuit before being injured badly. Although low-voltage ac also causes muscles to twitch, the reaction is less violent. As a result, the victims can't get free from the circuit, lengthening the time that they are subject to shock.

Direct current of the equivalent voltage and amperage of an ac line is more dangerous. Instead of making you twitch, a dc shock is like one long, constant jolt. Very low-voltage current can harm you indirectly, too. Even a slight shock that only makes you jump a little can prove fatal if you happen to be standing on a scaffold or a ladder. The current won't hurt you but the fall will.

You can hold onto a high-voltage line without injury if you hold it with one hand and are not grounded. The electricity simply will not flow through your body if the circuit (including you as part of the line) is not complete. Obviously, we don't think you should play the part of a switch in an electric circuit, but that is the reason why some electricians can handle electric lines with one bare hand and a smile. Don't do it. It's nothing but a grandstand stunt.

If you were to hold the same electric line with both hands, even though your feet were on dry pavement (you are not grounded), you would add a loop made by your body to the electric circuit. You would then get a big jolt of electricity that could kill you.

A related precaution is to never touch your electrode or electrode holder to a gas cylinder. The cylinder is metal and will conduct current. That's bad enough if somebody else is touching the cylinder, but the arc can damage the cylinder walls. What's more, even though you may be arc welding, it's not uncommon to have oxyfuel cutting equipment standing nearby. The sparks might ignite the gas inside the cylinder.

Always Ground Your Equipment

Never use a machine that has not been grounded. When you set up a job for arc welding, one of the first things you should do before turning on the power source is attach a ground cable to your workpiece with a ground clamp. That will complete the electric circuit when the machine's turned on and you close the circuit by striking an arc with an electrode. The cabinet of the welding machine also *must* be grounded to make sure that any unsafe or stray electric currents (caused, for example, by old wiring) go into the earth instead of into you. Most modern welding machines are built with the cabinet already set up for grounding, but read the installation instructions anyway and do what they say.

Grounding the metal machine cabinet also prevents the equipment from building up a static electric charge. The machine ground bleeds off any static charges that might build up.

Grounding the equipment also protects it (and you) from sudden excessive voltages from a short circuit, lightning, accidental contact between your machine and high-voltage power lines, and excess power surges. Electric utilities may momentarily send you extra electricity when they switch equipment at the power station—it's a problem that drives computer operators crazy because it ruins unshielded computer memories.

Any sudden surge of high voltage can penetrate conventional electrical insulation. Sometimes it can destroy the welding machine and hurt you too. A properly manufactured and grounded welding machine takes care of this problem. That's also why welding power sources have fuses that stop current overload before they go too far.

12-3
ARC VOLTAGES AND MACHINE RATINGS

As you can see from Fig. 12-1, arc-welding equipment has two basic types of voltages: the voltage into the machine, or the *input voltage,* and the *output voltage* that you use for welding. In the United States, the input voltages most often used for welding machines are either 220 or 440 V. In some New England areas, the input voltage can be as high as 575 V. You should never plug an arc welding machine into the wrong line. Call in an electrician to set up a new power line for your machine if one doesn't already exist.

Power-line plugs are built so

that you can't plug a 220-V machine into a 110-V circuit, or into a 440-V line. The only welding machines that plug into a conventional 110-V ac line are small utility arc welders designed to deliver around 100 A of ac for welding. They are portable and are used by hobby welders and for small maintenance jobs.

Of course you should never plug an ac outlet into a dc line, or vice versa. Electrical equipment manufacturers design plugs to make it just about impossible to do that. But, like gas welders who somehow find a method to attach the wrong regulators or hose onto a cylinder, some people create ways to get themselves into trouble.

Just like when you're taking medicine, *always read the label first.* The electrical requirements of every machine are printed on its permanent nameplate attached to the cabinet. Similarly, most input power lines are clearly marked for what they are, such as 220-V ac or 440-V ac current. You sometimes will find 550-V ac input lines used for very large, high-powered welding machines. But as we already mentioned, each type of line should have different kinds of plugs to keep you from plugging the wrong equipment into them.

Open-Circuit Voltages

The output, or rated, welding voltage for most generators and dc rectifiers used for stick-electrode welding is around 30 to 40 V, with 150- to around 600-A currents for very large machines. The OCV (no-load voltage or potential voltage difference when no welding is being done) for these stick-electrode machines is almost always 70-V dc to 80-V ac.

The power sources used for gas-metal arc and submerged-arc welding will have OCVs of 40 to 60 V and load voltage (the voltage while welding is taking place) of 35 to 50 V. The output amperage while welding may be anywhere from 1 A for light-duty TIG welding to 1000 A for very heavy duty flux-cored wire welding. Of course these amperages depend on the size of the welding machine as well as the job it's required to do.

Gas-tungsten arc welding (GTAW) power sources have about the same OCVs as other welding machines, but they tend to produce lower amperages when welding. Thus 1 A would be on the low side and 500 A would be more common for a TIG power source in use.

OCVs for a few GTAW machines are higher, around 150-V dc when welding, or about 70-V dc when on standby. By use of a special circuit, the OCV can be cut in half while the welding machine is idling. But the instant that the arc is struck, the full 150 V is again automatically available through the special circuit. Fortunately, since the circuit is designed for fusion welding and not to fuse you to the machine, it's in the open-circuit or standby condition that your machine presents the most danger to you.

Plasma arc-welding and arc-cutting power sources can be quite hot. A typical plasma dc-rectifier power source might have 400-V OCV and produce 200 V. Many plasma machines have even higher OCV. The voltage when a plasma machine is in use will normally be around 250 to 500 A of dc. Either way, this much voltage and amperage all in one place can make the conditions very hot if you grab the wrong thing or don't follow instructions. Be especially careful when you are working with plasma-arc welding or cutting.

Most plasma machines are very well insulated and protected from abuse and stupid practices by the operator. However, we want to paint as risky a picture as we can to imprint in your mind what can

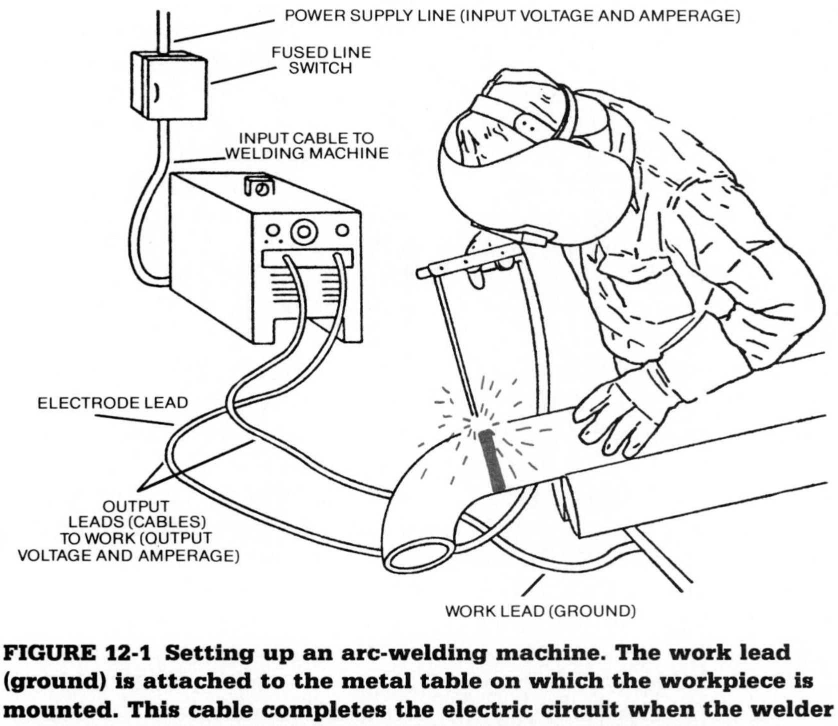

POWER SUPPLY LINE (INPUT VOLTAGE AND AMPERAGE)

FUSED LINE SWITCH

INPUT CABLE TO WELDING MACHINE

ELECTRODE LEAD

OUTPUT LEADS (CABLES) TO WORK (OUTPUT VOLTAGE AND AMPERAGE)

WORK LEAD (GROUND)

FIGURE 12-1 Setting up an arc-welding machine. The work lead (ground) is attached to the metal table on which the workpiece is mounted. This cable completes the electric circuit when the welder strikes an arc. It is connected to the welding machine's ground terminal. Inside the power-supply cabinet and the input cable from the welding machine is another protective ground cable that grounds the welding machine itself, so that you won't get shocked if stray current leaks to the cabinet or if the wiring is faulty.

happen to you if you don't follow instructions.

The most common type of ac arc-welding machine is the transformer with two sets of windings. One set is for the input circuit (the high-voltage end—which usually is at least 220 V and can be much higher) and the other winding is for the circuit that transforms or lowers the voltage to the output OCV (commonly 80-V ac or 70-V dc) while boosting the welding current out of the machine for delivery to your workpiece and electrode.

Unfortunately, some very small, low-priced transformer-type 110-V maintenance welding machines are still around called *autotransformers*. They only use a single winding. *Autotransformers are very unsafe.* Any failure in the winding may expose a welder to the full line voltage supplying the machine.

Any welding voltage can produce serious electrical shock. Even the output voltage from a small portable hobby welding outfit with a 110-V ac input line can hurt you. But machines with input voltages of 220 to 550 V or more are particularly dangerous if not properly installed and used. That input voltage range includes almost all the production welding machines used in industry. Almost all of these machines are made to be as safe to use as humanly possible. But it's still up to the welder to hook them up and use them correctly.

Changing Electrodes

When an arc is struck (especially on a machine suitable for stick electrodes), the open circuit voltage drops to the load or welding voltage, usually around 15 V to 35 V. Of course the welding current is boosted up pretty high because that's what produces the heat needed for welding. As low as these voltages may seem, there is enough current and voltage to give you a serious shock.

CAUTION: Never take off your gloves while changing electrodes or welding wire.

The time when the welder is most likely to be hurt is not while welding, but when the machine is on standby, with full OCV available as soon as the welding circuit is closed . . . maybe by the welder who is changing electrodes. Never take off your gloves while changing electrodes or welding wire.

Another problem is that the actual welding current may be higher than the meter reading on your machine if you are welding with short cables or are using low arc voltages. Particularly high overcurrents are likely on general-purpose welding machines when used for low-arc voltage processes such as air-carbon-arc cutting, oxygen-arc cutting, or gas-tungsten arc welding. That's why some extra safety features are added to some machines to make them suitable for these processes. These load voltages (when the machine is welding) can typically run around 30 V for stick and MIG machines and less than 15 V for TIG machines.

12-4
MULTIPLE MACHINES

Another potentially dangerous situation exists when several welders are welding on the same structure, close enough to each other that one welder might simultaneously touch both his or her electrode holder and one from a nearby welding machine. If the machines are not hooked up properly, you might be exposed to a voltage equal to the sum of the OCV of *both* machines. For example, if two nearby welding machines have an OCV of 80 V, the voltage difference between the two electrode holders on the separate machines can be 160 V, as shown in Fig. 12-2.

This hazard occurs when both machines are ac welders and are out-of-phase. Voltage, after all, is

the electrical pressure *difference* between two contact points in a circuit. Therefore, all ac machines anywhere near each other should be connected to the same phase of the power-supply circuit so that they will always have the same instantaneous polarity.

Hooking Up Several Machines

If the primary input circuits of two ac machines are connected in phase and the OCVs are the same, then each secondary welding circuit will reach its maximum OCV at the same time. As a result, the voltage difference between the two electrode holders will be zero. The ac welding current and voltage in both machines will rise equally and fall equally. Therefore, the worst the welder can get hit with (if he or she is grounded) is the normal OCV for either one of the machines.

Similarly, two nearby dc machines should be set up and used only with the same polarity. If one machine is set for DCRP (direct-current reverse-polarity) welding or DCSP (direct-current straight-polarity) welding, the other nearby machine also should be set up with the same polarity. In that way, if two electrode holders touch each other there won't be any difference in the voltage potential between them if both welding machines have equal settings. The worst a grounded operator can get hit with is the highest OCV from only one of the machines.

The literature that comes with each single-phase-transformer welding machine will provide you with details on how to connect them and how to avoid dangerous setups. Ask an electrician how to set up two nearby machines on a three-phase power line. If you have to, you also can check the potential (voltage) difference between two nearby electrode holders with a voltmeter.

You can use a test lamp to see if there is any voltage difference between the two holders. A test lamp is simply a light bulb with two insulated wire leads with alligator clips on their ends to hook it up to a circuit. It will light up only if there is a voltage difference between the two wires. No voltage means no current, which means that the test lamp won't light up . . . unless the bulb is burned out. Make sure the lamp works before using it.

12-5
MACHINE DUTY CYCLES
Standard arc-welding machines are designed and built to carry a current at a rated duty cycle. The duty cycle, as you learned in Chap. 11, is the amount of time a welding machine can operate at its rated current without overheating and is usually measured over a 10-min period. Most manufacturers put the duty cycle and rated current on the nameplate of the welding machine. Electrode holders and GMAW guns also have similar duty-cycle ratings.

If a machine or welding gun can be operated at its rated current for 4 out of 10 min without overheating, it has a 40 percent duty cycle. Power sources for stick electrodes usually have 20 percent to 60 percent duty cycles (9 out of 10 of these SMAW machines are rated at 60 percent duty cycles), simply because the welder doesn't need a higher duty-cycle machine for SMAW. After all, the welder has to stop welding periodically to change stick electrodes. While doing that the machine has a chance to cool off. The electrode is just like a switch in a light circuit. When the arc is not on, the machine's high amperage welding current is not flowing and the machine will cool off. Heat produced in any electrical device is a product of the current multiplied by itself then multiplied by the resistance in the circuit (or I^2R). No current means no heat.

Why not build SMAW power sources with 100 percent duty cycles just to be absolutely sure? Because the operator must stop frequently to change electrodes, and that lowers the overall effective duty cycle. Aside from the fact that higher duty cycle is totally unnecessary for stick-electrode welding, 100 percent duty cycle machines cost more. They have to have thicker copper wiring, heavier insulation, and more fan power to cool the transformer or rectifier. Some very big machines may even have water-cooled equipment combined with fan air cooling. They are designed to run on a production basis all day long if necessary.

A higher duty cycle may be required for certain stick machines if they are generating their own electricity with a gasoline or diesel engine (an engine-drive machine). The internal combustion engine produces its own heat along with that produced by electrical circuit heat losses. Water cooling may be required on very high amperage production arc-welding guns and plasma torches used for aluminum and other materials.

Machine Ratings
Machines that are rated for 100 percent duty cycle welding, such as those used for GMAW semiautomatic wire welding processes; machines for submerged-arc weld-

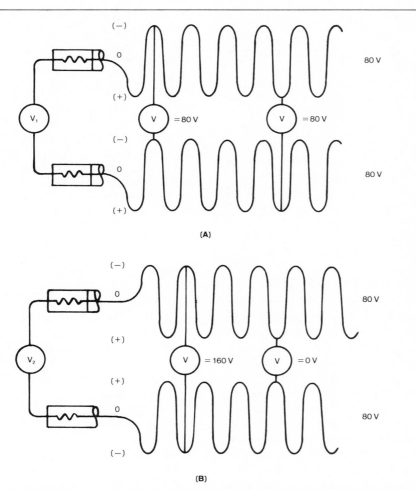

(A)

(B)

FIGURE 12-2 **This shows you why welding machines should be set up carefully if they are near each other. The gripper jaws (called electrode holders) on the left of the drawings carry current. If you happen to touch your electrode holder, and simultaneously touch the electrode holder of a welder next to you at the same time, the worst shock you can get is the open-circuit voltage of only one of the machines, if they are connected in-phase (A). If they are connected 180° out of phase (B), the jolt you get will be the open-circuit voltage of both machines added together.**

ing, carbon-arc welding, or plasma cutting or welding; or machines used for fully automatic welding cost more than conventional stick-electrode machines.

Manufacturers will tell you exactly what processes their machines are to be used for. Even two otherwise identical machine models, both with 100 percent duty cycle ratings, may not always be correct for all the same welding processes. Air-carbon arc cutting is just one example.

Standard welding machines are designed to carry their rated current at the rated duty cycle where the temperature of cooling air is below 104°F [41°C], and at altitudes below 3300 ft [1000 m]. At higher temperatures and altitudes, standard welding machines should be worked less than their rated duty cycles to give them more time to cool off unless the machines have been modified for these special conditions.

Most welding machines have built-in devices that will switch the machine off if it should overheat. If your machine gets unusually hot for any reason and starts shutting itself off while you are welding, slow down. Weld less and let the machine cool off more. You are not likely to see that happen in most plant jobs, but it can happen on a construction site during July while working on a very humid day.

Trust your machine if it turns itself off. It's trying to tell you something. Regardless of the type of machine and its duty cycle (even if it is rated at 100 percent), always use it according to the manufacturer's instructions, and don't exceed its rated duty cycle.

12-6
POWER-SUPPLY WIRING
Before you begin to work with any welding machine, you should make sure that it is properly connected to its power-supply lines and that the frame or cabinet of the machine is properly grounded.

The power input line to the machine should be in good working

condition and should be large enough to carry its rated current. As you know from the last chapter, the resistance of an electrical line is inversely proportional to the square of the wire diameter. A line with half the wire diameter will have four times the electrical resistance (and therefore it will heat up four times as much when carrying the same amount of current). That's due to the I^2R losses.

All connections to the welding machine should be strong enough so that they can't be pulled loose if accidentally jerked hard.

Sometimes it is necessary to use more current than one dc welding machine can supply. This sometimes occurs when using either plasma welding or air-carbon-arc gouging. Figure 12-3 shows how to hook up two machines in what is called a parallel circuit (machine outlets with the same polarity are hooked up together; the positive terminal to the positive terminal and negative terminal to the negative terminal). Don't do this kind of hookup unless you are (1) experienced, (2) check the power source manufacturer's literature (or phone them), or (3) have the help of an electrician who knows arc-welding equipment.

Where to Put the Cable
The power-supply cables should be connected to the nearest outlets available to keep the wiring as short as possible. Again, the reason is that the longer the conductor, the larger the resistance, and thus the higher the I^2R heating losses in the cable. If long runs of welding or power-line cable have to be used, hang the cable from insulated overhead hangers. Don't attach them directly to the walls or structural beams of a building. If the cable has a worn spot and the bare conductor touches a structural support or a metal wall, it will short out and might even start a fire.

When the cable can't be kept out of the way of passing forklift trucks, tractors, wheelbarrows, or

people with heavy feet, protect the cable by covering it with wooden planks. This also helps other workers because nobody wants to trip over your cables.

The welding machine also should be properly grounded. Pay special attention to safety ground connections on portable machines. If you are not sure about the safety of the ground connection (or can't find it), check with an electrician before you start to work. Don't fiddle with anything inside the machine yourself.

12-7
SELECTING WELDING CABLE
Welding cables and electrode holders are a welder's constant daily companions. Your safety depends upon how well you take care of this equipment.

The welding leads or cables are the ones on the outside of the machine. They form the circuit that carries the welding current to the work and back out again. One of the cables attached to the work is called a *ground cable* or *ground lead*. The other cable is connected to the electrode holder or welding

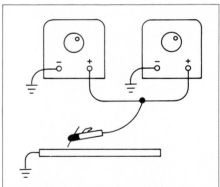

FIGURE 12-3 Two dc rectifiers can be connected in parallel (joining like-polarity terminals) to obtain much higher currents required for air-carbon-arc gouging than available when a single power source can deliver. Don't do that without using a skilled electrician, and checking with the welding machine supplier first.

gun, TIG holder, or plasma torch and is called the *electrode cable* or *lead*.

Any cables used for welding must be designed specifically for welding and for nothing else. They must be completely insulated, flexible enough to make it very easy to handle them, and large enough to handle the currents that they must carry. Just like welding machines, welding cables are rated for their maximum current-carrying capacity. But after the cable has been unpacked you might not be able to find the manufacturer's rating data. It's on the cable package, but usually not on the cable itself. See Chap. 11 to review any details you need on welding cable.

Use Cable Tables

Be sure that your welding cable is the right size for the amperage and the distance from your machine, as outlined in Tables 11-1 through 11-5. The reason is that each length of cable, the ground cable and the electrode cable, adds its own resistance to the welding circuit. *You have to account for the total resistance both cables add to the circuit.* If you are using two 25-ft cables, one for the ground and one for the electrode, it is just the same as using one 50-ft electrode cable and grounding the workpiece right at the machine. Therefore, if both your ground cable and electrode cable are 25 ft, select for a 50-ft cable length from Tables 11-1, 11-2, and 11-3.

The present practice for selecting aluminum cable, whose resistance is higher than copper cable, is to use two American Wire Gauge (AWG) sizes larger than shown for copper cable. Consequently, if you want to weld with 200 A at 50 ft from the welding machine (requiring a total of 100 ft of cable), you should use a size 3/0 aluminum cable instead of a size 1/0 copper cable. Personally, we prefer copper cable because it has about half the resistance of aluminum in equal sizes and also

has a higher melting temperature and greater mechanical strength. Most welding cable is copper.

Inspect Each Cable and Connection

Inspect your welding cables and their connections often. Always do that with the power OFF. Cables with damaged insulation or conductors must be replaced or repaired. Even the best welding cable can be damaged. For example, a heavy truck or a tractor can run over it. A heavy vehicle might not do much to the copper in the cable, but it can shear off a small patch of insulation to expose the bare conductor. We suggest that you inspect your cables every morning before hooking up and every evening after you shut down. It only takes a minute. You can save yourself a lot of trouble that way.

Minor repairs to damaged insulation can be made with approved electrical tape, provided that the insulating qualities of the repaired area are the same as or better than those of the original cable. If the cable is badly damaged, it is far safer to cut out the damaged section and splice in a section of good cable. When joining lengths of cable, use only insulated connectors specially designed for that purpose. Make sure that the cable connectors are tight-fitting. Loose connectors will flash and may start a fire. If you need help, ask for an electrician.

Cable splices must be strong enough to withstand normal handling and to carry the currents being used. Undersized splices will overheat. The first 10 ft [3 m] of cable on the electrode lead, measured from the electrode holder toward the welding machine, must be completely free of connectors, splices, or repairs. That's where your hands and body will be. Don't take chances with that end of the cable to save a few dollars in repairs.

Keep all welding cables dry and free of grease and oil. Grease and oil damage cable covers. They will

cause the cable cover to lose part of its insulating ability. Support long cable lengths out of the way, preferably on overhead insulated hangers. Otherwise, cover them with wooden planks or even put them under channel-iron covers.

CAUTION: Never coil up welding cable while in use. Don't wrap it around you.

Keep welding cable away from power-supply lines and other high-voltage conductors. Never coil up long lengths of welding cable when the cable is being used. Coiled cable will induce currents in itself, just like the wire turns inside a transformer. Coiled cable will heat up fast, damaging the insulation.

Never coil any cable about your body while you weld. The reason isn't electrical, it's practical. You may get pulled off your ladder or scaffold, or jerked into your arc, or pulled into moving machinery if somebody or something hooks onto your cable and gives it a hard tug. The worst that can happen if you are not tangled up in your own cable and someone pulls on it is that your electrode will be pulled away from the weld or out of your hands. But at least that will break the welding circuit without hurting you.

Check all cable connections and work-cable grounding clamps before you begin to weld and each time you change cables or reassemble any connections. Don't hook up to dirty, corroded connections. They are sources of greatly increased electrical resistance and will overheat. It doesn't take long to file down or wire brush a dirty or corroded electrical connection before you hook up the machine.

Always unplug the machine at the wall before changing cables or switching them from DCSP to DCRP or DCRP to DCSP. If you're using an engine drive, turn the engine off first. (Some machines have a simple switch you can use to change the polarity. Never switch polarity while welding.)

> **CAUTION: Unplug the machine before disconnecting cables.**

12-8
ELECTRODE HOLDERS

Electrode holders are used for clamping stick electrodes and holding them while you weld with the SMAW process. Only use electrode holders specifically designed for welding. Never use homemade holders. The handle part of the electrode holder must be insulated to withstand the maximum OCV you will use. You will be much safer if you use only fully insulated electrode holders. If the insulated handle is cracked, turn the holder in and get a new one. The same rule applies to GMAW guns, GTAW electrode holders, and plasma or air-carbon-arc electrode holders.

Hot Holders

If your electrode holder gets too hot while you are using it, never dip it in water to cool it off. (Believe or it not, some people have tried it.) You will get a big shock as a result. In fact, if your electrode holder is overheating, it probably means either that your wire connections inside the holder are not good or, more likely, you need a larger electrode holder with a higher electrical rating for the welding current you are using. First check the electrical connections that are inside the insulated handle . . . *after turning off the welding machine and disconnecting the welding cable from it.* If they look OK, get a bigger electrode holder. See Tables 11-7 and 11-8 for more details on electrode-holder sizes and amperage ratings.

When you insert welding electrodes into your holder, don't do it with bare hands. Always wear gloves. If you're doing gas tungsten-arc welding, always turn off the machine and also unplug it, and then disconnect the electrode holder from its power source before you change the permanent tungsten electrode in the holder.

12-9
GROUNDING THE WORKPIECE

There are certain safe practices that must be followed when grounding work to be welded. The safest and best way to ground your workpiece is to use a single cable running directly from the workpiece to the welding-machine ground connection to complete the electrical circuit as shown in Fig. 12-4. While you're at it, select a cable size that won't reduce your welding voltage very much if it has to be very long. Table 11-3 shows you how to select an electrode cable and ground cable to complete the circuit while getting a minimum voltage drop due to the resistance of the cable.

Sometimes it's not practical to attach the cable directly to the work, for instance when working on a welding positioner, a jig, or a welding fixture. In such cases, the cable can be attached to the positioner, jig, or fixture and the work can be grounded by contact with these work-holding devices, and through them to the proper ground connection.

Dangerous Grounds

Never use pipelines that carry gases or flammable liquids as grounding for welding circuits. If you have a poor connection, it will spark and cause a fire. You should never use conduits or pipes that contain electrical wiring for grounding because of the possibility of feeding high-voltage current back into your welding leads.

When a pipeline or steel building frame is used for a temporary welding ground, check all the joints between your ground connection and the earth to make sure that there is good electrical contact. If any loose joints along the way heat up or spark, the contact is not only poor but you also are asking for trouble . . . anything from a fire to hurting somebody. If you do hook up a ground to a structural section that may not be tightly connected to another section ultimately leading to the earth, you can use copper jumper cables across the loose joint to ensure adequate electrical contact. Inspect these connections frequently for corrosion and possible fire hazards. Use large jumper cables selected for size the same way that you select welding cable.

You should never use chains and wire rope; pipes with threaded connections; caulked, flanged, or bolted connections; cranes; hoists; or elevators for

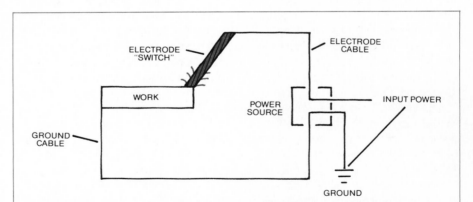

FIGURE 12-4 This is how a ground connection from a power-source cabinet works. The work cable (also often called the *ground*) completes the electrical welding circuit. But a separate ground line from the welding power source protects the cabinet (and you) from stray current. Never ground a power source or welding circuit to elevators, chains, cranes, plant piping, or anything else with loose or threaded connections. You can weld the loose parts of the equipment together. Solidly welding an overhead plant crane onto its rails is a bit expensive. Grounds attached to fuel-gas pipes start fires.

grounds. None of them have good connections. They will cause sparks and probably fires. A heavy ground current could weld an overhead crane to its tracks, or an elevator to its guide rails. Welders make good money, but not quite good enough to pay for a new overhead plant crane.

12-10
WATER IS HAZARDOUS
Moisture, as you already know, can increase the chance of your getting a heavy shock. Moisture from your sweaty gloves, from the weather, from a very humid work environment, or from any other source is a very real hazard. Wet hands, wet skin, wet clothing—any wet object—provide a path of low resistance to current flow. Where any wet condition exists, avoid contact between your bare skin or wet parts of your clothing and bare-metal current-carrying elements of your equipment and work.

Never stand in water or on a damp surface when you weld. Stand on dry insulating material such as a rubber mat or dry boards. Make sure that you are dry when you sit down; your shoes offer more insulation than the seat of your pants. When you have to weld in a prone or leaning position, make sure that your body is insulated from your work. Lean on a rubber mat, a fire-resistant blanket, dry canvas, or a wooden plank.

Always wear dry welder's gloves in good condition. Never use wet gloves. Leather welding sleeves and capes give you good protection from stray currents. Keep your clothes dry even if you have to stop work and change often. Don't weld in the rain. Use common sense and your new understanding of electricity to keep yourself safe.

12-11
SAFE MAINTENANCE
Here are a few good rules for safely operating and maintaining your welding equipment. Equipment that is maintained properly is safe to work with. Equipment that needs repair should never be used.

Do not work on any welding equipment unless you have been trained and authorized to do so. Further, any maintenance that you do should be performed strictly according to the manufacturer's instructions. *Always disconnect the power supply before you work on any welding equipment.*

Keep welding machines from overheating. Clean them from time to time. You can clean a machine by blowing it out with clean, dry compressed air. *Always disconnect the power supply before you do this.*

The commutators and brushes on motor-generator welding machines should be kept clean to prevent excessive flashing and pitting. Do not use gasoline or other flammable or toxic liquids to clean them. It's better and safer to use commutator polish or very fine sandpaper to clean the commutators, then blow them clean of dirt and grit with compressed air.

Protect Your Power Source
If a welding machine is used outdoors, you must protect it from rain, snow, ice, and dust by covering it with a tarpaulin or by putting it in a shed. Make sure that you do not block air circulation to the machine. If the machine is poorly ventilated it will overheat fast. Never use a wet welding machine without first drying it completely and testing the chassis against the power lines, ground cable, and electrode cable with a meter to detect a short circuit or poor grounding. If your electrical meter shows that a test current flows between the chassis and any of the electrical lines, the machine is still too wet to use or you need help from an electrician. Dry compressed air is a fast way to dry out a machine.

Always open the power-disconnect switch from the power-supply lines of any welding machine when you stop welding and leave the work area for any length of time. Disconnect the machine when you move the power source around and before you clean it, repair it, connect or disconnect the welding leads. Whenever you take a machine out of service, disconnect everything.

12-12
WORKING IN CONFINED SPACES
We've already cautioned you about working in confined places with apparently harmless shielding gases that actually can smother your. For safety's sake, whenever you are working in any confined space—inside a tank, a ship hull, a closed room, the base of a machine tool—use your head and follow these instructions.

Check the Air
■ Always check to make sure that the enclosure is well ventilated, or use an air mask with a trailing hose from a clean air supply. If the air in a nonexplosive atmosphere can't support the flame of a lit match, it can't support you. Gas masks and dust filters don't supply oxygen but air masks do. Remember that you need at least 16 percent oxygen in your air to sustain life.
■ If your air is pumped in from the outside, be sure that the air intake on the pump is not near the exhaust of a gasoline or diesel engine, or you will get carbon monoxide along with your air. For the same reason, if you work inside a plant with an engine-driven welding generator, vent the exhaust fumes to the outside air. Engine drives usually are used outside.

Check Your Equipment
■ Always check your gas-using equipment before using it to be sure that it doesn't leak.
■ Never bring gas cylinders with you into any confined space. If an inert-gas cylinder springs a fast leak, you can be asphyxiated before you can get out.

■ When you leave, turn off all your cylinders.

■ Post a sign in front of your work area, and another one in front of the inert-gas supply. The sign might read, "Inert gas may be in here. You can't breathe oxygen-deficient air and live. Be careful."

Look for Fire Hazards

Always watch out for possible fire hazards in a confined area. If you are arc-welding and leave for lunch or go home, remove the electrodes from all electrode holders. Place the electrode holders in a safe, insulated spot, where they cannot cause an accidental fire by arcing against some scrap metal or a steel floor. Also unplug your welding machine before you leave the job.

12-13
HOT METAL AND HIGH PLACES

It's obvious to you that welding makes workpieces hot. But people working around you might not know it. Worse, they might not know what you have just welded and what already has cooled. As you know, steel often remains hot enough to burn somebody long after it loses its red-hot color. Aluminum retains its original color even when it is so hot that it melts. Leave a sign such as "Don't touch—hot metal" if there is any chance that someone will pick up a hot workpiece.

Whenever you work in a high place—anything over a foot above the ground is a high place—take special precautions. Naturally, you know that falling off multistory buildings can flatten you. Also remember what it feels like when you are walking downstairs and miss a step, or when you take a step backward and fall off a curb. A fall from any height can hurt. Be careful.

If you are working on a platform or scaffold, protect yourself from falling by installing solid railings or using a harness or life-

line designed for quick release if you need to move fast, yet strong enough to hold you if you fall. OSHA requires them. So will your safety director. Safety nets are used for added protection against falls.

Many of the things you work with and take for granted can hurt you. Don't leave welding cable, hoses, and other equipment in hallways, stairways, hanging from ladders, or in passageways where someone can fall over them. When you are welding with stick electrodes, don't throw the stubs just anywhere. Underfoot they can roll and cause serious falls. Put your stubs in a container or in a place where no one will be liable to step on them and slip.

12-14
PROTECTING YOUR EYES

Besides the intense light, most welding processes produce invisible ultraviolet rays that can damage your eyes, burn your skin, and even blind you if you are not careful. These rays are the same ones that give you a sunburn on the beach. Ultraviolet rays reach out far enough to hurt people around you. They travel farther than infrared rays that produce heat, which is one of two reasons why eye protection is necessary even if you are not welding but just working near a welder.

The light given off by an open welding arc is very intense. It is more of a danger to your eyes than rays produced by oxyfuel welding. Ultraviolet rays can do more damage than infrared radiation partly because they travel farther and partly because they contain more energy. Infrared rays (heat) are also given off but are less dangerous.

The higher the amperage of the welding process, the more intense the ultraviolet radiation will be. Even the reflected arc radiation can seriously damage your eyes. The radiation is reflected back far more strongly from a bright aluminum or stainless-steel plate

than from the dull surface of hot-rolled steel. The same thing happens when sunlight reflects off water. You get a double dose of sunburn; one direct dose from the arc and one indirect dose reflected from the metal workpiece. The reflected radiation can sunburn you under your chin if your welding helmet doesn't fit properly.

The light from gas welding or flame-cutting is much less intense than that from arc welding. It has less ultraviolet and more infrared heat radiation in it. That is why different shades of tinted glass and different types of helmets or goggles are used with the two processes.

The radiation from the electric arc is so intense that it not only can burn your eyes but also your skin. The effect is exactly the same as a sunburn on your face, neck, or arms, which is why the arc-welder's helmet has a very dark glass that screens out all but enough light to see the arc (you can't see anything until the arc is struck). The helmet also protects your whole face and even keeps your ears from being sunburned.

If anyone wants to watch you work, including your helper, tell them to wear the proper eye protection. Welders often work behind screens for the same reason. If possible, do all your welding in screened-off areas, even if you have to erect portable screens yourself. The screens shield other people from your welding arc. Canvas screens (while not good) often are used. New fire-resistant semitransparent plastic screens that filter out dangerous radiation but also let the welder look out into the plant are now available.

GMAW and GTAW, using either argon or helium as the inert shielding gas, require special kinds of eye protection beyond that required by regular stick-electrode arc welding. The intensity of the ultraviolet radiation is many times greater when using these inert gases than when using covered stick electrodes. The filter glass in your welder's helmet should be several shades darker than what you use for stick-electrode welding. This glass is very easy to change.

You may also want to wear flash goggles equipped with side shields under your helmet. They will protect you from stray flashes from nearby welding equipment when you flip your mask up. By using a lighter lens shade in the flash goggles and a normal shade of filter glass in your helmet, you can get the darker shade required for GMAW or GTAW when you need it without changing the filter glass in your helmet.

Since you often will work where a lot of chipping and grinding is done, you must protect your eyes against chips and other flying particles. Some arc-welding helmets have hinged filter lens plates that can be flipped up to expose a clear glass shield that protects your eyes while you chip or grind, keeping your helmet down. Many welders always wear their flash goggles or regular safety glasses under their welding helmets for use when they are not welding.

Because they are so important to your eyesight, your helmet and goggles should be kept in excellent condition. Never wear a helmet that is cracked or broken; radiation can pass through the cracks and cause burns. If your lens filter gets cracked, replace it immediately. Also replace the clear glass cover plate if it gets cracked or becomes excessively pitted or covered with spatter.

Never lend someone else your goggles or helmet, and never use someone else's equipment without sterilizing it first. Eye infections can easily spread if you wear someone else's equipment.

The best way to avoid injuring other people's eyes, or bothering others while they work and you weld, is to weld in a closed bay or welding booth. If the type of work you do permits, you should always work in a welding booth that completely screens your arc from other people. Welding booths may

be made of rigid fire-resistant material or of firepoof canvas draped over a pipe framework. Both the rigid booth walls and the canvas should be painted or colored a dull, nonreflecting color inside and outside if needed to minimize reflected radiation. White paint using zinc oxide pigments provides very good eye protection because it absorbs ultraviolet radiation.

Good ventilation is essential in a welding booth. The partitions or curtains should end 18 to 24 in. [45 to 60 cm] above the floor and should also stop short of the ceiling to provide good draft air circulation, which will help remove welding fumes and dust. If you can't work in a booth or use screens, you must equip the people working near you with flash goggles.

Eye-Protection Standards
The most complete set of regulations on eye protection for all industries, including welding, is published by the American National Standards Institute (ANSI). This is ANSI Standard Z87.1 1975, *Practice for Occupational and Educational Eye and Face Protection*. Both the AWS and OSHA also provide suggestions

for the use of different kinds of goggles, filters, and lens shades. The two most important rules for eye protection are:

- Use the right kind of equipment.
- Use the right shade of tinted glass in your goggles or helmet.

This last item is the most likely to be ignored.

Table 12-1 lists the lens shades to be used with various welding processes and for certain applications of those processes. The lower the lens-shade number, the lighter the shade. (You'll find the oxyfuel half of this table in Chap. 4 on page 93.) In places where the table is not specific to your work, a good general rule is to pick the darkest shade that lets you see the welding operation. If spots begin to appear before your eyes as you work, stop welding and immediately remove your helmet or goggles and change the lens to a darker shade.

12-15
WELDING CURTAINS AND BLANKETS
Welding curtains should always be used whenever you are arc welding in an area where other

TABLE 12-1 Arc-welding lens shade selector chart*

Welding or Cutting Operation	Lens Shade Number
Reflected light from nearby welding arc (for flash goggles)	2
Shielded-metal arc welding (stick electrodes) for electrode diameters 1/16 through 5/32 in.	10
Gas-shielded arc welding of nonferrous metals, for electrode or TIG rod wire diameters from 1/16 through 5/32 in.	11
Gas-shielded arc welding of ferrous metals, for electrode or TIG rod wire diameters from 1/16 in. through 5/32 in.	12
Shielded-metal arc welding (stick electrodes) for 3/16 through 1/4 in. diameters	12
Shielded-metal arc welding (stick electrodes) for 5/16 through 3/8 in. diameters	14
Atomic-hydrogen arc welding	10–14
Carbon-arc welding and cutting	14
Plasma-arc welding and cutting	
Up to 300 A	9
300 to 400 A	12
400 A and over	14

* See Chap. 4, page 93, for data on oxyfuel lens shades.

people will see your arc. These curtains have two uses, one of which is to screen out ultraviolet radiation to protect other people working near you. The second is to deflect sparks away from areas where they might otherwise start a fire. Welding blankets are used on wooden floors for the same reason.

The ideal material for a welding curtain should be fireproof, not just fire-retardant. Fireproof means that the material won't catch on fire. Fire-retardant means that the material won't easily catch on fire, and if it does, it's not likely to continue to burn. That's not good enough for welding and flame cutting.

For many years asbestos was the only suitable material for welding curtains and blankets. Within the last decade or two, people have realized that asbestos particles in the air can produce lung cancer and respiratory ailments such as asbestosis (a disease similar to the silicosis that miners can get from breathing mineral dust) over long exposure periods.

Two new types of welding curtain materials have now replaced asbestos curtains. The first is a special translucent plastic produced as vertical, overlapping strips through which you can walk, while the material filters out ultraviolet light. The second material is fabric made from textured fiberglass yarn.

Studies by PPG Industries, Pittsburgh, have shown that the fiberglass fabrics actually are superior to asbestos cloth curtains, mats, and welding blankets. The textured fiberglass yarn fabrics have over 150 percent higher heat resistance; over three times more strength; are 50 percent more resistant to abrasion; and have twice the resistance to heat transfer (insulating values) than comparable commercial-grade asbestos fabrics that they replace. For example, Auburn Manufacturing Inc., headquartered in Maine, reports that the maximum continu-

FIGURE 12-5 The author's wife, Renée Smith, shows you how to hang a fiber glass welding curtain. She's using a conventional, heavy-duty expandable curtain rod she bought from a local store. You can even use two nails and suspend a wire between them, threaded through the grommets. Renée is demonstrating a key safety point—there's no excuse *not* to hang a welding curtain wherever you are welding or cutting, even if its in your own living room. (*Photo by Renée Byer.*)

ous service temperature of commercial-grade asbestos is 400°F [204°C] while the maximum continuous service temperature of the new textured fiberglass fabrics is 1000°F [538°C]. At 400°F for 24 hours of exposure, asbestos loses nearly half of its original strength while the fiberglass fabric retains around 98 percent of its original strength.

The weight of comparable thicknesses of the two fabrics is about 32 oz/yd^2 [1.1 kg/m^2] for the fiberglass textiles versus 40 oz/yd^2 [1.4 kg/m^2] for asbestos cloth. Therefore, the new woven fiberglass fabrics also weigh about 20 percent less than equal thicknesses of asbestos cloth.*

* You can get more information on these new products from Auburn Manufacturing Inc. by writing them at Walker Road, Mechanic Falls, Maine 04256. Auburn Manufacturing sells direct as well as through welding distributors.

In addition to welding curtains and blankets, the new fiber glass textiles are now used to make protective hoods, aprons, gloves, and various kinds of woven gaskets for heat-treating oven doors and woven flat tape. These new fabrics have such good insulating properties compared with asbestos blankets (roughly half the thermal conductivity of asbestos at room temperature) that they can be used as blankets to wrap hot weldments. As you already know from your oxyfuel welding work, some metals such as high-carbon steels and certain cast irons should be slowly cooled to room temperature after welding to make sure they don't crack. Sometimes sand is poured on a weldment, but more often the hot weld is wrapped with insulation. The new fiberglass blankets and tapes look like they will be widely used for this service in the future.

Welding curtains are easy to hang and take down after you use them. Many shops have permanent frames erected just for that purpose, creating individual welding booths. But if you move around a lot (as when doing maintenance work), you should build a temporary frame and hang the curtain from it (see Fig. 12-5). Most welding curtains are provided with metal rings (called *grommets*) through which you can place metal hooks to hold the curtain on the frame. Steel curtain rods make a better frame than wood, but you can build a temporary wood frame to hang your curtain on if you have nothing better to use. One way to do that is to erect a two-by-four frame around your temporary work area, and drive nails partway into the top sill, spaced the same distance apart as the grommets on the curtain. Hang the grommets on the nails. Don't drive the nails through the curtain fabric.

For very fast, temporary welding enclosures, hang the curtain on a rod suspended between two tall stepladders, or from any available overhead pipes. Don't use fire-retardant canvas for curtains. It won't hold up under spattering hot metal. What's more, the ultraviolet radiation from your welding arc will cause the canvas to fall apart sooner or later because canvas is made from cotton.

12-16
PROTECTIVE CLOTHING
There's really nothing to tell you about protective clothing for arc welding that you don't already know from studying Chap. 4. But we'll remind you that you should

wear woolen rather than cotton clothing when arc-welding. Wool does not burn readily. It protects you better than cotton. Cotton also falls apart when washed after exposure to ultraviolet radiation. Arc radiation has much less affect on wool. It has no effect on leather. Use protective leather sleeves, caps, and aprons when you need them. And always wear the best pair of gloves you can afford and have a second pair ready to switch to during the day.

Always wear a welder's skull cap to keep sparks out of your hair. Buy a fancy one while you are at it. Welding distributors carry them by the boxful. Don't wear cuffs or shirts with loose, open pockets; they collect live sparks. Use safety shoes or boots. Wear your pants outside the boots, don't tuck them into the boot tops. Be sure to wear steel metatarsal guards over the tops of your feet when you handle heavy workpieces.

The next chapter will start you off with a new way to cut any kind of metal using electricity instead of oxyfuel. It's called air-carbon-arc cutting. Do it safely.

REVIEW QUESTIONS
1. What is the maximum ambient (outside air) operating temperature of a welding power source that is not designed for very hot days? What is the lowest ambient operating temperature for the same power source if its insulation is not specially designed for cold weather?

2. Why is it unsafe to weld inside a flour mill?

3. Why should all arc-welding power sources be grounded?

4. Are you most likely to be hurt

while you are arc welding, or when your machine is on but you are not arc welding?

5. What should you check first before turning on any arc-welding power source?

6. Your electrode holder is getting too hot while you are working. What should you do?

7. Name three things that you should never connect your ground cable to when grounding your power source.

8. You have just stopped welding because the construction site is getting full of dust. What is the first thing you should do *before* blowing out the machine cabinet and circuitry with clean, dry compressed air?

9. What is the single biggest hazard in doing MIG welding (using an inert shielding gas) inside an enclosed water tank?

10. Which is more likely to give you a bad sunburn: welding polished, cold-rolled aluminum or welding hot-rolled carbon steel (assume that both materials will be welded with the same current when you answer this question).

11. Which is better for arc welding: a nylon shirt, a cotton shirt, or a wool shirt?

12. Should you tuck your pants legs inside your boots, or leave them out?

13. You have several different lenses for your welding helmet in your tool box. One has a No. 10 lens shade, another has a No. 12 lens shade, and the third has a No. 14 lens shade. Which lens is darkest?

14. Which lens shade should you use for GMAW work with ¹⁄₁₆-in.-diameter wire?

15. Which lens shade should you use for SMAW welding with ³⁄₈-in.-diameter electrodes?

Chapter 13

Cutting Metal with Arc and Air

Air-carbon-arc cutting (AAC) and arc gouging are widely used in foundries, fabricating shops, railroad repair shops, construction sites, and on maintenance jobs. The reason for the extensive use of this process is that it removes metal much faster than grinding, flame-cutting or flame-gouging, and pneumatic chipping hammers (Fig. 13-1). The process is commonly used to prepare grooved joints for welding moderately thick material. Very thick material would still be flame-cut because flame-cutting is faster on thick plates.

The AAC process also is relatively inexpensive and easy to do. A special AAC torch, a suitable power source, a compressed-air line, graphite gouging electrodes (usually called *rods* or *carbons*), and a few minutes of instruction are all that are needed to start work on almost any metal.

This chapter will explain how the process works. It will tell you how to use AAC and how to operate it safely.

13-1
HOW AAC WORKS

Air-carbon-arc cutting and arc gouging work by melting metal with an electric arc produced by a graphite electrode. High-velocity compressed air from holes in a special electrode holder then blows the molten metal out of the joint. It is arc-cutting's answer to the cold chisel and the pneumatic chipping hammer, and AAC is a lot easier to use.

Unlike oxyfuel flame-cutting, flame-gouging, or scarfing, the AAC process works on any kind of metal because the heat produced by the intense electric arc melts just about anything. Then the compressed air simply blows the molten metal away. It's not an oxidation process like oxyfuel flame-cutting.

Air-carbon-arc gouging is used not only to prepare the edges of plate or structurals for welding but also in foundries to remove excess metal left over in the casting process and to "wash" surface defects from castings. In that way

FIGURE 13-1 Air-carbon-arc cutting (AAC) is often used to gouge out bad welds, to take out tack welds before the final weld metal is deposited, to remove backing strips from submerged-arc weld joints, and sometimes to shape special edges for J-groove and U-groove joints prior to welding. The AAC process also is used to remove excess metal from castings. Unlike oxyfuel cutting, this process can cut both ferrous and nonferrous metals.

the process is used like oxyfuel flame scarfing, except that AAC is not limited to carbon and low-alloy steels.

Air-carbon-arc gouging also is widely used to remove bad weld metal and to cut the backside of welded plates down to good weld metal, before the reverse side of the plate is welded. Used this way the process removes weld-metal defects on the bottom of the root pass after it is made on the opposite side of the plate. That assures that the follow-up welding job on the back of the plate will penetrate right down to good solid weld metal already deposited on the other side.

Oxyfuel gas-cutting relies on the rapid oxidation and removal of low-melting-point slags. The AAC process, which doesn't have that restriction, can be used on metals that produce high-melting-point refractory slags as well as on carbon steels. It does a nice job on stainless steels, cast iron, aluminum, nickel, and copper alloys, which it cuts as easily as carbon steel.

13-2
GOUGING EQUIPMENT

Any arc-welding power source might be used to do double duty as a power source for air-carbon-arc gouging. Don't use any handy power source for this purpose unless the manufacturer's literature specifically says you can use it for gouging. Beyond that, AAC equipment (except for the correct selection of standard cable sizes) is unique to the process (see Table 13-1).

The Special Electrode Holder

AAC "torches" or electrode holders are designed either for manual, semiautomatic, or automatic use. Manual holders for gouging look very much like conventional heavy-duty SMAW electrode holders (Fig. 13-2). The holders for gouging, however, are equipped with air passages and holes in the

head of the holder to direct the high-velocity compressed-air stream in a direct line along the graphite electrode, aimed right at the base metal as it melts. A valve also is used to turn the compressed air on and off.

Another difference between a conventional SMAW electrode holder and an AAC torch is that the gouging torch has grooved copper buttons in the spring-operated jaws to get a better grip on the graphite electrode and to make a solid electrical contact.

Semiautomatic electrode holders for gouging are designed for mounting on track-mounted cutting-machine carriages. The operator manually feeds the graphite electrodes into the holder to produce the desired gouge, while the carriage moves the holder along the tracks on the work. The equipment is not at all complicated and will produce a gouge that is straighter and cleaner than most work done manually.

The graphite electrode is fed mechanically into fully automatic holders either by a feed motor activated by a spring-loaded device, or by a voltage-controlled automatic feed. The voltage-controlled automatic torch can produce consistent grooves with depth tolerances of $\pm$ 0.025 in. [0.6 mm]. Automatic holders usually are mounted on a torch carriage. They also can be mounted in a fixed position while the work rotates or slides past the stationary holder.

Carbon Electrodes

The so-called carbons or carbon electrodes actually are made of graphite, the crystalline form of carbon. There are three basic types of "carbons" used for different kinds of AAC operating currents. There are dc copper-coated graphite electrodes, dc plain or uncoated graphite electrodes, and ac copper-coated electrodes. DC reverse polarity (DCRP) setups are used for AAC with dc electrodes. The ac electrodes also can

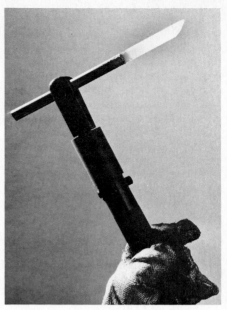

FIGURE 13-2 The AAC electrode holder is similar to a stick electrode holder except that it has compressed air passages in the head that are used to blow molten metal out of the cut.

be used with dc in a pinch, but dc electrodes can't be used with ac.

Copper-coated dc electrodes

The most commonly used electrodes are the dc copper-coated type of graphite electrodes, usually called *gouging rods*. The copper coating helps maintain the original shape of the electrode during gouging and this results in a uniform groove size and shape. The copper coating also reduces the rate of electrode wear and allows you to use higher currents so that the process goes faster. These dc electrodes are available in many convenient diameters from 3/16 in. up through 1 in. Special shapes such as half-rounds and flats also are used. A half-round gouging rod, for example, is very good for producing a half-round or U-groove joint in a plate.

Copper-coated dc graphite electrodes with flat or half round shapes also are available for special jobs such as "skimming" away excess weld metal and removing a hardfaced surface or

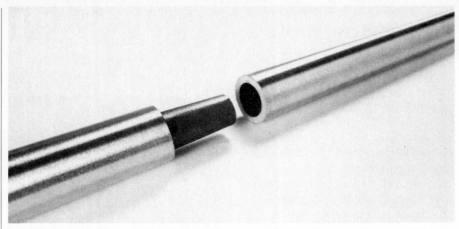

FIGURE 13-3 AAC electrodes are called *carbons*, or gouging rods. They are made of graphite. Many round gouging rods have tenon and socket joints to make up longer strings when the first section is worn down.

weld-crown reinforcements and backing strips after welding. Many round electrodes are manufactured with a tenon and socket that help you to join them together (Fig. 13-3). These jointed graphite electrodes can be used for either manual or automatic gouging to reduce stub losses. Without them, the last inch or so of the electrode would have to be thrown away. A variety of different kinds of gouging electrodes are shown in Fig. 13-4.

Copper-coated ac electrodes
Copper-coated graphite ac electrodes are manufactured with rare-earth-element additions to stabilize the ac arc during the rapid current and voltage swings characteristic of ac. They are used mainly for gouging stainless steels, ductile and malleable cast irons, and for nickel- and copper-based alloys.

Uncoated dc electrodes Plain dc uncoated graphite electrodes perform reasonably well in diameters less than ⅜ in. These uncoated electrodes are not used very often for gouging because they wear out rapidly and the hot ends change shape during gouging. Their only advantage is that they are less expensive than copper-coated electrodes. Considering the importance of accurate

grooves in any kind of welding, we suggest that you don't use them for AAC. The plain dc electrodes would only be used in scrap jobs where groove precision and tolerances don't matter. However, these uncoated dc electrodes do have another purpose for which they are better suited. They are used for carbon-arc-soldering and heating electrodes. Uncoated carbon-arc-welding, heating, and soldering electrodes are not fully discussed in this book as they are seldom used today.

AAC Power Sources
All standard welding power sources can, in theory, be used for air-carbon arc gouging. However, a minimum OCV of 60 V is necessary for good results with the larger-diameter electrodes. Arc voltages when operating may be as low as 35 V or as high as 56 V. Actual arc voltages and amperages are determined by the electrode size and the job to be done. Operating parameters are usually printed on each box of electrodes.

The output circuit of a power source used for AAC also should have overload protection to prevent damage to the power supply during high current surges that occur while gouging. Therefore, be sure to check the manufacturer's literature or call the supplier to

find out if your power source is suitable before using it for arc gouging.

Rectifier power sources are preferred where large-diameter electrodes are in common use, such as in foundries. Where the demand is constant for high arc currents, constant-potential rectifiers are used with all slope removed. A dc constant-current, drooper-type power source is required for automatic equipment using a voltage-controlled electrode feed.

Recommended power sources for air-carbon-arc gouging are listed in Table 13-1. Use a constant-current, drooping-voltage dc power source if you possibly can. It will give you the best results.

When dc rectifiers do not have sufficient output for gouging, two rectifiers can be connected in parallel. The output from both rectifiers will thus be combined to produce enough current for air-carbon-arc gouging. Figure 12-3

FIGURE 13-4 AAC electrodes come in many different round diameters; however, D-shaped rods and other special sections can be ordered for special kinds of arc-cutting work.

in the last chapter showed you how to make this two-machine setup. *But again we warn you, they should have the same OCV.* You don't want two machines near each other with a different OCV. The voltage *difference* between the machines could be dangerous if you happen to touch both of them at once.

Current Settings

Metal-removal rate is closely related to current. For example, at 1400 A a ¾-in. electrode removes twice as much metal per minute as it does at 1000 A. However, exceeding recommended current levels can overheat the electrodes, which will cause high copper-coating burnback, torch-holder damage, and too much heat reflected back against you.

NOTE: Do not exceed recommended current levels when gouging.

The best method to get very rapid metal-removal rates is to go to larger-diameter electrodes rather than increasing the amperage beyond the recommended maximum for the electrodes you already are using. Larger-diameter electrodes not only operate at higher current settings, but they don't wear down as fast. Larger electrodes also provide a smoother arc-cleaned finish because the arc current is spread over a larger area. Finish may also be improved by reducing the angle between the torch and workpiece to produce a more shallow but smoother cut.

When gouging U grooves for weld-joint preparation, electrode size and current settings are chosen to provide the desired groove width, depth, and contour. Groove width is generally ⅛ in. [3 mm] greater than the electrode diameter. Therefore a ¾-in. [19-mm] diameter electrode will produce a ⅞-in. [22-mm] wide groove. Groove depth and contour are controlled by your operating technique and the current setting you choose. You can always prac-

tice first on a piece of scrap until you get the best electrode diameter, amperage settings, and the technique you want for a given U-grooved joint.

Table 13-2 lists suggested current ranges for operating different kinds of air-carbon-arc-gouging electrodes.

Air Supply

Ordinary compressed air supplied at pressures from 80 to 100 psig [550 to 690 kPa] is all you need for air-carbon-arc gouging. Light-duty gouging torches sometimes can operate with as little as 40 psig [275 kPa] air flowing at 3 scf/min [0.001 m³/min], but most industrial torches require twice that much pressure with at least three times the flow rate or more. In a pinch you can use nitrogen gas for the job when you can't get com-

pressed air. Some operators use compressed oxygen (oxyarc gouging), but we feel that an inert gas is safer . . . as long as you have adequate ventilation for breathing air. Never use an inert gas in place of air for gouging if you work in an enclosed area.

Air pressure and volume Table 13-3 lists the compressed-air requirements for AAC cutting. The air stream must have enough volume, as well as pressure, to blow heavy, melted slag from the kerf cut by the graphite electrode. The holes in the AAC torch are designed to provide an adequate air stream for gouging.

Poor-quality gouging will result if the air pressure falls below the required 80 psig [550 kPa] minimum, or if the air holes in the electrode holder are restricted by

---

TABLE 13-1 Power sources for arc gouging (use drooping characteristic power sources when possible)

Type of Current	Type of Power Source	Remarks
DC	Variable-voltage generator, rectifier, or resistor-grid equipment	Recommended for all electrode sizes.
DC	Constant-voltage motor generator or rectifier	Recommended only for electrodes above ¼ in. diameter.
AC	AC transformer	Should be used only with ac electrodes.
DC	DC rectifier	DC supplied by three-phase transformer-rectifier is satisfactory. DC from single-phase power source is not recommended. AC from ac-dc power source is satisfactory if ac electrodes are used.

TABLE 13-2 Current ranges for arc gouging, in amperes

Electrode and Power Source	Electrode Size, in.					
	5/32	3/16	1/4	5/16	3/8	1/2
DC electrodes (DCRP)	90–150	150–200	200–400	250–450	350–600	600–1000
AC electrodes (ac)	—	150–200	200–300	—	300–500	400–600
AC electrodes (DCRP)	—	150–200	200–250	—	300–400	400–500

small-diameter hoses, fittings, or by slag blown back against the holder.

While gouges or cuts made with insufficient air may not always look particularly bad, they will be loaded with slag and carbon deposits. That is the reason why the air pressure should not fall below the minimum pressure specified for the type of torch being used. The inside diameter of all hoses and fittings always should be large enough to allow a high volume of air to reach the electrode holder.

Air hose and fittings Hoses and fittings with an inside diameter of ¼ in. are sufficient for small, light-duty torches. A minimum ⅜-in. inside diameter is recommended when gouging with electrode holders that are big enough to hold ⅜-in. diameter and larger electrodes. Automatic holders should be equipped with hoses and fittings with a minimum inside diameter of ½ in.

13-3
GOUGING TECHNIQUES

It only takes a few minutes to learn to use air-carbon-arc gouging, but the results will not be very good unless you pick up some operating skills. Gouging-rod suppliers often provide data on metal-removal rate in pounds per hour versus current settings for different sizes of gouging rods. If you can get this information for the gouging rods you will use, it will help you select the best rod

diameter and power source setting for the particular job you have to do. Here are some other procedures to help you do a good job.

ELECTRODE STICKOUT Electrode stickout is the distance between the torch and the hot end of the electrode. This distance should be no more than 6 in. [150 mm], or 4 in. [100 mm] when cutting aluminum. Any more than that and the compressed air will be too far from the groove and you won't blow out all the molten metal and slag. The electrode stub can be burned to within 1½ in. [38 mm] from the torch before repositioning the electrode or adding a new jointed rod.

NOTE: Turn the compressed air on before the arc starts.

START THE AIR FIRST It is very important that the compressed air be turned on before you strike an arc. When you do strike the arc, you have to touch the electrode to the work. Unlike SMAW, it is not necessary to pull back the electrode to establish an arc length. That makes the job a lot easier for you. The metal directly under the graphite electrode will immediately melt and be blown away by the compressed air. Electrode movement should begin as soon as the arc is started. Once the arc is established, the electrode should not touch the work or carbon from the electrode will be deposited on the base metal. You especially want to avoid getting any carbon into the base metal when welding

stainless steels, nonferrous metals, or low-carbon steels.

TORCH ANGLES The torch angle that you maintain between the electrode and the work determines your travel speed and depth of cut. A steep electrode angle will result in a deep gouge requiring you to travel slowly. Faster travel speeds and a shallower groove will result when a low electrode angle is used. The side of the electrode also can be used to wash down larger areas and to adjust the angle of a bevel joint on the edge of a square plate. You can judge the proper travel speed for any electrode angle simply by listening for, and maintaining, a steady hissing sound. If you hear a lot of intermittent spitting and spluttering noises, you are not making a good cut.

THE AIR STREAM Always remember that you must turn your air on before you strike the arc. The direction of the air stream from the electrode holder must be aimed directly at the molten metal. If you hold the electrode in the torch jaws so that the air stream is not parallel with the electrode, you will blow slag to one side or the other of the kerf and make a real mess because the air will not properly clean out the groove.

STEADY HANDS HELP You have to handle the electrode holder steadily. Move it smoothly forward (and of course, pointed away from you) as you work. Interrupted or shaky movements will result in rough edges on the workpiece with slag and carbon deposits in the base metal. Pointing the electrode in your own direction is dangerous. You'll get hit with all the slag. The best idea is to support the arm holding your torch either with your other arm and hand, or a steady rest. Also use both hands on the torch when possible. Sometimes you will want to support your arm or hand by putting an elbow on the

TABLE 13-3 Air consumption for arc gouging

Maximum Electrode Size, in.	Application	Air Pressure, psig	Air Volume, scf/min
¼	Intermittent-duty manual electrode holder	40	3
¼	Intermittent-duty manual electrode holder	80	9
⅜	General-purpose electrode holder	80	16
¾	Heavy-duty electrode holder	80–100	20
⅝	Semiautomatic or mechanized electrode holder	80–100	25

plate. Your elbow can get quite sore after a couple of hours resting on hard plate. A basketball elbow pad under your shirt will take care of that problem.

13-4
SOLVING GOUGING PROBLEMS

Although air-carbon-arc gouging is a very simple process to use, there are two areas that can give you problems: excessive *burnback* of your copper electrode coating, and adding carbon to certain metals (such as low-carbon steels) that you are cutting.

We mentioned earlier that all metals can be gouged or cut using the AAC process. However, different thermal and slag-forming characteristics require different techniques for some base metals. Table 13-4 gives you some operating suggestions for working on commonly gouged metals.

Excessive Burnback

Normally, a small portion of the copper coating on the electrode melts back from the tip when you use it, exposing the bare graphite material underneath. The exposed area may reach temperatures as high as 3000°F [1650°C]. However, when the burnback area extends to 5 or 6 in. [125 to 150 mm] the heat transmitted through the rod to the torch can cause extreme operating discomfort and damage torch jaws and insulators. Excessive burnback also oxidizes a significant portion of the electrode with the result that a ¾-in. [19-mm] diameter electrode may be tapered to only ⅝ in. [16 mm] at the arc.

High burnback (more than 3 or 4 in. [75 to 100 mm] of the copper coating has burned back from the electrode tip) can result from using an excessive current for the electrode diameter, insufficient air pressure, or a copper coating on the electrode that is too thin (you used a bad-quality electrode). Check your air pressure (it may be too low). Check your current for the electrode diameter you

picked for the same reason. . . too much current and heat for the electrode diameter you're using.

Added Carbon

Carbon from the gouging electrode can mix with the molten metal as the groove or cut is being made. With proper air velocity and torch movement this oxidized and carbonized slag will be blown right out of the kerf, leaving a clean base-metal edge free from carbon deposits. If the air-stream volume or pressure is too low, or the electrode-holder movement is irregular, carbon and slag deposits will be left in the kerf.

This is particularly bad if the material you are working on is to be welded. As you know, carbon will increase the strength of steel. As the base metal rapidly cools off, the steel will become very hard and brittle anywhere you accidentally added carbon. The added carbon will make the base

metal crack-sensitive along the welded joint.

You can add so much carbon to the steel base metal that it will crack right after you gouge it. Increasing the carbon content of low-carbon austenitic stainless steels also will greatly lower their corrosion resistance.

What do you do if you get carbon and slag in the kerf? All is not lost—you simply grind the edge afterward. But that, of course, adds extra labor cost to the job. However, if you do gouge a metal that can easily become crack-sensitive with small additions of carbon (such as high-strength, low-alloy steels, or one whose corrosion resistance depends on having a low-carbon content like many types of stainless steel), it's well worth finish-grinding the edge after gouging and before welding.

Since a groove or cut made with air-carbon-arc gouging moves very fast, little heat is put into

---

TABLE 13-4 Operating techniques for different base metals

Base Metal	Technique
Carbon steel and low-alloy steel	Use dc electrodes with DCRP, ac electrodes with ac. A transformer can be used, but ac is only 70% as efficient as dc. If you don't remember why, look back at Chap. 11 under "Effective (RMS) Voltage" of ac. It's only 0.707 as hot as the same dc voltage.
Stainless steels	Apply the same technique used for carbon steels.
Cast irons including malleable and ductile irons	Use ac electrodes with DCSP or with ac at the middle of the amperage range. DC electrodes can be used with DCRP at maximum amperage.
Copper alloys (Cu 60% and less)	Use dc electrodes with DCRP at maximum amperage.
Copper alloys (Cu 60% and more) or a large-size workpiece	Use ac electrodes with ac at maximum amperage.
Nickel alloys	Use ac electrodes with ac.
Magnesium alloys	Use dc electrodes with DCRP. Before welding, surface of groove should be wire-brushed.
Aluminum alloys	Use dc electrodes with DCRP. Wire brushing is mandatory prior to welding. Electrode extension (length of electrode between holder and work) must not exceed 4 in. [100 mm] for quality work.
Exotic metals (titanium, zirconium, hafnium, and their alloys)	At the present time these metals cannot be prepared for welding by the AAC process without subsequent cleaning. They can be cut for remelting.

the base metal. This results in very little distortion. It also means that the surface of the gouge, which is momentarily exposed to high temperatures, will cool rapidly. A rapid air-cooling rate will have no effect on non-hardenable steels, but it will cause a hard surface in the gouged area of a high-carbon steel or a cast iron. This hardened, heat-affected zone is very shallow (usually around 0.006 in. [0.15 mm] deep) and normally does not present problems unless the surface will be machined.

A slight amount of preheat will help reduce any objectionable surface hardness when gouging high-carbon steels or cast irons, especially if they will subsequently be machined. A light pass with a grinding wheel also will remove the hard surface.

13-5
ARC-GOUGING SAFETY

The arc-gouging process requires the same safety considerations as any arc-welding job. See Chap. 12 for general safety considerations for arc welding. The voltages used in air-carbon-arc gouging are relatively low. This reduces the chances of serious electrical shock. However, take precautions to wear proper clothing including leather aprons, sleeves, pants, and shoe covers (see Chap. 4). Remember that you will be spraying a lot of slag out in front of you. Some of it can either glance down on your shoes or bounce back at you. Be especially sure to protect yourself properly if you work overhead.

CAUTION: Arc gouging can throw molten metal and sparks over 20 ft [6 m] from where you're working.

Air-carbon-arc gouging produces a shower of sparks and molten metal that can easily travel 20 ft [6 m] or more. You have some control over the direction of the spray and should operate your electrode pointed away from other

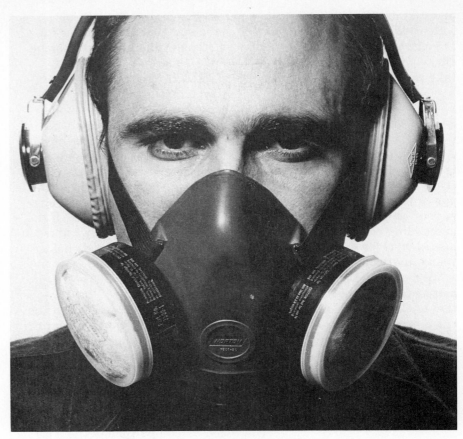

FIGURE 13-5 This is a filter mask. It does not supply oxygen, but only filters dust from the air. Supplied-oxygen masks are even safer.

workers and from flammable materials. Combustibles should be moved at least 35 ft [10.5 m] away from your work area. Don't work on a wooden floor without first laying down a fire-resistant blanket or sheet metal.

Splashing metal and sparks are actually easy to control if you set up a sheet-metal shield in front of your work. Deflect the shield slightly downward toward you and your base metal and the shield will make the sparks bounce back onto the floor, not up and over the slag barricade.

Protect yourself from arc radiation by using a standard welding helmet with at least a No. 12 lens shade. Use a No. 14 lens shade for larger-diameter electrodes. Be sure that your throat and the sides of your face and your hands above your gloves are covered so that you won't get sunburned or hit with sparks.

Noise is another problem with arc gouging. It stems directly from the use of compressed air. The noise level of the gouging process can be higher than a rock concert. Most plants have industrial sound-deadening earmuffs (see Fig. 13-5) that are provided for arc-gouging. These sound protectors are very comfortable. The safety director, plant nurse, or your welding supervisor should have a pair handy for you to use.

Use good ventilation above your work area. Most of the smoke and fumes that are produced by arc-gouging carbon steels are iron oxides and carbon mixed with trace elements of alloying elements. They usually are classed as nuisance particles because they are biologically inert, which means that they won't hurt your lungs and they are not toxic. But the fumes can be very annoying to work in and the dust is a nuisance, so get rid of them with good ventilation.

The smoke and fumes produced by stainless steels, high-alloy steels, zinc-coated galvanized steels, and copper and its alloys, however, are harmful. In fact, whenever you work on metals that contain chromium, zinc, copper, or nickel use a high ventilation rate and wear a good filter mask (Fig. 13-5). Zinc fumes, for example, will give you terrible but temporary headache and chills. A respirator with supplied oxygen is even better and we strongly recommend that you ask for one. *Never* gouge metals containing beryllium, cadmium, or lead (some copper alloys contain beryllium as an alloying element—they usually are referred to as *beryllium-coppers* so that you will know them). Beryllium, cadmium, and lead fumes are poisonous. (Cadmium is usually a silvery electroplated coating on high-strength steel bolts and some electrical equipment. Although it is dangerous, you are not very likely to run into it. Lead is easy to recognize. You can cut it with a knife, but lead-based paints should be removed before gouging.) If you are not sure, ask your welding supervisor, safety director, or welding engineer.

Just about all industrial companies are very concerned about worker health and safety. They will want to take care of you just as much as you want to protect yourself. Do not hesitate to ask questions and express your concerns when you work in any shop. The people you report to can specify safe procedures and improve working conditions, but it's up to you to let them know when you're having a problem.

REVIEW QUESTIONS

1. What process abbreviation is used on blueprints and operator instructions for the air-carbon-arc process?

2. Give an example of a material that the air-carbon-arc process can cut that oxyfuel flame cutting can't.

3. What material are gouging rods made of?

4. What metal are these electrodes often coated with? When they are coated, what is the purpose of the metal coating?

5. Coated air-carbon-arc electrodes are most often used. Sometimes uncoated (bare) dc electrodes are used. Why?

6. What affect does increasing the current have on metal removal rate in air-carbon-arc gouging? That is, does a higher current remove more metal, or less metal?

7. What current range (in amperes) would you set your welding machine at to run $\frac{3}{16}$-in.-diameter dc electrodes with a DCRP setup?

8. What air pressure would you use for gouging with a $\frac{3}{8}$-in. diameter electrode? What cubic feet per minute compressed-air volume (flow rate) would you need?

9. Should you strike an arc first, or start up the air first when gouging?

10. What type of electrode should you use for gouging aluminum alloys, ac or dc? If dc, what polarity, DCSP or DCRP?

11. You are getting excessive burnback of the electrode coating on a job. What should you check on your setup to solve the problem.

12. Roughly how far can arc gouging throw molten metal, slag, and burning sparks if you don't erect a barrier or work in a booth?

13. You are working on a zinc-coated (galvanized) steel. Should you worry about the fumes? If so, what should you do about them?

14. Should you gouge beryllium alloys, or copper alloys that contain lead, beryllium, or metal with a cadmium-electroplated coating? If not, why not?

LABORATORY

Getting Experience with Gouging

Your welding instructor may assign some projects for you to work on with arc-gouging equipment. By now you are familiar with grooved joints for welding. Try at least one or two of the projects we give here.

13-1-S
MATERIALS AND EQUIPMENT

You'll need a complete gouging outfit, electrode holder, compressed-air supply, and copper-coated gouging rods. You also will need some scrap pieces of steel ranging from $\frac{1}{4}$ in. to 1 in. thick. Select your gouging-rod diameters, if possible, to match the material thicknesses you'll work on.

1. Butt two pieces of 1-in.-thick plate together and clamp them into position. Then pass a gouging rod along the butted edges. You will find with a little practice that you can produce a double J-grooved joint (a U groove) with one pass of a gouging rod. With a little more practice you can even control the depth of the grooved joint to leave a land or flat nose on the bottom of the J groove without cutting all the way through the two plate edges.

2. The next job is to try gouging out a bad weld (one with surface porosity, for example—by now your class should have produced plenty of that stuff). Try cutting down to solid weld metal without removing the entire weld. Then try cutting out a section of a complete weld as if you were just notified by quality control that the weld had deep cracks in the root and it had to be removed and rewelded. If the bad weld section is thick, you will have to make more than one pass with the gouging electrode.

3. Also try working in the vertical position, first gouging up (if your welding instructor allows it). That will show you how far the process can throw sparks. Watch for those sparks and don't spray them all over the work area, in your hair, down your neck, or on other people. If you have to, put a sheet-metal lid on the top of your welding booth to protect other people in other booths. If you are producing too much slag spray, then simply gouge down instead of up.

4. After that, try making a J groove and then one side of a V groove on one piece of steel about 1 in. thick. Again, with a little practice you will be able to control the electrode so that you can either shave off a fairly smooth, straight edge on either side of a plate to make half of a V joint, or else produce a rounded J groove in the plate with a flat land on the base, ready to be fit up as a U-grooved butt joint.

Follow these guidelines in your shop work:

Cast irons, including gray, ductile, and malleable grades: Use ac electrodes with the ac at the middle of the amperage range. Direct-current electrodes can be used with DCRP at maximum amperage.

Copper alloys (Cu 60 percent and less): Use dc electrodes with DCRP at maximum amperage for maximum heat.

Copper alloys (Cu 60 percent and more): Use ac electrodes with the ac at maximum amperage.

Nickel alloys: Use ac electrodes with ac gouging rods.

Magnesium alloys: Use dc electrodes with DCRP. Before welding, surface of groove should be wire-brushed.

Aluminum alloys: Use dc electrodes with DCRP. Wire-brushing is mandatory prior to welding. Electrode extension (length of electrode between holder and work) must not exceed 4 in. for quality work.

Exotic metals (titanium, zirconium, hafnium, and their alloys): These metals cannot be prepared for welding by air-carbon-arc gouging without thorough subsequent cleaning. They can be gouged to produce scrap for remelting.

TIG Welding

You will enjoy working with TIG (tungsten-inert-gas) welding (Fig. 14-1). With the skills you have learned using gas welding and brazing, you will find gas-shielded tungsten-arc welding (GTAW) easy to learn. That is why we start you on TIG before working with stick electrodes.

Your welding instructor might not agree. Do what he or she wants. Most students traditionally learn stick-electrode welding (SMAW) as their first arc process. But the oxyfuel skills you have just learned are directly transferable to GTAW, and not to SMAW. That's why this chapter is where it is in this book.

14-1
WHAT IS TIG?

The GTAW process uses a tungsten electrode enclosed inside a gas-shielded electrode holder to create a welding arc (Fig. 14-2). The holder is very small; some of them are not much larger than a cigar, others are no bigger than a small welding torch. All TIG electrode holders have an insulated handle you hold in one hand while you work. The tungsten electrode holder is attached to the electrical power supply and shielding gas which flows through the torch, protecting the tungsten electrode and the work from oxidation. The tungsten inside the holder does not melt. It is essentially a permanent (but replaceable) part of the TIG torch. Filler metals, when used, are fed into the arc produced between the tungsten electrode and the workpiece. When filler metals are used, you hold the torch in one hand and feed the bare filler-metal rod into the arc and molten weld puddle with the other. Later on we will explain the process in much more detail.

However, by now you've probably guessed that GTAW and TIG are the same processes and the abbreviations are used interchangeably. How did this process wind up with two different names? TIG is the common shop

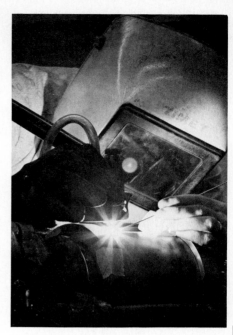

FIGURE 14-1 Tungsten-inert-gas welding is most often used to weld stainless steel, aluminum, copper, and (in this photo) titanium, although it can easily weld carbon steels, too. Note how this TIG welder feeds an uncoated filler-metal rod into the weld joint with a technique similar to oxy-fuel gas welding. TIG also can be used without a filler metal. Many TIG welders hold the filler metal without wearing gloves to get better control over the wire or rod. Safety directors might not like that, but TIG welders often do delicate work requiring very accurate positioning of the filler metal.

name for this welding process. It was the only name used for many years because GTAW was only one arc-welding process. Advances in welding technology finally expanded TIG into a number of related processes such as standard TIG welding, pulsed-current TIG welding, TIG spot welding, and hot-wire TIG welding.

The American Welding Society (AWS) decided it was time to give all these separate TIG processes one formal name, gas-shielded tungsten-arc welding, and one common abbreviation, GTAW. The same thing happened to MIG welding, which now is only one of several related GMAW or gas-shielded metal-arc-welding processes.

In the help-wanted sections of newspapers you will see other names used for TIG. These are manufacturers' trade names for the process or the equipment. The two names you are most likely to see in employment advertising are "Heliarc or Heliweld welders wanted."

The AWS name for the process, gas-shielded tungsten-arc welding, is a good description of all GTAW processes. First, GTAW is an arc-welding process. Second, the molten weld metal is shielded by an inert gas. Third, GTAW uses a permanent tungsten electrode to maintain an arc on a workpiece, instead of a consumable electrode that gets used up as filler metal during the welding process. Fourth, separate filler metals such as hand-held rod or automatically fed wire are optional, depending on the thickness of the workpiece and the joint strength desired. Thin sections usually don't require filler metal.

TIG Is Versatile

The GTAW process is the most flexible of all fusion-welding processes in terms of the large variety of metals it can join. GTAW can weld just about any metal. It will weld aluminum and magnesium and their alloys; carbon steels; alloy steels; stainless steels; copper and nickel alloys; titanium; tin, silicon, and aluminum bronzes; and even cast irons (although GTAW is seldom used for welding cast irons).

Unlike oxyfuel welding, the various GTAW processes can be operated manually or with automatic equipment. GTAW will join many materials with welds that are of equal or better quality than any other welding process.

It is this exceptional versatility that often makes GTAW the first choice for welding stainless steels and aluminum alloys, which are difficult to join. But if TIG can weld just about any metal that can be welded, why isn't it used everywhere to weld everything?

A lot of maintenance welding is still done with oxyfuel simply be-cause the equipment is much less expensive than any arc-welding process, including GTAW. For example, what you pay for one good arc-welding power source would buy enough complete gas-welding outfits for a small crew of people.

GMAW (which includes MIG and other welding processes) may not have the flexibility of GTAW in terms of the metals that can be joined, but since these other processes always use a continuous electrode wire, they can produce more pounds of weld metal per hour. In 1966, the continuous hot-wire GTAW process was introduced. That boosted the pounds-per-hour deposition rate of GTAW, and it has increased the use of GTAW in more production applications on carbon and high-strength, low-alloy steels than before. But human nature and economics have as much to do with the selection of a welding process as welding technology. Hot-wire TIG is still considered to be a specialized process. It requires two power sources (one for the TIG arc and one for the hot filler wire) and, therefore, the process is initially more expensive to use.

Any company that does welding work will tend to pick the process it knows the most about. More production people understand gas welding and stick electrodes than any of the other processes. They also are more likely to have the right equipment for these processes on hand. Obviously if you already have stick-electrode equipment and you know how to use it, you stay with it.

Less-skilled welders also have limited the use of GTAW to some extent. No matter how much a production manager wants to use a given welding process, he or she must consider what the work force does and doesn't know how to do. Many welders who know how to use stick electrodes do not make good TIG welders. Oxyfuel welders are easier to train as TIG welders, but there is a shortage of good oxyfuel welders, and therefore a shortage of TIG welders.

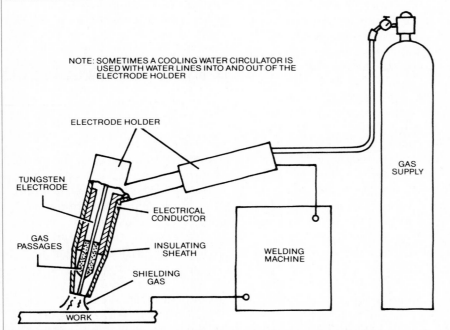

NOTE: SOMETIMES A COOLING WATER CIRCULATOR IS USED WITH WATER LINES INTO AND OUT OF THE ELECTRODE HOLDER

ELECTRODE HOLDER

TUNGSTEN ELECTRODE

ELECTRICAL CONDUCTOR

GAS PASSAGES

INSULATING SHEATH

SHIELDING GAS

WELDING MACHINE

GAS SUPPLY

WORK

FIGURE 14-2 This is a schematic of the equipment you will use in TIG welding. The TIG electrode holder is drawn much larger in proportion to the other equipment to help you understand the process. Actually the TIG holder is quite small. The tungsten electrode is not consumable. It should not melt, but instead simply helps to form the welding arc. TIG holders are either air- or water-cooled.

Why don't more welders understand GTAW? Because most arc welders only know one process, and that usually is stick-electrode welding. Also, since the GTAW process is so ideal for welding stainless steel and aluminum, it's most likely to be found in industries that use these materials. Examples are aerospace plants, fabricators of stainless-steel kitchen products, manufacturers of chemical equipment, and pipeline fabricators. Pipe-fabricating shops often make the critical root passes with GTAW. They then complete the joint with some other process with a higher deposition rate, like stick electrodes or submerged-arc welding.

GTAW requires continuous and efficient weld-metal shielding with moderately expensive inert gases. Backup inert-gas shielding on the opposite side of the weld also is needed in some applications, especially when welding titanium, and often when working on thin sections of easily oxidized metals. That makes job setup a little more difficult.

Most TIG welding is done indoors because of the gas shielding that is used. A high wind will blow the shielding gas away from the molten weld puddle. However, there are methods for solving that problem. Examples are working behind a canvas windscreen or in a portable welding booth that fits over the work. That is how some GTAW is used in field work.

Many of the industries that use GTAW extensively pay very good wages, which means that welders that know GTAW often get more money than other welders.

Reasons to Become a Skilled TIG Welder
Although TIG is a process that is not used as often as it should be, where it is used the wages are good. And *you* have an advantage over many arc welders: you have learned oxyfuel welding. This simplifies the learning process because TIG welding is similar in important ways to oxyfuel work.

The GTAW torch is operated in much the same way as a gas-welding torch (Fig. 14-3). You can use the same bare filler rods with GTAW that you use with gas welding and brazing. (Even the same AWS filler-metal specs apply to both processes.) You can feed the welding rod into the arc and the molten weld puddle using either the forehand or the backhand methods that you've already learned for gas welding. You can work in any position with GTAW including down-hand, horizontal, vertical, or overhead. And, just like oxyfuel gas welding, thin sections can be joined by GTAW without any filler rod at all. Even the welding arc is cone-shaped (somewhat like a gas flame), rather than cylindrical like many other arc-welding processes.

14-2
AN OVERVIEW OF TIG
We'll cover the GTAW processes in detail later on, but right now,

let's look back at Fig. 14-1 and imagine that you see the TIG welder at work. The operator is wearing a standard arc-welding helmet. The TIG operator is holding the torch in one hand and feeds a *cut length* or rod of filler metal into the molten weld puddle with the other hand.

There is a lot more equipment required than shown in Fig. 14-1. The electrical current is being supplied by an arc-welding power source that is compatible with GTAW process. Next to the power source is a cylinder of inert shielding gas, most likely either argon, helium, or an argon-helium mixture. The choice of shielding gas depends on what metal is being welded and what kind of weld joint penetration (shallow or deep) is desired.

Under the operator's workbench you may find a foot treadle used to control the welding amperage. Welding-current foot controls are used mostly for high-quality ac and dc GTAW, which means

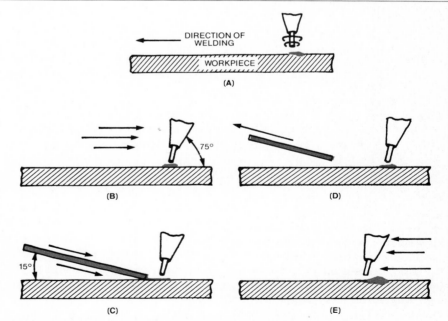

FIGURE 14-3 Feed GTAW filler metal just like you do when using oxyfuel welding. (A) First develop a small molten puddle. (B) Then move the TIG holder and arc back slightly, holding the holder so that the shielding gas continues to cover the molten area. (C) Add filler metal. (D) Then remove the rod, and (E) return the TIG arc to the leading edge of the weld puddle. You will repeat this process over and over as you develop a continuous TIG weld seam. The filler rod and arc can be used with either the forehand (shown here) or the backhand technique, just like oxyfuel welding.

that you have to work with one foot as well as both hands, and that adds a little more to the skills you need for this process over oxyfuel work. But there are other ways to control the welding current, as you'll find out shortly.

The inert shielding gas is fed through a conventional high-pressure gas regulator from a cylinder to the molten weld metal. However, the cylinder regulator may have a gas flowmeter attached to it, because shielding-gas flow rates in TIG welding often are critical. A separate shielding-gas line leads from the flowmeter or gas regulator through the TIG

FIGURE 14-4 This is a typical TIG holder. The long pointed cap on the upper right contains the tungsten electrode. The cap can be unscrewed so that you can push the tungsten farther into the holder if your tungsten tip (on the upper left) wears out. The tungsten tip extends from a ceramic gas-shield nozzle that slips over the holder. The ceramic nozzle is easily removed and replaced if you break it. The long handle also can be unscrewed in the middle so that you can wire up the electrical contacts to the holder, and then slide the lower part of the handle back in place and screw it tightly again.

holder and out through a ceramic cup called the *nozzle*.

Inside the nozzle (Fig. 14-4) is a small-diameter tungsten electrode that clamps into the holder and is attached to the electrical supply from the GTAW power source. The *tungsten,* as it's called, is usually between ¼ and 1/16 in. in diameter depending on your current and polarity requirements. Smaller tungstens are used with very light gauge metals. The tungsten is quite long and can be moved down into the nozzle as it wears out. That's why some TIG holders have a pointed cap on the back side, opposite the ceramic nozzle. The extended tungsten electrode (which is quite brittle) is protected by this cap.

TIG Power Sources

The power source you will use (Fig. 14-5), is probably very similar to the one you used for air-carbon-arc gouging, except that it may have some special high-frequency circuits built into it to make it easier to start or control the TIG arc. The power supplies used for GTAW include ac or dc motor generators, ac transformers, dc transformer-rectifiers, and ac or dc engine-driven power sources.

Alternating current If an ac welding machine is used, the power source must have special circuits to keep the arc "ignited" when the ac switches polarity. Without these special circuits it would be difficult to keep the arc going when the current waveform passes zero amperes 120 times per second (when using 60-Hz current) as it passes back and forth from straight polarity to reverse polarity.

One way to keep the TIG arc ignited is to superimpose a very high frequency ac waveform onto the regular 60-Hz ac welding current. Of course that means that an ac power source designed for use with GTAW will cost a little more than a regular ac welding machine for use only

FIGURE 14-5 This is a typical multipurpose GTAW power source containing both an ac transformer and a dc transformer-rectifier. You can select ac, dc straight polarity, or dc reverse polarity with the three-position handle on the right-hand side of the power source panel. Three current ranges (low, medium, and high) can then be selected with the left-hand handle. The welding current can then be fine-tuned with the round dials to give you exactly the current you want. Although you can't see it in this photo, every switch, dial and handle on the control panel is clearly marked to tell you what it does.

with ac stick electrodes, but an ac power source designed for GTAW can also be used for ac stick electrodes. That's why it's often better to buy a more versatile power source and get multi-use ac welding power.

Direct current Your workpiece for GTAW can be set up for either ac or dc welding. If dc is used, either DCSP (direct-current straight polarity) or DCRP (direct-current reverse polarity) can be used. Most GTAW work uses DCSP.

DCSP Straight polarity (electrode negative and workpiece positive) generates much more heat at the positive (work) end of the arc and less at the negative (electrode) end of the arc. Smaller-diameter tungsten electrodes can be used with DCSP without severely overheating the tungsten. For example a 1/16-in.-diameter tungsten is capable of carrying 125 A if DCSP is used. Twice the tungsten diameter would be needed for DCRP welding.

DCSP welding with the electrode lead attached to the negative power-source terminal and the ground lead attached to the positive power-source terminal produces a narrower molten weld pool and deeper weld penetration. The reason is that the negative electrons literally boil off the tungsten electrode and penetrate the base metal (Fig. 14-6). The result is a narrower heat-affected zone, deeper weld-metal penetration, less distortion, and faster welding.

DCRP DCRP is only used for special GTAW jobs. Reverse polarity with the electrode positive and work negative has one special advantage, surface cleaning of metals whose oxides cause problems in the welding operation. Examples of these metals are aluminum, magnesium, and titanium metal and their alloys. In DCRP welding the electrons boil off the negative workpiece, flow up and hit the positive tungsten electrode, keeping the surface oxides on the weld metal stirred up so that they do not form a solid refractory metal-oxide film on the weld puddle. This is called the *cleaning action* of the DCRP welding setup. Since the same cleaning action occurs when ac is used (because half of the time the polarity of the current is reversed), ac power sources often are used for GTAW of aluminum, magnesium, and other refractory metals.

If you were to set up a GTAW job with DCRP (electrode positive and work negative), you would find that the negative electrons striking the small tungsten electrode will cause it to overheat. Since part of the electron flow heats the electrode instead of heating the sheet or plate you are joining, less energy goes into the weld puddle and your molten weld metal will be shallow and wide. Therefore DCRP welding with TIG tends to produce shallow weld penetration, larger heat-affected zones, and more weld distortion than DCSP.

This excessive electrode heating limits the use of DCRP. Tungsten electrodes are fairly expensive. A 1/4-in.-diameter tungsten electrode has an allowable current-carrying capacity of about 125 A, but that is twice as big a tungsten electrode as you need for DCSP welding with the same 125 A of current.

Current Settings

When either ac or DCRP current is large enough, a pure tungsten electrode melts at the arc end and forms a tiny ball. Should the current be set too high, the molten ball of tungsten at the end of the electrode will fall into the weld puddle, producing a very bad inclusion in the weld metal. Welding current settings are quite important in GTAW.

Because of electrode effects like this, you have to choose between different kinds of tungsten-alloy electrodes. This chapter will tell you which ones to select for different kinds of work.

Your current setting is determined by the thickness of your weld joint and the type of metal you will join. Metals with high thermal conductivity like copper and aluminum require more heat (and therefore more amperage) than metals such as stainless steel. Thicker sections also require more heat (and amperage) than thinner sections of the same metal. Most of the time you will either be given the correct amperage settings, or you can look them up in a welding handbook or ex-

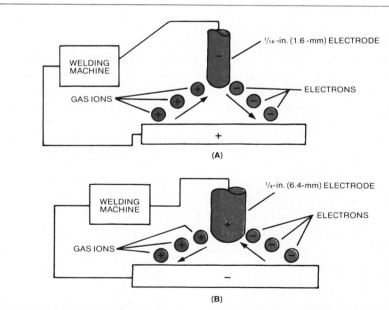

FIGURE 14-6 (A) Direct-current straight polarity (DCSP) shown schematically is the most commonly used dc current setting for TIG welding because it generates more heat in the positive workpiece and less heat in the negative tungsten electrode. (B) Direct-current reverse polarity (DCRP) is only used for special work (it does a better job of cleaning surface oxides off refractory metals), especially when ac (which does the same thing) is not available.

periment and see what settings give you the best-looking welds and highest welding speeds.

Shielding Gas

The shielding-gas flow through your TIG holder is set on your cylinder regulator, usually at around 20 to 25 scf/h [0.56 to 0.7 m³/h], depending on the properties of your shielding gas and the welding position in which you will work. Even smaller flow rates of only 5 scf/h [0.15 m³/h] may be used with very small TIG holders working on very thin gauge metals. Regulators with flowmeters often are used for GTAW because the shielding-gas flow rates are quite critical.

More shielding-gas flow is needed for lighter helium than heavier argon gas when working downhand. Working overhead, less helium than argon may be needed because helium is so light it rises up against the overhead weld metal. Again, the flow rates are usually determined for you in advance, or else you can experiment until you are sure you are getting adequate shielding with no discoloration of the cooled weld metal and tungsten electrode.

The shielding gas you use also affects the depth of penetration and the shape of the weld you produce, because helium and argon have different ionization rates in an electric arc. The ions subsequently return different amounts of heat energy when they regain their lost electrons. Therefore, in addition to current settings, you have to think about which shielding gas to use and how much to use to protect your weld metal.

On critical jobs requiring shielding-gas backup on the reverse side of your workpiece, you may have two different shielding-gas cylinders (or even two different shielding gases), and often different flow rates for the welding side versus the back side of your work.

The shielding gas sometimes is contained in a narrow movable hood that slides along the work. That's especially true when welding titanium alloys which are extremely sensitive to any degree of oxidation. Sometimes the workpiece is designed with a folded edge joint that functions as a shielding gas "tunnel" on the back of the weld. The tunnel contains the second shielding gas on the back of the joint to protect your metal from oxidation when you make full-penetration welds. (Another trick, however, is to use a stainless-steel backup strip instead of a shielding gas. It works very well on a lot of jobs when it's difficult to get backup gas up against the underside of the weld.) A pipe or tube often becomes its own backup protection because you can *lightly* plug both ends, purge the air from the inside, and inject shielding gas into the section before you start to weld on the outside. Don't plug the tube too tight. Remember that the gas inside will get hot and may expand enough to pop the "cork" and shoot it at you if the gas pressure can't be relieved.

Now that we've given you an overview of the GTAW process, let's get down to details.

14-3
GTAW POWER SOURCES

The welding a TIG arc can do depends very much on the amount of electrical power delivered by the welding machine. The electrical power in watts (amps times volts) produced by the machine is converted to heat by I^2R heating and by the ionized shielding gas at the arc. Once you have selected the shielding gas, it is the arc current (amps) and the arc length (which determines the arc voltage) that control the amount of heat produced with the GTAW process.

It takes some knowledge and even more manual skill to control these variables. You set the welding current on the power source, but you control the arc voltage by manipulating the length of the welding arc between the tip of the tungsten electrode and the workpiece. Automatic GTAW systems also use controlled arc current and arc length (arc voltage) to manipulate the heat input to the molten weld puddle.

GTAW Power and Arc Length

The electrical power (watts) converted to heat by a GTAW arc depends on the arc current and the arc voltage. The current to be set on the front panel of the welding machine is predetermined by the job, the metal thickness, the shielding gas used, and other factors.

Precise current settings often will be determined for you in advance. With practice, you can estimate what you need, strike an arc and try it on a piece of scrap, and then adjust your current settings according to how the arc, the molten weld metal, and your tungsten electrode look to you.

Arc length is not something you set on the machine. It depends on how you handle the GTAW electrode holder and how close or far away you hold it from the workpiece. Nevertheless, current and voltage (and therefore power and welding heat) are all related.

Imagine a dc constant-current (CC) GTAW power source set at about 100 A, using a dial mounted on the front panel of the power source. Suppose that your shielding gas is argon and that you are holding the GTAW torch and tungsten electrode just far enough away from the workpiece to produce an arc voltage of 15 V. This welding condition can be plotted on a graph (Fig. 14-7).

In this figure, the welding arc consumes 100 A × 15 V = 1500 W of electrical power. Most of this power is converted into useful heat which melts the workpiece and any filler metal you may add to the molten weld puddle. Some of the power is turned into heat and light which is radiated away from the workpiece and lost.

If you increase the distance of the TIG holder from the workpiece very slightly, the arc length becomes longer, increasing the arc voltage because it takes more volts of electrical pressure to push the current across the larger gap between the tungsten electrode and the workpiece. If you decrease the distance of the TIG holder slightly, the arc voltage becomes smaller and fewer volts are needed to make the current jump the gap between the tungsten electrode and the workpiece.

Let's assume that you continue to use the 100-A machine setting, but that you increase the arc length by $\frac{1}{16}$ in. [1.5 mm]. Your arc voltage will increase to 17 V in our example. The arc power will increase to 100 A × 17 V = 1700 W. The arc will become hotter because more power is being used.

If you lower the tip of the tungsten electrode by $\frac{1}{16}$ in. [1.5 mm] to put it closer to the workpiece, the arc length becomes shorter. Therefore, the arc voltage drops. That means that less power and heat are produced by the welding arc. Assume that the $\frac{1}{16}$ in. [1.5 mm] shorter arc length produces 13 V. The arc power is now 100 A × 13 V = 1300 W. The arc will run a little cooler because less power is being used.

This is a method used by TIG welders to change the amount of heat going into their weld metal when they don't have a foot pedal to control the amount of amps going through the arc (or they can't conveniently reach the amperage control foot pedal while

they work). It's very similar to changing the distance of a gas flame closer to or farther away from the workpiece. For example, the welder might increase the arc length slightly on a thicker section to get more heat and weld penetration with a given current setting. Then, when the operator comes to a thin section or has to control molten weld metal in the overhead position, the arc length is decreased slightly to reduce the power produced by the arc, which will lower the heat going into the section and avoid base-metal distortion and control the temperature of the molten weld metal. You also can avoid possible burn through on a very thin piece that way. In fact, GTAW can even be used to weld metal foil with a small enough TIG holder, a small-diameter tungsten electrode, lower amperage settings, and a suitable short arc length.

Most modern TIG machines include a foot pedal that you can use to increase or lower the amperage of the welding arc if you want it "hotter" or "colder."

Of course you, as the operator, won't see a meter that gives amps and volts and watts. You are behind an arc-welding helmet. About all you can see is the arc and some of the molten weld metal. But as a skilled welder you will know that raising or lowering a TIG holder has an effect very similar to controlling the heat into the weld metal by raising or lowering an oxyfuel gas flame.

The only thing is, you need a constant-current (CC) power source to make sure that the current won't change as you change the arc length. That is one reason why CC rather than constant-voltage (CV) power sources are almost always used for GTAW.

GTAW Power and Slope

Arc current will vary unless a CC power source is used. As you know, a perfect CC welding machine will have a zero slope. The graph of volts versus amps will be a perfectly flat horizontal line (as you can see in Fig. 14-7). No matter how the arc voltage changes, the amperage remains constant. In practice, all welding power sources have at least some degree of slope which is built into the machine, unless the machine has a special slope-control dial. In fact, one of the big differences between different types of welding machines for different processes is the amount of slope "built into" the machine.

If the power source has slope, the current and voltage will change together, depending on the volt-amp curve characteristic of the machine. Now look at Fig. 14-8. Imagine that this figure is the volt-amp graph of a welding machine with some slope built into it. The line sloping from left to right plots that slope relationship for many different individual current settings (the amps on the bottom of the graph) that you can make on your power source.

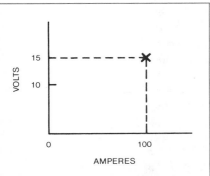

FIGURE 14-7 The perfect constant-current voltage/amperage curve for operation at 15 V and 100 A.

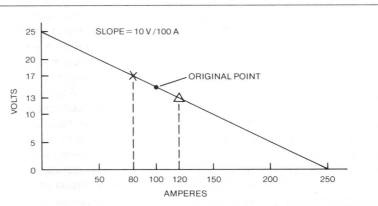

FIGURE 14-8 The effect of slope on the operating point of a GTAW power source.

How do you measure slope from this graph? Slope is the ratio of volts (on the left) to amps (on the bottom) at any point along the line. If you have ever done any surveying or had a little trigonometry, you would know that slope is the *tangent* of the angle at the right-hand corner of the graph.

The slope number (or tangent) will be the same no matter where you read it off the right-triangle in the graph. But let's make it simple. On the far left of the graph, the slope line cuts the volts side of the graph at 25 V. It also cuts the amps side on the bottom at 250 A. The ratio of volts to amps (which is what slope means) when the amperes are measured from the right-hand lower corner point of the triangle, and the volts are measured up from the bottom left-hand, right angle of the triangle, is 25 V/250 A = $\frac{1}{10}$, and that is the same as a slope of 0.1 or the tangent of a slope angle of about 5.7°.

Let's make one more slope measurement to clarify how you get the number. Although you won't have to calculate it when you are working in a welding shop, it helps you to understand a little bit more about what slope can do for you.

The slope at 150 A is measured from the right lower corner angle of the triangle starting with 250 A and subtracting 100 A for the base of the triangle. The voltage on the left, reading across where 150 A intersects the slope line, is 15 V. Therefore the slope is 15 V/150 A = $\frac{1}{10}$ = 0.1. Measured that way, the slope for the given volt-amp relationship will always be the same no matter what current setting you use to measure it. That is why a single slope setting can be applied to many different volt/ampere relations as long as their ratio remains the same.

Let's assume that you have just set the current on a welding machine with this 0.1 slope at the same 150 A you set on the CC,

no-slope machine graphed in Fig. 14-7. Read up to the sloping line (in Fig. 14-8) and over to the left and you will find that the arc voltage will be the same as before, 15 V. The power produced by the arc is 100 A × 15 V = 1500 W.

What happens when you set the welding current at 80 A? Reading up the chart from the 80 A spot to the slope curve (at the point marked X), and then over to the left to get volts, you will read 17 V on the left-hand side of the graph, which means that at a current setting of 80 A you will get 17 V. The power produced by your welding arc when the machine is set for 80 A will be 80 A × 17 V = 1360 W.

Next, imagine that we set the same welding-machine current to 120 A on the dial on the machine's front control panel. Reading up to the slope line (at the triangle mark) and over to the left, you get 13 V. The power produced at your welding arc will be 120 A × 13 V = 1560 W.

Using the welding machine with 0.1 slope, we can get at least 1360 W and up to 1560 W at the arc, depending on whether we set the amperage at 80 up to 120 A on the machine. That means we can get a cooler arc, a slightly hotter arc, and an even hotter arc, depending on how the slope angle and machine amperage is set.

Once a correct welding-current setting is made, you usually keep it there throughout a given job. If you start with a 100-A setting, you probably will stay with 100 A as long as the setting produces satisfactory welds. If you set the welding machine at 80 or 120 A, that's probably where you will keep it. Therefore, using either the pure CC machine or the second machine with slope, you can control the power, and therefore the actual heat produced by the arc, by changing the welding-current setting. The machine with adjustable slope gives you more control over the amperage power, and heat that you can use.

Arc Length and Slope

What would happen if you changed the arc length by increasing it, or decreasing it $\frac{1}{16}$ in. [1.5 mm], by moving your torch? In the pure CC machine, the current did not change when you changed the arc length. Only the arc voltage changed, and the amount of change produced was either +200 W or −200 W when you either lifted the torch $\frac{1}{16}$ in. [1.5 mm] or shortened the arc length by $\frac{1}{16}$ in. [1.5 mm].

However, changing the arc length on a machine with slope will not give you an equal change in the power (up or down). If you were to increase the arc length by $\frac{1}{16}$ in. [1.5 mm] on a machine with a 0.1 slope, you would actually *decrease* the power by 140 W. If you decrease the arc length, with the same slope, you would *increase* the power by 60 W.

The slope has reversed the direction of the power increase or decrease caused by changing the arc length. Instead of increasing the power by increasing the arc length (as in the CC machine), the arc power is decreased with an increase in arc length, at least on the machine with the 0.1 slope. What's more, the increase or decrease does not produce an equal amount of change in power, and heat, at the arc. The reason the change is not equal is that the slope relationship is a tangent function.

We have plotted all this data in Fig. 14-9 so you can see them all in one place. We show you these examples to make the point that a GTAW arc will behave differently when a welding machine has slope than when there is no slope.

Now consider a third welding machine. This one has a control that lets you select different degrees of slope, along with the welding current, when you set up the job. You can dial in zero slope (a pure CC machine), or a slope of 0.1 as in our example, or even more slope like 0.2 or 0.5, which would make the slope line on the volt-amp graph much steeper.

Adjustable volt-amperage slope control is an effective tool which you can use on more sophisticated GTAW power sources to change the arc power response to changes in arc length. Not all GTAW machines will have slope control. It requires extra circuits, and therefore slope-controlled GTAW machines cost more money.

There are limits to what slope can do for you. Let's look back again at Fig. 14-8. Look over to the far right of the slope curve, where it cuts the amperage line on the bottom of the graph at 250 A (and 0 V). What kind of a circuit could run 250 A of electrical current without any voltage? It's a short circuit with no resistance.

This would be how the 250 A of current would flow if you connected the positive and negative terminals of the power source together with a very large copper bar with almost no resistance (but please don't do that; it will melt the copper, probably burn you, and very likely ruin your power source).

If the slope were not 0.1, but 0.05 or 0.02 or even less, the slope line would get flatter and flatter as the slope angle gets smaller. The point at the bottom of the graph where the short circuiting current is would move farther and farther to the right for a given voltage. That means that the short-circuiting amps would become very large.

Too much short-circuit current is a problem with GTAW arcs that are started by the *scratch start* method (which we'll tell you about soon). Large short-circuit currents cause a big surge of electricity to flow through the tungsten electrode and damage it. A short-circuit current of from 1.5 to 2 times the operating current is generally considered safe to use for machines that are started with a scratch start. That puts a limit on just how much slope control will be built into any GTAW power source. That is a second reason why constant-voltage (CV) power sources are almost never used for GTAW.

14-4
STARTING YOUR ARC

Starting a GTAW arc is not as simple as it sounds.

Touch Starts

The most obvious way to start the arc is to touch the tungsten electrode (which sticks out slightly beyond the ceramic cup in the torch holder) to the workpiece. That will cause current to begin to flow. When you raise the electrode slightly from the workpiece, the current keeps flowing. The small spark created between the electrode and the workpiece grows into a full welding arc. You simply lift the electrode until you get the right arc length for the welding arc, power, and heat that you want.

This method is called *touch* or *scratch* starting. It has several advantages, but also a lot of drawbacks.

Touch Start Advantages
■ Simple to use. ■ Nonmechanical, nothing to fix.

Touch Start Problems
■ Tungsten can contaminate the weld metal. ■ Erosion of the tungsten electrode. ■ High initial arc current due to the short circuit of the tungsten on the workpiece. ■ Uses up expensive tungsten electrodes. ■ You're not 100 percent sure of starting at each touch. ■ Impractical to set arc length in advance. ■ Difficult to start remote arcs on automatic equipment.

Touch or scratch starting is only useful where you have a simple GTAW machine without special starting circuits, and when a high-quality weld is not essential. Since GTAW is usually used because it produces high-quality welds, it's better to start the arc some other way. That's true even if you have to buy a slightly more expensive GTAW machine with special high-frequency starting circuits.

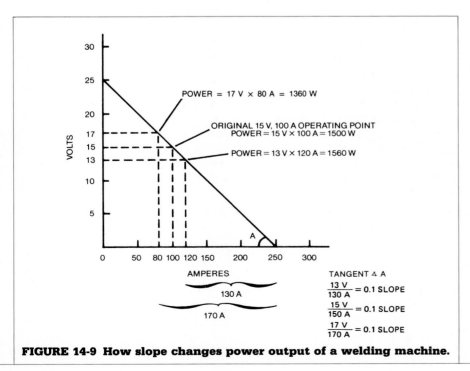

FIGURE 14-9 How slope changes power output of a welding machine.

High-Frequency Starts

High-frequency or HF starting is a method for igniting the GTAW arc that overcomes the problems of touch starting. A high-voltage (several thousand volts) but very low-amperage (a couple of amps) current is combined with a high frequency (up to a million or more hertz) and connected across the arc gap. The high voltage causes the gas in the arc gap to become ionized. Since an ionized gas with the electrons stripped off the outer orbits of the gas atoms carries a current (the gas in fluorescent lights is an example), the size of the gap that a starting spark can jump is much larger.

Properly adjusted HF circuits can cause a GTAW arc to start across a gap of 1/8 to 1/4 in. [3 to 6 mm] and more. Once the arc starts, it maintains the ionized gas column between the tungsten elecrode tip and the workpiece as long as dc is used.

If you are welding with ac, the voltage shuts itself off every time the waveform passes through zero voltage and zero current while it changes polarity. That happens 120 times a second with 60-Hz ac. Therefore the HF circuits automatically remain on as long as the ac welding machine is running.

HF starting currents with only a couple of amps of electricity will not affect your welding-current settings, and you won't notice any effect on your weld metal. But you will notice an effect on the ease of starting the arc. Several thousand volts of ac at up to a million or more hertz sounds terribly dangerous, even when the amperage is very low. Actually, without the high frequency it would be dangerous, but one characteristic of HF current is that it flows on the surface of conductors instead of through the conductor's cross section.

If you were to come in contact with the HF current for any reason, the energy would travel over your skin instead of through your body and vital organs. The most

you will feel is a slight shock or a tingling sensation. Other than being a little uncomfortable to touch, high-voltage, high-frequency, low-amperage current will not hurt a healthy person.

There are advantages and some disadvantages even with HF arc starting. They are:

HF Starting Advantages
■ Reliable, fast starting.
■ Very good for keeping ac arcs ignited.
■ No electrode or workpiece contamination.
■ HF arc starting is safe.
■ No starting short-circuit current surge.
■ HF can be used remotely for starting automatic GTAW equipment arcs.
■ HF starting is much easier to weld with than touch starting.

HF Starting Problems
■ HF causes radio and TV interference (it actually broadcasts radio signals).
■ HF circuits require maintenance.
■ HF arc-starting circuits increase the cost of the GTAW machine.

You can improve arc starting with a couple of simple techniques. For example, sharp-pointed tungstens start more easily than blunt ones with HF arc starting. Since the HF current travels on the surface of the tungsten electrode, grinding the arc end of the tungsten to a sharp point will concentrate the voltage and current and increase the pressure that forces electrons to jump off the tip and strike your workpiece. All you have to do is grind one end to a sharp point on a bench grinder and you're in business.

Very clean, bright electrodes in continuously flowing high-purity inert gas, after a few arc starts, become very difficult to restart.

Slightly oxidized electrodes actually start more easily than clean ones. A little oxide coating is good, not bad, on the tungsten. The electrons will jump off the surface at a lower electrical pressure. Even fingerprints on the tungsten make it more likely to give you a good start.

Don't polish your tungstens. (It reduces the HF arc-starting effect.) Select the correct tungsten for the job. We'll give you the data you need when we tell you how to select the correct tungsten electrode for your work.

GTAW equipment manufacturers have overcome many of the remaining problems with HF arc starting, especially radio interference. Some equipment even features a tiny pilot arc, much like a pilot light in a gas stove. The pilot arc is a small additional arc that creates a tiny ionized gas column. This column is maintained while you work and it helps the full welding arc reform almost instantly if it goes out.

14-5
ALTERNATING VERSUS DIRECT CURRENT

Both ac and dc can be used for GTAW. DCRP (electrode positive) heats the tungsten electrode much more than DCSP (electrode negative). As you already know, the positive tungsten in DCRP setups is constantly bombarded by hot electrons from the workpiece. In DCSP setups, the hot electrons flow from the tungsten to the work, actually pulling heat from the tungsten electrode and helping to keep it cool. The current-carrying capacity of a DCSP tungsten also is considerably higher than a DCRP tungsten.

What happens when you run the tungsten with ac, since ac is straight polarity half of the time and reverse polarity the other half?

You might guess that an ac tungsten electrode would heat up to some level between a DCRP tungsten and a DCSP tungsten. It does, but not halfway between them.

The tungsten electrode causes ac to become "unbalanced." It is easier to push current through the tungsten electrode and the arc in the straight-polarity direction (electrode negative—current flowing out of the electrode) than in the reverse-polarity direction (electrode positive and electrons flowing into the electrode). The same voltage applied to a straight-polarity arc will push more electrons or current through it than through a reverse-polarity arc, which means that half the time using ac, the voltage is higher than expected, and half the time it is lower than expected, to push the same amount of current through the electrode. The varying voltage causes uneven heating of the base metal. This produces an unbalanced ac waveform that looks like Fig. 14-10.

Balanced AC Machines
The unbalanced ac is allowed for in the design of GTAW-compatible ac power sources to make them fully suitable for TIG welding. Since not all ac welding machines have provisions for balancing the current, not all ac welding machines will work well with TIG welding. When you buy or use an ac machine, the manufacturer's literature will tell you if it has additional circuits and iron in the transformer to handle an unbalanced ac waveform. These specially modified machines are called *balanced* ac machines. If the manufacturer's literature doesn't use the term balanced, it will simply tell you that the

power source is "suitable" for GTAW.

You can use an unbalanced ac machine designed primarily for ac stick electrodes when you have to do TIG welding and have no other choice. But if you do use an unbalanced ac power source for GTAW with ac, never set the current over 50 percent of the machine's rated capacity. If you go above that the machine will overheat. Always use balanced ac machines for TIG when you can.

It's a good idea to buy an ac machine already designed for TIG, because any ac TIG machine will still operate ac stick electrodes. You will save yourself money and trouble by using a machine that can do either stick-electrode or GTA welding.

But fortunately, you have yet another choice of machine.

AC-DC Machines
A very popular kind of TIG power source is the ac-dc machine. It gives you a single welding machine designed to operate with either ac or DCSP and DCRP welding. Most of these machines have a simple switch that lets you change from ac to dc and back again. (But don't *ever* touch that switch when you have an arc on; only use it when your arc is off and you are not welding.)

CAUTION: Never use a switch to change dc polarity, or a switch to ac, while welding. If you change dc polarity by reversing the cables, unplug the machine first, or shut off the engine.

FIGURE 14-11 This is a production-sized power source designed to help you do either MIG, TIG, or stick-electrode welding. The wheeled trailer is an option that is often used for many different sizes of power sources. You can load your welding machine on the trailer, hitch it to your car or truck, and drive to a job.

Many of these machines also have a second switch that you can use to change the polarity of your dc welding setup from DCSP to DCRP and back again at the control panel, instead of having to reverse the electrode and ground leads at the machine's terminals. (Again, never switch polarity when your arc is on.) If one of these machines has been designed specifically for GTAW, you can use it for either ac or dc stick electrodes as well as ac or dc TIG welding.

Most GTAW machines with ac-dc capability (Fig. 14-11) have the following features in common.

- On-off provision using switches or contactors.
- Continuous-current adjustment from a dial on the front panel.
- HF arc starting.
- A switch for changing the dc polarity from straight to reverse and back again as desired.
- Switches for changing current ranges (from very low currents to rated current capacity for the machine, in several steps, with a fine-

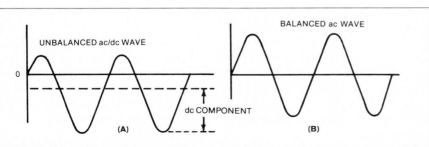

FIGURE 14-10 Comparison of (A) an unbalanced ac waveform and (B) a balanced ac waveform.

tune dial to select or adjust the exact current that's best after the general range has been selected).

■ Auxiliary controls such as gas and cooling water solenoids (a kind of electromechanical valve), amp meters, and maybe other dials and switches that help you control cooling water and gas-flow rates.

■ Timers that turn the shielding gas on and off before and after the arc starts, and for holding the current on for a predetermined time. These are very useful for spot TIG welding when you want to keep the gas flowing after the spot weld is made, to protect the molten metal and your tungsten electrode from oxidation for a short time until they cool down.

■ Some GTAW machines have a dial or a switch so you can select the slope you want to use.

■ Some GTAW power sources even have special controls so that you can change the shape of an ac waveform from a round water-wave shape to a square-wave shape with flat tops and bottoms. These special square-wave ac machines are often used when extremely accurate control is required (as it is in nuclear and aerospace work) and when you want to weld metal foil. They also are ideal for welding out of position.

■ Other auxiliary controls might even include a current-controlling foot pedal so that you can vary the current flowing into the weld metal (and thus the heat input) as you work. Other special pulsed-current dc controls sometimes can be added to existing dc machines to retrofit them as pulsed-current dc power sources.

As you can see, welding machines for GTAW work vary from simple ac or dc models to special high-production fabricating units that you might find in an aerospace plant. Unless you are completely familiar with all the dials, switches, and other gadgets on a machine, always read the operating instructions before you start work.

If you have the machine but not the instructions, phone the welding distributor who sold your company the power source, or call the manufacturer directly. (The name and address are right on the machine's name plate.) You often can get good instructions right over the phone, and a complete user's manual in the mail a few days later.

14-6
TYPES OF MACHINES FOR TIG WELDING

All welding power sources can be classified into one of four basic varieties by the way in which energy input to the power source is changed into electricity for welding. Within these four basic categories there are many subdivisions according to the volt-amp curves the machines produce such as CC or CV. (By now you should be getting a better feeling for what these terms mean.)

Additional subdivisions of each machine type might include how the output amperage is controlled (slope control, no slope control, pulsed current, square wave, and so on). Some machines are "dedicated" designs made for only one kind of welding work. Other machines are multipurpose; ac-dc machines are one example. There are some ac-dc welding machines that can do stick electrode welding, TIG, MIG, and TIG or MIG spot welding . . . all from one power source (Fig. 14-12).

Some of these versatile machines are not very expensive because they are designed for smaller shops (or smaller jobs in big plants) that do a lot of different kinds of work, but none of it in very high volume. Therefore these machines sacrifice high duty cycles and very high current capacities in exchange for increased process versatility. Lower duty cycles and welding current limits mean less copper and iron in the machine construction, keeping costs down.

Other machines may cost more because they are designed for one

FIGURE 14-12 This is a low-cost multipurpose power source designed for work on thinner sections (less than 1 in.) such as you will find in everything from auto body shops to plants making metal furniture. It offers stick-electrode power, MIG, or TIG welding in one package.

job, but they have 100 percent duty cycles, very high maximum amperage ratings, and maybe even special equipment for use with automated welding systems. These are usually very high production, dedicated welding machines.

The following descriptions will help you look for any kind of GTAW power source. What's covered here will be very useful to you when you get to other welding processes described in this textbook.

Transformers

You learned in Chap. 11 that welding transformers take in high-voltage, low-amperage current and output lower-voltage, higher-amperage current. This means that you can plug them into a high-voltage power line and get out low-voltage welding current.

The simplest type of ac transformer power source (Fig. 14-13) has a number of receptacles or taps that you can plug your electrode lead into to get ac welding current out. Each tap only gives you one current, which simplifies the design inside the machine, and thus the machine's cost. It also greatly limits its versatility.

As higher amperages are selected by using different taps, the open-circuit (no-load) voltage (OCV) decreases. It may go as low as 50 V. This decreased voltage at high amperage, plus the fact that

fine amperage adjustment is not possible, makes these tapped machines unsuitable for GTAW. They are used mostly for ac stick electrodes.

While tapped-reactor ac machines give you no fine-current control and they are not good for TIG welding, they are used for ac stick electrodes when you need a low-cost ac machine.

Movable-shunt ac-transformer power sources are a step up in sophistication (Fig. 14-14). They have a movable control inside the machine that slides along a transformer coil to provide a full range of current control. You can select any current within the machine's range by turning a crank on the outside front panel to operate the shunt. These machines are designed mostly for shielded metal-arc (stick electrode) welding. Some models may not provide adequate OCV for TIG welding throughout the machine's full current range.

If the ac machine has not been designed for GTAW (it doesn't have circuits for handling an unbalanced ac waveform), the maximum current you should use for GTAW should not exceed 50 percent of the machine's rated capacity. For example, a power source of this type with a 200-A rating should not be used with GTAW at over 100 A.

Transformer-Rectifiers

Transformer-rectifier machines that provide ac input and produce dc output for welding can be designed to produce either ac or dc by throwing a switch. A second switch often is included so that you can choose the dc polarity you want without disconnecting and switching the electrode and ground leads to the machine.

These ac-dc transformer-rectifier power sources (Fig. 14-15) may have mechanical amperage control (a moving shunt) or electrical current control (a dial). The ac-dc power source with electrical control is the type of transformer-rectifier machine most often de-

FIGURE 14-14 This is a movable-shunt ac-transformer. The crank handle lets you select any current within the machine's range. These machines have limited ac TIG welding capability.

FIGURE 14-13 Some of the least expensive ac power sources have receptacles called *taps* that you can plug your electrode lead cable into to get different levels of ac current out. This machine is even lower in cost. It has only one power level—100 A. It is used for small stick electrodes, mostly for light maintenance work. Simple tapped reactors are seldom suitable for TIG welding because they don't have fine-tuning dials for precise current adjustment after a current level is selected at a tap.

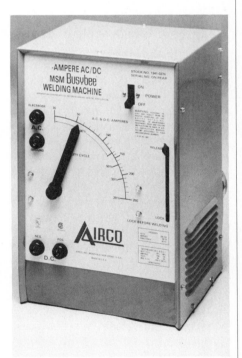

FIGURE 14-15 This is a low cost ac/dc transformer-rectifier with a moving shunt designed as a dial. If this machine also has high-frequency (HF) arc-starting circuits, it would be a good choice for TIG welding at lower current levels. Since this machine is rated for a maximum of 250 amps, a larger machine would probably be used for production work.

signed for use with GTAW (as well as for stick electrodes). If they are designed for GTAW, they almost always will have auxiliary controls for HF arc starting.

Transformer-rectifier machines that output only dc are usually intended for use with dc stick electrodes when they are drooping-characteristic machines, or for GMAW processes such as MIG when they are CV machines. Some tranformer-rectifier machine models are more adaptable than others for use with dc TIG welding. Consult the manufacturer's description of the machine for recommendations on its use by welding process.

Since all these units have transformers, their output is still at the mercy of the current sent into your plant by the local utility company. If the delivery line voltage and current vary going into the machine, the welding current will vary coming out of it.

The line voltage applied to the input side of the power source can vary as much as 5 to 10 percent during the day as the electrical load in your shop or in neighboring buildings on the same main power line varies. But this is not a major problem for most GTAW work because the variations in voltage at the welding arc probably won't be noticed. For those very precise jobs that require extremely accurate voltage control (like aerospace and nuclear work), auxiliary equipment to correct for line variation is available for installation in the power-line supply leading into your welding machine. This equipment is very expensive. It may cost several times the price of the TIG machine that the equipment is designed to protect.

Motor-Generators

The third basic type of power source used for GTAW is the motor-generator (Fig. 14-16), which has an electric motor used to drive an electric generator. AC generated within the rotor is converted to dc by a commutator and brushes. This power source has been used for many years for stick-electrode welding. Although it offers satisfactory characteristics for many types of GTAW, dc transformer-rectifier power sources are generally preferred.

Because the rotational speed of an electric motor is a function of the frequency, and the frequency does not vary, motor-generators will not be affected by variations in the line voltage as much as ac transformer and dc transformer-rectifier machines. The design of some motor-generator units is such that the OCV will decrease when the current is decreased. If such a unit is to be used, avoid trying to weld at a setting with low OCV, because arc starts may be difficult.

Engine-Driven Power Sources

The fourth basic welding power source is the engine-driven types (Fig. 14-17). They are essentially the same as the motor-generator types except that the electric motor is replaced with an engine using gasoline, diesel, or other fuel. Engine-driven power sources are used mostly for field work on pipelines and on construction jobs and farms where you don't have ready access to utility power supplies. Less advanced engine drives contain generators. Newer, more advanced types contain alternators instead.

The alternator, which is very much like a generator, can produce ac or dc, if it has the correct windings. One of the most important features of the alternator machines is that they weigh less and use much less gasoline or diesel fuel. That's an important consideration—just like buying a more fuel-efficient car.

14-7
SELECTING THE RIGHT TUNGSTEN

Now that you are more familiar with the various types of power sources used for TIG welding, let's take a look at the other end of the

FIGURE 14-16 According to the Lincoln Electric Company, motor generators are seldom used now in the U.S. They are used in other countries where utility line voltage variations are a more serious problem. You will find these machines in use in some old maintenance shops because motor-generators are nearly indestructable as long as the motor brushes are periodically replaced.

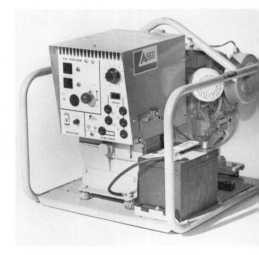

FIGURE 14-17 Engine-driven power sources for use in farms and in remote locations without power lines sometimes are designed to be suitable for TIG welding as well as for stick electrodes. This machine has taps to select different current ranges and a fine-tuning dial to select a specific welding current. It also provides an outlet for auxilliary power such as lights for night work, or for power tools. This machine has a gasoline engine. Some engine-driven power sources have diesel engines. Some have water cooling.

equipment, the tungsten electrode inside the TIG holder.

Tungsten and Its Alloys

There are several types of tungsten and tungsten-alloy electrodes. Each type of TIG electrode has advantages and drawbacks. For example, pure tungsten electrodes are the least expensive, but they shouldn't be used for critical work. The tungsten-alloy electrodes are much better for critical work, but they cost more. Here is a complete briefing on your choices.

Pure tungstens Pure-tungsten electrodes (grade EWP in the AWS specification A5.12) are used primarily for ac welding. For good arc stability with ac, pure-tungsten electrodes should produce a film of molten metal on the surface of the arcing end. Normally this film will appear as a well-rounded, bright, shiny surface. To obtain this condition, the electrode should be operated at the upper end of its recommended amperage range. If this shiny condition is absent and insufficient welding heat is being used, the arc may be unstable and will wander.

You will see this condition because the end of the pure-tungsten electrode will be rounded only on one side. If an excess current setting is used, the end of the electrode will tend to ball up or droop. Although this balling up of the tungsten electrode tip helps improve the stability of the arc by spreading the current density over a wider area than a pointed tungsten tip, especially for general purpose, less critical TIG welding jobs, it's frequently a disadvantage.

The reason for the problem is that the molten ball of tungsten will appear to pulse and move around. If more heat is developed in the pure tungsten, tungsten "spitting" will start and tungsten droplets will fall off the electrode and be included in your weld metal. This condition will get progressively worse. Tungsten spit-

ting also will cause higher than normal arc voltage and will seriously affect automatic GTAW.

Pure tungsten doesn't have very good electrical conductivity. In fact, its relatively high resistance (for a metal) is one reason why thin tungsten wires are used in incandescent light-bulb filaments. The conductivity of the pure-tungsten wire is very low (the resistance is high), and that causes the filament to heat up and emit light. That's not necessarily what you want when TIG welding.

The conductivity of TIG tungstens is improved by using one of several tungsten alloys. Some of the most common ones used for GTAW electrodes have a 1 percent addition of an oxide of thorium. These alloyed electrodes are called *thoriated tungstens*.

Tungsten-thorium alloys

Thoriated tungstens not only have higher current-carrying capacity than pure tungstens, but they also maintain their shape longer since they run cooler than pure tungstens and aren't as likely to melt, greatly reducing the chances of contaminating the weld metal with a drop of molten tungsten. Another advantage of thoriated tungstens is that accidentally touching the weld puddle is not as likely to cause contamination of the electrode.

Electrodes containing 2 percent thorium oxide (specified as AWS EWTh-2) also are available. They retain a pointed end even longer than the ½ percent (AWS EWTh-3) or 1 percent (AWS EWTh-1) thoriated electrodes. These 2 percent thoriated tungstens are used in the aircraft and missile industries for critical GTAW fabrica-

tion work on light-gauge metals. Although thoriated-tungsten electrodes normally are recommended for DCSP TIG welding, they also can be used for ac welding.

Tungsten-zirconium alloys

Tungsten electrodes containing a small amount of zirconium also are available. They are described in the AWS A5.12 specification as grade EWZr tungsten electrodes. Their performance generally is between that of pure-tungsten and thoriated-tungsten electrodes. However, for some applications, particularly joining aluminum while using ac and argon shielding gas, zirconium tungstens have been found very useful.

How to untangle the types of tungstens How do you know which is which if you mix up several different kinds of tungstens? The best procedure is never mix them up. Keep them in envelopes marked with the AWS tungsten grade, such as EWP or EWTh-2. The AWS tungsten electrode specification requires special color coding on the surface of each electrode. It may be a spot, a band, or some other colored mark. Table 14-1 lists the color codes for TIG tungstens from the AWS A5.12 specification, in case you mix them up. When you prepare your tungsten, don't grind off the color-coded end. Save that end for last so that your TIG tungsten grades won't get mixed up in your toolbox.

Surface finishes In addition to your choice of types of tungsten electrodes, you also have a choice of different surface finishes. The surface finish as well as the electrode grade will be marked on the

---

TABLE 14-1 GTAW tungsten electrode color codes

AWS Grade	Color Code	Type of Tungsten Electrode
EWP	Green	Pure tungsten
EWTh-1	Yellow	1% thorium
EWTh-2	Red	2% thorium
EWTh-3	Blue	½% thorium
EWZr	Brown	Up to 0.40% zirconium

package. The two most commonly used GTAW electrode finishes are *clean finish* and *centerless ground finish*.

Clean-finished electrodes are packaged as they are produced. They have a clean, smooth surface free from defects and imperfections. This grade of tungsten is satisfactory for most TIG welding jobs. Centerless-ground-finish electrodes are produced by passing clean-finish electrodes through a centerless grinder to produce an extremely smooth, mirrorlike surface. This grade of electrode finish is used to make sure you get the highest possible welding quality on critical jobs like aerospace welding, and occasionally on pipeline and nuclear-reactor piping work.

Electrode Sizes and Current Ranges

Table 14-2 gives you data for selecting the right tungsten electrode diameter and alloy for your work, whether you are using DCSP, DCRP, unbalanced-wave ac, or balanced-wave ac welding equipment and work setups. This table is taken from AWS A5.12, "Specification for Tungsten Arc-Welding Electrodes." This is the standard specification for GTAW electrodes.

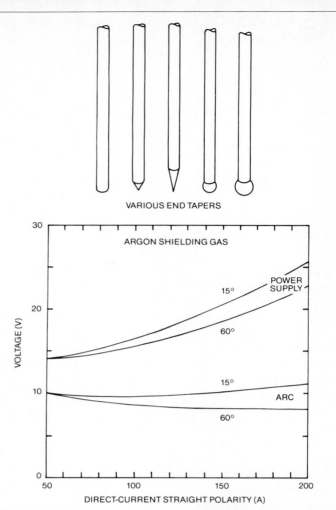

FIGURE 14-18 Voltage and current for two different tungsten electrode taper angles are graphed, as measured at the power supply and directly across the arc. The drawing shows a variety of different tapers that can be used for TIG tungstens.

TABLE 14-2 GTAW tungsten electrode current ratings*

Electrode Diameter, in.	DCSP EWP EWTh-1 EWTh-2 EWTh-3 Pure and Thoriated Tungsten	DCRP EWP EWTh-1 EWTh-2 Pure and Thoriated Tungsten	AC High-Frequency Unbalanced Wave			AC High-Frequency Balanced Wave		
			EWP Pure Tungsten	EWTh-1 EWTh-2 EWZr Zirconium and Thoriated Tungsten	EWTh-3 Thoriated Tungsten	EWP Pure Tungsten	EWTh-1 EWTh-2 EWZr Zirconium and Thoriated Tungsten	EWTh-3 Thoriated Tungsten
0.010	up to 15	—	up to 15	up to 15	—	up to 15	up to 15	—
0.020	5–20	—	5–15	5–20	—	10–20	5–20	10–20
0.040	15–80	—	10–60	15–80	10–80	20–30	20–60	20–60
1/16	70–150	10–20	50–100	70–150	50–150	30–80	60–120	30–120
3/32	150–250	15–30	100–160	140–235	100–235	60–130	100–180	60–180
1/8	250–400	25–40	150–210	225–325	150–325	100–180	160–250	100–250
5/32	400–500	40–55	200–275	300–400	200–400	160–240	200–320	160–320
3/16	500–750	55–80	250–350	400–500	250–500	190–300	290–390	190–390
1/4	750–1000	80–125	325–450	500–630	325–630	250–400	340–525	250–525

* Approximate current ranges for tungsten and thoriated tungsten electrodes (in amperes). Based only on the use of argon shielding gas; helium runs hotter. Use one-size-larger electrode for helium.

The E in the AWS specification means *electrode*. The W in the grade designation is the chemical formula for tungsten (it stands for wolfram, the principal tungsten ore). Th and Zr are the chemical symbols for thorium and zirconium metals.

Tungsten electrodes should not be used above the upper end of their current range. This range must be considered when selecting an electrode for a specific job. Because argon shielding gas normally is used with ac, the values shown in the table are for tungstens used in argon. The diameters of electrodes for use in helium gas shielding will be slightly larger, but this table gives you a pretty good starting point for selecting the electrode diameter that you need. (Also see Fig. 14-18.)

Now that you know all about TIG tungstens, you may wonder how they are sold. Almost all welding distributors carry them. They usually are packaged as long rods in diameters from a very narrow 0.010-in. up to ¼-in.-diameter tungstens. The standard length of ³⁄₁₆-in.-diameter tungstens fresh out of the package is 18 in. The lengths may be as short as 3 in. for 0.010-in.-diameter tungstens, which also are available in coils for very large users. Handle them carefully. They break easily.

14-8
CARE OF ELECTRODES

The tungsten electrode can be used with no end preparation, with the end beveled to a specific included angle (point), or with the electrode end "balled up" as shown in Fig. 14-18. These shapes will affect the weld-bead shape and size. For example, with DCSP at 90 A, welding on ¼-in. stainless steel plates, changing the end bevel or included angle from 30 to 120° (going from a pencil point to a blunt tip) will increase the width of the weld bead and decrease weld penetration. However, as the welding current is increased, the opposite is true. For

example, when welding stainless steel at 300 A DCSP, changing the electrode tip included angle from 30 to 120° will decrease the weld bead width by a factor of 2 and will increase the weld penetration by 45 percent. The degree of taper or beveling of the electrode tip also influences the tungsten's erosion or wear-out rate. Tapers over a length of two to three times the diameter will minimize electrode erosion.

If you are not given a specific electrode taper as part of your instructions and are using HF arc

starting, try experimenting with a couple of different tip shapes to see which gives the best results. You can experiment with scrap metal of the same composition and thickness as the job you will work on and quickly establish the best electrode tip shape.

Preparing Your Tip
When you need a new tungsten (Fig. 14-19), take a fresh rod out of the package. First put a sharp point on it to improve HF arc starting. Use an abrasive wheel to

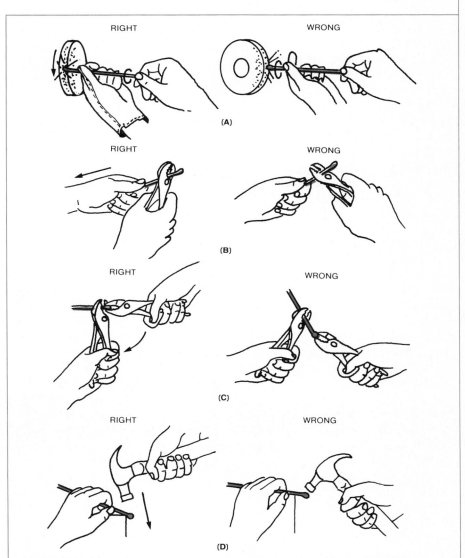

FIGURE 14-19 Right and wrong ways of tipping and redressing a GTAW tungsten electrode. (A) How to grind the ends of a tungsten electrode. (B) Cutoff ends of electrodes up to ¹⁄₁₆ in. [1.6 mm] in diameter. (C) How to break off ends of electrodes larger than ¹⁄₁₆-in. [1.6-mm] diameter but not larger than ⅛-in. [3.2-mm] diameter. (D) How to break off ends of electrodes ³⁄₃₂ in. (2.4 mm) and larger in diameter. Put electrode on edge of a bench.

grind one end of the tungsten to a sharp point. Never use a wheel that has ground something else besides TIG-welding tungstens. A dirty wheel will contaminate your electrodes.

Hold the electrode about 1 or 2 in. from the end and press the tip *lightly* against the side of the wheel, with one gloved hand, holding the tip pointing at the wheel axis. Be careful not to grind your fingers. Use a fine-grained, clean grinding wheel. Use your other hand to rotate the electrode while you grind it. Don't overheat the electrode when grinding, you'll burn it.

After pointing the electrode, cut a notch in it with a file or the wire cutters in the throat of a pair of pliers. The wire cutters will cut through a small tungsten. If the tungsten is larger, notch the electrode rod exactly where you want it to break off. Then hold one section with the jaws of a pair of pliers held very close to the notch and snap off your tungsten with a second pair of pliers. Or hold the notch close to the edge of a bench or the anvil on a vise and give the free end a whack with a hammer if the rod is $\frac{3}{32}$ in. in diameter or larger. Be sure to hold the tungsten right next to the notch. The material is brittle.

Another way to prepare a pure-tungsten electrode tip is to make it ball up using either ac or DCRP current. After cutting the tungsten to length, insert it in your TIG holder. Use the current settings and electrode diameters in Table 14-2 and initiate an arc with the HF arc-starting switch on. Start the arc on a water-cooled copper block. As soon as the arc is established, the electrode end will show a bright orange color. As the current is increased, the color will change to a brilliant white and the electrode tip will melt, forming a ball on the arc end. At that point, shut off the welding current and let the electrode cool in flowing shielding gas. That will give you

an electrode with a nice round ball on the end of it.

You can re-dress used tungstens the same way that you make them. Always check the end of any broken electrode for cracks. The break should be clean, smooth, and bright. Cracked electrodes will affect your welding.

No matter what type of electrode tip shape you use, the most important factor is to be consistent once a welding procedure begins. Since changes in electrode geometry have a significant influence on weld bead shape and size, electrode tip preparation is a welding variable that should be included in any GTAW procedure.

14-9
TIG HOLDERS
Of all the factors involved in selecting a TIG holder, its duty-cycle rating is the most important.

Duty-Cycle Ratings
TIG holders are used very much like a gas welding torch, but they are electrical devices and, like power sources, they have duty-cycle ratings. A specific amperage always is associated with the duty cycle. The amperage usually is the rated maximum current that the device will handle. The duty cycle also can be reported for different amperages, because lower amperages produce less heating.

A typical duty cycle for a TIG holder might be 100 percent at 100 A, 80 percent at 200 A, and 60 percent at 300 A maximum. In that case, the maximum current rating for the device is 300 A. You'll note that the TIG holder can run longer at lower amperage than at the maximum-rated amperage for the device.

The amperage rating for a TIG holder is based on using the maximum practical amperage without serious heating, when using the largest-diameter thoriated-tungsten electrode for which the holder is designed, using the long-

est nozzle and the largest diameter for the particular holder, and using DCSP with helium shielding gas (which runs hotter than argon gas). All of these are measured while making a weld on stainless steel in the flat position, which produces tough conditions for the electrode holder.

On a real welding job, if you use a TIG holder design that is rated for the largest-diameter thoriated-tungsten electrode it can use, operated at its maximum rated amperage with DCSP, then you should reduce the rating of the holder when using a pure tungsten with ac. The reason is the greater heating at the electrode caused by running with ac. With DCRP, the holder's rating would have to be reduced to the maximum allowable amperage for that diameter electrode with reverse polarity. Some holders are designed in such a way that the ratings for dc straight polarity and ac are the same.

The TIG holder has several parts (Fig. 14-20). The handle is insulated and holds electrical cable leading to the power source. The holder will also contain a shielding-gas tube or hose and nozzle, an electrode cap, and sometimes it will have two lines to deliver and return cooling water on larger-amperage models. Smaller TIG holders don't need water cooling.

Gas nozzles All TIG holders have gas nozzles of one kind or another (see Fig. 14-20) that protect the tungsten electrode and also direct the flow of shielding gas at the weld metal. Most TIG nozzles are now made of fairly tough ceramic materials, although they used to be made of glass. The ceramic material is usually an aluminum-oxide ceramic which is both heat resistant and an electrical insulator. These qualities are necessary because of the tremendous heat at this end of the holder and because the nozzle contains both an electrified tungsten and also hot, ionized shielding gases.

FIGURE 14-20 Compare this TIG holder with the holder shown in Fig. 14-21. This TIG holder doesn't have the long pointed cap to hold an extra length of tungsten. You would select this holder shape if getting at work in a tight space was more important to you than having an extra-long tungsten in the holder.

The replaceable ceramic nozzle not only protects the electrode inside, but it also protects you from getting hurt if you accidentally bump into the wrong end of the holder.

Electrode Caps
On the other end of the TIG holder is a cap that holds the tungsten electrode. Some holders have very long, pointed caps (Fig. 14-21) that can hold an entire tungsten electrode so you don't have to shorten your tungstens. It is more economical and faster to use a holder with a long electrode cap because then you don't have to break up as many tungsten electrodes and you can simply refurbish the arc end of the one you have from time to time.

On the other hand, if you will be working in confined areas on narrow joints, you will want a TIG holder with a very short cap.

Cooling Systems
One of the ways that TIG holders are classified is by their cooling system: air or water.

Air-cooled holders Air-cooled holders depend upon the surrounding air for cooling. Therefore, the size of the nozzle will affect the amperage rating of the holder. Many holders can use more than one length of nozzle. When a short nozzle is used, the holder rating must be reduced below the maximum value indicated. In addition, when the holder is being used in a confined space, as in a deep V joint, the rating must be reduced because of the greater amount of reflected heat in the weld joint, regardless of the nozzle size used.

Water-cooled holders Water-cooled TIG holders are designed with water passages in the head, or arc end, of the holder to permit the use of smaller equipment for a given amperage rating. Cooling water is brought into the holder in a hose and flows through passages in the head to remove heat. The heat from the nozzle may be transferred to the holder head, where it is removed by the water, or the cooling water may contact the upper portion of the nozzle to remove heat.

FIGURE 14-21 This TIG holder has just been used to weld a thin edge seam on a metal box. There is no clearance problem in this sheet-metal work, so the welder selected a TIG holder with a cap to contain an extra-long tungsten.

FIGURE 14-22 The TIG holder in the middle is a straight-headed pencil-shaped TIG holder. All three TIG holders are air-cooled models. A water-cooled holder (not shown) would look similar to an air-cooled TIG holder except that the handle and head would be a little thicker to make room for the cooling-water lines needed for high-amperage (up to 500 A) work.

The cooling water leaving the holder often flows in a hose which also serves as the insulation for the welding power cable. A cable of this type is known as a water-cooled power cable. It permits the use of a smaller cable for a given amperage rating. No attempt should be made to use a water-cooled holder as an air-cooled holder. Without the cooling water surrounding the welding power cable, the amperage-carrying capacity of the cable is reduced.

Guidance Methods
TIG holders can be guided either by hand or by machine. Their construction reflects the difference.

Manual Manually guided holders are available in many shapes and sizes for both air-cooled and water-cooled GTAW. Small, air-cooled pencil-size holders (Fig. 14-22) will accept electrode diameters from 0.020-in. to ¹⁄₁₆-in. diameter at currents up to 200 A, using either dc or ac. These very small holders are used for very

FIGURE 14-23 This air-cooled TIG holder has an angled head, and a short electrode cap. The neck swivels when the knob is turned so that it can get into difficult joints.

light duty GTAW such as joining delicate, thin-walled tubing, for making jewelry, and even for welding metal foil.

Larger air-cooled production holders usually have heads that are at an angle to the handle. Some newer TIG holders even have flexible necks between the handle and head so that you can adjust the head angle while you work (Fig. 14-23). These air-cooled units usually have current ratings between 200 and 300 A.

Larger manual water-cooled holders that use electrode diameters up to ¼ in. with dc ratings up to 500 A and 100 percent duty cycle are available.

Machine-guided holders
Machine holders are used when a job requires precise movement of the arc along a joint, when high travel speed is needed to obtain the desired joint type, when high production speeds are used, or when the number of similar parts makes it possible to use a welding robot.

The head or arc end of an automatic or machine TIG holder is essentially the same as a manual holder, but instead of having a handle, the machine holder has a short, straight barrel that fits into an automatic "seam follower" or similar device. GTAW machine holders are usually rated at 500 A with 100 percent duty cycles. Machine holders with even higher capacities are available.

14-10
WELDING CABLE
Welding and ground cable for TIG welding is selected just like cable for any other arc-welding process. See Chaps. 11 and 12 for details on selecting welding cable. Make sure that you use cable properly rated for your job. Also make sure that the cable you use is no longer than necessary. Remember that you lose power due to resistance heating of the cable, and your voltage drops when very long cables are used. What's worse, very long cables will get in your way. Reread pages 285 to 288 for cable safety procedures.

14-11
HOSE FOR GTAW
The shielding-gas pressure and flow rates you will use with GTAW will be much lower than those of fuel gas and oxygen. You don't always need a heavy-walled hose to counteract gas pressure. However, helium gas, especially, is very peculiar stuff. You may already be familiar with how rapidly a helium balloon shrinks and shrivels up after you buy it.

What's happening is that the helium atoms are so small that they diffuse right through the rubber wall of the balloon. They'll do the same thing in a conventional rubber hose. Be sure that your shielding gas and cooling-water hose is lined on the inside and on the outside with neoprene, not rubber. Also make sure that it is nonconductive.

Sometimes the wrong gas or water hose can be installed on a machine. When that happens, you may get a shock. Hoses with rubber inner tubes and rubber outer wraps are not suitable for TIG equipment. Grade M single-line hose with a neoprene inner tube and a neoprene outer wrap, or black neoprene braided hose should be used.

You can test the conductivity of your hose if you are concerned about it by using a Simpson model 260 multimeter to check the resistance of a piece of hose. The resistance measurement should be taken between the ends of a 2-ft [0.6 m] sample length with meter probes placed on the inner diameter surface of the hose. There should be no deflection of the meter when using the R × 10,000 ohm scale. When two different materials are used for the outside and the inside of the hose, perform the same test on the outside diameter.

A common problem with all forms of gas-shielded arc welding is contamination of the shielding gas by either air or condensed moisture inside the gas hose. Obviously, you should never use these hoses for anything but inert shielding gases.

NOTE: Always purge your lines and holder with very dry shielding gas before you use them.

If any air is in the hose, the oxygen in it will immediately oxidize your hot tungsten electrode. Even worse, the air will oxidize the weld metal and leave brittle oxide inclusions in critical welds in very expensive base metals.

Water vapor in your shielding-gas hose may produce porosity in your welds, and you will have a terrible time figuring out why. Water vapor in your shielding gas also will produce hydrogen embrittlement in high-strength-steel weld joints, and you may never know it until your welded workpiece is in use in somebody else's equipment. It is imperative that you purge your lines and TIG holder with very dry shielding gas before you use them.

14-12
SELECTING A SHIELDING GAS
The inert monatomic (one atom, no molecule) gases argon and he-

lium are used as shielding gases for GTAW. Mixtures of argon and helium are useful when you want some balance between the characteristics of individual gases. Additions of hydrogen and nitrogen have been used in special applications. Don't try using special mixtures without the advice of a welding engineer or knowledgeable welding distributor that specializes in shielding gases. Table 14-3 is a guide to selecting and using shielding gases for GTAW.

Not only can you buy welding-grade shielding gases in high-pressure cylinders from full-service welding distributors, but some of the larger distributors also carry mixed argon-helium shielding gases. Several welding equipment manufacturers who also are gas suppliers will list their shield-ing-gas grades and mixtures in their catalogs.

Inert shielding gases are extremely dry and very pure. They are vaporized from cryogenic liquids. Welding-grade argon and helium are supplied with a purity of 99.995 percent or greater. Dew points for these gases are around −70°F [−56.6°C] or lower. The dew point is the temperature at which a polished mirror would begin to fog up if the gas were blown on it. The lower the temperature of the dew point, the less moisture there is in the gas.

Argon is the most commonly used GTAW shielding gas. It provides a lower arc voltage than helium at any given current and arc length. Argon also provides a quieter, smoother-running arc, and arc starting in argon is easier

than in helium. Since argon is a heavy gas (it's 38 percent heavier than air), lower flow rates are required to provide good shielding . . . in the downhand position.

The reverse is true when welding overhead. Since helium is much lighter, it floats up against an overhead surface, while the argon will tend to sink as it cools. This difference is less, of course, when the gas is very hot, when it flows through the GTAW arc. The hot gas expands and gets lighter in weight. But keep the density of your shielding gas in mind when you think about your welding position.

Argon is used for welding lighter sections because it produces less heat and therefore less base-metal distortion with less chance of a burn through when

TABLE 14-3 How to select a shielding gas for GTAW

Metal	Welding Operation	Shielding Gas	Advantages and Applications
Aluminum and Magnesium	Manual	Argon	Better arc starting, cleaning action, weld quality, lower gas consumption.
		Argon-helium	High welding speeds are possible due to higher heat produced by helium.
	Machine	Argon-helium	Better weld quality, lower gas flow than required with straight helium.
		Helium (DCSP)	Deeper penetration and higher weld speeds than can be obtained with argon-helium.
Carbon Steel	Spot Welding	Argon	Generally preferred for longer electrode life. Better weld-nugget contour. Ease of starting, lower gas flows than helium.
	Manual	Argon	Better molten weld puddle control, especially for out-of-position welding.
	Machine	Helium	Higher speeds obtained than with argon.
Stainless Steel	Manual	Argon	Permits controlled penetration on thin-gauge material (up to 14 gauge).
	Machine	Argon	Excellent control of penetration on light-gauge materials.
		Argon-helium	Higher heat input, higher welding speeds possible on heavier gauges.
		Argon-hydrogen (up to 35% H₂)	Prevents undercutting, produces desirable weld contour at low current levels, requires lower gas flows.
		Argon-hydrogen-helium	An excellent selection for high-speed tube-mill operations.
		Helium	Provides highest heat input and deepest penetration.
Copper, Nickel, Cu-Ni Alloys	Manual	Argon	Easy to get molten puddle control, penetration, and good bead contour on thin-gauge sheets and strip.
		Argon-helium	Higher heat input to offset high conductivity of heavier gauges.
		Helium	Highest heat input for higher welding speeds on heavy metal sections.
Titanium	Manual	Argon	Low gas flow rate minimizes turbulence and air contamination in welds, improves heat-affected zone.
		Helium	Better penetration for manual welding of thick sections (inert-gas backing required to shield back of weld against contamination).
Silicon Bronze	Manual	Argon	Reduces cracking of this "hot short" metal.
Aluminum Bronze	Manual	Argon	Less weld penetration of base metal when welding thinner sections.

welding thinner material. Helium is used as a shielding gas in welding heavy sections where the gas's higher arc-voltage characteristics are an advantage.

The purpose of the shielding gas is not only to protect your molten weld metal from the air, but also to protect your hot tungsten electrode from oxidation. The shielding gas is delivered around the tungsten electrode though a ceramic nozzle. Stainless-steel nozzles sometimes are used on very large, high-amperage torches used with machine welding. Nozzle sizes must be adequate to provide coverage of the weld area and still allow you room to get into the weld joint.

Automatic welding setups may include supplemental shielding gas over the weld area, with leading or trailing gas shields that increase the coverage. Purge times (the time the purge is on and running) also must be set to allow the weld metal to cool for a longer period before it is exposed to the atmosphere.

Problems with Shielding

The most common causes of atmospheric contamination of weld metal are insufficient gas flow, excessive electrode extension (stickout beyond the nozzle), or insufficient postweld purge time. The flow rates will vary widely, depending upon the shielding gas used, nozzle size, and specific welding conditions.

Manual welding requires flow rates of at least 15 scf/h [0.4 m³/h] with argon. Helium requires slightly higher rates. The effectiveness of the gas shielding, even with adequate flow, can be greatly reduced with excessive electrode stick-out.

A rule of thumb for electrode extension is: *The electrode should extend beyond the nozzle a maximum of one times the diameter of the nozzle.*

When using a ⅜-in. OD nozzle, for example, the tungsten electrode should extend not more than ⅜ in. beyond the nozzle.

Also, the effectiveness of the gas shielding can be increased with the use of a gas "lens" or "gas straightener" attachment. Gas lenses often are provided as options with modern TIG holders. They fit on the front of the torch at the ceramic nozzle.

An insufficient shielding-gas flow rate during the postweld purge also can cause atmospheric contamination. The postflow rate should be sufficiently long to allow your tungsten to cool below its oxidizing temperature (so it won't glow red).

Water can be another source of electrode contamination. Water contamination results from leaks in a water-cooled torch caused by loose connnections, pinched O rings, defective seals, and so on. Water contamination also results from condensation of moisture in the welding torch head, even when there are no leaks in the torch. Condensation occurs when cold-water coolant is circulated through your torch during warm, humid days. This problem is easily corrected simply by using warmer cooling water.

When your tungsten is being contaminated, either by atmospheric gases, low flow rates, or by water, you will see the color of the cold tungsten electrode change. The color of a contaminated tungsten may range from a bright blue or purple to black, instead of the shiny silvery metallic color it should have. If the tungsten is blue-black colored, the dark-colored oxidized section of the tungsten electrode must be removed immediately and thrown away. You also will have to find out what's causing the problem and correct it before you continue welding.

14-13
FILLER METALS FOR GTAW

The filler metals used with GTAW are essentially the same as those used for oxyfuel welding when cut lengths or rods are used. Examples from the chapter on

oxyfuel welding are the RG 60 and RG 65 steel rods you used when learning to weld with gas. With killed steel (steel that has been heavily deoxidized when it was made) you have no trouble. But rimmed steel (high in oxygen) will give you porosity. Rimmed steels often are used to make cold-rolled sheet and strip because they give a very good surface finish. If you know that you are welding a rimmed steel, you must use filler metals that are higher in deoxidizers such as silicon to avoid porosity.

AWS Specifications

You also may find that some GTAW filler rods will have specs with a different grade designation than you've seen before, such as AWS A5.18, Grade 70S-1. These may be rods that have been produced to AWS filler-metal specifications for continuous-welding-wire processes, especially GMAW welding wires. The wires may have been straightened and cut to length for GTAW rods. More likely, such a specification refers to coiled wire for use with an automatic wire feeder (just like the ones used for GMAW that you will learn about in Chap. 17). When coiled, continuous filler-metal wires and automatic wire feeders are used for machine GTAW.

The 70 in the GMAW filler-metal continuous-wire designation stands for the tensile strength of the weld metal. The S means that the wire (or cut-length rod) is solid, as opposed to a flux-cored wire. The final 1 refers to a specific chemical composition for the filler metal that produces 70,000 psi [480 MPa] minimum tensile strength weld metal.

Table 14-4 lists the most commonly used AWS filler-metal specifications for GTAW. See Chap. 9 on gas welding, or look ahead to Chap. 17 on GMAW if you want more details on how these filler-metal grades are designated.

Cut lengths for GTAW are usually supplied either as straight lengths, normally 18 or 36 in. [45 to 90 cm] long, or occasionally on spools or reels for use with wire feeders. Welding-wire spools range from very small 2-lb [0.9-kg] spools up to 50- and 60-lb [22.5- to 27-kg] reels. The most commonly used wire diameters for both cut lengths and coiled wire range from 0.020 to ¼ in. The sizes most often used in GTAW are ¹⁄₁₆-, ³⁄₃₂-, and ⅛-in.-diameter cut lengths.

Filler-metal shapes other than round wire are usually in preformed ring inserts or rectangular wires. The preformed filler-metal inserts are of the same composition as other forms and are fused into the weld joint during the first welding pass. The advantages of preforms are uniformity of results and precise control over the chemical composition of the weld deposit. Pipe welding is a very good example of the use of preformed shapes used to fit the root of pipe joints.

Other examples of the use of preformed filler metal in GTAW are T rings and Y rings which are roll-formed and then machined to close tolerances for components that require precision welding, such as certain aerospace or nuclear-reactor welding jobs. Machined preforms are normally used with automatic TIG welding systems where precise control of the welding variables must be maintained to take the place of the skilled welder.

An important version of cut lengths for meeting nuclear and other highly critical ASME boiler and pressure vessel code welding work are called *flag-tagged* cut lengths. These are regular cut-length rods, except that they have met the special requirements of some part of the ASME Boiler and Pressure Vessel Code (see Chap. 23). Every rod will have a little paper flag on one end with information on the composition of the rod, such as type 304 stainless steel. Obviously, flag-tagged cut lengths will cost more than regular cut-length TIG rods.

Taking Care of TIG Rods

You must handle TIG filler rods with care. The surface must be kept absolutely clean. When the rods are produced, all wire-drawing lubricants have been cleaned off and any oxides have been removed to ensure good-quality welds. Once you get the rods, any dirt, grease, oil, marking pen, or anything else you get on them will result in a bad spot in your TIG weld metal. Since TIG welding will often be followed by x-ray testing to make sure that there are no unacceptable porosity or other defects, it's to your advantage to keep your rods clean.

Opened packages, dirty gloves, and greasy or dirty welding equipment are all poor shop practices that you must avoid as a competent TIG welder. Many TIG welders even use a "white-glove" test to inspect their own filler rods. If they can wipe the rod with a clean, white cloth and not leave any marks on the cloth, then they'll use the rod. If the rod will not pass the white-glove test, the operator will refuse to weld with it.

Good shop practice is to keep your filler rods in a clean, dry cabinet. Every rod package should be closed when not in use. Every package also should be clearly identified as to its diameter and chemical composition or grade. Simply writing "⅛-in.-diameter stainless steel" is not good enough. What grade of stainless steel is it . . . Type 304, Type 304L, Type 308, Type 347, or what?

If you ever work in a nuclear plant or pipe-fabricating shop, or in many shipyards and aerospace plants, you actually will have to sign out for your TIG rods and return what you don't use at the end of every shift. Once mixed up, different grades of TIG rods are very difficult to identify without chemical analysis. If you don't want to be responsible for your plant throwing away 10 lb of stainless-steel rod at $5 to $10 per pound, handle your rods with care and don't mix them up.

14-14
MANUAL GTAW TECHNIQUES

Manual TIG welding is, by far, the most commonly used GTAW process. We're going to take the techniques a step at a time and show you how to do them. After gas welding, brazing, and soldering, you'll find that your first experience with arc welding with TIG will be a pleasure. If you had started with stick-electrode welding, instead of oxyfuel, you would have a lot of trouble learning to do this work.

Basics

We'll start you through the basics of manual TIG welding a step at a time. The first step is to prepare a good joint.

TABLE 14-4 American Welding Society filler-metal specifications used with GTAW

AWS A5.2	Iron and Steel Gas-Welding Rods.
AWS A5.7	Copper and Copper-Alloy Welding Rods.
AWS A5.9	Corrosion-Resisting Chromium and Chromium-Nickel Steel Welding Rods and Bare Electrodes.
AWS A5.10	Aluminum and Aluminum-Alloy Welding Rods and Bare Electrodes.
AWS A5.13	Surfacing Welding Rods and Electrodes.
AWS A5.14	Nickel and Nickel-Alloy Bare Welding Rods and Electrodes.
AWS A5.16	Titanium and Titanium-Alloy Bare Welding Rods and Electrodes.
AWS A5.18	Mild Steel Electrodes for Gas-Metal-Arc Welding.
AWS A5.19	Mangesium and Magnesium-Alloy Welding Rods and Bare Electrodes.
AWS A5.21	Composite Surfacing Welding Rods and Electrodes.

Preparing the joint As you now know, you can't get a good weld with a bad joint. It's your job (or the responsibility of somebody in the machine shop, working closely with you) to prepare that joint. If the wrong method for preparing the edge bevels is used, you will pay the price in serious welding defects.

Starting the weld Once you have set up your equipment and are ready to weld, strike your arc

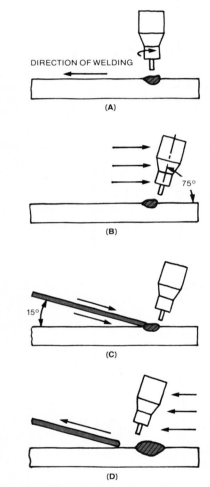

DIRECTION OF WELDING

(A)

75°

(B)

15°

(C)

(D)

FIGURE 14-24 Making welds in a butt joint in the flat position. (A) Develop the weld puddle in one spot. (B) Move the TIG electrode holder back slightly. (C) Add more filler metal. (D) Move the TIG holder back to the leading end of the weld puddle to allow weld metal and filler rod to cool under the shielding gas. Repeat the sequence until the weld is finished.

using one of the techniques discussed in Sec. 14-4. When the arc is started, hold the TIG torch with its tungsten electrode positioned at an angle of about 75° to the weld-pool surface.

To start welding, the arc is usually moved in a small circle until a suitable-size pool of molten metal is obtained. When adequate fusion is achieved at any one point, you can get up to speed and weld by slowly moving the electrode along the parts to be welded together, which will progressively melt the adjoining edges of the base metal.

Solidifying weld metal follows behind your GTAW-torch arc, completing the welding cycle. You almost never will have to use flux, or clean slag from the surface of a TIG weld. The argon or helium shielding gas will leave bright, shiny weld metal behind as long as you work slowly enough to let your hot weld metal solidify before the shielding gas leaves it.

Using filler metal The metal thickness and joint design, together with the weld-metal properties that you want, determine whether filler metal has be added to the joint. Filler metal, if added, is applied by you as you work. You feed the rod into the pool of molten metal near the arc, in much the same way that you fed filler rod into a weld during oxy-fuel welding. Figure 14-24 shows one of the most frequently used techniques for feeding filler rod into a TIG weld.

As shown in Fig. 14-24, use the TIG torch to start a weld puddle on the base metal. Then hold the filler rod at an angle 15° to the surface of the work and slowly feed the rod into the weld pool. Hold the TIG torch at a 75° angle while you dip the filler rod into the molten metal. The hot filler rod must not be completely removed from the protection of the inert-gas shield during welding. You continue in this manner, progressing along the weld seam until the joint is finished. After the

weld is finished, move the torch so that the shielding gas covers the weld metal and the end of the filler rod until *both are solid and cool enough not to oxidize* when exposed to air.

NOTE: Never expose hot weld metal or your rod to air until they are cool enough not to oxidize. Keep them shielded.

Another method frequently used in weld surfacing and in making large welds is to feed filler metal continuously into the weld pool by oscillating the filler rod and arc from side to side. The filler rod moves in one direction while the arc moves in the opposite direction, but the filler rod must be near the arc at all times as you feed it into the weld pool within the shielding-gas blanket.

Speed Arc penetration into the base metal is inversely proportional to your welding speed. The faster you travel across the weld, the less weld penetration you get.

Higher currents and fast welding speeds are preferred for welding high-conductivity metals such as copper, aluminum, and magnesium, to keep ahead of the heat conducted away from your molten weld puddle. Slower welding speeds are used with less conductive materials like carbon, alloy, and stainless steels. There's no magic to determining the correct welding speed. You will usually be able to see if you are going too fast or too slow simply by watching the size of your molten weld puddle.

Alloys such as silicon bronze and aluminum bronze, which are prone to thermal cracking, cannot be welded at high speeds, unfortunately, since the steep change in the heat curve away from your weld puddle will cause them to crack. Fairly slow weld speeds are needed to avoid this problem. Slow welding speeds are often used in combination with preheating the base metal to reduce thermal shock.

The size of your molten weld pool is directly influenced by the welding speed, and that's a good way to judge your speed as you work. Only a small weld pool can be carried when welding in positions other than the flat downhand position. Careful selection of your welding speed is required for out-of-position work. You have to learn to go more slowly when welding in the horizontal, vertical, or overhead welding positions.

You should tie your welds into the base metal by slowing down and pausing at the edges of the joint, just as you do with oxyfuel gas welding. You will need lower welding currents for working out of position, just like you needed to control your heat in welding overhead with a gas torch.

Pulsed-current dc welding controls are now available either as add-ons to existing TIG welding power sources or as part of the newer, more sophisticated machines. They will greatly help you in welding out of position. (This also is one of the purposes of a square-wave ac welding machine.) You can adjust the step-pulse frequency on your power source or its special control attachment to suit yourself.

Figure 14-25 shows what two different programmed welding pulse rates would look like used on fully controlled-penetration root passes for nuclear piping. Note that there are two current levels, upper and lower current settings, between which the welding current pulses. The high current melts the metal and the low current lets your weld metal cool a little to control it when working out of position. This is typical of pulsed-current dc methods used in GTAW and GMAW.

Metal as thick as ¼ in. (which is greater than the practical melt-through welding thickness range for GTAW without using a grooved joint) has been successfully welded with pulsed-current dc TIG. A pulsed-current welding program like this probably would be used on automatic equipment,

but pulsed-current dc TIG also can be used with manual welding. The equipment is expensive, and you may not get a chance to use it unless you work in a high-technology shop.

Thermal expansion and contraction from multiple-pass welding of heavy-walled base metals can cause distortions which you can reduce by selecting a suitable combination of welding and torch cross-seam oscillation speed. Adjusting your dwell time at the end of each stroke also provides the proper "wetting" action of filler metal to the wall of the base-metal joint and it eliminates any chance of "cold shuts," or unwelded areas. Also, proper selection of dwell time in the 2G (vertical) position is how you control your weld-bead sag during multiple-pass welding of pipes.

Arc position In welding seams, especially very long ones, position your welding arc precisely over the joint to be welded. Then use a steady rest or guide for your hand if you are welding manually. If automatic equipment is used, there are any number of "seam trackers" that you can use to follow the weld joint. There even are devices that oscillate the torch in a programmed fashion, while the equipment makes the weld. All you have to do is set up the equipment and monitor it while it runs.

Stopping the weld You stop welding either by gradually withdrawing the electrode from the workpiece or by tapering off the current (when you have equipment with a foot pedal). Gradually withdrawing the TIG electrode from the work is the most

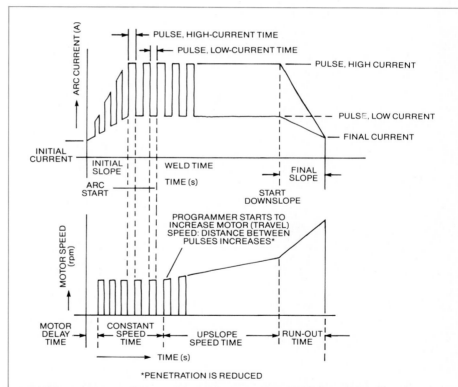

FIGURE 14-25 Automated, microprocessor-controlled, step-pulsed program. Advanced, pulsed GTAW welding machines with special controllers sometimes are used for aerospace and nuclear work. The top graph plots current from a machine sequence set for initial upslope, weld time, and final down-slope. The bottom graph shows how the controller varies the automated welding machine's drive-motor speed, operating the welding head to coincide with the electrical pulses of the welding current. Welding can be totally automated. But it's neither easy nor inexpensive to do so.

common way to stop when you are using dc welding.

You can stop simply by speeding up until you outrun the molten weld puddle, but that will leave some molten metal exposed to the air. One of the better methods for both starting and stopping is to use *runoff tabs* that will be removed from the work later on. You start or stop welding on these tabs, so that you are up to speed when you hit the base metal. Runoff tabs are commonly used in all kinds of welding work. Slope control also can be used to help you start and stop welding, if you have equipment with automatic remote slope control at your welding station.

The foot pedal that controls your welding current is far and away the best method for starting and stopping a TIG weld. You use it like the accelerator on the car. Pushing it down gives you more welding current. Lifting your foot up gradually slows you and your current down until the GTAW arc is gradually turned off.

You will most likely be using a foot pedal with ac, since the HF oscillators normally used with ac require on-off switching to control the starting spark. Foot controls for current and on-off switching are used for high-quality ac and dc GTAW.

Grinding wheels Grinding wheels should be cleaned and reserved exclusively for the material being welded. For example, soft materials such as aluminum become loaded with microsized abrasive particles when edges are ground on the wrong wheels. These abrasive, refractory particles from the grinding wheel become built-in weld defects along the interface between your weld metal and the base metal. Abrasive particles in the surface of the base-metal joint must be removed before you start welding. The ideal joint-preparation method is to use a cutting tool such as a lathe for round or cylindrical joints, or a milling cutter for lon-gitudinal joints. Another common solution, even after grinding, is to wire-brush the soft aluminum base-metal joint just before you start welding.

Cutting fluids You must be careful that your machine shop doesn't use the wrong cutting fluid (if any), too. Be sure they use new cutting lubricants that are free of machining particles. You must clean off the lubricants before you start to weld. Use safety-approved cleaning solvents that are free of residues. *Don't use chlorinated solvents that can make dangerous gases when you weld through them.* The base metal must be free of surface oxides, moisture, marking crayons, and any other foreign substances.

Joint tolerances The required joint tolerances vary a lot, depending upon whether you are setting up for manual or automatic GTAW. Automatic welding always requires tighter tolerances than manual work. Machines can't adjust like people to slight variations in fit-up. The particular tolerance for a given job depends on too many factors to tell you about here. However, you can tell your machine shop people that any good tolerance they give you will be adequate for manual welding. Welding tolerances are a lot looser than machining tolerances.

Bevel angles Not all joints need beveled edges. Edge beveling is not usually needed for butt joints 1/8 in. or less in thickness in carbon and low-alloy steels, stainless steels, and aluminum. For high-nickel alloys, you don't need beveling for material 3/32 in. thick or less. For greater thicknesses, the joint should be beveled to form a V, U, or J groove. Otherwise, erratic penetration will result, causing crevices and voids that are potential areas of accelerated corrosion, as well as built-in "cracks" in the finished weld joint. Notches resulting from erratic penetration act as mechanical stress raisers and cause early mechanical failure in weld joints.

Normally, a double-U or double-V joint is preferred for material over 1/2 in. thick. The added cost of preparation is justified by the decreased amount of weld metal and lower welding time needed to complete the joint. Also, less residual stress will be developed than with the single-groove joint designs.

As shown in Fig. 14-26, V-groove joints are usually beveled to provide a 60° groove for carbon, low-alloy, and stainless steels. An 80° groove is needed for high-nickel alloys. A 90° (right angle) groove is commonly used when welding aluminum with ac GTAW. The U-groove joints are beveled to 7 to 9° sidewalls for carbon, low-alloy, and stainless steels. A 15° sidewall is recommended for high-nickel alloys, and a 20 to 30° sidewall bevel is often required for U-groove joints in aluminum. Single bevels for T joints between dissimilar thicknesses of ferrous metals should have an angle of about 45°. An angle of as much as 60° is usually necessary for aluminum alloys.

Cleaning your base metal
Base-metal cleanliness is one of the most important and also most overlooked requirements for successful welding. That's never more true than in GTAW and GMAW. All sorts of things get on base metal before you ever see it. Paint, marking crayons, oil, grease, shop dirt, printing and stencil ink on plates marked for shipment by metal service centers, rust or other kinds of metal oxides, processing chemicals, cleaning solvents, machining lubricants . . . you name it. Every one of these contaminants will cause you trouble, especially when doing critical GTAW and GMAW work. It is absolutely necessary that you start work on a clean base-metal surface.

The entire joint surface must be cleaned. The minimum area away from the joint on either edge and on either side of the weld joint should be at least ½ in. [12.5 mm] away from the weld edges of the joint. The cleaned area should include the far and near edges of the workpiece and the interiors of hollow or tubular shapes.

The cleaning method depends largely on the substance to be removed. Shop dirt and materials with an oil or grease base can be removed by vapor degreasing or swabbing with nontoxic solvents. Paint and other materials not soluble in degreasing solvents may require the use of methylene chloride, alkaline cleaners, or special proprietary compounds.

If you are not sure what cleaner to use, first ask the nearest welding engineer. If you don't have

one, ask the plant engineer or somebody else in the shop who has had experience. Somebody in the plant finishing department also may have some good ideas, if you explain why you have to have very clean metal. They have similar problems cleaning metal surfaces before painting or electroplating them.

All oxides, of course, must be removed. Most metal oxides melt at much higher temperatures than the base metal. For some metals such as carbon and low-alloy steels, wire-brushing is sufficient. However, wire brushing may not be good enough for metals that have refractory oxides such as aluminum, stainless steels, titanium, and high-nickel

alloys. If you do use a wire brush on stainless steel, use one with stainless-steel wires. Carbon steel will leave behind low-carbon marks on the stainless steel where corrosion can start later on. Acceptable methods for removing metal oxides include grinding (with dedicated grinding wheels), abrasive blasting, machining, or pickling with acids (except for aluminum, which may require alkaline cleaners).

TIG Welding Joints
Your welding joint must leave room for you to work in it. The joint must be big enough for the arc, shielding gas, and your filler rod and electrode in the torch nozzle to reach the bottom. You have to have a joint wide enough for you to manipulate your electrode holder, and your joint must take into consideration the characteristics of the weld metal you are working with. For example, high-nickel alloys are very sluggish when molten. The weld metal does not spread around. You have to place the molten metal at the proper location in the joint. High-nickel-alloy joints must be more open than those for carbon or alloy steel to provide room for you to work. We'll give you other suggestions like this in the following sections covering the basic TIG joint designs.

The five basic joints (butt, lap, T, edge, and corner joints) are all used when TIG welding all metals. We remind you in Fig. 14-27 what these joints look like. Weld joints that require complete weld penetration through the section and high joint strength in thicker sections must have beveled edges. The three most common welding grooves for butt joints are used when the base-metal thickness is larger than sheet gauges.

While there are no fixed rules for using a particular joint design for any one metal, certain designs are commonly used and we'll describe them in the following sections and tell you how to put them to use.

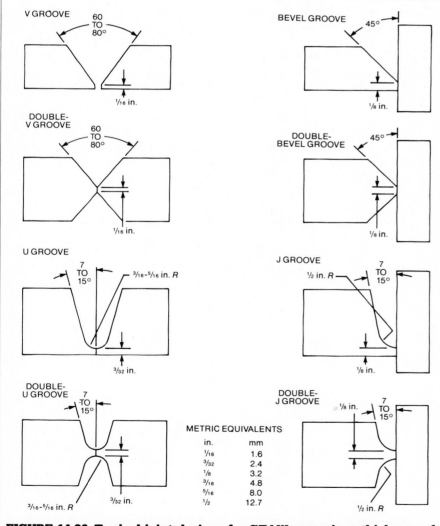

FIGURE 14-26 Typical joint designs for GTAW on various thickness of ferrous metals. Nickel, and especially aluminum included angles are wider.

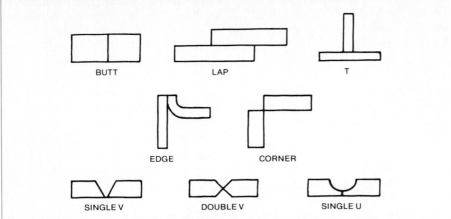

FIGURE 14-27 A reminder of the basic welding joints which are used for GTAW welding, as well as most other fusion welding processes.

Single-V butt joints The single-V butt joint, Fig. 14-28, is the most widely used of all joint designs when complete joint penetration is required from only one side.

The amount of root opening and thickness of the root face or land depends upon whether the GTAW process is to be manual or automatic, whether the filler metal is to be added during the root pass, and whether a backing strip is to be used. The amount of bevel-edge angle depends upon the thickness of the metal and the clearance you need for arc movement for your electrode holder to assure good fusion to both sides of the weld joint.

These welding variables are usually determined by welding test samples that have been prepared with different combinations of angles, gap openings, and other variables. In large shops with welding engineers, that's one thing they are supposed to do for you. Fig. 14-28 and similar drawings are only a guide. Actual joints will vary with the base metal, filler-metal costs, and the size and type of equipment you have available.

Double-V butt joints Double-V butt joints (Fig. 14-29) are used when thicker sections require complete joint penetration and a single-V butt joint would require too big an opening at the top of the V. Remember that the bigger the joint opening, the more filler metal and labor cost must go into

filling it. The double-V butt joint will use less filler metal and cost less to make on thicker sections than a single-V butt joint.

Single-U butt joints The single-U joint is really quite expensive joint to make. It's more difficult to cut or machine than straight-edged joints unless you can use a gouging rod. It requires more filler metal and more welding labor. However, certain metals such as titanium and aluminum are extremely fluid when molten. They are difficult to weld, especially when working in other than the flat downhand position. Sometimes you will be working on thick pipe and you can only weld from one side, even though you would prefer to have a double-V joint to join the thick material from both sides. Jobs like these are best handled with special joints. They may cost more initially, but they often make an otherwise impossible job practical. Figure 14-30 shows only two of many special joint designs.

The joint in Fig. 14-30A is used where metal fluidity creates problems in trying to get a good root pass. It's often used in welding thick-walled aluminum pipe or tubing to control root penetration when welding in all positions without internal backing or a consumable insert in the root gap.

The joint in Fig. 14-30B is used where two base-metal surfaces should be dead level after weld-

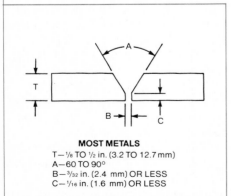

MOST METALS
T—⅛ TO ½ in. (3.2 TO 12.7 mm)
A—60 TO 90°
B—³⁄₃₂ in. (2.4 mm) OR LESS
C—¹⁄₁₆ in. (1.6 mm) OR LESS

FIGURE 14-28 The most commonly used V-groove joint for most metals in thicknesses from ⅛ to ½ in. [3 to 13 mm].

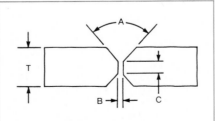

MOST METALS
T—½ in. (12.7 mm) AND OVER
A—60 TO 90°
B—³⁄₃₂ in. (2.4 mm) OR LESS
C—¹⁄₁₆ in. (1.6 mm) OR LESS

FIGURE 14-29 The basic double-side, double-V joint most commonly used for GTAW work on most metals over ½ in. thick.

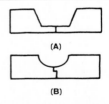

FIGURE 14-30 Special joints are sometimes used for TIG welding. (A) This joint is used where metal fluidity creates problems in obtaining a good root pass. (B) This joint can be used where locating surfaces are required for exceptionally accurate part alignment prior to welding.

ing. The interlock in the root of the joint makes sure that the surfaces of both sides of the joint will be absolutely flush and in line when the weld is completed. It's one way of handling workpiece distortion by designing a clamp right into the weld joint.

Many other special joint designs are used such as folded watertight seam joints with a fillet along the edge. Another type of joint is a plug weld, where you weld through a small hole in the plate. It is really just a small round fillet weld. Combinations of standard joints often are encountered, such as T joints blending into a fillet which might ultimately smooth out to a butt joint. Nevertheless, any joint you run into will be based on the same principles as the joints we've shown you here and in the previous chapters of this book.

TABLE 14-5 Typical welding procedure for manual GTAW of carbon steel

Material thickness, in.	1/16–1/8	1/8–1/4	1/4–1/2
Joint design	Square butt	Single-V groove	Double-V groove
Current, A	50–100	70 to 120	90 to 150
Polarity	DCSP	DCSP	DCSP
Arc voltage, V	12	12	12
Travel speed, in./min	As required	As required	As required
Electrode type	EWTh-2	EWTh-2	EWTh-2
Electrode size, in.	3/32	3/32	1/8
Filler-metal type	E70S-2	E70S-2	E70S-2
Filler-metal size, in.	1/16 or 3/32	3/32 or 1/8	3/32 or 1/8
Shielding gas	Argon	Argon	Argon
Shielding-gas flow rate, scf/h	20	20	25
Purging gas	Argon	Argon	Argon
Purging-gas flow rate, scf/h	5–7	5–7	5–7
Nozzle size	3/8	3/8	1/2
Nozzle-to-work distance, in.	1/2 maximum	1/2 maximum	1/2 maximum
Preheat, minimum	60°F [15.5°C]	60°F [15.5°C]	60°F [15.5°C]
Interpass temperature, maximum	500°F [260°C]	500°F [260°C]	500°F [260°C]
Postweld heat treatment	None	None	None
Welding position*	F, H, V, OH	F, H, V, OH	F, H, V, OH

* F = flat, H = horizontal, V = vertical, OH = overhead.

TABLE 14-6 Typical GTAW procedure for manual welding of stainless steel

Material thickness, in.	1/16–1/8	1/8–1/4	1/4–1/2
Joint design	Square butt	Single-V groove	Double-V groove
Current, A	50–90	70–120	100–150
Polarity	DCSP	DCSP	DCSP
Arc voltage, V	12	12	12
Travel speed, in./min	As required	As required	As required
Electrode type	EWTh-2	EWTh-2	EWTh-2
Electrode size, in.	3/32	3/32	3/32
Filler-metal type	ER-308	ER-308	ER-308
Filler-metal size, in.	1/16 or 3/32	3/32 or 1/8	3/32 or 1/8
Shielding gas	Argon	Argon	Argon
Shielding-gas flow rate, scf/h	20	20	25
Purging gas	Argon	Argon	Argon
Purging-gas flow rate, scf/h	5–7	5–7	5–7
Nozzle size	3/8	3/8	1/2
Nozzle-to-work distance, in.	1/2 maximum	1/2 maximum	1/2 maximum
Preheat, minimum	60°F [15.5°C]	60°F [15.5°C]	60°F [15.5°C]
Interpass temperature, maximum	500°F [260°C]	500°F [260°C]	500°F [260°C]
Postweld heat treatment	None	None	None
Welding position*	F, H, V, OH	F, H, V, OH	F, H, V, OH

* F = flat, H = horizontal, V = vertical, OH = overhead

14-15
WELDING SCHEDULES

Most TIG work that you will do in large plants will arrive along with a written *welding schedule* (often called a procedure). A welding schedule or procedure spells out most of the variables for each workpiece that you will weld. Table 14-5 is a general welding schedule for joining carbon steels. Table 14-6 does the same job for stainless steels. The next table, Table 14-7 on page 342, gives welding schedules for aluminum using DCSP, and Table 14-8 on page 342 gives you setup data for welding aluminum with ac.

If you don't have a welding schedule to work with, you can use these tables as guides to setting up your job and establishing your welding variables. Use these welding schedules for your lab work when learning how to weld with GTAW.

REVIEW QUESTIONS

1. Explain why GTAW is often used instead of oxyfuel welding.

2. Why is DCSP current used for most GTAW work?

3. What special advantage does DCRP current have when using GTAW?

4. Name two shielding gases that could be used with GTAW. Give one good reason for using each of these gases.

5. Describe scratch starting and give two advantages and two drawbacks for using this starting method.

6. Give two advantages and two problems associated with high-frequency starting.

TABLE 14-7 Typical procedure for manual DCSP GTAW of aluminum

Material Thickness, in.	Joint Design	Current* DCSP, A	Arc Voltage, V	Electrode† Diameter, in.	Helium‡ Gas Flow, scf/h	Travel, in./min	Filler Rod or Wire Diameter, in.	Number of Passes
0.010	Standing edge	10–15	10	0.020	20–50		None	1
0.020	Square butt	15–30	10	0.020	20–50		0.020	1
0.030	Square butt	20–50	10	0.020	20–50		0.020	1
0.032	Square butt	65–70	10	3/32	20–50	52	None	1
0.040	Square butt	25–65	10	3/64	20–50		3/64	1
0.050	Square butt	35–95	10	3/64	20–50		3/64	1
0.050	Square butt	70–80	10	3/32	20–50	36	None	1
0.060	Square butt	45–120	10	1/16	20–50		3/64	1
0.070	Square butt	55–145	10	1/16	20–50		1/16	1
0.080	Square butt	80–175	10	1/16	20–50		1/16	1
0.090	Square butt	90–185	10	1/8	20–50		1/8	1
1/8	Square butt	120–220	12.5	1/8	20–50	24	None	1
1/8	Square butt	180–200	12.5	1/8	25–60		1/8	1
1/4	Square butt	230–340	12.5	1/8	25–60	22	3/16	1
1/4	Square butt	220–240	12.5	3/16	25–60		None	1
1/2	60° V bevel, 1/4-in. land	300–450	13	5/32	25–60	20	3/16	1
1/2	Square butt	260–300	13	5/32	25–60		None	2
3/4	60° V bevel, 3/16-in. land	300–450	13	3/16	25–60		1/4	3 for single V
3/4	Square butt	450–470	10	3/16	40–60	6	None	4 for single V
1	60° V bevel 3/16-in. land	300–450	10	3/16	25–60		1/4	2 for double V
1	Square butt	550–570	10	1/4	40–60	5	None	2

* Automatic welding is required for the higher amperages. Manual welding is best done at the lower amperages.
† In lighter gauges of material, it is common to use larger-diameter electrodes than recommended and to taper the electrode tip.
‡ It is possible to substitute helium-argon mixtures. In automatic welding, the arc can be started in argon and the helium added to the shielding gas when welding begins. The best ratio of He to Ar is usually determined experimentally; also consult a welding shielding gas distributor.

TABLE 14-8 Typical procedure for manual ac GTAW of aluminum square and grooved joints

Material Thickness, in.	Edge Preparation	Joint Spacing, in.	Welding* Position	AC, A	Diameter of Electrode, in.	Argon Gas Flow, scf/h	Travel Speed, in./min	Filler Rod or Wire Diameter, in.	Number of Passes
1/16	Square butt	1/16	F, V, H	70–100	1/16–3/32	20	8–10	3/32	1
	Square butt	1/16	O	65–75	1/16	25	8–10	3/32	1
3/32	Square butt	3/32	F	95–115	3/32–1/8	20	8–10	1/8	1
	Square butt	3/32	V, H	95–110	3/32	20	8–10	3/32–1/8	1
	Square butt	3/32	O	90–110	3/32–1/8	25	8–10	3/32–1/8	1
1/8	Square butt	1/8	F	125–150	1/8	20	10–12	1/8–5/32	1–2
	Square butt	3/32	V, H	110–140	1/8	20	10	1/8	1–2
	Square butt	3/32	O	115–140	1/8	25	10–12	1/8–5/32	1–2
3/16	60° bevel	1/8	F	170–190	5/32–3/16	25	10–12	5/32–3/16	2
	60° bevel	3/32	V	160–175	5/32	25	10–12	5/32	2
	90° bevel	3/32	H	155–170	5/32	25	10–12	5/32	2
	110° bevel	3/32	O	165–180	5/32	30	10–12	5/32	2
1/4	60° bevel	1/8	F	220–275	3/16–1/4	30	8–10	3/16–1/4	2
	60° bevel	3/32	V	200–240	3/16	30	8–10	3/16	2
	90° bevel	3/32	H	190–225	5/32–3/16	30	8–10	5/32–3/16	2–3
	110° V bevel	3/32	O	210–250	3/16	35	8–10	3/16	2
3/8	60° V bevel	1/8	F	315–375	1/4	35	8–10	3/16–1/4	2
	90° V double bevel	3/32	F	340–380	1/4	35	8–10	3/16–1/4	2
	60° V bevel	3/32	V	260–300	3/16–1/4	35	8–10	3/16	3
	90° V double bevel	3/32	V, H, O	240–300	3/16–1/4	35	8–10	3/16	2
	90° V bevel	3/32	H	240–300	3/16–1/4	35	8–10	3/16	3
	110° V bevel	3/32	O	260–300	3/16–1/4	40	8–10	3/16	3

* F = Flat, V = Vertical, H = Horizontal, O = Overhead

7. Pure tungsten electrodes don't have very good electrical conductivity. What alloying element (as an oxide) is used to improve the conductivity of GTAW electrodes? What are these special electrodes (AWS grade) called?

8. What is the color coding for EWP AWS grade (pure tungsten) electrodes, and EWZr AWS grade electrodes?

9. What is the difference between a "clean finish" and a "centerless ground finish" tungsten electrode?

10. What duty cycle should your electrode holder (TIG torch) have for continous use in an automatic welding system?

11. Should you purge your TIG holder and gas lines with compressed air before you use them?

12. How far should your tungsten electrode extend beyond the nozzle of your TIG torch? A rule-of-thumb answer for this electrode extension is good enough.

13. You have to select a filler rod for welding aluminum with the GTAW process. What American Welding Society filler-metal specification should you read?

14. What AWS filler-metal specification should you check if you were going to use TIG welding on stainless steel?

15. After you start your arc, what's the best angle to hold your electrode and torch (measuring this angle between the molten metal and your electrode holder) while you weld? It is:
(a) 10° (holding the electrode almost flat over the molten metal)
(b) 50°
(c) 75°
(d) 90° (holding the electrode perpendicular to the molten metal)

16. How do you use your TIG holder to stop welding when you don't have a foot-pedal dc current control?

17. What should you do to the surface of your base metal around the weld area and on the joint surfaces before you start welding with GTAW? Why?

Learning to Weld with TIG

14-1-S
SHOP PRACTICE ON STEEL

The purpose of this work is to help you learn how to make GTAW fusion welds on low-carbon steel, first without using a filler rod, and then with a filler rod.

Materials

You will need ¹⁄₁₆- and ⅛-in.-thick sheet steel for this work. You also will need ¼-, ³⁄₁₆-, and ¾-in.-thick low-carbon steel plate or, preferably, Type 302 or Type 304 stainless steel if your school can afford it. Small 2- × 4-in. coupons are good to use when you are practicing this work, although your final welder qualification tests are best conducted on 6- × 6-in. coupons.

1. Make bead welds on the ¹⁄₁₆-in.-thick steel without filler metal. (Simply melt through the material, making a single good weld bead on the metal surface.)
2. Make bead welds on the ⅛-in.-thick steel without filler metal.
3. Make bead welds on the ¹⁄₁₆-in.-thick steel with filler metal.
4. Make bead welds on the ⅛-in.-thick steel with filler metal.

Bead-weld test There is no physical test for a bead weld. The only test is the appearance of the weld bead that you produce. The best weld will have a uniform width and an even, smoothly rippled surface with no pin holes, cracks, or gaps where you missed making the weld, and no holes burned through the material.

5. Make a weld on a corner joint on two pieces of ¹⁄₁₆-in.-thick steel sheet.
6. Make a weld on a corner joint on two pieces of ⅛-in.-thick steel.

7. Subject your completed corner welds to a hammer test. If you produced a satisfactory corner weld joint, you will be able to bend and then hammer the joint flat without cracking or breaking your weld.

8. Make a weld on a square-groove butt joint on ¹⁄₁₆-in.-thick steel in the flat position.
9. Make a weld on a square-groove butt joint on ⅛-in.-thick steel in the flat position.
10. Make a weld on a square-groove butt joint on ¹⁄₁₆-in.-thick steel in the vertical position.
11. Make a weld on a square-groove butt joint on ⅛-in.-thick steel in the vertical position.
12. Make a weld on a square-groove butt joint on ¹⁄₁₆-in.-thick steel in the overhead position.
13. Make a weld on a square-groove butt joint on ⅛-in.-thick steel in the overhead position.
14. Make a weld on a single-bevel butt joint on ³⁄₁₆-in.-thick steel plate in the flat position. Use a

45° edge-beveled joint for this and the following plate jobs.

15. Weld a single-bevel butt joint on ¼-in.-thick steel plate in the flat position.

16. Weld a single-bevel butt joint on ³⁄₁₆-in.-thick steel plate in the vertical position.

17. Weld a single-bevel butt joint on ¼-in.-thick steel plate in the vertical position.

18. Weld a single-bevel butt joint on ³⁄₁₆-in.-thick steel plate in the overhead position.

19. Weld a single-bevel butt joint on ¼-in.-thick steel plate in the overhead position.

Steel butt-joint test All butt welds should be tested by a guided root-bend test.

Making steel fillet and lap joints

20. Make a fillet-welded lap joint on ¹⁄₁₆-in.-thick steel in the flat position.

21. Make a fillet-welded lap joint on ⅛-in.-thick steel in the flat position.

22. Make a fillet-welded lap joint on ¹⁄₁₆-in.-thick steel in the vertical position.

23. Make a fillet-welded lap joint on ⅛-in.-thick steel in the vertical position.

24. Make a fillet-welded lap joint on ¹⁄₁₆-in.-thick steel in the overhead position.

25. Make a fillet-welded lap joint on ⅛-in.-thick steel in the overhead position.

26. Make a fillet-welded T joint on ¹⁄₁₆-in.-thick steel in the flat position.

27. Make a fillet-welded T joint on ⅛-in.-thick steel in the flat position.

28. Make a fillet-welded T joint on ¹⁄₁₆-in.-thick steel in the vertical position.

29. Make a fillet-welded T joint on ⅛-in.-thick steel in the vertical position.

30. Make a fillet-welded T joint on ¹⁄₁₆-in.-thick steel in the overhead position.

31. Make a fillet-welded T joint on ⅛-in.-thick steel sheet in the overhead position.

32. Make a fillet-welded lap joint on ³⁄₁₆-in.-thick steel plate in the flat position.

33. Make a fillet weld lap joint on ¼-in.-thick plate in the flat position. Use a 45° edge bevel.

34. Make a fillet weld on a lap joint on ³⁄₁₆-in.-thick steel plate in the vertical position.

35. Make a forehand fillet weld on a lap joint on ¼-in.-thick steel plate in the vertical position. Use a 45° edge bevel.

36. Make a forehand fillet weld on a lap joint on ³⁄₁₆-in.-thick steel plate in the overhead position.

37. Make a forehand fillet weld on a lap joint on ¼-in.-thick steel plate in the overhead position. Use a 45° edge bevel.

38. Make a forehand fillet weld on a T joint on ³⁄₁₆-in.-thick steel plate in the flat position.

39. Make a forehand fillet weld on a T joint on ¼-in.-thick steel plate in the flat position. Use a 45° edge bevel.

40. Make a forehand fillet weld on a T joint on ³⁄₁₆-in.-thick steel plate in the vertical position.

41. Make a forehand fillet weld on a T joint on ¼-in.-thick steel plate in the vertical position. Use a 45° edge bevel.

42. Make a forehand fillet weld on a T joint on ³⁄₁₆-in.-thick steel plate in overhead position.

43. Make a forehand fillet weld on a T joint on ¼-in.-thick steel plate in overhead position. Use a 45° edge bevel.

Testing steel fillet and lap joints

44. All of your lap joints are to be subjected to a hammer test, folding the joint 180° along the weld to make a flat sandwich with the welded joint running along one side. Then press the welded joint flat without any weld-metal cracking.

45. All of your fillet joints are to be subjected to a hammer test for T joints, folding the 90° leg of the T flat to one side of the joint, followed by hammering the joint without producing any weld-metal cracking.

14-2-S
WELDING COPPER

Copper-Welding Materials
You will need sheet copper samples, about 2 × 4 × ⅛ in. thick and ¼ in. thick. You also should get some standard-weight, 2-in.-diameter copper water pipe to work on. The sheet copper preferably should be a good selection of both deoxidized (OFC or oxygen-free copper) and electrolytic copper.

Select your filler rod from AWS A5.7, using a silicon-copper rod about ⅛ in. in diameter.

Making Welds on Copper
46. Make a weld on a butt joint of ⅛-in.-thick copper in the flat position.

47. Make a weld on a butt joint of ¼-in.-thick copper in the flat position.

48. Make a forehand weld on a butt joint on 2-in. copper pipe in the horizontal rolled position.

49. Make a forehand weld on a butt joint on 2-in. copper pipe in the horizontal fixed position.

50. Make a forehand weld on a butt joint on 2-in. copper pipe in the vertical fixed position.

Testing Your Copper Welds
Successful completion of each of these copper welding exercises will constitute qualification for this work.

14-3-S
WELDING ALUMINUM
GTAW is one of the two main processes used to join aluminum and its alloys. The other process most frequently used on aluminum is MIG welding, which you will learn shortly. Watch out for aluminum, as it melts without changing color.

Aluminum-Welding Materials
You should have 2- × 4-in. sheet and plate samples of aluminum to work on that are ¹⁄₁₆, ⅛, ³⁄₁₆, and ¼ in. thick. If you can obtain them, get some aluminum castings to work on. They should be about ¼ to ½ in. thick. If possi-

ble, you should practice welding on commercially pure 1100 series aluminum, and on at least one weldable aluminum alloy such as a silicon-aluminum grade.

You should select your aluminum TIG rod from AWS A5.10, using ER 1100 rods for the pure aluminum and ER 4043 rods for the silicon-aluminum alloy material.

Making Welds on Aluminum

51. Make bead welds on $\frac{1}{16}$-in.-thick aluminum sheets without filler metal. First try it with 1100 series aluminum, and then with a silicon-aluminum alloy such as 4043, using the correct filler rod for this material.

52. Make bead welds on $\frac{1}{8}$-in.-thick aluminum sheets without filler metal. First try it on commercially pure aluminum and then on the alloyed material.

53. Make welds on flanged-edge joints in the flat position on $\frac{1}{16}$-in.-thick aluminum sheets—no filler metal. Work with both 1100 and 4043 aluminum grades.

54. Make the same flanged-edge joints in the flat position, working on $\frac{1}{16}$-in.-thick sheets, but this time use a filler rod. Work with pure aluminum sheet and then with a weldable alloy grade.

55. Make welds on butt joints in the flat position on $\frac{1}{8}$-in.-thick 1100 series aluminum sheets. This time, use your filler rod.

56. Make welds on butt joints in the flat position on $\frac{1}{8}$-in.-thick silicon-aluminum sheets. Use filler rod.

57. Make welds on flanged-edge joints in the vertical position on $\frac{1}{16}$-in.-thick aluminum sheets—no filler metal.

58. Make welds on flanged-edge joints in the vertical position on $\frac{1}{8}$-in.-thick silicon-alloy aluminum sheets—no filler rod.

59. Make welds on butt joints in the vertical position on $\frac{1}{8}$-in.-thick 1100 aluminum sheets. Use filler rod.

60. Make welds on butt joints in the vertical position on $\frac{1}{8}$-in.-thick 4043 aluminum sheets. Use a silicon-aluminum filler rod.

61. Make welds on deep-flanged edge joints in the flat position on $\frac{1}{16}$-in.-thick aluminum sheets—no filler rod.

62. Make welds on deep-flanged edge joints in the flat position on $\frac{1}{16}$-in.-thick silicon-alloy aluminum sheets—no filler rod.

63. Make welds on butt joints in the flat position on $\frac{3}{16}$-in.-thick 1100 grade aluminum sheets.

64. Make welds on butt joints in the flat position on $\frac{3}{16}$-in.-thick 4043 silicon-aluminum alloy sheets.

65. Make welds on butt joints in the flat position on $\frac{1}{4}$-in.-thick 1100 grade aluminum sheets.

66. Make the same welds on butt joints in the flat position on $\frac{1}{4}$-in.-thick 4043 silicon-aluminum alloy plates.

67. Repeat the butt joints on $\frac{1}{4}$-in.-thick plate but this time work in the vertical position.

68. Make a butt joint on $\frac{1}{4}$-in.-thick plate in the overhead position. Watch out for the melting aluminum.

69. Not all cast-aluminum alloys are readily weldable. With luck, you'll get a chance to work on one that is. Use an AWS A5.10 ER 4043 TIG rod.

70. Turn your power source down to its lowest possible setting and then try welding aluminum foil. Your power source may not be able to do it, but with the right TIG machine and a very small tungsten, even foil thicknesses can be welded.

Testing Your Aluminum Welds

All aluminum butt welds on material $\frac{1}{4}$ in. and thicker should be submitted to the nick-break test. Test your flange welds and butt welds on thinner material with guided-bend tests. There is no test for bead-on-plate or bead-on-sheet welding except weld-bead appearance.

14-4-S
WELDING STEEL PIPE

This work is optional, but very important if you can possibly get it done. Use 4-in. steel pipe and an AWS A5.2 grade 45 or grade 60 welding rod about $\frac{1}{16}$ in. in diameter. Work in as many different pipe-welding positions as you can. Save the best samples for your pipe-welder operator qualification test which follows. This could be your preparation for nuclear power-plant pipe welding. A lot of that pipe is TIG welded, and the pay is very good.

71. Prepare a beveled pipe joint from two pieces of steel pipe about 4 in. in diameter. Bevel the outer edges to 37.5° before you start. Then tack-weld the two sections into position with four tacks spaced equally around the joint (at points about 0, 90, 180 and 270° around the pipe). Then weld the pipe sections together.

Keep the pipe centerline in the horizontal position while you work. It will help you to rest the pipe in a piece of channel iron or two angle irons. That way you can roll the pipe when you have to, welding a quarter of the section at a time, while always working in the downhand position. Pay special attention to getting a good solid root pass. That's what TIG is best suited for and often used for in pipe work. Make sure your cover passes are tied into the sides of the joint and into the root pass.

72. Repeat Exercise 71 with the pipe centerline in the vertical position (the pipe is standing upright) after you tack-weld it in line. This time the pipe joint is like making a circular weld in the horizontal position.

Testing Your Pipe Welds

Subject a strip cut from your pipe welds to guided root- and face-bend tests.

14-5-S
OPERATOR-QUALIFICATION TESTS FOR WROUGHT METALS

Test 1. Butt Welds

These tests will require six plates of metal (low-carbon steel, copper, or aluminum). In some exercises you will have to decide whether or not the material thickness requires edge beveling, and what angle to bevel the material. You also will do the required edge preparation before you start to weld. Your test material should be ½ in. [13 mm] thick or thicker for the single-sided V-groove butt welds.

The first test consists of making:

■ Single-V-groove butt joints in the flat position.
■ Single-V-groove butt joints in the vertical position.
■ Single-V-groove butt joints in the overhead position.

Each of the welded pieces you make will be cut into four strips. Discard the outside strips. The two inner strip specimens that are retained are to be machined flush all over as shown in Fig. 9S-2 in Chap. 9's "Shop Practice" section. One test specimen from each of the three welding positions (flat, vertical, and overhead) will be submitted to a guided face-bend test, and the other center strip will be submitted to a guided root-bend test. If any defects greater than ⅛ in. in any dimension are developed in these tests, your weld will be considered as failing. In addition to the guided face-bend and guided root-bend test, your weld will be examined for appearance. Sloppy-looking welds will be rejected.

Test 2. Fillet and Lap Welds

These tests require six plates, each of which is 6 × 6 in. × ¼ in. thick. One edge of each of the plates will be ground square to make a T joint. Also, one backup strip 6 × 1¼ in. × ¼ in. thick will be required for each joint.

Set up the test plates for a double fillet-weld T joint and make:

■ One set of fillet welds in the flat position.
■ One set of fillet welds in the vertical position.
■ One set of fillet welds in the overhead position.

After the fillet welds have cooled, cut off the leg of the T flush to the base and then make a third weld in the area marked A in Fig. 9S-3 in the Shop Practice Section at the end Chap. 9. This weld may be done in any position you want, regardless of the position in which the fillet welds were made. If any defects greater than ⅛ in. are developed in this test, the weld will be considered as failing.

Test 3. Pipe

Make several samples by welding two pieces of 4-in. standard steel pipe approximately 6 in. long. The centerline of the two pieces of pipe will be placed in a vertical plane and position-welded as a butt joint. This is a horizontal fixed-position pipe weld. You are not allowed to rotate it. You must work around the pipe. (If you think you're really good, put the axis of the pipe on a 45° angle, and weld that one *without rotation*.)

A minimum of four specimens should be taken from the completed weld, spaced at 90° intervals around the weld. These should be taken from the top, bottom, and opposite sides of the completed pipe weld. The top and one side specimen should be submitted to the guided face-bend test, and the bottom and other side specimen should be subjected to the root-bend test.

All of the above specimens should bend flat to 180° or the full capacity of the guided-bend-test jig without any cracks greater than ⅛ in. in any dimension resulting from bending the weld.

If you pass this pipe-welding test, you are well on your way to becoming a skilled TIG welder.

Welding and Cutting with a Plasma Arc

If you enjoyed TIG welding, wait until you see what you can do with plasma-arc welding and cutting, assuming, of course, that your welding school has the equipment (it's rather expensive). Whether it does or doesn't, we'll discuss both processes. They are becoming more important in industry, and any good welder should at least know how they work and what they can do.

Plasma-arc welding (or PAW as the American Welding Society codes it) will weld any weldable metal. PAW is very similar to GTAW (Fig. 15-1). It uses an electrode holder, a permanent tungsten electrode, the same filler rods and wire that are used for GTAW, and the same inert shielding gases. Plasma-arc cutting (abbreviated PAC) closely resembles the welding process, except for somewhat different cutting-torch head designs and much higher voltages and currents.

Let's take the two related processes one at a time. We'll start with an overview of PAW, then we'll give you more details on the welding process. After that, we'll finish this chapter with PAC.

We are going to start with PAW first because, operationally, it is very similar to the TIG process you just learned in Chap. 14. In fact, that's why this chapter immediately follows the one on TIG welding. The main difference between the GTAW and PAW processes as far as the operator is concerned is the shape of electrode holder nozzle and the shape of the plasma arc that flows through it.

The TIG plasma has an open plasma arc. Plasma welding and cutting equipment has a tightly restricted plasma arc (Fig. 15-2), which means that you have a longer arc length in the PAW process compared with the GTAW arc. The PAW arc also has a cylindrical shape rather than the bell shape that you see when working with GTAW. That's why the PAW arc is described as a constricted arc.

Like TIG, plasma welding and cutting torches are used with a conventional drooping-characteris-

FIGURE 15-1 Plasma-arc welding (PAW) can join any weldable metal. The PAW process is so similar to TIG welding that we've included it as your next lesson, even though it usually is considered to be an advanced welding process. PAW is used here for welding titanium.

tic power source (dc straight-polarity). A conventional TIG power source often can be used for PAW simply by adding a control box that monitors arc starting, the gases, and cooling-water flow, among other things. PAC often requires the same basic type of power source, but it must have much higher open-circuit voltage and should be able to produce much higher amperages than a TIG power source.

15-1
WHAT'S A PLASMA?

As you learned in Chap. 11, a plasma is ionized matter that carries an electrical current. It's the material that makes up the arc of any electric arc-welding process used in air or with a shielding gas. You've seen plasma in the lab work you did following the last two chapters. In fact, the only way you would ever see a welding arc without an associated gas plasma is if you welded in a vacuum. You can even have a plasma of electrons making up an "electron gas" if you welded in a vacuum—but vacuum welding methods certainly don't fit in a beginner's welding course, so we won't get into that subject.

How PAW Works

PAW gets its name because the plasma arc is much denser, a lot stiffer, and much hotter than the arcs of conventional shielded-gas welding processes such as TIG and MIG. Arc temperatures of 21,000°F [11,650°C] are not unusual in PAW and in plasma-arc flame-cutting (electrically very similar to plasma welding).

15-2
TWO TYPES OF ARCS

The plasma-arc torch is started by adjusting the flow of the plasma-forming and weld-metal shielding gases. Then, using HF arc starting just like TIG, a pilot arc is established between the tungsten electrode and the torch nozzle (called a *nontransferred arc*), or between the electrode and the workpiece (a *transferred arc*). The different arcs are used in different processes. We'll tell you more about them shortly.

The pilot arc begins as a non-transferred arc. It has a visible flame showing beyond the end of the electrode-holder nozzle. When the pilot arc is brought into contact with the work, it transfers over to the work, providing a nar-

row conductive path of ionized gas through which the full transferred welding arc is established against the workpiece.

The Constricted Plasma Arc

Remember when working with TIG welding that the GTAW arc was slightly cone shaped? The arc was wider at the bottom than at the top near the tungsten electrode. The GTAW arc plasma often has a cone shape. The arc you will see in PAW is much straighter, like a cylinder. The reason is that the plasma in PAW is constricted or squeezed as it passes through the ceramic nozzle on the end of the electrode holder (which, like TIG, is often called a torch). Look back at Fig. 15-2 and you will see how the ceramic nozzle constricts the plasma welding arc.

Constricting the arc plasma increases the arc voltage and current density for a given welding amperage. The increased current density means that there are more electrons per cubic inch of plasma (so to speak). Another way to put it is to say that there are more amps in each unit of plasma. That's why the plasma arc is so hot. The constriction of the PAW arc is achieved by forcing it to flow through the ceramic nozzle's central hole, which is rather small.

15-3
PLASMA WELDING'S ADVANTAGES

There are important operating advantages that PAW has over conventional GTAW.

■ The improved plasma-arc direction and stiffness give you easier control of heat input to the work without the arc wandering around.
■ Greater welding "standoff" distance (equivalent to a longer arc length) gives you fewer problems maintaining the correct torch position and distance from the work than with TIG.

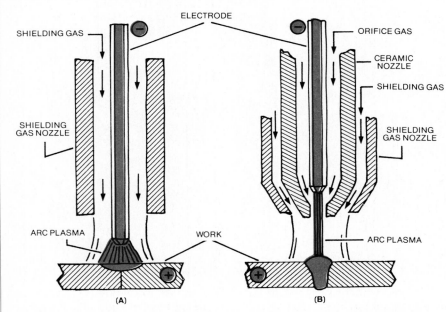

FIGURE 15-2 Comparison of (A) gas-tungsten-arc and (B) plasma-arc welding processes.

- The plasma weld-puddle size also is less sensitive to changes in arc length. Plasma welding is even more forgiving of slight variations in arc length as you move your hand, meaning that the welds you make will be even more uniform than TIG welds.
- You will find it easier to feed the filler rod or wire and handle torch manipulation because of the greater standoff distance of PAW versus GTAW. The standoff distance is usually about ³⁄₁₆ to ½ in. [4.7 to 12.5 mm] from the nozzle to the workpiece.
- You also can use larger-diameter filler wire or rod with minimum risk of contaminating the weld metal or the tungsten electrode. This is especially important in welding stainless steels, aluminum, and titanium in the critical jobs you will find in nuclear power and chemical-plant piping and in aerospace welding.
- Because the plasma arc is hotter than the GTAW arc, the welding speeds are higher and the heat-affected zone next to the solidified weld metal is narrower.
- The PAW process is less sensitive to joint mismatch and poor joint fit-up than GTAW. The main reasons are the higher heat input and the way the plasma arc makes a temporary "keyhole" right through your work which is surrounded by molten metal that flows back into the gap after the plasma arc passes by (as we'll explain shortly). That is why the PAW is often automated. It doesn't always require a skilled welder to manipulate the torch.
- There's much less chance of electrode or weld-puddle tungsten contamination because the tungsten electrode is recessed inside the ceramic gas nozzle on the working end of the torch (unlike GTAW).
- You can control weld-metal joint penetration of thinner sections by using a special keyhole welding technique.
- Like GTAW, PAW can be used in any welding position from flat to overhead.

15-4
PLASMA WELDING'S ONE DRAWBACK

With this long list of advantages, you would think that plasma welding would completely replace TIG welding. But there's one big drawback to the process. PAW equipment costs more than just about any other arc- or resistance-welding method. What's more, the ceramic nozzles on the end of the torch have a short service life due to the very high-temperature-arc plasma. The consumption of relatively expensive inert gas also is higher, because both a plasma-arc-forming gas and a shielding gas are needed.

An important fact to fabricators who want to use PAW is that there is a shortage of skilled plasma-arc operators, even worse than that of skilled TIG welders. That's good news for you. Also good is the fact that the kind of work often done with PAW is done under very clean conditions in modern plants such as aerospace, chemical, and sophisticated nuclear-reactor pipe-fabricating shops . . . places where the advantages of PAW outweigh the equipment cost.

In shops that do critical welding work, the cost of one bad weld can easily exceed the cost of an entire PAW outfit. Therefore the equipment cost is only a relative drawback. When you need the best welding process there is to do high-quality work, plasma welding is cheap. Even the apparent high cost of the equipment is usually paid off quickly when plasma welding is done with a "hot wire," which means using a current-carrying filler wire (or two) instead of a rod. The hot wire runs through a wire feeder (or two) and carries a second electrical current (similar to MIG welding). Deposition rates with hot-wire plasma can be as high as 40 to 70 lb/h [18 to 30 kg/h] of weld metal. That figure compares with the best deposition rates of submerged-arc welding. At those weld-metal deposition rates, the total cost of plasma welding is low. The productivity of the hot-wire equipment helps the fabricator write off the equipment cost rapidly.

What's more, the cost of a complete plasma-arc welding outfit is not far above that for some of the very sophisticated GTAW machines. Also, the add-on controls to convert one of the advanced GTAW machines into a plasma unit is a fraction of the cost of buying the GTAW power source to begin with, and since both GTAW and PAW are used for the same kinds of work (welding thin materials, nonferrous alloys like aluminum, and high-priced aerospace alloys), acceptance of PAW is increasing in industry.

Another reason for an increase in the use of PAW is that welding engineers are becoming more familiar with another closely related process that we've already briefly referred to—PAC. If you are plasma-arc-cutting nonferrous materials that you can't cut with oxyfuel flame-cutting, you already own a power source that probably can be used for plasma-arc-welding the same materials after you have finished cutting them.

15-5
PLASMA GASES

You will always have two separate gas flows, two gas cylinders, two gas regulators, and two gas hoses to contend with in PAW. One cylinder is for the plasma-forming gas. The other cylinder is for the plasma-shielding gas to protect your base metal from the atmosphere.

Plasma-Forming Gases
The stream of current (electrons and gas ions) in the arc-welding plasma-forming gas is chemically inert (in the sense that it won't oxidize your material or combine with it in any other fashion) as long as your plasma gas is an inert gas like argon. If you were plasma flame-cutting, you might have a plasma-forming gas that

also contains reactive gases such as oxygen, but the reactive gases are not there to oxidize away the metal when you cut it as much as to increase the temperature of the cutting-plasma jet.

Pure argon generally is used as the plasma-forming gas because it's less expensive than helium. Pure helium also is not used because it makes arc starting too difficult.

Similar to GTAW, your weld-metal penetration depth and, partly, your welding speed are dependent upon the arc current and shielding gas. However, the flow rate of the plasma-forming gas also has a significant effect on weld-penetration depth into the base metal. If the rate is too low, you will get poor base-metal penetration. The reason is that if you have a low flow rate on your plasma-forming gas, you are getting less plasma per second, which means less heat. As you know, less heat means less penetration into the base metal.

Of course, if you are welding metal foil, you don't need a large plasma, and therefore you want a low plasma-gas flow rate. If you are welding a thick section, or a section of aluminum or copper with high thermal conductivity, you want a higher plasma-gas flow rate to pump as much heat into the piece as possible. This fact will be reflected in the welding schedules you will use to set up your equipment to plasma-arc-weld different materials and thicknesses. What's most important right now is that you understand the reasons for the different welding schedules (or procedures if you prefer that word).

Plasma Shielding Gases

When you are plasma-welding, the column of plasma streaming from your electrode is too narrow to provide adequate shielding of the molten weld metal underneath it. Therefore a second, larger-diameter stream of shielding gas is provided that flows con-

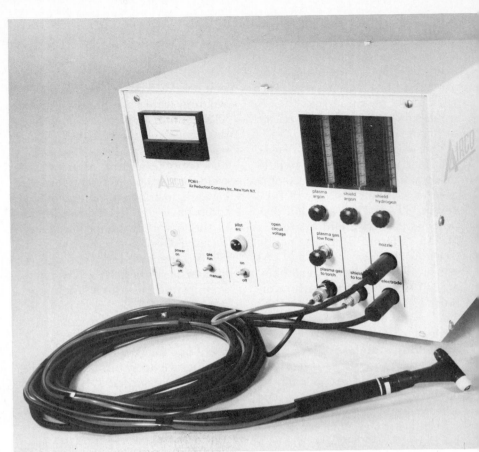

FIGURE 15-3 The plasma-arc welding torch looks very similar to a TIG electrode holder (another example of the close similarity among these two processes). The control box to which the gun is attached normally would sit on top of the plasma-arc power source.

centrically around the outside of the plasma-arc column and covers the base-metal area that you are working on. Very little of this second shielding gas will be ionized to form a plasma. Most of it remains as a conventional gas used primarily to protect your work from oxidation.

The same inert shielding gases (argon, or helium, or a mixture of the two gases) that you used in GTAW are used as shielding gases in the PAW. Argon, argon-hydrogen, or argon-helium mixtures are sometimes used for the plasma-shielding gas. (Note that the addition of hydrogen is an exception—hydrogen is hardly an inert gas—but not very much is required and what's there increases the temperature of the plasma arc for doing special welding jobs.)

15-6
PAW OUTFITS

Apart from the different nozzle and the doubled gas-flow equipment, the PAW outfit is very similar to a TIG-welding setup. You'll need a plasma-control device (usually in a separate box) to monitor and regulate the current as well as the cooling water, shielding-, and plasma-forming gas. The control box probably also will have preflow and postflow gas timers. You'll probably use a foot-pedal-operated current control, just like those used in more sophisticated GTAW operations. You'll also need shielding- and plasma-gas regulators and cylinders, and a cooling-water circulator pump and filter, just like those used for water-cooled TIG welding. Finally, you'll need a plasma torch.

Many plasma torches now include plasma-gas-flow controls right at the torch, along with a torch-mounted contactor control switch so that you can alter the welding current while you work. It's very much like adjusting the fuel and oxygen flow from a welding torch. This feature will help you make the special keyhole welds you will read about shortly.

The plasma-arc torch is the lightest, smallest welding torch you have ever used (see Fig. 15-3). Machine-mounted torches also are available, of course. Manual PAW torches have typical ratings of 100, 200, and 300 A. Larger torches up to 500 A are used with automated welding machines.

Your welding school may not have PAW equipment because of the cost. If not, spend your time after studying this lesson learning to be a good TIG operator. If you become really good at GTAW and fully understand this chapter, you probably could learn to operate PAW equipment and qualify to weld with plasma in just a day or two on the job.

15-7
PLASMA-WELDING POWER SOURCES

Most PAW power is direct-current straight-polarity (DCSP). One reason is that dc plasma arcs are easier to maintain than ac plasma arcs because the dc arc, once started, is constantly on while the ac arc goes off and has to be restarted electronically (with HF arc starting) every time the ac waveform passes through the zero point as it cycles from positive to negative polarity and back again. The most commonly used plasma-arc power supply is the constant-current (CC) or drooper-type power source.

The major difference between a dc plasma-arc power source and a conventional TIG-welding machine is the output voltage. The OCV of the plasma unit may be as high as 400 V. Actually, al-

most any dc drooper power source with a minimum of 80-V OCV and a primary contactor can be used to operate plasma-welding equipment. The minimum current you'll get depends on the power source selected. Most of the time, however, high voltage is supplied by specially designed PAW power sources.

Connecting Smaller Power Sources

Several TIG power sources also can be connected together, either in series to give very large load voltages for the upper range of plasma-arc work, or in parallel to produce high currents for welding thick sections and base metals that need more heat because they have very high conductivity (like pure aluminum or copper).

DC series connections If you don't have a PAW power source, you can ask an experienced electrician familiar with welding power sources to talk to the manufacturer about hooking up several TIG machines in series (Fig. 15-4). A dc *series connection* means that the positive terminal of one power source is connected to the negative terminal of the next one, whose positive terminal is connected to the negative ter-

minal of the first power source, or third unit, and so on. When any dc device is connected that way, the output voltages add. For example, three 100-V-output machines will produce one 300-V dc series circuit (or $3 \times 100 \text{ V} = 300 \text{ V}$ when welding). In the same way, you can hook up six 1.5-V dry cells in series to give you 9 V (because $6 \times 1.5 \text{ V} = 9 \text{ V}$).

Figure 15-5 shows the volt-amp curve of a typical specialized plasma power supply. Its rated output is 300 V at 250 A. In this case, the arc operates at more than ten times the voltage of a typical TIG arc, which might have a rated voltage of around 50 V at 250 A.

Figure 15-6 shows the composite volt-amp curves for several TIG power supplies. The TIG units are connected in series as shown in Fig. 15-4 to make their voltages add up. (Welding power sources should *never* be connected in series or parallel unless the manufacturer authorizes it and an *experienced* operator or electrician does it. That means you should never do it without expert advice.)

The voltage at each current on the curve for all three power sources is three times the voltage of one machine (assuming that all three power sources have identi-

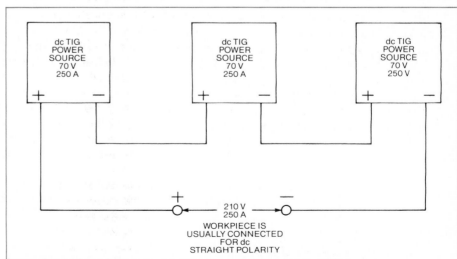

FIGURE 15-4 Series connection of three dc TIG welding power sources to produce high voltage for plasma work. Voltages are added when electrical devices are connected in series. The current remains unchanged.

cal ratings). However, the current rating for three machines connected in series is the same as that of any of the single power sources. That is typical of any series connection. *When electrical units are connected in series, the voltages add up, but the current in the line stays the same.* Three machines with 100-V loads would produce 300 V in the line.

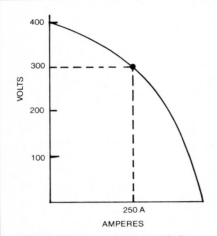

FIGURE 15-5 A specialized dc plasma-arc power source combines high voltage and high amperage. This voltage/amperage curve is for a machine rated at 300 V at 250 A which might be used for plasma cutting.

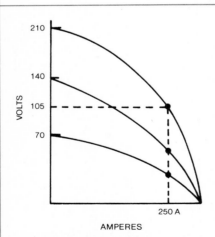

FIGURE 15-6 Composite voltage/amperage curves for three different 70-V, 150-A power sources connected in series. The rated output of the complete series circuit is 105 V at 250 A.

DC parallel connections Figure 15-7 shows another way to hook up welding (or any other electrical) equipment, even light bulbs. If you connect all the positive terminals together, and all the negative terminals together so that they are in parallel, rather than in series, the *rated load voltage of the parallel circuit will be unchanged, but the currents will add up.* Also, the power sources should all have the *exact same OCV* when the hookup is made, which will give you very high currents for plasma cutting as well as plasma welding. (They should all have the same OCV for safety reasons.) Again we repeat, this is not a job for beginners.

Figure 15-8 shows what the volt-amp curve for three power sources connected in parallel would look like. The rated OCV in the diagram is 70 V for all three machines. The rated load voltage of 50 V produces 250 A of current for each machine, or 750 A for all three machines connected in parallel. (Note again that all machines have the same OCV.) As a result, each machine adds all of its current to the overall parallel circuit. Three machines producing 100 A each at the rated voltage would result in 300 A in a parallel hookup. The machines rated at 400 A at some given rated voltage would produce a whopping 1200 A of current in a parallel connection. That's characteristic of many plasma-arc setups. They are *hot*.

Combined series and parallel connections A combination power source with series and parallel connections will provide both high voltage and high current. That is the principle used by manufacturers of specialized plasma-arc power sources. Now you can see why plasma-arc power sources cost more than regular power sources. They have a lot more wiring in them.

Figure 15-9 shows the family of volt-amp curves you would get from a plasma power source with

four separate 100-V open-circuit power lines. (Remember that OCV is the voltage supplied by the power source when you are not welding. It is the highest voltage output from the machine, while load voltage is the voltage in the circuit when you are welding.)

The circuits in this power source can be connected all in series, all in parallel, or two-by-two. (If you had the power source in front of you, you would select the combination you want with a dial.) When the circuits are series-connected, the combined out-

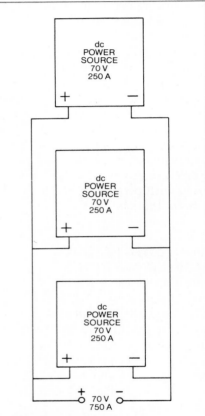

FIGURE 15-7 Direct-current power sources can be connected in parallel to produce high current while leaving voltage unchanged. Plasma power sources often combine connections together inside one machine cabinet to produce both high voltage and high current. The extra electrical connections and components required explains why plasma-arc power sources are more expensive than GTAW power sources.

put circuit will have a 250-A, 200-V load. When connected in parallel, the combined circuit will produce 1000 A and 50 V. The two-by-two connection provides 500 A at 100-V load. The OCV for the three hookups are 400 V, 100 V, and 200 V, respectively. Electricians use formulas to calculate these values.

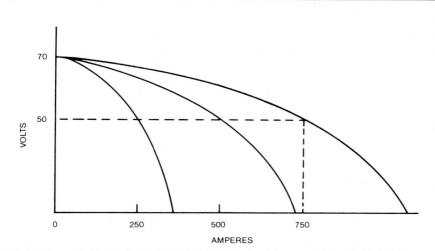

FIGURE 15-8 Voltage/amperage curve for three separate 70-V dc machines, each rated at 250 A and 50 V, connected in parallel. When used in a parallel circuit, the three combined machines produce 750 A at 50 V, or 3 times the rated amperage of each machine operated separately. The rated voltage of the combined circuit remains the same as for a single machine.

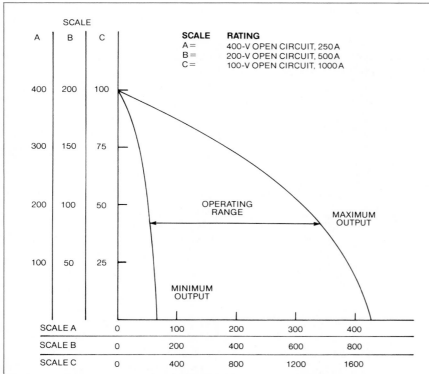

FIGURE 15-9 Voltage/amperage curves for a multirated power source. Scale settings similar to those shown on the graph often are included right on the front control panel of the power source.

Ratings

Some plasma-arc power supplies are rated in kilowatts instead of amperes. Instead of saying that the load condition is 250 A at 200 V, or 100 A at 75 V, for example, the rated voltage and amperage are multiplied together, giving watts. Since 1000 W = 1 kW, the power-source rating is in units of kilowatts. The load for our 250-A, 200-V machine is 50 kW (200 A × 250 V = 50,000 W), and the rating for our 75-V, 100-A machine is 7.5 kW.

Some plasma-arc machines are even calibrated in kilowatts at the rated voltage instead of amperes at the rated voltage. There is no difference between a power source rated in kilowatts and a machine rated in volts and amperes. The only difference for you is in how the indicator scale (or dial) on the power supply is labeled.

15-8
PLASMA-WELDING TORCHES

The nozzle is the most important part of the plasma torch. It narrows the arc because the nozzle bore is smaller than the arc diameter. The smaller the nozzle bore, the higher the temperature of the plasma arc. The longer and narrower the nozzle constriction, the more uniform the path of the plasma (the stiffer it is, in other words), the higher the temperature, and the more concentrated the electrical energy that the plasma delivers.

Plasma torches either have *nontransferred* arcs (the arc is between the electrode and the nozzle) or *transferred* arcs (the arc is passed between the electrode and the work). Figure 15-10 shows how they differ. (If you're interested in the kinds of electrodes used in the plasma torch, look back at Sec. 14-7—plasma torches use the same tungsten or tungsten-alloy electrodes as TIG torches for exactly the same reasons.)

The nontransferred arc projects

a cone-shaped jet of plasma which is cooler because the heat comes only from the plasma. It's also less concentrated because it is cone-shaped. Nontransferred plasma torches are very useful for metal spraying, where filler wire or powdered metal is fed through the arc where it melts into tiny droplets and sprays directly onto a workpiece, like paint from a spray gun (see Chap. 19).

These sprayed-metal coatings are very useful in maintenance work because high-Rockwell-hardness alloys can be applied directly onto parts after the surfaces are cleaned. Similarly, corrosion-resistant coatings such as zinc and aluminum, as well as different grades of stainless steel, can be sprayed directly onto a workpiece. However, torches with nontransferred arcs are not good for plasma welding.

Transferred arcs are used for welding. The reason is that the heat to the work comes from the arc and the plasma working together. The transferred plasma arc is cylindrical, as we saw back in Fig. 15-2. It usually is produced by DCSP. Very high energy reverse-polarity electrons from a negative workpiece jumping up against the electrode would eat away the tungsten electrode.

15-9
PLASMA-WELDING TECHNIQUES

The welding techniques used with the manual plasma process are similar to but not as difficult as those used with GTAW. Because of the longer standoff distance (arc length), there is no danger of contaminating your workpiece by touching the tungsten electrode to it, or contaminating the tungsten with the weld metal. You also will have more room to get your filler rod into the arc and weld puddle.

The Melt-in Technique

Most manual plasma welding is done in the puddle welding mode (often called the *melt-in* technique), which simply means that you stick the filler rod into the molten weld puddle or next to the plasma arc, just like you would with a TIG torch. It's very similar to the way you would handle your rod with an oxyfuel flame.

The Keyhole Technique

The other plasma-welding mode you can use is the *keyhole* technique. Keyhole welding is used on thinner sections such as sheet and strip, usually without using a filler rod. However, material up to

¼-in. [6-mm] thick can be welded with difficulty using higher current settings. One example of a good use for keyhole welding is making corner, edge, and lap welds to join two pieces of sheet metal.

To start a keyhole weld, hold the torch in a nearly vertical position. Increase the current with your foot pedal. Don't start moving your torch until you have burned a small keyhole through the workpiece. Once you have established a keyhole in the work (Fig. 15-11), you can start your torch motion to make a continuous fusion weld.

The weld metal will flow around the keyhole which is continuously being created by your plasma arc as you slowly advance your torch. If you don't travel too fast, the surface tension of the molten puddle will cause your weld metal to flow around the keyhole and close up behind it, filling and completing your fusion weld.

The most critical part of the manual keyhole weld is initially piercing the workpiece. During this time, be very careful to hold your torch perpendicular to the workpiece. The plasma arc must point directly at the base metal. If you angle the torch, the backwash from the plasma will cause your torch nozzle to crack due to thermal shock. Also make sure that the plasma-creating gas flow is at a higher level than when welding in the puddle mode. You will want to create even more heat than in conventional plasma puddle welding when you start keyholing.

When you come to the end of the weld, close the keyhole off by simultaneously lowering the welding current with your foot pedal to reduce the heat at the workpiece, and at the same time reduce the plasma-gas flow. Both of these methods, working together, will reduce the heat at the keyhole, allowing the remaining weld puddle to flow into and fill up the keyhole. That's a lot of things to do together, and it takes practice.

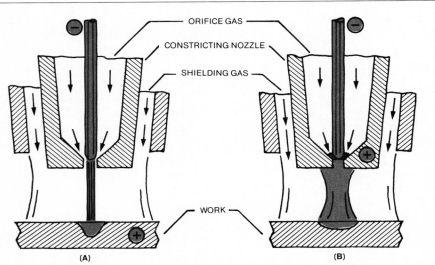

FIGURE 15-10 The difference between transferred and nontransferred plasma-arc processes. (A) A transferred arc will produce deep base-metal penetration. (B) A nontransferred arc will produce shallow base-metal penetration.

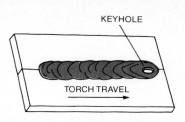

FIGURE 15-11 The keyhole welding mode. Keyhole through the base metal is carried along with the torch as molten metal flows into place behind the torch plasma. This is an important welding technique for plasma-arc welding.

Another reason why keyhole welding is more difficult than puddle welding is that you can burn out your torch nozzle, which can happen either during the keyhole welding or at the closure procedure at the end of the weld seam, if you reduce the plasma-gas flow too much while your welding current stays at the higher keyhole level. You have to be very careful with keyhole welding to make sure that your plasma-gas flow rates are compatible with the plasma-arc current at all times. It takes practice. It's not easy, but it's well worth learning how to make keyhole welds with a plasma-arc torch.

15-10 PLASMA-WELDING SCHEDULES

The American Welding Society publishes an excellent, detailed guide titled *Recommended Practices for Plasma-Arc Welding*. It is AWS publication C5.1. This lesson is only an introduction to the subject. For example, there are special hot-wire plasma welding techniques that combine the plasma process with additional current run through continuously fed electrode filler wires. The hot-wire plasma welding process (very much like hot-wire TIG) is used for high-deposition-rate automated welding.

You will most likely be given predetermined welding schedules for setting up PAW jobs whether they are manual or automatic. These schedules will be based on the metal thickness to be joined and will include data on your current setting, the torch-nozzle (orifice) size suitable for the given material thickness, and the shielding-gas and plasma-gas flow rates. The plasma-forming gas will always be argon. The shielding gas can be argon, argon-helium, or argon-hydrogen mixtures, but not 100 percent helium. A 100 percent helium shielding gas will make it difficult to start and maintain the arc.

A specific shielding gas is selected depending on the current used and the material thickness to be welded. Trial-and-error tests also are often made, but the following data will help you set up your own equipment if welding parameters have not already been worked out for you.

Since the plasma-gas flow rate is most critical for puddle or melt-in welding, we'll give you data on that first (Table 15-1) for several different current settings and argon flow rates. Look at this table. If you were to plot amperage against plasma-gas flow rate on a graph, you would see that high currents and low-plasma-gas flow rates would burn your torch nozzle.

If you choose too high an amperage and too high a plasma-gas flow rate, you would cut or gouge holes in your weld metal . . . it's like an excess keyholing effect, but that is just the settings used for PAC torches. The high plasma temperature melts and even vaporizes some metals, and then the high plasma-gas pressure does the plasma-arc cutting.

If your plasma-arc current is too low relative to the plasma-gas flow, your plasma arc will start to wander around and you will make bad welds, as it's no longer a stiff arc. If the current is very low, you will not only be producing heat too low to make good welds, you also will have trouble starting the arc.

In Table 15-1 you'll also find separate areas for five different filler-rod diameters, from 0.036-in.

TABLE 15-1 Conditions for plasma-welding Type 304 stainless steel with the melt-in (puddle) technique*

Plate Thickness, in.	Downhand Joint Type	Orifice Diameter, In.	Plasma-Gas Flow Rate, scf/h	Current, A	Voltage, V	Filler Wire Size, in.	Weld Speed, in./min
0.058	Lap	0.036	0.75	15	27	0.035	2.75
0.058	Butt	0.036	1.0	8	30	0.035	3
0.058	Fillet	0.036	0.75	15	28	0.035	3.25
0.093	Lap	0.036	0.75	25	26	0.045	3
0.093	Butt	0.036	0.75	25	26	0.045	5
0.093	Fillet	0.036	0.75	25	26	0.045	2.5
0.093	Lap	0.067	1.25	50	25	0.045	6.5
0.093	Butt	0.067	1.25	45	25	0.045	6.75
0.093	Fillet	0.067	1.25	50	25	0.045	4.25
0.118	Lap	0.067	1.25	70	27	1/16	7
0.118	Butt	0.067	1.50	70	28	1/16	8
0.118	Fillet	0.067	1.25	65	27	1/16	6
0.118	Lap	0.089	1.75	75	29	3/32	6
0.118	Butt	0.089	2.0	70	31	3/32	7.25
0.118	Fillet	0.089	1.50	75	27	3/32	5
3/16	Lap	0.089	1.25	100	32	3/32	4.25
3/16	Butt	0.089	2.0	100	34	3/32	3.75
3/16	Fillet	0.089	1.5	100	31	3/32	3.5

* Shielding Gas—argon plus 5 percent H_2 at 15 scf/h, filler wire—Type 304 stainless steel, torch rating—100 A.

rod at the top of the list to $^3/_{32}$-in. diameter filler rod at the bottom of the list. It's not surprising that the operating ranges in the tables for some of these different filler rods tend to overlap. The table only plots typical data for you to use. You still have to decide what machine settings and plasma-gas flow rates to use for an actual job.

Table 15-1 also gives you welding parameters for joining type 304 stainless steel with a 100-A-rated torch, argon plasma gas, and an argon plus 5 percent hydrogen shielding gas. You can use Type 304 stainless-steel filler wire for Type 304 stainless-steel base metal. These data are for the melt-in (nonkeyholing) welding technique. The joints are horizontal lap and flat-position butt and fillet joints. Plate thickness range from 0.058 in. to $^3/_{16}$ in.

Table 15-2 shows you a typical set of PAW conditions for manual keyhole welding with a 100-A torch on type 304 stainless steel. Note that the keyhole technique is *not* recommended for all plate thicknesses and all torch orifice sizes.

Table 15-3 gives data for welding Type 304 stainless steel using puddle welding, and a 100-A-rated torch. Table 15-4 gives the same data for a 200-A-rated plasma torch.

Table 15-5 also gives you typical welding parameters using plasma welding to join Type 304 stainless steel, Type 410 stainless steel, Inconel 62 (a high-nickel alloy), Waspalloy, and ASTM A 286 (both of which are high-strength, high-temperature iron-based alloys used in things like jet engines), commercially pure titanium, and a 70 percent copper–30 percent zinc brass.

The data in Table 15-6 are based on welding stainless-steel plate at the maximum rated current for a 100-A plasma torch and three nozzle diameters. Argon is the plasma-forming gas. Argon plus helium or argon plus hydrogen are the shielding gases. These flow rates are set at the highest shielding-gas flow rate that will not disturb the weld puddle.

The data show that higher flow rates are used with lower-density gas mixtures, particularly with 75 percent helium plus 25 percent argon mixtures. A 200-A torch would function similarly to the data in Table 15-6, with proportional increases in the flow rates, depending on the nozzle orifice size used.

TABLE 15-2 Typical conditions for plasma-welding Type 304 stainless steel with the keyhole technique

Plate Thickness, in.	High Plasma Gas Flow Rate, scf/h	Orifice Size, In.	Current, A	Voltage, V	Travel Speed, in./min
All	—	0.036	Not recommended		—
0.093	5	0.067	50	30	6.5
0.118	5	0.067	56	31.5	5.25
$^3/_{16}$	5	0.067	75	36	5.0
$^1/_4$	—	0.067	Not recommended		—
0.118	—	0.089	Not recommended		—
$^3/_{16}$	8	0.089	80	37	5.0
$^1/_4$	—	0.089	Not recommended		—

TABLE 15-3 Guide for selecting current and plasma-gas flow, 100-A torch, for Type 304 stainless steel

Nozzle Diameter, in.	Current Range, A	Plasma-Gas Flow Rate, scf/h	Ar + 5% H$_2$ Shielding Gas,* scf/h
0.036	1–25	0.05–1.5	15–20
0.067	2–75	0.5 –3.0	15–25
0.089	5–100	0.5 –4.0	15–30

* When welding ultrathin material, heat input can be lowered by reducing or eliminating the hydrogen in the shielding gas.

TABLE 15-4 Guide for selecting current and plasma-gas flow, 200-A torch, for type 304 stainless steel

Nozzle Diameter, in.	Current Range, A	Plasma-Gas Flow Rate, scf/h	Ar + 5% H$_2$ Shielding Gas,* scf/h
0.036	1–25	0.5–1.6	20–40
0.067	2–75	0.5–3.0	20–40
0.089	5–100	0.5–4.0	20–40
0.118	60–180	1–5	20–40
0.136	90–200	1–6	20–40

* When welding ultrathin material, heat input can be lowered by reducing or eliminating the hydrogen in the shielding gas.

15-11
PLASMA-ARC CUTTING
PAC can be used to cut any metal. Most applications are for carbon steel, aluminum, and stainless steels simply because they are the most common metals

TABLE 15-5 Plasma welding data for various alloys with the melt-in (puddle) technique

Material and Plate Thickness, in.	Joint Type	Orifice Diameter, in.	Argon Plasma-Gas Flow Rate, scf/h	Shielding Gas* Flow Rate, scf/h	Current, A	Maximum Nozzle-to-Work Distance, in.	Filler-Wire type and size, in.	Weld Speed, in./min	Welding Technique	Backup-Gas Type and Flow Rate, scf/h
304 SS 1/8	Vertical fillet	0.067	1.0	20	65	1/4	308 SS 3/32	—	Manual (Puddle)	—
304 SS 1/16	Vertical fillet	0.067	1.0	20	25	1/4	308 SS 1/16	—	Manual (Puddle)	—
304 SS 3/32	Vertical fillet	0.067	1.0	20	45	1/4	308 SS 1/16	—	Manual (Puddle)	—
304 SS 1/8	Bead on flat plate	0.067	1.0	20	65	1/4	—	—	Manual (Keyhole)	Argon 10
304 SS 1/8	Bead on flat plate	0.067	4.0	20	60	1/4	—	—	Machine	Argon 5
304 SS 0.022	Flat butt	0.036	1.0	30	15	1/4	—	5	Machine	Argon 5
304 SS 1/16	Flat butt	0.036	1.0	30	25	1/4	—	6.5	Machine	argon 5
304 SS 1/16	Flat butt	0.089	1.0	30	75	1/4	—	18	Machine	Argon 5
304 SS 1/16	Flat butt	0.136	1.0	30	125	1/4	—	30	Machine	Argon 5
304 SS 1/8	Flat butt	0.067	1.5	30	80	1/4	—	10	Machine	Argon 10
304 SS 1/8	Flat butt	0.136	5.0	30	180	1/4	—	18	Machine	Argon 10
304 SS 3/32	Flat butt	0.036	1.5	30	25	1/4	—	3	Machine	Argon 10
410 SS 1/8	Flat butt	0.067	2.5	25	55	3/8	410 SS 1/16	—	Manual (Keyhole)	Argon 25
410 SS 1/8	Flat butt	0.067	4.5	25	90	3/8	—	—	Manual (Keyhole)	Argon 25
Inconel 62 3/32	Flat butt	0.067	2.5	25	62	3/8	Inconel 62 1/16	—	Manual (Keyhole)	Argon 25
Waspalloy and A-286 0.045	Flat butt	0.036	0.75	25	45	3/8	—	—	Manual (Puddle)	Argon 10
Waspalloy and A-286 3/32	Flat butt	0.067	2.5	25	70	3/8	—	—	Manual (Keyhole)	Argon 25
70-30 brass 0.020	Flange butt	0.089	4.0	30	140	1/4	—	100	Machine	Argon 10
Titanium 50 A 0.025	Flat butt	0.067	5.0	35	50	3/16	Ti 50A 1/16	—	Manual (Keyhole)	Argon 25

* Shielding gas—Argon plus 5 percent H_2 at 15 scf/h, torch rating—100 A.

used in a welding shop. But the PAC process works equally well on any conductive material: copper, brass, and bronze; nickel and its alloys; zirconium metal—it's even been used to cut uranium.

Why PAC Is Used
PAC can be used for stack cutting, plate beveling, shape cut-

TABLE 15-6 Plasma-gas (argon) and shielding gas flow rates (scf/h) for various nozzle sizes

Nozzle Size, in.	Argon	He 75% + Argon 25%	Ar 75% + He 25%	Ar 92.5% + H_2 7.5%
0.036	0.75–1.0	5	1	1
0.067	1.25–1.5	7	2	1.5
0.089	1.5 –2.0	10	2	—

ting, and piercing. In fact, you'll find that it will do these jobs with less heat input into the base metal than an oxyfuel flame (even though the plasma is much hotter), partly because the plasma torch operates much faster than an oxyfuel flame and partly because it does not have to "burn" or oxidize its way through metal; instead it melts and sometimes even vaporizes metals out of the kerf. As a result there are fewer problems with base-metal distortion.

PAC torches are used most often on shape cutting and on high-speed ripping and squaring machines. Very little manual work is done because of the high currents and OCVs involved. The noise level of the high-speed plasma-gas jet also is very high, and the process can produce a lot of metal fumes.

The noise and fumes are very difficult to control with a manual torch. They are no problem for an automatic torch mounted on a suitable flame-cutting machine. The fumes, heat, and noise produced by the plasma torch are easily handled on a large cutting machine simply by cutting plate on a water-filled cutting table. The water just touches the bottom of the plate where it traps the fumes and slag (or dross as it is called when cutting nonferrous metals) as the plasma emerges from the bottom of the kerf. The water also muffles the sound that the high-speed plasma jet makes leaving the torch head. If necessary, you can wear "industrial-strength" earplugs or earmuffs. Your safety director can get a set for you.

Cutting Speeds
Using a suitable cutting machine (one that can take the high speeds of the plasma process without losing cut accuracy and tolerances) you can cut metals at speeds of 100 to 150 in./min [2.5 to 3.8 m/min] that would require speeds of only 20 to 25 in./min [0.5 to 0.63 m/min] using an oxyfuel torch.

FIGURE 15-12 Plasma-arc cutting on a highly automated, computer-controlled machine set up to cut openings in large-diameter aluminum cylinders used in electrical power plants.

Speeds up to 300 in./min [7 m/min] are used in cutting some thin materials. No manual operator can possibly keep up with the speed of an efficiently operated plasma-arc cutting torch.

Carbon-steel plate can be cut faster with the oxyfuel process than PAC if you are cutting material around 3 in. [75 mm] thick. However, for cutting thicknesses under 1 in. [25 mm], PAC is up to five times faster than the oxyfuel cutting process. The decision to use PAC for carbon steels that can also be cut by oxyfuel is made on the basis of the higher productivity of the PAC versus the much higher initial cost of the equipment.

The trade-off of higher speed versus higher equipment cost is why you most often will find PAC equipment used on high-speed flame-cutting machines designed to do a large amount of shape cutting. The equipment's speed and productivity help the fabricator pay off the much higher initial investment. It's also very common, when PAC is used, to find the equipment mounted on a cutting machine along with oxyfuel torches, which allows a high-volume parts fabricator to switch from oxyfuel to plasma and back again, depending on whether ferrous or nonferrous sheet and plate, and thick or thin material is being cut.

The economic advantages of PAC will show up most often where long, continuous cuts are made on a large number of workpieces. This type of high-volume cutting is most frequently seen in shipyards, tank fabrication plants, bridge-steel construction shops, and steel service centers. Figure 15-12 shows plasma-arc cutting in a computer-controlled, automated machine that produces complex, large-diameter cylinders made of aluminum with holes cut in them to insert electronic equipment.

15-12 PLASMA-ARC EQUIPMENT
Stripped to its bare essentials, PAC requires a cutting torch that operates very much like a plasma welding torch, a suitable power supply, and a supply of clean cooling water.

The Plasma Torch
The PAC torch consists of an electrode holder which centers the electrode tip in the orifice of the constricting plasma nozzle. The electrode and nozzle are water-cooled. Plasma gas is injected into the torch around the electrode and exits through the nozzle orifice. Nozzles with various orifice diameters are available for each model of torch used. The orifice diameter depends on the cutting current; larger diameters are required at higher currents. Nozzle design depends upon the type of PAC torch used and the metal being cut.

Both single- and multiple-port nozzles can be used for PAC. Multiple-port nozzles have the auxiliary shielding-gas ports in a circle around the main plasma-gas orifice. All of the arc plasma passes through the main orifice with a high plasma-gas flow rate per unit area. The plasma-gas speed is so high that it's usually supersonic (faster than the speed of sound), which explains why the plasma process makes so much noise. It's comparable to a tiny jet engine. These multiple-port nozzles produce better-quality cuts than single-port nozzles at equivalent travel speeds. However, cut quality (no matter what type of torch is used) decreases with increasing travel speed, just as with the oxyfuel process.

Plasma Cutting Controls
Control consoles for PAC contain special solenoid valves to turn gases and cooling water on and off. They usually have flowmeters for the various types of cutting and shielding gases used and a water-flow switch to stop the entire operation if the flow of cooling water falls below a safe limit. Controls for high-power automatic PAC also will include programming features for adjusting the upslope and downslope of the current and orifice gas flow.

Plasma Cutting Power Sources
Power sources for PAC are specially designed units with OCVs in the range of 120 to 400 V (compared with 70 to 85 V maximum for arc-welding power sources). A power source is selected on the basis of the design of the PAC torch to be used, the type and thickness of the work to be cut, and the cutting speed range. Constant-current, drooping-voltage-characteristic dc machines are used.

The plasma-cutting process operates on DCSP, electrode-negative current with a constricted, transferred arc. Heavy cutting requires high OCVs of 400 V for piercing material as thick as 2 in. [51 mm]. Low-current, manual plasma-cutting equipment uses a lower OCV of 120 to 200 V. Some power sources used for gouging have connections to change the OCV as required for specific jobs. Many welding equipment manu-

facturers would rather not supply the manual equipment because of the safety problems involved in handling OCVs that are about twice as high as those used in arc welding.

The output current from a plasma-arc power source can range from about 70 to 1000 A, depending upon the material to be cut, its thickness, and the cutting speed. These power sources also will have circuits for the pilot arc and HF power source arc starting.

15-13 PLASMA-ARC OPERATIONS
In the transferred-arc mode, an arc is struck between the electrode in the torch and the workpiece. The arc is initiated by a pilot arc between the electrode and constricting nozzle. The nozzle is connected to ground (positive) through a current-limiting resistor and a pilot-arc relay. The pilot arc is initiated by an HF generator connected to the electrode and the nozzle. The power source is designed to maintain this low-current arc inside the torch. Ionized gas from the pilot arc is blown through the constricting nozzle orifice, forming a low-resistance path to ignite the main arc between the electrode and the workpiece. When the main arc ignites, the pilot-arc relay may be opened automatically

to avoid unnecessarily heating the constricting nozzle.

Because the plasma-constricting nozzle is exposed to high temperatures (estimated to be between 18,000 and 25,000°F [9980 and 13,870°C]), the nozzle must be made of water-cooled copper. In addition, the torch should be designed to produce a boundary layer of gas between the plasma and the nozzle—otherwise the ultra-high-temperature plasma will melt or vaporize the nozzle walls as well as the workpiece kerf.

Plasma Process Modifications
Several PAC process variations are used to improve the quality of cuts. They are generally applicable to materials in the range of $\frac{1}{8}$ to $1\frac{1}{2}$ in. thick [3 to 38 mm]. Auxiliary shielding, in the form of gas or water, also is used to improve cut quality.

Dual-flow plasma cutting
Dual-flow plasma cutting provides a secondary gas blanket around the arc plasma, as shown in Fig. 15-13. The usual orifice gas is nitrogen. The shielding gas is selected for the material to be cut. For mild steel, it may be carbon dioxide or air and cutting speeds will be slightly higher than those for conventional PAC, but the cut quality is not satisfactory for many applications. Carbon dioxide is often used for shielding stainless steels. Argon-hydrogen

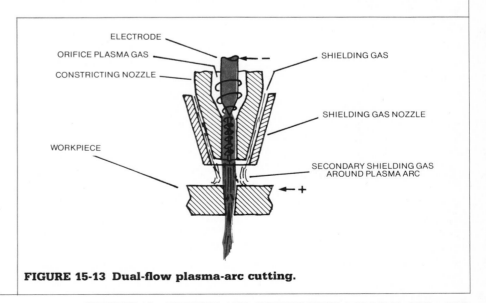

FIGURE 15-13 Dual-flow plasma-arc cutting.

shielding-gas mixtures are used for aluminum.

Water-shielded plasma cutting

This technique is similar to dual-flow plasma cutting. Water is used in place of the auxiliary shielding gas. Cut appearance and nozzle life are improved by the use of water for auxiliary shielding. Cut squareness and cutting speed are not significantly improved over conventional PAC.

Water-injection plasma cutting

This modification of PAC uses a symmetrical impinging water jet near the constricting nozzle orifice to further constrict the plasma "flame." The arrangement is shown in Fig. 15-14. The water jet also shields the plasma from turbulent mixing with the surrounding atmosphere. The end of the nozzle can be made of ceramic, which helps to prevent "double arcing." Double arcing results when the arc jumps from the electrode to the nozzle and then to the work, usually destroying the nozzle.

The water-constricted plasma produces a narrow, sharply defined cut at speeds above those of conventional PAC. Because most of the water leaves the nozzle as a liquid spray it cools the kerf edge, producing a sharp corner. When the orifice gas and water are injected tangentially, the plasma gas swirls as it emerges from the nozzle and water jet. This produces a high-quality perpendicular face on one side of the kerf. The other side of the kerf is beveled. In shape-cutting applications, the direction of torch travel must be selected to produce a perpendicular cut on the part, and the bevel cut on the scrap.

Plasma-Forming Gases

Plasma-forming gases are selected based on the material being cut and the cut surface quality desired. Most nonferrous metals are cut by using nitrogen or nitrogen-hydrogen or argon-hydrogen mixtures. Titanium and zirconium are cut with pure-argon plasma because these materials are extremely susceptible to embrittlement by reactive gases, especially hydrogen.

Carbon steels are cut by using compressed air (80 percent nitrogen and 20 percent oxygen) or pure nitrogen. Nitrogen is used with the water-injection method. Some systems use nitrogen for the plasma-forming gas with oxygen injected into the plasma downstream of the electrode. This arrangement prolongs the life of the electrode by not exposing it to oxygen. For some nonferrous cutting with the dual-flow system, nitrogen is used for the plasma gas with carbon dioxide used for the shielding gas. For better-quality cuts, argon-hydrogen plasma gas with nitrogen shielding is used.

15-14 SAFETY

PAC operators and people standing in the vicinity of the operation must be protected from arc glare, spatter, and fumes. Safety procedures are very similar to those for arc welding. Filter plates in protective helmets or goggles should be selected from the darkest shades available. See Chap. 12 on arc-welding safety for further details on lens shade selection.

Two of the most common safety accessories for PAC are the water table we already mentioned (it is simply a conventional cutting table filled with water up to the bottom surface of the plates being cut), and the water muffle. When the water table is used, high-speed gases emerging from the plasma jet produce turbulence in the water. Almost all of the fumes and particles produced during cutting are trapped in the water. The water acts like air-pollution-control devices called *scrubbers* that remove particles from smokestack gases.

The water muffler is a device that reduces noise. It is a nozzle attached to the torch body that produces a curtain of water below the nozzle of the torch. It is always used in connection with a water table. Water from the table is pumped through the nozzle. The water curtain at the top of the plate and the water contacting the plate bottom enclose the arc in a sound-deadening shield. Don't confuse the water muffler with water-injection or water-shielding PAC process variations. Detailed operator safety instructions are given in AWS publication A6.3, "Recommended Safe Practices for Plasma Arc Cutting." Most of the details in this publication follow those outlined in Chap. 12 on arc-welding safety. Some additional details also are usually included with any of the operating literature that comes with a plasma-arc power source.

15-15 OPERATING PROCEDURES

After selecting the proper process conditions for the metal you will cut, about the only thing you have to do in starting up an automatic shape-cutting machine using plasma arc is to push a button. These operating conditions usually are selected from data tables. We will give you examples of some of these data shortly. After the button has been pushed, the cutting control takes over and regulates the sequence of events. About the only thing you have to

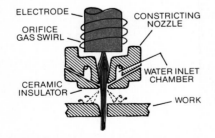

FIGURE 15-14 Water-injection plasma-arc cutting. Water injection into the cutting plasma produces smoother kerfs on cut surfaces.

worry about is the direction of cutting-head travel.

As we've already mentioned, most PAC torches swirl the orifice gas by injecting it through tangential holes or slots in the torch head. One feature of the gas swirl is a more efficient transfer of arc energy to one side of the kerf. With a clockwise swirl, for example, the right side of the cut (facing in the direction of travel) will be reasonably square, but the left side of the cut will be beveled. Therefore, the travel direction must be selected to place the scrap metal on the left, as illustrated in Fig. 15-15. Reverse-swirl components can be used if a square left side is desired—for example, when cutting opposite edges with two torches moving in the same direction.

For manual operation, you, as the operator, will select the orifice-gas-flow rate and current according to a list of recommended procedures. You will position the torch at the cut-start location on the workpiece, and then initiate the arc by pressing the torch contactor switch. Then you manually guide the torch across the workpiece at the desired cutting speed. The power and gas are automatically turned off when you release your torch trigger. The control

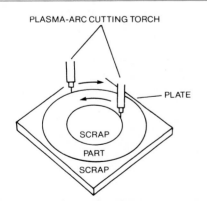

FIGURE 15-15 Relationship of torch travel direction to the part with clockwise swirl of the orifice gas. Get it wrong and you will either make the workpiece too big, or too small.

may reinitiate the pilot arc to allow immediate restarting for repetitive operation.

Note: Don't let the torch touch the work.

You must not allow the constricting nozzle of the torch to contact the work. That will cause nozzle damage if the arc goes from the electrode to the nozzle body and then to the work (double arcing), rather than from the electrode to the work through the nozzle orifice. Cutting-torch designs usually minimize double-arcing problems by insulating or recessing the nozzle.

Cut Quality

The factors that determine plasma-arc cut quality are surface smoothness, kerf width, parallelism of the cut faces, dross or slag formation on the bottom edge of the cut faces, squareness of the cut, and sharpness of the top edge. These factors are controlled by the type of material being cut, the equipment design and setup, and the operating variables.

Generally, high-quality cuts are obtained with moderate power and low cutting speeds. However, low-speed cutting may offset the otherwise economical features of PAC. Therefore, what constitutes your required cut quality (and therefore your cutting speed) should be defined before applying the process.

Plasma cuts in plates up to about 3 in. [76 mm] thick may have a surface smoothness very similar to that produced by oxyfuel gas cutting. Surface oxidation is almost nonexistent with modern automated equipment that uses water injection or water shielding. On thicker plates, slow travel speeds produce a rougher surface finish and discoloration. On very thick stainless steel (around 7 in. [178 mm] thick), the plasma-arc process has little advantage over oxyfuel gas cutting in terms of speed, although PAC is a lot cleaner.

Note: Kerf widths for PAC are 1½ to 2 times wider than the kerf widths for oxyfuel flame-cutting.

Kerf widths of plasma-arc cuts are 1½ to 2 times the width of oxyfuel gas cuts in plates up to 2 in. [51 mm] thick. For example, a typical kerf width in 1-in.-thick [25-mm] stainless steel is approximately ³⁄₁₆ in. [4.8 mm] compared with a ⅛-in. [3.2-mm] kerf for oxyfuel cutting. Kerf width increases with plate thickness using plasma cutting, just as with other cutting processes. A plasma cut in 7-in.-thick [178-mm] stainless steel made at about 4 in./min [102 mm/min] has a kerf width of 1⅛ in. [28.6 mm].

The plasma jet tends to remove more metal from the upper part of the kerf than from the lower part. This results in beveled cuts that are wider at the top than at the bottom. Because of this, a typical included angle of a cut in 1-in.-thick [25-mm] steel is 4 to 6° from a completely parallel cut edge. This bevel occurs on one side of the cut when orifice-gas swirl is used. The bevel angle on both sides of the cut tends to increase with cutting speed.

Dross (or slag) is the material that melts during cutting and adheres to the bottom edge of the cut face. With present automated equipment, slag-free or dross-free cuts can be produced on aluminum and stainless steel up to approximately 3 in. [76 mm] thick. With carbon steel, selection of speed and current are more critical, with slag tending to form on this material as cutting speed is increased.

Edge-rounding will result when excessive power is used to cut a given plate thickness, or when the torch standoff distance is too large. It may also occur in high-speed cutting of materials less than ¼ in. [6 mm] thick.

Typical operating recommendations are given for PAC of aluminum, stainless steels, and carbon steels in Tables 15-7, 15-8, and 15-9. More specific machine-operating conditions (which will probably vary somewhat from these tables) will be given to you when you operate shop equipment.

We won't list shop work in plasma-arc welding and cutting here. It's not too likely that your school will have the equipment. If they do, all you have to do is repeat the shop work in the last chapter on GTAW for experience with plasma-arc welding, or select some obvious nonferrous and ferrous cutting jobs, including edge beveling, if you want to do some work with plasma-arc cutting. Another reason for not recommending shop work with either process is the high voltages and currents used by the equipment. We would rather leave it up to your instructor to decide whether you know enough at this point to handle such electrically hot equipment.

TABLE 15-7 Plasma-arc cutting of aluminum*

Thickness, in.	Speed, in./min	Orifice Diameter, in.	Current DCSP, A	Power, kW
¼	300	⅛	300	60
½	200	⅛	350	70
1	90	5/32	400	80
2	20	5/32	400	80
3	15	3/16	450	90
4	12	3/16	450	90
6	8	¼	750	170

* Plasma-gas flow rates vary with orifice diameter from about 100 ft³/h for a ⅛-in.-diameter orifice to about 250 ft³/h for a ¼-in.-diameter orifice. The gases used are nitrogen and argon in this table, with hydrogen additions from 0 to 35 percent. The equipment manufacturer or a welding engineer should be consulted for correct setups for each application.

TABLE 15-8 Plasma-arc cutting of stainless steels*

Thickness, in.	Speed, in./min	Orifice Diameter, in.	Current DCSP, A	Power, kW
¼	200	⅛	300	60
½	100	⅛	300	60
1	50	5/32	400	80
2	20	3/16	500	100
3	16	3/16	500	100
4	8	3/16	500	100

* Plasma-gas flow rates vary with orifice diameter and the gas used from about 100 ft³/h for a ⅛-in.-diameter orifice to about 200 ft³/h for a 3/16-in.-diameter orifice. The gases used are nitrogen and argon with hydrogen additions from 0 to 35 percent. The equipment manufacturer or a welding engineer should be consulted for each setup.

TABLE 15-9 Plasma-arc cutting of carbon steels*

Thickness, in.	Speed, in./min	Orifice Diameter, in.	Current DCSP, A	Power, kW
¼	200	⅛	275	55
½	100	⅛	275	55
1	50	5/32	425	85
2	25	3/16	550	110

* Plasma-gas flow rates vary with orifice diameter from about 200 ft³/h for a ⅛-in.-diameter orifice to about 300 ft³/h for a 3/16-in.-diameter dual-flow orifice. The gases used are usually compressed air, nitrogen with up to 10 percent hydrogen additions, or nitrogen with oxygen added downstream from the electrode (dual flow). The equipment manufacturer or a welding engineer should be consulted for specific setups.

REVIEW QUESTIONS

1. The GTAW arc plasma is an open arc. How does this differ from the plasma arc used in PAW?

2. Is the plasma arc length longer or shorter than the GTAW welding arc?

3. Are the open-circuit voltages of plasma-arc welding and plasma-arc cutting generally higher or lower than gas-tungsten arc welding?

4. What is a plasma?

5. Explain the difference between a nontransferred plasma arc and a transferred plasma arc.

6. What part of the plasma torch is used to constrict the plasma arc?

7. Is the stand-off distance between the electrode and the molten metal that is used in most plasma-arc welding a little greater than, or a little less than that used in gas-tungsten arc welding?

8. What, in your opinion, is the one big drawback in using plasma-arc welding instead of gas-tungsten-arc welding?

9. What plasma-forming gas is most commonly used for PAW?

10. What welding shielding gases are most commonly used with the PAW process?

11. Most plasma-arc welding power sources are (select one):
AC machines
DC constant-current machines (droopers)

12. The polarity most frequently used with dc plasma-arc welding is (select one):
DC straight polarity (DCSP)
DC reverse polarity (DCRP)

13. Name the two most common plasma-arc welding techniques. Explain how they differ.

14. Name some metals plasma-arc cutting can cut that oxyfuel welding can't cut.

15. Which cuts faster, oxyfuel or plasma cutting?

16. What polarity is used for dc plasma-arc cutting? Is it DCSP or DCRP?

17. Why is water-injection plasma-arc cutting used?

18. What plasma-forming gases or gas mixtures are most often used for cutting nonferrous metals with the plasma-arc process?

19. What plasma-forming gas or gases are used for cutting carbon steels?

20. Are kerf widths for plasma cutting wider or narrower than those produced by oxyfuel flame cutting?

21. Approximately how fast can plasma-arc cutting cut on ¼-in.-thick aluminum? What about ¼-in.-thick stainless steel?

22. At what speed would you expect plasma-arc cutting to cut 1-in.-thick carbon steel plate?

CHAPTER 16

Welding with Stick Electrodes

16-1
WHAT IS SMAW?

About half of all welding filler metals used are stick electrodes (Fig. 16-1). Stick-electrode welding is the single most frequently used welding process. Stick-electrode welding is formally called *shielded-metal arc welding* by the American Welding Society, and the abbreviation you will often see in use is SMAW. The amount of stick-electrode welding done drops a little each year relative to other processes as the cost of welding labor goes up and manufacturers move to one or more of the semiautomatic or automatic processes to increase productivity. Nevertheless, SMAW still holds a large share of the total welding filler-metal business.

Had this book been written 10 years ago, we would have started the arc-welding section with the use of stick electrodes. For example, SMAW electrodes accounted for 61 percent of all the filler metals shipped in 1975. Only 5 years later they had dropped to 52 percent of shipments while GMAW and submerged-arc welding (SAW) filler metals took their place.

As the future welders of America (which is what you are), you have even chances of working with GMAW (MIG, for example), FCAW (flux-cored wire), GTAW (TIG), or even some other process such as submerged-arc welding (SAW) instead of using stick electrodes. Nevertheless, any arc welder needs to know how to handle stick electrodes because SMAW will be around for a long time to come.

Advantages of SMAW

Covered electrodes are used in some of the least sophisticated and some of the most sophisticated welding jobs done in industry. Using the process, job shops can handle most welding problems with a relatively small variety of electrodes in their inventory. A farmer can repair a tractor or a broken truck frame in the field; an iron worker can move around the high steel without lugging very much equipment; and the valves and piping on nuclear reactors will probably be welded with a flux-covered SMAW electrode (or TIG, or both together), as will pipelines.

So the first advantage of flux-covered stick electrodes (Fig. 16-2) is exceptional versatility. Unlike some of the other welding processes, there is a stick electrode available somewhere, either ferrous or nonferrous, that is designed to weld just about any common metal except low-melting-point metals such as lead and zinc.

SMAW equipment tends to be less expensive than that used for the semiautomatic wire-fed processes such as GMAW and SAW. For a little more than a hundred dollars you can buy a 100-A ac "buzz box," a welding helmet, gloves, electrode holder and cable, ground clamp and cable, and a few boxes of stick electrodes, and go into the light-duty repair welding business; or for a few thousand dollars you can buy a 600-A ac-dc power source for high-production SMAW and get a variety of attachments.

Power sources for SMAW include both ac and dc machines. They can be stationary (plug-in-the-wall) units, or engine drives that you can use anywhere as long as you have fuel for the engine. The machines range from motor-generators and alternators to high-performance ac transformers and dc rectifiers.

Welds can be made in any position with stick electrodes, from flat and horizontal to vertical and overhead (although soon you'll find that certain types of electrodes are better suited for some welding positions than others). There are heavy-coated "drag rod" electrodes with lots of iron powder in the flux coating that almost anybody can learn to use. Drag rods are used for making flat and hori-

FIGURE 16-1 While MIG, TIG, and submerged-arc welding have taken a large share of the applications that used to belong solely to stick electrodes, SMAW welding remains a major method not only for maintenance welding, but also for critical work such as nuclear pressure vessel welding (shown here being used to attach opening connections to a nuclear power plant reactor vessel).

FIGURE 16-2 This welder is joining two pieces of stainless-steel plate, welding in the vertical-up direction. There are hundreds of types of covered electrodes for welding different base metals. After this welder finishes the joint, she will back-chip the dark slag off the weld to reveal the bright, shiny weld metal underneath.

zontal tack welds, among other things, because they can touch the workpiece, making it easier for a beginner to use them.

There are SMAW electrodes with low-hydrogen flux coatings that are designed for welding in any position. These electrodes take a lot of skill to learn to use. Arc length (distance from the electrode tip to the work) is critical. Such skills can't be learned quickly, but after your experience with gas welding and GTAW you will have fewer problems than most beginning welders. That's one reason why we organized this book the way we did.

The SMAW process is technically suitable for joining metal from fairly thin sheet up to any plate or structural thickness you want to weld with it. In practice, most SMAW work is done on sections between ⅛ and ¾ in. [3.2 and 19 mm] thick. Below ⅛ in. you are probably better off with GTAW to help avoid base-metal distortion and burn-through. Above ¾ in. [19 mm] thick, it takes so long to fill up a grooved welding joint with stick electrodes that other processes such as FCAW or SAW often are used.

You'll find that fillet welds are particularly easy to make with SMAW, especially with heavily flux covered iron-powder electrodes. For that reason alone, a lot of people get their first jobs as tack welders using small-diameter SMAW electrodes to set up work for skilled arc welders to finish.

The typical current range used for SMAW is between 50 and 500 A, although some special electrodes are designed to be used with currents of 600 A or more, and others as low as 30 A. Most power sources used for SMAW are relatively simple compared with the machines used for GMAW and other wire-fed processes. They don't have to have 100 percent duty cycles, which is the main reason why power sources used only for SMAW cost less than other welding machines. Maintenance costs also are less because SMAW machines are very durable, relatively simple to operate, and easy to set up.

Disadvantages of SMAW

The weld-metal deposition rates for the SMAW process range between 1.5 and 17 lb/h [0.7 to 7.7 kg/h] in the flat position. However, a stick-electrode welder normally is not able to deposit more than 3.6 lb/h [1.6 kg/h] in all-position welding because smaller-diameter electrodes and lower welding currents are needed to work in the horizontal, vertical, and overhead positions. Some of the semiautomatic high-production processes (such as FCAW, SAW, and hot wire GTAW) have weld-metal deposition rates more than ten times greater than SMAW.

One of the reasons SMAW rates are so relatively low is that the operator has to stop and change electrodes every few minutes. Another reason is that you also have to clean slag after each pass with SMAW, and every minute your SMAW arc is not on and welding, you are costing somebody labor dollars for essentially nonproductive (but necessary) work. As a result, the hourly labor costs for using covered electrodes are high.

Material costs also are high since less than 60 percent of the weight of the average stick electrode winds up as weld metal. The rest of the electrode goes into slag and gas or remains unused in your holder. The fact that slag is produced when the flux coating of the electrode is melted also means more chances for slag inclusions in your finished weld.

Moisture in the air can enter your electrode flux on a humid day. When you weld, this slightly moist flux can produce everything from starting porosity in your welds at each point where you start a new electrode, to serious problems with hydrogen embrittlement when welding high-strength steels. There are ways of handling each of these problems, but there are other difficulties for

beginners with SMAW. You'll learn why when you strike your first arc and your electrode short-circuits and sticks to your plate.

16-2
SMAW ELECTRODES

SMAW owes most of its versatility to the wide variety of covered electrodes that are available. Your specific welding techniques also will be governed as much by the type of covered electrode you are using as any other welding variable. Therefore, let's start by looking at covered electrodes and find out how they are made, how they work, how they are specified, what they can do for you, and how each of the major types of SMAW electrodes are used.

The covered SMAW electrode is a short length of wire typically about 14 in. [0.36 m] long, although longer electrodes are available. The wire diameter normally ranges from 3/32 to 1/4 in. [2.4 to 6.4 mm]. This is the size of the electrode, even though the flux coating will increase the actual diameter.

The smaller-diameter electrodes are used with lower welding currents for joining sheet and for welding in all positions. Of

course, although smaller-diameter electrodes also produce less heat and cause less base-metal distortion, they also produce fewer pounds-per-hour of weld metal. The larger-diameter electrodes are designed for conducting higher welding currents to give you greater weld-metal deposition rates, especially for flat-position welding.

Anatomy of a SMAW Electrode

An electrode for welding steel or cast iron will have a mild-steel core wire (Fig. 16-3). An aluminum electrode would have an aluminum core wire. An SMAW electrode for welding copper or copper alloys will have a copper core wire. Similarly, an electrode for welding any other metal probably would have a relatively pure (as opposed to an alloy) core wire made out of that metal. There are several reasons why.

A high-alloy core wire would be very expensive to make. Many alloys can't even be made into wire; they aren't ductile enough. In addition, there's no need to buy small quantities of many different kinds of alloy-steel, alloy-aluminum, or bronze wires when the SMAW flux coating can be used to add the alloying elements.

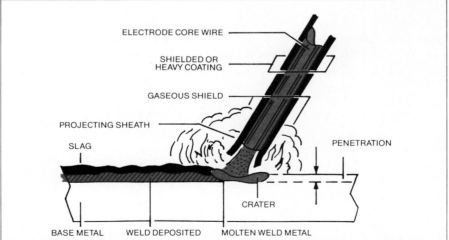

FIGURE 16-3 This shows schematically how a shielded-arc (SMAW) electrode works. Part of the flux coating vaporizes around the arc and molten metal, producing a welding shielding gas. The rest of the flux forms molten slag that floats on top of the weld metal, protecting it while it is hot. The digging action of the welding arc creates the small crater under the electrode.

When the welding arc forms between the electrode and the base metal, part of the flux on the hot working end of the electrode is vaporized, making a protective shielding gas that surrounds the hot weld metal, the heat-affected zone of the base metal next to the weld, and the molten end of the electrode wire. Other elements in the flux join with the molten core wire to make weld metal with the desired final properties.

For example, an austenitic stainless-steel electrode will use a mild-steel core wire, but the chromium and other elements like nickel that make stainless steel stainless, and austenitic, will be added to the weld metal from the flux.

The stick-electrode arc makes a molten puddle of weld metal under the electrode. The stronger the arc-current density and the larger the electrode size, the higher the welding-current setting for a given electrode will be. The type of polarity used (DCSP or DCRP), as well as the type of current (ac or dc), helps determine whether the electrode penetrates deep into the base metal or makes a shallow-penetration weld. So does the speed at which you move the electrode and the distance you hold it from the weld puddle.

SMAW Flux Coating

The molten puddle under the electrode has to be protected from the oxygen and nitrogen in the air. The gases produced by the vaporizing flux coating won't last long enough to protect the weld metal until it is fully solidified and cooled below a point where air won't hurt it. Therefore other elements are put into the flux coating to produce protective slag to keep the hot weld metal covered until it can be exposed to the air.

The molten slag is lighter in weight than the molten weld metal. It floats to the top of the weld puddle and hardens to protect the weld until it is cool. The slag also removes certain unwanted elements and impurities

from the molten weld metal and base metal. When the weld is cool, the slag is removed and a bright, shiny weld will be found underneath.

Still other elements are put into the flux coating to help control the stability of the arc under different conditions. Some flux additives are best for electrodes that operate with dc; other additives are best for electrodes that operate with ac; still other additives make electrodes that can operate on either ac, DCSP, or DCRP. These additives do other things, too. Some are deoxidizers that help remove any excess oxygen in the weld metal created by rust or scale on the steel (the oxide residue floats out of the weld metal and into the slag). If these additives are present, the catalog description of the electrode will tell you that it is "good for welding rusty or heavily scaled steel." There even are flux-coating additives that help keep the molten weld metal from becoming too fluid. These are put into the flux to make the electrode work better for out-of-position welding.

Even more elements can be added to the flux to increase the deposition rate of finished weld metal. These additives are excellent for making high-deposition-rate SMAW electrodes that are only used in the flat position.

Some electrodes even have additives in the flux coating that reduce the amount of moisture that the coating can pick up from humid air. These extra-dry, moisture-resistant electrodes are especially valuable for welding high-strength low-alloy and full-alloy steels that are subject to hydrogen embrittlement. Even a tiny amount of moisture in your electrode flux can pass through the welding arc and become oxygen and hydrogen atoms and ions. The hydrogen will make high-strength steels brittle.

And we're still not finished with fluxes. Binders are needed to hold the additives together so that the flux won't chip off the electrode,

and other chemical additives are needed to make it easy to extrude the flux coating onto the electrode wire.

How SMAW Electrodes Are Made

Electrode manufacturing is one of the most complicated trades in the welding industry. To begin with, the raw material requirements are very complex. There are over 100 different raw materials that can go into the production of welding electrodes. These materials consist of many minerals, ores, chemicals, and metallic elements that are carefully controlled by amount, chemistry, and physical properties.

These materials are made to very strict specifications. Expensive equipment is used to ensure that the specifications are consistently met. Some of the equipment used for quality control are x-ray diffraction, spectrophotometry, mass spectroscopy, analytical chemical instruments, and metallurgical inspection equipment including special metallurgical microscopes and tensile-testing machines.

Most of the raw materials for flux are packaged in small-lot containers so that no grinding takes place in shipment. The particle size of most raw materials is very important to the final operating characteristics of the finished SMAW electrodes.

Chemical analysis is made on every lot of materials that is received to make sure that the manufacturer's specifications are met. Such things as sulfur and phosphorus are poison to the electrode manufacturer, and to you, the welder, not so much because of what they can do to you, but because of what they do to weld metal. They make weld metal brittle and porous.

Electrodes are made in batch lots so that maximum control can be exercised in flux mixing. All batches are mixed by referring to code charts. These codes are set up to protect the formulations from getting into competitor's hands and also to ensure that the correct material gets into each batch of flux.

The dry materials in the batch are mixed first, and then the wet constituents are added after a pre-determined time. A heavy "mud" is formed, and after mixing, the mud is placed in a press to form a slug or cylinder. This cylinder is taken to the extrusion press for a production run. A usual production run amounts to about 40,000 lb [18,000 kg] of electrodes. Only a single electrode grade will be made in each batch.

The extrusion press is a machine that takes the mud and forms it around the cut-length core wire at very high speed. Fourteen-inch [0.36 m] electrodes are extruded from these machines at speeds up to 1200 electrodes per minute. The mud coating on the wire will become the flux coating on the finished electrode. Making electrodes is such a competitive business, and the unit price of electrodes is so small, that at least 40,000 lb [18,000 kg] of electrodes has to be extruded before any profit can be made by the manufacturer.

Flux-coated SMAW electrodes leaving the extrusion press are put on moving belts and sent to a drying oven. The oven is graduated in temperature so that the "green" electrodes are first sent into a low-temperature zone of the furnace, and then gradually raised in temperature until a pre-determined water content remains in the drying flux coating. Electrodes are never completely dry, and in fact many have controlled water content as part of their design. For example, AWS E6010 mild-steel electrodes will not operate properly if they are baked completely dry.

On the other hand, AWS E7018 low-hydrogen electrodes must have flux coatings that are as dry as possible to avoid hydrogen embrittlement in high-strength low-alloy weld metal. So you see, even the simplest rules have exceptions when it comes to making electrodes. That's also why you must follow the manufacturer's instructions when you use an electrode holding oven or rebake them to recondition them for future use.

After the electrodes come out of the drying oven they are sampled. Each lot has operation and control tests made on it to ensure that the product is made to specifications. Experienced welders will actually try welding with samples from each new lot to test the electrode's "operator appeal." The manufacturers know that no matter how well an electrode meets a customer's technical specifications, if the customer's welders don't like to use the electrodes, the electrodes won't be bought.

The core wire used to manufacture mild-steel electrodes and even low-alloy SMAW electrodes is, in most cases, the same. It usually is a "rimmed" steel (the surface is very smooth, pure iron while the insides of the wire are higher in carbon content). The wire drawn from the rimmed-steel ingot is used purposely to retain the pure-iron "skin" on the outside surface of the wire. A typical analysis of the full section of mild-steel core wire is iron plus

carbon (C)	0.15% maximum
sulfur (S)	0.04% maximum
phosphorus (P)	0.04% maximum
silicon (Si)	0.25% maximum
manganese (Mn)	0.30 to 0.60%

Sometimes a higher-carbon core wire is used in AWS E6020 covered electrodes, for example, for specific reasons, the main one being to give better spray-type transfer characteristics to the weld metal that passes through the arc.

The AWS specifications place no restrictions on the manufacturer as to the materials used in the flux coating. The usable welding position, electrical characteristics, and, for carbon steel and certain other types of electrodes, the mechanical properties of the weld-metal produced are the only things that are specified by the AWS. This gives the electrode-manufacturing industry a free hand to develop new concepts in electrode flux coatings.

The coatings of most competitive electrodes, however, are very similar. Don't let a hard-driving salesperson try to convince you that the special Super-Crack maintenance electrode is twice as good as anything else on the market (at three times the regular price) . . . unless the claim can be supported by hard facts, laboratory data (your lab, not the salesperson's), and shop experience with the "new, improved" electrode.

16-3
AWS SMAW ELECTRODES
The versatility of SMAW owes so much to the availability of different electrodes that you should start your study of this process by understanding the scope of this vast welding resource.

Proprietary electrodes are a mixed blessing. Because of them, individual manufacturers can compete with each other to make better electrodes for you. Because of that there are many brands of SMAW electrodes that you can select to solve almost any kind of metal or welding problem. But if you were faced with a list of 100 electrodes, each with a different name such as Superduperweld IV, Ferromanurium Extra, and Spatter and Splash Fast, you wouldn't have the slightest idea of what you were buying.

Thanks to the AWS, welders in the United States aren't driven crazy just trying to select a stick electrode. Every SMAW electrode can be grouped into some standard AWS classification. Once you understand the simple AWS code, you will know what kind of weld metal the electrode will produce and often whether it is an ac, dc, or dual-current (ac and dc) electrode. The code also will tell you the minimum strength of the weld metal the electrode will produce, and whether you should use it downhand or out of position. The AWS electrode codes even will give you a clue to other properties of the flux coating and will help you know a lot more about the general operating characteristics of the electrode you want to use.

The AWS specifications for SMAW electrodes even put limits on how far a manufacturer's core wire can deviate from a "¼-in. electrode," for example, and these valuable specifications list things like standard electrode lengths and other data to protect you, because just about any welding equipment or electrode holder you use will be compatible with just about anybody's electrode.

The ASTM (American Society for Testing and Materials) adopted the AWS system, and this joint AWS-ASTM electrode-coding system is used by almost all welding manufacturing codes, in one form or another, including very important ones like the ASME Boiler and Pressure Vessel Code, the various American Petroleum Institute (API) tank- and pipeline-fabricating codes, the Society of Automotive Engineers (SAE), and other important code-writing bodies for manufactured products.

First we'll explain the AWS-ASTM covered-electrode designations for carbon and low-alloy-steel SMAW electrodes, then we'll tell you about stainless-steel electrodes, and finally nonferrous SMAW electrodes for copper, nickel, and aluminum. We'll reserve our description of the special AWS hardfacing electrode designations for Chap. 19, "Hardfacing and Surfacing." Although many hardfacing jobs use covered-metal arc welding, filler-metal-selection problems (and AWS filler-metal codes) are entirely different. As just one example, hardness instead of strength is usually the primary requirement.

The AWS Electrode Classification System

The AWS-ASTM electrode filler-metal code is very simple, considering the amount of information each code number contains. Even

the most complex SMAW-electrode code names, those used for carbon and low-alloy steels, are easy to understand. We haven't asked you to memorize anything in this book so far, but it would be to your advantage to memorize the basic ideas behind the AWS system. The words for different electrodes that you will use, coded under this system, will be words (electrode names) that you will use as a welder almost every day.

Figure 16-4 is a general picture of how the carbon- and low-alloy-steel electrode code works (stainless steel and nonferrous electrodes are generally even simpler). In the illustration, the E stands for electrode. It simply means that we are going to talk about electrodes rather than a gas or TIG welding rod (which has an R in front of it).

The first two (or three) numbers are very important. They designate the minimum tensile strength of the weld metal that the electrode will produce if properly operated. For example, a steel electrode that produces weld metal with a minimum tensile strength of 60,000 psi [414 MPa] will be coded E60XX. An electrode that produces weld metal with a minimum tensile strength of 80,000 psi [552 MPa] would be coded as E80XX.

What if the electrode is used to weld submarine hulls and produces weld metal with a minimum tensile strength of 120,000

psi? Then the first part of the code designation would simply be E120XX.

The third digit (or sometimes the fourth digit for weld metal with a tensile strength of 100,000 psi [689 MPa] or more) tells you whether the electrode was designed to operate in all positions (1), only in the flat and horizontal positions (2), or only in the flat position (3). There are only these three possibilities: EXX1X, EXX2X, and EXX3X.

Let's stop and examine the most common carbon-steel electrode used today. It produces weld metal with a minimum tensile strength of 60,000 psi [414 MPa] and comes in three versions for each of the three groups of welding positions.

EXXXX	It's an electrode as opposed to a brazing or TIG rod
E60XX	that produces weld metal with a minimum tensile strength of 60,000 psi [414 MPa]
E601X	and is designed to operate in all positions
E602X	or works best only in the flat and horizontal positions (therefore it probably is designed to operate at higher welding currents and will give you more weld metal per hour than an E601X electrode)

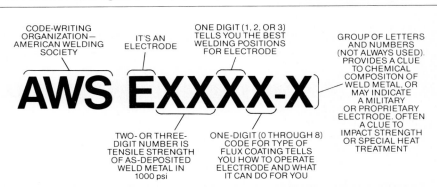

CODE-WRITING ORGANIZATION— AMERICAN WELDING SOCIETY

IT'S AN ELECTRODE

ONE DIGIT (1, 2, OR 3) TELLS YOU THE BEST WELDING POSITIONS FOR ELECTRODE

GROUP OF LETTERS AND NUMBERS (NOT ALWAYS USED). PROVIDES A CLUE TO CHEMICAL COMPOSITON OF WELD METAL, OR MAY INDICATE A MILITARY OR PROPRIETARY ELECTRODE. OFTEN A CLUE TO IMPACT STRENGTH OR SPECIAL HEAT TREATMENT

AWS EXXXX-X

TWO- OR THREE-DIGIT NUMBER IS TENSILE STRENGTH OF AS-DEPOSITED WELD METAL IN 1000 psi

ONE-DIGIT (0 THROUGH 8) CODE FOR TYPE OF FLUX COATING TELLS YOU HOW TO OPERATE ELECTRODE AND WHAT IT CAN DO FOR YOU

FIGURE 16-4 The American Welding Society filler-metal code for carbon steel and low-alloy steel-coated electrodes. Learn to read the code. Every arc welder should be able to look at the code printed on the electrode-holder end of a stick electrode and immediately know a great deal about the weld metal that the electrode will produce.

E60<u>3</u>X   or works best only in
the flat, downhand
position (it's a very
high volume produc-
tion electrode)

EXXX<u>X</u>   Now we'll tell you
about the last digit.
It's the only part of
this designation sys-
tem that takes any
real memory work.

The last digit in the main code
name tells the kind of flux coating
used, which, in turn, tells you a
great deal about how to operate
the electrode. The last digit can
be any one of these numbers.

EXXX0 = Cellulose-sodium, or
iron-oxide mineral
(We'll tell you how
to separately identify
them in a minute)
EXXX1 = Cellulose-potassium
EXXX2 = Titania-sodium
EXXX3 = Titania-potassium
EXXX4 = Iron powder–titania
EXXX5 = Low-hydrogen lime–
sodium
EXXX6 = Low-hydrogen lime–
potassium
EXXX7 = Iron oxide plus iron
powder
EXXX8 = Low-hydrogen lime–
iron powder
EXXX9 = A number left over for
peculiar coatings
that don't fit in any
standard category.
This number is al-
most never used.

Table 16-1 shows what these
flux-coating designations tell you.
So how do you tell whether an
electrode coded EXXX0 has a cellulose-sodium or an iron-oxide
flux coating? *Look at the next to
last number, just before the number that codes for the flux-coating
class, EXX?X.* If that number is a
1, as in AWS E6010, then the
coating is sodium-cellulose. Sodium-cellulose based-flux coatings
make electrodes that work in all
positions. If the number is a 2,
such as AWS E6020, or a 3, such
as in AWS E6030, then the coating is an iron-oxide type. Iron oxide is used in high-production
downhand welding electrodes.
High-production electrodes aren't
used for out-of-position work.

With experience you will easily
remember this difference when
you use a high-volume production
electrode. You will see and feel
how differently it operates. Iron
oxide is always put in electrode
flux coatings to increase the
downhand deposition rate. Both 2
and 3 indicate high-production,
downhand welding positions. Use
2 for flat and horizontal welding
and 3 for flat positions only. Let's
repeat that. Iron powder + iron
oxide = more iron in the weld =
faster deposition rates = flat or
horizontal welding because you
can't get high production when
working out of position.

Only three of these standard
flux-coating codes will give you an

"all-purpose" electrode that will
operate either with ac, DCRP, or
DCSP. They are EXX20 or
EXX30 iron-oxide electrodes, the
EXXX4 iron powder–titania electrodes, and the EXXX7 iron oxide–iron powder electrodes. This
leads to the obvious conclusion
that electrode flux coatings with
a lot of iron powder in them can
be used with ac, DCSP, or DCRP
welding current. To repeat, if the
last (flux) code number is a 0
(such as in EXX20 or EXX30) it's
a high-production, downhand electrode. If the last flux-code digit is
a 4 or a 7, the electrode is an all-
purpose, ac-dc electrode.

Before we go any further, let's
look at two very good common
electrodes and one that's so undesirable it doesn't exist. They are
AWS E6010, AWS E7018, and our
imaginary E11031 electrode.

E6010 AWS E6010 produces weld
metal with a minimum tensile
strength of 60,000 psi [414 MPa]
(E60XX). You can use it in all
welding positions (EXX1X), and it
has an iron-oxide flux coating
(EXX10) designed for use with
DCRP welding current. This electrode, and its high-production
downhand relatives, AWS E6020
and AWS E6030, are the most common carbon-steel electrodes in use.
You would probably use AWS
E6010 electrodes for carbon steel if
you had to weld in a variety of positions, but you would switch to
either AWS E6020 or AWS E6030
if you were working on a high-volume production job where all your
work was downhand welding.

E7018 AWS E7018 is an electrode that produces weld metal
with a minimum tensile strength
of 70,000 psi [483 MPa] (E70XX).
It also can operate in all positions
(EXX1X) and it has a low-hydrogen, iron-powder flux coating
(EXXX8) for use with either
DCRP or ac. Since DCSP is not
used very much, this electrode is
almost all-purpose. It happens to
be the most common electrode
used for welding high-strength
low-alloy steels.

TABLE 16-1 What flux-covered electrodes are covered with

Electrode	Flux Coating	Current Setting or Polarity		
		DCSP	DCRP	AC
EXXX0	Sodium-cellulose or iron oxide	No	Yes	No
EXXX1	Potassium-cellulose	No	Yes	Yes
EXXX2	Titania-sodium	Yes	No	Yes
EXXX3	Titania-potassium	Yes	No	Yes
EXXX4	Iron powder–titania	Yes	Yes	Yes
EXXX5	Low-hydrogen lime–sodium	No	Yes	No
EXXX6	Low-hydrogen lime–potassium	No	Yes	Yes
EXXX7	Iron oxide–iron powder	Yes	Yes	Yes
EXXX8	Low-hydrogen lime–iron powder	No	Yes	Yes
EXXX9	Proprietary or experimental coating. It's not officially listed but reserved for new ideas.	?	?	?

E11031 This electrode doesn't exist. All high-strength steel electrodes in use are AWS EXXXX5, EXXXX6, or EXXXX8 grades because low-hydrogen electrodes such as EXXXX8 are important for welding high-strength steels to prevent hydrogen embrittlement. E11031 would be a high-strength electrode with an organic cellulose flux coating that would produce hydrogen embrittlement in high-strength welds because all organic materials burn to produce water, and water contains hydrogen. That's why E11031 doesn't exist.

16-4 MILD-STEEL ELECTRODES

A very important AWS filler-metal specification covers SMAW electrodes for mild (that is, carbon) steel. It is listed in AWS specification A5.1 titled "Mild-Steel-Covered Arc-Welding Elec-

trodes." A second AWS specification, AWS A5.5, which is titled "Low-Alloy-Steel Covered Arc-Welding Electrodes," often goes hand-in-hand with AWS A5.1, since carbon steels and high-strength low-alloy steels are frequently used in the same products. We will describe the most frequently used electrodes in AWS A5.5 after we cover the mild-steel electrode grades in AWS A5.1. You will see that there will be some overlap (at the 70,000-psi tensile-strength level) in AWS A5.1 electrodes for mild steels and the lower-strength electrodes in AWS A5.5 for high-strength low-alloy steels.

The AWS A5.1 specification includes over twenty different classifications of mild-steel SMAW covered electrodes. This filler-metal specification not only explains the code system, it also gives detailed data on minimum mechanical properties of the weld

metal produced from each class of electrode (and a lot of other information). In the appendix that comes with every AWS filler-metal specification there are even suggestions where you can use these different electrodes to best advantage.

If you were to look into AWS A5.1 on carbon-steel electrodes, you would find these grades listed:

AWS E6010	AWS E7014
AWS E6011	AWS E7015
AWS E6012	AWS E7016
AWS E6013	AWS E7018
AWS E6020	AWS E7024
AWS E6027	AWS E7028

Selecting a Mild-Steel Electrode

Table 16-2 shows how to put this information to use. This table summarizes the characteristics of the most important SMAW electrodes used for carbon and high-strength low-alloy steels. It shows

TABLE 16-2 Mild-steel-coated electrode selection chart and performance ratings

	AWS A5.1 Specification Electrode Class Listed in Order of Flux Coating Designation (Rating Is Based on Same-Size Electrodes. 10 = Highest Rating, 1 = Lowest.)*										
Application	E6010	E6011	E6012	E6013	E7014	E7016	E7018	E6020	E7024	E6027	E7028
Groove-butt welds, flat downhand less than ¼ in.	4	5	3	8	9	7	9	10	9	10	10
Groove-butt welds, all positions, less than ¼ in.	10	9	5	8	6	7	6	Not	Not	Not	Not
Fillet welds, flat or horizontal, high production	2	3	8	7	9	5	9	10	10	9	9
Fillet welds, all positions, all-purpose	10	9	6	7	7	8	6	Not	Not	Not	Not
Welding current, polarity‡ ac, DCSP, or DCRP	DCRP	ac DCRP	ac DCSP	ac dc	ac dc	ac DCRP	ac DCRP	ac dc	ac dc	ac dc	ac DCRP
Thin material, distortion problems from excess heat	5	7	8	9	8	2	2	Not	7	Not	Not
Heavy plate, highly restrained joints	8	8	6	8	8	10	9	8	7	8	9
High-sulfur steel or off-analysis steel	Not	Not	5	3	3	10	9	8	5	Not	9
Deposition rate (high is better)	4	4	5	5	6	4	6	6	10	10	8
Depth of penetration into base metal	10	9	6	5	6	7	7	8	4	8	7
Weld appearance, good-looking beads, minimal undercutting	6	9	6	5	6	7	7	8	9	8	7
Ease of controlling weld-metal soundness	6	6	3	5	7	10	9	9	8	9	9
Ductility (high is better)	6	7	4	5	6	10	10	10	5	10	10
Low-temperature impact strength (high is better)	8	8	4	5	8	10	10	8	9	9	10
Low spatter loss (less likely to be cleanup problems)	1	2	6	7	9	6	8	9	10	10	9
Poor fit-up problems when sections mismatch	6	7	10	8	9	4	4	Not	8	Not	4
Welder appeal, good operating characteristics	7	6	8	9	10	6	8	9	10	10	9
Slag self-removing or easy removal	9	8	6	8	8	4	7	9	9	9	8

* This is a good general selection chart. It's based on electrodes of equal diameter. Of course your out-of-position ratings, for example, will improve if you use a smaller electrode and lower welding current. Similarly, this chart does not take into consideration some proprietary electrodes that, although they may fall within a given AWS electrode grade, will probably have better properties of one kind or another than indicated here relative to other electrodes in the same general AWS grade.

† No means not recommended for this use.

‡ AC = alternating current; DCRP = dc reverse polarity, electrode positive; DCSP = dc straight polarity, electrode negative.

you the effect of different flux coatings and what they will do for you (or against you) when you need an electrode to weld most different kinds of carbon steels.

The third and last letter, "L", on some of the electrode codes such as EXXXX-B2L and EXXX-B4L, has a special additional meaning. The L means that the weld metal has extra-low carbon. Among many other reasons, the low-carbon composition prevents the high-temperature material in a welded pipe from "heat-treating" the weld metal.

If you are welding thin-gauge material and want a weld-metal tensile strength of around 60,000 psi [414 MPa], check out an AWS E6013 electrode, or alternately an AWS E6012 grade. If your problem is poor joint fit-up, try an AWS E6012; this electrode is tolerant of wide gaps and bridges them very nicely. On the other hand, AWS E6010 should not be used for making fillet welds in the flat or horizontal position on a production basis because it doesn't have a very high weld-metal deposition rate, but AWS E6010 electrodes are fine if you are going to weld overhead for a while, then do some flat- or horizontal-position welding, and maybe some vertical welding. In other words, E6010 electrodes are an excellent choice for maintenance welding where you don't want to carry a large variety of electrodes in inventory.

Now that you're beginning to get a feeling for why there are different kinds of carbon-steel electrodes, let's talk a little more about the welding characteristics of each grade. While we're at it, we'll give you some tables of typical weld-metal properties and power-source settings for each AWS grade. But you need to know two things. First, the characteristics of individual electrodes manufactured as a given AWS grade, such as AWS E6010, may vary a little from what we tell you because each electrode manufacturer is constantly trying to

improve the welding characteristics of the proprietary grades they sell; and second, the welding voltage and current settings we've listed here are only approximate.

The best way to use any electrode is to follow the manufacturer's instructions, which isn't hard since the recommended current (with recommended dc polarity) and voltage settings are almost always printed right on the label of the electrode package.

More than your power source, more than your electrode holder, more than anything else you have to work with except your own basic skill as a welder, these electrodes are your most important tools for SMAW. The more you know about the characteristics of each grade, the better welder you will be.

AWS E6010 All-position AWS E6010 electrodes are used for DCRP (electrode-positive) welding. They are best suited for making vertical and overhead welds and also for some sheet-metal applications in any position. The molten weld metal sprays through the welding arc something like a miniature paint gun. (The physics of transferring molten metal through an arc is discussed in the next chapter, so bear with us on this terminology until then.) This spray transfer helps you weld in the vertical and overhead positions. Globs of weld metal are less likely to fall on your head while the spray transfer blasts weld metal into out-of-position joints.

This type of metal transfer also tends to dig into your base metal. As a result, AWS E6010 electrodes give you deep weld penetration, which means that you have to be careful in handling the electrode to minimize spatter. AWS E6010 electrodes have a tendency to undercut your welds. Reduce your current setting if that happens.

Fillet welds produced by AWS E6010 electrodes are relatively flat in profile. They have a rather coarse, unevenly spaced surface

ripple. Where the surface quality of your weld deposit is of prime importance, particularly on multi-pass weld joints in vertical and overhead positions, and where radiographic requirements such as x-ray quality are necessary, these electrodes are highly recommended.

Most of the AWS E6010 electrodes used today are for mild steel. However, they also can do a nice job on galvanized steel sheet. The AWS E6010 grade is excellent for temporarily tacking a weldment together because of the weld metal's good ductility and deep penetration. Its mechanical properties are generally excellent, and when properly applied, weld deposits produced by AWS E6010 electrodes will meet some very tough inspection standards.

The thickness of the flux coating on AWS E6010 electrodes is held to a minimum to make it easier to weld in the vertical and overhead positions, but the coating will give you enough shielding for high-quality weld deposits. The electrode flux coating is high in cellulose, usually exceeding 30 percent cellulose by weight. Other materials in the coating generally include titania (titanium dioxide, TiO_2), various types of magnesium and aluminum silicates, metallic deoxidizers such as ferromanganese, and liquid sodium silicate.

Some AWS E6010 electrode flux coatings have a small amount (less than 10 percent by weight) of iron powder in them to improve their arc characteristics. Because of the coating composition, these are generally classified as high-cellulose sodium-type electrodes.

The maximum welding current that can be used with the larger diameters of AWS E6010 electrodes is somewhat limited compared with other electrode classifications due to the high spatter loss that occurs with high currents. Table 16-3 presents typical welding currents, voltage settings, and mechanical properties for AWS E6010 and AWS E6011 electrodes of different diameters.

AWS E6011 AWS E6011 electrodes are almost identical to AWS E6010 electrodes except that they operate on ac as well as dc. Their performance is very similar. However, AWS E6011 electrodes perform equally well with either ac or DCRP (electrode-positive) power settings. These electrodes have a forceful digging arc action that results in deep base-metal penetration.

While the flux coating is slightly heavier than that of AWS E6010 electrodes, the resulting slag and weld profiles of AWS E6011 stick electrodes are quite similar to those of E6010 electrodes. The coating is high in cellulose and is designated as a high-cellulose potassium type. (Potassium rather than sodium makes these electrodes work well with ac.)

As is true of AWS E6010 electrodes, AWS E6011 electrodes in sizes larger than 3/16-in. [4.8-mm] diameter are not usually used for all-position welding. You need smaller-diameter electrodes to weld horizontally, vertically, and overhead.

The current and voltage ranges recommended for ASW E6010 and E6011 electrodes are usually identical. High current settings will produce a lot of spatter. One difference between these electrodes is that the ductility of the deposited weld metal from an AWS E6011 electrode is higher than that obtained with AWS E6010. When the electrodes are used in the vertical or overhead position, currents nearer to the lower limit of each range should be used.

AWS E6012 AWS E6012 stick electrodes are used for all-purpose welding in all positions, using either DCSP (electrode-positive) or an ac power source. Although AWS E6012 is considered to be an all-purpose electrode, it is used much more frequently in flat and horizontal positions than in vertical or overhead welding.

AWS E6012 electrodes are specifically recommended by most manufacturers for horizontal and flat or slightly downhill welding. They are especially recommended for single-pass, high-speed, high-current, horizontal fillet welds. Ease of handling, good fillet-weld profile, and the ability to withstand high currents as well as to bridge gaps when you have to deal with poor fit-up problems are all characteristic of AWS E6012 electrodes.

When used for vertical and overhead welding, the electrode diameter is frequently selected to be one size smaller than you might use for an E6010 or E6011 electrode. Since the arc is highly stabilized, welds have a good appearance and are relatively free from undercutting. Fillet welds usually have a convex weld-bead profile and a smooth even ripple in the horizontal or vertical-down positions.

AWS E6012 electrodes have also been used to advantage on many low-alloy steels, particularly the higher-carbon grades. This is probably due to the fact that the penetration obtained, although adequate, is by no means the deep penetration achieved with an AWS E6010 or E6011 electrode. As a result, the pickup of alloying ingredients from the base metal is not as great, which means that there is less chance that your welds will crack. That's also why this electrode is used with some medium-carbon steels. It won't pick up as much carbon in the weld metal, which would rapidly quench on cooling to produce a harder, less ductile weld deposit with a strong possibility of cracking.

AWS E6012 electrodes have a rather quiet arc. This means that although you get medium base-metal penetration, you also get a lot less spatter. Good weld-bead buildup and no excess base-metal penetration make these electrodes very nice to use under poor joint fit-up conditions. Since the arc is highly stabilized, your welds will look good and will be relatively free from undercutting.

The flux coating of AWS E6012 electrodes usually is high in titanium dioxide, exceeding 35 percent by weight, which is why these electrodes are often called *titania-* or *rutile-coated* grades. (Rutile is the name of the titanium dioxide mineral used in the flux coating.) The flux also contains other minerals such as feldspar (the pink or white mineral in granite rock), clay minerals, and small mounts of cellulose and sodium silicate as the binder.

Ferromanganese (iron-manganese metallics produced in an electric furnace) also may be added as a deoxidizer along with calcium compounds such as lime-

TABLE 16-3 AWS E6010 and AWS E6011		
3/32	40–80	23–25
1/8	75–125	24–26
5/32	110–170	24–26
3/16	140–215	26–30
7/32	170–250	26–30
1/4	210–320	28–34
5/16	275–425	28–34
	As-welded	**Stress-relieved**
Tensile strength, psi	62,000–70,000	60,000–67,000
Yield strength, psi	52,000–58,000	47,000–54,000
Elongation in 2 in., %	22–30	28–35
Charpy V-notch impact	20 ft·lb at −20°F [−29°C]	

stone to produce a satisfactory arc on straight polarity. Some AWS E6012 electrodes have a little iron powder added to improve arc characteristics and to increase their downhand weld-metal deposition rate. The slag coverage is complete and slag removal is easy. While single-pass welds may meet radiographic requirements, multipass welds with AWS E6012 electrodes probably will fall far short of what you want.

When using a dc power source, straight-polarity (electrode-positive) current is preferred. AWS E6012 electrodes are often used where the appearance of the weld bead and high deposition rates are far more important than maximum weld-metal ductility. For example, AWS E6012 electrodes make very good welds in sheet metal.

The weld metal deposited by AWS E6012 electrodes is lower in ductility and higher in yield and tensile strength than weld metal made by either AWS E6010 or E6011 electrodes. Table 16-4 gives you typical welding currents, voltage settings, and mechanical properties for AWS E6012 electrodes of different diameters.

AWS E6013 Although AWS E6013 electrodes are similar to E6012 filler metals, they have some very important differences. AWS E6013 electrodes are designed for use in all welding positions, using either ac or dc (either straight or reverse polarity). AWS E6013 electrodes produce minimum spatter and only a small tendency to undercut your weld beads. The beads have a fine ripple and superior surface finish.

Slag removal is somewhat better and the arc can be established and maintained more easily than with AWS E6012 electrodes, particularly when you are working with the smaller $1/16$-, $5/64$-, and $3/32$-in. [1.6-, 2.0-, and 2.4-mm] sizes. AWS E6013 electrodes are very good for working at lower amperages. As you know, lower amperages mean less heat input

into the base metal, and that means less distortion. Hence, AWS 6013 electrodes are ideal for welding thin metals. Mechanical properties are slightly better than the AWS E6012 electrodes, too, as are your chances for making acceptable x-ray-quality welds.

Originally, AWS E6013 electrodes were designed only for sheet-metal work. However, the larger diameters are now used for many jobs previously welded with AWS E6012 electrodes. While the operation of AWS E6013 electrodes is quite similar to that of AWS E6012 electrodes, the arc is quieter, the bead smoother, and the ripple finer.

Changing from one manufacturer's AWS E6013 electrode to another can result in a change in the way the molten metal is transferred through the arc, which can affect the amount of spatter you get and whether you can use the electrodes out-of-position or not.

Some manufacturers make AWS E6013 electrodes so that they transfer weld metal through the arc in a fine spray. Others compound the flux coating so that the weld metal passes through the arc as globules. Spray transfer is preferred for vertical or overhead welding. Globular transfer is for downhand welding. There are

other differences between individual brand-name electrodes.

Some AWS E6013 electrodes, recommended for sheet-metal work, will give you good welds working in the vertical position, welding from the top down (known as vertical-down welding). Other AWS E6013 electrodes have a more fluid slag and are partially replacing AWS E6012 electrodes for horizontal fillet welds and for other general-purpose welding.

Rather than the convex contour characteristics of AWS E6012 electrodes, the E6013 filler metals produce a flat fillet weld similar to that of the AWS E6020 electrode. They are also readily usable for making groove welds because of the concave (depressed center) bead shape and easily removed slag.

In addition, the weld metal of E6013 electrodes is definitely free of slag and oxide inclusions compared with that of AWS E6012 weld metal, meaning that x-ray-quality welds are easier to make. In fact, AWS E6013 electrodes often are specified when very high quality carbon-steel welds must be made that will pass x-ray inspection.

The AWS E6013 flux coating is very similar to that used with the AWS E6012 electrode class. It

TABLE 16-4 AWS E6012

Electrode Diameter, in.	Current, A	Voltage, arc volts
$1/16$	20–40	17–20
$5/64$	25–60	17–21
$3/32$	35–85	17–21
$1/8$	80–140	18–22
$5/32$	110–190	18–22
$3/16$	140–240	20–24
$7/32$	200–320	20–24
$1/4$	250–400	20–24
$5/16$	300–500	22–26

	As-welded	Stress-relieved
Tensile strength, psi	62,000–78,000	60,000–75,000
Yield strength, psi	52,000–65,000	50,000–60,000
Elongation in 2 in., %	17–22	22–27
Charpy V-notch impact	Not required	

contains rutile, silicate-bearing minerals, cellulose, ferromanganese deoxidizers, and liquid silicate binders. An important difference, however, is that easily ionized materials are used in the coating so that you can establish and maintain your arc at lower welding currents and low OCVs. Some manufacturers also introduce small quantities of iron powder into their AWS E6012 flux coating to give you better arc control and a slightly higher weld-metal deposition rate.

The high welding currents possible with AWS E6012 electrodes cannot be used for AWS E6013 electrodes. When welding in a vertical or overhead position, however, the current and voltage settings for both types of electrodes will be very similar.

Table 16-5 lists typical welding currents, voltage settings, and mechanical properties for AWS E6013 electrodes of different diameters.

AWS E7014 As the first two numbers of this electrode designation indicate, weld metal produced by this electrode has a 70,000-psi [483-MPa] tensile strength. At one time there was an AWS E6014 electrode, but since no manufacturer ever made one for this classification, and most did make electrodes for the AWS E7014 class, the E6014 grade was dropped.

AWS E7014 electrode flux coatings are similar to those of the AWS E6013 filler metals. However, the coating of this electrode is considerably thicker since it contains substantial amounts of iron powder (30 percent or more of the coating weight). The amount of coating and the percentage of iron powder in the flux are usually less than those found in the AWS E7024 electrodes we'll tell you about shortly.

The welding characteristics of AWS E7014 electrodes are a compromise between the AWS E6013 and the AWS E7024 classifications. The deposition rate is higher than AWS E6013 electrodes, but not as high as AWS E7024 electrodes. Iron powder in a flux coating lets you use higher welding currents and that means higher deposition rates and higher welding speeds.

While AWS E7014 electrodes are classified for all-position welding, the thicker coating does not make them suitable for out-of-position work on thin-gauge materials. However, they will perform satisfactorily on the occasional job to which they are suited. The performance characteristics of AWS E7014 electrodes do make them suitable for production welding of irregularly shaped products where you will run into some out-of-position welding by the very nature of the job.

AWS E7014 electrodes are good for welding both mild and low-alloy steels. Typical weld beads are smooth with fine ripples, and penetration into the base metal is about the same as you get with AWS E6012 electrodes, which is an advantage when you are welding over gaps caused by poor fit-up. Fillet-weld profiles made by AWS E7014 electrodes tend to be flat or slightly convex (bowed out in the middle).

Mechanical properties of AWS E7014 weld metal are superior to those of AWS E6012 and E6013 electrodes. Slag removal is very easy, at times almost self-cleaning. Good base-metal penetration and rapid solidification make AWS E7014 electrodes well suited for handling poor fit-up and working out of position. The usable currents for AWS E7014 electrodes are higher than for AWS E6012 or E6013 filler metals because of the extra iron powder in them.

Table 16-6 gives typical welding currents, voltage settings, and mechanical properties for AWS E7014 electrodes of different diameters.

AWS E7015 Electrodes made to AWS E7015 standards are commonly referred to as *low-hydrogen* electrodes. The last digit in the code, 5, tells you that. (The other flux coatings that are low-hydrogen are coded by a 6 or an 8.) The last digit, 5, also tells you that this electrode is a low-hydrogen sodium flux coating.

All low-hydrogen electrodes have a lot of calcium carbonate (the mineral in limestone) or calcium fluoride (a mineral called fluorite) in them. They sometimes are called lime-ferritic, or basic-type electrodes. Materials such as cellulose, clays, asbestos, and other minerals that contain water in the crystal lattice are not used

TABLE 16-5 AWS E6013

Electrode Diameter, in.	Current, A	Voltage, arc volts
$1/16$	20–40	17–20
$5/64$	25–60	17–21
$3/32$	45–90	17–21
$1/8$	80–120	18–22
$5/32$	105–180	18–22
$3/16$	150–230	20–24
$7/32$	210–300	21–25
$1/4$	250–350	22–26
$5/16$	320–430	23–27

	As-welded	Stress-relieved
Tensile strength, psi	67,000–72,000	65,000–71,000
Yield strength, psi	55,000–60,000	50,000–56,000
Elongation in 2 in., %	17–20	25–30
Charpy V-notch impact	Not required	

to ensure that the electrode flux coating has a very low hydrogen content (because water is made of hydrogen and oxygen). In addition, low-hydrogen electrodes are baked at higher temperatures after the flux has been extruded onto them to ensure that all the water possible has been driven out of the coating. As you'll learn shortly, you must handle all kinds of low-hydrogen electrodes with care to keep them absolutely dry when you are using them.

Low-hydrogen electrodes (with stainless-steel weld-metal composition) are used to weld stainless steels. Low-hydrogen electrodes (with special alloying elements) are used to weld all high-strength steels.

AWS E7015 electrodes were the first SMAW filler metals designed to operate on DCRP to make all-position welds on high-sulfur and high-carbon steels. They are included in the high-strength low-alloy steel electrode category because the weld metal they deposit (not necessarily the steels that they weld) are of high-strength low-alloy steel composition. The base metals such as high-carbon steels welded by high-strength low-alloy E7015 electrodes are materials that are prone to cracking under the weld bead. High-carbon steels often are found in

springs and very hard parts in machine tools. High sulfur is put into certain "free-machining" steels to make them easier to cut at high speeds. Both materials are very difficult to weld without underbead cracking and porosity.

Underbead cracks occur in the base metal, usually just below the weld metal, and are caused by hydrogen ions being absorbed from the welding-arc atmosphere into the weld metal and from there into the base metal. The hydrogen ions are very small. In fact, they are protons and enter steel very easily. When they enter the steel, the hydrogen ions can combine with oxygen to reform into water . . . but there's no room in the steel for the water molecules. The result is very high stresses in the surface of the base metal. If the base metal is not very ductile, as is the case in most higher-strength steels, the metal acts like it has suddenly become very brittle. That's what hydrogen embrittlement is. The result is a crack right under the weld bead.

Low-hydrogen electrodes are not normally needed for joining regular mild steels, but they are critical to the successful welding of high-strength low-alloy steels used in bridges and buildings and for other low-alloy steels such as the chromium-molybdenum

grades used in critical steam pipes and many pressure vessels.

When an electrode that is not a low-hydrogen grade is used for high-sulfur steels, those steels containing 0.10 percent to 0.25 percent sulfur will have weld deposits that are badly honeycombed. Other free-machining steels contain selenium. This element also is a free-machining additive and makes it very difficult to weld free-machining steels without producing porosity.

AWS E7015 electrodes can be used to weld these steels with fewer problems, but free-machining steels are still difficult materials to weld.

The flux coating on AWS E7015 electrodes is high in limestone (containing calcium) and other ingredients that have been very carefully selected to have as little hydrogen in them as possible. There must be no moisture, no organic materials, and nothing else that might have hydrogen atoms in it, to prevent hydrogen embrittlement. For the same reason, you have to keep your electrodes as dry as possible at all times.

The arc produced by AWS E7015 electrodes is moderately penetrating into the base metal. The slag is heavy and very brittle so it crumbles and is easily removed. The deposited weld metal lies in a flat bead or may even be slightly concave.

AWS E7015 electrodes are used in all positions up to and including the larger 5/32-in.-diameter [4-mm] sizes. The larger diameters are useful for fillet welds and for making horizontal and flat-position welds. Welding currents are somewhat higher than recommended for the same diameters of AWS E6010 electrodes.

When you use low-hydrogen electrodes, keep your arc-to-weld-metal distance as short as possible to reduce the tendency for underbead cracking. Using a "short-arc" technique also will improve the quality of your as-welded deposit and will somewhat reduce

TABLE 16-6 AWS E7014

Electrode Diameter, in.	Current, A	Voltage, arc volts
3/32	80–125	17–21
1/8	110–160	18–22
5/32	150–210	19–23
3/16	200–275	20–24
7/32	260–340	21–25
1/4	330–415	22–26
5/16	390–500	23–28

	As-welded	Stress-relieved
Tensile strength, psi	72,000–78,000	68,000–74,000
Yield strength, psi	60,000–66,000	55,000–61,000
Elongation in 2 in., %	17–21	21–27
Charpy V-notch impact	Not required	

the need for preheating and post-heating of difficult-to-weld steels.

AWS E7015 electrodes were originally designed for welding hardenable steels (those with medium- to high-carbon content). They also have been found useful for welding malleable cast irons, spring steels, and the mild-steel sides of plates clad on the other side with stainless steel. These electrodes are commonly used for making small welds on heavy weldments since they are less susceptible to cracking than electrodes that are not low-hydrogen grades.

Another extensive use made of AWS E7015 electrodes has been in welding steels that are subsequently to be porcelain-enameled. Porcelain-enameled irons and steels are used for the exposed surfaces of refrigerators, sinks, bathtubs, the insides and outsides of home water heaters, the decorative colored curtain walls on buildings, colored steel road signs, some permanent store advertising signs, and even the deep-blue A. O. Smith "Harvester" silos that wealthy farmers often have. Porcelain enameling usually uses special very low carbon steels, and the coating is a type of ceramic material that is very hard and corrosion-resistant.

The success of AWS E7015 electrodes led to the development of AWS E7016 and E7018 low-hydrogen electrodes; two of the most important low-hydrogen electrodes in a welder's bag of tricks. You'll note in the data in Table 16-7 that both AWS E7015 and E7016 electrodes have the same operating characteristics and weld-metal mechanical properties. Also note the higher ductility (measured by percent elongation) in the weld metal produced by these electrodes, while giving you the same tensile strength as AWS E7014 electrodes.

AWS E7016 AWS E7016 electrodes have all the welding characteristics of AWS E7015 electrodes. The core wire and flux coatings are very similar except for the use of small amounts of potassium silicate and other potassium compounds in the coating. Potassium is easily ionized and makes these electrodes suitable for use with ac as well as DCRP current. Everything else we told you about AWS E7015 electrodes also is true of AWS E7016 electrodes, so use Table 16-7 for details.

AWS E7018 AWS E7018 low-hydrogen, iron-powder electrodes are to welding low-alloy steels what

AWS E6010 electrodes are to welding carbon steels. They are the fundamental "tonnage" low-hydrogen grade used today. Like AWS E7016 electrodes, this all-position, ac or DCRP low-hydrogen electrode has 25 to 40 percent iron powder in the flux coating. It has all the desirable low-hydrogen characteristics including producing sound welds on difficult-to-weld high-sulfur and high-carbon steels. Its primary job is welding high-strength low-alloy steels within its tensile-strength range. If you were told that you could only carry around two kinds of steel electrodes, your best choices would be AWS E6010 and AWS E7018.

Like other low-hydrogen electrodes, AWS E7018 is not quite as easy to work with as AWS E6010. You have to keep a short arc at all times, which is not easy to do until you have become skilled at handling arc-welding electrodes.

The fillet welds made with AWS E7018 filler metal in the horizontal or flat position have slightly convex bead profiles. The bead has a smooth, finely rippled surface. These electrodes are characterized by a smooth, quiet arc, low penetration into the base metal, very low spatter, and they can be used at high welding speeds.

The minerals in the low-hydrogen coatings are limited to inorganic compounds (no cellulose, for example). The minerals include calcium fluoride, calcium carbonate, magnesium-aluminum silicate, various ferroalloys, and such binding agents as sodium and potassium silicate. AWS E7018 electrodes are sometimes referred to as lime-ferritic electrodes because of their general use of lime (that is, calcium) compounds in the flux coating.

Since the coating of AWS E7018 electrodes is heavier than normal, vertical and overhead welding is usually limited to the smaller electrode diameters. Currents used are somewhat higher

TABLE 16-7 AWS E7015 and AWS E7016

Electrode Diameter, in.	Current, A	Voltage, arc volts
3/32	65–110	17–22
1/8	100–150	18–22
5/32	140–200	20–24
3/16	180–225	21–25
7/32	240–320	23–27
1/4	300–390	24–28
5/16	375–475	24–28

	As-welded	Stress-relieved
Tensile strength, psi	72,000–78,000	68,000–74,000
Yield strength, psi	60,000–66,000	55,000–60,000
Elongation in 2 in., %	22–28	28–34
Charpy V-notch impact (see specs for further details)	20 ft·lb at −20°F [−29°C]	

than those for AWS 6010 electrodes of the same size.

There have been many proprietary developments in low-hydrogen electrodes. One of the most important started with AWS E7018 electrodes and has now been extended to other, even stronger low-hydrogen SMAW electrodes. Airco Welding Products first developed MR or moisture-resistant grades that stay dry much longer than conventional low-hydrogen electrodes. Now other suppliers like Hobart have them.

The flux coating of most electrodes tends to pick up moisture after several hours of use. As you know, that leads directly to hydrogen embrittlement of the low-alloy-steel base metal. Normally, all low-hydrogen electrodes are shipped in hermetically sealed, moisture-proof aluminum or steel cans. In contrast, AWS E6010 electrodes often are shipped in cardboard boxes.

Once the seal is broken and the electrodes are in use, the flux coating tends to pick up small amounts of moisture from the air. After a predetermined time, usually 4 h or less depending upon the relative humidity of the air in which you work, you have to return the electrodes that you haven't used yet to an electrode baking oven, where they will be "cooked" for an hour or more to dry them out again.

Low-hydrogen MR electrodes also pick up moisture from the air (such as in a humid Gulf Coast shipyard), but at a rate many times slower than conventional low-hydrogen electrodes. Oftentimes a welder drops off MR electrodes for rebaking only at lunch time, or even after carrying them around all day long. Newer, Super MR electrodes are now being produced by Airco that remain below the allowable flux-coating moisture limits for several days.

The American Welding Society has placed strict limits on just how much moisture a low-hydrogen electrode can contain and still be used before rebaking. Only 0.04 percent moisture in the flux coating is allowed for AWS E7018-X and AWS E8018-X electrodes. The even stronger AWS E9018-X, AWS 10018-X, and AWS E11018-X low-hydrogen electrodes are not allowed to contain more than 0.2 percent moisture before they must be returned for rebaking. The ASME Boiler and Pressure Vessel Code also puts very strict limits on how wet a low-hydrogen electrode can be.

(The X after each of these electrode designations refers to additional chemical requirements that will affect weld-metal properties in the various AWS specifications. The significance of these additional codes is discussed in Sec. 16-6, "Low-Alloy-Steel Electrodes.")

Military specifications require even tighter moisture control. The military specification MIL-E-0022200/ID requires that the moisture content of AWS E9018-M, E10018-M, E11018-M, and E12018-M must be less than 0.15 percent or they must be returned for rebaking. Some of these military grades, such as AWS E12018-M and even higher-strength welding electrodes, are used for welding submarine hulls. A little underbead cracking in one small weld in the hull of a large vessel could cause the loss of an entire boat crew. That's why so many critical welds are 100 percent x-rayed.

The tensile properties (including impact strength at low temperatures) of any low-hydrogen electrode are superior to those of AWS E6010 and electrodes that deposit weld metal of similar overall composition. The use of a low-hydrogen electrode often reduces the need for preheating and postheating. Nevertheless, any time you are told to preheat and postheat a weldment, do it.

Table 16-8 gives typical properties and operating data for AWS E7018 electrodes. Consult with the manufacturer (or simply read the label on the electrode can) for further details. Remember that in addition to tensile and yield strength and ductility data (measured by percent elongation), weld metal produced by low-hydrogen electrodes also may have all sorts of additional requirements, the most important usually being low-temperature impact strength.

AWS E6020   Now we'll give you a rundown of the more important high-production mild-steel electrodes designed primarily for use in downhand welding. The first is AWS E6020.

AWS E6020 electrodes are de-

TABLE 16-8 AWS E7018

Electrode Diameter, in.	Current, A	Voltage, arc volts
3/32	70–100	17–21
1/8	115–165	18–22
5/32	150–220	20–24
3/16	200–275	21–25
7/32	260–340	22–26
1/4	315–400	23–27
5/16	375–470	23–28

	As-welded	Stress-relieved
Tensile strength, psi	72,000–78,000	68,000–74,000
Yield strength, psi	60,000–66,000	55,000–61,000
Elongation in 2 in., %	22–28	28–33
Charpy V-notch impact (see specs for further details)	20 ft·lb at −20°F [−29°C]	

signed to produce high-quality, horizontal fillet welds at high welding speeds using either ac or DCRP current. In the flat position, AWS E6020 electrodes can be used with ac, or dc of either polarity.

These electrodes are characterized by a forceful spray-type arc and heavy slag which completely covers the weld deposit and may be easy to remove. Penetration into the base metal is medium at normal welding speeds, but high currents will give you deep penetration. Weld metal deposits are usually flat or even slightly concave in profile and they have a smooth, even-rippled appearance. The x-ray qualities of these welds are excellent (if you are a skilled welder and can do radiographic-quality work), but the weld bead shows a medium spatter and a tendency to undercut if you are not careful.

The AWS E6020 electrodes are mineral-coated with high percentages of iron oxide, manganese compounds, and silicates in the flux, along with sufficient deoxidizers to give your weld deposit the desired composition. The slag coverage is so extensive, and the slag-metal reaction is of such a nature, that these electrodes generally do not depend on the usual carbon-dioxide shielding-gas cloud formed by many other electrodes in this strength category.

The flux coatings also are such that an iron oxide, manganese oxide, silica slag is usually produced. Other materials such as aluminum, magnesium, or sodium may be present in the flux to modify this slag. Ferromanganese is used as the main deoxidizer; sodium silicate is used as a binder.

The quantity of basic oxide, acid silica, and silicates and deoxidizers must be carefully controlled to produce satisfactory operating characteristics and good weld metal. The heavy slag produced will be well honeycombed with bubbles on the underside (but the weld metal will be sound if you do your work correctly).

This slag porosity is the main reason that you can very easily remove it from your welds.

In general, AWS E6020 electrodes are recommended for horizontal fillet and flat-position welds where radiographic (x-ray or gamma-ray) requirements must be met. Fillet welds tend to have a flat or concave profile and a smooth, even ripple. In many instances the surface of the deposit is dimpled.

The more restricted the opening in which you deposit the metal, the greater the tendency toward surface dimples. So expect dimpled weld metal on practically all your weld beads when making the first few passes in deep-groove welds using AWS E6020 electrodes. As the welding nears completion, this tendency decreases.

Many welders have noted that ac tends to promote this dimpled condition more than dc. No undesirable mechanical or physical defects are associated with this dimpled surface condition.

High weld-metal deposition rates can be made in the flat position on heavy plate. Usually AWS E6020 electrodes are not used on thin sections due to their higher heat input, which will cause warping. For making horizontal fillet welds using conventional techniques, current values near the lower end of the range should be used if undercutting of your welds

is to be avoided. If you make deep-penetration fillet welds, higher currents should be used. Table 16-9 gives you the details.

AWS E6027 With a 50 percent iron-powder flux coating AWS E6027 electrodes have arc characteristics that are very similar to those of AWS E6020 electrodes. AWS E6027 electrodes are designed to produce satisfactory fillet or groove welds in flat positions with either ac, or dc of either polarity. They will produce flat or slightly concave horizontal fillet welds with either ac or DCSP.

AWS E6027 electrodes have a spray-type metal transfer through the arc and they deposit weld metal at high speed. The penetration into the base metal is medium and spatter loss is very low. In fact, the arc characteristics of AWS E6027 electrodes closely duplicate those of AWS 6020 electrodes. The slag, though very heavy and honeycombed on the underside, crumbles for easy removal. AWS E6027 electrodes are particularly well suited for multipass, deep-groove welding.

Welds produced with AWS E6027 electrodes have a flat to slightly concaved profile with a smooth, fine, even ripple and with good metal wash-up at the joint sides. The weld metal is apt to vary in radiographic quality and

TABLE 16-9 AWS E6020

Electrode Diameter, in.	Current, A	Voltage, arc volts
3/32	70–100	17–21
1/8	115–165	18–22
5/32	150–220	20–24
3/16	200–275	21–25
7/32	260–340	22–26
1/4	315–400	23–27
5/16	375–470	23–28

	As-welded	Stress-relieved
Tensile strength, psi	72,000–78,000	68,000–74,000
Yield strength, psi	60,000–66,000	55,000–61,000
Elongation in 2 in., %	22–28	28–33
Charpy V-notch impact	Not required	

be somewhat inferior to that made from AWS E6020 electrodes.

High welding currents are used to melt the high iron-powder flux coating of AWS E6027 electrodes. These electrodes are well suited for welding heavy sections.

In many respects, the AWS E6027 electrodes produce high-quality weld metal having physical properties closely duplicating those of AWS E6010 electrodes. Operating characteristics make this electrode slightly harder to handle than AWS E6024 electrodes. However, properly deposited weld beads will have a neater surface appearance. See Table 16-10 for details.

AWS E7024 AWS E7024 electrodes are ideally suited for production fillet welding. They are designed for horizontal fillet and flat positions using either ac or dc power sources.

The AWS E7024 electrodes, although generally used on mild steels, also produce satisfactory welds on many low-alloy, medium-, and high-carbon steels. These electrodes are particularly good for making fillet welds on mild steel. The welds are slightly convex in profile with a smooth surface and an extremely fine ripple. AWS E7024 electrodes are characterized by low spatter, low penetration into the base metal,

and they can be used at high welding speeds.

The iron-powder flux coatings of AWS E7024 electrodes help produce deposition rates and welding speeds considerably higher than those of AWS E6012, AWS E6013, or AWS E7014 electrodes, which have similar performance characteristics. Other than the high percentage of iron powder, the flux coating is very similar to the coatings commonly used in the AWS E6012 and AWS E6013 electrode classifications. See Table 16-11 for details.

AWS E7028 AWS E7028 are the last of the SMAW electrodes used for mild steels that we'll cover. They have a low-hydrogen coating containing 50 percent iron powder. Since they produce 70,000-psi [483-MPa] tensile-strength weld metal, you know that they are high-strength low-alloy steel electrodes that are used (in this instance) for mild steels. In many ways these electrodes are very much like AWS E7018 electrodes, although they differ in several respects.

AWS E7018 electrodes are designed for use in all positions. AWS E7028 electrodes are for high-volume production welding in downhand positions, making horizontal fillet and flat-position welds, only. The flux coating of AWS E7028 electrodes is much

thicker than that of AWS E7018 electrodes because it has a higher iron-powder content. Therefore, on horizontal fillet and flat welding, AWS E7028 electrodes have much higher deposition rates than AWS E7018 electrodes for any given electrode diameter. The flux coating of an AWS E7028 electrode actually equals about 50 percent of the total weight of the electrode.

The way that molten metal is transferred through the arc is different from AWS E7018 electrodes, too. While AWS E7018 electrodes have a spray-type transfer, AWS E7028 electrodes have a globular-type molten weld-metal transfer through the arc, which is the primary reason why AWS E7018 electrodes can be used in overhead and vertical welding while AWS E7028 electrodes are used for welding downhand at high production speeds. Apart from these differences, the mechanical properties of AWS E7028 and AWS E7018 electrodes are just about equal, like AWS E6010 and AWS E6011 electrodes and AWS E7015 and AWS E7016 electrodes. See Table 16-11, again, for details.

16-5 LOW-ALLOY-STEEL ELECTRODES

The list of electrodes and operating characteristics we just gave you only covers what are generally considered to be mild-steel SMAW electrodes, as described in the AWS A5.1 specification titled "Mild Steel Covered Arc-Welding Electrodes." This specification also includes the AWS E7015, E7016, E7018, and E7028 electrodes, because the boundary between mild steels and high-strength low-alloy steels isn't that clear-cut.

AWS A5.5 includes more than forty different classifications of low-alloy-steel electrodes. If you were to look into the AWS A5.5 specification on low-alloy steels, you would find one more important addition to the electrode grade designation code, a letter,

TABLE 16-10 AWS E6027

Electrode Diameter, in.	Current, A	Voltage, arc volts
3/16	175–250	22–26
7/32	225–325	23–27
1/4	250–350	24–29
5/16	325–460	24–30
3/8	450–600	26–32

	As-welded	Stress-relieved
Tensile strength, psi	62,000–70,000	60,000–67,000
Yield strength, psi	52,000–58,000	47,000–54,000
Elongation in 2 in., %	22–30	28–35
Charpy V-notch impact (see specs for further details)	20 ft·lb at −20°F [−29°C]	

or a letter and a number, attached to each basic electrode class with a hyphen. For example, the following lists only eight of the forty-two electrodes you can find in this filler-metal specification for welding high-strength low-alloy steels.

AWS E7018-A1
AWS E7020-G
AWS E8018-C3
AWS E9015-B3L
AWS E9018-D1
AWS E10018-M
AWS E11016-D2
AWS E12018-M

Note that the highest-strength electrodes in AWS A5.1 covering mild steel are the E70XX grades. These are the lowest-strength electrodes in AWS A5.5 covering high-strength low-alloy-steel electrodes.

Why do you find AWS E70XX electrodes listed in both specifications? Simply because the same basic electrode grade, E70XX, is used for welding both carbon steels (that is, mild steels). Also higher-strength steels require at least the same weld-metal strength produced by AWS

E70XX electrodes but with special restrictions on the alloying elements supplied by these electrodes so that they will produce very specific properties above and beyond a minimum weld-metal tensile strength of around 70,000 psi [483 MPa] . One frequently used example out of AWS A5.5 is the high-strength low-alloy-steel electrode E7018-A1.

The weld metal produced by an AWS E7018-A1 will not only equal the tensile strength of an AWS E7018 electrode, but it may have somewhat better low-temperature impact strength, depending on whether the steel is stress-relieved or not. At the very least, these extra mechanical-property requirements are more or less assured, because they have been made part of the filler-metal specification in the case of AWS E7018-A1, but not in the case of E7018.

The companion specification to AWS A5.1 on mild-steel electrodes, as we've already mentioned, is AWS A5.5, "Low-Alloy-Steel Covered Arc-Welding Electrodes." It begins at the weld-metal strength level of 70,000 psi

[483 MPa], includes AWS A7018 electrodes and six other E70XX electrodes, and also has six E80XX (80,000-psi [552-MPa] minimum tensile strength) electrodes, six E90XX (90,000-psi [620-MPa] minimum tensile strength) electrodes, six electrodes that produce steel weld metal with at least 100,000-psi [689-MPa] minimum tensile strength in the E100XX category, three electrodes in the E110XX category, and three more high-strength-steel electrodes in the E120XX category that produce weld metal with a minimum 120,000-psi [872-MPa] tensile strength. Table 16-12 lists them, with an "-X" hanging on the code name for the electrode. The X is a letter or two and a number. This tail is important. It identifies the specific chemical composition of the weld metal the electrode will produce.

The AWS is not trying to make life hard for you. What these letters and numbers indicate is that there are slightly different chemical compositions available for special jobs of equal tensile strength. This problem doesn't crop up very often when welding mild steels. However, it's often a consideration when welding high-strength low-alloy steels.

No welder would memorize all these differences. The alloying elements in each electrode are not anywhere near as important as the mechanical properties that these chemical additives produce in finished weld metal. That's where the differences are important. To find what the mechanical property differences are, you simply look them up in the appropriate AWS filler-metal specification.

The AWS gives many details on the mechanical properties of these different electrodes in the as-welded versus the stress-relieved condition because many of these higher-strength steels may be subjected to special stress relief or other heat treatments after welding. Very often, the tail on the end of the electrode name implies

TABLE 16-11 AWS E7024 and AWS E7028

Electrode Diameter, in.	Current, A	Voltage, arc volts
3/32	100–145	20–24
1/8	140–190	21–25
5/32	180–250	22–26
3/16	230–305	23–27
7/32	275–365	23–28
1/4	335–430	24–29
5/16	400–525	24–30

	As-welded	Stress-relieved
Tensile strength, psi	72,000–78,000	68,000–72,000
Yield strength, psi	60,000–66,000	55,000–61,000
Elongation in 2 in., % for AWS E7024	17–22	22–27
Elongation in 2 in., % for AWS E7028	22–28	27–32
Charpy V-notch impact for AWS E7024	Not required	
Charpy V-notch impact for AWS E7028 (see specs for further details)	20 ft·lb at 0°F [−18°C]	

an important difference in ductility or low-temperature impact strength, even when the weld-metal tensile strength and the type of flux coating is essentially the same.

You won't have much choice among which of these grades you will use; that will be determined in advance for you, often by a welding or metallurgical engineer, and sometimes by the welding code you are working with.

Filler metals that have an M attached to them, as in AWS E10018-M or AWS E12018-M, are a special military grade for welding submarine hulls and tank turrets. The G tail attached to the E7020-G electrode is typical of many AWS specifications. It lists only a very limited number of chemical, physical, or mechanical requirements beyond simple tensile strength and a few restrictions on things like sulfur and phosphorus. The G grades are mostly in the specifications to allow for proprietary electrodes, because many large companies will not buy welding electrodes unless they can meet some kind of AWS grade designation. The AWS created G grades so that everybody has a chance to compete, even if a new kind of electrode is developed. A lot of pipeline steels are welded in the field with electrodes ending in G.

Table 16-13 explains more about the tails on the electrode specifications.

Each one of these electrodes, including our now familiar AWS E7018-X, has this "tail" attached because many of these electrodes are quite special and have different mechanical properties depending on whether or not, and how, the weld metal and the base metal are stress-relieved after welding. That's also true with conventional E7018 electrodes (without the tail), but AWS A5.5 covering low-alloy-steel electrodes takes these differences into consideration, while AWS A5.1 covering mild-steel electrodes doesn't because these differences simply aren't that critical when welding most carbon steels.

We will not give you too many details (other than what you see in Table 16-13) on these grades because many of them are for pressure piping, ship hulls, special pressure vessels, steam piping, and other jobs that beginning welders are not likely to get involved with. But you should know that each of these grades has a recommended preheat and interpass temperature for welding them. They also have different maximum allowable moisture contents in the flux coating that cannot be exceeded. If these electrodes are exposed to the atmosphere for too long, they must be returned to an electrode baking oven for reconditioning.

These grades often are tied into very detailed specifications in addition to those written by the AWS. Examples are the ASME Boiler and Pressure Vessel Code, the American Petroleum Institute, the United States Bureau of Ships, and military specifications for everything from welding tanks and submarines to armored personnel carriers and howitzers. However, with what you are learning about the common SMAW electrodes, you will have no trouble catching up when you're ready for advanced work.

16-6
STAINLESS STEELS
So far we have not mentioned much about the different kinds of stainless steels. The welding metallurgy of stainless steel is complex. You could learn it if you need it, and sooner or later you probably will, but for the time being, let's just say that there are four basic kinds of stainless steels.

Kinds of Stainless Steels
The four kinds of stainless steels are *ferritic, austenitic, martensitic,* and *precipitation-hardening.* The different designations refer to the crystal structure of the iron that makes them. If you want to refer back to Chap. 8 on mechanical properties, we described some of these steel microstructures. The

TABLE 16-12 Weld-metal chemistry codes

EXXXX-A1	Carbon-molybdenum-steel electrodes with Mo from 0.40 to 0.65%
EXXXX-B1	Chromium-molybdenum-steel electrodes with Cr and Mo from 0.40 to 0.65%
EXXXX-B2	Chromium-molybdenum-steel electrodes with Cr from 1.00 to 1.50% and Mo from 0.40 to 0.65%
EXXXX-B2L	Chromium-molybdenum-steel electrodes with Cr from 1.00 to 1.50% and Mo from 0.40 to 0.65% and a low-carbon content of 0.05% (that's what the L signifies)
EXXXX-B3	Chromium-molybdenum-steel electrodes with Cr from 2.00 to 2.50% and Mo from 0.90 to 1.20%
EXXXX-B4L	Chromium-molybdenum-steel electrodes with Cr from 1.75 to 2.25% and Mo from 0.40 to 0.65% and a low-carbon content of 0.05%
EXXXX-B5	Chromium-molybdenum-steel electrodes with 0.40 to 0.60% Mo, 1.00 to 1.25% molybdenum, and a 0.05% trace of vanadium (V)
EXXXX-C1	Nickel-steel electrodes with 2.00 to 2.75% Ni
EXXXX-C2	Nickel-steel electrodes with 3.00 to 3.75% Ni
EXXXX-C3	Nickel-steel electrodes with 0.80 to 1.10% Ni, 0.15% Cr, 0.35% Mo, 0.05% V
EXXXX-D1	Manganese-molybdenum-steel electrodes with 1.25 to 1.75% Mn and 0.25 to 0.45% Mo
EXXXX-D2	Manganese-molybdenum-steel electrodes with 1.65 to 2.00% Mn and 0.25 to 0.45% Mo
EXXXX-G	All other low-alloy-steel electrodes . . . at least 1.00% Mn plus various amounts of nickel, chromium, molybdenum and vanadium
EXXXX-M	Special military grades

TABLE 16-13 The low-alloy electrodes*
As-Welded Mechanical Properties

AWS Classification	Minimum Tensile Strength, psi	Minimum Yield Strength, psi	Minimum % Elongation in 2 in.	Minimum Charpy V-Notch Impact Strength
Carbon-Molybdenum-Steel Electrodes				
E7010-A1	70,000	57,000	22	Not required
E7011-A1	70,000	57,000	22	Not required
E7015-A1	70,000	57,000	25	Not required
E7016-A1	70,000	57,000	25	Not required
E7018-A1	70,000	57,000	25	Not required
E7020-A1	70,000	57,000	25	Not required
E7027-A1	70,000	57,000	25	Not required
Chromium-Molybdenum-Steel Electrodes				
E8016-B1	80,000	67,000	19	Not required
E8018-B1	80,000	67,000	19	Not required
E8015-B2L	80,000	67,000	19	Not required
E8015-B4L	80,000	67,000	19	Not required
E8016-B2	80,000	67,000	19	Not required
E8016-B5	80,000	67,000	19	Not required
E8018-B2	80,000	67,000	19	Not required
E8018-B2L	80,000	67,000	19	Not required
E9015-B3L	90,000	77,000	17	Not required
E9015-B3	90,000	77,000	17	Not required
E9016-B3	90,000	77,000	17	Not required
E9018-B3	90,000	77,000	17	Not required
E9018-B3L	90,000	77,000	17	Not required
Nickel-Steel Electrodes				
E8016-C1	80,000	68,000	19	20 ft·lb at −75°F [24°C]
E8016-C2	80,000	68,000	19	20 ft·lb at −100°F* [38°C]
E8016-C3	80,000	68,000	24	20 ft·lb at −40°F [4°C]
E8018-C1	80,000	68,000	19	20 ft·lb at −75°F [24°C]
E8018-C2	80,000	68,000	19	20 ft·lb at −100°F* [38°C]
E8018-C3	80,000	68,000	24	20 ft·lb at −40°F [4°C]
E9015-D1	90,000	77,000	17	20 ft·lb at −60°F [15°C]
E9018-D1	90,000	77,000	17	20 ft·lb at −60°F [15°C]
E10015-D2	100,000	87,000	16	20 ft·lb at −60°F [15°C]
E10016-D2	100,000	87,000	16	20 ft·lb at −60°F [15°C]
E10018-D2	100,000	87,000	16	20 ft·lb at −60°F [15°C]
Manganese-Molybdenum-Steel Electrodes				
E9015-D1	90,000	77,000	17	20 ft·lb at −60°F [15°C]
E9018-D1	90,000	77,000	17	20 ft·lb at −60°F [15°C]
E10015-D2	100,000	87,000	16	20 ft·lb at −60°F [15°C]
E10016-D2	100,000	87,000	16	20 ft·lb at −60°F [15°C]
E10018-D2	100,000	87,000	16	20 ft·lb at −60°F [15°C]
Other Low-Alloy-Steel Electrodes				
EXX10-G	All other data depend upon chemical composition of weld metal produced. These electrodes often are made for welding pipelines and other uses. Many of them are proprietary welding electrodes.			
EXX11-G				
EXX13-G				
EXX15-G				
EXX16-G				
EXX18-G				
E7020-G	70,000			
Typical Military Electrodes				
E9018-M	90,000	78,000	24	20 ft·lb at −60°F [15°C]
E10018-M	100,000	88,000	20	20 ft·lb at −60°F [15°C]
E11018-M	110,000	98,000	20	20 ft·lb at −60°F [15°C]
E12018-M	120,000	108,000	18	20 ft·lb at −60°F [15°C]

* See AWS specifications for further details on individual grades and how they differ.
† Data are for stress-relieved weld metal.

way the iron and carbon are arranged in the crystal lattice of the steel has a great deal to do with how it is hardened and strengthened. That's never more true than in the complex field of stainless steels.

Ferritic stainless steels The first kind of stainless steels have so little carbon in them that they are irons not steels that contain a lot of chromium. Chromium, when it's over 12.5 percent in iron or steel, makes it a full stainless steel instead of a chromium-alloy steel.

Ferritic stainless steels do not act like carbon and low-alloy steels. They are not hardened by heat treatment. Rather they work-harden by being rolled at the steel mill or in some forming process such as deep drawing or forging. These steels are included in an American Iron and Steel Institute category called the "400 series" stainless steels, along with martensitic stainless steels, which we'll mention in a minute. AISI Type 405 is a typical ferritic stainless-steel grade. Type 405 stainless steels not only have good corrosion resistance but also good resistance to high-temperature environments. Furnace parts often are made from them.

AISI steel designations are numbers that signify chemical compositions. Specific mill forms (pipe, tubing, plate, sheet, strip, structurals, wire) are specified by separate ASTM (American Society for Testing and Materials) specification numbers.

One example is ASTM specification A 514 covering welded stainless-steel mechanical tubing, that is, tubing made to very high precision dimensions (unlike pipe) and produced to specific strength levels. It has a welded seam instead of being extruded like seamless tubing. There are literally thousands of ASTM materials specifications for everything from steel pressure-vessel plate to copper trolley wire and the graphite used to make motor brushes and plastics used to make toys.

All ASTM specifications that begin with the letter A are specifications for ferrous materials. Nonferrous ASTM specifications begin with the letter B. Testing procedures, such as how to make a Rockwell hardness test, begin with the letter D, and so on through everything from paint to rubber. The entire ASTM *Book of Standards* has over 30 large volumes containing detailed specifications.

Unlike ASTM specifications, which take mill form into consideration, AISI steel specifications are chemical codes telling you a little about what elements are in the steel. The AISI ferritic stainless steels include AISI Types 405, 409, 429, 430, 430F (a free-machining grade), 430FSe (another free-machining grade with selenium added as well as sulfur for extra machinability), 434, 436, 442, and 446. These nonhardenable (by heat treatment) ferritic stainless steels develop their maximum ductility and corrosion resistance in the annealed condition.

Austenitic Stainless Steels

Austenitic stainless steels are composed of iron that is in a face-centered cubic crystal lattice instead of the body-centered cubic crystal lattice of ferritic iron. Austenitic stainless steels which have around 8 percent nickel (or more) in them, along with lots of chromium, also are not hardened by heat treatment but work-harden to high strengths by cold-rolling. Like ferritic stainless steels, most of them have very little carbon in them. As a group, they often are called 18-8 stainless steels because most of them have around 18 percent chromium and 8 percent nickel. They are grouped together in an AISI category called the 300 series stainless steels.

Some austenitic stainless steels have more manganese and less nickel in them and they are in the AISI category called the 200

series stainless steels. The 18-8, AISI Type 300 series stainless steels and their AISI 200 cousins are usually the most corrosion resistant of the stainless steels. Chemical plants make wide use of them. Stainless-steel kitchenware and architectural siding are other uses. The four most important grades are AISI Types 302, 304, 308, and 316. There are other austenitic stainless steels such as AISI 201 and 202 that contain less nickel, which is replaced by more manganese, making them less expensive than the 300-series austenitic grades.

The nonhardenable (by heat treatment) chromium-nickel austenitic stainless steels include AISI grades 201, 202, 205, 301, 302, 302B, 303, 303Se, 304, 304L (an extra-low-carbon grade), 304N, 305, 308, 309, 310, 310S, 314, 316, 316F, 316L, 316N, 317, 317L, 321, 329, 330, 347, 348, and 384. These stainless steels are essentially nonmagnetic in the annealed condition. Cold-working (rolling, forging, wire drawing) develops extremely high tensile strengths at which point they may become very slightly magnetic. These steels often will show a wide gap between the yield strength and tensile strength. That is, they often are very ductile. (Remember what the stress-strain curve looks like? Take a look at Figs. 8-4, 8-5, and 8-6 in Chap. 8, on pages 182, 183, and 184 if you've forgotten.)

When austenitic stainless steels are annealed (heated to high temperatures) and then cooled very rapidly, instead of being quenched and hardened like steels containing more than a tiny amount of carbon in them, these steels actually get soft and very ductile. As a group they are exceptionally corrosion resistant.

AISI Type 304 stainless steel, even in the soft-annealed condition, will have a tensile strength of 84,000 psi [579 MPa] but a yield strength of only 42,000 psi [290 MPa]. Hand in hand with such a high-tensile-strength and

low-yield-strength is the material's very high ductility. Annealed Type 304 sheet often will have an elongation in 2 in. of 55 percent. It is extremely formable and "stretchable" and often is made into complex shapes by stamping, deep drawing, or forging.

Martensitic Stainless Steels

Martensitic stainless steels act very much like carbon steels and low-alloy steels because they do contain carbon—sometimes a lot of it. When these steels are rapidly cooled from certain high temperatures where austenite likes to exist, the austenite iron atom stacking (face-centered cubic) wants to change to ferrite (body-centered cubic) crystals. However, the carbon jams up the crystal lattice, and instead of forming ferrite, a very strong, hard steel microstructure called *martensite* forms. It's the martensite that makes carbon, alloy, and stainless steels with significant carbon content become very hard and strong when they are quenched from high temperatures.

The martensitic stainless steels are included in the AISI series called the 400 stainless steels. Like all the other stainless steels, each individual grade has a three-digit AISI number that identifies it. One example is Type 440A stainless steel. It, and its relatives Types 440B and 440C, are used where a heat-treatable, high-carbon martensitic stainless steel is needed to make good-quality cutlery, expensive hunting knives, and the blades that cut bark off trees in a paper pulp mill.

The AISI martensitic stainless steels include the standard AISI grades 403, 410, 414, 416, 416Se, 420, 422, 431, 440A, 440B, 440C, 501, and 502. With the use of different kinds of heat treatment, these steels develop an extraordinary range of mechanical properties and hardness. It's not unusual to see heat-treated martensitic stainless steels with yield strengths of 150,000 to over 200,000 psi [1035 to 1380 MPa] and tensile strengths as high as 280,000 psi [1930 MPa]. Martensitic stainless steels offer high strength, high hardness, and good corrosion resistance, but nowhere near the formability of ferritic or austenitic stainless steels.

Precipitation-Hardening Stainless Steels

Last but not least is a group of very special stainless steels often called *precipitation-hardening alloys*. These chromium-nickel alloys have additional elements that make them hardenable by a very special heat treatment called *solution treating and aging*. This complex type of heat treatment normally is used on aluminum alloys. The precipitation-hardening stainless steels get very hard and strong by this type of heat treatment for entirely different reasons than carbon steels. They have names such as 17-4 PH, 17-7 PH, 15-5 PH, PH 15-7 Mo, and PH 14-8 Mo, as well as names such as AM 362 and AM 363.

The precipitation-hardening stainless steels are essentially aerospace alloys or specialty steels. They are used for things like boat propeller shafts and the landing gears of large jet airliners. The precipitation-hardening high-alloy steels include martensitic, ferritic, and austenitic grades.

Many of these alloys, and other special very high chromium stainless steels (often with molybdenum added to them), are included in an AISI 500 series of steels called high-strength high-temperature alloys. Two key examples are AISI Type 501 and Type 502. As we already mentioned, these high-strength high-temperature grades are martensitic stainless steels. They are especially useful for parts and piping that operate at temperatures of 500 to 1000°F [260 to 538°C] and higher.

Table 16-14 lists the standard AISI stainless steels and tells you something about them. With this very brief introduction to stainless steels, let's look at the elec-

trodes most often used for welding them.

Stainless-Steel Electrodes

Stainless-steel electrodes and electrodes for other very high chromium ferrous metals are grouped together in AWS specification A5.4 titled "Corrosion-Resisting Chromium and Chromium-Nickel Steel Covered Arc-Welding Electrodes." We'll cover the major types of stainless-steel electrodes.

The information here on which type of stainless-steel weld metal to use to weld each type of stain-

---

TABLE 16-14 The standard AISI stainless steels

Ferritic Stainless Steels

AISI Types 405, 409, 429, 430, 430F, 430FSe, 434, 436, 442, 446

Type 430 accounts for 75 percent of all the straight-chromium nonhardenable (by heat treatment) ferritic stainless steels used. Type 430F is the equivalent free-machining grade. So is Type 430FSe. Ferritic stainless steels are often used for their excellent corrosion resistance in strong acids. They often are called the *straight chromium* stainless steels because they are essentially chromium-bearing irons with very little or no carbon or nickel in them. Grades other than Type 430 are used for special reasons ranging from higher corrosion resistance to increased weldability or higher strength.

Austenitic Stainless Steels

AISI Types 201, 202, 205, 301, 302, 302B, 303, 303Se, 304, 304L, 304N, 305, 308, 309, 310, 310S, 314, 316, 316F, 316N, 316L, 317, 317L, 321, 329, 330, 347, 348, 384

As a group, the austenitic stainless steels are hardenable by work-hardening (cold-rolling or forging for example). They are generally the most corrosion resistant of the stainless steels. They contain a lot of nickel as well as chromium. The general designation 18-8 stainless steels (18 percent Cr and 8 percent nickel) fits many of them. Some have even higher chromium, nickel, and other alloying elements.

The grades followed by an L or an ELC (for extra-low carbon) are used when welding might otherwise create carbide precipitation and loss of corrosion resistance in the heat-affected zone. Types 321, 347, and 348 also are used to prevent carbide precipitation. The carbides in these grades are stabilized with columbium or other elements. Type 301 has the fewest alloying elements and is the least expensive of the 300 series. Types 303 and 303Se are most important free-machining grades and they are not very weldable. Types 302 and 304 are most commonly used.

Martensitic Stainless Steels

AISI Types 403, 410, 414, 416, 416Se, 420, 422, 431, 440A, 440B, 440C, 501, 502

The martensitic stainless steels are hardenable by heat treatment. They all contain carbon, ranging up to very high carbon grades (up to 1.20 percent C) such as Types 440A, 440B, and 440C. As a group, they tend to be the least costly of the stainless steels. They also have lower corrosion resistance as a group than ferritic and austenitic stainless steels. They can be hardened and strengthened by heat treatment to very high levels.

The free-machining grades of the martensitic stainless steels such as Types 416 and 416Se contain sulfur and/or selenium additives that make them very difficult to weld. The very high carbon grades such as Types 440A, B, and C also are not considered to be weldable with conventional processes. Types 501 and 502 are typical high-strength high-temperature alloys.

Precipitation-Hardening Stainless Steels

Precipitation-hardening stainless steels are mostly used in aerospace applications and where very high strength and corrosion resistance are needed. They are hardened by special and rather complex heat-treating procedures that strengthen them without causing much distortion.

less-steel base metal also applies to GMAW with MIG, which we cover in the next chapter, as well as to the bare stainless-steel welding rods used in TIG welding. The primary difference is that MIG uses continuous bare welding wire and a shielding gas so you don't have to worry about flux. GTAW uses bare-wire filler metals cut to length.

Unlike mild-carbon and low-alloy-steel electrodes, there are basically only two types of electrode flux-coating designations to worry about with stainless-steel SMAW electrodes. They are flux coatings that make the electrodes suitable for ac welding, and coatings that make them work for dc welding. The electrode designations are much simpler, too. The type of stainless steel that the electrode will deposit is right in its name. Let's look at two of them.

AWS E308-15 is an electrode with dc flux coating. It will produce AISI Type 308 stainless-steel weld metal. AWS E308-16 is an electrode with a flux coating for ac and dc welding. It also produces AISI Type 308 austenitic stainless-steel weld metal. That is,

EXXX-15 is a stainless-steel dc electrode only.
EXXX-16 is a stainless-steel ac-dc electrode.
The XXX gives the AISI grade designation for the weld metal.

DC flux coatings on stainless-steel electrodes are basically lime (calcium) flux coatings. The electrodes that use this type of coating have excellent chracteristics for out-of-position welding. These are the only stainless-steel electrodes that can be used for *vertical-down* welding and they are almost always used with DCRP.

Proprietary stainless-steel electrodes often work on both ac and dc. Many proprietary electrodes are available because manufacturers make a better profit on them than on carbon-steel electrodes, so they can invest more money in research and development.

The arc characteristics of EXXX-15 lime coatings are much less stable than those you'll get with the next ac coating we'll talk about. The weld beads are more irregular in appearance than the ac flux-coated electrodes will produce, but the quality of the weld deposit is superior to that produced by the ac stainless-steel electrodes. Lime-coated EXXX-15 stainless-steel electrodes are preferred for use when high-restraint joints are to be welded. They are much less crack sensitive than the titania-coated ac electrodes.

Stainless-steel electrodes with the designation EXXX-16 were originally intended to be ac electrodes. Most commercial EXXX-16 electrodes now operate either on ac or dc. They have high percentages of titania in the flux coating to help stabilize the arc for ac operation. Other materials such as potassium are added for the same reason.

These EXXX-16 flux-coated stainless-steel electrodes will perform out-of-position welding for you *except for vertical-down welding*. The slag is much more fluid than that of the lime-coated dc electrodes. The EXXX-16 ac electrodes with a high-titania flux coating will give you flat-contoured weld beads and a very smooth, good-looking bead. This fact makes the flux-coated EXXX-16 electrodes operated with ac an excellent choice for fabricating food-processing equipment, consumer products, and other things where a superior surface finish is needed on the weld bead. The welds produced by ac-type EXXX-16 stainless steel electrodes are much easier to grind to a smooth-blending contour to the base metal. The EXXX-16 stainless steel electrodes are, by far, the most popular stainless steel electrodes. However, because of their flat bead contour and the high-titania content in the flux coating, these electrodes are more prone to cracking than the lime-coated dc EXXX-15 stainless-steel electrodes.

Problems of Welding Stainless Steels

Stainless steels present a lot of new welding problems for you. These problems are not so much a function of the welding process you use as simply a result of the unusual microstructure of stainless steels and the very difficult conditions such as corrosive fluids and high temperatures that they are expected to operate in.

Carbide precipitation One of the problems you get when welding stainless steels with carbon-bearing electrodes is *carbide precipitation*. You can't see it. You won't know it's happened until the stainless weldment is put into service in an extremely corrosive environment. The reason it occurs is well worth explaining because some of the stainless-steel electrodes you can choose are designed specifically to prevent it.

Ordinarily, the chromium in an austenitic stainless steel remains dissolved in the solid face-centered-cubic iron crystal lattice that differentiates austenite from ferrite. The body-centered iron atoms in ferritic iron are at the corners of a cube with one iron atom left over in the middle of the cube, like stacked cannon balls. When the iron atoms rearrange themselves to form austenite, they are still at the corners of a cube, but there is no iron atom in the center of the cube. Instead, other iron atoms take up places in the six faces of the cubic crystal lattice. The chromium atoms in stainless steels take up some of the spaces where iron atoms would normally sit in a pure iron crystal lattice.

When your austenitic stainless-steel base metal has some carbon in it (it doesn't take much), the chromium will combine with the carbon to make chromium carbides when the base metal is heated or cooled within the temperature range of 800 to 1550°F [425 to 843°C]. When part of the chromium that normally keeps the stainless-steel stainless is tied up by carbon to make chromium carbide, the area where this occurs becomes much less stainless. That's when severe hairline corrosion can start in the heat-affected zone of your weld when it is put into service.

Carbide precipitation occurs in the base metal, not in the weld metal. It occurs along the narrow heat-affected zone in the base metal right near your weld, because that's the part of your weldment that will most likely be held at the 800 to 1550°F [427 to 843°C] temperature long enough to let the carbon atoms grab up the available chromium and "take it out of solution" as chromium carbides.

As a result, you sometimes will see good-looking stainless-steel parts with a red rust line running like a knife edge on either side of good-looking stainless-steel weld metal. The effect is called *intergranular corrosion,* because the corrosive medium attacks the chromium-depleted steel along the crystal grain boundaries. A corrosion-resistant pipe or tube, or a reactor vessel in a chemical plant, can literally come apart at the seams (or at least right next to them) when carbide precipitation has occurred.

There are four ways to solve that problem. One is to make the weld and base metal cool very fast through the dangerous 1550 to 800°F [843 to 425°C] zone so that the chromium doesn't have time to hook up with very much carbon. A water quench will do the job nicely. The second solution is to use an extra-low-carbon (ELC) austenitic stainless-steel base metal. AISI Type 304L—the L standing for extra-low carbon—is an example.

The third solution is to use austenitic stainless-steel base metal that has special elements in it to keep the chromium from making chromium carbides. One example is AISI Type 347, which is a stainless steel whose carbides are stabilized by columbium (Cb) and tantalum (Ta). Carbon would

much rather form columbium or tantalum carbides than chromium carbides. Therefore, the extra Cb and Ta use up the leftover carbon in solution, so that the chromium remains in solid solution in the weld metal and base metal to keep your steel corrosion-resistant.

The fourth way to avoid the problem is to reheat the entire weldment to about 1900°F [1040°C] and hold it at that temperature for a length of time that depends on the thickness of the base metal. This *solution-anneal* heat treatment will redissolve the chromium carbide and put the chromium back into solution where it belongs. Then, of course, the entire weldment must be cooled quickly through the dangerous 1550 to 800°F [843 to 425°C] temperature range so that the problem won't occur again. That last solution can be pretty difficult if the weldment is a very large chemical tank which won't fit in a heat-treating furnace.

The most common alloy additive used to stabilize both your weld metal and the base metal to keep the carbon from combining with the chromium, is the chemical element columbium. Its chemical symbol is Cb. AISI Type 309Cb stainless steel is an example. Columbium works (as we mentioned) because it has a bigger desire to make carbides than chromium does. Therefore the carbides that form are columbium carbides, leaving the chromium free.

The problem for you is that you have to match your stainless-steel filler metal to the base metal. There are several different stainless-steel welding electrodes that are designed specifically for use with austenitic stainless-steel base metals to prevent carbide precipitation.

Temper embrittlement A problem with martensitic stainless steels that you might have to watch out for is called *temper embrittlement.* When these higher-carbon stainless steels are hardened by heat treatment and rapid

quenching, they must be reheated (tempered) to a lower temperature to get back some ductility. But tempering them in the range of 750 to 1050°F [400 to 565°C] is not recommended because it produces low-impact properties and you lose some of the corrosion resistance. The reasons for temper embrittlement are too complicated to explain here. The result, however, can be a serious welding problem.

The selection of the correct stainless-steel electrode for any given stainless-steel base metal depends on a lot more than what we've told you. Stainless steels are an extremely complicated subject. Trained metallurgists or welding engineers often are required to advise you on how to work with them. You not only have to pick a stainless-steel welding electrode that will produce weld metal matching the mechanical properties of the base metal, but the corrosion resistance of the base metal too; and corrosion is an even more complex subject than welding.

Welding Different Stainless Steels Together

You may have to weld two different kinds of stainless steel together that don't have the same properties. Compound that with the problem of matching the color of the base metal with the weld metal, because stainless steels usually are not painted and they often are used where appearance as well as corrosion resistance is important. (You may think that all stainless steels look alike; but you'll find that they don't, and you often can see a color difference between the weld metal and the base metal.) Then try solving the common problem of joining two different kinds of stainless steel together with a filler metal that has to work well with both of them.

Table 16-15 won't solve all your problems, but it does give a fairly complete list of stainless-steel welding electrodes and the stainless-steel base metals most likely

TABLE 16-15 Stainless-steel filler-metal selection chart

Base-Metal Alloy	201, 202, 301, 302, 302B, 303ᵃ, 304, 305, 308	304L	309 309S	310 310S 314ᵃ,	316	316L	317	317L
201, 202 301, 302 302B, 303ᵃ 304, 305 308	E308	E308	E308	E308	E308	E308	E308	E308
304L		E308L	E308	E308	E308	E308	E308	E308
309, 309S			E309	E309	E309	E309	E309	E309
310, 310S 314ᵃ				E310	E316	E317	E308	E310
316					E316	E316	E316	E316
316L						E316L	E316	E316L
317							E317	E317
317L								E317L
321, 347 348								
330ᵃ								
403, 405 410, 414 416, 420								
430, 430F 431, 440A 440B, 440C								
446								
501, 502								
505								

Note: Use this chart for welding any stainless-steel base metal to any other. The stainless-steel base metals are at the top and left side. Suggested stainless-steel electrodes (or wires) for welding them are in the chart.

Notes: Grades shown are those most selected for most applications; other combinations may be used. Wherever possible, recommendation is based upon the most-available and lowest-cost filler metal. Filler-metal designations are those appearing in AWS Specification A5.4 for covered electrodes, A5.9 for bare filler wire, and A5.22 for flux-cored wire.

ᵃ These alloys are sensitive to weld cracks and fissures; for this reason, E312 filler metal is a frequently recommended alternative. It is preferred especially when thick sections or highly restrained joints are required. Buttering these metals with Type 312 before joining is often desirable. Use EXXX-15 electrodes for SMAW process.

to be joined by them. It's a starting point to every stainless-steel welding problem. It would take a book this size to cover all possible problems and solutions.

16-7 COPPER-ALLOY ELECTRODES

You may recognize the SMAW filler-metal designations for electrodes used for copper and its alloys. Remember your work with brazing and braze-welding copper and bronze gas rods in Chap. 7? Add an E to the copper-alloy filler-metal designation instead of an R for a gas (or TIG) rod and you'll have a covered SMAW copper or copper-alloy electrode.

AWS A5.6, "Specification for Copper and Copper-Alloy Arc-Welding Electrodes," includes eight classes of these SMAW electrodes. We'll list them for you in Table 16-16 along with the type of weld metal they produce and the minimum tensile strength of that as-deposited weld metal. They are used in much the same ways you used them for gas welding. Some can join steels and cast irons as well as copper and its alloys.

321 347 348	330[a]	403, 405 410, 412 414, 420[c]	430, 430F 431, 440A 440B, 440C[c]	446[d]	501 502[c,d]	505[c,d]	Carbon Steels[c,d]	Low-Alloy Chromium Molybdenum Steels[c,d]
E309	E309	E309[f]	E309	E310	E309	E309	E309	E309
E309	E309	E309	E309	E310	E309	E309	E309	E309
E309	E309	E309	E309	E310	E309	E309	E309	E309
E309	E309	E309	E309	E310	E309	E309	E309	E309
E308[b]	E309Mo	E309	E309	E310	E309	E309	E309	E309
E316L	E309Mo	E309	E309	E310	E309	E309	E309	E309
E308[b]	E309Mo	E309	E309	E310	E309	E309	E309	E309
E308L	E309Mo	E309	E309	E310	E309	E309	E309	E309
E347	E309	E309	E309	E310	E309	E309	E309	E309
	E330	E309	E309	E310	E312	E312	E312	E312
		E410	E430[e]	E410[e]	E502[e]	E505[e]	E410[e,f]	E410[e]
			E430	E430	E502[e]	E505[e]	E430[e,f]	E430[e]
				E446	E502[e]	E502[e]	E430[e,f]	E430[e]
					E502	E502[e]	E502[e,f]	E502[e]
						E505	E505[e,f]	E505[e]

[b] E16-8-2 is preferred to lower embrittlement danger in elevated temperature service.
[c] When joining an austenitic steel, alternate choice is to butter carbon or chromium steel with E309 and join with E308 or with filler metal similar to austenitic base metal. E307 is also commonly used for welds between austenitic stainless steel and either carbon or low-alloy steels.
[d] ENiCrFe3 is preferred for elevated temperature service, except when sulfur compounds are present.
[e] If austenitic weld metal is acceptable for service conditions, E309 or E310 is often employed.
[f] E7018 and mild steel wires are frequently acceptable if preheating is used and hydrogen in weld zone is kept as low as possible.

Copper metal, generally speaking, comes in three types. There is deoxidized copper, oxygen-free copper, and electrolytic tough-pitch copper. We've listed the three kinds of copper metal in order of increasing electrical conductivity. Deoxidized copper has the lowest conductivity and electrolytic tough-pitch copper (very pure copper) has the highest.

AWS ECu (Copper)
Electrolytic tough-pitch copper is not often welded. Welding will add elements that will lower the material's conductivity. Mechanical fasteners or high-conductivity brazing alloys often are used to join it. The material also has low tensile strength compared with other copper alloys. It's mostly used in electrical conductors. The deoxidized coppers, on the other hand, can be welded to produce joints of maximum strength. That's where the high-copper-content AWS ECu covered electrodes are used.

Covered electrodes for manual SMAW are available for fabricating silicon bronzes whenever it is not economical or practical to use MIG welding, which is the preferred method. Silicon-bronze electrodes (the AWS ECuSi class) are also used to weld copper, dissimilar metals, and some iron-base metals. Silicon-bronze weld deposits are seldom used for surfacing bearings but they are used to surface areas subject to corrosion.

AWS ECuSi electrodes are used with DCRP (electrode-positive) and are operated very much like mild-steel electrodes. You should follow the supplier's instructions for volt and amperage settings on your power source.

AWS ECuSn (Copper-Tin) Bronze

AWS ECuSn are copper-tin electrodes for joining phosphor bronzes of similar composition. They also are useful for joining brasses and, in some cases, for welding cast iron and mild steel. These covered electrodes are usually designed for operation with DCRP. The weld metal produced by phosphor-bronze electrodes has a tendency to flow sluggishly. You often will have to preheat your workpiece to at least 400°F [205°C] and keep it there while you work, especially if you have to work on thick sections. Post-weld heat treatment may not be

necessary, but to get maximum weld-metal ductility, you should use post-weld heat treatment, such as a stress-relieve anneal. This is even more important if the copper has been cold-worked. Like austenitic stainless steels, copper gets harder and less ductile when cold-worked. Any moisture on the workpiece and in your electrode flux coatings must be avoided. You should bake your electrodes at 250 to 300°F [121 to 150°C] before using them.

The two basic copper-tin (bronze) electrode grades, ECuSn-A and ECuSn-C, differ slightly in that ECuSn-A electrodes are used mostly to weld plates of similar bronze composition. They may also be used for welding copper if the reduction in electrical conductivity isn't important.

The AWS ECuSn-C electrodes have higher tin content than ECuSn-A electrodes. These higher-tin electrodes produce harder weld metal with higher tensile and yield strength than AWS ECuSn-A electrodes. Tin is a strengthening element for copper. In fact, the warriors in the Bronze Age used copper-tin alloy shields and swords. After all, copper-tin is bronze, and some bronze

alloys can be produced to higher strengths than many carbon steels.

AWS ECuNi (Copper-Nickel)

AWS electrodes in the ECuNi class are very important. They are used mostly for welding copper-nickel alloys. Many copper-nickel alloys are resistant to corrosion in seawater. Therefore, you will find them used in everything from boatyards and shipyards to seawater desalting plants. AWS ECuNi electrodes are operated with DCRP with a medium to short arc. Although the covered electrodes may be used in all positions, welding is best done in a flat position. In general, you won't have to preheat the weld because copper-nickel base metal has lower thermal conductivity than alloys with more copper.

AWS ECuAl (Copper-Aluminum) Bronze

There are three different AWS ECuAl electrodes (copper-aluminum electrodes). Two of them (ECuAl-A2 and ECuAl-B) also have iron in them.

Iron-free ECuAl Iron-free AWS ECuAl-A1 electrodes are used primarily for welding annealed aluminum-bronze plate, sheet, and strip and for repairing castings of similar composition. They also are used to surface bearings and for making corrosion-resistant surfaces.

Iron-bearing ECuAl The iron-bearing AWS ECuAl-A2 electrodes produce higher yield and tensile strength weld metal than the iron-free AWS ECuAl-A1 electrodes, but the ductility is lower. They are used for welding high-strength copper-zinc alloys, silicon bronzes, manganese bronze, some nickel alloys, many ferrous metals and alloys, and combinations of dissimilar metals. The weld metal is also suitable for surfacing products that need good bearing strength, and wear and corrosion resistance. AWS ECuAl-A2 elec-

TABLE 16-16 Copper-alloy-covered electrodes

AWS Classification	Type of Weld Metal	Minimum As-Welded Tensile Strength, psi
ECu	98% copper	25,000
ECuSi	Silicon-bronze	50,000
ECuSn-A	Phosphor-bronze	35,000
ECuSn-C	Phosphor-bronze	40,000
ECuNi	Copper-nickel	50,000
ECuAl-A1	Aluminum-bronze	55,000
ECuAl-A2	Aluminum-bronze	60,000
ECuAl-B	Aluminum-bronze	65,000
For Cast-Iron Welding		
ECuSn-A	Phosphor-bronze	Depends on how much dilution of the weld metal occurs. AWS ECuAl-A2 produces highest weld strength. It's normally your best choice.
ECuSn-C	Phosphor-bronze	
ECuAl-A2	Aluminum-bronze	

trodes operate on DCRP. From the wide variety of applications you might guess that these electrodes are good for general-purpose maintenance. You're correct.

The iron-bearing AWS ECuAl-B electrodes are used to join aluminum bronzes of similar composition, high-strength copper-zinc alloys, silicon bronzes, manganese bronzes, some nickel alloys, many ferrous metals and alloys, and combinations of dissimilar metals. The as-deposited weld metal has higher yield and tensile strength but is less ductile than that produced by AWS ECuAl-A2 electrodes. The aluminum-bronze weld metal produced by these electrodes may have ultimate and tensile and yield strengths almost double those of the copper-tin bronze electrodes, plus higher ductility. The weld metal also is suitable for surfacing bearings, and wear- and corrosion-resisting surfaces. All of these electrodes operate on DCRP.

It's not hard to see why AWS ECuAl-A2 and AWS ECuAl-B electrodes are very useful, nearly all-purpose covered electrodes. They're both good choices for repairing broken cast-iron parts, among other things. They are good maintenance arc-welding electrodes, but you have to use them only in the flat position, which can be a drawback for maintenance welding. You also will have to preheat iron castings and steel base metals to 200 to 300°F [93 to 150°C] and silicon bronzes from 300 to 400°F [150 to 205°C], and brasses from 500 to 600°F [260 to 315°C] if you use these electrodes. Preheating will avoid base-metal or weld-metal cracking and other problems.

16-8
NICKEL ELECTRODES
There are twelve nickel-based covered electrodes listed in the AWS A5.11 specification titled "Nickel and Nickel-Alloy Covered Arc-Welding Electrodes." They are listed in Table 16-17. They

have many uses when welding different kinds of nickel-based alloys, but we won't get into that. You are not likely to run into nickel-based metals as a beginning welder. What you very frequently will have to do is repair cast iron. That's where some of these AWS nickel-covered electrodes really shine.

In fact, if you were to turn to another AWS specification, AWS A5.15, "Welding Rods and Covered Electrodes for Welding Cast Iron," you would find a slug of nickel electrodes listed, along with gas-welding rods for welding cast irons that we already have described in the chapter on oxy-fuel welding.

The nickel electrodes in Table 16-17 that are used to weld cast irons include AWS ENi-CI, ENiFe-CI, and ENiCu (nickel-copper). Each of these SMAW electrodes has different amperage and voltage settings that also depend on the diameter of the electrode used, so your best bet is to look up the correct operating settings when you use them. Very often, you will find the data printed right on the electrode package.

Don't forget that most of the practices we recommended for gas

welding cast iron apply equally well to arc welding this material, including cleaning the base metal before welding and especially preheating the weld, and postweld heating after you are finished.

16-9
ALUMINUM AND ALUMINUM-ALLOY ELECTRODES
You won't find a separate AWS specification for covered-titanium or covered-magnesium arc-welding electrodes. These materials are welded either by the GMAW (MIG) or GTAW (TIG) processes (or maybe something more exotic like PAW). There are no SMAW electrode specifications for these materials. Electrodes for these materials are bare wires that will be shielded by inert gases. You also won't find flux-covered stick electrodes used for zinc (galvanized or zinc-coated steel, yes, but not zinc metal) for the same reason. The same is true for lead. All of these low-melting-point alloys are either gas-welded, or brazed or soldered.

You can, however, find a very small AWS specification for aluminum SMAW electrodes, even though almost all aluminum

TABLE 16-17 Nickel electrodes

AWS Classification	Minimum As-Welded Tensile Strength, psi	Percent Elongation
ENi-1	60,000	20
ENiCu-1	70,000	30
ENiCu-2	70,000	30
ENiCuAl-1	100,000	15
ENiCu-4	70,000	30
ENiCr-1	80,000	20
ENiCrFe-1	80,000	30
ENiCrFe-2	80,000	30
ENiCrFe-3	80,000	30
ENiMo-1	80,000	10
ENiMoCr-1	80,000	10
ENiMo-3	80,000	10

For Cast-Iron Welding	
ENiCu-A ⎫ ENiCu-B ⎬ ENi-CI ⎪ ENiFe-CI ⎭	Depends on how much dilution of the weld metal by the cast-iron base metal occurs. Hardness can be as high as 350 Brinell.

welding is also done either by the MIG or TIG or plasma processes. The name of the specification is AWS A5.3, "Specification for Aluminum and Aluminum Alloy Arc-Welding Electrodes." It only has two electrodes listed. One of the covered-aluminum electrodes, AWS EA1-2, is used for welding commercially pure (1100-series) aluminum. The other one, AWS EA1-43, is used for welding certain high-strength aluminum alloys.

The Aluminum Association, a trade association for the aluminum industry like the American Iron and Steel Institute is for steel producers, names aluminum alloys. Commercially pure aluminum is called 1100 aluminum. Silicon is used as an alloying element to make some aluminum alloys much stronger. One of the most important aluminum-silicon alloys is 4043 alloy aluminum. The two covered electrodes in the aluminum covered-electrode specification are for use on both of these materials, the pure aluminum and the alloyed material.

Don't think that that's all there is to welding aluminum. There are many aluminum filler-metal grades for use with dozens of different aluminum base metals. The aluminum filler-metal selector chart for them would look as complex as the one we gave you for stainless-steel electrodes (see pages 452–453). These aluminum filler metals are bare wires used either for MIG welding or as cut lengths for TIG. Table 16-18 lists the two basic SMAW aluminum electrodes.

Don't be surprised by the low tensile strengths of as-deposited aluminum weld metal. Aluminum

is quite different from steel. It is hardened and strengthened either by cold-working or by special heat-treating processes very similar to some used with precipitation-hardening stainless steels. In other words, your weld metal may be strengthened later on, after you have finished welding (as is the case of the AWS EA1-43 electrodes).

On the other hand, the commercially pure 1100-series aluminums are inherently low in strength compared with steel, which is also a reason why AWS A5.3 doesn't even list the ductility of the as-deposited weld metal. It's very much subject to change depending on what's done with the metal later on. What's more, corrosion resistance is often more important than tensile strength when selecting an aluminum-alloy base metal. Aluminum welding metallurgy, in fact, is as complicated as that of stainless steels.

The AWS EA1-2 covered electrodes produce weld metal with about 99 percent aluminum and just a trace of copper, manganese, zinc, silicon, and iron in it. These are the common alloying elements used to strengthen different kinds of aluminum.

The AWS EA1-43 electrodes, for example, have from 4.5 to 6.0 percent silicon in them because the 4000-series aluminum alloys are aluminum-silicon grades. Although the as-deposited weld metal may not show it, 4300-series aluminums can be made to be quite strong with the right kind of heat treatment. AWS EA1-43 electrodes are quite versatile. They can be used for welding many different kinds of aluminum alloys. But MIG, TIG, and PAW, not SMAW stick electrodes, are the primary methods used to join aluminum alloys.

You can, of course, use oxyfuel welding to join aluminum, magnesium, and zinc, in fact, almost any metal except titanium. (Titanium and its alloys are such difficult materials to weld that all sorts of special cleanliness precautions

must be taken. For example, even a fingerprint left on a titanium section will affect the weld metal.)

Special kinds of stick electrodes are used for hardfacing and making corrosion-resistant surface deposits. But as we said earlier, that is the subject of Chap. 19. It's about time we got around to telling you how to arc weld with the more important electrodes we've already described. We do that in the next section of this chapter. However, since this introduction to SMAW has been quite tough, let's take a midlesson break and turn to Review Questions 1 through 27 at the end of the chapter. Answer the questions.

16-10
WELDING WITH STICK ELECTRODES

You needed the information on stick electrodes that we gave in the first half of this chapter for a very important reason—knowing how to select stick electrodes is equally as important as knowing how to use them. Later on, when you become experienced, you won't have any trouble selecting electrodes.

Now that you know all these details on stick electrodes, let's learn how to use them.

Starting Your Arc
We shouldn't have to remind you by now to make sure that you have your helmet down. As you already know from GTAW and maybe plasma work, it'll mean that you can't see what you are doing. You should be used to that. You won't get used to being blind.

CAUTION: Never strike an arc while your helmet is up.

There are two methods for starting an arc with stick electrodes: the scratch method and the tap method. In both methods, your arc is formed by short circuiting the welding current between your electrode and the workpiece surface. The surge of

---

TABLE 16-18 Aluminum electrodes

AWS Classification	Minimum As-Welded Tensile Strength psi
EA1-2	12,000
EA1-43	14,000

current causes both the end of your electrode and a small spot on the base metal beneath your electrode to instantly melt. That's really all there is to getting your electrode started.

The scratch start When you use a scratch start (Fig. 16-5), you will bring your electrode down to the work as if you were striking a match. As soon as the bare metal tip of the electrode touches the work surface, you raise the electrode very slightly to establish a nice arc. *The arc length or gap between the end of your electrode and the work should be about equal to the diameter of the electrode wire at its uncoated end.*

When you have the proper arc length, you'll hear a sharp, crackling sound. The correct arc length for your electrode will sound something like sizzling bacon on a frying pan. If you get a spitting noise, you are too far away. If your electrode gets stuck to the plate, you are too close. We'll tell you what to do about that in a minute.

The tap start Using the tapping start (Fig. 16-6), your electrode is held in a position vertical to the work surface. You establish your arc by quickly lowering your electrode, tapping it or bouncing it on the work surface, and then slightly raising the electrode a short distance to establish the correct arc length. The correct arc length for most electrodes is, again, about the diameter of the electrode wire at the uncoated

end. And again, when you hear the sizzling sound, you know you have the correct arc length. If you hear spitting and snapping, you are too far away. The arc is constantly making and breaking itself.

Now, here's where the fun begins for beginning welders. Your first few attempts with either method will probably result in freezing your electrode to the workpiece. Don't panic, even though your electrode will start getting very hot. Just give the electrode a sharp twist of your wrist to snap the end of the electrode loose from the plate.

If that fails, you have welded your electrode to your workpiece. Remain calm. Remove the electrode holder from the electrode, or stop the welding machine and use a light blow with a chisel to free your electrode from the base metal.

How to Start Welding

After your arc is running, droplets of metal will fall off the end of the electrode and will be fed into the molten crater on the base metal. This crater is a little different from gas welding. You'll see the weld puddle in the crater, but the arc is more forceful than the gas flame. It depresses the middle of the weld puddle.

You have to keep the electrode moving to prevent digging an even bigger crater in your base metal. Meanwhile, your electrode is melting itself off and becoming shorter and shorter. You have to gradually adjust for the shortening of your electrode by moving it slowly closer to the weld puddle or else your arc length will grow longer. The secret of any good welding job is to keep the arc length constant.

If you feed your electrode gradually into the weld puddle and simultaneously move it with a short, constant arc length, back and forth or side to side, weaving it while you move the electrode forward at a constant average rate, you will start making a weld

bead. Before advancing your arc, hold it for a short time at the starting point to "burn in" the weld metal and get good fusion into the base metal. That also will build up the weld bead slightly. Good arc welding depends upon good arc control as you weave the electrode back and forth or side to side a fraction of an inch, move it forward at just the right speed, and also adjust your arm for the length of the electrode as it slowly melts off.

You will have trouble at first doing all of those things at once. But think of what you have to do driving in heavy traffic. You have to work both hands, at least one foot (two if you have a stick shift), watch front and back and both sides, and maybe talk to someone. Arc welding is actually a lot easier than driving, but just like learning to drive, you have to do it until it becomes second nature. The name of the game is *practice*.

Breaking Your Arc

It's very easy to stop arc welding. Just lift your electrode up and break the arc. That's also the wrong way to do it. Remember that crater that holds your weld metal? It will freeze just the way you leave it. What's worse, it may chill so fast that you will get a nasty welding defect called a *crater crack*. It's often a star-shaped tiny bunch of cracks running along the bottom of the crater. You may not see them. We'll tell you how to find them in Chap. 20 on welding inspection. But let's

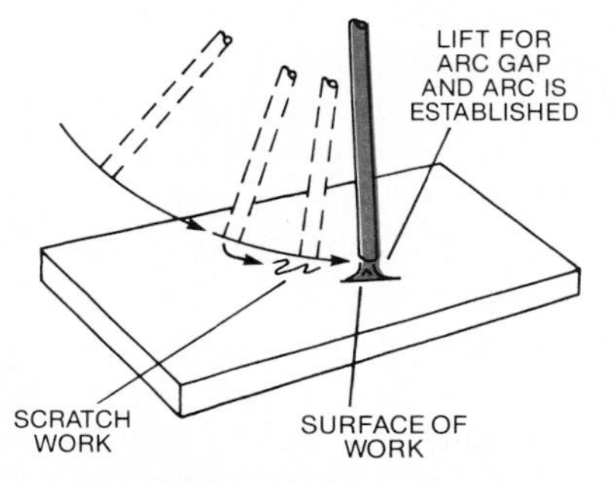

FIGURE 16-5 Striking or brushing method for scratch starting the arc.

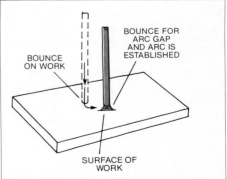

FIGURE 16-6 Tapping method for starting the arc.

prevent them right from the start.

When you are finished welding, or when you are coming to the end of your electrode and have to change it, withdraw your electrode slightly and fill up the crater that you normally would leave behind if you simply jerked the electrode away from the base metal. Now you can pull your electrode away and change the stub for a new electrode. That should eliminate both the crater and the crack in the bottom of it. The same action will also tie the end of your weld into the base metal so you won't get a "cold lap" of weld metal covering, but not penetrating into, the base metal.

Clean Slag between Passes

If you stop your weld at the end of a pass, you will have a small layer of slag on top of the weld crater that you filled in when you stopped welding. You'll first have to clean out that slag in the crater and make sure that your weld is penetrating into your base metal at that point before you continue making a weld bead. If you aren't careful, you will build a weld defect into your work at every point where you restart your electrode. Use your chipping hammer and wire brush to clean the slag off your weld. Make sure that you don't leave flakes of slag in the crevices between weld beads. While your chipping hammer will remove slag from the tops of weld beads, your wire brush (often attached to your chipping hammer) is best for cleaning the last remaining slag particles from your weld.

Cleaning your weld is very important if you are using an electrode with very sticky slag that won't easily remelt. Chip the slag away from your starting point in the weld crater with your chipping hammer. If your electrode has a refusible slag that will remelt easily, you can clean out the crater with your arc (with your helmet down, of course).

Restarting Your Arc

Every time you use up a stick electrode, you have to stop welding, lift your mask, remove the electrode stub from your electrode holder, throw the useless stub away, insert a new electrode, drop your mask down, and restart your weld with the fresh electrode. That's the primary reason that stick-electrode welding is gradually being replaced by semiautomatic and automatic welding processes. A lot of welding defects can creep in at the point where you change electrodes if you don't restart your electrode correctly.

First, drop your mask. Then restart your electrode by either tapping or scratching it. Strike the tip of your new electrode at the forward (cold) end of the crater. Move the arc backward over the crater to remelt the slag and make sure that you are picking up your base-metal penetration where you left off (otherwise you will leave a built-in crack in the workpiece between your stopping and starting points). Then move the tip of the electrode forward again to continue the weld. This procedure fills in the crater and prevents porosity and slag inclusions from occurring where you stopped the weld to change electrodes.

16-11
WELDING FLAT

The types of weld joints and welding positions for arc welding are the same as for oxyfuel gas welding and GTAW. The techniques, of course, are somewhat different. The best way to learn arc welding is first to learn to make a simple bead-on-plate. You will first make simple, straight-line weld beads. They're often called *stringer* beads.

You will use stringer beads later on when welding the root passes of pipe and tube joints. A modification of the stringer-bead technique often is used in hardfacing and weld surfacing. Stringer beads are often all you

need to hold thin sheet together. They also are usually what's used in making the first pass on pipelines during field welding. So learning to make stringer beads correctly is an important exercise.

Flat-Bead Welds

When you start, make sure you pause at the starting point for a moment to build up the bead a little and make sure that your arc is penetrating the base metal. That gives your weld good fusion with the base metal. If you travel at the right welding speed, this fusion into the base metal will continue along your work.

You'll have to determine the best welding speed for making a straight-line weld. Your speed will vary with the deposition rate of your electrode and the current setting you have made on your machine. Only practice will show you how to judge what's best.

For making a straight-line bead, the electrode (in theory) should be held at a 90° angle to the base metal as shown in Fig. 16-7A. However, you won't be able to see your weld puddle through your helmet's filter glass that way.

You can still make a good weld bead and see what you are doing by tipping your electrode slightly to an angle between 5 and 15° from the vertical (shown in Fig. 16-7B), leaning in the direction you are welding. If you keep a short arc of the proper length, you won't be able to see it, as much as hear that it is correct. Listen for the sound of bacon frying on a hot skillet.

A properly made bead weld should leave little spatter on the surface of your work (see Fig. 16-8A). The arc crater or depression in the bead, when the arc is broken, should be about $\frac{1}{16}$ in. [1.6 mm] deep, varying with the size of the electrode and the plate thickness (Fig. 16-8B). The bead metal should be built up slightly (see Fig. 16-8C), but without any weld metal overlapping onto the base plate at the top surface that would indicate poor fusion. (That creates a cold shut, in effect a self-created crack.) The depth of your crater at the end of the bead (Fig. 16-8B) can be used as an indication of penetration into the base metal (or the lack of it).

You should start with bead-on-plate welds and not try anything else until you can make good-looking stringer beads.

Refer back to Table 16-2 on page 371 on the performance of different kinds of carbon-steel electrodes. Follow your welding instructor's directions. Your instructor may want you to experiment with different kinds of electrodes. Try, for example, using a cellulose-coated electrode, a heavy-coated iron-powder–iron-oxide electrode, and a low-hydrogen electrode. Try welding with electrodes that work in all positions (such as AWS E6010) and electrodes only designed for flat and horizontal welding such as AWS E6020. Then compare them to working with AWS E6030, which is only for flat-position welding.

Compare the amount of spatter you get with an AWS E6010 and an AWS E6020 electrode. Next, weld with an E6010 electrode, using DCRP, and then compare it with an E6011 electrode operated on ac. You'll quickly notice a difference in the sound of the arc and the way the electrode operates, especially compared with an E6020 electrode.

Then try working with an E7018 electrode. You'll find that low-hydrogen electrodes are more difficult to work with. You really have to control your arc length to do a good job.

You'll soon see that all that electrode information we gave you will start making a lot more sense. It will no longer be words in a book, but the sound and the feel of your electrode and its arc as you use it.

Flat-Position Butt Joints

Butt joints often are made in the flat position. Sometimes you can make them without a backing strip, and sometimes it's better to make the joint with a strip of metal of the same composition as your base metal to "hold" the first pass of molten weld in the bottom of joint in place. Later on, the backing strip will be removed. Now we'll tell you how to make butt joints in the flat position with and without backing strips.

A butt joint, as you know, is used to join two pieces of plate with the top and bottom surfaces in approximately the same plane. We've already shown you most of the possible butt joints in the gas-welding chapter. This time we'll show you what they look like again, with the weld metal added. Look at Fig. 16-9.

Plates ⅛ in. [3.2 mm] thick can be welded in one pass with no special edge preparation. Plates from ⅛ to 3⁄16 in. [3.2 to 4.8 mm] thick can be welded with no special edge preparation by making a

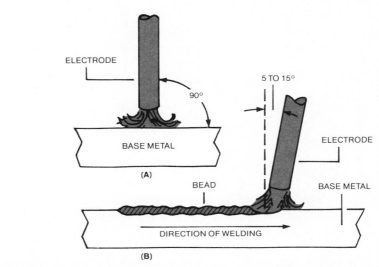

FIGURE 16-7 How to hold a stick electrode when making a bead in the flat position.

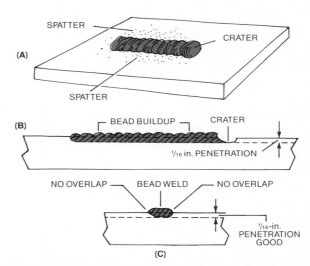

FIGURE 16-8 Properly made weld bead produced in the flat position. The amount of spatter you get depends on the type of electrode you use and your power source settings. Heavy spatter is objectionable and should be avoided. Spatter not only means extra clean-up work for you, but also wasted filler metal which is sometimes quite expensive.

bead weld on both sides of the joint. Tack welds should be used to keep the plates aligned. The electrode should be selected to give you deep-penetration welds. Your electrode motion should be about the same as making a bead weld on top of the plate.

In welding ¼-in.-thick [6.4-mm] plates and heavier, the edges of the plates should be prepared by oxy-fuel flame beveling or by making J, U, or V grooves with flame-cutting, air carbon-arc gouging, or by machining them, whichever you think will work best. Single or double bevels or grooves can be used, depending on the thickness of the plate being welded. The first bead should be deposited to seal the space between the two plates and to weld the root of the joint. That pass is called the *root pass*. It's the most important pass you can make because any welding defects you create will be deeply hidden inside the weld, right where they do the most damage when your weldment is put to work.

The root pass must be thoroughly cleaned by chipping and wire-brushing (or even by grinding if necessary) to remove all slag before the second layer of metal is deposited. There's no guarantee that slag remaining in the root pass will be remelted when you make the second pass. Any slag left in the root pass becomes a brittle notch where a crack can start deep inside the metal.

When you make multipass welds, as shown in Fig. 16-10, the second, third, and fourth layers of weld metal are deposited with a weaving motion of your electrode, and each layer must be cleaned of slag before additional layers are deposited. Any of the weaving motions illustrated in Fig. 16-11 can be used, depending on the type of weld joint and size of the electrode you are using.

Learn from the start to make your weaving motions so that the electrode oscillates or moves uniformly from side to side, with a slight hesitation at the end of each oscillation. Just like in oxy-fuel welding, that momentary hesitation helps tie the weld metal into the side of the joint. As in bead welding, the electrode should be inclined 5 to 15° in the direction of welding. If you don't make your weaving motion correctly, you'll get undercutting of your base metal as shown in Fig. 16-12. Welding too fast also will cause undercutting and poor weld fusion at the edges of your weld bead.

Flat Butts with Backup
Backing strips can be used when you weld plate ³⁄₁₆ in. [4.8 mm] thick or heavier. They will help you get complete fusion at the root pass of the weld and give you

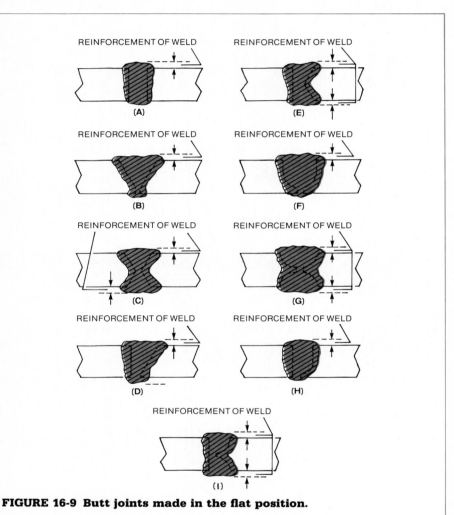

FIGURE 16-9 Butt joints made in the flat position.

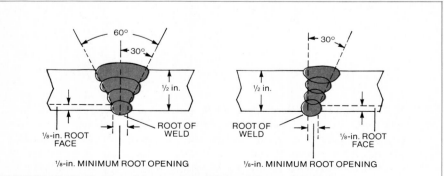

FIGURE 16-10 Butt welds made with multipass beads.

better control of your arc and weld metal when you make your root pass. The edges of your plates are prepared in the same way as making joints for joining without backing strips.

For plate up to ³⁄₈ in. [9.5 mm] thick, the backing strips should be about 1 in. wide and ³⁄₁₆ in. thick [25 mm × 4.8 mm]. They should be made of the same metal you are welding. For plate over ½

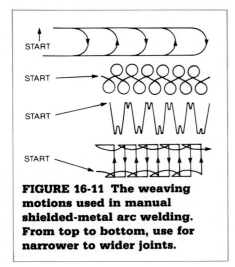

FIGURE 16-11 The weaving motions used in manual shielded-metal arc welding. From top to bottom, use for narrower to wider joints.

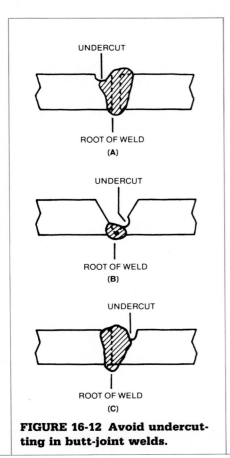

FIGURE 16-12 Avoid undercutting in butt-joint welds.

in. thick, your backing strips should be 1½ in. wide and ¼ in. thick [3.8 mm × 0.6 mm]. These strips are tack welded to the base of your joint as shown in Fig. 16-13. They act like cushions for the root pass. Your weld should penetrate right into the backing strip.

First, you will complete your joint making as many passes as necessary to fill the gap between the plates. Make sure that you clean each pass of all slag after making it. Also make sure that each pass is tied into the base metal or into all the weld metal below it, and on either side. You will remove the backing strip after you finish welding.

After your joint is completed, the backing strip can be "washed" off or cut away from the finished plate with a cutting torch or air carbon-arc electrode. If necessary, you may want to add what amounts to a "sealer bead" along the bottom surface, along the root joint where the backing strip was. That's usually simply a light stringer bead or capping pass made to cover the area where you removed the backing strip.

Don't forget that it will not always be possible to use a backing strip. Just because backing strips make it easier to weld thicker

plate or work out of position, that doesn't mean that you can always rely on using one. Welding pipe is one example where you often can't get a backing strip into the section. You must learn to run a good root pass and get good penetration in any position, without forming icicles of weld metal hanging down below the bottom of your root pass.

16-12 HORIZONTAL WELDING

As an inexperienced welder, you will find working in the horizontal position somewhat difficult. The main reason for your problem will be that you have no "shoulder" of previously deposited weld metal to hold your molten metal in place as you progress along the weld seam.

Horizontal Bead-on-Plate Welds

Your best bet is to start making bead-on-plate welds in the horizontal position until you get good at them (Fig. 16-14), then progress to more difficult joints. Practice and more practice is the only way you will learn to do a good job. Making a bead-on-plate weld in the horizontal position is very much like making it in the flat

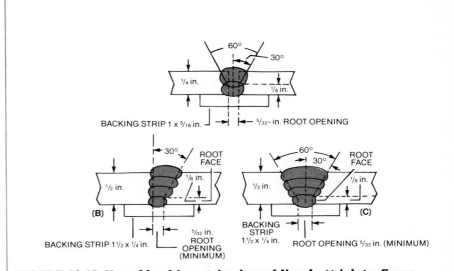

FIGURE 16-13 Use of backing strips in welding butt joints. Some backing strips may even be water-cooled. The backing strip is removed by air-carbon-arc gouging or flame-cutting after the top side of the weld is finished. A final cover or capping pass may then be applied to the back of the weld to finish the job.

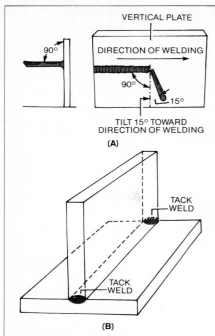

FIGURE 16-14 (A) How to hold an electrode while making a horizontal weld bead on a vertical plate. (B) How tack welds can be used to hold a T joint in place until it is fully welded.

position, except that your electrode should be held at an angle of 90° to the surface and inclined about 15° in the direction you are welding. When you deposit weld metal, move a little more slowly than you would welding down-hand. It also helps to reduce your welding current a little if you find that your weld metal is too fluid. (Reducing your current, of course,

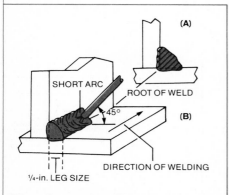

FIGURE 16-15 Position of electrode and fusion zone of fillet weld on a T joint.

reduces the heat input into the weld metal and allows it to solid-ify faster.) You can steady your electrode with one (gloved) hand while you work if the electrode has not burned down too far.

Horizontal T Joints

Once you have learned to make acceptable bead-on-plate welds in the horizontal position, learn how to make a T joint in the horizon-tal position (see Fig. 16-15A). In making horizontal T joints, the two plates are located at right an-gles to each other in the form of an inverted T. The edge of the vertical plate may be tack-welded to the surface of the horizontal plate as shown in Fig. 16-14B.

Horizontal Fillets

A fillet weld is used in making the T joint and a short arc is nec-essary to provide good fusion at the root and along the edges of the weld (see Fig. 16-15A). Your electrode should be held at an angle of 45° to the two plate sur-faces (Fig. 16-15B) and inclined about 15° in the direction you are welding.

Light plates can be welded with a fillet weld in one pass with little or no weaving of your electrode. Welding heavier plates will re-quire making two or more passes, in which the second pass or layer is made with a semicircular weav-ing motion as shown in Fig. 16-16. You should make a slight pause at the end of each weave to obtain good fusion between the weld metal and base metal with-out undercutting.

A fillet-welded T joint on ½-in.-thick [13-mm] plate or heavier can be made by depositing stringer beads in the sequence shown in Fig. 16-17, in which a ½-in. [13-mm] fillet weld is illus-trated.

Chain-intermittent or stag-gered-intermittent fillet welding, as shown in Fig. 16-18, is used for making long T joints. Fillet welds of these types are used where high weld strength is not required so you don't have to make a full-

length weld. However, the short welds are arranged so that the finished joint is equal in strength to a fillet weld along the entire length of a joint from one side only. Also, you avoid base-metal warping and distorting by balanc-ing the heat input and stresses using staggered or chain-intermit-tent welding.

Horizontal Lap Joints

Lap joints have two overlapping plates tack-welded in place so you can work on them (Fig. 16-19). You can deposit a fillet weld in the horizontal position along the lap joint if the plate is in the posi-tion shown in the figure.

The procedure for making this fillet weld is similar to that used for making fillet welds in T joints. Your electrode should be held so that it forms an angle of about 30° from the vertical and is tilted 15° in the direction you are weld-ing. The position of your electrode in relation to the plates is shown in Fig. 16-20. The weaving motion you should use is the same as for T joints except that the pause at the end of the top plate must be sufficiently long to make sure that you get good fusion and no undercutting. Lap joints on ½-in.-thick [13-mm] plate or heavier are satisfactorily made by deposit-ing a sequence of stringer beads, shown in order of deposition in Fig. 16-20.

In making lap joints on plate of different thicknesses, as shown in

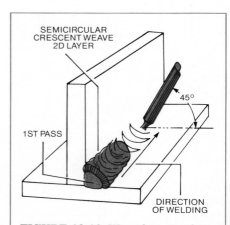

FIGURE 16-16 Weaving motion for multipass fillet welds.

Fig. 16-21, you will have to hold your electrode so as to form an angle of between 20 and 30° from the vertical or you will undercut the thinner plate edge. By tipping your electrode, you direct the force of the arc more along the surface of the base metal and less directly into the thinner section, which puts less heat on the thinner section and avoids undercutting it. You must use your arc to control your weld metal. You can wash up the molten metal to the edge of this thinner plate without burning through it or cutting under it. The idea behind this technique is to put more heat on the thicker section and less heat on the thinner one.

16-13
VERTICAL WELDING

Welding on a vertical surface is progressively more difficult than welding in the flat or horizontal position because, due to the force of gravity, your molten weld

metal tends to flow downward. We'll give you the techniques you need to make different kinds of welds in the vertical position using either mild-steel electrodes or high-strength low-hydrogen electrodes, but reading about them won't help you learn to make them. The only way you will get good is by constant practice.

In SMAW in the vertical position, your welding-current settings become very important. They should be less than those used for the same electrode in the flat position, and the currents used for welding upward on vertical plate are slightly higher than those used for welding downward on the same plate.

While you're learning to weld in the vertical position, let's give you two common shop terms. They are welding *vertical-up* and *vertical-down*. In the first one, you start at the bottom of the weld and work up. In the second, you weld from the top and work your way down.

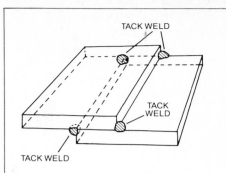

FIGURE 16-19 Tack-welding a lap joint in preparation for welding. Note that tack welds are not full-penetration welds.

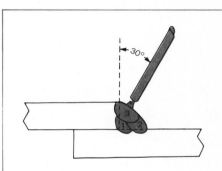

FIGURE 16-20 Position of electrode when making a lap joint.

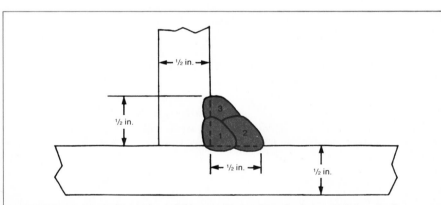

FIGURE 16-17 Order of making stringer beads for a T joint in heavy plate.

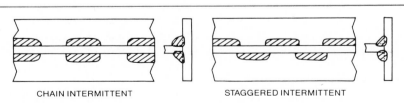

CHAIN INTERMITTENT   STAGGERED INTERMITTENT

FIGURE 16-18 The use of intermittent fillet welds saves weld metal when maximum joint strength is not required. Intermittent welds can either be made in chains or staggered. Staggered spacing is best to control base-metal distortion while balancing built-in welding stresses.

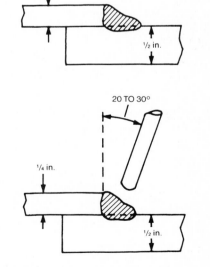

FIGURE 16-21 Lap joints made on plates of different thicknesses.

Vertical Stringer Beads

Use smaller-diameter electrodes to restrict the size of your molten weld puddle when making vertical stringer-bead welds. A 3/16-in.-diameter [4.8-mm] E6010 or E6011 electrode is about the maximum size you can use for vertical and overhead welding and for butt welding in the horizontal position. If you let the weld puddle get too big, gravity will cause it to sag and run out of your weld. You have to use an even smaller 5/32-in. [4.0-mm] low-hydrogen electrode such as E7018 to keep your weld metal in place.

The proper angle between your electrode and the base metal also is very important to help you make a good bead weld and control it in the vertical position. In welding vertical-up, your electrode should be held at 90° to the vertical, as shown in Fig. 16-22A. When welding upward and weaving is necessary, your electrode should be oscillated as shown in Fig. 16-22B. Since the weld bead will be larger on thicker material, it's best to stick to welding vertical-up on anything but sheet metal.

You can weld vertical-down on sheet metal using an E6010 or E6011 electrode. The outer end of your electrode should be inclined downward about 15° from the horizontal, with the arc pointing upward toward the deposited weld metal, as shown in Fig. 16-22C. When welding vertical-down and a weaving bead is required, your electrode should be held at right angles to the vertical, as illustrated in Fig. 16-22C, and tilted about 15° toward the direction you are heading so as to provide you with a better view of your welding arc and weld crater. Use the weaving pattern shown in Fig. 16-22D.

Vertical Butt Joints

The edges of butt joints on plates in the vertical position are prepared for welding in the same way as those required for butt joints in the flat position. To get good fusion and penetration into your base metal with no undercutting, you'll need to use a short arc and carefully control the motion of your welding arc and electrode.

Butt joints on beveled plates 1/4 in. [6.3 mm] in thickness can be made by using a triangular weaving motion, as shown in Fig. 16-23A. Welds on 1/2-in. [12.7-mm] plate or heavier should be made

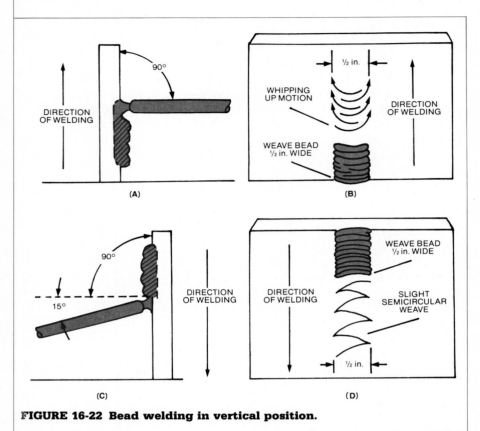

FIGURE 16-22 Bead welding in vertical position.

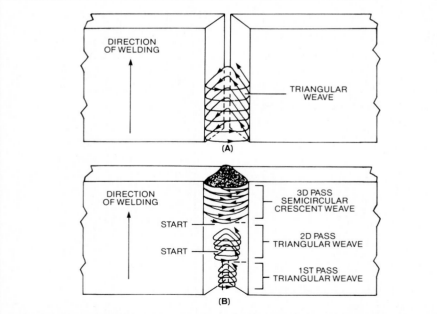

FIGURE 16-23 Electrode motion for butt-welding a joint in the vertical position.

in several passes, as shown in Fig. 16-23B. The last pass should be deposited with a semicircular weaving motion, with a slight "whip-up" and a pause of your electrode at the edge of the bead if you are using an E6010 or E6011 electrode. Welds with a backing strip should be made in the same way.

The triangular weave and the crossover and hook-up patterns both have a major flaw in welding certain kinds of work, especially with high-strength low-hydrogen electrodes (and with MIG welding, too). The force of the arc moving across the outer edge of the puddle tends to deposit molten metal on top of molten metal, thus causing slower cooling of the weld puddle and more chance of slag forming in the center of the bead, possibly creating cold laps of weld metal at the edges.

Pipe welders work vertical-up sweeping their electrode over the top edges of the puddle and pause at the sides, placing the force of the arc on the base metal, or on previously solidified weld beads, which does a better job of cleaning out slag at the toes of the other welds. This allows the outer edge of the weld puddle to cool faster. When changing over from stick welding to MIG welding, you will find that it is essential to use this welding pattern to prevent cold laps (layers of weld metal that have not fully bonded to each other). By learning this method, you can produce sound welds using any SMAW electrode, or you can apply the same vertical-up pattern to MIG welding and not have to change techniques each time you change electrodes or your welding process.

When welding butt joints in the horizontal direction on vertical plates, a short arc is necessary at all times and the metal is deposited in multipass beads (see Fig. 16-24). The first pass is made with your electrode held at 90° to the vertical plates as shown in Fig. 16-24A. The second, third, and all following passes are made

with the electrode held parallel to the beveled edge opposite the edge from which your weld bead is being deposited as shown in Figs. 16-24B and C.

Vertical Fillet Welds

When you make fillet welds in either T or lap joints in the vertical position, the electrode should be held at 90° to the plates or not more than 15° off the horizontal for proper molten-metal control. The arc should be held short to obtain good fusion and base-metal penetration.

In welding T joints in the vertical position, you should start the joint at the bottom and weld upward, and your electrode should be moved in a triangular weaving pattern, as shown in Fig. 16-25A, to control your weld metal. A slight pause in the weave at the points indicated will improve the sidewall penetration you get and give you good fusion at the root of the joint.

If the weld metal should overheat and start to sag, shift your electrode away quickly from the crater without breaking the arc, as shown in Fig. 16-25B. That lets the molten metal solidify without running down the plate. Your electrode should be returned immediately to the crater of the weld to maintain the desired size of your weld and avoid a crater crack.

When more than one pass is necessary to make a T weld, either of the weaving motions

shown in Fig. 16-25C or D can be used. A slight pause at the end of your weave will develop good fusion without undercutting at the edges of your plates.

To make welds on lap joints in the vertical position, your welding electrode should be moved in the triangular weaving motion shown in Fig. 16-25E. The same procedure as outlined above for the T joint is used, except that the electrode is directed more toward the vertical plate marked G. The object is to tie the weld into the base plate G while you travel up the joint. Your arc should be held short, and the pause at the surface of plate G should be slightly longer, again to tie the weld into the plate. (Naturally, if plate G were considerably thinner than the overlapping plate H, you would direct your arc at the thicker plate—in the drawing, both plates are of nearly equal thickness.) Take care not to undercut either of the plates and be careful not to allow your molten metal to overlap at the edges of the weave.

Lap joints in the vertical position on heavy plate require more than one layer of weld metal. The bead that you deposit should be thoroughly cleaned and all following beads should be deposited as shown in Fig. 16-25F. The precautions outlined above to ensure good fusion and uniform weld-metal deposit in T joints apply also to lap joints made in the vertical position.

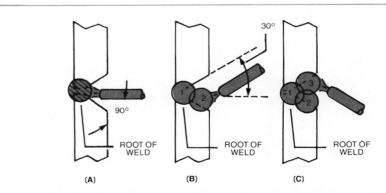

FIGURE 16-24 Butt-joint welding in the horizontal direction on vertical plates when making multipass welds.

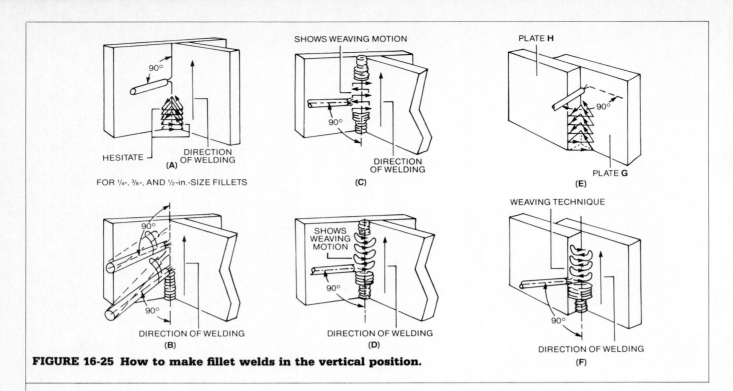

FIGURE 16-25 How to make fillet welds in the vertical position.

16-14
OVERHEAD WELDS

Overhead welding is the most difficult position of all to work in. You need to constantly maintain a very short arc to retain complete control of your weld metal. As in vertical welding, the force of gravity tends to cause your molten weld metal to drip down or sag on the plate. If you hold your arc length too long, you will have difficulty transferring metal from your electrode to the base metal.

This problem is increased if you use an electrode with the type of flux coating that produces globular drops instead of spraying the weld metal into the joint. You can fight the force of gravity by first shortening and then lengthening your arc at intervals. That will let your weld metal cool a little and stay in place. However, care should be taken not to carry too large a pool of molten metal in the weld.

In the following sections we'll tell you how to make overhead bead welds, butt joints, and fillet welds.

Overhead Stringer Beads

When you make stringer beads in the overhead position, your electrode should be held at an angle of 90° to the base metal as shown in Fig. 16-26A, or you can tilt it to about 15° in the direction of welding as shown in Fig. 16-26B, if this will help give you a better view of arc and crater on your weld.

Weave beads can be made in the overhead position by using the motion illustrated in Fig. 16-26C. You'll have to use a rather rapid motion at the end of each semicircular weave to control your metal deposit. Excessive weaving should be avoided be-

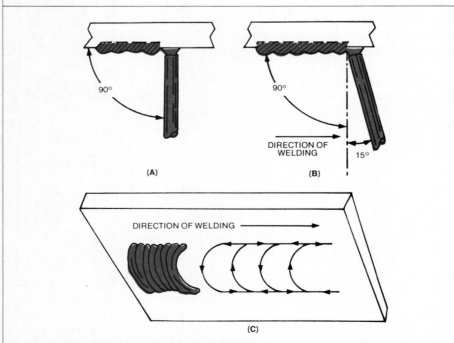

FIGURE 16-26 Electrode positions and weaving motions for overhead welding.

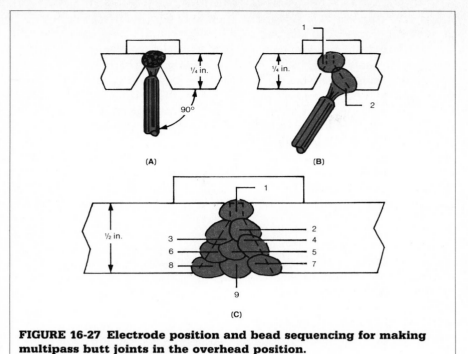

FIGURE 16-27 Electrode position and bead sequencing for making multipass butt joints in the overhead position.

chipped out before the following pass is deposited.

The positions of your electrode and the order to be followed in depositing beads in ¼- and ½-in.-thick [6.3- and 13-mm] plates are shown in Figure 16-27B and C. The first pass should be made with your electrode held at 90° to the plate as shown in Fig. 16-27A.

Another thing you can do to help yourself is to use a smaller-diameter electrode than you would use welding the same joint in the flat position. This will help you hold a short arc and still get good penetration at the root of your joint. Excessive current will create a very fluid puddle that will fall on your head. *Use low currents for welding overhead.*

cause this will cause overheating of your weld deposit and the formation of a large pool of molten metal that will be hard to control.

Overhead Butt Joints
Your plates for overhead butt joints are prepared the same way as in the flat position. You'll get the best results with backing strips, if you can use them. You must remember, however, that

you will not always be able to use a backing strip, so learn to make these joints without one. If your plates are beveled to a feather edge and no backing strip is used, the root pass will tend to burn through repeatedly unless you use extreme care.

For overhead butt welding, stringer-bead rather than weave welds are best. Each bead should be cleaned and the rough areas

Overhead Fillet Welds
Use a short arc when making overhead fillet welds in either T or lap joints. Don't weave your electrode. The electrode should be held approximately 30° to the vertical plate and moved uniformly in the direction of welding as shown in Fig. 16-28B. Control your arc motion to get good penetration to the root of your weld and good fusion with the side-

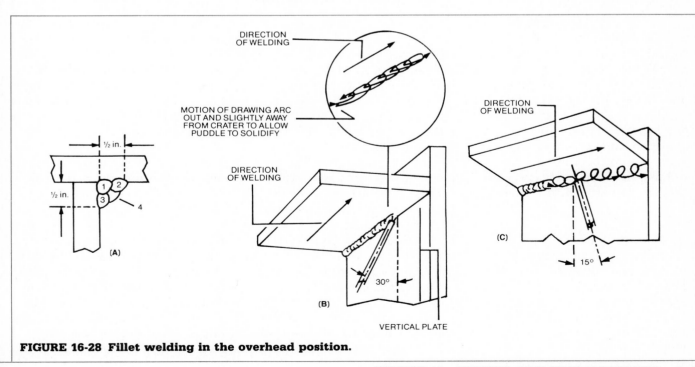

FIGURE 16-28 Fillet welding in the overhead position.

walls of your vertical and horizontal plates.

If the molten metal becomes too fluid and tends to sag, whip your electrode away quickly from the crater and ahead of your weld without breaking your arc so as to lengthen your arc and allow the metal to solidify. Your electrode should then be returned immediately to the crater where you can continue welding.

Fillet welds for either T or lap joints on heavy plate in the overhead position require several passes or beads to make the joint. The order in which these beads are deposited is shown in Fig. 16-28A. The root pass is a stringer bead with no weaving motion of your electrode. The second, third, and fourth passes are made with a slight circular motion of your electrode with its top tilted about 15° in the direction you are welding, as shown in Fig. 16-28C.

This weaving motion with your electrode gives you greater control and better distribution of the weld metal you are depositing. All slag and oxides must be removed from the surface of each pass by chipping or wire-brushing before you add more beads into the joint.

16-15
SOLVING PROBLEMS WITH SMAW

Now that we've given you basic SMAW techniques, let's help you stay out of trouble when welding, and tell you how to solve some problems if they do occur. Many of these techniques will help you solve the same problems when you run into them using other welding methods, such as GTAW, GMAW, or hardfacing.

Reducing Spatter

A common problem that both beginning and advanced welders have with certain electrodes is spatter—the tiny droplets of molten metal that are thrown out of the weld and stick to the base metal.

Spatter often has to be cleaned up when it occurs. You can't always eliminate it, but you can modify your welding techniques to reduce it or select a welding electrode that will produce less spatter. (Spatter occurs most often with lightly flux-coated electrodes such as AWS E6010. It's much less of a problem with heavily coated electrodes such as AWS E6020.) Some ways to reduce spatter are

■ Use the correct welding current (and the correct dc polarity) for your electrode.
■ Keep the arc length at the proper length (shorter is better).
■ Take precautions to prevent arc blow (which we'll tell you about next).

If spatter continues to be a problem even when using proper techniques, areas next to the weld can be painted with water-soluble whitewash before you weld. The whitewash keeps the little balls of molten weld metal from sticking to the plate and you can wash or wire-brush them off afterward. There also are proprietary coatings that come in a spray can that you can use on your base metal. One common one is called Spattercat. It's a colorless liquid that keeps the spatter from sticking.

Arc Blow

Arc blow is a nasty problem you will run into when working with dc welding, no matter what process you use. Arc blow is caused by magnetic fields induced by dc that turn your base metal near the arc into an electromagnet. The local magnetic field in the workpiece will force the arc to blow away from the point where you direct it. Most often, the arc will wander back and forth in the joint. Sometimes your arc will wander from one side of the joint to the other.

Unless you can control arc blow, it will cause problems for you in controlling the molten weld puddle and slag. The result will be excessive spatter, incom-

plete fusion, reduced welding speed, porosity, and lower weld quality. Sometimes arc blow will even force your arc into the sides of the joint and gouge out craters in the sidewall.

You'll recognize arc blow if, for example, you are welding on a wide-flange beam and are approaching a corner where the joint will make an abrupt turn, and all of a sudden your arc tends to blow away from your travel direction.

Backward arc blow occurs when you are welding *toward* your ground connection, the end of a joint, or into a corner. Forward arc blow occurs when you are welding *away* from your ground connection or at the start of a joint (Fig. 16-29). Forward arc blow is especially troublesome with iron-powder or other electrodes that produce lots of slag, where the effect is to drag the heavy slag on your welding crater forward and under the arc. The statement of the problem also gives you one solution. Use two ground cables instead of one and put them on opposite sides of your work. The forward and backward arc blows will tend to oppose each other and cancel out the problem.

In addition to the visible action of your arc, backward arc blow is indicated by increased spatter; intermittent undercutting; narrow, high beads, usually with intermittent undercutting; an increase in base-metal penetration; and surface porosity at the finished end of welds on sheet metal. Forward arc blow is indicated by an intermittent wide bead, wavy spots in the bead, intermittent undercutting, and an intermittent decrease in base-metal penetration.

Arc blow is worse on high-nickel steels because they are inherently more magnetic than straight carbon steels. Aluminum and copper alloys don't show arc blow because they don't become magnetic. Nickel alloys will give you even more problems than iron alloys. They become very magnetic.

Arc blow also is more common in heavy plates than on thin material simply because they make a larger local magnetic field. Working on thick plates, you'll run into the problem more often when your electrode is deep inside a joint, because the electrode is surrounded by more metal (and more magnetic field).

Arc blow also occurs when working on the inside fillets of H beams for the same reason; your arc is surrounded by more metal. Arc blow on thick plates can get very bad if the plates were handled by a magnetic crane. Magnetic cranes are often used to handle plates, but if you have a high-nickel steel, for example, the crane can turn your plate into a permanent magnet that you can't weld without severe arc blow. Simply ask the crane operator to handle the plate some other way, such as by using slings.

Arc blow can start at any time. If it does, put the ground cable in the direction of the arc blow, or add a connection to the ground cable so that you have one ground lead but two ground connections clamped onto your workpiece on opposite sides. Sometimes you also can weld toward the direction of arc blow (that is, go with the flow). Holding a short arc helps. So does reducing your welding current.

You can clamp steel blocks on the base metal near the weld joint to alter the magnetic path around the arc, or you can change the angle of your electrode somewhat while you work. Some welders even try wrapping the welding ground cable around the work to induce a more balanced magnetic field in the workpiece. It often works, but don't weld into your cable.

If a magnetic crane has turned your workpiece into a giant permanent magnet, it may be necessary to have the plate stress-relieved or annealed. When steel is heated over 1414°F [768°C], it can no longer be magnetic. That is called the *curie point* of the metal. Pure nickel has a curie point of about 700°F [370°C].

The curie point of a metal sounds very theoretical, but it's very practical when you are in a steel mill and hot steel bar stock or hot-rolled sheet goes out of control on the rolling mill and starts piling up at one of the rolling stands, while more stock is rushing into the line. A magnetic-crane operator will desperately try to lift the stuff out of the way and won't succeed, until the metal cools below 1414°F [768°C].

The simplest solution to arc blow, if you have an ac-dc power source and can use electrodes suitable for your work that can operate on ac, is simply to switch to ac. That's especially useful if you are operating with dc above 250 A. While ac won't completely eliminate serious arc-blow problems, it will greatly reduce them. Remember that ac rapidly switches polarity 60 times a second (if you are using 60-Hz current), meaning that the local magnetic field in your workpiece is rapidly switched back and forth, somewhat neutralizing the problem.

Arc blow also can occur when you are running automatic equipment with two electrodes (usually continuous GMAW wires and not stick electrodes) operated with dc in the same joint, one behind the other (Fig. 16-30). If the two wires get within 4 in. [102 mm] of each other, their induced magnetic fields will affect each other. If the dc polarity of one wire is positive and the other is negative, the two arcs will be forced away from each other. If the polarity of both wires is the same, the two arcs will be drawn toward each other. The solution is to run one electrode wire on dc and one on ac.

With experience you will learn which of these corrections are more likely to produce satisfactory results under various conditions.

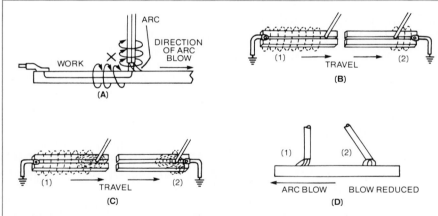

FIGURE 16-29 (A) Arc blow occurs on particularly magnetic alloys, and also is a problem on automated machines using two or more simultaneous, nearby arcs (one for root and one for cover-pass welding, for example). The magnetic flux set up by the ground current combines with the flux around the electrode to create a high flux concentration. This concentrated magnetic field blows the arc away from the direction of the ground cable connection as if it were an extremely strong wind. That is why the effect is called arc "blow." (B) The magnetic flux set up by the ground current is shown both behind the arc and ahead of the arc. The direction in which the arc blow occurs is a major clue to recognizing and solving the problem. (C) Superimposed magnetic fields. Magnetic blow at the finish end of the joint is reduced because the two flux fields tend to offset one another. At (B) the two fields are additive and cause a strong back blow. (D) Arc blow, can sometimes be corrected by angling the electrode to compensate for the effect and weld in the direction of arc flow.

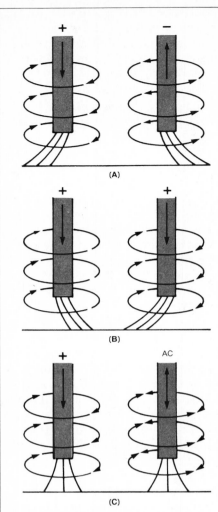

FIGURE 16-30 Reactions of the magnetic fields when two arcs are close together. (A) Arcs of different polarities. (B) Arcs with the same polarity; the magnetic fields oppose each other and the arcs blow inward. (C) One of these arcs is direct current and the other is alternating current. There is very little magnetic arc blow as a result. That's one way to solve arc blow problems when welding with multiple arcs.

16-16
PREHEATING AND POSTHEATING

Remember how we stressed the importance of preheating and postheating after gas welding gray cast iron in Chap. 9? Well, here we go again, except this time it's even more important for two reasons. First, the welding arc, as hot as it is, doesn't pump as much heat into the workpiece as a gas flame simply because the arc is a lot smaller than the oxyfuel flame. Second, arc welding often is used to weld some higher-alloy steels that are even more prone to cracking than cast irons.

Heat treatments (other than simple preheating and postheating for welding) are seldom required for low-carbon or structural steels, except occasionally when used to stress-relieve weldments, to prevent warping, or to assure low hardness for machining. Medium- and high-carbon steels, however, are something else entirely. So are thick sections of any metal, typically massive gray iron castings that are huge heat sinks, copper alloys that consume heat like a sponge, and any metal joined outdoors during cold weather.

Let's make some distinctions here about heat treating. Preheating a weld to warm it up to about 300°F [150°C] is technically heat treating. So is postweld heating to let your weldment cool down slowly. But these are entirely different from high-temperature heat treatments such as annealing, normalizing, quenching, and tempering, which are critically important in terms of time and temperature and are used to alter the strength and hardness of metals.

For example, when you weld medium- and high-carbon steels, or lower-carbon steels with lots of alloying elements in them, there is a danger that the weld deposit and heat-affected zone will contain a high percentage of martensite. You'll remember that martensite is the very hard, brittle microstructure in steel with almost no ductility. It forms when steel is rapidly cooled from austenite. Up to 0.8 percent carbon the more carbon in the steel, the harder the martensite will be. Beyond 0.8 percent carbon, you start forming iron carbides. Figures 16-31 through 16-35 show you what various kinds of steel look like on a polished section when seen under a special metallographic

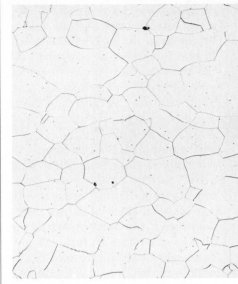

FIGURE 16-31 Ferrite is the nearly pure form of iron that makes up most of the crystals in a low-carbon steel at room temperature. This photomicrograph was taken on a polished specimen of low-carbon ferritic steel at 500 power (500×) magnification.

microscope.

Not much martensite will show up when you work on low-carbon steels, but the more carbon (and alloying elements) in the metal, the more likely that martensite will form on fast cooling. That's what makes a steel knife blade very hard and strong when it's quenched, but also very brittle. Reheating it "tempers" the martensite, making it more ductile and less likely to crack. That's the basic idea behind quenching and tempering techniques that have been used for centuries by everybody from the village blacksmith to tool-and-die makers.

Also remember that your molten weld pool is a tiny hot spot on a very big piece of otherwise cold steel. In between that cold base metal and your molten weld is your heat-affected zone. The heat-affected zone can take a lot of thermal punishment, but don't push it too far or martensite is likely to form when a medium- to

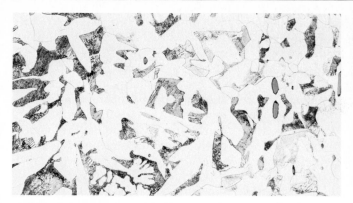

FIGURE 16-32 As the carbon content of steel increases, very little of it can be dissolved in ferritic iron. The excess carbon that the ferrite can't dissolve forms another, stronger structure called pearlite. In this photomicrograph taken at 500× magnification, the slow-cooled ferrite and pearlite structure is clear. The white grains (or crystals) are ferrite while the darker grains are pearlite. If this steel had been cooled quickly, martensite would have formed instead. At about 0.8 percent carbon, the steel (if slow-cooled) would be 100 percent carbon, and the microstructure would show pearlite mixed with a very hard carbide called cementite. Note that the pearlite grains have striations or lines on them. That's typical of pearlite.

FIGURE 16-34 This high-carbon steel shows pearlite grains surrounded by the iron carbide called cementite. Cementite commonly forms around grain boundaries. Cementite forms when the steel has so much carbon in it that even the pearlite can't use it all up. While cementite is the material that makes high-carbon steels (such as tool steels) extremely hard and abrasion-resistant, the connected cementite grain boundaries made of this extremely hard, brittle material also are paths for micro-cracks that can run right around the pearlite grain boundaries. High-carbon steels like tool steels often are forged or rolled many times while still hot to break up these connected grain boundaries to increase the overall resistance of the steel to cracking. You would have great trouble welding this steel. Fast cooling would produce untempered martensite *and* cementite.

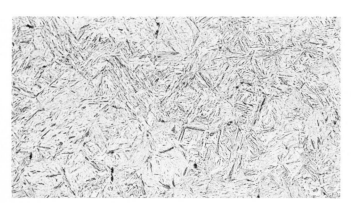

FIGURE 16-33 When a carbon or alloy steel is cooled very quickly (quenched), a hard, strong microstructure called martensite forms. Here you see untempered martensite in a steel that contains 0.4 percent carbon. Untempered martensite makes steel very strong and hard, but also very brittle. If the steel is reheated to a lower tempering temperature, it loses very little of the strength of the martensite, but it gains enough ductility so that it can be used without risk of cracking. Martensite is a much finer grained structure than ferrite or pearlite. These long bladelike crystals were photographed at 800× magnification. Because so much untempered martensite forms on fast cooling, this steel would be difficult to weld unless you used preheating and slow cooling after welding to reduce the amount of untempered martensite that otherwise would form in the heat-affected zone of your base metal.

FIGURE 16-35 Austenite is the form that iron or steel takes when heated to high temperatures (below the melting point). Austenite can dissolve much more carbon than ferrite can, so that a ferritic, pearlitic, or martensitic steel at room temperature usually will become all austenite at high temperatures. Certain alloying elements (like lots of nickel or manganese) make the austenite stable even at room temperature. That's why certain stainless steels (such as Type 304) contain 18 percent chromium (to make it stainless) and 8 percent nickel (to make it austenitic, even at room temperature or cooler). This happens to be a photomicrograph of Type 304 austenitic stainless steel at 600× magnification.

high-carbon-steel weldment cools too fast. That's where your cracks are most likely to occur when welding on these more difficult-to-weld materials.

The purpose of preheating and postheating your base metal, when they are required, is to keep the martensite content of your weld and heat-affected zone to a minimum.

The martensite is actually formed during the cooling of your weld and heat-affected zone. The slower your weldment cools off, the less martensite forms. Below a characteristic cooling rate for each metal, no martensite forms at all. Instead, individual grains of a more ductile microstructure called *pearlite* will be produced. Therefore, heating your steel before welding it, keeping it fairly hot between passes, and controlling the cooling rate after welding all slow down the amount of martensite that can form, greatly reducing the chances of weld and base-metal cracking.

For example, a preheat and postheat treatment of 300°F [150°C] will reduce the cooling rate of 700°F [370°C] weld metal to that of 400°F [205°C] weld metal with no prior preheating. You can actually "fool" the base metal into not making martensite by cooling it more slowly. It will make pearlite instead of martensite. Whether you should use preheat and postheat treatments depends on the amount of carbon and other alloying elements in the steel being welded.

Big cast-iron weldments require preheating and postheating for the same reason. Gray cast iron can turn into glass-hard and very brittle white iron that is full of very hard martensite and iron carbides simply by being cooled too fast.

In the case of nonferrous metals such as copper and aluminum that are very thermally conductive, preheating is used on very thick sections mostly just to keep the base metal hot enough to pre-vent chilling your weld metal.

Whether these heat treatments should be used for irons and steels depends upon the amount of carbon and other alloying elements in the metal being welded. If test welds without heat treatment are found to have too low ductility or too high hardness, you had better take a look at modifying your welding procedure to include preheating and postheating. Many welding codes will, in fact, specify exactly what you should do when working on certain metals, certain thicknesses, and even certain ambient temperatures in your welding shop or in the field. "Ambient" simply means the temperature of the air where you are.

Preheating

There is a simple way to determine if you need to preheat a steel without making test welds. The weldability of a steel is closely related to its carbon content plus the content of certain other alloying elements. The approximate amount of the other alloying elements which will produce the same hardness as 1 percent carbon is well known. Thus, the amount of all the important elements in your base metal that will affect hardness can be put on the same basis as the amount of carbon in your base metal. It's called *equivalent carbon content* and there's a simple formula for calculating it.

Equivalent carbon content
$$= C + \frac{\% \text{ Mn} + \% \text{ Si}}{6}$$
$$+ \frac{\% \text{ Cr} + \% \text{ Mo} + \% \text{ V}}{5}$$
$$+ \frac{\% \text{ Cu} + \% \text{ Ni}}{15}$$

This formula shows that manganese (Mn) and silicon (Si) are only about one-sixth as powerful in hardening effect as carbon. Similarly, chromium (Cr), molybdenum (Mo), and vanadium (V) are only one-fifth as powerful as carbon in their hardening effect, while copper (Cu) and nickel (Ni) are only one-fifteenth as effective in causing an increase in hardness. Alloying elements can take the place of a certain amount of carbon to help you reduce welding problems associated with higher-carbon steels, while keeping the steels strong and tough.

This formula is only valid when the percentages of these alloying elements are less than what we show here (and you can get that chemistry data for your deposited weld metal directly from the AWS filler metal specification for the electrode you use):

% carbon less than 0.50 percent
% manganese less than 1.60 percent
% silicon less than 0.50 percent
% chromium less than 1.00 percent
% molybdenum less than 0.60 percent
% vanadium less than 0.20 percent
% copper less than 1.00 percent
% nickel less than 3.50 percent

The formula doesn't work well for full-alloy steels and it's useless for stainless steels. Similarly, it wouldn't apply well to the high-manganese austenitic steels used for wear plates on construction equipment, and the carbon range of the equation is far too low for cast irons. Also, stainless steels have much more chromium in them (at least 12.5 percent Cr), but many of them are quite weldable. You'll also find exceptions to the equation for some high-strength low-alloy steels.

There are other carbon-equivalent formulas, but this one is very good for your purposes, especially when you have to work with conventional carbon steels, and more carbon steels are used than anything else. It's only on plain carbon steels and most low-alloy steels that you should use this formula. What's important is that a quick calculation of the carbon equivalent of a steel will give you a good clue as to whether or not you can weld it without cracking, unless you use preheating and probably postheating.

Tied in with the carbon-equivalent formula, the following data give you a good rule-of-thumb guide for deciding how much you need to preheat carbon steels with different ranges of *equivalent* carbon content.

Equivalent Carbon (%)	Preheating Required
Up to 0.45	Preheat optional
0.45 to 0.60	Preheat to 200 to 400°F [93 to 205°C]
Over 0.60	Preheat to 400 to 700°F [205 to 370°C]

Some steels, particulary those having *equivalent* carbon greater than 0.60 percent, may require postheating as well as preheating. This is particularly true when welding heavy sections. However, for most plain carbon steels and alloy steels, only preheat is needed, if any treatment is required at all.

Preheating from 250 to 300°F [121 to 150°C] is also generally used when multipass welding of sections greater than 1 to 2 in. [25 mm to 51 mm] thick to reduce the joint's susceptibility to weld cracking. Table 16-19 will give you more details for commonly used carbon, high-strength low-alloy and full-alloy steels.

Postheating

Postheating, as we'll use it here, means heating your weldment *immediately* after the weld is made. It is very different from other heat treatments made after the weld cools off, such as stress relieving, tempering, normalizing, and annealing, or solution heat

TABLE 16-19 Suggested temperatures for preheating metals before welding

Material	Alloying Elements, %*						Preheat Temperature °F[°C]
	C	Mn	Ni	Mo	Cr	Cu	
Plain carbon steels	Below 0.20						Up to 200 [93]
	0.20–0.30						200–300 [93–150]
	0.30–0.45						300–500 [150–260]
	0.45–0.80						500–800 [260–425]
Carbon-molybdenum-alloy steels	0.10–0.20			0.50			300–500 [150–260]
	0.20–0.30			0.50			400–600 [205–315]
	0.30–0.35			0.50			500–800 [260–425]
Manganese steels							
Silicon structural	0.35	0.80					300–500 [150–260]
Medium manganese	0.20–0.25	1.0–1.75					300–500 [150–260]
AISI 1330	0.30	1.75					400–600 [205–315]
AISI 1340	0.40	1.75					500–800 [260–425]
AISI 1350	0.50	1.75					600–900 [315–482]
High-manganese wear-resistant steels	1.25	12.0					Not required
High-strength low-alloy steels							
Manganese-molybdenum	0.20	1.65		0.35			300–500 [150–260]
Chromium-copper nickel	0.12 max	0.75	0.75			0.55	200–400 [93–205]
Chromium-manganese	0.40	0.90	0.40				400–600 [205–315]
(There are many exceptions to these treatments for HSLA steels)							
Nickel-alloy steels							
AISI 2015	0.15		0.50				Up to 300 [150]
AISI 2115	0.15		1.50				200–300 [93–150]
AISI 2315	0.15		3.50				200–500 [93–260]
AISI 2320	0.20		3.50				200–500 [93–260]
AISI 2330	0.30		3.50				300–600 [150–315]
AISI 2340	0.40		3.50				400–700 [205–370]
Nickel-chromium-alloy steels							
AISI 3115	0.15		1.25		0.60		200–400 [93–205]
AISI 3125	0.25		1.25		0.60		300–500 [150–260]
AISI 3130	0.30		1.25		0.60		400–700 [205–370]
AISI 3140	0.40		1.25		0.60		500–800 [260–425]
AISI 3150	0.50		1.25		0.60		600–900 [315–482]
AISI 3215	0.15		1.75		1.00		300–500 [150–260]
AISI 3230	0.30		1.75		1.00		500–700 [260–370]
AISI 3240	0.40		1.75		1.00		700–1000 [370–538]
AISI 3250	0.50		1.75		1.00		900–1100 [482–593]
AISI 3315	0.15		3.50		1.50		500–700 [260–370]

* Not all alloying elements are shown. All steels, for example, contain manganese. To keep the table simple, we included only those elements that will help you identify the steel.

(Continued on page 410)

TABLE 16-19 Suggested temperatures for preheating metals before welding (*Continued*)

Material	Alloying Elements, %*						Preheat Temperature °F[°C]
	C	Mn	Ni	Mo	Cr	Cu	
AISI 3325	0.25		3.50		1.50		900–1100 [482–593]
AISI 3435	0.35		3.00		0.75		900–1100 [482–593]
AISI 3450	0.50		3.00		0.75		900–1100 [482–593]
Molybdenum-alloy steels							
AISI 4140	0.40				0.20	0.95	600–800 [315–425]
AISI 4340	0.40		1.75		0.35	0.65	700–900 [370–482]
AISI 4615	0.15		1.80		0.25		400–600 [205–315]
AISI 4630	0.30		1.80	0.25			500–700 [260–370]
AISI 4640	0.40		1.80	0.25			600–800 [315–425]
AISI 4820	0.20		3.50	0.25			600–800 [315–425]
Low-chromium molybdenum steels	Up to 0.15			0.5	2.0		400–600 [205–315]
	0.15–0.25			0.5	2.0		500–800 [260–425]
	Up to 0.15			1.0	2.0		500–700 [260–370]
	0.15–0.25			1.0	2.0		600–800 [315–425]
Medium-chromium molybdenum steels	Up to 0.15			0.5	5.0		500–800 [260–425]
	0.15–0.25			0.5	5.0		600–900 [315–482]
	0.15 max			1.0	8.0		600–900 [315–482]
Straight-chromium stainless steels							
Type 410	0.10				13.0		300–500 [150–260]
Type 430	0.10				17.0		300–500 [150–260]
Type 446	0.10				26.0		300–500 [150–260]
Chromium-nickel austenitic stainless steels							
Type 304	0.07		8.0		18.0		Usually do not require preheat but it may be desirable to heat to 32°F [0°C]
Type 309	0.07		8.0		25.0		
Type 310	0.10		8.0		25.0		
Type 316	0.07		8.0		18.0		
Type 317	0.07		8.0	2.5	18.0		
Type 347	0.07		8.0	3.5	18.0		
Cast irons							
Gray cast iron							700–900 [370–482]
Ni-bearing iron							500–1000 [260–538]
Non-ferrous metals							
(Preheat only required on thick sections)							
Aluminum							500–700 [260–370]
Monel							200–300 [93–150]
Nickel							200–300 [93–150]
Inconel							200–300 [93–150]
Copper alloys							500–800 [260–425]

* Not all alloying elements are shown. All steels, for example, contain manganese. To keep the table simple, we included only those elements that will help you identify the steel.

treatments done on precipitation-hardening stainless steels and hardenable aluminum alloys.

Postheating does the same job as preheating. It holds the temperature of your work at a sufficiently high level so that the weld will cool slowly. As with preheating, the result is greater ductility in your weld and the heat-affected zone. Postheating is rarely used alone; it is almost always used with preheating.

Postweld heating is most often used on the highly hardenable steels, but it sometimes is used on less-hardenable steels if adequate preheat is difficult because of the size of the sections. For this reason, Table 16-20 can help you a lot, *but it only is useful if you use postweld heating right after you finish making each welding pass, while the metal is still hot.*

Stress Relieving

Following a welding operation on a big workpiece, or one that is likely to have large built-up stresses in it, or when you are having trouble with the workpiece warping out of shape, stress relief is required. You already know that very high stresses can build up because of the cooling and contraction of your weld metal. Sometimes these stresses are high enough to tear the steel apart. If they are above the yield strength, you get distortion. If they are close to the yield or tensile strength of your weldment and get locked into the cooling weldment, when the part is used, it doesn't take much additional stress to cause the part to break when put into service.

The purpose of stress relieving is to eliminate or greatly reduce built-up stresses caused by welding.

The thermal stress-relief treatment improves ductility and raises low-temperature impact values even though it may slightly reduce the tensile and yield strength of your base metal and weld. Many welding codes allow greater design stresses if stress relieving is to be used on a weldment. Typically, stress relieving consists of heating your workpiece to about 1100°F [593°C] and hold-

TABLE 16-21 Stress-relief times and temperatures for carbon steels

Metal Temperature, °F [°C]	Time per Inch of Section Thickness h
1100 [593]	1
1050 [565]	2
1000 [538]	3
950 [510]	5
900 [482]	10

TABLE 16-20 Suggested postheat time and temperature for typical hardenable steels*

AISI Steel	Postheat Temperature, °F [°C]	Postheat Time, Min	Typical Hardness, R_c	Maximum Hardness, R_c
1019	—	—	19	42
1050	600 [315]	10	48	62
	1000 [538]	1	28	
10B60	600 [315]	25	50	65
	1000 [538]	1	36	
1335	700 [370]	10	39	59
	1000 [538]	20	18	
2160	600 [315]	10	48	65
	1000 [538]	1	24	
2260	600 [315]	25	50	64
	1000 [538]	1	25	
2340	600 [315]	35	46	60
	1000 [538]	5	21	
2512	700 [370]	5	36	42
	900 [482]	75	20	
3140	700 [370]	25	48	60
	1000 [538]	15	21	
3330	600 [315]	25	43	57
	800 [425]	20	38	
	1100 [593]	10 h	23	
4037	700 [370]	10	41	56
	900 [482]	5	25	
4130	700 [370]	10	44	56
	800 [425]	5	37	
4340	600 [315]	50	48	62
	700 [370]	5 h	40	
4360	500 [260]	10 h	54	64
	650 [343]	50 h	48	
	1200 [649]	75	26	
4615	700 [370]	10	38	45
	900 [482]	10	26	
4640	600 [315]	25	48	60
	900 [482]	25	23	
5140	600 [315]	15	50	62
	1000 [538]	25	26	
6145	600 [315]	25	51	61
	1000 [538]	25	33	
86B20	700 [370]	5	40	45
	900 [482]	15	26	
8630	600 [315]	15	46	53
	900 [482]	25	27	
8660	500 [260]	5 h	52	64
	800 [425]	50 h	34	
8745	600 [315]	25	49	61
	800 [425]	100	30	
9260	600 [315]	100 h	50	65
	1200 [649]	1	35	
9440	600 [315]	50	50	60
	800 [425]	10	33	

* You must apply heat immediately after your weld is completed and before it cools. Get advice from a welding engineer, metallurgist, or heat treater if you plan to use preheating and postheating. Note the long times (hours) required at lower temperatures.

ing it at that temperature for one hour per inch of thickness. The work is then slowly cooled in still air (furnace-cooled) to below 600°F [315°C]. If temperatures as high as 1100°F [593°C] are not practical, lower temperatures and longer heating times can be used. Table 16-21 lists typical stress-relief times and temperatures for carbon steels.

16-17
PIPE WELDING

The real art of the skilled SMAW welder is in joining pipe and tubing. Your best bet, if your welding instructor wants you to do some work on pipe and tubing, is to re-read Chap. 10 on oxyfuel pipe welding before you do anything else. Review especially the description of the various pipe-welding positions. The same basic principles and pipe-welding positions that apply to oxyfuel pipe welding also apply to arc welding pipe.

Don't try welding pipe or tube with SMAW until you have had extensive practice welding plate and structural out-of-position. When you and your welding instructor think you're ready to try it, repeat all the work we described in Chap. 10, except do it with stick electrodes instead of oxyfuel.

Welding Small Pipe
Figure 16-36 shows you two joints commonly used to weld smaller-diameter butt joints on pipe. You will have to make the bevels very

accurately with a cutting torch or pipecutter, or you'll have terrible fit-up problems. You also will have to carefully align the work before welding. Use jigs or fixtures as holding devices. A piece of angle iron makes a good jig for small-diameter pipe, while a section of I beam or H beam is suitable for larger-diameter pipe.

You can use tack welds to hold the joint in place. The number of tack welds will depend on the diameter of the pipe. You will need two tack welds for $\frac{1}{2}$-in.-diameter [13-mm] pipe. Put them directly opposite from each other. Pipe over 1 in. [25 mm] in diameter will probably need four oppositely placed tack welds. Be sure that the tack welds are not any longer than twice the pipe thickness. Also don't make the tack welds any thicker than one-half of the pipe's wall thickness. Tack welds should be made with the same electrode you will use to finish the pipe weld.

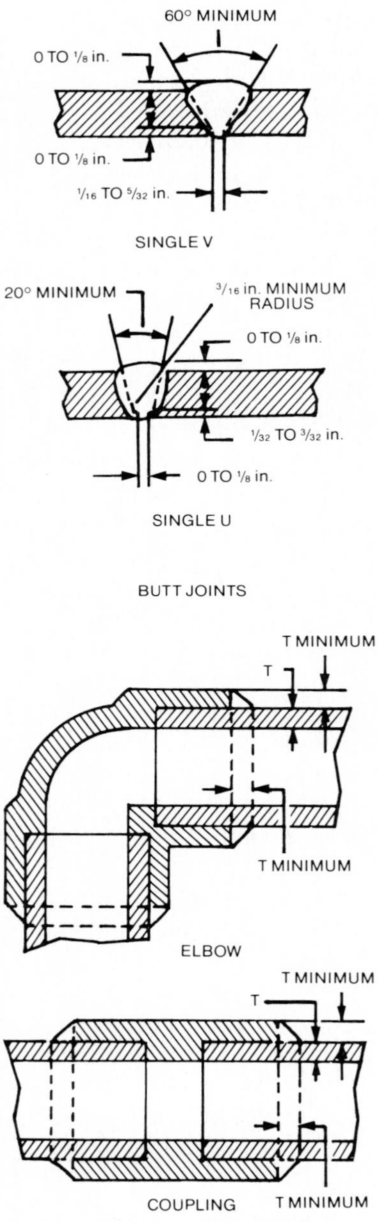

In addition to tack welds, spacers may be needed to maintain the proper joint alignment. Spacers are accurately machined pieces of metal that conform to the dimensions of the pipe joint. In pipe welding, spacers sometimes are called "chill rings" or "backing rings." Spacers serve a number of purposes. They provide you with a method for maintaining the specified root opening; spacers provide a convenient location for tack welds, and spacers aid you in securing the proper alignment of the pipe centerline. In addition, pipe spacers prevent weld spatter and the formation of slag or icicles inside the pipe. Spacers are tack-welded so that the parts being joined are rigidly locked together.

Never weld pipe when it is wet, raining, or snowing, or when the pipe is cold. Also never weld pipe in a high wind (it cools the pipe) unless you have it protected by a canvas windbreak. If the temperature where you are working is between 32 and 0°F [0 and −18°C], preheat the pipe with a heating torch over an area at least 3 in. [76 mm] on either side of the joint before welding. The pipe should be heated until it's warm to your hand before you start welding.

Welding Line Pipe
A large book could be written just on welding line pipe. You need a few years' basic welding experience before you can even apprentice in the fine art of cross-country pipeline welding. Besides, 48-in.-diameter [1.2-m] line pipe isn't something your local welding supplier will have in stock, anyway. But if you are interested in more detail, the American Petroleum Institute specifications for line-pipe welding would be the first and best place to read more about it. Meanwhile, we'll give you some pipeliner's terms here that will help you, and we'll show you what a cross-country pipeline joint might look like.

High-pressure gas pipelines are only one kind of pipeline work. There are hundreds, maybe several thousand shops and plants in the United States where structural tubing, steam pressure piping, petrochemical pipe and tubing, and similar materials are fabricated with stick electrodes and other welding methods, too. But by the nature of cross-country pipeline work, covered-metal arc welding is still the main method for joining pipe. If the work can't come to the welders, the welders go to the work.

Your first term to know is *double jointing*. Half of every pipeline is prefabricated under shop conditions by joining two lengths of large-diameter pipe into a single double joint before it is shipped to the pipeline where the double-length sections will be welded into a completed pipeline by field welders working for pipeline contractors.

One of the largest pipeline contractors is Brown & Root, Inc., a subsidiary of the Halliburton Company. Brown & Root is headquartered in Houston. Their pipeline division works out of Belle Chasse, Louisiana. They work around the world. For example, Brown & Root's pipeline division has installed more than 8000 mi [13,000 km] of submarine pipelines in the Gulf of Mexico alone. These submarine pipelines carry natural gas, oil, and other petroleum products to the Gulf states and beyond. Pipelines range from 4 to

FIGURE 16-36 Butt joints and socket-fitting joints used in pipe welding.

48 in. [100 mm to 1.2 m] in diameter. Anything over 30-in. [0.76 m] diameter is a large job.

Line-Pipe Specifications

Most pipeline field welding work follows American Petroleum Institute (API) Specification 1104 titled "Welding Pipelines and Related Facilities." This very detailed specification covers welding procedures, repair procedures, x-ray inspection, and other radiographic and test methods, and much more.

One contractor seldom if ever fabricates a complete pipeline. Contract awards are usually let for only a certain section of the job, sometimes only a mile or two if work is done in each section; sometimes one contractor will work on a section crossing several states. Each unit section of a pipeline is called a *spread*.

The pipe itself will have an API designation such as API 5LX70 or API 5LX52. The last two digits refer to the strength of steel from which the pipeline is fabricated. For example, API 5LX52 is made of carbon steel with a 52,000-psi [358-MPa] minimum yield strength, with specified low-temperature impact properties. API 5LX70 pipe has a minimum yield strength of 70,000 psi [483 MPa]. Charpy V-notch impact properties also are very important for pipelines, and for the weld metal used to fabricate them.

Line-Pipe Procedures

Larger-diameter pipe joints usually require two welders working on each side of the joint to keep the welding stresses in balance. Welders use internal line-up clamps on pipe of 16 in. [406 mm] diameter and larger. Fit-up usually must be within $\frac{1}{16}$ in. [1.6 mm]. They also have to preheat the pipeline steel when the outside (ambient) temperature drops to less than 40°F [4°C]. API X60 and API X65 pipe grades require extra precautions to avoid cracking. They often are preheated to 300°F [150°C] if the pipe and outside temperature is less than 70°F

[21°C]. In very cold weather or in high wind that would tend to chill the weld metal, welders work inside canvas huts that may completely surround the pipeline joint.

A typical pipe joint for API 5LX material (the X means that the wall is extra thick) is a 30° V-groove joint with a $\frac{1}{16}$-in. [1.6-mm] gap between the lands of the joint at the root face. The lands also are $\frac{1}{16}$ in. [1.6 mm] thick (Fig. 16-37). The first pass made in the root is a stringer pass, usually made by the most experienced welders on the crew. Working on opposite sides of the previously positioned pipe, two welders (more for larger-diameter pipe) start at the top and drag electrodes down to the bottom, filling in the root. After completing the stringer bead, they remove the line-up clamps.

The stringer pass is followed by what is called the *hot pass*. The hot pass must use enough current to melt out any external undercuts (they call them "wagon tracks") and any slag that just might remain. The hot passes are made within 5 min of finishing the stringer pass to prevent any chance of cracks. During that 5 min the complete stringer pass must be cleaned with a power brush until bare metal without any slag is left. Each subsequent pass is fully cleaned before others are made.

The hot pass is followed by *filler passes* to fill up the weld joint. Filler passes are made with a slight weave, enough to fuse the sides of grooves and previously

deposited weld metal to the new deposit. The filler passes continue in a previously established sequence until the pipe groove joint is filled almost to the top.

The final passes at the surface of the joint are called *cover passes* or *capping passes*. Welders run the cover pass all around to fill the groove to within $\frac{1}{32}$ to $\frac{1}{16}$ in. [0.8 to 1.6 mm] above the pipe surface so that it overlaps the outside surface of the pipe by $\frac{1}{16}$ in. [1.6 mm] on each side of the joint.

Electrodes of $\frac{3}{16}$-in. [4.8-mm] diameter are often used for the cover and filler passes, while $\frac{5}{32}$-in. [4-mm] diameter electrodes are used for the hot pass and root-pass stringer bead. All these passes are made in sequence, with the two welders working across from each other on opposite sides of the pipe.

Electrodes are usually burned down to stubs no shorter than 4 in. [100 mm] even though more of the electrodes could be used. Four-inch stubs are considered normal for cross-country pipeline welding.

All pipeline work requires x-ray-quality welding. Every inch of every weld will be fully inspected by radiographic techniques to detect the smallest imperfections. Every pipeline inspector knows what crew and what welders worked on each foot of pipe. If welders produce more welding imperfections than are allowed, they seldom are warned, they are simply fired from the job. Except for welder's helpers, there are no beginners welding cross-country pipelines.

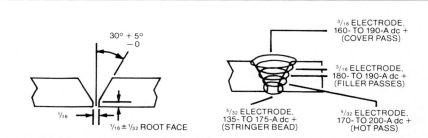

FIGURE 16-37 Typical line-pipe welding procedure and joint design for large-diameter pipe.

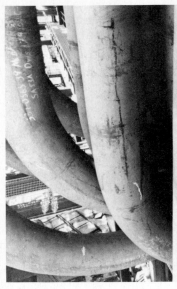

FIGURE 16-38 This is only 13 of the 56 pipes, each 22 in. in diameter, that carry pulverized coal from coal crushers to the burners of a 450-MW electric utility boiler. Every joint in this network of large-diameter, heavy-walled pipes was manually welded with SMAW electrodes and then inspected both visually and by x-rays. Most of the work was done on site. Only highly skilled (and very well paid) welders can do this kind of work.

If you think you can weld high-strength low-alloy steel from ½ to 1 in. thick or more, with stick electrodes, while standing or lying on your back in the mud or ice, with a 60-mi/h [96 km/h] wind blowing over you, while working continuously from the flat to the vertical-down to the overhead position, in exact sequence with another welder on the opposite side of the pipe, while making nothing but x-ray-quality welds, all the while working on joints preheated to 300°F [150°C] or more, then nobody will argue with you that you really are a skilled welder (Fig. 16-38).

REVIEW QUESTIONS

1. Name three advantages of SMAW over MIG and TIG welding.

2. Name three disadvantages of stick-electrode welding.

3. List two things that the flux coating on SMAW electrodes do.

4. You are holding an AWS E6010 electrode. Tell as much as you can about it just from the AWS designation, including the mechanical properties of the weld metal it produces, the type of flux coating it has, the positions that you can operate it in, and the correct current and polarity settings for using it.

5. Can you operate an AWS E6010 electrode on ac current and get good results?

6. Without looking it up, what is the minimum tensile strength of the steel weld metal produced by AWS E11018 electrodes?

7. Recommend a mild-steel SMAW welding electrode that will give your weld metal maximum ductility. The electrode also should produce welds with a minimum tensile strength of 70,000 psi [483 MPa].

8. You are welding a mild-steel weldment (60,000-psi [414-MPa] tensile strength or less) and have a serious fit-up problem because somebody did a poor flame-cutting job on the parts. What electrode should you use to help solve the fit-up problem?

9. What mild-steel electrode (60,000-psi [414-MPa] tensile strength weld metal) would you recommend for welding groove-butt welds in the flat position on material less than ¼ in. [6.4 mm] thick?

10. What high-strength low-alloy steel electrodes would you suggest for welding a pressure vessel with very good low-temperature impact strength?

11. On this mild-steel welding job, appearance is everything. What mild-steel electrode is most likely to give you very good looking beads with minimal undercutting?

12. You are depositing mild-steel weld metal but are getting very bad spatter that you must clean up after welding. What electrode do you recommend to solve the problem?

13. You have been asked to weld a part made of free-machining steel (with a high sulfur content). Normally high-sulfur steels produce very poor welds, but the part must be repaired by welding. What electrode do you suggest?

14. The steel sheet you are welding is giving you a lot of problems with distortion when you weld it with AWS E7018 electrodes. What AWS electrode grade do you suggest as a replacement for this job?

15. You need a general-purpose, all-position carbon-steel electrode for general maintenance welding mostly involving fillet welds and groove welds on material less than ¼ in. [6.3 mm] thick. What one electrode do you recommend?

16. Your shop needs a high-strength low-alloy steel electrode (70,000-psi [483-MPa] tensile) with a maximum deposition rate for flat position downhand welding. What should they buy?

17. You have to go to the storage area and select AWS E8016 electrodes for a job requiring as-deposited weld metal with a 20 ft·lb impact strength at −80°F. The storekeeper has three cans containing AWS E8016-C1, E8016-C2, and E8016-C3 electrodes and wants to know which can of electrodes you want. Which can do you want?

18. Your supervisor, just out of welding engineering school, claims that all AWS E70XX-A1 electrodes must be accompanied by minimum Charpy V-notch impact strength data when they are purchased. Is the new engineer right or wrong?

19. What group of stainless steels are hardenable by quenching and tempering heat treatments: ferritic, austenitic, or martensitic grades?

20. Name three austenitic stainless steels. What is the first number in the AISI designation code for each of the grades you list?

21. You are holding an AWS E308-16 stainless-steel electrode. Can you run it on ac as well as dc current?

22. What stainless-steel electrode would you recommend for welding Type 304L (low-carbon) stainless-steel parts together?

23. You have been asked to join Type 430 stainless-steel brackets to a part made from Type 440A stainless steel. What electrode should you use?

24. What AWS electrode specification covers arc-welding electrodes for copper-nickel tubing used in sea-water desalinization plants?

25. Name two arc-welding electrodes that are good for welding gray cast iron.

26. You have to weld the commercially pure (1100 series) aluminum sheets onto the sides of trailer trucks. Suggest an SMAW electrode.

27. What SMAW electrode grade would you first investigate for welding 4300-series aluminum-silicon alloys?

28. What is the difference between a tap start and a scratch start?

29. If you stop welding by rapidly lifting your electrode from the workpiece to break the arc, what welding defect are you likely to leave behind?

30. Why should you clean slag between passes when using most SMAW electrodes?

31. What is a good, quick way to remove the backing strip from a finished weld?

32. You are welding 9 percent nickel steel plate and your welding arc keeps jumping over to the base-metal. At other times the arc seems to wander around as if it

had a mind of its own. What is the most likely problem that is causing this arc action?

33. Should high welding currents be used for overhead welding to make sure the weld metal stays in place, or should low welding currents be used?

34. Why are some welds preheated and postheated after welding?

35. You are supposed to weld some 2-in. [51-mm] diameter pipe out in the pipe storage yard. Since it is raining heavily you can use a poncho or tarp to keep the rain off yourself. You have dry planks to stand on so you have no safety hazard to worry about. Should you go ahead and weld the wet pipe, or should you do something first? If so, what?

Welding Metal with Stick Electrodes

This shop work will help you learn how to make fusion welds on low-carbon steel with the SMAW process.

16-1-S
MATERIALS

You will need $\frac{1}{16}$-in.- and $\frac{1}{8}$-in.-thick sheet steel for this work. You also will need $\frac{1}{4}$-in.-, $\frac{3}{16}$-in.-, and $\frac{3}{4}$-in.-thick low-carbon steel plate. Small 2- × 4-in. coupons are OK to use when you are practicing this work, although your final welder qualification tests are best conducted on 6- × 6-in. coupons (low-cost metal test pieces).

You will have to select the correct filler metal and electrode diameter for each piece of steel. Make sure that your welding machine is set at the correct polarity for the electrode you are using (unless we specify otherwise). If you think that the material thickness and joint you are making will require an edge bevel, it's your responsibility to determine the bevel angle for your joint, and make it. Then complete your weld.

16-2-S
MAKING BEAD-ON-PLATE WELDS

First use an AWS 6010 electrode, and then compare it with other

types of electrodes such as AWS 6020 (in welding positions where it can be used). Then try an AWS 7018 electrode. Also try unplugging your power source and then switching the polarity on an electrode designed, for example, for dc reverse-polarity to dc straight-polarity current. We want you to gain a feel (and hearing) for how an electrode reacts when the wrong polarity is used.

In addition to testing the operating characteristics of different electrodes, learn to adjust your power source on the basis of the way your SMAW electrode sounds. Use the electrode supplier's operating settings for what they are—approximate guides. Experiment with machine settings that are purposely a little too low and too high for your electrode and its diameter. You will quickly get a feeling for the correct operating conditions based on nothing more than the sound your electrode makes when you weld.

1. Make bead welds on the $\frac{1}{16}$-in.-thick steel in the flat position.

2. Make bead welds on the $\frac{1}{8}$-in.-thick steel in the horizontal position.

3. Make bead welds on the $\frac{1}{16}$-in. steel in the vertical position.

4. Make bead welds on the $\frac{1}{8}$-in. steel in the overhead position.

Bead-Weld Test

There is no physical test for a bead weld. The only test is the appearance of the weld bead that you produce. The best weld bead will have a uniform width and an even, smoothly rippled surface with no pinholes, cracks, or gaps where you missed making the weld, and with no holes burned through the material. The more you practice making good-looking bead-on-sheet or bead-on-plate welds, the easier it will be to make the more difficult welds you will do next.

16-3-S
MAKING CORNER WELDS

5. Make a weld on a corner joint on two pieces of 1/16-in.-thick steel sheet.

6. Make a weld on a corner joint on two pieces of 1/8-in.-thick steel.

Corner Weld Test

7. Subject your completed corner welds to a hammer test. If you produced a satisfactory corner weld joint, you will be able to bend and then hammer the joint flat without cracking or breaking your weld. (Cracks in your base metal are acceptable.)

16-4-S
MAKING SQUARE-GROOVE JOINTS

These next exercises will teach you how to make square-groove butt joints on different thicknesses of material. You will start with thin-gauge steel and work up to thicker joints. You also will work gradually into more difficult welding positions.

Square-Groove Butt Joints on Sheet Steel

8. Make a weld on a square-groove butt joint on 1/16-in.-thick steel in the flat position.

9. Make a weld on a square-groove butt joint on 1/8-in.-thick steel in the flat position.

10. Make a weld on a square-groove butt joint on 1/16-in.-thick steel in the vertical position.

11. Make a weld on a square-groove butt joint on 1/8-in.-thick steel in the vertical position.

12. Make a weld on a square-groove butt joint on 1/16-in.-thick steel in the overhead position.

13. Make a weld on a square-groove butt joint on 1/8-in.-thick steel in the overhead position.

14. Make a weld on a butt joint on 3/16-in. steel plate in the flat position.

15. Weld a butt joint on 1/4-in. steel plate in the flat position.

16. Weld a butt joint on 3/16-in. steel plate in the vertical position.

17. Weld a butt joint on 1/4-in. steel plate in the vertical position.

18. Weld a butt joint on 3/16-in. steel plate in the overhead position.

19. Weld a butt joint on 1/4-in. steel plate in the overhead position.

Steel Butt-Joint Test

All butt welds should be tested by a guided root-bend test.

16-5-S
MAKING FILLET AND LAP JOINTS

Now you are ready to try making fillet and lap joints on steel. As before, you will start with thinner-gauge material and work up in thickness, using progressively more difficult welding positions.

20. Make a fillet-welded lap joint on 1/16-in.-thick steel in the flat position.

21. Make a fillet-welded lap joint on 1/8-in.-thick steel in the flat position.

22. Make a fillet-welded lap joint on 1/16-in.-thick steel in the vertical position.

23. Make a fillet-welded lap joint on 1/8-in.-thick steel in the vertical position.

24. Make a fillet-welded lap joint on 1/16-in.-thick steel in the overhead position.

25. Make a fillet-welded lap joint on 1/8-in.-thick steel in the overhead position.

26. Make a fillet-welded T joint on 1/16-in.-thick steel in the flat position.

27. Make a fillet-welded T joint on 1/8-in.-thick steel in the flat position.

28. Make a fillet-welded T joint on 1/16-in.-thick steel in the vertical position.

29. Make a fillet-welded T joint on 1/8-in.-thick steel in the vertical position.

30. Make a fillet-welded T joint on 1/16-in.-thick steel in the overhead position.

31. Make a fillet-welded T joint on 1/8-in.-thick steel sheet in the overhead position.

32. Make a fillet-welded lap joint on 3/16-in.-thick steel plate in the flat position.

33. Make a fillet weld on a lap joint on 1/4-in.-thick plate in the flat position.

34. Make a fillet weld on a lap joint on 3/16-in.-steel plate in the vertical position.

35. Make a fillet weld on a lap joint on 1/4-in.-thick steel plate in the vertical position.

36. Make a fillet weld on a lap joint on 3/16-in.-steel plate in the overhead position.

37. Make a fillet weld on a lap joint on 1/4-in.-steel plate in the overhead position.

38. Make a fillet weld on a T joint on 3/16-in.-steel plate in the flat position.

39. Make a fillet weld on a T joint on 1/4-in.-steel plate in the flat position.

40. Make a fillet weld on a T joint on 3/16-in.-steel plate in the vertical position.

41. Make a fillet weld on a T joint on 1/4-in.-steel plate in the vertical position.

42. Make a fillet weld on a T joint on 3/16-in.-steel plate in the overhead position.

43. Make a fillet weld on a T joint on 1/4-in.-steel plate in the overhead position.

Testing Fillet and Lap Joints

44. All of your lap joints are to be subjected to a hammer test, folding the joint 180° along the weld to make a flat sandwich with the welded joint running along one edge. Then pound the welded joint

flat without any weld metal cracking (if you can, you pass the test).

45. All of your fillet joints are to be subjected to a hammer test for T joints, folding the 90° leg of the T joint flat to one side of the joint, followed by hammering the joint without producing any weld-metal cracking.

16-6-S
WELDING CAST IRON

Welding cast iron is a very common, and very important repair job. Learn how to do it well.

Cast-Iron Welding Materials

You should have some samples of gray cast iron about ¼ in. thick, ½ in. thick, and ¾ in. thick for this work. You also should have a small cast-iron part to repair. A quick trip to a junkyard with a sledgehammer could soon get you all the cast-iron you need. A foundry also should be glad to donate leftover risers and other scrap castings for you to work on. Select your cast-iron filler rod from AWS A5.15. A rod that is ⅛ in. square will be a good size to work with.

Making Cast-Iron Welds

46. Make a weld on a butt joint on ¼-in.-thick cast iron in the flat position. Repeat the job welding in the vertical position.

47. Make a weld on a butt joint on ½-in.-thick cast iron in the flat position. Repeat the job welding in the vertical position.

48. Make a weld on a butt joint on ¾-in.-thick cast iron in the flat position. Repeat the job welding in the vertical position.

49. Repair a fractured gray cast-iron part by stick-electrode welding. The bigger the part, the better. A worn-out auto engine block from a junkyard is useful.

Testing Cast-Iron Welds

Although in commercial practice a repair weld cannot be broken and tested to see if the weld metal is sound, you should know by the way you made your weld whether or not you got sound weld metal. However, if your practice castings are small enough, they should be broken to determine the quality of the welds you made. Each weld and the adjacent base metal should be tested with a file for hardness. If the metal is too hard to be cut by the file, either you let the casting cool too fast after welding or you didn't preheat it enough beforehand, or you made both errors. Cast-iron welds are often machined to a smooth finish after welding. But a file-hard weld in a casting is a machinist's nightmare.

16-7-S
WELDING COPPER

This work will help you learn how to weld copper with the SMAW process.

Copper Welding Materials

You will need sheet copper samples, about 2 × 4 in. and ¹⁄₁₆ to ¼ in. thick. Also practice on standardweight, 2-in.-diameter copper water pipe. The sheet copper preferably should be a good selection of both deoxidized (OFC, or oxygen-free copper) and electrolytic copper. Scrap copper used in electrical parts will probably be made of the latter material. You will have to select the proper SMAW electrode flux and wire diameter for this job.

Making Copper Welds

50. Make a butt joint in the flat position on sheet.

51. Make a butt joint in the vertical position on sheet.

52. Make a butt joint in the overhead position on sheet.

53. Repeat the work in the three jobs above using copper pipe or tubing.

Testing Copper Welds

All your welds should be subjected to a guided-bend test. Successful completion of each of these copper welding exercises will constitute qualification for this work. There are many tests for copper welding, but they vary so much with the material to be tested and the application for the product that simply making a decent weld on this highly conductive material is proof enough that you can weld this material as far as your shop course is concerned.

16-8-S
WELDING ALUMINUM

You must select an SMAW electrode that is compatible with the aluminum base metal grade you will weld. If a beveled joint is required, you have to decide what the bevel angle should be.

Aluminum Welding Materials

You should have a 2- × 4-in. sheet and plate samples of aluminum to work on that are ¹⁄₁₆ in., ⅛ in., ³⁄₁₆ in., and ¼ in. thick. If you can obtain them, get some aluminum castings to work on. They should be about ¼ to ½ in. thick. If possible, you should practice welding on commercially pure 1100-series aluminum, and on at least one weldable aluminum alloy such as a silicon-aluminum grade like 4043.

Making Aluminum Welds

54. Make bead welds on ¹⁄₁₆-in.-thick aluminum sheets. First try it with 1100 series aluminum, and then with a silicon-aluminum alloy such as 4043, using the correct SMAW electrodes for this material.

55. Make bead welds on ⅛-in.-thick aluminum sheets. First try it on commercially pure aluminum, and then on the alloyed material.

56. Make welds on flanged edge joints in the flat position on ¹⁄₁₆-in.-thick aluminum sheets. Work with both 1100 and 4043 aluminum grades.

57. Make welds on butt joints in the flat position on ⅛-in.-thick 1100 series sluminum sheets.

58. Make welds on butt joints in the flat position on ⅛-in.-thick silicon-aluminum sheets.

59. Make welds on flanged-edge joints in the vertical position on $\frac{1}{16}$-in.-thick aluminum sheets.

60. Make welds on flanged-edge joints in the vertical position on $\frac{1}{8}$-in.-thick silicon-alloy aluminum sheets.

61. Make welds on butt joints in the vertical position on $\frac{1}{8}$-in.-thick 1100 aluminum sheets.

62. Make welds on butt joints in the vertical position on $\frac{1}{8}$-in.-thick 4043 aluminum sheets.

63. Make welds on deep flanged-edge joints in the flat position on $\frac{1}{16}$-in.-thick aluminum sheets.

64. Make welds on deep flanged-edge joints in the flat position on $\frac{1}{16}$-in.-thick silicon-alloy aluminum sheets.

65. Make welds on butt joints in the flat position on $\frac{3}{16}$-in.-thick 1100 grade aluminum sheets.

66. Make welds on butt joints in the flat position on $\frac{3}{16}$-in.-thick 4043 silicon-aluminum alloy sheets.

67. Make welds on butt joints in the flat position on $\frac{1}{4}$-in.-thick 1100 grade aluminum sheets.

68. Make the same welds on butt joints in the flat position on $\frac{1}{4}$-in.-thick 4043 silicon-aluminum alloy plates.

69. Repeat the butt joints on $\frac{1}{4}$-in.-thick plate, but this time work in the vertical position.

70. Make a butt joint on $\frac{1}{4}$-in.-thick plate in the overhead position.

Testing Aluminum Welds

All aluminum butt welds on material $\frac{1}{4}$ in. and thicker should be submitted to the nick-break test. Test your flange welds and butt welds on thinner material with guided-bend tests. There is no test for bead-on-plate or bead-on-sheet except weld-bead appearance.

16-9-S
WELDING STEEL PIPE

This work is optional, but very important if you can possibly get it done. Using 4-in. steel pipe, work in as many different pipe welding positions as you can or

have equipment to operate in. Save the best samples for your pipe welder operator qualification test (Test 3), which follows.

71. Pipe welding is very difficult but well-paying work. Prepare a beveled pipe joint from two pieces of steel pipe about 4 in. in diameter. Bevel the outer edges to 45° before you start. Then tack-weld the two sections into position with four tacks spaced equally around the joint (at points about 0, 90, 180, and 270° around the pipe). Then weld the pipe sections together.

Keep the pipe centerline in the horizontal position while you work. It will help you to rest the pipe in a piece of channel iron or two angle irons while you work. That way you can roll the pipe when you have to, welding a quarter of the section at a time, while always working in the downhand position. Pay special attention to getting a good solid root pass. Make sure your cover passes are tied into the sides of the joint and the root pass.

72. Repeat Exercise 71 with the pipe centerline in the vertical position (the pipe is standing upright) after you tack-weld it in-line. This time welding the pipe joint is like making a circular weld in the horizontal position. It's tough to do.

Testing Pipe Welds

Subject a strip cut from your pipe welds to guided-root-bend and guided-face-bend tests.

16-10-S
OPERATOR-QUALIFICATION TESTS

The purpose of this shop practice has been to develop your ability to make butt and fillet welds under conditions approximating those you will find in industry.

It's time to take some operator-qualification tests very similar to those you will have to take to get a job. In these tests, you will select your own equipment and filler metal and set up your ma-

chine. You also will conduct your own tests on your own material. In a real operator qualification test, you would make the welds but someone else would test your work. The objective of these tests is to show you where you are having problems. If you have problems, solve them with extra practice.

If you don't pass one of these tests, it shows that you need more work. For example, you may be able to make very good welds in the flat and horizontal positions, but you may be having trouble making sound welds in the vertical and overhead positions. Or you may be able to make good welds in any position on steel, but aluminum is giving you trouble. Or you may not have any difficulty with wrought materials, but cast iron is giving you a hard time. These tests will show you where you need extra shop practice.

Test 1. Butt Welds

These tests will require six plates of metal (low-carbon steel, copper, or aluminum). You will have to decide whether or not the material thickness requires edge beveling, and at what angle to bevel the material. You also will do the required edge preparation before you start to weld.

The first test consists of making:

■ Single-V-groove butt joints in the flat position
■ Single-V-groove butt joints in the vertical position
■ Single-V-groove butt joints in the overhead position

Each of the welded plates you make will be cut into four strips as shown in Fig. 9S-1, Shop Practice section, Chap. 9. The two specimens that are retained are machined flush all over as shown in Fig. 9S-2, Shop Practice section, Chap. 9. One test specimen from each of the three welding positions (flat, vertical, and overhead) will be submitted to a guided-face-bend test, and the other center strip will be submitted to a guided-root-bend test. If any defects greater than ⅛ in. in any dimension are developed in these tests, your weld will be considered as failing. In addition to the guided-face-bend and guided-root-bend test, your weld will be examined for appearance. Sloppy-looking welds, whether they pass the mechanical tests or not, will be rejected.

Test 2. Fillet and Lap Welds

These tests require six plates each of which is 6- × 6- × ¼-in. thick. One edge of each of the plates will be ground square. Also, one backup strip 6- × 1¼- × ¼-in. will be required for each joint. Set up the test plates for a double-fillet-weld T-joint and make:

- One set of fillet welds in the flat position
- One set of fillet welds in the vertical position
- One set of fillet welds in the overhead position

After the fillet welds have cooled, cut off the leg of the T flush to the base and then make a third weld in the area marked "A" in Fig. 9S-3, Shop Practice section, Chap. 9, in the position where the leg of the T joint previously stood. This third weld may be done in any position you want, regardless of the position in which the fillet welds were made. However, you must use the 1¼-in.-wide × ¼-in.-thick backup strip behind the final weld in area A, when you make it.

When welding is completed, the backup strip shown on the bottom of Fig. 9S-3, Shop Practice section, Chap. 9, should be removed flush with the underside of the plates. For best results, this backup strip should be machined off. However, if a suitable grinding machine or other machine tool is not available, this strip can be removed by a hand-cutting torch. You must in this case be very careful not to cut into the root of the weld. That would tend to fuse in or cover up any root defects and would not permit a true or fair test of your ability.

The plates are then cut into four strips as shown in Fig. 9S-4, Shop Practice section, Chap. 9. The two specimens retained are machined flush all over as shown in Fig. 9S-2, Shop Practice section, Chap. 9. One specimen from each of three welding positions is submitted to the guided-face-bend test and the others are submitted to the guided-root-bend test.

If any defects greater than ⅛ in. in any dimension develop in this test, you will fail this test.

Test 3. Pipe Welding

Make several sample joints by welding two pieces of 4-in. standard steel pipe approximately 6 in. long. The two pieces of pipe will be placed in a horizontal line and position-welded as a butt joint. This is a horizontal fixed-position pipe weld.

A minimum of four specimens should be taken from the completed weld, spaced at 90° intervals around the weld. These should be taken from the top, bottom, and opposite sides of the completed pipe weld. The top and one side specimen should be submitted to the guided-face-bend test, and the bottom and other side specimen should be subjected to the root-bend test.

All of the above specimens should bend to 180° or the full capacity of the guided-bend test jig without any cracks greater than ⅛ in. in any dimension, either as a result of bending or existing prior to welding.

If you pass this pipe welding test, we congratulate you. If you don't pass, work on your pipe welding technique.

MIG and Cored-Wire Welding

Gas-metal arc welding (GMAW) uses an externally supplied shielding gas to protect the arc and molten weld metal from oxygen in the air. The filler metal is a continuous bare electrode wire fed through a wire feeder and a welding "gun." Low-carbon steels, high-strength low-alloy and stainless steels, aluminum and magnesium alloys, copper alloys, and other metals can be deposited. The gun delivers both the shielding gas and the electrode filler-metal wire to the weld. The various GMAW processes can be manually operated as in Fig. 17-1, or they can be fully automatic as shown in Fig. 17-2.

The reason behind the development of GMAW remains its single most important advantage. You can work continuously without stopping to change electrodes, or pick up a new welding rod. Since about 85 percent of the cost of manual welding is labor, doubling the time that you have your arc on cuts the total cost of welding by over 40 percent. As a result, MIG and other forms of GMAW, and the closely related flux-cored-wire welding (FCAW), are usually thought of as high-production welding processes. However, they are so versatile that GMAW also is extensively used in plant and farm maintenance work, and occasionally even in making welded works of art and fixing cars.

Welding equipment makers have recently developed relatively small, low-cost power sources, wire feeders, and guns that use the GMAW process. These outfits are sold as integral welding packages so that farmers, metal sculptors, automotive body shops, and plant maintenance people can afford the advantages of MIG and FCAW without investing in high-production equipment.

17-1
THE GMAW PROCESSES

MIG stands for metal inert-gas arc welding. The present process is very much like the original late 1940s invention of the Airco Welding Products Division of Airco, Inc. However, the equipment used for the first MIG welding outfits compares with present-day equipment about as much as the Wright brothers' first airplane compares with a modern jet fighter. Nevertheless, any welder reading Airco's first patents would immediately recognize the MIG process:

U.S. Patent No. 2,504,868—"New concepts of a combination of factors including welding current density, gas shielding, filler wire speeds and direct current, reverse polarity . . ."

U.S. Patent No. 2,544,801—"Apparatus to feed a wire electrode continuously to an arc of the gas-shielded type . . ."

Other companies that have made major contributions to GMAW technology are the Linde Division of Union Carbide, the Lincoln Electric Company, Hobart and Miller Electric (major welding equipment manufacturers), and numerous producers of the welding wires that are used with the GMAW processes.

Many Processes from One Concept

A large array of related welding processes has developed from this one semiautomatic welding concept. Airco, Inc., alone holds over 100 additional patents on related processes and equipment. For example, using ultra-high-speed motion pictures, Airco researchers saw for the first time that, under certain conditions, the new process sprayed droplets of molten metal across the arc. Before that, nobody knew exactly how the weld metal in any arc-welding process got from the solid filler-metal electrode to the base metal and molten weld puddle. Many other welding processes did not spray molten metal.

Studies of different high-speed molten-metal-transfer processes soon led to the development of a variety of major welding pro-

FIGURE 17-1 MIG welding is an extremely versatile wire-fed welding process. It can be used for maintenance and repair work, or for high production. Unlike stick electrodes, the process makes continuous welds nonstop, but it also can be used for tack welding, like the job shown here.

FIGURE 17-2 Ten MIG guns, power sources, wire feeders (encased in plastic covers to keep the wire clean) simultaneously weld steel-hat sections to a base plate to make rigid decking for a bridge. Compare this to Fig. 17-1, and you see two extremes of the MIG process.

cesses. A few of the more important ones are *short-circuit* GMAW welding (which doesn't spray metal) used for thin materials and out-of-position work; conventional solid-wire MIG; pulsed-spray arc welding, extends the MIG process to thinner materials with less distortion and to more difficult out-of-position work; high-deposition-rate flux-cored-wire welding with filler wire containing a core of flux instead of a solid wire; and now a cored-wire welding process that uses powdered metals and combines the advantages of solid and cored-wire welding.

Equipment innovations include automatic wire feeders and welding machines that can join steel plate several inches thick while climbing up the side of a pressure vessel or a ship hull. These innovations produced two more welding processes called *electroslag* and *electrogas* welding (see Appendix). Most of the work done by welding robots uses one of these wire-fed processes because robots won't pay off quickly enough if stick electrodes are used. (A welding robot would also have a terrible time remaining productive while trying to manipulate a stick electrode, anyway.)

Further development work has shown that the relatively expensive and inert gas argon, which was originally used in MIG welding, is not always needed. Mixtures of carbon dioxide and argon produce weld-metal spray transfer through the arc while reducing shielding gas costs (pure carbon dioxide, however, does not produce spray transfer).

Other gases mixed with argon give different welding characteristics. It was even found that 100 percent CO_2 shielding gas, which is less expensive than argon, can be used if the continuous solid or flux-cored filler wire has some deoxidizing elements in it. The wires now used with CO_2-shielded welding processes do contain special deoxidizing elements (which is a clue to you that not all

GMAW wire is right for all processes).

The deoxidizers in the wire make up for the fact that CO_2 shielding gas isn't fully inert. CO_2 is in fact highly oxidizing when it passes through an arc and is ionized. The gas ions turn back into CO_2 quickly after passing through the welding arc. The reconstituted gas continues to protect the molten weld metal, but it also can cause oxide porosity in welds if welding wire containing deoxidizers isn't used.

Other researchers found that certain kinds of flux-cored wires could be run without any exterior gas shielding. Enough flux of the right kind could be put into the wire core to produce a relatively protective (but not totally inert) shielding gas when the flux inside the welding wire was vaporized by the arc. The idea is like using a continuous inside-out flux-coated stick electrode. These special unshielded flux-cored electrodes are often called *air wires* in the shop.

Meanwhile, "pulsed-spray" MIG welding was developed for welding thin metals and metals sensitive to the less expensive but more reactive carbon-dioxide shielding gases. The result was a new process that was easier to use for out-of-position welding. It also works on thin sheets by reducing heat input and distortion. The pulsed welding current idea was later adopted for GTAW and is called *pulsed TIG*.

There even is a modification of the MIG process called *MIG-spot* arc welding because the results resemble resistance spot welding (see Appendix for resistance welding). MIG-spot welding is used to join sheet metal just like resistance welding does. It was even found that a solid MIG wire with high current and very high speed could be used to cut steel instead of weld it. What's more, the lower-melting-point filler metals used in braze-welding can be used in a MIG outfit to do continuous MIG arc braze-welding (a newer

development that is becoming more frequent in automotive production).

What's in a Name?
The AWS finally had to abandon the use of the name MIG for all of these diverse processes. The AWS now calls all processes using exterior gas shielding (whether argon or argon mixed with other gases, or straight CO_2 shielding) GMAW processes. The original solid-wire process with inert argon gas (or argon plus an additive shielding gas) continues to be called MIG by most shop people. But now it's simply a branch of the much larger family of GMAW processes. The pulsed-current processes are handled simply by calling them *pulsed GMAW*. Examples are *pulsed MIG* similar to *pulsed GTAW* in the case of TIG welding. You can lump all of them together with the name *pulsed-current* welding. This chapter will tell you a lot about solid-wire MIG and cored-wire welding. The other related processes are easy to understand and operate once you become skilled at using these two basic processes. The one other important welding process that uses continuous wire filler metals is submerged-arc welding. We'll cover that process in the next chapter.

The Metals Welded by GMAW
The various GMAW processes can join just about any metal in almost any commercial thickness from light sheet to heavy plates and structurals. The only major technical limitation is using filler metals that can be drawn into wires. Cast iron can't be made into flexible filler-metal wires because its ductility is far too low. If cast iron is welded by some GMAW process, some other filler metal than cast iron must be used.

However, there are numerous grades of continuous, solid GMAW filler wires used to weld carbon steels, high-strength low-alloy steels, and even high-strength alloy steels. Many

GMAW wires also exist for welding stainless steels, aluminum alloys, copper and copper-based alloys, nickel and its alloys, and some hardfacing alloys.

The flux-cored wire processes similarly offer quite a variety of steel weld-metal grades. The low-cost carbon-steel wire used for the outer part of the continuous wire electrode can be alloyed with various chemical elements put into the flux (or powdered metal) core to make many different kinds of ferrous weld metal. FCAW is used almost exclusively on ferrous metals—irons and steels.

When to Use Each Process
Roughly speaking, the high-deposition-rate flux-cored-wire processes are reserved for welding carbon steels and high-strength low-alloy and full-alloy steels. Solid MIG welding wire grades are far more numerous and can be used for welding carbon steels, high-strength low-alloy steels, full-alloy steels (even submarine hull plate), and for welding almost all of the weldable grades of aluminum, copper, and nickel-based alloys. In other words, MIG welding is more versatile than flux-cored-wire welding in terms of the metals that it can weld. But flux-cored-wire welding, reserved for use on steel, can lay down weld metal a lot faster than solid-wire MIG welding.

Both GMAW processes, MIG and cored-wire welding, offer much higher weld-metal deposition rates than most stick electrodes. However, stick electrodes still give you the widest choice of weld-metal grades and properties, lower initial equipment cost in many cases, and portability. A stick electrode welder can get almost anywhere he or she can crawl or climb. A GMAW welder is tethered to the wire feeder.

A GMAW welder is usually limited to moving about 15 to 20 ft [4.5 to 6 m] from the power source and wire feeder because of problems with push-feeding wire greater distances. The recent development of extended-reach wire feeders has changed this restriction. These special feeders let you operate MIG equipment up to 100 ft [30 m] from your power source when using steel wire and over 50 ft [15 m] when welding with softer aluminum filler-metal wires. These extended-reach units use two feeders: one near the power source is a conventional push-type feeder, and a much smaller secondary pull-type feeder is carried along by the welder. The small wire feeder pulls on the wire because you can pull wire much farther than you can push it through a service cable and welding gun.

The choice among these different processes is not always easy to make. New process and equipment developments continue to be made every year. We'll look at some of the differences among the most important GMAW processes in this lesson.

GMAW Equipment

No matter what type of GMAW is done, MIG or flux-cored-wire welding, pulsed MIG, or some other GMAW process such as FCAW, the basic equipment and supplies you need are almost always the same (Fig. 17-3). Essentially the same equipment list will be needed for submerged-arc welding (which we'll cover in the

next chapter), except that granulated welding flux is used in place of a shielding gas in the submerged-arc process, along with some kind of device to spread the flux on the base metal ahead of the advancing welding wire and arc. Therefore, much of what you learn about the wires and equipment used for GMAW welding will also apply to the submerged-arc process. In general, you need the following:

■ A suitable welding power source to provide current to melt the wire and base metal being joined.
■ A wire feeder, contactor (switch), and controller to deliver the filler wire at the required speed.
■ A welding gun to direct the wire and the shielding gas at the workpiece.
■ A shielding-gas cylinder and high-pressure regulator with hose.
■ A spool or coil of electrode wire of specified type and diameter for the work to be done—solid wire for MIG welding and cored wire for FCAW.
■ Proper welding techniques for the given GMAW or FCAW process and base metal.

One characteristic of MIG welding is that welders who have already had some experience usually learn enough about the process in a week or less to go into full production. You certainly

qualify on that score. You also will find that some equipment items may be combined. A small wire spool may be included with the welding gun. A larger wire feeder and reel sometimes are installed right inside the power source, along with the wire feed and speed controls. High-production equipment will have separate wire feeders and power sources.

Before we tell you about each of these basic elements, let's look at the different ways that filler metal is transferred from the wire to the weld metal through the GMAW arc. That will help explain almost everything else about GMAW processes and equipment.

17-2 MOVING METAL THROUGH AN ARC

MIG welding is almost always done with dc, and almost always with reverse-polarity hookups at the power source terminals. The thickness and type of base metal and the diameter and feed rate of the wire will govern the welding machine settings. The voltage, current, and type of shielding gas determine the manner (or *mode* as it's called) in which metal is transferred from the wire to the molten weld pool.

There are three different ways in which the molten weld metal moves through an arc. They are *dip transfer* (also called *short circuiting* or *short arc*), *globular transfer,* and *spray transfer*. To get a better feeling for them, think of the electrode wire first getting just hot enough to melt at a fairly low current (dip transfer). Then increase the welding current until the weld metal starts to ball up on the end of the electrode wire (globular transfer). Turn the welding current up even higher and tiny droplets of the weld metal will start to spray off the end of the welding wire (but only when you have the right kind of shielding gas).

Each transfer mode has a major effect on welding characteristics.

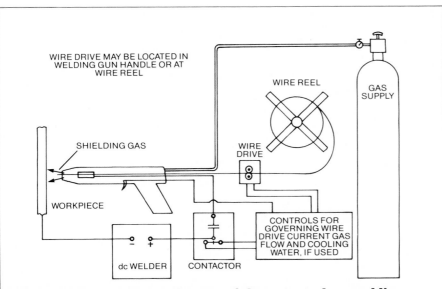

FIGURE 17-3 A schematic diagram of the gas-metal-arc welding process. (Courtesy Linde Division, Union Carbide Corp.)

WIRE DRIVE MAY BE LOCATED IN WELDING GUN HANDLE OR AT WIRE REEL

WIRE REEL

GAS SUPPLY

SHIELDING GAS

WIRE DRIVE

WORKPIECE

dc WELDER

CONTACTOR

CONTROLS FOR GOVERNING WIRE DRIVE CURRENT GAS FLOW AND COOLING WATER, IF USED

They can help you get the work done or make it difficult. You can control and select the transfer mode by the machine settings, and the shielding gas. Each of these weld-metal transfer modes works best in certain welding positions, too. In fact, selecting the right transfer mode is the major decision you must make when welding out-of-position with MIG wire. Choose the right transfer mode (spray transfer) and you'll get cleaner welds with less spatter. Choose the wrong one for overhead welding (dip transfer or globular transfer) and you'll have problems keeping the overhead weld metal in place.

17-3
DIP-TRANSFER WELDING

This transfer mode is generally used on sheet metal and for welding in all positions. This process also is called *short-circuiting, short-arc, microwire,* and *pinch-arc* welding. We'll call it dip-transfer welding because that's a good description of how it works.

Dip-transfer MIG welding deposits electrode metal by contact between the tip of the electrode

and the molten weld metal. Contacts are made regularly between the electrode wire and workpiece at a rate of from forty to hundreds of times per second.

Spatter is practically non-existent when dip-transfer welding if the settings are correctly made. Drops are not transferred across a space between the electrode and the work. The molten-metal transfer is made only when the electrode wire touches the weld puddle.

The dip-transfer process uses low welding currents and voltage with electrodes of 0.030 in., 0.035 in., and 0.045 in. [0.76, 0.89, and 1.1 mm] diameter, which compares with the larger-diameter 0.0625-in. wire (1/16-in. [1.6-mm] wire) frequently used in spray-transfer MIG welding.

Dip transfer is a dc welding process used for joining materials up to several inches thick, as well as for welding thin-gauge metals. Reverse polarity is used for almost all jobs. Dip-transfer welding is especially useful for welding all kinds of steels. Aluminum can be welded with the dip-transfer MIG process, but the results

are not very good. The reason is that dip transfer is a relatively low heat process. The high thermal conductivity of aluminum causes the weld puddle to freeze rapidly. The rapid freezing traps gases in the aluminum weld puddle, creating porosity. The low thermal conductivity of steel delays the weld puddle's solidification long enough for gas bubbles to rise to the surface of the molten metal and escape.

The dip-transfer welding current and arc-voltage is low (see Table 17-1), and therefore the heat input also is low. That's why dip transfer is excellent for welding thin-gauge steel that otherwise would be distorted by a high heat input.

Another advantage of the low heat input is that dip-transfer welding can bridge gaps where the joint fit-up is poor. Since it's a low-heat-input welding process with relatively fast puddle freezing (contrasted to spray-transfer MIG welding), dip transfer also is useful for welding in all positions. When needed, the heat of the arc can be increased by using special shielding-gas mixtures and by "tuning in" the power supply.

TABLE 17-1 Typical arc voltages for GMAW of different metals*

Metal	Droplet Transfer (1/16-in. or 0.625-in.-diameter electrode wire)					Dip Transfer (0.035-in.-diameter electrode wire)			
	Argon	Helium	25% Ar, 75% He	Argon + Oxygen (1–5% O_2)	CO_2	Argon	25% Ar, 75% He	Argon + Oxygen (1–5% O_2)	CO_2
Aluminum	25	30	29	—	—	19	—	—	—
Magnesium	26	—	28	—	—	16	—	—	—
Carbon steel	—	—	—	28	30	17	18	19	20
Low-alloy steel	—	—	—	28	30	17	18	19	20
Stainless steel	24	—	—	26	—	18	19	21	—
Nickel	26	30	28	—	—	22	—	—	—
Nickel-copper alloys	26	30	28	—	—	22	—	—	—
Nickel-chromium-iron alloys	26	30	28	—	—	22	—	—	—
Copper	30	36	33	—	—	24	22	—	—
Copper-nickel alloys	28	32	30	—	—	23	—	—	—
Silicon bronze	28	32	30	28	—	23	—	—	—
Aluminum bronze	28	32	30	—	—	23	—	—	—
Phosphor bronze	28	32	30	23	—	23	—	—	—

* Plus or minus approximately 10 percent. The lower voltages are normally used on light material and at low amperage; the higher voltages with high amperage are used on heavier sections.

We'll tell you more about tuning in in a moment. First we'll give you a better description of the dip-transfer welding process.

The Dip-Transfer Process

The dip-transfer method uses the lowest range of welding currents and smallest electrodes wires of any GMAW process. Typical welding currents are 50-A minimum to 150-A maximum for 0.030-in.-diameter [0.76-mm] steel wire; 75 to 175 A for 0.035-in. [0.89 mm] wire; and 100 A to 225 A for 0.045-in.-diameter [1.1-mm] steel wire, using DCRP (electrode-wire positive) setups.

Dip-transfer welding is characterized by repeated short-circuiting of the electrode wire against the workpiece (Fig. 17-4). When the electrode tip touches the weld puddle the arc goes out for a moment as the voltage and amperage rise rapidly (Fig. 17-4D). A liquid-metal bridge forms between the tip of the electrode wire and the molten weld puddle. When that happens, part of the molten electrode-wire tip is transferred to the weld puddle, and then the molten-metal bridge is pinched off (Fig. 17-4A). When the molten metal separates from the electrode wire, the amperage drops rapidly, and the voltage dips. When the droplet breaks from the electrode wire, the welding arc resumes (Fig. 17-4B). The arc amperage drops and the voltage falls. Meanwhile, the electrode wire continues to feed in toward the molten weld puddle (Fig. 17-4C). As soon as the wire hits the weld metal (Fig. 17-4D), the voltage and amperage rise again and the cycle is repeated. The individual drops of molten filler metal that are transferred to the puddle are very small. It takes a great many drops per second to supply enough metal to make a weld bead.

For example, assume that each drop is 1/32 in. [0.79 mm] in diameter, and that there are 120 drops transferred per second (120 short circuits per second). That is equiv-alent to a deposition rate for steel of only about 2 lb/h [0.9 kg/h]. If the wire electrode were also 1/32 in. [0.79 mm] in diameter, the electrode-wire feed rate would be about 150 in./min [3.8 m/min].

The important thing to remember about dip-transfer MIG welding is that the molten weld metal is only transferred when the electrode wire touches the weld puddle. The welding action is completed by the arc during the time between short circuits.

The average frequency of short circuits (metal transfers) will vary from 20 to 200 times a second, depending upon welding conditions. Whatever the frequency of "dipping" happens to be, it's not constant. The normal differences in materials, equipment, and power sources will cause the dip-ping frequency to vary from one second to the next. For example, there might be 108 shorts for one second, then 132 shorts for the next second. Over a period of a minute or so, however, the *average frequency* of dipping might be about 120 shorts per second. At some other current and wire feed speed, the average dipping frequency might be 70 or 150 shorts per second.

You will not be able to see the short-circuiting dips occurring. The frequency is faster than the frame speed in a motion picture film. The arc looks like it is on all the time just like a motion picture film makes the picture on the screen look like it is in continuous motion. However, if you had a device that could produce a stop-action picture, you would see what

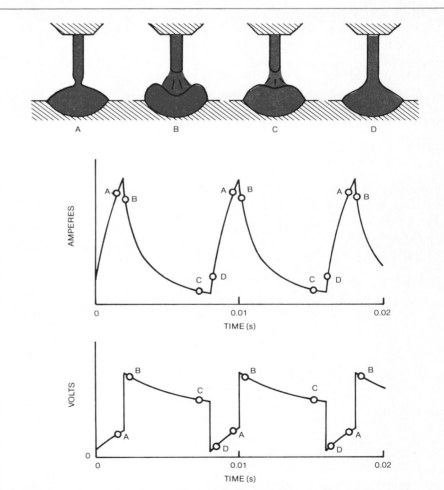

FIGURE 17-4 Dip-transfer welding. The two curves below the diagram show how amperes and volts vary with the rapid melt-off cycle of the electrode wire.

was really happening. There is just such a device called an *oscilloscope*. The lower drawings displaying voltage and amperage variations in Figure 17-4 are exactly what you would see on the screen of an oscilloscope.

Controlling Dip Transfer

The most stable dip-transfer arc is one with a low wire feed rate versus a high frequency of short circuiting. In other words, very short sections of electrode wire would get melted off very fast to produce the most stable dip-transfer arc. But since the wire feed rate often is fixed in advance, you can control the process by increasing the frequency of the dips. You can do that by altering the open-circuit (no-load) voltage (OCV) which controls the arc (load) voltage and the arc length.

If your OCV is too low, you won't get a high enough short-circuiting frequency as the electrode wire dips into the molten weld puddle. The arc will act erratically as it goes on and off with each relatively slow dip into the molten weld pool.

Open-circuit voltage If your OCV is too high, the dipping frequency also drops off. Too much voltage causes the arc gap to get too long for shorts to occur at all. Metal will be transferred without any short-circuiting, and the dipping frequency will drop to zero. Therefore, somewhere in between there must be an optimum OCV for every dip-transfer welding job. You could find the best OCV with an oscilloscope, but nobody does. Most operators simply tune the voltage settings by watching how the weld metal and arc act. They usually adjust the arc voltage up and down around an approximate setting until they find a welding condition that they like. The voltage setting that an experienced welder will pick will almost always be very close to the optimum welding conditions.

Wire feed rates An experienced operator also might adjust the wire feed rate to find the best operating conditions. Too low a wire feed rate and the arc becomes long. Large drops form causing the short-circuiting frequency to drop off or even go to zero.

With medium feed rates, the arc becomes shorter, the current rises, and small drops are formed. The short-circuiting frequency rises toward the optimum. When feed rates are too high, the wire feeds too fast. The arc becomes too short and the wire begins to stub on the workpiece. The short-circuiting frequency becomes erratic and drops lower.

Optimum dip-transfer conditions Obviously, there also must be some *combination* of feed rates and OCVs that produces optimum conditions for dip-transfer welding. When all the variables are juggled properly it is possible to achieve these optimum conditions. However, there is an easier way to tune in your power source than trying to guess which combinations of voltage and feed rates will produce the best results.

Figure 17-5 is a graph of arc volts plotted against wire feed rate. The zone of good short-circuiting or dip-transfer arc welding conditions is about 2 V wide (inside the box shown in the drawing). A welding engineer might actually plot these data and tell you to operate within a certain range of wire feed rates and welding voltages. The welding engineer frequently will get these data by measuring the conditions used by an experienced operator who set the initial conditions and tuned the power source by "eye" and experience.

What do you do when you are not a welding engineer; don't have curves like the one shown in Fig. 17-5 for specific welding equipment, welding wire, and base metal; or an electrical measuring device such as an oscilloscope? One way is simply to increase your wire feed speed at the wire feeder control box until the arc begins to "bump" and "stub out" on the workpiece. Then decrease the wire-feed rate slightly until the erratic bumpy action disappears. The frequency ratio of voltage to wire feed speed will then be just about optimum.

Instead of changing the wire feed speed, you can gradually change the voltage output of the power source. When the slope is set (fixed at one value), merely decrease the voltage setting until the arc begins to bump and stub out. Then increase the voltage about 2 to 3 V. This method also will give you an optimum frequency ratio of voltage to wire feed speed.

You can tune your power source a third way if you have one with a separate slope-control knob. When the slope is increased (becomes steeper), the arc will begin to stumble and stub out. Decreasing the slope (making it flatter) will eliminate the stubbing. Avoid taking out too much slope or you will start producing a lot of spatter.

Some GMAW power sources made for use with dip-transfer welding will have one knob instead of two. The wire feed and the power supply adjustment knobs are interconnected mechanically. Adjusting the one knob automatically selects the feed rate and arc voltage to place your

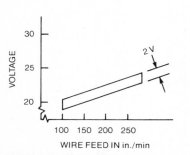

FIGURE 17-5 Typical zone of good short-circuit (dip-transfer) welding conditions. The power source settings and wire feed speed must be tuned to operate in a narrow zone for the process to work well.

operating conditions inside the optimum zone.

17-4
STRAIGHT CO_2-SHIELDED BURIED-ARC WELDING

A GMAW arc shielded by pure carbon-dioxide gas produces either short-circuiting transfer if the current is very low, or globular transfer if the current is higher. A pure carbon dioxide shielding gas will never produce spray transfer, which means you will get lots of spatter if the arc is open (above the weld joint and not buried in the gap). You won't get much spatter, however, if the arc is inside the weld joint. This fact led to the idea of welding with a "buried arc," using the arc slightly below the surface of the base metal, inside the joint.

Carbon dioxide–shielded, buried-arc welding is used mostly for automatic, high-speed welding jobs with high currents. One reason for the use of the process is that carbon dioxide costs less than argon per unit weight of deposited weld metal. In CO_2-shielded buried-arc welding, approximately the same equipment is required as for the other GMAW methods. To get the best results, you must use high current, minimum arc length, the correct electrode extension, and the proper filler metal. A dc constant-current power source modified for the process (with slope control, for example) also helps.

In CO_2-shielded welding with an open-arc technique, the weld-metal transfer is either by dip or globular transfer and there is a lot of spatter. With the buried-arc technique, which is a short-circuiting metal-transfer technique, arc instability and spatter are minimized. Too short an arc length produces a high-crowned weld bead. An increase in arc voltage will bring about a better looking, flatter weld contour. If the arc voltage is further increased, it will be accompanied by increasing spatter levels. Therefore there is an optimum arc volt-age for each buried-arc wire and travel speed.

Advantages

When buried-arc welding is done on single-pass joints for both groove and fillet welds in plate up to 1¼ in. [32 mm] thick, it usually is more economical than any other GMAW process because the cost for pure carbon dioxide is low. There is no flux to damage fabricated parts by abrasion or moving parts in welding fixtures as can happen with submerged-arc welding. You usually also don't have to spend much (or any) time cleaning the slag off the workpiece, thereby increasing your productivity.

Another advantage of the buried-arc process is the excellent joint penetration that you get. The weld metal is narrow and deep and the heat-affected zone is small, which reduces the need for special joint preparation and extra welding fixtures.

Filler Wires

Since CO_2 is highly oxidizing after passing through the welding arc, much of the silicon and manganese in the filler wire is not transferred into the weld metal. It is used up controlling the oxygen produced from the dissociation of CO_2 in the arc. Unless the solid filler wire contains enough deoxidizing alloying elements to compensate for the oxygen produced, porosity and reduced mechanical properties of the weld metal will result.

More highly deoxidized filler wires are used for welding mild steel with the CO_2 buried-arc process than with the regular CO_2-shielded GMAW process because these wires contain a sufficient amount of silicon and manganese to allow for losses of alloying elements in the arc. Welds are sound and have good mechanical properties.

Some steel wires contain aluminum or titanium additions. These elements are excellent deoxidizers. Wires with these deoxidizing elements are often preferred for making fillet welds. The tendency toward porosity is more pronounced on fillet welds because of the longer arc lengths required, which cause the CO_2 shielding gas to produce more ions as it passes through the longer welding arc.

Humans versus Machines

Thin material can be welded automatically with speeds in excess of 100 in./min [2.6 m/min] with the buried-arc process. Although the majority of these jobs are mechanized, the process also is used for the manual welding of mild steel. Manual applications are generally limited to base metal with a minimum of ⅛-in. [3.2-mm] thickness because of the inability of any welding operator to maintain the high speeds required to avoid burn-through of thin sheets using buried-arc GMAW. (You would have to travel at about 9 ft/min [2.7 m/min] to keep up.) Lap and horizontal fillet joints are welded easily with manual buried-arc CO_2-shielded welding.

17-5
GLOBULAR AND SPRAY TRANSFER

In GMAW droplets of molten metal are transferred through the arc to the work from a continuously fed electrode wire. As shown in Fig. 17-6, for a given diameter of electrode wire within the shield of a substantially inert gas, the amount of welding current determines both the size of the droplets and the number of them that are detached from the electrode per unit of time.

For current in the range from the lowest value at which an arc can be maintained up to an amperage value called the *transition point*, the drops grow to a diameter that is several times that of the electrode before they are detached. The transfer rate with low current is only several drops per second.

As the drops grow on the end of the electrode wire they wobble around and disturb the arc plasma so much that the arc also moves around the work. The heat-affected zone in the work is narrow, the weld penetration is small, and the weld-metal deposit is irregular. Drops may grow large enough to cause a short circuit (they touch the weld pool before leaving the electrode), resulting in blasts of spatter that you have to clean up after welding. If you are welding in the vertical position with globular transfer, some of the molten-metal drops will never get to the weld. They will be deflected by gravity and become weld spatter falling onto the floor, your shoes, or your arms (Fig. 17-7).

At the welding current's transition point, the size of the detached drops abruptly decreases (Fig. 17-6 and Fig. 17-7) becoming equal to

or less than the diameter of the electrode wire. Simultaneously, the rate of drop detachment suddenly increases to several hundred droplets per second. This point is associated with a specific current level for each electrode diameter (wire size).

Above the transition point, the rate of drop detachment continues to increase as the current increases, but very gradually compared with the rate before you reached the transition current. (The right half of the curves in Fig. 17-6 show this effect.) The arc cone stabilizes. Your base-metal penetration increases because the increased current is concentrated at your work. The molten metal droplets are small enough to pass through the arc plasma without disturbing its

shape or position. It's a very steady arc.

From Globular to Spray Transfer

The molten-metal transfer mode *below* the transition current is called *globular transfer*. The transfer mode above the transition current is called *spray transfer* because the molten metal passing through the arc at the electrode tip acts like the spray of water from a nozzle. *You can actually see the difference through the lens in your welding helmet* so that you will immediately know when it happens.

The direction in which the droplets move relative to the electrode wire axis changes when the transfer mode changes from globular to spray for any welding position other than working flat. Gravity is the reason.

In globular transfer, the droplets detach when their weight exceeds the surface tension of the molten metal that holds the drop on the electrode tip. The electromagnetic force that acts in a direction to detach the drop is small relative to the force of gravity when the current is in the globular transfer range. Consequently, molten metal droplets in globular transfer fall downward regardless of the direction in which the electrode wire is pointed.

In spray transfer, the electromagnetic force is strong enough to eject the droplets from the electrode tip in line with the electrode axis. That happens no matter which direction you point the electrode wire.

Now you can understand why the globular-transfer mode is a terrible choice for welding overhead, or in the horizontal or vertical position. Instead of the weld metal spraying into the joint, some of it falls out of the joint and on the floor. On the other hand, the spray-transfer mode is good for out-of-position welding because the weld metal literally sprays into the joint like paint out of a spray can. Both transfer

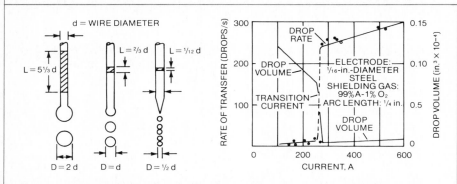

FIGURE 17-6 Metal transfer in the spray mode of the pulsed GMAW welding process. Special pulsed-power welding machines are required for this process.

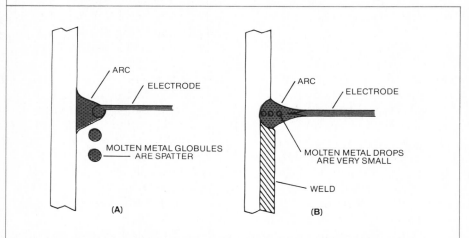

FIGURE 17-7 Horizontally held electrode wires are shown producing globular and spray transfer during gas-metal-arc welding.

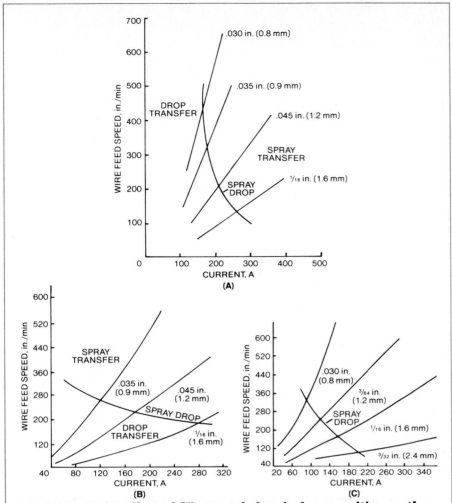

FIGURE 17-8 The effect of filler metal chemical composition on the spray transfer critical current also varies with the electrode wire diameter. Typical curves for (A) mild steel, (B) stainless steel, (C) aluminum.

modes work well for downhand welding. Whether they work equally well is a question that we'll answer later on.

The Transition Point

Different globular-spray transfer welding-current transition points exist for all kinds of electrode wires. Figure 17-8 shows curves for several different kinds of electrode wires. Similar charts are available for many kinds of welding wire. The minimum current setting for spray-transfer welding depends on several factors. Other variables also affect the results you get, so first let's look at these basic MIG welding variables in Fig. 17-9, and then we can talk about them. For a given diameter of electrode wire these factors are:

■ The chemical composition of the electrode wire (carbon steel versus stainless steel versus aluminum, for example) strongly affects the transition current. We show that difference in Table 17-2 and Fig. 17-8.

■ The electrode polarity (whether it is DCRP with the electrode positive, or DCSP with the electrode negative) has a big affect on transition current and other operating variables.

■ The composition of the shielding gas, and the length of the electrode *stick-out* distance between the arc and the contact tip both

TABLE 17-2 Globular transfer to spray transfer transition currents

Wire Electrode Material	Wire Electrode Diameter, in.	Shielding Gas	Minimum Spray Arc Current, A
Mild steel	0.030	98% argon + 2% oxygen	150
Mild steel	0.035	98% argon + 2% oxygen	165
Mild steel	0.045	98% argon + 2% oxygen	220
Mild steel	0.062	98% argon + 2% oxygen	275
Stainless steel	0.035	99% argon + 1% oxygen	170
Stainless steel	0.045	99% argon + 1% oxygen	225
Stainless steel	0.062	99% argon + 1% oxygen	285
Aluminum	0.030	Argon	95
Aluminum	0.045	Argon	135
Aluminum	0.062	Argon	180
Deoxidized copper	0.035	Argon	180
Deoxidized copper	0.045	Argon	210
Deoxidized copper	0.062	Argon	310
Silicon bronze	0.035	Argon	165
Silicon bronze	0.045	Argon	205
Silicon bronze	0.062	Argon	270

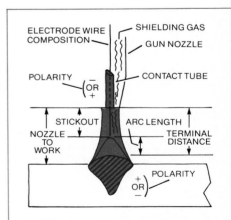

FIGURE 17-9 The working end of the MIG welding system and its basic operating variables. You will learn to control these variables and put them to use.

affect operating variables. The stick-out distance is quite important. The contact tip is the point where the electrode wire picks up its current from the power source. (The contact tip is the copper tube inside the front-end nozzle of the welding gun.)

Let's see how these factors might affect the transition current you would use. (You can get many more machine-setting recommendations than we give you in this chapter. They are provided for almost all GMAW power sources, usually in users' manuals or process manuals that go with each machine. Simple slide-rule charts also are provided by many manufacturers.)

Chemical composition of electrode wire A ¹⁄₁₆-in.-diameter [1.6-mm] aluminum electrode wire has a transition current of 160 A, while a carbon-steel electrode of the same diameter has a transition current of around 275 A, as shown in Fig. 17-8. Therefore, *the spray-transfer mode for steel welding wire occurs at a much higher transition current than for aluminum wire.*

Electrode polarity In a shielding gas with 99 percent argon and 1 percent oxygen, there is a transition current for DCRP welding, but none exists for DCSP. There is no spray-transfer mode. Only globular transfer occurs during DCSP, *which is one reason why DCRP is almost always used for MIG welding.* Another reason is illustrated in Fig. 17-10. If you set up for GMAW welding with dc straight polarity, 70 percent of the heat produced by the arc is in the welding wire and only 30 percent is in the base metal. On the other hand, a dc reverse polarity setup will put 70 percent of the heat in the base metal where it belongs, and only 30 percent in the welding filler metal. The net result is that DCRP welds so much faster than DCSP polarity setups that DCSP is almost never used in the MIG process. How-

ever, DCSP has uses in other GMAW processes.

Shielding-gas composition With a positive electrode, if carbon dioxide is substituted for the 99 percent argon plus 1 percent oxygen gas used above, the transition current is so high that, for practical purposes, it does not exist and you remain in the globular transfer mode. *This is why a pure carbon dioxide gas shield limits you to downhand (flat) welding and a pure CO_2 gas shield produces a lot of spatter.* On the other hand, CO_2 shielding gas is very inexpensive. If you don't mind weld spatter, and you intend to work exclusively in the flat position, carbon dioxide may be a good shielding-gas choice.

Electrode stick-out As the stick-out of a ¹⁄₁₆-in.-diameter [1.6-mm] carbon-steel electrode is increased from the lowest possible value (³⁄₈ in. [9.5 mm]) without melting the contact tube, to a distance of 3 in. [76 mm], the transition current changes from 280 to 200 A. The minimum wire stick-out distance for MIG welding is ³⁄₈ in. [9.5 mm] because a shorter wire stick-out will ruin the contact tube. *The transition current from the globular to the spray*

mode decreases as the stick-out distance increases.

17-6 SPRAY-TRANSFER WELDING

When all these conditions are set so that spray transfer is produced, weld spatter is negligible and you get a weld-metal deposition efficiency as high as 99 percent (almost all of your filler metal becomes weld metal). Since spray transfer is achieved in a gas shield made mostly of inert argon gas, the harmful effects of a non-inert (reactive) gas such as carbon dioxide on the weld metal are avoided. The inert gas dilutes much of the reactive gas in the mixture. Meanwhile a small addition of a reactive gas improves the welding characteristics, especially in the arc, that the pure-argon inert gas does not have. For example, your welding arc is much more stable with a gas mixture than it would be in pure argon.

Spray transfer is associated primarily with the use of inert gases. Either pure argon or argon-rich (0.5 percent to 5 percent oxygen added) shielding gas is used. With these gas mixtures, true axial spray transfer is possible with

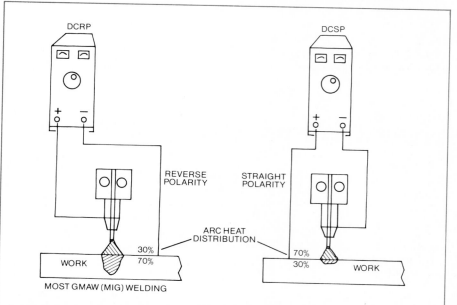

FIGURE 17-10 About 70 percent more heat is dumped into the base metal using direct-current reverse-polarity setups than using direct-current straight-polarity setups.

DCRP welding. Spatter and weld-metal porosity, however, can result even from spray-transfer welding if the molten electrode metal is allowed to bridge too much of the space between the electrode wire and the molten weld pool, causing a rapid release of gases. (Porosity is caused by absorption of nitrogen from the air due to arc instability.) *Increasing the arc length eliminates this problem.* In other words, hold the welding gun a little farther back from the base metal if you are having weld-porosity problems.

Conventional Spray-Transfer Welding

Spray-transfer welding is widely used. There is, however, a metal thickness below which the process cannot be used unless the arc is pulsed because, unlike dip transfer, spray-transfer welding is a higher heat-input process. (This fact led to the development of pulsed MIG welding in 1968.) Also, when used on aluminum and its alloys, conventional spray-transfer welding cannot be performed in the vertical or overhead positions in thicknesses greater than 1/8 in [3.2 mm]. In these greater thicknesses, the aluminum cooling rate is too fast. The reason is that aluminum has much higher thermal conductivity than steel. This is even truer of copper.

Spray-transfer welding with steady instead of pulsed dc uses voltages and currents that are relatively high. It produces deep weld penetration and high weld-metal fluidity (the molten weld metal is very hot). This high fluidity makes it harder for you to work when welding steel overhead versus welding in the flat and horizontal positions. One of the ways of overcoming this disadvantage is by the use of pulsed-spray welding.

17-7
PULSED-SPRAY WELDING

The limits on the use of spray-transfer welding have been greatly extended by the development of pulsed-spray welding. In pulsed-spray welding, the welding current switches automatically from a low level to a higher level and back again, back and forth between the spray-transfer and globular-transfer current ranges. The power source is adjusted so that the lower level of current (known as the *background current*) is set below the transition current for spray transfer. Your maximum welding current pulses are set well above the spray-transfer level. The result is the best of both worlds.

Actual weld-metal transfer through the arc is restricted to spray transfer. Globular transfer is suppressed by not allowing the weld metal enough time to transfer in the globular mode. Plenty of time is allowed for spray transfer to occur. The average heat input into the weld is much lower than for constant-spray transfer. In effect, the base metal "sees" a current level (and therefore I^2R heating) that is the average of the globular and spray settings as shown in Fig. 17-11.

As a result, you get all the advantages of spray transfer welding without putting much extra heat into the base metal, which makes it much easier to weld out-of-position. You get less base-metal distortion. Therefore, you can weld thinner sections than are possible with constant-spray transfer. You can also use larger-diameter electrodes that cost less than fine wires and have fewer feeding problems, too.

Large-diameter wires also have more volume for a given surface area. Since there is less surface area per unit volume of wire in larger wires, surface impurities

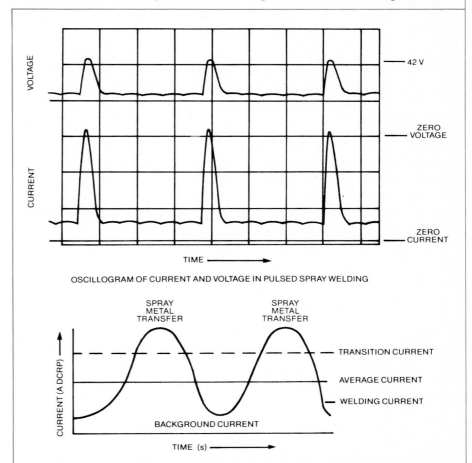

OSCILLOGRAM OF CURRENT AND VOLTAGE IN PULSED SPRAY WELDING

FIGURE 17-11 Pulsed-spray GMAW makes the workpiece "see" only the average current and (by I^2R heating) the total heat input is less. That produces much less base-metal distortion and makes possible the welding of thinner gauge materials.

such as residual drawing lubricants and oxides on the wire are less likely to wind up in your weld metal, which means fewer welding defects per foot of weld. Pulsed-spray GMA welding is also essentially spatter-free.

The pulsing current is produced by special welding machines that, in effect, have two separate power sources—one for the background current and one for the pulsing current. And that's the one drawback to the pulsed process. Pulsed-spray power sources don't cost twice as much as a single power source, but they are more expensive. The extra expense often is worth the benefits, but sometimes it isn't. An economic analysis is required to determine the tradeoffs, to decide which way to go.

Many high-volume production welding shops that make aluminum, copper, and steel articles use a lot of pulsed-spray equipment. Truck fabricators use it, for example, to build the big aluminum cargo boxes on truck trailers. This work combines welding relatively thin aluminum sheet with out-of-position welding.

17-8
NARROW-GAP WELDING
Narrow-gap welding is a modification of conventional GMAW where square grooves or V grooves with extremely small included angles are used in thick sections of ferrous metals. The small included angles mean that less filler metal is needed to make the joint. On large welding jobs on thick plate, this can result in a substantial saving in materials costs.

Root openings of about ¼ to ⅜ in. [6.4 to 9.5 mm] are used in narrow-gap GMAW, regardless of the material thickness. This process requires the use of specially designed welding-gun accessories including water-cooled contact tubes and nozzles that introduce shielding gas from the plate surface. All positioned welds are made from one side of the plate. For thicknesses over 2 in. [5.1 mm], this process is competitive with other automatic arc-welding processes used for welding heavy plate.

Welding is usually done with two small-diameter wire electrodes running in tandem, one behind the other, fed through two contact tubes (Fig. 17-12). Axial spray transfer with DCRP is the most commonly used metal transfer mode. Each electrode is oriented so that a weld bead is directed toward each sidewall.

The travel speeds used in narrow-gap welding must be high, resulting in low heat input and small weld puddles which are easily controlled in out-of-position welding. Heat-affected zones in the base metal are narrow. The necessary use of low heat input in thick materials can cause sidewall or interbead lack of fusion, the problem most often encountered with this process. Careful attention is required to every detail of the process, particularly wire electrode placement. Interpass cleaning of all slag spots also is necessary if slag entrapment in the finished weld metal is to be avoided.

Nevertheless, narrow-gap welding has advantages when operated by experienced welders in shops that have had plenty of experience with the process. These advantages are

■ Lower residual stresses and distortion
■ Improved as-welded joint properties
■ Better (lower) weld-metal consumption on thick plates
■ High productivity and low labor costs per pound of weld metal deposited.

17-9
FLUX-CORED ARC WELDING (FCAW)
FCAW uses a tubular electrode with powdered flux inside and a shielding gas of carbon dioxide. Flux-cored wires are used for welding either carbon or high-strength low-alloy steels, high-strength alloy steels, or stainless steels. The wire is usually made of a strip of carbon steel rolled up to make a tube with the powdered flux inside. The flux has alloying elements that modify the molten weld metal to produce the desired composition and mechanical properties when the finished weld metal freezes and cools.

The Flux Core
As well as containing alloying elements that modify the molten weld metal, the flux in the wire also provides deoxidizers and other scavenger elements that remove extra dissolved gases. The

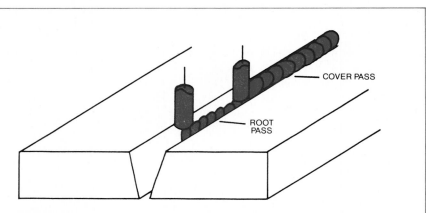

FIGURE 17-12 Tandem narrow-gap welding heads on an automatic welder. The heads are slightly offset to tie each pass into the opposite walls of the final joint. Narrow V-gap design saves filler metal costs. The process operates at high speeds because less filler metal is deposited into smaller joint volumes. Dual welding heads double production speed because two passes are made at the same time.

flux also contains arc stabilizers, slag-forming elements, and other things to make the arc operate properly and to produce good sound weld metal. Table 17-3 lists some of this information for you.

The carbon-dioxide shielding gas used in flux-cored GMAW serves mostly to protect the weld pool from the harmful effects of the oxygen and nitrogen in the air, including oxidation of the weld metal, oxide inclusions in the weld, and porosity caused by nitrogen bubbles. The gas-flow rate is very important in FCAW.

Too little gas flow, unfavorable electrode-holder position, or an improper gas nozzle on the welding gun can result in weld-metal porosity and poor gas shielding.

Advantages of FCAW

FCAW has several advantages over stick-electrode welding, in addition to the obvious ones of semiautomatic welding or continuous fully automatic welding. Two of the most important of these advantages are deeper base-metal penetration and higher weld-metal deposition rates.

Deeper penetration For a given size of fillet-weld leg, a fillet weld that is made by FCAW can have a greater throat depth and therefore greater strength than a fillet weld that is made by SMAW. You can see that this is true in Fig. 17-13. Because of the deep root penetration that can be obtained in FCAW, a weld of a given strength can be smaller than a weld made by SMAW. This is graphically illustrated in Fig. 17-14.

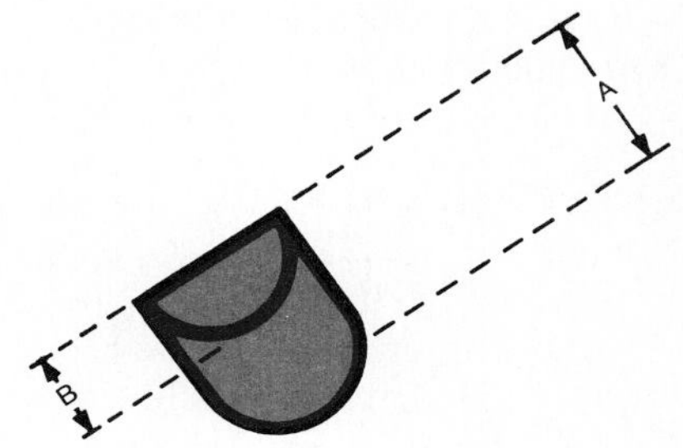

TABLE 17-3 Common core elements in flux-cored electrodes

Element	Usually present as	Purpose in weld metal
Aluminum	Metal powder	Deoxidize and denitrify
Calcium	Minerals such as fluorspar (CaF_2) and limestone ($CaCO_3$)	Provide shielding and form slag
Carbon	Element in ferroalloys such as ferromanganese	Increases hardness and strength
Chromium	Ferroalloy or metal powder	Alloying to improve creep resistance, hardness, strength, and corrosion resistance
Iron	Ferroalloys and iron powder	Alloy matrix in iron-based weld-metal deposits, alloy in nickel-based and other nonferrous weld-metal deposits
Manganese	Ferroalloy such as ferromanganese or as metal powder	Deoxidize, prevent hot shortness by combining with sulfur to form MnS, increase hardness and strength, form slag
Molybdenum	Ferroalloy such as ferromolybdenum	Alloying to increase hardness, strength, and in austenitic stainless steels to increase resistance to pitting-type corrosion
Nickel	Metal powder	Alloying to improve hardness, strength, toughness, and corrosion resistance; an austenite former as in 18% Cr–8% Ni austenitic stainless steels
Potassium	Minerals such as potassium-bearing feldspars and silicates	Stabilize the arc and form slag
Silicon	Ferroalloys such as ferrosilicon manganese; mineral silicates such as feldspar	Deoxidize and form slag
Sodium	Minerals such as sodium- and potassium-bearing feldspars and silicates	Stabilize the arc and form slag
Titanium	Ferroalloys such as ferrotitanium; in the mineral, rutile (TiO_2)	Deoxidize and denitrify, form slag, stabilize carbon to prevent the formation of chromium carbides which lead to intergranular corrosion in some stainless steels
Zirconium	Oxide or metal powder	Deoxidize and denitrify
Vanadium	Oxide or metal powder	Increase strength and hardness

FIGURE 17-13 (A) A flux-cored wire fillet often has a greater throat depth (and joint strength) than (B) a covered-electrode fillet weld, even when the two processes produce the same leg size on the fillet weld.

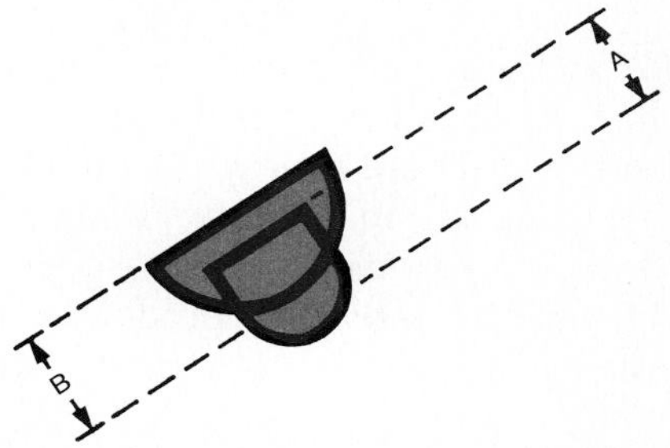

FIGURE 17-14 (A) A semiautomatic flux-cored wire fillet often will have the same throat depth (and joint strength) as (B) a covered-electrode fillet weld. The cored-wire weld has a smaller leg size, and needs much less weld metal. Deposition rates are much higher than those of manual SMAW stick electrodes. Labor costs per pound of weld metal deposited are much lower.

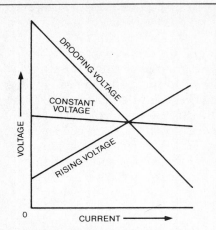

FIGURE 17-15 Power source voltage/amperage characteristics. Note the small CV slope.

Faster deposition The deposition rate of FCAW electrodes can be boosted by increasing either the welding current or the electrode stick-out distance. In the horizontal position, the weldmetal deposition rate is limited to about 18 lb/h [8 kg/h]. The rates are limited by the puddle size, bead contour, and possible undercutting. Deposition efficiency is about 90 percent, which means that 90 percent of the welding wire you start with ends up as finished weld metal, and 10 percent goes off as smoke, slag, or spatter.

There now are some metal-cored wires that have powdered metal instead of granular flux in the core. These new wires have high deposition rates, very high deposition efficiencies (well over 95 percent), and they produce very little smoke and almost no slag.

In fact, the metal-cored wires are usually run in 75 to 80 percent argon and 25 to 20 percent CO_2 shielding gas instead of the 100 percent CO_2 shielding gas used for conventional flux-cored wires. The slight increase in shielding gas costs is far offset by the improved weld-metal deposition efficiency and higher deposition rates produced by these wires, and by the improved welding conditions and better weld quality. The number of steel

grades in which metal-cored wires are available, however, is limited.

No edge preparation Another advantage of FCAW is that full-penetration welds can be made without edge preparation. Such welds, for example, can be made on double-welded T joints in ⅝-in.-thick [16-mm] steel. In the case of butt welds, full-penetration welds also can be obtained on square-edged plates up to ¾ in. [19 mm] thick by making one pass from each side. Plates thicker than ¾ in. [19 mm] are usually joined with single- or double-V grooves with 30° included angles. These grooves need less weld metal than those normally used in solid-wire welding, which provides weld-metal savings and increased welding speed.

17-10
GMAW POWER SOURCES
All GMAW processes use only dc power. Most of the time DCRP (electrode positive) is used. Most dc MIG welding machines also are constant-voltage (CV) machines. Constant potential (CP) means the same as CV.

Differences among the Three DC Power Sources
There are three basic types of dc power sources for GMAW and FCAW. The differences among

them are based on their *static volt-amperage curves.* You've already seen these curves in Chap. 11. The volt-amp curves for dc power sources are either drooping-voltage, constant-voltage, or rising-voltage. Each type of GMAW dc power source (Fig. 17-15) is built to react automatically and uniquely to compensate for variations in electrical loads encountered during welding.

Volt-amp curves If the power source is the drooping-voltage type, the voltage will drop as the current increases. If it is the rising-voltage type, the voltage will rise as the current increases. The current stays steady as voltage changes in constant-current welding machines. The static volt-amperage curves for drooping-type dc power source with adjustable open-circuit voltage, with and without slope control, are shown in Fig. 17-16.

The vertical line at the left of the figure indicates voltage, and the points where the curves intersect the vertical voltage line indicate the OCV setting on the machine (when no welding is taking place). The right ends of the curves where they intersect the horizontal amperage line tell you that the electrode wire is touching against the grounded workpiece to

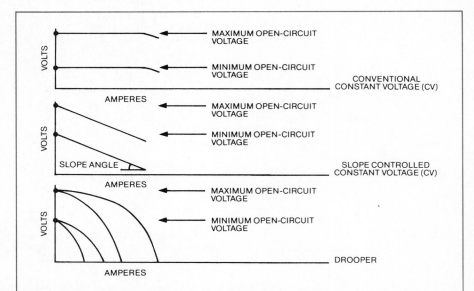

FIGURE 17-16 Various types of dc welding power sources with their volt-ampere curves are shown here for comparison.

start the arc (it is short circuiting). Those are the extreme ends of the welding conditions you will run into, ranging from no load (no welding) to a complete short circuit (when the electrode touches the workpiece or weld pool).

In between these two extremes are the conditions that occur most of the time during actual welding. With a drooping-voltage power source, when the electrode wire is touched to the work to start the arc the voltage drops to near zero, but the current increases very rapidly to melt the wire and start the arc instantaneously (see Fig. 17-17, on the left).

As the arc increases in length, its resistance increases, which causes the welding current to decrease and the voltage to rise to the operating range you set on the power source with the voltage-control knob. If your power source has a knob for setting the slope of the volt-amp curve, you also must set that.

With a rising-voltage power source, the effects of striking an arc are about the opposite from a drooping-characteristic power source (see Fig. 17-17, on the right). Rising-voltage power sources are seldom used because of the wide use of constant-voltage (CV) power sources.

Using the CV power source (by far the most commonly used type of machine for MIG and flux-cored-wire welding), the voltage remains relatively constant and droops only slightly with an increase in current. This system assures a self-regulating arc based on a fixed rate of wire feed. You only have to set the wire speed when you set up the job. The arc length is controlled when you set the voltage on the power source (see Fig. 17-17, in the center). The welding current is controlled by adjusting the wire feed speed.

After you set the wire feed speed, you don't have to worry about readjusting it while you work, unless you change the welding parameters on the machine and wire feeder. Changes in wire

feed speed that will occur when you move your welding gun toward or away from the work are compensated for by the machine and wire feeder, which momentarily change the current and melt-off rate of the wire until equilibrium conditions (the welding parameters that you initially set on the machine) are reestablished.

Arc length The effects on arc length of each of the three types of power sources in response to

changes in contact tip to work distance are shown in Figs. 17-17 and 17-18. Increased resistance from a greater electrode wire stick-out moves the operating conditions to the left on the curve. Decreased resistance from a closer contact tip to work distance moves the operating conditions to the right on the curve in Fig. 17-18 and results in changes in voltage and arc length that correspond to the volt-amp curves for each type of power source.

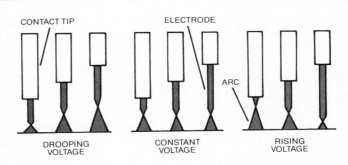

FIGURE 17-17 Power supply characteristics affect arc length and wire feed speed differently. Note that the arc length of a constant-voltage (CV) power source is constant with different electrode feed speeds; the arc length of other power sources varies with electrode wire feed rate. Therefore, constant voltage (CV) machines are preferred for most GMAW welding.

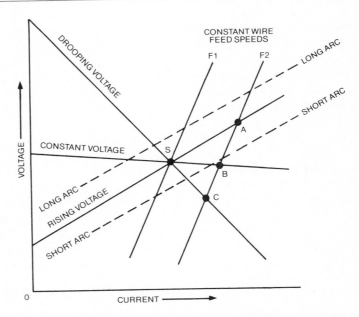

FIGURE 17-18 The static volt/ampere characteristics of constant-voltage, drooping- and rising-voltage power sources. The change in voltage and current that is produced from an increase in wire feed speed from F1 to F2 is plotted for each power source.

Wire feed speed When a change is made in the electrode wire feed speed, each of the three types of power sources reacts differently. The differences can be best understood by looking again at Fig. 17-18. In this figure, the volt-amp curves for the welding arc and the three types of power sources are superimposed. The curves for arc length and electrode wire feed are typical of those that you would get by holding the contact tip to work distance constant and measuring the resulting arc length, voltage, and current values at two different wire feed speeds.

Referring to Fig. 17-18, assume that welding is being done under a stable condition designated by point S. For that condition, the wire feed speed is represented by the line F1. If the wire feed speed were to be increased to F2, the new operating conditions would be at point A for the rising-voltage power source, at point B for the CV power source, and at point C for the drooping-voltage power source.

Since the volt-amp curve for the rising-voltage power source satisfies the requirement for higher arc voltage at high wire speeds, the change in arc length is hardly noticeable. However, the increase in current (and therefore melt rate) is greater than what occurs with the other two power sources, and therefore rising-voltage machines adjust for slight changes in arc length even faster than the other machines. This ability makes rising-voltage machines suitable for highly automated, high-speed GMAW and for jobs where crater filling is a problem because you lifted the electrode from the weld too fast (which also is often a problem with automated systems).

With the CV power source, the arc becomes slightly shorter, and although the change in current is significant, it is less than the change that results with the rising voltage power source. But the change in current is fast enough that these machines are ideal for

almost all solid-wire GMAW. That's why these dc CV machines are the most commonly used welding power sources for MIG welding.

The greatest reduction in arc length and the least amount of current increase occurs with drooping-voltage power sources. Since extremes in arc length are limited by stubbing the electrode wire against the work at a short arc length and melting the contact tip when the arc is long, the rising-voltage power source permits the greatest change of wire feed speed for a given setting of the power source controls. The drooping-voltage power source permits the least change in feed speed, and the CV source permits a change somewhere between the changes of the other two power sources. As you already know, droopers are used mostly for dc stick-electrode welding.

The arc length is self-regulating with the rising-voltage power source and also, but to a lesser extent, with the CV power source. When you make moderate adjustments in the wire feed speed, you do not normally have to adjust the controls to maintain the original arc length. With the drooping-voltage power sources, however, you must adjust the feed speed and controls to maintain the original arc length. This is called *tuning* the power source. That's yet another reason why CV power sources are used mostly for GMAW welding.

CV Power Sources

We mentioned that the CV power source is the most widely favored for GMAW because it can be used for all the modes of weld-metal transfer and because it offers other advantages as well. Because its volt-amp curve slopes only very slightly, the CV power source can produce extremely high short-circuit currents to clear the short when the electrode wire is touched to the work.

Another feature of CV power sources is that "burnback" (fusion

of the welding wire to the contact tip in the end of the welding gun due to excessive lengthening of the arc) is all but eliminated. Since the current decreases as the arc length increases, the arc will extinguish itself before it can grow long enough to reach back and melt the contact tip.

CV power sources operated with DCRP are used for GMAW. When DCSP current is used, the arc is erratic and produces an inferior weld. Therefore, DCSP current is almost never used for GMAW.

DCSP, however, can be used for submerged-arc welding and FCAW. CV power is normally not used with ac. It also can be used for electroslag welding (a welding method we describe in the Appendix) but it is not popular.

⚙ **CAREFUL: Don't use CV machines for stick electrodes.**

CV power sources should not be used for SMAW. SMAW may overload and damage the power source by drawing too much current for too long a time. CV power sources can be used for carbon-arc cutting and gouging with small electrodes, but if you are in doubt, ask a welding engineer or the power-source supplier or distributor before using any kind of power source for jobs that its instruction manual doesn't describe.

Slope and inductance controls

For dip-transfer welding, CV power sources are often furnished with slope and inductance controls.

SLOPE A little additional droop in the voltage-amperage curve, or slope, is required for dip-transfer welding with a CV power source because that type of welding machine produces extremely high currents to clear a short circuit. For dip-transfer welding, the short-circuit current must be high enough to clear the short by melting the wire; but if it is too high, it will result in excessive spatter.

With very little slope in the welding circuit, as in a CV power source, the short-circuit current will be very high. If too much slope is present (as with a drooping-voltage power source), the short-circuit current will be too low to clear the short, which will result in stubbing of the electrode wire against the work (Fig. 17-19). For comparison, a typical drooping-voltage power source for stick-electrode welding droops at a rate of 15 to 20 V/100 A, whereas the controlled-slope power source for GMAW droops at a rate of only 2 to 7 V/100 A.

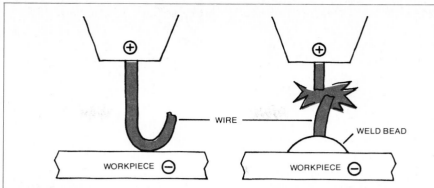

FIGURE 17-19 Too much slope can be too much of a good thing when MIG welding, as shown here. The wire stubs and then shorts out.

INDUCTANCE All electrical devices have some inductance in the circuit. This electrical phenomenon is especially important in CV power sources used for short-circuit (dip-transfer) GMAW. Inductance (Fig. 17-20) controls the rate of current rise that occurs during a short circuit to prevent violent current fluctuations, which would cause weld spatter.

Inductance is found in GMAW dc power supplies because no electrical device can respond instantly to rapid load changes. The current takes a measurable time to reach a new level every time the arc is turned on and off. Inductance in the circuit is responsible for this time lag. Since GMAW wire-fed welding processes involve rapid on and off sequences of the welding arc, the effect of inductance, or volt-ampere time lag, is important.

Curve A in Fig. 17-20 shows a typical current-time curve with inductance present as the current rises from zero to a final value. Curve B shows the path which the current would have taken if there were no inductance in the circuit. The maximum amount of current attainable during a short circuit is determined by the slope of the welding power source. However, the *rate* at which that current is reached is controlled by the inductance in the circuit. The rate can be controlled by the operator (dialed in the machine's control panel) on many power sources used for dip-transfer welding so

that the short circuit may clear with a minimum of weld spatter.

Inductance in the circuit also stores energy. It returns this stored energy to the arc after the short circuit has cleared. In short-circuiting (dip transfer) welding, an increase in inductance increases the arc-on time. This, in turn, makes the weld puddle more fluid, resulting in a flatter, smoother weld bead. The opposite is true when the inductance is decreased. In spray-transfer welding, the addition of some inductance to the power supply circuit will produce a better arc start. Too much inductance will result in erratic welding. When conditions of both correct short circuiting and correct rate of current rise (inductance) exist, spatter is minimal.

As a general rule, both the amount of short circuit current and the amount of inductance needed for ideal operation are increased as the electrode diameter is increased. Therefore, larger-diameter wires may be more difficult to work with than small-diameter wires unless you can directly dial in increased inductance on the panel of the power source.

Settings vary with wire diameter Slope and inductance adjustments are normally based on the electrode wire diameter. Both the amount of inductance and the short-circuit current (less slope) should be increased as the diameter of electrode wire in-

creases. The instruction manual that comes with your welding machine usually will give recommended settings for different wire diameters, so all you have to learn are the ideas, not the details.

Rising-Voltage Power Sources
The rising-voltage power source finds its major application in spray-transfer welding in which crater filling is an important consideration. Craters, as you know from SMAW, occur at the end of a weld bead when you lift your electrode too fast. They also are a bigger problem (along with the cracks that form on the bottom of them, called crater cracks) when welding higher-strength steels.

Drooping-Voltage Power Sources
The drooping-voltage power source can be used for spray-transfer welding. At one time it was very popular because it could also be used for stick-electrode welding. Drooping-voltage dc power sources often are referred to as "droopers" in a welding shop.

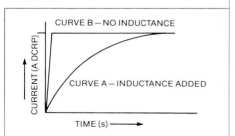

FIGURE 17-20 Change in current rise owing to inductance.

The main problem with using a drooping-voltage power source for GMAW involves arc starting. Since the volt-amp curve of a drooping-voltage power source slopes down sharply to the right, the amount of current available to melt the wire and clear the short circuit during arc starting is limited. This makes for extremely poor arc starting in comparison with the other types of power sources. When drooping-voltage power sources are to be used for GMAW, either a scratch-start kit or a slow wire run-in control may be added to overcome the arc-starting problems.

Scratch-start kit The scratch-start kit is normally an option that is mounted in the electrode wire-feeder control. It works like this. When you depress the GMAW gun trigger, that turns on the power source so that voltage is available to start the arc, but the wire feed does not start immediately.

You then hold the gun close to the weld area and scratch the end of the electrode wire that is sticking out of the contact-tube tip against the workpiece, much as if you were striking a match. Since not as much current is required to strike the arc for a nonmoving electrode wire as for a wire being fed, the arc is easily started. As soon as the arc is established, the wire begins feeding automatically at the rate you preset.

17-11
OTHER GMAW EQUIPMENT
GMAW equipment is significantly different from the equipment used for the other welding and cutting processes that we've already studied. Now that we've looked at the power sources, let's take a closer look at the rest of the equipment you will use.

Wire Feeders and Controls
Wire feeders and controls go together. As the names imply, the wire feeder holds a spool or coil of

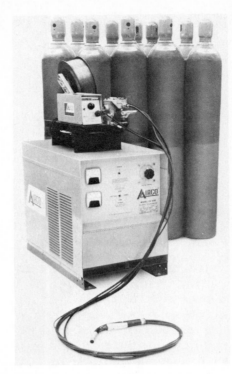

FIGURE 17-21 The main components of a production GMAW welding outfit (MIG in this case) include the power source, a separate wire feeder (on top of the power source in this photo) with spooled wire, a welding gun, and shielding-gas cylinders.

wire. The controller that goes with it feeds the wire at a preset speed through a service line into the welding gun, through the arc, and into the weld metal.

Types of wire feeders There are three basic types of wire feeders. They are push-type feeders, pull-type feeders, and push-pull feeders. The difference is mostly in how they get wire off a spool or coil and through the cable to the welding gun. The reason that three basic types of feeders exist is that steel wires are stiffer than nonferrous wires, especially aluminum. If you remember what we said in Chap. 8 on mechanical properties of materials, steel has three times the modulus of elasticity of aluminum, which means that a steel wire of a given diameter is three times stiffer than an aluminum wire of the same diam-

eter. But even steel wire is not all that stiff. Therefore, there are limits to how far you can be from a wire feeder and still weld. Those limits are much longer than they are from a stick-electrode power source.

If you use a big wire feeder, it probably will be hung from a boom not too far away from where you are working. If you don't have a very long welding cable and wire and shielding gas conduit (15 feet [4.5 m] is usual), the wire feeder may sit right on top of the power source (Fig. 17-21). Some power sources include wire feeder and controls as a single package, with the feeder inside the welding machine. In that case, you simply plug your service cables directly into the connections provided for them on the machine's front panel. High-production GMAW outfits, however, will have the feeder and controls as a separate unit from the welding power source.

Push-type feeders are used for steel wires which often are called *hard wires*. Pull-type feeders are used for aluminum wires which are frequently called *soft wires*. Very small diameter wires also are fed with pull-type feeders. Larger-diameter copper wires can usually be fed through either push or pull feeders, depending on the distance. For very long distances, an auxiliary wire feeder may be used for either aluminum, copper, or steel. For example, an extended-reach wire feeder is an auxiliary pull-type feeder that often works with a main push-type feeder up to 100 ft [30 m] away (Fig. 17-22).

The welding wire itself is mounted on anything from 1-lb [0.45-kg] spools on up to large payoff packs, which are drums with a wire guide on the top that can carry up to several thousand pounds of welding wire. At the other extreme are the 1-lb [0.45-kg] spools often used in "spool-on-guns," which include the wire feeder as an integral part of the welding gun (Fig. 17-23).

(A)

(B)

FIGURE 17-22 (A) MIG power source, wire feeder, large spool of wire, and shielding gas on the left feed welding wire to an intermediate extended-reach wire feeder (on top of the pressure vessel in middle) and on to the operator at the right. Extended-reach wire feeders can supply an operator working up to 100 ft from the wire spool and power source. (B) Close-up of the primary wire feeder and the extended-reach intermediate feeder.

FIGURE 17-23 This combination MIG gun and wire feeder contains a tiny spool of wire inside the gun's case (the round section on the right). These "spool-on-guns" often are used to feed fine wire which otherwise would be difficult or impossible to feed through long service lines.

wire such as 0.030-in.-diameter [0.76-mm] aluminum welding wire.

Not all wire feeders automatically feed all wire diameters. The feed rolls often must be changed to match the range of wire diameters being fed. For example, flux-cored wires often use knurled (bumpy) drive rolls to make sure that you get positive feeding from these stiff, larger-diameter wires. Very small, smooth drive rolls are used for fine, small-diameter, soft aluminum welding wire. Changing rolls is very easy to do. The instructions with each wire feeder will either tell you how, or at least tell you what wire diameters the feeder is designed to handle.

Each model will specify its wire feed-speed range, wire diameter (for both hard and soft wires), and whether wire can be fed from spools or from larger coils. Following is a description of a typical high-production GMAW wire feeder (Fig. 17-24):

■ Push feeder with solid-state speed control and totally enclosed dc motor to maintain speed at ±2 percent from zero to full load and for a line voltage variation of 10 percent. It also features an enclosed motor and drive stand mounted on a turntable with a boom-mounting hook for swinging over work area.

■ Wire feed speed, 45 to 600 in./min [1.1 to 15 m/min]

Wire feeder drive rolls The welding wire on all but the large payoff packs is mounted on a post from which it feeds wire as the spool turns. Drive rolls inside the wire feeder do the pushing or pulling. A push-type feeder for hard (steel) wires will strip the wire from the reel and push it through a service conduit and into

the gun and at the workpiece and weld metal.

Some welding guns other than spool-on-guns have an extra pull-type wire feeder built into them that works with a larger wire feeder and spool usually located at a distance, near the welding power source. These guns are usually used for soft, small-diameter

FIGURE 17-24 This is a typical production-sized push-type wire feeder. The wire reel attaches to the spindle on the upper right. Wire is threaded through the wire drive rolls and into an attached MIG gun. Wire feed speed is set on the control panel on the left. The wire feed rolls are changed to accommodate different welding wire diameters. In use, this wire feeder might sit on the floor, be on top of the power source, or even be hung from a hook from a small boom over the work area.

- Wire diameter
    Hard wire, 0.025 to ¹⁄₁₆ in. [0.64 to 1.6 mm]
    Cored wire, 0.045 to ¹⁄₁₆ in. [1.1 to 1.6 mm]
    Soft wire, ³⁄₆₄ to ¹⁄₁₆ in. [1.2 to 1.6 mm]
- Wire supply, 12- and 14-in.-diameter [305- to 356-mm] spools and 60-lb [27-kg] wire reels.

Auxiliary functions The principal job of the wire-feeder control is to maintain the feed speed you set. Many high-production feed controls include sophisticated electrical circuitry that will maintain your preset wire feed speed within ± 2 percent.

However, wire feeders and controls do more than just feed the wire to the welding gun. They start and stop the flow of shielding gases (and cooling water if a high-amperage water-cooled gun is used). They also feed signals from the welding gun back to the feed-speed control that tell the control whether to speed up the wire feed, slow it down, or start or

stop it altogether.

For example, a rotating spool of heavy wire will continue to turn when you stop welding, jamming your wire liner, gun, and wire feeder like a backlash on a fishing reel. This mess is usually called a "bird's nest" by very annoyed welders who have to cut it out with wire cutters. Bird's nests are usually avoided by using wire feeders that have dynamic braking controls that stop the reel from rotating when you stop welding.

Once you have dialed in your desired wire feed speed, all you have to do to signal the wire feeder and control is to pull the trigger on the welding gun. The wire feeder and its control take care of everything else.

Wire feeder accessories Wire-feeder control accessories include things like short-cycle weld timers that allow the equipment to be used for making spot welds. They control the entire spot-welding cycle from pre- and postpurge of shielding gas and cooling water (if it is used).

Other accessories may include slow wire run-in, stub burn-off, dual schedule (two separate current levels) of wire feed, scratch-start kits (which we've already mentioned), turntables to mount the wire feeder on to follow the direction in which you work. A wire feeder and control mounted on a turntable will follow you in a 360° circle if you walk completely around the unit. Special spool holders are available for fast changing of your welding wire.

Wire feeder hangers often come with wire feeders so that the feeders can be mounted on booms over the area in which you are working. Canvas and plastic covers for the spooled wire protect it from shop dust and moisture (see Fig. 17-2).

All of these welding accessories are valuable additions to your welding outfit. In many instances they are not only nice to have; they are essential for good GMAW.

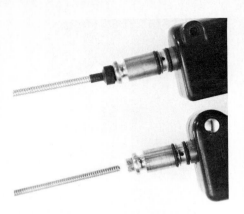

FIGURE 17-25 The wire liner inside the Neoprene service cable of a welding gun runs between the gun and wire feeder rolls. Steel wire liners (like this) are used for steel welding wire. Plastic or nylon liners are used for aluminum welding wires. Different welding wire diameters require different-sized wire liners, which are easily changed. Cable ends plugs in fast.

The service cable The wire fed from the wire feeder to the gun travels through a flexible cable that contains a specific wire liner selected to match the type and size of welding wire you use. Wire liners, like the drive rolls on wire feeders, are easy to change to match the given type of wire and diameter you will use. Changing wire liners only takes a couple of minutes.

The wire conduit and its liner (Fig. 17-25) are connected between the welding gun and the drive rolls in the wire feeder. Together they direct the welding wire to the gun and into the contact tube. Uniform feeding of the wire is absolutely necessary for arc stability. When not properly supported by the conduit and liner, the welding wire may jam somewhere between the drive rolls and the contact tube.

The wire liner can either be an integral part of the conduit or supplied separately. In either case, the inner diameter and the material that the liner is made of are very important. When using steel-wire electrodes, a steel spring liner is recommended. Nylon and other plastic liners should be used for aluminum wire because the spring-steel wire liner might scratch the soft aluminum-wire electrode. The literature supplied with each welding gun lists the recommended conduits and liners for each wire size and type of filler metal you will use.

One purpose of the wire liner (which usually is a metal or plastic tube that looks like a long, tight spring with a small hole down the middle) is to help keep the long welding wire from buckling as it is pushed or pulled through the service cable. Wire liners should be changed from time to time and inspected and cleaned frequently. If you are not using clean, bright welding wire, or even if you are, wire liners tend to accumulate oxides and shop dust after a while. If they clog up the slightest bit, you will have wire feeding problems.

The service cable leading to the welding gun also includes neoprene tubes for the shielding gas and larger-diameter cable for the welding current from the power source, as well as separate tubes for cooling water supply and re-

turn if you use a water-cooled gun. Many times these separate lines are enclosed and protected from wear and tear in a single protective outer rubber or neoprene jacket.

A good service cable also will have quick-disconnect couplings so that you can hook up the cable to the wire feeder or change guns quickly (see Fig. 17-25).

The Welding Gun

The GMAW gun guides the wire and shielding gas into the weld zone. It also brings the current to the wire so you can weld. Figure 17-26 shows a cross section of a typical water-cooled, straight-front-end welding gun with the following necessary components:

■ Contact tube (or tip)
■ Shielding gas nozzle
■ Wire liner from service cable
■ Wire conduit assembly
■ Gas hose from service cable
■ Water hose from service cable
■ Power cable
■ Gun switch with cable to wire feeder control

Contact tube The wire guide tube (Fig. 17-27), also called the

contact tube, is made of copper and is used to bring welding current into the wire as well as to direct the wire to the work. Contact tubes (often called *contact tips*) frequently become clogged and dirty, but they are very easy to replace. They simply screw into the front end of the gun. Contact tips are expendable items, but they are very important.

Keep your contact tips clean. They have a tendency to get spatter in them. If you are using a poor grade of welding wire, the contact tip will become clogged by surface dirt, dust, or lubricant from the wire.

The welding gun (and guide-contact tube) is connected to the welding power supply by the power cable. Because the wire must feed easily through the contact tube and also make good electrical contact with it, the inside bore diameter of the contact tube is important. The instruction book supplied with each welding gun will list the correct internal tube diameter for each wire size you use. The tube must be firmly screwed into place and centered in the shielding-gas nozzle.

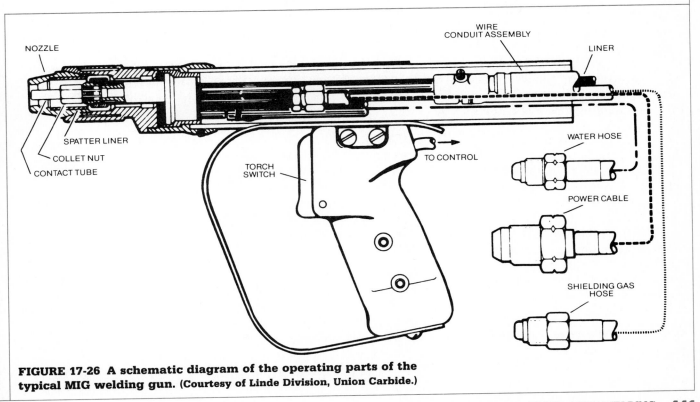

FIGURE 17-26 A schematic diagram of the operating parts of the typical MIG welding gun. (Courtesy of Linde Division, Union Carbide.)

FIGURE 17-27 A close-up of the business end of a GMAW gun. The welding wire exits from a copper contact tube in the middle of the gun's nozzle. The contact tube delivers welding current to the wire. The nozzle directs the flow of shielding gas around the wire and the weld metal when you strike an arc.

Shielding-gas nozzle The shielding-gas nozzle (Fig. 17-28) directs a protective mantle of gas to the welding zone. Large nozzles are used for high-current welding where the weld puddle is large. Smaller cups are used for lower-current welding.

To start welding, you simply depress the trigger on the welding gun, which signals the control box to start shielding gas flowing and then to start feeding you wire. To stop welding, you simply release the trigger on the gun.

Shielding gas will continue to flow for a predetermined purge time to make sure that your weld metal is shielded until it has cooled. That's done with the purge-time indicator on the control box at the wire feeder (sometimes it's simply preset if you have a less sophisticated wire feeder control).

Rating GMAW guns There are dozens of different kinds of GMAW guns. Some are shaped like a pistol, some look a little

more like a flexible-headed fire hose, some include a small wire reel (the spool-on-guns), and some have special miniature auxiliary pull-type wire feeders built right into the gun for feeding fine wires and very soft aluminum wires. Some GMAW guns are water-cooled for use at high amperages. Many guns are air-cooled and so inexpensive that they have almost become throwaway items.

All of these guns have one thing in common. They are rated by their manufacturers, just like power sources, for their maximum welding amperage at 100 percent duty cycle. Here, for example, is the rating on a typical water-cooled GMAW gun:

- Air-cooled manual welding gun with one-quarter turn heavy-duty nozzle and quick-disconnect coupling to the wire feeder. A 360° rotatable gooseneck gets into those hard-to-reach joints.
- Rating: 400 A DCRP in CO_2
              200 A DCRP in argon
              100 percent duty cycle
- Welding wire diameters:

Hard	Soft
0.030 in.	0.035 in.
0.035 in.	3/64 in.
0.045 in.	1/16 in.
1/16 in.	
5/64 in.	

17-12
GMAW WIRE
Before we learn about specific types of solid and flux-cored GMAW wire, let's look at some very important characteristics that all such wires should have if you want to avoid welding problems.

Feedability and Cleanliness
Let's start with the problems of feeding wire from a coil or reel into a wire liner and out through a straight contact tube. Since the wire starts on a coil it will have a permanent set in it. If the set is too curved, the wire will certainly

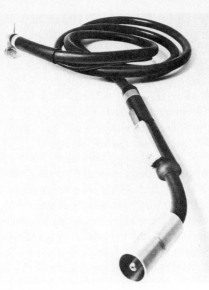

FIGURE 17-28 The welding gun nozzle in the foreground slips easily off the welding gun. Nozzles can be changed for different welding conditions, or replaced if they become too dirty or damaged. Similarly, the contact tube in the middle of the gun nozzle can be unscrewed and replaced with a clean contact tube. The other end of the gun service cable has a quick connect/disconnect device for attachment to the wire feeder.

make contact with the tube in the gun nozzle and pick up the current it needs, but it will also wander around as it feeds out of the gun. This can happen if the wire is fed off a tightly wound small reel. On the other hand, if the wire is too straight, it may not make good contact inside the bore of the short, straight contact tube, which sometimes happens when the wire is fed out of a large, loose coil from a big barrel.

Cast and helix One of the first things you look for in any continuously fed welding wire is *cast and helix*. The best way to understand these commonly used terms is to cut a length of welding wire about 15 ft [4.5 m] long and throw it loosely on the floor (Fig. 17-29). It will not lay straight. It will tend to coil up. The cast is the diameter of the single loose wire coil. The helix is the spring effect; it's the vertical distance between the two ends as they lift away from each other on the floor like the first winding of a very large helical spring.

The AWS only requires a minimum cast of 15 in. [380 mm] for welding wire. But a cast close to 5 ft (60 in.) [1.5 m] will give you much better feeding characteristics through your liner and especially through long service cables. Excessive helix (over an inch [25 mm] high) will create an annoying wandering arc as the wire emerges from the welding gun and contact tip. The ends of a wire with no helix at all would lie flat on the floor. You're not likely to ever get a welding wire that has no helix, but if you ever do, don't use it! Figure 17-29 tells you why.

Another way to check the feedability of a wire is to run it through a welding gun held at a 45° angle to a flat surface (with no current on). Pull the trigger of the welding gun. The wire should coil off the surface into a spring. Feedability improves with wire of higher tensile strength, larger cast, and smooth finish.

New wire feeders, guns, and wire liners usually provide good support over the full path of wire travel. As equipment ages, unsupported areas develop and a support gap somewhere inside the equipment will allow a wire to buckle. Generally, the higher a wire's tensile strength, the better its resistance to buckling under compression between the drive rolls and cable inlet guide.

Cleanliness Clean welding wire is critical. The lack of an absolutely clean surface often results in welding defects. When welding wire is produced, the wire has to withstand the pressure and heat of being cold-drawn through dies. It must be lubricated to make that happen. The lubricants used to make welding wire are difficult for the manufacturer to clean off after the wire is made, but wire cleanness is critical to any welding job.

The wire-drawing compounds or lubricants left on the wire surface create welding problems. Porosity, unstable arcs, humped welds, and even cracking are often traceable directly to dirty welding wire. You normally can't see the difference between clean and dirty wire. The white glove test is still the best way to check wire cleanliness. If the wire doesn't pass inspection (makes the glove dirty), don't use it without complaining first. Otherwise you might be blamed for the welding defects instead of the bad wire you are using.

Many steel GMAW wires have a thin copper coating to increase the conductivity of the surface as it touches the contact tip at the business end of the welding gun. If the copper coating is not properly applied by the wire manufacturer, small flakes of copper will break off. These copper flakes will jam up your contact tips, your wire liner, and everything in between. If you note tiny flakes of copper in your wire liner or con-

tact tip and are having wire feeding problems, tell somebody quickly and get the problem corrected, by getting a better wire (or a better wire supplier).

Selecting Your Electrode Wire
One easy way to approach GMAW electrode wires is by type, solid or flux-cored. We will start by looking at the solid wires used for steel, then we'll learn about the solid wires used for other metals. After that, we'll study the different kinds of flux-cored wires. The AWS will help us a lot here. They have specifications for each kind of electrode wire. These specifications make electrode wire selection fairly easy, once you know the kind of base metal you will join. We'll list those specifications for you.

Solid wires Let's take a look at AWS specification, AWS A5.18 "Specification for Mild Steel Electrodes for Gas Metal-Arc Welding." This lists quite a few different classifications of carbon-steel

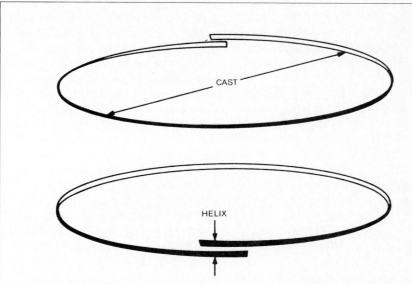

FIGURE 17-29 Cast, an important property of GMAW welding wire, is the diameter of the circle formed by a length of filler wire that has been thrown loosely on the floor. Cast close to 5 ft enhances feedability through the wire liner inside the service cable to the MIG run. The same wire thrown on the floor also shows the amount of helix in the wire coil from which it is taken. Excessive helix (more than 1 in.) creates annoying arc wander when you weld, as the wire emerges from the welding in varying directions. No helix at all also will create problems because the wire won't be in constant contact with the electrical contact tube inside the MIG gun's nozzle.

and low-alloy steel GMAW electrode wires. What does it describe?

In the filler-metal specifications described in AWS A5.18, the prefix ER designates electrode (in this instance a welding wire). The number 70 indicates a minimum of around 70,000-psi [483-MPa] tensile strength for weld metal produced by that electrode wire, and S tells you that the electrode wire is a bare, solid wire and not a flux-cored wire. Weld metal with a 70,000-psi [483-MPa] tensile strength will join many high-strength low-alloy steels as well as carbon steels.

■ ER70S-1 class electrode wires contain the lowest silicon content of all the solid-steel wire electrodes. These wires usually are run only under argon shielding gas, sometimes with a little bit of oxygen added. They have very few deoxidizers. Welding wires of this class can be used with CO_2 shielding-gas mixtures, of course, but only if weld quality isn't important and your only objective is to get the lowest possible cost. Weld metal produced by ER70S-1 steel wires does not require a minimum Charpy V-notch impact strength test according to the AWS A5.18 specification.

■ ER70S-2 covers multi-ply-deoxidized steel filler wire. It contains a combined total of 0.20 percent zirconium, titanium, and aluminum deoxidizers, plus silicon, manganese, and 0.07 percent carbon by weight. If you flip through the AWS specification, you will see that this wire is designed primarily for operation with argon shielding gas with from 1 to 5 percent oxygen added to it, or for use with CO_2 gas shielding.

■ ER70S-3 is by far the most popular electrode wire grade for welding carbon and high-strength low-alloy steels. It is used for about two-thirds of all MIG solid-wire welding. With a carbon content of 0.06 to 0.15 percent for a higher-tensile-strength weld metal, this electrode wire has

enough deoxidants in it to provide porosity-free welds under normal welding procedures. This electrode wire is designed for use with argon-oxygen and CO_2 shielding gases. A minimum Charpy impact strength of of 20 ft-lb [27 J] at 0°F [−18°C] is required for as-deposited weld metal.

■ ER70S-4 contains slightly more manganese and silicon than the previous electrode classification, up to 1.50 and 0.85 percent, respectively. This electrode wire produces a sound weld deposit of higher tensile strength than the previously described wires. ER70S-4 wire is designed to operate with CO_2 gas shielding. No minimum impact strength is specified.

■ ER70S-5 includes 0.50 to 0.90 percent aluminum, in addition to manganese and silicon, for powerful deoxidation. It permits you to weld on oily steel without porosity, and where the base metals contain high carbon and/or high sulfur. This is a wire for very tough welding conditions. CO_2 gas shielding can be used. No minimum impact strength is specified.

■ ER70S-6 has the highest combination of manganese and silicon, up to 1.85 and 1.15 percent, respectively, and is frequently recommended for use when good weld appearance is a requirement—a smooth weld bead and contour with low spatter. You can use it with CO_2 gas shielding. In addition to minimum tensile strengths, this wire includes a minimum of 20 ft·lb [27 J] Charpy impact strength at −20°F [−29°C] for deposited weld metal.

■ ER70S-7 has higher manganese content than the other wires, about 2 percent or more, to give you better wetting action and weld appearance with slightly higher tensile and yield strengths, and it may even let you increase welding speeds a little. It can be used with CO_2 gas shielding.

■ ER70S-G does not specify chemical composition or ranges because they are subject to, as the AWS says, "agreement between sup-

plier and purchaser." What that means is that the AWS has made room for the specification of proprietary solid GMAW wires that don't fall into any of the other categories. This is typical of many AWS filler-metal specifications—and a far-sighted idea, too. It allows welding filler-metal producers to innovate wires that often exceed standard specifications in one or more ways while also meeting the demands of many companies that require their welders to buy materials manufactured to some kind of AWS specification.

■ ER80S-D2 is a low-alloy electrode wire that produces weld metal with a tensile strength over 80,000 psi [552 MPa]. It contains between 0.40 and 0.60 percent molybdenum as an alloying element to add tensile strength to the weld deposit, and it is recommended by many fabricators for welding low-alloy steels such as AISI 4130.

Other AWS welding specifications cover GMAW solid welding wires for stainless steel, aluminum, magnesium, nickel and copper and their alloys, as well as titanium. These other specifications are mostly the same ones you would use to select a GTAW (TIG) cut-length welding rod or an oxyfuel gas welding rod for these same base metals.

Unlike the solid-steel electrode classifications, electrode wires for most other materials are identified by the Type of weld metal produced, such as Type 308 stainless steel (produced by an ER308 electrode filler wire), or Type 1100 aluminum (pure aluminum rather than an aluminum alloy), which would be welded by an ER1100 solid welding wire. An ERCuSi welding wire would be used to join copper-silicon (silicon bronze) base metal.

Flux-cored wires Flux-cored electrode wire (used only for welding carbon, high-strength low-alloy, and stainless steels) has two separate AWS specifications. One is for low-carbon steels and one is for high-chromium and chromium-nickel steels (mostly stainless steels). The two flux-cored electrode-wire specifications are AWS A5.20 for mild-steel flux-cored electrode wires and AWS A5.22 for corrosion-resisting chromium and chromium-nickel flux-cored electrode wires used for stainless steels.

The flux-cored-wire classes are very similar to those for solid wires. The only difference is that they include the letter T. For example, the flux-cored equivalent of a solid ER70S-3 welding wire is E70T-3. The flux-cored equivalent of an ER308 solid stainless-steel welding wire is E308T. Whenever you see a T in an electrode specification you know that it's for flux-cored welding.

Why do solid wires have both an E (for electrode) and an R in their AWS designation? Because if you cut a bare solid electrode wire into a short length you could use it as a welding rod for gas or TIG welding. You can't do that with flux-cored electrode wires.

Why the T designation for a flux-cored wire? Probably because F had already been used and the AWS was running out of letters. As you can see, picking the correct electrode wire is like fishing in alphabet soup.

There is no standard set of fluxing materials inside each cored-wire grade. Every wire producer is allowed to use available metallurgical talent to develop flux-cored wires that will produce weld metal meeting a given AWS specification. We've listed some of the materials that can be put into a flux-cored wire in Table 17-3, just to give you an idea of what that flux core might contain.

AWS Filler-Metal Specifications

AWS filler-metal specs are very detailed. They run up to 50 pages each and there are over 20 separate booklets that contain them. But if you sit down and look at one of the specifications carefully, you will find a wealth of practical information and advice on which electrodes or rods or wires to use with which base metals and shielding gases, and so on. That's in addition to details on mechanical properties, available diameters, packaging, testing requirements, and other information.

You also can get very good electrode selection booklets from any filler-metal producer that supplies arc-welding electrodes and wire. Most of the electrodes that meet various AWS specifications will be cross-indexed so that a product like SPATTER-ARC 300 can be traced back to some AWS standard classification. You will find, however, that in addition to arc-welding electrodes meeting the AWS specifications, many filler-metal producers also make a large variety of proprietary grades. Many of these proprietary electrodes exceed AWS requirements or are designed for special work.

Here is a list of the AWS filler-metal specifications that will contain data to help you select either solid or flux-cored GMAW electrode wires.

AWS A5.3	Aluminum and Its Alloys (solid wires).
AWS A5.6	Copper and Its Alloys (solid wires).
AWS A5.9	Stainless Steels and High Chromium Alloy Steels (solid wires).
AWS A5.14	Nickel and Its Alloys (solid wires).
AWS A5.16	Titanium and Its Alloys (solid wires).
AWS A5.18	Mild Steel and High-Strength Low-Alloy Steel (solid wires).
AWS A5.19	Magnesium and Its Alloys (solid wires).
AWS A5.20	Mild Steel and High-Strength Low-Alloy Steel (flux-cored wires).
AWS A5.22	Stainless Steels (flux-cored wires).

17-13 SHIELDING GASES

The main job of your shielding gas is to displace the air in the weld zone and thus prevent contamination of the weld metal by nitrogen, oxygen, and water vapor. These gases cause a variety of welding defects and put impurities into the hardened weld metal. The impurities range from metal oxides to slag inclusions, dissolved hydrogen atoms that cause high-strength weld metal to crack, as well as producing other problems such as weld porosity. The porosity is in the form of tiny bubbles that you may see on the surface of the solidified weld metal. More often, however, the bubbles are deep inside the weld where you won't see them without x-ray inspection.

A very small amount of porosity may be acceptable in lower-quality welds if the bubbles are not lined up to initiate a crack. But any time you see weld metal with bubbles in it, tear out the weld, correct the problem that's giving you the porosity, and lay in good, solid weld metal. A lot of porosity and other defects result from improper gas shielding, especially if you don't purge your shielding-gas lines first or when the gas has some water vapor in it (it has a high dew point). A very small amount of water vapor can cause a lot of bad welding.

Your shielding gas also has an effect on your welding arc's characteristics, on how much or how little spatter is produced during welding, and on how deep or shallow your penetration into the base metal will be. No single shielding gas can be used with every welding process or every metal. Each gas has only certain applications (Table 17-4). What's more, different welding shops will use different gases or gas mixtures for the same job. Sometimes it's a matter of individual economics. Sometimes a shielding-gas mixture is used to solve welding problems that are encountered with a single shielding gas.

The Three Principal Shielding Gases

Argon (A), helium (He), and carbon dioxide (CO_2) are the principal shielding gases. Selecting the best shielding gas or mixture for a given job initially depends upon the welding process (mode of weld-metal transfer). Certain gases are not compatible with certain modes of molten-metal transfer through the welding arc. For example, spray transfer will not occur with a shielding gas of pure CO_2 or pure helium. Spray transfer requires a large amount of argon in the shielding gas, usually above 85 percent. Where a choice of shielding gas is possible for given welding process, cost and welding characteristics are the main considerations.

A pure inert shielding gas (argon or helium) will protect the arc and weld metal from the atmosphere, but it is suitable only for steel, thus it cannot be used for all GMAW jobs. The controlled mixing of quantities of "reactive" gases such as CO_2 and O_2 with the inert gases produces arc stability and substantially spatter-free welding at a lower total gas cost. Filler metals that have been adjusted with additional deoxidizers such as manganese to balance any oxidative reaction that may occur due to the reactive gas additives also must be used.

Let's look at the inert gases first. Although pure argon is fine for conventional spray and pulsed-spray welding of aluminum and copper, the pure argon-shielded arc is not stable enough for welding steel, stainless steel, or other ferrous metals. The arc wanders and produces an irregular welding bead. This condition is overcome by using a small amount of O_2 or CO_2 mixed in with the argon. This small addition of reactive gas (or gases) not only stabilizes your arc's characteristics, but also improves weld metal wetting of the base metal and gives you reproducible welding results. These small percentages of O_2 or CO_2 also minimize undercutting of the base metal next to the weld bead and improve the appearance of the weld bead. The recommended amount of additives will range from 0.5 to 5 percent O_2 and from 2 to 25 percent CO_2 in the argon, depending on the joint being welded, the composition of the steel, and the welding technique.

Carbon dioxide additions to argon generally do not exceed 10 percent when spray or pulsed-spray transfer through the arc is desired. More CO_2 may be added for dip transfer. In this case, the amount of gas added depends upon the metal being welded. Stainless steel can tolerate far less CO_2 (about 2.5 percent maximum) than carbon or low-alloy steels, where up to 25 percent CO_2 often is used.

Before we go on to take a look at the shielding gases that are used in GMAW (MIG) and FCAW (cored-wire) welding, let's look at the physical characteristics of shielding gases that help make the gas selection job easier.

Physical Characteristics

A deeper understanding of why shielding gases work the way they do will help you understand why they are selected for use with different metals and different welding processes.

Ionization potential The first idea that will help you is your gas's *ionization potential*. That's the voltage needed to remove an electron from an atom of the shielding gas, turning the atom into an ion. (You might want to reread the appropriate sections in

TABLE 17-4 Selecting a shielding gas for GMAW

Base Metal	Spray-Transfer Welding	Dip-Transfer Welding	Pulsed-Spray Welding	Flux-Cored or Buried-Arc Welding
Stainless steels	Argon + 0.5% oxygen Argon + 1% oxygen Argon + 2% oxygen	90% helium 7.5% argon 2.5% CO_2	Argon + 0.5% oxygen Argon + 1% oxygen Argon + 2% oxygen	Not used
Carbon and low-alloy steels	Argon + 1% oxygen Argon + 2% oxygen Argon + 5% oxygen Argon + 5% CO_2 Argon + 8% CO_2	CO_2 Argon + 25% CO_2 Argon + 8% CO_2 Argon + 5% CO_2	Argon + 1% oxygen Argon + 2% oxygen Argon + 5% oxygen Argon + 5% CO_2 Argon + 8% CO_2	CO_2 Argon + 25% CO_2
Aluminum and magnesium	Argon + 25% helium Argon + 75% helium Argon + 0.15% oxygen		Argon Argon + 25% helium Argon + 75% helium Argon + 0.15% oxygen	Not used
Copper	Argon + 1% oxygen Argon + 75% helium Argon + 0.3% oxygen		Argon + 1% oxygen Argon + 75% helium Argon + 0.3% oxygen	Not used
9% nickel steel	Argon + 0.03% oxygen Argon + 25% helium Argon + 75% helium		Argon + 0.03% oxygen Argon + 25% helium Argon + 75% helium	Not used

Chaps. 11 and 15.) A cloud of ions in an electrically charged gas is called a *plasma*. You used a gas plasma when you did PAW. You also used a gas plasma when you did GTAW. And you'll use it in any gas-shielded GMAW process. Whether a shielding gas has a high or a low ionization potential matters very much in the welding arc.

Take argon as an example. It requires a low ionization potential of 15.7 electron volts (eV) to remove the first electron from the argon atom, and only 27.6 eV to remove the second electron. In GTAW with argon, the argon easily forms a plasma, which works like an electrically charged path that encourages the welding current to travel from the electrode to the workpiece. That's one reason why the argon arc in GTAW is tight and constricted and why your puddle control is good.

Arc stability depends on your shielding gas's ionization potential, too. The low ionization potential of argon turns atoms into ions easily, which helps to sustain a smooth, even arc. Helium, with its higher ionization potential of 24.5 eV to remove one electron from the atom, and 54.4 eV to remove the remaining electron from the atom, is very high compared with argon, which only needs 15.7 eV to remove one electron. Therefore, helium-shielded welding produces a less stable arc. Arc and puddle control are difficult when using pure helium shielding gas compared with argon or an argon-helium, argon-CO_2, or argon-O_2 gas mixture.

The high ionization potential of helium also explains why welding with helium requires a high arc voltage, usually 75 percent higher than that for argon using the same arc length. The high arc voltage of helium gives you an intensive arc that has the high heat input needed for rapid welding speeds and a high-flowability (hot) weld puddle that aids degasification and produces welds of good

integrity. Helium is insoluble in molten metal. Helium is not used as a pure shielding gas in GMAW welding; it is used in mixtures with other gases.

Thermal conductivity *Thermal conductivity* is another important idea that applies to gases (and to your workpiece). The thermal conductivity measures the ability of a substance to conduct heat. A gas that has good thermal conductivity helps to conduct the heat to the workpiece. The degree of thermal conductivity of your shielding gas influences the shape of the weld bead (Fig. 17-30) and the conduction of the metal next to the weld in the heat-affected zone.

Helium, for example, has excellent thermal conductivity. In GTAW, or as an additive in GMAW work, with helium, the arc column broadens due to heat and spreads the heat to make a wide, smooth bead. Argon has lower thermal conductivity than helium. It makes a narrower arc column and a narrower weld bead. But since the heat is concentrated in a narrower arc column, the argon-shielded arc "digs in" more and produces a deeper weld with a finger-shaped root.

Carbon dioxide, with its thermal conductivity in between that of argon and helium, produces a weld bead between these two shapes—as you might expect. Pure carbon dioxide also produces a lot of spatter when you use it in GMAW. That's why carbon dioxide often is mixed with argon for GMAW. (It's not used for GTAW because carbon dioxide is not totally inert. The gas will oxidize the tungsten electrode.) Carbon dioxide often is used in GMAW shielding mixture, and as 100 percent gas shielding in FCAW because it is very inexpensive.

Gas density Density is the weight per unit volume of the gas. A heavy shielding gas will provide better cover, at an equal flow rate, than will a light or low-density gas . . . most of the time. Argon, which is 1½ times as heavy

as air and 10 times heavier than helium, requires one-half to one-third the flow rate of helium to do the same work (see Chap. 3).

The density of helium is 0.01 lb/ft³ [0.16 kg/m³] at 30°F [−1°C] and one atmosphere pressure (14.7 psi [469 Pa]). The density of air is 0.08 lb/ft³ [1.3 kg/m³] under the same conditions. Because it is so light, helium does not shield as well as other gases at the same flow rate. It is an excellent gas shield at high flow rates. To equal the protection of argon, flow rates for helium must be two to three times higher. One exception is when you are welding overhead, where the tendency of helium to rise helps you out.

Dew point The shielding gas you use for any gas-shielded welding process must have a very low dew point. You may remember from Chap. 3 that the dew point of a gas is a temperature, like 0°F or

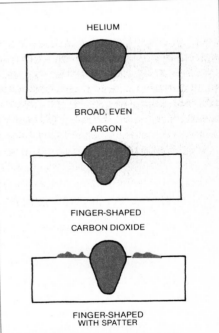

HELIUM

BROAD, EVEN

ARGON

FINGER-SHAPED

CARBON DIOXIDE

FINGER-SHAPED WITH SPATTER

FIGURE 17-30 This shows how shielding gases affect the shape of a weld bead and the penetration depth into the base metal. A pure carbon dioxide gas shield, sometimes used on carbon steel, will produce excessive spatter which not only wastes filler metal but also increases labor costs for cleanup.

−60°F [−18 or −51°C], at which moisture in the gas would condense out as water vapor if you held a mirror at that temperature in front of the gas flowing from the cylinder. Therefore the dew point of the gas is equivalent to stating how much moisture it has in it; it's a measure of its purity.

Shielding gases for welding must be very dry (have very low dew points) because any moisture (water) in the gas will turn into hydrogen and oxygen when it passes through the welding arc. The hydrogen, especially, can be very destructive. Small amounts of moisture also will cause weld-metal porosity. Shielding gases should have dew points of at least −40°F [−40°C] and lower if possible. Better-grade shielding gases often have dew points of −80°F [−62°C]. How much or how little moisture is that? Let's show you.

Dew point	% moisture
0°F [−18°C] =	0.059
−20°F [−29°C] =	0.021
−40°F [−40°C] =	0.0065
−60°F [−51°C] =	0.0048
−80°F [−62°C] =	0.0030

As you can see, we are talking about fractions of 1 percent moisture. Since 1 percent is the same as one part per hundred, a very "wet" gas with a dew point of 0°F [−18°C] or 0.059 percent moisture still only has 59 parts of moisture per 100,000 volumes of gas. A very dry gas with a −80°F [−62°C] dew point would have only 3 volumes of moisture per 100,000 volumes of gas. The expression other than dew point most commonly used for dry gases is parts per million or ppm, so that the gas with a dew point of −80°F [−62°C] would have only 30-ppm water vapor. That gives you an idea of how dry a good welding shielding gas should be.

Now that you understand how important certain physical characteristics of shielding gases are, let's see how they are used in our most common shielding gases.

Helium

Helium additives in argon allow you to make GMAW joints with narrow weld gaps, which translates into savings in filler metal. Helium favors heat transfer from the arc to the work. High heat input and current density make helium ideal for welding thick sections and for materials with high melting temperatures or high thermal conductivities.

This gas finds most of its use in high-speed automatic GTAW of light-gauge metals such as in tube mills and for heavy-section welding, where it gives deep penetration welds of high quality. If GMAW or GTAW is used for metal over 1/8-in. [3.2-mm] thick, helium may be mixed with argon to improve weld-metal penetration into the workpiece. Voltage-control equipment can be a big help in controlling the sensitive helium arc.

The cost of helium is higher than that of other shielding gases. When looking at the cost of your shielding gas, it is well to consider the cost of the total operation. Helium costs more per unit volume than does argon or carbon dioxide, but it allows fast welding with narrow gaps. Savings in labor and filler metal may offset the higher gas cost. Another point: Heat input with a helium gas mixture may be enough to avoid preheating, especially with copper and other high-heat-conductivity materials.

Argon

Argon is the most common shielding gas used for both GMAW and GTAW to join aluminum and stainless steels. The reasons for its popularity are easy to figure out. It costs less than helium and it makes welding easy. Argon has a lower arc voltage than helium, so it lets you vary the arc length with less variation in the current and bead shape. Another reason for argon's popularity is that with ac TIG welding, argon gives you good cleaning action, removing surface oxides from aluminum and stainless steel. The high density means that you can reduce flow rates, which also keeps arc turbulence low. Argon used in DCRP welding in atmospheres over 75 to 85 percent Ar also produce spray transfer, and that's a big factor in its use in GMAW.

Not a high-energy-input gas, argon makes a weld that freezes quickly. If the metal is not molten long enough to wet out to the weld toe, undercutting results. For ferrous materials, additions of from 1 to 5 percent oxygen, which superheats the metal, allows the molten weld metal shielded by argon to flow out to the toes of the weld and help avoid undercutting. The argon-oxygen mixture also flattens the weld-bead reinforcement.

The relatively narrow and deep argon-shielded weld may actually promote the trapping of gases in narrow-groove welds. These gases come mostly from the weld metal or base metal because even solid metal can trap some gases which are released when the metal is melted. The argon itself has very low solubility in molten weld metal, and even less argon will stay in solid weld metal. But argon is useful for getting deep penetration at a low current.

The solution to entrapped gases in narrow and deep weld joints is to use helium-argon mixtures. The broader helium-argon shielded weld improves the chances for gases dissolved in the weld metal to escape, and helium sometimes is used specifically for this purpose in place of pure-argon gas shielding.

Carbon Dioxide

CO_2 is an active gas, not an inert gas. Its great advantage is that it has a very low cost.

Pure CO_2 can be used with the short-circuiting process and naturally is used with the buried-arc CO_2 welding process. CO_2 is an inexpensive shielding gas when compared to argon or argon with mixtures of other gases such as helium. It is used in mixtures with inert gases, especially argon, and throughout the welding industry for joining carbon and high-strength low-alloy steels. CO_2 shielding should not be used with stainless steel, aluminum, or other metals such as titanium that are especially sensitive to oxidation.

In the welding arc, CO_2 momentarily breaks down into carbon monoxide and oxygen. The oxygen burns, mostly recombining with the carbon monoxide, and gives off plenty of heat that adds to the arc energy. The CO_2 gas shield is oxidizing and is about equal to a mixture of 91 percent argon and 9 percent oxygen. The advantage of CO_2 shielding is that it is much less expensive than truly inert argon or helium.

The reaction of CO_2 in the arc produces a harsh arc that causes excessive weld spatter in anything but short, low-voltage arc processes. It also produces a heavily oxidized weld bead. Welding fabricators who want to cut shielding gas costs get around weld porosity and the oxidation by using filler metals that contain deoxidizers and by welding with a short arc where possible. Because CO_2 promotes oxidation, it can put lots of metal oxide particles into the welding atmosphere as fumes or smoke, reducing your visibility.

Spray transfer cannot be attained with CO_2 or with argon-gas mixtures containing more than 15 percent CO_2. A common way to improve the qualities of CO_2 is to mix it with a little argon (much less than you need for spray-transfer welding). The argon stabilizes and quiets the arc, at a small increase in shielding gas cost. Power sources modified to maintain a short arc help too.

CO_2 works well with flux-cored wires because it provides a good shield at high currents and gives good weld-metal penetration. CO_2 is not suitable for GTAW because of its oxidating nature. Spatter also is a problem with CO_2 because it reduces the deposition efficiency of your filler metals (the pounds of filler metal used versus the pounds of weld metal produced). A common solution in GMAW is to use a mixture of 75 percent argon and 25 percent CO_2 for your MIG welding gas. That will give you spray-transfer MIG welding and also lower the cost of the gas shielding.

Under a straight CO_2 shield, about 50 percent of the manganese and 60 percent of the silicon in the welding wire are converted to oxides when passing through the welding arc. These deoxidizers are added to the welding wire specifically for this purpose, which is why welding wires (either solid or flux-cored) must be formulated for use with CO_2 shielding gas.

In the heat of the arc, CO_2 breaks down into its component parts (CO and O). At normal arc lengths (the distance between the workpiece and the welding wire contact tube) about 7 percent of the total volume of the gas shield dissociates to carbon monoxide. At excessive arc lengths the quantity reaches 12 percent. That is why CO_2-shielded welding is normally used with a short-arc technique. Keep in mind, of course, that the gases recombine almost instantly after leaving the welding arc as the gas temperature drops very rapidly at short distances away from the arc. There is no worker hazard from carbon-monoxide gas. You would literally have to stick your bare nose in the welding arc to get near enough to pick up some carbon monoxide before it reconverts to carbon dioxide.

Carbon dioxide gas used for blanketing the weld puddle is termed *welding grade* and is packaged in special cylinders. The cylinders are marked with the proper grade designation. Cylinders of welding-grade CO_2 are tested to maintain high purity and a low dew point of at least $-40°F$ $[-40°C]$. Since even small amounts of moisture in the CO_2 would result in porosity in the weld deposit, or hydrogen embrittlement in higher-strength steels, it is important that only welding-grade CO_2 be used.

The commercial cylinder for welding-grade CO_2 is roughly 9 in. [230 mm] in diameter, 51 in. [1.3 m] long, and empty normally weighs 105 lb [47 kg]. Lightweight cylinders are in use in some areas that weight 90 lb [41 kg]. Both types of cylinders contain 50 lb [23 kg] of CO_2, which is equivalent to 436 scf [12.2 m³] of gas.

A full cylinder at 70°F [21°C] will contain approximately 45 lb [20 kg] of CO_2 in the liquid state. The liquid level on a "full" cylinder is about two-thirds up from the bottom. Space for gas over the liquid is necessary. The liquid in the cylinder absorbs heat from the atmosphere and "boils." Gas is formed until the pressure in the cylinder raises the boiling temperature above the temperature of the liquid. Boiling, and consequently gas production, ceases until either the atmospheric temperature increases or some gas is drawn off, lowering the pressure in the cylinder and allowing boiling to continue, producing more gas.

Since the rate of withdrawal of CO_2 is limited by the heat transfer through the cylinder walls, termed the *wetted surface,* the maximum discharge rate of a single CO_2 cylinder is from 3 to 4 lb/h [1.3 to 1.8 kg/h] (25 to 35 scf/h [0.7 to 0.9 m³/h]). For heavier flows connect several cylinders together with a manifold. Generally, at least two cylinders are suggested on a manifold for best results with the GMAW or FCAW processes.

Because CO_2 cylinders contain liquefied CO_2, the pressure under normal conditions will be maintained until all the liquid is vaporized. After the last bit of liquid is vaporized the cylinder pressure will begin to fall. Gas content at that point is still about 130 scf [3.6 m³] per cylinder, which allows approximately 3 h arc time at a 40 scf/h [1 m³/h] flow rate before the cylinder is empty. Cyl-

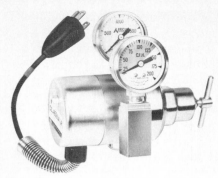

FIGURE 17-31 This electrically heated gas regulator is used on carbon-dioxide shielding-gas cylinders when high gas withdrawal rates otherwise would cause the CO_2 to turn into dry ice when pressures drop rapidly. If you see your CO_2 regulator freeze, you know you need a heated regulator. Ice from moisture in the air will actually freeze up the unheated regulator and stop it from working because dry ice forms at $-109°$ F.

inders should be changed when pressures drop to about 150 psig [1030 kPa]. Typical CO_2 cylinder pressures for various temperatures are:

100°F [38°C]	1450 psig [10,000 kPa]
70°F [21°C]	835 psig [5760 kPa]
30°F [-1°C]	476 psig [3280 kPa]
0°F [-18°C]	290 psig [2000 kPa]

Solid CO_2 particles will clog the flow control orifice of the gas regulator causing sharp fluctuations in flow rates, which result in improper welding shielding and porosity if special regulators are not used. Flow can also be restricted or reduced on applications other than welding, therefore welding regulators for CO_2 are in general use.

The problem of "freezing up" a regulator is caused by the expansion of the gas through an orifice from a high inlet pressure to a low outlet pressure; the very same technique used to produce solid dry-ice snow.

At cylinder pressures above 950 psig [6500 kPa] (cylinder temperatures above 79°F [26°C]), CO_2 becomes what is called a supercriti-

cal fluid. Such a fluid has some properties of a gas and some of a liquid but has a lower heat content than that of a gas, and thus can be solidified much more easily than gaseous CO_2. When this supercritical fluid is expanded through an unheated regulator orifice, the subsequent cooling causes the dry-ice formation and freeze-up. The problem is magnified when high flows are required, since increasing the flow rate causes more dry ice to be formed.

Regulator manufacturers provide electrically heated regulators for CO_2 (Fig. 17-31). They convert the frozen particles back to gaseous CO_2 as it passes through the throttling nozzle in the regulator. Any dry ice that does form also is thawed as it travels through the heated regulator passages.

Heated regulators normally require a 115-V, 1.3-A ac circuit producing 50 W of power to the heaters (available from a standard ac wall circuit). These regulators have control thermostats and a safety cutout. Normally a thermostat cycles the heater to maintain the regulator at body temperature between the control limits of 95 and 120°F [35 and 49°C]. Under conditions of maximum heat demand, that is, inlet pressures over 950 psig [6500 kPa], and at flow rates over 60 scf/h [1.7 m³/h], the heaters will run continuously. The regulator surfaces may even begin to frost up. The presence of this frost is normal and does not indicate a regulator freeze-up. Flowmeters often are attached to the regulator outlet to monitor CO_2 flow rates precisely.

Mixing Your Own Shielding Gases

You can buy GMAW shielding gases premixed in cylinders from many welding distributors. You also can mix your own. Several companies produce gas proportioners for mixing large volumes of shielding gas. Most likely, you will not have one. Here's how to mix a shielding gas if all you have are two gases in their sepa-

rate cylinders and two flowmeters.

Let's assume that you want to weld with a mixture of 75 percent argon and 25 percent CO_2. The job requires a total flow rate of 40 standard cubic feet per hour (scf/h) [1.1 m³/h] through your GMAW gun. Your flowmeter settings are made as follows:

Argon flowmeter setting: $^{75}/_{100} \times 40$ scf/h = 30 scf/h [0.8 m³/h] CO_2 flowmeter setting: $^{25}/_{100} \times 40$ scf/h = 10 scf/h [0.3 m³/h]

The control valve on the CO_2 flowmeter is adjusted until the center of the ball float in the glass tube is opposite the 10-scf/h [0.3 m³/h] graduation which is on the *low flow* scale. When reading the low flow scale, the scale selector valve must be turned *in*.

The control valve on the argon flow meter is adjusted until the center of the ball float in the glass tube is opposite the 30-scf/h [0.8 m³/h] graduation which is on the high flow scale. When reading the *high* flow scale, the scale selector valve must be turned all the way *out*.

When the gas from the two cylinders is combined, the result will be 10 scf/h [0.3 m³/h] of CO_2 and 30 scf/h [m³/h] of argon, or 40 scf/h [1.1 m³/h] of your 75 percent argon and 25 percent CO_2 shielding gas mixture.

Here's another example of setting up your flowmeter to produce 120 scf/h [3.4 m³/h] of a mixed gas consisting of 10 percent helium and 90 percent argon. The meter settings are determined as follows:

Helium setting: $^{10}/_{100} \times 120$ scf/h = 12 scf/h [0.3 m³/h] Argon setting: $^{90}/_{100} \times 120$ scf/h = 108 scf/h [3.1 m³/h] Total flow = 120 scf/h [3.4 m³/h]

The control valve of your helium flowmeter is adjusted until the center of the ball float is opposite the 12-cf/h [0.3 m³/h] graduation in the glass cylinder which is on the low flow scale. When reading the *low* flow scale, the scale selector valve must be turned *in*.

The control valve on the argon flowmeter is adjusted until the center of the ball float is opposite the 108-scf/h [3.1-m³/h] graduation which is on the high flow scale. When reading the *high* flow scale, the scale selector valve must be turned all the way *out*.

You also can increase the capacity of a given flowmeter if you have more than one of the same type. For example, hooking the two flow meters to two cylinders of argon will double the flow through the line when the gas flows are combined into one stream. This is equivalent to combining two arc-welding power sources in parallel. Always keep in mind that a pressure drop and reduced flow rate will occur in long shielding-gas hoses (just like electricity). Keep your shielding-gas lines as short as possible.

Your Base Metal Determines the Shielding Gas

Different shielding gases are used on different base metals, depending both on the cost of the shielding gas and the quality of the weld that you want.

Carbon steels Most plain carbon steel over ⅛ in. [3.2 mm] thick is GMA welded with some kind of CO_2 shielding, often a mixture (see Table 17-4 for more details). The shielding gas used depends a lot on the type of weld-metal transfer through the arc that you want.

Where spatter can be tolerated, straight CO_2 is the choice. Where you can't have spatter on your weldment, use argon plus CO_2 to quiet the arc, reduce spatter, and to give you a flat weld bead with smooth transition to the base metal. For low-alloy steels, a mix of 75 percent argon and 25 percent CO_2 works well, as it does on mild steel that requires only shallow penetration. If fit-up is poor, a mix of 92 to 94 percent argon with the remainder CO_2 gives you a wide arc that makes gap-bridging easier and helps keep spatter low.

If a high-quality weld in your carbon steel is the overriding consideration, argon with 0.5 percent to 2 percent oxygen are possible choices. These mixtures produce spray transfer at a low arc current, which allows you to use a larger-diameter wire for a given job. The weld will have little porosity and a wide, flat bead.

Stainless steels The stainless steels weld best under argon with ½ to 2 percent oxygen. The oxygen improves metal transfer through the arc. The austenitic stainless steels do well in a 90 percent helium, 8 percent argon, 2 percent CO_2 mixture. This is a short-arc welding mixture that gives you a flat weld that economizes on filler wire and weld finishing.

In dip-transfer welding of stainless steel, the preferred shielding gas is a mixture of three gases, argon, helium, and CO_2. The gas is combined in the following percentages: 90 percent helium, 7.5 percent argon, and 2.5 percent CO_2. The helium produces the heat necessary for good wetting of the base metal and the argon and CO_2 are necessary to produce additional arc stability and to improve weld-metal wetting of the base metal even more. The small percentage of CO_2 does not harm the corrosion resistance of the stainless-steel weld metal.

Aluminum Certain facts limit shielding gas selection for aluminum.

■ Hydrogen dissolves in molten aluminum.
■ Aluminum readily forms oxides.
■ Aluminum has high electrical and thermal conductivity.
■ Aluminum needs a gas to provide cleaning action to break up the aluminum oxides in the weld zone.

A shielding gas for aluminum must not contain hydrogen or oxygen. This leaves only argon and helium, and they must be moisture-free. Investigators have tried small (1 percent and less) oxygen

additions to argon. These gave improved joint penetration, but small oxide inclusions in the aluminum weld metal ruled out even this small amount of oxygen.

Argon and argon plus helium mixtures are the gases best suited for the spray and pulsed-spray welding of aluminum, aluminum alloys, and magnesium. Pure argon is used for light-gauge work (⅜ in. [9.5 mm] thick and less). A 75 percent helium and 25 percent argon mixture is preferred for heavier gauges of aluminum and magnesium to improve weld penetration and reduce porosity, especially when welding heavy plate from both sides.

Since the argon penetration broadens and deepens to a maximum at 75 percent helium and 25 percent argon, this mixture is common for getting sound MIG welds in heavy aluminum plate welded from both sides. Argon-helium mixtures also are used on thick sheet metal. Because more spatter is produced with helium than with argon, helium's use for nonferrous jobs has been small compared with argon. When making an argon-shielded weld from two sides of a plate, the weld placement must be carefully controlled to assure overlap of the two passes at the center.

Helium makes good aluminum welds. The high energy input from the high-voltage arc offsets the metal's high thermal conductivity. Where argon is used, helium additions make for faster welding and fewer defects. Helium additions to argon also increase the deposition rate of aluminum filler metals. (See Table 17-5.)

Copper and copper alloys

Argon is usually used for welding light-gauge copper up to about ⅛ in. [3.2 mm] thick. Helium, argon-helium, or argon-oxygen mixtures are generally preferable for welding heavier-gauge copper or copper alloys. Argon is generally suitable for welding most copper alloys up to about ⅜ in. [9.5 mm]

TABLE 17-5 Selecting Filler Metals for Aluminum Alloys

Each base-alloy column is rated for the characteristics **W S D C T M** (in that order). Shaded diagonal cells show the filler-metal names.

Base Alloys	Filler Alloys	1060, EC (W S D C T M)	1100 (W S D C T M)	2014, 2036 (W S D C T M)	2219 (W S D C T M)	3003, ALCLAD 3003 (W S D C T M)	3004 (W S D C T M)	ALCLAD 3004 (W S D C T M)	5005, 5050 (W S D C T M)	5052, 5662 (W S D C T M)
319.0, 333.0, 354.0, 355.0, C355.0, 380.0	2319			B A A A A A	B A A A A A					
	4043	B A A A A A	B A A A A A	C C B C A A	C C B C A A	B B A A A A	B B A A A A	B B A A A A	B B A A A A	A A A A A A
	4145	A A B A A A	A A B A A A	A B C B A A	A B C B A A	A A B A A A	A A B A A A	A A B A A A	A A B A A A	
413.0, 443.0, 444.0, 356.0, A356.0, A357.0, 359.0	4043	A A A A A A	A A A A A A	B B A A A A	B B A A A A	A A A A A A	A A A A A A	A A A A A A	A A A A A A	A B A A A A
	4145	A A B B A	A A B B A	A A B A A	A A B A A	A A B B A				
	5356									B A B B A
7005, 7021, 7039, 7046, 7146, A712.0, C712.0	4043	A A C A A	A A C A A	B B A A A	B B A A A	A B C A A	A D C B A	A D C B A	A B C B A	B D C B A
	4145			A B A A	A A B A A					
	5183	B A B A A	B A B A A			B A B A A	B A B A A	B A B A A	B A B A A	A A B A A
	5356	B A A A A	B A A A A			B A A A A	B B A A A	B B A A A	B A A A A	A B A A A
	5554						C C A A A A	C C A A A A	C A A A A A	B C A A A A
	5556	B A B A A	B A B A A			B A B A A	B A B A A	B A B A A	B A B A A	A A B A A
	5654						C C A A B	C C A A B	C A A A A	B C A A A
6061, 6070	4043	A A C A A	A A C A A	B B A A A	B B A A A	A B C A A	A D C A A	A D C A A	A B C A A	A D C A A
	4145	A A D B A	A A D B A	A A B A A	A A B A A	A A D B A	B C D B A	B C D B A	A B D B A	
	5183	B A B A	B A B A			B A B A	B A B A	B A B A	B A B A	B A B C B
	5356	B A A A	B A A A			B A A A	B B A A	B B A A	B A A A	C C A B A B
	5554									
	5556	B A B A	B A B A			B A B A	B A B A	B A B A	B A B A	B A B C B
	5654									C C A B A
6005, 6063, 6101, 6151, 6201, 6351, 6951	4043	A A C A A	A A C A A	B B A A A	B B A A A	A B C A A	A D C A A	A D C A A	A B C A A	A D C A A
	4145	A A D B A	A A D B A	A A B A A	A A B A A	A A D B A	B C D B A	B C D B A	A B D B A	
	5183	B A B A	B A B A			B A B A	B A B A	B A B A	B A B A	B A B C B
	5356	B A A A	B A A A			B A A A	B B A A	B B A A	B A A A	B B A C A
	5554									C C A B A B
	5556	B A B A	B A B A			B A B A	B A B A	B A B A	B A B A	B A B C B
	5654									C C A B A
5454	4043	A B C C A	A B C C A		A A A A A	A B C C A	A D C C A	A D C C A	A B C C A	A D C C A
	5183	B A B B A	B A B B A			B A B B A	B A B B A	B A B B A	B A B B A	A A A B A
	5356	B A A B A	B A A B A			B A A B A	B B A B A	B B A B A	B A A B A	A B A B A
	5554	C A A A A A	C A A A A A			C A A A A A	C C A A A A	C C A A A A	C A A A A A	C C A A A A
	5556	B A B B A	B A B B A			B A B B A	B A B B A	B A B B A	B A B B A	A A B B A
	5654									B C A B B
514.0, A514.0, B514.0, F514.0, 5154, 5254	4043	A B C C	A B C C		A A A A	A B C C	A D C C	A D C C	A B C C	A D C C
	5183	B A B B A	B A B B A			B A B B A	B A B B A	B A B B A	B A B B A	A A B B B
	5356	B A A B A	B A A B A			B A A B A	B B A B A	B B A B A	B A A B A	A B A B A
	5554	C A A A A	C A A A A			C A A A A	C C A A A	C C A A A	C A A A A	C C A A B
	5556	B A B B A	B A B B A			B A B B A	B A B B A	B A B B A	B A B B A	A A B B B
	5654	C A A A B	C A A A B			C A A A B	C C A A B	C C A A B	C A A A B	B C A A A
5086, 5356	4043	A B C B	A B C B		A A A A	A B C B	A C C B	A C C B	A B C B	
	5183	A A B A	A A B A			A A B A	A A B A A	A A B A A	A A B A A	A B A A A
	5356	A A A A A	A A A A A			A A A A A	A B A A A	A B A A A	A A A A A	C C A A A
	5554									
	5556	A A B A A	A A B A A			A A B A A	A A B A A	A A B A A	A A B A A	A A B A A
	5654									B C A A B
5083, 5456	4043	A B C B	A B C B		A A A A	A B C B	A C C B	A C C B	A B C B	
	5183	A A B A A	A A B A A			A A B A A	A A B A A	A A B A A	A A B A A	A A B A A
	5356	A A A A A	A A A A A			A A A A A	A B A A A	A B A A A	A A A A A	A B A A A
	5554									C C A A A
	5556	A A B A A	A A B A A			A A B A A	A A B A A	A A B A A	A A B A A	A A B A A
	5654									B C A A B
5052, 5662	4043	A B C A A	A B C A A	A A A A A	A A A A A	A B C A A	A B C A A	A C C A A	A B C A A	A D C B A
	5183	B A B A	B A B A			B A B A	B A B A	B A B A	B A B A	A A B C B
	5356	B A A A	B A A A			B A A A	B A A A	B B A A	B A A A	A B A C A
	5554									C C A A A B
	5556	B A B A	B A B A			B A B A	B A B A	B A B A	B A B A	A A B C B
	5654									B C A B A
5005, 5050	1100	C B A A A A	C B A A A A			C C A A A A				[1100]
	4043	A A C A A	A A C A A	B B A A A	B B A A A	A B C A A	A B C A A	A B C A A	A B D A A	[4043]
	4145	B A D B A	B A D B A	A A B A A	A A B A A	B B D B A				[4145]
	5183	C A B B	C A B B			C A B C B	B A B A	B A B B A	B A C A	[5183]
	5356	C A B B	C A B B			C A B C B	B A A A	B A A B A	B A B A	[5356]
	5556	C A B B	C A B B			C A B C B	B A B A	B A B B A	B A C B	[5556]
ALCLAD 3004	1100	D B A A A	D B A A A			C C A A A			[1100]	
	4043	A A C A A	A A C A A	B B A A A	B B A A A	A B C A A	A D D A A	A D D A A	[4043]	
	4145	B A D B A	B A D B A	A A D A A	A A D A A	B B D B A			[4145]	
	5183	C A B C B	C A B C B			C A B C A	B A C C A	B A C C A	[5183]	
	5356	C A B C B	C A B C B			C A B C A	B B B C A	B B B C A	[5356]	
	5554						C C A B A A	C C A B A A	[5554]	
	5556	C A B C B	C A B C B			C A B C A	B A C C A	B A C C A	[5556]	
3004	1100	D B A A A A	D B A A A A			C C A A A A		[1100]		
	4043	A A C A A	A A C A A	B B A A A	B B A A A	A B C A A	A B D A A	[4043]		
	4145	B A D B A	B A D B A	A A B A A	A A B A A	B B D B A		[4145]		
	5183	C A B B	C A B B			C B C A	B A C C A	[5183]		
	5356	C A B B	C A B B			C A B C A	B B B C A	[5356]		
	5554						C C A B A A	[5554]		
	5556	C A B B	C A B B			C B C A	B A C C A	[5556]		
3003, ALCLAD 3003	1100	B B A A A A	B B A A A A			B B A A A A	[1100]			
	4043	A A B A A	A A B A A	A A B A A	A A B A A	A A B A A	[4043]			
	4145	A A C B A	A A C B A	A A B A A	A A B A A	A A C B A	[4145]			
2219	2319			B A A A A A	A A A A A A	[2319]				
	4043	B A A A A	B A A A A	B C B C A	B C B C A	[4043]				
	4145	A A B A A	A A B A A	A B C B A	A B C B A	[4145]				
2014, 2036	2319			C A A A A A	[2319]					
	4043	B A A A A	B A A A A	B C B C A	[4043]					
	4145	A A B A A	A A B A A	A B C B A	[4145]					
1100	1100	B B A A A A	B B A A A	[1100]						
	4043	A A B A A	A A B A A	[4043]						
	5356			[5356]						
1060, EC	1100	B B A A A B	[1100]							
	1260	C C A A A A	[1260]							
	4043	A A B A A	[4043]							

Guide to Choice of Filler Alloy for General-Purpose Welding — ratings given as W S D C T M for each base-alloy pairing.

Base Alloys	Filler Alloys	5083, 5456 (W S D C T M)	5086, 5356 (W S D C T M)	514.0, A514.0, B514.0, F514.0, 5154, 5254 (W S D C T M)	5454 (W S D C T M)	6005, 6063, 6101, 6151, 6201, 6351, 6951 (W S D C T M)	6061, 6070 (W S D C T M)	7005, 7021, 7039, 7046, 7146, A712.0, C712.0 (W S D C T M)	413.0, 443.0, 444.0, 356.0, A356.0, A357.0, 359.0 (W S D C T M)	319.0, 333.0, 354.0, 355.0, C355.0, 380.0 (W S D C T M)
…9.0, 333.0 …4.0, 355.0 …55.0, 380.0	2319	A A A A A	A A A A A	A A A A A	A A A A A A	B B A A A	B B A A A	B B A A A	B B A A A	B A A A A
	4043					A A B A A	A A B A A	A A B A A	A A B A A	A B B A A
	4145									
…3.0, 443.0 …4.0, 356.0 …6.0, A357.0 359.0	4043	A B B A A	A B B A A	A B B A A	A B B A A A	A B A A A	A B A A A	A B B A A A	A B A A A A	
	4145					A A B B A	A A B B A	A A B B A	A A B B A	
	5356	A A A A A	A A A A A	A A A B A	A A A B A			A A A A B		
…005, 7021 …039, 7046 …46, A712.0 C712.0	4043					A D C B A	A D C B A	B D C B A		
	4145									
	5183	A A B A A	A A B A A	A A B A A	A A B A A	A A B A A	A A B A A	A A B A A		
	5356	A B A A A	A B A A A	A B A A A	A B A A A	A B A A A	A B A A A	A B A A A		
	5554			B C A A A	B C A A A A	B C A A A A	B C A A A A	B C A A A		
	5556	A A B A A	A A B A A	A A B A A	A A B A A	A A B A A	A A B A A	A A B A A		
	5654	B C A A A	B C A A A	B C A A A	B C A A A	B C A A A	B C A A A	B C A A A		
6061 6070	4043	A D C A	A D C A	A D C A	A D C B A	A C B A A	A C B A A			
	5183	A A B A A	A A B A A	B A B C B	B A B C A	B A A C A	B A A C B			
	5356	A B A A A	A B A A A	B B A C A	B B A C A	B A A C A	B A A C A			
	5554	B C A A A	B C A A A	C C A B B	C C A A A A	C B A B B A	C B A B B B			
	5556	A A B A A	A A B A A	B A B C A	B A B C A	B A A C A	B A A C B			
	5654	B C A A B	B C A A B	C C A B B	C C A B B	C B A B B	C B A B B			
6065, 6063, 6101, 6151, 6201, 6351, 6951	4043	A B C A	A B C A	A B C A	A B C B A	A C B A A				
	5183	A A B A A	A A A A A	B A B C A	B A A C A	B A A C A				
	5356	A A A A A	A A A A A	B A A C A	B A A C A	B A A C A				
	5554	B A A A A	B A A A A	C A A B A	C A A A A A	C B A B B A				
	5556	A A B A A	A A B A A	B A B C A	B A B C A	B A A C A				
	5654	B A A A B	B A A A B	C A A B B	C A A B B	C B A B B				
5454	4043									
	5183	A A B B A	A A B B A	A A B B A	A A B B A					
	5356	A B A B A	A B A B A	A B A B A	A B A B A					
	5554	B C A A A	B C A A A	B C A A A	B C A A A A					
	5556	A A B B A	A A B B A	A A B B A	A A B B A					
	5654			B C A A B	B C A B B					
514.0 A514.0 B514.0 F514.0 5154 5254	4043									
	5183	A A A A A	A A A A A	A A B B B						
	5356	A B A A A	A B A A A	A B A B A						
	5554	B C A A A	B C A A A	B C A A A						
	5556	A A B A A	A A B A A	A A B B B						
	5654	B C A A B	B C A A B	B C A A A						
5086 5356	4043									
	5183	A A B A A	A A B A A							
	5356	A B A A A	A B A A A							
	5554									
	5556	A A B A A	A A B A A							
	5654									
5083 5456	4043									
	5183	A A B A A								
	5356	A A A A								
	5554									
	5556	A A B A A								
	5654									
5052 5662	4043									
	5183									
	5356									
	5554									
	5556									
	5654									

Notes

1. A, B, C, and D are relative ratings in decreasing order of merit. The ratings have relative meaning only within a given block.

2. Combinations having no rating are not usually recommended.

3. Ratings do not cover these alloys when heat treated after welding.

Source: Aluminum Co. of America (February 1981).

How To Use the Chart

First select the base alloys to be joined, one from the column at far left and the other from the row running along the top of the chart. Then, from the base alloy in the left column, move to the right until you reach the block directly under the base alloy from the top row. Or, from the base alloy in the top row, move down until you reach the block directly across from the base alloy in the left column.

This intersecting block contains horizontal lines of letters (A through D) which represent the filler alloys directly across from them in the color tinted box at the ends of each row. The letters in each line give the A to D rating of the characteristics listed at the top of each column—W, S, D, C, T, and M (see Legend for explanation).

You will find that by choosing the different filler alloys in each block, you can vary the characteristics of the weld; that is, "trade-off" one characteristic for another until you find the filler that best meets your needs.

Example: When joining base alloys 3003 and 1100, find the intersecting block. Now, note that filler alloy 1100 provides excellent ductility (D), corrosion resistance (C), performance at elevated temperatures (T), and color match after anodizing (M), with good ease of welding (W) and strength (S). However, if ease of welding and shear strength are utmost in importance, and ductility and color match can be sacrificed slightly, filler alloy 4043 can be used advantageously.

Legend

Filler alloys are rated on the following characteristics:

Symbol	Characteristic
W	Ease of welding (relative freedom from weld cracking).
S	Strength of welded joint (as-welded condition). Rating applies particularly to fillet welds. All rods and electrodes rated will develop presently specified minimum strengths for butt welds.
D	Ductility. Rating is based upon free bend elongation of the weld.
C	Corrosion resistance in continuous or alternate immersion in fresh or salt water.
T	Recommended for service at sustained temperatures above 150°F (65°C).
M	Color match after anodizing.

thick. Above that thickness, argon-helium mixtures usually are used because the helium gives greater depth of penetration.

Nickel and its alloys Argon is generally preferable to other shielding gases for welding nickel and most nickel alloys up to ⅜ in. [9.5 mm] thick. Above that thickness, argon-helium mixtures usually are desirable to get maximum base-metal penetration.

Magnesium and its alloys The arc characteristics of helium and argon are somewhat different in welding magnesium than they are with other metals. The burn-off rates of magnesium wires are equal when shielded either by argon or helium, but the depth of penetration is greater with argon-helium mixtures. Argon is recommended for most work on magnesium and its alloys because of the excellent surface-cleaning action provided by this gas. It breaks up the magnesium oxide coating that forms on this material, just as the argon gas cleans the surface of aluminum and titanium. The argon to helium mixtures might be preferred in multipass welding where the rounded type of penetration pattern produced by this mixture is more desirable.

Refractory metals Titanium and zirconium, ultrasensitive to hydrogen, oxygen, and all other contaminants, weld best in a high-purity argon arc. These materials require a second, special trailing shield to protect the weld metal as it cools. All difficulties considered, it's sometimes easiest to weld these materials in an enclosed glove box or chamber flooded with inert gas. Argon breaks up the surface oxides that form on all refractory metals, and its low flow rate makes for a quiet weld puddle.

17-14 GMAW TECHNIQUES

Up to now we have deluged you with details on why GMAW works. You may think at this point that the process is terribly complicated to operate. Just the reverse is, in fact, the truth. GMAW is very easy to learn. With the background you now have, you will quickly pick up the techniques you need.

Gun Position

Your manual GMAW gun can be operated with the forehand or backhand technique just like gas welding. You also can operate the gun pointing it vertically toward the base metal (Fig. 17-32). The backhand method means that the welding gun is pointed so that the wire feeds opposite to the direction of arc travel. The filler metal is fed into the weld metal previously deposited. For the forehand method, the gun is angled so that the electrode wire is fed in the same direction as your arc is traveling. Now the filler metal is being deposited, for the most part, directly onto the workpiece. You also can point your welding gun straight down into the workpiece and move it along the weld seam. Most welders prefer the backhand technique because it yields a more stable arc and less spatter.

For maximum base-metal penetration, hold the gun at an angle of 25° to the work and use the backhand welding technique. The forehand welding technique has one important advantage over backhand welding. The weld penetration into the base metal is less. That's what you want when you are welding thin materials and need low penetration to prevent burnthrough as well as to reduce welding distortion. You can do GMAW spot welding simply by holding the gun perpendicular to the workpiece and briefly making a small spot weld to join two thin sheets.

When welding in the vertical position, you can either operate your gun by welding up the seam, or by welding down. Most people find it easier to weld vertical-up because the previously deposited weld metal acts as a dam, supporting the new weld metal as you work up the weld seam.

Even though the position of your welding gun has some effect on the depth of penetration into the base metal, welding current

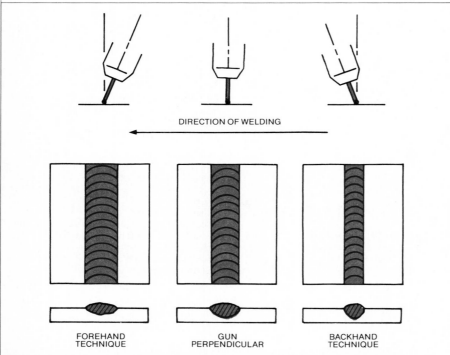

DIRECTION OF WELDING

FOREHAND TECHNIQUE

GUN PERPENDICULAR

BACKHAND TECHNIQUE

FIGURE 17-32 The effect of electrode position and welding techniques using the forehand, vertical, and backhand techniques with the GMAW process.

has the greatest effect. Weld penetration is closely related to your current setting. An increase in welding current increases penetration. Decreasing the current decreases weld penetration. However, that welding current can be varied without changing the wire feed speed. All you have to do is change the tip-to-work distance. The effect of your tip-to-work distance is the opposite of changing the welding current. Increasing your spacing reduces penetration. Decreasing the distance increases penetration. Experienced welders change the tip-to-work distance to prevent burnthrough when there are changes in thickness (from a thick member to a thin one) almost without thinking about it.

Using welding voltage to adjust weld-metal penetration is not good practice. There generally is an optimum voltage setting for each welding wire you will use. Moving to either side of that recommended voltage will generally decrease penetration. But it also affects your arc stability.

Travel Speed

The effects of travel speed are similar to that of changing welding voltage. Base-metal penetration is maximum at a certain value and decreases as the arc speed is varied. When your travel speed is too low, too much weld metal will be deposited. At high speeds, the heat generated by the arc hasn't sufficient time to substantially melt the base metal.

Different wire diameters (and different wire compositions) have different optimum welding characteristics including machine settings and travel speed. The supplier's literature usually gives these details, so you don't have to worry about setting up these variables. Simply follow recommended directions, which often are printed right on the filler-metal package label.

Deposition Rates

The deposition rate describes how much useable weld metal you can deposit in one hour of actual arc-on welding time. Because the various GMAW welding processes are very efficient, only a very small amount of weld metal is lost as spatter. The deposition rate for any wire is calculated by the equation:

$$\text{Deposition rate} = \frac{\text{wire feed speed (in./min, or m/min)} \times 60 \text{ min/hr}}{\text{inches of wire per lb (or m/kg)}}$$

The deposition rate for a given wire diameter and wire feed rate often is supplied directly by the filler-metal manufacturer.

Increased deposition rate goes hand-in-hand with higher wire feed speeds. You can increase your wire feed speed by increasing your tip-to-work distance slightly, which increases the CV welding current and wire feed speed. Increasing the welding current setting on your power source also increases the deposition rate (but stay within recommended ranges).

Long tip-to-work distances and high wire feed speeds are used for high-speed welding of thin materials because the welding current can be kept low to avoid overheating and burnthrough or base-metal distortion. As you already know, the forehand welding technique is used with thin materials. Increasing the deposition rate in this way will also have a slight effect on weld penetration. Because more weld metal is being deposited at a given welding current, the penetration will be reduced. This results from a cushioning of the arc force by the extra weld metal deposited.

Bead Shape

The two most important weld-bead characteristics for you to control are the bead height and bead width. These are important to assure that that the weld joint is properly filled with a minimum of welding defects, especially in multipass weldments. In this case, if bead height is too great, it becomes very difficult to make subsequent weld passes that will have good fusion.

You can alter both the weld bead size and shape. To change the size of your weld bead, the weight of weld metal per linear foot (or meter) of the weldment must be changed. Welding current and travel speed are the welding parameters primarily used to control the size of your weld bead. For instance, when the current is decreased, the weld bead will become smaller. When the current is increased, the weld bead will become larger.

The weld bead size can also be changed by varying the arc travel speed. The two are inversely related. A high speed means a smaller bead. A low travel speed means a bigger, wider bead.

Both the welding current and travel speed have little effect on the shape of the weld bead. The bead width and height increase or decrease together.

Arc voltage (arc length) is used to control the shape of your weld bead. As the arc voltage increases, the bead height decreases and bead width increases. The weld flattens out. The total size of the weld bead (the amount of weld metal in it) remains the same. Only the shape and contour of the bead surface is changed. By increasing the bead width, the bead height becomes taller and the weld metal wets the base metal more efficiently. Fusion to the workpiece is improved.

Wire stick-out and the welding technique you use (backhand or forehand welding) also affect these characteristics, but only to a limited extent. When long stick-outs are used to increase deposition rates, the bead height will increase to a greater extent than the bead width. Although larger, the weld bead comes more peaked. The backhand welding technique will also produce high, narrow weld beads. Decrease the lagging gun angle to decrease the bead height and increase the width. The forehand technique yields the flattest, widest weld beads.

WELDING VARIABLES TO CHANGE	DESIRED CHANGES: PENETRATION ▲	PENETRATION ▼	DEPOSITION RATE ▲	DEPOSITION RATE ▼	BEAD SIZE ▲	BEAD SIZE ▼	BEAD WIDTH ▲	BEAD WIDTH ▼
CURRENT AND WIRE FEED SPEED	↑	↓	↑	↓	↑	↓	*	*
VOLTAGE	+	+	*	*	*	*	↓	↑
TRAVEL SPEED	+	+	*	*	↓	↑	↑	↓
STICKOUT	↓	↑	↑	↓	↑	↓	↑	↓
WIRE DIAMETER	↓	↑	↓	↑	*	*	*	*
SHIELD GAS % CO_2	↑	↓	*	*	*	+	↑	↓
GUN ANGLE	BACK-HAND TO 25°	FORE-HAND	*	*	*	*	BACK-HAND	FORE-HAND

* NO EFFECT
+ LITTLE EFFECT
↑ INCREASE
↓ DECREASE

FIGURE 17-33 The effects of all GMAW parameters and techniques are summarized in this one chart. (Courtesy Linde Division of Union Carbide.)

We have given you a great deal to think about. Now let's put it all in one summary figure so you can easily refer back to it when you need it. Figure 17-33 wraps up all these GMAW welding adjustments, and their effects, in one place.

Welding Gun Motions

The biggest help we can give to get you started with the following shop work is to show you some methods for handling your welding gun in different positions. As you gain experience, these gun motions will be second nature, and you will use whatever technique is required for a given joint without thinking about it. With enough practice in your school's welding shop, GMAW welding motions will be as automatic to you as driving a car or riding a bike.

Flat position The recommended weaving patterns, gun positions, and bead sequence for flat-position welding are shown in Fig. 17-34. A slight back-stepping motion is used for single-pass butt joints. Gapped root passes are made with a small back-and-forth weave pattern. For fill-and-cover passes, the same weave, with an adjustment for the desired bead width, is used. Take care to pause at the sidewalls to obtain adequate tie-in to the base metal.

Horizontal position Recommended weaving patterns, gun positions, and bead sequences for horizontal weld joints are shown in Fig. 17-35. A circular motion is recommended for fillet welds. For butt-weld root passes and fill passes, an in-line, back-and-forth motion is used with bead width adjustments as required. A slight pause is used at the tie-in to the previous weld bead.

Vertical and overhead position Recommended weaving patterns and gun positions for vertical-up and vertical-down welding are shown in Fig. 17-36. With vertical-up, for a square-edge joint, an in-line, back-and-forth weave is used. For a beveled mutlipass joint a U pattern is used for the root. The fill-and-cover passes are made using a side-by-side weave with a backstep at the walls. The length of the backstep is on the order of a wire diameter. For a vertical-up or overhead fillet a "Christmas tree" pattern is used with pauses at the sidewalls (Fig. 17-37).

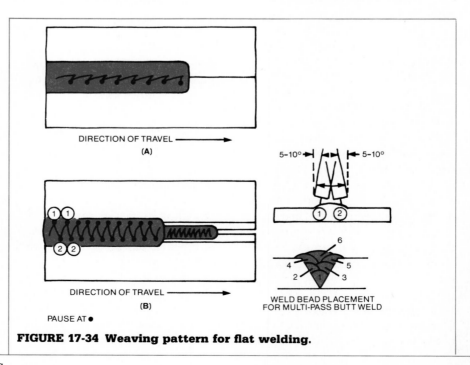

FIGURE 17-34 Weaving pattern for flat welding.

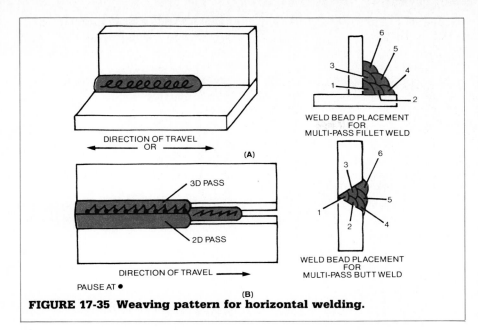

WELD BEAD PLACEMENT
FOR
MULTI-PASS FILLET WELD

WELD BEAD PLACEMENT
FOR
MULTI-PASS BUTT WELD

DIRECTION OF TRAVEL
OR

3D PASS

2D PASS

DIRECTION OF TRAVEL

PAUSE AT ●

(A)

(B)

FIGURE 17-35 Weaving pattern for horizontal welding.

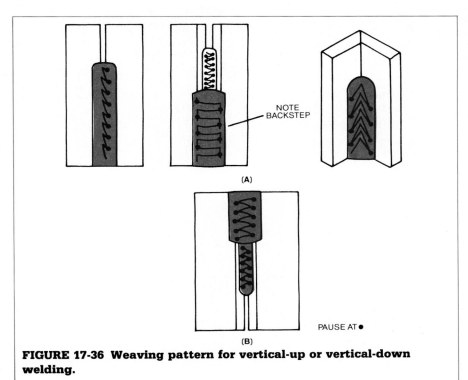

NOTE
BACKSTEP

(A)

PAUSE AT ●

(B)

FIGURE 17-36 Weaving pattern for vertical-up or vertical-down welding.

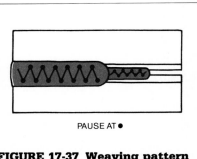

PAUSE AT ●

FIGURE 17-37 Weaving pattern for vertical-up or overhead welding.

REVIEW QUESTIONS

1. What percent, on the average, does the welding labor cost contribute to the total cost of welding a product?

2. What does the American Welding Society process abbreviation, GMAW, mean?

3. What does FCAW mean?

4. What happens to weld metal in an argon-shielded GMAW arc when the dc welding current is increased above the critical transition current? That is, how is the molten metal transferred to the workpiece?

5. Name four process variables that affect the transition current of a given type of steel welding wire.

6. What is the minimum electrode stick-out distance beyond the nozzle for GMAW welding wire?

7. If the stickout distance is increased, does the transition current from globular to spray transfer decrease, or increase?

8. Give a brief description of the difference between conventional spray transfer and pulsed-spray transfer GMAW.

9. Name three reasons why pulsed-spray GMAW might be used instead of conventional spray-transfer GMAW.

10. Name one reason why conventional spray-transfer GMAW equipment would be used instead of pulsed-spray equipment.

11. What type of material (thickness) is dip-transfer (or short-circuiting transfer) GMAW welding generally used for, and what common welding problem does the dip transfer process control?

12. What shielding gas is used for buried-arc welding?

13. Why is the buried-arc process used? Would buried-arc GMAW most likely be used for maintenance welding, or high-volume, high-deposition-rate production work?

14. How does the welding wire of the FCAW process differ from that of other GMAW welding methods?

15. Name several advantages of the FCAW process over other kinds of GMAW work.

16. What kind of power source is used most commonly for solid-wire GMAW welding? Is it ac, or dc? And if dc, is it a constant-voltage or a constant-current machine?

17. Do you think you understand slope control? If you don't, reread this chapter or better, discuss it with your welding instructor.

18. What are the three basic types of wire feeders as defined by the way they feed the welding wire?
19. What is a contact tube and why do you have to keep it clean?
20. What is the difference between cast and helix, why does it matter, and how can you measure them?
21. Describe a simple way to test the surface cleanliness of solid GMAW welding wire using nothing more than a white cloth.
22. What is ER70S-1?
23. What is the solid welding wire most commonly used for joining carbon and high-strength low-alloy steels?
24. Name an AWS solid-wire electrode grade that will give you steel weld metal with a minimum tensile strength of at least 80,000 psi [552 MPa].
25. What type of weld metal would an E308T electrode produce, and in what welding process would you use it?
26. What AWS filler-metal specification would you use to select solid GMAW welding wires for mild and high-strength low-alloy steels?
27. What AWS filler-metal specification would you use to select a flux-cored welding wire for joining stainless steel?
28. What shielding gas is often added to carbon dioxide to reduce weld spatter?
29. Why is the dew point of a shielding gas important? What substance in the gas does it measure? Should the dew point be high or low?
30. What equipment do you need to mix your own shielding gases?

Becoming Skilled at GMAW

Start your shop practice with downhand welding. Don't try welding out-of-position until you are making good-quality downhand welds. Work first with carbon steels before you try welding aluminum and stainless steel. Start your work with MIG, then move to one or more of the other GMAW processes. Reserve pipe and tube welding in different positions for last. As usual, it's the toughest work to do.

Your welding school may not have all the possible types of GMAW welding power sources we have talked about. That's not terribly important. Most welding schools use equipment and GMAW processes similar to the companies in your area that will hire you. What's more, once you have learned basic MIG welding, the other GMAW processes will be easy to pick up during on-the-job training.

While you are learning on the job, you also will have to learn how to weld specific metals. Even different grades of steel (low-carbon steel versus high-strength low-alloy steels versus quenched and tempered full-alloy steels, or different grades of aluminum or stainless steel) have somewhat different welding properties and require different techniques.

You can look up the best methods for welding each type of material in a standard welding handbook . . . after you have learned the basics. Even an experienced welding engineer would have to do that.

It's far more important that you can do a good job MIG welding downhand on carbon steel than welding fancy stainless steel or aluminum grades out-of-position, poorly. And the only way you'll ever learn to be a good GMAW welder is to practice, practice, practice.

17-1-S
MIG WELDING STEEL
This shop work will help you learn how to make fusion welds on low-carbon steel with the GMAW process.

Materials
You will need $\frac{1}{16}$-in.- and $\frac{1}{8}$-in.-thick sheet steel for this work. You also will need $\frac{1}{4}$-, $\frac{3}{16}$-, and $\frac{3}{4}$-in.-thick low-carbon steel plate.

Small 2- × 4-in. coupons are OK to use when you are practicing this work, although your final welder qualification tests are best conducted on 6- × 6-in. coupons.

You will have to select the correct filler wire and wire diameter for each piece of steel. One steel welding wire grade and diameter will probably be suitable for almost all of the work. Make sure that your contact tip and wire liner is clean, and that your shielding gas is properly selected and adjusted for the material and GMAW process you will use. If you think that the material thickness and joint you are making will require an edge bevel, it's your responsibility to determine the bevel angle for your joint, and make it. Then complete your weld.

17-2-S
MAKING GMAW BEAD-ON-PLATE WELDS
1. Make bead welds on the $\frac{1}{16}$-in.-thick steel in flat position.
2. Make bead welds on the $\frac{1}{8}$-in.-thick steel in horizontal position.
3. Make bead welds on the $\frac{1}{16}$-in. steel in vertical position.

4. Make bead welds on the ⅛-in. steel in overhead position.

Bead-Weld Test

There will be no standard physical test for a bead weld. The only test is the appearance of the weld bead that you produce. The best weld bead will have a uniform width and an even, smoothly rippled surface with no pinholes, cracks, or gaps where you missed making the weld, and with no holes burned through the material. The more you practice making good-looking bead-on-sheet or bead-on-plate welds, the easier it will be to make the more difficult welds you will do next.

17-3-S
MAKING CORNER WELDS

5. Make a weld on a corner joint on two pieces of 1⁄16-in.-thick steel sheet.

6. Make a weld on a corner joint on two pieces of ⅛-in.-thick steel.

Corner Weld Test

7. Subject your completed corner welds to a hammer test. If you produced a satisfactory corner weld joint, you will be able to bend and then hammer the joint flat without cracking or breaking your weld. (Cracks in your base metal are acceptable.)

17-4-S
MAKING SQUARE-GROOVE BUTT JOINTS

8. Make a weld on a square-groove butt joint on 1⁄16-in.-thick steel in the flat position.

9. Make a weld on a square-groove butt joint on ⅛-in.-thick steel in the flat position.

10. Make a weld on a square-groove butt joint on 1⁄16-in.-thick steel in the vertical position.

11. Make a weld on a square-groove butt joint on ⅛-in.-thick steel in the vertical position.

12. Make a weld on a square-groove butt joint on 1⁄16-in.-thick steel in the overhead position.

13. Make a weld on a square-groove butt joint on ⅛-in.-thick steel in the overhead position.

14. Make a weld on a butt joint on 3⁄16-in. steel plate in the flat position.

15. Weld a butt joint on ¼-in. steel plate in the flat position.

16. Weld a butt joint on 3⁄16-in. steel plate in the vertical position.

17. Weld a butt joint on ¼-in. steel plate in the vertical position.

18. Weld a butt joint on 3⁄16-in. steel plate in the overhead position.

19. Weld a butt joint on ¼-in. steel plate in the overhead position.

Steel Butt-Joint Test

All butt welds should be tested by a guided-root-bend test.

17-5-S
MAKING FILLET AND LAP JOINTS

20. Make a fillet-welded lap joint on 1⁄16-in.-thick steel in the flat position.

21. Make a fillet-welded lap joint on ⅛-in.-thick steel in the flat position.

22. Make a fillet-welded lap joint on 1⁄16-in.-thick steel in the vertical position.

23. Make a fillet-welded lap joint on ⅛-in.-thick steel in the vertical position.

24. Make a fillet-welded lap joint on 1⁄16-in.-thick steel in the overhead position.

25. Make a fillet-welded lap joint on ⅛-in.-thick steel in the overhead position.

26. Make a fillet-welded T joint on 1⁄16-in.-thick steel in the flat position.

27. Make a fillet-welded T joint on ⅛-in.-thick steel in the flat position.

28. Make a fillet-welded T joint on 1⁄16-in.-thick steel in the vertical position.

29. Make a fillet-welded T joint on ⅛-in.-thick steel in the vertical position.

30. Make a fillet-welded T joint on 1⁄16-in.-thick steel in the overhead position.

31. Make a fillet-welded T joint on ⅛-in.-thick steel sheet in the overhead position.

32. Make a fillet-welded lap joint on 3⁄16-in.-thick steel plate in the flat position.

33. Make a fillet weld on a lap joint on ¼-in.-thick plate in the flat position.

34. Make a fillet weld on a lap joint on 3⁄16-in.-thick steel plate in the vertical position.

35. Make a fillet weld on a lap joint on ¼-in.-thick steel plate in the vertical position.

36. Make a fillet weld on a lap joint on 3⁄16-in.-thick steel plate in the overhead position.

37. Make a fillet weld on a lap joint on ¼-in.-thick steel plate in the overhead position.

38. Make a fillet weld on a T joint on 3⁄16-in.-thick steel plate in the flat position.

39. Make a fillet weld on a T joint on ¼-in.-thick steel plate in the flat position.

40. Make a fillet weld on a T joint on 3⁄16-in.-thick steel plate in the vertical position.

41. Make a fillet weld on a T joint on ¼-in.-thick steel plate in the vertical position.

42. Make a fillet weld on a T joint on 3⁄16-in.-thick steel plate in the overhead position.

43. Make a fillet weld on a T joint on ¼-in.-thick steel plate in the overhead position.

Testing Fillet and Lap Joints

44. All of your lap joints are to be subjected to a hammer test. Fold the joint 180° along the weld to make a flat sandwich with the welded joint running along one edge. Then pound the welded joint flat without any producing any weld-metal cracking.

45. All of your fillet joints are to be subjected to a hammer test for T joints. Fold the 90° leg of the T joint flat to one side of the joint. Hammer the joint without producing any weld-metal cracking.

17-6-S
FLUX-CORED WIRE WELDING

First, select a cored-wire grade and diameter suitable for your steel and for the welding position you will work in. Next, select an appropriate shielding gas for the cored wire you will use. After you have done the following, test the joint without producing any weld-metal cracking.

46. Make a weld on a square-groove butt joint on ⅛-in.-thick steel in the flat position.

47. Make a weld on a square-groove butt joint on ⅛-in.-thick steel in the vertical position.

48. Make a weld on a square-groove butt joint on ⅛-in.-thick steel in the overhead position.

49. Weld a butt joint on ¼-in.-thick steel plate in the flat position.

50. Weld a butt joint on ¼-in.-thick steel plate in the vertical position.

51. Weld a butt joint on ¼-in.-thick steel plate in the overhead position.

52. Make a fillet weld on a lap joint on ¼-in.-thick steel plate in the flat position.

53. Make a fillet weld on a lap joint on ¼-in.-thick steel plate in the vertical position.

54. Make a fillet weld on a lap joint on ¼-in.-thick steel plate in the overhead position.

55. Make a fillet weld on a T joint on ¼-in.-thick steel plate in the flat position.

56. Make a fillet weld on a T joint on ¼-in.-thick steel plate in the vertical position.

57. Make a fillet weld on a T joint on ¼-in.-thick steel plate in the overhead position.

17-7-S
WELDING STEEL PIPE

This work is optional, but very important if you can possibly get it done. Using 4-in. steel pipe, work in as many different pipe welding positions as you can or have equipment to operate in. Use MIG welding. Save the best sam-

ples for your pipe welder operator-qualification test (Test 3), which follows.

58. Pipe welding is very difficult but well-paying work. Prepare a beveled pipe joint from two pieces of steel pipe about 4 in. in diameter. Bevel the outer edges to 45° before you start. Then tack-weld the two sections into position with four tacks spaced equally around the joint (at points about 0, 90, 180, and 270° around the pipe). Then weld the pipe sections together.

Keep the pipe centerline in the horizontal position while you work. It will help you to rest the pipe in a piece of channel iron or two angle irons while you work. That way you can roll the pipe when you have to, welding a quarter of the section at a time, while always working in the downhand position. Pay special attention to getting a good solid root pass. Make sure your cover passes are tied into the sides of the joint and the root pass.

59. Repeat the previous exercise with the pipe centerline in the vertical position (the pipe is standing upright) after you tack-weld it in-line. This time welding the pipe joint is like making a circular weld in the horizontal position. It's tough to do.

Testing Pipe Welds

Subject a strip cut from your pipe welds to guided-root-bend and guided-face-bend tests.

17-8-S
OPERATOR-QUALIFICATION TESTS

The purpose of this shop practice has been to develop your ability to make butt and fillet welds under conditions approximating those you will find in industry.

It's time to take some operator-qualification tests very similar to those you will have to take to get a job. In these tests, you will select your own equipment, filler wire, wire diameter, and shielding

gas. You also will conduct your own tests on your own material. In a real operator qualification test, you would make the welds, but someone else would test your work. The objective of these tests is to show you where you are having problems. If you have problems, solve them with extra practice. Try to qualify first with the MIG process. Then, if you have time, repeat the qualification test (on steel) with the flux-cored wire process.

If you don't pass one of these tests, it shows that you need more work. For example, you may be able to make very good welds in the flat and horizontal position, but you may be having trouble making sound welds in the vertical and overhead positions. Or maybe you can make good welds in any position on steel, but aluminum is giving you trouble. Or you may not have any difficulty with wrought materials, but cast iron is giving you a hard time. These tests will show you where you need extra shop practice.

Test 1. Butt Welds

These tests will require six plates of metal (low-carbon steel, copper, or aluminum). You will have to decide whether or not the material thickness requires edge beveling, and what angle to bevel the material. You also will do the required edge preparation before you start to weld.

The first MIG welding test consists of making:

■ Single-V-groove butt joints in the flat position.
■ Single-V-groove butt joints in the horizontal position.
■ Single-V-groove butt joints in the vertical position.
■ Single-V-groove butt joints in the overhead position.

Each of the welded plates you make will be cut into four strips as shown in Fig. 9S-1, Shop Practice section, Chap. 9. The two specimens that are retained are machined flush all over as shown in Fig. 9S-2, Shop Practice section, Chap. 9. One test specimen from each of the three welding positions (flat, vertical, and overhead) will be submitted to a guided-face-bend test, and the other center strip will be submitted to a guided-root-bend test. If any defects greater than ⅛ in. in any dimension are developed in these tests, your weld will be considered as failing. In addition to the guided-face-bend and guided-root-bend test, your weld will be examined for appearance. Sloppy-looking welds, whether they pass the mechanical tests or not, will be rejected.

Test 2. Fillet and Lap Welds

These tests require six plates, each of which is 6- × 6- × ¼-in. thick. One edge of each of the plates will be ground square. Also, one backup strip 6 × 1¼ × ¼ in. will be required for each joint.

Set up the test plates for a double-fillet-weld T joint and make:

■ One set of fillet welds in the flat position.
■ One set of fillet welds in the horizontal position.
■ One set of fillet welds in the vertical position.
■ One set of fillet welds in the overhead position.

After the fillet welds have cooled, cut off the leg of the T flush to the base and then make a third weld in the area marked "A" in Fig. 9S-3, Shop Practice section, Chap. 9, in the position where the leg of the T joint previously stood. This third weld may be done in any position you want, regardless of the position in which the fillet welds were made. However, you must use the 1¼-in.-wide × ¼-in.-thick backup strip behind the final weld in area A when you make it.

When welding is completed, the backup strip shown on the bottom of Fig. 9S-3, Shop Practice section, Chap. 9, should be removed flush with the underside of the plates. For best results, this backup strip should be machined off. However, if a suitable grinding machine or other machine tool is not available, this strip can be removed by a hand-cutting torch. You must in this case be very careful not to cut into the root of the weld. That would tend to fuse in or cover up any root defects and would not permit a true or fair test of your ability.

The plates are then cut into four strips as shown in Fig. 9S-4, Shop Practice section, Chap. 9. The two specimens retained are machined flush all over as shown in Fig. 9S-2, Shop Practice section, Chap. 9. One specimen from each of the three welding positions is submitted to the guided-face-bend test and the others to the guided-root-bend test.

If any defects greater than ⅛ in. in any dimension are developed in this test, the weld will be considered as failing.

Test 3. Pipe Welding

Try this test with the MIG process. Make several samples by welding two pieces of 4-in. standard steel pipe approximately 6 in. long. The two pieces of pipe will be placed in a horizontal line and position-welded as a butt joint. This is a horizontal fixed-position pipe weld.

A minimum of four specimens should be taken from the completed weld, spaced at 90° intervals around the weld. These should be taken from the top, bottom, and opposite sides of the completed pipe weld. The top and one side specimen should be submitted to the guided-face-bend test, and the bottom and other side specimen should be subjected to the root-bend test.

All of the above specimens should bend to 180° or the full capacity of the guided bend test jig without any cracks greater than ⅛ in. in any dimension, either as a result of bending or existing prior to welding.

If you pass this pipe welding test, we congratulate you. If you don't pass, work on your MIG pipe-welding technique.

Now repeat exercises 1 through 45 using aluminum instead of steel.

CHAPTER 18

Sub-merged-Arc Welding

Now that you are an expert at gas-shielded arc welding we're going to take away your gas. Submerged-arc welding (SAW is the proper AWS abbreviation—"sub-arc" is a common nickname in welding shops) uses an electric current passing through a continuous filler-metal wire to produce an arc. The arc melts the wire and the base metal, just like in GMAW welding, but no shielding gas is used. The tip of the welding wire, the arc, and the welding joint in the workpiece are covered by a layer of granulated minerals known as SAW flux.

The flux protects the arc, the molten filler metal, and the hot base metal from oxygen, nitrogen, and water vapor in the air. There is no visible arc, no sparks, no spatter, and no smoke. Here's an overview of why this process is used. The drawing in Fig. 18-1 will give you a better feeling for the overall process as we explain it.

In addition to learning about SAW in this chapter, you'll also learn more about joint designs and edge preparation for very thick plates. You'll learn, once again, that plate edge preparation is as important as the welding process. In relation to the cost of heavy steel plates, the time and expense required to prepare them for welding is very small.

18-1
THE SAW PROCESS

Very high welding speeds and weld-metal deposition rates are characteristic of SAW. A ¼-in. [6.4-mm] steel wire operated at 1400 A can deposit around 45 lb [20.4 kg] of weld metal per hour. Welding currents up to 4000 A, ac or dc, on a single welding wire have been used (although 4000 A is an extreme case). Welding speeds up to 200 in./min [5 m/min] are possible using high-amperage single-wire welding. With more than one electrode wire in the same weld joint, even higher speeds and very high deposition rates are possible.

Submerged-arc welding is used mostly on thick steel plates and heavy structural sections. One-pass welding of plate or structurals up to 3 in. [76 mm] thick is possible. Multipass welding of sections of any thickness can be done. The practical minimum thickness for SAW is 18-gauge sheet metal.

You can use a low-carbon-steel filler-metal wire (for example) and produce high-strength low-alloy (HSLA) steel weld metal with the sub-arc process. Or you can deposit hardfacing layers by selecting a flux that adds the alloying elements you want to the melting low-carbon electrode wire to produce one of many possible as-deposited weld-metal chemistries. The flux sprinkles on the workpiece ahead of the welding wire and the electrode filler wire plows through the flux, making the weld.

Submerged-arc welding is fast and welding costs are low. So why isn't this process used everywhere and on all kinds of products? The reason is that SAW is limited to carbon, HSLA, and special alloy steels for pressure vessels, although some hardfacing flux and wires also are available. The very large weld puddle produced at the high currents that are used in SAW tends to limit the process to heavy steel plate.

You are also limited to flat-position welding when using SAW. If the workpiece slopes too much, the flux will fall off (although there are a few limited ways to get around that problem, mostly by using welding positioners or special weld-backing devices). The process also uses high welding currents that produce great heat input into the weld zone, which produces a large pool of liquid metal. Not only will the flux fall off if you tried to work out-of-position, but the puddle would spill out of the weld and run down your workpiece. Unlike inert-gas-shielded welding, you have to clean slag and excess flux from your finished weld.

Submerged-arc welding is limited to work that slants no more than 15° from the horizontal. It is possible to cheat a little and make individual passes on smaller joints in the vertical or nearly vertical position, but that requires support for the flux pouring out ahead of the welding arc and some method to dam up the large volume of filler metal until it solidifies. So for all practical purposes, SAW is strictly a downhand, flat-position, high-production welding process.

Because of its high deposition speed, sub-arc is mostly used for automatic welding. A single welder can't run fast enough to keep up with the process without using special semiautomatic or automatic equipment. Some manual equipment is used. The welding gun looks like a GMAW gun

except for the flux hopper. We expect to see more manual SAW done in the future because of the low cost of the process and the fact that flash, smoke, and spatter are not produced. In days when everything is becoming automatic, it will be a switch to see an automatic process loan its technology to manual operation.

18-2
SAW EQUIPMENT

You will recognize most of the equipment used in SAW welding, like the power source and wire feeder, but this time you will have some new equipment to deal with and some new types of controls and accessories.

Components of the SAW System

An SAW "head," or assembly, containing a hopper, a funnel, and a flux feed tube is used for feeding the granular flux to the joint

ahead of the welding wire. The head has some controls, often a seam follower to guide the welding wire, and various other attachments. A common attachment is a vacuum sweeper that sucks up excess flux that didn't get melted into slag. (Submerged-arc fluxes are expensive because they have a lot of alloying elements in them, and therefore unused flux is recovered and reused.)

Another piece of equipment that is attached to the SAW head is the wire feeder and a wire reel for the filler metal. The welding head and wire feeder with its controls often ride along a track or on a welding manipulator carriage that runs along the workpiece or weld seam. If you are welding a wide joint, the welding head usually will move automatically from side to side to tie the weld into the face of the welding joint. The controls on the welding head often can be set so that the

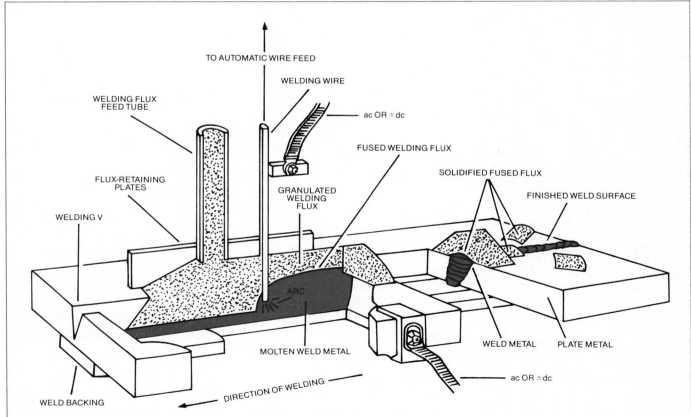

FIGURE 18-1 Cutaway view of submerged-arc welding zone. The Linde Division of Union Carbide, which supplied information for this chapter, is the major U.S. supplier of submerged-arc welding supplies.

wire will "dwell," or stop momentarily, at each side of the joint to ensure that the weld metal is strongly tied into the base metal.

The welding head can be suspended from a boom on the welding manipulator or it can ride directly on tracks laid down on the workpiece. A power source suitable for use with SAW often is mounted on the manipulator along with all the other equipment. The manipulator rides along the workpiece, but circumferential (around the pipe) welds also can be made in pipe, large-diameter tubing, or a pressure vessel by rotating the work on rollers while the sub-arc head stays stationary over the top of the rotating joint. Figure 18-2 shows you a typical welding-head manipulator complete with all its submerged-arc equipment. Figure 18-3 shows you what a portable submerged-arc welding outfit looks like.

Weld backing (discussed in great detail later on in this chapter) often is used if the joint is not structure-backed by its design. The weld backing keeps the flux and molten metal from leaking out of the bottom gap in the weld joint. Since sub-arc often is used to weld thick plates, the joint designs used vary from single-V to double-V or other, wider joint designs. The pieces to be welded are usually placed in position for welding and are tack-welded by some other process such as stick-electrode welding or MIG, before the first SAW pass is made. Thinner workpieces often have to be clamped or held in jigs to maintain the desired gap width and keep the work from distorting due to the large heat input from the high-amperage submerged arc.

Each element of the submerged-arc system has an effect on the finished weld. The values for welding voltage and current, the composition and diameter of the welding wire, wire feeds and speeds, and the flux are usually selected from tables of SAW parameters that you will be given before you start a job. We list some of these data at the end of this chapter.

It's your responsibility to prepare the weld joint and clean it up before welding. You also will set up the equipment, tack weld or jig up the workpiece, adjust the welding head over the work, check the system to make sure everything is ready and operating, and monitor the work as the SAW system produces a weld. You also may have to adjust the welding parameters if the equipment isn't performing as it should.

Other work for you to do is make sure that the flux hopper is full and that the welding wire isn't running low on the feeder reel. After the weld is complete you may have to follow up with a pneumatic chipping tool and maybe an air-driven wire brush to clean the slag off the finished weld. However, you'll find that most submerged-arc fluxes produce very brittle, glassy slags that not only protect the weld metal until it is cool but also detach themselves from the weld by thermal contraction when the slag cools. Deslagging submerged-arc welds usually is not very difficult, and you won't have any weld spatter to take off around the weld joint.

Another point about the process that may surprise you at first is that you don't need a welding helmet and dark-glass filter while you run the equipment . . . as long as the arc is fully buried in the flux. However, you should have a hand-held face mask with a dark lens ready when you need it. You also should wear safety glasses or flash goggles at all times.

Submerged-Arc Fluxes

The fluxes used in SAW are very similar to those used in making steel. The job of the flux is to protect molten weld metal and nearby hot base metal from oxy-

FIGURE 18-2 Large automated welding manipulator used for either high-production MIG, FCAW, or submerged-arc.

FIGURE 18-3 Small submerged-arc unit on wheeled carriage welds floor plate for huge oil-storage tank in Valdez, Alaska.

gen in the air until the metal cools. The flux also lowers the melting point of certain impurities in the metal and floats them out of the weld as slag. Alloying additions also are made with the flux, so that a low-cost carbon-steel wire can produce high-strength low-alloy steel weld metal if used in combination with the correct flux.

An SAW flux is chosen for the type of welding job you have. A certain flux might be best for welding mill-scaled plate or rusty joints. Another flux might be used to control weld-metal reinforcement and penetration through the joint. A third flux might help you bridge joints with poor fit-up.

Still another flux might help give you weld metal with high impact toughness at low temperatures. One or more of the fluxes might give you slag that is almost self-peeling or slag that is very difficult to remove. Some other sub-arc fluxes may be better if you switch from dc to ac welding.

The choice of a flux to cope with factors peculiar to a given welding job is generally made by the welding shop based on prior experience and the guidance and recommendations of the flux supplier. In other words, submerged-arc fluxes are one of those areas in welding where there are a lot of proprietary products. You get a recommendation from a reputable supplier and try it out. If it doesn't do the job, you either try another flux or another supplier.

AWS specifications Does that mean that there is no AWS specification for submerged-arc fluxes? No. There is a general flux specification. It's the same one that covers SAW wires. The name of the American Welding Society (AWS) submerged-arc consumables specification for carbon steels is AWS A5.17, "Bare Mild Steel Electrodes and Fluxes for Submerged-Arc Welding." The AWS spec for low-alloy steels is AWS A5.23, "Low Alloy Steel Electrodes and Fluxes for Submerged-Arc Welding."

AWS A5.17, the carbon-steel specification, classifies SAW fluxes on the basis of the tensile strength of the weld metal they will produce when used with specific SAW wires. AWS A5.23 does something very similar for low-alloy steels. The point is *you have to specify the flux and the filler wire together*.

The flux designations in AWS A5.17 begin with the letter F and have a number, such as 60 or 72, after them. The number refers to the tensile strength of the weld metal the flux will produce. For example, F60 is one AWS grade of SAW flux. Another one is F74. There are 10 AWS flux grades in the AWS A5.17 specification on carbon steels alone, ranging from F60 to F74. We'll list them for you shortly. The flux designations in A5.23 on low-alloy steels work the same way, but there are even more flux–electrode wire combinations.

First we want you to understand that an AWS submerged-arc flux designation never stands alone. It is always followed by the AWS (or a proprietary) submerged-arc filler-wire designation. An example would be AWS F60-EXXXX, where EXXXX is the AWS A-5.17 filler-wire designation.

Without specifying the wire, all the AWS spec tells you is that flux grades from F60 through F64 will produce weld metal with a tensile strength somewhere between 62,000 and 80,000 psi [427 to 551 MPa] with a minimum yield strength of at least 50,000 psi [345 MPa]. It also says that AWS flux grades from F70 through F74 will produce weld metal with a tensile strength somewhere between 72,000 and 95,000 psi [496 to 655 MPa] and a minimum yield strength of at least 60,000 psi [431 MPa]. That's a pretty vague specification.

The specification becomes more meaningful when you fill in the

EXXXXs for the AWS filler metal, which is the main point to learn here. You never select a submerged-arc flux and a wire separately from each other.

Table 18-1 gives you some of the flux data from AWS A5.17.

Note that the mechanical property data are still rather wide. That's partly a reflection of the fact that SAW consumables often are proprietary; data on wire and flux combinations from individual producers will be far more specific. The vagueness in the data also reflects the fact that specific weld-metal mechanical properties are strongly affected by the plate thickness, joint design, and welding parameters. For example, a heavy plate section can cause considerable weld-metal dilution. These metallurgical complications are one reason why we left a lot of data out of Table 18-2 under "Low-Alloy Steels." Explaining the data and its use is simply beyond your present knowledge. It requires substantial metallurgical training. But, as you will soon see, there are practical solutions to the problem.

Submerged-Arc Filler Wires AWS A5.17 also lists designations for submerged-arc filler wires for welding carbon and HSLA steels. The AWS uses a classification system that groups the steel submerged-arc wires into low-manganese (L), medium-manganese (M), and high-manganese (H) classes. They are listed in Table 18-2.

It's difficult to give general rules on selecting these filler metals and fluxes without getting into a lot of metallurgy; but the following rules of thumb will be helpful for most jobs.

General guidelines for selecting SAW flux and wires Low-alloy steels are much more complicated to work with than carbon steels. You'll have to look closely at detailed materials specifications to see what to do with them. Generally, low-alloy steels are used to get higher strength than carbon steels, or for higher strength at high temperatures, or for increased toughness at low temperatures. You'll find them used in everything from boilers to submarines.

Yield, tensile strength, and hardness will increase with increasing carbon, manganese, and silicon content. Silicon has the least effect and carbon has the strongest effect on tensile strength. The best notch toughness will be from weld metal with low carbon, manganese, and silicon content. Electrodes containing high manganese, such as EH14, promote weld-metal soundness by reducing the tendency for cracking and porosity.

Electrodes with larger amounts of silicon, such as EL8K, EM5K, and EM13K, increase the fluidity of the molten weld metal and its ability to wet the base metal, which improves the shape of weld-metal reinforcing and produces more regular edges and sounder welds at maximum welding speeds. What's more, the way the flux is made also may affect the final mechanical properties and welding characteristics that you'll get. Some fluxes are fused (the grains look like black glass). Others are mechanically mixed (they look gray) or bonded. Bonded flux looks like the grinding material in a grinding wheel crushed into a fine granular powder.

High-current, single-pass SAW welds have much larger mixtures of base metal in them than lower-current multipass welds where much of the previously deposited weld metal is remelted with only minimum penetration into the base metal. Large single-pass welds solidify and cool more slowly than the individual beads of multipass welds.

Furthermore, each pass of a multipass weld is subject to annealing (heat softening) due to the heat input of the succeeding passes, which is one reason why mechanical property tests often are made on sample welds designed to be as similar to production welds as possible when an actual flux–electrode wire combination is selected.

You can get all kinds of flux–electrode wire combinations by putting the filler wire and the fluxes together. For example, your best weld metal for the job at hand might be made with F60-EH14, F71-EM12K, or F62-EL8K flux and wire.

There are dozens (maybe hundreds) of proprietary submerged-arc fluxes and welding wires. The AWS A5.17 carbon steel and AWS A5.23 low-alloy steel sub-arc materials specifications are purposely written with wide ranges on the numbers to allow suppliers and users to work together to select the best flux–electrode wire combination for a given problem.

The actual weld metal that any flux-wire combination will produce is evaluated by welding standard test specimens and measuring their mechanical properties, then by looking at the appearance of the weld metal. AWS A5.17 and AWS A5.23 give complete details for running these tests and for evaluating the mechanical properties of the result-

TABLE 18-1 Mechanical properties of AWS flux and electrode classes

AWS flux class	Tensile strength, psi	Minimum yield strength at 0.2% offset, psi	Minimum elongation % in 2-in., %	Charpy V-notch impact energy, ft·lb
AWS A5.17. Carbon Steels				
F60-EXXXX				Not required
F61-EXXXX				20 at 0°F [−18°C]
F62-EXXXX	62,000–80,000	50,000	22	20 at −20°F [−29°C]
F63-EXXXX				20 at −40°F [−40°C]
F64-EXXXX				20 at −60°F [−51°C]
F70-EXXXX				Not required
F71-EXXXX				20 at 0°F [−18°C]
F72-EXXXX	72,000–95,000	60,000	22	20 at −20°F [−29°C]
F73-EXXXX				20 at −40°F [−40°C]
F74-EXXXX				20 at −60°F [−51°C]
AWS A5.23. Low-Alloy Steels				
F7XX-EXXX-X	70,000–95,000	58,000	22	The impact properties vary with details in the specification, but can range from 75 ft·lb at + 70°F [21°C] to 20 ft·lb at a very cold − 150°F [−101°C]. The final, lone X at the tail end of each spec is used to designate details like that.
F8XX-EXXX-X	80,000–100,000	68,000	20	
F9XX-EXXX-X	90,000–110,000	78,000	17	
F10XX-EXXX-X	100,000–110,000	88,000	16	
F11XX-EXXX-X	110,000–130,000	98,000	15	
F12XX-EXXX-X	120,000–140,000	108,000	14	

TABLE 18-2 AWS submerged-arc electrode wires

AWS Class	Chemical composition, %			
	Carbon	Manganese	Silicon	Other Elements*
AWS A5.17. Low-Carbon Steels				
Low-Manganese				
EL8	0.06	0.30–0.55	0.05	
EL8K	0.10	0.30–0.55	0.10–0.20	
EL12	0.07–0.15	0.35–0.60	0.05	
Medium-Manganese				
EM5K	0.06	0.90–1.40	0.40–0.70	*
EM12	0.07–0.15	0.85–1.25	0.05	
EM12K	0.07–0.15	0.85–1.25	0.15–0.35	
EM13K	0.07–0.19	0.90–1.40	0.45–0.70	
EM15K	0.12–0.20	0.85–1.25	0.15–0.35	
High-Manganese				
EH14	0.10–0.18	1.75–2.25	0.05	
AWS A5.23. Low-Alloy Steels†				
Carbon Steels				
EL12	Carbon steels included from A5.17 because they are used as backup welds for low-alloy sub-arc welds.			
EM12K				
Carbon-Molybdenum Steels				
EA1				
EA2	See AWS A5.23 for detailed chemical composition and other data.			
EA3				
EA4				
Chromium-Molybdenum Steels				
EB2				
EB2H				
EB3	See AWS A5.23 for detailed chemical composition and other data.			
EB5				
EB6				
EB6H				
Nickel Steels				
ENi1				
ENi2	See AWS A5.23 for detailed chemical composition and other data.			
ENi3				
ENi4				
Other Low-Alloy Steels				
EF1				
EF2				
EF3				
EF4				
EF5	See AWS A5.23 for detailed chemical composition and other data.			
EF6				
EM2				
EM3				
EM4				
EW				
EG				

* See AWS A5.17 for further details.
† See AWS A5.23 for more details.

ing weld metal. The detailed procedures are complex and we won't include them here. New proprietary flux-wire combinations also are evaluated the same way. While the evaluation tests are being made, welding parameters such as current, voltage, and wire speed are tested at the same time for a given plate material and joint design.

Low-alloy-steel flux-wire combinations often are dictated by the ASME Boiler and Pressure Vessel Code. We repeat, low-alloy steels are much more complicated than carbon steels, but flux-wire combinations would still be tested in the shop before they are used.

This may sound like a lot of work to do before production welding gets started, but once all the variables are nailed down, the setup may be used to weld the entire hull of a ship or a dozen large pressure vessels.

To help give you a better feeling for SAW electrodes, we'll list the standard electrode diameters for you in fractions and decimal inches.

$\frac{1}{16}$ in. (0.063 in.) $\frac{3}{16}$ in. (0.188 in.)
$\frac{5}{64}$ in. (0.078 in.) $\frac{7}{32}$ in. (0.219 in.)
$\frac{3}{32}$ in. (0.094 in.) $\frac{1}{4}$ in.   (0.250 in.)
$\frac{1}{8}$ in.   (0.125 in.) $\frac{5}{16}$ in. (0.312 in.)
$\frac{5}{32}$ in. (0.156 in.) $\frac{3}{8}$ in. (0.375 in.)

SAW Power Sources

Power sources for SAW can be either ac or dc machines. The dc machines can be either constant-current (CC) or constant-voltage (CV) units. Since spatter is no problem with SAW (the molten slag surrounding the arc traps the spatter), it is possible to use currents that are two to three times as high as those used in MIG welding. Currents of 1500 A on a single electrode wire are not unusual. The weld puddles are large and the power released as heat in the arc zone is high. The sub-arc system is ideal for use on heavy steel plate. It also can be used on thin materials, but sub-arc's prime use is on heavy steel weldments.

Because there is no great problem with spatter in sub-arc welding, there's no need to use slope to control the short-circuit current. Constant-voltage power supplies without slope control are less expensive than those with slope. Therefore, most dc CV welding power sources have very flat volt-amp curves. In addition to holding down spatter, the flux makes it easier to run an arc. Gases released in the weld zone produced by selecting the appropriate flux make it possible to run ac as well as dc arcs.

A large amount of inductance is recommended, but not essential, for smoothing out the large current swings that would occur if inductance were not present. Large current swings tend to cause puddle turbulence and inductance helps to quiet the arc action.

Direct current The requirements for a dc CV power source for SAW are not very complicated. Practically any machine that can provide enough current and voltage will do the job.

You pick a dc CV power source by telling the manufacturer or welding distributor that you want to do submerged-arc welding. Every machine suitable for this process will say so in its literature. You do have to make sure that you are getting a very high amperage machine because high operating currents are characteristic of the process.

The same is true of a dc CC machine or an ac machine suitable for SAW. The normal OCV for conventional ac SAW power sources is about 80 V. Constant-voltage ac power sources are not normally used for SAW because they have lower OCVs and you will have problems keeping the arc going and even getting it started.

Alternating current Alternating current is preferred whenever you expect to run into arc blow. Arc blow is caused by the interaction of the magnetic field of the arc and magnetism induced in the steel plate. It causes weird, erratic metal transfer and an irregular bead shape. When dc is used, the arc is deflected in one direction by the induced magnetic field, usually to one side or the other of the workpiece joint. The arc can even gouge out small holes in the side of the joint because the dc arc is attracted by the magnetic field set up in the workpiece (see pages 404–406).

Alternating current also induces magnetic fields in the workpiece, but since the polarity of ac alternates from positive to negative and back again about 60 times a second, it breaks down magnetic fields in the steel plate as fast as they are set up. Therefore, ac tends to even out the effects of arc blow while neutralizing the induced magnetic field.

High-nickel steels that are used for low-temperature and cryogenic pressure vessels and liquefied gas containers are especially subject to arc-blow problems because nickel is even more magnetic than iron. High-nickel steels are typical of materials that are welded with ac power sources.

Two SAW wires sometimes are run in tandem to get very high deposition rates on thick joints. Arc blow is a common problem when two dc arcs (two wires) are run in the same arc zone because their fields interact. They produce mutual arc blow. When two ac arcs of the same phase are operated in the same arc zone they also cause mutual arc blow.

The solution is simply to use two ac arcs (and two separate power sources) that are out of phase with each other. When one of the two ac arcs is at its maximum magnetic strength (maximum current) at the top of the current peak, the other one can be set to be at zero current. They are said to be 90° out of phase, which cancels out the effects of the mutual magnetic fields and the arc blow that results from running two wires.

One phase is usually called the *leading* electrode phase and the other is called the *trailing* electrode phase. The two electrodes operate one behind the other in the same weld pool. The ability to adjust the current separately in each electrode is a big advantage when doing SAW.

No matter what the technical reasons are, the important thing for you to remember is that *ac power sources are used to control arc blow when you are doing SAW.*

Wire Feeder Controls

Submerged arc produces very high welding speeds and high weld-metal deposition rates. There is no way you could work fast enough to do it manually and maintain control of the weld. Therefore, just about everything on a high-production SAW system operates automatically.

The automatic control and power supply system used in SAW operate to maintain a constant voltage and constant current. The welding voltage is proportional to the length of the current path between the welding wire and the workpiece. Therefore:

■ If the distance between the wire and the workpiece increases, the welding voltage increases.
■ If the distance between the wire and workpiece decreases, the welding voltage decreases.
■ If the distance between the wire and workpiece remains constant, the welding voltage will remain constant.

With a CV power source, the arc voltage is maintained by the power supply. Arc current is controlled by the wire feed speed with increased feed speed producing increased current. Therefore, the wire feed system is simplified to a constant-speed device and arc control is performed by the power source.

With a CC power source, if, for any short period of time, the current flowing through the welding zone melts off the electrode wire at a faster rate than it is being fed, the distance between the electrode wire and work will increase and the welding voltage will increase. Conversely, if, for any short period of time, wire is feeding faster than it melts off, the distance between the wire and the work will decrease and welding voltage will decrease.

A constant welding voltage can be maintained with a CC power source if a control unit is used that will automatically vary the rate of electrode wire feed with changes in the welding voltage.

18-3
WELD-METAL BACKING
Submerged-arc welding creates a large volume of molten weld metal that remains fluid for quite a while, so you have to support the molten metal and contain it until it has solidified. If you don't, the weld metal will leak out the bottom of the gap in the weld joint.

There are five ways of supporting molten weld metal in a joint. There are:

- Nonfusible backing
- Submerged-arc flux backing
- Root backing
- Weld backing
- Fusible metallic backing

The first two methods use temporary backing that is removed after the weld is completed. In the other three, the backing becomes part of the completed weld joint. Let's take a close look at each of these five flux-support systems.

Nonfusible Backing
Copper backing with water cooling frequently is used as a nonfusible backing when welding steel, as shown in Fig. 18-4. Water-cooled copper backing strips are used when the workpiece is not massive enough to provide adequate weld-metal support or when you must get complete weld penetration in one pass. Water-cooled copper backing is especially useful for welding thinner-gauge steel.

Because pure copper is an extremely good heat conductor it cools the molten weld metal rapidly and provides support without being fused with the joint. People have tried using copper alloys, aluminum, aluminum alloys, and ceramic materials to replace pure copper backing strips, but the results have been invariably inadequate in both welding and service quality.

The poor service results from the fact that these substitutes have lower thermal conductivity than pure copper and they fuse at lower temperatures. Thus, they cannot withstand high welding temperatures and they deteriorate rapidly in use.

You must make absolutely sure that the water-cooled copper backing strip is held tightly against the bottom of the joint to prevent the weld metal from flowing out between the backing strip and the underside of the workpiece. The

water-cooled copper backing strip can be grooved or recessed to make it easier for weld metal to penetrate at the root of the joint. In practice, grooving is generally not used on sheet below about 10-gauge in thickness because the full chill effect of the copper is required for support. For material above 10 gauge, grooved backing strip with groove dimensions varying from 0.02 to 0.06 in. [0.5 to 1.5 mm] deep and 1/4 to 3/4 in. [6.4 to 19 mm] wide is used with the groove dimensions and strip width increasing with the workpiece thickness (Fig. 18-5).

The groove corners can be rounded or squared. The groove is larger for the thicker plates because the full chill effect of the water-cooled copper would prevent complete penetration, and because you'll want to allow some weld metal to flow through and give additional reinforcement to the underside of the joint. Wider

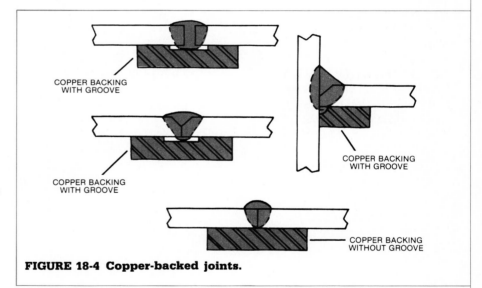

FIGURE 18-4 Copper-backed joints.

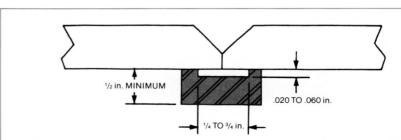

FIGURE 18-5 Recommended groove dimensions for copper backing bars.

grooves let you get by with greater misalignment of your plates, which is convenient when you are welding long joints.

The life of your water-cooled copper backing strip depends upon how well you take care of it. Guard it against mechanical damage and against overheating due to poor joint fit-up. If the contact surfaces of the backing strip become grooved or pitted, they can be made serviceable again by re-machining. A backing strip must not be machined too often or else it will not provide the required chill effect because it will be too thin.

The water-cooled copper backing strip will tend to warp and become less abrasion resistant if excessively heated. One of the most common causes of excessive heating is poor joint fit-up. Poor fit-up allows too much of your molten metal to contact the copper bar. When a copper backing "shoe" is used, the welding heat is applied continuously to the same section of copper. With a long backing bar the heat is applied progressively along its entire length. When a short backing shoe is used (or when the same copper bar is used constantly without allowing it to cool) the full chill effect is reduced.

You may want to add extra cooling to the backing strip by running circulating water through flattened copper tubes inserted into grooves machined in the bottom of the backing strip (Fig. 18-6). When water cooling is used, moisture condensation or water leaking into the joint must be avoided. The presence of moisture will lower the quality of the weld metal by introducing hydrogen embrittlement and porosity.

Warpage of the backing strip can be minimized by mounting the strip in a restraining jig or fixture as shown in Fig. 18-7.

When copper backing is used for single-pass fillet welds (Fig. 18-8), the corners should be chamfered to at least ⅛ in. [3.2 mm] to allow some metal to fuse through the joint and produce a small fillet at the root. This small fillet gives your weld added strength and avoids a "nick" effect that can cause weld failure that sometimes occurs when fillet welds are made without adequate weld-metal penetration.

Flux Backing

Although all grades of sub-merged-arc flux are sometimes used for backing, a special backing flux is recommended. Two methods for using flux backing

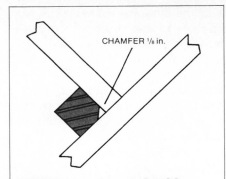

FIGURE 18-8 Copper backing bar for fillet weld. Note position of chamfer on backing bar.

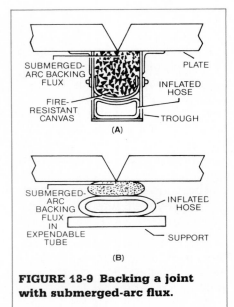

FIGURE 18-9 Backing a joint with submerged-arc flux.

are shown in Fig. 18-9. The backing flux is held uniformly against the bottom of the joint by an air- or water-inflated fire hose. The flux can be laid on top of the hose (Fig. 18-9A) or it can be contained in a tube of cotton, glass fiber, paper, or other material (Fig. 18-9B).

Looser joint fit-up can be tolerated when using backing flux than with other backing methods because the granular flux shifts and accommodates a lot of mis-alignment.

Pressure exceeding that needed to hold the flux against the bottom of the joint tends to produce a concave weld bottom, which is highly undesirable because it

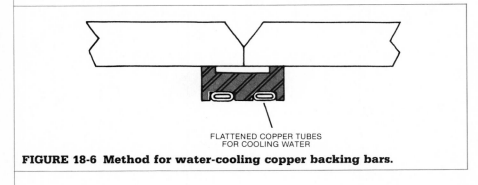

FIGURE 18-6 Method for water-cooling copper backing bars.

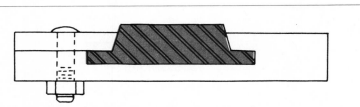

FIGURE 18-7 Method for mechanically restraining warping in copper backing bars.

makes a weak joint (the section thickness is less than it should be), so be careful not to use too much pressure.

Root Backing and Weld Backing

Root backing and weld backing are the most widely used methods of obtaining weld-metal support under a submerged-arc joint.

In a root-backed joint, the root face is thick enough to support the incompletely penetrated first pass of the weld. This method is used for butt welds (either grooved or square edge), for fillet welds (lap or T joints), and for plug or slot welds. Extra backing or chilling sometimes is used. It is most important that your joint edges be tightly butted at the point of maximum penetration of the weld.

In a weld-backed joint, the first pass (made by submerged arc, gas metal-arc, or shield metal-arc) forms the backing for subsequent passes (Fig. 18-10). These passes will be made either from the opposite side of the plate or from the same side.

Manual-Weld Backing

Manual welds often are used as backing for submerged-arc welds (Fig. 18-11) when it is not convenient to use other backing methods. This often happens when you can't get at the joint easily with backing strip, your joint preparation or fit-up is poor, or you're having problems turning the welded plate over to work on the other side. Your manual weld should be removed later by arc or flame gouging or chipping, or it may be machined out after the weld has been made. When the manual weld is removed, it is replaced by a permanent submerged-arc weld.

If shielded-metal-arc (stick-electrode) welding must be used for submerged-arc weld backing, a low-hydrogen electrode such as AWS E7016 or E7018 electrode is recommended. Also use E7016 or E7018 for overhead or vertical welding because, although you could use E6010 or E6011 electrodes, the low-hydrogen electrodes will make sure you don't get weld porosity. SMAW electrodes E6012 and E6013 and their iron-powder equivalents are not recommended because they tend to cause porosity in the finished submerged-arc weld.

It is very important that you make your manual weld backup of good-quality, nonporous weld metal that is free from entrapped slag. Otherwise, the submerged-arc finishing weld also will contain these defects since it penetrates down into and re-fuses a portion of your manual weld. The root gap must be maintained and weld "bridging" must be avoided across a joint. Gas-metal-arc backing welds are preferred since previous slag-generated defects will not occur.

Fusible Metallic Backing

When using fusible metallic backing, the weld metal penetrates into and fuses with the backing material which, either temporarily or permanently, becomes an integral part of the weldment.

Fusible backing can be made with steel backing strips, preferably of the same material being welded (Fig. 18-12A). Otherwise a joint can be located so that it is already backed up against another part of the weldment (Fig.

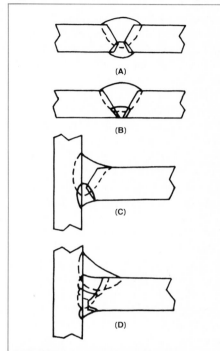

FIGURE 18-11 Manual welds used as weld backing.

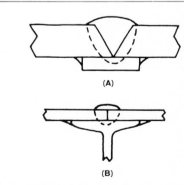

FIGURE 18-12 Fusible metal backing for submerged-arc welding. Sometimes the structure itself is part of the fusible backing. When a backing strip is used, it is removed after welding with flame cutting or carbon-arc gouging, followed by a final surface finishing pass.

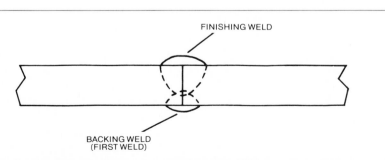

FIGURE 18-10 Root backing and weld backing for submerged-arc welding. Root backing is used for first weld pass. Weld backing is used for finishing pass.

18-12B). It is important that your joint contact surfaces be clean and close together or else porosity will occur and your molten weld metal will leak out of the joint.

18-4
PREPARING TO WELD

Now that you have some of the basics of SAW and have chosen a backing system, let's take a look at the next two steps, edge preparation and precleaning your joint so you'll be ready to start sub-arc welding.

Edge Preparation

The way you prepare the edges of your work for SAW is very important. Your edge beveling or grooving work will be done to control the amount of weld reinforcement you get on top of the weld bead (Fig. 18-13). The amount of weld penetration you get also is affected by changes in bevel depth and volume. For butt, fillet, and multipass welds, the maximum weld width should be at least somewhat wider than the weld depth. A weld-width-to-weld-depth ratio of 1.25 to 1.5 is preferred to reduce the possibility of center weld cracking (caused when the weld metal cools and shrinks). Proper joint bevels and joint design will help you get set up.

Beveling or grooving is especially desirable for submerged-arc butt joints thicker than 5/8 in [16 mm]. Beveling is sometimes used for material as thin as 1/4 in. [6.4 mm], where it can help you in tracking the weld seam with knife-edged wheel followers. In positioned T joints or corner-joint fillet welding where you want complete joint penetration, bevel the abutting section if the depth of fusion you want for each weld exceeds 3/8 in.

The unbeveled root face should be thick enough for the weld metal to fuse down into, but not through, the butted root faces or "nose" of the joint. If the nose is not thick enough, there will not be enough metal mass to absorb the heat of your molten weld metal and it will spill out of the bottom of the joint. Figure 18-14 shows good and bad practice.

There are several methods for preparing the edges of a submerged-arc joint. You are already familiar with most of them, but let's go over them again. We'll tell you what's special about them (if anything) when you do SAW.

Flame-cutting edges You can prepare your weld joints with manual or machine flame-cutting. All loose scale and slag resulting from the cutting operation must be removed before assembly and welding. It is *not* necessary to remove the oxide film formed on cooling. In fact, it is good practice to let this dark oxide film remain on the edges if the plates are to be stored for some time before welding since the thin oxide coating will help protect the plates from rusting.

Machined edge preparation Your plate edges can be machined with gate shears, planers, lathes, boring mills, shapers, and so on, depending upon the availability of the equipment, the type of cut you need, the size, and handling problems the machinist may have with your plate.

If gate shears are used for edge preparation, any heavy rust must be removed to prevent the rust from becoming pressed into and embedded in the plate edges. Rust that has been embedded in the steel by shearing cannot be removed by a wire brush. The edges must be machined or ground off before welding. All cutting fluids and oils or greases must be removed before welding. A chlorinated solvent or some other degreaser that will leave no residue upon evaporation should be used.

Manual chipping and grinding It's lots of hard work, but manual chipping and grinding sometimes are used to prepare plate edges for SAW. The accuracy of the edge preparation depends on your skill. Do it accurately and you will eliminate a lot of problems when you weld.

Precleaning the Joint

After the joint edges have been prepared, you still have the most important preparation job ahead of you. The joint must be cleaned up. Any material that will evolve gases (such as cleaning solvents, oil, grease, water, paint, rust, or scale) when subjected to the heat of welding must be removed. Mill scale or even crayon marks can cause you trouble. Cleanliness is particularly important when you are going to weld thin material at high speeds. Here are some important cleaning methods.

Flame cleaning An oxyfuel flame from a heavy-duty heating torch with a large tip often is used to remove rust, scale, and moisture. Flame cleaning is one of the most effective means of eliminating porosity in SAW. Since there is no radiant-arc heat preceding the weld, and speeds are usually quite fast, flame cleaning is even more important than with an open-arc process.

Sometimes a heating torch is mounted directly on the SAW car-

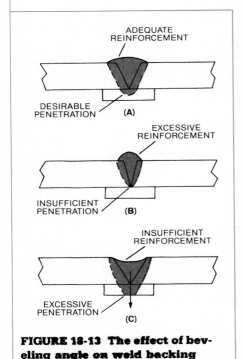

FIGURE 18-13 The effect of beveling angle on weld backing and joint reinforcement.

ADEQUATE REINFORCEMENT

DESIRABLE PENETRATION **(A)**

EXCESSIVE REINFORCEMENT

INSUFFICIENT PENETRATION **(B)**

INSUFFICIENT REINFORCEMENT

EXCESSIVE PENETRATION **(C)**

riage. The hottest section of the flame should hit directly on the weld seam. The immediate weld area should be heated to more than 400°F [205°C] to avoid condensing additional moisture from the oxyfuel gas combustion products in the joint.

Grinding Grinding can be used to remove mill scale or rust. Hand-operated grinders or wheels mounted on the welding machine can be used. Protective glasses must be worn at all times when grinding and the proper type of grinding wheel should be selected for the speed at which the wheel will operate. If you're not sure which grinding wheel to use, ask a machinist or your welding supervisor, or read the directions that come with the wheels. (Chapter 22 gives you useful information on selecting grinding wheels.)

Wire brushing Wire brushing with an air-operated brush wheel is useful only for removing light rust and dirt. Some types of paint can be removed by wire brushing, but a paint remover, flame cleaning, or grinding is necessary to remove heavy paint.

Paint removers Paint removers should be applied to painted surfaces and the loose paint scraped off. Then brush the plate with a wire brush and wash it with a chlorinated solvent or other volatile solvent. You have to recognize some risks here. Be careful with flammable or explosive solvents. Take precautions and work in a well-ventilated area. Paint can sometimes be removed by grinding, but surfaces should be carefully inspected to be sure that all paint has been removed. Paint also clogs some grinding wheels.

Sand or shot blasting Shot or sand blasting readily removes heavy rust, mill scale, and paint. Sheared edges are not satisfactorily cleaned by this method. You must be sure to work in a well-ventilated area and use eye and face protection and possibly a filter mask. See Chap. 22 for more details on abrasive shot blasting.

Pickling Pickling is the chemical cleaning of mill scale and oxides with acids. It can be used to remove even the heaviest mill scale and rust. The pickling bath should contain inhibitors and a neutralizing rinse should be used to prevent hydrogen absorption into the steel which will result in porosity and hydrogen embrittlement. Most welding shops that do pickling use proprietary solutions. They usually contain hydrochloric acid. If you do this work, make sure that you work in a well-ventilated area and that you and your clothes are fully protected from splashing acid. Also remember to "do like you aughter . . . add the acid to the water."

Degreasing There are a lot of proprietary degreasing solutions that are used in welding shops. They are especially useful for cleaning cold-formed parts. Chlorinated solvents are suitable as degreasers. Although the solvents will evaporate in time, you should *make sure that they are completely removed before welding.* The arc will turn any residual chlorinated solvents into a poisonous gas called phosgene.

18-5
TYPES OF SAW JOINTS
A number of different types of joints are commonly used for SAW. In the following sections we'll tell you how they will affect your work. Even if you forget the details (you can always look them up later on), the important thing to learn about SAW is that joint design and preparation are critical.

Remember, the SAW process goes too fast for a manual operator to keep up with it. It's essentially an automated process. Without a human being guiding the electrode wire, the joint design, preparation, and fit-up have to be perfect. A machine can't make adjustments for poor conditions like you can. The automatic SAW head is a machine, not a skilled welder.

Butt Joints
As you know, there are many different kinds of butt joints. Not all of them are set up the same way when you weld them with the sub-arc process. See Fig. 18-15A and B.

Square-groove butt joints The square-groove butt joint (Fig. 18-15A) is good for single-pass submerged-arc welds up to $5/16$ in. [7.9 mm]. It has no root opening and requires suitable backing. Reinforcement on top of the SAW bead tends to become excessive for thicker welds but it can be controlled by adjusting the root opening. Variations in root opening,

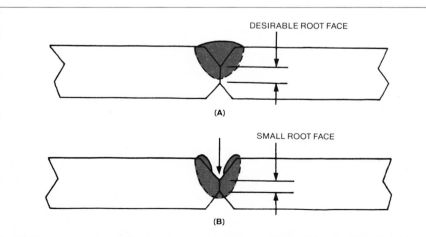

FIGURE 18-14 The effect of root-face dimension on root fusion and weld-pass bead shape. In (B) the weld leaked through the root.

alignment of welding wire with the joint, and the amount of weld metal required limit the thickness of this type of weld to ¾ in. [19 mm].

Two-pass welds up to ⅝ in. [16 mm] thick are made without root openings, but it's essential in two-pass welding that the edges be closely butted since weld backing is *not* used. The maximum permissible root opening is 1/32 in. [0.79 mm] unless the joint is sufficiently backed to prevent weld metal from flowing through the root opening. With such support greater root openings can be used.

When the root opening exceeds 1/16 in. [1.6 mm], however, the opening should be packed with SAW flux ahead of the weld. The maximum root opening is about ⅛ in. [3.2 mm] because of the difficulty of re-fusing the welding flux at the base of the first-pass weld. Plate up to ¾ in. [19 mm] thick can be welded with square butt joints if the root gap is maintained constant for the entire length of the seam.

Your first weld is a backing weld made on the reverse side of the joint. Next turn the work over and finish the weld. The finishing weld penetrates down into and re-fuses a portion of the backing weld to make sure that you get a continuous weld structure all the way through the plate thickness.

A good way of getting your best weld-metal penetration without excessive reinforcement is to back-gouge a groove about ⅛ to 5/16 in. [3.2 to 7.9 mm] deep into the top of the joint after the backing weld has been made. When gouging is used, no additional preparation or cleaning is needed except to remove any adhering slag.

The advantage of the square-groove butt joint (Fig. 18-15A) is that it requires minimum edge preparation while it produces good-quality welds with adequate base-metal penetration.

Single-V-groove butt joint with root face The single-V-groove butt joint with a root face (Fig. 18-15B) is used with nonfusible backing for single-pass butt welds of 5/16 in. [7.9 mm] or greater thickness. For most industrial jobs, the maximum thickness is around 1¼ to 1½ in. [32 to 38 mm] for this joint design. The root face gives you several advantages.

The square edges of the single-V-groove joint with a root face simplify assembly. You can get excellent penetration and weld reinforcement. Normal, practical variations in voltage, current, welding speed, and edge preparation will cause minimum damage to the backing. Relatively small quantities of electrode wire are used because the V joint gives you good base-metal penetration without excessive current. The volume of the V space is considerably smaller than that required by other welding methods. Therefore you'll use less filler metal to complete the job.

Using nonfusible backing, the root-face dimension should be ⅛ to 3/16 in. [3.2 to 4.8 mm]. Your root gap should not exceed 1/16 in. [1.6 mm]. Fusible metallic backing also is used with this joint design with a root gap of at least ⅛ in. [3.2 mm].

The single-V-groove butt joint with root face and without external backing is also used for two-pass welds where the plate thickness exceeds ⅝ in. [16 mm]. The first weld, usually the larger one, is the backing weld made in the V side of the joint. The work is then turned over and the finishing weld is made on the flat side. The finishing weld penetrates down into the first weld and re-fuses a portion of this weld backing to make sure that you get complete penetration in the joint. Figure 18-16 shows a single-V-groove butt weld with two overlapping welds. This joint could be made in

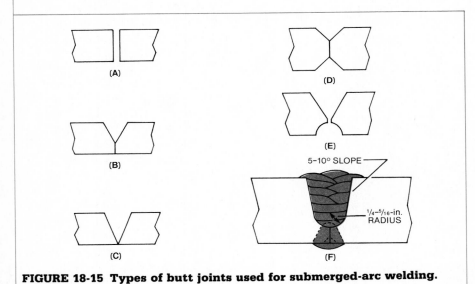

FIGURE 18-15 Types of butt joints used for submerged-arc welding.

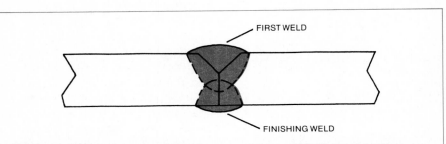

FIGURE 18-16 Two-pass single-V-groove butt weld. Note that there are overlapping fusion zones.

one pass with two welding heads on the bottom inside and then the top outside of a cylinder (see Fig. 18-28).

The root face is about ⅜ in. [9.5 mm] for all commercially welded plate thicknesses. The plate edges must be tightly butted together (¹⁄₃₂ in. [0.79 mm] is your maximum allowable gap between the plates), just like the square-butt joint. When welding flux and weld metal are retained by a support below the joint, slightly greater root gaps are permissible; if the root gap exceeds ¹⁄₁₆ in. [1.6 mm], SAW flux should be tamped into the gap ahead of the weld.

Single-V groove without root face The single-V groove butt joint without a root face (Fig. 18-15C) is used for nearly all plate thicknesses when using submerged arc backing flux. It is not commonly used below ⅜-in. [9.5-mm] thickness since you can get adequate penetration for these thicknesses without beveling the edge of your plate.

The single-V groove butt joint without a root face must always have backing since the mass of metal at the joint root is not sufficient to provide support for the weld metal. Reasonable misalignment in fit-up and variations in the root gap can be tolerated when using submerged-arc backing flux because the granular material will shift to accommodate them. Copper backing is not recommended because of the tendency of the weld metal to fuse to the backing piece. Fusible metallic backing is acceptable if you have no objections to its remaining on the work as part of the completed weldment.

Double-V-groove butt joint The double-V-groove butt joint (Fig. 18-15D) is the basic design for two-pass submerged-arc welds. It is commonly used for welding section thicknesses up to 2 in. [51 mm]. Even greater thicknesses have been welded successfully with this joint design. For welding thicknesses above 2 in. [51

mm], however, the joint for multi-pass welding (shown in Fig. 18-15F) should be used.

The double-V-groove joint is designed with a large root face to provide adequate support for the initial weld, which will be used as backing for the opposite-side weld. The maximum possible fit-up misalignment is 25 percent of the root-face thickness. You should try to keep misalignment much less than that.

The root faces must be closely butted along their entire length. The maximum permissible root gap is ¹⁄₃₂ in. [0.79 mm]; any larger root gap, and molten metal or slag will flow through the gap. Several methods are used to stop metal flowing through the gap if it's too big. A small stringer bead can be laid manually in the bottom of the V in which the finishing weld is to be made. A length of wire can be tacked into the finishing-weld V. Submerged-arc backing flux can be tamped into the gap ahead of the weld. The stringer bead, wire, or backing flux should be removed before making the finishing pass if you expect to get a high-quality weld joint. You simply cannot leave these materials in the joint and expect to get a superior weld.

To make absolutely sure that you get full penetration and that all slag or porosity has been removed from the bottom of the backing weld, the finishing weld should penetrate into the backing weld and re-fuse it to a depth of ³⁄₁₆ to ⁵⁄₁₆ in. [4.8 to 7.9 mm].

Since this joint is widely used in pressure-vessel fabrication, you should know about one limitation the double-V groove butt joint has. When making circumferential welds (around the curved section), the ratio of wall thickness to cylinder diameter must be at least 1:25. Otherwise, the large pool of molten weld metal will tend to run, causing unstable welding action and an undesirable weld shape.

Manual weld backing sometimes is used with double-V-

groove butt joints when the joint has a small (⅛-in. [3.2 mm] maximum) root face and a root gap of about ⅛ in. [3.2 mm]. If conditions require that you do manual welding on material thicker than ⅜ in. [9.5 mm], however, the joint shown in Fig. 18-15E is better.

Single- and double-U-groove butt joints The single-U-groove butt joint (Fig. 18-15F) is often used for multipass submerged-arc welds. Any thickness of material can be welded using this joint design.

A small manual backing weld often is made from the reverse side of the joint. If the manual weld is not made, the root faces must be closely butted (¹⁄₃₂-in. [0.79-mm] maximum root gap).

Double-U groove butt joints can be used for extremely thick material. These are essentially two single-U-groove butt joints with a common root. If a SMAW weld is used to back the first pass, you may want to remove it later on if maximum weld quality is required. The use of gas metal-arc (GMA) manual welds will eliminate the need for removing this first weld pass. Because of the internal and external slag-free nature of GMA welds, subsequent submerged-arc welds of excellent quality can be produced.

Lap Joints

The main advantage of single- or double-fillet lap joints (Fig. 18-17A to E) is their simplicity during fit-up and the minimum amount of edge preparation necessary before welding. These joints must be clean and the lapping surfaces must be clean and dry.

Single-fillet lap joint The single-fillet lap joint is used primarily where the underside of the weld is not accessible, or for service where little joint strength is required and tightness is the primary function of the joint. Double-fillet lap joints are used for structural strength. (Fig. 18-17A).

Joggled and through-welded lap joints The joggled lap joint (Fig. 18-17B) makes fit-up easy and requires little edge preparation. In addition, it provides one flush surface that the ordinary lap joint does not have. The joint must be clean and the lapping surfaces must be tightly fitted and dry. The joggled lap joint is used with a single-fillet weld when fabricating small fuel-gas tanks. The double-fillet-welded joggled lap joint is used extensively in shipbuilding to join the last of a series of butted plates to make fit-up a lot easier.

The continuous through-welded lap joints (Fig. 18-17C, D, and E) are used widely for welding sheet up to 11 gauge. These joints are useful for joining sheets or for attaching sheets (either singly or simultaneously) to a backing member. The lapping surfaces must be clean, dry, and in contact.

Fillet Joints

Submerged-arc welding produces deep penetration, which can save a lot of money when making fillet welds by reducing the size of the fillet (and the filler metal required) without reducing the strength of the joint. Design calculations of fillet-weld strengths are based on the throat dimension of the fillet (Fig. 18-18).

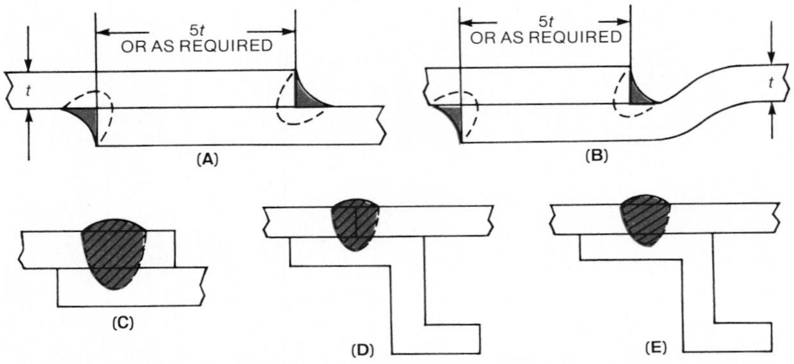

For conventional arc-welded fillets, the throat dimension is obtained by multiplying the minimum leg dimension by half the square root of 2 (which is about 0.707), since the root of the fillet joint is rarely penetrated by the weld metal. The deeper weld-metal penetration of sub-arc welding produces an effective throat depth about 20 to 30 percent greater than with ordinary arc welds. Therefore, the size of your submerged-arc-welded fillet joints can be made smaller and you will still get the same joint strength as a conventional arc weld of greater size. Almost any reduction in the fillet size saves money because less weld metal is used (since the volume of deposited weld metal varies directly as the square of the leg dimension). If you reduce the leg of the fillet weld by a factor of 2, you will use four times less weld metal to make the joint.

Horizontally welded T fillets Single-pass fillet welds up to $5/16$-in. [7.9-mm] leg dimension (equivalent to $3/8$-in. [9.5-mm] welds made by other processes) are used for making T joints in the horizontal position. If the web thickness is no greater than $3/8$ in. [9.5 mm], two $5/16$-in. [7.9 mm] fillets will penetrate into each other at the root (Fig. 18-19A). T joints that need deeper penetration or larger fillet welds can be made with multipass welding (Fig. 18-19B).

The size limits of single-pass horizontal fillet welds are determined not by the capacity of the SAW equipment, but by the volume of the molten metal that will remain in position without flowing off the workpiece.

Flat-position-welded T fillets Fillets of equal leg dimension can be made by positioning the joint at a 45° angle (with the weld surface flat). The depth of penetration can be increased still further by increasing the angle of the "stem" or bottom leg of the T by as much as 60° from the hori-

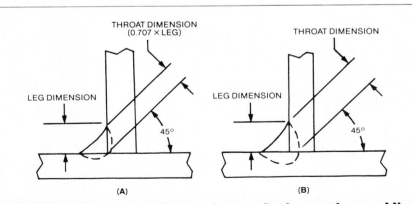

FIGURE 18-17 Types of lap joints used with submerged-arc welding. (A) Double fillet-welded lap joint. (B) Double fillet-welded joggle lap joint. (C) Continuous through-welded lap joint. (D) Continuous through-welded lap joint joining adjacent sheet edges to backing structure. (E) Continuous through-welded lap joint attaching a single sheet to a backing structure.

FIGURE 18-18 Comparison of manual arc and submerged-arc welding of fillet welds. Deep-throat penetration of fillet weld during submerged-arc welding is similar to that of flux-cored arc welding.

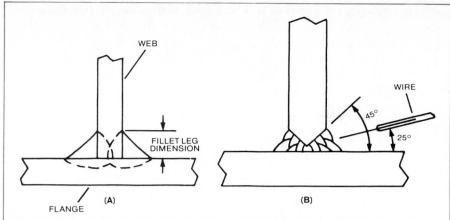

FIGURE 18-19 T-joint fillet welds in the horizontal position. (A) Horizontal fillet weld. (B) Double-V-groove weld.

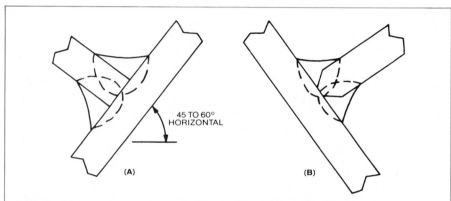

FIGURE 18-20 T-joint fillet welds in the flat position. (A) Flat fillet weld. (B) Preparation for double-V T-joint in flat position for joint thicknesses greater than ¾ in.

zontal and offsetting the wire toward the stem or leg side of the T joint (Fig. 18-20A). If the bottom leg or stem thickness of the T joint exceeds ¾ in. [19 mm] and complete penetration is desired, the edges should be beveled (Fig. 18-20B).

Corner Joints

It's pretty obvious that corner joints are used whenever two pieces of metal come together at the corner of a part. What's not so obvious is that corner joints often can be designed right out of a part, saving extra welding costs, simply by using an extra sheet or plate that "wraps around" at least one of the corners on the part, even if it ends at another corner. That is, it's easier to make a box shape from two wide side sheets than four narrower ones, and it cuts the welding work in half. Figure 18-21A to E shows some types of corner joints.

Square-groove corners The square-groove corner joint (Fig. 18-21A) is recommended for joint thicknesses up to ½ in. [13 mm]. A submerged-arc fillet weld is first made on the inside corner of the joint and then a square-groove butt weld is made on the reverse side of the joint. The butt weld should penetrate down into and

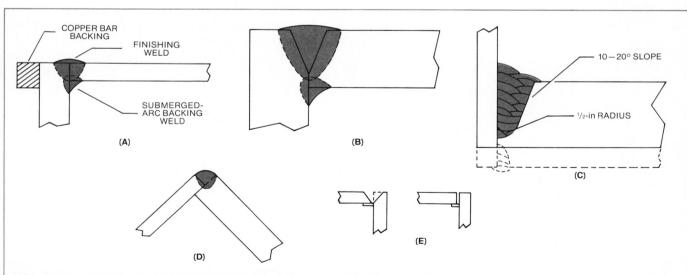

FIGURE 18-21 Types of corner joint welds. (A) Square-groove corner joint. (B) Single-V-groove corner joint. (C) J-groove corner joint (single- or double-J). (D) Corner fillet joint. (E) Single-V-groove and square-groove corner joints with fusible or nonfusible backing.

re-fuse a portion of backing fillet weld.

The joint can be positioned for the fillet weld if larger fillets are desired. Backing is not generally required if the edges are tightly butted. A copper backing bar sometimes is needed for chilling the outer corner of the vertical member to prevent excessive fusion while the finishing butt weld is being made.

V-groove corners For thicker joints, single-V-groove joints are recommended to get penetration without excessive weld-metal reinforcement. The V groove is preferred over the single-bevel groove as it will give you full weld-metal penetration into the joint with an optimum weld shape. The fillet weld is made first and then the butt weld is made on the reverse side. If the fillet weld is submerged-arc-welded, a deep root face must be used as shown in Fig. 18-21B.

For processes other than SAW, the preparation should provide a narrow root face and a root gap. These fillet welds can be made with the edge preparation shown, or with a single-bevel grooved edge. A copper chill bar can be useful with this joint as with the square butt joint.

J-groove corners The J-groove joint (Fig. 18-21C), either with a single-J groove or a double-J groove, is used for multipass welding of even larger thicknesses than those for which the V-groove corner joint can be used. The welding technique is similar to that used in the U-groove butt joint. A small backing weld often is made before the multipass submerged-arc weld can be made. If no backing weld is used, the maximum permissible root gap is $\frac{1}{32}$ in. [0.79 mm]. Since only one face of the joint is prepared, the edge slope and radius of the J's curvature should be as specified in Fig. 18-21C to give you sufficient working area to deposit the welds at the root of the joint.

Corner Fillet Joints
The outside single-fillet-welded corner joint (Fig. 18-21D) is useful for many jobs. The strength of the joint can be increased by adding a second fillet at the interior angle of the joint to form a double-fillet-welded corner joint. The corner joint has the advantage of requiring no additional backing other than that provided by the structure itself. The butting surfaces must be clean, dry, and fitted tightly together.

Corner fillets with backing
Either square-groove or single-V joints can be used with backing to give you complete weld-metal penetration when welding from one side of the joint (Fig. 18-21E). The requirements of this method are the same as for butt-joint welds with backing.

Plug-weld joints Plug welds (Fig. 18-22) are used to join two pieces by welding through a hole in one of them to secure a bond and fill the hole with weld metal. It is important that the hole be large enough to prevent arcing between the electrode wire and the upper member. Unless the hole is beveled or tapered, its diameter should be no less than the thickness of the upper member of the joint.

If you can use sufficient welding current and can tolerate excess weld metal on the surface, weld penetration through the upper member of a section can be obtained without the hole. You simply weld right through the section and into the underlying piece (very much like making a spot weld). This is only practical, however, for attaching a thin section to a heavier piece underneath.

The size of the hole that can be used for a completely fused plug weld made without changing electrode position during welding will be determined by the current that can be applied. A large hole may require you to move the electrode during welding to get complete root fusion.

18-6 POSITIONING YOUR WORK

Unlike other welding processes where the operator can control the position he or she works in, when the work can't be positioned for them, sub-arc-welded parts must be positioned for downhand welding. You can't weld with SAW in the vertical and overhead positions without the flux falling out of the weld. You also have to position the weldment correctly before you start the process, be-

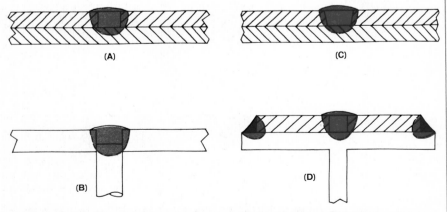

FIGURE 18-22 Applications of plug welds. (A) Plug weld joining two plates. (B) Plug weld used to attach a stray bolt in pressure vessel. (C) Plug weld hole beveling when plate thickness is over 1 in. and hole diameter is less than 1 in. (D) Plug weld reinforcing joint primarily attached by fillet welding.

cause you have so little manual control once the process gets started. Figure 18-23 shows the effect of work angle on flat-position SAW.

In Chap. 22 we will tell you about welding positioners, manipulators, and turning rolls that are used for positioning work. They are very productive machines that are used for all kinds of welding work, not just SAW.

Most SAW is done in the flat position to keep the large molten metal puddle from running off the work (Fig. 18-23A). Sometimes you can weld with your work slightly inclined. For example, in high-speed welding of 18-gauge steel, a better weld results when the work is inclined 15 to 18° and the welding is done downhill. Inclined welding is also done on many curved, preformed plate sections such as the bow and stern

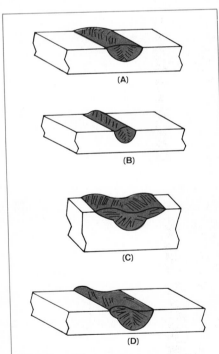

FIGURE 18-23 Effect of work inclination on submerged-arc weld profile. (A) Level weld on ½-in. plate. (B) Uphill (1½-in.- per-ft incline) weld on ½-in.-thick plate. (C) Downhill (1½-in.-per-ft incline) weld on ½-in.-thick plate. (D) Lateral (side) slope (⅝-in.-per-ft) weld on ½-in.-thick plate.

plates on ships. The angle of maximum inclination decreases as the plate thickness increases. All circumferential welding on the convex (curved out) surface of a weldment rotated on a horizontal axis under the welding is downhill welding, although control of the molten weld metal determines the puddle position.

Uphill welding affects the fusion zone contour and the weld surface as illustrated in Fig. 18-23B. The force of gravity causes the weld puddle to flow back and lag behind the welding wire. The edges of the weld lose metal which flows to the middle of the weld bead. As the angle of inclination increases, buildup of weld metal in the center of the bead increases, and the width of the weld decreases. Also, the bigger the weld puddle, the greater the penetration and center buildup. The limiting angle of inclination is about 6°, or a drop of about 1¼ in. [32 mm] for each foot [305 mm] of plate length. When higher welding currents are used, the maximum workable angle decreases. Greater workpiece inclination than that given above makes the weld uncontrollable and run-outs of molten weld metal are likely to occur.

Down welding affects the weld as shown in Fig. 18-23C. The weld puddle tends to flow toward the welding wire and preheats the base metal, particularly at the surface. This produces an irregularly shaped fusion zone, called a *secondary wash*. As the angle of declination increases, the middle surface of the weld is depressed, penetration decreases, and the width of the weld increases. Note that these effects are exactly opposite those produced by uphill welding.

Lateral (sideways) inclination of the workpiece produces the effects shown in Fig. 18-23D. The limit for a lateral slope is approximately 3°, or ⅝ in./ft [53 mm/m]. Permissible lateral slope varies somewhat, depending upon the size of the weld puddle.

Wire Position

Submerged-arc welding heads can be angled in different directions to point the electrode wire at the workpiece. You have to consider three things when you decide what angle to position the welding head and the wire. They are:

■ The alignment of welding wire in relation to the joint
■ The angle of tilt in the lateral (sideways) direction transverse to (across) the joint
■ The direction the welding wire points—forward or backward. Forward is the direction of travel of the sub-arc welding head. The wire makes an angle with the workpiece less than 90°. A backward-pointing wire makes an angle with the workpiece larger than 90°.

In general, a backward-pointing electrode wire produces greater, more uniform weld-metal penetration. Increased height and decreased width of the weld reinforcement also are likely. A forward-pointing electrode wire produces less weld-metal penetration and the weld will be wide and flat.

Here is how to set up your electrode wire for each of the several types of welding joints.

Butt joint wire angle See Fig. 18-24 for wire alignment for butt-joint welding. There should be no lateral tilt at all. You can angle the wire so that it's pointed forward or backward. You'll get good weld puddle and arc stability with a vertical welding wire when working on thick material (½ in. [13 mm] and thicker). However, for welding thin sections (14- to 16-gauge steel) a backward-pointing wire angled at 25 to 45° from the vertical will be needed for voltage stability.

Horizontal fillet wire angle See Fig. 18-25 for the best welding alignment when making a horizontal fillet weld. The centerline of the wire should not be on the joint centerline, but down from it toward the horizontal piece a distance equal to one-half to one-fourth of the wire diameter. The greater distance is used when making fillet welds of a larger size (about a ⅜-in. [9.5-mm] leg). Careless or inaccurate alignment will cause an unsatisfactory weld.

The lateral tilt for making horizontal fillet welds should be between 20 and 40° from the vertical. The exact angle is determined by either or both of the following factors:

■ Clearance for the nozzle or jaw assembly, especially when weld-ing structural sections to plate as shown in Fig. 18-26.

■ The relative thickness of the members forming the joint. (If the possibility of burning through one of the members exists, it will be necessary to aim the welding wire toward the thicker member.)

Wire that points forward, backward, or vertically downward can be used equally well for horizontal welding. When making a large fillet weld, the effects of changing the direction that the electrode wire points are relatively minor. Only in making small, high-speed fillet welds are the effects important. For making high-speed fillet welds in thin (14- to 16-gauge) steel, a backward-pointing wire with an angle of 25 to 45° from the vertical will improve voltage stability.

Flat fillet wire angle The electrode wire alignment for flat-position fillet welding is shown in Fig. 18-27. The lateral tilt for the electrode wire normally is vertical (zero lateral tilt angle). Occasionally in making positioned fillet welds, the wire is tilted slightly off the vertical as shown in Fig. 18-27B.

The use of forward- or backward-pointing electrode wire is the same for fillet welding in the flat position as it is for the horizontal fillet-welding position.

Circumferential rotated welds When welding on rotated assemblies (Fig. 18-28), the welding wire is aligned as in normal butt welding on a horizontal surface. In addition, the wire is usually positioned slightly ahead of the top point of the rotating work to avoid objectionable downhill and uphill effects. Sometimes when welding thin materials and when making fillet welds, these downhill and uphill effects will assist you in producing desirable weld shapes. Your best weld wire positioning must be determined by trial and adjustment.

Circumferential welding on rotated sections usually does not require any lateral tilt of your welding wire. A backward-pointing electrode wire is used on outside circumferential welds as shown in Fig. 18-28A; a forward-pointing wire is used on inside welds as shown in Fig. 18-28B.

18-7 STARTING THE WELDING ARC

The method used to start the weld in a particular position depends upon such factors as the time required for starting relative to the total setup and welding time, the number of pieces to be welded, the type of power source, and the importance of starting the weld on a

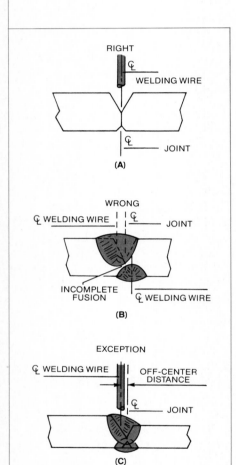

FIGURE 18-24 Effect of alignment in butt-joint welding. (A) Welding wire directly over the joint centerline. This alignment results in a centered fusion weld. (B) Wire not held to centerline results in incomplete fusion. (C) Off-center alignment sometimes used when butt-welding dissimilar metals, or plates of different thicknesses. ℄ symbol means "centerline."

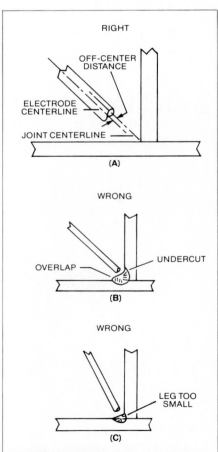

FIGURE 18-25 Effect of alignment during horizontal fillet welding. (A) Normal welding wire alignment for horizontal fillet welding. (B) Welding wire is too close to vertical surface. (C) Welding wire is too far from vertical surface.

particular spot on the workpiece. There are several SAW starting methods.

Note in the following descriptions that the actual controls and switches that you will use to start welding are not always named since these controls will differ depending upon the equipment. The instructions furnished with each power source and sub-arc system will tell you what controls to use to position the wire, close the contactor, start the travel carriage, jog the wire, and start welding.

Using a dc CC power source for SAW is very much like doing MIG welding with a dc CC machine. The arc length (voltage) is maintained by variations in the wire speed. A reference arc length (voltage) is set on the welding control which compares it to the actual arc voltage at the end of the welding wire. The wire feed rate adjusts itself automatically to maintain the arc voltage at the same reference voltage. The current is set by adjusting the power supply.

Retract Arc Start

There are several ways to start the welding arc. One way is called *retract starting*. Retract starting is possible only when the welding equipment is specifically adapted

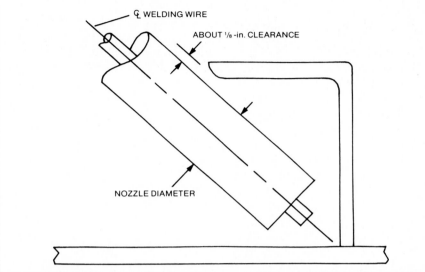

FIGURE 18-26 Lateral (side) tilt determined by nozzle clearance. Here the welding wire angle depends on the projection of the flange over the weld, and the diameter of nozzle or jaw assembly used. The ⅛-in. clearance allows for minor lateral adjustments of the welding wire.

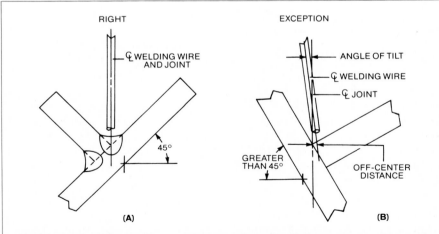

FIGURE 18-27 Alignment of fillet welds and submerged-arc electrode wire in flat position. (A) Normal alignment for fillet welds in flat position. Work is positioned at a 45° angle. Welding wire is centered in fillet corner. (B) When more than the usual amount of penetration is required, the work may be positioned at other than a 45° angle with the horizontal. The welding wire is positioned so that its centerline intersection with the joint is near the center of the joint. The wire may have to be tilted to avoid undercutting.

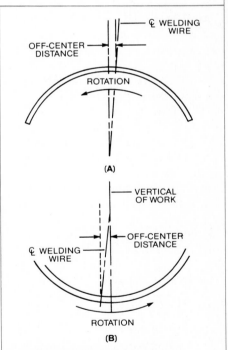

FIGURE 18-28 Welding wire position for submerged-arc welds made on rotated assemblies. (A) Outside circumferential welding. Incline welding wire so that it is in-line with radius of curvature of workpiece. (B) Inside circumferential welding. Incline welding wire so that it is in-line with radius of curvature of workpiece.

to this procedure. It is used when frequent starts have to be made over a short time, and when the starting position is particularly important.

With this method the electrode wire is jogged or inched down into the work. Make sure that you have good electrical contact with the workpiece and the welding wire. Then cover the end of the wire with SAW flux and turn the welding current on.

As soon as the welding contactor (power source switch) closes, the wire retract starting device in the equipment will momentarily reverse the wire feed motor, preventing the welding wire from fusing to the workpiece. As the wire retracts, it breaks contact with the work. It's like opening a switch when the current is flowing. When the weld circuit is broken by pulling the wire back from the work, a spark flows in the arc gap. The spark grows into a big welding arc. As soon as the arc is started, the wire feed direction is changed automatically to feed the wire into the workpiece and normal welding begins.

The starting conditions become critical if your workpiece is light-gauge metal. The wire must make only the lightest possible contact consistent with good electrical conductivity. Also, the wire size must be chosen to permit operation at high current densities since the higher the current density, the easier the start.

Steel-Wool Start

Another way to start a submerged-arc is called the *steel-wool start*. In this method, a small piece of steel wool is rolled into a $\frac{3}{8}$-in.-diameter [9.5-mm] ball. The steel wool is placed on a clean spot on the workpiece. Then the electrode is jogged down (without welding power) until it just touches the ball of steel wool, compressing it to half size.

The weld zone, wire electrode, and steel wool are all covered with flux and then the power supply and the wire feeder are

started simultaneously. Welding current flows through the steel-wool ball and "fuses" it away. The arc starts in the space left by the melted steel wool. The steel-wool start is one of the most frequently used starting methods, particularly when the starting time is only a small part of total setup and welding time.

Flying Starts

If the welding head is in motion before the power and wire feeder are on, the wire will drag across the base metal. A frequently used starting method is to use a starting strip or piece of extra metal temporarily welded to the leading edge of the joint, where the dragging electrode wire will strike an arc as soon as the power and wire feeder are on. This is called a *flying start* and it's often used to start a sub-arc weld, leaving any welding defects that might occur at the start of welding behind on the temporary starting strip.

A rather rigid mechanical welding-head carriage is needed for this type of start. Otherwise the feeding electrode will merely lift the carriage instead of starting the arc by scratching the workpiece.

In addition to starting on a starting strip, flying starts are used when the starting position is not important, and when frequent starts must be made such as making girth welds on small tanks or pipe. For such work, flying scratch starts can be made by normally feeding the wire down to the rotating workpiece through the layer of granulated submerged-arc flux, *provided* that fairly high current density is used.

You also can make a flying scratch start on a workpiece that initially is not rotating or moving. You start by inching the welding wire down until it lightly contacts the work. Then apply the submerged-arc flux. Start the carriage, and then apply the welding current. Because of the motion of the workpiece the welding wire

will not fuse to the workpiece and the weld will start.

Pointing the welding wire also helps with scratch starting because it increases the current density in the tip. Take a pair of bolt cutters and cut the welding wire end to a point and then inch the wire down until the end just contacts the workpiece. Apply the submerged-arc flux, start the welding current, and begin welding.

Molten-Metal Starts

Molten-metal starts are very simple. Whenever there is a molten puddle of submerged-arc flux, a weld can be started by simply running the electrode wire into the puddle and turning on the welding current and wire feed.

When two or more welding wires are separately fed into one weld puddle, as is sometimes the case in multiple-electrode welding, it is only necessary for you to start one wire by some other method to establish the puddle. The second wire will then start when they both are fed into the molten puddle.

High-Frequency Starts

High-frequency starts require special equipment. Using this method you don't have to do anything except close the starting switch. High-frequency starting is particularly useful for intermittent welding, or for welding at high production rates where a large number of starts is required.

When the welding wire approaches to within about $\frac{1}{16}$ in. [1.6 mm] above the workpiece, a high-frequency, high-voltage generator connected to the welding wire causes a spark to jump from the wire to the workpiece. This spark produces an ionized path through which the welding current can flow, and the welding action begins.

A CV submerged arc is usually started in one of two ways. One way is to run the electrode into the plate at the welding speed and let the short-circuit surge fuse off the wire, just as in MIG welding. Using this type of start, the current can rise to several thousand amps in a fraction of a second.

Another way to start the arc with a dc CV machine is to use the flying start. In this method, the welding-head carriage, which moves the wire feed mechanism along the weld seam, is turned on before the wire feed motor is started. Then, when the electrode runs into the weld zone the motion of the carriage causes the electrode to scratch the workpiece. The scratching action helps start the arc just like starting a shielded stick electrode.

You also can use pointed-wire starts and scratch starts with dc CV equipment.

18-8
VARIABLES AFFECTING WELD QUALITY

You have to set up several variables in SAW before you start if you want good quality welds. These variables, in approximate order of importance, are

- Welding current, wire diameter, and burn-off rates
- Welding voltage
- Welding speed
- Flux width and depth
- Mechanical adjustments

Welding Wire and Current

Welding current is the most influential variable in sub-arc welding. It controls the rate at which welding wire is burned off, the depth of fusion, and the amount of base metal fused. If the current is too high, the depth of fusion will be too great and the weld may melt through the backing. In addition to this, the higher heat developed may extend the heat-affected zone of the adjacent plate too much. Too high a current also means a waste of power and a waste of expensive welding wire in the form of excessive welding metal reinforcement.

If the current is too low, there is not enough weld penetration into the workpiece and not enough weld reinforcement on the joint.

Welding Voltage

Next in importance to welding current is the welding voltage. This is the electrical potential difference between the tip of the welding wire and the surface of the molten weld metal. The welding voltage varies with the length of the gap between the welding wire and the molten weld metal. If the gap increases, the welding voltage increases; if the gap decreases, the welding voltage decreases.

The welding voltage has little effect on the amount of welding wire deposited. Weld-metal deposition rates are determined mainly by your welding current. The voltage determines the shape of the fusion zone and weld reinforcement. High welding voltages produce wider, flatter, less deeply penetrating welds than low welding voltages.

Welding Speed

With any combination of welding current and voltage, the effects of changing the welding speed conform to a general pattern.

If the welding speed is increased:

- Power or heat input per unit length of the weld is decreased.
- Less welding wire is used up per unit length of weld.
- Consequently, there is less weld reinforcement.

If the welding speed is decreased:

- Power or heat input per length of weld is increased.
- More welding wire is used up per unit length of weld.
- Consequently, there is more weld reinforcement.

In addition to this pattern, welding speed may have another effect on the finished weld. Normally only welding current affects the penetration of the weld into the workpiece. However, if the welding speed is decreased beyond a certain point, the penetration also will decrease. This happens because a good portion of the molten weld puddle will be beneath the welding wire and the penetrating force of the arc will be cushioned by the puddle. Conversely, if the speed is increased beyond a certain point, the penetration of the weld into the joint will increase since the welding wire moves ahead of the weld puddle.

Width and Depth of Flux

The width and depth of the layer of flux will affect the appearance and soundness of the finished weld as well as the welding action itself. If the sub-arc flux layer is too deep, a rough, ropey weld is likely to result. The gases generated during welding cannot easily escape through high volumes of slag, and the surface of your molten metal will be irregular, distorted, and full of porosity.

If the sub-arc flux layer is too shallow, your welding zone will not be entirely submerged. Arc flashing and spattering will occur. The weld will have a bad appearance and may be porous or oxidized.

It is seldom that too narrow a layer of flux is applied. The safest procedure is to apply a layer that is three times the width of the fused portion. In large welds, a greater allowance for the width of the heat-affected zone has to be made. A layer of flux that is too narrow messes up your normal sideways flow of weld metal, resulting in reinforcement that is narrow, steep-sided, and poorly "faired" (smoothed out) into the baseplate or edges of the weld joint.

An optimum flux depth exists for any set of welding conditions. This depth can be established by slowly increasing the flux layer until the welding action is submerged and arc flashing no longer occurs. The gases will then puff up quietly around the welding wire. Sometimes you can see the gases burning in the air.

The unfused flux can be removed a short distance behind the welding zone where the fused slag has solidified. However, under certain conditions, you may not want to disturb the material until the heat has become better distributed throughout the section.

The fused submerged-arc slag should not be forcibly loosened while the weld metal is at a high temperature. Let it cool first. If allowed to cool, the slag will probably detach itself or you can knock it off with little effort. Sometimes you may want to knock off a tiny bit of the hot slag with a hammer just to get a quick look at the weld surface. That's OK on a solid weld.

It's very important that no foreign material gets into your flux. Submerged-arc welding flux that has not fused into slag is picked up with a broom, or often with a vacuum sweeper-like device that follows the welding head. To prevent impurities from getting into your reclaimed flux, a space about a foot wide should be cleaned on either side of your welding zone before you lay the flux down. The recovered flux should be passed through a screen with openings no larger than ⅛ in. [3.2 mm] before you reuse it to get rid of fused pieces of slag.

Submerged-arc flux is thoroughly dry when it is shipped to you. If it becomes damp or wet, it should be dried in an electrode baking oven at 400°F [205°C] or higher for several hours before being used. The smallest amount of moisture in the flux can cause porosity or even hydrogen embrittlement in your weld.

Mechanical Adjustments
The position of your welding wire must be maintained to control the shape of the weld and depth of penetration. The wire may be guided mechanically or manually adjusted as the weld moves along.

While the welding is going on, you should inspect the backing to see whether it is tight up against the underside of the joint. If it isn't, too much metal may flow into the space, resulting in reduced weld reinforcement on the top side of the joint, undercutting, and a ruined weld.

Typical Operating Conditions
Table 18-3 gives you the generally accepted current ranges for commonly used SAW wire diameters.

The contact points in the welding head that put current into the sub-arc wire must be clean and in good condition if you are going to get these current values. Poor current transfer between your welding wire and contacts will cause the wire to heat irregularly above the welding zone. This condition should be corrected by

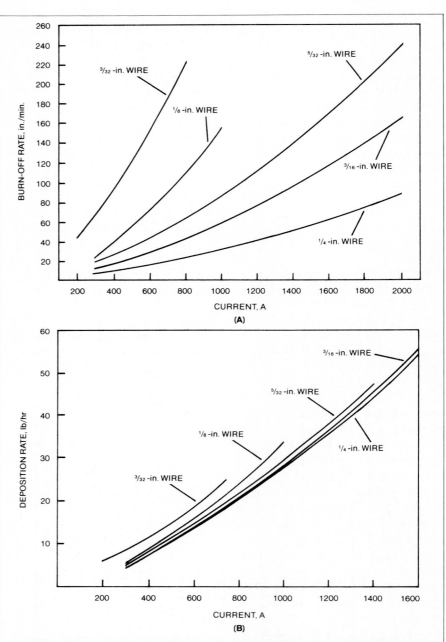

FIGURE 18-29 (A) Alternating-current burn-off rate and (B) deposition rate for different submerged-arc wires. Notes: For DCRP subtract 10 to 15 percent from the ac burn-off rate. For DCSP add 10 to 15 percent to burn-off rate. Electrode extensions are set at 8 times wire diameter. Burn-off rates will vary by ±10 percent with different fluxes.

TABLE 18-3 Welding current versus wire diameter

Wire Diameter, In.	Current range, A
3/32	120– 700
1/8	220–1100
5/32	340–1200
3/16	400–1300
1/4	600–1600
5/16	1000–2500

cleaning or remachining the contact points.

In Fig. 18-29A we've plotted the relationship between wire feed speed in inches per minute versus ac amps for several wire sizes. Note that when using dc reverse polarity (electrode positive) the wire feed speed should be *lowered* by 10 to 15 percent. When using dc straight polarity, the wire feed speed should be *increased* 10 to 15 percent for a given amperage.

In Fig. 18-29B we've plotted weld-metal deposition rate versus ac amps for several wire sizes. The same changes from the graphed data hold. When using dc reverse polarity, the wire feed speed should be *lowered* by 10 to 15 percent. When using dc straight polarity, the wire feed speed should be *increased* 10 to 15 percent.

Wire extension referred to in Fig. 18-29B refers to the distance between the welding contact tip and your plate, just like in GMAW. Generally this extension distance is set at eight times your wire diameter. Therefore if you are using 1/8-in. [3.2-mm] wire, your electrode extension should be 1 in. [25 mm].

Increasing the extension of the wire out of the contact tip increases I^2R resistance heating of the wire as it passes from the contact tip to your plate. This also increases your weld-metal deposition rate per ampere of current. However, excessive extensions will give you bad wire positioning

and unstable arcs, causing irregular weld beads, reduced and uneven weld penetration, and poor-quality welds.

When normal SAW conditions are used, there is a maximum welding current for every vessel girth diameter. This current is the highest that can be used without having a weld-metal run-out. It is affected by the welding speed and fluidity of the weld metal and its alloy composition. Figure 18-30 graphs the most likely maximum current versus the outside diameter of a round pressure vessel or cylinder made of carbon or HSLA steel.

Finally, Figs. 18-31 through 18-44 give you welding conditions for making single- and multiple-pass submerged-arc welds in different joints.

18-9
MULTIPASS WELDING
In Fig. 18-41, we show you the best way to handle multipass welding with SAW. When plate thickness exceeds the limits of two-pass techniques, or where you can't provide accurate joint fit-up so you can't use high welding currents, multipass SAW should be used.

Where possible a split-pass procedure should be used, as shown in Fig. 18-41A. This procedure makes it easy to remove flux and prevents weld cracking. Each weld pass should be slightly convex (bowed up) as shown on the left in the figure (Fig. 18-41A) to help you in slag removal and to prevent your welds from cracking. In certain base metals that are sensitive to heating affects, you will want to use multipass welding to reduce the heat input into the heat affected zone of the base metal.

You also may have to preheat and use postweld heating to slowly cool some of the steels you will weld with submerged-arc. The preheating is particularly useful because of the thickness of the sections joined with this process. Remember that a massive weldment is a heat sink. It will rapidly chill your weld metal if the base metal is not preheated first. Chilled weld metal will have a tendency to crack because the rapid cooling hardens it and reduces ductility.

Preheating from 250 to 300°F [120 to 150°C] is generally used when making multipass welds in sections greater than 1 to 2 in. [25 to 51 mm] thick. To help you

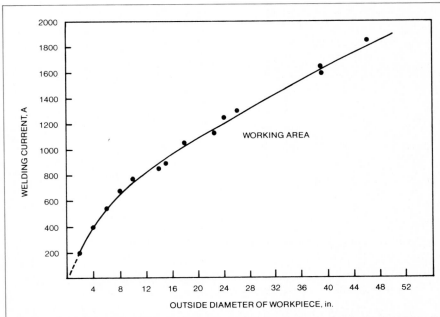

FIGURE 18-30 Carbon steel welding parameters. Maximum welding current for circumferential seam welding.

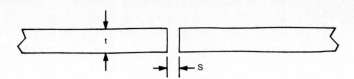

Gauge/ US.S	t, in.	S Root Opening, in.	Welding Wire		Welding Flux, lb/ft (approx.)	Welding Current*		Speed, in./min
			Diam., in.	lb/ft (approx.)		Amperes	Volts	
16	0.062 (1/16)	0	3/32	0.02	0.02	250–350	22–24	100–150
14	0.078 (5/64)	0	3/32	0.02	0.02	325–400	24–26	100–150
12	0.109 (7/64)	0	3/32	0.03	0.03	350–425	24–26	75–100
10	0.140 (9/64)	0–1/16	3/32	0.06	0.05	400–475	24–27	50–80
8	0.172 (11/64)	0–1/16	1/8	0.07	0.06	500–600	25–27	40–70
7	0.187 (3/16)	0–1/16	1/8	0.10–0.13	0.09	575–650	25–27	35–45
3	0.250 (1/4)	0–3/32	5/32	0.14–0.23	0.12–0.20	750–850	27–29	30–35
0	0.312 (5/16)	0–3/32	3/16	0.25–0.30	0.21–0.26	800–900	26–30	26–29

* AC is used successfully for light-gauge welding, but reversed-polarity DC is preferred for high-speed production welding.

FIGURE 18-31 Carbon steel welding parameters. Typical production preparations and welding conditions for x-ray quality sub-arc welds in steels conforming to ASME specifications SA-201, SA-204, and SA-212.

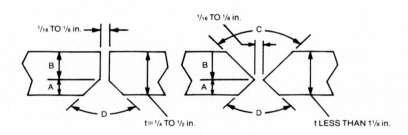

t Thickness, in.	Single Machine Pass Min. ID, in.	Finishing Weld					Welding Wire		Manual Weld Backing	
		B, in.	C, degrees	Min., amperes	Max., volts	Speed, in./min	Diam., in.	lb/ft (approx.)	A, in.	D
3/8	14	3/16*	60	700	33	18	5/32	0.26	3/16	90
1/2	14	1/4*	60	850	33	16	5/32	0.32	1/4	90
9/16	15	5/16*	60	900	35	16	3/16	0.42	1/4	90
5/8	15	5/16	90	1000	35	14	3/16	0.63	5/16	90
11/16	16	3/8	90	1000	35	13	3/16	0.81	5/16	90
3/4	18	3/8	90	1050	35	12	3/16	0.86	3/8	90
13/16	20	7/16	70	1100	35	13	3/16	0.77	3/8	90
7/8	20	1/2	70	1150	35	12	1/4	0.93	3/8	90
1	22	5/8	70	1250	35	11	1/4	1.32	3/8	90
1 1/8	30	3/4	70	1350	36	10	1/4	1.76	3/8	90

* Chip B side for submerged-arc weld after manual weld is installed. Good overlap between manual and submerged-arc welds is needed to avoid pinholes.
NOTE: 1. C—±5°.
    2. Speed—to be varied slightly to secure desired reinforcement.
    3. Check accuracy of meters frequently.
    4. Gas metal-arc preferred.

FIGURE 18-32 Carbon steel welding parameters. Double-V-groove welds with manual weld backing (3/8 in. to 1 1/8 in.).

understand the importance of preheating, if your weld metal has solidified and already has cooled to 700°F [370°C], it will have about 625°F [330°C] more to go to cool to room temperature. A preheat of 300°F [150°C], for example, will reduce the cooling rate to one equivalent to a 400°F [205°C] section. The drop in temperature from 700 to 75°F [370 to 24°C] will be much faster than from 700 to 300°F [370 to 150°C]. Since slower cooling avoids the formation of brittle, low-ductility martensite, preheating avoids cracking. The higher the carbon or alloy content of the steel, the more likely it is that preheating will be needed.

If you want to put preheating in another perspective, just one thick plate is worth a lot more than a new car. One big crack is more than a little embarrassing. A little preheating could keep you employed.

Large weldments also will have to be stress-relieved. Most likely the procedures will be spelled out by some fabricating code such as

(Continued on page 490)

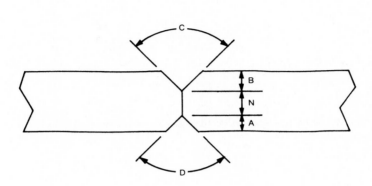

			Finishing Weld								Backing Weld				Rod
Thickness, in.	Min. ID, in.	B, in.	C, deg.	Min., amperes	Max., volts	Speed, in./min	Wire Diam. in.	N, in.	A, in.	D, deg.	Amperes	Max., volts	Speed, in./min	Wire Diam. in.	lb/ft (approx.) per Joint
3/8	14	1/8	Chip*	600	33	20	5/32	3/8	0	0	550	33	22	5/32	0.40
1/2	14	1/8	Chip*	900	35	16	5/32	3/8	1/8	90	650	35	18	5/32	0.59
9/16	16	1/8	Chip*	1000	35	16	3/16	3/8	3/16	90	700	35	18	3/16	0.62
5/8	18	3/16	90	1050	35	14	3/16	1/4	3/16	90	750	33	16	3/16	0.65
11/16	20	1/4	90	1100	35	13	3/16	1/4	3/16	90	800	33	16	3/16	0.73
3/4	21	1/4	90	1150	35	13	1/4	5/16	3/16	90	850	33	16	1/4	0.80
13/16	22	5/16	90	1200	35	13	1/4	5/16	3/16	90	900	33	16	1/4	0.88
7/8	24	5/16	90	1250	35	12	1/4	5/16	1/4	90	950	34	15	1/4	0.95
1	24	3/8	90	1300	36	11	1/4	5/16	5/16	90	1000	34	15	1/4	1.10
1 1/8	24	3/8	90	1350	36	11	1/4	3/8	3/8	60	1050	34	14	1/4	1.44
1 1/4	30	1/2	70	1450	36	10	1/4	3/8	3/8	60	1100	35	13	1/4	1.60
1 3/8	32	1/2	70	1500	37	10	1/4	7/16	7/16	60	1200	35	11	1/4	1.77
1 1/2	36	5/8	70	1600	37	9	1/4	7/16	7/16	60	1300	35	10	1/4	1.96

* Chip vee in second side after backing weld is made.

Recommended Tolerances:
1. Gap between root faces—1/32 in. (use flux backing to exceed 1/32 gap).
2. N—plus 1/16 in. minus zero.
3. Misalignment between edges of root faces—1/8 in.
4. C or D—allowance must be made in edge preparation for distortion due to rolling.
5. Amperes used in first pass will vary according to fit-up and should be maximum joint can carry.
6. Speed—to be varied slightly to secure reinforcement.
7. Check accuracy of meters frequently.

FIGURE 18-33 Carbon steel welding parameters. High-quality butt welds in high-quality steels. Two-pass double-V-groove welds (3/8 in. to 1 1/2 in.).

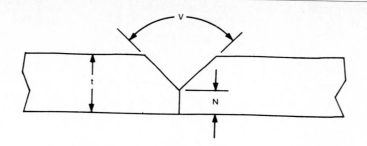

t Thickness, in.	Vee*, degrees	N*, in.	Amperes	Volts	Speed, in./min	Welding Wire	
						Diameter, in.	lb/ft (approx.)
3/16	60	1/8	500–575	28–31	29–50	5/32	0.11–0.13
1/4	60	1/8	725–825	29–32	28–45	5/32	0.14–0.23
5/16	60	1/8	775–900	30–33	26–40	3/16	0.25–0.30
3/8	60	1/8	900–1000	32–36	24–27	3/16	0.28–0.35
7/16	60	1/8	1000–1100	32–36	22–25	3/16	0.33–0.40
1/2	60	3/16	1075–1175	34–37	20–23	1/4	0.38–0.45
5/8	45	3/16	1150–1250	35–38	16–19	1/4	0.50–0.58
3/4	45	3/16	1200–1300	36–39	13–14	1/4	0.60–0.75

NOTE: The weight of submerged-arc flux consumed (amount fused) is 85 to 100 percent of the weight of wire deposited.
* Recommended maximum tolerances:
1. Gap between root faces—1/32 in.
2. N—± 1/32 in.
3. V—± 5°.
4. Check accuracy of meters frequently.

FIGURE 18-34 Carbon steel welding parameters for structural-quality welds—that is, welds with 90 percent penetration. Preparations and conditions for structural-quality welds may be used when there is no need for x-ray quality. Shown here are specifications for single-V-groove welds with copper backing (3/16 in. to 3/4 in.).

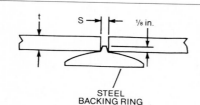

STEEL BACKING RING

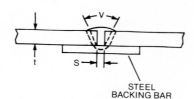

STEEL BACKING BAR

t Thickness, in.	V Edge Preparation	S Min. Root Opening, in.	Min. Thickness Backing Bar, in.	Amperes	Volts	Speed, in./min	Welding Wire	
							Diameter, in.	lb/ft (approx.)
3/16	Square	1/16	3/16	750	27	28–40	5/32	0.13
1/4	Square	1/8	1/4	850	27	22–30	5/32	0.19
5/16	Square	1/8	1/4	875	28	20–30	3/16	0.22
3/8	Square	1/8	1/4	900	28	18–30	3/16	0.25
7/16	30° Vee	3/16	3/8	950	30	12–20	3/16	0.49
1/2	30° Vee	3/16	3/8	975	30	12–20	3/16	0.55
5/8	30° Vee	3/16	1/2	1100	30	11	1/4	0.83
3/4	30° Vee	3/16	1/2	1200	30	9	1/4	1.10

FIGURE 18-35 Carbon steel welding parameters for structural-quality welds. These welds are most often used in structural welding, but x-ray quality can be obtained if close control is held over variables. Shown here are specifications for square groove and single-V-groove welds with steel backing (3/16 and 3/4 in.).

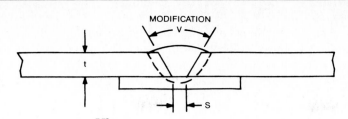

t Thickness, in.	V Edge Preparation	S Min. Root Opening, in.	Min. Thickness Backing Bar, in.	Amperes	Volts	Speed, in./min	Welding Wire Diameter, in.	lb/ft (approx.)
¼	45° Vee	⅛	¼	800	30	18	5⁄32	0.27
5⁄16	45° Vee	⅛	¼	800	30	16	3⁄16	0.30
⅜	45° Vee	⅛	¼	800	30	12	3⁄16	0.40
½	45° Vee	3⁄16	⅜	960	30	9.5	3⁄16	0.61
⅝	45° Vee	3⁄16	⅜	{1000*	33	10	3⁄16	0.93
				800	35	12	3⁄16	0.93
¾	45° Vee	3⁄16	⅜	{1000*	33	10	¼	1.24
				1000	36	9.5	¼	1.24

NOTE: Root opening should be uniform.
* First pass of two-pass weld.

FIGURE 18-36 Modified V-groove butt-joint welding parameters for carbon steel structural-quality welds.

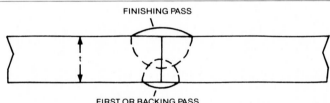

t, in.		Amperes	Volts	Speed, in./min	Welding Wire Diameter, in.	lb/ft (approx.)	Preparation before Finishing Pass
¼	BP*	400	32	28–45	3⁄32 or ⅛	0.10	None
	FP†	500	30	27–45	3⁄32 or ⅛	0.12	
5⁄16	BP*	420	32	28–40	3⁄32 or ⅛	0.11	None
	FP†	550	30	26–40	⅛ or 5⁄32	0.15	
⅜	BP*	500	32	28–32	⅛ or 5⁄32	0.14	None
	FP†	650	32	24–32	⅛ or 5⁄32	0.18	
7⁄16	BP*	600	33	24	⅛ or 5⁄32	0.20	None
	FP†	700	33	22–27	⅛ or 5⁄32	0.25	
½	BP*	650	33	22	⅛ or 5⁄32	0.23	None
	FP†	750	35	20–25	⅛ or 5⁄32	0.27	
9⁄16	BP*	700	33	20	⅛ or 5⁄32	0.25	None
	FP†	800	35	18	⅛ or 5⁄32	0.32	
⅝	BP*	725	33	18	5⁄32 or 3⁄16	0.28	Gouge
	FP†	850	35	16	5⁄32 or 3⁄16	0.38	
11⁄16	BP*	850	38	12	5⁄32 or 3⁄16	0.40	Gouge
	FP†	1100	42	12	3⁄16	0.50	
¾	BP*	960	38	12	3⁄16	0.45	Gouge
	FP†	1100	42	12	3⁄16	0.50	

* BP = Backing Pass
† FP = Finishing Pass
Flame-gouging a groove ⅛ to 5⁄16 in. deep assures adequate penetration of the finishing weld without excessive reinforcement.

FIGURE 18-37 Carbon steel welding parameters for structural-quality welds. Shown here are specifications for two-pass square and single-V-groove welds (¼ to ¾ in.).

the ASME Boiler and Pressure Vessel Code. See Chap. 16 on SMAW for more information on preheating and postheating of thick weldments and hardenable steels (see pages 408 to 411).

An example of the use of multipass welding is shown in Fig. 18-43. For pressure-vessel circumferential welds such as head-to-head and shell butts, this double-bevel plate preparation with semiautomatic or automatic GMAW (MIG) used to handle varying fit-up in the root area is an excellent combination procedure. Fill passes are then welded with SAW to provide consistent quality, low-cost welds. The GMAW process is the best choice for manual or automatic root welding or for welding the first pass because the resulting weld metal is free of slag, which means that the critical root pass will be free of welding defects caused by slag particles that might otherwise be included in the weld metal.

Multipass procedures must be used for welding plates over 2 in. [51 mm] thick. Figures 18-41 to 18-44 give you examples of joint preparation and both single- and two-sided welding conditions for thick plates.

t, in.		Amperes	Volts	Speed, in./min	Welding Wire		Total Vee Angle, degrees
					Diameter, in.	lb/ft	
9/16	BP*	850	33	20	5/32	0.38	75
	FP†	650	33	22	5/32	0.21	
5/8	BP*	900	33	18	3/16	0.44	75
	FP†	700	33	22	3/16	0.21	
3/4	BP*	950	33	16	3/16	0.50	60
	FP†	750	33	20	3/16	0.25	
7/8	BP*	1100	35	14	3/16	0.60	45
	FP†	800	35	18	3/16	0.32	
1	BP*	1200	35	12	1/4	0.75	45
	FP†	850	35	18	1/4	0.34	

* BP = Backing pass
† FP = Finishing pass
Flame-gouging a groove 1/8 to 5/16 in. deep assures adequate penetration of the finishing weld without excessive reinforcement.

FIGURE 18-38 Carbon steel welding parameters for structural-quality welds. Shown here are specifications for two-pass square and single-V-groove welds (9/16 to 1 in.).

Normal Fillet Size L, in.	Amperes	Volts	Speed, in./min	Welding Wire	
				Diameter, in.	lb/ft (approx.)
1/8	400	25	36–65	3/32	0.04
3/16	500	25	32–40	1/8	0.07
1/4	650	27	28–35	5/32	0.11
5/16	650	27	22	5/32	0.17
3/8	750	29	18	3/16	0.25
1/2	900	32	16	3/16	0.43
5/8	1050	32	13	1/4	0.66
3/4	1150	32	11	1/4	0.95

FIGURE 18-39(A) Carbon steel fillet welds in flat position (1/8- to 1½-in. leg).

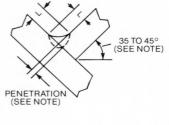

NOTE:
INCLINATION OF 45° GIVES NORMAL PENETRATION (0.4 L TO 0.5 L). DEEPER PENETRATION (0.55 L) OBTAINED WITH INCLINATION OF 35°

(A)

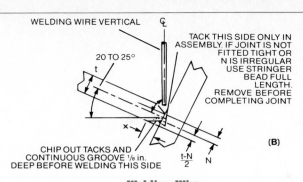

t, in.	N, in.	X, in.	Bevel a, degrees	Welding Wire Diameter, in.	lb/ft (approx.) per Joint	Weld	Amperes	Max. Volts	Speed, in./min
⅝	¼	⅜	26.5	⁵⁄₃₂	0.56	First	750	30	9
						Final	800	30	7
¾	⅜	½	21.0	³⁄₁₆	0.72	First	950	30	8.5
						Final	1050	30	8.5
1	⁷⁄₁₆	⅝	24.0	³⁄₁₆	1.33	First	1050	30	8
						Final	1150	30	7.5
1¼	⁷⁄₁₆	¾	28.5	¼	2.75	First	1100	30	7
						Final	1150	30	7
1½	⁷⁄₁₆	⅞	31.0	¼	3.00	First	1150	30	6.5
						Final	1200	30	6

FIGURE 18-39(B) (*Continued*) Welds in flat position (⅝- to 1½-in. leg).

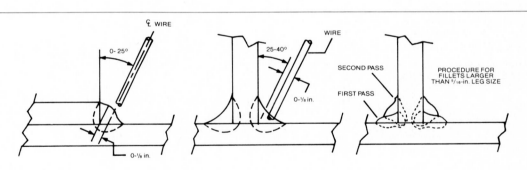

Dimension of Vertical Leg, in.	Manual Fillet Size (approx.) for Equiv. Strength, in.		Amperes	Volts	Speed, in./min	Welding Wire Diameter, in.	lb/ft (approx.)
⅛	⅛–³⁄₁₆		400	25	30–65	³⁄₃₂	0.06
⁵⁄₃₂	³⁄₁₆–¼		450	27	26–55	⅛	0.07
³⁄₁₆	¼		500	27	22–40	⅛	0.09
¼	⁵⁄₁₆		550	28	20–30	⅛	0.14
⁵⁄₁₆	⅜		650	28	18–25	⁵⁄₃₂	0.20
⅜	½		700	28	15–20	⁵⁄₃₂	0.29
⅜		1st pass	520	30	22	⅛	0.31
		2d pass	520	30	22	⅛	
½		1st pass	650	33	22	⁵⁄₃₂	0.41
		2d pass	750	35	20	⁵⁄₃₂	
⅝		1st pass	725	33	18	⁵⁄₃₂	0.575
		2d pass	850	35	16	⁵⁄₃₂	
¾		1st pass	800	35	9	⁵⁄₃₂	1.12
		2d pass	820	33	9	⁵⁄₃₂	

FIGURE 18-40 Carbon steel fillet welds in horizontal position (⅛- to ¾-in. vertical leg).

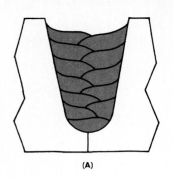

(A)

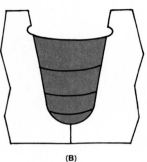

(B)

FIGURE 18-41 Multipass welding on thick plates with wide joints. (A) Procedure with overlapping weld passes is preferred to (B) procedure with thick beads. The previously deposited weld metal under each bead in (A) is further refined by the heat of overlapping passes, increasing the toughness of the finished weld metal.

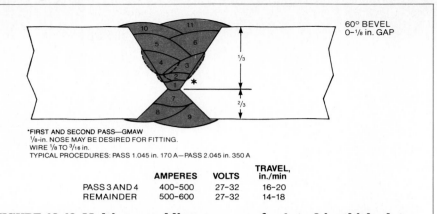

*FIRST AND SECOND PASS—GMAW
1/8-in. NOSE MAY BE DESIRED FOR FITTING.
WIRE 1/8 TO 3/16 in.
TYPICAL PROCEDURES: PASS 1.045 in. 170 A—PASS 2.045 in. 350 A

	AMPERES	VOLTS	TRAVEL, in./min
PASS 3 AND 4	400–500	27–32	16–20
REMAINDER	500–600	27–32	14–18

FIGURE 18-43 Multipass welding sequence for 1- to 2-in.-thick plates.

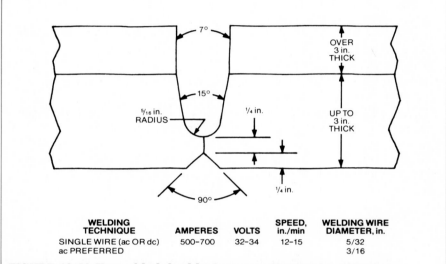

WELDING TECHNIQUE	AMPERES	VOLTS	SPEED, in./min	WELDING WIRE DIAMETER, in.
SINGLE WIRE (ac OR dc) ac PREFERRED	500–700	32–34	12–15	5/32 3/16

FIGURE 18-44 Two-sided double-J-groove, V-groove weld joint.

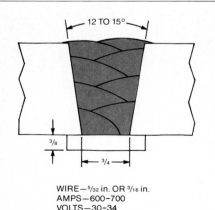

WIRE—5/32 in. OR 3/16 in.
AMPS—600–700
VOLTS—30–34
TRAVEL—12 TO 15 in./min

FIGURE 18-42 Multipass welding with overlapping passes on 2-in. plates and thicker ones as well.

REVIEW QUESTIONS

1. What material and section thicknesses is the SAW process mostly used on?

2. What is the practical *minimum* thickness for steel to be welded with the SAW process?

3. Why is sub-arc used?

4. What welding position limitations does submerged-arc welding have? Why?

5. Why are different fluxes used for SAW?

6. What as-deposited weld metal tensile strength would you expect to get if you used an AWS F70 grade flux and an EL12 wire?

7. Name at least one AWS welding wire grade that you might use for joining nickel-steel plate. What AWS filler metal specification would you use to look up the mechanical properties of the weld metal your wire probably will produce?

8. Are only ac, or only dc, or both ac and dc welding power sources used for submerged-arc welding?

9. If the distance between the electrode wire and workpiece increases, does the welding voltage increase, decrease, or remain the same?

10. Name the five kinds of weld-metal joint backing used for SAW. Which ones are temporary, and which backing methods become a part of the final weldment? Why?

11. List six ways to preclean a weld joint before using the SAW process. Which method or methods can present health or safety hazards to you if you are not careful? Why?

12. What is the typical current range, in amps, for these three steel SAW welding electrode wire diameters?

a. $\frac{3}{32}$ in.

b. $\frac{3}{16}$ in.

c. $\frac{5}{16}$ in.

Welding with Submerged Arc

Most welding schools will not have submerged-arc equipment. Sub-arc is a production welding process that often requires welding positioners, manipulators, and turning rolls (which we discuss in Chap. 22), as well as the sub-arc equipment to do the job. The plate samples (over 1 in. [25 mm] thick) that show the process to its best advantage also are very expensive.

Follow your welding instructor's advice. If your school owns or can develop a small sub-arc system, first try making sub-arc bead-on-plate welds. Then try making several passes on grooved butt joints on plate over 1 in. thick, using different backing methods. If you have the equipment, try making pipe welds with your sub-arc system using turning rolls. Use the Shop Practice for FCAW and steel pipe welding in Chap. 17 as a guide to your SAW Shop Practice, modifed by the data we provide in this chapter.

If your welding school does not have the equipment, then make sure that you fully understand this chapter. Nevertheless, if you get assigned to a submerged-arc welding job, you will find that your manual welding skills with other arc welding processes, plus the information in this chapter, will make it easy to learn submerged-arc welding.

Applying Metal Surfaces

With what you know now, you can quickly learn to use MIG welding wire to put a stainless-steel liner on a carbon-steel reactor vessel for a paper mill to protect the vessel from corrosion. There is a variant on the MIG process that uses stainless-steel strip instead of wire, because the strip will lay down a wider bead much faster for jobs like this. Or maybe your problem is on a construction site and you have to put down a hardfacing layer of wear- and abrasion-resistant metal on the dipper teeth of a dragline with special hardfacing stick electrodes (Fig. 19-1). You might even want to build up the worn surfaces of a truck axle while you are at it, to make the axle as good as new. Oxyfuel gas metallizing (a metal-spraying process) might be best for that.

Perhaps you have been asked to fix a hard stamping die, put a new edge on some metal shears, and resurface a stack of rock-worn rock drills. Your TIG outfit can do all of that—with the correct alloys.

The point we're making is that surfacing is a very large subject. It can use oxyfuel processes, SMAW electrodes, MIG or flux-cored wire, and even GTAW to lay down a useful surface on a previously fabricated workpiece. Table 19-1 lists them.

Surfacing includes hardfacing, which usually means maintenance welding, but it also includes the fabrication of original equipment (like surfacing our low-cost carbon-steel reactor vessel with the far more expensive stainless-steel lining).

Obviously, one chapter in one book could never cover the entire subject. (The subject is worth a book in itself.) We won't even try. What we'll do is give you a broad overview of some of the more important things you should know about the key filler metals that are used for surfacing and hardfacing and explain how one new process that we haven't covered yet, metal spraying, works.

19-1 HARDFACING TERMINOLOGY

Whatever surfacing work you do, whatever hardfacing process you use, you'll run across the same words used over and over again. Let's tell you what they mean.

The first term is *hardness*. You already know a lot about measuring hardness from the lesson on mechanical properties. Hardness, for lack of a better definition, is a measure of the resistance to deformation (like trying to press a dent in a workpiece with a portable Rockwell or Brinell hardness tester).

Hot Hardness

Hot hardness is hardness at high temperatures. As you know, almost everything gets softer as it gets hotter, except graphite. Hot hardness is a very important property of certain weld metals. It refers to metals and alloys that remain *relatively* hard at high temperatures, compared with conventional engineering metals such as carbon steel. Hot-hard metals and alloys are used, for example, for everything from surfacing jet engine parts to fly-ash hoppers in fossil-fuel boilers. Tungsten and molybdenum are probably the most important alloying elements in ferrous metals that have hot hardness.

Tungsten and molybdenum react with carbon to make tungsten carbides and molybdenum carbides. These carbides are very hard. The atoms of tungsten and molybdenum also are very large relative to carbon atoms and they don't move around very much in hot solid metals. As a result their carbides remain as very small grains in the alloy, only going into solid solution very slowly, thus contributing a lot to the material's hardness at high temperatures.

At temperatures up to 1100°F [593°C], the as-deposited Rockwell hardness of weld metal containing these alloying elements falls off very slowly, typically from Rockwell C 60 at room temperature to Rockwell C 47 (448 Brinell) at 1100°F [593°C]. It's one of the properties you may want to use when you do hardfacing or other surfacing work.

At higher temperatures the hardness falls off a lot faster. At about 1200°F [650°C] the maximum Rockwell hardness of a typical surfacing-alloy weld metal containing tungsten or molybdenum (or both of them) drops to about Rockwell C 30 (283 Brinell). Nothing stays hard forever. Some metals just stay hard longer (when they get hotter) than other materials.

Impact Resistance

As you know, impact strength normally refers to how many foot-pounds (ft·lb) or joules (J) of energy a piece of metal can absorb without breaking when it's at a certain temperature. You already know from other chapters that weld metal often will give you a minimum of 15 ft·lb [20 J] impact at some specified low temperature. Some as-deposited weld metals will give you a lot more than that, even at very low temperatures. Austenitic stainless steels are an example.

FIGURE 19-1 This rock is harder than the steel from which this Caterpillar D-10 bulldozer's ripper tooth (in the rear) and blade (in front) are made. The D-10 Cat couldn't last long ripping up solid rock layers and moving them with the blade without special, extremely abrasion- and impact-resistant, weld-metal overlays. The Stoody Company of Industry, California (whose products protect this bulldozer from wear), specializes in developing hardfacing filler metals and corrosion-resistant surfacing alloys which are marketed through welding distributors across the United States.

TABLE 19-1 Surfacing methods and materials

Process	Mode of application	Hardfacing		Cladding and buildup		Remarks
		Filler metal form	Applicable filler metals	Filler metal form	Applicable filler metals	
Welding Processes						
Oxyfuel gas	Manual and automatic	Bare cast or tube rod	Co, Ni, and Fe base; tungsten-carbide composites	—	—	—
	Manual	Powder	Co, Ni, and Fe base; tungsten-carbide composites	—	—	—
Bare metal arc	Manual and semiautomatic	Bare solid or tube wire	Austenitic manganese steel	—	—	Limited use
Shielded metal arc	Manual	Flux covered cast rod, wire, or tube rod	Co, Ni, and Fe base; tungsten-carbide composites	Flux covered wire	Stainless steel, Ni, Cu, and Fe base	—
Flux cored electrode (self-shielded)	Semiautomatic	Flux cored wire	Fe base	Flux cored wire	Stainless steel, Fe base	—
	Automatic	Flux cored wire	Fe base	Flux cored wire	Stainless steel, Fe base	—

Source: American Welding Society

(Table 19-1 continued)

TABLE 19-1 Surfacing methods and materials (*Continued*)

Process	Mode of application	Hardfacing Filler metal form	Hardfacing Applicable filler metals	Cladding and buildup Filler metal form	Cladding and buildup Applicable filler metals	Remarks
Welding processes (*Continued*)						
Gas tungsten arc (GTA)	Manual	Bare cast or tube rod	Co, Ni, and Fe base; tungsten-carbide composites	Bare solid wire or rod	Stainless steel, Ni, Cu, and Fe base	—
	Automatic	Bare tube wire; extra long bare cast rod; tungsten-carbide powder with bare solid wire	Co, Ni, and Fe base; tungsten-carbide composites	Bare solid wire	Stainless steel, Ni, Cu, and Fe base	—
Gas metal arc (GMA)	Semiautomatic, automatic			Bare solid wire	Stainless steel, Ni, Cu, and Fe base	—
Submerged arc						
Single wire electrode	Semiautomatic	Bare solid or tube wire	Fe base	—	—	Limited use
	Automatic	Bare tube or solid wire	Fe base	Bare solid or tube wire	Stainless steel, Ni, Cu, and Fe base	Limited use with Cu base alloys
Multiple electrode	Automatic	Bare tube or solid wire	Fe base	Solid or tube wire	Stainless steel, Ni, Cu, and Fe base	Limited use with Cu base alloys
Series arc	Automatic	Solid wire	Fe base	Solid wire	Stainless steel, and Ni base	Used primarily for cladding large vessels
Strip electrode and auxiliary strip	Automatic	—	—	Bare strip	Stainless steel, and Ni base	Used primarily for cladding large vessels
Auxiliary powder	Automatic	Bare solid or tube wire with metal powder	Fe and Co base	Bare solid wire with metal powder	Stainless steel and Ni base	—
Plasma arc (transferred arc)						
Powder	Automatic	Powder with or without tungsten-carbide granules	Fe, Co, Ni base; tungsten-carbide composites	Powder	Stainless steel, Ni, and Cu base	Used primarily for production hardfacing
Hot wire	Automatic	Bare solid or tube wire	Fe base	Bare solid or tube wire	Stainless steel, Ni, Cu, and Fe base	Used primarily for cladding large vessels and related components
Electroslag	Automatic	—	—	Plate or wire	Stainless steel, Ni, and Fe base	Use on heavy sections only
Coating processes						
Flame spray (metallizing)	Semiautomatic and automatic	Powder, rod	Fe, Co, Ni base; tungsten-carbide composites	Bare solid wire	Stainless steel, Ni, Cu base, Al. Zr. etc.	Some hardfacing alloys may be fused after spraying to produce a fused bond
Detonation gun plating	Automatic (proprietary)	Powder	Tungsten-carbide with selected matrices; selected oxides	—	—	—
Plasma spraying (metallizing)	Semiautomatic and automatic	Powder	Fe, Co, Ni base; tungsten-carbide composites, ceramics	—	—	—
Brazing		Wire, sheet, powder	Fe, Co, Ni, Cu base	Wire, sheet, powder	Ag, Au, Ni, and alloys	Limited use

Source: American Welding Society.

Impact resistance, however, refers to a slightly different (but closely related) idea. Impact resistance measures the amount of force required to make a welded surface crack. A practical example is the bottom of an off-highway truck bed with a big coal-stripping shovel dumping rock into it. If the truck bed can hold up without cracking, it has impact resistance. Some surfacing alloys have much higher impact resistance than others, even though they may have lower initial hardness. High-manganese austenitic alloy steels used as wear plates are an example. The high-manganese steel work-hardens on impact or deformation, just like austenitic stainless steels. (Manganese, like nickel, can make the austenitic iron crystal structure stable at room temperature instead of at high temperatures.)

Oxidation Resistance

A material may have good hot hardness but can't stand up to oxidation at high temperatures. Certain weld-metal deposits that contain molybdenum will oxidize very rapidly at high temperatures if they are exposed to air. Nevertheless, they have high hot hardness and high strength at high temperatures. An alloy containing lots of chromium and molybdenum might give you both high hot hardness and high oxidation resistance. Furnace parts exposed to the air need high-temperature oxidation resistance as well as high-temperature strength.

Corrosion Resistance

Corrosion caused by chemical reactions is an enormously complicated subject. A material can be almost completely corrosion resistant in one medium (such as pure water), but if you add the smallest amount of table salt to the water the metal will start to rust very fast. That's even true of many stainless steels, which is why very special alloys have to be used in seawater. Copper-nickel alloys often are used for salt-

water and brine services in coastal desalting plants. Aluminum has high corrosion resistance to acids, but bases (such as lye or sodium hydroxide) will gobble it right up. Stainless steels similarly have super corrosion resistance in some materials, and very low corrosion resistance in others.

What's more, the same material may have high corrosion resistance if the corrosive fluid is not moving, or not full of dissolved oxygen, or not too hot, but terrible corrosion resistance if you change the conditions of the corrosive medium very slightly, like making it flow through a sharp-angled bend in a pipe. The corrosion resistance of materials should always state exactly what the test conditions were when the corrosion test was made. The most common method of measuring corrosion resistance is by weight loss over a given period of time, or by the amount of thickness a surface loses to corrosion in a given time.

You'll understand the weight-loss units when you see them. The loss of section thickness due to corrosion will use a unit of measure you haven't seen before. It's called a *mil*. A mil is simply $1/1000$ in. (or 0.001 in.). Therefore, 50 mils is 0.050 in. and 100 mils is 0.100 in. or $1/10$ of an inch. The metric equivalent of $1/1000$ in. is 0.025 mm.

A common system for rating the corrosion resistance of all metals, not just welding overlays, is based on thickness loss measured in mils. Corrosion losses of 5 mil/year [0.13 mm/year] are considered excellent. From 5 to 20 mil/year [0.13 to 0.50 mm/year] corrosion loss is rated good. A corrosion loss of 20 to 50 mil/year [0.50 to 1.3 mm/year] is rated acceptable. More than 50 mil/year [1.3 mm/year] corrosion loss is bad.

Corrosion resistance is a metal property that is very important in chemical plants, paper mills, and petroleum refineries. It often is combined with a need for oxidation resistance and sometimes for hot hardness. Corrosion is a very

complicated subject. We know a lot more about why things corrode than what to do about it when they do.

Chemical and paper mill vats, pressure vessels, piping, valves, pumps, stirrers, and even industrial fans used in humid environments are examples of products that need exceptional corrosion resistance. Obvious consumer products that need good corrosion resistance are washing machines, sinks, hot-water heaters, and exposed metal surfaces on buildings.

Abrasion Resistance

Hardness, impact resistance, and abrasion resistance are related, but they are not the same. Abrasion means sandpaper-like wear (as opposed to other kinds of wear). A weld-metal deposit that gives you high abrasion resistance probably has high hardness, but that's not necessarily so. For example, the special high-manganese austenitic steels we already mentioned have excellent abrasion-resistant applications, like a gravel chute or the inside of the ball mill in a cement factory, even though they are not hard when installed.

These austenitic steels don't start out very hard but the manganese makes them work harden. Up to a point, the more abrasion that occurs on the steel, the more resistant to abrasion the steel becomes. Sooner or later, however, the metal will reach its maximum hardness for the given alloy. Abrasion resistance is a property that often is needed in surfacing alloys used for hardfacing. Quartz sand is an example of an abrasive material. Mining and construction equipment need abrasion resistance, as do many types of farm equipment.

Erosion Resistance

Combine abrasion with corrosion, especially when the corrosive medium is flowing, and you get erosion. Erosion has both chemical and mechanical aspects. A warm dilute solution of sulfuric acid with sand particles in it flowing through a pipe would be very erosive. Where would you find such an awful combination? In mine water.

Metal-to-Metal Wear
Now we have another kind of wear. Almost any tool steel you can name will have very high metal-to-metal wear resistance. Some welding alloys can be deposited that have 65 percent or more of the metal-to-metal wear resistance of a high-carbon, high-alloy tool steel. That's very high. Alloys with high metal-to-metal wear resistance also often have low coefficients of friction, and the ability to take a polish and retain their hardness at elevated temperatures. They also often have high compressive (as opposed to tensile) strength because metal-to-metal wear often involves compression. An example is a machine tool slide. Another example is a bearing. As you know from Chap. 8 on mechanical properties, gray cast irons have high compressive strength even though they have low tensile strength. That's why they make good machine-tool bases and frames.

The dies on machine tools need to resist metal-to-metal wear. So do the valves in piston engines (when the lubrication isn't all that good). Certain types of bearings also need as much metal-to-metal wear resistance as they can get. The machine slides on machine tools often require metal-to-metal wear resistance.

Machinability
It doesn't do you any good to resurface a worn shaft with a hardfacing alloy if the machinist can't work the shaft to the final correct diameter after you have built up a worn spot. While the machinist may not be able to turn (machine) the shaft on a lathe because your surfacing deposit is too hard, maybe he or she can grind it to final dimensions. Machinability often is a trade-off you have to make between one surfacing alloy that's very hard, and another one that's not quite so hard, but machinable.

Heat Treatment
Some of the alloys you may lay down as a surfacing layer may not be all that hard until they are heat-treated later on. That, in fact, is one way out for your friend the machinist. If you can select an alloy that can be machined, and then heat-treated to final hardness, you and the machinist are both ahead of the game.

There are other properties besides these that some people may want from time to time in a weld-surfacing alloy, but this list is enough to give you an idea of some of the things you will read and hear about when you do weld-surfacing work. Heat treatment is a very complicated subject. We can't possibly cover it in detail in this book, although we've explained some of the ideas behind it from time to time. You already know more about heat treatment than you might think. Stress-relieving a weldment is one example.

Cladding
When you put a thick layer of some weld metal such as stainless steel onto a carbon- or low-alloy-steel plate to change the nature of the surface, you are doing a cladding operation. The cladding does not have to be stainless, but that's the most common material used.

The benefit of cladding is that you can create a relatively low-cost, corrosion-resistant surface without paying a fortune for thick stainless-steel plate. Another advantage is that you can combine a high-strength material such as low-alloy steel for the backing, with some other material (such as stainless steel or a nickel-based alloy or a bronze) for the surface. As a general rule, the strength of the cladding material is not included in the design of the component.

Cladding often is used to make chemical, paper mill, petroleum refining, and nuclear power plant vessels that cost less but do the required job longer than unclad vessels. Beer is corrosive, too. It's made in special copper-lined reactors. Food plants make extensive use of stainless steel for the same reason.

Hardfacing
Hardfacing usually means adding weld metal over an existing surface to give the surface of a part properties such as abrasion resistance, surface hardness, impact resistance, hot hardness, erosion resistance, and so on. As with cladding, the strength of the hardfacing layer is not included in the design of the part. Hardfacing also can mean hardening the surface of a part strictly by the use of heat treatment.

When we talked about the use of heating torches in Chap. 6, we mentioned that some steels could be surface-hardened simply by heating them and cooling (quenching) them very quickly. Hardfacing sometimes combines both processes. The heat treatment can be used to harden the weld-metal surface layer already deposited on a part.

Figure 19-2 shows you the relative hardness and abrasion resistance of common hardfacing materials. While the information is fairly general (specific alloys will vary a lot in hardness), this drawing will give you a good feeling for the relative hardness of different surfacing alloys.

Construction equipment, including bulldozer blades, scraper blades, and rock chutes; coal and ore mining equipment; and textile equipment last longer when hardfaced. (See Fig. 19-3 and 19-7.)

Buildup
Parts that wear out often can be rebuilt. Let's assume that you have a shaft with a groove worn into it after several years of use. The groove can be filled up with some weld metal, then the shaft remachined to its original dimensions and you have done a buildup job on the worn surface. Similarly, the teeth of dragline buckets and the edges of bulldozer

blades and scrapers get worn down. Rather than buying a new dragline bucket or dozer blade, a new part is simply built up to approximately the original shape from the worn section. Repairing broken gear teeth with bronze braze-welding alloys is another example of building up a part. You did that in the Shop Practice in Chap. 7, when you learned how to braze-weld.

Sometimes weld-metal buildup with a tough, impact-resistant alloy can be combined with a final pass of a more brittle, but much higher-hardness abrasion-resist-ant hardfacing alloy. The result may be better than the original product. The strength and other mechanical properties of the built-up deposit also are important because the material has to replace some of the original part that has been worn away.

Buttering

Some alloys can't be welded to each other with good results, but they both can be welded to an intermediate alloy. Other times, welding one metal directly to another (such as a high-alloy stainless steel to a carbon steel) will dilute the stainless steel with carbon steel, reducing its corrosion resistance and other properties. But you can weld a high-nickel or nickel-chromium transition deposit to the carbon steel and then weld the high-alloy stainless steel to the other side of the nickel-chromium metal. This is an example of *buttering*. Buttering usually refers to putting one or more layers of weld metal onto the face of a joint or surface that will be welded later on to something else.

Dilution

Probably the single biggest difference between welding a joint and depositing surfacing is that the base metal and the weld overlay will have very different chemical compositions. The base metal might be a low-carbon steel and the weld overlay could be stainless steel, or even contain tungsten carbide. Obviously, one of them (usually the base metal) will dilute the other material (usually the surfacing deposit). Therefore the surfacing deposit will not have the properties that you want it to have.

A good example is surfacing low-alloy steel with stainless-steel using SMAW electrodes. This process normally gives 15 to 50 percent weld-metal dilution when the first two or three beads in the surface layer have been made. Thus, if the stainless-steel surfacing deposit is laid down with an E308 stick electrode that produces austenitic stainless-steel weld metal with 19 percent chromium and 9 percent nickel, the first layer of the surfacing weld metal will give you only 12 percent Cr and 6 percent Ni. That's because of base-metal dilution. That will result in poor corrosion resistance and lower mechanical properties (low ductility).

On the other hand, if you had selected an E309 SMAW stainless-steel electrode with 25 percent Cr and 12 percent Ni for the same job, the as-deposited weld metal, complete with base-metal dilution, would have about 16 per-

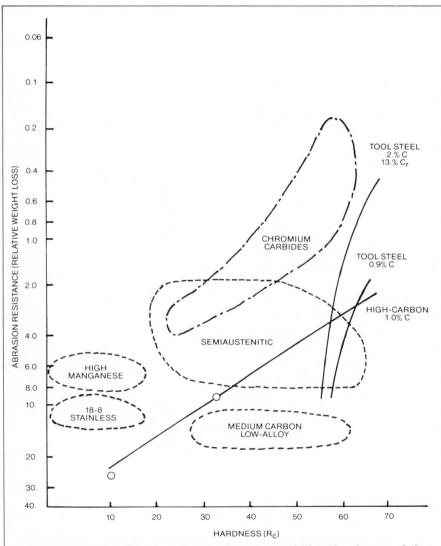

FIGURE 19-2 The Lincoln Electric Company of Cleveland, one of the leading arc-welding equipment and filler metal suppliers, developed this graph comparing the Rockwell C hardness of different types of weld overlays versus abrasion resistance. It gives you an overview of the different abrasion-resistant filler-metal families you will learn about in this chapter.

cent Cr and 8 percent Ni. This surfacing deposit is right on target. It will have good corrosion resistance and better mechanical properties. If you had been able to hold dilution to the low end of the range (about 15 percent), the weld-metal corrosion resistance would have been even better.

Many hardfacing and surfacing techniques are set up specifically to keep weld-metal dilution as small as possible. For example:

■ Increasing the amperage (current density) increases dilution. The arc becomes stiffer and hotter, penetrating more deeply and melting more base metal. Set your amperage on the low side to reduce dilution.
■ DCSP (electrode positive) gives less penetration and, hence, lower dilution than DCRP. AC dilutes weld metal to a level intermediate between DCSP and DCRP.
■ The smaller the electrode, the lower the amperage as a rule, and therefore the lower the dilution. For a given amperage, however, the larger the electrode (the lower the current density), the lower the dilution.
■ A long electrode stickout for consumable electrode processes such as GMAW, FCAW, and SAW lowers dilution by increasing the melting rate of the electrode (I^2R heating) and by diffusing the energy of the arc as it strikes the base metal. Conversely, short electrode extensions increase dilution, within limits.
■ Tight bead spacing (more overlap) reduces dilution because some of the previously deposited bead is melted, rather than a lot of the base metal. Wider bead spacing (less overlap) increases dilution.
■ The more you oscillate your electrode from side to side the less dilution you will get. A stringer bead gives maximum dilution. Hesitating at each side of the weld bead is usually a good idea when making a deep joint because it "burns" the weld metal into the base metal. However, weaving

your electrode with a pause at each side of the pass increases weld-metal dilution. Similarly, weaving your electrode in a slight curve will increase dilution a little more than making straight zig-zag lines at a constant speed as you work down the weld (Fig. 19-3). This straight back-and-forth line made with a constant-velocity weaving motion is best for surfacing and hardfacing.*
■ A slow travel speed decreases the amount of base metal melted and increases the amount of filler metal added, which lowers dilution. The reason is that slow travel speeds change the bead shape and thickness, and the arc force is directed into the weld puddle and not dug into the base metal.
■ The longer the weld puddle stays under the arc to shield the base metal from the arc force, the less dilution you get. Therefore, your welding position has an effect on weld-metal dilution. In order of decreasing dilution (most dilution at the top of the list and the least dilution at the bottom of the list), the effect of welding positions is as follows:

5. Vertical-up (the most dilution)
4. Horizontal
3. Uphill
2. Flat
1. Downhill (the least dilution)

Uphill and downhill welding can be done by inclining the part to be clad or hardfaced or by placing the arc off center if you are working on a rotating cylindrical part while using a welding positioner.
■ Arc shielding has a significant effect on weld-metal dilution. It influences the fluidity and the surface tension of the weld pool. These, in turn, determine the ex-

* The Stoody Company has a 144-page manual titled *Stoody Hardfacing Guidebook* that gives you much more detail on specific patterns and filler metals than we can in this chapter. Write for a free copy to: The Stoody Company, 16425 Gale Avenue, Industry, California 91749.

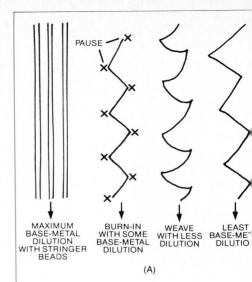

MAXIMUM BASE-METAL DILUTION WITH STRINGER BEADS   BURN-IN WITH SOME BASE-METAL DILUTION   WEAVE WITH LESS DILUTION   LEAST BASE-METAL DILUTION

PAUSE

(A)

(B)

(C)

(D)

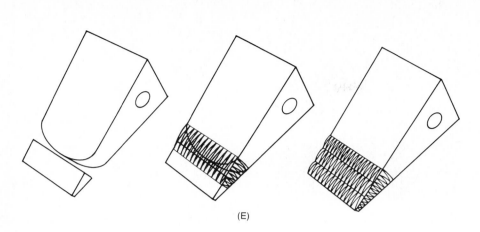

(E)

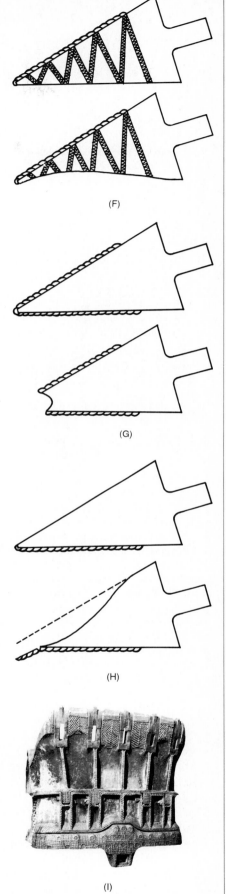

(F)

(G)

(H)

(I)

FIGURE 19-3 These weld overlay patterns, recommended by the Stoody Company are only a few of the patterns they have developed for different kinds of work. Unlike corrosion-resistant coatings, abrasion-resistant overlays don't always have to cover the entire work area of a part. That saves expensive filler metals and extra welding labor and time to deposit the overlay. **(A)** The effect of different electrode or wire-manipulation techniques on dilution of the filler metal by the base metal. It's best to hardface shovel teeth and other products subject to extreme wear when they are new—before putting them into service. The weld-bead pattern that you use can have a marked effect on service life. **(B)** For example, the Stoody Company recommends that dipper or dozer teeth working primarily in rock should be hardfaced with beads running the length of the tooth. This allows the rock to ride on the hard metal beads. **(C)** When working with dirt, clay, or sand, run the beads across the flow of material. The dirt will fill cavities between the weld beads, giving the base metal added protection. **(D)** Another effective pattern is the waffle or cross-hatch pattern used for combination rock and dirt. Apply hardfacing to the top and sides of the tooth 2 in. [50 mm] up from the cutting point. Extra-hard tungsten-carbide-bearing weld metal applied as stringer beads on the cutting edge of the tooth will prolong service life even more. **(E)** Shovel teeth can be repointed using austenitic manganese-steel bars welded to the carbon-steel teeth with austenitic Type 308 stainless-steel SMAW electrodes or GMAW welding wire. The bars can then be hardfaced with an appropriate wear-resistant alloy. **(F)** According to the Stoody Company, the best procedure for surfacing shovel teeth is to make them self-sharpening. Self-sharpening occurs when the unprotected bottom wears faster than the top or sides. After initial wear on the bottom, hard weld metal on the sides will retard continued wear of the tooth and will result in a self-sharpening bucket tooth. The top of the teeth should always be protected as described above, using an appropriate pattern for dirt, clay and sand, or hard rock. **(G)** It is not recommended that you hardface both the top and bottom of the tooth. After the hard metal is worn away on the cutting edge, the tooth will cavitate as shown and the hardfacing will chip off, producing a dull tooth. **(H)** Hardfacing just the bottom of the tooth does not produce best life, either. The hardfacing overlay will chip off through lack of support as the unprotected top surface wears away. **(I)** Herringbone, waffle, and stringer bead patterns were used on this bucket to fit wear conditions to bucket contours.

(Continued on page 502)

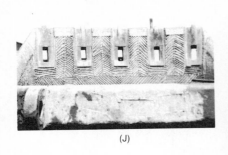

(J)

(L)

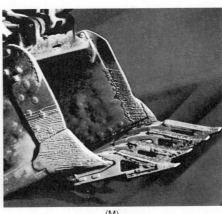

(K)

(M)

FIGURE 19-3 (*Continued*) (J) The inside view of bucket (I). The lips between tooth mounting areas were completely overlaid with a wash pass for maximum protection, while the space between was overlaid with a herringbone pattern. (K) These are the same patterns for hardfacing dipper teeth. The waffle pattern works best in wet sand and dirt as it does with bucket teeth. The dirt or sand packs between the hard-metal weld beads to provide additional base-metal protection. (L) The Stoody Company suggests that you use a dot pattern which works best on hard-to-face base metals that should not be overheated (such as austenitic high-manganese abrasion-resistant steels). Spacing of the dots should be varied—by trial and error—to determine the best dot pattern. (M) Hardfaced buckets working in abrasive slag and rock should be protected with beads running parallel to the flow of the abrasive material to let it ride on the hard weld-metal overlay, and not on the buckets' base metal.

tent to which the weld metal will wet the base metal and feather in well along the edges of the bead. The shielding medium also has a significant effect on the type of welding current that you can use. The following list ranks, in general, the different shielding media in order of decreasing dilution.

5. Helium (highest dilution)
4. Carbon dioxide
3. Argon
2. SAW fluxes, no alloy additions
1. SAW fluxes with alloy addition (lowest dilution)

■ Auxiliary filler metals reduce dilution. Some processes such as

SAW, GMAW, and hot-wire GTAW can add a second or third electrode wire to the weld pool (if you have the equipment). The extra metal also may be added separately as powder, wire, strip, or with the flux. Increasing the amount of filler metal added to the weld puddle reduces the amount of dilution caused by the base metal.

Contamination

Some alloying elements, even in small amounts, can affect the usefulness of a surfaced deposit. These contaminating elements can cause cracking, reduce corrosion resistance, or reduce strength, ductility, and toughness. Carbon, for example, can reduce the corrosion resistance of austenitic stainless-steel cladding. (Put carbon into an austenitic stainless steel and you're likely to get a martensitic stainless steel that is harder and stronger, but has less corrosion resistance.) The carbon can easily get into the stainless surface from the carbon-steel base metal through dilution. It will reduce the austenitic alloy's resistance to intergranular corrosion. (See Chap. 16, pages 386–387, "Problems Welding Stainless Steels.")

Also, the diffusion of carbon from ferritic base metals into austenitic stainless-steel cladding can create hard, brittle carbides in the cladding and in components that must be heat-treated after cladding or that operate at high temperatures for long times. Such carbides can lead to mechanical failure of the part or surface during use.

Lead, phosphorus, and tin must be controlled in nickel-base alloys. Copper must be controlled to prevent cracking in ferritic materials. Low-dilution surfacing techniques, adding extra filler metal or flux, or even the use of a buttered transition layer of a different (or intermediate) material can help solve these problems. But it may require a welding engineer or a metallurgist to decide what to use.

Surface Appearance

Cladding usually produces the finished surface of a part or vessel or weldment. Subsequent machining is not always possible or even economically feasible. For this reason, the as-deposited weld metal frequently must be smooth enough to meet the requirements of the job. Surface smoothness may be very important because even the "valleys" between weld beads can be areas where corrosion will start faster than on top of the weld bead.

Warping and Cracking

Thermal stresses build up fast in a hardfacing and surfacing operation. Residual stresses from welding may add to or oppose stresses encountered in the surface. For example, if the weld metal is in compressive stress and the applied stresses on the part are tensile, the weld surface will actually help reduce the effect of the service stresses on the part. This effect was discussed in Chap. 8, pages 197–198, "Metal Fatigue."

It can also work the opposite way. That's when you get cracks. Hardfacing thin material is especially difficult because the wide weld area will cause all sorts of warping. Also, hardfacing alloys seldom are stress-relieved after they are applied. Therefore, you may have to use a lot of jigging or very strong fixtures on some parts. However, that can lead to cracking your weld metal instead of bending the base metal. The solution is to use hardfacing or surfacing alloys with relatively high ductility and toughness for these jobs.

Among the surfacing alloys, the austenitic stainless steels are the most ductile and the toughest. They have high ductility but moderate yield strength. They contrast somewhat with martensitic stainless steels that have lower ductility but higher strength. The high-carbon cast irons are relatively strong in compression, but they are brittle and weak in ten-sion. Surfacing with these high-carbon irons is likely to crack the weld metal when it is on base metals that are not extraordinarily hard but are thick enough to resist distortion.

19-2 WHICH WELD-SURFACING PROCESS?

Surfacing and hardfacing can be applied with welding, brazing, and metal spraying. Most processes involve fusion welding, where the filler metal and surface of the base metal melt join and then resolidify.

The oxyfuel process produces slower heating and cooling of the base metal. Since things don't happen so fast, you'll have better control over your work. The equipment also is very portable and less expensive than arc welding.

Arc welding, on the other hand, is faster, less expensive overall if a lot of work has to be done, and you don't have to be as skilled to do it as with gas welding. The very fast heating and cooling cycles in arc welding also mean that thermal stresses are much higher on the base metal and overlay and therefore they are more susceptible to cracking.

For small jobs and field use, your choice is usually between some oxyfuel process and one of the arc-welding processes like SMAW. For plant production use, you're more likely to get involved with a process that will be semi- or fully automatic such as GMAW or SAW.

19-3 FILLER-METAL FORMS

So many different welding processes can do surfacing and hardfacing that the filler metals come in many different forms such as rods, alloy wire and flux-filled tubes, strip, granular alloy-containing flux, or even powders that are melted and sprayed on the work. Fluxes can be in the central core of the tubes or wire, on the surface as in covered stick electrodes, or mixed in with powdered alloying elements as in SAW or metal spraying (which often uses oxyfuel or plasma-arc heat).

For gas welding, rods and powders are commonly used. Solid rods may be drawn into wires (for ductile alloys) or your welding rod may be a casting (gray cast-iron-rods are an example). The usual sheath around the wire or tube will be mild steel and the tube will be filled with flux and alloying elements. An important composite type of electrode wire or cut-length rod has granular tungsten carbide as 60 percent of the filling. All types of rods should be bare and clean for oxyfuel or TIG arc welding.

Many tubular electrode wires are used for surfacing. These flux-cored surfacing electrodes contain enough alloying elements to produce stainless-steel weld overlays. When you select one, don't forget to adjust the as-deposited weld-metal composition to allow for base-metal dilution.

Powders are very versatile. Practically any metal or alloy can be used. The powder alloys are usually arc melted after precise formulation and then atomized, or cast and finely granulated. The alloy powders can be simple mechanical mixtures as well as pre-fused alloys. (Prefused alloys are alloys that have been made by melting and cooling, before being turned into a powder.) Composition control of mechanical alloy mixtures, usually by a metering device and a hopper next to your work station or even attached to your metallizing spray gun, is less precise than using prefused alloys.

19-4 USING STICK ELECTRODES

SMAW is used extensively for hardfacing, buttering, buildup, and cladding. Surfacing of carbon and low-alloy steels and many nonferrous metals can be done with the SMAW process. Base-metal thickness can range from below ¼ in. [6.4 mm] to 18 in. [0.46 m] and more. The filler metals you can use include both low- and high-alloy-steel stick electrodes, stainless steels, nickel-base alloys, cobalt-base alloys, and copper-base alloys.

Keeping in mind the methods for avoiding base-metal dilution we already listed, hardfacing with stick electrodes is just the same as any other welding job. The main differences will be the characteristics of your individual electrodes, not the overall process. You should be able to deposit about 1 to 4 lb [0.45 to 1.8 kg] of weld metal per hour at dilution levels of 30 to 50 percent.

Every manufacturer who makes stick electrodes has dozens of proprietary filler metals for hardfacing. It's tough to figure out what's really what. But the AWS has a very good specification that will help you out. AWS A.5.21 is titled "Specification for Composite Surfacing Welding Rods and Electrodes" and it's the most important filler-metal specification you will use for hardfacing alloys. The specification includes four major groups of filler metals:

- High-speed steel gas rods and electrodes coded with RFe5-A and RFe5-B, and EFe5-A, EFe5-B and EFe5-C class designations. These are tool-steel alloys for high hardness.
- Austenitic manganese-steel electrodes coded with EFeMn-A and EFeMn-B class designations. These are work-hardening alloys for abrasion resistance.
- Austenitic high-chromium iron gas rods and electrodes coded with RFeCr-A1 or EFeCr-A1 class designations. These are low-carbon stainless steels used for corrosion resistance.
- Tungsten carbide gas rods and electrodes coded with RWC-XX and EWC-XX (there are 13 grades of tungsten carbide rods and electrodes; since W is the chemical symbol for tungsten (one of its ores is a mineral called wolfram)

and C is the chemical symbol for carbon, the WC simply means tungsten carbide. The XX will be explained later. These filler metals are used for abrasion resistance as well as hardness.

The AWS A5.21 specification will give you a lot of detail on the mechanical properties of the weld metal that each filler-metal grade should be able to give you. You should consult with a supplier for more information if you need it.

High-Speed Rods and Electrodes

The RFe5 gas rods and EFe5 electrodes are very popular for use where hardness is required at service temperatures up to 1100°F [593°C], and where good wear resistance and toughness also are required. These filler metals produce weld metal that essentially is high-speed tool steel.

The three classifications Fe5-A, Fe5-B, and EFe5-C are more or less interchangeable except that Fe5-A and Fe5-B (with highest carbon) are more suitable for cutting and machining (edge-holding) jobs while EFe5-C (lower carbon) is most suitable for hot working (such as hot forging dies) and for jobs requiring toughness. Typical surfacing applications are cutting tools, shear blades, reamers, forming dies, shearing dies, cable guides, ingot tongs, for fixing broaches, and similar tool repair jobs.

The Rockwell hardness of the undiluted Fe5 filler metal in the as-welded condition is in the range of C 55 to C 60. Where a machining operation is required, the hardness of the weld deposit can be reduced to Rockwell C 30 by an annealing heat treatment, followed by machining, and then quenching and tempering to bring the weld metal back up to its final hardness.

The hot hardness of these alloys is very important. Tungsten and molybdenum are the most important elements in keeping the weld metal fairly hard at high temperatures. At temperatures up to

1100°F [593°C], the Rockwell C 60 as-deposited hardness falls off very slowly to about Rockwell C 47 (448 Brinell). At higher temperatures, it falls off a lot faster. At about 1200°F [667°C], the maximum Rockwell hardness you can expect is C 30.

The impact resistance of the as-deposited Fe5 filler metals is only medium. A hard impact will make them crack. If the weld metal is correctly tempered (reheated) after deposition it will be a lot tougher.

Deposits of Fe5 filler metals, because of their high molybdenum content, will oxidize readily at high temperatures. When heat treatments are required, a nonoxidizing furnace atmosphere should be used, or else use a salt-bath furnace or a borax coating to prevent decarburization (loss of carbon during heat treatment).

The corrosion resistance of Fe5 filler metal is not too bad in the atmosphere but the weld metal will not hold up well in moisture or in corrosive liquids. You should use one of the chromium-nickel grades for these conditions.

The high-stress abrasion resistance of these filler metals, as-deposited, at room temperature is much better than that of low-carbon steel. But they are not considered to be high-abrasion-resistant alloys. Resistance to deformation at high temperatures (up to 1100°F [593°C]) is their outstanding feature and this may help you when you need hot abrasion resistance, such as in resurfacing the inside of the combustion zone of a boiler where the fly ash is both hot and abrasive.

Deposits of Fe5 filler metals are well suited for metal-to-metal wear, especially at elevated temperatures. They have a low coefficient of friction and the ability to take a high polish and retain their hardness at high temperatures. The compressive strength of the weld metal also is very good, which means these materials are a good choice for repairing hot-work forming dies and for fixing machine-tool slides, for example.

The machinability of the Fe5 surfacing filler metals is not good in the as-deposited condition. They are too hard. Either ceramic or other very hard carbide or nitride cutting tools, or even diamond-cutting tools or grinding will be needed to machine these metals, or else you will have to have the part annealed. After annealing and machining it will have to be rehardened by heat treatment.

One way to identify weld metals produced by the Fe5 alloys in the hardened or as-deposited condition is that they are very magnetic (they contain ferrite and martensite instead of austenite which is not magnetic). When spark tested they give off a very small, thin stream of sparks about 60 in. [1.5 m] long. Close to the grinding wheel, you will see that the spark is red. At the far end of the spark it will be a straw color.

The heat treatment of Fe5 weld metal is outlined here.

■ Preheat your workpiece before welding. Preheat is usually necessary, although in some instances you can get away without it. When a preheat is used, the minimum preheat temperature of the workpiece you will repair is 300°F [150°C].
■ Annealing is only necessary if machining is required. The annealing temperature is 1550 to 1650°F [843 to 899°C]. Get advice from a heat treater or a metallurgist before you try it. The time the work is held at this temperature depends on the thickness of the workpiece. The cooling rate after annealing also is important.
■ Hardening is needed only if your weld metal already has been annealed. Preheat to 1300 to 1500°F [704 to 815°C]. Harden at 2200 to 2250°F [1204 to 1232°C] and use an air or oil quench. Double temper at 1025°F [552°C] for 2 h, then air cool to room temperature. Then reheat to 1025°F [552°C] for 2 h, air cool to room temperature again. (That's the

double temper.) Get the help of a heat treater. These alloys are subject to decarburization in the heat-treating furnace if not properly handled.
■ You should preheat your base metal to at least 300°F [150°C] before welding with these alloys. If you don't, the weld metal may crack after it cools. It also helps to peen each weld bead immediately after deposition to reduce built-up stresses in the weldment. You can use a peening hammer, but shot blasting may be more efficient.

Austenitic Manganese Electrodes

The two classes of EFeMn electrodes are substantially equivalent, except that the yield strength of EFeMn-B weld deposits is higher than that of EFeMn-A. For track work, the higher yield is a considerable asset. These electrodes are most commonly used to build up worn parts.

For surfacing with EFeMn electrodes, most jobs are those dealing with metal-to-metal wear and impact resistance, and where the work-hardening quality of the weld deposit is a major asset. These steels get harder the more they are hammered or battered about. Equipment used in crushing soft rock and in transporting limestone, dolomite, or shale, for example, can also benefit from surfacing by EFeMn electrodes. Because these steels work harden rapidly and also because they are brittle at high temperatures, you should immediately peen each weld bead with a pneumatic hammer or a machinist's ball peen hammer.

You should peen each bead immediately after depositing one (or even one-half) electrode of the material to avoid cracking problems later on. In no case should a weld bead longer than 9 in. [230 mm] be left without immediate peening. The weld metal is weakest when it is hot and it also easily deforms when red or yellow

hot. Since cracking is most likely to occur above 1500°F [815°C], you should peen the bead as quickly as possible.

Very severe abrasion from rock with quartz particles is solved by using these manganese-steel surfacing electrodes to produce the base layer, using a buttering technique. Then cover the weld metal with a final deposit of a hard, martensitic cast iron. The buttered underlay of EFeMn weld metal will make the surface tough. The very hard cast iron, which is inherently brittle, still has extremely high abrasion resistance. The combination gives you the best of both materials.

One way to avoid cracking when overlaying carbon steels that normally would add extra carbon to the FeMn5 weld metal (making the austenite convert to ferrite or martensite while increasing its brittleness and tendency to crack) is to butter the carbon steel with a layer of austenitic stainless steel. Then overlay the stainless steel with the austenitic high-manganese weld metal. The stainless steel will act as a transition barrier between the high carbon content of the carbon steel and the austenitic high-manganese steel on top.

Under very high stress, like in a jaw crusher, you'll find that all wear-resistant metals except manganese steel are too brittle. Surface protection then becomes a matter of replacing worn metal with more EFeMn filler metal. Railway frogs and crossing switches also are reclaimed in this way. Extensive areas, as in crushers and power-shovel parts, are usually protected with a combination of weld deposits and filler bars. Filler bars are flats and rounds of high-manganese steel that are welded in place with EFeMn alloys. Such protection may be applied up to 3 in. [76 mm] thick. That is the upper thickness limit for common surface-protection methods.

The initial as-deposited hardness of EFeMn weld metal is

about 170 to 230 BHN, but this is misleading since these materials will work-harden very rapidly up to 450 BHN to 550 BHN. Reheating EFeMn weld metal above 500 to 600°F [260 to 315°C] may cause serious embrittlement. Thus you don't want to use these alloys where hot hardness is needed.

The impact resistance of as-deposited EFeMn electrodes is outstanding. They often are used for heavy-impact service.

The oxidation and corrosion resistance of EFeMn weld metal is similar to that of ordinary carbon steels. Weld metal produced by EFeMn electrodes is not considered to be oxidation or corrosion resistant.

The metal-to-metal wear resistance of EFeMn weld metal is generally excellent. The yield strength in compression is low, but any compressive force rapidly raises the strength of the metal just like work hardening increases the hardness. For this reason, EFeMn weld metal is sometimes used for battering, pounding, and bumping applications.

Machining EFeMn weld metal is very difficult with ordinary tools and equipment. The finished surfaces usually are ground instead of being cut.

Because of their unusual response to heating (EFeMn weld metal gets brittle over 500°F [260°C]), correct identification is very important. A small magnet and a grinding wheel will do the job. A clean, ground surface of the austenitic manganese-steel weld deposit will be nonmagnetic. Grinding sparks will be plentiful in contrast to austenitic stainless steels, which also are nonmagnetic.

Deposits of EFeMn weld metal are usually not heat treated, since the filler metals are formulated to be air-hardening. However, it sometimes may be advisable to heat-treat a weldment to restore the toughness of the manganese base metal after it has been embrittled by too much heat (over 500°F [260°C]). Water quenching after heating the workpiece to 1850°F [1010°C] for 2 h will do the job. The weld deposit should be free of cracks if you do this. Otherwise oxidation or corrosion in the cracks will cause a lot of damage.

Austenitic High-Chromium Irons

FeCr-A1 as-deposited filler metal is a very high carbon, high-chromium austenitic cast iron with up to 5 percent carbon, up to 8 percent manganese, and up to 32 percent chromium.

The RFeCr-A1 welding rods and EFeCr-A1 electrodes have proved to be very popular for surfacing agricultural equipment, machinery, and parts. Arc welding is used on heavy materials and large areas. Oxyfuel welding is used for thin sections. Fixing plowshares and combine cutters are typical applications because these filler metals flow well enough to produce a thin-edged deposit, and the wear conditions in sandy soil are typically those of erosion or low-stress scratching abrasion. FeCr-A12 filler metals are not suitable for use in very rocky soil because of the impact that occurs. Typical industrial uses include coke chutes, cable reel guides, and sandblasting equipment.

The as-welded hardness of iron-chromium deposited by oxyfuel welding will vary with the carbon content. The average Rockwell hardness will be around C 56 with a range from C 51 to C 62. That represents a carbon content ranging from 4.3 to 5.2 percent. These hardness values will increase slowly as the carbon content increases. The Rockwell hardness readings measure the hardness of hard carbides in the metal which are in a softer austenitic steel matrix. Therefore the hardness readings are not a reliable guide to abrasion resistance. Small abrasive particles will eat out the soft matrix and leave the hard carbides behind.

Since little dilution is expected in normal oxyfuel welding, your chief problem will be with carbon pickup from the flame adjustment. With a 3× feather-to-cone reducing flame, a pickup of 0.4 percent carbon is possible if the welding rod is on the low side of the carbon range. On the high side of the carbon range, a neutral flame can slightly decarburize the weld deposit. You're better off using a neutral or slightly oxidizing flame (see pages 51–54).

The austenitic matrix can be work-hardened somewhat under impact, but since that can lead to cracking, impact service should be avoided.

The hot hardness of Fe-Cr weld metal falls slowly with increasing temperatures up to about 800 to 900°F [425 to 482°C] and thereafter it falls rapidly and also becomes strongly affected by creep stresses. At 900°F [482°C] the instantaneous Rockwell hardness is about C 43, and 3 min under load will cause an apparent drop to nearly C 37. At 1200°F [650°C] the instantaneous hardness value may be no higher than Rockwell C 5, and the apparent loss due to high-temperature creep may be as much as C 45. However, the loss of hardness due to tempering is negligible in comparison with many martensitic alloys, and the drop in hardness shown by hot testing is practically reversed on cooling to ordinary temperatures. Very little is known about these alloys' resistance to thermal shock and thermal fatigue.

The impact resistance of FeCr-A1 weld-metal deposits is low. They will withstand very light impact without cracking, but cracks will form readily if blows deform the metal. These filler metals are seldom used under conditions of medium impact and are unsuitable for heavy impact where cracking is objectionable. High-compression stresses should be avoided, too.

The high chromium content of EFeCr filler metals and weld metal gives them excellent oxidation resistance up to 1800°F [982°C] and they can be considered for hot-wear applications, where their tendency to be slightly plastic at hot temperatures is not objectionable. They will not hold a precise shape at high temperatures.

The corrosion resistance of the chromium in the weld-metal deposit is comparatively low and these alloys are not very effective in providing resistance to liquid corrosion. The as-deposited weld metal will rust in moist air. It is not stainless. However, its corrosion resistance is better than ordinary iron and steel.

The resistance of these alloys to low-stress scratching abrasion is outstanding. It is related to the volume of hard carbides that they contain. Deposits of FeCr-A1 weld metal will wear only about one-eighteenth as much as soft carbon steel, even against rounded quartz grains and sharp angular flint fragments. As stress on abrasion increases, their performance declines. As deposited, the wear resistance of FeCr-A1 is only mediocre under high-stress grinding abrasion. Its use is usually not correct for this kind of service.

Low-stress abrasion produces a good polish on FeCr-A1 weld metal with a resulting low coefficient of friction. Where the polish is produced by metal-to-metal contact, the performance also is good. Resistance to galling (sticking) is considered to be better than for ordinary hardened steel because tempering from frictional heat is negligible.

The hard carbides in the metal matrix can stand up from the surface and cause excessive cutting wear on a mating surface. Therefore metal-to-metal service such as a machine slide or dry bearing should be considered carefully. Surfacing of rolling-mill guides is a typical application for these alloys.

In compression, the deposited FeCr-A1 filler metals are expected to have a yield strength (at 0.1 percent offset) of between 80,000

to 140,000 psi [552 to 965 MPa] with an ultimate compressive strength from 150,000 psi [1034 MPa] to as high as 280,000 psi [1930 MPa]. Like other cast irons, their tensile strength is low, and tensile stresses should be avoided when you use them.

The machinability of FeCr-A1 weld metal deposits is very low. These alloys are generally not machinable and they are even difficult to grind with aluminum-oxide wheels. Diamond or boron-nitride wheels have to be used.

The austenite matrix of the weld metal is stabilized to keep it from becoming martensite by dissolved manganese in the metal. Therefore, these materials do not harden by heat treatment.

Welding with oxyfuel in the flat position with a 3× feather-to-cone reducing flame is recommended. The coefficient of thermal expansion of FeCr-A1 alloys is about 50 percent greater than that of carbon steels and irons. You can expect a lot of thermal expansion and contraction. Contraction stresses are prone to crack the weld deposit and, while the cracks will do no harm to a thin-welded surface layer, they can be minimized by preheating the workpiece and postheating after welding to produce slow cooling. The use of a flux is helpful in dealing with dirt, scale, and other undesirable surface contamination that can add carbon to your weld-metal deposit.

It isn't necessary to add flux to clear, bright metal surfaces that have been produced by grinding. A good bond can be produced on all iron-base materials, providing that the base metal is not damaged by the high temperature of your welding and weld-cooling procedures. In arc welding, the procedures for applying FeCr-A1 filler metals are similar to those for other surfacing electrodes.

Tungsten Carbide Rods and Electrodes

Tungsten carbide gas-welding rods and arc-welding electrodes are usu-

ally sold as steel tubes containing 60 percent carbide granules by weight. Lower percentages are available for certain applications. The same carbide material can be purchased in bulk as loose granules and applied directly on your base metal, like a sub-arc flux.

Overlays containing tungsten carbide make weld-metal deposits whose abrasion resistance surpasses that of any other available hardfacing material. These WC electrodes and rods are used on the cutting edges of rock drills; the wearing surfaces of mining, quarrying, digging, and earth-moving equipment; and on hundreds of other parts where the roughness of the weld deposit (as it wears and exposes the tungsten carbide grains) is not a problem, but where extremely high abrasion resistance is needed.

The weld metal isn't just a softer matrix with hard tungsten carbide particles. Part of the tungsten and the carbon dissolve in the weld-metal matrix and produce an extremely hard tungsten steel with the rest of the extra-hard tungsten-carbide particles in it. This matrix material has characteristics that range from those of air-hardening tungsten steel to cast iron containing tungsten. It is similar to armor plate.

The balance of the tiny tungsten-carbide particles is cemented in this extremely wear-resistant matrix. The size of these particles is the basis for the eight RWC gas-rod grades and the five EWC electrode grades. Two typical AWS grade designations are RWC-12/30 and EWC-12/20. The number code at the end means that the tungsten-carbide particles in the filler metal will pass through a standard #12 Tyler screen mesh but will not pass through a #30 or #20 Tyler screen mesh.

The surface hardness and abrasiveness of the weld deposit depends upon how much of the tungsten carbide has dissolved in the matrix and how much is left as cemented particles. Small carbide particles dissolve faster than large ones, so that the AWS grade system, based on screen mesh sizes of the particles, makes good sense.

The hardness of a good-quality cast tungsten carbide is extremely high. The Vickers pyramid number is about 2400. This material falls within 90 to 95 on the Rockwell A scale. It's about as hard as silicon carbide or all but the hardest type of grinding wheel.

The tungsten gives the weld deposit hot hardness up to 1000°F [538°C], which is considerably better than ordinary hardened steels. The higher temperatures of the welding arc will let more tungsten carbide go into solution in the matrix metal than oxyfuel welding does. So arc-welded deposits will be harder than weld metal layed down with an oxyfuel process.

Both the carbide granules and the weld deposits are relatively brittle and vulnerable to cracking under sudden tensile stress. However, they have extremely high compressive strength. They can withstand light impacts that do not produce compression above the yield strength of the weld metal (which may reach 200,000 psi [1379 MPa]). Impact blows faster than 50 ft/s [15 m/s] should be avoided.

Tungsten carbide has a low resistance to oxidation. Exposed granules in the weld metal will oxidize to form a yellow tungsten oxide at temperatures above 1000°F [538°C].

Although the carbide granules can have good corrosion resistance in many fluids, the matrix that the granules are in is as vulnerable to corrosion as ordinary steel.

Don't select these materials to resist corrosion. If you have to use them in a corrosive environment because of their extreme high hardness and abrasion resistance, make a test to see how long some of the weld metal will last in the corrosive environment.

These composite tungsten carbide weld-metal deposits are very good for resisting low-stress scratching and high-stress grinding abrasion. In either type of service the matrix metal will abrade more rapidly than the tungsten carbide particles. As long as the matrix wear does not undermine the carbide particles so they fall out, this shouldn't be a problem.

Tungsten-carbide weld-metal deposits are not used for metal-to-metal wear. This material will wear out almost any metal that it rubs against. The tiny carbide particles sticking out of the surface make the weld overlay act like a grinding wheel.

As you might guess, tungsten-carbide weld-metal deposits are practically unmachinable. The deposits are finished, when absolutely necessary, with silicon carbide, boron nitride, or diamond grinding wheels.

You can identify WC weld metal because it is nonmagnetic, has a very high density, is insoluble in most acids, and readily forms a yellow oxide coating when heated red hot in air. These materials have such a high melting point that they are practically impossible to melt with an oxyacetylene or MAPP Gas flame. Either oxyhydrogen (which we recommend you don't use) or an arc-welding process is needed to deposit the tungsten-carbide surface.

Tungsten carbide weld deposits seldom are heat-treated. It is possible to boost the carbon content of the matrix metal by carburizing it, followed by quenching and tempering, but that's not often done.

The welding conditions usually used for WC filler metals are listed in Table 19-2.

The usual overlay thickness is about ⅛ in. [3.2 mm]. Skilled arc welders can make thinner deposits, and thicker ones are possible by using several layers, but tension cracks in the overlay are likely.

19-5 SURFACING WITH GMAW, GTAW, AND SAW

We won't give you details on surfacing methods with other welding processes, except to tell you that a lot of cladding work is done on large pressure vessels with GMAW or SAW. Most of this involves laying down layers of stainless steel inside carbon-steel or low-alloy-steel vessels to make them corrosion resistant on the inside. This work is rather advanced for a beginning welding course and it would require extensive knowledge of the metallurgy of stainless steels to pick the correct grade for the service conditions. However, the basic welding techniques are those you have already learned. Other alloys, such as cobalt-base materials, sometimes are put down with GTAW, which also can apply stainless steels when the work area is not too large.

We will spend the rest of this lesson telling you about metal spraying, which is an important maintenance welding and repair method. With metal spraying you can do a lot of things that would be difficult to do with conventional welding electrodes and gas rods.

19-6 SPRAYING MOLTEN METAL

Metal spraying, also called metallizing, is used to spray metal coatings on fabricated workpieces. The coating metal is initially in wire or powder form. It is fed through a special gun or oxyfuel torch and melted by an oxyfuel gas or plasma-arc flame, then atomized by a blast of compressed air. The air and combustion gases

TABLE 19-2 Tungsten-carbide welding parameters

Electrode Diameter	Current, in A	
	DCRP	AC
⅛ in.	100–125	110–135
3/16 in.	125–150	135–160

spray the atomized molten metal onto a prepared surface where the coating is formed. The process is shown schematically in Fig. 19-4.

Metallized coatings are used to repair worn parts, to salvage mismachined components, or to provide special properties to the surface of original equipment.

Metallized coatings are used for improving the bearing strength of surfaces, for adding corrosion resistance, increasing surface lubricity, improving thermal and electrical conductivity, and for producing decorative coatings.

Corrosion-resistant coatings such as aluminum and zinc are applied to ship hulls, bridges, storage tanks, and canal gates, for example. Hardfacing alloys are put on shafts, gear teeth, amd other machine components, as well as on mining equipment, ore chutes, hoppers, track, and rail. Coatings with combined bearing and lubricity properties such as bronze and babbitt metal are used to improve the surface life of machine shafting, slides, and ways.

Metallized Coating Characteristics

The mechanical properties of some sprayed metal coatings are listed for you in Table 19-3. The physical properties are often quite different from steels, and you can do a lot of things with metallized coatings that would be impossible to do with conventional welding.

As-sprayed metal coatings are not solid. The first molten droplets from the metallizing gun hit the substrate metal and flatten out. Subsequent particles overlay the first deposit, building up a porous coating. Particle bonding is essentially mechanical, although some metallurgical bonding also occurs.

The small pores between the droplets soon become closed as the coating thickness increases droplet by droplet. These microscopic pores can hold lubricants and are one of the reasons why metallized coatings are used to increase the lubricity of wear surfaces.

The tensile strengths of sprayed coatings are high for the relatively low melting points of most of the metals used, but ductility is uniformly low. Therefore, parts must be formed first and then sprayed. Thin coatings of low-melting-point metals such as sprayed zinc on steel are a minor exception to this rule. They can withstand limited forming.

Workpiece Limitations

Metallizing is not limited to any particular size workpiece. The work can vary from a crane boom to a small electrical contact. Metallizing can be done on a production line or by hand, in a plant or in the field.

The shape of your workpiece, however, does have an important influence on the process. Cylindri-

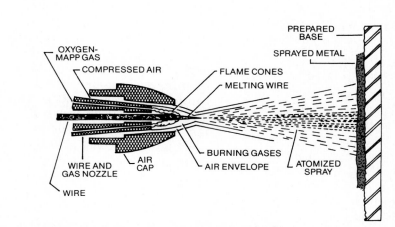

FIGURE 19-4 This schematic drawing of the wire spray process shows you how wire metallizing works. Powders as well as wires can be fed through certain types of equipment, and some metallizing guns use plasma arc instead of oxyfuel to melt the filler metal. Acetylene or MAPP Gas are the most common fuel gases used. Propane and natural gas do not have high enough flame temperatures to melt most of the alloys used in this work.

TABLE 19-3 Mechanical properties of sprayed coatings

Coating Metal	Rockwell Hardness	Tensile Strength, psi	Elongation, in./in.
1100 aluminum	H 72	19,500	0.23
Aluminum-bronze (90-9-1)	B 78	29,000	0.46
Babbitt (89% Sn)	H 58	Very soft	—
Tin bronze (60–40)	B 50	26,500	0.50
Copper	B 33	15,500	—
Molybdenum	C 38	7,500	0.30
Monel	B 39	—	—
Nickel	B 49	—	—
Steels:			
AISI 1010	B 89	30,000	0.30
AISI 1025	B 90	34,700	0.46
AISI 1080	C 36	27,500	0.42
AISI type 202	B 88	30,000	0.50
AISI type 304	B 78	30,000	0.27
AISI type 420	C 29	40,000	0.50
Zinc	H 46	13,000	1.43

cal parts such as shafts, driers, and press rolls that can be rotated in a lathe, on turning rolls, or in a headstock-tailstock welding positioner are ideal for spraying with a machine-mounted setup. The curved parts of the inside of a small pump casing, or threaded shafts, are difficult to coat because the backdraft or splash of the metal spray can be hard to control. Small-diameter holes, bores of any depth, or narrow grooves are also difficult or impossible to coat because of bridging of the sprayed-metal coating.

Metallizing Materials

As you probably suspect, there is a wide range of materials that can be flame sprayed. Most of them include metals, but refractory oxides in the form of either powder or rods also can be applied. Wires for flame spraying include the entire range of alloys and metals from lead (which melts at 618°F [325°C]) to molybdenum (with a melting point of 4730°F [2610°C]). Higher-melting-point materials, even tungsten, also can be sprayed, but a plasma-arc spray gun is required.

Between the extremes of lead and molybdenum are common metal coatings such as zinc; aluminum; tin; copper; various brasses, bronzes, and carbon steels; stainless steels; and nickel-chromium alloys. Spray coatings can be combined on one workpiece. For example, molybdenum or nickel aluminide often is used as a thin transition coating on steel parts to increase bond strength with a top coating. Then the second metal is applied to build up the deposit. This is the metallizing equivalent of buttering a transition layer between two other alloys with stick electrodes.

Surface Preparation

Surfaces for metallizing must be very clean. They also require roughening to make sure that you get a good mechanical bond between the workpiece and the coat-ing. Grease, oil, and other contaminants are removed with any suitable solvent. Cast iron or other porous metals should be preheated at 500 to 800°F [260 to 425°C] to remove entrapped oil or other foreign matter. Sandblasting can be used to remove excessive carbon resulting from preheating cast iron. Chemical cleaning may be necessary prior to preheating.

Shaving off the original surface with a lathe often is necessary on shafts and similar surfaces to permit a uniformly thick buildup on the finished part. The depth of undercutting depends on the diameter of the shaft and on service conditions. If the undercut surface becomes oxidized or contaminated, it should be cleaned before roughing and spraying.

Roughing the workpiece usually is the final step before flame spraying. Various methods are used (see Table 19-4), ranging from rough threading or threading and knurling to abrasive blasting. Details are given in the AWS report, "Recommended Practices for Metallizing Shafts and Similar Objects." The main idea is to make the surface rough enough to hold the first layer of sprayed metal.

Thin molybdenum or nickel-aluminide spray coatings are often applied to the roughened surface to improve the bond strength of subsequent coatings. Applications that require only a thin coating of sprayed metal often eliminate the roughing step and go directly to a bonding coat. The surface is then built up with some other metal.

Coating Thickness

Cost and service requirements are the basis for determining the practical maximum coating thickness you will want to apply for a particular job, such as building up a worn machine part. Total metallizing cost includes your cost of preparation, oxygen, fuel gas and materials, application time, and finishing operations. If repair costs are too high, it may be more economical to buy a replacement part.

The total thickness for the as-sprayed coating on shafts is determined by the maximum wear allowance, the minimum coating thickness that must be sprayed, and the amount of stock required for the finishing operation. The minimum coating thickness depends upon the diameter of the shaft and is given in Table 19-5. For press-fit sections, regardless of diameter, a minimum 0.005-in.-thick [0.13 mm] coating is required.

Variations in the thickness of a deposit depend on the type of surface preparation you use. The

TABLE 19-4 Typical spray-bond strength versus surface preparation in pounds per square inch (blanks indicate that data are not available or coating not used)

Substrate	Type of Test	Grit Blast	Groove and Roughen	Rough Thread	Molybdenum Bond Coat
1100 aluminum	Tensile	1,150	2,610	990	2,150
	Shear	3,270	6,500	5,530	5,200
Commercial bronze	Tensile	1,170	4,290	1,780	—
	Shear	5,890	13,580	10,650	—
Steel drill rod (R_c 67)	Tensile	2,330	—		—
	Shear	2,105	—	—	9,820
Hardened steel (R_c 52)	Tensile	1,720	—	—	2,250
	Shear	4,710	—	—	9,850
Cold-finished steel	Tensile	2,660	4,580	—	2,150
	Shear	6,970	14,600	—	8,850
Gray iron	Tensile	2,510	3,720	—	2,250
	Shear	4,110	15,660	—	10,200

thickness of a deposit over a threaded surface varies more than that of a deposit over an abrasive-blasted surface or a smooth surface prepared by spray bonding a molybdenum or nickel-aluminide coating. In general, the total variation in thickness that you can expect for routine production spraying with mounted equipment is 0.002 in. [0.05 mm] for deposits from a metallizing wire.

Coating Shrinkage

You must also consider the effects of the shrinkage of your sprayed metal because it affects the thickness of the final deposit

(Table 19-6). For example, deposits on the inside diameters of parts must be held to a minimum thickness to conform with the shrinkage stresses; thick coatings will separate from the workpiece because of excessive stresses and inadequate bond strength.

Table 19-6 gives shrinkage values for the metals commonly used for spray coatings. However, thicker coatings can be deposited with metals that have lower shrinkage. Metals with high shrinkage rates may tend to crack when very thick coatings are made.

All sprayed-metal coatings are stressed in tension to some degree except those where the substrate material has a high coefficient of expansion and is preheated before spraying. The stresses can cause cracking of thick metal coatings with a high shrinkage value. The austenitic stainless steels are in this category.

The susceptibility to cracking of thick austenitic stainless steels can be prevented by first spraying a martensitic stainless steel on the substrate, then depositing austenitic stainless steel over it to get the required coating thickness. The martensitic stainless steel produces a strong bond with the carbon-steel substrate, has good strength in the as-sprayed condition, and gives you an excellent surface for your austenitic stainless-steel top coat.

Spray Speeds

Both acetylene and MAPP Gas are used for oxyfuel metallizing. MAPP Gas produces higher spraying speeds, up to 25 percent higher than acetylene, on all but the highest-melting-point alloys. The reason is that MAPP Gas has a much higher heat of combustion than acetylene (2406 Btu/ft^3 versus 1470 Btu/ft^3 for acetylene), which means that your workpiece will come up to the correct temperature must faster and the wire will melt off quicker. Acetylene is used for applying cobalt and some tungsten-based spray coatings because these materials have very high melting temperatures and the oxyacetylene flame temperature is a couple of hundred degrees higher than that of MAPP Gas. Tungsten is so difficult to melt and spray that you may need a plasma-arc spray gun.

Table 19-7 compares the spraying speeds of MAPP Gas and acetylene for ten metal coatings.

Metallizing Processes

According to the Wall Colmonoy Corporation of Detroit (the leading equipment supplier for metallizing processes) there are four basic methods for spraying molten metal. They are the spray without fusion, spray and fuse, plasma spray (done under ambient conditions or in a soft vacuum), and the detonation gun process. All

TABLE 19-5 Minimum coating thickness on shafts

Shaft Diameter, in.	Coating Thickness, in.
1 or less	0.010
1–2	0.015
2–3	0.020
3–4	0.025
4–5	0.030
5–6	0.035
6 or more	0.040

TABLE 19-6 Shrinkage values for sprayed coatings

Coating Metal	Shrinkage, in./in. thickness
Ferrous Metals	
0.10% carbon steel	0.008
0.25% carbon steel	0.006
0.80% carbon steel	0.0014
Type 304 austenitic stainless steel	0.012
Type 420 martensitic stainless steel	0.0018
Nonferrous metals	
1100 aluminum	0.0068
Al-Si (4–6% Si)	0.0057
Aluminum-bronze	0.0055
Manganese-bronze	0.009
Phosphor-bronze	0.010
Molybdenum	0.003
Zinc	0.010

TABLE 19-7 Spray rates of MAPP Gas versus acetylene

Material	Wire Diameter, in.	Spray Rate, lb/h	
		MAPP Gas	Acetylene
Molybdenum	1/8	5.8	4.9
Zinc	1/8	43	34.4
Aluminum	1/8	11.5	9.4
0.10% carbon steel	1/8	12.5	8.5
18-8 stainless	1/8	12.5	8.8
Zinc	3/16	80	68
Aluminum	3/16	19.5	16.2
Aluminum	3/16	18.6	16.9
0.10% carbon steel	3/16	20.8	15.8
18-8 stainless	3/16	28.5	19.2
Aluminum-bronze	3/16		

four processes are line-of-sight spray methods which means that powdered metal is heated to a plastic or molten state and accelerated straight at a workpiece. The powdered and heated metal hits the substrate workpiece and sticks to it. On impact, the hot metal particles form a coating consisting of many layers of overlapping spots or "splats" that build up the coating thickness. Since all four processes are line-of-sight, you can't coat around corners or underneath some area that you can't see. Keep that in mind when using metallizing.

Spray and fuse Metal spraying without follow-up fusion is just like the first step in the spray and fuse process. The plain spray process deposits low-melting-point materials such as aluminum. The spray and fuse process requires more know-how. It also produces stronger, less porous coatings. The spray and fuse process is a two-step process in which powdered coating metals are first deposited with a normal thermal spray method, using either a combustion gun, an oxyfuel torch, or a plasma-spray gun, and then the coating is fused with a heating torch or fused in a heat-treating furnace. Spray and fuse coatings can be low-melting-point alloys such as aluminum, but they most often are made of nickel or cobalt "self-fluxing" alloys to which tungsten carbide particles have been added for increased wear resistance. Coatings range from 0.020 to 0.080 in. [0.51 to 2.0 mm] thick. The coatings are built up a layer at a time at a rate of 0.005 to 0.030 in. [0.13 to 0.75 mm] per pass. Typical metal deposition rates are 8 to 12 lb [3.5 to 5.5 kg] per hour.

The equipment used by the spray, or spray and fuse methods, differs primarily in the way the powder is fed into the flame (Fig. 19-5). Compressed air can be used as a propellant for the powder which is injected directly into the flame through a central orifice in

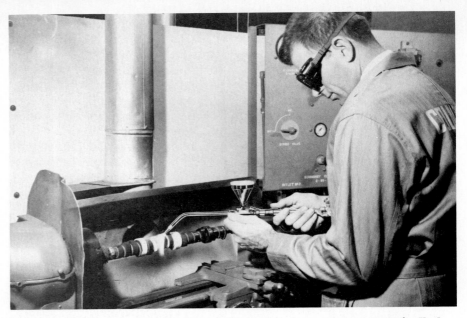

FIGURE 19-5 Powdered metals can be sprayed on a base-metal (called the substrate) with special oxyfuel torches like this one. These powder-metal alloys are mostly used for abrasion and wear resistance. The deposited overlay is made more dense (less porous) by reheating the coated surface with a bushy-flamed heating torch as soon after the overlay is deposited as possible. (Courtesy of Wall Colmonoy Corporation.)

the gun nozzle or through a torch tip. Another system feeds the powder into an aspirating (sucking) gas ahead of the nozzle and prior to the flame. In still another system, gravity feeds the powder directly into the flame. The molten powder is then fed directly to the workpiece by the force of the combustion gases.

Oxygen and fuel gases (either acetylene or MAPP Gas) are used for their higher flame temperatures, with acetylene being the most likely fuel to use for high-melting-point alloys. Propane only has limited use for low-melting-point alloys. The powder also can be enclosed in a thin plastic tube and fed through wire-type spray equipment. Alternatively, a plasma-spray torch rather than a combustion gun can be used. In all cases, the powdered alloy becomes molten or semimolten as it passes through the flame or plasma arc, making an initial bond on the substrate prior to the final fusion step. Fusion after spraying is accomplished using an oxyacetylene or oxy–MAPP Gas heating torch with a multiflame

tip. An alternative to using a heating torch is to fuse the part in a furnace with a vacuum or reducing atmosphere, or with electrical induction heating equipment.

Many metals can be coated using the spray and fuse process without special precautions, while others require preheating or slow cooling procedures to prevent cracking of the metal coating caused by differences in the thermal expansion and contraction rates of the coating and base-metal substrate. In general, carbon steels and conventional alloy steels can be coated with ferrous metals without special precautions if the carbon content of the coating and substrate is less than 0.25 percent. When the carbon content of the steel coating or substrate is above 0.25 percent, the part requires a 500 to 700°F [260 to 370°C] preheat prior to fusing the sprayed coating, and a slow cooling after fusion. Cast irons and copper also present no problems. But the austenitic AISI Type 300 series stainless steels have higher coefficients of thermal expansion. If austenitic stainless steel is spray-coated for corrosion resistance (for example) the substrate should be preheated to 600 or 700°F [315 to 370°C] prior to spraying. This will expand the part sufficiently to prevent the subsequent expansion or contraction of the coating from breaking the mechanical bond between the sprayed metal and the part prior to final fusion. The fusion step for these coatings is best done immediately after spraying.

Air-hardening tool steels and the AISI 400 series martensitic stainless steels (except Type 414 and 421) can be coated easily by the spray and fuse process. However, these metals require preheating prior to fusing the overlay, and slow cooling after fusion, to prevent the overlay from cracking. Certain stainless steels (such as AISI Type 303) have sulfur, lead, phosphorus, or selenium additives to make them free-machining. These free-machining additives produce porous overlays due to outgassing during fusion. Therefore, these materials are not good overlay metals. Titanium- or aluminum-bearing alloys (such as AISI Type 321 stainless steel) also are not good candidates for metal spraying because of the highly stable oxide layers they form on substrate metals which interfere with proper bonding of the overlay during fusion.

SURFACE PREPARATION Undercutting of the surface to be coated is usually necessary to produce an even thickness of overlay metal on an unevenly worn surface, on parts not originally built to include an overlay, or to provide room for sufficiently thick overlay to take the expected wear. The depth of undercutting should be determined by the amount of wear permitted in service. It is best to undercut at least 0.010 in. [0.25 mm] because this is considered to be the minimum thickness for a sound overlay. When undercutting a shoulder, the machinist that does the job should "feather" up the shoulder at a 30° angle

from the surface. The reason is that the corner of right-angle turns are difficult to coat (remember that metal spraying is a straight line-of-sight process). If an external corner is going to be sprayed, it should have a radius of at least 1/32 in. [0.8 mm]. If rough threading is used to prepare the surface, the shoulder angle should be reduced to 15°.

After roughing or undercutting, the surface to be sprayed should be cleaned of all electroplated coatings, carburized surface layers, nitrided surfaces, or other surface treatments. Oil or grease must be removed with degreasing solutions. Abbrasive shot blasting also can be used to clean surfaces. The surface to be coated usually must be roughened up, anyway, and grit blasting is the preferred method. When the surface hardness is less than 30 Rockwell C, it can be grit blasted with crushed angular chilled-iron grit. Grit size can range from SAE No. 12 to No. 16; some manufacturers also recommend mixtures of No. 25 and No. 40 grit. Silicon-carbide or aluminum-oxide grit can be used as well, particularly for harder surfaces. It's important to grit-blast more area than you will spray, and to blast surfaces around external corners and beyond the shoulder of undercut areas.

Rough threading by a machine tool (usually a lathe) sometimes is used to prepare the surface, but that is the second-best method. Threading should be done with 32 to 40 threads per inch (cut to not over 0.008 in. [0.2 mm] deep). Metal slivers are removed by running a clean, new file over the threaded surface.

MASKING You often will want to stop off certain areas so that they do not get coated. There are stop-off chemicals available from Wall Colmonoy that you can paint or spray on the metal substrate to prevent coating adhesion in areas where you don't want the spray coating to stick. These stop-off coatings can be removed after

metal spraying by wire brushing and polishing.

If there are holes, keyways, or slots in the workpiece that are not to be coated, they can be masked (for example, plugged with wood) during grit-blasting. Graphite is another good plugging material because it not only withstands high temperatures, but graphite also is soft and easy to machine or carve with a knife into the plug shape you want. The top surface of a masking plug should be flush with the height of the finished overlay. If the mask rises above the part surface, it will cast an uncoated shadow when you spray the part.

SPRAYING The best distance to hold the metal spray gun is about 7 to 12 in. [180 to 300 mm] from the work surface, measured as the gun tip to work distance. The relative gun-to-part surface speed is usually about 30 to 50 surface feet per minute [9 to 15 surface meters per minute]. Multiple spray passes are best with the application of about 0.005- to 0.030-in. [0.127- to 0.762-mm] coating per pass. When a plasma gun is used for depositing a sprayed coating, follow the manufacturer's suggested operating procedures. When calculating the required thickness of the sprayed overlay about 20 percent extra must be allowed for coating shrinkage upon follow-up fusion. In addition, allow at least 0.010 in. [0.25 mm] per side for finishing the part if that is necessary. When spraying heavy sections or internal diameters, or when laying down heavy surface coatings, preheat the part to 400 to 500°F [205 to 260°C]. That will help prevent cracking of the as-deposited mechanical bond.

FUSION Fusion should be done immediately after spraying. The idea is to bring the overlay and base metal surface up to a temperature where the overlay will wet and bond to the base metal without the coating losing shape or running. Depending upon the coating metal you use, the fusion temperature will be in the range of 1850 to 2050°F [1010 to 1120°C].

Heating with an oxyacetylene or oxy–MAPP Gas torch is the best fusion method. Use a multi-flame tip with a soft, bushy flame adjusted to a neutral flame setting. Hold the torch at a 90° angle to the workpiece at the proper distance so that the outer flame cone is close to, but not touching, the part. Play the heating torch flame on the base metal adjacent to the overlay coating until the base metal reaches a dull red (a minimum of 1300°F, or 705°C).

If the overlay does not extend to the end of the workpiece, heating should begin about 2 in. [50 mm] from the overlay. If the overlay extends to both ends of the part, begin heating at the center of one end, or on the inside if the part is a hollow cylinder. The heat should then be concentrated on the overlay. As the overlay reaches its bonding temperature, the rough-sprayed surface becomes molten (looks glassy), and it reflects the light of the torch flame. Then move the flame along at a constant rate, slow enough to heat the overlay to its bonding temperature but fast enough to prevent the overlay from sagging or running.

Fusion also can be done in a vacuum furnace or in a controlled-atmosphere furnace with a reducing atmosphere. The exact temperature and degree of control required in the furnace depends on the atmosphere being used. That's a job for a heat treater to handle, not a welder. Furnace settings are not your problem. The furnace fusion method is best used on parts with irregular cross sections where the torch fusion method is not practical. Another fusion method you can use is induction heating. This is particularly well suited for high production where the induction process is automated. Again, induction heating is for heat treaters, and

(A)

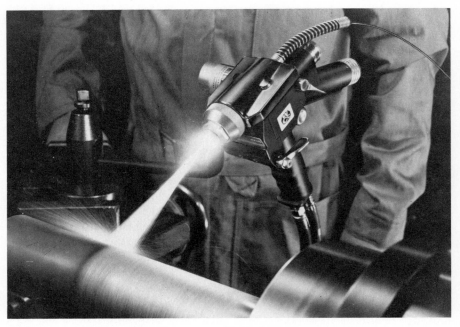

(B)

FIGURE 19-6 (A) Plasma-arc spray equipment can deposit powdered metals that oxyfuel processes can't melt, as well as (B) wire-fed filler metals such as aluminum, carbon steel, stainless steel, or copper. Plasma coatings often don't require a follow-up fusion pass with a heating torch. Powdered metals can produce weld overlays with chemical compositions that can't be produced by using wire filler metals because the coating material can't be drawn into a wire, but it can be made into a powder. Wire-fed coatings, however, usually cost less because they are commonly used metals such as aluminum or stainless steel. But the powder-fed overlays often have superior properties such as extremely high hardness. The two types of coating materials can be combined. For example, a carbon-steel wire-fed process can be used to build up a worn part, and a powder-fed process can then apply an extremely hard and abrasion-resistant surface coating.

you don't have to worry about how to do it.

FINISHING Sprayed and fused coatings of alloys harder than 55 Rockwell C cannot be machined. They must be ground. Surfaces softer than 55 Rockwell C can be machined with a carbide-tipped tool. Most sprayed and fused alloy deposits can be ground with green silicon-carbide grinding wheels of the H, I, or J hardness (see Chap. 22 for details on grinding wheels). Wet grinding is recommended. For roughing, use No. 24 grit wheels. For finishing, use a #60 grit or finer-grained wheel.

Plasma spraying The plasma-spraying process (Fig. 19-6) uses a plasma gas stream that propels powder particles onto a substrate. Because the plasma temperatures are extremely high, this process can be used to apply refractory coatings that cannot be applied by the spray and fuse process. For example, parts can even be glass-coated.

Although plasma spray coatings have inherent porosity and are mechanically bonded to the substrate, they generally have higher density and better adhesion than is possible with single-step flame spraying. In most applications, coating thickness is in the range of 0.005 to 0.015 in. [0.127 to 0.381 mm], but depending upon the coating and the application, much thicker coatings can be applied by the plasma-spray process.

The plasma-spray process is a one-step process, unlike the spray and fuse method. The substrate usually is maintained below 300°F [150°C] which makes things simpler and results in much less distortion and no dilution of the coating by the substrate. Very smooth coatings can be applied, and finishing often is not necessary. The drawback is that plasma spray equipment costs much more than oxyfuel spray equipment.

PLASMA EQUIPMENT Plasma spraying is similar in many ways to plasma-arc welding. We won't go into the details of the process, because it is not much different from plasma-arc welding or cutting. Wall Colmonoy or other suppliers can tell you what is needed, if you need specific equipment details. The powder carrier gases also are similar to plasma-arc welding (the primary plasma-forming gas is argon, or argon plus nitrogen). The choice of powder feed rate and size, however, is important.

PLASMA POWDERS Plasma spray powders must be suitable for rapid melting and deposition. Most of the conventional plasma metallizing guns used up to the mid 1970s produced powder velocities of 400 to 1000 ft/s [122 to 305 m/s]. Higher-velocity torches are now available that produce higher coating densities and make some bonds with strengths in excess of 10,000 psi [6900 kPa].

According to Wall Colmonoy, the plasma powder for the highest coating density should be between 44 μm (micrometers) (325 mesh) and 10 μm in size. The lower limit ensures free powder flow. Does that mean that you have to count particles and screen them for your work? Of course not. Just make sure that you use powders suitable for plasma spraying. Work with a reputable supplier and you won't have problems.

PLASMA SUBSTRATES Almost any metal substrate that can be adequately roughened can be plasma-coated. The surface roughness should normally exceed 150 microinch (μin.) rms (rms means root-mean-square—it's a way of measuring the surface smoothness of a part and we won't explain it here). Another fact about plasma spraying is that the coating needs to be supported; that is, the base metal must be strong enough to resist deformation (such as bending) when used in service. On the other hand, plasma can spray just about any material that can be powdered and fed through the gun.

The substrate shape may limit the stand-off distance (the distance from the nozzle or front piece of the plasma gun to the workpiece) if you have to coat the inside of a cylinder. Unlike the oxyfuel spray process, you don't have the advantage of a torch tip; the plasma gun is a rigid piece of equipment. Plasma torches which allow you to apply plasma coatings on the inside of cylinders as small as 1¼ in. [32 mm] in diameter are available from Wall Colmonoy. Coating quality is partly a function of the workpiece shape. Coating complex parts, particularly those with narrow grooves or sharp angles, is difficult because the angle between the axis of the plasma jet should be 90° (although some torches can be used at angles of 60°).

SURFACE PREPARATION The surface to be coated must be free of oils, lubricants, oxide or scale, and other foreign matter. After the surfaces are cleaned, they must be roughened. The preferred method is grit blasting. Chilled-steel grit is often satisfactory for relatively soft substrates, while aluminum (aluminum oxide) or silicon-carbide grit will give better cutting action on harder surfaces. The surface roughness should normally exceed 150 μin.

MASKING There are many masking techniques that can be used for plasma spraying. In most cases, it is cheaper to mask a part before plasma spraying than to try to grind off a coating after spraying. Many types of tape and oxide-loaded paints or stop-off lacquers are satisfactory for low-velocity, long stand-off plasma torches. For high-velocity torches with short stand-off distances, more substantial masking is required, such as glass-fiber reinforced high-temperature tape, adhesive-backed steel or aluminum foil, or sheet metal.

FINISHING Plasma coatings are often used as deposited, but a variety of finishing techniques also can be used if necessary. These include wire brushing to produce a nodular surface, machining (suitable for some softer metal coatings), grinding (for harder coatings, usually done with silicon-carbide or diamond wheels), and lapping to produce surface roughnesses of less than 2 μin. rms. The best surface finish that can be obtained is a function not only of the finishing technique, but also of the coating composition and the deposition conditions. Recommendations for machining, grinding, and lapping techniques should be obtained from the coating supplier. Great care is required in finishing plasma coatings to avoid damaging the coating through heat checking, pullout, or edge chipping.

PLASMA COATING ADVANTAGES The plasma spray process has several advantages. Plasma coatings have higher density and better adhesion than those achieved with single-step flame spraying. Usually, the substrate is maintained below 300°F [150°C] so that there is little or no warping. Also, very smooth coatings can be obtained and finishing may not be required. The drawbacks to the process (in addition to higher equipment costs) are that the plasma sprayed coating is porous, and the process (like all metallizing processes) is a straight line-of-sight process so that it will not coat sharp corners. Also, the bond is mechanical, not metallurgical, and therefore plasma coatings can be chipped by high-impact forces. For applications in corrosive environments up to about 350°F [177°C], it is possible to seal the plasma-sprayed coating. For high-temperature applications, densification can be promoted by mechanically working the coating surface (for example, by shot peening) and by a high-temperature sintering treatment (usually done in a furnace).

Vacuum plasma spraying In the vacuum variation of the plasma process, the plasma gun and the workpiece are both en-

closed in a vacuum tank. Since welders can't be enclosed in vacuum tanks and survive, the low-pressure or vacuum plasma coating process (which also can be performed in a low-pressure 50-torr or 0.02 atmosphere) is not a process requiring operator skills. Very fine powder is used. If high-purity coatings are desired, then the powder must be manufactured under inert gas. The advantage is that higher bond strengths can be achieved because higher substrate preheat temperatures and higher operating temperatures can be used without detrimental oxidation of the substrate and coating. Excellent dimensional control on the coating thickness also is possible. Beyond that, we won't go into the vacuum or low-pressure inert gas techniques because they are specialized processes and, as we already mentioned, do not require operator skills.

Detonation guns The detonation gun is markedly different from the flame and plasma-spray equipment. It is used primarily to deposit very hard, wear-resistant materials such as metal oxides and carbides on a substrate. The extremely high particle velocities achieved by detonation guns result in coatings with higher density, greater internal strength, and superior bond strength than can be achieved by conventional plasma and oxyfuel spray processes.

Detonation gun coatings have been successfully applied to critical areas of precision components made from all commercial alloys. The typical coating thickness is less than 0.015 in. [0.04 mm], but many applications require only 0.003 in. [0.008 mm] or less of finished coating.

DETONATION EQUIPMENT The detonation gun has a water-cooled barrel several feet long with an inside diameter of about 1 in. [25 mm] and associated gas and powder metering equipment. In operation, a mixture of oxygen and acetylene is fed into the barrel

along with a charge of metal powder. The gas is ignited and the detonation wave accelerates the powder to about 2400 ft/s [731 m/s], while heating the powder close to or above the material's melting point. The distance that the powder travels and accelerates through the gun's tube is much longer than that in a plasma or flame spray device, which accounts for the high particle velocity.

After the powder leaves the gun barrel, a pulse of nitrogen purges the gun for the next "shot." The cycle is repeated many times a second. Because of the gases used in the detonation gun, the powder can be exposed to either an oxidizing, a carburizing, or an essentially inert environment. Coating deposition is closely controlled by fully automated equipment to achieve uniform coating thickness and to minimize substrate heating and coating residual stress.

POWDERS Suitable powders for detonation guns include pure metals, alloys, oxides, carbides, composites, and mechanical blends of two or more components. Detonation guns have proven particularly good at applying wear-resistant carbide and oxide coatings.

SUBSTRATE The substrate surface is seldom heated above 300°F [150°C] during the coating process; therefore, there are no distortion problems. The detonation gun is a line-of-sight process so that the angle between the axis of the gun and the surface of the workpiece must be 90° to achieve maximum coverage and highest density and bond strength. The detonation speed of the particles, however, means that in many cases the coatings can be applied at angles as low at 45° without loss of surface-coating properties.

SURFACE PREPARATION Clean surfaces are required, as they are by all metallizing processes. However, grit blasting is not always necessary prior to the application of detonation gun coatings. Because the detonation gun accelerates the powder particles to such a high velocity, the particles are actually driven into the surface of some metal surfaces. This results in substantial surface roughening. Some substrates such as titanium do not require grit blasting to achieve adequate bonding. With harder substrates, grit blasting is usually used to improve surface roughness.

MASKING Sheet metal masking is used with detonation guns to limit the deposition area to the required section of the part. Other masking methods don't work because of the high particle velocity of the process.

COATINGS The detonation gun deposits a circle of coating metal with each blast. The coated circles are about 1 in. [25 mm] in diameter and a few μm (ten thousandths of an inch) thick. Each circle of coating is composed of many overlapping splats (corresponding to the individual powder particles). The placement of the circles is closely controlled to build up a smooth coating and to minimize substrate heating and coating residual stress. Detonation-gun coatings have higher density and bond strength than conventional plasma coatings.

FINISHING The surface roughness of most as-deposited detonation-gun coatings fall in the range of 125 to 250 μin. rms. In some applications the coating is used as is. In at least one application, a tungsten carbide plus cobalt coating was grit-blasted to further roughen the surface for better gripping action. Frequently the coating surface is finished before it is placed in service. Finishing techniques vary from wire brushing (to produce a nodular surface) to machining (used on softer metallic coatings) and grinding (used on harder coatings), or lapping to produce surfaces with very low roughness and high polish (as low as 2 μin. rms).

19-7 APPLICATIONS AND SERVICE CONDITIONS

Surfacing and hardfacing problems can be divided into approximately five classes of service conditions. These conditions are not absolute, but they'll help you think about the most important kind of problems you have to solve when working on a piece of equipment. This in turn will help you select the best alloy for the job, since most hardfacing alloys and many surfacing alloys are described in terms of service conditions very similar to the ones we are going to list for you. We'll use our own unofficial class designations here so we won't have to repeat each condition when we show you where it's most likely to occur. These service conditions are

- Class I. Severe impact
- Class II. Very severe abrasion
- Class III. Corrosion, often with abrasion, and frequently at high temperature
- Class IV. Severe abrasion with moderate impact
- Class V. Abrasion with moderate to heavy impact

Now we'll show you some products that are commonly surfaced or hardfaced, and we'll attach a service condition label to each part. The hatch marks show where the hardfacing or surfacing alloy is most likely to be needed. Note in many of the drawings in Fig. 19-7 that you don't always

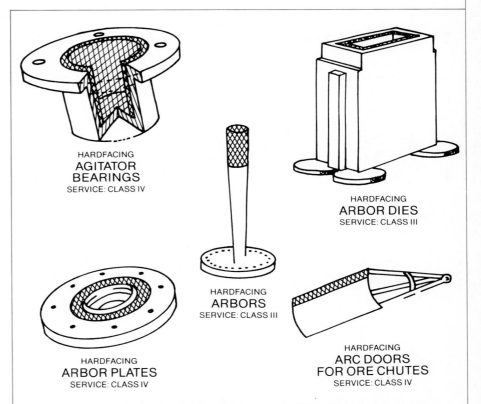

HARDFACING
AGITATOR BEARINGS
SERVICE: CLASS IV

HARDFACING
ARBOR DIES
SERVICE: CLASS III

HARDFACING
ARBORS
SERVICE: CLASS III

HARDFACING
ARBOR PLATES
SERVICE: CLASS IV

HARDFACING
ARC DOORS FOR ORE CHUTES
SERVICE: CLASS IV

FIGURE 19-7 Commonly hardfaced parts are listed here by service classes which are explained in the text. Note the hatch-marked areas on these drawings where the overlay weld metal is applied. Only service conditions and prior experience can assure you that you have the best hardfacing alloy and deposition pattern and that you have applied the material in the right place. Nevertheless, common sense (and these drawings) will guide you to selecting and applying your filler metal and overlay pattern correctly for most industrial applications. A phone call to weld overlay equipment or filler-metal suppliers will give you immediate direction to the most probable alloys and processes to use, based on their years of experience in this kind of work. (Continued on page 518)

HARDFACING
ASPHALT
MIXER PADDLES
SERVICE: CLASS IV

HARDFACING
BULLDOZER
BLADES
SERVICE:
CLASS II, IV, OR V

HARDFACING
CLASSIFIER
FLIGHT SECTIONS
SERVICE: CLASS IV

HARDFACING
COLTERS
SERVICE: CLASS II OR IV

HARDFACING
AUGER BITS
SERVICE: CLASS II

HARDFACING
BULLDOZER
TRUNNIONS
SERVICE: CLASS V

HARDFACING
CLUTCH LUGS
SERVICE: CLASS V

HARDFACING
CONVEYOR
SCREWS
SERVICE: CLASS III
(i.e., FOR SERVICE INVOLVING
TEMPERATURES OVER 800°F)
OR IV

HARDFACING
BAFFLE PLATES
SERVICE: CLASS V

HARDFACING
CENTRIFUGAL DRYER
ANGLE IRONS
SERVICE: CLASS IV

HARDFACING
COAL RECOVERY
AUGERS
SERVICE: CLASS IV

HARDFACING
CRUSHER
HAMMERS
(MANGANESE STEEL)
SERVICE: CLASS I OR IV

HARDFACING
BEARINGS
(KILN FEED SCREW)
SERVICE: CLASS III

HARDFACING
CRUSHER
HAMMERS
(FLEX-TOOTH)
SERVICE: CLASS I OR IV

HARDFACING
BLISTER BAR
TONG BITS
SERVICE: CLASS III

HARDFACING
CLAMSHELL
BUCKET LIPS
SERVICE: CLASS IV

HARDFACING
COAL RECOVERY
CORE BARRELS
SERVICE: CLASS IV AND II

HARDFACING
CRUSHER JAWS
SERVICE: CLASS I OR IV

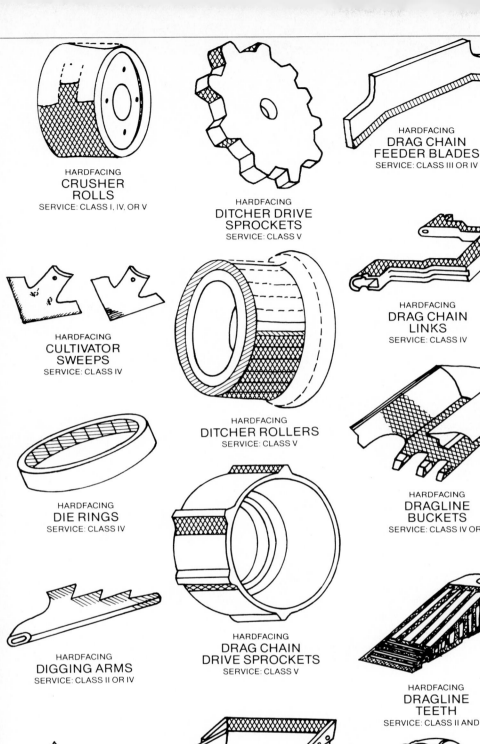

HARDFACING
**CRUSHER
ROLLS**
SERVICE: CLASS I, IV, OR V

HARDFACING
**DITCHER DRIVE
SPROCKETS**
SERVICE: CLASS V

HARDFACING
**DRAG CHAIN
FEEDER BLADES**
SERVICE: CLASS III OR IV

HARDFACING
**DREDGE PUMP
CUTTERS**
SERVICE: CLASS II OR IV

HARDFACING
**CULTIVATOR
SWEEPS**
SERVICE: CLASS IV

HARDFACING
DITCHER ROLLERS
SERVICE: CLASS V

HARDFACING
**DRAG CHAIN
LINKS**
SERVICE: CLASS IV

HARDFACING
**DREDGE PUMP
HEAD LINERS**
SERVICE: CLASS IV

HARDFACING
DIE RINGS
SERVICE: CLASS IV

HARDFACING
**DRAGLINE
BUCKETS**
SERVICE: CLASS IV OR I

HARDFACING
**DREDGE PUMP
IMPELLERS**
SERVICE: CLASS IV

HARDFACING
DIGGING ARMS
SERVICE: CLASS II OR IV

HARDFACING
**DRAG CHAIN
DRIVE SPROCKETS**
SERVICE: CLASS V

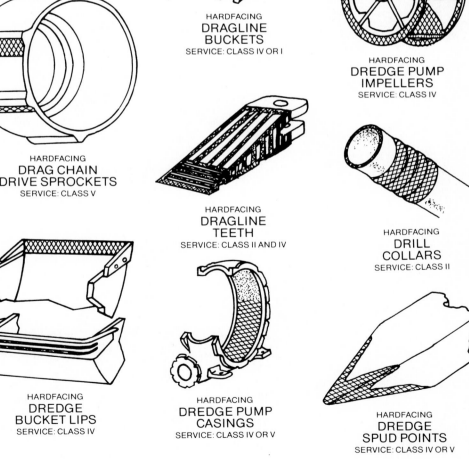

HARDFACING
**DRAGLINE
TEETH**
SERVICE: CLASS II AND IV

HARDFACING
**DRILL
COLLARS**
SERVICE: CLASS II

HARDFACING
**DITCHER DRIVE
SEGMENTS**
SERVICE: CLASS V

HARDFACING
**DREDGE
BUCKET LIPS**
SERVICE: CLASS IV

HARDFACING
**DREDGE PUMP
CASINGS**
SERVICE: CLASS IV OR V

HARDFACING
**DREDGE
SPUD POINTS**
SERVICE: CLASS IV OR V

(Continued on page 520)

HARDFACING
DREDGE PUMP
SHOULDERS
SERVICE: CLASS IV

HARDFACING
ENSILAGE
KNIVES
SERVICE: CLASS II

HARDFACING
FLAPPER
VALVES
SERVICE: CLASS III

HARDFACING
GYRATORY
CRUSHER MANTLES
SERVICE: CLASS IV

HARDFACING
DUCK BILLS
SERVICE: CLASS IV

HARDFACING
EXHAUST
VALVES
SERVICE: CLASS III

HARDFACING
GRADER
BLADES
SERVICE: CLASS II OR IV

HARDFACING
IMPACT
BREAKER ROTORS
SERVICE: CLASS IV

HARDFACING
ELEVATOR
BUCKET LIPS
SERVICE: CLASS I AND V

HARDFACING
FAN BLADES
(CEMENT)
SERVICE: CLASS II OR IV

HARDFACING
GEAR TEETH
(PINION GEARS)
SERVICE: CLASS V

HARDFACING
GYRATORY
CRUSHER LINERS
SERVICE: CLASS IV

HARDFACING
ELEVATOR
GRADER DISCS
SERVICE: CLASS II

HARDFACING
FEEDER
BLADES
SERVICE: CLASS II

HARDFACING
INDUCED
DRAFT FANS
SERVICE: CLASS II

HARDFACING
GRIZZLIES
SERVICE: CLASS IV

HARDFACING
KILN SCOOP
SHOVELS
SERVICE: CLASS II

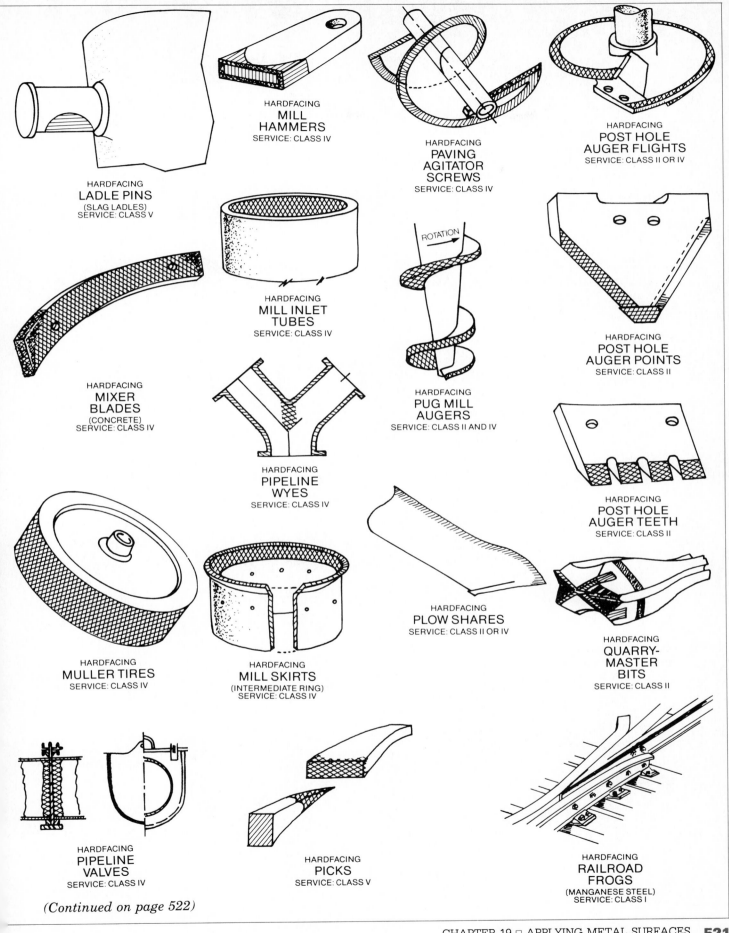

HARDFACING
LADLE PINS
(SLAG LADLES)
SERVICE: CLASS V

HARDFACING
MILL
HAMMERS
SERVICE: CLASS IV

HARDFACING
PAVING
AGITATOR
SCREWS
SERVICE: CLASS IV

HARDFACING
POST HOLE
AUGER FLIGHTS
SERVICE: CLASS II OR IV

HARDFACING
MIXER
BLADES
(CONCRETE)
SERVICE: CLASS IV

HARDFACING
MILL INLET
TUBES
SERVICE: CLASS IV

ROTATION

HARDFACING
PUG MILL
AUGERS
SERVICE: CLASS II AND IV

HARDFACING
POST HOLE
AUGER POINTS
SERVICE: CLASS II

HARDFACING
PIPELINE
WYES
SERVICE: CLASS IV

HARDFACING
POST HOLE
AUGER TEETH
SERVICE: CLASS II

HARDFACING
MULLER TIRES
SERVICE: CLASS IV

HARDFACING
MILL SKIRTS
(INTERMEDIATE RING)
SERVICE: CLASS IV

HARDFACING
PLOW SHARES
SERVICE: CLASS II OR IV

HARDFACING
QUARRY-
MASTER
BITS
SERVICE: CLASS II

HARDFACING
PIPELINE
VALVES
SERVICE: CLASS IV

HARDFACING
PICKS
SERVICE: CLASS V

HARDFACING
RAILROAD
FROGS
(MANGANESE STEEL)
SERVICE: CLASS I

(Continued on page 522)

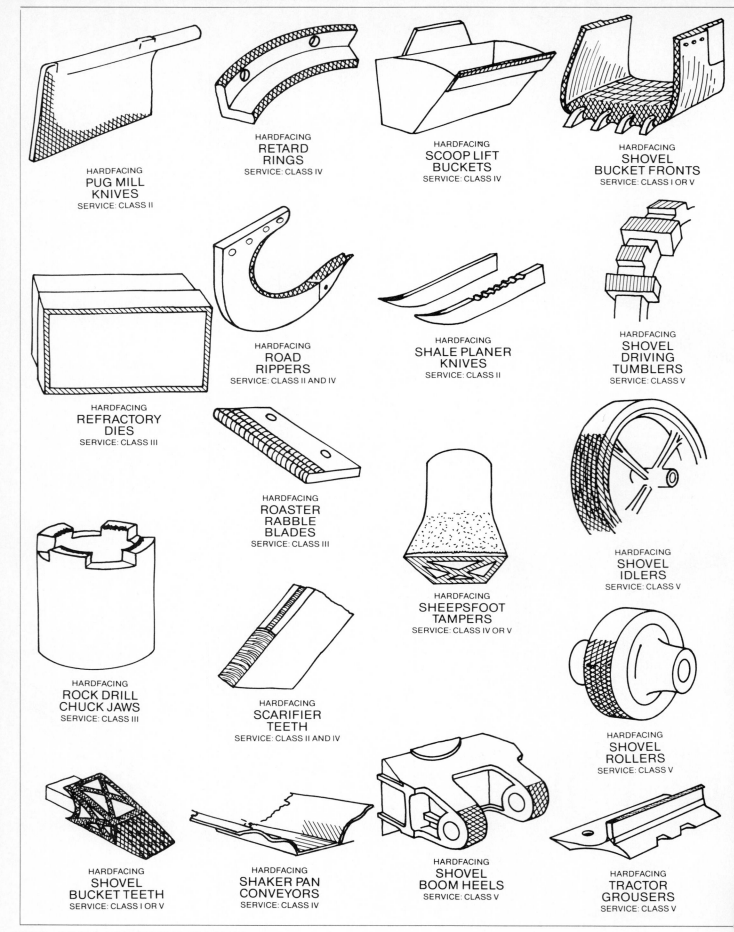

HARDFACING
PUG MILL
KNIVES
SERVICE: CLASS II

HARDFACING
RETARD
RINGS
SERVICE: CLASS IV

HARDFACING
SCOOP LIFT
BUCKETS
SERVICE: CLASS IV

HARDFACING
SHOVEL
BUCKET FRONTS
SERVICE: CLASS I OR V

HARDFACING
REFRACTORY
DIES
SERVICE: CLASS III

HARDFACING
ROAD
RIPPERS
SERVICE: CLASS II AND IV

HARDFACING
SHALE PLANER
KNIVES
SERVICE: CLASS II

HARDFACING
SHOVEL
DRIVING
TUMBLERS
SERVICE: CLASS V

HARDFACING
ROASTER
RABBLE
BLADES
SERVICE: CLASS III

HARDFACING
SHEEPSFOOT
TAMPERS
SERVICE: CLASS IV OR V

HARDFACING
SHOVEL
IDLERS
SERVICE: CLASS V

HARDFACING
ROCK DRILL
CHUCK JAWS
SERVICE: CLASS III

HARDFACING
SCARIFIER
TEETH
SERVICE: CLASS II AND IV

HARDFACING
SHOVEL
ROLLERS
SERVICE: CLASS V

HARDFACING
SHOVEL
BUCKET TEETH
SERVICE: CLASS I OR V

HARDFACING
SHAKER PAN
CONVEYORS
SERVICE: CLASS IV

HARDFACING
SHOVEL
BOOM HEELS
SERVICE: CLASS V

HARDFACING
TRACTOR
GROUSERS
SERVICE: CLASS V

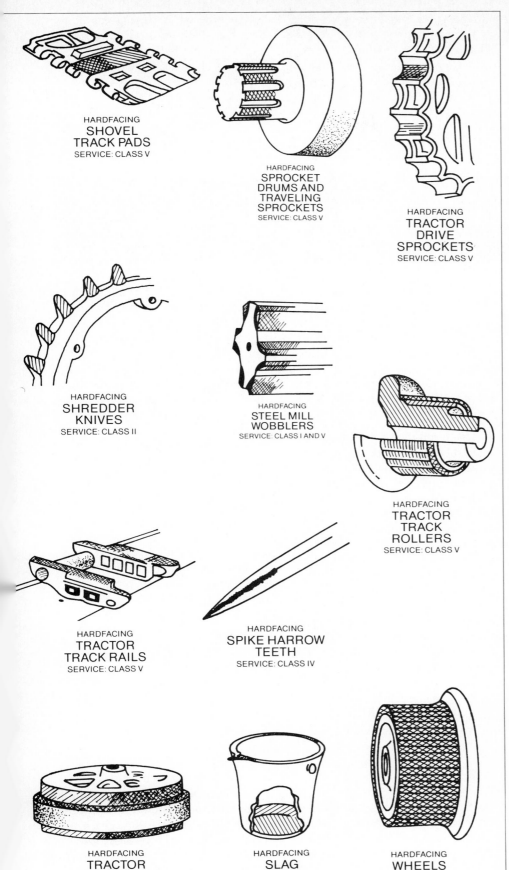

HARDFACING
**SHOVEL
TRACK PADS**
SERVICE: CLASS V

HARDFACING
**SPROCKET
DRUMS AND
TRAVELING
SPROCKETS**
SERVICE: CLASS V

HARDFACING
**TRACTOR
DRIVE
SPROCKETS**
SERVICE: CLASS V

HARDFACING
**SHREDDER
KNIVES**
SERVICE: CLASS II

HARDFACING
**STEEL MILL
WOBBLERS**
SERVICE: CLASS I AND V

HARDFACING
**TRACTOR
TRACK
ROLLERS**
SERVICE: CLASS V

HARDFACING
**TRACTOR
TRACK RAILS**
SERVICE: CLASS V

HARDFACING
**SPIKE HARROW
TEETH**
SERVICE: CLASS IV

HARDFACING
**TRACTOR
IDLERS**
SERVICE: CLASS V

HARDFACING
**SLAG
LADLES**
SERVICE: CLASS III

HARDFACING
WHEELS
(MINE-CAR, SKIP,
CRANE, ETC.)
SERVICE: CLASS V

have to surface the entire part, or even completely cover a given area with hardfacing alloy. Some well-placed stringer beads may be all that the part needs.

This was your last chapter on how to weld. The next chapters will tell you how to know how good or bad your welds are, how to read welding blueprints, and how to use several types of large machinery such as turning rolls and welding manipulators, which you will find in most welding shops. And finally, the last chapter tells you about getting certified as a skilled welder.

REVIEW QUESTIONS

1. Name five products that you might be asked to hardface.

2. Name two applications that you think might benefit from a corrosion-resistant weld-metal overlay.

3. What is hot hardness?

4. What two alloying elements in combination might give you a combination of both high hot hardness and high oxidation resistance?

5. What is the difference between abrasion, erosion, and corrosion?

6. What is buttering? Why is it used?

7. Dilution of surfacing or hardfacing weld metal is a serious problem. Can you name and describe 10 ways you can reduce it?

8. Which welding method gives the most weld-metal dilution, welding vertical-up, or downhill?

9. You have to select an SMAW electrode for hardfacing steel plate. Which AWS filler-metal specification do you need to select an electrode?

10. What do you think the principal wear-resistant alloy is in a gas rod meeting one of the AWS RWC-XX grades?

11. What oxyfuel gas rod or SMAW electrode grade would you most likely select for producing high hardness at service temperatures up to 1100°F where good wear resistance and toughness also are required?

12. Reheating weld metal produced by an EFeMn electrode to above 500 to 600°F may cause what serious weld-metal problem?

13. What is the Vickers pyramid hardness number for a good quality tungsten carbide weld overlay? What Rockwell hardness and what Rockwell scale does this Vickers pyramid hardness correspond to? Do you think this is a very hard, moderately hard, or relatively soft deposit?

14. Briefly, what should you do to a surface before metallizing it?

15. Which metallized coating is likely to shrink the most after deposition: zinc or Type 420 martensitic stainless steel?

Learning Cladding, Hardfacing, and Buildup

This shop will give you valuable hands-on experience in cladding, hardfacing, and buildup work. Use steel samples at least ¼ in. thick (½ in. thick is better) to avoid distortion problems while you learn to use the various methods.

19-1-S
CLADDING

1. Try applying a uniform, porosity-free layer of austenitic stainless-steel weld metal completely covering a steel-plate coupon with parallel, slightly overlapping weld beads. If you can successfully produce a sound, uniform coating layer, run a second layer over the first one with the parallel weld bead at a right angle to the layer underneath.

19-2-S
HARDFACING

2. Start with parallel bead-on-plate welds. Select from the following filler metals and gas or arc processes. Then learn to produce different surfacing patterns. Work in all positions. Select your filler metals (and welding process) from the following list, but be sure to learn to do the work with SMAW electrodes.

■ High-speed steel gas rods and electrodes coded with RFe5-A and RFe5-B, and EFe5-A, EFe5-B, and EFe5-C class designations.
■ Austenitic manganese-steel

electrodes coded with EFeMn-A and EFeMn-B class designations.
■ Austenitic high-chromium iron gas rods and electrodes coded with RFeCr-Al or EFeCr-Al class designations.
■ Tungsten-carbide gas rods and electrodes coded with RWC-XX and EWC-XX class designations.

3. Use a small steel plate or thick sheet, about 12 in. [305 mm] square or less. This is a part that must be surfaced with a suitable welding alloy to prevent wear. Use stick electrodes. Put a layer of the wear-resistant alloy of uniform thickness over the entire surface of the wear plate. Lay down parallel, slightly overlapping beads until the plate is covered. Make the deposits as equal in thickness over the entire area as you can. Assume that it will subsequently be used as part of a bulldozer blade.

4. Test the surfaces you produced in this exercise for Brinell or Rockwell hardness. Also test the hardness with a machinist's file.

19-3-S
BUILDUP

5. If you can get a scrap steel or cast-iron gear, break off one tooth (or melt it off with a torch). Then build up a new tooth on the gear with SMAW welding. If you can't get an old gear, build new teeth on a piece of round bar. Make the gear tooth you build up slightly

oversized, assuming that it will be machined to the correct size later on. Work downhand. This is a typical maintenance job often using SMAW electrodes.

6. Start with a steel bar. Grind a groove around the circumference, as if it were a worn axle. Then fill up the groove with steel filler metal, leaving only a little excess metal to be machined off. Now grind the excess metal smoothly flush to match the surface of the bar.

7. Fix a broken gear tooth (or create one on a round bar). If you have a used gear from a junkyard, you can knock off some teeth with a sledgehammer and then build them up again using a process other than SMAW.

19-4-S
SPRAYING MOLTEN METAL

8. If your welding school has metal spray equipment, learn to apply a thin coating of aluminum to steel. Try coating a round bar, first.

9. Now coat a pipe or piece of square structural tubing.

10. Coat a section of structural steel with a protective aluminum coating. Make sure that the section is completely covered, and that none of the base-metal is exposed. Also be sure that the coating is uniform. Don't let it build up in one place and become too thin in another.

PART 4

Skills for All Processes

Many skilled welders find better-paying jobs checking the work of other welders. That's why we devoted a special chapter to inspection methods.

Rather than duplicate other textbooks that concentrate solely on blueprint reading, we will simply introduce you to the subject. If your instructor desires, you can gain further understanding of welding blueprints from one of these specialized textbooks.

Now that you are fully trained as a basic welder, it's time to learn about different kinds of equipment that will save you time and effort in real shop situations, if not in a welding class. We will tell you about welding positioners, welding robots, and other tools of the trade.

Then we finish your formal training as a skilled welder by telling you about welder qualification tests so you will be ready to take them to get a good job.

The appendix of this book describes other welding methods that don't necessarily require high welding skills, but are processes you should know about. Here is where we introduce you to resistance welding, stud welding, electron-beam welding, and laser welding. Another part of the appendix will help you apply for a job.

Welding Defects and NDT

There are a lot of ways to check whether a weld is sound and meets specifications. One of the more important ways is simply to look at the weld. By now you already have a good idea of what bad welds look like.

In this chapter we will tell you what to look for when inspecting your weld metal. We'll also tell you about other methods that are used to "look" inside your welds to see things below the surface. This could be your start toward a career as a welding inspector, which is a very good job, but it requires everything you now know, plus more training in welding metallurgy, detailed knowledge of standard welding codes, and skills in various nondestructive testing (NDT) methods. We'll give you a preliminary look at some NDT techniques after we look at the common defects found in bad welds. The American Welding Society has a very good book on this subject. It's titled "Welding Inspection" and was first printed in 1980. Get a copy whether or not you get serious about being a welding inspector, because every welder should be very serious about inspecting welds. The knowledge in this AWS book could get you a job (or a promotion).

20-1
WELDING DEFECTS

Distortion
We don't have to tell you about the effect of heat on a welded structure. Distortion can be corrected by proper jigging, preheating or post-weld slow cooling or stress-relief heat treatment, and sometimes by peening the weld metal and heat-affected zone if your manufacturing specification allows it. Sequencing your welding also may balance stresses and eliminate the problem. Reducing amperage to lower heat input and using smaller electrodes are done for the same reason. Pulsed-current GMAW is another way to lower heat input. If the weldment is out-of-shape, out-of-alignment, or not dimensionally correct, it's simply a useless piece of metal.

Bad Weld Profiles
Unacceptable weld profiles have a considerable effect on the performance of your weld under load. If one pass of a multipass weld has a bad profile it can cause incomplete fusion or slag inclusions, even though the next weld passes will partially remelt the first pass. A bad weld profile on one pass can cause subsequent cover passes to be out of size, out of contour, or create many other weld defects that will be covered up by follow-up passes, making the problems impossible to see without special test procedures. The problem is most severe when the bad pass is the first, or root pass of the weld. That is why very

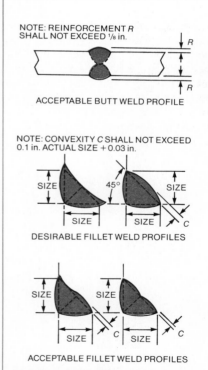

NOTE: REINFORCEMENT *R* SHALL NOT EXCEED ⅛ in.

ACCEPTABLE BUTT WELD PROFILE

NOTE: CONVEXITY *C* SHALL NOT EXCEED 0.1 in. ACTUAL SIZE + 0.03 in.

DESIRABLE FILLET WELD PROFILES

ACCEPTABLE FILLET WELD PROFILES

FIGURE 20-1 These are examples of some desirable and some not as desirable but acceptable weld profiles. Welding specifications often state what the allowed reinforcement and weld size should be. Always try to produce welds that are better than simply acceptable.

experienced welders are often used to make the root passes on pipelines and less experienced welders follow them up making the added passes needed to fill the joint.

To understand what bad weld profiles are, you need to look first at what's normally considered to be good (or at least adequate) weld profiles. These are shown in Fig. 20-1).

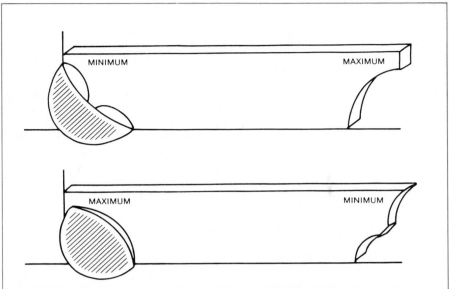

FIGURE 20-2 You can make your own weld profile gauges out of heavy sheet metal and carry them with you when you work. Profile gauges (often called templates) are worth making if you will be producing a lot of welds of the same size. This gauge is designed to check fillet welds. You can design a similar template for testing butt welds by cutting the minimum and maximum weld-bead profiles on the long edge of the template instead of on the ends. If you have several weld templates for different sizes of welds, you should scribe or stamp them with the weld size for which they are used.

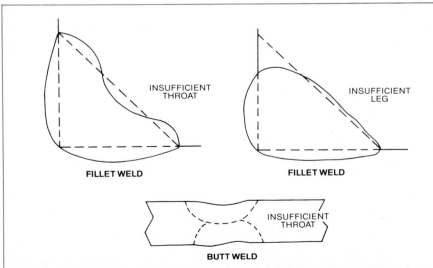

FIGURE 20-3 These are examples of unacceptable fillet-weld and butt-weld profiles. In every case the weld-metal section is too small, and therefore the weld will not be strong enough. The right profile (insufficient leg) would probably also produce undercutting of the base metal, and therefore a notch where a crack can start.

Incorrect Weld Size

If your weld is not the size required by the blueprints or your instructions, you are in trouble. Small welds will be too weak. Extra-large welds waste expensive filler metals. If you make a lot of welds of the same size, make up a sheet-metal template on the two ends of a piece of metal strip that has cutouts to match your correct maximum and minimum weld size (Fig. 20-2).

For example, if your fillet welds are supposed to have ½-in. [13-mm] legs, the weld bead should be at a 45° angle to the base-metal T, and your bead is not allowed to have more than a certain amount of convexity or concavity, make one end of the template to match the maximum dimensions of your weld and the other end to match the minimum dimensions. Check your work as you go. Figure 20-3 shows some examples of incorrect weld-size profiles that your profile gauge can help you find.

Excess concavity Excess concavity can occur with all types of welds (Fig. 20-4). The actual strength of these welds is considerably less than that of correctly made welds since the throat of the weld is smaller than the design requires. Excess concavity is a bad problem in cover-pass welds and root passes on pipe or tubing. In the case of fillet welds, the throat of the weld is less than normal as measured by the length of the leg. This condition occurs most often in the flat position in fillet welding and in the 5G and 6G pipe-welding positions. It's caused by excess welding current.

Excess convexity Excessive weld reinforcement (excessive weld convexity) also is undesirable (Fig. 20-5). It tends to stiffen the section and makes notches that create stress concentrations. Excess reinforcement also uses extra weld metal and adds to the cost of welding without gaining any worthwhile strength. Excess convexity is harmful in the case of an intermediate pass in a mul-

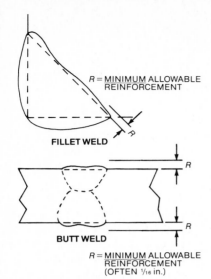

FIGURE 20-4 Adding too little weld reinforcement reduces joint strength. Therefore, most welding specifications state a specific minimum amount of weld-metal reinforcement that you can safely use. That minimum allowable reinforcement often is 1/16 in. [1.6 mm]. Unless the specification says otherwise, always use 1/16 in. as your minimum reinforcement on fillet and butt welds (unless you are joining very thin material).

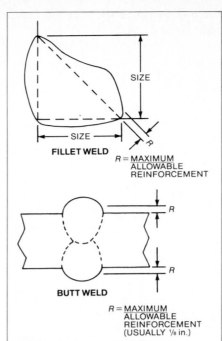

FIGURE 20-5 Too much weld-metal reinforcement not only wastes expensive filler metal but can actually reduce the strength of your weld joint, rather than increase it. Unless your welding specifications state otherwise, assume your maximum allowable reinforcement to be no more than 1/8 in. [3.2 mm].

Inadequate penetration

Inadequate penetration (Fig. 20-7) is another obvious problem, but it's harder (in fact just about impossible) to see. It occurs most often in the root of a weld and it also becomes a built-in crack ready to start running through the base metal, weld metal, or heat-affected zone when your weldment is put under stress. (Certain NDT methods we'll tell you about shortly are used to find problems you can't see like inadequate base-metal penetration.)

Undercutting Undercutting (Fig. 20-8) is another common problem. Instead of overlapping your base metal, your welding arc cuts into it. In addition to reducing the section thickness of the base metal, undercutting can create a notch that will someday start a crack if it's not corrected. Undercutting the base metal is mostly due to bad welding techniques. You may not be holding your electrode steady when you work. Your welding current also may be too high and your arc is acting like a gouging rod. Sometimes undercutting results from arc blow. If you have arc blow, switching to ac and using ac electrodes will correct it. (See the section on arc blow in Chap. 16 for more ways to solve the problem.)

tilayer weld because incomplete fusion or slag inclusions can occur unless prevented by grinding or back-gouging prior to each new pass. The problem is usually due to insufficient welding current or incorrect welding techniques.

Overlap Overlap (Fig. 20-6) is the condition in which weld metal sticks out beyond the bond line at the toe of a weld. This problem tends to produce notches in your weldment where stress will build up and start a crack at the sharp point where the weld metal and base metal come together at the overlap area. These built-in cracks also are places where corrosion often starts and isn't seen until it begins to undermine the base metal.

Overlap is usually caused by the use of either incorrect welding technique or by improper electri-

cal conditions. Overlap can also occur at the toe of the reinforcement of a complete weld groove.

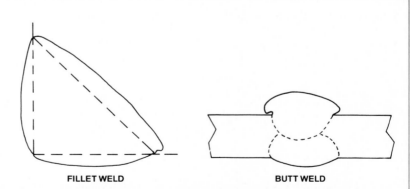

FIGURE 20-6 Look closely at the edges of these weld beads and you will see examples of weld-metal overlap onto the base metal. The notch effect produced by overlap can initiate a crack when the part is heavily stressed. Equally important, corrosive attack often begins inside these narrow gaps and eats into the weld and base metal while leaving an almost invisible trace on the surface of the part. Many failures due to corrosive attack start in hidden crevices such as these.

Interruptions

When you weld, you can't possibly avoid leaving some things in your weld that shouldn't be there. They are interruptions of what should be continuous weld metal. Examples are porosity, slag inclusions, and incomplete fusion. Not the existence of, but the size, amount, and distribution of these defects are what's important to making an acceptable weld as opposed to an unacceptable one.

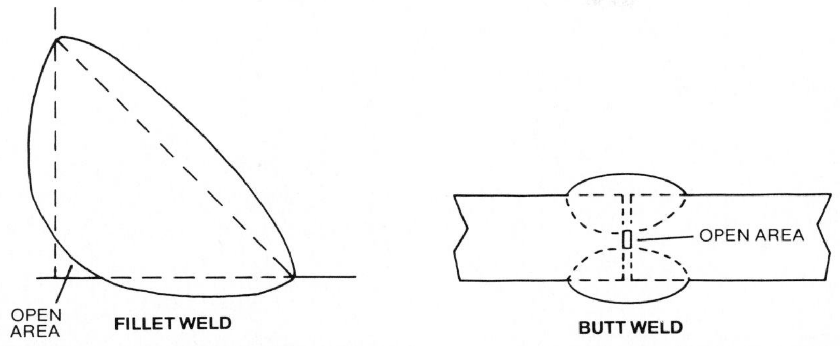

Porosity The main reasons you get porosity (bubbles) in your welds are dirt, rust, and moisture on the surface of the base metal; moisture in the flux coating of your welding electrodes or in your shielding gases; or moisture in your welding equipment (hoses, wire drive rolls). Porosity is caused by gases produced by these substances that are trapped in the solidifying weld metal.

The single best way to avoid porosity is to use perfectly clean, dry welding equipment and electrodes that either have just been baked in an electrode-holding oven (the times and temperatures differ with different electrode flux coatings—300°F for 4 h is often used) or you can use the newer moisture-resistant SMAW stick electrodes with hydrophobic (water-hating) flux coatings.

Always purge your shielding-gas lines before welding. Excessive current and arc lengths that are too long also should be avoided. High welding currents will burn up the deoxidizing elements in SMAW electrodes at such a high rate that not enough deoxidizers are left to combine with the weld metal before it starts to cool. Cooling weld metal drives gases out of solution, and then the weld hardens with gas bubbles in it.

Uniformly scattered porosity looks like you have scattered voids or tiny bubbles in your weld (Fig. 20-9). They look like black spots on an x-ray negative. The individual voids that you will see on an x-ray can range in size from almost microscopic to ⅛ in. [3.2 mm] or more in diameter. Whether uniformly scattered porosity is acceptable or not depends on the welding code and the quality-control requirements that you are working with.

Cluster porosity usually is worse (Fig. 20-10). Frequently the pores occur in clusters separated by con-

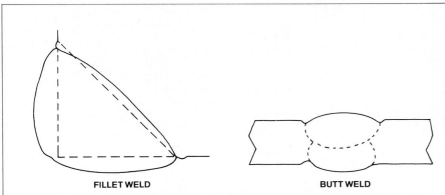

FIGURE 20-7 Incomplete fusion in the root zone of welds is one of the most serious welding defects that you can produce. The open area is a notch inside the weld at its most critical point, where cracks are most likely to start. The open area inside the weld lowers the actual throat depth or thickness of the weld metal, reducing its true strength. And these defects are impossible to detect without special nondestructive test methods.

OPEN AREA   FILLET WELD   BUTT WELD   OPEN AREA

FIGURE 20-8 Notches on the base metal next to the weld bead are examples of undercutting. Like overlaps, undercutting can turn into a source for cracks that can ultimately cause a part to fail.

FILLET WELD   BUTT WELD

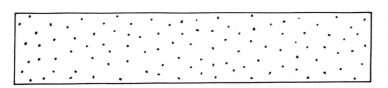

FIGURE 20-9 There are several kinds of weld-metal porosity. This shows you what scattered porosity looks like on an x-ray. Welding specifications often state just how much scattered porosity can be allowed (if any) before a part is not acceptable. A little porosity that may be perfectly acceptable in welds used to join furniture could be disasterous if the same amount of porosity were found in welds on a pipeline or pressure vessel.

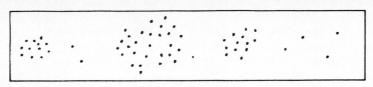

FIGURE 20-10 Cluster porosity would look like this drawing if you saw it on an x-ray.

siderable lengths of porosity-free weld metal. Such clusters are often associated with changes in welding conditions (for example, changes in arc settings when you start or stop welding).

Starting porosity occurs when you are using stick electrodes and get cluster porosity spaced at dis-

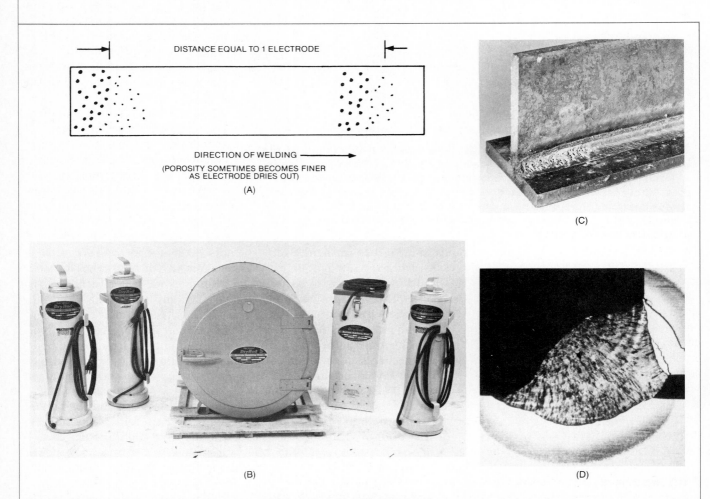

DISTANCE EQUAL TO 1 ELECTRODE

DIRECTION OF WELDING →
(POROSITY SOMETIMES BECOMES FINER
AS ELECTRODE DRIES OUT)
(A)

(B)

(C)

(D)

FIGURE 20-11 (A) If you saw an x-ray that looked like this drawing, you would know that your problem is starting porosity. This pattern occurs when using SMAW electrodes that have too much moisture in the flux coating. Note how the porosity drops off in the direction of welding. That's because the electrode is heating up as it is used, and the remaining flux on the electrode is drying out. When a new electrode is used, the pattern repeats itself. (B) This is one of the main reasons why many electrodes should be baked or temporarily stored in the special electrode ovens shown in this photo before use. Most filler metals that require rebaking will list the recommended oven heating time and temperature for the flux coating on the electrode package. (C) This photo shows an extreme case of starting porosity produced by using SMAW electrodes in a very humid environment without rebaking them first to dry out the flux coating. (D) Rebaking also is required with most high-strength steel SMAW electrodes because even small amounts of moisture that would not produce starting porosity will still create hydrogen embrittlement cracks in the high-strength base metal next to the weld metal.

tances just about the equivalent of the weld metal that one stick electrode would lay down (Fig. 20-11). Your electrode flux coatings probably contained moisture when you started welding. As the electrode heated up, the moisture evaporated, but you left behind porosity at every point where you started a new electrode.

Linear porosity (Fig. 20-12) occurs most frequently in the root pass of welds and is often regarded as a special case of inadequate joint penetration. Any defects that line up in a weld are bad. When they occur in the root of the weld, they are worse. You have almost guaranteed that the weldment will start cracking along the line of porosity (frequently in the root zone, where the crack often can't be seen) as soon as the workpiece is put in use and heavily stressed. Linear (in-line) porosity often is defined in welding codes by the number and size of the pores and their linear distribution with respect to the axis of the weld. The lined-up pores can become connected together very quickly by short cracks to become one big disastrous failure when the welded part is heavily stressed during usage. A very little bit of it may be acceptable, but probably not on any critical x-ray-quality weldments.

Slag inclusions The most common cause of slag inclusions is inadequate cleaning of your weld metal between passes; as a result you have permanently welded some slag particles into your work (Fig. 20-13).

During SMAW, slag also can be formed and forced below the surface of your molten metal by the stirring action of your arc. Slag may also flow ahead of your arc, causing the metal to be deposited over the slag. Also, with some types of electrodes, slag in crevices of previously deposited weld metal will not remelt and will be trapped in the weld. That usually occurs because you didn't clean each weld pass thoroughly before starting the next one.

Slag also can be present in the molten weld metal for other reasons such as high-viscosity (stiff) weld metal that is too cool to flow properly, rapid solidification, or too low a preheat temperature that prevents the slag from floating to the top of the weld before the weld metal turns solid. Gas "dents" or pock marks that collect slag also can occur on the weld surface when gas is produced by chemical reactions between the slag and the molten weld metal.

Always clean your weld metal thoroughly between passes. Use electrodes that produce refusible slag when you can. If you have an electrode that produces a tough slag that is difficult to remove, grind down or back-gouge the surface of each pass before making another pass. (See Chap. 13 on arc-gouging or Sec. 22-12 for more advice.)

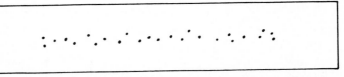

FIGURE 20-12 This is what linear porosity would look like on an x-ray. The porosity (tiny gas bubbles) are lined up in a row. Even one tiny crack that connects two of these bubbles can continue to the next bubble and on along the line until you have a giant crack in the workpiece. Linear porosity is usually seen in the root pass and often is a borderline example of inadequate root penetration. This is an example of why root-pass welding is so critical. Scattered porosity that might be perfectly acceptable in cover passes would cause weld failure if lined up in the root pass.

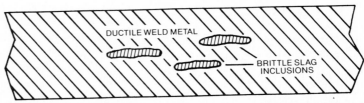

FIGURE 20-13 Slag inclusions are bad because slag is brittle, not strong and ductile like weld metal. Each slag inclusion is a ready-made defect that may soon initiate an unseen crack inside the weld metal. Always clean your welds completely between passes when using a slag-producing process such as SMAW and SAW. Some SMAW electrodes are said to have a "refusible slag" which is easily remelted on follow-up passes and is supposed to float to the surface of the next layer of weld metal. Don't trust that your weld is free of slag inclusions simply because you are using electrodes that produce refusible slag, unless you have thoroughly tested the electrodes first. Some electrodes produce slag that will peel up off the weld as the metal cools. These are very handy electrodes that make slag cleaning easier. Other electrodes produce slag that is tightly adherent to the weld metal. If in doubt, thoroughly clean all weld metal between passes.

Tungsten inclusions When you are careless with your tungsten while using GTAW, pieces of tungsten metal will get into your weld. Never touch the tungsten to the weld metal. Sometimes tungstens break off while you are welding. If you see it happen, flick the tungsten particle out with your filler rod if you are using one. You may want to switch to a better grade of tungsten if you are having this problem. Thoriated tungstens or zirconium-tungsten electrodes are used with DCRP to reduce or eliminate tungsten inclusions (see Sec. 14-10).

Incomplete fusion Incomplete fusion (Fig. 20-14) is very bad, and a direct result of either sloppy welding or incorrect weld settings. Your weld metal has not fully joined with the base metal or with prior passes. You have made a built-in crack in the weld. Incomplete fusion can be caused by not preheating your base metal (or previously deposited weld metal) when working on thicker plates. It may be caused by your failure to use proper fluxes when gas welding, welding too fast, or even worse, not watching what you are doing.

Cracks
Of all the things that can go wrong with welds, cracks are the worst. Cracks occur because the stress at that point in the weldment exceeds the ultimate tensile strength (pull-apart strength) or ultimate shear strength (slice-apart strength) of the base metal or weld metal. Cracks can't occur where the metal is under compression because it's being squeezed together instead of being pulled apart. That's one of the reasons why peening a weld and your heat-affected zone will reduce surface cracking. Peening the weld puts the surface of the metal under compression. However, peening is not always allowed by welding codes.

Tensile cracks are by far the most common (Fig. 20-15). They can occur either in your weld metal or base metal. Shear cracks, when they occur, often occur at an angle to tensile cracks. They also may appear in line running off at an angle from the direction of stress. After your weld metal has cooled, cracks are most likely to occur if the weld metal or base metal is hard and brittle. In other words, it has low ductility. The metal can't stretch very much to release the built-up stresses. The only thing left for it to do is crack. That is why steels with over about 0.30 to 0.35 percent carbon in them are very difficult to weld. These medium- to high-carbon steels have high strength but low ductility.

Preheating a weldment before you work on it and post-weld slow cooling after you have finished often are specified for thicker sections or for base metal that is prone to cracking. The preheating and post-weld heating and stress-

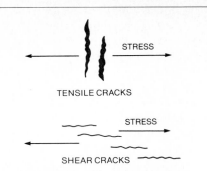

FIGURE 20-15 Tensile cracks are the most common type of crack you will see. They can occur either in weld metal or in the base metal. Shear cracks often occur at staggered intervals. If you were to measure the angle of stagger or overlap, it would probably be close to 60°. Both types of cracks show you the direction of maximum stress when the cracks occurred. Tensile cracks are perpendicular to the maximum stress while shear cracks are parallel to the maximum stress. The direction of stress often is a clue to solving the problem.

relieve heat treatment help reduce residual stresses that otherwise could start a crack. Whenever your specification calls for preheating the base metal before you weld, you *must* do it.

There's something else that you know about a crack. Once it starts, it will continue right through the weld and the base metal until the stress that causes the cracking is relieved (meaning that the stress falls below the ultimate tensile or yield strength of the metal). Most often, instead of relieving the stress, the crack makes the structural section smaller so that the unit stress per section area becomes even larger (Fig. 20-16). Guess what that nightmare means? It means that the crack, once started, can continue right through the weldment. Ships have been known to split in two because a single crack started at a hatch cover corner. A crack in a high-pressure pipeline can

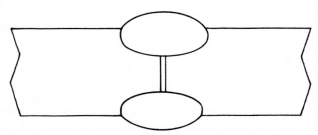

FIGURE 20-14 Incomplete fusion results from welding too fast or using incorrect weld settings on your power source (or a gas-welding tip that is too small for the section). Incomplete fusion is just about the worst welding defect you can make.

literally race at the speed of sound for miles from one pumping station to another. Cracks are no joke, even tiny ones.

In multilayer welds, cracking occurs most frequently in the first layer of the weld metal, in the root zone of the weld joint. That's why so much attention is paid to making a good, solid first pass in a large joint. If cracks are found in the weld by using a nondestructive method, the weld metal will have to be gouged or ground out and the weld remade.

When you run into cracks and there doesn't seem to be any metallurgical reason for them (like working on a higher-carbon steel or resulfurized free-machining steel), here are some things you can do.

■ Change the way you are manipulating your welding gun or electrode to improve the contour of your weld metal or change to a filler metal that gives more ductile weld metal.

■ Slow down your travel speed. Slowing down increases the thickness of your deposit to provide more weld metal to resist stress.

■ Use preheat to reduce thermal stresses. This is especially important if your workpiece is cold. Welding anything in cold weather without preheating invites trouble.

■ Use low-hydrogen electrodes when doing SMAW so that you won't get hydrogen embrittlement, which reduces the ductility of your weld and base metal.

■ Sequence your welds to balance shrinkage stresses when the weld metal solidifies. Intermittent welding sequences followed by welding in the missing spots often is used. Many times in fabricating shops you will be asked to weld alternately on different sides of a structure. That's also a good way to avoid distortion and for exactly the same reason. It balances stresses.

■ Avoid letting weld metal or base metal cool too fast, especially if it is subject to cracking. That's why post-weld heating is used. If your ferrous weld metal or base metal is very hot and it cools too quickly, you can form brittle martensite and that's the beginning of lots of problems.

Weld metal cracks There are three kinds of cracks that occur in weld metal. They are transverse cracks, longitudinal cracks, and crater cracks.

Transverse weld-metal cracks are cracks that run across the weld perpendicular to the weld axis (Fig. 20-17). This is a common kind of crack and, in some cases, it extends beyond the weld metal and into the base metal. Transverse cracks commonly occur in weld joints that are very tightly restricted so that the structural section or the weld metal can't shift to relieve the built-up stresses. These often are shrinkage cracks produced when the weld metal is cooling and contracts. Another kind of crack called *lamellar tearing* occurs in certain restricted joints under high stress. We'll describe it later on.

Longitudinal weld-metal cracks, usually running through the center of the weld metal (Fig. 20-18), often are extensions of crater cracks formed at the ends of a weld (discussed below). Longitudinal cracks at the surface of your weld metal also can be extensions of a crack that started deeper down in the joint, probably in the root zone. That's another reason

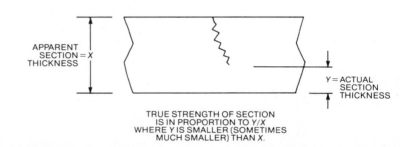

FIGURE 20-16 **Cracks that you can't even see may run deep into the section. What you think is a 1-in.-thick section could be only a fraction of an inch where the crack is. The true strength of a section is in proportion to the actual section thickness divided by the apparent section thickness (Y/X in the drawing). For example, if the actual section thickness is ¼ in. and the apparent section thickness is 1 in., the ratio of Y/X is 0.25. If the strength of the 1-in.-thick plate is 40,000 psi, then the actual strength of the section is 40,000 psi multiplied by 0.25 which equals only 10,000 psi.**

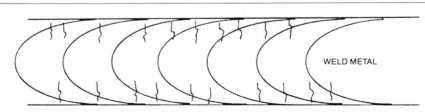

FIGURE 20-17 **Transverse weld cracks run across the weld bead. They may imply that your weld joint is too tightly clamped or held by the rest of the workpiece and that the weld metal can't stretch enough to relieve the stress (its ductility is too low). You may need more ductile weld metal, or you may need to preheat and postheat the weld to reduce built-in stresses.**

why root passes must be made very carefully.

Crater cracks can occur whenever you interrupt your welding (Fig. 20-19). These cracks are usually star-shaped and progress only to the edge of the crater. However, crater cracks can be the starting point for longitudinal cracks, particularly when they occur in the crater formed at the end of your weld.

When you make a crater somewhere else in the weld pass (for instance, when you change electrodes), the crater cracks are usually welded up when you start again, but not always. Sometimes fine, star-shaped cracks are seen at various locations in a weld. Crater cracks are frequently found in weld metal that has a high coefficient of thermal expansion; the prime example is austenitic stainless steel.

You can prevent crater cracks by starting each weld properly, correctly restarting your electrode, and by filling craters that you do make to a slightly convex shape prior to breaking your arc.

Base-metal cracks Cracks that occur in your base metal usually occur in the heat-affected zone and they run longitudinally (parallel) to your weld. They almost always occur because the heat-affected zone has become hard and brittle (probably because you didn't preheat your weldment or properly cool a thick section with post-weld heating).

Transverse base-metal cracks perpendicular to your welding bead but in your base metal (Fig. 20-20) are usually associated with welding high-hardenability steels such as tool steels and certain full-alloy steels. You often can't see these cracks until your weld metal has cooled to room temperature. NDT methods may be needed to find them.

Longitudinal base-metal cracks parallel to your weld but in your base metal (Fig. 20-21) may be extensions of bond-line cracks due to poor weld-metal penetration into the sides of your joint (you didn't properly tie the weld metal into the faces of the groove weld joint). For fillet welds, longitudinal base-metal cracks can be divided into two groups; those that start at the toe of the weld, and root cracks which proceed from the root of the fillet weld into the base metal and sometimes even out the other side of the joint.

In the case of groove welds, cracks in your base metal are more likely to show up in the heat-affected zone or at the very edge of your weld in the fusion zone between the base metal and the weld metal. Cracks in the heat-affected zone or the fusion zone are associated with high-hardenability steels when the filler metal and base metal are of entirely different composition. The combination of different alloys sometimes can produce unpredictable results in the fusion zone,

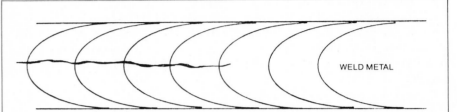

WELD METAL

FIGURE 20-18 Longitudinal weld-metal cracks, running along the weld bead (and sometimes along the base metal next to the weld), may show up on the surface as extensions of cracks that started in the root zone of the weld. Cracks that occur primarily in the base-metal heat-affected zone and not in the weld metal have other causes, usually resulting from improper (or lack of) preheating and postheating treatment of high-strength steels. When you see longitudinal cracks in the base metal and not in the weld, it means that the weld metal is more ductile than the base metal and residual welding stresses are tearing the base metal up. The higher the strength of the base metal, the more likely you are to have this problem. Longitudinal cracks in the base-metal heat-affected zone of high-carbon steels are just about a sure thing to occur. That's why high-carbon steels are so difficult to weld.

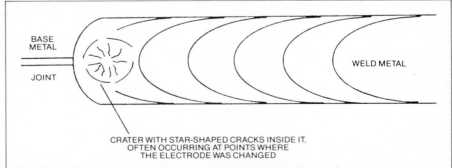

BASE METAL

JOINT

WELD METAL

CRATER WITH STAR-SHAPED CRACKS INSIDE IT,
OFTEN OCCURRING AT POINTS WHERE
THE ELECTRODE WAS CHANGED

FIGURE 20-19 Crater cracks are easy to recognize. They are star-shaped cracks radiating from the center of the depression and most often will be seen only at the end of a weld bead, or less often inside a weld bead where you stopped welding to change electrodes. Crater cracks also occur during TIG welding, when you lift the arc off the work too fast. If you see crater cracks forming while you change electrodes, fill in the crater before continuing to weld. And don't lift your arc off the work so quickly.

even though that may be the only way to weld a high-hardenability steel. One solution is to make a transition joint by buttering up the joint face with a very ductile electrode and then using a different electrode to finish the weld.

Steels most likely to crack

Given the same cooling rate, the low-carbon steels will harden considerably less than the medium-carbon steels, and high-carbon steels will harden even faster than medium-carbon steels. That's the order in which you are most likely to find cracks in your base metal.

Low-alloy steels show a much wider variation in their hardening characteristics. Some may react similarly to low-carbon steel. Others may act like medium-carbon steel. It all depends on the alloying elements they contain.

High-alloy steels also show many differences. They range from the austenitic and ferritic stainless steels and austenitic wear-resistant manganese steels to martensitic stainless steels and special high-strength, high-temperature ferrous metals.

Martensitic stainless steels behave a lot like medium-carbon and low-alloy steels, except that the martensitic stainless steels harden to a greater degree with a given cooling rate.

Neither the austenitic stainless steels (the most commonly used grades contain 18 percent Cr and 8 percent Ni) nor the ferritic stainless steels (usually low-carbon straight-chromium steels containing 12 percent or more chromium) harden upon quenching from a high temperature. However, in general, the ferritic (straight chromium) stainless steels may become brittle (but are not hardened) when they are welded.

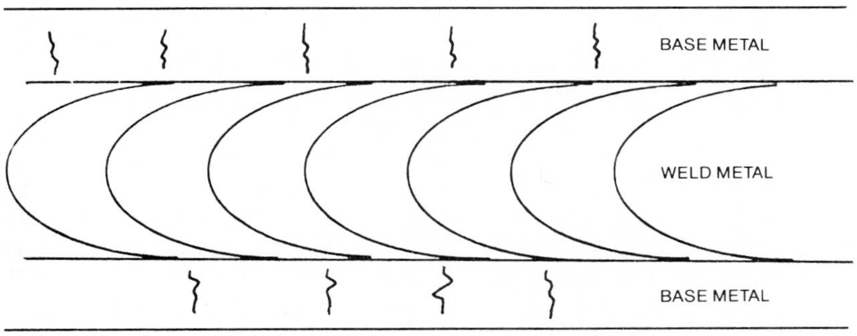

Other Defects

That's not all that can happen to a weld; a sloppy operator can do a lot to help get a workpiece rejected. Good weld appearance is valuable not only in consumer products like refrigerators, cans, and pots and pans but also on pressure vessels and turbines.

Workmanship We know of a major turbine failure that shut down the entire pulp operation of a major paper mill. Looking inside the equipment to find out why it failed (costing both the paper mill and the turbine manufacturer millions of dollars), the first thing the inspectors saw was weld spatter. The first thing they thought of was bad welding.

The engineers (who know a lot about welding because it's used in paper mills) knew that good welders make good-looking welds. That alone was enough to start them looking for anything that they could use to pin the failure of the turbine on the manufacturer.

The weld spatter, alone, was enough to start court action in a law suit that ran for 2 years. It ultimately turned out that the turbine failed because the paper mill operators ran it incorrectly, but that weld spatter left by a sloppy operator cost his company a fortune in consulting and legal fees and for getting nondestructive tests taken on every welded part of the turbine to prove that the welds were structurally sound. Ultimately, the failure resulted in a complete reorganization of the manufacturer's welding shop. A lot of welders lost their jobs because of a little weld spatter.

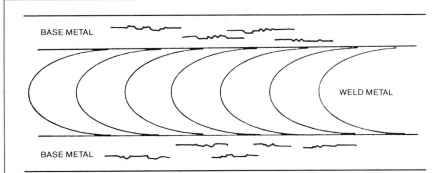

FIGURE 20-20 Transverse base-metal cracks are most often seen when higher-strength steels are not properly welded. These cracks frequently show up only after the weld metal has cooled. You must use sufficiently high preheat temperatures and postweld heat treatment if necessary.

FIGURE 20-21 Longitudinal base-metal cracks run parallel to weld beads, and usually show up in the heat-affected zone. There are numerous reasons why they occur, but every reason means trouble that must be solved before you continue welding.

FIGURE 20-22 Arc strikes are caused by sloppy welders. Don't touch your electrode to the base metal except where you want to weld. If you accidently create an arc strike on a workpiece, grind the imperfection down to give it a smooth contour and check for soundness (no cracks) before you continue welding or put the part into use.

Bad weld-surface appearance indicates a bad welder was at work, whether or not the surface defects are structurally important. You already know what to look for, because by now you have produced a lot of bad-looking welds. Obvious variations in the width of the weld bead, humps in the surface, occasional depressions, nonuniform ripples, and changes in the height of the weld reinforcement are all examples. Poor weld appearance is covered in most welding specifications under the heading of "workmanship." You can even flunk a welder qualification test when all your work passes the mechanical tests and dimensional requirements if your welds simply look bad.

Arc strikes are operator-caused defects. Whenever you accidentally touch an electrically hot electrode to any area of the base metal you don't intend to weld, you get an arc strike (Fig. 20-22). The AWS Structural Welding Code requires that all arc strikes must be ground to a smooth contour and checked for soundness before the weldment is put into use.

High-low is a problem area in a plate or pipe joint where the edges simply don't match up (Fig. 20-23).

Start-stop will usually show up on an x-ray as an approximately circular area extending irregularly into the weld metal (Fig. 20-24). It's caused when you abruptly lift your electrode from the workpiece instead of smoothly tapering off your weld.

Excess penetration also crops up from time to time. Instead of incomplete penetration, your arc digs too deeply into the section. You usually see it as excess weld metal on the back side of your joint (Fig. 20-25). It's most often a problem that will occur when you are making root passes. Are you using a big enough root-face land or nose in the bottom of the joint?

Defects in the base metal

Sometimes steel from a mill will have defects in it when it arrives. Some of these defects are not the fault of the steel producer. They may depend on how the steel was stored. (Rusty steel is the responsibility of the warehouse or the people who store it at your plant). There's a whole array of awful things a steel mill can do to your material. We won't bother giving you the complete list. We'll just mention one problem that can be very serious for welders.

Edge laminations are lined-up inclusions that are rolled or otherwise shaped into the steel when it is made. It probably exists to some degree in all plate, pipe, and tubing, especially if the steel is made in ingots instead of being continuously cast.

These imperfections may have started out as blow holes in the ingot casting or other inclusions in the molten steel that got lined up in the material when it was finish-rolled or forged. The ASTM and the ASME have criteria that tell you how much edge lamination (and other defects) is acceptable. While defects in the base metal may not be your direct responsibility, they can become your problem. If you see something that doesn't look right, tell somebody immediately.

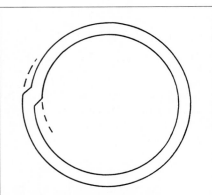

FIGURE 20-23 High-low is a problem you most often have when welding large diameter pipe and rolled cylinders. The curved parts don't match up. It also can result from trying to butt-weld wavy plate. One solution is to use better (or more) tack welds. Another solution, if the base-metal mismatch is caused by heat distortion, is to use welding jigs and clamps or fixtures that will hold the base metal in place until you finish the weld.

Lamellar tearing is a problem most often occurring with rolled structural-steel sections and heavy plate used in large structures and big pressure vessels. The cracks in the base metal have a characteristic stair-step shape (Fig. 20-26). The problem most often occurs in large buildings, bridges, and in pressure vessels where the weld joints are under extreme stress. It looks like a structural section was trying to pull itself out of the joint.

The stair-step cracks usually start as microscopic brittle inclusions in the steel and tear through the ductile metal until another brittle inclusion is reached, where the crack changes direction and continues. That's why the crack has a stair-step shape. Lamellar tearing usually goes completely through the section.

Lamellar tearing occurs most often in tightly restrained structural joints. It's often a welding-design problem and probably not your fault. There are, however, methods for handling it in the welding shop and in the field (such as buttering the connection's joint face with a more ductile electrode, followed by completion of the joint with another type of electrode). If you don't follow these procedures when they are specified, you may be responsible for lamellar tearing (and maybe for an entire bridge or building falling down).

20-2
PROOF TESTS

Now that we have given you a short list of all the sins to which welds (and welders) are subject, how are they to be found? One way is to subject your weldment to stresses in excess of those it will operate under when actually used. Technically, these are not considered to be nondestructive tests because the product can fail if it doesn't pass the test.

Proof tests usually are conducted with stresses above the operating levels the work will be used at but below the yield strength of the metal. Proof tests usually are combined with visual inspection. The details of the test are often specified by the drawings, specifications, or contract for the weldment.

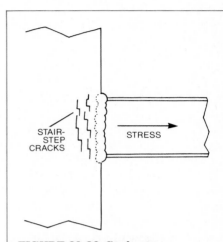

FIGURE 20-26 Stair-step–shaped cracks in the base-metal joint of large welds such as you will find on structural steel in buildings and bridges may occur long after the structure is in use. This is called *lamellar tearing* and it is produced by highly restrained weld joints under stresses so high that the structural section is trying to pull itself right out of the joint. Lamellar tearing problems have been reduced by extensive studies that have shown welding designers how to design better welds. Newer, cleaner steel-making methods that reduce the number of brittle sulfide and oxide microinclusions in the steel also have reduced this problem. You can help avoid lamellar tearing by strictly following any special welding procedures you are given for joints where this problem may occur. For example, using more than one kind of electrode, first to make a buttered more ductile-metal pass, and then to complete the weld with a higher-strength electrode.

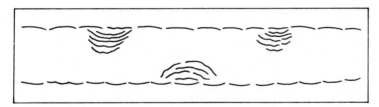

FIGURE 20-24 Stop-start is a defect that usually shows up on x-ray film. It has a circular area extending irregularly into the weld metal. It could be a type of crater crack that was not properly filled when tying weld metal into the walls of a thick joint.

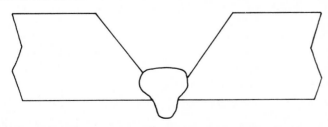

FIGURE 20-25 Excess penetration is often called *icicles.* It either means that your root-face land or nose on a groove-welded joint is not wide enough, or that you are using way too much current or a gas tip that is far too big for the section. It also occurs when the root gap is too wide. Do you need a backing bar?

Pressure Testing

High-pressure liquids or gases are usually used for proof testing closed containers. For example, a vessel may be pumped full of high-pressure water with the test pressure set higher than the working pressure of the design. The vessel is then visually inspected to see whether there are any leaks or bulges in the vessel walls or whether the valves are holding up. *Proof tests are conducted with strict safety rules, just in case the product fails destructively; never stand close to any vessel while it is being proof-tested.*

Proof testing is used on storage tanks for liquids and pressure vessels of all sorts such as oil-cracking stills and tanks for compressed gases. Water-pressure (hydrostatic) testing often is used because if the vessel fails there is less risk of a serious explosion. Pneumatic testing using high-pressure gases, including air or nitrogen, can be used, but if the container fails, it's likely to do so explosively. As a result, pneumatic testing is more risky.

Open containers also are tested hydrostatically and visually examined for leaks. Examples are water- or oil-storage tanks. Watertight open vessels such as ship hulls are often tested by fully loading them after they are launched. Leaks are noted and sealed before the vessel is put into service.

Overloading

Load-bearing structural sections fabricated in a shop sometimes are tested by loading the section with sand bags or bricks to stress levels higher than the section will handle in service. The section's dimensions are then checked to make sure that any deflections in it are less than those required by the design. The load is then removed to see whether any part of the weldment has a permanent set. Something as simple as a welded steel door frame (or as complex as a bay of a building) can be tested this way. If the section holds up, the fabricator sometimes may go ahead and build more sections like it without testing the rest of them.

Overspinning

Weldments that are used with rotating machinery, parts, or equipment can be proof-tested by overspinning the work and permitting centrifugal force to provide the excess test load. Sometimes you will see statements such as "tested to 150 percent of design stress" on such parts. After stress testing, the dimensions of the part are rechecked to make sure that the excess load did not cause too much permanent deformation. Manufacturers of grinding wheels and tires selectively test their products that way.

20-3
LEAK TESTING

Leak tests are often handy methods for inspecting pressure vessels, piping, hydraulic lines, and open containers. Proof tests and leak tests often overlap. The primary difference is that leakage is the only thing you look for in a leak test, not necessarily dimensional changes in the part. Water, oil, compressed air, helium, or nitrogen can be used in leak tests. The test fluid might be used at atmospheric pressure or at higher pressures. Before leak testing, all welds should be completely finished and finish grinding (if required) should be done. Leak testing often is done just before the workpiece is painted and readied for shipment.

Not all test fluids are equally sensitive to leaks. A light oil may leak at a lower pressure than water because the water has a higher surface tension. However, when you add a household detergent to the water, the detergent not only lowers the surface tension of the water (so that it can penetrate smaller cracks or holes), but the leak also will show up as "soap" bubbles on the outside surface of the part.

Fluorescent dyes are sometimes added to the water to make a small leak more visible. Ultraviolet light is used to make the leaking liquid show up better. A "black light" lamp will make the fluorescent dye stand out so that the leak is hard to miss.

When a hydrostatic leak test is used, all high points in the vessel, where possible, should have vents to let air out so that the water can fill the space. Hydrostatic leak testing has the advantage of being relatively safe compared with the use of compressed gases (pneumatic testing) if all fittings are securely fastened and air entrapped during filling is minimized. If the workpiece does fail during hydrostatic testing the energy release is much less than in pneumatic testing. You may get wet but you won't be seriously hurt.

When air or an inert gas is used, leaks often are located by bubbles that rise from the work when it is submerged in a tank of water. It's the same idea as finding a leak in a tire. Of course you can't get very big workpieces into a tank of water, but you can coat the seams of the workpiece with a wet soap film and look for soap bubbles.

Helium gives you a chance to find very small leaks. As you probably know, helium atoms will even leak right through a solid rubber balloon. They are so small that they can get through tiny holes and cracks that will stop other gases. Helium is even used to test for leaks in laboratory glassware that has been sealed with a brazing or welding torch. Helium leak detectors using electronic methods can find as little as 1 part helium per 200,000 parts air.

Leak Test Specifications

The ASTM (American Society for Testing and Materials) publishes ASTM Standard E432 (plus whatever the latest revision date may be) titled "Selection of a Leak Testing Method." ASTM E432 will tell you more about leak testing than you ever thought possible. You also might want a copy of ASTM E425, which is titled "Standard Definition of Terms Relating to Leak Testing." Section V of the ASME Boiler and Pressure Vessel Code gives you the latest requirements for leak testing the products it covers in Article 10.

API Standard 620, "Design and Construction of Large Welded Storage Tanks," covers hydrostatic and pneumatic testing. API Standard 650, "Welded Steel Tanks for Oil Storage," includes information on inspection, testing, and repairs of these steel tanks.

AWS D5.2 (with the latest revision date) titled "Standard for Welded Steel Elevated Tanks, Standpipes and Reservoirs for Water Storage," includes information on inspection and testing them. If you repair any container that previously carried combustible or explosive liquids or gases (such as fixing a leak in an automotive gas tank), first look at AWS A6.0, "Safe Practice for Welding and Cutting Containers That Have Held Combustibles."

20-4 NONDESTRUCTIVE TESTING

Nondestructive testing (not even counting leak testing and proof testing) is a big subject and a very important one for welders. NDT, as it's called, can be used to locate and identify just about any kind of welding defect without tearing the weldment apart to find it.

If NDT methods have any drawback at all, it's that a trained welding inspector is frequently needed to interpret what the tests

TABLE 20-1 Summary of NDT methods

Equipment	Applications	Advantages	Limitations
Leak Testing			
Exact equipment depends upon the method used (see text). Generally needs equipment capable of inducing a pressure differential and some form of detection device capable of sensing the leak. Some applications require special fluids such as helium.	Welds that have defects extending through the weld volume.	Many components can be inspected in "real time" along with a proof test. Some applications require very little operator training. Test results are usually obtained fast. Certain processes can be automated, for example, soft drink cans tested at 350–400 per min.	Some methods require special facilities and time-consuming inspection. Applications requiring high levels of sensitivity usually are uneconomical and may require personnel with special training.
Visual Inspection			
Magnifiers; projectors; other measuring equipment such as rulers, micrometers, optical comparitors; light source including ultraviolet light.	Welds with surface defects.	Economical and fast. Needs relatively little training and little equipment for most jobs.	Limited to external or surface conditions only. Limited by inspector's eyesight.
Gamma Radiography			
Gamma-ray source (radioactive isotopes), gamma-ray camera, projectors, film holders, films, lead screens, film processing equipment, film viewers, exposure facilities, radiation monitors.	Most weld discontinuities including cracks, porosity, lack of fusion, incomplete penetration, slag, as well as corrosion and fit-up defects, wall thickness, dimensions.	Permanent record so that other people can use data later on. Gamma-ray source can be put inside an accessible weld such as pipes and vessels for unusual radiographic techniques. Energy-efficient source requires no outside energy to produce gamma rays.	Radiation is a safety hazard. Requires special facilities or area where personnel are monitored. Gamma source must be replaced periodically. Wavelength of radiation source cannot be adjusted. Gamma source must be licensed. Requires highly skilled operator and data interpreter.
X-Radiography			
X-ray sources (from high-voltage x-ray machine), electrical power source for x-ray machine, same general equipment used with gamma-ray sources.	Same applications as above.	Adjustable energy levels, generally produces higher quality radiographs than gamma radiation. Offers permanent film record as does gamma radiation.	High initial cost for equipment. Not generally considered portable. Radiation hazard. Skilled operator and interpreter needed.

Continued on page 540

TABLE 20-1 Summary of NDT methods (*Continued*)

Equipment	Applications	Advantages	Limitations
Ultrasonic Testing			
Pulse-echo instrument capable of exciting a piezoelectric material and generating ultrasonic energy within a workpiece, and a suitable cathode-ray tube oscilloscope capable of displaying the magnitude of received sound energy.	Most weld discontinuities including cracks, slag, lack of fusion, lack of bond, thickness. Some mechanical property data can be calculated using this test.	Most sensitive to planar (flat) defects. Test results are known quickly and equipment is portable. Most ultrasonic flaw detectors do not require electrical power outlet. They can use batteries. High penetration capability.	Surface condition must be suitable for coupling with transducer. Coupling liquid needed. Welds may be difficult to inspect. Reference required. Skilled operator and data interpreter needed.
Magnetic-Particle Testing			
Prods, yokes, coils suitable for inducing magnetism into the test workpiece. Power source, magnetic powders; some applications require special equipment.	Most weld discontinuities open to the surface and some large voids slightly below the surface. Most suitable for cracks.	Economical and fast. Inspection equipment is portable. Unlike dye penetrants, magnetic particle can detect some subsurface defects.	Must be applied to magnetic materials. Parts must be cleaned before and after inspection. Coatings will mask rejectable defects. Some applications require parts to be demagnetized after inspection. Magnetic-particle inspection requires electrical outlets for most jobs.
Liquid-Penetrant Testing			
Fluorescent or dye penetrant, developers, cleaners (solvents and emulsifiers). Suitable cleaning gear. Ultraviolet light source if fluorescent dye is used.	Weld discontinuities open to the surface such as cracks, porosity, lack of fusion.	Can be used on all nonporous materials. Portable, inexpensive equipment gives fast results. Data are easy to interpret. Requires no electrical energy except for ultraviolet light source. Indications can be further examined visually.	Surface films such as coatings and scale will hide rejectable defects. Bleed out from porous surface can mask indications. Parts must be cleaned before and after inspection.
Eddy-Current Testing			
An instrument capable of introducing electromagnetic fields within a test workpiece and sensing the resulting electrical currents (eddies) induced with a suitable probe or detector. Calibration standards.	Weld discontinuities open to surface such as cracks, porosity, fusion problems, as well as subsurface inclusions. Alloy content, heat-treatment variations, wall thickness.	Fast, low-cost method. Automation possible for symmetrical parts. No coupling fluid needed. Probe does not have to be in intimate contact with test piece.	Limited to conductive materials. Shallow depth of penetration. Indications may be masked by part geometry. Reference standards required.
Acoustic-Emission Testing			
Emission sensors, amplifying electronics, signal-processing electronics including frequency gates, filters. A suitable output system for evaluating the acoustic signal (audio monitor, visual monitor, counters, tape recorders, X-Y axis recorder).	Internal cracking in welds during cooling, crack initiation, and growth rates.	Real time and continuous surveillance inspection. May be inspected remotely. Portable equipment.	Requires the use of transducers coupled on the test-part surface. Part must be "in-use" stressed. Ductile materials yield low-amplitude emissions. Noise must be filtered out of the inspection system.

show and determine whether the welding defects uncovered are minor or are serious enough to require repairing the weld.

The main NDT techniques that we'll tell you about in this lesson include:

- Visual inspection
- Radiography
- Ultrasonic testing
- Magnetic-particle testing
- Liquid-penetrant testing
- Eddy-current testing
- Acoustic-emission testing

Table 20-1 (shown on pages 539–540) lists the equipment, applications, advantages, and limitations of each of these methods. The rest of this chapter will give you more details on each test except visual inspection, which we covered in the first part of this lesson.

Radiography

Let's look at x-ray and gamma-ray testing together. They are very similar in how they work and what they do.

An x-ray photograph of the inside of a weld can be taken just like the x-rays you get from a doctor or dentist. X-rays show differences in the densities of materials. Your bones are the most dense material in your body, so they show up clearest on the x-ray photo. The same thing is true of radiographs of a weld.

Very intense, short-wavelength radiation (about one ten-thousandth the wavelength of visible light or less) will penetrate metals. Very short wavelength x-rays are produced by a machine. Even shorter wavelength gamma rays are more penetrating than x-rays, and they are produced by the radioactive decay of isotopes such as cobalt 60, cesium 137, and iridium 192. Radium would do the same thing, but this material is very rare and quite expensive.

Special photographic film sensitive to very short wavelength radiation is used to detect the radiation that "shines" through the metal. Variations in the density of your weld metal are detected on the film. A hole (porosity) in your weld will let more radiation travel through the metal and hit the film than solid, dense metal without any porosity.

A slag inclusion will also pass more radiation than your weld metal, as will a crack. Since the film used is usually developed as a negative (a positive print is a waste of time), the light areas on the negative are spots where less radiation hit the film and the darker areas are places where more radiation passed through (Fig. 20-27).

The radiation source, whether it is an x-ray machine or a radioactive isotope, is put on one side of the weldment. The film, in a light-tight envelope, is taped to the other side of the weld. Workers are protected from back-reflected or back-scattered radiation

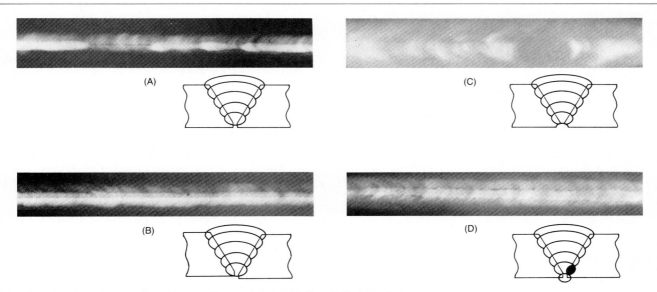

(A)

(B)

(C)

(D)

FIGURE 20-27 X-rays of weld metal require special training and experience to fully interpret. However, many welders soon learn to recognize welding defects on x-rays, especially when the x-ray defects show up on their own work. Here are some examples. (A) Inadequate penetration of the weld root. (B) Inadequate penetration of the root with high-low (poor fit-up). (C) Burn-through where excess weld metal penetration and arc force has caused the weld puddle to be blown away from the interior root section of a pipe weld. (D) Elongated slag inclusions entrapped in the weld metal or between the weld metal and base metal. Elongated slag inclusions (continuous or broken slag lines) are called wagon tracks by pipe welders. They are usually found at the fusion zone of a weld.

Continued on page 542

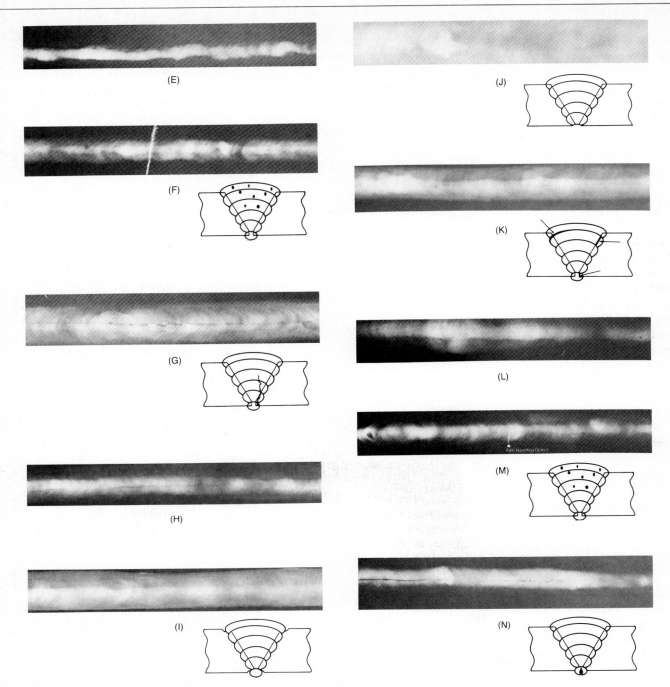

(E) Isolated slag inclusions are irregularly shaped inclusions that may be found anywhere in the weld. (F) Piping or wormhole porosity is an elongated hole that results when the gas rising from the solidifying weld metal does not escape before the weld metal becomes solid. (G) With the exception of very shallow cracks, no weld containing cracks, regardless of size or location, is acceptable. Even very tiny surface cracks may not be permitted in most applications. (H) Inside undercutting on this pipe weld shows up in this x-ray as a line. (I) Outside undercutting in this pipe joint has notched the wall of the base-metal. (J) Internal bead concavity in this pipe weld differs from burned-through areas which are associated with intermittently deposited weld metal. (K) Incomplete fusion or bond at the root of the joint or at the top of the joint between the weld metal and base metal. (L) Elongated porosity or gas pockets occurring in the weld metal. (M) Spherical porosity or gas bubbles distributed throughout the weld metal. (N) A hollow weld bead in the root zone.

from the surface of the metal by heavy sheet-lead screens and lead-containing gloves, aprons, and masks.

The film will show slag inclusions, undercutting, porosity, incomplete fusion and lack of penetration, cracks, and excess penetration or burn-through. However, the interpretation of the picture sometimes is difficult. The ideal angle to take the radiographic photo of most welds is from the side, at right angles to the weld. That's usually impossible because the base metal is in the way. Most radiographs of welds, therefore, are taken from the top surface of the weld and the radiograph is interpreted by a specialist who figures out what it shows, and whether the defect is minor and can be tolerated or serious and not allowed by the welding code or company quality-control policy.

The angle that the flaw makes with the incoming radiation is another complication. A crack shown edge-on may be harder to see than one shown at an angle. On the film, the edge-on crack may only look like a thin black hairline.

The quality of the resulting picture also depends on the intensity of the radiation source, the angle used, the type and thickness of the metal, and a lot of other details we won't cover here. Nevertheless, a skilled radiographer can find and pinpoint every internal welding flaw that you can create and then tell you not only where on the weldment the problem is but also what it is, how severe it is, how deep it is, and what side of the weldment you should start on to get at and remove the defect. That helps you fix your welds without taking out too much good, solid weld metal when you do it.

You should never do radiography by yourself. Always work with a qualified specialist. The work is not only risky for an inexperienced operator because of the radiation hazard but you also won't be able to understand a lot of what you're looking at without someone to help you interpret it. Nevertheless, we have seen experienced aircraft welders standing in line at the radiographer's office to "see their x-rays." After a while, if you work in a shop that does a lot of radiographic test work, you will become very good at interpreting most of the results yourself.

Ultrasonic Testing

Ultrasonic testing is like solid-metal sonar (Fig. 20-28). High-frequency sound waves penetrate your weld metal and base metal and bounce back from welding defects (as well as from the back

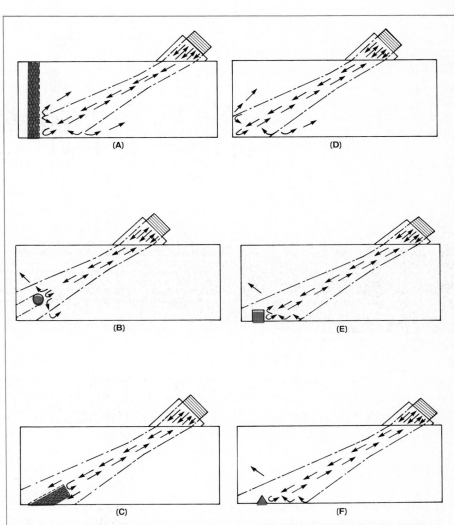

FIGURE 20-28 Ultrasonic inspection locates weld defects that reflect high-frequency sound waves (much like solid-metal radar). The signals require even more training than x-ray work to interpret. These drawings show how the high-frequency sound reflects from various types of defects. Part of the problem with ultrasonic inspection is that different defects (and some things that aren't defects at all) give similar signals. On the other hand, the signals often can be calibrated to precisely locate the depth of the defect, which is not usually possible with x-ray inspection. (A) Vertical defect reflections. (B) Horizontal defect reflections. (C) End-on defect reflections. (D) Corners and faces of the workpiece also reflect sound which adds "noise" to the signal, making interpretation more difficult. (E) Normal keyway in a workpiece. (F) A notch defect in the workpiece. (E) and (F) give almost the same signal.

edge of the plate). The intensity of the reflected sound can be interpreted to give a very good idea of the kind of welding defect, where it is in relation to the surface of the weldment, and how deep in the section it may be. The results usually are displayed on a TV-like screen called a cathode-ray tube (CRT), usually in the form of an electrician's calibrated oscilloscope.

The frequency of the ultrasonic sound used to test welds is most often betweeen 1 and 6 MHz (1 MHz equals 1 million cycles per second). A device called the *transducer* contains a crystal that vibrates at a natural frequency in this range and produces the ultrasonic sound.

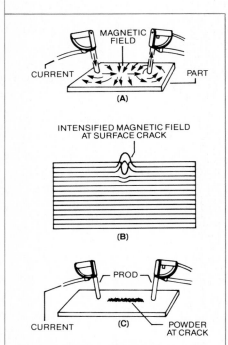

FIGURE 20-29 Magnetic particle inspection uses an induced magnetic force (produced by the two handheld prods). Defects near the surface, especially cracks and slag, distort the shape of the magnetic field. Iron powder sprinkled on the surface shows the shape of the magnetic field over the defect by lining up with the crack or imperfections such as slag inclusions. However, magnetic particle detection does not reach very far into a workpiece.

Crystals that vibrate in sequence with an alternating electric current are called *piezoelectric* materials. (Quartz is the most common material of this kind.) When subjected to an alternating current, a piezoelectric material will vibrate. An ultrafast alternating current produces ultrafast crystal vibrations and, therefore, ultra-high-frequency sound.

You can't hear ultra-high-frequency sound in the range of 1 to 6 MHz. In fact, sound in this frequency range won't even travel very far through the air. That's why liquids called *coupling materials* are put on your metal to get as much of the sound as possible into the work. Typical coupling compounds are water, grease, glycerine and cellulose-gum powder in water, and proprietary materials.

The smallest flaw size that ultrasonic testing will find is about one-half the sound wavelength. This is very small and therefore this test method will identify and locate cracks, lack of fusion, incomplete penetration, slag inclusions, and even fine porosity. Just like radiography (but even more so), this test method requires an experienced operator to interpret the results. Also just like radiography, the quality of your results depend on the angle of the transducer to the flaw and many other variables.

Magnetic-Particle Testing
The shape of a magnetic field is strongly affected by changes in the density and shape of the magnetized metal that carries the field. These changes will show up if you sprinkle a magnetic powder like iron filings on the surface of the metal. Magnetic-particle inspection is very good at detecting surface flaws and other problems that are on or slightly below the surface of the weld metal or base metal, even when they are too small for you to see.

Magnetic-particle testing will show fine hairline surface cracks of all kinds, laminations caused

by chemical or physical differences in the base metal, incomplete fusion and undercutting, subsurface cracks, and inadequate joint penetration. But the technique is *not* a substitute for radiography or ultrasonics. Rather, it offers special advantages over these other NDT techniques when locating tight cracks and surface defects that the other methods have trouble spotting.

Magnetic-particle inspection, of course, is limited to ferrous metals, nickel and cobalt alloys, and even then only when these materials are magnetic. It won't work on aluminum, copper, zinc, austenitic stainless steel, or austenitic high-manganese steels because these materials are nonmagnetic. (Some austenitic stainless steels will become slightly magnetic after severe cold working, but not enough to matter much.)

The method is affected by the surface condition of your iron or steel so that testing must be done on clean, dry surfaces free of oil, water, excessive slag, and other contaminants. Wire brushing or sandblasting the surface before testing helps a lot and should be done if the surface is not very smooth.

The magnetizing force is delivered by hand-held devices called *prods* (Fig. 20-29). Two prods are placed on the metal and a magnetic field is set up between them. The prods are nothing more than coils with a soft iron core. They are small electromagnets to concentrate the magnetic field. The magnetic field in the weldment will be strong between the two prods. It also will line up with any flaws in or near the surface of your metal. When iron powder is sprinkled on the metal a crack, for example, will be clearly "painted" by the way the iron powder lines up in the local magnetic field over the crack.

Magnetic-particle inspection can be used on all types of heavy weldments as long as the materials are magnetic. It usually is used to inspect finished welds. On heavy, multipass welds, however, it is most often used to inspect each pass right after it is made because the deeper the defect is inside the metal, the weaker the pattern made by the magnetic force will be on the surface. Magnetic-particle inspection is excellent for detecting crater cracks. The magnetic test also is used to inspect root passes that have been back-chipped down to what you hope will be sound weld metal without cracks.

Magnetic-particle inspection is used after stress-relief heat treatment, not before. The subsurface cracks it detects will open up more after stresses are relieved.

Many steel weldments used in the aircraft industry, such as landing gears, are tested by magnetic-particle inspection after the part has been in use for a long time. The magnetic-particle inspection technique is excellent for detecting the beginning of the tiny fatigue cracks that always start on the surface.

Magnetic-particle inspection also is used to test the edges of thick steel plate prior to welding to detect cracks, laminations, nonmetallic inclusions, and alloy segregations in the steel.

There are two basic methods used. They are the dry method (dry magnetic powder is sprinkled on the part) and the wet method, which uses magnetic particles suspended in a liquid. The magnetic particles in the wet method are much smaller than in the dry method and the wet process is far more sensitive to fine surface defects. The liquid can be water, but often it is a dye penetrant, so that magnetic-particle inspection can be combined with the next test method we'll tell you about.

Liquid-Penetrant Testing

Certain types of liquids have much lower surface tension than water, which means that these liquids can penetrate into cracks that even water can't get into. Once they have entered these sur-

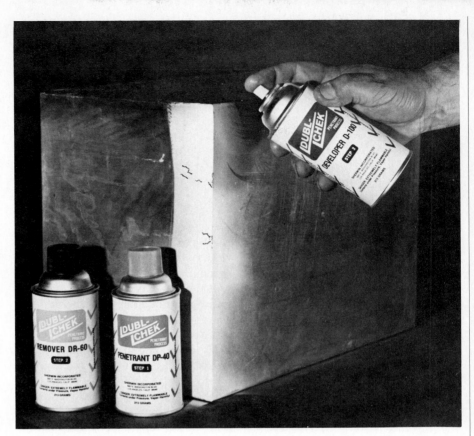

FIGURE 20-30 Mr. Amos Sherwin, president of Sherwin, Inc., Los Angeles, a producer of dye penetrant and fluorescent crack-detection compounds, found these surface cracks in less than 4 min using nothing more than spray cans of Sherwin DP-40 dye penetrant liquid, DP-60 remover compound, and can of Sherwin D-100 developer. The liquid dye penetrant has a low surface tension that makes the liquid rapidly penetrate tight cracks that even water normally would not enter. Even after the rest of the penetrant compound is removed by a towel dampened with Sherwin's remover compound, the penetrant liquid remains in the cracks. A special developer sprayed on the surface of the metal pulls the test penetrant liquid in the cracks to the surface and gives it a bright red color that strongly contrasts with the white developer. The tiny cracks stand out in bold, red relief against the white background.

face cracks, the rest of the liquid can be washed away but liquid in the cracks remains there. Now imagine what you would have if that penetrating liquid could be made to show up very strongly with a bright, easily seen color (Fig.20-30). That's the idea behind liquid-penetrant testing.

There are two basic kinds of liquids. One type is a dye (the process is called *dye-penetrant testing*). The other kind of liquid fluoresces under an ultraviolet or black-light bulb. The fluorescent method is exceptionally good at resolving very fine cracks. The

dye-penetrant method is almost as good and has the advantage that you don't have to carry an ultraviolet light source around with you. The dye is usually a vivid red or yellow color that contrasts sharply with the white background color of the second, white developer chemical.

Both methods are simple to use, easy to interpret, very portable, and subject only to minor restrictions such as thoroughly cleaning the surface of the metal by wire brushing or sandblasting away slag, mill scale, and paint, and removing grease and oil with chlorinated petroleum solvents or acetone. (Vapor degreasers work fine, but you should wait a short time to let the cleaning solution evaporate before applying the dye or fluorescing compound.)

After you apply the compound, wait for 15 to 30 min to let it really sink in. You can apply the compound with a spray can or a paintbrush. Just make sure that your work surface is dry before you apply the penetrant. After waiting for a while, wipe or wash off excess liquid (most of these compounds are water-soluble) and apply the developer. Then shine the ultraviolet light on the area (if it's a fluorescent material). When you wash off the material, don't use water over 110°F [43°C] because hot water will wash these compounds out of the surface defects.

If you have removed all the excess penetrant liquid and followed the manufacturer's instructions, and if you have any kind of surface cracks in your weld metal, they will stick out like a sore thumb. If necessary, you can use a Polaroid Land camera with color film to take a picture of the defect to have a permanent record. If necessary (or required), you can remove all the penetrant liquid after you are through using solvents specified by the manufacturer.

There are only a few false clues to watch out for. Excess penetrant liquid in the depressions between your weld beads may look like cracks when they are not. If you run into that problem, lightly grind the weld surface smooth and then retest it.

Be especially on the lookout for very fine lines that indicate tight cracks. Also look for any kind of markings that seem to line up. They could be tiny surface cracks that will join up later on to make one big crack. Deep cracks will show fine lines at first but will continue to bleed out penetrant, and the colored line will get larger as you wait.

Liquid-penetrant testing is just as effective for catching surface imperfections as the magnetic-particle inspection method, and it works on any kind of metal, whether or not it is magnetic. On the other hand, if the defect is just below the surface but the surface itself is solid, liquid penetrants won't show the flaw while magnetic-particle inspection will.

Eddy-Current Testing

Eddy-current testing is a nondestructive test method based on the principle that an electric current will flow in any conductor subject to a changing magnetic field. That's also the principle behind the generation of ac electricity. Eddy-current testing is used to check welds in both magnetic and nonmagnetic materials. It's especially useful in automatic systems that check bars, billets, and pipe and tubing at high speeds.

Changes in the electric or magnetic field and the resistance of the material are measured. The current used can range from 50 Hz to 1 MHz depending on the type and thickness of material being tested. The theory behind eddy-current testing is rather technical and we won't go into it here, except to say that the workpiece runs through an electrical coil which measures changes in the current induced by differences in the material. The data are usually recorded on a strip chart, so the process is not exactly portable.

The process will not only detect weld flaws on or near the surface of the workpiece such as cracks, porosity and lack of fusion, but it also will pick up shallow subsurface defects. Eddy currents also can be used to check the thickness of electroplated coatings on metals, help gauge the alloy content of a metal, the wall thickness of a pipe or tube as it is being produced, and a lot of other things. The instruments used must be calibrated before the readings can be

interpreted. Calibration standards are used for this purpose.

Eddy-current testing is most often used in large metal-producing plants. For example, welded tubing can be checked for minute cracks and flaws as it travels through a tubing mill at 300 ft/min [91 m/min].

Acoustic-Emission Testing

Acoustic-emission testing is sometimes used with other test methods, especially proof testing. When metals deform plastically or start cracking, they make a lot of noise characteristic of the type of action taking place. This noise can be detected, recorded, and interpreted.

A pressure vessel will grunt and groan as it's loaded in a proof test. A sound, properly welded vessel will stop making noise when the stress is reduced and does not make further bursts of noise until the previous load level is exceeded. A growing crack emits a continuously increasing sound level as it's loaded and it keeps on making noise when the load is reapplied, even though the reloading is below the first load level.

You can listen to the test being conducted without getting near the test piece by using a microphone taped to the vessel and a tape recorder to collect the sounds. Therefore, acoustic-emission testing is one way you can know how well a proof test is going without hanging around too close to the workpiece—just in case it happens to blow up, or as the engineers like to say, "fail in a destructive mode." Let's hope that all your errors will be discovered nondestructively.

REVIEW QUESTIONS

1. Name two kinds of bad weld profiles.

2. What is overlap? What usually causes it?

3. Name four kinds of weld-metal porosity patterns that you might see in an x-ray. Where or when do they most often occur?

4. How can you avoid slag inclusions in your welds?

5. How can you avoid leaving tungsten inclusions in your welds when using GTAW?

6. Name three conditions that can produce cracking in the weld metal or base metal.

7. Discuss six ways to reduce or eliminate cracking that doesn't seem to be associated with the metallurgy of your weld metal or base metal.

8. What kinds of steels are most likely to crack?

9. List four kinds of proof tests and give examples of where or when they might be used.

10. What are the seven basic kinds of NDT tests? Briefly describe each one, especially the inspection source used in the test.

11. What types of weld defects will radiography most likely find?

12. Will magnetic-particle inspection tests find porosity deep inside the root of a thick weld?

13. What type of weld defects is liquid penetrant testing most likely to be used on?

14. When might you use eddy-current testing?

15. Describe what sonic testing is and how you might be able to use it. Can you think of any simple ways to amplify the sounds made in the test?

Developing an Eye for Defects

Most welding schools have limited NDT facilities. If your welding instructor can provide you with a liquid penetrant test kit, learn to use it. A few schools will have radiographic test equipment. If your school has it, learn to interpret X radiographs until you can recognize various welding defects and not only determine what they look like, but also learn how to judge how severe the problems are.

One idea several schools have tried is to develop a collection of radiographs of both good and bad welds. This study collection can be used to learn elementary film interpretation. Then a second set of film that you haven't seen before can be used to test your skill at interpreting unknown radiographs. If you have a local code-welding shop in town, your instructor may even be able to invite an NDT inspector over to show you a variety of test results. (Some firms may let your instructor borrow film and make prints.)

Some welding instructors have collections of visually detectable defective welds which can be studied in class. A collection of defective welds is certainly easy to make in a welding school. (By now you should have made just about every quality-control horror that welders are capable of producing.)

Obviously, we can't outline a detailed shop session on NDT and quality control for you. It depends too much on available equipment and supplies. But by putting this chapter near the end of the book, we think your own prior shop experience will remind you of many of the problems we have discussed.

Reading Blueprints

All that you have learned to do as a skilled welder isn't worth much if you can't follow instructions. Those instructions can range from a supervisor saying "fix it" to detailed drawings produced by design engineers. These drawings often come to you in the form of blueprints. This lesson won't tell you everything you need to know about reading blueprints, which really should be a separate shop course, but we will give you the basics of reading welding blueprints.

There is a lot to know about reading blueprints. For example, the blueprints on a piping job use special symbols that are quite different from those used on ship-hull or truck-frame blueprints or a wiring diagram. But all of these drawings use the same symbols for welding, brazing, soldering, and nondestructive testing. You will learn these special welding symbols in this chapter.

The AWS has the best reference work on welding, brazing, soldering, and nondestructive testing symbols. It's called AWS A2.4, "Symbols for Welding and Nondestructive Testing." AWS A2.4 is a very useful guide to welding symbols of all kinds, including symbols for many welding processes we don't cover in this book because they are automated or machine processes and don't require skilled operators.

21-1
WHY WELDING SYMBOLS ARE USED

Let's look at a typical welding symbol in Fig. 21-1. It may look to you like the instructions for making the putting green on a golf course, but this one symbol tells you nine things: (1) what welding process to use (GMAW), (2) what kind of joint to prepare (a) 30° single-bevel fillet-weld with a ⅛-in. [3.2-mm] root gap), (3) how to finish the joint after welding (G and the slanting line mean grind after welding to a flat surface finish), (4) which side of the part to weld, (5) instructions to weld all around the joint (that's the circle in symbol), (6) the type and size of the welded joint (the triangle means a fillet weld—the ⅜ means it's a fillet weld with a ⅜-in. [9.5-mm] leg), (7) the size of the groove (¼ in. [6.4 mm]), (8) the "effective throat" of the weld (the ¹⁄₁₆ in. or its metric equivalent, 1.6 mm, in parentheses), (9) where you are supposed to do the work (the golf flag means field welding). And we haven't even shown you the part that you are supposed to work on. The arrow will either point right at the place on the part where the weld will be put or the arrow will point at the side of the part just opposite where the weld goes.

Now compare the space the symbol takes up with the space the above paragraph takes to describe it. You can see at a glance that one symbol can say a lot more than the nearly 200 words we needed in the above paragraph. What's more, there's much less chance for error when reading the symbol than when reading the paragraph. Nevertheless, even more information could have been put into the one welding symbol in Fig. 21-1 if it were necessary.

For example, if the request were not to "weld all around," then a number would be used to tell you how long to make the fillet weld. The symbol can even be used to tell you to make intermittent fillet welds, each one of which is 6 in. [152 mm] long, and even more things that you will learn shortly. Like most subjects that look complex when you start, welding symbols are actually easy to read after you learn how. Equally important in helping you learn them, the symbols make sense. They usually resemble the work to be done, which means that you don't have to work too hard to memorize what they mean. The best way to learn how to understand welding symbols is to take the symbols apart and look separately at each part and what it means.*

21-2
TAKING SYMBOLS APART

All welding symbols have a minimum of three basic parts. These are a reference line, an arrow, and a stylistically feathered tail on the reference line opposite the arrow that contains the welding process code or a basic welding symbol. (The feathered tail of the arrow and a welding symbol on the reference line are often combined into one symbol.) A welding symbol also can have four more parts: dimensions and other data, finishing symbols, and supplementary symbols to make the instructions easier to understand. A complete welding symbol might contain eight different kinds of information. But let's strip the welding symbol of everything but the mandatory reference line, arrow, and feathers on the tail. Now let's see what just these three elements can tell us.

Reference Lines, Arrows, and Tails

We show you a stripped-down welding symbol in Fig. 21-2. Note that the reference line is always horizontal on all drawings and that the arrow is always broken (points either up or down from the reference line). That means that the reference line has two sides. One is *the arrow side* and the other side is called (as you might guess) *the other side*.

The arrow points to a spot on the drawing where the welding is supposed to be done. Many times it's difficult to put the arrow on the same side as the actual spot where the weld is to be made, so the two sides of the reference line are put to use. Any welding instructions on the arrow side of the

reference line are to be made on the same side of the part as the arrow points to. Any welding instructions on the other side of the reference line apply to the opposite side of the part. The arrow side is always *under* the reference line (no matter which way the arrow points). The other side is always *on top of* the reference line.

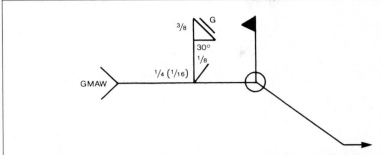

FIGURE 21-1 A typical welding symbol. At the end of this chapter you will understand everything this symbol is saying to you. For example, the golf flag means "do this work in the field, not the shop" while the circle means "weld all around the joint on the opposite side of the part from where the arrow is pointing." The triangle means "this is a fillet weld with a 3/8-in. leg that should be ground to a flat surface after welding." The symbol just below the fillet weld tells you that the other weld is a single-groove weld with a 30° included angle, a 1/8-in. gap width, and a 1/4-in.-deep throat. You should make sure that your weld penetrates another 1/16 in. deep below the bottom of the bevel-weld joint. And the process to use for both welds is GMAW. If that sounds incredible, just read this chapter and you soon will be reading welding symbols as easily as you read this book.

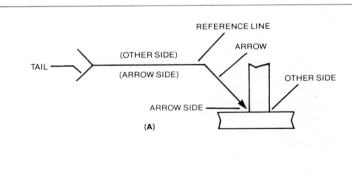

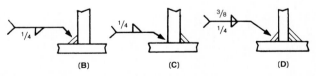

FIGURE 21-2 First learn the basics of welding symbols, especially which side of a part you should weld when the symbol is on the arrow side or on the other side of the reference line. The arrow side is on the bottom and the other side is on the top of the reference line. The arrow points to the part, so that a symbol on the lower arrow side means weld the side you can see in the drawing that the arrow points to. When the symbol is on top of the reference line, it means that you should weld the side of the part on the other side, just opposite where the arrow points.

---

* Gene Wolfe, Senior Editor of the trade magazine *Plant Engineering*, was the first to use this approach to explain welding symbols. We all owe him a debt of gratitude for much of the following explanation.

Just think "same side down" and you'll keep it straight. This is shown in Fig. 21-2A. We put these ideas to use in Fig. 21-2B, C, and D.

As shown in Fig. 21-2D, welding on both sides of a part still conforms to the same rules. The instructions on the arrow side apply to the side of the part the arrow is on. Instructions on the other side apply to the opposite side of the part. In Fig. 21-2, the part happens to be a fillet weld. The triangle is the symbol for fillet weld because it looks like a fillet weld. The ¼ and ⅜ refer to the size of the fillet-weld legs in inches (unless the units are otherwise specified somewhere else on the drawing).

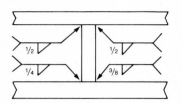

FIGURE 21-3 These are fillet welds with arrow-side welding required on the joint. The arrowheads point to the weld location. The numbers tell you the size of the required fillet-weld leg. The dimensions will normally be in inches, unless otherwise specified in a note on the drawing.

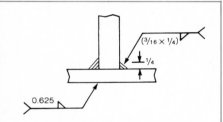

FIGURE 21-4 Decimals as well as fractions of an inch (or other units) are used on welding symbols. The ³⁄₁₆- × ¼-in. fillet weld has uneven legs. The drawing shows you which one is the ¼-in. leg. Also note where the arrow-side and other-side welds are actually located on the part.

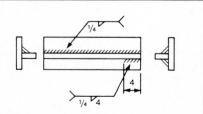

FIGURE 21-5 This is one way to indicate the length of a weld along the part, as well as its size.

Dimensioning the Weld

No matter how the arrow is positioned on the drawing, the straight side of the triangle must be on the left as shown in Fig. 21-3. The size of the fillet-weld symbol as drawn on the arrow need not be related to the size of the fillet weld desired, and its legs need not be uneven when a fillet weld with uneven legs is required, as shown in Fig. 21-4. This figure also shows the use of decimal dimensions.

The size of the fillet may be given to the left of the fillet-weld symbol. (When there are several welds of the same size on a drawing, it is completely acceptable to omit the dimension from the arrows and give it in a note such as: "Unless otherwise specified, all fillet welds are to be ¾ in.") Dimensions are normally given in inches on welding blueprints in the United States, but no inch symbol (″) is used.

A dimension may be expressed as a decimal (such as 0.75) instead of a fraction (such as ¾). If the decimal form is used, a zero should precede the decimal point. The reason is that decimal points and dirt can often be confused, but the leading zero makes it clear what dimensions the designer wants and where the decimal point is located. If a fraction is used, the line indicating division should be horizontal, not diagonal like the type in the book.

When they are used, metric dimensions are often indicated by a note on the blueprint. If feet instead of inches is used, dimensions on the parts drawing or a separate note will tell you so.

When a fillet weld with uneven legs is needed, the dimensions of the legs are given in parentheses to the left of the fillet symbol. The drawing (not the arrow) should make it clear to you which leg of the fillet is to be longer.

The length of the weld should be indicated on the right of the fillet symbol, as shown in Fig. 21-5, on the bottom drawing. This ¼-in. [6.4-mm] fillet weld should be 4 in. [102 mm] long. If no length is specified (as in the top drawing in Fig. 21-5), you have to assume that the weld is to run the full length of the part.

Continuing to use our fillet weld as the example of how almost all types of welds are dimensioned in welding symbols, look at Fig. 21-6. The upper welding symbol shows you a method of dimensioning a fillet weld with unequal legs. The drawing shows that the short ¼-in. [6.4-mm] leg is the vertical leg and that the longer ⅜-in. [9.5-mm] leg is the horizontal leg. The lower drawing shows that an equal-legged fillet weld is required. Both welds are dimensioned on the lower (arrow) side of the horizontal reference line so that both symbols apply to welds on the same side of the part as the welding symbol. The top plate in the drawing is ¼ in. [6.4 mm] thick.

Figure 21-7 shows three different ways that the length of a weld (in this case a fillet weld) can be indicated. First we show a number to the right of the fillet-weld triangle. This is the length of the weld in inches (unless other units are specified).

In Fig. 21-8 we show another way to get across the same information. This time the part itself is dimensioned with the welding symbol. On the left in the I-beam drawing, the blueprint shows that two ½-in. [12.7-mm] fillet welds are to run for 60 in. [1.5 m] along the connection between the web and the top and bottom

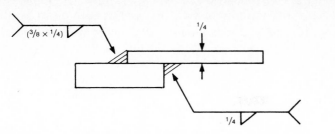

FIGURE 21-6 The top fillet weld has unequal legs. The drawing shows which leg is which. The bottom symbol is for an equal-legged fillet.

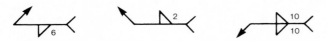

FIGURE 21-7 The lengths of these three fillet welds are 6 in., 2 in., and 10 in. along a part (not shown here). Weld length is always shown to the right of the symbol, not only for fillet welds but for other weld symbols. Most of these rules for positioning dimensions will carry over to other welding symbols, as you'll see shortly. In this example, the fillet-weld leg size is not specified.

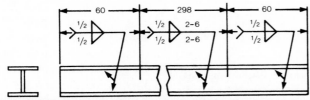

FIGURE 21-8 Six fillet-welded areas are required. Two ½-in. fillets are needed on each end of the I beam, each running for 60 in. The fillet welds in the middle of the section are ½-in. intermittent fillets, each 2-in. long, placed on 6-in. centers (6 in. between the center-to-center distance of each 2-in.-long weld which leaves 4 in. between each intermittent weld). You calculate the distance between weld bead simply by taking the center-to-center distance and subtracting the weld length, or in this example 6 − 2 = 4 in. between welds.

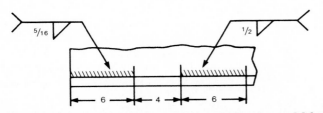

FIGURE 21-9 Another way to indicate the length of a weld is by showing the length on the drawing, not on the weld symbol.

flanges. The right-hand side of this sketch shows that the other end of the I-beam should be welded for the same distance.

The middle section of the I-beam drawing shows an additional, but very simple idea. It asks for 2-in. [51 mm] intermittent fillet welds followed by 4-in. [104-mm] space, followed by 2-in. fillet welds, followed by another 6-in. [152-mm] space until 298 in. [7.6 m] of the section have been welded.

Figure 21-9 shows a third way to indicate the length of a weld. On the left a ⁵⁄₁₆-in. [7.9-mm] equal-legged fillet weld is needed for 6 in. [152 mm]. On the right a ½-in. [13-mm] equal-legged fillet weld is needed. Leave 4 in. [102 mm] of space unwelded between the right and left ends of the section.

Figure 21-10 shows another way that welds can be dimensioned to make the instructions clear for you. Weld A is to run the full length of the joint. Note the use of multiple arrows to make clear what full length means. Weld B should be only 1¾ in. [44 mm] long, and that information is given to the right of the fillet-weld symbol.

Dimensioning intermittent welds Intermittent welds are used when a joint does not require the strength of a full-length fillet weld. They are relatively short welds separated by spaces, and their use can save both labor and materials (but only if a full-strength weld is not needed). Intermittent welds also are used to reduce the chances of warping. You can run a series of intermittent welds around a workpiece to keep weld stresses balanced and then fill in the gaps later on.

Intermittent-weld symbols on a drawing include both the length of the weld beads (called *increments*) and the pitch, or *center-to-center distance,* between the welds. In other words, if you see an intermittent welding dimension symbol that reads 6–10, it means put down 6 in. [150 mm] of weld and leave a 4-in. [102-mm] gap between the welds (10 − 6 = 4). It would be ridiculous to have intermittent welds that overlap or butt up against each other, so that the distance between increments is easy to calculate as we have done. It's the distance of the pitch minus the distance of the weld increment. Obviously, the pitch must be longer than the weld increment or else the welds will overlap. If the increment and the pitch were equal, the welds would butt up against each other.

As a rule of thumb, the increments should be at least four times the size of the fillet-weld leg and not shorter than 1 in. [25 mm]. The pitch should not be greater than 12 in. [305 mm]. On the welding symbol, the increment length is given to the right of the fillet symbol, and the pitch is given to the right of the increment, which is shown in Fig. 21-11.

Chain intermittent welds (Fig. 21-12) are interrupted fillets placed opposite each other on opposite sides of the joint. Because the welds are to be made on both the arrow side and the other side of the joint, the information required to make them appears on both sides of the reference line.

A drawing often will use the reference line as part of a larger dimension line (Fig. 21-13). Note in Fig. 21-13 that the weld is to be 3 in. [76 mm] long, not ¼ in. [6.4 mm]. The ¼-in. [6.4-mm] dimension refers to the length of the fillet leg.

Intermittent welds are specified by giving the length of the increments (welds) and the center-to-center distance between increments to the right of the weld symbol with the dimensions separated by a minus sign (which reminds you how to calculate the pitch between the increments). But a drawing showing an intermittent weld often will only have one weld symbol for the entire joint. It will be clear by reading the blueprints that the same intermittent weld continues along the length of the joint. Sometimes the design engineer will add extra dimensions to make sure that the welders in the shop don't make mistakes (Fig. 21-14).

Unless the drawing specifies otherwise, you have to assume that an intermittent weld begins and ends with an increment (weld metal). That is, the outside edge

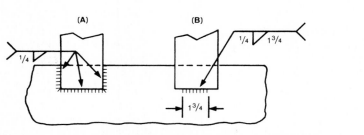

FIGURE 21-10 The ¼-in. fillet weld on the left goes on three sides of the joint. The unequal-legged fillet on the right only runs for 1¾ in. on one side of the joint.

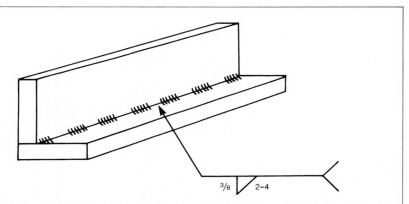

FIGURE 21-11 This one symbol plus the part drawing shows you that ⅜-in. fillet welds are required only on the arrow side of the part, and that they are to be 2 in. long, with a 4-in. center-to-center spacing which gives you 4 − 2 = 2 in. between each weld.

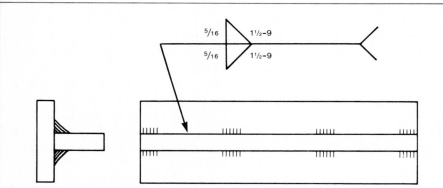

FIGURE 21-12 Chain intermittent fillet welds are required on both sides of this part. Weld size and spacing is the same for both sides in a chain intermittent weld. The ⁵⁄₁₆-in. fillet welds will be short (only 1½ in. long), and they will be made on 9-in. center-to-center distances, or 9 − 1½ = 7½ in. between each weld bead.

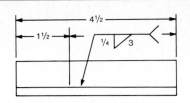

FIGURE 21-13 This symbol uses the weld arrow reference line as part of a dimension line for the overall part. Since 4½ − 1½ = 3 in., the arrow-side fillet weld must run for 3 in. on the right side of the 4½-in.-long part.

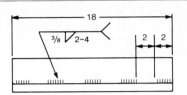

FIGURE 21-14 This part drawing has added dimensions to help you understand the symbol, and your welding job. The two 2-in. dimensions on the right side of the part are excess information, but still very helpful.

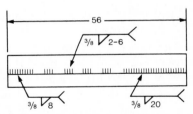

FIGURE 21-15 All welds should begin at the outside edge of a part whenever possible, unless you are otherwise specifically told to do something else (which is almost never the case). A long joint that does not end in a weld will be weak. The contact between the sections of the joint can initiate a crack that can run right through the weld metal. When intermittent welds are made between continuous welds, leave the specified center-to-center distance between the continuous welds (6 − 2 = 4 in. in this example) before you start the intermittent welding.

of a part should start with a weld. However, when an intermittent weld is used between continuous welds, the intermittent weld is assumed to begin and end with a space between itself and the continuous weld as shown in Fig. 21-15. Note that the total length of the intermittent weld itself (28 in. [710 mm] in this case) is not specified.

The symbol shown in Fig. 21-16 specifies an intermittent weld on the arrow side of the joint and a continuous weld on the other side. The fillets on each side will have the same cross section. The intermittent fillets will be 1½ in. [38 mm] long and there will be 7½ in. [190 mm] of unwelded space between them.

Dimensioning staggered intermittent welds When a part is only lightly loaded but may be stressed from either side, chain intermittent welds can provide welds that are too widely separated from each other. Staggered intermittent welding provides increments that alternate on opposite sides of the joint (Fig. 21-17). The symbol for staggered intermittent welds in this drawing actually is staggered above and below the reference line, just like the welds on the part. Both welds on opposite sides of the section

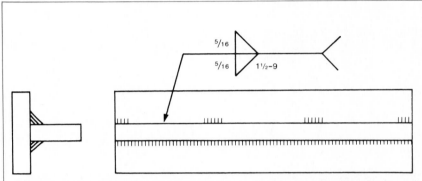

FIGURE 21-16 This one symbol calls for intermittent welding on the arrow side and continuous welding on the other side of the symbol and the part.

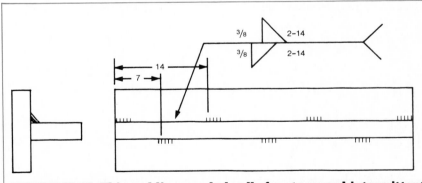

FIGURE 21-17 This welding symbol calls for staggered intermittent welding instead of chain welding (where the welds on the opposite side of the part would also be opposite each other). Each weld should have ⅜-in. legs and should be 2 in. long, spaced on 14-in. center-to-center distances. Therefore the welds will be 14 − 2, or 12 in., apart. The second weld from the left on top should start 14 in. from the left side of the section. The first weld on the top left is flush left with the end of the part. The first weld on the bottom left should start 7 in. from the left end of the part.

will be the same size, have the same increment, and the same pitch. The beginning of each staggered welding sequence is shown on the drawing. Note that although the fillet-welding symbols are staggered, the dimensions are not. They are in-line in vertical columns, one above the other.

Like intermittent welding and chain intermittent welding, staggered intermittent welding should be dimensioned so that there is an increment (weld metal) at each end of the joint. These increments on the end of the section may be on the same side of the joint (as in Fig. 21-17) or on opposite sides of the joint (as in Fig. 21-18). Always start at least one side of the workpiece with a weld-metal increment unless you are instructed to do otherwise (by a note on the drawing for example). If the ends

of the workpiece are not tied down, you may have created a built-in stress raiser or crack that can cause the entire weld to fail.

Multiple arrows sometimes are used to specify multiple welds with the same welding symbol. The I beam in Fig. 21-19 will have staggered intermittent welds on both joints. All the increments will be 2 in. [51 mm] long, and the space between increments will be 5 in. [127 mm]. Omitting the numbers to the right of the weld symbols might have resulted in continuous fillet welds on both sides of both joints if you didn't note that the fillet-weld triangles are offset from each other. Offset weld symbols mean offset welds.

The International Standards Organization has standardized a different staggered intermittent welding symbol. It is shown in

Fig. 21-20. Although it is not good practice to use this symbol in the United States, you may find it on drawings of products designed or made outside of the United States.

Weld All Around
Very often a weld must be made all around a joint. There are sev-

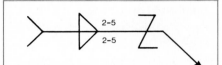

FIGURE 21-20 This is another symbol for staggered intermittent welding that you may see on European blueprints and drawings.

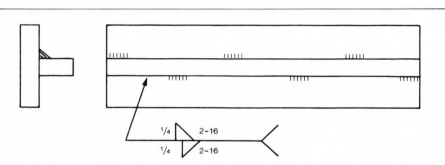

FIGURE 21-18 Here's another example of staggered intermittent fillet welding. Note that one weld on each side of the joint ends at the edge of the workpiece. That's much better practice than *not* ending a section with weld metal on at least one side of the joint.

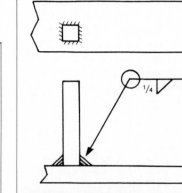

FIGURE 21-21 This symbol tells you to make a ¼-in. fillet weld all around a column.

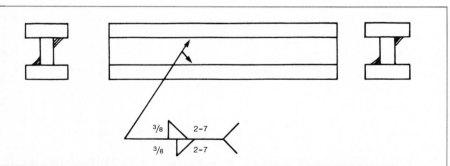

FIGURE 21-19 Now you are to make two staggered intermittent ⅜-in. fillet welds; each weld is 2 in. long and they are to be 7-in. center-to-center distances (7 − 2 = 5 in. between each weld). It's up to you to make sure that the section has at least one weld starting on each end to avoid creating an open "crack" where the joint surfaces meet.

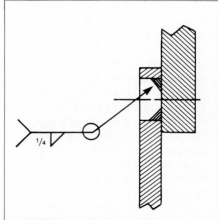

FIGURE 21-22 This symbol tells you to make a ¼-in. fillet weld all around the inside of a hole.

eral ways to make that clear to you. The design engineer might use three or four arrowheads coming out of one horizontal reference line, or two or three welding symbols scattered over the weld seam. But there is a much better way of showing you that you are supposed to weld all around a joint. A special symbol is used. It is a circle drawn around the point at which the arrow joins the reference line (Fig. 21-21). There is a round spot-welding symbol that appears on the reference line (not at the joint between the reference line and the arrowhead) that we will describe later. There also is an obsolete field-welding symbol, a black dot instead of the golf flag. Don't confuse them with this open-circle weld-all-around symbol.

Figure 21-22 shows another way to use the weld-all-around symbol inside holes and slots as well as on the outside of the parts. The weld-all-around symbol can even be used for a weld that curves in three dimensions, like many pipe connections do. The weld-all-around symbol can tell you a lot in a very little space.

Weld Contour and Finish

Welding symbols can also specify that the surface of our fillet weld (the crown) is to be finished flat, concave, or convex, as shown in Fig. 21-23. The symbols for these surfaces resemble the desired work to be done. They are even placed right on top of the hypotenuse (the longest leg) of the triangle that would be the surface of the weld on the part. If no finishing symbol is added, you have to assume that you should weld the fillet with the contour shown or use established shop practices where you work for finishing welds. If a letter code is added it will tell you that after welding you should finish the weld with a final process. The letter codes used for those final processes (Fig. 21-24) are:

G = grinding
C = chipping
M = machining
R = rolling
H = hammering

Any other letter codes can be used. For example, if you saw a straight line (flat finish) with a P,

look over the blueprint. You will probably find a note saying "peen weld surface with glass beads," or "polish weld to match base metal." The designer is free to choose any code for finishing that he or she wants to use. However, the engineer who made the blueprints should leave you a note on what the made-up code means. Letters denoting surface finishing should never be used without a contour symbol. *The same weld-finishing symbols are used on all types of weld joints, not just on fillet welds.*

21-3
PUTTING THE SYMBOLS BACK TOGETHER

By now you should be feeling pretty good about welding symbols. As we said at the start, they really aren't that complicated. Before we go on with welding symbols for other kinds of welds than fillet welds, let's show you a group of typical fillet-weld symbols along with drawings of the welds that would be produced if the instructions were followed correctly.

Figure 21-25 shows a variety of fillet welds and the weld joints that would result. Included are instructions for making single-fillet welds, double-fillet welds of equal size, double-fillet welds of unequal size, fillet welds with unequal legs, a continuous fillet weld, and a fillet weld with a specified length.

Figure 21-26 shows various kinds of intermittent fillet welds. They include intermittent welding only on one side, chain intermittent welding, and staggered chain intermittent welding.

Figure 21-27A shows how fillet-weld symbols can be used along with a groove weld in a T joint with a specified root opening. (We haven't told you about the groove-weld symbols, but we will shortly.) The letters P, R, and S simply stand for dimensions that have not been filled in on the diagram. Sometimes dimensions are

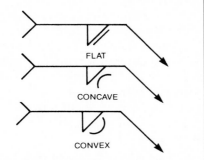

FIGURE 21-23 **These symbols tell you how the surface of your fillet weld should be contoured. They apply to any type of weld symbol, not just fillet welds. You'll soon learn that most of the dimension practices we have already covered also apply generally to all types of weld joint symbols whenever possible.**

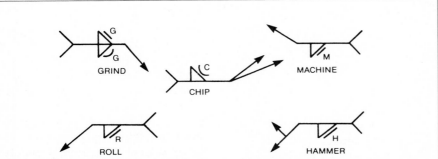

FIGURE 21-24 **These symbols tell you how to finish your weld metal, as well as what the contour of the surface of the weld should be.**

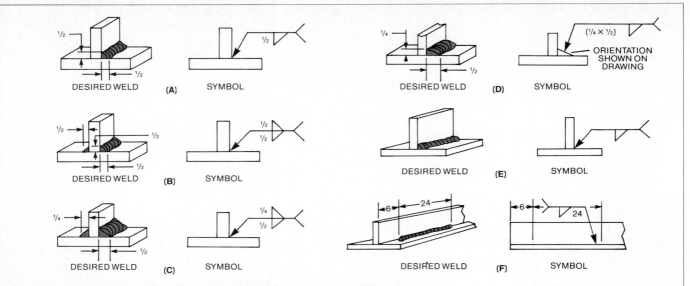

FIGURE 21-25 A variety of fillet-weld symbols are shown here along with the resulting parts.

referred to in notes and not directly on the main body of the drawing, which is useful when a family of parts will have the same design but different dimensions depending on the model. It would be silly to make separate blueprints for each model.

The fillet weld in a hole or slot (Fig. 21-27B) shows you how cross sections of parts are sometimes shown. The arrows pointing at A refer to cross section A-A. If you were to cut the part shown in the plan (looking down) view in the middle of Fig. 21-27B along the

section marked by the arrows, you would see that the hole does not go all the way through the part. You often will see cross sections like this on engineering drawings.

Figure 21-27C shows how to dimension both sides of a part in section or end view, and in an elevation or plan view, with the resulting weld.

Figure 21-27D is a little more complicated, but it shows four equal-legged fillet welds produced from just two symbols. If you count the number of fillet-weld triangles, of course, you will find that four fillet welds are called for and dimensioned. Finally, Fig. 21-27E puts all these ideas together and generally summarizes how all the welding symbols we'll learn about are used.

Skewed Joints
Figure 21-28 shows you how fillet welds that are not at right angles can be symbolized on a blueprint. These are called *skewed joints*. Your first attempts to make fillet welds probably resulted in a lot of unintentional skewed joints, but sometimes they are required to make a part. When they are, you have to know the angle that the two sections of the joint make with each other.

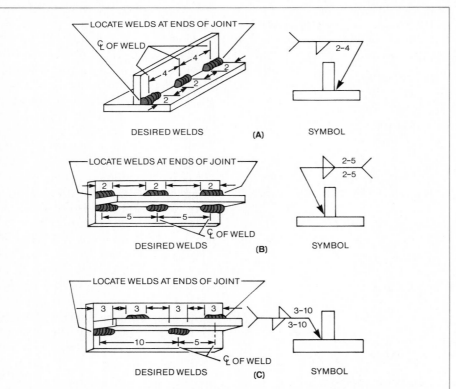

FIGURE 21-26 Intermittent, chain intermittent, and staggered intermittent fillet-weld symbols and the resulting parts are shown here.

A skewed joint has characteristics between a fillet weld and a beveled joint. When two pieces are at right angles you have a standard fillet joint. When they are lying flat on each other you also have a standard fillet joint. But when the pieces meet at some angle between vertical and horizontal, the point where they meet forms an angle that looks more like a groove weld than a fillet

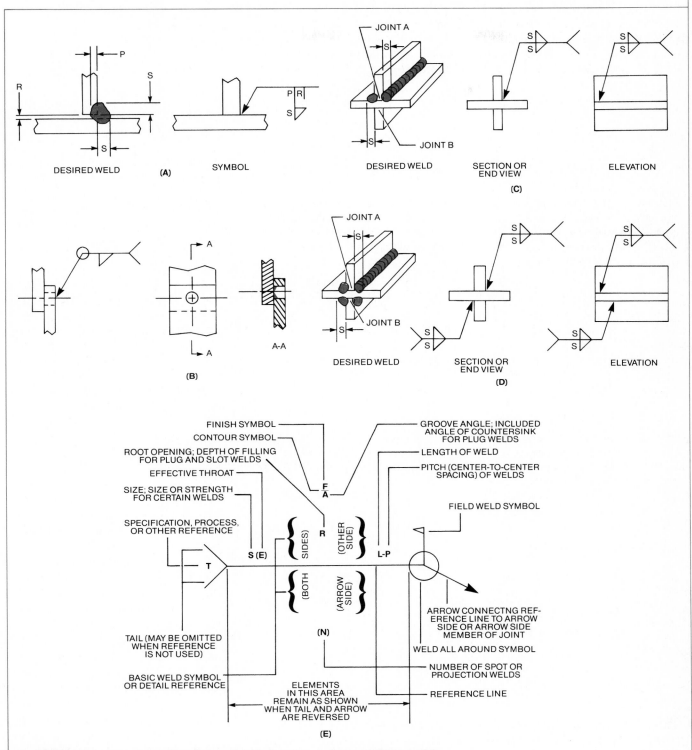

FIGURE 21-27 Drawings (A) through (D) show a variety of fillet welds for many different parts, and (E) gives you a grand summary of how welding symbols, in general, are put together. All that remains for you to learn are the specific symbols for joints other than fillet welds, and the rest of what you have just learned can be applied to these other types of welding joints.

weld. The AWS welding symbols take that into consideration.

In Fig. 21-28 (left) the vertical section makes an angle with the horizontal section that is at least 60°. It can be more right up to 90°, which is a standard fillet weld. When the angle is between 60° and 90°, the fillet-weld symbol is used, as shown in Fig. 21-28.

The vertical section in the workpiece on the right in Fig. 21-28 tilts over to make an angle between 30° and 60° with the horizontal section. The gap between the sections looks and acts more like a groove weld than a fillet weld, so that the welding symbol used is for groove welds. If the top piece were tipped even farther

over so that it made an included angle less than 30°, the other side of the joint where you would weld it looks and acts more like a fillet weld where the vertical leg is not very high. The blueprint symbol would then go back to using the fillet-weld symbol.

To help you out, the AWS includes both angles made by the top section with the bottom section on each diagram. The angle on the side of the piece is given, along with the angle on the other side of the piece, which is put in a special kind of bracket. Since the process is easier to see than to describe, look closely at Fig. 21-28 and you will see how the system works. Just keep in mind that the angle in broken brackets is the angle for the other side of the joint. In practice you are not likely to run into many skewed joints, but we want to equip you to recognize them on a blueprint if you ever have to make them.

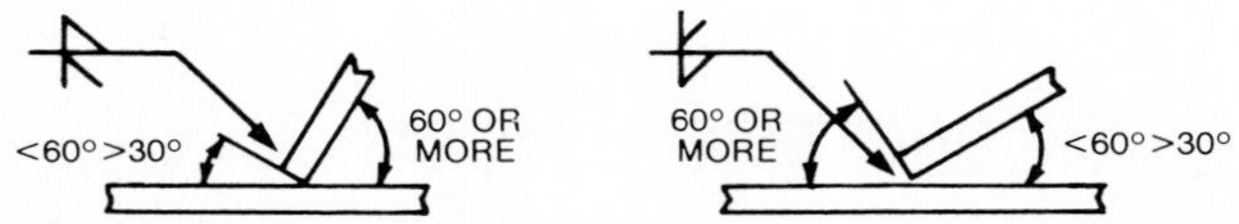

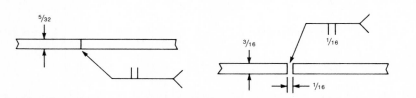

FIGURE 21-28 You need a fillet weld on one side and groove weld on the other side of this oddly angled section. (We will give you details on groove-weld symbols shortly.) This type of tilted "fillet-groove" joint is not often seen. The angle in broken brackets is the angle for the other side of the joint.

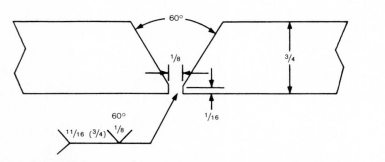

FIGURE 21-29 This is how square-butt welds are indicated by welding symbols. The butt joint is tight in the left drawing and has a ¹⁄₁₆-in. gap in the right drawing. Note how the gap width is shown in the square-butt-joint symbol.

21-4
GROOVE WELDS

As you already know, there are about 10 basic types of grooves. They are square-butt, single-bevel, V-, U-, J-, flare-V, flare-bevel, plug- and slot-, edge-flange and corner-flange joints. Except for the square-butt edge and the flange joints, the others can be single-sided or double sided (on both sides of the section). The best joint depends on how the groove will be made and how the weld will be made and stressed, as well as on the design of the joint and the size and thickness of the workpiece. All of these joints can be symbolized and dimensioned with just a few symbols and number codes. Most of what you've already learned about fillet welds also applies to groove-weld symbols, including the reference line, the arrow, the tail, and the dimensions and finishing details we've already talked about.

Figure 21-29 shows the use of welding symbols for a simple

FIGURE 21-30 This 60° included-angle V-groove butt joint has a ⅛-in. root gap, and a ¹⁄₁₆-in.-deep land. The section is ¾ in. thick. Therefore, the throat depth of the V joint is ¾ − ¹⁄₁₆ = ¹¹⁄₁₆ in. All of this information is shown on the one welding symbol. Note that the full section thickness (¾ in.) is in parentheses on the symbol. In Fig. 21-33 we will put that information to work when you see a weld symbol that does not require a full-penetration weld.

square-groove butt joint. It's two vertical lines (which look like a butt joint). In practice it's often all you need because the drawing will show the thickness of the material being joined. If a gap is required between the butted sections, the distance is shown by a small dimension placed between the vertical lines. The CP symbol you will often see in the tail of the arrow simply means "complete penetration."

The V-groove symbol (Fig. 21-30) is always drawn as a right angle no matter what the actual angle of the V groove is. The angle is shown above the V symbol. The gap (if any) between the plates in the root of the joint is shown as a fraction inside the V symbol. This distance is usually about $\frac{1}{32}$ or $\frac{1}{16}$ in. [0.8 or 1.6 mm] on all but thicker sections. We show a $\frac{1}{8}$-in. [3.2-mm] root gap in the $\frac{3}{4}$-in. [19 mm] plate drawing.

The depth of penetration is given to the left of the groove symbol and it's always in parentheses. In the case of Fig. 21-30, the required depth of penetration is $\frac{3}{4}$ in. [19 mm]. Since the plate is $\frac{3}{4}$ in. [19 mm] thick, the specification requires complete penetration. The depth of penetration also is called the *size of the weld* or the *effective throat* of the weld. Putting the weld size number in parentheses makes the symbol work just like it did for a fillet weld (or for any other welding symbol).

The same number in parentheses can be found on all other groove-weld symbols, whether it is a butt-joint or something very fancy like a J groove. If the joint were to be welded from both sides, the depth of penetration on one weld would be less than the plate thickness, and a second welding symbol would be used for welding on the other side of the joint. However, depth of penetration or effective throat-thickness numbers, added together, would equal or exceed the section thickness

(unless a gap were left in the weld joint, which is very bad practice).

The vertical dimension of the V groove is called the *depth of preparation* or bevel or chamfer depth. It is given to the left of the depth of penetration. In Fig. 21-31 the depth of penetration on the top 90° V joint is $\frac{3}{8}$ in. [9.5 mm]; the depth of preparation is less, of course. It's $\frac{1}{4}$ in. [6.3 mm]. The gap or root opening is $\frac{1}{8}$ in. [3.2 mm] wide. Figure 21-31 shows that two separate welds on either side of the double-V joint can be symbolized with one arrow.

The second V joint under the plate is a 60° angle V joint with a $\frac{1}{2}$-in. [13-mm] depth of penetration and $\frac{3}{8}$-in. [9.5-mm] depth of preparation. The dimensions on

the plate show the distance from the top and bottom surface of the $\frac{5}{8}$-in.-thick [16-mm] plate to the root of the weld. That data will help you prepare the joint.

Figure 21-32 shows three different bevel joints. One on the left is a single-bevel joint. The other two welds make up a double-bevel joint. All the bevels are 45°. The bevel symbol is a "semi-V" shape, just like the bevel joint itself. Since depth of penetration and depth of preparation of the joint are not shown in Fig. 21-31, you must assume that the designer wants complete penetration and a feathered edge on the weld because that's simply good welding practice. Just as the vertical side of the fillet-weld symbol always faces left, the vertical side of the

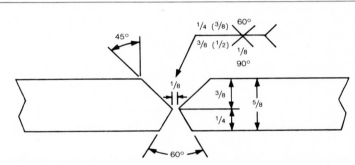

FIGURE 21-31 Here is a complex 60 to 90° double-V bevel joint without a land. The section drawing contains dimensions that are reflected by the data contained in the welding symbol. In fact, if we gave you the welding symbol without the drawing, you should be able to reconstruct the drawing that the symbol points to, and do it quickly.

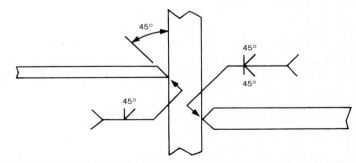

FIGURE 21-32 A single 45° bevel weld on the left, and a double 45° bevel weld on the right are symbolized in two single welding symbols. Note the use of the other side (not the arrow side) weld bevel symbol on the left. Compare it with the location of the bevel joint in the part drawing.

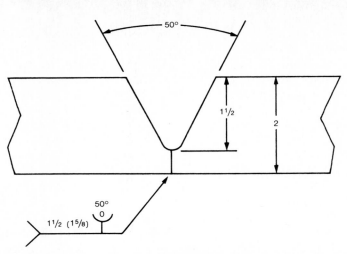

FIGURE 21-33 This U-bevel joint with a 50° included angle has a throat depth of 1½ in. The depth of penetration (in parentheses) is 1⅝ in., but the full section thickness shown in the drawing is 2 in. Therefore, the weld is not a full penetration joint. Most likely, another butt-joint welding symbol would be on the right side of the drawing in a real blueprint because a section of this thickness should have a full penetration weld. If you see this design on a real project, and the other side of the weld joint does not indicate additional welding, you should ask your welding superintendent if something is missing. Welding engineers who design parts have been known to leave out some details.

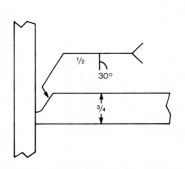

FIGURE 21-34 This J-groove weld has a 30° included angle and a ½-in. vertical throat depth. The full section thickness for the member on the right is ¾ in. Therefore, the depth of the flat land where the joint butts against the vertical section must be ¾ − ½ = ¼ in. Engineering drawings often leave off some dimensions that can easily be recovered with a little arithmetic. Since no depth of penetration is indicated on the left of the symbol for this joint, you should assume that the weld is a full-penetration weld through the remaining ¼ in. of the section.

bevel-angle symbol also always faces left, even if the vertical side of the joint is actually on the right in the real workpiece or on the engineering drawing.

The U-groove joint symbol is a U (Fig. 21-33). The U groove in Fig. 21-33 requires a 50° bevel angle, 1⅝-in. [41-mm] penetration, and is cut 1½ in. [38 mm] deep (depth of preparation) into the 2-in.-thick [51-mm] plate.

The J-groove joint in Fig. 21-34 has a 30° bevel angle and it's cut ½ in. [13 mm] deep into the ¾-in.-thick [19-mm] plate. The vertical line in the J-groove symbol is to the left, as it should be. The J-groove symbol is a sort of J-shape. The bevel angle is shown with each symbol, just as it is for all groove joints except the simple square-butt joint, which has no bevel angle (unless you count the parallel faces as an angle of 180°, which is the angle of a straight line).

Flare-V and flare-bevel groove joints (Fig. 21-35) are used when one or both sections are bent and

must be joined. The symbols, two back-to-back curved surfaces, look very much like the weld joints they describe. The double-flare-V-groove joint is used in drawing (A) where two round pieces are to be joined. Depth of preparation (in effect) is the radius of the smaller piece. Flare bevel in the right drawing (B) describes the bend angle of the plate as a radius of 1 in. [25 mm]. The depth of penetration of the weld metal and the depth of preparation of the joint (or the effective throat) are the same in these designs.

21-5
PLUG AND SLOT WELDS
Plug and slot welds are welding's answer to the rivet. The names plug and slot weld clearly describe the joint they make. A plug weld plugs up a round hole with weld metal. A slot weld does the same thing for a slot-shaped hole. There usually is a second piece of metal underneath the plug or slot hole so that the weld metal bonds the top section (the one with the hole in it) to the bottom section.

The welding symbol for a plug or slot weld is a rectangular box (the symbol looks like the cross section of the weld). Just like other welding symbols, the box will be on the arrow side (underneath the reference line) when the weld is to be made on the same side of the part where the arrow points (Fig. 21-36), and on the other side (on top of the reference line) when the weld is to be made on the opposite side of the part from where the arrow points (Fig. 21-37).

The dimensions of plug welds are shown on the same side of the reference line as the welding symbol (Fig. 21-38). Three items of information are given in these sketches. You see two numbers, one on either side of the box, a third number inside the box, and an angle on each drawing. Let's consider what the left-hand number means first.

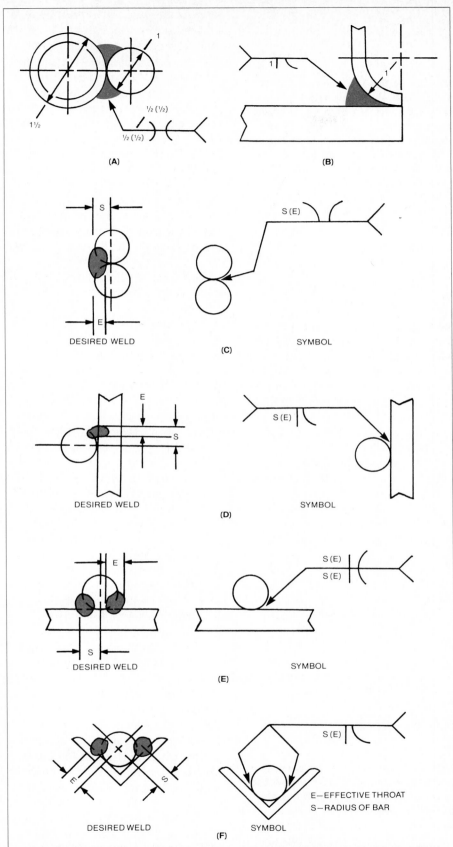

(A)

(B)

DESIRED WELD

S (E)

SYMBOL

(C)

DESIRED WELD

S (E)

SYMBOL

(D)

DESIRED WELD

S (E)
S (E)

SYMBOL

(E)

DESIRED WELD

S (E)

E—EFFECTIVE THROAT
S—RADIUS OF BAR

SYMBOL

(F)

FIGURE 21-35 Flare-V and flare-bevel groove weld symbols are shown on the right with the desired weld bead for the same part shown on the left in (C) to (F). Each welding symbol looks similar to the actual joint. The flare-V symbol has two facing curves, and the flare-bevel has one straight line and one curve.

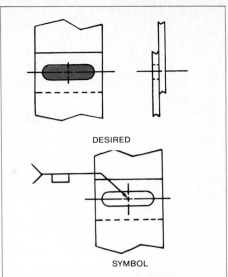

DESIRED

SYMBOL

FIGURE 21-36 Plug and slot welds are both shown by the same boxlike welding symbol. This symbol shows you that the weld is on the arrow side of the part and not on the other side.

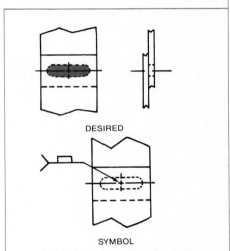

DESIRED

SYMBOL

FIGURE 21-37 This plug or slot weld is on the other side, not the arrow side of the part, but it can still indicate where the weld is required, even though the actual hole on the other side of the part in the drawing would not be visible in the actual part. The designer shows that the hole is not visible on the side of the part (and drawing) facing you by sketching the outline of the hole with a dashed line in the drawing. These dashed lines are called "hidden lines." They help you understand what parts look like by giving you x-ray vision (in effect).

FIGURE 21-38 The slot or plug-weld symbol on the left tells you that the weld is on the arrow side (even though the part drawing is missing). The left-hand symbol also tells you that the hole to be filled with weld metal is 2 in. across, that the weld metal should be filled only ½ in. deep in the hole, and that there must be several of these welds made because their center-to-center distance is 6 in. The lip or edge of the hole has a 45° chamfer angle. The symbol on the right tells you that 1-in.-diameter slot or plug welds are to be filled ⅜ in. deep with weld metal; each weld should be on 4-in. center-to-center distances, and that each hole is chamfered or countersunk to a 30° angle. The next figure will make this clear.

Plug-Weld Dimensions

The size of a plug weld (the distance across the hole) is shown by the number on the left. If the number is 2, it means that the hole is 2 in. [51 mm] across. The number on the right is the spacing between the center-to-center distances of plug welds when more than one plug weld is used. If the number is 6, it means that the plug welds are 6 in. [152 mm] apart, measured center-to-center. (Of course if metric units were being used, the numbers would

refer to millimeters or centimeters or whatever scale units are used on the blueprint.)

The number inside the box tells you how much the hole should be filled up with weld metal. The number inside the box usually (but not always) is the same as the depth of the hole.

Holes sometimes are countersunk, just like countersinking a screw hole so that the head of the screw doesn't stick up above the surface of the part being joined. Figure 21-39 shows you a counter-

sunk plug weld with a 45° bevel on the countersunk hole on top. In this figure, the welding symbol tells you that the hole (at the bottom) is 1 in. [25 mm] in diameter and opens up with a chamfer to the top of the countersunk hole; the angle of chamfer on the hole is 45°, the hole is to be filled to ⅜ in. [9.5 mm] (which is less than the ½-in. [13-mm] thickness of the top plate), and that there will be other plug welds just like it spaced on 8-in. distances center-to-center from this plug weld (the pitch). The welding symbol makes it very easy to say that—much easier than trying to explain what's required with the words in this paragraph.

The same surface-finishing symbols used on fillet welds and groove welds also are used on plug welds. We won't bother repeating them here.

Slot-Weld Dimensions

Once you understand plug welds, you understand slot welds. Precisely the same symbols and dimensioning units are used . . . with one additional, helpful detail often added. You frequently will see a note such as DET A or DWG 7 or some such notation telling you to look for a separate detail or drawing that will give you the precise dimensions of the slot. That's needed simply because a slot is harder to explain than a simple circular hole.

Figure 21-40 shows how slot welds might be detailed on a welding blueprint. Figure 21-40A shows you completely filled slot welds and Fig. 21-40B shows partially filled slot welds. Note in these drawings that the center-to-center pitch of the slots is given on the drawing instead of on the welding symbol. Either system can be used. The best way to do it depends entirely on which method makes the job easier to understand for the person who has to do the work.

The same surface-finishing symbols are used on slot welds as on plug, groove, and fillet welds.

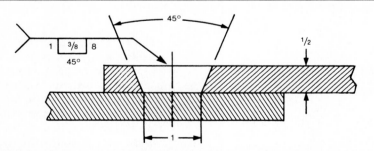

FIGURE 21-39 This part drawing will make plug and slot weld symbols easier to understand. The hole is 1 in. across and you are to fill it ⅜ in. deep with weld metal. There are several of these holes to make, spaced on 8-in. centers. Each hole has a 45° chamfer. Since the section thickness is ½ in., and the required weld metal depth in the hole is only ⅜ in., the weld metal will be ½ − ⅜ = ⅛ in. from the top of the hole when you finish welding. It's obviously much easier to measure down into the hole to your weld metal than to try to guess if the weld in a hole is deep enough. Therefore, instead of trying to guess whether you have made a weld ⅜ in. deep, fill the hole to within ⅛ in. of the top and you get the same result.

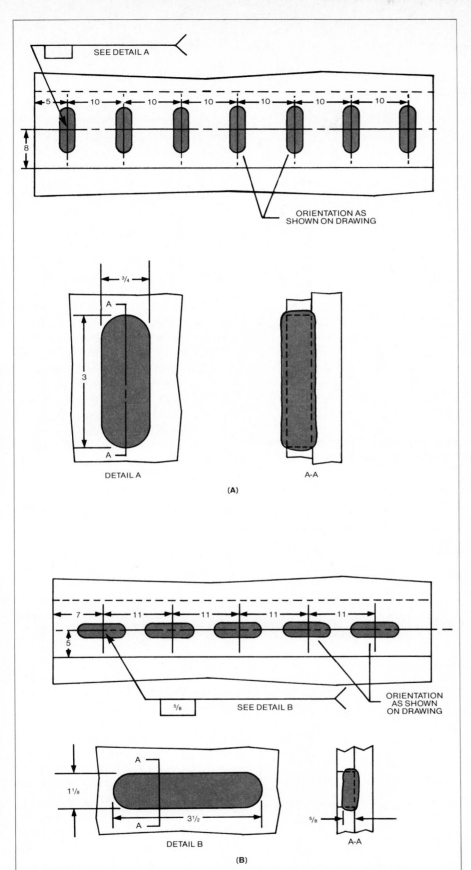

FIGURE 21-40 Here are a variety of plug and slot welds with their welding symbols. The designer has asked you to see a special detail drawing in (A) and (B) for more information. The symbol for a cross section of a part also is shown and the cross section is coded as A-A.

Whenever a drawing does not include surface-finish requirements, it probably means that you are to use your own judgment or follow some standard shop practice for finishing welds. It also could mean that the engineer making the blueprint didn't know that much about welding, or forgot to tell you how the weld was to be finished. Always ask what's required if any welding symbol is not perfectly clear or leaves out some details that you consider important.

21-6
FLANGE WELDS
When light-gauge-metal joints are made by joining flanges that have been bent before welding to make corner and edge welds, another set of welding symbols is used. These edge-flange and corner-flange symbols look like the work to be done. They follow the same basic scheme used for other welding symbols. These are common welding symbols used on sheet-metal work.

The edge-flange symbol is shown in Fig. 21-41. Fig. 21-41A shows the arrow-side use of the edge-flange symbol, while Fig. 21-41B shows how the other-side edge-flange symbol is used.

The corner-flange weld symbol is shown in Fig. 21-42. Fig. 21-42A shows the arrow-side use of the corner-flange symbol, while Fig. 21-42B shows how the other-side edge-flange symbol works.

Dimensioning Flange Welds
Flange-weld symbols will show the radius and the height above the point where the curved sheet or sheets first contact each other (their point of tangency). The height of the weld will be on the right and the radius of the flange curve will be on the left. They are separated by a + sign. The penetration depth of the weld is always shown above or below the other two numbers. Figure 21-43 will make it clear to you how the dimensioning system works.

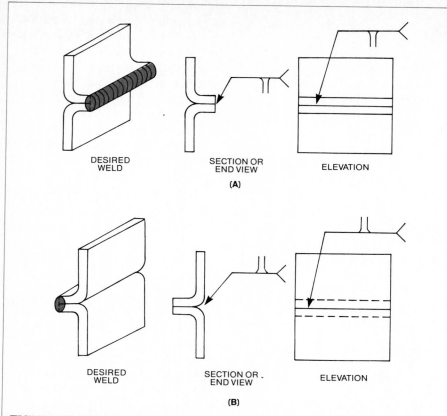

DESIRED WELD

SECTION OR END VIEW

(A)

ELEVATION

DESIRED WELD

SECTION OR END VIEW

(B)

ELEVATION

FIGURE 21-41 Flange welds are easy to understand. Here are several undimensioned flange weld symbols on drawings, with the desired welds on the left.

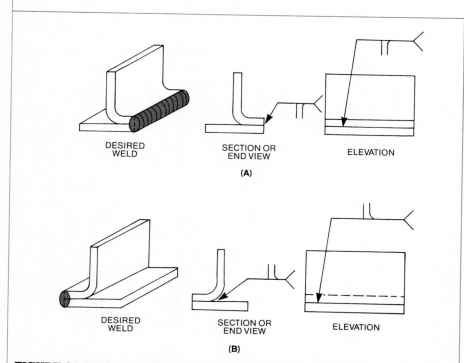

DESIRED WELD

SECTION OR END VIEW

(A)

ELEVATION

DESIRED WELD

SECTION OR END VIEW

(B)

ELEVATION

FIGURE 21-42 Corner-flange weld symbols are quite similar to conventional flange weld symbols, except that one side of the joint is flat and the other side is curved. So is the symbol.

21-7
MELT-THROUGH AND WELD-BACKING SYMBOLS

When welds are made from one side only and complete penetration is required with weld-metal reinforcement on the opposite side, a special melt-through symbol is used. A similar symbol is used when the weld joint has a backup strip. We'll describe the melt-through symbol first.

The melt-through symbol is used in combination with square-butt, single-V-bevel, double-V-bevel, J-groove, and U-groove joints. It looks like the cap of a mushroom that is filled in solid with colored (or black) ink. Figure 21-44 shows what this symbol looks like when used with several different types of weld joints, and with directions for finishing the contour of the melt-through weld.

The melt-through symbol is always placed on the opposite side of the weld-joint symbol. No dimensions other than height are used on the melt-through symbol, and the height of the weld reinforcement often is not included.

As with the other welding symbols, melt-through that is to be made flush by mechanical means such as grinding or machining will have a letter code combined with the finishing symbol. If the melt-through weld is to be contoured, it will have the contoured arc-finishing symbol combined with it. All of these possibilities also are shown in Fig. 21-44.

When a backing strip is used (often the case in flux-cored-wire welding), a very similar mushroom-cap symbol is used except that the cap is open instead of solidly filled with ink. If there is a gap between the backing strip and the weld (to be used for weld-metal reinforcement on the opposite side of the joint from the welding), the distance of the gap (the thickness of weld-metal reinforcement) between the backup strip and the joint is shown to the left of the backup symbol. It also may be indicated separately on

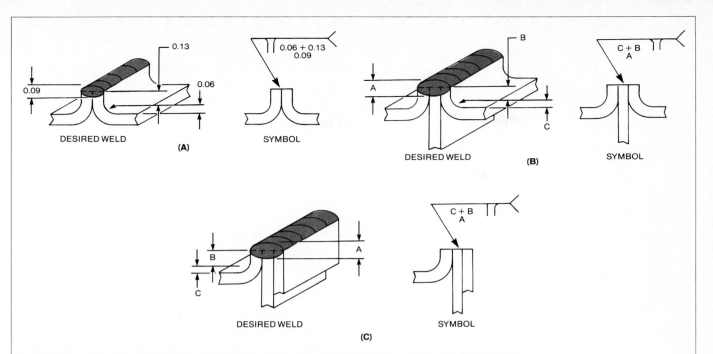

FIGURE 21-43 Here are several flange weld and corner-flange weld drawings. The desired weld joint is shown to the left for each drawing on the right. The height of the center of the weld above the tangent point where the two sections come together is on the right (B) and the radius of the flange curve on the left (C), separated by a plus sign. The figure underneath (C + B) is the thickness of the weld bead (A) and is always shown either above or below the other two dimensions. In the top left drawing, the weld is 0.09 in. thick and the radius of curvature of the flange is 0.06 in.

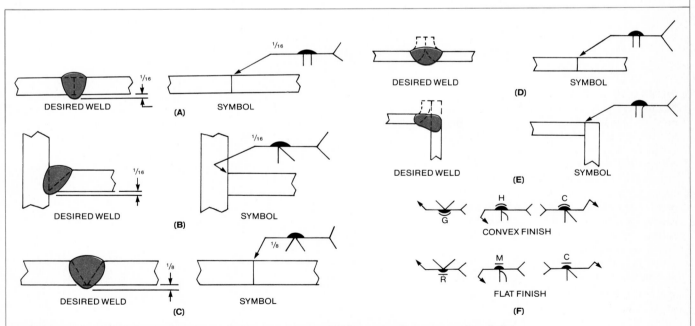

FIGURE 21-44 Melt-through symbols for a variety of joints are shown on the right with the resulting welds shown on the left in drawings A through E. Finishing symbols for full-penetration melt-through welds are shown in drawing F. The solid melt-through symbol is always placed on the opposite side of the reference line from the weld symbol to which it applies. The melt-through symbol is always solid (filled in) so don't confuse it with the next symbol shown in Fig. 21-45.

the engineering drawing. See Fig. 21-45.

21-8
SEAM, SPOT, AND PROJECTION WELDS

Electrical-resistance spot welding is used to join sheets of steel, aluminum, or other metals. Most people think of electrical-resistance welding when they think of spot welds. Resistance welding (see Appendix) is not normally a job for skilled welders, even though it is widely used in industry.

The skill in doing spot, seam, or projection welding comes in engineering the system and determining the operating parameters for producing thousands of spot-welded joints before the water-cooled copper resistance-welding electrodes have to be cleaned up. The machine operator seldom has to know much more about the process than how to stick the work into the machine. The oper-

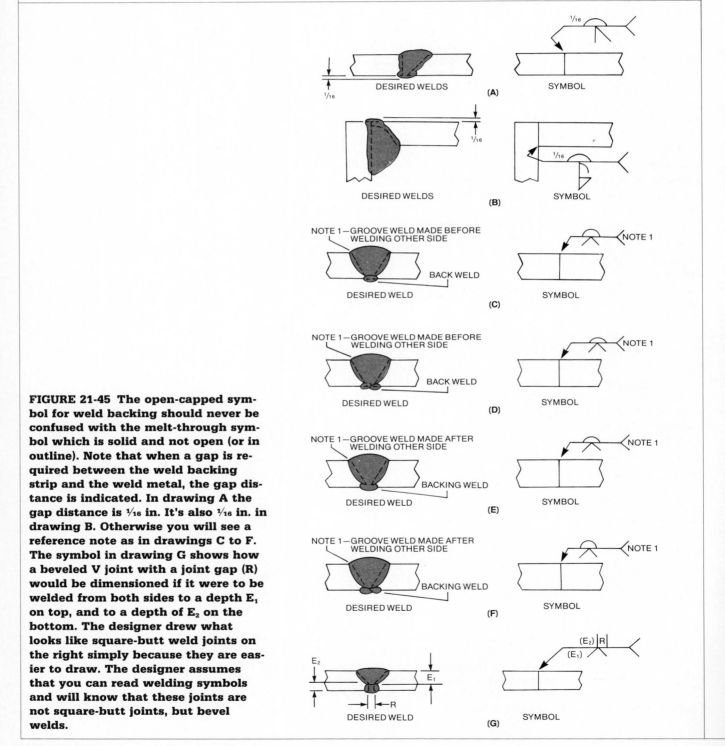

FIGURE 21-45 The open-capped symbol for weld backing should never be confused with the melt-through symbol which is solid and not open (or in outline). Note that when a gap is required between the weld backing strip and the weld metal, the gap distance is indicated. In drawing A the gap distance is $\frac{1}{16}$ in. It's also $\frac{1}{16}$ in. in drawing B. Otherwise you will see a reference note as in drawings C to F. The symbol in drawing G shows how a beveled V joint with a joint gap (R) would be dimensioned if it were to be welded from both sides to a depth E_1 on top, and to a depth of E_2 on the bottom. The designer drew what looks like square-butt weld joints on the right simply because they are easier to draw. The designer assumes that you can read welding symbols and will know that these joints are not square-butt joints, but bevel welds.

ator usually does very little more than punch a button and maybe clean the copper electrodes from time to time. Look at the sheet metal on a steel office desk or the handles on pots and pans and you will see small indentations here and there where spot welds have been made.

A lot of the electrical-resistance spot and seam welding (a continuous spot weld, in effect) is done by fully automated equipment in the automotive, appliance, and furniture industries. But spot welding doesn't have to be done with electrical resistance. Spot welds also can be made with fusion-welding processes. When spot welds are not made with electrical-resistance welding, they most often are made with GTAW and sometimes with GMAW. Electron-beam welds (a process which is very specialized and again requires very little operator skill) also can produce spot or seam welds. There is no great mystery to making most spot or seam welds with a fusion-welding process.

Just think of a fusion spot weld as a plug weld without a hole. That is, you can join two thin sheets of metal simply by welding all the way through one of them and into the other one. In other words, you make a complete penetration weld through two sheets instead of one. Obviously, spot welding is limited to sheets. You can't get complete penetration through two pieces of overlapping thick plate by spot welding.

Projection Welds
If plug welds are the welding equivalent of rivets, projection welding is the welding equivalent of the automatic nail guns carpenters often use. One major form of projection welding uses a small resistance-welding gun that welds studs, screws, bolts, and other parts directly onto some object. The weld is made by the heat produced between the parts by the electrical resistance at the joint (the I^2R heating you learned about when we studied electricity

is at work again). The main skill (other than being able to pull the trigger of the gun) is simply putting the fastener on straight instead of at an angle.

The projection-welding symbol will have an arrow-side or an other-side position for a circle. Spot and seam-welding symbols usually don't make this distinction. Spot, seam, and projection welds have very simple symbols on welding blueprints (Fig. 21-46). The symbol for a spot weld is simply a circle that can be on the arrow side, the other side, or placed right on the reference line. When the circle is on the reference line, it is *centered* on the line and not at the point where the line is broken to form the arrow.

Don't confuse the spot-welding symbol with the weld-all-around symbol. (That's not likely to happen; you certainly will know what kind of equipment you are using.) Sometimes you will see a number in parentheses above or below the spot-weld circle, such as (6) or (15). This is the number of individual spot welds to be made.

How do you tell resistance spot welding from GTAW spot welding, GMAW spot welding, or projection welding? If there is any doubt, the process letter symbol will be placed inside the tail of the arrow, just like it is (when required) for other welding processes.

Figure 21-47 shows you how weld parameters are handled. Spot welds can be dimensioned either by size or by shear strength (the force parallel to the metal required to pull the two lap-welded sheets apart). The size of a spot weld (its diameter) is shown to the left of the circle. The strength of the spot weld, if that is used, also will be on the left of the circle. For example, a ⅛-in.-diameter spot weld will have a ⅛ on the left of the circle. A spot weld that produces a minimum shear strength of 600 psi [4.1 MPa] per spot will have a 600 to the left of the circle. If you see a number to the right of the circle it refers to

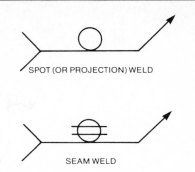

FIGURE 21-46 Spot and projection welds are indicated by simple circles that can be on top of, overlapping, or on the bottom of the reference line. Since spot and projection welds are inherently one-side welds, there is no important distinction to be made between arrow-side and other-side symbols. Seam welds work the same way, except that the symbol is a circle with tracks running through it to indicate a continuous weld instead of a spot weld.

the pitch (center-to-center spacing) of the spot welds. This is the same practice used for all other welding symbols.

The surface-finish and weld-contour symbols used for other processes also apply to spot welds. A straight line over or under the spot-weld circle means "finish the weld flush with the exposed surface of the workpiece." How you are supposed to do that usually is covered in a separate note on the drawing.

Figure 21-48 shows several examples of the use of the spot welding symbol for GTAW, electron-beam, and resistance spot welding. The S will be a dimension for the weld diameter or strength.

Seam Welds
A seam weld is the same thing as a spot weld except that it is continuous. You keep the complete-penetration weld moving from the beginning of the seam to the end of it. If electrical-resistance welding is used, the copper electrodes are round wheels that roll along

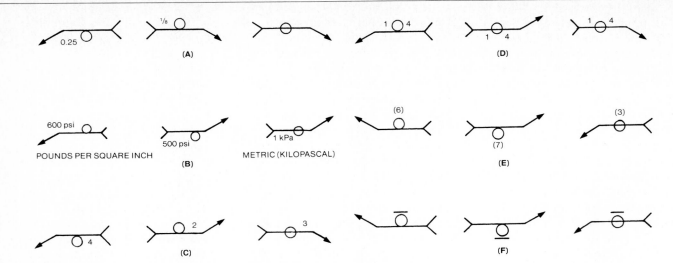

FIGURE 21-47 (A) The diameter of the spot weld nugget is 0.25 in. or ⅛ in. in these examples. Note the right side drawing has a spot-weld symbol that is centered on the reference line. There's no special meaning attached to that fact. (B) Spot welds are often described by their minimum tear-part or peel-apart (shear) strength, which is 600 psi in one example, 500 psi in another, and 1 Pascal of force in another (don't worry about what that means). (C) These drawings tell you that the welds are on 4-, 2-, or 3-in. centers. (D) The center-to-center spacing of the spot weld is 4 in. The weld nugget of each spot weld is to be 1 in. in diameter. (E) The numbers in parentheses simply tell you how many spot welds are needed. (F) These welds are to be finished flat to the base-metal surface.

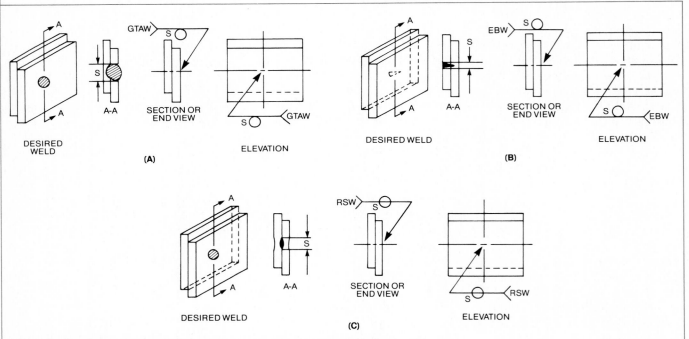

FIGURE 21-48 Spot and seam welds can be produced by several different processes. The welding process to be used (if you don't already know) is tucked into the feathered end of the arrow. GTAW, as you know, means TIG welding while EBW means electron-beam welding. RSW means resistance spot welding. The S in our examples is a dimension. On a real drawing it should give you the diameter of the spot-weld nugget or the shear strength of the weld with a number.

the seam making a continuous weld instead of a spot. Again, seam welding is a highly automated process. Special seam-welding machines using either electrical-resistance methods or GTAW are sold just for this kind of work. One example of the use of seam welding is joining the overlapping edges of galvanized-steel culvert pipe to form a watertight joint.

The seam-weld symbol (Fig. 21-49) shows its close relationship to spot welding. The symbol is a spot-welding circle centered on the reference line, except that the circle has two "railroad tracks" running through it. Just like spot welding, the seam-welding circle with two horizontal lines on it can be on the bottom arrow side of the reference line, the top other side of the reference line, or right in the middle of the reference line. The reason (just like spot welding) that the symbol can be right on the reference line is that seam welding is a complete-penetration process. Except for getting the

equipment on the seam, it doesn't matter too much on which side of the seam the weld is made. You usually won't have a choice because of the shape of the part.

The process used to make the weld is identified by a symbol tucked inside the feathers of the arrow. If electrical-resistance welding will be used, the symbol inside the arrow feathers will be RSW (for resistance-spot welding). If GTAW is to be used, the process symbol in the tail feathers of the arrow will be GTAW, and so on. The shear strength (or the width of the weld seam) is given on the left of the symbol, just like spot welding. The pitch or center-to-center distance between seams is on the right as shown in Fig. 21-50.

21-9 SURFACING SYMBOLS

The surfacing or hardfacing welding symbol resembles the two weld beads on the surface of the

part. Since surfacing can only be done on one surface (it's not a full-penetration process and doesn't make a weld joint), there is no arrow side or other side to the symbol's reference line. The arrow simply points to the surface on which the weld will be deposited.

The dimensions used with hardfacing and other surfacing processes are always shown on the same side of the reference line as the two weld beads of the surfacing symbol. This number is always placed to the left of the symbol. The "size" of a surface built up by welding is the minimum thickness of the welded deposit. When only a portion of the surface area of a part is to be built up by welding, the extent, location, and orientation of the area and the stringer beads or passes to be used are indicated on the blueprint (Fig. 21-51). If there are any special requirements for the pattern of weld metal to be laid down, or the minimum hardness

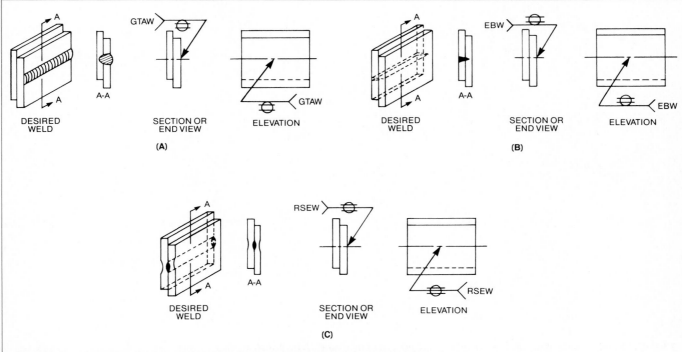

FIGURE 21-49 This is how different seam-welding processes would be indicated for GTAW (TIG), electron beam (EBW), and electrical resistance seam welding (RSEW). Electrical resistance welding is not a skilled welding process so we have described it in the appendix of this book along with electron-beam welding. Both processes are primarily automated machine-welding methods.

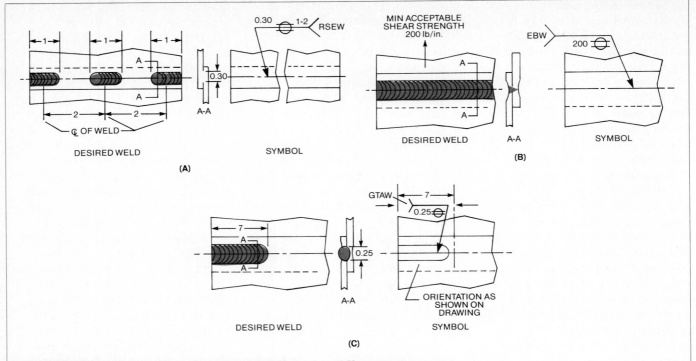

FIGURE 21-50 Seam welds using several different welding processes are dimensioned in the drawings on the right and shown as finished welds on the left.

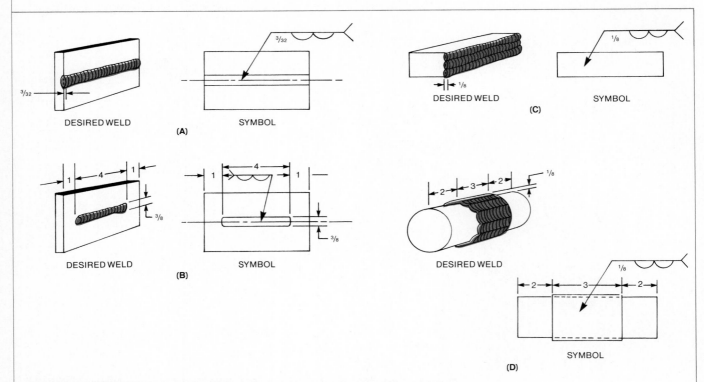

FIGURE 21-51 Surfacing or hardfacing are shown by a welding symbol that represents two side-by-side weld beads, since parallel stringer beads are commonly used to surface parts. The fraction on the left under the surface welding symbol is the minimum thickness of the built-up surface. Other details, if any, usually are given directly on the part drawing. The desired weld overlays are shown on the left and their drawings are on the right.

of the weld deposit is specified, they are covered in a separate note on the drawing.

21-10
BRAZING SYMBOLS

The symbols used for brazed joints are very simple and closely resemble those used for welding processes. The process symbol most frequently used is TB, which means "torch braze." Another common symbol is FB, which means "furnace braze." If no special joint preparation other than very careful cleaning is required, only the arrow is used with the brazing process indicated. Figure 21-52 shows you how joint clearances are indicated, how the length of the overlapping parts is shown on a drawing, the fillet size of the brazed joint, and how several different brazing joints would be indicated on a drawing.

21-11
NDT SYMBOLS

The standard symbols for nondestructive testing are very simple (Fig. 21-53). They follow the same basic method that is used for welding symbols. They have an arrow with a reference line which has an arrow side and other side. If you have to test all around the part, the same circle on the point where the reference line and arrow join that means weld all around also means test all around. The arrow symbol will include code letters for the NDT tests required.

Sometimes a number in parentheses is used. It will always be a whole number and not a fraction. It is the number of tests you have to conduct. Another number to the right of it will tell you the length of the section to be tested. The arrow, as always, points to the location on the part to be tested.

The symbols in Table 21-1 are used with the arrow to tell you what type of nondestructive test to make. Note that you are not likely to ever confuse them with any welding process symbols. All the test process symbols end in T.

21-12
SUMMING UP

Figure 21-54 lists all of the information we gave you in this chapter and also includes a few symbols for a few welding processes such as flash-butt welding (a special friction-welding process for joining large shafts) that we only cover in the Appendix of this book. Like the other processes we include in the Appendix, it's a machine operation with little operator skill required. Figure 21-54 is a good reference for when you work with welding blueprints.

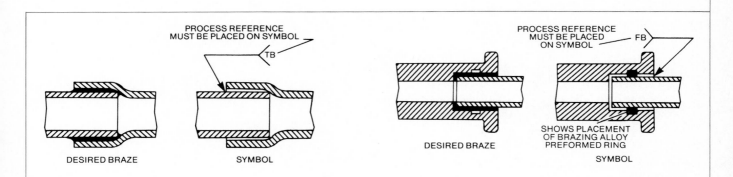

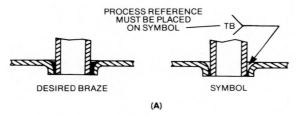

FIGURE 21-52 (A) Brazing symbols often include the process to be used, as in these drawings where TB stands for torch brazing and FB for furnace brazing.

Continued on page 572

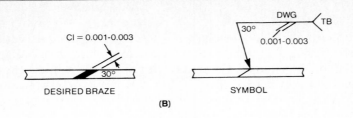

CI = 0.001–0.003

30°

DESIRED BRAZE

DWG

30°

0.001–0.003

TB

SYMBOL

(B)

TABLE 21-1 Symbols used in NDT arrow tails

Test	Symbol
Acoustic emission	AET
Eddy current	ET
Leak	LT
Magnetic particle	MT
Neutron radiographic	NRT
Penetrant	PT
Dye penetrant	DPT
Fluorescent penetrant	FPT
Proof test	PRT
Radiographic	RT
Ultrasonic	UT
Visual	VT

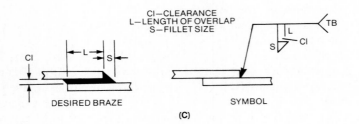

CI—CLEARANCE
L—LENGTH OF OVERLAP
S—FILLET SIZE

CI

L

S

DESIRED BRAZE

S

L

CI

TB

SYMBOL

(C)

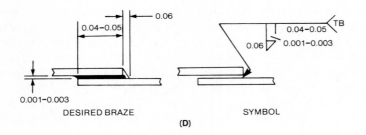

0.06

0.04–0.05

0.001–0.003

DESIRED BRAZE

0.04–0.05

0.06

0.001–0.003

TB

SYMBOL

(D)

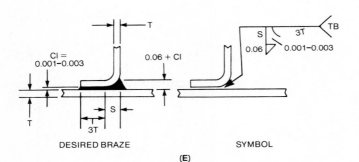

T

CI = 0.001–0.003

0.06 + CI

S

3T

T

DESIRED BRAZE

S

0.06

3T

0.001–0.003

TB

SYMBOL

(E)

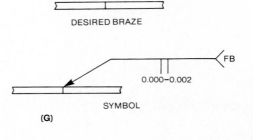

DESIRED BRAZE

0.000–0.002

FB

SYMBOL

(G)

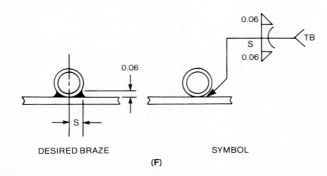

0.06

S

DESIRED BRAZE

0.06

S

0.06

TB

SYMBOL

(F)

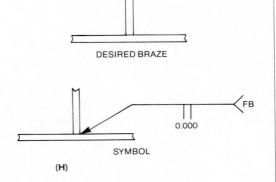

DESIRED BRAZE

0.000

FB

SYMBOL

(H)

FIGURE 21-52 (*continued*) (B) Brazing symbols may include the part clearance. This drawing also shows how a brazed scarf joint would be dimensioned. (C) These drawings show how clearance, length of overlap, and fillet size are indicated on a brazed lap joint with a fillet. The drawings in D through H give you specific examples of typical brazed joints with actual dimensions and the resulting part.

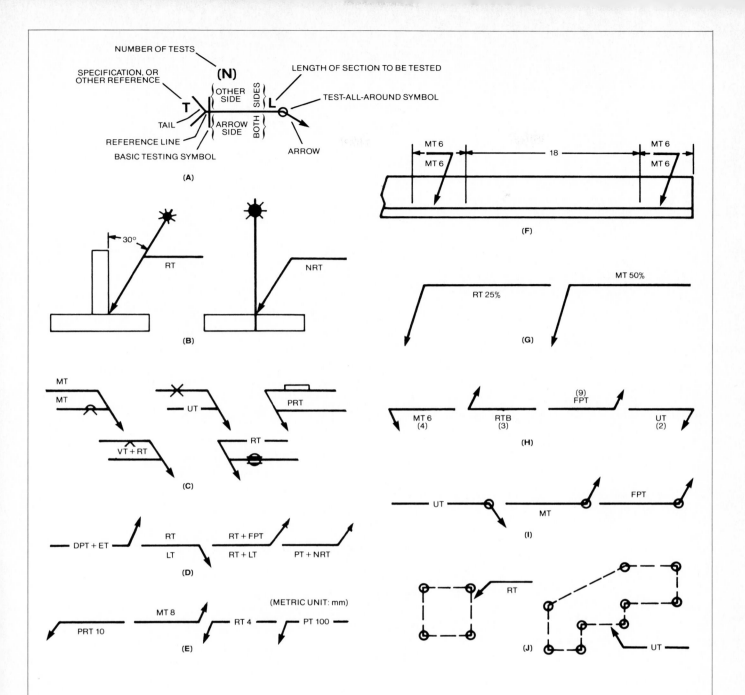

FIGURE 21-53 Nondestructive testing symbols are shown here (also refer to Table 21-1 for process symbols). (A) The general location of data for all NDT symbols. (B) X-radiographic testing (RT) and neutron-radiographic testing (NRT), including the angle of the beam to the workpiece. (C) These symbols refer to tests to be performed on specific types of welds (all of which you should recognize by now). MT means magnetic-particle inspection. UT means ultrasonic testing. PRT calls for a proof test of some kind. VT + RT means visual inspection plus radiographic testing. (D) The arrows simply point to different weld areas and tell you what to do. DPT means dye penetrant testing. LT is leak testing. FPT means fluorescent penetrant testing, and PT is penetrant testing, in general. (E) The numbers tell you how much of the weld length should be tested, such as 10, 8, and 4 in., and 100 mm in our examples. (F) How an NDT test would be shown on a part diagram. (G) The percentage of the welds on a part that should be tested. (H) The number of tests to conduct, such as (4), (3), (9), or (2) in our example. Note the number of tests to be conducted is always in parentheses, just as the number of welds to be made is in parentheses in symbols for spot welding. (I) The test-all-around symbol is just like the weld-all-around symbol. (J) Testing all over a given area of a part (enclosed in the dashed lines) is shown in this drawing.

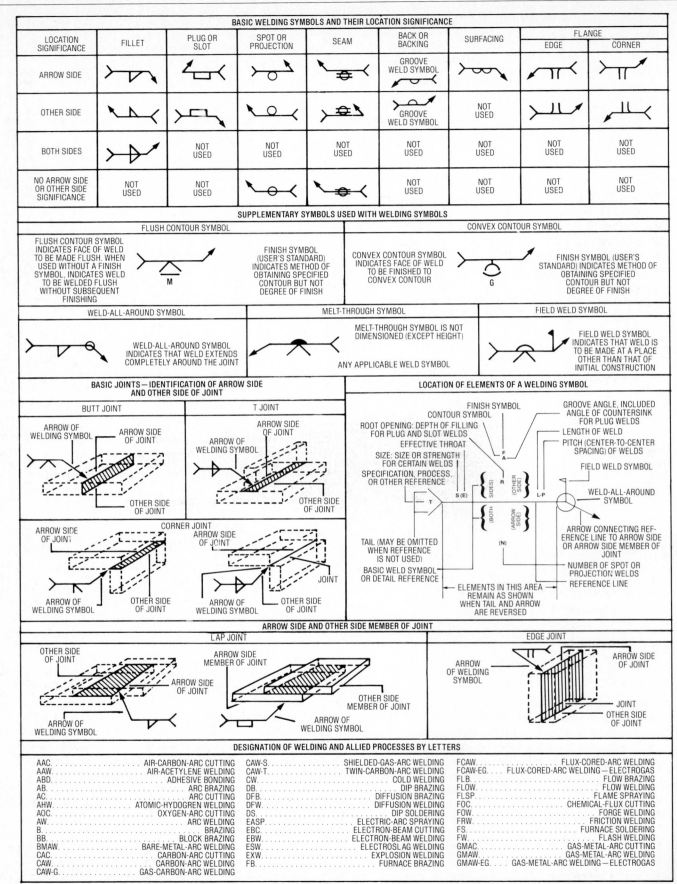

FIGURE 21-54 These big charts sum up everything you have learned in this chapter except the symbols for nondestructive testing, which are covered in Fig. 21-53 and Table 21-1.

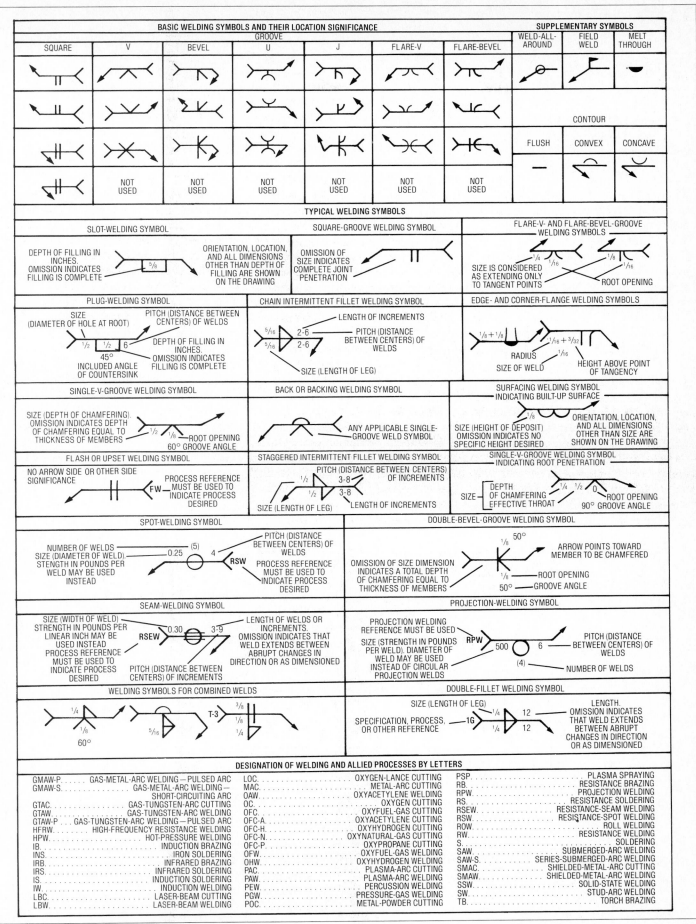

BASIC WELDING SYMBOLS AND THEIR LOCATION SIGNIFICANCE

GROOVE

SQUARE	V	BEVEL	U	J	FLARE-V	FLARE-BEVEL

SUPPLEMENTARY SYMBOLS

WELD-ALL-AROUND	FIELD WELD	MELT THROUGH

CONTOUR

FLUSH	CONVEX	CONCAVE

	V	BEVEL	U	J	FLARE-V	FLARE-BEVEL
	NOT USED	NOT USED	NOT USED	NOT USED	NOT USED	NOT USED

TYPICAL WELDING SYMBOLS

SLOT-WELDING SYMBOL
DEPTH OF FILLING IN INCHES. OMISSION INDICATES FILLING IS COMPLETE — 5/8 — ORIENTATION, LOCATION, AND ALL DIMENSIONS OTHER THAN DEPTH OF FILLING ARE SHOWN ON THE DRAWING

SQUARE-GROOVE WELDING SYMBOL
OMISSION OF SIZE INDICATES COMPLETE JOINT PENETRATION

FLARE-V- AND FLARE-BEVEL-GROOVE WELDING SYMBOLS
1/4 1/16 1/8 1/16 SIZE IS CONSIDERED AS EXTENDING ONLY TO TANGENT POINTS ROOT OPENING

PLUG-WELDING SYMBOL
SIZE (DIAMETER OF HOLE AT ROOT) 1/2 1/2 6 PITCH (DISTANCE BETWEEN CENTERS) OF WELDS DEPTH OF FILLING IN INCHES. OMISSION INDICATES FILLING IS COMPLETE 45° INCLUDED ANGLE OF COUNTERSINK

CHAIN INTERMITTENT FILLET WELDING SYMBOL
5/16 2-6 5/16 2-6 LENGTH OF INCREMENTS PITCH (DISTANCE BETWEEN CENTERS) OF WELDS SIZE (LENGTH OF LEG)

EDGE- AND CORNER-FLANGE WELDING SYMBOLS
1/8 + 1/8 1/16 + 3/32 RADIUS 1/16 SIZE OF WELD HEIGHT ABOVE POINT OF TANGENCY

SINGLE-V-GROOVE WELDING SYMBOL
SIZE (DEPTH OF CHAMFERING). OMISSION INDICATES DEPTH OF CHAMFERING EQUAL TO THICKNESS OF MEMBERS 1/2 1/8 ROOT OPENING 60° GROOVE ANGLE

BACK OR BACKING WELDING SYMBOL
ANY APPLICABLE SINGLE-GROOVE WELD SYMBOL

SURFACING WELDING SYMBOL INDICATING BUILT-UP SURFACE
1/8 SIZE (HEIGHT OF DEPOSIT) OMISSION INDICATES NO SPECIFIC HEIGHT DESIRED ORIENTATION, LOCATION, AND ALL DIMENSIONS OTHER THAN SIZE ARE SHOWN ON THE DRAWING

FLASH OR UPSET WELDING SYMBOL
NO ARROW SIDE OR OTHER SIDE SIGNIFICANCE FW PROCESS REFERENCE MUST BE USED TO INDICATE PROCESS DESIRED

STAGGERED INTERMITTENT FILLET WELDING SYMBOL
PITCH (DISTANCE BETWEEN CENTERS) OF INCREMENTS 1/2 3-8 1/2 3-8 SIZE (LENGTH OF LEG) LENGTH OF INCREMENTS

SINGLE-V-GROOVE WELDING SYMBOL INDICATING ROOT PENETRATION
1/4 1/2 0 SIZE — DEPTH OF CHAMFERING EFFECTIVE THROAT ROOT OPENING 90° GROOVE ANGLE

SPOT-WELDING SYMBOL
NUMBER OF WELDS (5) SIZE (DIAMETER OF WELD). STENGTH IN POUNDS PER WELD MAY BE USED INSTEAD 0.25 4 RSW PITCH (DISTANCE BETWEEN CENTERS) OF WELDS PROCESS REFERENCE MUST BE USED TO INDICATE PROCESS DESIRED

DOUBLE-BEVEL-GROOVE WELDING SYMBOL
1/8 50° ARROW POINTS TOWARD MEMBER TO BE CHAMFERED OMISSION OF SIZE DIMENSION INDICATES A TOTAL DEPTH OF CHAMFERING EQUAL TO THICKNESS OF MEMBERS 1/8 ROOT OPENING 50° GROOVE ANGLE

SEAM-WELDING SYMBOL
SIZE (WIDTH OF WELD) STRENGTH IN POUNDS PER LINEAR INCH MAY BE USED INSTEAD PROCESS REFERENCE MUST BE USED TO INDICATE PROCESS DESIRED RSEW 0.30 3-9 LENGTH OF WELDS OR INCREMENTS. OMISSION INDICATES THAT WELD EXTENDS BETWEEN ABRUPT CHANGES IN DIRECTION OR AS DIMENSIONED PITCH (DISTANCE BETWEEN CENTERS) OF INCREMENTS

PROJECTION-WELDING SYMBOL
PROJECTION WELDING REFERENCE MUST BE USED SIZE (STRENGTH IN POUNDS PER WELD). DIAMETER OF WELD MAY BE USED INSTEAD OF CIRCULAR PROJECTION WELDS RPW 500 6 (4) PITCH (DISTANCE BETWEEN CENTERS) OF WELDS NUMBER OF WELDS

WELDING SYMBOLS FOR COMBINED WELDS
1/4 1/8 60° 5/16 T-3 3/8 1/8 1/4

DOUBLE-FILLET WELDING SYMBOL
SIZE (LENGTH OF LEG) 1/4 12 1/4 12 LENGTH. OMISSION INDICATES THAT WELD EXTENDS BETWEEN ABRUPT CHANGES IN DIRECTION OR AS DIMENSIONED SPECIFICATION, PROCESS, OR OTHER REFERENCE 1G

DESIGNATION OF WELDING AND ALLIED PROCESSES BY LETTERS

GMAW-P. . . . . . . GAS-METAL-ARC WELDING — PULSED ARC	LOC. . . . . OXYGEN-LANCE CUTTING	PSP. . . . . PLASMA SPRAYING
GMAW-S. . . . . GAS-METAL-ARC WELDING — SHORT-CIRCUITING ARC	MAC. . . . . METAL-ARC CUTTING	RB. . . . . RESISTANCE BRAZING
GTAC. . . . . GAS-TUNGSTEN-ARC CUTTING	OAW. . . . . OXYACETYLENE WELDING	RPW. PROJECTION WELDING
GTAW. . . . . GAS-TUNGSTEN-ARC WELDING	OC. . . . . OXYGEN CUTTING	RS. . . . . RESISTANCE SOLDERING
GTAW-P . . . GAS-TUNGSTEN-ARC WELDING — PULSED ARC	OFC. . . . . OXYFUEL-GAS CUTTING	RSEW. . . . . RESISTANCE-SEAM WELDING
HFRW. . . . . . . HIGH-FREQUENCY RESISTANCE WELDING	OFC-A. . . . . OXYACETYLENE CUTTING	RSW. . . . . RESISTANCE-SPOT WELDING
HPW. . . . . HOT-PRESSURE WELDING	OFC-H. . . . . OXYHYDROGEN CUTTING	ROW. . . . . ROLL WELDING
IB. . . . . INDUCTION BRAZING	OFC-N. . . . . OXYNATURAL-GAS CUTTING	RW. RESISTANCE WELDING
INS. . . . . IRON SOLDERING	OFC-P. OXYPROPANE CUTTING	S. . . . . SOLDERING
IRB. . . . . INFRARED BRAZING	OFW. . . . . OXYFUEL-GAS WELDING	SAW. . . . . SUBMERGED-ARC WELDING
IRS. . . . . INFRARED SOLDERING	OHW. . . . . OXYHYDROGEN WELDING	SAW-S. . . . . SERIES-SUBMERGED-ARC WELDING
IS. . . . . INDUCTION SOLDERING	PAC. PLASMA-ARC CUTTING	SMAC. . . . . SHIELDED-METAL-ARC CUTTING
IW. . . . . INDUCTION WELDING	PAW. PLASMA-ARC WELDING	SMAW. . . . . SHIELDED-METAL-ARC WELDING
LBC. . . . . LASER-BEAM CUTTING	PEW. PERCUSSION WELDING	SSW. . . . . SOLID-STATE WELDING
LBW. . . . . LASER-BEAM WELDING	PGW. PRESSURE-GAS WELDING	SW. . . . . STUD-ARC WELDING
	POC. . . . . METAL-POWDER CUTTING	TB. . . . . TORCH BRAZING

REVIEW QUESTIONS

1. You are going to leave a note for another welder to do some work. In the note you will leave are welding symbols. Sketch the following welding symbols for your note, using an arrow with a reference line and a feathered tail. You do not have to include dimensions, process notes, or finishing details, nor indicate arrow-side and other-side information when answering this first question. All we want are the basic welding symbols with whatever additional detail you need on them to complete your weld symbol so that the instructions are perfectly clear to the other welder.

(a) Weld all around this joint after the part is in the field

(b) Make a single-fillet weld

(c) Make a double-fillet weld

(d) Make a single-groove butt weld

(e) Make a single V-groove butt weld with a 60° included angle

(f) Make a double V-groove butt weld with a 60° included angle on one side and a 45° included angle on the other side

(g) Make a single-bevel weld with 45° bevel angle

(h) Make a double-bevel weld (no angle specified)

(i) Make a U-groove weld joint

(j) Make a J-groove weld joint

(k) Make a corner-flange weld

(l) Make a flare-V weld

(m) Make a flare-bevel weld

(n) Make a plug weld

(o) Make a square-groove butt weld with melt-through

2. Contour these three single-bevel fillet-weld surfaces (three different weld symbols are required):

(a) Flush to the base metal

(b) Convex to the base metal

(c) Concave to the base metal

3. Draw a symbol to tell another welder to surface (or hardface) an area of a workpiece.

4. Tell another welder with a symbol to finish this single-bevel weld with a convex surface by (three symbols required):

(a) Grinding

(b) Machining

(c) Hammering

5. Write the blueprint symbol for "centerline." (It's an overlapping combination of a C and an L; if you don't know it, look through the illustrations in this chapter until you find it).

6. Show the following symbols for fully dimensioned welds using only the arrow symbols (you don't have to draw the weldment):

(a) Five intermittent ¼-in. [6.4-mm] arrow-side fillet welds, each 10 in. [0.25 m] long, on 30-in. [760 mm] centers (use only one weld symbol)

(b) Single 60° V-groove butt weld with ⅛-in. [3.2-mm] root gap but no land, using a full melt-through GMAW welding process on ¾-in. [19-mm] thick plate while making the weld from the other side (not the arrow side) of the joint

(c) Torch braze this joint

(d) Spot weld this joint

(e) Test this joint all around with dye-penetrant testing

7. If your welding school has any real blueprints of actual welded parts (preferably of a large, complex structure), look at the drawings and discuss them with your instructor and class. How much can you tell about the job from reading the blueprints? Could you direct other welders to do the work indicated?

Other Equipment Large and Small

Now that you know how important welding in different positions is, did you ever wonder how heavy weldments are turned over so you can get at them? What about peening weld beads? Do you really think that big plants have a dozen welders at a time pounding on welds with ball-peen hammers to stress-relieve weldments mechanically? How do you know when the plate is preheated to 300°F [150°C] and not 325°F [163°C]? How in the world do they bend a 1-in.-thick [25-mm] plate into a curved section for a large-diameter pressure vessel or water tank? And how do you pick a grinding wheel?

Large welding shops have special equipment to perform these jobs and to make your life much easier. This chapter will tell you about some of that equipment.

22-1
WHY WELDING POSITIONERS ARE USED

Gravity has a nasty effect on welding. It makes welding more expensive and a lot harder for you. You have to work more slowly out of position than downhand. Your weld metal wants to sag instead of staying in place when you work overhead. You can solve these problems, but the solutions will cost your company money because your weld-metal deposition rates will be a lot less when working out of position.

There are ways to solve vertical and overhead welding problems. If you are working with MIG or TIG, you can use a pulsed-power welding machine that produces less heat. You can use lower current and voltage settings that reduce your burn-off rate and help reduce the fluidity of your weld metal while welding with SMAW. But that also reduces your weld-metal deposition rate. Or you can use smaller-diameter electrodes with good out-of-position characteristics. But that still means lower deposition rates. Or you can stop and wait for a crane to come over and turn your workpiece when you want it in a new position, but few welders have their own personal cranes (and crane operators).

Working in the vertical position with SMAW electrodes you can expect to get about 3 lb/h [1.4 kg/h] instantaneous deposition rates if you are a good welder. We say "instantaneous" rates because you still have to stop from time to time to change electrodes or scratch your head, and that lowers your average deposition rate.

Even a good welder can keep the arc on only about 25 percent of the time when working overhead. Among other things, your arms get very tired. Down-hand, you will easily get 10 to 15 lb/h [4.5 to 6.8 kg/h] of weld metal deposited. That's at about five times the rate you'll get while your arc is on when working overhead.

Replace the stick electrode with flux-cored wire or submerged arc and weld down-hand and your productivity will at least double or triple. Productivity can even increase by 10 times what stick electrodes will produce if you use a high-deposition-rate process. Welding positioners turn your work so that you are always welding in the down-hand position.

Positioning Machines

The Aronson Machine Company in Arcade, New York, makes more welding positioners, manipulators, and turning rolls than anybody else. One of their smaller gadgets is the Universal Balance positioner. The Universal Balance (Fig. 22-1) lets you turn a workpiece by hand, even when it weighs up to 4000 lb [1800 kg], so you can assemble and jig your work and weld it downhand all the time.

The Aronson Universal Balance can rotate a workpiece 360° in two axes through the part's center of gravity, or it can rotate the entire assembly 360° around the positioner's column. Only fingertip pressure is needed to move a bal-

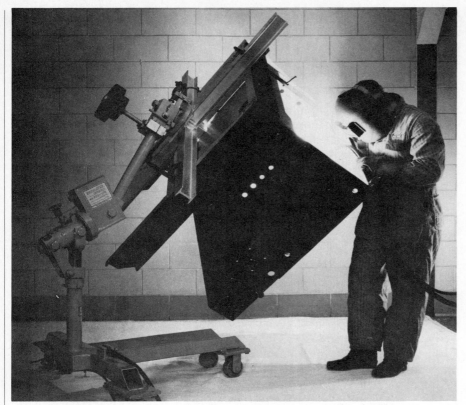

FIGURE 22-1 The Aronson Machine Company's Universal Balance can rotate a workpiece 360° around the positioner's column, and also 360° around two other axes through the part's center of gravity. Only finger-tip pressure is needed to move a workpiece weighing up to 4000 lb [1800 kg] into a new position. You can turn any odd-shaped workpiece that will fit on this machine into any position that will help you weld downhand.

anced weldment to a new position. You don't need to break your arc or lift your helmet while you move your work around.

Turning rolls If your workpiece is cylindrical, anything from a piece of pipe to a large pressure vessel (Fig. 22-2), you can use turning rolls to turn your work steadily into the downhand position.

Boilers, pressure vessels, silos, axles, rocket casings, and both small- and large-diameter pipe and tubing are examples of welding applications using turning rolls. Rolls also can be used to manipulate these products during x-ray and sonic testing. The rolls can turn the work while it is being cleaned, painted, or inspected, too. Even some things that aren't round can be rolled on turning rolls. Railroad cars are an example. If the railroad car is mounted in support rings, the car rides in the rings and the rings ride on the rolls.

Turning rolls are simple, low-cost positioners with many productive options (Fig. 22-3). They can be powered to turn at almost

(A)

(B)

FIGURE 22-2 Turning rolls come in dozens of shapes and sizes. They can turn anything from a small pipe (A) to a huge pressure vessel (B) if you have the right model. Turning rolls with several roll sets will have driven rolls that rotate the workpiece (operated by electric motors) and idler rolls that turn with the workpiece and simply provide support. The object in all cases is to be able to weld downhand while the workpiece turns round and round.

FIGURE 22-3 Turning rolls can be powered to operate at any speed you would need for welding. They also come in an enormous range of options, such as this Aronson Machine Company tractor-tread turning roll with support belts made of metal designed to hold up a thin-walled rocket casing. Even steel silos are assembled on turning rolls.

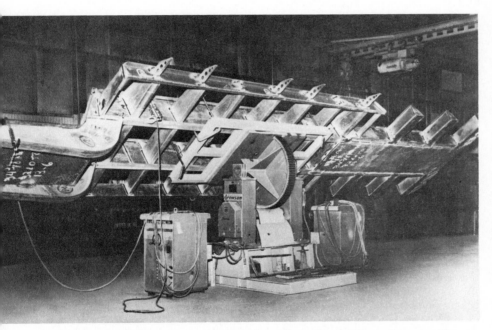

FIGURE 22-4 Positioners can be designed to tilt weldments as well as rotate them. This Aronson tilting positioner also can be mounted on railroad tracks to move a workpiece from one end of a plant to another. In fact, most positioners can be track-mounted to move work around as well as rotate it. Tilting-turning rolls (not shown) would look something like this geared tilting machine except that the rolls could also rotate a pipe or cylinder while it was being tilted.

any speed you want. You can operate the roll-drive motor with a foot treadle while you sit down comfortably and weld pipe. Turning rolls can even be mounted on a tilting frame to make tilting-turning rolls (Fig. 22-4) that not only rotate weldments but also tilt them for you. Turning rolls can also be mounted on tracked carriages so that large pressure vessels, for example, can move

from one end of a plant to another, turning all the while. These are only examples of the many kinds of turning rolls that exist. You often will find turning rolls in use in pressure-vessel shops and pipe-fabricating plants.

Turning rolls range in capacity from individual rolls that will handle 5000-lb [2000-kg] work loads per roll set, to rolls that will handle over 1 million lb [450,000 kg] per roll set. Turning capacities range from 150,000 to 3.6 million lb [68,000 to 1,600,000 kg] with multiple idler rolls. Idlers are rolls that are not powered. They often are combined with drivers that have motors in them. Driver speeds range from as little as 0.4 in./min [10 mm/min] up to 80 in./min [2 m/min].

Gear-driven positioners Some products don't fit on a Universal Balance and they aren't cylindrical, so you can't use turning rolls very easily. Take a dish-shaped pressure-vessel head, for example; positioning a piece like that is a job for an Aronson gear-driven positioner.

Gear-driven positioners (Fig. 22-5) are the most common type of welding positioner used in fabricating plants. Most of them have tables on which work can be mounted and rotated from a horizontal position downward through a 135° angle, so that the table and the mounted workpiece hang down toward the floor. The table also rotates on its own central axis so that the work can be loaded when the table is in the horizontal position. The weldment can then be turned nearly upside down while it's also rotating 360° to help you get at any welds that you want to make. Figure 22-6 shows you what you would have to do if you didn't have a geared positioner.

Aronson makes many different kinds of gear-driven positioners for work loads weighing from as little as 250 lb [110 kg] up to monster machines that can handle workpieces that weigh up to 4

FIGURE 22-5 Geared positioners are the most common type of welding positioner. Most Aronson models have rotating tables on which you can mount the workpiece. The table not only rotates but also tilts (or tilts and rotates at the same time) to move very heavy work into position so that you can weld it. Geared positioners are available in a variety of models and we will discuss them in this chapter. This photo shows the author of this book welding a temporary support bracket to a large workpiece. The support bracket is bolted to the positioner table. That's an example of how heavy workpieces are fastened to positioner tables. Aronson's large positioning manual gives many more details on positioner use than we can cover in this chapter.

million lb [1,800,000 kg]. You would need a special foundation under the plant floor to hold one of these big ones.

Gear-driven positioners are rated by load torque measured in pound·inches (or if you prefer, inch-pounds). The metric equivalent is kilogram-meters. Torque is a calculation of the force operating at a distance from the center of rotation of the tilt axis to the center of gravity of the load, when the table is in the vertical position and the workpiece sticks out in space (Fig. 22-7). It's similar to calculating the force on a lever

FIGURE 22-6 This is what you have to do if you don't have a geared positioner, a Universal Balance, a turning roll, or some other type of welding positioner to move your work while you weld. Welders without positioners spend more time climbing over work than they do welding it. Even more important, temporary supports for heavy weldments are not safe. If somebody accidentally kicks a sawhorse, the welder can be seriously injured.

arm. For example, if the distance from the center of gravity of the work to the center of the positioner table (the lever arm distance) is 10 in., and the work (the force on the end of the lever arm) weighs 10,000 lb, then the load torque required to lift the workpiece is 10 in. × 10,000 lb = 100,000 in.·lb. Rotating torque is measured about the same way.

The maximum capacity of the largest 135° gear-driven positioners is 700,000 lb [300,000 kg], but there are much bigger gear-driven positioners than that. A similar type of gear-driven positioner, called a 45°-90° positioner, is just about unlimited in the size of work it can handle. The bigger units are usually anchored to the floor or to their own special foundations. They can lift and rotate entire submarine hulls, parts for blast furnaces, the entire base for a coal-stripping dragline, and other big things (Fig. 22-8) so that you can work around the part (or inside it) easily.

A variation on the gear-driven positioner puts a 135° tilting-table

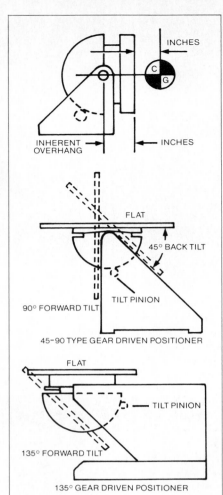

FIGURE 22-7 Geared positioners are rated by the distance of the center of gravity of a workpiece from their rotation axis, measured in inches. This distance requires two measurements: the distance of the workpiece CG to the positioner table (which is a variable) and the distance of the table to the axis (which is a machine constant for each model and is called the *inherent overhang*). You will learn in this chapter to use this information to calculate the force required to turn large workpieces. The middle drawing shows a 45° backward, 90° forward-tilting geared positioner. The bottom drawing shows a 135° forward-tilting positioner. In both machines the work is loaded when the table is in the flat position, and then the work is tilted into a position where you can weld downhand. The tables also rotate the work if you want them to.

(A) (B) (C)

FIGURE 22-8 (A) The 45°–90° tilting/turning gear-driven positioners have nearly unlimited load capacity in the largest sizes. These very big machines are often custom-fabricated to handle specific parts ranging from pressure vessels to submarine hulls and the bases for large construction equipment. (B) How a pressure vessel can be mounted on the positioner when the huge weldment is even bigger than the large positioner. (C) A huge pressure vessel with a ring of preheating flames around it produced by a circumferential natural-gas pipeline. The author and a welding superintendent are in the cherry picker in the upper left (that gives you an idea of how big this machine is). They are watching a welder inside the steel bowl surfacing it with stainless-steel weld overlay for corrosion resistance. Although you can't see the operator inside the steel vessel, you might like to know that she is wearing wooden shoes and an air-conditioned heat-resistant suit to protect her from being cooked in the steel pot while she deposits the stainless-steel overlay.

positioner on legs, then provides means for lifting the entire unit up the leg posts by rack-and-pinion gearing. These are Aronson's patented Geared Elevation positioners. Geared Elevation models range in work-load capacity from 2500 to 350,000 lb [1000 to 159,000 kg].

The most common application for Geared Elevation positioners (Fig. 22-9) is mounting weldments onto a horizontal table when headroom in your plant is low. The positioner is lowered so that the work can be loaded on the table. Once the weldment has been mounted on the table, the entire positioner table and its load can be raised on its posts to bring the weldment up where you can work on all sides of it. Then the work can be rotated, tilted, and lifted all at once.

Headstock-tailstock positioners Some products are elongated but not cylindrical. They are too big for a Universal Balance and too ungainly for a geared positioner or turning rolls. Products like that are probably just right for headstock-tailstock positioners (Fig. 22-10). These positioners work like giant lathes to rotate truck bodies, railroad hopper cars, or even much smaller things like pipe and tubing with side connections and elbows into position so that you can work on them.

Headstock-tailstock positioners range in capacity from 5000 to 240,000 lb [2300 to 110,000 kg] (that's the maximum load weight between the head and tail). They have powered rotary turntables mounted perpendicular to the axis of rotation through the center of

gravity of the workpiece. You can control the rotation speed of these units while you work, and even make them back up if you missed part of the weld.

Some headstock-tailstock positioners can also incorporate the powered elevation feature so that then they not only can rotate work but also lift and lower it.

Rotational torque is very important in headstock-tailstock positioners. If your work is very heavy and rotating, the rotation-torque rating for the machine and workpiece must be calculated. A high torque drive is required not only to rotate the weldment but also to stop it from turning without it twisting itself right off the mounting table.

Floor turntables Floor turntables (Fig. 22-11) are sometimes

FIGURE 22-9 An Aronson geared-elevation positioner is an unusual modification of a 135° tilting/turning gear-driven positioner that also has legs. The frame of the positioner can be driven up the legs to raise or lower large workpieces, as well as tilting and turning them. This unit, the Aronson Titan Model GE3500, can lift a weldment weighing up to 175 tons and measuring up to 20 × 20 × 20 ft when the positioner is fully raised on its elevating legs.

(A)

used in pressure vessel shops simply to turn work around and around. The load capacities of turntables range from 1000 to 400,000 lb [450 to 180,000 kg].

In addition to using them for welding, floor turntables are used to burn circular shapes in steel plate (the table rotates and the torch stays put), and to burn or grind the risers off castings. They're also ideal for x-ray inspection of large weldments when it's easier to move the weldment than the x-ray source.

Another example of the use of a turntable is welding flanges on extralong, extra-large-diameter pipe. The flange is simply put on the turntable and the pipe is suspended vertically on top of it. Then the turntable rotates while the welder stands by and welds the flange to the pipe without lifting anything heavier than a weld-

(B)

FIGURE 22-10 The headstock-tailstock positioner is a type of gear-driven positioner that uses the same principle as the metalworking lathe to rotate long workpieces that would be difficult to rotate on a conventional tilting/turning positioner. (A) A small headstock-tailstock positioner with a weldment mounted between the two ends. One end is usually a driven end (the headstock) and the other is usually an idler (the tailstock). (B) A headstock-tailstock positioner that also uses Aronson's geared-elevation principle. The headstock and tailstock can lift a wide workpiece high off the floor before rotating it into positions where a welder can reach the work easily.

FIGURE 22-11 Floor turntables are sometimes used in pressure vessel shops simply to turn work round and round. A large cylinder can be mounted vertically on a floor turntable and welding equipment, or machine tools can work on the pressure vessel walls. When a new position is required, or a new part of the cylinder must be turned, the table rotates the work. Horizontal weld seams also can be made continuously around the circumference of a cylindrical workpiece as it rotates steadily on a floor turntable. The turntable in this photo was designed by Merrick Engineering Company. They left room to slip a nuclear pressure vessel onto the turntable between the equipment, so that the inside and outside of the big cylinder can be welded at the same time.

ing electrode. There also are tilting turning tables that rotate and tilt weldments at the same time.

Manipulators Positioners bring the work to the welder. Manipulators bring the welding equipment to the work. They usually mount automatic equipment, often with multiwire SAW, flux-cored wire, or MIG-wire setups (Fig. 22-12).

Manipulators are best used for repetitive welding sequences, for surfacing large areas, or for making long welds where high travel speed pays off in high productivity. For example, on a long weldment two or more sets of power sources, wire feeders, and guns can be mounted in tandem on one welding manipulator to produce both the root pass and cover passes in one pass. Manipulators usually have long booms on which the welding equipment hangs.

In one shop a manipulator was placed between two rows of pressure vessels and rode on railroad tracks down the center of a plant. As the manipulator rode down the tracks it welded the vessels on both sides of the aisle simultaneously with equipment hanging from two separate booms. The welder rode in a comfortable chair on the manipulator while monitoring the control panel.

Manipulators can be produced with swinging booms and traveling booms, booms that reach out 40 ft [12.2 m] and booms that reach out in all directions. Some track-mounted welding manipulators with traveling cars have rapid traverse speeds of up to 200 in./min [5 m/min]. Their equipment load includes one or more welding power sources and wire feeders, welding heads, control panels, seam followers, flux hoppers and flux recovery units for SAW, and the operator's chair.

Utility positioners If you scale down all these different kinds of big positioners you would have a whole new family of special utility positioners. Most of them are small enough to be mounted on a workbench. They are used for light-duty machining, grinding, inspection, and assembly, as well as welding.

Some Aronson utility positioners are no more complicated than a rotating vise. Others can be small but very sophisticated. For example, there are bench-mounted power-driven positioners that will give you tilt, drive, and rotation while enclosed in a "glove box" containing a con-

FIGURE 22-12 Positioners bring the work to the welder. Manipulators bring the welding equipment to the work. Welding positioners range from simple columns with horizontal masts from which welding equipment such as submerged-arc flux containers and guns can hang, to fully automated machines that carry an enormously complex array of welding equipment, power sources, and controls to totally automate the welding process. A large, automated welding positioner with its full array of equipment can outweld any known welding robot. However, manipulators tend to be dedicated machines capable of doing only one thing, while robots can be programmed to do many things and, therefore, have wider application in general shop welding, as well as high-volume production work.

trolled atmosphere (Fig. 22-13). These positioners are used for welding titanium, zirconium, hafnium, and other metals used in nuclear and aerospace work.

Center of Gravity
If you are going to use a welding positioner, you need to know

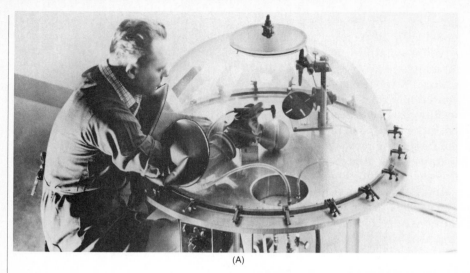

(A)

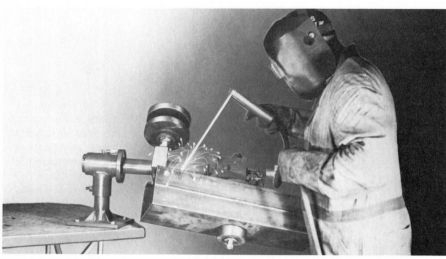

(B)

(C)

FIGURE 22-13 Welding positioners don't have to be big machines. (A) Many of the positioners you have just read about also can be made in miniature models small enough to work inside glove boxes containing inert atmospheres. They are used for welding nuclear fuel components made of zirconium, aerospace parts, instrument components, and other extremely critical work (usually done with GTAW). (B) Small utility positioners also are mounted directly on a workbench to help you weld faster and safer. (C) A geared positioner can be turned into a pipe positioner with a small outboard turning roll (on the left).

about three properties that all moving weldments have. They are center of gravity (CG), momentum, and torque. We'll tell you about CG first.

The idea behind CG Every weldment has a CG. The CG is an imaginary balance point at which the mass of an object is considered to be concentrated. It is the point that makes it possible to use a welding positioner. It's also the balance point that you use when putting slings on a workpiece so a crane can lift it. For example, if you put two slings on a cylinder and they were both on the same side of the CG, the pipe would fall out of the slings.

The best way to understand CG is to look at the steel box in Fig. 22-14. Whether the box is solid steel or hollow, as long as all six sides are equally thick and of the same dimensions, the CG will be located in the center of the box. The CG of any solid cube is midway between any of its faces. If the cube is hollow, the CG is still midway between the faces, as long as each face weighs the same.

A cylindrical object like a pipe section has its CG or balance point located at the geometrical center on the axis of the cylinder (Fig. 22-15). A dish-shaped pressure-vessel head attached to one end of the cylinder like a dome would move the CG along the central axis toward the attached dome, while something else, like a smaller inlet pipe extending from the side of the cylinder, would move the CG from the central position on the long axis of the cylinder over toward the side of the attached object.

CG is really a simple idea. You use it every day, because you also have a CG where all your mass seems to be concentrated. If your CG extends beyond your toes when you lean over, you will fall down.

Calculating the CG Most welding work is symmetrical around at least one axis. Very few things are so complicated that it's difficult to estimate where their CGs are. You often can locate the CG of a weldment simply by looking at it.

Even when a workpiece shape gets complicated, two very fast shop methods can be used to locate the CG exactly (see Fig. 22-16). All you need is a chain sling, a plumb bob, and some chalk. Lift the work (or have a crane do it) and hang the plumb bob from the top. Mark a chalk line along the line formed by the string of the plumb bob. Lift the

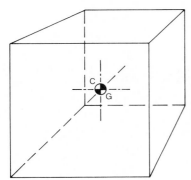

FIGURE 22-14 When working with the idea of center of gravity (CG), simply think where the weight would be centered in an object if all of that weight were in one place. In this cube, the weight would be centered right in the middle of the cube. Even if the cube were a hollow box, the weight of the box would act as if it were still at the center of the cube. As long as all sides of the box were of the same material and the same thickness, the CG of the system would remain centered in the middle of the cubic shape. If it helps you, think the way a juggler trying to balance a box does. The juggler doesn't care about the size of the box, only the point where all the weight seems to be centered—the middle of the box. As long as the juggler's finger is balancing the box on an axis that runs through the center of the box (where the CG is located), the box will be in balance.

work again (you can often use a piece of pipe and teeter-totter the work on it so that you don't have to lift it at all) and hang the plumb line from a different point. Then make a second chalk mark along the same face of the workpiece. The point where the two chalk marks intersect is one point of an imaginary axis through the workpiece that also runs through the middle of the part's CG.

Now turn the work over and repeat the above process. The CG is on an axis that passes through the two points where the two chalk lines on each face of the work intersect.

It's possible to calculate the exact location of the CG of a very complex workpiece right from the engineering drawings, even before the piece is built. But since that won't be your job, we won't go into details on how to do it. It's a complex process and would take up several pages of this book.

Knowing how to locate the CG of a workpiece, even approximately, is a very practical tool. Sometimes it can be a way to protect yourself. Remember that no work is going to fall over if its CG is placed over its base. As the CG gets closer and closer to extending beyond the base of the workpiece, the more likely the object is to fall over, whether it's a large pressure vessel, a tractor being driven up a hill, or a compressed gas cylinder.

Moment and Torque

When a heavy load is put on a welding positioner the engineers will say that a large *load moment* has been put on the machine. The force supplied by the positioner to rotate the load so you can stand in one place and weld it is referred to as *torque*. All positioners have maximum allowable load moments and maximum torque loads. Both terms are very simple to understand. They both are different kinds of levers.

You can move a 1000-lb [450-kg] object with a steel bar 11 ft

[3.4 m] long and do the work with only 100 lbs [45 kg] of force if 10 ft [3 m] of the bar is on one side of the balance point (called a *fulcrum*) and 1 ft [0.3 m] is on the other side. Put the short end of the bar under the object, and the long arm of the bar will multiply the force you apply on it by its length, relative to the short arm of the bar. What we're talking about, for example, is a crowbar, a hammer, a crane boom, a baseball bat, or anything else that uses a lever arm to multiply the force applied to the other end of it.

A simple way to think about load moments is the example of a teeter-totter in a park (Fig. 22-17 or 22-18). We will use the teeter-totter to put the ideas of CG and load moments together.

A 200-lb [91-kg] man on one end of a plank with his CG 48 in. or 4 ft [1.2 m] from the fulcrum point (measured along the plank) will apply a load moment of 200

FIGURE 22-15 The center of gravity of a cylinder also is located in the geometric center of the cylinder. You can locate that CG by imagining a long axis running through the center of the cylinder's circular ends, and any other axis halfway down that long axis and at right angles that intersects the long axis. Any one of these shorter, perpendicular axes would be a diameter of the circular section through the cylinder. The drawing explains the idea much more simply than words can.

(A)

(C)

(B)

(D)

FIGURE 22-16 There are two practical shop methods for finding the center of gravity of a symmetrical workpiece without calculating it with complicated formulas. (A) Balance the work on a round bar until the work rocks back and forth easily (it is in balance) on the bar. Draw a vertical line on the side with a plumb bob and chalk down to the round balance bar. (B) Turn the work into another position repeat the process. You have located the center of gravity of the *side* of the workpiece where the two chalk lines intersect. If the right and left sides (or front and back sides as in this photo) are the same, you don't have to repeat this work on the opposite side of the weldment. Simply measure the location to the first work *surface* CG and repeat it on the opposite side. These two opposite points are the ends of an axis that runs *through* the full workpiece. The CG of the entire workpiece will be located somewhere on this axis. If both sides of the work are the same, the full workpiece CG will be halfway between these two end points. (C) In the second method, hang the work from an overhead hoist. Drop a plumb bob along one surface and draw a chalk line. (D) Reposition the work on the hoist, drop a second plumb bob, and make a chalk line on the same surface. The CG of the surface will be at the intersection of the two chalk lines, just as in the first method. (If you repeated this process many times, all the lines would intersect at the same point.) The CG of the entire workpiece will be on an axis that runs through the part whose end points are defined by the two workpiece surface CGs. Again, if the work is symmetrical, the full workpiece CG will be on this through-the-workpiece axis at a distance halfway between the two surface points. While you are at it, you can use a scale in the second method to weigh the workpiece while you find the CG. There are several types of crane scales or force measurement systems that attach to the hook of a crane to weigh much larger workpieces than we show you here.

lb × 48 in. or 9600 lb·in. [13 kJ] of force on his end of the plank. Note the units used for moment are pound-inches or kilogram-meters (that is, distance × force). When that force is due to the weight of an object, it is measured vertically down through the CG of the object exerting the force right onto the lever arm where the force is applied.

We need an equal and opposite moment (not necessarily an equal force because the lever arm distance is also involved) to balance the man. If we don't, the plank will rotate around the balance point or fulcrum and he'll fall off. Let's say that his daughter is on the other end of the plank, on the opposite side of the fulcrum, as shown in the drawing in Fig. 22-18.

The daughter weighs 100 lb [45 kg], or half as much as the man. But she can exert an equal and opposite force around the fulcrum or balance point simply by making the lever arm 2 × 48 in. = 96 in. [2.4 m] long. The girl's opposing force is now 100 lb × 96 in. or 9600 lb·in. [13 kJ], except that the moment is working in the opposite direction. It's forcing the plank to rotate the opposite way from how it would rotate if the man alone were on the teeter-totter. The two 9600-lb [4400-kg] loads are equal but they work in opposite directions. They cancel each other out. And the father and daughter are perfectly balanced. Only a very small additional force is needed to make them go up and down. That's how welding positioners work. Given the right lever arm, a relatively small force can move and rotate a very large object once the object is balanced on the positioner.

Torque is a special case of load moments applied to things that rotate (or are trying to do so). It's the rotational force caused by an applied load. All welding positioners that rotate (and all of them do in one direction or another) have torque ratings for the rotary force they provide. Torque, just like

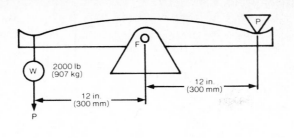

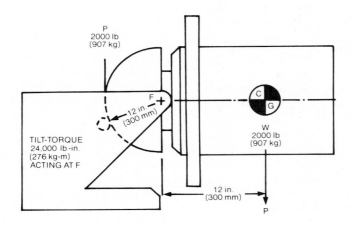

any other load moment, is measured in pound·inches.

Now let's look at both load moment and torque. The man applied a load moment of 9600 lb·in. [13 kJ] on his side of the teeter-totter plank. The girl applied a load moment of 9600 lb·in. on her (other) side of the plank. Since both people are applying an equal torque or rotational force to the teeter-totter plank, both torques cancel out because they work in opposite directions . . . unless the plank breaks, of course.

Here's another example of torque. Pulling on a 6-in.-long wrench handle with 100 lb of force will produce 600 lb·in. [800 J] of torque (100 lb × 6 in.) on a stuck nut. The nut doesn't have to move for torque to be there. Since the nut isn't moving because it is stuck, maybe we should apply more torque. We can't get any stronger, but we can use a 12-in.-long wrench, which doubles the moment of the lever arm and doubles the torque trying to force the nut to move. It's now 100 lb × 12 in. = 1200 lb·in. [1.6 kJ] of torque. The force you apply to the wrench handle is the same, but the torque has doubled because the lever arm is twice as long.

The weight of a workpiece multiplied by the distance of its CG from the fulcrum (balance) point determines the load moment and the torque. That's as true of weldments and positioners as it is of fathers and daughters on teeter-totters.

All gear-driven positioners have tables that provide geared rotation or tilting (which is just another kind of rotation). A number of models provide both motions. Therefore, rotation moments and tilt-load moments must be considered when getting ready to use a welding positioner.

Inherent overhang The drawing in Fig. 22-19 shows a typical gear-driven welding positioner in the 90° tilt position. This position

FIGURE 22-17 An old-fashioned weighing scale or a teeter-totter in a playground use the same principle as a geared positioner (and all other positioning equipment). In the top drawing we have two weights (think of them as forces acting through two separate CG's). One force is on one side of the balance point, F, and the other force is on the other side of F. In this example the forces are equal and the lever arm distances from the balance point also are equal. If the 2000-lb load on the right is to be kept in balance when it is 12 in. from the fulcrum (F), the load (P) on the left also must equal 2000 lb. The only difference is that the two forces operate in *opposite* directions. The right-hand force is trying to rotate the balance beam down and to the left (clockwise). The force W is trying to rotate the balance beam down and to the right (counterclockwise). Now look at the bottom drawing of a geared positioner. The fulcrum or balance point is still at F, which is still 12 in. from the CG of the 2000-lb horizontal cylinder attached to its table on the right. The tilt-torque gear on the left also is 12 in. from the fulcrum or rotational axis of the positioner (F), but on the opposite side from the load. To keep a 2000-lb load in balance, the gears must produce an equal 2000 lb of force working in the *opposite* direction. Torque is a way of describing these lever-arm forces for *any* system. Torque is simply force multiplied by the lever-arm distance through which the force operates. In both the balance in the top drawing and geared positioner in the bottom drawing, the two opposing torques are trying to rotate the lever arm in equal but opposite directions. Another way to say that is that one torque is positive and the other is negative (which is which doesn't matter as long as you are consistent). Therefore, 2000 lb × 12 in. = 24,000 lb·in. of torque on the left which equals the 2000 lb × 12 in. = 24,000 lb·in. of torque on the right. If you add them together, you get +24,000 lb·in. −24,000 lb·in. which equals 0 lb·in. The total system has zero net torque. it must be in balance.

puts the greatest stress on the positioner's gear train when a workpiece is mounted on the table and tilted. The symbol used for CG is shown on the drawing and marked C/G. It indicates where the workpiece's CG is located relative to the surface of the positioner's table, which presently is vertical. The F in the drawing is the fulcrum or point of rotation of the machine's drive axis.

The distance from the drive axis or F point to the surface of the table is obviously fixed by the machine model. The only distance that can vary is the distance from the work table to the CG of the workpiece. The distance from the axis at F to the surface of the table is called the machine's *inherent overhang*. You are stuck with it, as long as you use that machine.

Inherent overhang must always be used in making calculations about the tilt moments when loading work on the positioner table. You add the distance of the workpiece's CG from the positioner table to the inherent overhang to get the total distance for moment calculations. This is the distance that the weight of the workpiece acts through to put stress on the positioner's gears. (See Fig. 22-20 for a more complex example.)

Tilt torque Now we'll mount a workpiece on the gear-driven positioner, as shown in Fig. 22-19. In this sketch we show a 2000-lb weldment (W) with a CG 12 in. from the tilt axis (F) of the positioner. The total moment on this lever arm in the positioner is 2000 lb × 12 in. or 24,000 lb·in. [33 kJ]. To hold this workpiece, the positioner's gear drive must produce at least 24,000 lb·in. [33 kJ] of rotational torque (not even counting a safety factor which always is included).

Welding positioners have name plates on the side that tell how much weight the positioner can tilt when the CG of the workpiece is at different distances from the surface of the table. Figure 22-21 is an example. This illustration shows a plate of typical capacity that is attached to a small gear-driven positioner. The plate shows that the inherent overhang of the

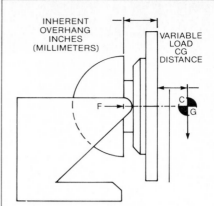

FIGURE 22-19 You must always take the inherent overhang of a positioner into consideration when calculating torque. The lever arm from the rotation point (F) is *not* the distance from the table to the workpiece CG. It is the distance from the table to the workpiece CG added to the distance from the rotation point (F) to the table. That last distance is called the *inherent overhang*.

positioner is 8.75 in., and that the machine can tilt up to 50,000 lb [23,000 kg] if the workpiece CG is no more than 6 in. from the surface of the work-holding table. If the distance of the workpiece CG is 60 in. from the surface of the table, the machine can, at best, only tilt a 12,000-lb [54,000-kg] workpiece. If you're interested, 12,000 lb × 60 in. from the workpiece table plus 8.75 in. of inherent overhang = 825,000 lb·in. of tilt torque. With a workpiece CG at 12 in. from the surface of the positioner's table, it can provide 40,000 lb × (12 in. + 8.75 in.) = 830,000 lb·in. [1.1 MJ] of tilt torque.

How big a workpiece could this positioner handle if the workpiece CG were 12 in. from the machine's table surface?

$$\frac{\text{Tilt capacity}}{\text{Inherent overhang} + \text{CG distance to table}} = \text{Maximum load}$$

or

$$\frac{830,000 \text{ lb·in.}}{8.75 \text{ in.} + 12 \text{ in.}} = 40,000 \text{ lb}$$

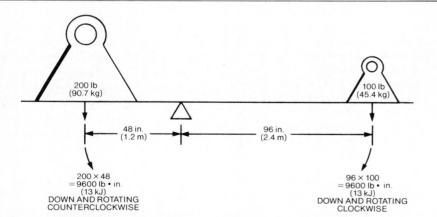

FIGURE 22-18 Here are two loads that are not equal, and two lever arms that are different. But this torque calculation still tells us that the system is in balance. The torque on the left is 200 lb × 48 in. = 9600 lb·in. The torque on the right is 100 lb × 96 in. = 9600 lb·in. They operate in opposite directions so that their signs are different; one is positive and the other is negative. If you add them together you get +9600 lb·in. − 9600 lb·in. = 0 lb·in. The system is in balance even though the lever arms and forces on each side of the fulcrum are different. This example shows the power of making torque calculations.

The heaviest object that this positioner can tilt with a workpiece CG 12 in. from the surface of the table weighs 40,000 lb [18,000 kg]. The Aronson Machine Company makes gear-driven positioners with tilting capacities up to 90 million lb [41 million kg].

There is much more to know about welding positioners, manipulators, and turning rolls than we can possibly cover in this book. Fortunately, Aronson has a 100-page manual titled *The New Handbook of Positioneering* that will tell you all about welding po-

sitioners, manipulators, and turning rolls. You can get a free copy of the manual by writing to the Aronson Machine Company, Arcade, NY 14009.

22-2
WELDING ROBOTS

You're smart enough to know that a welding robot doesn't walk around with a torch in one hand, a helmet thrown back on its tin head, with a light in its gleaming eyes that may remind you of your

welding instructor. Welding robots don't look like science fiction robots (Fig. 22-21). For one thing, they have no head or legs and usually look more like a ditch-digging machine with one boom (arm) than anything else. But if there is a welding robot in your future, is it likely to take your job away from you or make life easier? The honest answer is probably neither of the above.

Welding robots (and all the other kinds of industrial robots) only pay off if they are dedicated to doing one thing over and over and over again. They are actually pretty good at very routine, boring assembly work. They are great at spot welding auto bodies. They can be "trained" to do automatic welding—if the job is not too complicated.

Welding robots are fairly new. Automatic welding machines that do a lot of the things that robots are doing now have been around for years, and neither the new robots nor the older automated welding machines can climb around inside a ship hull, work on the structural steel of a high-rise building, or squat in cold mud and repair bad welds on a pipeline job in the middle of a swamp.

For one thing, industrial robots and automated welding machines can't walk around like you can (Fig. 22-22). If they are programmed to do one kind of work, they have to be reprogrammed for something else, even if the job is simply welding the same part one size larger. Robots, automatic welding machines, and computers aren't very smart.

Robots are very efficient, however, if a job requires making thousands of the same thing because they don't get bored. They also are expensive, and the only way they can earn their keep is to do lots and lots of repetitive jobs. A welding supervisor can take you off one job and put you on another in one minute. A robot takes a lot longer. It has to be reprogrammed, or retrained to do a new part, and when it's not

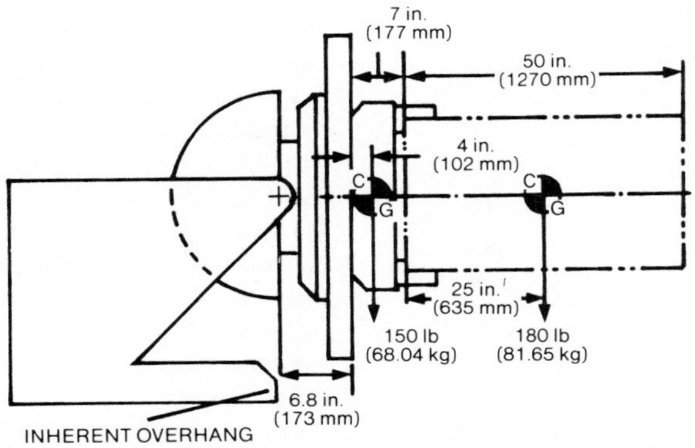

FIGURE 22-20 Now we have two loads at different distances from the positioner table on the right. The first load of 150 lb is the weight of a welding fixture. The second load farther outboard is the 180-lb weight of the workpiece. One way to handle this problem is to treat the loads separately. First calculate the torque of the fixture, then the torque produced by the load, and add them together to get the torque of the fixture-workload system. The fixture CG is 6.8 in. (inherent overhang) plus 4 in. (distance from the table surface to the fixture CG) which equals 10.8 in. Now 10.8 in. × 150 lb (fixture weight) = 1620 lb·in. (fixture torque). The workpiece is 6.8 in. (inherent overhang) + 7 in. (from the table surface to the base of the workpiece) + 25 in. (from the base of the workpiece to the workpiece CG), which is a total of 38.8 in. And 38.8 in. × 180 lb (workpiece weight) equals 6984 lb·in., which is the workpiece torque. The total workpiece plus fixture torque is 6984 lb·in. + 1620 lb·in. = 8604 lb·in. We arbitrarily assign the workpiece and fixture system a positive torque since both forces are on the same side of the fulcrum (they both are trying to turn the lever arm in the same direction). Therefore, the geared positioner must be able to produce at least 8604 lb·in. of torque *in the opposite direction* to rotate the workpiece and the fixture up from the horizontal position, or to keep the workpiece and fixture from falling downward from the horizontal position. What good does that information do for you? Figure 22-21 explains how to put this calculation to work to determine whether your positioner can do the job. If it can't, you could overload the machine and possibly break it.

welding it is a very expensive toy to have around.

As we have said, robots are very powerful production tools for making repetitive parts. They are the ideal assemblyline workers. Humans are not. Just as human beings thrive on diversity, robots are at their best at the most boring possible work you can find. An industrial robot will do the same thing over and over day and night if necessary. They are perfect for spray painting parts and welding auto bodies, appliances, and other high-volume products in mass production plants. In that kind of work, a robot will usually pay for itself in less than 2 years.

However, industrial robots have difficulty operating in the less structured manufacturing environments of general industry. It's hardly a problem for you to pick up a square block that has been turned slightly to one side so that it's not perfectly aligned. It can be a major task for a robot. Automated equipment doesn't cope well with randomly oriented work.

Advances in controls, joints, and gripper "hands" have improved industrial robot performance since their introduction in the early 1960s. However, most industrial robots are still deaf, blind, have no sense of touch, and no creativity.

Present robotics research is aimed at developing a new breed of robot adaptable enough to operate effectively in general manufacturing. These advanced robots usually have computer "intelligence" for limited decision-making capability, sophisticated programming languages for readily accepting complex instructions, and electronic sensors to detect events and conditions in the world around them. Thus, rather than performing as blind slaves, future robots may serve as active partners with humans in manufacturing processes.

The task of making robots effective for work other than very routine assemblyline projects is formidable. Let's take sight as an example. You might think that

FIGURE 22-22 Robots don't have arms and legs and walk around. At least not the ones that do welding. The machine in the upper right is a welding robot. The screen in the middle protects the operator from arc-flash. The lower half of the machine is an ESAB brand welding positioner designed to work with welding robots. The operator loads a new workpiece into the fixture while the robot welds one that already has been fixtured. When the robot is finished welding, the entire robot positioner system swings 90° and the robot starts welding the newly fixtured part while the operator unloads the completed weldment.

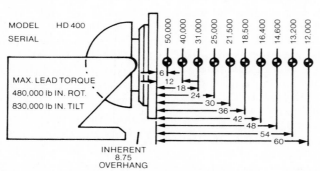

FIGURE 22-21 Aronson geared positioners always have name plates that let you read torque data directly as maximum load weight at a given CG from the table. In other words, Aronson does most of the arithmetic for you. Here is data for an Aronson HD 400 positioner. This drawing shows that the Aronson HD 400 can handle a total load of 50,000 lb if its center of gravity is no more than 6 in. from the positioner table; or it can handle a total load of up to 25,000 lb if the load CG is no more than 30 in. from the tabletop; or the machine will handle a load no greater than 12,000 lb if the distance from the tabletop to the load's CG is no more than 60 in. from the tabletop. Therefore, as long as you understand how the load and torque calculations are made, and you know how to weigh a workpiece and determine its CG, you don't have to know exactly what the torque load is . . . as long as it's *less than the capacity of the positioner.*

TV or some such electronic sensors can be used for a robot's "eyes." You are right and such systems already exist. But think of what a little smoke and weld spatter would do to a TV lens. Think about getting a robot to simply move to the other side of a dragline bucket. You can walk around the bucket. The robot needs special welding positioners that can position the bucket for the robot and turn the workpiece so that the robot can get at it.

Robots are very expensive, and an equal investment may be required for programmable welding positioners tied into the same computer that controls the robot's motions. This is enough money to pay for a whole lot of skilled welders who can do more than one kind of work on a moment's notice, and the human welders will go to the work if the work can't be moved to them.

So just how much might robots affect your job future? All told, about 20,000 employees of all kinds in industry can expect to lose their jobs during the decade of the 1980's to robots. This is a much smaller number of people than uninformed estimates have projected. The Society of Manufacturing Engineers in Dearborn, Michigan made this and many other predictions about robots and work. In a detailed study titled "Industrial Robots: A Delphi Forecast of Markets and Technology," first published in 1982, the SME forecast (with 90 percent probability of occurrence according to the industry experts who were surveyed by the Delphi prediction method, first developed by the RAND Corporation) that only 5 percent of welders and burners will be displaced by robots by 1985. That compares with 10 percent of production assembly-line painters and 3 percent of packers and wrappers in the shipping departments of industrial plants.

The SME study predicts that the vast majority of affected workers, about 70,000 to 100,000 people, will be employed in new positions within their companies. Of these people, 50 percent will be retrained as robot programmers and technicians, and 25 percent more will be transferred within the plant without retraining. That leaves only a little more than 1 percent of all industrial welders and burners who will lose jobs to robots by 1985. Skilled welders will be the first people to be considered in most plants for additional training in robot programming. And most skilled welders who don't learn to work with robots will be transferred to other

welding work within a plant.

Meanwhile, the installation of industrial robots for all types of applications is projected to decrease factories' accident rates by as much as 11 percent by 1985, by 24 percent by 1990, and by as much as 41 percent by 1995. Machine-loading robots are expected to be responsible for the greatest reduction in accidents. Robots in assembly and painting applications are projected to foster a decrease in shift lengths among 5 percent of US factories by 1995.

You probably are beginning to realize that just like any new technology, nothing does everything. Each development finds its own niche and fills it. That's as true of industrial robots as it is of computers and automobiles. Automobiles were once considered a disaster for workers because they would put so many blacksmiths, carriage makers, and stable cleaners out of work, while replacing the faithful horse.

Of course, if you really want to protect your investment in welder training, learn computer technology and programming, too. Welding robots are going to create a huge demand for welder–computer operators, just as numerically controlled and computer-controlled machine tools created high-paying jobs for skilled machinist-programmers. But many industrial robots can be operated without any programming skills on your part. (See Fig. 22-23.)

Teaching by Showing
The easiest way to program robots is by manual teaching. You literally take the robot's "hand" and make it do what it's supposed to do. You only have to show a robot how to do something once and it will repeat the task endlessly. The robot simply loads all the motions into its microprocessor computer memory. Of course, if that task is welding, the teacher had better be a good welder, and the task had better not vary from the

set pattern, or the robot will have to be retrained all over again.

It's often only necessary to show the robot the end points of a motion, such as a circular curve. The robot will fill in the rest of the path. This can be very effective for materials handling and loading and unloading parts, but for applications such as arc welding and spray painting, the entire path of the robot arm has to be input into the machine's memory.

Teaching programs require relatively little training for simple robot jobs. Almost anybody can do it, but the teach mode works best for jobs where the speed of the robot's arm and the exact path of the motion is unimportant. As far as complex welding jobs go, it's a big program for robots and an easy job for you.

Manual teaching programming has many deficiencies that may cause problems in all but very routine plant work. Manual teaching is time-consuming and error-prone for complex jobs, and production facilities usually must be tied up during programming. In addition, modifying some of the steps in the program to accommodate design changes or new tooling may require that the entire program be retaught.

Off-Line Programming
Many of the deficiencies of manual teaching programming are being overcome with off-line programming (Fig. 22-24), a relatively new approach in which the user describes sequences of movements through a computer instead of with robot hardware. Advanced robot languages permit the user to specify these sequences at a keyboard terminal connected to a computer. With the most recently developed languages, you don't have to be a computer programmer to program the robot.

Off-line programming defines and documents robot instructions better than manual teaching. Software (computer program) aids available with many of the newer programming languages used for

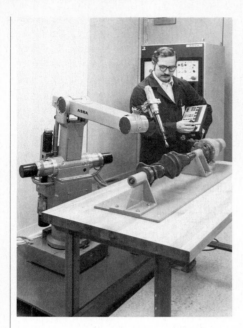

FIGURE 22-23 Robots are now very easy to program. You don't have to be a computer expert to do it. This small ASEA robot is holding a deburring tool in a lab demonstration of robot deburring of crankshaft oil holes. The operator holds a programming module (think of it as a large hand-held calculator that signals to you in a little window with plain English, Spanish, German, or Swedish words at your choice). The ASEA SII robot controller tells you (or often asks you) what you want to do (e.g., "Swing the robot to the right, or to the left?" "How much?" "How fast?"). The operator in this photo also has a joy stick like those you see in computer games in arcades with which he can move the robot around, make it go faster or slower, and so on. Therefore, the operator can either enter data on the "calculator" or operate the robot with the joy stick. Once the operator is satisfied that the robot is doing everything right, he punches a button and the complete set of complex robot motions are entered in a permanent computer program. When instructed, the robot will repeat that program over and over until it is told to stop. The control panel behind the operator can store up to 999 separate operating programs which can be recalled at any time (or even mixed together, or changed at will) to make many kinds of parts or do many different kinds of work.

robots make programming faster and more accurate. For example, subroutines (pieces of a program that are written to do part of a bigger job) can describe frequently repeated steps just once, and the program will keep referring back to them when they are needed. Very large and complex programs often can be built up from a handful of subroutines.

In addition, programs are readily modified at the keyboard by using editing routines and symbolic data references, and because programming is done off line, shop-floor production is not held up while the program is being debugged.

Talking to computers Off-line programming languages are classified broadly as either explicit or implicit. Explicit languages tell the robot exactly what to do. Programming commands in explicit control languages are often no more complex than "open," "move," and "pick" for controlling the robot's gripper hand. Implicit languages are those in which the user describes the task to be performed rather than detailed robot motions.

One example of a real computer program instruction for a rather advanced industrial robot is "place interlock on bracket such that interlock hole is aligned with bracket hole." The computer program then selects the grip points on the part, the approach path, and the motion required to assemble the parts. Implicit languages are far more advanced and more costly than explicit languages. They require so much more active memory in the computer that more powerful computers are required to run them.

Some researchers are developing speech recognition systems that may eventually allow operators to give robots spoken instructions. These voice data-entry systems are already in use in some computer-aided-design systems. Experts predict their use on industrial robots by the middle or

the end of the 1980s. People are also working on robotic speech synthesizers, creating a generation of robots able to converse with humans. One of the nice things about a robot that talks back too much is that if you disagree with it you can always pull its plug.

Robots That See and Feel
Research in robotic sensory capabilities is directed at developing and refining sight and touch. Providing these senses for robots requires on effective closed-loop system interconnecting with the sensor, computer, robot manipulator arm, and maybe the cooperating welding positioner that holds the part. Each element of such a loop already exists, but researchers have several obstacles to overcome before these elements can be linked together in a feedback loop comparable with that of a human brain. Feedback means that you not only see (or sense) the problem, but you realize you are doing something about it as you do it and can adjust your movement accordingly.

These feedback sensors must be sufficiently sensitive to provide accurate information without overloading the computer with a flood of data. Also, the computer must process these data rapidly enough for the robot arm and grippers to react almost instantaneously. What's more, the robot arm and gripper hand must be sufficiently agile to carry out the instructions accurately. And finally, a coding technique must be devised so that all of these signals circulating in the loop are comprehensible to all of the elements in the loop: the robot computer, often the welding positioner's computer, and sometimes a supervising human being, too.

Robot vision In robot vision systems, a television camera feeds images to a computer. The computer converts the images to digital computer code. This digital representation is then compared with other images stored in the computer memory. The part type is determined when two of the codes are matched. The computer sends a command to the robot arm based on the recognized part. For example, Consight-1 is a robot at General Motors that can recognize, grasp, reorient, and transfer each of five different types of parts on a moving conveyor.

Most vision systems can recognize only a few characteristic shapes that have been placed in their memories. However, several computer specialists predict that future robots may be able to recognize objects they have never seen before. Computers for these advanced robots would use deduction to determine the meaning of shapes in the same manner as the human brain reaches conclusions about new objects.

This area of artificial intelligence is in the early stages of development. However, a very powerful computer language called LISP, developed by MIT, closely models some things that the human brain can do. It will probably be used (in one version or another) for programming smarter industrial robots. Researchers believe that robots with vision and deductive capabilities will appear in laboratories by the mid-1980s and could be operating on the shop floor by 1990.

Experts also predict that robots with 3D stereo vision also could move onto the factory floor by that time. Unlike present 2D systems that can distinguish only silhouettes, 3D systems could give robots the required depth perception to navigate across the factory floor or perform complex assembly tasks.

Unfortunately, for 3D vision to be practical—and it's not yet developed—the software required must be able to provide an immense data base for reference and comparison. That data base will take up enormous amounts of computer memory. The more memory a computer requires, the

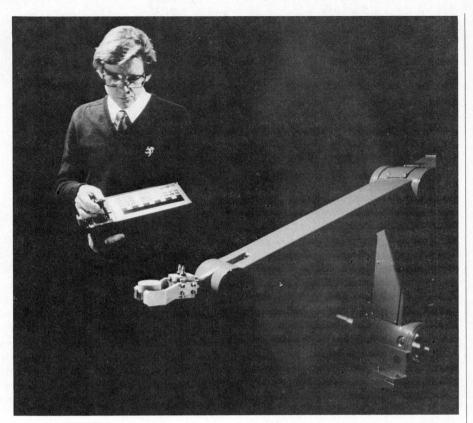

FIGURE 22-24 Off-line programming means programming a robot before it does the work. On-line programming means changing the robot's program while it is working. This is often called *hot editing*. This robot is being programmed off-line to pick up little round workpieces and put them somewhere else. Note the robot's gripper hand. These hands can be easily changed to do many different kinds of work. The problem with industrial robots is that they can't move around very easily (unless they are mounted on tracks like some welding positioners and turning rolls are). It will be a very long time before an industrial robot can walk around a plant. There is nothing really very mysterious about robots. They are simply another kind of machine tool, but one that is capable of doing many things. Nevertheless, people are still needed to tell them what to do. A poor welder will make a robot produce terrible welds. A skilled welder will make the robot make good welds. The skilled welder is still needed, robots or no robots.

more expensive it is. In contrast, most human beings don't need to use much brain power to walk across the floor.

Robot touch Researchers are also developing artificial tactile (feeling) senses for robots. Whereas vision may guide the robot arm through many manufacturing operations, it is the sense of touch that will allow the robot to perform delicate gripping and assembly work. Tactile sensors will provide position data for contacting parts more accurately than that provided by vision. In

some cases, the robot may be able to "feel" its way through a task in regions obstructed from view.

Most robot grippers (Fig. 22-25) have switches to indicate the presence or absence of an object. Also, grip-controlling sensors detect how tightly an object is grasped. However, sensitive tactile systems are being developed for handling parts more delicately.

In one advanced system, skin-like arrays of pressure-sensitive piezoelectric materials (producing tiny electrical currents under pressure—quartz crystals are one example) on the robot gripper generate small electric currents when stressed. This current is applied to a built-in microprocessor and analyzed to yield tactile information about the object shape and applied grip pressure. Thousands of these tiny sensors detect subtle pressure changes, and a multifingered gripper hand accurately responds to computer commands based on the tactile feedback. This type of system operates in a manner similar to that of the human system, in which electrical nerve impulses are routed on a feedback loop or network between fingers and the brain.

Real robot specifications

The total number of companies making either industrial (and welding) robots or equipment and software for use with them probably numbers several hundred.*

It's not possible to describe in detail all the different kinds of industrial robots now on the market, but we will describe one line of robots simply because it's considered to be the "Cadillac" of the robot industry.

ASEA is one of the leading U.S. producers of microprocessor-controlled manufacturing robots designed to help industry improve productivity and reduce manufacturing costs. The robot division maintains engineering centers in Troy, Michigan, New Berlin, Wisconsin, and White Plains, New York. A robot-manufacturing plant in New Berlin, Wisconsin, near Milwaukee, produces the ASEA units.

ASEA robots are used for welding, parts assembly, materials handling, adhesive bonding, cleaning, polishing, gauging, inspection, and in thermally hot or radioactive environments too dan-

* Several of the leading industrial and welding robot makers are Unimation, Inc., Shelter Rock Lane, Danbury, CT, 06810; Cincinnati Milacron, Inc., 4701 Marburg Avenue, Cincinnati, OH 45209; DeVilbiss Co., Box 913, Toledo, OH 43692; and ASEA Inc.'s Robot Division, P.O. Box 372, Milwaukee, WI 53201; 1176 East Big Beaver Road, Troy, MI 48084; and 4 New King St., White Plains, NY.

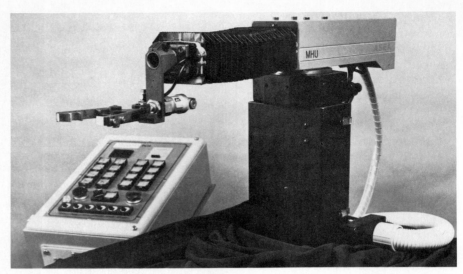

FIGURE 22-25 This is an ASEA MHU Minior robot. It is typical of a "pick-and-place" or material-handling robot. It can handle a workpiece weighing up to 2 lb [1 kg]. It is pneumatically operated instead of electrically operated. A robot like this is often used to feed small parts into presses and assembly machines (jobs that are dangerous for people to do because of the risk of the press operating on the person's hand instead of the workpiece). Especially note the different type of grippers (hand) on this robot. Robot grippers are often custom-made for handling specific parts. They can be changed very quickly when different parts are to be handled.

FIGURE 22-26 This is an ASEA IRb 6/2 robot loading work into a machine and unloading finished parts. The work-load capacity of this robot is 13 lb [6 kg].

gerous for a human operator. ASEA robots are assembling automobiles and trucks in Detroit, Europe, and Japan. They have a "top of the line" reputation among industrial robot users primarily because of their sophistication, pre-

cision, durability, and the manufacturing productivity increases they offer.

The company offers two families of industrial robots as of 1982, the ASEA IRb family and MHU family. These robots are used for welding, machining, and grinding; parts assembly; precision gauging; and quality control. The three models in the IRb family have maximum load-handling capacities of 13 to 200 lb [5.9 to 91 kg], with up to seven axes of motion. The ASEA IRb models are mini-computer-controlled and have smooth, precise electric drives. The three models in the MHU family have maximum load-handling capacities of 2 to 33 lb [0.9 to 15 kg] and up to seven axes of motion. Each MHU model has an increasingly longer reach. The MHU robots have pneumatic drives and are microprocessor-controlled. They are primarily smaller, parts-handling robots rather than welding robots (an important distinction in robot technology).

The members of the ASEA IRb robot family are:

■ Industrial Robot IRb 6/2 (Fig. 22-26). This is an all-electric robot with a 13-lb [5.9-kg] handling capacity. Position repeatability is better than ±0.008 in. [±0.20 mm]. The most frequent applications include arc welding, deburring, polishing, gluing, machine tending, and inspection.
■ Industrial Robot IRb 60/2 (Fig. 22-27). An all-electric model with position repeatability better than ±0.004 in. [±0.10 mm]. Typical applications are for cleaning castings, grinding, polishing, materials handling, and machine tending.
■ Industrial Robot IRb 90S/2 (Fig. 22-28). An all-electric robot specially designed and equipped for spot welding. It copes with handling weights up to 198 lb [90 kg] with five axes and 132 lb [60 kg] with six axes of motion. Position repeatability is better than ±0.004 in. [±0.10 mm].

ASEA's three parts-handling robots are:

■ MHU Senior (Fig. 22-29). Pneumatically driven with modular construction this robot is for use in heavy-duty materials handling. It can handle weights up to 33 lb

FIGURE 22-27 On the left is an ASEA IRb 60/2 robot with a workpiece-load capacity of 132 lb [60 kg]. It is picking up work from the box on the far left and putting the parts on an assembly-line conveyor belt on the right. It will do this incredibly boring job day in and day out until it either runs out of workpieces or is turned off. If you had to work at this assembly station, the job would soon drive you crazy or put you to sleep.

FIGURE 22-28 This is a large ASEA 90S/2 spot-welding robot which is frequently used in the automobile industry to spot-weld car bodies. This robot can handle spot-welding equipment weighing up to 198 lb. You can't, and if you ever worked on a spot-welding line in an auto plant, you wouldn't want to, anyway. There are few jobs that are more boring and unrewarding.

[15 kg] with repeatability better than ±0.004 in. [0.10 mm].
■ MHU Junior (Fig. 22-30). Pneumatically driven with modular construction, the MHU Junior handles weights up to 11 lb [5 kg]. The robot can work with up to three arms, which provides high handling speed. Position repeatability is better than +0.004 in. [±0.10 mm].
■ MHU Minior (Fig. 22-31). The Minior is the smallest of the ASEA robots. It has modular con-

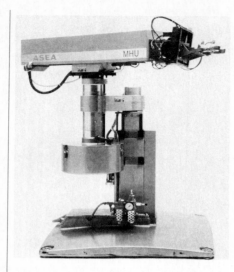

FIGURE 22-29 This ASEA MHU Senior robot is pneumatically (compressed-air) driven and is used for heavy-duty materials handling. It can handle parts weighing up to 33 lb [15 kg]. Its positioning accuracy is ±0.004 in. [±0.10 mm].

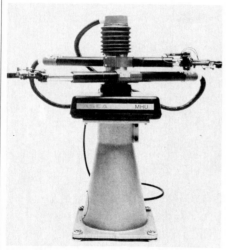

FIGURE 22-30 The ASEA MHU Junior handles work weighing up to 11 lb [5 kg]. It comes with either one, two, or three arms, all of which can put work in position with an accuracy of ±0.004 in. [±0.10 mm].

struction, is pneumatically driven, and handles weights up to 2.2 lb [1 kg]. The robot can work with one or two arms.

While materials-handling robots are not used for welding, they can be used for assembling parts for brazing and soldering, including electronic circuits and circuit boards. They also are good for the uncomfortable job of putting parts into, and taking them out of, furnaces and molten-salt brazing tanks and dip-solder baths.

ASEA industrial robots are programmed through teach pendants. (Fig. 22-23). The robot is taught its work by running it through its paces under manual control. Once the robot has learned the job (which usually takes just one run-through), it tirelessly repeats the task, without making mistakes, through many thousands of production cycles.

ASEA robot-operating programs can be modified through the teach pendant at any time. In some cases, ASEA industrial robots are also equipped with sensing devices that enable them to see and feel the presence or absence of objects. Adaptive control programs permit them to pick up a part that may be slightly out of position or sense a missing part and take appropriate action.

A new control system has now been introduced for these robots. Its main advantages are in programming. Programming takes place as a simple display dialogue between the human operator and the robot control system. The operator can choose the language, English, French, German, or Swedish.

The new ASEA plain-language programs can be used with the pendant-controlled positioning program previously used on ASEA robots. The new programming feature includes main programs and subroutines which save lots of time when programs are being prepared for different parts of a workpiece. Another time-saving feature is the possibility of running the robot around a defined working point (such as a tool center or the tip of a welding electrode). Up to 10 different working points can be programmed. The robot will then interpret the points as locations on straight lines or curves for the operator.

A joy stick (Fig. 22-24), like those used on arcade computer games, also can be used to direct the robots. Together with the dialogue routine for entering instructions, the joy stick helps in most cases to reduce programming time by 25 percent compared with programming using a set of push buttons. In addition, it lets the human operator concentrate on the operation being programmed. The alphanumeric (letters and numbers) display with each robot gives the operator continuous information about the options available and the meaning of various function buttons.

Other features incorporated into the control system include the operator's choice of coordinate systems, the possibility to adjust positions while a program is running, self-diagnosis of faults in the system, and operator security functions. No prior computer programming or robot experience is required to operate any of the programs.

More than 5000 ASEA robots have been built as of 1984. They are currently in use in over 20 countries. By the time you read this book, many more industrial robots will be in use.

22-3
PRESS-BRAKE FORMING

A press brake is a big, wide machine with an open space in which you can stick a plate or sheet that you want to bend. When you are ready, a big punch is pushed down by a ram onto the sheet or plate and the material bends down, folding up the sheet or plate on either side of the punch (Fig. 22-32). If you don't have a die underneath the material, the process is called *air bending*. If you want to give the material a special shape, use a female die under the ram-driven punch so that the material is pushed into the die and takes on a special

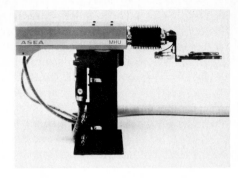

FIGURE 22-31 This is the ASEA MHU Minior. It is the smallest of the ASEA robots. It can mount either one or two arms. It handles work weighing up to 2.2 lb [1 kg]. Like the other MHU robots, this is a typical pick-and-place model.

shape like a jog or a U-shaped corner flange.

Press brakes are very good for making simple V bends or round-cornered bends on long sheets or plates. You can make the sides of a box by using a press brake to bend a sheet in the middle into a right angle to make two sides of the box. Bend a second sheet the same way, and all you have to do is weld two opposite corners to complete the sides of the box.

Operations such as blanking, piercing, lancing, shearing, straightening, embossing, beading, flattening, corrugating, and flanging can also be done in a press brake. Press brakes also are handy for straightening out distorted work.

Press-brake forming is most widely used for producing simple shapes from ferrous- and nonferrous-metal sheet and plate that is ¼ in. [6 mm] thick or less. When the material gets over ¼ in. [6 mm] thick the load on the machine gets pretty big. Plates up to 1 in. [25 mm] thick or more have been bent by very large machines.

Press-brake forming is more or less limited to long or narrow parts that won't fit into a conventional metalworking press. They also are used for small numbers of parts that don't warrant special tooling. For example, you can make your own angle iron with a

press brake from narrow plates. The length of the plate or sheet that can be bent is limited only by the width of the opening or "throat" of the machine. For instance, a 600-ton [540-tonne] press brake can bend a 10-ft [3-m] length of ¾-in.-thick [19-mm] low-carbon-steel plate to a 90° angle, with an inside bend radius equal to the thickness of the plate.

Any metal that can be bent can be shaped in a press brake without the outside edge of the metal stretching until it cracks, as long as you don't exceed the capacity of the machine or the bending limit of the metal and the given thickness. Press brakes are used on low-carbon steels, high-strength low-alloy steels, aluminum alloys, and copper alloys. High-carbon steels and titanium alloys are less frequently formed this way because they are more difficult to bend without special procedures.

The bending limit of your metal is related directly to its ductility and thickness. The machine capacity needed to do the bending job is a function of the yield strength of the material and its thickness. A steel sheet with a yield strength of 60,000 psi [413 MPa] takes twice as much power to bend as sheet with a 30,000-psi yield strength. A ½-in.-thick [13-mm] plate takes twice as much power to bend as a ¼-in. [6-mm] plate.

Most press-brake forming is done at room temperature. You can, however, reduce the yield strength of your material and increase its ductility by heating it and bending it in a press brake while it is still hot. In that case, the punch and die also should be heated to keep from chilling the metal at the critical part of the bend. Make sure that heating and cooling the metal won't permanently change its mechanical properties. That's usually only a problem with high-strength steels with a high carbon content, and then only if you heat it to too high a temperature. Keep the heating below 500°F [260°C].

The thickness of material that you can bend successfully to a given angle also is a function of the material's yield strength because materials with higher yield strengths will tend to "spring back" more than low-yield-strength materials.

In the wiping type of bending operation typical of a press brake, the metal is bent into position but when the punch is raised the metal attempts to return to its original shape. This movement, called *springback,* is seen in all metals and it increases with the yield strength of the metal.

The amount of springback you will get when forming a soft metal like 1100 aluminum is negligible. However, higher-strength aluminum alloys like 2024 have significant springback. In general, low-carbon steels show more springback than do aluminum or copper alloys, and you can expect still more springback from stainless steels.

The usual solution to springback is to overbend the material a number of degrees so that when springback occurs the material will go back to exactly the shape and angle that you want. Several trials will soon tell you just how much overbending is needed. Tables for springback have been developed for specific metals.

One of the big problems with press-brake forming is that sheet and plate tend to crack when bent parallel to the final rolling direction. This happens because the crystals inside the material tend to line up in that direction. So do inclusions that make weak zones in the material. Therefore there are two directions in most materials as far as bending is concerned. One bending direction is the "easy" way (across the material's final rolling direction in the steel mill). The other is bending it "the hard way" (parallel to final rolling direction). Hot bending is one solution when you have to bend something the hard way. But hot bending only reduces the problem; it doesn't eliminate it.

FIGURE 22-32 A Verson press-brake which is very useful for folding and forming sheet metal into various shapes. Stamping dies can be added to the ram to cut shapes in sheet metal. However, most press brakes are used primarily to bend sheet or, in larger models, plate. The operator is not chained to keep him at work. The chains attached to the operator's wrists keep the machine from working if the operator's hands are inside the throat of the machine.

22-4
SHEARING AND SLITTING

In addition to flame-cutting, metal can be sheared or slit (Fig. 22-33). These processes are the same as cutting paper with a pair of scissors except that the blades of the machine are made of tool steel. Shearing machines look a lot like press brakes except that they have two slightly offset straight-sided blades instead of a punch and die.

A large pair of shears can cut low-carbon steel up to 1¼ in. [32 mm] thick. Using the same amount of force, stainless steel or full-hard high-carbon strip up to 1 in. thick [25 mm] can be sheared as well as aluminum alloys up to 2 in. [50 mm] thick. Conventional shears are used mostly to make straight-edged cuts. There are other devices such as rotary shears that can cut circles. Certain kinds of steel strip are made from much wider sheet with slit-

ting machines that have many shears set in a line. (These shears have a row of tool-steel blades instead of a single cutting edge.) Metal service centers use them to make slit sheet.

Most shearing machines are rated according to the section thickness of low-carbon steel that they can cut. They are usually limited to shearing steels with tensile strengths less than 75,000 psi [517 MPa]. The ductility of your material, measured by percent elongation, can affect the machine's capacity. Annealed copper, for example, because of its high elongation, requires as much shearing effort as low-carbon steel even though the copper has considerably lower tensile strength. The copper tends to stretch instead of shearing cleanly apart.

One of the drawbacks of using sheared edges for weldments is that the shearing action tends to wipe a burr on each edge of the

material where the blades finish the cut. That can make problems when you fit up the material. A bigger problem is that very large machines are needed to slit plate over ½ in. [13 mm] thick. Most of this material is produced on flame-cutting machines. Nevertheless, a lot of the thinner-gauge metal that you will use will be slit instead of flame cut.

22-5
BENDING ROLLS

If you ever saw an old-fashioned washing machine with rollers on the wringer you would have a good idea of what two-thirds of a plate-bending machine looks like. Add a third roll in front of the other two that you can adjust up or down and your plate-bending machine will be complete (except for the frame, a big motor, and, of course, very large hardened rolls). A related type of machine is a tangent bender that can form thick sections into complex shapes (Fig. 22-34).

When flat plate is put through a bending machine, the first two rolls pinch the plate and drive it forward. The leading edge of the plate rides up on the third bending roll, putting a slight curve in the plate. If you pass the plate back into the machine, adjust the third bending roll up a little more, and run the plate forward through the roll bender again, the plate will curve up more.

After a few passes through the roll bender, your flat plate will be a smoothly curved plate. You can even produce a complete cylinder of steel or aluminum on a roll-bending machine. It will be ready for you to complete the final longitudinal weld seam to put it together.

The limitations that apply to press brakes [bending the easy way across the grain (the steel-mill rolling direction) versus the hard way parallel to the grain, and the fact that high-yield-strength materials are more difficult to bend than low-yield-

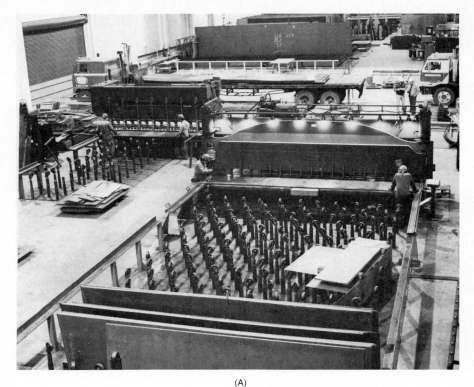

(A)

(B)

FIGURE 22-33 (A) A machine in use at Sea-Port Steel, Seattle, a large steel-service center. This machine is a hydraulic shear. It is very similar in operation to a press brake, but it cuts steel plate up to ½ in. thick and 20 ft wide with one stroke. On the left is a smaller ¾-in. × 12-ft mechanical shear. Press brakes can save welding time by forming parts instead of joining them. Press sheers can be used to cut long sections quickly instead of flame-cutting them. Both press brakes and sheers are common machine tools seen in large weldment fabricating shops. (B) A Wean-United slitting line turning wide sheet into narrow strip at high speed at Universal Steel Company's Cleveland, Ohio, plant. Universal Steel, like Sea-Port Steel, is a metal-service center that processes material from steel and aluminum mills, and then resells it to fabricating shops. Metal-service centers probably handle about 25 percent of all the metal used in the United States at one time or another before it is fabricated into useful products.

strength materials] also apply to roll benders. Nevertheless, the bend angle produced by a roll bender is much shallower and more smoothly curved than that produced by a press brake. Big roll benders also are very massive, rigid, squat machines that can bend very thick plate.

22-6
SHOT PEENING

Shot peening is a cold-working process used primarily to increase the fatigue life and prevent stress corrosion cracking of metal parts. It also can be used to stress-relieve weldments (if allowed by a code), to straighten pieces that have been deformed by built-up welding stresses, and to work-harden weld-metal overlays made of austenitic manganese steel or certain austenitic stainless steels. It can be used on very complicated shapes and will do the job far faster than you can with a hammer.

In shot peening (Fig. 22-35), the surface of the finished part is bombarded with round steel shot in special machines under controlled conditions. When the surface has been peened all over by millions of hard impacts from the tiny steel shot (nonferrous metals are peened with glass beads), the resultant residually stressed surface layer, which is in compression, prevents the formation of cracks. Cracks in the heat-affected zone of a weld usually occur because it is initially in tension after the weld metal cools, solidifies, and contracts. A crack will not develop in a compressed layer.

Nearly all fatigue and stress-corrosion failures originate at the surface of a part. The layer of compressively stressed surface induced by shot peening produces a tremendous increase in the life of the part. The maximum compressive residual stress produced at or near the surface of a shot-peened part is at least as great as half of the ultimate tensile strength of

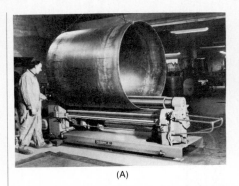

(A)

(B)

(C)

FIGURE 22-34 There are various types of machines that are designed to form heavy sheet and plate into useful shapes before you weld them. (A) A small roll bender. Flat plate is fed into it, and it rolls the plate into a cylinder. That is how almost all cylindrical workpieces are fabricated. Roll bends range from this size up to machines that can bend 8-in.-thick plate. (B) The Taylor-Winfield Company of Warren, Ohio, makes an enormous range of metalworking machines. One example is a tangent bender that produces formed parts such as this from flat sheet and plate, ready for welding. (C) The tangent bender machine.

the metal. If the original built-in tensile stresses in the surface of a part are making the part bend, shot peening will eliminate those stresses and the part will straighten out again.

The Metal Improvement Company, a wholly owned subsidiary of the Curtis Wright Corporation, has 28 service centers in the United States, Europe, and Canada that perform special peening services. They peen critically stressed parts such as gears,

shafts, aircraft components, and weldments. Parts that are too large to transport are stress-relieved or hardened on site. The company's peening services are not only used to stress-relieve or work-harden surfaces, but also to form sheet metal into complex shapes (such as forming the metal skin onto the wing structure of a Boeing 747).

Metal Improvement has developed methods to measure the deformation intensity of a shot-

peened surface (there are no non-destructive tests to do that). The company has found that the deformation intensity of a surface is a function of the time it is exposed to the process. The surface hardness continues to increase as the part is shot peened until the hardness reaches a "saturation" level at which point the hardness levels off. You can use a test sample or two in the shot-peening machine to determine how long to run production parts to reach any given level of surface hardness. If a flat piece of your sample sheet or plate is clamped to a solid block, you can actually plot a curve of hardness versus shot-peening time. Standard samples also are available from the Metal Improvement Company for determining or calibrating your shot-peening operation. These standard samples are called Almen strips.

22-7
DETERMINING WELD TEMPERATURE
Moving from very big things to very small ones, let's tell you about crayons, paints, and pellets that will tell you the temperature of your weld.

Temperature-indicating materials are easily applied chemical compounds in liquid and solid forms that melt or change color in a known way when exposed to certain temperatures. They are thermometers with (usually) one scale reading. When you are asked to preheat a weldment to 300°F [150°C], how do you know when 300°F [150°C] is reached? A temperature-indicating crayon set exactly for 300°F [150°C] will melt; a dab of indicating liquid set for the same temperature will change color; the pellet will droop and turn into a liquid blob.

You don't need to calibrate these "instruments." They usually are accurate to within 1 percent. You can carry them in your pocket. They are inexpensive. Several marks or broad coatings

FIGURE 22-35 We have mentioned shot-peening several times in this book. This is a view inside a typical shot-peening machine cabinet. The nozzles project the shot at high velocity against a workpiece (such as these gears) and peen the surface to improve mechanical properties. The Metal Improvement Company, Inc., of Paramus, New Jersey, has offices and plants all over the United States that specialize in shot-peening work for other people. Shot-peening work after welding, for example, substantially increases a part's fatigue strength. That's one reason why these gears are being treated.

can show the temperature of a large surface all at once.

Crayons, paints, sprays, and melting pellets are available for a wide range of temperatures. You can get them for use from 100°F [38°C] (in temperature increments of 3 to 50°F [1.7 to 27.7°C]) up to 2500°F [1370°C] in crayons and paints, and up to 3000°F [1650°C] in units of 100°F [55.5°C] in pellets. The most well known manufacturer of temperature-indicating materials is a company called Tempil°, a division of Big Three Industries, Inc., South Plainfield, NJ 07080.

Special crayons (Fig. 22-36 on page 604) are the best tools for spot checking easily accessible parts. A touch of the stick to the surface tells you instantly whether the part is above or below the rated temperature of the indicator stick. By watching a mark made by one or more crayons change when the work is heated you can accurately gauge the temperature of the workpiece.

Paints are more durable. Many are corrosion-resistant, weatherproof coatings that also protect the surface. Fast-drying liquids are available for use on highly polished surfaces. These may be thinned out and sprayed on the part before heating. A single paint that goes through several color changes (as many as four at four specific temperatures) can indicate heating progress and temperature distribution over a very large surface area.

Pellets come into their own in furnace applications. They resist evaporation caused by extended heating, and they melt upon reaching their rated temperature. A drawback with pellets, of course, is that the furnace has to have a viewing port and be adequately lit so that you can see the pellet melting.

22-8 SCALES, TAPES, AND MARKERS

Of all the measuring tools you will use, the most useful will be a long roll of string coated with chalk. If you tightly fasten one end of the string to an edge of a wide part (such as a steel plate or sheet), then get on the other side of the work and pull the string tight, you can mark a very straight line with no effort at all.

Simply lift the string up and let it snap back. When it hits the surface of the work, the string will leave a chalk line. That's all you have to do if you want a guide line to follow when getting ready to flame cut a straight or beveled edge on a plate, for example. Carpenters use the same marking strings or chalk lines in their work. You can buy chalk lines at almost any hardware store.

A material called soapstone is another common marking device. You can buy it at almost any welding distributor's store. Soapstone has a soapy feel to it. It is made of soft minerals such as talc and chlorite, and often has a little

bit of a harder, magnetic mineral called magnetite (Fe_3O_4) in it. Soapstone is the welder's all-purpose pencil for marking metals. It leaves a white or yellow mark that you can easily wash off.

A hard tool-steel scribe (which looks something like a thin ice pick or sharply pointed dentist's tool) can be used to draw lines right in the surface of metals. A tool-steel punch is a good marker for locating holes to be drilled or burned out.

Other markers including marking crayons that leave more permanent marks also are available, including ink and paints. Be careful what you mark with markers that contain organic materials (which is most paints, inks, and grease pencils). If a part is made of steel and will be heat treated later on, the surface marking will turn to carbon. If the heat-treat temperature is high enough, the carbon will diffuse into the steel and carburize (add carbon to) the material. As you know, carbon increases the strength of steel. The result will be a hard, relatively brittle, narrow surface line where you marked the steel.

You also should be careful when marking aluminum and stainless steel. Some marking materials will cause corrosion to begin where the mark was put on the metal. If you have any doubts (or are working on critical materials or any special code-welding job), ask your welding supervisor first, before using a marking material that is not approved.

Of all the measuring tools you will use, the simplest and most common is the steel scale. This is usually a steel ruler 6 to 12 in. [152 to 305 mm] long, although other lengths are available. Many welders carry a tiny 6-in. steel ruler around with them in their pocket. It's good for measuring the dimensions of weld beads.

Many steel rulers now have English units in fractions of an inch and inches on one side and metric units in millimeters on the other. The best small steel ruler you can buy is one with English fractions of an inch down to $\frac{1}{32}$ in. (or even 64ths of an inch) on one edge, decimal equivalents of inch fractions on the other edge, and metric units on the reverse side of the scale. Machinists often use these small pocket scales, so if you don't know where to get one, ask a machinist in your plant.

Folding rules usually are 2 to 6 ft [0.6 to 1.8 m] long. Folding wooden rulers are less useful than steel measuring tapes. The tapes are more accurate. They can be from 6 to 100 ft [1.8 to 30.5 m] long. Shorter tapes are made with curved cross sections so that they are flexible enough to be rolled up, but remain rigid when extended. Long, flat tapes require support over their full length when measuring, or the natural sag in the tape will cause an error in the reading. Even when a long tape is supported, you must make sure that you pull the tape tight before you read the scale.

Metal tape measures should be handled carefully and kept lightly oiled to prevent rust. Do not kink tapes; pull them straight out from their spring-operated metal cases. Don't ever bend a metal tape backwards. It will either take a permanent set or, more likely, will break. Don't let tapes snap back violently into their cases (something which many people do allow). You can put a kink in the tape, twist it, or break off the ring or hook at the end.

When using any linear measuring tool, make sure that the end is not worn. An old wooden ruler, for example, may have as much as $\frac{1}{16}$ in. or more worn off the end. It's always best to start measuring at some unit within the scale, such as the first inch, and subtract the starting point from the final reading. For example, if you start at the 1-in. mark and measure 9.75 in., the length you are measuring is really 8.75 in.

There are many special linear measuring devices such as small hole gages, depth gages, depth mi-

crometers, inside and outside dimension micrometer calipers, and other devices. Most of these other linear measuring devices are used more often by machinists than welders. One linear measuring device now used more frequently in plants to measure very long distances, or to measure distances very accurately, is a laser. Don't use laser-measuring equipment without supervision. A laser can burn out the retinas in your eyes and blind you.

22-9
ANGLE-MEASURING DEVICES

The most simple angle-measuring device is a cheap protractor that you can buy in a stationery store, usually for use by schoolchildren. There are, however, many kinds of precision steel protractors used in plants. They often are attached to linear measuring tools such as steel rulers.

Right angles are best measured by a simple steel square. Unlike the common ruled carpenter's square (which you can often use), you may find your tool crib will have a right-angle steel square that consists of a rectangular handlebar with a thinner blade attached to it at right angles. It may not have any markings on the blade.

Sometimes the blade is removable or adjustable so that it projects a greater or lesser distance from the handlebar. Extra blades are often supplied with this type, with the ends ground to commonly used angles such as 45° for octagons (eight-sided figures) and 60° for hexagons (six-sided figures). The device may have other angles ground on it for measuring the cutting angle of a drill, or the angle of a countersunk hole. You can make your own tools like this if you need something special. Just be sure that your tool is accurately made. Check for the desired angle with a good protractor *after* you put it together. Remem-

ber that any measuring tool that is out of alignment will produce nothing but errors whenever you use it.

Like linear measuring tools, there are many special angle measuring tools ranging from combination squares and angle protractors with steel rulers on them, to bevel protractors.

Never try borrowing tools from experienced machinists. No good machinists will give anybody else the precision equipment they are responsible for. Instead, check the tool room.

22-10
HACKSAWS

Although you are now an expert at flame cutting, you don't always want to flame cut every piece of metal you will work with. Alloy steels that will get a hardened edge from flame cutting (by air quenching), or thin tubing are examples. And, sometimes, a hacksaw is simply faster to pick up and use.

There are two parts to a hand-held hacksaw—the frame and the blade—and most frames are adjustable to take 8-, 10-, or 12-in. blades. Older styles have a straight handle in line with the blade, but newer types have a full pistol grip, which is easier and less fatiguing to use. Some manufacturers now combine the convenience of a pistol grip with the advantage of the straight handle—where the applied force on the forward stroke is delivered in direct line with the blade—by making an inverted handle.

All adjustable hacksaw frames are made so that the blade can be used in either a vertical or horizontal position. On some types of hacksaws, the blade can be positioned at four or six different angles. This is an advantage when the saw must be used in places where there would not be sufficient clearance for the conventional saw with only two blade positions.

To install a blade in a frame, adjust the frame for the length of the blade with sufficient adjustment remaining to permit the blade to be tightly stretched. Place the blade on the pins with the teeth pointing away from the handle (point the teeth toward the front end of the saw), and tighten the adjusting nut until the blade is rigid.

In starting a cut, use the thumb of your left hand (if you are right handed) against the upper surface of the blade to guide the blade until the cut is started. (Keep your thumb away from the cutting edge). Use sufficient pressure in starting the cut so that the saw begins to bite into the metal immediately. If the teeth do not bite into and cut the metal, the rubbing action will dull the blade. Relieve the pressure on the return stroke, but do not lift the blade off the work when the cut is being started. When the kerf is deep enough to guide the blade, the saw can be lifted slightly on the back stroke.

As far as possible, use the entire length of the blade on every stroke, except when starting, and keep the blade moving in a straight line to avoid any twisting or binding action. Try to keep at least three teeth in contact with the work at all times. If the work is thin, use a fine-tooth blade and, if necessary, tilt the saw up or down to increase the length of the kerf. For thin sheet stock, clamp the work between two pieces of wood and cut through the wood–metal sandwich.

Keep enough pressure on the blade to prevent it from being pinched or jammed, as this often breaks some of the teeth or breaks the blade. If a blade breaks and the operation must be finished with a new blade, always start a new cut with the new blade if possible. If the work is round, rotate it and start the new cut in line with the first one. If the work is flat, start the new cut from the other edge. This is because a new blade has more *set*

than a worn one, and will usually jam in the narrow kerf you have already made.

The set in a saw refers to how much the teeth are pushed out in opposite directions from the sides of the blade. Set assumes that the kerf will be slightly wider than the thickness of the blade and thus provide clearance to prevent sticking. Blades are made with 14, 18, 24, and 32 teeth per inch. The 18-tooth blade is most commonly used for general work, and the 32-tooth blade is used for thin sheet and tubing.

In addition to hand-held hacksaws, many shops have machine-driven (sometimes called automatic) hacksaws that will chop through steel bar stock at high speed. The same principles apply to operating these hacksaws as the hand-held tools you will use. However, an electric motor does the work for you.

Protect the teeth of your hacksaw from contact with hard materials such as slag. Cover the blade when not in use, or keep your hacksaw in a tool box. Wipe the blade with an oily rag after use, unless you are concerned about contaminating a surface that you will subsequently weld.

22-11
CHISELS AND CHIPPING HAMMERS

In the hands of a skilled worker a chisel can be made to do almost anything a milling machine can do, although perhaps less accurately and at greater expense in time and money. Chisels are made from tool steel by heating the end of a bar and hammering it into the desired shape. The bar is usually octagonal, but may be round, square, rectangular, or hexagonal. After forging, the chisel must be hardened to give it the ability to cut metal, and the edge must be tempered or drawn to prevent chipping. The head end opposite the cutting edge is usually chamfered to reduce the amount of mushrooming that oc-

curs from repeated hammering.

The type most commonly used by welders is the flat cold chisel, which you can use to cut rivets, split nuts, chip castings and forgings, remove bad weld metal, and cut thin metal sheets when you don't want to do this work with a cutting torch or a gouging rod. There are other types of chisels with specially-shaped cutting ends and edges for cutting keyways, narrow grooves, and square corners (the cape chisel). There also are chisels for cutting semicircular grooves and for chipping a fillet on the inside corners of parts (round-nose chisels). Diamond-point chisels are used for cutting V-grooves and sharp corners.

As with any other tool, there is a correct technique for using a chisel. To start with, use a chisel that is big enough for the job. The width of the cutting edge denotes the size, but it is sometimes necessary to use an extra-light or extra-heavy tool for special jobs. Also be sure to use a hammer that matches the chisel—the larger the chisel, the heavier the hammer. A heavy chisel will absorb the blows of a light hammer like an anvil and will do virtually no cutting.

A chisel is usually held in your left hand (if you are right handed) with your thumb and first finger about 1 in. [25 mm] from the top. Hold the chisel steady, but not tightly. Your finger muscles should be relaxed, so that if the hammer does strike your hand, it will slide down the tool and lessen the pain. Keep your eyes on the cutting edge of the tool, not on the head, and swing the hammer in the same plane as the body of the chisel. If you are doing a lot of work, you can slip a piece of rubber hose over the chisel to make an easier grip and lessen the shock on your hand and arm.

When you are chipping, the depth of your cut is controlled by the angle at which you hold the chisel. Don't try to make too deep a cut at one time. For rough cuts, $\frac{1}{16}$ in. [1.6 mm] is sufficient; for

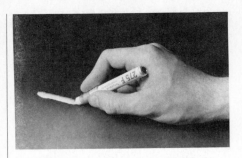

FIGURE 22-36 The Tempil° Division of Big Three Industries, with division headquarters in South Plainfield, New Jersey, is well known to most skilled welders for its temperature-indicating materials. These indicators are marketed through most welding distributors. For example, this Tempstik° crayon is calibrated to melt at 275°F. The operator is applying a mark with the material before heating the work. At the specified temperature, the mark will melt and become glassy-looking. You can buy an entire test kit of these crayons, each calibrated to be within 1 percent of the specified temperature (accuracy traceable to the National Bureau of Standards). The 20 crayons in each kit are systematically spaced between 125°F [52°C] and 800°F [427°C]. The marking crayons are also certified to be free of lead and sulfur so that they will not harm your heated metal. Temperature indicators such as these (paints and other materials also are available) are used for brazing, soldering, and heat-treating, as well as for welding temperature indication. Some indicators melt while others change color at the specified temperature.

finishing cuts, take 1/32 in. [0.8 mm] or less. Always wear safety goggles when chipping, and see that anybody around you is protected by a screen from the flying chips. Always chip away from yourself. If the work is held in a vise, chip toward the stationary jaw.

A chisel must be sharp for effective, fast cutting. Sharpen your chisel by holding the edge against a grinding wheel with only light pressure to avoid overheating the chisel's cutting edge, and dip the edge in water frequently enough to keep it cool. You don't want to heat the edge so much on the grinder that you temper the hardness right out of the tool steel.

Try to maintain the original angle of the chisel's cutting edge by grinding only a small amount at a time from each surface of the cutting edge. For best results the wide faces of the cutting edge should not be straight, but should be ground to a slight radius; higher at the center than at the ends.

If the temper on your chisel has been drawn by overheating during grinding, or if the tool has been reground so often that the hardened section at the cutting edge has been worn away, the tool will no longer hold an edge and it must be rehardened. This job should be given to an experienced heat treater or a good blacksmith. Grind the head end off smooth with a slight chamfer along the edge. At every grinding, check the top end for ragged edges or mushrooming. Mushroomed chisel heads are dangerous. Not only can you cut your hand on them, but a piece of steel can break off when you hit the chisel, and the flying metal chip will hit you.

Chipping hammers (Fig. 22-37) are lovely devices that you will enjoy using. They are compressed-air-operated tools with replaceable chisel cutting heads. For heavy cutting, and for gouging out bad weld metal when you don't want to use (or don't own) an air carbon-arc cutting outfit, you can plow through solid steel and imbedded slag with a chipping hammer like pushing a plow through wet snow.

Air-operated wire brushes are hand-held tools that beat slag off welds like a whip (Fig. 22-38).

22-12 GRINDING WHEELS AND DISKS

Grinding wheels are available in an enormous variety of shapes and materials. Fortunately, welders don't have to worry about the

FIGURE 22-37 The ARO Corporation, Bryan, Ohio, makes a wide range of air-operated tools for welders. This is an air-operated chisel that can remove bad weld metal or smooth the contour of a weld bead. It sure beats hand chiseling with a heavy hammer.

FIGURE 22-38 ARO Corporation makes pneumatic (air-operated) wire brushes that will beat the most tenacious slag off a workpiece at high speed, even in narrow spaces such as this fillet weld inside a box. Their welder's tools are available through many welding distributors as off-the-shelf items that you can pick up when you visit a welding store.

(A)

(B)

(C)

FIGURE 22-39 ARO Corporation's pneumatic grinders come in many shapes, sizes, and wheel types for everything from removing slag to finishing weld beads to use as cutting wheels to slice through light-gauge sections. (A) A large disk grinder. (B) A small grinder. Notice that both operators have eye protection. The operator with the big grinder is wearing a transparent full-face shield. The operator with the small grinder has safety glasses with side shields. (C) Another kind of air-operated grinder. This is a cone-grinder used for heavy metal removal, especially on rounded corners and in hard-to-reach areas where you can't get a disk grinder. This operator is wearing a full-face shield attached to his hard hat.

vast array of wheels and materials that are now available. Most of your grinding work will simply be removing slag and bad weld metal (Fig. 22-39). However, there are some very important things you should know about abrasive tools, which include not only grinding wheels but also abrasive belts. Whatever shape they come in, grinding wheels and belts are not toys. They are cutting tools and should be treated with care (Fig. 22-39).

Formerly, each grinding wheel manufacturer used a different system for indicating the nature of a grinding wheel. This led to considerable confusion, so a standard marking system was set up, sponsored by the Grinding Wheel Institute.

The marking consists of six parts: (1) abrasive, (2) grain size,

(3) grade, (4) structure, (5) bond, and (6) the manufacturer's record or private marking code. The code will be found on every grinding wheel you use. A typical marking on a wheel might read:

51 A 36 L 10 V Δs

The first number (51 for example) is the manufacturer's symbol indicating the exact kind of abrasive used in the wheel. This number is a prefix to the standard code, and it is optional.

The first standard part of the code is always a letter (A in our example) code for the type of abrasive used in the wheel. As a welder, you will want to know what the abrasive is, as we'll explain shortly. The abrasive code for grinding wheels commonly used by welders is:

A = Aluminum oxide
C = Silicon carbide
D = Diamond

Other code letters designate special grinding wheel materials. One newer material is boron carbide, which is harder than aluminum oxide. Of course, diamond remains the hardest abrasive. Diamond wheels can cut anything, including stone.

The third code (36 in our example) tells you the grain size of the abrasive in the wheel. That's also important for welders to know.

Coarse-grained wheels will have numbers ranging from 10 to 24. Medium-grained wheels will have numbers from 30 to 60 (our example is 36; therefore the wheel has a medium-grained abrasive). Fine-grained abrasive wheels will have numbers from 70 to 180. Very fine grained abrasive wheels will have grain-size numbers from 220 to 600. Therefore, the larger the grain-size number, the finer the abrasive grain size. The smaller the grain-size number, the larger, more coarse-grained the wheel abrasive is.

The code letter for grade (L in our example) can range from A to Z. Very soft wheels will have letters at the beginning of the alphabet, like C or D. Medium-grade wheels will have letters in the middle of the alphabet, like L in our example, or O or P. Very hard wheels will have letter codes at the end of the alphabet, like W, X, Y, or Z.

Structure refers to the relative grain spacing, which can range from a porous wheel (lots of space between grains) to a very dense wheel (closely packed grains). Wheels with close grain spacing (relatively dense wheels) have low numbers, such as 0, 1, 2, 3, 4, 5. Wheels with more open structure, or wide grain spacing, are designated by higher numbers, such as 8, 9, 10, 11, or 12. The available structure depends somewhat on the grain size, grade, and bond. Our example has a moderately open structure (10).

The type of bond (V in our example) is one of several possible letters. The bonding material is denoted by the following code letters:

V = Vitrified
S = Silicate
E = Shellac or elastic
R = Rubber
B = Resinoid (various plastics)
O = Oxychloride

The sixth and final position in the grinding-wheel code is reserved for the private use of the wheel manufacturer, usually to indicate a particular bond modification. The last code entry on a grinding wheel label might mean anything.

Here is another typical marking for a grinding wheel:

+ A 46 M 5 V BE

This code tells you that the wheel is made of aluminum oxide abrasive (A) with a 46 grit (grain size), M grade (the wheel is of approximately medium hardness), with a number 5 (low–medium) structure number, and a vitrified (V) bond of the manufacturer's BE type.

That's an overview of the standard grinding-wheel code. Now we'll give you some additional information that will help you work with grinding wheels.

Classification by Bonding Material

The American National Standards Institute (ANSI), working with the Grinding Wheel Institute, defines an abrasive wheel as a cutting tool consisting of abrasive particles held together by various bonding materials. The wheels are further categorized into two types: organic wheels (bonded by an organic material such as resin, rubber, shellac, or similar agents), and inorganic wheels (bonded by inorganic materials such as clay, glass, porcelain, sodium silicate, magnesium oxychloride, or metal). Wheels bonded with clay, glass, porcelain, or related ceramic materials are called *vitrified bonded wheels*.

Classification by Use

ANSI also describes the different operations for which abrasive wheels are used. Wheels in different size ranges are required depending on the intended use (something that machinists have to worry about more than you do). The major categories include:

■ Cutting off—for slicing or slotting parts. This operation is usually performed with a thin abrasive wheel, usually one with an organic bond. Cut-off wheels are very useful to welders because they often can slice up steel much faster than a hacksaw.
■ Cylindrical, between centers—for grinding the outside surface of a part that rotates about its own center. That's mostly a job for machinists.
■ Centerless—where the work is rotated by a revolving roll. Machinists use them.
■ Internal—for grinding the inside bore of a workpiece. Again, mostly work for machinists.
■ Offhand grinding—where the work is held in the operator's hand. That normally means wheels on small bench grinders.
■ Saw gumming—sawtooth sharpening and sharpening with a grinding wheel.
■ Snagging—for removing relatively large amounts of material without regard for close tolerances or surface-finish requirements. Typically, snagging removes surface defects from billets and excess metal or slag from welds. Most of the wheels you use will be snagging wheels.
■ Surfacing grinding—where the workpiece is flat. Again, this work is mostly for precision grinding. However, you might use a portable surface grinder on your work.
■ Tool grinding—for grinding and sharpening the edges of tools such as drills, taps, and reamers. From time to time you will want to use tool grinders to sharpen chisels.

Different Surfaces for Different Jobs

Standard grinding wheels are made to cut either with the periphery (narrow edge) of the wheel or with the side face. You would use the edge of a grinding wheel as a cut-off wheel and probably remove excess weld metal with the side face. However, you would gouge out bad weld metal with the periphery or edge of a wider wheel than you would use for cutting off sections.

ANSI has a detailed classification for grinding wheels by their shape. For example, a simple peripheral cutting wheel is ANSI type 1. ANSI type 6 is a grinding wheel with a flat face. Type 6 wheels often are reinforced around their periphery with wire or metal straps. You don't have to worry about ANSI wheel-shape designations. You can certainly pick a wheel just by looking at your workpiece and matching the wheel shape to the job. Only machinists would have to get into the details of the ANSI coding system. Complete details, if you ever need them, are found in ANSI publication B74.2.

Selecting Wheels by What You Grind

Workpiece material has the greatest impact on your choice of a grinding wheel. High-strength steels, tool steels, alloy steels, and hardened steels of different kinds are generally ground with aluminum oxide wheels. Soft materials and metals that are both hard and brittle such as cast iron are ground with silicon carbide wheels. These materials include cast iron, brass, bronze, aluminum, stainless steels, low-carbon steels, and sintered carbides. The tiny abrasive particles in silicon carbide wheels break down more easily than aluminum oxide particles and keep forming new sharp edges for rapid metal removal.

There is a vast array of new grinding-wheel materials available for special jobs. Examples are cubic boron nitride wheels which are second only to diamond in hardness, diamond wheels, and zirconium oxide wheels. Most of these wheels are considered to be superabrasives and are used mostly for extremely hard materials and for high-production grinding.

You won't be faced with many decisions on the selection of a grinding-wheel material, but you will have to keep a very important fact about grinding wheels in mind. *Don't use the same wheel on different materials, especially if grinding stainless steel after you have used a wheel on carbon steel.* A wheel with particles of low-carbon steel in it will deposit the particles on the stainless steel. These low-carbon-steel particles will become points where corrosion will begin on otherwise highly corrosion resistant stain-

less steel. Similar problems occur with aluminum.

CAUTION: Wear a clear face mask when grinding, unplug the grinder after you use it, and use common sense.

Also, use fine-grained wheels for finishing and coarse-grained wheels for rapid metal removal. Always wear goggles that fully wrap around your face or eyes (a full transparent face shield is even safer). And, of course, use common sense when grinding. You want to remove metal, not your fingers. So don't jam the work against the wheel and hold a portable grinder steady when you use it. Always pull the plug on the grinder when you finish.

Remember that portable grinders have triggers (even though they also have trigger guards). You can start one accidentally by depressing the trigger with your shoe. Grinders can go through your feet even faster than they go through steel. And, of course, never change grinding wheels while the grinder is plugged in.

Most grinding is done dry. That is, you don't use the cutting lubricants that a machinist would use. These lubricants will contaminate your workpiece with organic compounds and produce hydrogen embrittlement, weld-metal porosity, and all sorts of other troubles. If a lubricant is absolutely necessary, you often can use water—but make sure that you thoroughly dry your workpiece with a torch if you continue welding.

The literature that comes with different grinding wheels will give you recommendations on the materials they are best for and the speeds at which they should be used. The best advice we can give you is keep separate wheels available for different classes of materials. Use your stainless-steel wheel only on stainless steels. Use your aluminum wheel only on aluminum, and reserve the carbon-steel wheel for carbon steels and cast irons. If you are

not sure what to use, ask a machinist or your welding supervisor. A welding distributor who sells abrasive wheels also can advise you.

Before any wheel is put into service, you should examine it carefully for cracks. A cracked wheel can fly apart when rotating at high speed. The best way to inspect a wheel for cracks is to hang the wheel on a hook and tap it lightly with a hammer. A sound wheel will give a clear ring. If it sounds cracked (clunk instead of ring), don't use the grinding wheel. Return it to the tool crib for more careful inspection.

REVIEW QUESTIONS

1. Why are welding positioners used?

2. Name the best kind of welding positioner for making butt welds in 20-ft lengths of 12-in.-diameter pipe.

3. What kind of welding positioner do you think would be good for working on a square steel box frame that weighs less than 1300 lb and is 4 ft on a side?

4. Do you think a vise on a workbench is a kind of welding positioner?

5. Show how to calculate the tilt torque of the workpiece in Fig. 22-20. Did you remember to add in the inherent overhang in your calculation?

6. You will be making a steel silo 40 ft high and 10 ft in diameter. Its wall thickness is uniform from top to bottom and it has no openings or connections on it. Indicate exactly where the center of the gravity is located.

7. Where is the center of gravity of a football (ignoring the slight affect of the laces on the open seam)?

8. Estimate approximately where the center of gravity of your body is. If you are standing up straight, about how high off the floor would your CG be?

9. Two straight pipes of equal length have been butt-welded together. However, one section has a wall thickness twice that of the other (it's twice as heavy). The CG of the connected pipe sections will be somewhere on their centerline. Will it be located toward the heavy-walled section or the lighter-walled section? Tell exactly where the center of gravity is located.

10. What welding process, or processes, are welding-head manipulators most likely to be used for?

11. A 51-ft-long board is balanced horizontally on a sharp edge underneath it, like a teeter-totter. Fifty feet of the board extends out on one side, and only 1 ft on the other. But the 1-ft end that sticks out carries a 50-lb weight. How much weight is needed on the long end of the board to balance the big weight on the short end? Ignore the weight of the board itself when you answer.

12. Does the increasing use of welding robots worry you?

13. Name a method for processing metals that will increase their fatigue life. (Sometimes the same process can help prevent stress-corrosion cracking of stainless-steel parts.)

14. Name four simple products that can be used to test the temperature of solid metal. Are there any ways you can estimate the temperature of very hot steel if you didn't have any special products or instruments?

15. Name a piece of equipment that can roll 2-in.-thick plate into a cylinder.

16. What kind of machinery would you use to bend long, narrow metal sections into U-shaped channels?

17. Look at the grinding wheels in your shop and find out what wheel grades are used.

18. After you are through using a hand grinder, what's the first thing you should do with it?

19. Grinding wheels used for carbon steels should not be used afterward for stainless steels or aluminum? Why?

20. If you have used a grinding lubricant on a weldment (even water), what should you do before you weld over the area?

Welding Codes and Getting Certified

There is no such thing as a welder certified for all types of work. There's even a good chance that you may not have to be certified at all. It all depends on where you get a job and what kind of work you do. However, getting some kind of welder certification will go a long way toward getting you a good job.

23-1
THE PERFECT WELDER DOESN'T EXIST

One of the reasons that there is no general certification for welders is that every welder who becomes certified does it under different welding codes. The different codes either spell out exactly what you have to do to be certified for the work they cover or at least specify the kinds of materials and filler metals that you can use for that work. Table 23-1 lists the key groups that write these codes and specifications, along with other organizations that you'll find it useful to know about in your career as a welder.

Some of the most important code-writing groups are the American Society of Mechanical Engineers (ASME) which writes the Boiler and Pressure Vessel Code (for boilers, pressure vessels, and pressure piping including nuclear power generation reactors), the American Welding Society (AWS) Structural Welding Code (it covers the welding of bridges and buildings), and the American Petroleum Institute (API) standards (for welding gas and oil line pipe and large open petroleum and gasoline storage tanks). Shipbuilding is covered variously by U.S. Coast Guard (USCG) requirements, the requirements of the American Bureau of Shipping (ABS), and, for naval vessels, the U.S. Navy.

For example, you may be certified to do certain kinds of work governed by the ASME Boiler and Pressure Vessel Code, but that does not mean that you are certified to do all kinds of ASME code work. If you get certified to do certain kinds of ASME Boiler and Pressure Vessel Code work with SMAW electrodes on a specific kind of steel, you are only certified for that particular job. If the steel, the welding process, or even the plate thickness, the diameter of the electrode, or welding position is changed, you probably are no longer certified for the new work and you will have to requalify.

Fortunately, passing a test for the most difficult ASME work will probably qualify you for most ASME (or AWS) code jobs that use the same welding process but are considered to be easier to do. For example, if you can weld thick-walled high-alloy-steel pipe in any position with stick electrodes, you'll probably be qualified for welding the same material with stick electrodes while working on thinner plate in the flat position.

It doesn't matter how many years you have been a welder, you still have to requalify when the nature of the work changes. (What's considered a change in procedure and what's not are defined by the codes.) That's one reason why the ASME Boiler and Pressure Vessel Code is so tough and so important. You'll find the same thing is true of the API and AWS codes.

In addition to these very detailed fabricating and welder qualification codes, individual companies usually have welder qualification tests. Shipyards, for example, will test you on anything from tack welds to your ability to weld nuclear-powered-submarine hull plate, depending on the job you apply for. But if you are not qualified, or can't pass a more advanced company test, there's still a very good chance that they will want you.

You can start in a union training program, or work on the job as a welder's helper until you improve your skills and can demonstrate the ability to pass a more

TABLE 23-1 Organizations you should know about*

AES	Abrasive Engineering Society 1700 Painters Run Road Pittsburgh, PA 15243 (412) 221-0900	ASTM	American Society for Testing and Materials 1916 Race Street Philadelphia, PA 19103 (215) 299-5400
AA	Aluminum Association 818 Connecticut N.W. Washington, DC 20006 (202) 862-5100	ASQC	American Society for Quality Control 161 West Wisconsin Avenue Milwaukee, WI 53203 (414) 272-8575
ABS	American Bureau of Shipping 65 Broadway New York, NY 10006 (212) 440-0300	AWWA	American Water Works Association 6666 West Quincy Denver, CO 80235 (303) 642-7090
AFS	American Foundrymen's Society Golf and Wolf Roads Des Plaines, IL 60016 (312) 824-0181	AWS	American Welding Society 550 Northwest LeJeune Road Miami, FL 33126 (305) 443-9353
AGA	American Gas Association 1515 Wilson Boulevard Arlington, VA 22209 (703) 841-8400	AAR	Association of American Railroads 1920 L Street N.W. Washington, DC 20036 (202) 293-4000
AIAA	American Institute of Aeronautics and Astronautics 1290 Avenue of the Americas New York, NY 10019 (212) 581-4300	AISE	Association of Iron and Steel Engineers Suite 2350 Three Gateway Center Pittsburgh, PA 15222 (412) 281-6323
AIME	American Institute of Mining, Metallurgical and Petroleum Engineers 345 E. 47th Street New York, NY 10017 (212) 644-7695	CISC	Canadian Institute of Steel Construction 201 Consumers Willowdale, Ontario M2J 4G8 (416) 487-5415
AIPE	American Institute of Plant Engineers 3975 Erie Avenue Cincinnati, OH 45208 (513) 561-6000	CWB	Canadian Welding Bureau 254 Merton Street Toronto, Ontario M4S 1A9 (416) 487-5415
AISC	American Institute of Steel Construction, Inc. 400 North Michigan Avenue Wrigley Building Chicago, IL 60611 (312) 670-2400	CWDI	Canadian Welding Development Institute 391 Burnhamthorpe Road East Oakville, Ontario L6J 4Z2 (416) 845-9881
AISI	American Iron and Steel Institute 1000 16th Street N.W. Washington, DC 20036 (202) 452-7100	CGA	Compressed Gas Association 500 Fifth Avenue New York, NY 10036 (212) 354-1130
ANSI	American National Standards Institute 1430 Broadway New York, NY 10018 (212) 354-3300	CDA	Copper Development Association Chrysler Building 405 Lexington Avenue New York, NY 10017 (212) 953-7300
API	American Petroleum Institute 2101 L Street N.W. Washington, DC 20037	CMAA	Crane Manufacturers Association of America 1326 Freeport Road Pittsburgh, PA 15238 (412) 782-1624
ASME	American Society of Mechanical Engineers 345 E. 47th Street New York, NY 10017	CSA	Cryogenic Society of America 3 Rockinghorse Road West Rancho Palos Verdes, CA 90274 (213) 832-4848
ASM	American Society for Metals Route 87 Metals Park, OH 44073 (216) 338-5151	FMA	Fabricating Manufacturers' Association 7811 North Alpine Road Rockford, IL 61111 (815) 654-1902
ASNT	American Society for Nondestructive Testing 3200 Riverside Drive Columbus, OH 43221	GWI	Grinding Wheel Institute 712 Lakewood Center North Cleveland, OH 44107 (216) 226-7700
ASSE	American Society of Safety Engineers 850 Busse Highway Park Ridge, IL 60068 (312) 692-4121		

* This table is courtesy of *Welding Design & Fabrication* magazine, Penton/IPC Inc. Publishing Corporation, 1111 Chester Avenue, Cleveland, OH 44114, (216) 696-7000.

HMI	Hoist Manufacturing Institute 1236 Freeport Road Pittsburgh, PA 15238 (412) 782-1624		SME	Society of Manufacturing Engineers One SME Drive P.O. Box 930 Dearborn, MI 48128
IAPA	Industrial Accident Prevention Association 100 Front Street West Royal York Hotel Arcade Toronto, Ontario M5J 1E3		SPFA	Steel Plate Fabricators Association 2901 Finley Road Suite 103 Downers Grove, IL 60515 (312) 629-3630
IIW	International Institute of Welding 2501 North West 7th Street Miami, FL 33125 (305) 642-7090		SSCI	Steel Service Center Institute 1600 Terminal Tower Cleveland, OH 44113 (216) 644-6610
IOMA	International Oxygen Manufacturers Association P.O. Box 16248 Cleveland, OH 44116 (216) 228-2166		STI	Steel Tank Institute 111 East Wacker Drive Chicago, IL 60601 (312) 644-6610
ISE	Industrial Safety Equipment Association Suite 501 1901 North Moore Street Arlington, VA 22209 (703) 525-1695		TMHI	The Material Handling Institute 1326 Freeport Road Pittsburgh, PA 15238 (412) 782-1624
NASA	National Aeronautics and Space Administration 400 Maryland Avenue S.W. Washington, DC 20546 (202) 755-2320		TWI	The Welding Institute P.O. Box 5268 Hilton Head Island, SC 29928 (803) 785-3417
NAAD	National Association of Aluminum Distributors 1900 Arch Street Philadelphia, PA 19103 (215) 564-3484		UIA	Ultrasonic Industry Association 481 Main Street New Rochelle, NY 10801 (914) 235-4020
NACE	National Association of Corrosion Engineers P.O. Box 986 Katy, TX 77450 (713) 492-0535		DOD	U.S. Department of Defense Procurement Information The Pentagon Washington, DC 20301 (202) 697-1481
NEMA	National Electrical Manufacturers Association Suite 300 2101 L Street, N.W. Washington, DC 20037 (202) 457-8400		DOL	U.S. Department of Labor 200 Constitution Avenue N.W. Washington, DC 20210 (202) 523-8165
NLPGA	National LP-Gas Association 1301 West 22nd Street Oakbrook, IL 60521 (312) 986-4800		DOT	U.S. Department of Transportation 400 Seventh Street N.W. Washington, DC 20590 (202) 426-4000
NSC	National Safety Council 444 North Michigan Avenue Chicago, IL 60611 (312) 527-4800		WRC	Welding Research Council Room 801 345 East 47th Street New York, NY 10017 (212) 644-7956
NWSA	National Welding Supply Association 1900 Arch Street Philadelphia, PA 19103 (215) 564-3484		WSTI	Welded Steel Tube Institute 522 Westgate Tower Cleveland, OH 44116 (216) 333-4550
OSHA	Occupational Safety and Health Administration 200 Constitution Avenue N.W. Washington, DC 20210 (202) 523-8151			
RWMA	Resistance Welder Manufacturer's Association 1900 Arch Street Philadelphia, PA 19103 (215) 564-3484			
SAE	Society of Automotive Engineers 400 Commonwealth Drive Warrdendale, PA 15096 (412) 776-4841			

difficult test. The company will want you to advance just as much as you want to improve your skills and your earning power. They need skilled welders just as badly as you need a job. Therefore, most companies who take beginning welders will bring them up to speed as fast as possible.

23-2
RELY ON YOUR WELDING SCHOOL
The job market in every part of the country is different. It not only varies from place to place but also from time to time. Your welding instructor or placement counselor will know where the jobs are and what type of skills prospective employers are looking for at any given moment.

About a month or two before you graduate, you should start training yourself for the skills that the companies in your area will most likely be looking for. If the demand is for SMAW welders, concentrate on your stick-electrode welding skills while you continue to learn the rest of the major welding processes. If you live near an aerospace or chemical company, work hard on MIG, TIG, and plasma-arc welding.

Many welding students come into school after hours to prepare themselves for specific jobs. Some people even buy their own portable welding equipment and work at home and on weekends to get ready for the first interview and the first test. It doesn't matter how good you think you are, you'll only get one job. Assume that everybody else in your class is after that same job. Your job is to be better than they are.

23-3
GETTING CERTIFIED
Qualified and certified are different. You may have all the skills you need to do a job; that means you are qualified for the work. Certification means that somebody else has given you a special

TABLE 23-2 Metallurgical labs that certify welders and specialize in weld testing

Engineers Testing Laboratories, Inc.
Phoenix, AZ
(602) 268-1381
Metals Engrg. & Testing Lab.
Phoenix, AZ
(602) 272-4571
Accurate Metallurgical Services, Inc.
Santa Fe Springs, CA
(213) 693-6201
Accurate Weld Testing Lab.
Southgate, CA
(213) 564-5870
Acoustic Emission Technology Corp.
Sacramento, CA
(916) 927-3861
Anamet Lab. Inc.
Berkeley, CA
(415) 841-5771
Atlas Testing Laboratories, Inc.
Los Angeles, CA
(213) 722-8810
Borchardt & Assoc. Inc.,
T. J. Baldwin Park, CA
(213) 962-3568
BTC Laboratories Inc.
Ventura, CA
(805) 656-6074
Durkee Testing Lab., Inc.
Gardena, CA
(213) 321-9800
General Dynamics
San Diego, CA
(714) 277-8900
Grandia Laboratories
Costa Mesa, CA
(714) 645-9080
Kenneth H. Holko, Inc.
San Diego, CA
(619) 270-8043
Metallurgical Associates
Walnut Creek, CA
(415) 934-1161
Nutech
San Jose, CA
(408) 629-9800
Testing and Controls
Mountain View, CA
(415) 967-6982
Testing Engineers Inc.
Oakland, CA
(415) 835-3142
Testing Services & Inspection, Inc.
National City, CA
(619) 474-6645
Commercial Testing Lab.
Denver, CO
(303) 825-3207
Hauser Lab.
Boulder, CO
(303) 443-4662
Mangone Lab., Inc.
Golden, CO
(303) 279-2768
Ponderosa Assoc., Ltd.
Louisville, CO
(303) 666-8112
Flecher Technology
Windsor, CT
(800) 243-8417
Hartford Steam Boiler Insp. & Ins.
Essex, CT
(203) 767-2113
Stanford Technology Corp.
Glenbrook, CT
(203) 348-4080

Lehigh Testing Labs., Inc.
Wilmington, DE
(302) 655-7358
Alvine Assoc. Inc.
William Casselburry, FL
(305) 339-3473
Applied Research Laboratories, Inc.
Miami, FL
(305) 624-4800
Industrial Testing Services
St. Petersburg, FL
(813) 577-0247
Pensacola Testing Laboratories, Inc.
Pensacola, FL
(904) 477-5100
Applied Technical Services, Inc.
Marietta, GA
(404) 952-8705
AT & E Consultants
Norcross, GA
(404) 448-6644
Northern Testing Lab.
Boise, ID
(208) 377-2100
CONAM Inspection Div.
Itasca, IL
(312) 773-9400
Hunt Co., Robert W.
Chicago, IL
(312) 922-0872
Kawin Co., Inc., Charles C.
Broadview, IL
(312) 865-0400
Magnaflux Corp.
Chicago, IL
(312) 867-8000
Magnaflux Quality Services
Chicago, IL
(312) 867-8000
Magnetic Insp. Lab., Inc.
Rosemont, IL
(312) 678-5415
Packer Engineering Associates, Inc.
Naperville, IL
(312) 355-5722
Pennies Industrial X-Ray Co.
Chicago, IL
(312) 722-2750
Taussig Associates, Inc.
Chicago, IL
(312) 676-2100
Time National Laboratories, Inc.
Chicago, IL
(312) 646-1200
Calumet Testing Services, Inc.
Highland, IN
(312) 474-5860
Jimona, Inc.
Mishiwaka, IN
(219) 259-8548
Sherry Lab.
Muncie, IN
(317) 747-9000
Patzig Testing Laboratories Co. Inc.
Des Moines, IA
(515) 266-5101
Metallurgical Services Co.
Louisville, KY
(502) 966-8701
Mechanical Testing Laboratory
Kenner, LA
(504) 466-0864
Minco Inc.
Westwego, LA
(504) 773-1013

Owensby & Kritikos Inc.
Gretna LA
(504) 368-3122
Teleshak Metallurgical Lab.
New Orleans, LA
(504) 945-3255
Energy Materials Testing Lab.
Biddleford, ME
(207) 282-5911
EMV Associates, Inc.
Rockville, MD
(301) 948-7400
Penniman & Browne, Inc.
Baltimore, MD
(301) 825-4131
Reliance Testing Lab. Inc.
Timonium, MD
(301) 252-6030
Arnold Greene Testing Laboratories, Inc.
Natick, MA
(617) 653-5950
Arthur D. Little, Inc.
Cambridge, MA
(617) 864-5770
Briggs Engineering Inc.
Norwell, MA
(617) 773-2780
Dirats Laboratories
Westfield, MA
(413) 568-1571
Factory Mutual Research Corp.
Norwood, MA
(617) 762-4300
Massachusetts Materials Research, Inc.
West Boylston, MA
(617) 835-6262
Skinner & Sherman Lab., Inc.
Waltham, MA
(617) 890-7200
Cleveland X-ray Inspection, Inc.
Jackson, MI
(517) 750-1558
Detroit Testing Laboratory, Inc.
Oak Park, MI
(313) 398-2100
Engrg. & Testing Lab.
Grand Rapids, MI
(616) 247-0515
National Testing & Res. Lab., Inc.
Detroit, MI
(313) 834-7500
Spectrographic Testing Laboratory
Detroit, MI
(313) 366-2933
Braun Engineering Testing
Minneapolis, MN
(612) 941-5600
R. T. Picha Company
New Brighton, MN
(612) 636-2968
Facts Inc.
St. Louis, MO
(314) 645-1066
General Testing Laboratories, Inc.
Kansas City, MO
(816) 471-1205
Industrial Testing Laboratories, Inc.
St. Louis, MO
(314) 771-7111
St. Louis Testing Laboratories Inc.
St. Louis, MO
(314) 531-8080

Source: *Metal Progress*, American Society for Metals.

Nebraska Testing Laboratories, Inc.
Omaha, NB
(402) 331-4453

Omaha Engineering and Testing Consultants, Inc.
Omaha, NB
(402) 341-5181

Omaha Nondestructive & Metallurgical Testing, Inc.
Omaha, NB
(402) 341-5181

Omaha Testing Laboratories, Inc.
Omaha, NB
(402) 341-5181

Fairfield Testing/Labtech Inc.
Fairfield, NJ
(201) 575-8665

International Testing Lab., Inc.
Newark, NJ
(201) 589-4772

Martin Metallurgical Inc.
Roebling, NJ
(609) 499-3200

Physical Acoustics Corp.
Princeton, NJ
(609) 452-2510

PSE&G Research Corp.
Maplewood, NJ
(201) 761-1906

Ramball Testlab
Pennsauken, NJ
(609) 488-0515

Spectrum Laboratories, Inc.
Piscataway, NJ
(201) 752-1400

ATL Engrg. Services
Albuquerque, NM
(505) 268-4537

Academy Testing Laboratories, Inc.
Farmingdale, NY
(516) 420-8666

American Standards Testing Bureau, Inc.
New York, NY
(212) 943-3156

Certified Testing Laboratories, Inc.
Bronx, NY
(212) 824-1616

Consolidated Testing Laboratories, Inc.
New Hyde Park, NY
(516) 746-3705

Heidi Testing Laboratories, Inc.
Burnt Hills, NY
(518) 399-3689

Lucius Pitkin Inc.
New York, NY
(212) 233-2737

Product Safety Laboratories, Inc.
New York, NY
(212) 233-2558

Law Engineering Testing Co.
Charlotte, NC
(704) 523-2022

Soil & Material Engineers, Inc.
Raleigh, NC
(919) 872-2660

Bowser-Morner Testing Lab., Inc.
Dayton, OH
(513) 253-8805

Central Testing Laboratories, Inc.
Cleveland, OH
(216) 475-9000

Col-X Corp.
Columbus, OH
(614) 267-1201

Conam Inspection
Columbus, OH
(614) 491-3000

Herron Testing Laboratories, Inc.
Cleveland, OH
(216) 524-1450

Metcut Research Associates, Inc.
Cincinnati, OH
(513) 271-5100

Midwest Testing Laboratories, Inc.
Piqua, OH
(513) 773-1013

Nutting Co., H. C.
Cincinnati, OH
(513) 321-5816

Quality Testing Inc.
Cleveland, OH
(216) 433-4424

Toledo Testing Lab., Inc.
Toledo, OH
(419) 241-7175

EMTEC Corp.
Norman, OK
(405) 321-1154

Metlab Testing Services, Inc.
Tulsa, OK
(918) 664-7767

Midstates Analytical Lab., Inc.
Tulsa, OK
(918) 622-6030

Oklahoma Testing Laboratories
Oklahoma City, OK
(405) 232-4666

Standard Testing & Engineering Co.
Oklahoma City, OK
(405) 528-0541

Koon-Hall Testing Corp.
Albany, OR
(503) 928-1668

MEI-Charlton Inc.
94Portland, OR
(503) 228-9663

Conam Inspection
Folcroft, PA
(215) 237-1500

Energy Consultants Inc.
Pittsburgh, PA
(800) 242-1767

Forney Inc.
Wampum, PA
(412) 535-4341

Gilbert Associates, Inc.
Reading, PA
(215) 775-2600

Industrial Testing Laboratory Services Corp.
Pittsburgh, PA
(412) 963-1900

Pennsylvania Instrument Service Corp.
Spring City, PA
(215) 495-5244

Pittsburgh Testing Laboratory
Pittsburgh, PA
(412) 922-4000

SPS Laboratories
Jenkintown, PA
(215) 572-3562

EG & G Sealol
Warwick, RI
(401) 781-4700

Applied Engineering Company
Orangeburg, SC
(803) 534-2424

Industrial NDT Company, Inc.
Charleston, SC
(803) 744-7412

Soil Consultants, Inc.
Charleston, SC
(803) 723-4539

An-Tech Laboratories, Inc.
Houston, TX
(713) 644-7501

Automation Industries, Inc.
Dallas, TX
(214) 351-5381

Bryan Lab., Inc.
Houston, TX
(713) 521-9595

Engineering Metals, Inc.
Houston, TX
(713) 492-1440

Hackney, Inc.
Dallas, TX
(214) 631-4420

Hurst Met. Research Lab., Inc.
Euless, TX
(817) 267-3421

Metallon, Inc.
Houston, TX
(713) 923-7761

NDE-AIDS, Inc.
Benbrook, TX
(817) 249-4760

REMSCO
Houston, TX
(713) 641-0436

Shilstone Engineering Testing Lab. Inc.
Houston, TX
(713) 224-2047

Southwestern Laboratories
Ft. Worth, TX
(817) 332-5181

Southwestern Laboratories
Houston, TX
(713) 692-9151

Texas Testing Lab., Inc.
Dallas, TX
(214) 428-7481

Trinity Engineering Testing Corp.
Austin, TX
(512) 926-6650

Associated Piping & Engineering Materials Testing Lab
Clearfield, UT
(800) 453-2170

Terra Tek Systems
Salt Lake City, UT
(801) 583-6186

Artech Corp.
Falls Church, VA
(703) 560-3292

Froehling & Robertson, Inc.
Richmond, VA
(804) 264-2701

Newport News Industrial
Newport News, VA
(804) 380-7821

Boeing Technology Services
Seattle, WA
(206) 394-3333

Northwest Laboratories of Seattle, Inc.
Seattle, WA
(206) 622-0680

Magnaflux Quality Services
Milwaukee, WI
(414) 771-3060

Technimet Corp.
Milwaukee, WI
(414) 483-0054

examination, inspected your work, tested it according to some previously established procedure (usually spelled out in the ASME, AWS, or API codes), and that you passed. The people who certify you have to be certified themselves just to test you.

The company that will certify you may be a contractor doing code work that requires certified welders. It might be a union working with industry to supply skilled welders (the pipeliner's union out of Tulsa is just one example), or it could be an independent testing organization.

Where to Get Certified

Herron Testing Laboratories in Cleveland, Ohio, tested 1500 welders in 1979. In 1980 the number of welders they tested doubled to 3000, and the next year it doubled again. This growth occurred because more manufacturers are being required to fabricate products to code rules to assure that their fabricated products will serve dependably and safely. Many companies simply don't have the time to test all the welders they need. They also may not know all the detailed code requirements for testing and certification because the details are very complicated and they change periodically. Independent test labs often do the testing work for them.

Arnold Greene Testing Laboratories in Natick, Massachusetts, tests welders for nearby shipyards, as well as for boiler and pressure-vessel fabricators and structural-steel erectors in their area. This company is typical of the many test labs in the United States that also certify welders.

Most of these special metallurgical service labs will certify welders to various welding-code requirements (see Table 23-2). Note, however, that they charge for their services. Try to get an employer (or a prospective employer) to pay for your certification. (Also note that many welding jobs don't require certification at all.) If you do want to become certified, have a specific code in mind. Get a copy of the code certification requirements and make sure that you can pass it before you try for official certification. Many codes won't let you try again for several months after you fail the first time.

To find a good local welder certification test lab, first talk it over with your school's welding instructor, then try the telephone Business Yellow Pages. Look under the category "Laboratories—Testing." Read the ads. Not all test labs certify welders. In fact, not all test labs even work with metals. A few phone calls, however, should get you some recommendations for a test lab that certifies welders. If you find a good test lab, be sure to ask what the certification test will cost. Also ask which tests are most in demand by local companies.

Also call nearby welding distributors. Use the Business Yellow Pages again, and simply look under "Welding Equipment and Supplies." The inside salesperson who answers the phone will probably know a lot about what's going on in his or her area. Inside sales personnel might not only know a local welder certification test lab but they also might know if any of their customers are looking for a good beginning welder.

What to do To get tested by a lab (once you've found one), first call them and tell them what you can do and ask them what tests would be best for you. They will set up a test procedure that meets the qualifications of whatever code you are interested in. They also will provide an impartial third-party endorsement of your work by actually witnessing that you did it.

If you pass the test, good for you. If you don't, it means that you need more training. Any independent test lab will charge you for their services. Testing can be expensive. If you pass, it can be

well worth the cost. There's nothing like walking into a prospective employer holding a signed and witnessed statement from a reputable independent test lab that you have already passed a welder certification test for some part of a major fabricating code important to the company.

23-4
JOBS WITHOUT CERTIFICATION

A lot of welding work is not governed by any codes. If you work in the production or maintenance departments of many large metal-working companies, and they make steel or aluminum, construction equipment, railroad cars or automobiles, refrigerators, or bicycles, you may only have to prove that you can do the job they want you to do. If you work in a company's maintenance department, all they want to know is that you can help fix broken machinery, install pipes, or help put up a new addition to the plant building.

Working with your welding instructor and your school's placement director, pick out a few target companies that you would like to work for. Either you or your school can call their personnel department a month or more before you graduate and find out what types of qualification tests they require of new welders. If they don't require tests, find out what kind of skills they need and concentrate on getting good at what they want.

Also remember that there are jobs for people who know a lot about welding but simply can't do it very well because they don't have very good manual skills. You may be well-qualified to sell welding products. You might prefer to work as a welding inspector, or maybe you should get a job in the quality-control or testing department. A hands-on course in welding also is ideal before entering college to (or afterward) study welding engineering.

On the other hand, not all welding jobs require the ability to join all kinds of metals with many different processes. You might be best operating a flame-cutting machine. In addition to fabricating plants, metal service centers need cutting machine operators. There are over 1000 metal service centers in the United States and they are often large companies. Or maybe you are a farmer or artist and simply want to be able to repair your own equipment or use welding to create metal sculpture. Alternatively, imagine what you can do as a steamfitter with a little more training. Since computers are used heavily in welding and manufacturing now, maybe you should follow up this course with one on computer programming (while you earn a living as a welder). That should make you very special to some companies.

You can practice for a welder certification test before you take it. If the personnel department of the company you have targeted doesn't have details on what kind of testing is performed, call one of the welding supervisors in the plant. Many supervisors who get a call from a new welder wanting to prequalify for a job before making an official application will take down your name and ask you to call back just as soon as you get out of school. Good people are hard to find. A résumé of your experience and a good cover letter helps to get you that first interview. It also helps you get the attention of a union training manager. See the Appendix for examples and advice on getting interviews.

23-5
THE ASME CODE

About 1 million boilers and pressure vessels are fabricated every year in the United States. Every one of them (with a few exceptions noted below) are made to ASME code requirements. You can't work on one of these products unless you have been certified. A company can't make any of these products and sell them in the United States without being certified by the code, too. Producers of steel, aluminum, and copper pipe and tubing (and other metals) can't even supply materials for code-fabricated equipment without meeting certain ASME code requirements.

Even welding filler-metal producers can't sell electrodes and gas rods for ASME code work unless the products and their companies meet code requirements. A materials supplier or a fabricator is issued a code stamp certifying that they meet the requirements. (See page 616 for further information on code stamps.)

The complete ASME Boiler and Pressure Vessel Code is terribly complicated. It would take you 2 months or more to read it. It's about a foot thick, expensive, and it comes in a dozen volumes of closely mimeographed and printed pages. The details of the code also change several times each year. The ASME code is accepted as law in almost every state, in all the provinces of Canada, and in a number of cities in the United States.

A volunteer army of some 750 engineers, almost all of whom hold down full-time jobs in industry, keep the code up to date. No group is more familiar with the code than these people. They are the engineers and government people who have to use it; the companies that make the materials; the companies that use the pressure vessels, piping, and other products covered by the code; and even insurance companies that insure pressure vessels.

To the best of our knowledge, no one has ever been killed by a boiler or pressure-vessel explosion due to poor design or faulty construction practice where the ASME code has become law, if the code requirements were followed. What fatalities there have been have always been the result of improper operation of the boiler or pressure vessel.

The ASME code covers the design and construction of power-generation boilers, heating boilers, nuclear power plant components, and any pressure vessel that will operate at a pressure of at least 15 psig [103 kPa]. (An open-top oil storage tank would be covered by the API codes.) The ASME code even covers the design and construction of fiberglass-reinforced plastic pressure vessels.

The code does not cover gas and oil transmission pipelines (that's handled by the API), or railroad tank cars (but some people think it should). The rules governing the fabrication of nuclear submarines also do not fall into the ASME code category; that's covered by military specifications under the U.S. Department of Defense (DOD).

How the ASME Code Is Divided
The ASME code has 11 sections. Five of them are construction-code sections pertaining to power boilers, nuclear power plant components, heating boilers, pressure vessels, and fiberglass-reinforced plastic pressure tanks.

Three other sections cover requirements for materials, nondestructive testing and examination, and welding. The welding sections include specific and very detailed instructions for certifying welders. These sections are referenced by the five ASME code construction sections.

There also are two sections on recommended rules for the care of boilers plus a final section on rules for the in-service inspection of nuclear power plant components.

If you do any work that must meet ASME code requirements, you will often hear people talking about different parts of the code. Even your filler metals will have to meet certain requirements of one or more of these sections.

Section by section Let's take a short tour through the ASME code sections.

SECTION I. POWER BOILERS This is a construction code covering power, electrical, and miniature boilers, and high-temperature boilers used in stationary service. This section includes power-generation boilers used in locomotives.

SECTION II. MATERIALS SPECIFICATIONS This is a very important section for you. It is divided into three parts (A, B, C) and contains specifications on code-accepted ferrous materials; nonferrous materials; and welding rods, electrodes, and other filler metals such as brazing alloys.

SECTION III. NUCLEAR POWER PLANT COMPONENTS Because of the ever-increasing amount of information on nuclear power plant components, section III of the ASME code has been subdivided into seven subsections. They are (1) general requirements, (2) class 1 components, (3) class 2 components, (4) class 3 components, (5) class MC components, (6) component supports, and (7) core support structures.

SECTION IV. HEATING BOILERS This is another construction code covering the design, fabrication, installation, and inspection of steam-heating, hot-water-heating, and hot-water-supply boilers which are directly fired by oil, gas, electricity, or coal.

SECTION V. NONDESTRUCTIVE EXAMINATION Covered here are such NDT methods accepted for use in the code as radiography, ultrasonics, liquid-penetrant, magnetic-particle, eddy-current, visual-inspection, and leak testing methods.

SECTION VI. RECOMMENDED RULES FOR CARE AND OPERATION OF HEATING BOILERS This section is directed toward the owners of steel and cast-iron heating boilers and is a guide to their operation, maintenance, and repair. It covers such areas as boiler-room accessories and facilities, automatic fuel-burning equipment and controls, and water treatment.

SECTION VII. RECOMMENDED RULES FOR CARE OF POWER BOILERS Similar to section VI, this section covers the operation and maintenance of stationary, portable water, and traction-type power boilers.

SECTION VIII. PRESSURE VESSELS—DIVISION 1 Basic rules for the construction, design, fabrication, inspection, and certification of pressure vessels are covered in division 1. These rules have been formulated on the basis of design principles and construction practices for vessels designed for pressures up to 3000 psi [20.7 MPa]. Stamping and coding of the vessels are also covered in this big section.

SECTION VIII. PRESSURE VESSELS—DIVISION 2 Division 2 provides an alternative to the minimum construction requirements for pressure vessels outlined in division 1. The rules here are more restrictive in the choice of materials that may be used, but they permit higher design stress in the range of temperatures over which the design stress is controlled by the ultimate or yield strength. More complete testing and inspection is called for. The rules in this section cover vessels to be installed at stationary locations.

SECTION IX. WELDING AND BRAZING QUALIFICATIONS This section covers the qualification of welders and welding operators and the procedures you must follow in order to comply with the code. *This is the most critical section of the ASME code for you.* Under procedure qualifications, each process is listed and the essential and nonessential variables of each process are spelled out for use in writing and in conducting welder certification tests. Welding performance qualifications also are included.

SECTION X. FIBERGLASS-REINFORCED PLASTIC PRESSURE VESSELS This section establishes general specifications for the glass and resin used to fabricate pressure vessels. It sets limits on the service conditions and sets rules under which fabricating procedures are qualified. It also outlines the requirements for stamping and marking the vessels.

SECTION XI. RULES FOR IN-SERVICE INSPECTION OF NUCLEAR POWER PLANT COMPONENTS These rules constitute requirements to maintain a nuclear power plant in a safe manner and to return the plant to service following plant outages.

ASME Code Stamps

If a company wants to make pressure vessels, it doesn't just write the ASME and ask them to send a code stamp. The company must write to ASME and tell them it is interested in obtaining one. But before a company even writes to the ASME, it should have a quality-control manual. Any company that you might work for that makes products to code requirements already has compiled one. These quality-control manuals sometimes are several thousand pages long. They must conform to ASME code requirements, and they tell how the company will meet the requirements of the code.

After writing a quality-control manual, the next step in getting a code stamp is a visit from a two-person review team to inspect the plant. This team consists of a representative from the inspection agency and a representative from the local governmental area. If there is no local government representative who is suitable, or in instances where the local government group also is the inspection agency, then someone from the national ASME board will come. For nonnuclear work this team will spend at least a day and a half at the company's plant to determine whether the company is, indeed, adhering to its own procedures outlined in the quality-control manual. If the company does not have a quality-control manual, the supervisors from one of the insurance companies that insure boilers and pressure vessels will help the company provide what's needed.

The procedure for getting an N stamp for nuclear work is much more severe. Here the ASME assigns a four- to six-person survey team that will probably spend at least 3 days inspecting the company's operations.

Once the fabricator passes the quality-control or quality-assurance requirements, the ASME will issue a certificate of authorization that is good for 3 years. When the 3-year period is up, the company must be recertified.

All stamps for use on ASME work are issued by ASME. Table 23-3 lists the types of code stamps available. Any company, no mat-

TABLE 23-3 ASME code stamps

Code	Covers the Fabrication of
A	Field assembly of power boilers and steel plate heating boilers
H	Steel plate and cast-iron sectional heating boilers
HLW	Lined potable (drinkable) water heaters
L	Locomotive boilers
M	Miniature boilers
N	Nuclear vessels
NPT	Nuclear vessel parts
NA	Nuclear installation
NV	Nuclear vessel safety valves
PP	Pressure piping
RP	Reinforced plastic pressure vessels
S	Power generation boilers
U	Pressure vessels (Division 1)
U2	Pressure vessels (Division 2)
UM	Miniature pressure vessels
UV	Pressure vessel safety valves
V	Boiler safety valves
R	Repair work on existing structures

ter how big or important, can lose its ASME stamp if it is not careful (that's why you have to help your company retain its code stamp by following exactly every welding procedure you are given).

23-6
QUALIFICATION TESTS

To see how good you are, let's give you several hypothetical welder qualification tests. They are very similar to the welder qualification requirements in AWS D1.1, "Structural Welding Code for Steel." Tests similar to these are found in the ASME Boiler and Pressure Vessel Code and in API standards, too. However, since beginning welders seldom start on large, thick-walled pressure vessels made from high-alloy steels or on high-pressure line pipe, we think that the AWS test will be fairer than the various tests in the ASME or API codes.

We'll use some of the procedures from the AWS qualification tests for structural steel welding (without some of the details the actual tests require) because they are close to what you will most likely have to do when you first get a job, even in a company that does not make products covered by some national standard. The detailed AWS tests can be found in AWS D1.1, "Structural Welding Code—Steel," available from the American Welding Society. Note that *passing only one test, welding pipe in the 6G position (described below), would qualify you for all the other tests that we describe here.* But if you can pass any of these tests you should be proud. You are on your way to becoming a skilled welder. If you do pass these tests, you should take the test details with you, "certified" by your welding instructor, when you go for a job interview. These are not to be considered official certification tests. Certified welding tests can only be given by companies that are qualified to conduct the test-

ing. But if you can pass any of these tests you should be proud. Keep your certification paperwork with you when you look for a new job. Even though you may have to recertify at a new company, the old welder certification test may get you hired on the spot.

Test 1 Tack Welder
This test will show that you can make sound tack welds.

Test samples Prepare four samples as shown in Fig. 23-1. Make four complete samples for testing to show what you can do working in four different welding positions (flat, horizontal, vertical, and overhead).

Material The test shall be conducted with one or more of the following materials. Materials may be combined in one test sample if desired. The candidate must choose the correct filler metal to match the yield strength and ductility of the base metal, or else the person giving the test may specify the correct filler-metal grades for you from the following:

■ ASTM A-36 specification for structural carbon steel with a minimum yield strength of 36,000 psi[250 MPa].
■ ASTM A-588 specification for high-strength low-alloy structural steel up to 4 in. thick with 50,000-psi [340-MPa] minimum yield point.
■ ASTM A-572 grade 50 specification for high-strength low alloy columbium-vanadium steel of structural quality with a 50,000-psi [340-MPa] yield strength.
■ ASTM A-514 specification for high-yield strength, quenched and tempered alloy steel plate, suitable for welding, with nominal yield strength of around 100,000 psi [689 MPa].
■ Alternate nonferrous materials may be used in additional tests with appropriate welding processes and filler metals.

Sample preparation Eight pieces of ½- × 4-in. steel are to

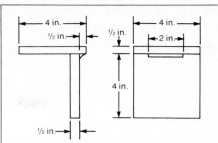

FIGURE 23-1 Make four tack-welder test samples like these. Make the first sample in the downhand flat (1F) position. Make the second sample in the horizontal (2F) position. The third sample should be welded in the vertical-up (3F) position. The last sample is to be welded in the overhead (4F) position. Mark each sample with a note indicating the position in which you welded it. These samples will be used to test your ability to make tack welds in the four different positions. Each weld is ¼ × 2 in.

be flame cut to size by the candidate to prepare four tack-welded samples for testing.

Positions Weld one sample in each of these four positions (see Fig. 23-2):

Flat (1F position)
Horizontal (2F position)
Vertical-up (3F position)
Overhead (4F position)

Tack weld size ¼-in. [6.4-mm] maximum leg by approximately 2 in. [50 mm] long.

Filler metal SMAW electrodes of ⁵⁄₃₂ in. diameter shall be used for qualifying tack welders. Select one or more electrodes from Table 23-4 to make samples. The candidate tack welder or person providing the test should select and match the correct filler-metal tensile strength to the strongest material to be used in each sample.

Additional tests may be conducted with oxyfuel gas welding, GTAW or GMAW if suitable filler metals are substituted for the SMAW electrodes (Table 23-4). The candidate also may wish to qualify on alternate nonferrous base metals of equal thickness.

Test method All four samples are to be tested as shown in Fig. 23-3 until rupture occurs. The force may be applied downward by any means to fracture the test weld. The surface of the weld and of the fracture shall be examined visually for defects.

Test results

1. Each tack weld shall present a reasonably uniform appearance and shall be free of overlap, cracks, and excessive undercut. There shall be no porosity visible on the surface of the tack welds.
2. The fracture surface of the tack welds shall show fusion to the root, but not necessarily beyond, and shall exhibit no incomplete fusion to the base metal or any inclusion or porosity larger than $3/32$ in. [2.4 mm] in its greatest dimension.
3. A tack welder who passes the fillet-weld break test shall be eligible to tack weld all types of joints for the welding process, welding positions, and steel for which he or she has qualified. Successful completion of high-strength low-alloy steel test samples will qualify the tacker to work with both high-strength low-alloy steels and carbon steels. Similarly, successful completion of welds made on quenched and tempered ASTM A 514 steel with a 100,000-psi [689-MPa] yield strength will qualify the tack welder to work on high-strength quenched and tempered alloy structural steels, high-strength low-alloy structural steels, and structural carbon steels.

Retest In case of failure to pass the above test, the tacker may make one retest without additional training.

Qualification period A tacker who passes this test shall be considered eligible to perform tack welding indefinitely in the positions and with the process and materials for which he or she is qualified, unless there is some specific reason to question his or her ability. In such a case, the tacker shall be required to demonstrate the ability to make sound tack welds by again passing the same tack-welding test.

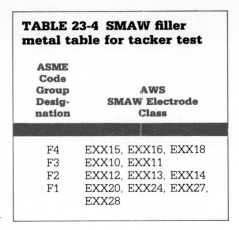

TABLE 23-4 SMAW filler metal table for tacker test

ASME Code Group Designation	AWS SMAW Electrode Class
F4	EXX15, EXX16, EXX18
F3	EXX10, EXX11
F2	EXX12, EXX13, EXX14
F1	EXX20, EXX24, EXX27, EXX28

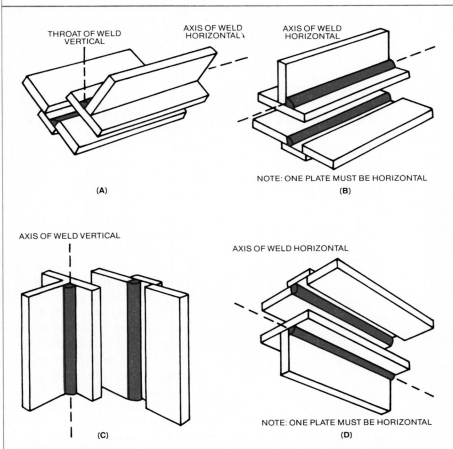

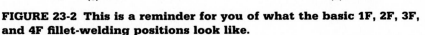

FIGURE 23-2 This is a reminder for you of what the basic 1F, 2F, 3F, and 4F fillet-welding positions look like.

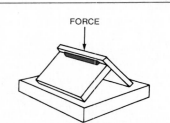

FIGURE 23-3 Test all four fillet-weld samples you produce in the four basic welding positions by smashing them flat like this. Apply the force straight down until the sample welds crack, or until the welded sample is squashed flat. Then examine the work as described in the text. A couple of heavy-duty welding clamps will do the job if you don't have a press. Keep them turning evenly to uniformly deform the work. If you really have confidence in your work, try a sledgehammer as a weight for pressing the pieces flat. Smacking them with a sledgehammer will cause cracking due to impact that might not occur if you smoothly and slowly flatten the test specimens.

Records Records of these test results shall be kept by the candidate, the manufacturer or contractor for whom the tests are conducted, and by the welding school conducting the test. These records shall be available to those authorized to examine them.

Test 2 Fillet Welder

The candidate welder shall make four sample fillet welds as shown and dimensioned in Fig. 23-4; one sample for each welding position in which he or she will be tested.

Material The same materials shall be used as described in the tack welder test. Note that nonferrous materials of equal thickness to the steel test sample may be substituted for additional testing if suitable filler metals and an appropriate welding process are selected.

Sample preparation The candidate welder shall prepare the steel for his or her test by flame-cutting the parts to the dimensions shown in Fig. 23-4.

Positions The same positions shall be used as for the tack welder test, namely the flat (1F position), horizontal (2F position), vertical-up (3F position), and overhead (4F position).

Filler metal Use ⁵⁄₃₂-in.-diameter [4.0-mm] SMAW electrodes. The candidate welder will select one or more SMAW electrodes from the same electrode selection table as in the tack welder test (Table 23-4), matching the correct filler metal to the strongest material to be used in each sample.

For additional testing with other processes, suitable oxyfuel, GTAW, or GMAW filler metals may be used.

Test method and results

1. The surfaces of all four samples produced in different welding positions are to be examined visually for defects. Then 1 in. [25 mm] will be cut from both ends of the fillet weld and the interior face of the weld-metal cross section will be macroetched with a suitable acid to show that no undercutting, overlapping, or cracking has occurred. The fillet welds shall have fusion to the root of the joint but not necessarily beyond. The minimum leg size shall meet the specified size of the fillet weld. The end of the macroetch test specimen shall be smooth for etching.

2. After etching and visual examination of the etched surface, the center section of the test sample shall pass a bend test if it can be bent flat upon itself. If the fillet weld fractures, the fractured surface shall show complete fusion to the root of the joint and shall exhibit no inclusion or porosity larger than ³⁄₃₂ in. [2.4 mm] in greatest dimension. The sum of the greatest dimensions of all inclusions and porosity shall not exceed ³⁄₈ in. [9.5 mm] in the 6-in.-long [150-mm] bend test specimen.

3. A welder who passes the macroetch and fillet-weld break test shall be eligible to fillet weld all types of joints for the welding process, welding positions, and steel for which he or she has qualified. Successful completion of high-strength low-alloy test samples will qualify the welder to work with both high-strength low-alloy steels and carbon steels. Similarly, successful completion of welds made on quenched and tempered ASTM A-514 steel of 100,000-psi [689-MPa] yield strength shall qualify the welder to work on high-strength quenched and tempered alloy structural steels, high-strength low-alloy structural steels, and structural carbon steels.

4. If alternate materials and welding processes are used in additional testing, more difficult welding conditions qualify the welder for all less difficult work.

Retest In case of failure to pass the above test, the candidate welder may make one retest without additional training.

Qualification period A welder who passes this test shall be considered eligible to perform fillet welding indefinitely under the same conditions described in the tack welder test.

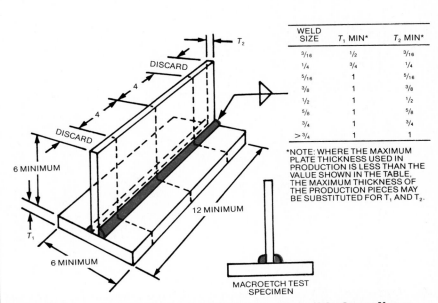

WELD SIZE	T_1 MIN*	T_2 MIN*
³⁄₁₆	¹⁄₂	³⁄₁₆
¹⁄₄	³⁄₄	¹⁄₄
⁵⁄₁₆	1	⁵⁄₁₆
³⁄₈	1	³⁄₈
¹⁄₂	1	¹⁄₂
⁵⁄₈	1	⁵⁄₈
³⁄₄	1	³⁄₄
> ³⁄₄	1	1

*NOTE: WHERE THE MAXIMUM PLATE THICKNESS USED IN PRODUCTION IS LESS THAN THE VALUE SHOWN IN THE TABLE, THE MAXIMUM THICKNESS OF THE PRODUCTION PIECES MAY BE SUBSTITUTED FOR T_1 AND T_2.

DISCARD
DISCARD
4
4
6 MINIMUM
T_1
6 MINIMUM
12 MINIMUM
T_2
MACROETCH TEST SPECIMEN

FIGURE 23-4 Make four fillet-weld test samples with these dimensions. Use the table to guide you in making the correct fillet-weld size for the material thickness you will weld. If you are going to qualify for making fillet welds in all four basic positions, make four separate samples. Then prepare each sample as indicated for further testing.

Records Records of these test results shall be kept as described in the tack welder test.

Test 3 Plate Welder
The joint design, shown in Fig. 23-5A and 23-5B, shall use a ⅜-in.- thick [9.5-mm] plate with a single-V groove, 45° included angle, ¼-in. [6.3-mm] root opening with backing. The backing must be at least ⅜ in. thick by 3 in. [9.5 by 76 mm] wide if radiographic testing is used without removal of the backing. It must be at least ⅜ in. by 1 in. [9.5 by 25 mm] for mechanical testing or for radiographic testing after the backing has been removed. The minimum length of the welding groove should be at least 7 in. [178 mm]. See Fig. 23-6 for cut-

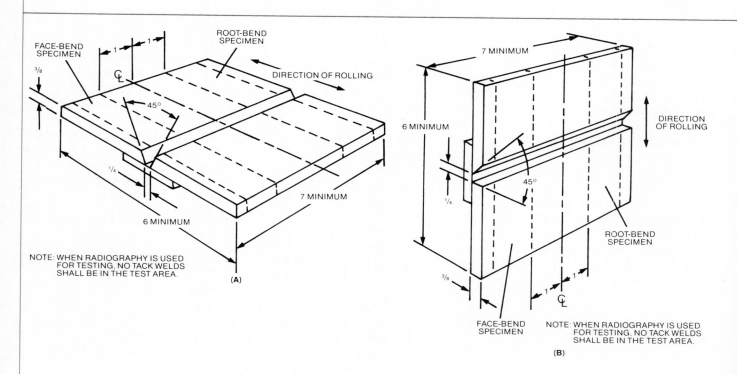

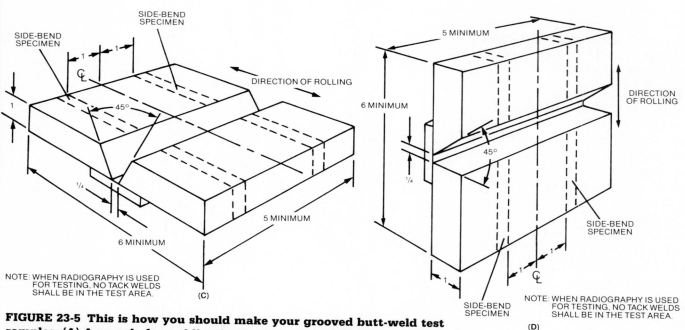

FIGURE 23-5 This is how you should make your grooved butt-weld test samples. (A) A sample for welding in the 1G position. (B) A sample for the 2G position. (C) The way to make your sample for the 1G position if you use 1-in.-thick plate. (D) The sample you should make for the 2G groove-weld test if the plate is 1 in. thick. (E) How these weld samples should be prepared for further testing if you want to use 1-in.-thick plate and qualify for welding unlimited plate thicknesses. If you want

ting bend-test samples. To try a more difficult test with unlimited plate thickness, follow the details in Fig. 23-5C, 23-5D, and 23-5E.

Materials The same materials shall be used as in the tack welder test. Alternate nonferrous materials of equal thickness may be used for additional testing with appropriate filler metals and alternate welding processes. Alternate processes may be selected from oxyfuel braze-welding, oxy-fuel welding, SMAW, GMAW, GTAW, or the submerged-arc (SAW) process.

Sample preparation The candidate welder shall prepare four samples for welding (eight plates)

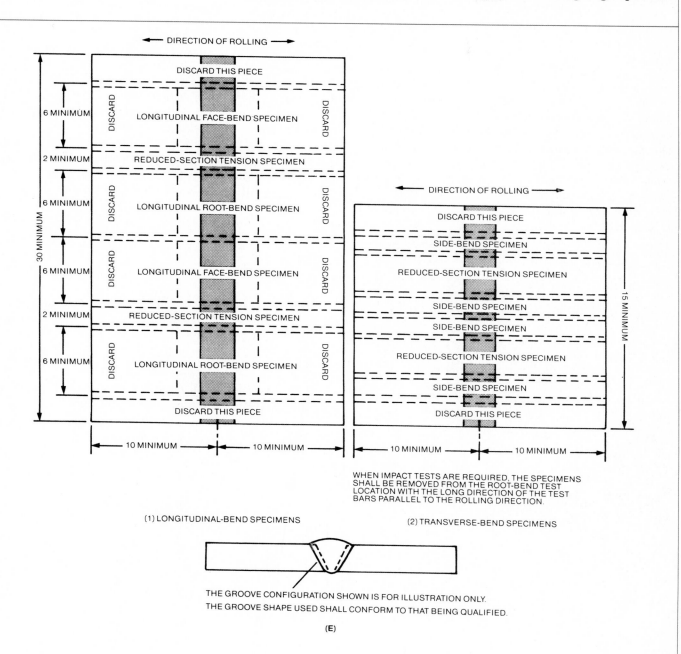

(1) LONGITUDINAL-BEND SPECIMENS

WHEN IMPACT TESTS ARE REQUIRED, THE SPECIMENS SHALL BE REMOVED FROM THE ROOT-BEND TEST LOCATION WITH THE LONG DIRECTION OF THE TEST BARS PARALLEL TO THE ROLLING DIRECTION.

(2) TRANSVERSE-BEND SPECIMENS

THE GROOVE CONFIGURATION SHOWN IS FOR ILLUSTRATION ONLY.
THE GROOVE SHAPE USED SHALL CONFORM TO THAT BEING QUALIFIED.

(E)

to be tested with more than one welding process (GTAW and GMAW as well as SMAW, for example), you will have to prepare separate samples for *each* process that you use. Part E in this set of drawings may resemble Fig. 23-6, but E should be used *only* if you plan to qualify for welding plate of *any* thickness. (For example, you might make your samples of 1-in.-thick plate and test them as shown in E, but use Fig. 23-6 if you only want to be tested on ⅜-in.-thick plate.)

in four separate positions using oxyfuel or plasma-arc cutting or gouging to prepare the beveled edges. (See Figs. 23-5 and 23-6.)

Positions The groove-weld plate test shall be conducted in the 1G (plates horizontal), 2G (plates vertical, axis of weld horizontal), 3G (plates vertical, axis of weld vertical), and 4G (overhead welding with plates horizontal and weld groove opening down).

Filler metals For SMAW, the candidate welder may select from the same filler metals used in the tacker test, matching the electrode to the base metal. If GMAW, GTAW, or SAW is used, or if nonferrous plate is welded

with oxyfuel, SMAW, GMAW, or GTAW processes, filler metals that are compatible with the welding process and material selected should be used.

Test method and results Before preparing mechanical test specimens, the qualification test plates shall be nondestructively tested for soundness as follows:

1. Either radiographic or ultrasonic testing shall be used. For an acceptable qualification test, the weld shall meet the minimum inclusion density and size established by the person giving the test. (AWS D1.1, the "AWS Structural Steel Welding Code," spells out these requirements in detail.) No cracks may be allowed.

2. On visual inspection the samples shall show no surface cracking; all craters shall be filled to the full cross section of the weld and the face of the weld shall be at least flush with the outside surface of the plate, and the weld shall merge smoothly with the base metal. Undercutting shall not exceed $1/64$ in. [0.4 mm]. Weld reinforcement shall not exceed $3/32$ in. [2.4 mm].

3. The root of the weld shall be inspected visually. There shall be no evidence of cracks, incomplete fusion, or inadequate joint penetration. A concave root face is not allowed if it is deeper than $1/16$ in. [1.6 mm].

4. A side-bend and a face-bend test specimen will be cut from

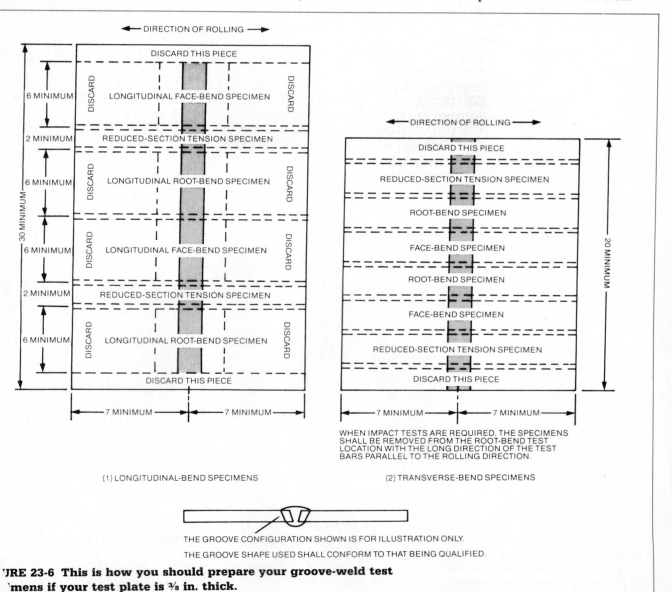

(1) LONGITUDINAL-BEND SPECIMENS   (2) TRANSVERSE-BEND SPECIMENS

THE GROOVE CONFIGURATION SHOWN IS FOR ILLUSTRATION ONLY.

THE GROOVE SHAPE USED SHALL CONFORM TO THAT BEING QUALIFIED.

FIGURE 23-6 This is how you should prepare your groove-weld test specimens if your test plate is $3/8$ in. thick.

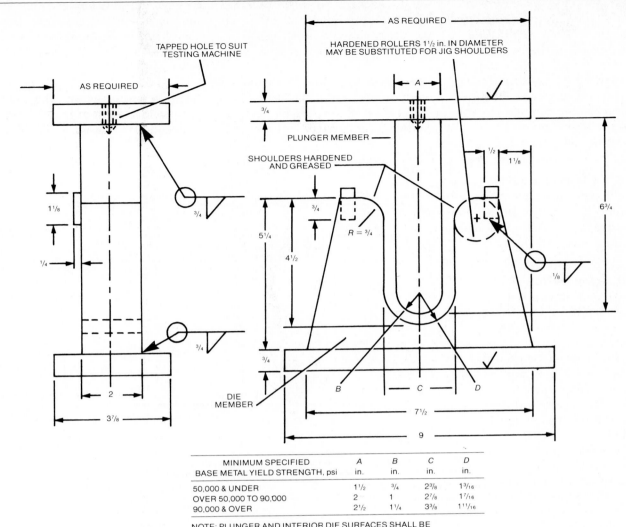

MINIMUM SPECIFIED BASE METAL YIELD STRENGTH, psi	A in.	B in.	C in.	D in.
50,000 & UNDER	1½	¾	2⅜	1³/₁₆
OVER 50,000 TO 90,000	2	1	2⅞	1⁷/₁₆
90,000 & OVER	2½	1¼	3⅜	1¹¹/₁₆

NOTE: PLUNGER AND INTERIOR DIE SURFACES SHALL BE MACHINE-FINISHED.

(A)

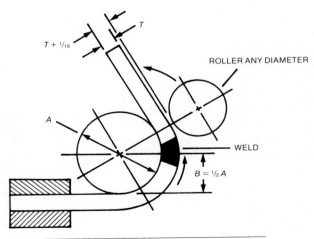

MINIMUM SPECIFIED BASE METAL YIELD STRENGTH, psi	A in.	B in.
50,000 & UNDER	1½	¾
OVER 50,000 TO 90,000	2	1
90,000 & OVER	2½	1¼

(B)

FIGURE 23-7 (A) In case your welding school does not have a bend-test machine, this diagram shows you how to build one. (B) How the bend test should be conducted for plate materials of different minimum-yield strength and different thicknesses.

each sample plate (as shown in Fig. 23-8) and subjected to guided bend tests. (See Fig. 23-7). The convex surface of the bend test plate with a crack or other opening exceeding ⅛ in. [3 mm] measured in any direction will fail the test. Cracks occurring on the corners of the specimen during testing shall not be considered as a failure.

Retest An immediate retest may be conducted consisting of two test welds of each type on which the welder failed. All retest specimens shall meet all the specified requirements.

Qualification period The welder's qualification shall be considered as remaining in effect indefinitely unless (1) the welder has not done welding in the process for which he or she is qualified for a period exceeding 6 months, or (2) there is some specific reason to question the welder's ability, in which case the welder must be requalified.

If the welder qualified on 1-in.-thick or thicker material instead of ⅜-in.-thick [9.5-mm] plate, he or she will be qualified to work on plate of any thickness using the qualified welding processes and materials.

Records Records of these test results shall be kept as described in the tack welder test.

Test 4 Pipe Welder

The joint design, shown in Fig. 23-9, shall be single-welded pipe butt joint in ⅜-in.-thick [9.5-mm] round pipe or square structural tubing with a single-V groove, 45° included angle, a ¼-in. [6-mm] root opening and a ⅛-in. [3-mm] maximum root face with backing (see Fig. 23-9D). Thicker pipe walls may be used if desired. The backing must be at least ⅜ in. [9 mm] thick by 3 in. [75 mm] wide if radiographic testing is used without removal of the backing. It must be at least ⅜ in. [9.5 mm] by 1 in. [25 mm] for mechanical testing or for radiographic testing af-

ter the backing has been removed. The minimum length of the welding groove should be at least 7 in. [180 mm].

Materials The same materials shall be used as in the tack welder test. Alternate nonferrous materials of equal thickness may

be used for additional testing with appropriate filler and either oxyfuel braze-welding, oxyfuel welding, SMAW, GMAW, or GTAW processes.

Sample preparation The candidate welder shall prepare three pipe samples (six pipes) for mak-

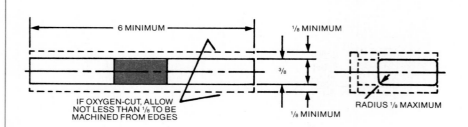

WHEN t EXCEEDS 1½, CUT ALONG THIS LINE. EDGE MAY BE OXYGEN-CUT.

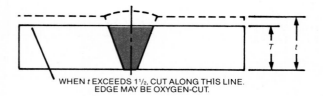

t, in.	T, in.
⅜ to 1½	t
>1½	See Note 2

NOTES:
1. A LONGER SPECIMEN LENGTH MAY BE NECESSARY WHEN USING A WRAPAROUND-TYPE BENDING FIXTURE OR WHEN TESTING STEEL WITH A YIELD POINT OF 90 kips/in² OR MORE.
2. FOR PLATES OVER 1½ in. THICK, CUT THE SPECIMEN INTO APPROXIMATELY EQUAL STRIPS WITH T BETWEEN ¾ AND 1½ in. AND TEST EACH STRIP.
3. t = PLATE OR PIPE THICKNESS.

(A)

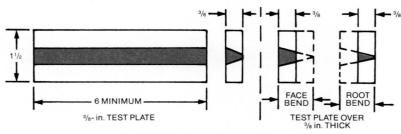

(1) LONGITUDINAL BEND SPECIMENS

FIGURE 23-8 (A) Detailed procedures for preparing plate and pipe samples cut from test pieces for a side-bend test. (B) Further details for preparing a specimen for a longitudinal bend test. (C) How to prepare a face-bend test specimen. (D) This is used to make a transverse root-bend test specimen.

ing welds in three separate positions using oxyfuel or plasma-arc cutting or gouging to prepare the beveled edges (Fig. 23-10).

Positions The groove-weld pipe test shall be conducted in the 1G (pipe horizontal and rolled) position and qualifies the welder for flat-position groove welding of pipe, tubing, and plate; flat- and horizontal-position fillet welding of pipe and tubing; and flat and horizontal fillet welding of plate (Fig. 23-10).

Certification in the 5G (pipe horizontal and fixed) position qualifies the welder for flat, vertical, and overhead position fillet welding of pipe, tubing, and plate.

Certification in the 6GR (inclined pipe axis with the pipe fixed) position qualifies the welder for all-position groove and all-position fillet welding of pipe, tubing, and plate. *This one test, therefore, certifies the welder for all of the preceding work.* (See Fig. 23-11).

Filler metals For SMAW, the candidate welder may select from the same filler metals used in the tack welder test, matching the electrode to the base metal.

If GMAW or GTAW is used, or if nonferrous plate is welded with oxyfuel, GMAW, or GTAW processes, suitable matching filler metals should be used that are compatible with the welding process and material selected.

Test method and results

1. Before preparing mechanical test specimens, the test pipes or tubes shall be nondestructively tested for soundness as described in the groove-welded plate test using either radiographic or ultrasonic testing.

2. On visual inspection the samples shall show no surface cracking; all craters shall be filled to the full cross section of the weld and the face of the weld shall be at least flush with the outside surface of the plate and the weld shall merge smoothly with the base metal. Undercutting shall not exceed $1/64$ in. [0.4 mm]. Weld reinforcement shall not exceed $3/32$ in. [2.4 mm].

3. The root of the pipe weld shall be inspected visually. There shall be no evidence of cracks, incomplete fusion, or inadequate joint penetration. A concave root face is not allowed if it is deeper than $1/16$ in. [1.5 mm].

4. Side-bend and a face-bend test specimens will be cut from each sample pipe (as shown in Fig. 23-8) and subjected to guided bend tests. The convex surface of the bent test plate with a crack or other opening exceeding $1/8$ in. [3 mm] measured in any direction will fail the test. Cracks occurring

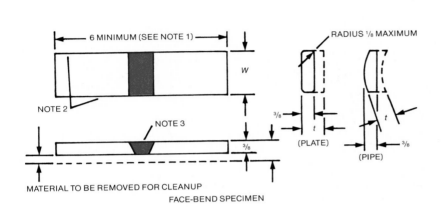

FACE-BEND SPECIMEN

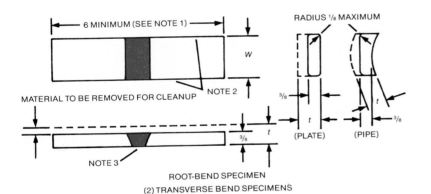

ROOT-BEND SPECIMEN
(2) TRANSVERSE BEND SPECIMENS

DIMENSIONS	
TEST WELDMENT	TEST SPECIMEN WIDTH, in. (W)
PLATE	$1\frac{1}{2}$
TEST PIPE 2 in. AND 3 in. IN DIAMETER	1
TEST PLATE 6 in. AND 8 in. IN DIAMETER	$1\frac{1}{2}$

NOTES:
1. A LONGER SPECIMEN LENGTH MAY BE NECESSARY WHEN USING A WRAPAROUND-TYPE BENDING FIXTURE OR WHEN TESTING STEEL WITH A YIELD STRENGTH OF 90 kips/in²
2. THESE EDGES MAY BE OXYGEN-CUT AND MAY OR MAY NOT BE MACHINED.
3. THE WELD REINFORCEMENT AND BACKING, IF ANY, SHALL BE REMOVED FLUSH WITH THE SURFACE OF THE SPECIMEN. IF A RECESSED BACKING IS USED, THIS SURFACE MAY BE MACHINED TO A DEPTH NOT EXCEEDING THE DEPTH OF THE RECESS TO REMOVE THE BACKING; IN SUCH CASES, THE THICKNESS OF THE FINISHED SPECIMEN SHALL BE THAT SPECIFIED ABOVE. CUT SURFACES SHALL BE SMOOTH AND PARALLEL.
4. t = PLATE OR PIPE THICKNESS.

(B)

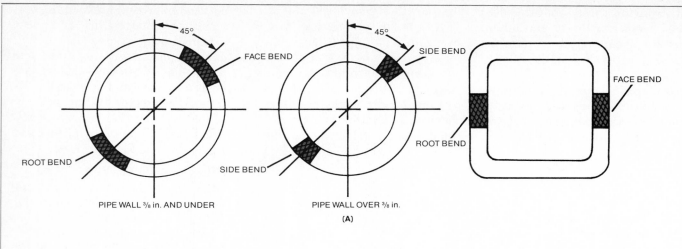

(A)

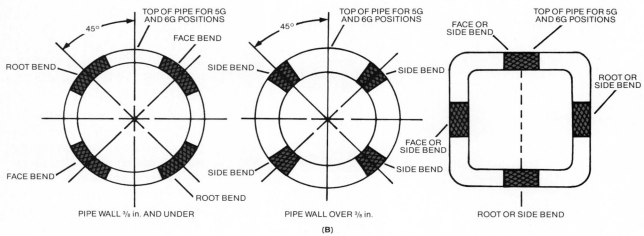

(B)

FIGURE 23-9 For a pipe welder test, make a butt joint in pipe as outlined here. Thicker pipe walls may be used if desired (or if your instructor will let you use up expensive thick-walled pipe). Follow the instructions in the text. **(A)** This shows you where to select samples from both round and square pipe or tube for different tests and different wall thicknesses. The hatched areas will be cut from your sample for testing. **(B)** Then prepare your test samples as outlined here. **(C)** This drawing should be used as a guide for making V-joint test samples if you are going to weld extra-thick-walled pipe or tubing without a backing bar. **(D)** A guide for making V-joint test samples on extra-thick-walled pipe or tubing with a backing bar.

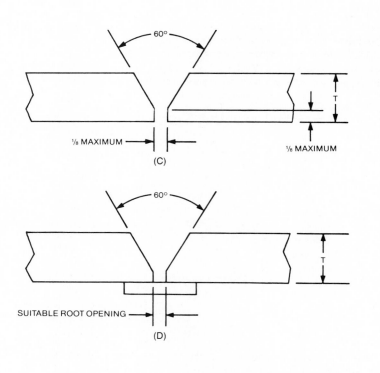

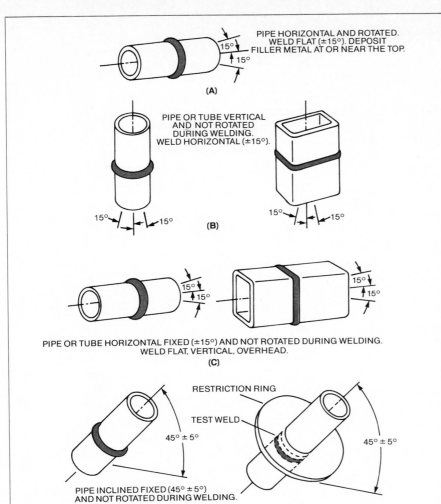

FIGURE 23-10 This is a reminder for you of what the basic pipe-welding positions look like. (A) The 1G position. (B) The 2G position. (C) The 5G welding test position. (D) The 6G position. (E) The 6GR (the R stands for restricted) test position for qualifying to make T, K, Y, and other pipe connections. See Fig. 23-11 for more details on this most difficult 6GR pipe-welding test.

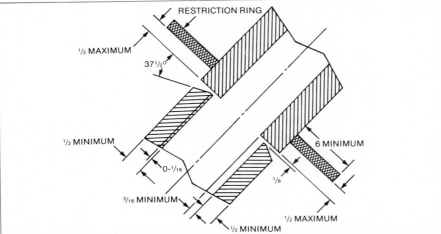

FIGURE 23-11 If you successfully complete this difficult 6GR test, you would be qualified to weld pipe in any position with any kind of connections, and you would also be qualified to weld plate or structurals in any position.

on the corners of the specimen during testing shall not be considered as a failure.

Retest An immediate retest may be conducted consisting of two test welds of each type on which the welder failed. All retest specimens shall meet all the specified requirements.

Qualification period The welder's qualification shall be considered as remaining in effect indefinitely unless (1) the welder has not done welding in the process for which he or she is qualified for a period exceeding 6 months, or (2) there is some specific reason to question the welder's ability, in which case the welder must be requalified.

If the welder qualifies on 1-in.-thick [25-mm] or thicker pipe, instead of ⅜-in.-thick [9.5-mm] pipe, he or she will be qualified to work on pipe or plate of any thickness using the qualified welding processes and materials (see Fig. 23-12).

Records Records of these test results shall be kept as described in the tack welder test.

23-7
WHAT'S IT WORTH TO YOU?

Your skills are worth whatever you can get. The law of supply and demand works in the job market just like any other market. However, one rule of thumb was proposed in the American Welding Society's technical magazine, *The Welding Journal*. It's certainly not official, but it's not too far off, either. The editorial said:

"One of the reasons for a career in welding is to have high earnings. Earnings depend on productivity; productivity depends on experience; and experience is a function of training, effort, skill and time.

"So that potential welding students understand not only 'what's a welder,' but also 'what's a proficient welder,' and how their earn-

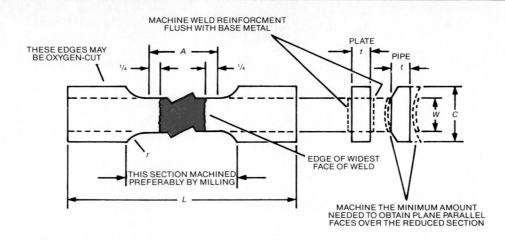

THESE EDGES MAY BE OXYGEN-CUT

MACHINE WELD REINFORCMENT FLUSH WITH BASE METAL

THIS SECTION MACHINED PREFERABLY BY MILLING

EDGE OF WIDEST FACE OF WELD

MACHINE THE MINIMUM AMOUNT NEEDED TO OBTAIN PLANE PARALLEL FACES OVER THE REDUCED SECTION

DIMENSIONS

	TEST PLATE			TEST PIPE	
	$Tp \leq 1$ in.	$1 < Tp < 1\frac{1}{2}$ in.	$Tp \geq 1\frac{1}{2}$ in.	2 in. AND 3 in. DIAMETER	6 in. AND 8 in. DIAMETER OR LARGE JOB SIZE PIPE
A—LENGTH OF REDUCED SECTION	WIDEST FACE OF WELD + $\frac{1}{2}$ in., $2\frac{1}{4}$ MINIMUM			WIDEST FACE OF WELD + $\frac{1}{2}$ in., $2\frac{1}{4}$ MINIMUM	
L—OVERALL LENGTH, MINIMUM (NOTE 2)	AS REQUIRED BY TESTING EQUIPMENT			AS REQUIRED BY TESTING EQUIPMENT	
W—WIDTH OF REDUCED SECTION (NOTES 3, 4)	$1\frac{1}{2} \pm 0.01$	1 ± 0.01	1 ± 0.01	$\frac{1}{2} \pm 0.01$	$\frac{3}{4} \pm 0.01$
C—WIDTH OF GRIP SECTION, MINIMUM (NOTES 4, 5)	2	$1\frac{1}{2}$	$1\frac{1}{2}$	1 APPROXIMATELY	$1\frac{1}{4}$ APPROXIMATELY
t— SPECIMEN THICKNESS (NOTES 6, 7)	Tp	Tp	Tp/n (NOTE 7)	MAXIMUM POSSIBLE WITH PLANE PARALLEL FACES WITHIN LENGTH A	
r— RADIUS OF FILLET, MINIMUM	$\frac{1}{2}$	$\frac{1}{2}$	$\frac{1}{2}$	1	1

NOTES:
1. Tp = THICKNESS OF THE PLATE.
2. IT IS DESIRABLE, IF POSSIBLE, TO MAKE THE LENGTH OF THE GRIP SECTION LARGE ENOUGH TO ALLOW THE SPECIMEN TO EXTEND INTO THE GRIPS A DISTANCE EQUAL TO TWO-THIRDS OR MORE OF THE LENGTH OF THE GRIPS.
3. THE ENDS OF THE REDUCED SECTION SHALL NOT DIFFER IN WIDTH BY MORE THAN 0.004 in. ALSO, THERE MAY BE A GRADUAL DECREASE IN WIDTH FROM THE ENDS TO THE CENTER, BUT THE WIDTH AT EITHER END SHALL NOT BE MORE THAN 0.015 in. LARGER THAN THE WIDTH AT THE CENTER.
4. NARROWER WIDTHS (W AND C) MAY BE USED WHEN NECESSARY. IN SUCH CASES, THE WIDTH OF THE REDUCED SECTION SHOULD BE AS LARGE AS THE WIDTH OF THE MATERIAL BEING TESTED PERMITS. IF THE WIDTH OF THE MATERIAL IS LESS THAN W, THE SIDES MAY BE PARALLEL THROUGHOUT THE LENGTH OF THE SPECIMEN.
5. FOR STANDARD PLATE-TYPE SPECIMENS, THE ENDS OF THE SPECIMEN SHALL BE SYMMETRICAL WITH THE CENTERLINE OF THE REDUCED SECTION WITHIN 0.25 in. EXCEPT FOR REFEREE TESTING, IN WHICH CASE THE ENDS OF THE SPECIMEN SHALL BE SYMMETRICAL WITH THE CENTERLINE OF THE REDUCED SECTION WITHIN 0.10 in.
6. THE DIMENSION T IS THE THICKNESS OF THE TEST SPECIMEN AS PROVIDED FOR IN THE APPLICABLE MATERIAL SPECIFICATIONS. THE MINIMUM NOMINAL THICKNESS OF $1\frac{1}{2}$ in.-WIDE SPECIMENS SHALL BE $\frac{3}{16}$ in. EXCEPT AS PERMITTED BY THE PRODUCT SPECIFICATION.
7. FOR PLATES OVER $1\frac{1}{2}$ in. THICK, SPECIMENS MAY BE CUT INTO THE MINIMUM NUMBER OF APPROXIMATELY EQUAL STRIPS NOT EXCEEDING $1\frac{1}{2}$ in. IN THICKNESS. TEST EACH STRIP AND AVERAGE THE RESULTS.

(A)

---

ings can be increased, the following progressive steps, equated to potential earnings, are suggested (where Class 9 is worth over three times as much per year as Class 1):

"Class 1. Three months training. Minimum starting wage

"Class 2. Class 1 plus passing the 1G downhand manual carbon steel ASME Code test plate. Class 1 plus 50%

"Class 3. Class 2 plus passing the 2G horizontal and 3G vertical low-carbon steel ASME Code test plates. Class 2 plus 30%

"Class 4. Class 3 plus can weld thick (1 in. and over) carbon as well as stainless steel to make x-ray quality welds. Class 3 plus 20%

"Class 5. Class 4 plus can efficiently operate automatic welding equipment—sub-arc, GTA, and GMA. Class 4 plus 10%

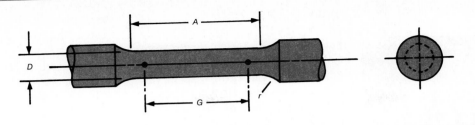

	DIMENSIONS		
	STANDARD SPECIMEN	SMALL-SIZE SPECIMENS PROPORTIONAL TO STANDARD	
NOMINAL DIAMETER	0.500 in. ROUND	0.350 in. ROUND	0.250 in. ROUND
G—GAGE LENGTH	2.000 ± 0.005	1.400 ± 0.005	1.000 ± 0.005
D—DIAMETER (NOTE 1)	0.500 ± 0.010	0.350 ± 0.007	0.250 ± 0.005
r — RADIUS OF FILLET, MINIMUM	3/8	1/4	3/16
A — LENGTH OF REDUCED SECTION (NOTE 2), MINIMUM	2 1/4	1 3/4	1 1/4

NOTES:
1. THE REDUCED SECTION MAY HAVE A GRADUAL TAPER FROM THE ENDS TOWARD THE CENTER, WITH THE ENDS NOT MORE THAN 1 PERCENT LARGER IN DIAMETER THAN THE CENTER (CONTROLLING DIMENSION).
2. IF DESIRED, THE LENGTH OF THE REDUCED SECTION MAY BE INCREASED TO ACCOMMODATE AN EXTENSOMETER OF ANY CONVENIENT GAGE LENGTH. REFERENCE MARKS FOR THE MEASUREMENT OF ELONGATION SHOULD BE SPACED AT THE INDICATED GAGE LENGTH.
3. THE GAGE LENGTH AND FILLETS SHALL BE AS SHOWN, BUT THE ENDS MAY BE OF ANY FORM TO FIT THE HOLDERS OF THE TESTING MACHINE IN SUCH A WAY THAT THE LOAD SHALL BE AXIAL. IF THE ENDS ARE TO BE HELD IN WEDGE GRIPS, IT IS DESIRABLE, IF POSSIBLE, TO MAKE THE LENGTH OF THE GRIP SECTION GREAT ENOUGH TO ALLOW THE SPECIMEN TO EXTEND INTO THE GRIPS A DISTANCE EQUAL TO TWO-THIRDS OR MORE OF THE LENGTH OF THE GRIPS.

(B)

FIGURE 23-12 These are details on how to prepare tensile test specimens for (A) pipe sections, sheet and plate and (B) round bar stock or thick plate with enough section to make a round tensile test bar.

"Class 6.   Class 5 plus can operate burning equipment and read blueprints.

Class 5 plus 5%

"Class 7.   Class 6 plus can act as a working leadman.

Class 6 plus 5%

"Class 8.   Class 7 plus can act as a welding inspector.

Class 7 plus 5%

"Class 9.   Class 8 plus can act as a welding technician.

Class 8 plus 5%

"Class 10.   Class 9 plus a degree in welding engineering to start. (Note that Class 10 is worth 344% of the value of Class 1 . . . to start with.)

Class 9 plus 10% to start. Up from there to who knows where. You might wind up owning your own company.

"Class 11.   Class 10 plus an MBA.

Go start your own company today."

Your course is finished. Go make a good living as a skilled welder.

A-1
OTHER WELDING PROCESSES

It would take a much larger textbook to cover all 98 welding processes, and even then we'd have to leave important information out. What we decided to leave out, instead, were welding and cutting processes that don't require highly skilled operators, processes that are not widely used, and processes that are so specialized or highly automated that you aren't likely to run into them. One group of processes that you are likely to see in use are the resistance-welding processes. Although they don't require skilled operators, they are commercially important.

Resistance Welding

Resistance spot welding (RSW), roll-seam welding (RSEW), and projection welding (RPW) are electric welding processes that do not use an open arc to produce heat and melt the base metal. Instead, heat is provided by I^2R resistance heating between two pieces of metal in contact with each other. No filler metal is required. A short burst of low-voltage, high-current density, high-amperage ac current does the job.

You will find resistance welding being done just about anywhere that large quantities of sheet metal are joined on a production basis. Examples are automotive body assembly, and the making of appliances, pots and pans, some architectural metal products, culverts and drainage products, and metal furniture. A couple of the resistance-welding processes also are used to make butt joints in wire products such as display racks and supermarket shopping baskets, and to put fasteners on structural steel.

In the United States and Canada, the most frequently used resistance-welding currents are 60-Hz ac with the voltage between 1.0 and 25 V. That is a much lower voltage than you are accustomed to using for arc welding.

However, resistance-welding currents range from 1000 to more than 100,000 A. Sometimes very high frequencies of up to 450,000 Hz are used. Direct current can be used, but its use is infrequent.

Resistance welding requires time as well as current. The actual amount of heat produced in resistance welding depends upon how long the current is on, that is, how long the I^2R heating continues. You might say that all resistance-welding processes are heat-producing processes where heat $(H) = I^2Rt$. The time factor, t, is usually less than a minute.

Most resistance-welding processes require still another process variable, pressure. The pressure or squeezing force is combined with the momentary burst of high heat at the contact point between the overlapping sheets to complete the weld.

The final process variable is the resistance of the base metals and the resistance of the contact point where they meet. As you can see from the equation $H = I^2Rt$, a base material with high resistance (or a contact point with high resistance) will produce heat faster than a low-resistance joint. That explains why most resistance welding is done on steel,

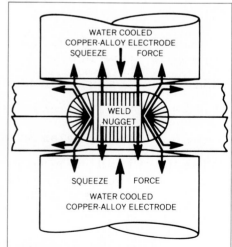

FIGURE A-1 Heat flow in a spot weld. Most of the heat is drawn off by the water-cooled copper electrodes. The weld nugget remains inside the weld joint below the surface.

rather than on copper or aluminum which have much lower resistivity (higher conductivity). It also explains why some types of resistance-welding processes use very small contact points between the metals being joined, because the smaller the contact point, the larger the resistance (actually, the higher the current density).

These main process variables, voltage, current, time, force, and contact area, are controlled by part design, welding schedules, and machine settings used to operate the process. Once the settings are correct, the operator only has to feed material into the welding machine and pull it out when welded. Many high-production resistance-welding machines include materials-handling devices (or even robots) to do that. Now you can see why we didn't include resistance welding in the main chapters of this book.

Resistance-spot and -seam welding The big difference between resistance-spot and resis-tance-seam welding is the shape of the weld bead. In this case, the bead doesn't even have that name, since its buried inside the work. It's called a weld nugget or button (Fig. A-1). If a set of big water-cooled copper-alloy electrodes are used, the process is called *resistance-spot welding* (RSW). If the permanent electrodes are two copper wheels on either side of the joint, the process is called *resistance–roll-seam welding* (RSEW). Figure A-2 shows the difference.

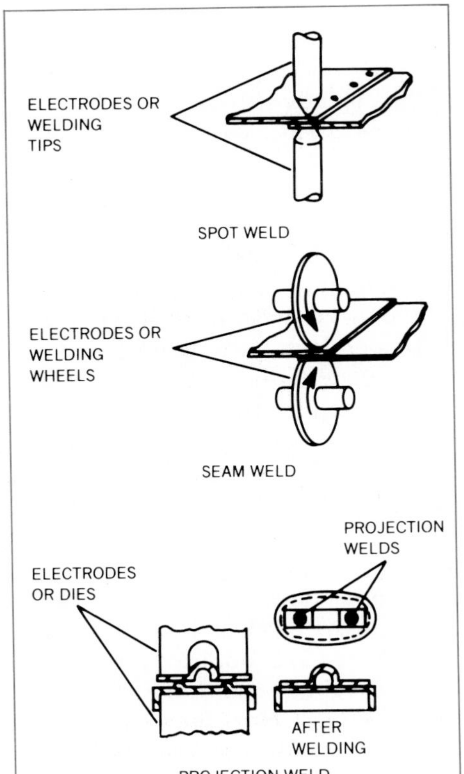

You can see in Fig. A-2 that resistance welding produces lap joints. While a few different joint designs are used, all resistance-spot and resistance-seam welds are variations of the basic lap joint. That, alone, should explain why these processes are limited to welding material less than 1/4 in. [0.6 mm] and usually less than 1/8 in. [0.3 mm] thick. On the other hand, these processes have been used to join material as thin as 0.001 in. [0.025 mm]. Metal foil can be resistance-welded.

If you want to see a resistance weld, the easiest place to find one is on a metal desk. If you look closely at the surfaces of a desk or filing cabinet, you may see tiny indentations spaced a short distance apart. The joint may be hidden so that it's not easy to find, but if you look for it, you will see the "spots," even though they are painted over. These spots appear because of the pressure of the copper welding electrodes (Fig. A-3) squeezing down on the joint after the metal inside the joint has melted. A spot weld is very similar to making an internal rivet in a workpiece.

The strength of the joint is measured much the same way that joint strength for rivets and soldered and brazed lap joints is measured. Instead of pulling the weld metal apart to see when it breaks, the joint is peeled apart. This measures the shear strength of the joint. (We talked about shear strength in Chap. 8.) When the initial machine settings are made, lap-joint samples of a part are welded and their shear strengths are determined. The machine variables are then adjusted until the process is producing a desired shear strength. Then the machine goes into full production, with only an occasional part tested to make sure that the welds are remaining within specification.

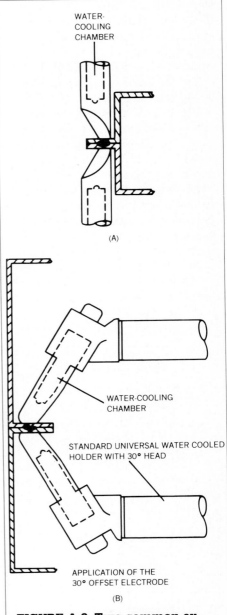

FIGURE A-2 Comparison of spot, seam, and projection welding.

ELECTRODES OR WELDING TIPS

SPOT WELD

ELECTRODES OR WELDING WHEELS

SEAM WELD

ELECTRODES OR DIES

PROJECTION WELDS

AFTER WELDING

PROJECTION WELD

FIGURE A-2 Comparison of spot, seam, and projection welding.

WATER-COOLING CHAMBER

(A)

WATER-COOLING CHAMBER

STANDARD UNIVERSAL WATER COOLED HOLDER WITH 30° HEAD

APPLICATION OF THE 30° OFFSET ELECTRODE

(B)

FIGURE A-3 Two common examples of many different electrode shapes and arrangements used in resistance-spot welding. (A) Spot welding electrode face offset at angle to axis. (B) Spot welding electrode with angled head.

After a while (maybe 4 h of continuous spot welding), the copper electrodes begin to mushroom out of shape under the heat and pressure. If galvanized steel is being joined, the zinc coating will slowly alloy with the copper electrodes, turning their contact surfaces into brass. Brass has lower conductivity than copper; therefore, the preset welding current will start producing smaller weld nuggets than it should. When that happens, the welding operation is stopped and the copper electrodes are filed down to remove the contaminated surfaces before the shear strength of the weld is reduced too much.

These are examples of resistance-welding procedures. The details can be quite complex. The weld force, current, time, and other factors often are combined into complex welding schedules for different parts. The point, however, is that all these variables are controlled by the welding machine, and very little is controlled by the machine operator.

Since no filler metal is required, and since the weld nugget is formed inside the joint, the position of the weld (flat, horizontal, vertical, or overhead) hardly affects the process. That makes resistance welding much easier to automate. In fact, resistance welding was the first process to be applied extensively with welding robots.

Projection and stud welding
Imagine that you want to add a bolt or stud to a piece of structural steel. If the bolt or stud has a couple of little projections on it to increase the resistance and current density at the point of contact with the structural section, you can use a special welding process called *resistance-stud welding* (SW). The stud is put into a handheld gun that operates in a similar way to a carpenter's nail gun, and the stud or welding fastener is quickly resistance-welded to the steel plate or structural section (Fig. A-4).

In resistance-projection welding (RPW) a flange with small projections can quickly be resistance-welded to a sheet-metal surface (like part of a filing cabinet). Special weld fasteners also can be attached to parts (Fig. A-5). Again, there is not much operator skill involved. Most of the work is in the design of the connection and the type of machine used, not in how the operator does the work.

Flash and upset welding
Imagine, now, that you had two pieces of large-diameter bar stock (or even a large machine shaft) that you wanted to butt-weld together. With the right kind of equipment, you can put the two sections end to end, press them tightly together, create a large resistance-welding current between them to heat up the joint until the interface melts, and then suddenly jam the parts together tightly, forcing molten metal out of the joint. Hold the joint motionless briefly while the molten metal at the interface solidifies, and you will have created a flash-butt–welded joint.

Flash-butt welding (FW) is used to join shafts, bars, and tubes end-to-end. A closely related process, upset welding (UW), is very similar, except that the force to press the parts together is applied before the heating starts, and remains steady throughout the heating period. After the joint is made (using either process), the extruded "flash" or extra upset weld metal squeezed out of the weld may be machined off to make a smooth joint.

The most noticeable difference between flash-butt welding and upset welding is that there is no violent expulsion of molten metal during the upset welding process. Upset welding is used mostly to join wire and smaller-diameter tubes. Upset welding can even make butt welds in wire rope.

As with other resistance-welding processes, special machinery is required and no filler metal is used. Minimal operator skills are required, but machine and welding cycle parameters are very important.

High-frequency resistance welding
Resistance welding at

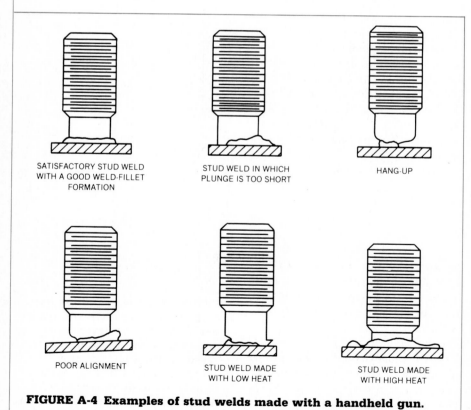

SATISFACTORY STUD WELD WITH A GOOD WELD-FILLET FORMATION

STUD WELD IN WHICH PLUNGE IS TOO SHORT

HANG-UP

POOR ALIGNMENT

STUD WELD MADE WITH LOW HEAT

STUD WELD MADE WITH HIGH HEAT

FIGURE A-4 Examples of stud welds made with a handheld gun.

very high ac frequencies reduces the amount of burning that sometimes occurs in welding thin-gage metal. This is one reason that HF resistance welding is used to join welded tubing at high speeds in automated welding lines. A set of rolls first bend strip into the shape of a tube, then two electrode contacts on either side of the continuous seam produce the current that joins the strip together, edge-to-edge, into a welded tube (Fig. A-6). The speed of the tube welding line is usually a function of the gage of the material being welded and the current used. The diameter of the tube is only a factor on very small tubes of ⅝-in. OD [16-mm OD] and less, or where the diameter is small compared with the tube wall thickness. High-frequency re-

sistance tube welding is obviously a highly automated process used in tube mills. We mention it here because, since you will often work with welded tubing, you should know how it is made.

Friction Welding

Another way to join solid bars into a shaft is by friction welding (FRW). If the two shafts are carefully aligned and one or both are spun at high speed with a gap between them, suddenly jamming them together will create enormous frictional heat. No electricity is used. The solid butt weld that is produced is made by converting the mechanical energy of the spinning shafts into heat energy produced by friction (Fig. A-7).

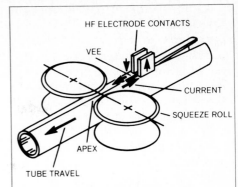

FIGURE A-6 High-frequency resistance welding of tube and piping.

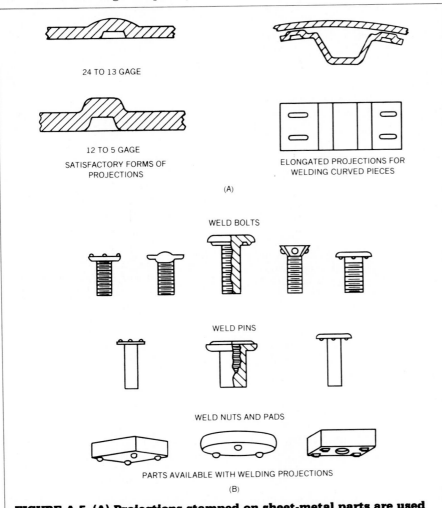

24 TO 13 GAGE

12 TO 5 GAGE

SATISFACTORY FORMS OF PROJECTIONS

ELONGATED PROJECTIONS FOR WELDING CURVED PIECES

(A)

WELD BOLTS

WELD PINS

WELD NUTS AND PADS

PARTS AVAILABLE WITH WELDING PROJECTIONS

(B)

FIGURE A-5 (A) Projections stamped on sheet-metal parts are used in projection welding. (B) Weld fasteners with special projections are commercially available (just like any other fasteners) for projection welding.

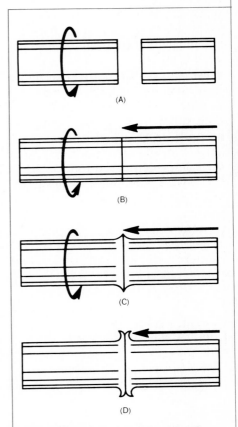

(A)

(B)

(C)

(D)

FIGURE A-7 Basic process of friction welding. (A) Rotating member brought up to desired speed. (B) Nonrotating member is advanced to meet the rotating member and pressure is applied. (C) Heating phase—pressure and rotation are maintained for a specified period of time. (D) Forging phase—rotation is either maintained or increased for a specified period of time. Total welding time—2 to 30 s.

An important characteristic of the friction-welding process is that the weld-joint interface is so narrow, and the temperature produced is sometimes below the melting point of the base metal, that the weld bond is actually a solid-state bond achieved by the diffusion of metal atoms across the joint. In that sense, the process is related to hammer welding (blacksmith welding). This solid-state bond makes it possible to join dissimilar metals that normally would never alloy together if they were melted.

Friction welding is not limited much by the diameter of the round sections being joined. Very large shafts can be welded together as long as the interface between them is clean, machined flat before they are oppositely spun, and then rammed together. As you can guess from this description, friction welding requires an especially rigid machine designed to hold large shafts in perfect alignment, spin them, and then slam them together to make the weld. It's obviously a specialized welding process that uses special machinery, but friction welding sometimes is the only way to economically join two large round sections. Machine and part setup, not operator skills, are the things that count.

Thermit Welding
Thermit welding is an old process that was used before arc welding was invented. When powdered aluminum metal is mixed with powdered iron oxide, and then heated with an oxyfuel torch to its kindling temperature (or by igniting the mix with a strip of burning magnesium metal), the result is a very hot fire. The iron oxide provides the oxygen needed to burn the aluminum powder to form aluminum oxide. The chemical reaction has a temperature of around 6500°F [3600°C]. That's a high enough temperature to melt just about anything you will ever see. Imagine putting this metal-powder mixture between the

joints of two butting railroad rails and igniting it. That's the way rails were originally welded together.

One advantage of this old process is that there is very little contraction of the joint on slow cooling. The welded section, even if highly restrained (as railroad rails often are), is put under very little stress. There is one big drawback. Powdered metals are very dangerous because of their extremely high surface area. When they burn, they burn fast and furiously.

The message to you is *Don't try it.* You can start a terrible fire simply by trying to make the powdered metals, let alone igniting them together.

Electroslag Welding
Electroslag welding (EW) is an offshoot of submerged-arc welding. It is used to join extremely thick plates by literally making a casting between the two butting joints. This is a wire-fed process, similar to submerged-arc welding, except that a heavy molten slag is produced on top of the weld-metal casting which the wire (or wires) penetrate to feed the liquid weld puddle underneath.

Both sides of the joint are dammed up with water-cooled copper shoes that slide up the joint as the weld metal cools and solidifies. This process does not use the same kind of granulated slag used for submerged-arc welding. Instead the slag is much thicker, and also is electrically conductive, even though it offers enough resistance to current flow to add extra heat to the process. This process obviously is used only in the downhand (flat) position because the weld puddle and molten slag layer can be very large. Electroslag welding is used occasionally to join very thick plate sections in one pass (sometimes the joints are more than a foot thick). Electroslag is a specialized process you are not likely to see in use. It has some limitations in use on heavy-walled

pressure vessels because the weld metal it produces does not have the high-impact properties produced by other welding processes.

Electrogas Welding
Imagine making a solid-wire MIG weld in the vertical position, except that the wire is feeding directly into the joint from above. In effect, the joint is in the vertical position but the weld puddle is in the flat position inside the joint. Again, both sides of the joint are dammed up by water-cooled copper shoes that slide up the plate slowly as the weld metal solidifies.

Further imagine that the wire feeder is mounted on a platform that creeps up the joint. You are sitting on the platform, monitoring the process. One wire may be feeding vertically down into the flat weld puddle, or maybe more than one wire will be in use. The setup resembles electroslag welding except that an inert shielding gas is used instead of slag. As the weld progresses, the equipment, platform, wire feeder, and power sources rise slowly up the side of a ship hull or the walls of a large-diameter water tank, pulled up by two chains that hang from the top of the structure. That's an example of electrogas welding.

Electrogas welding is obviously a variation of GMAW. It's highly automated, and it requires special equipment (although the basic power sources, wire and welding head, by themselves, are not that unusual). Among other special equipment in the weld joint would be a device to move the welding wire back and forth across the weld puddle and timing devices to give the wire a pause at the end of each pass to tie the weld into the plate. Electrogas welding is a specialized process that is occasionally used in shipyards, to build water tanks, and to make sections of offshore drilling rigs.

Now we will describe one welding process that requires good operator skills and three others that don't.

Atomic Hydrogen Welding

When molecular hydrogen (H_2) is passed through an open arc, it dissociates into two hydrogen ions (H^+ and H^+). When these ions recombine forming hydrogen gas again, the reaction releases a great deal of heat. Additional heat is generated by burning molecular hydrogen with oxygen to form water vapor. This ionization, recombination, and combustion process effectively shields the molten weld metal from oxidation and the formation of nitrides from the surrounding air. The temperature of an atomic hydrogen arc-flame is about 10,000°F [5540°C]. That's extremely high . . . in fact, it is so high that the process will even weld tungsten and molybdenum metal. Its use, however, is less exotic. Atomic hydrogen welding is used to join tungsten-bearing tool steels, and tungsten-carbide cutting inserts on drill heads such as those used in oil exploration when a brazed joint will not be strong enough.

In operation, two permanent electrodes create the arc and the gas flows between them. The process probably would remind you of GTAW with a funny-looking welding torch. Filler metal, if required, is fed into the arc the way it is in TIG welding. The big drawback is the use of hydrogen gas, both from a safety point of view and because of the possibility of hydrogen embrittlement of the base metal.

Laser Welding and Cutting

Now that powerful lasers are available, some welding is done, especially in the aerospace industry, using the energy produced by intense light whose wavelengths are very narrow and almost all the same, and whose waves are almost fully in phase with each other. This very high power, narrow-band light beam can produce enough energy and power to either weld or cut metals. A laser beam can even cut a diamond. The process can be used to join material up to about $1/32$ in.

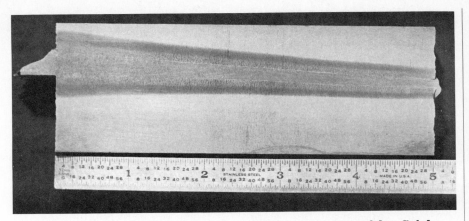

FIGURE A-8 A 5-in.-thick steel plate welded in one pass with a Sciaky 100-kW electron-beam (EB) welder. Plate thicknesses up to 10 in. have been welded in one pass with EB welding using a concentrated beam of high-velocity electrons that literally bore a hole with the leading edge of the beam. The molten steel through which the full section is melted then flows back together to solidify as a weld as the electron beam passes through the section.

[0.8 mm] thick with a power output of 1000 W or more.

When laser welding was first introduced, it received a lot of publicity. It had all the elements of science fiction—welding and cutting with a ray gun. In fact, the process is limited to rather thin material and requires a lot of equipment. Unlike a ray gun, the process isn't exactly portable. It might become so someday, but is not now. Laser welding (and especially cutting) has one big advantage: The kerf (or the weld zone) is very thin. In addition, the process is not limited by the base metal used. However, the equipment is expensive, and the process is limited to rather thin sections.

A word of warning about lasers. They are becoming used more frequently in industry. One of their major uses has nothing to do with welding. They are ideal as instruments to judge the distance or level surface of large workpieces. They have replaced conventional surveying instruments on many construction sites, too. *Never look into a laser beam, even with special goggles.* The beam can do the same thing to your eyes that it can do to even the most exotic high-strength, high-temperature metals.

Electron-Beam Welding

Like laser welding, electron-beam welding (EBW) generates a lot more excitement than it does weld metal. The idea behind electron-beam welding is to create an enormously powerful beam of electrons with a high power density in watts per unit area. Power concentrations as high as 1 million to 10 million W per square centimeter are routinely achieved, and even higher power densities are possible. Electron-beam welding is generally performed at voltages between 20,000 and 150,000 V, with the higher voltages producing the higher power densities. These high-powered electron beams require much of the same kind of operator protection as large x-ray machines. In fact, when a beam of high-energy electrons hits a workpiece, x-rays are produced. Now you can see why you won't find an EB welder in everybody's backyard.

One important characteristic of electron-beam welding is that it produces extremely deep base-metal penetration with a weld joint that can be knife-edge thin (Fig. A-8). As a result, a single pass with an EB welder can join thick material with a conventional butt joint, without special beveled edges (or filler metals).

Most EB welding is done in an extremely high vacuum. Otherwise the air would alter the characteristics of the beam and cause it to disperse and lose energy. You can't get much of a workpiece inside a vacuum chamber and keep fabricating costs down. EB welding processes have been developed that don't require the workpiece itself to be inside of a high vacuum, although almost all of the electron beam travels through a vacuum column before it exits and strikes the workpiece. However, when the EB welding is done in the air, the workpiece has to be less than about ⅝ in. [16 mm] from the exit nozzle of the EB welding gun or else the beam will be scattered by molecules in the air. That doesn't exactly make EB welding very portable or very versatile from a practical shop point of view. Like many other welding processes, EB welding is quite specialized, and is used mostly in the aerospace industry where the high capital costs for the equipment can be tolerated. Operator skill is minimal.

Explosive Welding

From time to time people would like to be able to produce low-cost plain-carbon steel plate, tubing, or heavy sheet with a layer of almost anything else on top of it as a cladding. Examples are stainless-steel–clad carbon steel, aluminum-clad steel, titanium-clad steel, and so on. There is a process that will join these highly dissimilar metals with a strong bond over a wide surface area in a very short time. Simply blow the two metals together with explosives.

If you put the cladding layer on top of the base metal with a uniform small air gap between them, then put a uniform layer of explosives on top of that, you are ready for explosive bonding. When the explosives are ignited, the detonation wave literally drives the metals together. The force of the explosion produces a momentary overpressure of 100,000 atm (more than 1.5 million psi or 10,000 MPa). This is more of a mechanical than a welded joint. It has a wavelike appearance if you looked at it under a microscope. However, the bond that is formed by the explosive-bond process is as strong as the metals themselves.

Since your average fabricating shop is not really equipped to perform explosive bonding (the work requires skilled blasters, special explosives, and always is done outdoors in remote areas), you are not ever likely to do this kind of work. You don't have to, because there are special firms that will do the work for you, cladding just about anything to anything else . . . mostly in flat shapes but occasionally in other shapes such as tubing. One interesting use of explosive bonding is to create special transition joints such as steel to aluminum. You can buy these transition joints and weld steel to one side and aluminum to the other without ever trying to figure out how to fuse weld steel to aluminum (which you can't do) without creating brittle iron-aluminum intermetallic compounds with just about zero ductility.

Now that we've given you a sampling of some of the other 98 welding and cutting processes that we haven't covered in the main part of the textbook, you can see why we stopped where we did with your training. Most of the other welding and cutting processes are either very specialized or very exotic. And almost all the ones we didn't cover in this textbook require very few operator skills.

Index

A

A-286 alloy, plasma welding data for, 357
AAC (*see* Air-carbon-arc cutting)
Abrasion resistance in surfacing alloys, 497, 504
Acetone in acetylene cylinders, 62–63
Acetylene:
  Btu/ft^3 when burned, 34
  burning velocity in oxygen, 49
  chemical formula, 61
  cutting data, 117
  cutting tips, 103–104
  density, 31
  explosive limits in air and oxygen (chart), 49
  flame adjustments, 51–54
  flame properties (table), 44
  flame temperature, 43
  full cylinder pressure, 62
  odor of leak, 61
  preheat flame adjustments, 107–110
  withdrawal rates from cylinder, 62–63
Acids, pickling, brazing and, 161
Acoustic-emission testing:
  applications (table), 540
  test methods, 546
Aerospace industry, 11–12
Air:
  composition of, 110
  density of, 447
  specific gravity of, 32
Air bending, 596–97
Air-carbon-arc cutting (AAC), 12, 304–312
  added carbon, problem of, 309–310
  air hose and fittings, 308
  air pressure and volume, 307–308
  air supply, 302
  burnback problems, 309
  current ranges by electrode size (table), 307
  current settings, 307
  methods and techniques, 308–309
    for different base metals (table), 309
  power sources for, 287, 306–307
  process description, 304–305
  safety: distance from combustibles, 310
    fumes, 310–311
    lens shades, 310
    noise, 310
Aircraft hydraulic tubing, 235
Almen strips, 600

Alternating current (ac), 273
  amplitude of, 278
  average voltage of, 278
  cycles, 277
  frequency, 277–278
  out-of-phase, 279
  peak height, 278
  period of, 277
  root-mean-square: amperage, 278
    voltage, 278
  in submerged-arc welding, 468
  three-phase, 278–79
Alternators, 272
Aluminum alloys:
  bending, 145
  cleaning prior to welding, 222
  ductile-brittle transition temperature, 194
  elastic modulus of, 185
  feeders for welding wires, 438
  filler metals for (table), 452–453
    GMAW specifications, 445
    oxyfuel, 205–206
  GMAW shielding gases for, 451
  grinding wheels for, 607
  melting temperature range, 222
  oxyfuel welding of, 221–225
  pipe joints, thick-walled, 241
  plasma arc cutting (table), 362
  SMAW electrodes for, 391–392
  soldering of, 162
Alloy steels, bending of, 145
Aluminum-silicon brazing alloys (table), 165
American Bureau of Shipping (ABS), 609, 610
American Iron and Steel Institute, 235
American Petroleum Institute, 235, 609, 610
American Society of Mechanical Engineers (ASME), 235, 236, 610
  code stamps, 616–617
  Boiler and Pressure Vessel Code of, 235, 609, 610, 614–616
  Structural Welding Code of, 609, 617
American Society for Testing and Materials, 235–237
Amperage, GMAW:
  bead shape and, 455
  weld parameters and, 456
Amperes, defined, 259, 264
Ampere-twins, 271
Amplitude of alternating current, 278
Angle-measuring, 602
Annealing, 505

Appearance:
   of surfacing alloys, 503, 505
   weld-surface, 536
Arc-blow, 404–406
Arc length, GMAW:
   bead shape and, 455
   type of power source and, 435–436
Arc strikes, 536
Arc voltage (*see* Arc length)
Architectural ironwork, 7
Argon:
   cylinders, 56–58
   density of, 31, 447
   ionization potential, 447
   regulator flow rates, 73
   shielding, 446–448
   specific gravity, 32
   weight in cylinder, 57
Asbestos welding curtains, 302
ASME (*see* American Society of Mechanical
   Engineers)
Atmosphere (*see* Pressure, gas)
Atomic hydrogen welding, 635
   lens shades for (table), 301
Atoms, 256
   in solids, 257
Austenite in carbon steel, 407
Austenitic high chromium irons, 504, 506–
   507
Austenitic manganese steel, 505–506
Austenitizing temperature, 148
Autotransformers, danger in use of, 296

B

Backfires, 46–49
   procedures in case of, 98
   (*See also* Combustion; Flashbacks; Fuel
     gases)
Background current in pulsed spray GMAW,
   431
Backhand method:
   in braze-welding, 171, 173
   in GMAW, 454
   in oxyfuel welding, 207–208
   in submerged-arc welding, 469–471
Backing:
   in submerged arc welding: flux, 470–471
     fusible metallic, 471
     nonfusible copper, 469–470
     root design, 471
     weld metal, 471
   weld symbol for, 566
Barometric pressure, 31
Bars, bending of, 143–144
Batteries, lead-acid, 263
Bead shape, GMAW:
   control of, 455–456
   weld parameters and, 456

Bend elongation, 196
Bend radius, minimum, 196
Bend tests, 196–197
Bending:
   of bars, 143–144
   "easy way" method, 597
   "hard way" method, 597
   of rolls, 598–599
   of structural sections, 142
Beryllium coppers, fumes from, 311
Bevel cutting:
   angle-calibrated tips for, 132–133
   with machines, 132–133
   pipe joints, 244
Blacksmith welding, 147–148
Blankets, welding, 302
Blueprints (*see* Welding symbols)
Boiler and Pressure Vessel Code, ASME, 609,
   610, 614–616
Boiling range of liquefied gases, 39–40
Brass:
   bending of, 145
   70-30, plasma welding data on, 357
   yellow, oxyfuel welding of, 220–221
     pipe, 252
Braze-welding, 169–174
   applications, 152
   filler metals, 170–171
   joints, 170, 171
Brazing/soldering:
   alloys (table), 165–166
   applications, 151
   definition of, 152
   fit-up tolerances, 151
   heat conduction in, 159
   heat sources for, 153
   master process chart, 21
   precleaning methods, 154–155
   safety, 161
   symbols for, 571, 572
   torch methods, 153–161
Brinell hardness, 187–188
   BHN numbers, 187
   tensile strength and, 187
British thermal units (Btu):
   conversion to calories, 268
   use of, 33
Bronze brazing-welding alloys, 170–171
Brushes, wire, 604
Buildup, surfacing alloys and, 498–499
Buried-arc GMAW, 427
Butt joints:
   brazed, 168
   in submerged-arc welding, 473–475
   welding positions, 239
Butt-lap joints, brazed, 169
Butt welds, oxyfuel, thin aluminum, 222–
   223
Buttering, 499

C

Cable, welding:
 American Wire Gauge sizes (table), 285
 ground cable, 293
 resistivity versus size, 269–270
 selection of, 296–297
 sizes for welding machines (table), 282
 splices and repairs, 297
 types of, for power sources, 293
 voltage drops (tables), 286
Cadmium, fumes from, 311
Cadmium-silver solders (table), 168
Cadmium-zinc solders (table), 164
Calcium carbide, 61–62
Calcium hydroxide (slack lime), 62
Calories and kilocalories:
 applications of, 33
 definition of, 34
Capacitors, electric, 260–261
Capillary force in brazing, 152–153
Caps for welders, 94
Carbon-dioxide:
 CO_2-shielded buried-arc welding, 427
 cylinder pressures, 450
 cylinder sizes, 449
 cylinder withdrawal rate, 449
 density, 31
 freezing temperature (dry ice), 56–57
 frozen regulators, 450
 liquid, 449–450
 shielding in GMAW and FCAW, 450
 specific gravity, 32
 and spray transfer, 449
Carbon steel:
 carbon content of, 121, 532
 cracking of, 119–122, 535
 Curie point of, 405
 elastic modulus, 185
 flame cutting: kindling temperature for,
  100
   problems with, 119–122
 GMAW filler-metal specifications, 445
 GMAW shielding gases for, 451
 grinding wheels for, 607
 low-alloy, strength range for, 182
 plasma-arc cutting schedule (table), 362
 in submerged-arc welding: filler metals for
  (table), 466, 467
   schedules (tables), 486–492
 in surfacing alloys, 504–506
Carburizing flames (see Flame adjustment)
Cast aluminum, oxyfuel welding of, 224–225
Cast iron:
 braze-welding methods, 171
 carbon content of, 121, 218
 copper-alloy electrodes for, 390
 elastic modulus, 185
 flame cutting methods, 125–126

Cast iron (Cont.):
 nickel alloy electrodes for (table), 391
 oxyfuel welding of: broken castings, 220
  cooling rates, 219
  groove joints, 219–220
  preheating/postheating temperatures,
   218–219
 pipe, ASTM specifications (table), 237
 surfacing with, 503
CC power sources, 284, 434–435
Center of gravity, 583–585
 calculation of, 584–589
  shop method, 586
 explanation of, 584, 585
Chain intermittant welding, 552, 556
Charge, electrical, 259–260
Charpy impact test, 192–193
Check values, 48, 75, 83–84
Chemical elements, 16–17
Chemical properties of metals, 178
Chemical valence bonds for acetylene, 61
Chipping hammers, 603–604
Chisels, 603–604
Cladding, 498
Circuit breakers, 281
Circuit diagram, 275
Clothing:
 for arc welding, 303
 for general work, 94
Cobalt brazing alloy (table), 166
Combustion, 42–46
 flash points, 43
 heat distribution in flames, 43–44
 kindling temperature, 42–43
 oxygen consumption (table), 45
  by fuel gases, 44–45
 supplied oxygen (neutral flames), 45
Compressive forces in metal fatigue, 198
Compressive stress, definition of, 180
Concavity, excess of, in weld profile, 527
Conduction, thermal, 41–42, 447
Conductors, electrical, 257–258
Consumable inserts in pipe welding, 240–241
Contamination:
 surfacing alloys and, 502
 (See also Dilution)
Convection, thermal, 42
Convexity, excess of, in weld profile, 527–528
Constant current (CC) or constant potential
  (CP) power sources, 284, 285, 434–435
Constant voltage (CV) power sources, 283–
  285, 434–435
Construction equipment maintenance, 10
Construction industries, 8–10
Contact tubes for GMAW guns, 441
Contactors in power sources, 280–281
Copper alloys:
 bending of, 145
 brazing alloys (table), 165

Copper alloys (*Cont.*):
  elastic modulus of, 185
  filler metals: GMAW, 445
    oxyfuel, 206
  GMAW shielding gases for, 451–452
  oxyfuel welding techniques, 221
Corner joints, submerged-arc welding of, 477–478
  J-groove, 478
Corner-flange joint, welding symbol for, 564, 565
Corrosion and metal fatigue, 198
Corrosion resistance, 497
Coulombs, 264
Coupling compounds, ultrasonic testing of, 544
Coupling distance:
  in braze-welding, 174
  in flame cutting, 111–112
CP power sources, 284, 434–435
Cracks:
  crater, 534
    in GMAW, 393
  from fatigue, 599
  hydrogen embrittlement, 500
  longitudinal: base metal, 534–535
    weld metal, 533
  shear, 532
  strength loss from, 533
  in surfacing alloys, 503
  tensile, 532
  transverse: base metal
    weld metal, 533
Creep rupture, 198–199
Culvert pipe, corrugated, 235
Curie point, 405
Current, electrical (*see* Alternating current;
  Direct current)
Curtains, welding, 301–303
Cutting machines:
  bridge models, 133
  microprocessor controlled, 135
  motorized torches for, 131
  numerically controlled, 135
  right- and left-handed, 133
  ripping and squaring, 117
  shape control methods, 133–134
  shape cutters, 117, 133
  speed calibration of, 133
  tips for, 130–132
  torches for, 130–132
  track-mounted, small, 132
Cutting speed, oxyfuel, 112–113
CV power sources, 283–285, 434–435
Cycles, alternating current, 277
Cylinder pressure, 29–30
Cylinder wrenches, 63, 67, 91–92
Cylinders, 56–68
  acetylene, 61–63

Cylinders, acetylene (*Cont.*):
  filling of, by weight, 64
  carbon dioxide, 449–450
  CGA-code connections for, 60–63
  handling and storage of, 65–68
  for high pressure gases, 57–58
  hydrogen, 67
  for liquefied gases, 57–58
  liquid, 60–61
  LPG gases, 64–65
  MAPP Gas, 64–65
  valves of, 58–59
    caps, 58
    oxygen, 60
    stem and tang, 59

D

Deep flange edge joints, 223
Degreased steels, flame cutting of, 106
Delivery pressure from cylinders, 30
Dents, removal of, 145–47
Deposition rates:
  GMAW: and current, 455
    and gravity, 577
    and welding parameters, 456
    and wire feed speed, 455
    and wire stickout, 455
  in submerged-arc welding, 462
  and welding positions, 477
Dew point of gases, 32–33
  conversion to percent moisture, 32
  cryogenic gases, 33
  cylinder gases, 33
  fuel gases, 33
  shielding gases, 447–448
Dewars, safe handling of, 60
Diamond pyramid hardness, 189
Dilution in surfacing, 499–502
  arc shielding, 500–501
  bead spacing, 500
  current density, 500
  electrodes, 500
  filler metals, 499–500, 502
  travel speed, 500
  welding position, 500
Diode in direct current power sources, 279
Dip-transfer GMAW, 423–427
  applications of, 424
  control by open-circuit voltage, 426
  electrode wire diameters for, 424
  inductance control, 437
  schedules for (table), 424
  short-circuiting frequency of, 425
  slope control, 436–437
  tuning in power source for, 424, 425
  typical welding currents for, 425
  wire feed speed for, 426

Direct current (dc), 273–274
    constant current (CC), in submerged-arc
        welding, 468
    constant voltage (CV), in submerged-arc
        welding, 468
    reverse polarity (dcrp), 276–277
    ripple, filtering of, 279
    straight polarity (dcsp), 276–277
Distortion, control of, 526
Divergent bore tips for flame cutting, 132
Drag lines in flame cutting, 112–113
Drooping voltage power sources (droopers):
    applications of, 437–438
    definition of, 284
    volt-amp curves, 434–435
Drop cuts in flame cutting, 110
Dross in plasma-arc cutting, 361
Duty cycle:
    calculation of, 282–283
    defined, for power sources, 281–283
    multiple welding machines on 1000-A
        power line (table), 287
Dry ice, 56–57
Ductility:
    measurement of, 184–185
    temperature effects of, 185
Dye-penetrant testing, 545–546
    symbols for (table), 572, 573

E

Eddy-current testing, 540, 546
    symbols for, 572, 573
Edge laminations, 536
Edge starts in flame cutting, 110
"Effective throat," 548
Elastic range, 182, 184
Elasticity, definition of, 181
Electrical conduction in metals, 256, 258
Electrical equipment builders, 14
Electrical grounds, 298, 299
Electrical hazards, 291–292
Electrical shock, 291–292
Electricians, use of, 291
Electricity:
    static, 260
    for welders, 256–288
Electrode holders:
    AAC, 305
    SMAW, 295
        hazards of, 298
        sizes of (table), 288
Electrodes:
    AAC: alternating current, coated, 306
        direct current, coated, 305
        graphite, 306
        uncoated, 305, 306
    FCAW, 444–445

Electrodes (*Cont.*):
    GMAW, 442–445
        cast and helix, 442–443
        cleanliness of, 443
        solid wires for steel, 443–444
    submerged-arc, wire sizes for (table), 467
    (*See also* Shielded-metal-arc welding)
Electrogas welding, 421, 634
Electromotive force (emf), 265
Electron-beam welding, 635–636
Electrons as charge carriers, 256
Electroplated steels, flame cutting of, 106
Electroslag welding (EW), 421, 634
    power sources for (table), 287
Elements, chemical, 16–17
Elongation:
    bend, 196
    total, 184
Embrittlement:
    low-temperature, 179
    (*See also* Hydrogen embrittlement)
Endurance limit, 198
Energy, definition of, 268
Engine-driven power sources, 326
Engineering properties of metals, 179
Equivalent carbon content, calculation of,
    408–409
Erosion resistance of surfacing alloys, 497–
    498
Eutectic point, 166–167
EW (*see* Electroslag welding)
Eyes, protection of, in arc welding, 300–302
    lens shade selector chart, 301
Explosive welding, 636

F

Fabricated metal products, 6
Fatigue, metal:
    peening and, 198, 599
    tests for, 197
Faying surface:
    in blacksmith welding, 147
    in braze-welding, 172, 173
Feather on oxyfuel flame tip, 43–44
Feeders (*see* Wire feeders)
Ferrite:
    in carbon steels, 406
    in stainless steels, 383–384
Filler metals:
    for buried-arc GMAW, 427
    in metal spraying, 510
    oxyfuel, 202–206
        aluminum, 205–206
        cast iron, 205
        copper and copper alloys, 206
        mild and low-alloy steel, 203–205
    SAW, 465–467

Filler metals (*Cont.*):
  SMAW, 287–288
    (*See also specific types of electrodes
      under* Shielded-metal-arc welding)
Fillet joints:
  in submerged-arc welding, 476–479
  welding position codes, 239
  welding symbols for, 549–558
Finish and metal fatigue, 198
Fit-up:
  in brazed joints, 155–156
  tolerances vs. strength, 155–156
Flame adjustment, 51–54
  braze welding, 171
  brazing/soldering, 159
  carburizing: acetylene, 45
    MAPP Gas, 46
  neutral: acetylene, 45–46
    MAPP Gas, 46
  oxidizing: acetylene, 45–46
    MAPP Gas, 46
  slightly carburizing, backhand method,
    207–208
Flame cutters and burners (*see* Jobs)
Flame cutting:
  applications of, 99–100
  bars and pipes, 124
  bevel cuts, 114–115
  blowholes, avoidance of, 119
  cast iron, methods, 125–126
  chemistry of (table), 100–101
  coated steels, 105–106
  cracking problems, 119, 121–122
  drilling, 123
  edges, types of, 124–125
  gouging, 122–123
  high-low oxygen switch, 109
  kerf width measurement, 116–117
  marking plate for, 118
  neutral preheat flames, 109
  oxyfuel ratios (table), 103
  by oxygen lancing, 126–127
  oxygen pressure settings, 107
  preheat flame adjustments, 51–54, 107–110
  preparing to start work, 107–108
  by removing rivets, 123–124
  by scarfing, 124
  with slightly carburizing preheat flames,
    108–109
  with slightly oxidizing preheat flames, 109
  by stack cutting, 118
  steels, by carbon content, 121
  strongly carburizing preheat flames, 108
  tip lead angles, 114–115
  troubleshooting using kerf surface, 119–
    122
  underwater, 110, 128–129
Flame drilling, 123
Flame gouging, 122

Flame hardening, 148
Flange edge joints, 223
  welding symbol for, 564, 565
Flare-V joint, welding symbol for, 361
Flash welding, 632
Flashbacks, 47, 48
  procedures upon occurence of, 98
  (*See also* Combustion; Backfires; Fuel gases)
Flat position, GMAW, 456, 457
Flowmeters, 76–78
  calibration of, 77–78
  GMAW use of, 450–451
  operation of, 76–77
Fluorescent liquid testing, 545–546
Fluorescent penetrant testing, 572, 573
Flux, 156–158
  application of, 157
  in braze-welding, 172, 173
  choice of, 157–158
  quantity needed, 158
  reason for use of, 151
  removal of: in braze welding, 173
    in brazing/soldering, 160
  in submerged-arc welding, 464–465
Flux-cored arc welding (FCAW), 421
  core wire composition, 432–433
  deposition rates, 434
  edge preparation, 434
  elements in flux core (table), 433
  penetration and throat depth, 433
  power sources for (table), 287
Forehand welding:
  braze-welding, 171, 172
  GMAW, 454
  oxyfuel, 171, 172, 207
Free-bend test, 195, 196
Frequency, alternating current (ac):
  calculation of, 278
  definition of, 277–278
Friction welding, 633–634
Fuel gases:
  boiling temperatures of (table), 40
  burning velocity of, 49
  industrial uses of, 27–28
  oxygen-to-fuel ratios, 46
  relative use of, by process, 26–27
  (*See also* Gases; *and specific gases, for
    example:* Acetylene)
Fuel-line tubing, ASTM specifications (table),
  237
Fumes, flame-cutting, mask for, 106
Furnace brazing, 158
Fuses in power sources, 281
Fusible solders (table), 164
Fusion:
  incomplete, 532
  lack of, as weld defect, 529
Fusion welding, 20
  master chart of processes, 23

G

Galvanized steel:
  pipe welding, 235
  welding, 251–252
Gamma radiography (table), 539
Gas cylinders (*see* Cylinders; *and specific gases, for example:* Carbon dioxide)
Gas hoses (*see* Hoses, gas)
Gas manifolds, 78–79
  regulators, 79
Gas-metal-arc welding (GMAW):
  buried-arc, 427
  electrogas, 421, 634
  narrow-gap, 432
  shielding gases in, 445–454
    carbon steels, 451
    copper alloys, 451–452
    magnesium alloys, 454
    mixing methods, 450–451
    nickel alloys, 454
    refractory metals, 454
    selection chart, 446
    stainless steels, 451
  (*See also* Flux-cored-arc welding; Metal-inert-gas welding)
Gas pressure (*see* Pressure, gas)
Gas regulators, 68–76
  carbon dioxide, 450
  creep problems, 73
  cylinder end points, 71
  high-pressure gases, 69–70
  low-pressure gases, 70
  manifolds, 79
  maximum working pressure, 50
  operating details, 74–75
  oxygen, seat fires, 61, 74
  pressure settings (table), 227
  ratings by flow capacity, 72–73
  safety devices, 73–74
  single-stage, 70–71, 75, 76
  two-stage, 70–71, 75
  by type of gas, 68–69
Gas torches (*see* Torches, oxyfuel)
Gas-tungsten-arc welding (GTAW), 313–346
  applications, 314–315
  arc length, 318–319
    effect of slope on, 320–321
  current settings: DCRP, 317
    DCSP, 316–317
  cut length rods for, 315
  electrode sizes (table), 328
    current ranges, 328–329
  equipment for, 314
  filler metals for, 334–335
    specifications (table), 335
  grinding wheels, 338
  high frequency starting, 320–322
  hose for shielding gases, 332

Gas-tungsten-arc welding (GTAW) (*Cont.*):
  joint tolerances, 338
  joints for, 339–341
    bevel angles, 338
    cleaning before welding, 338–339
    double-V butt, 340
    single-U butt, 340–341
    single-V butt, 340
  methods, 335–339
    filler-metal feeding, 315, 336
    starting the weld, 336
  power sources for, 318–326
    ac-dc machines, 323–324
    ac machines, 316–317, 323–324
    auxiliary controls, 323–324
    dc, 316–317
    selection of (table), 287
    slope in, 319–321
  process overview, 313–316
  pulsed current, dc, 337
  schedules for: aluminum alloys, 342
    carbon steel, 341
    stainless steel, 341
  scratch starts, 321
  seam trackers, 337
  shielding gases, 332–334
    selection of (table), 333
  stopping the weld, 337–338
  systems components of, 316
  TIG holders: caps over electrodes, 331
    cooling systems, 331
    duty cycle ratings, 330
    gas nozzles, 330–331
    machine-guided, 332
  tungsten-electrode care, 329–330
  tungsten selection, 326–329
    color coding (table), 327
    surface finish, 327–328
    types of tungstens, 327
Gas welding (*see* Oxyfuel welding)
Gas welding rods (*see* Filler metals)
Gases:
  diffusion rates: vs. density, 50
    multiplying factors (*n*) for, 82
  expansion and contraction of, 35–38
  high-pressure (*see specific gases, for example:* Carbon dioxide)
  leak detection, 50
  liquefied, boiling range of, 39–40
  low-pressure (*see* Natural gas)
  medium-pressure (*see* Acetylene; MAPP Gas)
  pressure-volume changes, 37–38
  specific gravity of, 32
  standard conditions, 38, 39
  temperature-pressure-volume changes, 38
  temperature-volume changes, 36–37
Gauge length in tensile testing, 180, 181
Gear-driven positioning, 579–582

Geared Elevation® positioning, 581, 582
General industrial machinery, 10–11
Generators, 272
Globular transfer, 427–430
Glove box, 583, 584
Gloves, 93
GMAW (see Gas-metal-arc welding)
Goggles:
    flash, lens shades for, 301
    for oxyfuel welding, 92–93
Gold brazing alloys (table), 165
Gouging with flame cutting, 122
Gouging rods:
    for air-carbon-arc cutting, 305–306
    copper-coated, 306
    uncoated, 306
Grinding, 604–607
Grinding wheels:
    for aluminum, 607
    classification of, 606–607
    codes for, 605–606
    for stainless steel, 607
Grooved joints, braze-welding of, 173
Grounds, electrical, 261–262
Guiding devices in flame cutting, 130
GTAW (see Gas-tungsten-arc welding)
Gun angle and bead shape, 455

H

Hacksaws, 602–603
Hammers, chipping, 603–604
Hard wires, wire feeders for, 438
Hardening surfacing alloys, 505
Hardfacing, 498
    applications of, 494
    methods (table), 495–496
    (See also Surfacing)
Hardness:
    average, 186
    conversion tables for, 190–191
    definition of, 186–187
    hot, 494–495, 504
    measurement of, 187–191
        Brinell, 187–188
        Rockwell, 187–189
        scleroscope, 189–190
        Vickers, 189
    microhardness, 186
    of surfacing alloys, 499, 504
Headstock-tailstock positioning, 581, 582
Heat-affected zone, ductility of, 179
Heat conduction in brazing/soldering, 159
Heat treatment of surfacing alloys, 498
Heating boilers, ASME code on, 615–616
Heating tips, types of, 138–139
Heating torches, types of, 138–139
Heavy construction, 8

Helium:
    atom, 256
    cylinders, 56–58
    density, 31, 447
    ionization potential, 447
    liquid, 194
        boiling point of, 60
        in dewars, 60–61
    shielding, 446–448
    specific gravity of, 32
Hertz units of ac frequency, 227
High-low pipe welding defect, 536
High-quality cut, definition of, 105
High-speed cut, definition of, 105
High-speed steels, 504–505
Horizontal position, GMAW weaving patterns
        for, 456, 457
Hoses, gas, 79–83
    gas pressure drop, 81–83
    purging new hose, 97–98
    repair and connections for, 83
    selection of, 80–81
    sizes of, 80–81
Hot bending, 143–145
Hot hardness, 494–495
    of surfacing alloys, 504
Hydraulic tubing, 235
Hydrocarbon gases, 39
Hydrocarbons as gas impurities, 32
Hydrogen:
    density, 31
    regulator flow rate, 73
    weight in cylinder, 57
Hydrogen embrittlement, 530
    during SMAW welding, 367
    prevention of, 533

I

ICC (Interstate Commerce Commission), 65
Icicles (weld defect), 537
Impact resistance of surfacing alloys, 495,
        497
Inclusions:
    slag, 531
    tungsten, 532
Inconel, plasma-arc-welding data on (table),
        357
Increments in intermittent welds, 552–554
Indium solders (table), 164
Induced current, 271
Inductance control in GMAW, 437
Infrared radiation, 41
Inherent overhang, 587–589
Inspection of brazed joints, 160, 173–174
Insulators, electrical, 256–257
Intermittent welds, 551–556
Interruptions (weld defect), 529

Interstate Commerce Commission (ICC), 65
Ionization potential of shielding gases, 446–447
Ions as electrical conductors, 258, 262–263
Iron gas rods (*see* Filler metals)
Iron powder in flame cutting, 127–128
Ironworkers, 9–10
Izod impact test, 193–194

J

J-groove joint, welding symbol for, 560
Jobs:
    aricraft, missiles and aerospace industry, 11–12
    architectural and ornamental ironwork, 7
    with certification, 614
    construction equipment maintenance, 10
    construction industries, 8–10
    electrical equipment builders, 14
    fabricated metal products, 6
    general industrial machinery, 10–11
    heavy construction, 8
    by industry (table), 6
    ironworkers, 9–10
    metal service centers, 14
    pipe fittings and valves, 7
    prefabricated structures, 7
    primary metals, 12–13
    repair services, 13–14
    shipbuilding, 12
    specialty contractors, 8
    steamfitters, 9
    structural metal products, 7
    transportation equipment, 11–12
    value of training, 627–629
    through welding schools, 612
    wholesale trade, 14
    (*See also* Welders)
Joules, 268

K

Kerf:
    control when starting cut, 110
    diagram of, 113
    location on part, 115
    providing for width of, 115
    use of surface in troubleshooting, 119–122
    width measurement, 99, 116–117
Keyhole technique in plasma-arc welding, 354–355
Kilowatts, 267
Kindling temperature:
    carbon steel, 42, 100
    recognition by color, 101

L

Labor unions, 15–16
    "pipeliners," 3
    with welders (table), 16
Lack of fusion from flame cutting, 114
Lamellar tearing, 533, 537
Lap joints:
    brazed, 168–169
        overlap for, 168
    submerged-arc: joggled, 478
        single-fillet lap, 475
        through-weld lap, 478
Laser:
    cutting with, 635
    linear measurement with, 602
    welding with, 635
Lead, fumes from, 311
Lead-acid batteries, 263
Lead-coated steels, flame cutting of, 106
Lead-silver solders (table), 168
Lead-tin solders (table), 163
Leak testing, 539
    hydrostatic, 538
    pneumatic, 538
    specifications for, 538–539
    symbols for (table), 572, 573
    weldments, 538
Leatherware, 93–94
Leg, insufficient, weld profile of, 527
Lens shades:
    arc-welding, 301
    oxyfuel welding, 92–93
Line pipe, 412–414
    API specifications (table), 413
    capping pass, 413
    cover pass, 413
    double jointing, 412
    filler pass, 413
    hot pass, 413
    joint design, 413
Line tracers, 134
Liquefied fuel gases, cylinders for, 40
Liquid-penetrant testing, 540, 545–546
Liquidus, 163–166
    definition of, 166
Load on structures, 179
Load moment, 585–587

M

Machinability of surfacing alloys, 498
Magnesium alloys:
    brazing (table), 166
    as GMAW filler metals, 445
    shielding gases for, 454
Magnetic field, 270
Magnetic flux, 274

Magnetic-particle testing, 540, 544–545
  symbol for, 572, 573
Magnetism, 270–275
Manipulators (*see* Positioners, welding)
MAPP Gas:
  Btu/ft³ when burned, 34
  bulk storage tanks, 65
  burning velocity in oxygen, 49
  composition of, 26
  cutting data, 116
  cutting tips for, 103–104, 105
  density, 31
  explosive limits in air and oxygen, 49
  flame adjustments, 51–54
  flame properties (table), 44
  flame temperature, 43
  heavy cutting data, 118
  preheat cutting flame adjustments, 107–110
  underwater cutting data, 129
  underwater cutting depth, 128
  vapor pressure, 40
  withdrawal rates, 65
Markers for metals, 601
Martensite:
  carbon steels, 407–408
  stainless steel, 384
Materials specifications, ASME code, 615
Mechanical properties of metals, 178–179
Melt-in technique in plasma-arc welding, 354
Melt-through, weld symbol for, 565
Metal-cored FCAW wires, 434
  shielding gases for, 434
Metal fatigue:
  peening and, 198, 599
  tests for, 197
Metal-inert-gas (MIG) GMAW, 422
  air wires, 422
  dip-transfer (*see* Dip-transfer GMAW)
  equipment for, 423
  extended-reach wire feeder, 423
  globular transfer, 423
  lens shades for (table), 301
  MIG-arc brazing, 422
  MIG-spot welding, 422
  power sources for (table), 287
  process overview, 421–422
  pull feeder, 423
  pulsed-current, 276, 422
  pulsed-spray, 421, 431–432
  push feeder, 422
  short circuiting transfer, 423
  spray transfer, 423
Metal service centers, 14
Metal spraying, 508–517
  applications of, 508–509
  bond strength (table), 510
  coating characteristics (table), 509
  coating shrinkage (table), 511

Metal spraying (*Cont.*):
  coating thickness, 510–511
    on shafts (table), 511
  methods, 511–517
    detonation guns, 516–517
    plasma spraying, 514–516
    spray and fuse, 512–515
    vacuum plasma, 518
  preparation of surface, 510
  spray speeds, 511
Metallizing (*see* Metal spraying)
Microwire GMAW (*see* Dip-transfer GMAW)
MIG welding (*see* Metal-inert-gas GMAW)
Mitering in pipe cutting, 244
Modulus of elasticity, 185–186
Moisture content of gases (table), 33
Motor generators, 326
Muntz metal, 170–171

N

Narrow-gap GMAW, 432
  root openings for, 432
  with tandem electrodes, 432
Natural gas:
  Btu/ft³ when burned, 34
  cutting data, 118
  explosive limits in air and oxygen, 49
  flame properties (table), 44
  flame temperature, 43
  liquefaction temperature, 191
Neutral flames (*see* Flame adjustment)
Nickel alloys:
  Curie point of, 405
  as filler metals: brazing, 166
    GMAW specifications, 445
  as surfacing alloys, 502
Nitrogen:
  cylinders, 56–58
  density, 31
Nondestructive testing (NDT), 526–546
  ASME code on, 615
  methods (table), 539–540
  symbols for, 571–573
Nontransferred plasma arcs, 353–354
Notch sensitivity in toughness tests, 192
Notches (base-metal weld defect), 529
Nozzle, shielding-gas, 442
Nuclear power plant components, ASME code on, 615, 616

O

OFC (*see* Flame cutting)
OFW (*see* Oxyfuel welding)
Ohms, definition of, 259, 265–266
Ohm's law, equations for, 266
Oil-country goods, 235

Open-circuit voltage (OCV), 284
  multiple machines, hazards of, 294–295
Operator qualification test:
  GMAW: butt welds, 460–461
    fillet welds, 461
    pipe welding, 461
  GTAW: butt welds, 345–346
    fillet and lap welds, 346
    pipe welding, 346
  OFW: butt joints, 231
    cast iron, 232
    fillet joints, 231–232
    lap joints, 231–232
    pipe and tubing welding, 232, 254
  SMAW: butt welds, 418–419
    fillet and lap welds, 419
    pipe welding, 419
  (See also Welder qualification tests)
Optical tracers of cutting machines, 134
Ornamental ironwork, 7
Oscilloscope, 278
OSHA, 9–10, 300
Ovens for electrode drying, 530
Overhead positions, GMAW, weaving
    patterns for, 456, 457
Overlap (weld defect), 528
Overlay patterns in weld surfacing, 500–502
Overload in weld testing, 538
Overspin in weld testing, 538
Oxidation resistance of surfacing alloys, 497
Oxides of iron, 100
Oxidizing flames (see Flame adjustment)
Oxyfuel welding (OFW), 206–225
  aluminum alloys, 205–206, 221–225
    bead welds, 221–222
    butt welds, 222–223
    melting temperature, 222
    sheet, 222–223
  applications of, 201–202
  backhand method, 207–208
  cast aluminum, 224–225
  cast iron, 218–220
  copper, 221
  corner joints, 211–212
  fillet and lap joints: backhand, 217–218
    forehand, 215–216
  fluxes, 209
  forehand method, 171, 172, 207
  gas rod manipulation, 210–211
  pipe and tubing, 233–254
  plate: backhand method, 214–215
    forehand method, 213–214
  puddle welding, 207
  ripple welding, 207
  sheet metal, 212–213
  steel, bead-on plate, 209–210
  T-joints, 216
  tack welds, 211, 212
  thick sections, 209

Oxyfuel welding (OFW) (Cont.):
  yellow brass, 220–221
Oxygen (gas):
  cylinders, 56–61
    handling and storage, 67–68
    pressure changes, 58
    safety, 58–59
    weight, 57
  density, 31
  purity of, for flame cutting, 32
  supplied during combustion (table), 103
  volume of, in air, 42
  weight of, in air, 42
Oxygen (liquid):
  handling and storage of, 67–68
  storage temperature, 57
Oxygen lancing, 126–127
Oxyhydrogen welding, 223–224
Ozone (O_3), 42

P

PAC (see Plasma-arc cutting)
Parallel connections, dc, 352
Parts-per-million, 32
Pascal, definition of, 29
Pasty range, 166
  of solders (table), 163–164
PAW (see Plasma-arc welding)
Peak height, ac, 278
Pearlite in carbon steels, 407–408
Peening:
  compressive stress from, 599–600
  and metal fatigue, 198, 599
  shot, 198, 599–600
  of weld metal, 532
Penetration:
  excess, 536, 537
  weld parameters and, in GMAW, 455–456
  inadequate (weld defect), 528
Percent elongation, 184
Percent reduction of area, 184
Period, ac:
  definition of, 277
  formula for, 278
Phase diagrams, 166
  lead-tin, 166, 167
  silver-copper, 167–168
Physical properties of metals, 178
Pickling acids, brazing and, 161
Piercing starts in flame cutting, 110–111
Piezoelectricity, 544
Pinch-arc GMAW (see Dip-transfer GMAW)
Pipe fittings and valves, 7
Pipe and tubing:
  bending problems, 142–145
  beveling, 238, 244
  cascade-welding method, 246, 247

Pipe and Tubing (*Cont.*):
  consumable inserts, 240–241
  differences between, 233–234
  galvanized, 251–252
  jigs for, 248
  joints: bevel edge cuts, 238, 244
    changing flow direction, 242–246
    edge preparation, 241–242
    multilayer, 246–247, 250–251
    small sections, 241, 411–412
    steel, 241
    T-, K-, J-connections, 240
  oxyfuel welding and cutting of, 233–254
    backhand and forehand methods, 245
    wall thickness, maximum, 201, 233–234
    welding positions, 239, 240, 245
  SMAW techniques for, 411–414
    line pipe, 412–414
    small pipes, 411–412
  specifications for (table), 235–237
  stress-relief methods, 247–248
  template cutting, 243
  turning rolls, 237–238, 248–249, 578, 580,
    582, 584
"Pipeliners," 3, 4
Pitch of intermittent weld, 552
Plasma-arc cutting (PAC), 13, 101, 358–362
  applications, 356–358
  arc temperature, 359
  controls for, 359
  cut quality, 360
  dual-flow, 359–360
  kerf width, 361
  lens shades (table), 301
  open-circuit voltage for, 359
  plasma-forming gases, 360
  power sources for, 359
  safety, 360
  speed of, 358
  thickness cut, 361
  torches for, 358–359
  travel direction, 361
  water-injection, 360
Plasma-arc welding (PAW):
  advantages of, 348–349
  applications of, 347–348
  arc temperatures, 348
  constricted arc, 348
  drawbacks of, 349
  keyhole technique, 354–355
  lens shades for (table), 301
  melt-in technique, 354
  nontransferred arcs, 348, 353, 354
  open-circuit voltage for, 293, 351–353
  outfits for, 350–351
  plasma-forming gases, 349–350
  power sources for, 287, 351–353
    kilowatt ratings, 353
    parallel connection, dc, 353

Plasma-arc welding (PAW) (*Cont.*):
    series connection, dc, 353
    series-parallel connections, dc, 351–352
  schedules: A-286 alloy, 357
    Inconel, 357
    titanium, 357
    waspalloy, 357
    70-30 brass, 357
    304 stainless steel, 355–357
  shielding gases for, 350
  torches for, 348, 353–354
Plasmas:
  electrical, 262, 263
  nature of, 348
Plastic range, 184
Plate, maximum OFW thickness of, 200
Plug welds:
  in submerged-arc welding, 478
  symbol for, 560–562
Porosity of weld metal, 529–531
  cluster, 529–530
  linear, 531
  starting, 530–531
  uniformly scattered, 529
Positioners, welding:
  floor turntables, 581–583
  gear-driven:
    calculation of torque, 586
    Geared Elevation®, 581, 582
    45°–90° type, 580
    135° type, 579–580
  headstock-tailstock, 581, 582
  manipulators, 583
  tilting capacity of, 588–589
  torque of, 586–590
  turning rolls, 578–579
  Universal Balance®, 577–578
  utility, 583
Positioning parts in brazing/soldering, 158
Postheating, suggested temperatures for
  (table), 411
Potential difference, electrical 265
  (*See also* Volts)
Powder cutting in OFW, 101, 127–128
Power:
  calculations of, 267–268
  equation for, 266–267
  loss of, to electrical resistance, 268–269
  units of, 267
Power boilers, ASME code on, 615, 616
Power sources:
  altitude and, 290–291
  ambient temperature limits, 291
  constant current (CC), 284, 285, 434–435
  constant voltage (CV), 283–285, 434–435
  GMAW, 434–438
  grounding of, 292, 293
  hazardous operating conditions, 290–291
  maintenance for safety, 299

Power sources (*Cont.*):
   manufacturers' machine ratings, 296
   multiple machine hookups, 294
   parallel connections, 296
   types of, by welding process (table), 287
Precipitation hardening of stainless steels, 385
Prefabricated structures, 7
Preheat and postheat, 406–411
Preheating:
   based on equivalent carbon content (table), 409
   crack prevention, 532–533
   submerged-arc welds, 485–487
   suggested temperatures for (table), 409–410
   surfacing alloys, 505
Prepainted steels, flame cutting of, 106
Press brakes, 196
   application of, 597, 598
   forming with, 596–598
Pressure, cylinder:
   definition of, 29–30
   delivery and, 30
Pressure, gas, 28–31
   barometric, 31
   measurement of, 29–31
   psi and psia, 30–31
Pressure testing, 538
Pressure vessels, ASME Code on, 616
Primary metals, 12–13
Prods in magnetic particle testing, 544
Products, welded (table), 3
Profiles, weld, 526–527
   gauges for, 537
Projection welding, 632, 633
   joints, welding symbol for, 566–567
Proof testing:
   symbol for, 572, 573
   weldments, 537–538
Propane:
   Btu/ft^3 when burned, 34
   explosive limits in air and oxygen, 49
   flame properties (table), 44
   vapor pressure, 40
Proportional limit, 182
Propylene:
   Btu/ft^3 when burned, 34
   explosive limits in air and oxygen, 49
   flame properties (table), 44
   flame temperature, 43
   vapor pressure, 40
Psi, 29–31
Psia, 29–31
Psig, 29–31
Pulsed-current MIG, 276, 422
Pulsed-spray GMAW, 431–432
Purging of gas lines, 75–76
Purity of gases, 32

Q

Qualification tests (*see* Operator qualification tests; Welder qualification tests)
Quenching to remove flux, 160

R

Radiation, 40–41
Rectifier:
   breakdowns, 280
   mercury-pool, 280–281
   silicon-controlled (SCR), 279–280
   selenium, 280
Reducing flame (*see* Flame adjustment, carburizing)
Regulators, gas (*see* Gas regulators)
Reinforcement in weld profile, 526
Relays in power sources, 280–281
Repair services, 13–14
Resistance, electrical, 258–259
Resistance-seam welding, 631–632
Resistance-spot welding, 631–632
Resistance welding, 630–631
   high-frequency, 632–633
Resistivity, cable size and, 269
Restricted bend test, 195, 196
Reverse bend test in blacksmith welding, 148
Rig welders, 14
Rivet removal, 123–124
Robots, welding, 583, 589–596
   application of, 589–591
   communication with, 592
   grippers for, 593
   programming, 591–592
   specifications for, 594–596
   tactile sensors of, 593–594
   teaching of, 591
   vision of, 592–593
Rockwell hardness, 187–189
   and tensile strength, 187
   tests for thin sections, 189
Root-mean-square amperage (RMS), 278
Root pass in braze-welding, 173
Rotor in electrical power source, 272
Rupture strength, 199

S

Safety:
   air-carbon-arc cutting, 310–311
   arc welding, 290–303
     arc voltages and machine ratings, 292–294
     confined spaces, 299–300
     electrical hazards, 291–292
     electrode holders, 298
     eye protection, 300–301

Safety: arc welding (*Cont.*):
        grounding of workpiece, 298–299
        high places, 300
        hot metal, 300
        machine duty cycles, 295–296
        maintenance of equipment, 299
        multiple machines, 294–295
        power-supply wiring, 296
        protective clothing, 303
        selecting cable, 296–297
        water hazards, 299
        welding curtains, 301–303
        welding machines, 290–291
    brazing/soldering, 161
    plasma-arc cutting, 360
SAW (*see* Submerged-arc welding)
Scales, 601–602
Scarf joints, brazed, 169
Scarfing, 13
    methods, 124
Scleroscope hardness, 189–190
Scratch start, SMAW, 393
Seam trackers, 337
Seam weld joint, symbol for, 567–569
Selenium rectifiers, 280
Self-jigging parts, 158
Semiconductors, 257
Series connection, dc, 351–352
Service cable, GMAW, 440–441
Shearing and slitting, 598, 599
Shielded-metal-arc welding (SMAW), 364–419
    advantages of, 364–366
    alloying with flux coating, 367
    aluminum alloy electrodes, 391–392
        selection (table), 392
    copper alloy electrodes, 388–391
        selection (table), 390
    disadvantages of, 366
    electrode classification system, 369–370
    electrode flux coatings, 367, 370
    electrode manufacture, 367–368
    electrode operation, 366–367
    electrode selection, 368–392
    hydrogen embrittlement, 367
    lens shades for (table), 301
    low-alloy steel electrodes, 380–382
        mechanical properties of (table), 383
        weld-metal chemical codes (table), 382
    methods: backup strips in butt joints, 397
        breaking the arc, 393–394
        flat-position bead welding, 394–395
        flat-position butt joints, 395–397
        horizontal fillets, 398
        horizontal lap joints, 398–399
        horizontal T joints, 398
        overheat butt joints, 403
        overhead fillet joints, 403–405
        overhead stringer beads, 402–403
        pipe welding, 411–414

Shielded-metal-arc welding (SMAW), methods (*Cont.*):
        restarting the arc, 394
        scratch start, 393
        starting the arc, 392–393
        spatter removal, 404
        tap start, 393
        undercutting in butt joints, 397
        vertical butt joints, 400–401
        vertical fillet joints, 401
        vertical stringer beads, 400
        weaving motions, 397, 400
    mild-steel electrodes, 371–380
        selection (table), 371
    nickel alloy electrodes, 391
        selection (table), 391
    power sources for (table), 287
    process overview, 364
    slag cleaning, 394
    stainless steel electrodes, 382–388
        selection (table), 388–389
    in surfacing and hardfacing, 503–508
    typical current range, 365
    weld thickness, 365
Shielding gases (*see* Gas-metal-arc welding,
        shielding gases in; Gas-tungsten-arc
        welding, shielding gases in)
Shipbuilding, 12
Short-arc welding (*see* Dip-transfer GMAW)
Short circuiting (*see* Dip-transfer GMAW)
Short circuits, 264
Shot peening, 198, 599–600
Shrink fitting, 148
Silhouette tracers, 134
Silicon-controlled rectifiers (SCR), 279–280
Silver brazing alloys (table), 165
Silver solders, 152
Skewed joints, welding symbol for, 556–558
Slack lime, 62
Slag:
    during flame cutting, 114
    in plasma-arc cutting, 361
Slitting and shearing, 598, 599
Slope:
    CV power sources, 284
    dip-transfer GMAW, 436–437
    GTAW power sources, 319–321
    power source voltage regulation, 283–285
Slot weld joint, 560
    welding symbol for, 562–563
SMAW (*see* Shielded-metal arc welding)
Society of Automotive Engineers, 235
Soft wires, GMAW feeders for, 438
Solder alloys:
    AWS list of (table), 163–164
    selection of, 161–165
        by melting temperature, 161–162
Soldering:
    definition of, 152
    (*See also* Brazing/soldering)

Solenoid, 130
Solid-state welding, 19, 20
    master chart of process, 22
Solidus, 163–166
    definition of, 166
Solution treating and aging of stainless
        steels, 385
Spark lighters, 90
    electric, 131
Spatter, 535
Specialty contractors, 8
Specific gravity of gas, 32
Spray transfer GMAW:
    effect of increased arc length, 430–431
    minimum weldable thickness, 431
    operating schedules for, 424
    transition currents for, 429
Spot weld joint, 566–567
Springback, 597
Square-butt joint, welding symbol for, 558
Square-groove corner joints, 477–478
Square-wave current for TIG, 276
Stack cutting, 129–130
Staggered intermittent welding, 553–554, 556
Stainless steel, 194
    austenitic, 194, 384
    carbide precipitation, 386–387
    ferritic grades, 383–384
    filler-metal specifications, 445
    GMAW shielding gases for, 451
    grinding wheels for, 607
    martensitic grades, 384–385
    plasma-arc welding schedule, 362
    precipitation-hardening grades, 385
    SMAW electrodes for, 385–386
        selection (table), 388–389
    solution annealing of, 387
    standard AISI grades (table), 385
    surfacing with, 503
    temper embrittlement of, 387
Start-stop weld defect, 536, 537
Static electricity, 260
Stator in electrical power source, 272
Steamfitters, 9
Steel (see Carbon steel; Stainless steel)
Steel tubing, ASTM specifications for (table),
        236–237
Steel wires, feeders for, 438
Stick-electrode welding (see Shielded-metal-
        arc welding)
Still tubes, ASTM specifications for (table),
        237
Strain, definition of, 179
Stiffness of sections, 185–186
Stress:
    definition of, 179
    and metal fatigue, 198
    relief of: in submerged-arc welding, 487–
        489

Stress (Cont.):
        times and temperatures (table), 411
        after welding, 410–411
    units of, 180
Stress-strain curves, 182, 183
Structural metal products, 7
Structural sections, bending of, 145–147
Structural steel tubing, 235
    ASTM specifications for (table), 237
Stud-arc welding (SW), 63
    power sources for (table), 287
Submerged-arc welding (SAW), 462–493
    advantages of, 462–463
    arc voltages, 483
    burnoff rate, 484–489
    butt joints, 473–475
    carbon-steel schedules, 486–492
    corner joints, 477–478
    current, 483
    edge preparation, 472–473
    equipment, 463–464
    filler metals for, 465–467
    fillet joints, 476
    flux width and depth, 483–484
    fluxes for, 464–465
    lap joints, 475–476
    limitations of, 462–463
    multipass methods, 485–492
        preheating, 485–487
        stress relief, 487–489
    positioning for, 478–480
    power sources for, 287, 467–468
    process overview, 462–463
    travel speed, 483
    weld-metal backing, 469–471
    weld start-up, 480–482
        flying start, 482
        molten metal start, 482–483
        retract start, 481–482
        steel-wool start, 482
    wire feeder control, 468–469
    wire position for, 479–480
    wire size in, 467, 483, 485
Surfacing, 494–524
    applications, 494, 517–523
    filler-metal forms, 503
    filler metals for: high-speed steels, 504
        tungsten carbide, 507–508
    GMAW, 508
    GTAW, 508
    metallizing (see Metal spraying)
    methods (table), 495–496
    process selection, 403
    SAW, 508
    service conditions, 512–523
    SMAW, 504–508
    welding symbols for, 569–571
SW (see Stud-arc welding)
Sweat soldering/brazing, 160

Switches in power sources, 280–281
Symbols (*see* Welding symbols)

T

Tack welds:
  in braze welding, 172
  pipe, 241–242, 245
Tapes, linear measurement, 601–602
Tearing, lamellar, 533, 537
Temperature:
  for braze-welding, 172
  color of steel and, 101
  definition of, 33–34
  and metal fatigue, 198
  hazardous levels, 290
  indicators of, 600–601
  measurement of, 33–35
Tensile stress, definition of, 179, 180
Tensile test bar:
  standard flat bar, 181
  standard round bar, 180
Tensile testing, 179–181
Thermal conduction, 41–42
  of shielding gases, 447
Thermal convection, 42
Thermal cutouts in power sources, 281
Thermal expansion/contraction of metals,
    139–142
  coefficients of (tables): large, 140
    small, 156–157
Thermit welding, 634
Thoriated tungstens, 327
Three-phase ac, 278–279
Throat of weld:
  effective size of, 559
  insufficient, weld profile of, 527
TIG welding (*see* Gas-tungsten-arc welding)
Tilt torque, 588, 590
Tin-antimony-lead solders (table), 163
Tin-antimony solder (table), 163
Tin-zinc solders (table), 163
Tinning:
  in braze welding, 172, 174
  prior to soldering, 163
Tip cleaners, 91
Tips:
  oxyfuel flame-cutting, 89, 101–105
    angle calibrated for bevel cutting, 132–
      133
    for cutting machines, 130–132
    for difficult conditions, 104–105
    divergent bore, 105, 132
    dual-type, 133
    MAPP Gas, 103–105
    one-piece, 102
    seat designs, 102
    sizes for manual cutting (table), 104

Tips: oxyfuel flame-cutting (*Cont.*):
    skirted type, 102–103
    straight-bore, 105
    two-piece, 102–103
  oxyfuel welding, 226
    regulator settings (table), 227
Titanium alloys:
  ductile-brittle transition temperature, 194
  glove box welding of, 583, 584
  GMAW filler-metal specifications, 445
  GMAW shielding gases for, 454
  plasma-arc welding schedule, 357
Torches:
  flame-cutting, manipulation of: carbon
      steel, 114
    cast iron, 114, 126
    stainless steel, 114
  oxyfuel, 84–90
    air-fuel type, 87–88
    combination, 89–90
    for cutting machines, 130–132
    flame scarfing, 124
    injection type, 64, 87
    packaged outfits, 90–91
    positive-pressure mixers, 85–87
    rated by gas-flow, 84–85
    tip-mix type, 85
    tips for (*see* Tips)
    underwater, 110, 128–129
  in plasma-arc welding, 348, 353–354
Torque:
  calculation of, 586–589
  gear-driven positioners and, 580
  load moment and, 585–587
  tilt, 588, 590
Total elongation, 184
Toughness of metals, 190–195
  temperature and, 191–192
Tracer control cutting machine, 133, 134
Transferred plasma arcs, 353–354
Transformer-rectifiers, 325–326
Transformers:
  ac and dc, 274–275
  moveable shunt, 325
  tapped reactor, 325
Transition joint, soldering, 163
Transition points, globular-spray-transfer,
    427–430
Transition temperature, ductile-brittle, 194
Transportation equipment, 11–12
Transverse bend tests, 197
Tube trailer, 31
Tubing (*see* Pipe and tubing)
Tungsten carbide surfacing alloys, 507–508
Tungsten-inert-gas (TIG) welding (*see* Gas-
    tungsten-arc welding)
Turning rolls, 578–579
  in pipe welding, 237–238, 248–249, 578,
    580, 582, 584

U

Ultimate tensile strength, 182
Ultrasonic testing, 540, 543–544
    symbol for, 572, 573
Undercutting (weld defect), 528
Underwater cutting, 110, 128–129
United States Coast Guard, 609
United States Navy, 609
Universal Balance®, 577–578
Upset welding, 632

V

V-groove butt joint, welding symbol for, 558–
    560
V-groove corner joint, 478
V-joint, braze-welding, 173
Vapor pressure, 40
Vaporizers for cryogenic gases, 61
Vertical position, GMAW weaving patterns
    for, 456, 457
Vickers hardness, 189
Visual inspection:
    symbol for, 572, 573
    of welds (table), 539
Volt-amp curves:
    CC and CV, 283–285
    (See also Slope)
Voltage:
    average, ac, 278
    GMAW power sources, 434–438
    input to power source, 292
    output from power source, 292
    and penetration depth, 455
    variation in supply, 276
Volts, definition of, 259, 265

W

Warping:
    of plate, cutting and, 131
    of surfacing alloys, 503
    (See also Thermal expansion/contraction)
Waspalloy, plasma-arc welding (table), 357
Watts, 267
Watt-second (see Joules)
Waveforms, electrical, 259
    ac, 276, 277
    dc, 275–276
Weld defects, 211, 527–532
Weld size, incorrect, 527
Welded products (table), 3
Welder Certification Tests, 612–614
    laboratories for, 612–614
Welder qualifications:
    ASME Code on, 616
    tests, 617–629

Welders:
    employment, by state (table), 5
    shortage of, 4–5
    (See also Jobs)
Welding blankets, 302
Welding cable (see Cable, welding)
Welding curtains, 301–303
Welding and cutting processes, master chart
    of, 19
Welding distributors, 94
Welding guns, GMAW, 441
    ratings for, 442
Welding positioners (see Positioners, welding)
Welding positions:
    butt joints, 239
    fillet joints, 239
    pipe joints, 239, 240, 245
Welding symbols, 548–576
    chain intermittent welds, 552, 556
    components of, 549–557
    contour and finishing method, 555
    dimensioning of, 550–555
    fillet welds, 555–556
    flange weld joint, 563–564
    flare-V joint, 561
    groove joints, 559–560
    melt-through, 565
    plug weld joint, 560–562
    slot weld joint, 560, 562–563
    staggered intermittent welds, 553–554, 556
    summary of, 574–575
    U-bevel joint, 560
    weld-all-around, 554–555
Wetting of metal, 151
    soldered or brazed basemetal, 162
Wire brushes, 604
Wire feed speed, GMAW:
    and bead shape, 455
    and weld parameters, 456
Wire feeders:
    accessories, 440
    auxiliary functions, 440
    controls of, in SAW, 468–469
    CV and CC power sources (table), 288
    extended reach, 438–439
    GMAW, 438–440
    pull type, 438, 439
    push type, 438, 439
Wire liners:
    for aluminum wires, 441
    in GMAW service cable, 440–441
    for steel wires, 440
Wire metallizing oxyfuel, 509
Work hardening, 182
Workmanship, 535–536
Workpiece limitations in metallizing, 509–
    510
Wrenches, cylinder, 63, 67, 91–92
Wrinkle bending, 142–143

Wrought iron, bending of, 145
Wrought-steel pipe, ASTM specifications
    (table), 236

X

X-radiography, 539, 541–543
    symbol for, 572, 573

Y

Yield points, 182
Yield strength, 182–184

Z

Zinc-aluminum solder (table), 168
Zinc-coated steels, flame cutting of, 106
Zinc fumes, 311
Zirconium alloys:
    glove-box welding of, 583, 584
    GMAW shielding gases for, 454
    tungstens, 327